The Little Oxford Dictionary & Thesaurus

Edited by
Sara Hawker

with
Chris Cowley

OXFORD UNIVERSITY PRESS

Oxford University Press, Walton Street, Oxford OX2 6DP

Oxford New York
Athens Auckland Bangkok Bombay
Calcutta Cape Town Dar es Salaam Delhi
Florence Hong Kong Istanbul Karachi
Kuala Lumpur Madras Madrid Melbourne
Mexico City Nairobi Paris Singapore
Taipei Tokyo Toronto

and associated companies in
Berlin Ibadan

The publishers are grateful to the Reader's Digest Association Limited for permission to use the respelling pronunciation system shown in this dictionary

Oxford is a trade mark of Oxford University Press

Published in the United States
by Oxford University Press Inc., New York

First published as the Oxford Minireference Dictionary & Thesaurus, 1995

British Library Cataloguing in Publication Data
Data available

Library of Congress Cataloging in Publication Data
The little Oxford dictionary & thesaurus / edited by Sara Hawker, with Chris Cowley.
p. cm.
1. English language—Dictionaries. 2. English language—Synonyms and antonyms. I. Hawker, Sara. II. Cowley, Chris.
423'.1—dc20 PE1628.08645 1996 95-45277

ISBN 0-19-863151-0

10 9 8 7 6 5 4 3

Printed in Great Britain by
Clays Ltd.
Bungay, Suffolk

Contents

Preface

The *Little Oxford Dictionary & Thesaurus* combines the information you would expect to find in a conventional dictionary and a thesaurus in a single handy volume. Within an individual entry it offers a guide to the spelling and meaning of a word, together with lists of synonyms from which an alternative word can be selected.

Suitable for use on many occasions, from writing a letter to solving a crossword puzzle, the *Little Oxford Dictionary & Thesaurus* is a convenient and compact quick-reference book.

S. J. H.

How to use the Little Oxford Dictionary & Thesaurus

The 'entry map' below explains the different parts of an entry.

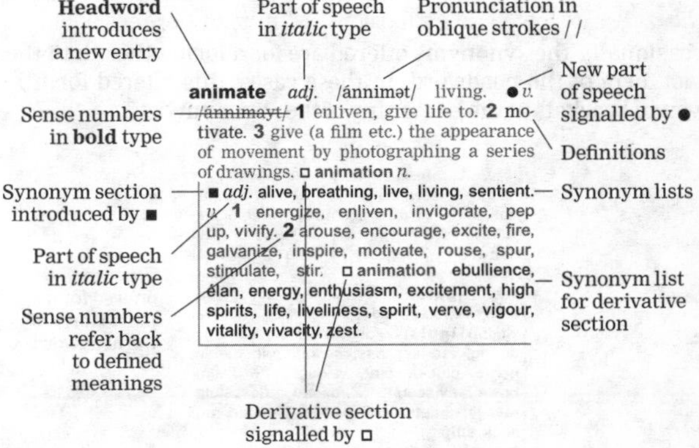

Arrangement of synonym sections

Synonyms are offered for many of the words defined in the *Little Oxford Dictionary & Thesaurus*. The sense numbering in the synonym sections follows the pattern of the defining sections, so that users can easily see which synonyms are appropriate to which meaning of the headword. Synonyms are not always offered for every dictionary sense and in some cases a single list of synonyms is offered for several senses. The sense numbering may then take the form of a list, as, for example, **1, 3, 4**. Wherever a single synonym list covers all the meanings explained in the defining section, sense numbers are dispensed with and the synonym list is simply introduced by the symbol ■.

Within each separate sense synonyms are arranged alphabetically, but a list may be divided by a semicolon to indicate a different 'branch' of meaning.

hire *v.* engage or grant temporary use of, for payment. ● *n.* hiring. □ **hire purchase** system of purchase by paying in instalments. **hirer** *n.*

■ *v.* charter, lease, rent; employ, engage, take on. *n.* lease, rental.

semicolon indicates new 'branch' of meaning.

Occasionally, the synonyms offered are for a form other than the exact form of the headword. In these cases, this altered form is given in brackets at the beginning of the synonym list.

gut *n.* **1** intestine. **2** thread made from animal intestines. **3** (*pl.*) abdominal organs. **4** (*pl., colloq.*) courage and determination. ● *v.* (**gutted**) **1** remove guts from (fish). **2** remove or destroy internal fittings or parts of.

■ *n.* **3** **(guts)** bowels, entrails, *colloq.* innards, *colloq.* insides, intestines, viscera, vitals. **4** **(guts)** boldness, bravery, courage, daring, fearlessness, *colloq.* grit, mettle, nerve, pluck, spirit, valour. *v.* **1** disembowel, eviscerate. **2** destroy, devastate, ravage; empty, loot, pillage, plunder, ransack, strip.

synonyms offered for a plural form of the headword.

Key to the Pronunciations

This dictionary uses a simple respelling system to show how words are pronounced. The following symbols are used:

a, á	*as in*	**pat** /pat/, **pattern** /páttərn/
aa, áa	*as in*	**palm** /paam/, **rather** /ráathər/
air, áir	*as in*	**fair** /fair/, **fairy** /fáiri/
aw, áw	*as in*	**law** /law/, **caught** /kawt/, **caution** /káwsh'n/
awr, áwr	*as in*	**warm** /wawrm/, **warning** /wáwrning/
ay, áy	*as in*	**gauge** /gayj/, **daily** /dáyli/
ch	*as in*	**church** /church/, **cello** /chéllō/
e, é	*as in*	**said** /sed/, **jealous** /jélləss/
ee, ée	*as in*	**feet** /feet/, **recent** /réess'nt/
er, ér	*as in*	**fern** /fern/, **early** /érli/
érr	*as in*	**ferry** /férri/, **burial** /bérriəl/
ə	*as in*	**along** /əlóng/, **pollen** /póllən, **lemon** /lémmən/, **serious** /séeriəss/
g	*as in*	**get** /get/
i, í	*as in*	**pin** /pin/, **women** /wímmin/
ī, ī́	*as in*	**time** /tīm/, **writing** /rī́ting/
īr, ī́r	*as in*	**fire** /fīr/, **choir** /kwīr/, **desire** /dizī́r/
írr	*as in*	**lyrics** /lírriks/
j	*as in*	**judge** /juj/
kh	*as in*	**loch** /lokh/
N	*as in*	**en route** /ON rṓot/
ng	*as in*	**sing** /sing/, **sink** /singk/
ngg	*as in*	**single** /síngg'l/, **anger** /ánggər/
o, ó	*as in*	**rob** /rob/, **robin** /róbbin/
ō, ṓ	*as in*	**boat** /bōt/, **motion** /mṓsh'n/

ö, ő	*as in*	**colonel** /kőn'l/
oo	*as in*	**unite** /yooní̄t/
o͝o, ő͝o	*as in*	**wood** /wo͝od/, **football** /fő͝otbawl/
o͞o, ő͞o	*as in*	**food** /fo͞od/, **music** /myő͞ozik/
oor, óor	*as in*	**cure** /kyoor/, **jury** /jóori/
or, ór	*as in*	**door** /dor/, **corner** /kórnər/
ow, ów	*as in*	**mouse** /mowss/, **coward** /kóward/
oy, óy	*as in*	**boy** /boy/, **noisy** /nóyzi/
r, rr	*as in*	**run** /run/, **fur** /fur/, **spirit** /spírrit/
sh	*as in*	**shut** /shut/
th	*as in*	**thin** /thin/, **truth** /tro͞oth/
th̲	*as in*	**then** /th̲en/, **mother** /múth̲ər/
u, ú	*as in*	**cut** /kut/, **money** /múnni/
ur, úr	*as in*	**curl** /kurl/, **journey** /júrni/
úrr	*as in*	**hurry** /húrri/
y	*as in*	**yet** /yet/, **million** /mílyən/
z̲h̲	*as in*	**measure** /méz̲h̲ər/, **vision** /víz̲h̲'n/

Consonants

The consonants *b, d, f, h, k, l, m, n, p, s, t, v, w, z* are pronounced in the usual way. A doubled consonant indicates that the preceding vowel is short, as in **robin** /róbbin/.

Stress

The mark ´ that appears over the vowel symbol in words of more than one syllable indicates the part of the word which carries the stress.

Abbreviations

abbr. abbreviation
adj. adjective
adjs. adjectives
adv. adverb
attrib. attributively
Austr. Australian
colloq. colloquial
conj. conjunction
Dec. December
derog. derogatory
esp. especially
fem. feminine
Fr. French
int. interjection
Ir. Irish
Jan. January
joc. jocularly
n. noun
N. Engl. northern England
Nov. November
ns. nouns
orig. originally
[P.] proprietary term
pl. plural
poss. possessive
pref. prefix
prep. preposition
pron. pronoun
rel.pron. relative pronoun
S. Afr. South African
Sc. Scottish
sing. singular
sl. slang
US United States
usu. usually
v. verb
v. aux. auxiliary verb

Abbreviations that are in general use (such as ft, RC) appear in the dictionary itself.

Proprietary terms

This dictionary includes some words which are, or are asserted to be, proprietary terms or trade marks. Their inclusion does not mean that they have acquired for legal purposes a non-proprietary or general significance, nor is any other judgement implied concerning their legal status. In cases where the editor has some evidence that a word is used as a proprietary name or trade mark this is indicated by a letter [P.], but no judgement concerning the legal status of such words is made or implied thereby.

Abbreviations

Proprietary terms

Aa

a *adj.* **1** one, any. **2** in, to, or for each.

aback *adv.* **taken aback** disconcerted.

abacus *n.* (*pl.* **-cuses**) frame with balls sliding on rods, used for counting.

abandon *v.* **1** leave without intending to return. **2** give up. ● *n.* careless freedom of manner. □ **abandonment** *n.*

■ *v.* **1** evacuate, leave, quit, vacate, withdraw from; desert, *sl.* ditch, forsake, jilt, leave behind, leave in the lurch, maroon, strand, wash one's hands of. **2** cede, discontinue, disown, *sl.* ditch, drop, forfeit, forgo, give up, relinquish, renounce, resign, surrender, waive, yield.

abandoned *adj.* **1** deserted. **2** unrestrained.

abase *v.* humiliate, degrade. □ **abasement** *n.*

abashed *adj.* embarrassed, ashamed.

abate *v.* make or become less intense. □ **abatement** *n.*

abattoir /ábbətwaar/ *n.* slaughterhouse.

abbess *n.* head of a community of nuns.

abbey *n.* building occupied by a community of monks or nuns.

abbot *n.* head of a community of monks.

abbreviate *v.* shorten.

■ abridge, condense, cut, edit, précis, reduce, shorten, summarize, truncate.

abbreviation *n.* shortened form of word(s).

ABC *n.* **1** alphabet. **2** alphabetical guide. **3** rudiments (of a subject).

abdicate *v.* renounce the throne. □ **abdication** *n.*

abdomen *n.* part of the body containing the digestive organs. □ **abdominal** *adj.*

abduct *v.* kidnap. □ **abduction** *n.*, **abductor** *n.*

■ carry off, kidnap, seize.

aberrant *adj.* showing aberration. □ **aberrance** *n.*

aberration *n.* **1** deviation from what is normal. **2** distortion.

abet *v.* (**abetted**) encourage or assist in wrongdoing. □ **abettor** *n.*

abeyance *n.* **in abeyance** not being used for a time.

abhor *v.* (**abhorred**) detest. □ **abhorrence** *n.*

■ abominate, despise, detest, execrate, hate, loathe, shudder at.

abhorrent *adj.* detestable.

■ abominable, detestable, execrable, hateful, loathsome, obnoxious, repellent, revolting.

abide *v.* tolerate. □ **abide by** **1** keep (a promise). **2** accept (consequences etc.).

■ accept, bear, endure, put up with, stand, stomach, suffer, tolerate. □ **abide by** accept, adhere to, comply with, conform to, heed, honour, keep (to), obey, observe, pay attention to, stick to.

abiding *adj.* lasting, permanent.

ability *n.* **1** power to do something. **2** cleverness.

■ **1** capacity, means, power, resources, scope. **2** aptitude, bent, brains, capability, cleverness, competence, expertise, flair, genius, gift, intelligence, knack, know-how, knowledge, proficiency, prowess, skill, strength, talent, training, wit.

abject *adj.* **1** wretched. **2** lacking pride. □ **abjectly** *adv.*

ablaze *adj.* blazing.

able *adj.* **1** having power or capacity. **2** talented. □ **ably** *adv.*

■ **1** allowed, at liberty, authorized, available, capable, eligible, equipped, fit, free, permitted, prepared, willing. **2** accomplished, adept, capable, clever, competent, effective, efficient, experienced, expert, handy, intelligent, masterly, practised, proficient, skilful, skilled, talented.

ablutions *n.pl.* process of washing oneself.

abnegate *v.* renounce.

abnormal *adj.* not normal. □ **abnormally** *adv.*, **abnormality** *n.*

■ aberrant, anomalous, atypical, bizarre, curious, deviant, eccentric, exceptional, extraordinary, funny, idiosyncratic, irregular, *colloq.* kinky, odd, peculiar, perverted, queer, singular, strange, unnatural, unusual, weird.

aboard *adv.* & *prep.* on board.

abode *n.* home, dwelling place.

abolish *v.* put an end to. □ **abolition** *n.*

■ annul, destroy, dispense with, do away with, eliminate, end, eradicate, get rid of, liquidate, nullify, overturn, put an end to, quash, suppress, terminate.

abominable *adj.* very bad or unpleasant. □ **abominably** *adv.*

■ abhorrent, appalling, atrocious, awful, *colloq.* beastly, despicable, detestable, disgusting, dreadful, execrable, foul, hateful, heinous, horrible, loathsome, nasty, obnoxious, odious, repellent, repugnant, repulsive, revolting, terrible, vile.

abominate *v.* detest. □ **abomination** *n.*

aboriginal *adj.* existing in a country from its earliest times.

aborigine /ábbəríjini/ *n.* aboriginal inhabitant.

abort *v.* **1** (cause to) expel a foetus prematurely. **2** end prematurely and unsuccessfully.

abortion *n.* **1** premature expulsion of a foetus from the womb. **2** operation to cause this.

abortionist *n.* person who performs abortions.

abortive *adj.* **1** causing an abortion. **2** unsuccessful.

■ **2** fruitless, futile, ineffective, unsuccessful, vain.

abound *v.* be plentiful.

about *adv.* & *prep.* **1** near. **2** here and there. **3** in circulation. **4** approximately. **5** in connection with. **6** so as to face in the opposite direction. □ **about-face, -turn** *ns.* reversal of direction or policy. **be about to** be on the point of (doing).

above *adv.* & *prep.* **1** at or to a higher point (than). **2** beyond the level or understanding of. □ **above board** without deception.

abracadabra *n.* magic formula.

abrasion *n.* **1** rubbing or scraping away. **2** injury caused by this.

abrasive *adj.* **1** causing abrasion. **2** harsh. ● *n.* substance used for grinding or polishing surfaces.

abreast *adv.* side by side.

abridge *v.* shorten by using fewer words. □ **abridgement** *n.*

■ abbreviate, condense, cut, edit, précis, reduce, shorten, summarize.

abroad *adv.* away from one's home country.

abrogate *v.* repeal, cancel. □ **abrogation** *n.*

abrupt *adj.* **1** sudden. **2** curt. **3** steep. □ **abruptly** *adv.*, **abruptness** *n.*

■ **1** hasty, hurried, precipitate, quick, rapid, sudden, unexpected, unforeseen. **2** blunt, brusque, curt, discourteous, impolite, rude, short, terse, ungracious. **3** precipitous, sharp, sheer, steep.

abscess *n.* collection of pus formed in the body.

abscond *v.* go away secretly or illegally.

abseil *v.* descend by using a rope fixed at a higher point. ● *n.* such a descent.

absence *n.* **1** being absent. **2** lack.

absent[1] /ábs'nt/ *adj.* **1** not present. **2** lacking, non-existent. □ **absent-minded** *adj.* forgetful, inattentive.

■ **1** away, elsewhere, gone, missing, off, out, playing truant. **2** lacking, missing, non-existent. □ **absent-minded** careless, forgetful, inattentive, preoccupied, scatterbrained, thoughtless.

absent[2] /əbsént/ *v.* **absent oneself** stay away.

absentee *n.* person who is absent from work etc. □ **absenteeism** *n.*

absinthe *n.* a green liqueur.

absolute *adj.* **1** complete. **2** despotic. □ **absolutely** *adv.*

■ **1** categorical, complete, downright, out and out, perfect, pure, sheer, thorough, total, unadulterated, unconditional, unmitigated, unqualified, unreserved, utter. **2** autocratic, despotic, dictatorial, totalitarian, tyrannical.

absolution *n.* priest's formal declaration of forgiveness of sins.

absolutism *n.* principle of government with unrestricted powers. □ **absolutist** *n.*

absolve *v.* clear of blame or guilt.

■ acquit, clear, excuse, exonerate, forgive, pardon.

absorb *v.* **1** take in, combine into itself or oneself. **2** reduce the intensity of. **3** occupy the attention or interest of. □ **absorption** *n.*, **absorptive** *adj.*

■ **1** assimilate, consume, digest, imbibe, incorporate, soak up, take in. **2** cushion, deaden, lessen, soften. **3** captivate, engage, engross, enthral, fascinate, interest, occupy, preoccupy.

absorbent *adj.* able to absorb moisture etc.

abstain *v.* **1** refrain, esp. from drinking alcohol. **2** decide not to use one's vote. □ **abstainer** *n.*, **abstention** *n.*

abstemious *adj.* not self-indulgent. □ **abstemiously** *adv.*

■ ascetic, frugal, moderate, restrained, self-denying, temperate.

abstinence *n.* abstaining esp. from food or alcohol.

abstract *adj.* /ábstrakt/ **1** having no material existence, theoretical. **2** (of art) not representing things pictorially. ● *n.* /ábstrakt/ **1** summary. **2** abstract quality or idea. **3** piece of abstract art. ● *v.* /əbstrákt/ **1** take out, remove. **2** make a summary of. □ **abstraction** *n.*

■ *adj.* **1** academic, conceptual, intangible, metaphysical, notional, philosophical, theoretical. ● *n.* **1** outline, précis, résumé, summary, synopsis.

abstruse *adj.* hard to understand, profound.

■ complex, cryptic, deep, difficult, enigmatic, esoteric, mysterious, obscure, profound, recondite.

absurd *adj.* not in accordance with common sense, ridiculous. □ **absurdly** *adv.*, **absurdity** *n.*

■ crazy, daft, farcical, foolish, idiotic, illogical, incongruous, irrational, laughable, ludicrous, nonsensical, outlandish, paradoxical, preposterous, ridiculous, senseless, silly, stupid, unreasonable.

abundant *adj.* plentiful. □ **abundant in** having plenty of. **abundantly** *adv.*, **abundance** *n.*

■ ample, bountiful, copious, generous, lavish, liberal, luxuriant, plentiful, profuse. □ **abundant in** full of, overflowing with, rich in, teeming with.

abuse *v.* /əbyŏŏz/ **1** ill-treat. **2** make bad use of. **3** attack with abusive language. ● *n.* /əbyŏŏss/ **1** ill-treatment. **2** incorrect use. **3** abusive language.

■ *v.* **1** damage, harm, hurt, ill-treat, injure, maltreat, molest, wrong. **2** exploit, misapply, misuse, pervert, take advantage of. **3** berate, be rude to, curse (at), defame, insult, libel, malign, rail at, revile, slander, swear at, vilify, vituperate. ● *n.* **1** ill-treatment, maltreatment, molestation. **2** misapplication, misuse, perversion. **3** curses, insults, invective, vilification, vituperation.

abusive *adj.* using harsh words or insults. □ **abusively** *adv.*

■ censorious, critical, defamatory, derogatory, disparaging, insulting, libellous, offensive, opprobrious, pejorative, rude, scurrilous, slanderous, vituperative.

abut *v.* (**abutted**) **1** border (upon). **2** touch or lean (against).

abysmal *adj.* very bad.

abyss *n.* very deep chasm.

acacia /əkáyshə/ *n.* flowering tree or shrub.

academic *adj.* **1** of a college or university. **2** scholarly, intellectual. **3** of theoretical interest only. ● *n.* academic person. □ **academically** *adv.*

■ *adj.* **2** bookish, brainy, clever, erudite, highbrow, intellectual, learned, scholarly, studious. **3** abstract, hypothetical, speculative, theoretical. ● *n.* don, highbrow, intellectual, lecturer, professor, scholar, thinker.

academician *n.* member of an Academy.

academy *n.* **1** school, esp. for specialized training. **2** (**Academy**) society of scholars or artists.

acanthus *n.* plant with large thistle-like leaves.

accede /akseed/ *v.* agree (to).

accelerate *v.* **1** increase speed (of). **2** (cause to) happen earlier. □ **acceleration** *n.*

■ **1** go faster, hasten, pick up speed, quicken, speed up. **2** expedite, forward, hasten, precipitate, speed up, spur on, stimulate.

accelerator *n.* pedal on a vehicle for increasing speed.

accent *n.* /áks'nt/ **1** particular regional, national, or other way of pronouncing words. **2** emphasis on a word. **3** mark showing how a letter is pronounced. ● *v.* /aksént/ emphasize.

accentuate *v.* **1** emphasize. **2** make prominent. □ **accentuation** *n.*

accept *v.* **1** say yes (to). **2** tolerate. **3** take as true. **4** acknowledge. □ **acceptance** *n.*

■ **1** accede (to), acquiesce (in), agree (to), consent (to). **2** bear, come to terms with, put up with, resign oneself to, stomach, submit to, suffer, tolerate. **3** believe, credit, find credible, have faith in. **4** acknowledge, admit, concede, recognize.

acceptable *adj.* **1** worth accepting, pleasing. **2** adequate. □ **acceptably** *adv.*, **acceptability** *n.*

■ **1** agreeable, gratifying, pleasant, pleasing, suitable, welcome. **2** adequate, admissible, passable, satisfactory, tolerable.

access *n.* **1** way in. **2** right to enter. **3** right to visit.

accessible *adj.* able to be reached or obtained. □ **accessibly** *adv.*, **accessibility** *n.*

■ at hand, attainable, available, close, convenient, handy, to hand, within reach.

accessory *adj.* additional. ● *n.* **1** additional fitment. **2** person who helps in a crime.

■ *adj.* added, additional, extra, supplementary. ● *n.* **2** abettor, accomplice, assistant, associate, collaborator, confederate, conspirator, helper, partner.

accident *n.* **1** unexpected event, chance. **2** unfortunate event.

■ **1** coincidence, fluke; chance, fate, fortune, luck, serendipity. **2** catastrophe, disaster, misadventure, mishap, mischance; collision, crash, pile-up.

accidental *adj.* happening by accident. □ **accidentally** *adv.*

■ casual, chance, coincidental, fortuitous, inadvertent, lucky, random, unexpected, unforeseen, unintended, unintentional, unlooked-for, unlucky, unplanned, unpremeditated.

acclaim *v.* welcome or applaud enthusiastically. ● *n.* shout of welcome, applause. □ **acclamation** *n.*

■ *v.* applaud, extol, hail, honour, praise, salute, welcome.

acclimatize *v.* adapt to a new climate or conditions. □ **acclimatization** *n.*

accolade *n.* **1** bestowal of a knighthood. **2** praise.

accommodate *v.* **1** provide lodging or room for. **2** adapt.

■ **1** billet, harbour, house, lodge, put up, quarter, shelter. **2** adapt, adjust, fit, harmonize, modify, reconcile.

accommodating *adj.* obliging.

■ adaptable, amenable, complaisant, compliant, cooperative, easygoing, flexible, helpful, kind, obliging.

accommodation *n.* place to live.

■ domicile, digs, home, house, housing, lodgings, quarters, residence, rooms, shelter.

accompany *v.* **1** go with. **2** be done or found with. **3** play an instrumental part supporting (singer(s) or an instrument). □ **accompaniment** *n.*, **accompanist** *n.*

■ **1** chaperon, conduct, escort, go with, guide, partner, usher.

accomplice *n.* partner, esp. in a crime.

■ abettor, accessory, associate, collaborator, confederate, conspirator, helper, partner.

accomplish *v.* succeed in doing or achieving.

■ achieve, attain, bring off, carry out, complete, consummate, do successfully, effect, finish, fulfil, realize, succeed in.

accomplished *adj.* **1** skilled. **2** having many accomplishments.

■ able, adept, expert, gifted, practised, proficient, skilful, skilled, talented.

accomplishment *n.* **1** useful ability. **2** thing achieved.

■ **1** ability, gift, skill, talent. **2** achievement, attainment, deed, exploit, feat, *tour de force*.

accord *v.* be consistent. ● *n.* consent, agreement. □ **of one's own accord** without being asked.

accordance *n.* conformity.

according *adv.* **according to 1** as stated by. **2** in proportion to. □ **accordingly** *adv.*

accordion *n.* musical instrument with bellows and a keyboard.

accost *v.* approach and speak to.

account *n.* **1** description, report. **2** statement of money paid or owed. **3** importance. **4** credit arrangement with a bank or firm. ● *v.* **account for 1** explain. **2** kill, overcome. □ **on account of** because of.

■ *n.* **1** commentary, description, explanation, history, log, narration, narrative, record, report, statement, story, tale. **2** bill, *US* check, invoice, statement. **3** consequence, importance, significance, use, value, worth.

accountable *adj.* obliged to account for one's actions. □ **accountability** *n.*

accountant *n.* person who keeps or inspects business accounts. □ **accountancy** *n.*

accoutrements /əkóōtrəmənts/ *n.pl.* equipment, trappings.

accretion *n.* **1** growth. **2** matter added.

accrue *v.* accumulate.

accumulate *v.* **1** acquire more and more of. **2** increase in amount. □ **accumulation** *n.*

■ **1** amass, collect, gather, heap up, hoard, pile up, stockpile, store up. **2** accrue, build up, collect, gather, grow, increase, multiply, pile up. □ **accumulation** collection, heap, hoard, mass, stack, stockpile, store.

accumulator *n.* **1** rechargeable electric battery. **2** bet on a series of events with winnings restaked.

accurate *adj.* **1** free from error. **2** careful and precise. □ **accurately** *adv.*, **accuracy** *n.*

■ **1** correct, exact, faultless, flawless, perfect, right, true, unerring. **2** careful, meticulous, precise, scrupulous.

accuse *v.* lay the blame for a crime or fault on. □ **accusation** *n.*, **accuser** *n.*

■ blame, censure, charge, denounce, impeach, indict. □ **accusation** allegation, charge, denunciation, impeachment, indictment.

accustom *v.* make used (to).

ace *n.* **1** playing card with one spot. **2** expert. **3** unreturnable service in tennis.

acerbity *n.* sharpness of manner.

acetate *n.* synthetic textile fibre.

acetic acid clear liquid acid in vinegar.

acetone *n.* colourless liquid used as a solvent.

acetylene *n.* colourless gas burning with a bright flame.

ache *n.* dull continuous pain. ● *v.* suffer an ache. □ **achy** *adj.*

■ *n.* discomfort, pain, pang, soreness, throbbing, twinge. ● *v.* hurt, smart, sting, throb.

achieve *v.* **1** accomplish. **2** reach or gain by effort. □ **achievement** *n.*, **achiever** *n.*

■ **1** accomplish, bring off, carry out, complete, conclude, do successfully, effect, engineer, execute, finish, fulfil, realize, succeed in. **2** acquire, attain, earn, gain, get, obtain, reach, win.

Achilles heel vulnerable point. □ **Achilles tendon** tendon attaching the calf muscles to the heel.

acid *adj.* sour. ● *n.* any of a class of substances that contain hydrogen and neutralize alkalis. □ **acidly** *adv.*, **acidity** *n.*

■ *adj.* sharp, sour, tangy, tart, vinegary.

acknowledge *v.* **1** admit the truth of. **2** confirm receipt of. □ **acknowledgement** *n.*

■ **1** accept, admit, allow, concede, confess, grant, own, recognize.

acme /ákmi/ *n.* highest point.

acne /ákni/ *n.* eruption of pimples.

acolyte *n.* person assisting a priest in a church service.

acorn *n.* oval nut of the oak tree.

acoustic *adj.* of sound. ● *n.pl.* qualities of a room that affect the way sound carries in it.

acquaint *v.* **acquaint with** make aware of. □ **be acquainted with** know slightly.

■ □ **acquaint with** apprise of, inform of, familiarize with, make aware of, notify of, tell of.

acquaintance *n.* **1** slight knowledge. **2** person one knows slightly.

acquiesce *v.* assent. □ **acquiescent** *adj.*, **acquiescence** *n.*

acquire *v.* get possession of.

■ buy, come by, earn, gain, get, obtain, pick up, procure, purchase, secure.

acquisition *n.* **1** acquiring. **2** thing acquired.

acquisitive *adj.* eager to acquire things. □ **acquisitiveness** *n.*

acquit *v.* (**acquitted**) declare to be not guilty. □ **acquittal** *n.*

■ absolve, clear, declare innocent, discharge, excuse, exonerate, free, release, reprieve, set free.

acre *n.* measure of land, 4,840 sq. yds (0.405 hectares).

acreage /áykərij/ *n.* number of acres.

acrid *adj.* bitterly pungent.

acrimonious *adj.* angry and bitter. □ **acrimony** *n.*

■ biting, bitter, caustic, cutting, harsh, sarcastic, scathing, sharp, spiteful, tart, testy, venomous, virulent, waspish.

acrobat *n.* performer of acrobatics.

acrobatic *adj.* involving spectacular gymnastic feats. ● *n.pl.* acrobatic feats.

acronym *n.* word formed from the initial letters of others.

acropolis *n.* upper fortified part of an ancient Greek city.

across *prep.* & *adv.* **1** from side to side (of). **2** on the other side (of).

acrostic *n.* poem in which the first and/or last letters of lines form word(s).

acrylic *adj.* & *n.* (synthetic fibre) made from an organic substance.

act *n.* **1** thing done. **2** law made by parliament. **3** item in a circus or variety show. **4** section of a play. ● *v.* **1** perform actions, behave. **2** have an effect. **3** play the part of. **4** pretend (to be).

■ *n.* **1** accomplishment, achievement, action, deed, exploit, feat, operation, step. **2** bill, decree, edict, law, measure, regulation, statute. **3** performance, routine, sketch, turn. ● *v.* **1** behave, carry on, conduct oneself. **2** be effective, function, operate, take effect, work. **3** appear (as), enact, perform, play, portray, represent. **4** fake, feign, pose (as), pretend (to be), sham, simulate.

action *n.* **1** process of doing something. **2** thing done. **3** battle. **4** lawsuit.

■ **1** activity, movement, performance, practice. **2** act, deed, exploit, feat, step, undertaking. **3** battle, combat, conflict, encounter, engagement, fight, fray, skirmish.

actionable *adj.* giving cause for a lawsuit.

activate *v.* make active. □ **activation** *n.*, **activator** *n.*

active *adj.* **1** doing things, energetic. **2** working. □ **actively** *adv.*

■ **1** animated, brisk, bustling, busy, dynamic, energetic, hyperactive, lively, *colloq.* on the go, tireless, vigorous, vivacious. **2** functioning, in operation, operative, working.

activist *n.* person adopting a policy of vigorous action in politics etc. □ **activism** *n.*

activity *n.* action, occupation.

actor, actress *ns.* performer in play(s) or film(s).

actual *adj.* existing, current.

■ authentic, bona fide, factual, genuine, material, real, tangible, true, verifiable; current, existent, existing.

actuality *n.* reality.

actually *adv.* in fact, really.

actuary *n.* insurance expert who calculates risks and premiums. □ **actuarial** *adj.*

actuate *v.* **1** activate. **2** be a motive for. □ **actuation** *n.*

acumen *n.* shrewdness.

acupuncture *n.* pricking the body with needles to relieve pain. □ **acupuncturist** *n.*

acute *adj.* **1** sharp, intense. **2** (of illness) severe for a time. **3** quick at understanding. □ **acute accent** the accent (´). **acutely** *adv.*, **acuteness** *n.*

■ **1** excruciating, exquisite, intense, keen, penetrating, piercing, severe, sharp, sudden, violent. **2** critical, dangerous, grave, serious, severe. **3** alert, astute, canny, clever, discerning, incisive, intelligent, penetrating, perceptive, sharp, shrewd.

ad *n.* (*colloq.*) advertisement.

adamant *adj.* not yielding to requests.

Adam's apple prominent cartilage at the front of the neck.

adapt *v.* **1** alter or modify. **2** make or become suitable for new use or conditions. □ **adaptable** *adj.*, **adaptation** *n.*, **adaptor** *n.*

■ **1** alter, amend, change, correct, edit, modify, revise, rewrite. **2** adjust, attune, fit; become acclimatized *or* accustomed *or* habituated *or* inured.

add *v.* **1** join as an increase or supplement. **2** put together to get a total. **3** say further. □ **add up** find the total of.

■ **1** affix, annex, append, attach, combine with, join (on to), tack on (to), unite with. **2** add up, count up, reckon, total, *colloq.* tot up.

addendum *n.* (*pl.* **-da**) section added to a book.

adder *n.* small poisonous snake.

addict *n.* one who is addicted, esp. to drug(s).

addicted *adj.* doing or using something as a habit or compulsively. □ **addiction** *n.*, **addictive** *adj.*

addition *n.* **1** act or process of adding. **2** thing added.

■ **1** adding-up, calculation, computation, reckoning, *colloq.* totting-up. **2** addendum, appendage, appendix, postscript, supplement; annexe, extension, wing.

additional *adj.* added, extra. □ **additionally** *adv.*

■ added, extra, further, increased, more, new, other, spare, supplementary.

additive *n.* substance added.

addle *v.* **1** (of an egg) become rotten. **2** muddle, confuse.

address *n.* **1** details of where a person lives or where mail should be delivered. **2** speech. ● *v.* **1** write the address on. **2** speak to. **3** apply (oneself) to a task.

■ *n.* **2** discourse, harangue, lecture, oration, sermon, speech, talk. ● *v.* **2** accost, greet, hail, salute; give a speech to, lecture, speak to, talk to. **3** concentrate on, focus on.

addressee *n.* person to whom a letter etc. is addressed.

adduce *v.* cite as proof.

adenoids *n.pl.* enlarged tissue at the back of the throat. □ **adenoidal** *adj.*

adept *adj.* & *n.* skilful (person).

■ *adj.* able, accomplished, adroit, clever, competent, deft, dexterous, expert, gifted, proficient, skilful, talented.

adequate *adj.* **1** satisfactory but not excellent. **2** enough. □ **adequately** *adv.*, **adequacy** *n.*

■ **1** acceptable, all right, average, competent, fair, middling, *colloq.* OK, passable, satisfactory, tolerable. **2** enough, sufficient.

adhere *v.* **1** stick. **2** continue to give one's support. □ **adherence** *n.*, **adherent** *adj.* & *n.*

■ □ **adherent** *n.* admirer, devotee, disciple, fan, follower, supporter.

adhesion *n.* process or fact of sticking to something.

adhesive *adj.* sticking, sticky.

ad hoc for a specific purpose.

adieu /ədyoó/ *int.* & *n.* goodbye.

ad infinitum for ever.

adjacent *adj.* **1** lying near. **2** next to.

adjective *n.* descriptive word. □ **adjectival** *adj.*

adjourn *v.* move (a meeting etc.) to another place or time. □ **adjournment** *n.*

adjudge *v.* decide judicially.

adjudicate *v.* **1** act as judge (of). **2** adjudge. □ **adjudication** *n.*, **adjudicator** *n.*

adjunct *n.* thing that is subordinate to another.

adjure *v.* beg or command.

adjust *v.* **1** alter slightly so as to be correct or in the proper position. **2** adapt oneself to new conditions. □ **adjustable** *adj.*, **adjustment** *n.*

■ **1** adapt, alter, amend, change, correct, modify, put right, rearrange, rectify, regulate, reorganize, reset, tailor, tune. **2** acclimatize, accommodate oneself, accustom oneself, adapt, reconcile oneself.

adjutant *n.* army officer assisting in administrative work.

ad lib 1 as one pleases. **2** improvise(d).

administer *v.* **1** manage (business affairs). **2** give or hand out.

■ **1** administrate, conduct, control, direct, manage, organize, oversee, preside over, regulate, run, supervise. **2** dispense, distribute, give out, hand out, mete out.

administrate *v.* act as manager (of). □ **administrator** *n.*

administration *n.* **1** administering, esp. of public or business affairs. **2** government in power. □ **administrative** *adj.*

admirable *adj.* **1** worthy of admiration. **2** excellent. □ **admirably** *adv.*

■ **1** commendable, creditable, estimable, laudable, meritorious, praiseworthy, worthy. **2** excellent, great, first-class, first-rate, marvellous, splendid, superb, wonderful.

admiral *n.* naval officer of the highest rank.

admire *v.* regard with approval, think highly of. □ **admiration** *n.*

■ approve of, esteem, honour, idolize, like, look up to, love, respect, revere, value, venerate.

admissible *adj.* able to be admitted or allowed. □ **admissibility** *n.*

admission *n.* **1** process or right of entering. **2** statement admitting something.

■ **1** access, admittance, entrance, entry. **2** acknowledgement, concession, confession, declaration, disclosure, revelation.

admit *v.* (**admitted**) **1** allow to enter. **2** accept as valid. **3** state reluctantly.

■ **2** accept, acknowledge, allow, concede, grant, recognize. **3** confess, disclose, own up (to), reveal.

admittance *n.* admitting, esp. to a private place.

admittedly *adv.* as an acknowledged fact.

admixture *n.* **1** thing added as an ingredient. **2** adding of this.

admonish *v.* **1** exhort. **2** reprove. □ **admonition** *n.*

ad nauseam to a sickening extent.

ado *n.* fuss, trouble.

adobe /ədóbi/ *n.* sun-dried brick.

adolescent *adj.* & *n.* (person) between childhood and maturity. □ **adolescence** *n.*

adopt *v.* **1** take (esp. a child) as one's own. **2** choose. **3** take and use. **4** approve (a report etc.). □ **adoption** *n.*

■ **2** choose, embrace, select. **3** appropriate, borrow, take over. **4** accept, approve, back, endorse, ratify, sanction, support.

adorable *adj.* very lovable.

adore *v.* love deeply. □ **adoration** *n.*

■ be in love with, dote on, idolize, love, revere, worship.

adorn *v.* **1** decorate. **2** be an ornament to. □ **adornment** *n.*

■ **1** beautify, decorate, embellish, garnish, ornament, trim.

adrenal /ədreén'l/ *adj.* close to the kidneys.

adrenalin /ədrénnəlin/ *n.* stimulant hormone produced by the adrenal glands.

adrift *adj.* & *adv.* **1** drifting. **2** loose.

adroit *adj.* skilful.
■ able, accomplished, adept, clever, deft, dexterous, expert, proficient, resourceful, skilful.

adsorb *v.* attract and hold (a gas or liquid) to a surface.

adulation *n.* excessive admiration or praise.

adult *adj.* & *n.* fully grown (person etc.). □ **adulthood** *n.*
■ *adj.* fully grown, grown-up, mature, of age.

adulterate *v.* make impure by adding substance(s). □ **adulteration** *n.*
■ contaminate, corrupt, debase, doctor, pollute, taint.

adultery *n.* sexual infidelity to one's wife or husband. □ **adulterer** *n.*, **adulterous** *adj.*

advance *v.* **1** move forward, make progress. **2** put forward. **3** promote. **4** lend (money). ● *n.* **1** forward movement, progress. **2** loan. **3** increase in price. □ **advancement** *n.*
■ *v.* **1** forge ahead, make headway, make progress, proceed, progress. **2** present, propose, put forward, submit, suggest. **3** aid, assist, boost, forward, further, help, improve, promote. **4** lend, loan. ● *n.* **1** breakthrough, development, headway, improvement, progress, progression.

advanced *adj.* **1** well ahead. **2** not elementary.

advantage *n.* **1** beneficial feature, favourable circumstance. **2** benefit. □ **take advantage of 1** make use of. **2** exploit unfairly.
■ **1** asset, bonus, convenience, feature, good point, plus; superiority, upper hand. **2** benefit, gain, profit, service, use, usefulness. □ **take advantage of 1** benefit from, build on, make (good) use of, profit from, use. **2** abuse, exploit, impose on, manipulate, trick.

advantageous *adj.* profitable, beneficial.
■ beneficial, favourable, helpful, profitable, useful, valuable, worthwhile.

advent *n.* **1** arrival. **2** (**Advent**) season before Christmas.

adventure *n.* exciting or dangerous experience. □ **adventurer** *n.*, **adventurous** *adj.*
■ deed, escapade, event, experience, exploit, happening, incident; risk, undertaking, venture. □ **adventurous** audacious, bold, brave, courageous, daredevil, daring, enterprising, intrepid, venturesome.

adverb *n.* word qualifying a verb, adjective, or other adverb. □ **adverbial** *adj.*

adversary *n.* opponent, enemy. □ **adversarial** *adj.*
■ antagonist, attacker, enemy, foe, opponent.

adverse *adj.* **1** unfavourable. **2** harmful. □ **adversely** *adv.*, **adversity** *n.*
■ **1** disadvantageous, hostile, inimical, inauspicious, unfavourable, unpropitious. **2** damaging, deleterious, detrimental, harmful, hurtful, injurious.

advert *n.* (*colloq.*) advertisement.

advertise *v.* make publicly known, esp. to encourage sales.
■ announce, broadcast, make known, make public, proclaim, promulgate; *colloq.* plug, promote, publicize.

advertisement *n.* **1** advertising. **2** public notice about something.

advice *n.* opinion given about what should be done.
■ counsel, guidance, opinion, recommendation, suggestion, tip, view; admonition, warning.

advisable *adj.* worth recommending as a course of action. □ **advisability** *n.*
■ expedient, judicious, prudent, recommendable, sensible, wise.

advise *v.* **1** give advice to, recommend. **2** inform. □ **adviser** *n.*
■ **1** counsel, guide; advocate, recommend, suggest, urge; caution, warn. **2** apprise, inform, make known to, notify, tell.

advisory *adj.* giving advice.

advocacy *n.* speaking in support.

advocate *n.* /ádvəkət/ person who speaks in court on behalf of another. ● *v.* /ádvəkayt/ recommend.
■ *v.* advise, counsel, favour, recommend, support, urge.

aegis /eéjiss/ *n.* protection, sponsorship.

aeon /eé-on/ *n.* immense time.

aerate *v.* **1** expose to the action of air. **2** add carbon dioxide to.

aerial *adj.* **1** of or like air. **2** existing or moving in the air. **3** by or from aircraft. ● *n.* wire for transmitting or receiving radio waves. □ **aerially** *adv.*

aerobatics *n.pl.* spectacular feats by aircraft in flight.

aerobics *n.pl.* vigorous exercises designed to increase oxygen intake. □ **aerobic** *adj.*

aerodynamics *n.* science dealing with forces acting on solid objects moving through air. □ **aerodynamic** *adj.*

aerofoil *n.* aircraft wing, fin, or tailplane giving lift in flight.

aeronautics *n.* study of the flight of aircraft. □ **aeronautical** *adj.*

aeroplane *n.* power-driven aircraft with wings.

aerosol *n.* container holding a substance for release as a fine spray.

aerospace *n.* earth's atmosphere and space beyond this.

aesthete /éess-theet/ *n.* person who appreciates beauty.

aesthetic /eess-théttik/ *adj.* **1** of or showing appreciation of beauty. **2** artistic, tasteful. □ **aesthetically** *adv.*

aetiology /éetiólləji/ *n.* study of causes, esp. of disease.

affable *adj.* polite and friendly. □ **affably** *adv.*, **affability** *n.*

affair *n.* **1** thing to be done. **2** (*colloq.*) thing or event. **3** temporary sexual relationship.

■ **1** business, concern, matter. **2** episode, event, happening, incident, occurrence. **3** intrigue, liaison, relationship, romance.

affect *v.* **1** pretend to have or feel. **2** have an effect on. **3** touch the feelings of.

■ **1** counterfeit, fake, feign, imitate, pretend, sham, simulate. **2** alter, change, have an effect on, have an impact on, influence, transform. **3** move, stir, touch, trouble, upset.

affectation *n.* pretence, esp. in behaviour.

affected *adj.* full of affectation.

affection *n.* love, liking.

■ fondness, friendship, goodwill, liking, love, tenderness, warmth.

affectionate *adj.* loving. □ **affectionately** *adv.*

■ caring, devoted, doting, fond, kind, loving, tender.

affidavit *n.* written statement sworn on oath to be true.

affiliate *v.* connect as a subordinate member or branch. □ **affiliation** *n.*

affinity *n.* **1** close resemblance. **2** attraction, natural liking.

■ **1** closeness, correspondence, likeness, resemblance, similarity, similitude. **2** attraction, fondness, like-mindedness, liking, rapport, sympathy.

affirm *v.* **1** state as a fact. **2** declare formally and solemnly. □ **affirmation** *n.*

■ assert, avow, declare, maintain, proclaim, state, swear, testify.

affirmative *adj.* & *n.* saying 'yes'.

affix *v.* /əfíks/ **1** attach. **2** add (a signature etc.). ● *n.* /áffiks/ **1** thing affixed. **2** prefix, suffix.

afflict *v.* distress physically or mentally.

■ beset, burden, distress, oppress, rack, torment, trouble.

affliction *n.* **1** distress, suffering. **2** cause of this.

■ **1** distress, grief, hardship, illness, misery, misfortune, pain, sorrow, suffering, torment, torture, tribulation.

affluent *adj.* rich. □ **affluence** *n.*

■ □ moneyed, prosperous, rich, wealthy, *colloq.* well-heeled, well-off, well-to-do.

afford *v.* **1** have enough money or time for. **2** provide.

■ **2** furnish, give, provide, supply, yield.

afforest *v.* **1** convert into forest. **2** plant with trees. □ **afforestation** *n.*

affray *n.* public fight or riot.

affront *v.* & *n.* insult.

afloat *adv.* & *adj.* **1** floating. **2** on the sea.

afoot *adv.* & *adj.* going on.

aforesaid *adj.* mentioned previously.

afraid *adj.* **1** frightened. **2** regretful.

■ **1** alarmed, anxious, apprehensive, *colloq.* chicken, cowardly, faint-hearted, fearful, frightened, intimidated, *colloq.* jittery, nervous, panicky, panic-stricken, scared, terrified, timid, timorous, trembling, *colloq.* yellow. **2** apologetic, regretful, sorry.

afresh *adv.* anew, with a fresh start.

Afrikaans *n.* language of S. Africa, developed from Dutch.

Afrikaner *n.* Afrikaans-speaking white person in S. Africa.

aft *adv.* at or towards the rear of a ship or aircraft.

after *prep., adv., & adj.* **1** behind. **2** later (than). **3** in pursuit of. **4** concerning. **5** according to. ● *conj.* at a time later than. □ **after-effect** *n.* effect persisting after its cause has gone.

afterbirth *n.* placenta discharged from the womb after childbirth.

aftermath *n.* after-effects.

afternoon *n.* time between midday and evening.

afterthought *n.* thing thought of or added later.

afterwards *adv.* at a later time.

again *adv.* **1** another time, once more. **2** besides.

against *prep.* **1** in opposition or contrast to. **2** in preparation or return for. **3** into collision or contact with.

age *n.* **1** length of life or existence. **2** later part of life. **3** historical period. **4** (*colloq.*, usu. *pl.*) very long time. ● *v.* (**ageing**) **1** grow old, show signs of age. **2** cause to do this. □ **of age** old enough, adult.

■ *n.* **2** maturity, seniority; old age, senescence. **3** epoch, era, period, time(s). ● *v.* **1** get on, grow older; mature, mellow, ripen.

aged *adj.* **1** /ayjd/ of the age of. **2** /áyjid/ old.

ageism *n.* prejudice on grounds of age.

ageless *adj.* **1** not growing old. **2** not seeming old.

agency *n.* **1** business or office of an agent. **2** means of action by which something is done.

agenda *n.* list of things to be dealt with, esp. at a meeting.

agent *n.* **1** person acting for another. **2** person or thing producing an effect.

■ **1** broker, delegate, envoy, executor, go-between, intermediary, mediator, middleman, negotiator, proxy, representative, surrogate.

agent provocateur /áazhoɴ prəvókkətőr/ person employed to tempt suspected offenders into overt action.

aggrandize *v.* make seem greater. □ **aggrandizement** *n.*

aggravate *v.* **1** make worse. **2** (*colloq.*) annoy. □ **aggravation** *n.*

■ **1** exacerbate, increase, intensify, make worse, worsen. **2** annoy, exasperate, infuriate, irk, irritate, needle, nettle, vex.

aggregate *adj.* /ágrigət/ combined, total. ● *n.* /ágrigət/ **1** collected mass. **2** broken stone etc. used in making concrete. ● *v.* /ágrigayt/ **1** collect into an aggregate, unite. **2** (*colloq.*) amount to. □ **aggregation** *n.*

aggression *n.* **1** unprovoked attack. **2** hostile act(s) or behaviour.

■ **1** attack, assault, onslaught, invasion. **2** belligerence, hostility, pugnacity, truculence.

aggressive *adj.* **1** openly hostile. **2** forceful. □ **aggressively** *adv.*

■ **1** antagonistic, bellicose, belligerent, hostile, militant, offensive, pugnacious, quarrelsome, truculent, warlike. **2** assertive, brash, forceful, *colloq.* pushy.

aggressor *n.* one who begins hostilities.

aggrieved *adj.* having a grievance.

aghast *adj.* filled with horror.

agile *adj.* nimble, quick-moving. □ **agility** *n.*

■ acrobatic, active, lissom, lithe, nimble, quick-moving, sprightly, spry, supple, swift.

agitate *v.* **1** shake or move briskly. **2** cause anxiety to. **3** stir up public concern. □ **agitation** *n.*, **agitator** *n.*

■ **1** churn, shake, stir. **2** alarm, discomfit, disconcert, disturb, fluster, perturb, ruffle, trouble, unsettle, upset, worry. **3** campaign, fight.

agnostic *adj.* & *n.* (person) holding that nothing can be known about the existence of God. □ **agnosticism** *n.*

ago *adv.* in the past.

agog *adj.* eager, expectant.

agonize *v.* **1** cause agony to. **2** suffer agony, worry intensely.

agony *n.* extreme suffering.

■ anguish, distress, pain, suffering, torment, torture.

agoraphobia *n.* abnormal fear of open spaces or public places.

agrarian *adj.* of land or agriculture.

agree *v.* **1** hold or reach a similar opinion. **2** consent. **3** be consistent with. □ **agree with** suit the health or digestion of.

■ **1** be of one mind, be unanimous, concur; accept, admit, allow, concede, grant. **2** accede, acquiesce, assent, consent. **3** accord, correspond, square, tally.

agreeable *adj.* **1** pleasing. **2** willing to agree. □ **agreeably** *adv.*

■ **1** congenial, delightful, enjoyable, gratifying, nice, pleasant, pleasing, satisfying. **2** accommodating, acquiescent, amenable, compliant, willing.

agreement *n.* **1** act or state of agreeing. **2** arrangement agreed between people.

■ **1** accord, concord, conformity, consensus, unanimity, unity. **2** arrangement, bargain, compact, contract, covenant, deal, pact, settlement, treaty, truce, understanding.

agriculture *n.* large-scale cultivation of land. □ **agricultural** *adj.*, **agriculturalist** *n.*

agronomy *n.* soil management and crop production.

aground *adv.* & *adj.* (of a ship) on the bottom in shallow water.

ahead *adv.* further forward in position or time.

ahoy *int.* seaman's shout to call attention.

aid *v.* & *n.* help.

■ *n.* assistance, backing, cooperation, help, relief, succour, support. ● *v.* abet, assist, back, cooperate with, facilitate, help, promote, relieve, succour, support.

aide *n.* **1** aide-de-camp. **2** assistant.

aide-de-camp *n.* officer assisting a senior officer.

Aids *abbr.* (also **AIDS**) acquired immune deficiency syndrome, a condition developing after infection with the HIV virus, breaking down a person's natural defences against illness.

ail *v.* make or become ill.

aileron *n.* hinged flap on an aircraft wing.

ailment *n.* slight illness.

aim *v.* **1** direct towards a target. **2** intend, try. ● *n.* **1** purpose, intention. **2** action of aiming.

■ *v.* **1** direct, focus, level, point, train. **2** intend, mean, plan, propose, seek, strive, try, want. ● *n.* **1** ambition, aspiration, design, end, goal, intention, object, objective, plan, purpose, target.

aimless *adj.* without a purpose. □ **aimlessly** *adv.*, **aimlessness** *n.*

■ erratic, haphazard, pointless, purposeless, random.

air *n.* **1** mixture of gases surrounding the earth, atmosphere overhead. **2** light wind. **3** impression given. **4** manner. **5** melody. ● *v.* **1** expose to air, dry off. **2** express publicly. □ **air-bed** *n.* inflatable mattress. **airbrick** *n.* perforated brick for ventilation. **air-conditioned** *adj.* supplied with **air-conditioning**, system controlling the humidity and temperature of air. **air force** branch of the armed forces using aircraft. **air raid** attack by aircraft dropping bombs. **on the air** broadcasting by radio or television.

■ *n.* **1** atmosphere, ether. **2** breeze, draught, wind, *poetic* zephyr. **3** atmosphere, aura, feeling, impression, sense. **4** appearance, aspect, bearing, demeanour, look, manner. **5** melody, song, strain, tune. ● *v.* **1** aerate, dry off, freshen, ventilate. **2** broadcast, declare, express, give vent to, make known, make public, vent, voice.

airborne *adj.* **1** carried by air or aircraft. **2** (of aircraft) in flight.

aircraft *n.* machine capable of flight in air.

airfield *n.* area with runways etc. for aircraft.

airgun *n.* gun with a missile propelled by compressed air.

airlift *n.* large-scale transport of supplies by aircraft. ● *v.* transport thus.

airline *n.* company providing air transport service.

airliner *n.* passenger aircraft.

airlock *n.* **1** stoppage of the flow in a pipe, caused by an air-bubble. **2** airtight compartment giving access to a pressurized chamber.

airmail *n.* mail carried by aircraft. ● *v.* send by airmail.

airman *n.* member of an air force.

airport *n.* airfield with facilities for passengers and goods.

airship *n.* power-driven aircraft that is lighter than air.

airstrip *n.* strip of ground for take-off and landing of aircraft.

airtight *adj.* not allowing air to enter or escape.

airworthy *adj.* (of aircraft) fit to fly.

airy *adj.* (**-ier, -iest**) **1** well-ventilated. **2** light as air. **3** careless and light-hearted.

aisle /īl/ *n.* **1** side part of a church. **2** gangway between rows of seats or shelves.

ajar *adv.* & *adj.* slightly open.

akimbo *adv.* with hands on hips and elbows pointed outwards.

akin *adj.* related, similar.

alabaster *n.* translucent usu. white form of gypsum.

à la carte (of a meal) ordered as separate items from a menu.

alacrity *n.* eager readiness.

alarm *n.* **1** warning sound or signal. **2** fear caused by expectation of danger. ● *v.* cause alarm to.

■ *n.* **1** red light, siren, tocsin, warning. **2** anxiety, consternation, dismay, dread, fear, fright, nervousness, panic, terror, trepidation, uneasiness. ● *v.* dismay, frighten, panic, *colloq.* put the wind up, scare, shock, startle, terrify, worry.

alarmist *n.* person who raises unnecessary or excessive alarm.

alas *int.* exclamation of sorrow.

albatross *n.* long-winged seabird.

albino *n.* (*pl.* **-os**) person or animal with no natural colouring matter in the hair or skin.

album *n.* **1** blank book for holding photographs, stamps, etc. **2** set of recordings.

albumen *n.* white of egg.

albumin *n.* protein found in egg white, milk, blood, etc.

alchemy *n.* medieval form of chemistry, seeking to turn other metals into gold. □ **alchemist** *n.*

alcohol *n.* **1** colourless inflammable liquid, intoxicant in wine, beer, etc. **2** liquor containing this.

alcoholic *adj.* of alcohol. ● *n.* person addicted to drinking alcohol. □ **alcoholism** *n.*

alcove *n.* recess in a wall or room.

alder *n.* tree related to birch.

ale *n.* beer.

alert *adj.* watchful, observant. ● *v.* rouse to be alert.

■ *adj.* attentive, awake, careful, heedful, observant, on one's guard, on one's toes, on the lookout, vigilant, watchful, *colloq.* wide awake.

alfresco *adv.* & *adj.* in the open air.

alga *n.* (*pl.* **-gae**) water plant with no true stems or leaves.

algebra *n.* branch of mathematics using letters etc. to represent quantities. □ **algebraic** *adj.*

algorithm *n.* step by step procedure for calculation.

alias *n.* (*pl.* **-ases**) false name. ● *adv.* also called.

alibi *n.* **1** proof that one was elsewhere. **2** excuse.

alien *n.* **1** person who is not a citizen of the country where he or she lives. **2** a being from another world. ● *adj.* **1** unfamiliar. **2** foreign.

■ *n.* **1** foreigner, newcomer, outsider, stranger. ● *adj.* **1** peculiar, odd, outlandish, strange, unknown, unfamiliar. **2** exotic, foreign, imported, overseas.

alienate *v.* cause to become unfriendly. □ **alienation** *n.*

alight[1] *v.* **1** get down or off. **2** descend and settle.

alight[2] *adj.* on fire.

align *v.* **1** place or bring into line. **2** join as an ally. □ **alignment** *n.*

alike *adj.* like one another. ● *adv.* in the same way.

■ *adj.* akin, comparable, similar; identical, indistinguishable.

alimentary *adj.* of nourishment.

alimony *n.* money payable to a divorced or separated spouse.

alive *adj.* **1** living. **2** lively. □ **alive to** aware of.

■ **1** animate, breathing, extant, live, living. **2** active, animated, brisk, energetic, lively, spirited, sprightly, vibrant, vigorous, vivacious.

alkali *n.* (*pl.* **-is**) any of a class of substances that neutralize acids. □ **alkaline** *adj.*

alkaloid *n.* a kind of organic compound containing nitrogen.

all *adj.* whole amount, number, or extent of. ● *n.* all those concerned. □ **all but** almost. **all-clear** *n.* signal that danger is over. **all in 1** exhausted. **2** including everything. **all out** using maximum effort. **all right 1** satisfactory, satisfactorily. **2** in good condition.

allay *v.* lessen, alleviate.

allegation *n.* thing alleged.

allege *v.* declare without proof.

■ assert, avow, claim, contend, declare, maintain, profess, state.

allegedly *adv.* according to allegation.

allegiance *n.* support given to a government, sovereign, or cause.

■ constancy, devotion, faithfulness, fidelity, loyalty.

allegory *n.* story symbolizing an underlying meaning. □ **allegorical** *adj.*, **allegorically** *adv.*

allegro *adv.* & *n.* (passage to be played) briskly.

allergen *n.* substance causing an allergic reaction.

allergic *adj.* **1** having or caused by an allergy. **2** (*colloq.*) having a strong dislike.

allergy *n.* unfavourable reaction to certain foods, pollens, etc.

alleviate *v.* lessen (pain or distress). □ **alleviation** *n.*

■ allay, assuage, diminish, ease, lessen, lighten, mitigate, moderate, palliate, reduce, relieve, soften, soothe, subdue.

alley *n.* **1** narrow street. **2** long enclosure for tenpin bowling.

alliance *n.* association formed for mutual benefit.

■ association, bloc, cartel, coalition, confederation, consortium, federation, league, partnership, syndicate, union.

allied *adj.* **1** of an alliance. **2** similar.

alligator *n.* reptile of the crocodile family.

alliteration *n.* occurrence of the same sound at the start of words. □ **alliterative** *adj.*

allocate *v.* allot. □ **allocation** *n.*

allot *v.* (**allotted**) distribute officially, give as a share.

allotment *n.* **1** share allotted. **2** small area of land for cultivation.

allow *v.* **1** permit. **2** give a limited quantity or sum. **3** admit, agree. **4** add or deduct in estimating.

■ **1** agree to, authorize, consent to, give the go-ahead to, *colloq.* give the green light to, grant permission for, let, permit, sanction; put up with, stand (for), tolerate. **2** allocate, allot, assign, give, grant. **3** acknowledge, admit, agree, concede, confess, grant, own.

allowance *n.* **1** amount or sum allowed. **2** deduction, discount. □ **make allowances for** be lenient towards or because of.

■ **1** allocation, portion, quota, ration, share. **2** deduction, discount, rebate, reduction.

alloy *n.* mixture of metals. ● *v.* **1** mix (with another metal). **2** spoil or weaken (pleasure etc.).

allude *v.* **allude to** refer briefly or indirectly to.

■ mention, refer to, speak of, touch on.

allure *v.* entice, attract. ● *n.* attractiveness.

■ *n.* appeal, attractiveness, charisma, charm, magnetism, pull, seductiveness.

allusion *n.* statement alluding to something. □ **allusive** *adj.*

■ hint, innuendo, insinuation, mention, reference.

alluvium *n.* deposit left by a flood. □ **alluvial** *adj.*

ally *n.* /állī/ country or person in alliance with another. ● *v.* /əlī́/ join as an ally.

■ *n.* accomplice, associate, collaborator, colleague, comrade, confederate, helper, partner, supporter. ● *v.* combine, join, join forces, team up, unite.

almanac *n.* calendar with astronomical or other data.

almighty *adj.* **1** all-powerful. **2** very great.

almond *n.* **1** kernel of a fruit related to the peach. **2** tree bearing this.

almost *adv.* very little short of, as the nearest thing to.

■ about, all but, approximately, around, as good as, nearly, not quite, practically, virtually.

alms *n.* money given to the poor.

almshouse *n.* charitable institution for the poor.

aloe *n.* plant with bitter juice.

aloft *adv.* **1** high up. **2** upwards.

alone *adj.* **1** without company or help. **2** lonely. ● *adv.* only.

■ *adj.* **1** by oneself, single-handed, solo, unassisted, unaccompanied. **2** desolate, forlorn, forsaken, friendless, isolated, lonely, solitary.

along *adv.* **1** through part or all of a thing's length. **2** onward. **3** in company with others. ● *prep.* beside the length of.

alongside *adv.* close to the side of a ship or wharf etc.

aloof *adv.* apart. ● *adj.* showing no interest, unfriendly.

■ *adj.* chilly, cold, cool, distant, frosty, haughty, indifferent, reserved, reticent, standoffish, supercilious, unapproachable, undemonstrative, unforthcoming, unfriendly, unsociable, unsympathetic, withdrawn.

aloud *adv.* audibly.

alpaca *n.* **1** llama with long wool. **2** its wool. **3** cloth made from this.

alpha *n.* first letter of the Greek alphabet, = a.

alphabet *n.* letters used in writing a language. □ **alphabetical** *adj.*, **alphabetically** *adv.*

alphabetize *v.* put into alphabetical order.

alpine *adj.* of high mountains. ● *n.* plant growing on mountains or in rock gardens.

already *adv.* **1** before this time. **2** as early as this.

Alsatian *n.* German shepherd dog.

also *adv.* in addition, besides.

■ additionally, besides, furthermore, in addition, moreover, too.

altar *n.* table used in religious service.

alter *v.* make or become different. □ **alteration** *n.*

■ adapt, adjust, amend, change, convert, modify, reform, remodel, reorganize, reshape, revise, transform, vary. □ **alteration** adjustment, amendment, change, correction, modification, reorganization, revision, transformation.

altercation *n.* noisy dispute.

alternate *adj.* /awltérnət/ first one then the other successively. ● *v.* /áwltərnayt/ place or occur alternately. □ **alternately** *adv.*, **alternation** *n.*

alternative *adj.* **1** usable instead of another. **2** unconventional. ● *n.* any of two or more possibilities. □ **alternatively** *adv.*

although *conj.* though.

altimeter *n.* instrument in an aircraft showing altitude.

altitude *n.* height above sea level or above the horizon.

alto *n.* (*pl.* **-os**) highest adult male voice.

altogether *adv.* **1** entirely. **2** on the whole. **3** in total.

■ **1** absolutely, completely, entirely, fully, perfectly, quite, thoroughly, totally, utterly, wholly. **2** by and large, in general, on the whole.

altruism *n.* unselfishness. □ **altruist** *n.*, **altruistic** *adj.*

aluminium *n.* light silvery metal.

always *adv.* **1** repeatedly, often. **2** for ever. **3** at all times. **4** whatever the circumstances.

■ **1** constantly, continually, perpetually, repeatedly, usually; frequently, often. **2** eternally, evermore, for ever, unceasingly. **3** consistently, every time, invariably, without exception.

alyssum *n.* plant with small yellow or white flowers.

a.m. *abbr.* (Latin *ante meridiem*) before noon.

amalgam *n.* **1** alloy of mercury. **2** soft pliable mixture.

amalgamate *v.* mix, combine. □ **amalgamation** *n.*

■ blend, combine, compound, consolidate, fuse, integrate, join, merge, mix, synthesize, unite.

amaryllis *n.* lily-like plant.

amass *v.* heap up, collect.

amateur *n.* person who does something as a pastime not as a profession.

■ dabbler, dilettante, layperson, nonprofessional.

amateurish *adj.* lacking professional skill.

amatory *adj.* of or showing love.

amaze *v.* fill with surprise or wonder. □ **amazing** *adj.*, **amazement** *n.*

■ astonish, astound, awe, bowl over, dumbfound, *colloq.* flabbergast, shock, stagger, startle, stun, stupefy, surprise. □ **amazing** astonishing, astounding, breathtaking, extraordinary, fabulous, marvellous, miraculous, phenomenal, prodigious, remarkable, sensational, staggering, startling, *colloq.* stunning, stupendous, surprising, wonderful.

amazon *n.* fierce strong woman.

ambassador *n.* diplomat representing his or her country abroad.

amber *n.* **1** hardened brownish-yellow resin. **2** its colour.

ambergris *n.* waxy substance found in tropical seas, used in perfume manufacture.

ambidextrous *adj.* able to use either hand equally well.

ambience *n.* surroundings, atmosphere.

ambiguous *adj.* having two or more possible meanings. □ **ambiguously** *adv.*, **ambiguity** *n.*

■ confusing, equivocal, indefinite, obscure, puzzling, uncertain, unclear, vague.

ambit *n.* bounds, scope.

ambition *n.* **1** strong desire to achieve something. **2** object of this desire.

■ **1** determination, drive, eagerness, energy, enterprise, enthusiasm, initiative, motivation, zeal. **2** aim, aspiration, desire, dream, goal, hope, intention, object, objective, target, wish.

ambitious *adj.* full of ambition.

■ determined, eager, enterprising, enthusiastic, go-ahead, go-getting, keen, motivated, *colloq.* pushy, zealous.

ambivalent *adj.* with mixed feelings towards something. □ **ambivalence** *n.*

amble *v.* & *n.* walk at a leisurely pace.

ambrosia *n.* something delicious.

ambulance *n.* vehicle equipped to carry sick or injured people.

ambuscade *v.* & *n.* ambush.

ambush *n.* surprise attack from a concealed position. ● *v.* attack thus.

■ *n.* ambuscade, trap. ● *v.* ambuscade, ensnare, intercept, pounce on, surprise, trap, waylay.

ameliorate *v.* make or become better. □ **amelioration** *n.*

amenable *adj.* accommodating, responsive. □ **amenably** *adv.*

■ accommodating, adaptable, agreeable, biddable, complaisant, compliant, cooperative, docile, persuadable, responsive, tractable, willing.

amend *v.* **1** make minor alterations in. **2** improve. □ **make amends** compensate for something. **amendment** *n.*

■ **1** adjust, alter, correct, emend, modify, polish, refine. **2** change, improve, make better, mend, put right, reform.

amenity *n.* pleasant or useful feature of a place.

American *adj.* **1** of America. **2** of the USA. ● *n.* American person.

Americanism *n.* American word or phrase.

Americanize *v.* make American in character.

amethyst *n.* **1** purple or violet semi-precious stone. **2** its colour.

amiable *adj.* likeable, friendly. □ **amiably** *adv.*, **amiability** *n.*

■ affable, agreeable, amicable, charming, cordial, friendly, genial, good-natured, kind, kind-hearted, kindly, likeable, pleasant.

amicable *adj.* friendly. □ **amicably** *adv.*

amid *prep.* (also **amidst**) in the middle of, during.

amino acid organic acid found in proteins.

amiss *adj.* & *adv.* wrong(ly).

ammonia *n.* **1** strong-smelling gas. **2** solution of this in water.

ammonite *n.* fossil of a spiral shell.

ammunition *n.* bullets, shells, etc.

amnesia *n.* loss of memory. □ **amnesiac** *adj.* & *n.*

amnesty *n.* general pardon.

amniotic fluid fluid surrounding the foetus in the womb.

amoeba /əmeébə/ *n.* (*pl.* **-bae** or **-bas**) simple microscopic organism changing shape constantly.

amok *adv.* **run amok** be out of control and do much damage.

among *prep.* (also **amongst**) **1** surrounded by. **2** in the category of. **3** between.

amoral *adj.* not based on moral standards.

amorous *adj.* showing or feeling sexual love.

■ amatory, ardent, impassioned, loving, lustful, passionate.

amorphous *adj.* shapeless.

amount *n.* total of anything, quantity. ● *v.* **amount to** be equivalent to in number, size, significance, etc.

■ *n.* aggregate, bulk, extent, lot, mass, number, quantity, sum, total, volume. ● *v.* add up to, come to, make, total; be equal *or* equivalent to.

ampere *n.* unit of electric current.

ampersand *n.* the sign & (= and).

amphetamine *n.* stimulant drug.

amphibian *n.* amphibious animal or vehicle.

amphibious *adj.* able to live or operate on land and in water.

amphitheatre *n.* semicircular unroofed building with tiers of seats round a central arena.

ample *adj.* **1** plentiful. **2** quite enough. **3** large. □ **amply** *adv.*

■ **1** abundant, bountiful, copious, generous, lavish, liberal, plentiful, profuse, unstinting. **2** adequate, enough, sufficient. **3** capacious, commodious, large, spacious, substantial.

amplify *v.* **1** increase the strength or volume of. **2** add details to (a statement). □ **amplification** *n.*, **amplifier** *n.*

■ **1** augment, boost, heighten, increase, intensify, magnify, make louder. **2** add to, broaden, develop, elaborate on, embellish, embroider, enlarge on, expand on, expatiate on, lengthen, make fuller, supplement.

amplitude *n.* **1** breadth. **2** abundance.

amputate *v.* cut off by surgical operation. □ **amputation** *n.*

amulet *n.* thing worn as a charm against evil.

amuse *v.* **1** cause to laugh or smile. **2** make time pass pleasantly for. □ **amusing** *adj.*

■ **1** cheer (up), delight, tickle. **2** absorb, beguile, divert, engross, entertain, interest, please. □ **amusing** comical, droll, enjoyable, entertaining, funny, hilarious, humorous, pleasing, witty.

amusement *n.* **1** being amused. **2** thing that amuses.

an *adj.* form of *a* used before vowel sounds other than long 'u'.

anachronism *n.* thing that does not belong in the period in which it is placed. □ **anachronistic** *adj.*

anaemia /əneémiə/ *n.* lack of haemoglobin in blood.

anaemic /ənéemik/ *adj.* **1** suffering from anaemia. **2** lacking strong colour or characteristics.

> ■ **2** colourless, pale, pallid, pasty, sallow, sickly, unhealthy, wan, washed out.

anaesthesia /ánniss-théeziə/ *n.* loss of sensation, esp. induced by anaesthetics.

anaesthetic /ánniss-théttik/ *adj.* & *n.* (substance) causing loss of sensation.

anaesthetist /ənéess-thətist/ *n.* person who administers anaesthetics.

anagram *n.* word formed from the rearranged letters of another.

anal *adj.* of the anus.

analgesic *adj.* & *n.* (drug) relieving pain. □ **analgesia** *n.*

analogous *adj.* similar in certain respects.

analogue *n.* analogous thing.

analogy *n.* partial likeness between things.

> ■ correlation, correspondence, likeness, parallel, relation, resemblance, similarity.

analyse *v.* **1** make an analysis of. **2** psychoanalyse. □ **analyst** *n.*

analysis *n.* (*pl.* **-lyses**) detailed examination or study.

> ■ breakdown, critique, dissection, evaluation, examination, inquiry, interpretation, investigation, review, scrutiny, study.

analytical *adj.* (also **analytic**) of or using analysis.

anarchism *n.* belief that government and law should be abolished. □ **anarchist** *n.*

anarchy *n.* **1** total lack of organized control. **2** lawlessness. □ **anarchical** *adj.*

anathema /ənáthəmə/ *n.* **1** formal curse. **2** detested thing.

anatomist *n.* expert in anatomy.

anatomy *n.* **1** bodily structure. **2** study of this. □ **anatomical** *adj.*

ancestor *n.* person from whom another is descended. □ **ancestral** *adj.*

> ■ antecedent, forebear, forefather, forerunner, precursor, predecessor, progenitor.

ancestry *n.* line of ancestors.

anchor *n.* heavy metal structure for mooring a ship to the sea bottom. ● *v.* **1** moor with an anchor. **2** fix firmly.

anchorage *n.* **1** place for anchoring. **2** lying at anchor.

anchovy *n.* small strong-tasting fish.

ancient *adj.* **1** of long ago. **2** very old.

> ■ **1** antediluvian, antiquated, archaic, bygone, past, prehistoric, primeval, primitive, primordial. **2** aged, elderly, hoary, old, venerable.

ancillary *adj.* helping in a subsidiary way.

and *conj.* connecting words, phrases, or sentences.

anecdote *n.* short amusing or interesting true story.

anemone *n.* plant with white, red, or purple flowers.

aneurysm *n.* excessive swelling of an artery.

anew *adv.* **1** again. **2** in a new way.

angel *n.* **1** messenger of God. **2** kind person. □ **angelic** *adj.*

angelica *n.* **1** candied stalks of a fragrant plant. **2** this plant.

anger *n.* extreme displeasure. ● *v.* make angry.

> ■ *n.* annoyance, displeasure, exasperation, fury, indignation, ire, irritation, rage, resentment, vexation, wrath. ● *v.* annoy, *colloq.* aggravate, displease, drive mad, enrage, exasperate, gall, incense, inflame, infuriate, irritate, madden, outrage, provoke, vex.

angina *n.* (in full **angina pectoris**) sharp pain in the chest.

angle[1] *n.* **1** point of view. **2** space between two lines or surfaces that meet. ● *v.* **1** present from a particular point of view. **2** place obliquely.

> ■ *n.* **1** approach, perspective, point of view, slant, standpoint, viewpoint.

angle[2] *v.* **1** fish with hook and bait. **2** try to obtain by hinting. □ **angler** *n.*

Anglican *adj.* & *n.* (member) of the Church of England. □ **Anglicanism** *n.*

Anglicism *n.* English idiom.

Anglicize *v.* make English in character. □ **Anglicization** *n.*

Anglo- *pref.* English, British.

Anglo-Saxon *n.* & *adj.* **1** (of) English person or language before the Norman Conquest. **2** (of) person of English descent.

angora *n.* **1** long-haired variety of cat, goat, or rabbit. **2** yarn or fabric made from the hair of such goats or rabbits.

angostura *n.* aromatic bitter bark of S. American tree.

angry *adj.* (**-ier**, **-iest**) feeling or showing anger. □ **angrily** *adv.*

> ■ annoyed, *colloq.* apoplectic, choleric, cross, enraged, exasperated, fuming, furious, heated, in a bad temper, incensed, indignant, infuriated, irate, irritated, *colloq.* livid, *colloq.* mad, outraged, raging, raving,

resentful, seething, smouldering, sore, vexed, wrathful.

angstrom *n.* unit of measurement for wavelengths.

anguish *n.* severe physical or mental pain. □ **anguished** *adj.*

■ agony, anxiety, distress, grief, heartache, misery, pain, sorrow, suffering, torment, torture, woe.

angular *adj.* **1** having angles or sharp corners. **2** forming an angle.

aniline *n.* oily liquid used in making dyes and plastics.

animal *n.* & *adj.* (of) a living thing that can move voluntarily.

animate *adj.* /ánnimət/ living. ● *v.* /ánnimayt/ **1** enliven, give life to. **2** motivate. **3** give (a film etc.) the appearance of movement by photographing a series of drawings. □ **animation** *n.*

■ *adj.* alive, breathing, live, living, sentient. ● *v.* **1** energize, enliven, invigorate, pep up, vivify. **2** arouse, encourage, excite, fire, galvanize, inspire, motivate, rouse, spur, stimulate, stir. □ **animation** ebullience, élan, energy, enthusiasm, excitement, high spirits, life, liveliness, spirit, verve, vigour, vitality, vivacity, zest.

animated *adj.* lively, vigorous.

■ active, alive, brisk, bubbly, ebullient, energetic, enthusiastic, excited, exuberant, high-spirited, lively, spirited, sprightly, vibrant, vigorous, vivacious.

animosity *n.* hostility.

■ acrimony, antagonism, animus, antipathy, aversion, bitterness, dislike, enmity, hate, hatred, hostility, ill will, loathing, malevolence, malice, rancour, resentment, spite, unfriendliness, venom, vindictiveness, virulence.

animus *n.* animosity.

aniseed *n.* fragrant seed of a plant (**anise**), used for flavouring.

ankle *n.* joint connecting the foot with the leg.

anklet *n.* chain or band worn round the ankle.

annals *n.pl.* **1** narrative of events year by year. **2** historical records.

anneal *v.* toughen (metal or glass) by heat and slow cooling.

annex *v.* **1** take possession of. **2** add as a subordinate part. □ **annexation** *n.*

■ **1** appropriate, conquer, occupy, seize, take over, usurp.

annexe *n.* additional building.

annihilate *v.* destroy completely. □ **annihilation** *n.*

■ destroy, eliminate, eradicate, exterminate, extinguish, extirpate, obliterate, raze, wipe out.

anniversary *n.* yearly return of the date of an event.

annotate *v.* add explanatory notes to. □ **annotation** *n.*

announce *v.* **1** make known publicly. **2** make known the presence or arrival of. □ **announcement** *n.*

■ **1** advertise, broadcast, declare, disclose, divulge, give notice of, give out, make public, proclaim, promulgate, publicize, publish, reveal, state. **2** introduce, present. □ **announcement** bulletin, communiqué, declaration, proclamation, report, statement.

announcer *n.* person who announces items in a broadcast.

annoy *v.* **1** cause slight anger to. **2** be troublesome to. □ **annoyance** *n.*

■ **1** *colloq.* aggravate, anger, bother, *sl.* bug, displease, drive mad, exasperate, gall, infuriate, irk, irritate, madden, needle, nettle, offend, pique, provoke, *colloq.* rile, ruffle, upset, vex. **2** badger, harass, harry, *colloq.* hassle, molest, pester, *colloq.* plague.

annoyed *adj.* slightly angry.

■ cross, disgruntled, displeased, exasperated, indignant, irritated, nettled, offended, *colloq.* peeved, piqued, *colloq.* shirty, sore, upset, *colloq.* uptight, vexed.

annual *adj.* yearly. ● *n.* **1** plant that lives for one year or one season. **2** book published in yearly issues. □ **annually** *adv.*

annuity *n.* yearly allowance provided by an investment.

annul *v.* (**annulled**) make null and void. □ **annulment** *n.*

annular *adj.* ring-shaped.

Annunciation *n.* announcement by the angel Gabriel to the Virgin Mary that she was to be the mother of Christ.

anode *n.* electrode by which current enters a device.

anodize *v.* coat (metal) with a protective layer by electrolysis.

anodyne *n.* something that relieves pain or distress.

anoint *v.* apply ointment or oil to, esp. ritually.

anomaly *n.* something irregular or inconsistent. □ **anomalous** *adj.*

anon *adv.* (*old use*) soon.

anon. *abbr.* anonymous.

anonymous *adj.* of unknown or undisclosed name or authorship. ▫ **anonymity** *n.*

anorak *n.* waterproof jacket with hood attached.

anorexia *n.* loss of appetite. ▫ **anorexia nervosa** obsessive desire to lose weight by refusing to eat. **anorexic** *adj.* & *n.*

another *adj.* **1** one more. **2** a different. **3** any other. ● *pron.* another one.

answer *n.* **1** thing said, written, needed, or done to deal with a question, accusation, etc. **2** solution to a problem. ● *v.* **1** make an answer or response (to). **2** be suitable for. **3** correspond (to a description). ▫ **answer for 1** take responsibility for. **2** vouch for.

> ■ *n.* **1** acknowledgement, reaction, rejoinder, reply, response, retort, riposte. **2** explanation, solution. ● *v.* **1** acknowledge, reply (to), rejoin, respond (to), retort. **2** fulfil, meet, satisfy, serve, suffice for, suit. **3** correspond to, fit, match, tally with.

answerable *adj.* having to account for something.

ant *n.* small insect that lives in highly organized groups.

antacid *n.* & *adj.* (substance) preventing or correcting acidity.

antagonism *n.* active opposition, hostility. ▫ **antagonistic** *adj.*

> ■ animosity, antipathy, conflict, enmity, friction, hostility, opposition, rivalry, strife.

antagonist *n.* opponent.

antagonize *v.* rouse antagonism in.

Antarctic *adj.* & *n.* (of) regions round the South Pole.

ante *n.* stake put up by a poker player before receiving cards.

ante- *pref.* before.

anteater *n.* mammal that eats ants.

antecedent *n.* preceding thing or circumstance. ● *adj.* previous.

antedate *v.* **1** put an earlier date on. **2** precede in time.

antediluvian *adj.* **1** before the Flood. **2** antiquated.

antelope *n.* animal resembling a deer.

antenatal *adj.* **1** before birth. **2** of or during pregnancy.

antenna *n.* **1** (*pl.* **-ae**) insect's feeler. **2** (*pl.* **-as**) radio or TV aerial.

anterior *adj.* coming before in position or time.

ante-room *n.* room leading to a more important one.

anthem *n.* piece of music to be sung in a religious service.

anther *n.* part of a flower's stamen containing pollen.

anthology *n.* collection of poems, stories, etc.

anthracite *n.* form of coal burning with little flame or smoke.

anthrax *n.* disease of sheep and cattle, transmissible to people.

anthropoid *adj.* & *n.* human-like (ape).

anthropology *n.* study of the origin and customs of humankind. ▫ **anthropological** *adj.*, **anthropologist** *n.*

anthropomorphic *adj.* attributing human form to a god or animal. ▫ **anthropomorphism** *n.*

anti- *pref.* **1** opposed to. **2** counteracting. ▫ **anti-aircraft** *adj.* used against enemy aircraft.

antibiotic *n.* substance that destroys bacteria.

antibody *n.* protein formed in the blood in reaction to a substance which it then destroys.

anticipate *v.* **1** deal with in advance. **2** foresee. **3** look forward to. ▫ **anticipation** *n.*

> ■ **1** forestall, intercept, preclude, pre-empt, prevent. **2** envisage, forecast, foresee, foretell, predict, prophesy. **3** await, expect, look forward to, wait for.

anticlimax *n.* dull ending where a climax was expected.

anticlockwise *adj.* & *adv.* in the direction opposite to clockwise.

antics *n.pl.* absurd behaviour.

anticyclone *n.* outward flow of air from an area of high pressure, producing fine weather.

antidote *n.* substance that counteracts the effects of poison.

antifreeze *n.* substance added to water to prevent freezing.

antigen *n.* foreign substance stimulating the production of antibodies.

antihistamine *n.* substance counteracting the effect of histamine.

antimony *n.* brittle silvery metallic element.

antipathy *n.* strong dislike.

> ■ abhorrence, aversion, disgust, dislike, hatred, loathing, repugnance, revulsion.

antiperspirant *n.* substance that prevents or reduces sweating.

antipodes /antippədeez/ *n.pl.* places on opposite sides of the earth, esp. Australia and New Zealand (opposite Europe).

antiquarian *adj.* of the study of antiques. ● *n.* person who studies antiques.

antiquated *adj.* very old-fashioned.

■ **antediluvian, archaic, dated, obsolete, old, old-fashioned, outmoded, out of date, unfashionable.**

antique *adj.* belonging to the distant past. ● *n.* antique interesting or valuable object.

antiquity *n.* **1** ancient times. **2** object dating from ancient times.

antirrhinum *n.* snapdragon.

anti-Semitic *adj.* hostile to Jews.

antiseptic *adj.* & *n.* (substance) preventing things from becoming septic. □ **antiseptically** *adv.*

antisocial *adj.* destructive or hostile to society.

antistatic *adj.* counteracting the effects of static electricity.

antithesis *n.* (*pl.* **-eses**) **1** opposite. **2** contrast. □ **antithetical** *adj.*

antitoxin *n.* substance neutralizing a toxin. □ **antitoxic** *adj.*

antivivisectionist *n.* person opposed to making experiments on live animals.

antler *n.* branched horn of a deer.

antonym *n.* word opposite to another in meaning.

anus *n.* opening at the excretory end of the alimentary canal.

anvil *n.* iron block on which a smith hammers metal.

anxiety *n.* **1** state of being anxious. **2** anxious desire.

■ **1 agitation, apprehension, concern, consternation, disquiet, distress, dread, fear, foreboding, misgiving, nervousness, panic, solicitude, tension, unease, uneasiness, worry. 2 desire, eagerness, longing, yearning.**

anxious *adj.* **1** mentally troubled. **2** eager. □ **anxiously** *adv.*

■ **1 afraid, agitated, apprehensive, concerned, distressed, edgy, fearful,** *colloq.* **fraught, fretful,** *colloq.* **jittery, nervous, panicky, perturbed, restless, solicitous, tense, troubled, uneasy, upset, worried. 2 dying, eager, keen, longing, yearning.**

any *adj.* **1** one or some from a quantity. **2** every.

anybody *n.* & *pron.* any person.

anyhow *adv.* **1** anyway. **2** not in an orderly manner.

anyone *n.* & *pron.* anybody.

anything *n.* & *pron.* any item.

anyway *adv.* whatever the truth or possible outcome is.

anywhere *adv.* & *pron.* (in or to) any place.

aorta *n.* main artery carrying blood from the heart.

apart *adv.* **1** separately, so as to become separated. **2** to or at a distance. **3** into pieces.

apartheid /əpaártayt/ *n.* former policy of racial segregation in S. Africa.

apartment *n.* **1** set of rooms. **2** (*US*) flat.

apathy *n.* lack of interest or concern. □ **apathetic** *adj.*

■ **indifference, lassitude, lethargy, listlessness, passivity, torpor.** □ **apathetic cool, half-hearted, impassive, indifferent, languid, lethargic, listless, passive, phlegmatic, sluggish, torpid, unconcerned, unenthusiastic, uninterested, unmoved.**

ape *n.* tailless monkey. ● *v.* imitate, mimic.

aperitif *n.* alcoholic drink taken as an appetizer.

aperture *n.* opening, esp. one that admits light.

apex *n.* **1** highest point. **2** tip.

■ **1 crest, crown, peak, pinnacle, point, summit, top; acme, climax, culmination, height, zenith.**

aphid *n.* small insect destructive to plants.

aphorism *n.* pithy saying.

aphrodisiac *adj.* & *n.* (substance) arousing sexual desire.

apiary *n.* place where bees are kept. □ **apiarist** *n.*

apiece *adv.* to or for or by each.

aplomb /əplóm/ *n.* dignity and confidence.

apocalypse *n.* **1** violent event. **2** revelation, esp. about the end of the world. □ **apocalyptic** *adj.*

Apocrypha *n.pl.* books of the Old Testament not accepted as part of the Hebrew scriptures.

apocryphal *adj.* **1** of doubtful authenticity. **2** invented.

apogee *n.* point in the moon's orbit furthest from the earth.

apologetic *adj.* expressing regret. □ **apologetically** *adv.*

■ **conscience-stricken, contrite, penitent, regretful, remorseful, repentant, rueful, sorry.**

apologize *v.* make an apology.

apology *n.* **1** statement of regret for having done wrong or hurt. **2** explanation of one's beliefs.

apoplexy *n.* **1** a stroke. **2** (*colloq.*) rush of extreme emotion, esp. anger. □ **apoplectic** *adj.*

apostasy *n.* abandonment of one's former religious belief.

apostate *n.* person who is guilty of apostasy.

Apostle *n.* any of the twelve men sent forth by Christ to preach the gospel. □ **apostolic** *adj.*

apostrophe /əpóstrəfi/ *n.* the sign ' used to show the possessive case or omission of a letter.

apothecary *n.* (*old use*) pharmacist.

appal *v.* (**appalled**) fill with horror or dismay. □ **appalling** *adj.*

■ disgust, dismay, distress, horrify, outrage, revolt, scandalize, shock, sicken. □ **appalling** abominable, atrocious, awful, deplorable, dreadful, ghastly, grim, grisly, gruesome, hideous, horrible, horrifying, outrageous, revolting, shocking, terrible.

apparatus *n.* equipment for scientific or other work.

■ equipment, gear, implements, instruments, machinery, paraphernalia, tackle, tools, utensils; appliance, contraption, device, gadget, machine.

apparel *n.* clothing.

apparent *adj.* **1** clearly seen or understood. **2** seeming but not real. □ **apparently** *adv.*

■ **1** blatant, clear, conspicuous, discernible, evident, manifest, marked, noticeable, obvious, patent, perceptible, plain, unconcealed, unmistakable, visible. **2** ostensible, outward, seeming.

apparition *n.* **1** ghost. **2** thing appearing, esp. of a startling or remarkable kind.

■ ghost, hallucination, phantom, spectre, spirit, *colloq.* spook, wraith.

appeal *v.* **1** make an earnest or formal request. **2** apply to a higher court. **3** be attractive or of interest. ● *n.* **1** act of appealing. **2** attractiveness.

■ *v.* **1** beg, pray, solicit, supplicate; (**appeal to**) ask, beseech, entreat, implore, petition, plead with. **3** (**appeal to**) allure, attract, fascinate, please, interest. ● *n.* **1** application, cry, entreaty, petition, plea, prayer, request, supplication. **2** allure, attractiveness, charisma, charm, fascination, pull, seductiveness.

appear *v.* **1** be or become visible or evident. **2** seem. **3** take part in a play, film, etc.

■ **1** arrive, *colloq.* show up, turn up; arise, come out, crop up, emerge, materialize, surface. **2** look, seem. **3** act, perform, play a role *or* part in.

appearance *n.* **1** act of appearing. **2** look, semblance.

■ **1** advent, arrival, coming, emergence. **2** air, aspect, bearing, demeanour, look, manner, mien; guise, impression, semblance, show.

appease *v.* soothe or conciliate, esp. by giving what was asked. □ **appeasement** *n.*

■ assuage, calm, conciliate, humour, mollify, pacify, placate, quiet, satisfy, soothe.

appellant *n.* person who appeals to a higher court.

append *v.* add at the end.

appendage *n.* thing appended.

appendicitis *n.* inflammation of the intestinal appendix.

appendix *n.* **1** (*pl.* **-ices**) section at the end of a book, giving extra information. **2** (*pl.* **-ixes**) small blind tube of tissue attached to the intestine.

■ **1** addendum, addition, codicil, postscript, supplement.

appertain *v.* be relevant.

appetite *n.* desire, esp. for food.

■ craving, desire, eagerness, enthusiasm, hankering, hunger, inclination, keenness, liking, longing, passion, predilection, preference, relish, stomach, taste, thirst, yearning, *colloq.* yen.

appetizer *n.* thing eaten or drunk to stimulate the appetite.

appetizing *adj.* stimulating the appetite.

applaud *v.* **1** express approval (of), esp. by clapping. **2** praise. □ **applause** *n.*

■ **1** clap, cheer, give a person an ovation. **2** acclaim, commend, compliment, congratulate, eulogize, extol, hail, pay tribute to, praise, salute.

apple *n.* fruit with firm flesh.

appliance *n.* device, instrument.

■ apparatus, contraption, device, gadget, implement, instrument, machine, tool, utensil.

applicable *adj.* **1** appropriate. **2** relevant. □ **applicability** *n.*

applicant *n.* person who applies for a job.

■ candidate, job-hunter, job-seeker, interviewee.

application *n.* **1** act of applying. **2** formal request. **3** sustained effort, diligence. **4** relevance.

■ **2** appeal, request, petition, submission. **3** assiduity, assiduousness, attention, dedication, diligence, effort, industry, perseverance. **4** applicability, bearing, relevance, pertinence.

applied *adj.* put to practical use.

appliqué /apleékay/ *n.* piece of fabric attached ornamentally.

apply *v.* **1** put or spread on. **2** bring into use or action. **3** be relevant. □ **apply to** or **for** make a formal request for something to be done, given, etc. **apply oneself** give one's attention and energy.

■ **1** put on, rub in, spread on. **2** employ, exercise, implement, practise, use, utilize. **3** be relevant, have a bearing, pertain, relate. □ **apply to** appeal to, make an application to, petition, solicit.

appoint *v.* **1** choose (a person) for a job, committee, etc. **2** decide on (a time, place, etc.).

■ **1** assign, choose, co-opt, delegate, depute, designate, detail, elect, name, nominate, select. **2** arrange, decide on, determine, establish, fix, ordain, set, settle.

appointee *n.* person appointed.

appointment *n.* **1** job. **2** arrangement to meet.

■ **1** job, office, place, position, post, situation. **2** arrangement, assignation, *colloq.* date, engagement, meeting, rendezvous.

apportion *v.* divide into shares.

apposite *adj.* appropriate.

apposition *n.* juxtaposition, esp. of syntactically parallel words.

appraise *v.* estimate the value or quality of. □ **appraisal** *n.*

appreciable *adj.* **1** perceptible. **2** considerable. □ **appreciably** *adv.*

appreciate *v.* **1** value. **2** understand. **3** rise in value. □ **appreciation** *n.*, **appreciative** *adj.*

■ **1** admire, enjoy, esteem, prize, rate highly, respect, think highly of, treasure, value. **2** comprehend, know, realize, recognize, see, understand.

apprehend *v.* **1** seize, arrest. **2** understand. **3** expect with fear or anxiety. □ **apprehension** *n.*

■ **1** arrest, capture, catch, collar, *sl.* nab, *sl.* nick, seize, take. **2** *colloq.* catch on to, comprehend, grasp, perceive, understand.

apprehensive *adj.* fearful, anxious. □ **apprehensively** *adv.*

■ afraid, anxious, concerned, edgy, fearful, *colloq.* jittery, nervous, tense, uneasy, worried.

apprentice *n.* person learning a craft. □ **apprenticeship** *n.*

■ beginner, learner, novice, probationer, pupil, starter, trainee.

apprise *v.* inform.

approach *v.* **1** come nearer (to). **2** set about doing. **3** go to with a request or offer. ● *n.* **1** act or means of approaching. **2** way of dealing with a person or thing.

■ *v.* **1** catch up (with), gain on, move towards, near; advance, loom. **2** address, begin, buckle down to, embark on, set about, tackle. **3** contact, make a proposal to, make overtures to, proposition. ● *n.* **1** advance, advent, arrival; access, drive, entrance, path, way. **2** manner, method, mode, procedure, style, system, technique, way.

approachable *adj.* easy to talk to.

approbation *n.* approval.

appropriate *adj.* /əprṓpriət/ suitable, proper. ● *v.* /əprṓpriayt/ **1** take and use. **2** set aside (money) for a special purpose. □ **appropriately** *adv.*, **appropriation** *n.*

■ *adj.* apposite, apt, befitting, correct, deserved, felicitous, fit, fitting, germane, pertinent, proper, relevant, right, suitable, suited. ● *v.* **1** commandeer, make off with, pocket, purloin, seize, *colloq.* snaffle, snatch, steal, take. **2** allot, apportion, earmark, set aside.

approval *n.* **1** act of approving. **2** consent. □ **on approval** (of goods) returnable if not satisfactory.

■ acceptance, agreement, approbation, assent, authorization, backing, blessing, consent, endorsement, go-ahead, *colloq.* green light, leave, *colloq.* OK, permission, sanction.

approve *v.* agree to, sanction. □ **approve of** say or think (a thing) is good or suitable.

■ accede to, accept, agree to, allow, assent to, authorize, consent to, endorse,

■ give the go-ahead to, permit, ratify, rubber-stamp, sanction.

approximate *adj.* /əpróksimət/ almost but not quite exact. ● *v.* /əpróksimayt/ be or make almost the same. □ **approximately** *adv.*, **approximation** *n.*

■ *adj.* estimated, imprecise, inexact, rough. ● *v.* be similar to, border on, come near to, resemble. □ **approximately** about, approaching, around, close to, in the region of, more or less, nearly, roughly, round about.

après-ski /áprayskeé/ *adj.* & *n.* (of or for) the social activities following a day's skiing.

apricot *n.* **1** orange-yellow peach-like fruit. **2** its colour.

apron *n.* garment worn over the front of the body to protect clothes.

apropos /áprəpṓ/ *adv.* concerning.

apse *n.* recess with an arched or domed roof in a church.

apt *adj.* **1** suitable. **2** having a tendency. **3** quick to learn.

■ **1** apposite, appropriate, befitting, felicitous, fitting, relevant, suitable, well-chosen. **2** disposed, given, inclined, liable, likely, predisposed, prone, ready. **3** able, bright, capable, clever, intelligent, quick.

aptitude *n.* natural ability.

■ ability, bent, capability, flair, gift, knack, skill, talent.

aqualung *n.* portable underwater breathing-apparatus.

aquamarine *n.* **1** bluish-green beryl. **2** its colour.

aquarium *n.* (*pl.* **-ums**) tank for keeping living fish etc.

aquatic *adj.* **1** living in or near water. **2** done in or on water.

aquatint *n.* a kind of etching.

aqueduct *n.* artificial channel on a raised structure, carrying water across country.

aqueous /áykwiəss/ *adj.* of or like water.

aquifer *n.* water-bearing rock or soil.

aquiline *adj.* **1** like an eagle. **2** (of a nose) hooked.

Arab *n.* & *adj.* (member) of a Semitic people of the Middle East.

arabesque *n.* **1** dancer's posture with the body bent forward and leg and arm extended in line. **2** decoration with intertwined lines etc.

Arabian *adj.* of Arabia.

Arabic *adj.* & *n.* (of) the language of the Arabs. □ **arabic numerals** the symbols 1, 2, 3, etc.

arable *adj.* & *n.* (land) suitable for growing crops.

arachnid *n.* member of the class to which spiders belong.

arachnophobia *n.* fear of spiders.

arbiter *n.* **1** person with power to decide what shall be done or accepted. **2** arbitrator.

arbitrary *adj.* based on random choice. □ **arbitrarily** *adv.*

■ capricious, chance, erratic, inconsistent, irrational, random, subjective, unpredictable, whimsical.

arbitrate *v.* act as arbitrator. □ **arbitration** *n.*

■ adjudge, adjudicate, decide, judge, referee, umpire.

arbitrator *n.* impartial person chosen to settle a dispute.

■ adjudicator, arbiter, intermediary, judge, mediator, negotiator, peacemaker, referee, umpire.

arboreal *adj.* of or living in trees.

arboretum *n.* place where trees are grown for study and display.

arbour *n.* shady shelter under trees etc.

arc *n.* **1** part of a curve. **2** luminous electric current crossing a gap between terminals.

arcade *n.* **1** covered walk between shops. **2** place with pin-tables, gambling machines, etc.

arcane *adj.* mysterious.

arch[1] *n.* curved structure, esp. as a support. ● *v.* form into an arch.

arch[2] *adj.* consciously or affectedly playful. □ **archly** *adv.*

archaeology *n.* study of civilizations through their material remains. □ **archaeological** *adj.*, **archaeologist** *n.*

archaic *adj.* belonging to former or ancient times.

archaism *n.* (use of) an archaic word or phrase.

archangel *n.* angel of the highest rank.

archbishop *n.* chief bishop.

archdeacon *n.* priest ranking next below bishop.

archer *n.* person who shoots with bow and arrows. □ **archery** *n.*

archetype /áarkitīp/ *n.* **1** prototype. **2** typical specimen. □ **archetypal** *adj.*

■ **1** master, original, pattern, prototype. **2** classic, epitome, exemplar, ideal, model, paradigm, standard.

archipelago *n.* (*pl.* **-os**) **1** group of islands. **2** sea round this.

architect *n.* designer of buildings.

architecture *n.* **1** designing of buildings. **2** style of building(s). □ **architectural** *adj.*

architrave *n.* moulded frame round a doorway or window.

archive /áarkīv/ *n.* (usu. *pl.*) historical documents.

archivist /áarkivist/ *n.* person trained to deal with archives.

archway *n.* arched entrance or passage.

Arctic *adj.* **1** of regions round the North Pole. **2** (**arctic**) very cold. ● *n.* Arctic regions.

ardent *adj.* full of ardour, enthusiastic. □ **ardently** *adv.*

■ avid, burning, eager, enthusiastic, fervent, fervid, hot, impassioned, intense, keen, passionate, vehement, warm, zealous.

ardour *n.* great warmth of feeling, enthusiasm.

■ desire, eagerness, enthusiasm, fervour, intensity, heat, passion, vehemence, warmth, zeal.

arduous *adj.* needing much effort.

■ demanding, difficult, exhausting, formidable, gruelling, hard, herculean, laborious, onerous, severe, strenuous, taxing, tiring, tough.

area *n.* **1** extent or measure of a surface. **2** region. **3** range of a subject etc.

■ **1** acreage, expanse, extent, measure, size, space. **2** district, locality, neighbourhood, part, precinct, quarter, region, sector, territory, vicinity, zone. **3** compass, range, scope, sphere.

arena *n.* **1** level area in the centre of an amphitheatre or sports stadium. **2** scene of conflict.

argon *n.* an inert gas.

argot /áargō/ *n.* jargon.

arguable *adj.* **1** able to be asserted. **2** not certain. □ **arguably** *adv.*

argue *v.* **1** exchange views or opinions, esp. angrily. **2** maintain by reasoning.

■ **1** bicker, debate, disagree, dispute, fall out, fight, quarrel, row, *colloq.* scrap, spar, squabble, wrangle. **2** assert, claim, contend, hold, maintain, reason.

argument *n.* **1** (angry) exchange of views. **2** reason put forward, reasoning process.

■ **1** altercation, clash, conflict, controversy, debate, disagreement, disputation, dispute, fight, fracas, quarrel, row, *colloq.* scrap, squabble, tiff, wrangle. **2** case, contention, reasoning, thesis.

argumentative *adj.* fond of arguing.

aria *n.* solo in opera.

arid *adj.* dry, barren. □ **aridly** *adv.*, **aridity** *n.*

■ barren, desert, dry, lifeless, parched, sterile, unproductive, waste, waterless.

arise *v.* (**arose**, **arisen**) **1** come into existence or to people's notice. **2** rise.

■ **1** appear, come up, crop up, develop, emerge, materialize, occur, surface.

aristocracy *n.* hereditary upper classes. □ **aristocratic** *adj.*

■ □ **aristocratic** blue-blooded, elite, noble, titled, upper class, *colloq.* upper crust.

aristocrat *n.* member of the aristocracy.

arithmetic *n.* calculating by means of numbers.

ark *n.* Noah's boat in which he and his family and animals were saved from the Flood. □ **Ark of the Covenant** wooden chest in which the writings of Jewish Law were kept.

arm[1] *n.* **1** upper limb of the human body. **2** raised side part of a chair.

arm[2] *v.* **1** equip with weapons. **2** make (a bomb) ready to explode. ● *n.pl.* weapons.

armada *n.* fleet of warships.

armadillo *n.* S. American burrowing mammal with a plated body.

Armageddon *n.* final disastrous conflict.

armament *n.* **1** military weapons. **2** process of equipping for war.

armature *n.* wire-wound core of a dynamo.

armchair *n.* chair with raised sides.

armistice *n.* agreement to stop fighting temporarily.

■ ceasefire, peace, truce.

armlet *n.* band worn round an arm or sleeve.

armour *n.* protective metal covering, formerly worn in fighting. □ **armoured** *adj.*

armourer *n.* maker, repairer, or keeper of weapons.

armoury *n.* arsenal.

armpit *n.* hollow under the arm at the shoulder.

army *n.* **1** organized force for fighting on land. **2** vast group.

aroma *n.* smell, esp. a pleasant one. □ **aromatic** *adj.*

■ bouquet, fragrance, odour, perfume, redolence, scent, smell.

aromatherapy *n.* use of fragrant oils etc., esp. in massage.

arose *see* **arise.**

around *adv.* & *prep.* **1** all round, on every side (of). **2** approximately.

arouse *v.* rouse.

■ awaken, rouse, wake, wake up, waken; encourage, excite, inspire, kindle, provoke, stimulate, stir up, whip up.

arpeggio *n.* (*pl.* **-os**) notes of a chord played in succession.

arraign *v.* **1** indict, accuse. **2** find fault with. □ **arraignment** *n.*

arrange *v.* **1** put into order. **2** settle the details of. **3** adapt.

■ **1** align, array, categorize, classify, display, dispose, group, lay out, marshal, order, organize, position, put in order, range, sort (out), systematize. **2** decide on, determine, fix, organize, plan, prepare, see to, settle. **3** adapt, orchestrate, score.

arrangement *n.* **1** act or process of arranging. **2** settlement between people. **3** (*pl.*) plans.

■ **1** classification, disposition, grouping, organization, planning. **2** agreement, bargain, contract, settlement, understanding. **3** (**arrangements**) measures, plans, preparations; itinerary, programme, schedule.

array *v.* **1** arrange in order. **2** adorn. ● *n.* imposing series, display.

■ *v.* **1** arrange, assemble, display, dispose, lay out, range. **2** adorn, attire, clothe, deck, decorate, drape, dress, garb, robe. ● *n.* arrangement, collection, display, exhibition, panoply, parade, range, series, show, variety.

arrears *n.pl.* **1** money owed and overdue for repayment. **2** work overdue for being finished.

arrest *v.* **1** stop (a movement or moving thing). **2** seize by authority of law. ● *n.* legal seizure of an offender.

■ *v.* **1** block, check, delay, halt, hinder, impede, interrupt, obstruct, restrain, retard, slow, stem, stop. **2** apprehend, capture, catch, collar, detain, *sl.* nab, *sl.* nick, seize, take into custody. ● *n.* apprehension, capture, detention, seizure.

arrival *n.* **1** act of arriving. **2** person or thing that has arrived.

■ **1** advent, appearance, approach, entrance; disembarkation, landing, touchdown. **2** immigrant, newcomer, visitor.

arrive *v.* **1** reach a destination. **2** (*colloq.*) establish one's reputation or success. **3** (of time) come.

■ **1** appear, come, make one's appearance, *colloq.* show up, turn up; disembark, land, touch down. **2** *colloq.* make it, make the grade, succeed.

arrogant *adj.* proud and overbearing. □ **arrogantly** *adv.*, **arrogance** *n.*

■ boastful, bumptious, cavalier, cocksure, cocky, conceited, contemptuous, disdainful, egotistical, haughty, high-handed, imperious, lofty, overbearing, patronizing, pompous, presumptuous, proud, scornful, self-assertive, self-important, *colloq.* snooty, supercilious, superior.

arrow *n.* **1** straight shaft with a sharp point, shot from a bow. **2** line with a V at the end, indicating direction.

arrowroot *n.* edible starch made from a W. Indian plant.

arsenal *n.* place where weapons are stored or made.

arsenic *n.* **1** semi-metallic element. **2** strongly poisonous compound of this.

arson *n.* intentional and unlawful setting on fire of a building.

art *n.* **1** creative skill, works such as paintings or sculptures produced by this. **2** (*pl.*) subjects or activities concerned with creativity (e.g. painting, music, writing). **3** aptitude or knack.

■ **1** artistry, craftsmanship, creativity, imagination, inventiveness. **3** aptitude, knack, skill, talent, technique, trick.

artefact *n.* man-made object.

arterial *adj.* of an artery. □ **arterial road** main trunk road.

artery *n.* blood vessel carrying blood away from the heart.

artesian well a well that is bored vertically into oblique strata so that water rises naturally with little or no pumping.

artful *adj.* crafty. □ **artfully** *adv.*

■ crafty, cunning, deceitful, designing, devious, scheming, sly, tricky, wily.

arthritis *n.* condition in which there is pain and stiffness in the joints. □ **arthritic** *adj.*

arthropod *n.* animal with a segmented body and jointed limbs (e.g. an insect or crustacean).

artichoke *n.* plant with a flower of leaf-like scales used as a vegetable. □ **Jerusalem artichoke** sunflower with an edible root.

article *n.* **1** particular or separate thing. **2** piece of writing in a newspaper etc. **3** clause in an agreement. □ **definite article** the word 'the'. **indefinite article** 'a' or 'an'.

> ■ **1** commodity, item, object, thing. **2** editorial, essay, feature, item, leader, piece, story.

articulate *adj.* /aartíkyoolət/ **1** spoken distinctly. **2** able to express ideas clearly. ● *v.* /aartíkyoolayt/ **1** say or speak distinctly. **2** form a joint, connect by joints. □ **articulated lorry** one with sections connected by a flexible joint. **articulation** *n.*

> ■ *adj.* **1** clear, coherent, comprehensible, eloquent, intelligible, lucid, understandable. **2** coherent, eloquent, fluent. ● *v.* **1** enunciate, express, pronounce, say, speak, utter, vocalize, voice.

artifice *n.* **1** trickery. **2** device.

artificial *adj.* **1** not originating naturally, not real. **2** insincere. □ **artificially** *adv.*, **artificiality** *n.*

> ■ **1** bogus, fabricated, fake, false, imitation, man-made, manufactured, mock, plastic, simulated, synthetic. **2** affected, feigned, insincere, *colloq.* phoney, pretended, sham, studied, unnatural.

artillery *n.* **1** large guns used in fighting on land. **2** branch of an army using these.

artisan *n.* skilled workman.

artist *n.* **1** person who produces works of art, esp. paintings. **2** one who does something with exceptional skill. **3** artiste. □ **artistry** *n.*, **artistic** *adj.*

artiste /aartéest/ *n.* professional entertainer.

artless *adj.* guileless, ingenuous. □ **artlessly** *adv.*

> ■ childlike, frank, guileless, honest, ingenuous, innocent, naive, natural, simple, sincere, unaffected, unsophisticated.

arty *adj.* (**-ier**, **-iest**) affectedly or pretentiously artistic.

as *adv.* & *conj.* **1** in the same degree, similarly. **2** in the form or function of. **3** while, when. **4** because. □ **as for, as to** with regard to. **as well** in addition.

asbestos *n.* **1** soft fibrous mineral substance. **2** fireproof material made from this.

asbestosis *n.* lung disease caused by inhaling asbestos particles.

ascend *v.* go or come up.

> ■ climb, come up, go up, mount, move up, rise, scale, soar.

ascendant *adj.* rising. □ **in the ascendant** rising in power or influence.

ascension *n.* ascent, esp. (**Ascension**) that of Christ to heaven.

ascent *n.* **1** ascending. **2** way up.

ascertain *v.* find out by enquiring. □ **ascertainable** *adj.*

> ■ confirm, determine, discover, establish, find out, learn, make certain, make sure, verify.

ascetic *adj.* not allowing oneself pleasures and luxuries. ● *n.* ascetic person. □ **asceticism** *n.*

> ■ *adj.* abstemious, austere, celibate, frugal, puritanical, rigorous, self-denying, self-disciplined, severe, spartan.

ascorbic acid vitamin C.

ascribe *v.* attribute. □ **ascription** *n.*

asepsis *n.* absence of harmful bacteria. □ **aseptic** *adj.*

asexual *adj.* without sex. □ **asexually** *adv.*

ash¹ *n.* tree with silver-grey bark.

ash² *n.* powder that remains after something has burnt.

ashamed *adj.* feeling shame.

> ■ abashed, conscience-stricken, contrite, embarrassed, humiliated, mortified, penitent, remorseful, repentant, shamefaced, sheepish, sorry.

ashen *adj.* **1** pale as ashes. **2** grey.

ashore *adv.* to or on shore.

ashram *n.* (orig. in India) retreat for religious meditation.

ashy *adj.* (**-ier**, **-iest**) **1** ashen. **2** covered with ash.

Asian *adj.* of Asia or its people. ● *n.* Asian person.

Asiatic *adj.* of Asia.

aside *adv.* to or on one side, away from the main part or group. ● *n.* words spoken so that only certain people will hear.

asinine /ássinīn/ *adj.* silly.

ask *v.* **1** call for an answer to or about. **2** seek to obtain. **3** invite.

> ■ **1** enquire (of), interrogate, question, quiz. **2** appeal for, apply for, beg, beseech, demand, entreat, implore, petition, plead

for, request, seek, solicit, supplicate. **3** invite, summon.

askance *adv.* **look askance at** regard suspiciously.

askew *adv.* & *adj.* crooked(ly).

asleep *adv.* & *adj.* in or into a state of sleep.

asp *n.* small poisonous snake.

asparagus *n.* plant whose shoots are used as a vegetable.

aspect *n.* **1** look or appearance. **2** feature of a problem or situation. **3** direction a thing faces.

■ **1** air, appearance, bearing, countenance, demeanour, expression, face, look, manner, mien. **2** attribute, characteristic, detail, element, facet, feature, ingredient, quality, respect, side. **3** direction, outlook, prospect, view.

aspen *n.* a kind of poplar tree.

asperity *n.* harshness.

■ acerbity, astringency, bitterness, gall, harshness, severity, sharpness.

aspersion *n.* **cast aspersions on** defame, criticize.

asphalt *n.* **1** black substance like coal tar. **2** mixture of this with gravel etc. for paving.

asphyxia *n.* suffocation.

asphyxiate *v.* suffocate. □ **asphyxiation** *n.*

aspic *n.* clear savoury jelly.

aspidistra *n.* house plant with broad tapering leaves.

aspirant *n.* person who aspires to something.

aspirate *n.* /áspirət/ sound of h. ● *v.* /áspiráyt/ pronounce with an h.

aspiration *n.* desire, ambition.

■ aim, ambition, desire, dream, goal, hope, longing, object, objective, wish, yearning.

aspire *v.* **aspire to** have a strong ambition to achieve.

■ aim for, desire, dream of, hanker after, hope for, strive for, want, wish for, yearn for.

aspirin *n.* **1** drug that reduces pain and fever. **2** tablet of this.

ass *n.* **1** donkey. **2** stupid person.

assail *v.* attack violently.

assailant *n.* attacker.

assassin *n.* person who assassinates another.

assassinate *v.* kill (an important person) by violent means. □ **assassination** *n.*

assault *n.* & *v.* attack.

■ *n.* attack, battering, battery, beating; charge, offensive, onset, onslaught, raid, strike. ● *v.* assail, attack, beat up, go for, hit, *colloq.* lay into, pounce on, set about, strike.

assay *n.* test of metal for quality. ● *v.* make an assay of.

assemble *v.* **1** gather together, collect. **2** put or fit together.

■ **1** collect, congregate, convene, forgather, gather, get together, meet, rally; bring together, marshal, round up. **2** build, construct, erect, fabricate, make, manufacture, put together.

assembly *n.* assembled group.

■ collection, company, congregation, crowd, flock, gathering, group, meeting, rally, throng; committee, congress, convention, convocation, council.

assent *v.* express agreement, consent. ● *n.* approval, consent.

■ *v.* accede, accept, acquiesce, agree, comply, concur, consent. ● *n.* acceptance, acquiescence, agreement, approbation, approval, concurrence, consent, permission, sanction.

assert *v.* **1** state, declare to be true. **2** use (power etc.) effectively. □ **assertion** *n.*

■ **1** affirm, allege, attest, claim, contend, declare, insist, maintain, profess, protest, state, swear, testify.

assertive *adj.* **1** forthright and self-assured. **2** dogmatic. □ **assertiveness** *n.*

■ **1** certain, confident, decided, definite, emphatic, firm, forceful, forthright, insistent, positive, self-assertive, self-assured, self-confident. **2** *colloq.* bossy, dogmatic, opinionated, peremptory, *colloq.* pushy.

assess *v.* estimate the size, value, or quality of. □ **assessment** *n.*, **assessor** *n.*

■ appraise, calculate, consider, estimate, evaluate, gauge, judge, rate, *colloq.* size up, weigh up, work out.

asset *n.* **1** useful or valuable thing or person. **2** (usu. *pl.*) property with money value.

■ **1** advantage, attraction, benefit, plus, resource, strength, talent. **2** (**assets**) capital, effects, estate, funds, holdings, means, money, possessions, property, resources, savings, valuables, wealth.

assiduous *adj.* diligent and persevering. □ **assiduously** *adv.*, **assiduity** *n.*

assign *v.* **1** allot. **2** designate to perform a task.

■ **1** allocate, allot, apportion, dispense, distribute, give out, hand (out). **2** appoint, delegate, designate, detail, name, nominate.

assignation *n.* arrangement to meet.

assignment *n.* task assigned.

■ chore, duty, errand, job, mission, project, task.

assimilate *v.* absorb or be absorbed into the body or a group etc., or into the mind as knowledge. □ **assimilation** *n.*

assist *v.* help. □ **assistance** *n.*

■ aid, back (up), cooperate with, help, lend a hand, support; advance, expedite, facilitate, further. □ **assistance** aid, backing, collaboration, cooperation, help, reinforcement, relief, succour, support.

assistant *n.* **1** helper. **2** person who serves customers in a shop.

■ **1** acolyte, aide, helper, right-hand man *or* woman; abettor, accessory, accomplice, collaborator, henchman; adjutant, aide-de-camp, auxiliary, deputy, subordinate, underling.

associate *v.* /əsṓshiayt/ **1** join or combine. **2** connect in one's mind. **3** mix socially. ● *n.* /əsṓshiət/ **1** companion, partner. **2** subordinate member.

■ *v.* **1** affiliate, ally, combine, join (together), unite. **2** connect, link, put together, relate. **3** consort, fraternize, hobnob, mingle, socialize. ● *n.* **1** accomplice, ally, colleague, companion, comrade, crony, fellow worker, friend, mate, partner.

association *n.* **1** act of associating. **2** group organized for a common purpose. **3** connection between ideas.

■ **1** amalgamation, combination, union. **2** alliance, cartel, club, coalition, confederation, consortium, federation, fellowship, group, guild, league, organization, society, syndicate, union. **3** connection, interconnection, link, relationship.

assonance *n.* partial resemblance of sound between syllables. □ **assonant** *adj.*

assorted *adj.* of different sorts.

assortment *n.* collection composed of several sorts.

■ array, collection, group, hotchpotch, jumble, medley, miscellany, mixture, potpourri, range, selection, variety.

assuage /əswáyj/ *v.* soothe, allay.

assume *v.* **1** take as true. **2** take or put upon oneself. **3** simulate.

■ **1** believe, fancy, imagine, presume, suppose, take it for granted, think. **2** accept, take on, undertake. **3** affect, fake, feign, pretend, simulate.

assumption *n.* **1** thing assumed to be true. **2** act of assuming.

■ **1** belief, conjecture, guess, hypothesis, postulation, premiss, presumption, presupposition, supposition, surmise, theory.

assurance *n.* **1** solemn promise or guarantee. **2** self-confidence. **3** life insurance.

■ **1** guarantee, oath, pledge, promise, vow, undertaking, word (of honour). **2** aplomb, confidence, poise, self-assurance, self-confidence.

assure *v.* **1** convince. **2** tell confidently.

assured *adj.* **1** sure, confident. **2** insured.

assuredly *adv.* certainly.

aster *n.* garden plant with daisy-like flowers.

asterisk *n.* star-shaped symbol *.

astern *adv.* **1** at or towards the stern. **2** backwards.

asteroid *n.* any of the tiny planets revolving round the sun.

asthma /ásmə/ *n.* chronic condition causing difficulty in breathing. □ **asthmatic** *adj.* & *n.*

astigmatism *n.* defect in an eye, preventing proper focusing. □ **astigmatic** *adj.*

astonish *v.* surprise greatly. □ **astonishment** *n.*

■ amaze, astound, bowl over, confound, dumbfound, *colloq.* flabbergast, shock, stagger, startle, stun, stupefy, surprise, take aback, take by surprise.

astound *v.* shock with surprise.

astrakhan *n.* dark curly fleece of lambs from Russia.

astray *adv.* & *adj.* away from the proper path.

astride *adv.* with one leg on each side.

astringent *adj.* **1** causing tissue to contract. **2** harsh, severe. ● *n.* astringent substance. □ **astringency** *n.*, **astringently** *adv.*

astrology *n.* study of the supposed influence of stars on human affairs. □ **astrologer** *n.*, **astrological** *adj.*

astronaut *n.* space traveller.

astronautics *n.* study of space travel and its technology.

astronomical *adj.* **1** of astronomy. **2** enormous in amount. □ **astronomically** *adv.*

astronomy *n.* study of stars and planets. □ **astronomer** *n.*

astute *adj.* shrewd, quick at seeing how to gain an advantage. □ **astutely** *adv.*, **astuteness** *n.*

■ acute, clever, discerning, intelligent, observant, perceptive, perspicacious, quick, sagacious, sharp, shrewd, wise; artful, canny, crafty, cunning, sly, wily.

asunder *adv.* apart, into pieces.

asylum *n.* **1** refuge. **2** (*old use*) mental institution.

■ **1** haven, refuge, retreat, sanctuary, shelter.

asymmetry *n.* lack of symmetry. □ **asymmetrical** *adj.*

at *prep.* having as position, time of day, condition, or price.

atavism *n.* resemblance to remote ancestors. □ **atavistic** *adj.*

ate *see* **eat**.

atheist *n.* person who does not believe in God. □ **atheism** *n.*

athlete *n.* person who is good at athletics.

athletic *adj.* **1** of athletes. **2** muscular and physically active. □ **athletically** *adv.*, **athleticism** *n.*

■ **2** energetic, fit, lithe, muscular, powerful, sinewy, strong, supple, vigorous, wiry.

athletics *n.pl.* sports, esp. running, jumping, and throwing.

atlas *n.* book of maps.

atmosphere *n.* **1** feeling conveyed by an environment or group. **2** mixture of gases surrounding a planet. **3** unit of pressure. □ **atmospheric** *adj.*

■ **1** air, ambience, aura, feeling, mood, spirit, tone. **2** air, ether, heavens, sky, stratosphere.

atoll *n.* ring-shaped coral reef enclosing a lagoon.

atom *n.* **1** smallest particle of a chemical element. **2** very small quantity or thing.

■ **2** bit, crumb, grain, iota, jot, mite, morsel, scrap, speck, spot.

atomic *adj.* of atom(s). □ **atomic bomb** bomb deriving its power from atomic energy. **atomic energy** that obtained from nuclear fission.

atomize *v.* reduce to atoms or fine particles. □ **atomizer** *n.*

atonal /áytṓn'l/ *adj.* (of music) not written in any key. □ **atonality** *n.*

atone *v.* make amends. □ **atonement** *n.*

■ compensate, make amends, make reparation, redeem oneself.

atrocious *adj.* **1** very bad. **2** wicked. □ **atrociously** *adv.*

■ **1** abominable, abysmal, awful, bad, dreadful, *colloq.* frightful, *colloq.* ghastly, horrendous, *colloq.* horrible, *colloq.* lousy, *colloq.* shocking, terrible. **2** appalling, barbaric, brutal, cruel, diabolical, evil, fiendish, heinous, horrific, horrifying, inhuman, savage, sickening, vicious, vile, villainous, wicked.

atrocity *n.* **1** wickedness. **2** cruel act.

■ **1** cruelty, enormity, evil, inhumanity, iniquity, villainy, wickedness. **2** crime, enormity, outrage.

atrophy *n.* wasting away, esp. through disuse. ● *v.* **1** cause atrophy in. **2** suffer atrophy.

attach *v.* **1** fix or join to something else. **2** attribute, be attributable. □ **attachment** *n.*

■ **1** add, affix, append, connect, couple, fasten, fix, hitch, join, link, pin, secure, tie, unite. **2** ascribe, assign, attribute, impute.

attaché *n.* person attached to an ambassador's staff. □ **attaché case** small rectangular case for carrying documents.

attached *adj.* devoted.

attack *v.* **1** attempt to hurt or defeat. **2** criticize adversely. ● *n.* **1** act or process of attacking. **2** sudden onset of illness. □ **attacker** *n.*

■ *v.* **1** assail, assault, beat up, go for, jump, lash out at, *colloq.* lay into, mug, pounce on, set about; charge, rush, storm. **2** abuse, berate, censure, criticize, denounce, *sl.* knock, *colloq.* lambaste, *colloq.* lay into, malign, revile, *sl.* slag (off), *sl.* slam, *colloq.* slate, vilify. ● *n.* **1** ambush, assault, battery, blitz, bombardment, charge, foray, incursion, invasion, offensive, onset, onslaught, raid, rush, sortie, strike. **2** bout, fit, outbreak, paroxysm, seizure, spasm, *colloq.* turn. □ **attacker** aggressor, assailant, mugger; critic, detractor.

attain *v.* achieve. □ **attainable** *adj.*, **attainment** *n.*

■ accomplish, achieve, arrive at, fulfil, gain, get, grasp, make, obtain, reach, realize, secure, win.

attempt *v.* make an effort to do. ● *n.* such an effort.

■ *v.* endeavour, have a go, make a bid, seek, strive, try. ● *n.* bid, effort, endeavour, essay, go, shot, *colloq.* stab, try.

attend *v.* **1** be present at. **2** escort. □ **attend to** give attention to. **attendance** *n.*

■ **1** be present (at), go (to), *colloq.* show up (at), turn up (at). **2** accompany, chaperon, conduct, escort, usher. □ **attend to** concentrate on, heed, listen to, mark, mind, note, pay attention to, take notice of, watch; deal with, do, handle, look after, see to, take care of.

attendant *adj.* accompanying. ● *n.* person present to provide service.

■ *n.* assistant, chaperon, escort, helper, servant, steward, stewardess, usher, valet.

attention *n.* **1** consideration, care. **2** applying one's mind. **3** erect attitude in military drill.

■ **1** care, consideration, heed, notice, regard, thought. **2** attentiveness, concentration.

attentive *adj.* paying attention. □ **attentively** *adv.*

■ alert, awake, concentrating, heedful, intent, observant, vigilant, watchful.

attenuate *v.* make slender, thin, or weaker. □ **attenuation** *n.*

attest *v.* **1** provide proof of. **2** declare true or genuine. □ **attestation** *n.*

attic *n.* room in the top storey of a house.

attire *n.* clothes. ● *v.* clothe.

■ *n.* apparel, clothes, clothing, costume, dress, finery, garb, garments, wear.

attitude *n.* **1** position of the body. **2** way of thinking or behaving.

■ **1** pose, position, posture, stance. **2** approach, opinion, outlook, perspective, point of view, standpoint, thought, view, viewpoint.

attorney *n.* (*US*) lawyer.

attract *v.* **1** draw to oneself or itself. **2** arouse the interest or pleasure of. □ **attraction** *n.*

■ **1** catch, capture, draw, invite, pull. **2** allure, appeal to, captivate, charm, enchant, entice, fascinate, interest, lure, tempt.

attractive *adj.* **1** capable of attracting. **2** pleasing in appearance. □ **attractively** *adv.*

■ **1** alluring, appealing, captivating, enchanting, engaging, enticing, fascinating, interesting, inviting, pleasing, seductive, tempting, winning. **2** beautiful, *Sc.* bonny, comely, desirable, glamorous, good-looking, *colloq.* gorgeous, handsome, personable, pretty, *colloq.* stunning, taking.

attribute[1] /ətríbyōōt/ *v.* **attribute to** regard as belonging to or caused by. □ **attributable** *adj.*, **attribution** *n.*

■ ascribe, assign, impute.

attribute[2] /átribyōōt/ *n.* characteristic quality.

■ characteristic, feature, property, quality, trait.

attrition *n.* wearing away.

attune *v.* **1** adapt. **2** tune.

atypical *adj.* not typical. □ **atypically** *adv.*

aubergine /óbərzheen/ *n.* **1** deep-purple vegetable. **2** its colour.

aubrietia *n.* perennial rock-plant.

auburn *adj.* (of hair) reddish-brown.

auction *n.* public sale where articles are sold to the highest bidder. ● *v.* sell by auction.

auctioneer *n.* person who conducts an auction.

audacious *adj.* bold, daring. □ **audaciously** *adv.*, **audacity** *n.*

■ adventurous, bold, brave, courageous, daring, fearless, intrepid, plucky, valiant; daredevil, foolhardy, rash, reckless. □ **audacity** boldness, bravery, courage, daring, *colloq.* guts, nerve, pluck, valour; rashness, recklessness, temerity.

audible *adj.* loud enough to be heard. □ **audibly** *adv.*

audience *n.* **1** group of listeners or spectators. **2** formal interview.

■ **1** assembly, congregation, crowd, gathering, listeners, spectators, turn-out, viewers.

audio *n.* (reproduction of) sound. □ **audio-visual** *adj.* using both sight and sound.

audit *n.* official examination of accounts. ● *v.* make an audit of.

audition *n.* test of a prospective performer's ability. ● *v.* test or be tested in an audition.

auditor *n.* one who audits accounts.

auditorium *n.* part of a building where the audience sits.

auditory *adj.* of hearing.

augment *v.* increase. ▫ **augmentation** *n.*, **augmentative** *adj.*

■ add to, amplify, boost, build up, enlarge, increase, multiply, step up, supplement, swell.

augur *v.* bode.

■ bode, foreshadow, herald, portend, predict, presage, promise, prophesy.

augury *n.* divination, omen.

august *adj.* majestic.

auk *n.* northern seabird.

aunt *n.* sister or sister-in-law of one's father or mother.

au pair young person from overseas helping with housework in return for board and lodging.

aura *n.* atmosphere surrounding a person or thing.

aural *adj.* of the ear. ▫ **aurally** *adv.*

aureole *n.* halo.

auscultation *n.* listening to the sound of the heart for diagnosis.

auspice *n.* **1** omen. **2** (*pl.*) patronage.

auspicious *adj.* showing signs that promise success.

■ bright, encouraging, favourable, hopeful, promising, propitious, rosy.

austere *adj.* **1** severely simple and plain. **2** stern. ▫ **austerity** *n.*

■ **1** modest, plain, simple, spartan, stark, unadorned, unostentatious. **2** cold, dour, forbidding, hard, harsh, rigorous, serious, sober, stern, strict.

Australasian *adj.* & *n.* (native, inhabitant) of Australia, New Zealand, and neighbouring islands.

Australian *adj.* & *n.* (native, inhabitant) of Australia.

authentic *adj.* genuine, known to be true. ▫ **authentically** *adv.*, **authenticity** *n.*

■ actual, bona fide, genuine, legitimate, real, true, valid; authoritative, reliable, truthful.

authenticate *v.* prove the truth or authenticity of. ▫ **authentication** *n.*

■ certify, confirm, corroborate, prove, substantiate, validate, verify.

author *n.* **1** writer of a book etc. **2** originator. ▫ **authorship** *n.*

■ **1** columnist, dramatist, journalist, novelist, playwright, poet, scriptwriter, writer. **2** creator, designer, father, founder, inventor, maker, originator.

authoritarian *adj.* favouring complete obedience to authority. ▫ **authoritarianism** *n.*

■ autocratic, *colloq.* bossy, despotic, dictatorial, dogmatic, domineering, high-handed, strict, tyrannical.

authoritative *adj.* **1** recognized as true or reliable. **2** having authority. ▫ **authoritatively** *adv.*

■ **1** authentic, definitive, dependable, reliable, true, valid. **2** approved, lawful, legitimate, official, sanctioned.

authority *n.* **1** power to enforce obedience. **2** delegated power. **3** person(s) with authority. **4** influence. **5** person with specialized knowledge.

■ **1** command, control, dominion, jurisdiction, power, prerogative, right, supremacy. **2** authorization, licence, permission. **3** (**authorities**) the establishment, government, officials. **4** hold, influence, sway, weight. **5** *colloq.* buff, connoisseur, expert, specialist.

authorize *v.* give permission for. ▫ **authorization** *n.*

■ agree to, allow, approve, consent to, endorse, give the go-ahead to, *colloq.* give the green light to, legalize, license, *colloq.* OK, permit, rubber-stamp, sanction.

autism *n.* mental disorder characterized by self-absorption and withdrawal. ▫ **autistic** *adj.*

auto- *pref.* self-.

autobiography *n.* story of a person's life written by that person. ▫ **autobiographical** *adj.*

autocracy *n.* despotism.

autocrat *n.* person with unrestricted power. ▫ **autocratic** *adj.*, **autocratically** *adv.*

autocross *n.* motor racing on dirt tracks.

autograph *n.* person's signature. ● *v.* write one's name in or on.

automate *v.* control by automation.

automatic *adj.* **1** (of a machine etc.) working by itself. **2** done without thinking. ● *n.* automatic machine or firearm. ▫ **automatically** *adv.*

■ *adj.* **1** automated, mechanical, robotic. **2** impulsive, instinctive, involuntary, knee-jerk, mechanical, reflex, spontaneous, unconscious, unthinking.

automation *n.* use of automatic equipment in industry.

automaton *n.* (*pl.* **-tons** or **-ta**) robot.

automobile *n.* (*US*) car.

automotive *adj.* concerned with motor vehicles.

autonomous *adj.* self-governing. □ **autonomy** *n.*

autopilot *n.* device for keeping an aircraft on a set course automatically.

autopsy *n.* post-mortem.

autumn *n.* season between summer and winter. □ **autumnal** *adj.*

auxiliary *adj.* giving help or support. ● *n.* helper. □ **auxiliary verb** one used in forming tenses of other verbs.

■ *adj.* additional, ancillary, extra, secondary, subsidiary, supplementary, supporting.

avail *v.* be of use or help (to). ● *n.* effectiveness, advantage. □ **avail oneself of** make use of.

available *adj.* **1** ready to be used. **2** within reach. □ **availability** *n.*

■ **1** at one's disposal, free, unoccupied. **2** accessible, at hand, handy, obtainable, *colloq.* on tap, to hand, within reach.

avalanche *n.* mass of snow pouring down a mountain.

avant-garde /ávvon-gaárd/ *n.* group of innovators. ● *adj.* progressive.

avarice *n.* greed for gain. □ **avaricious** *adj.*, **avariciously** *adv.*

■ □ **avaricious** covetous, grasping, greedy, mercenary, rapacious, selfish.

avenge *v.* take vengeance for.

avenue *n.* **1** wide street or road. **2** way of approach.

average *n.* **1** standard regarded as usual. **2** value arrived at by adding several quantities together and dividing by the number of these. ● *adj.* **1** of ordinary standard. **2** found by making an average.

■ *adj.* **1** commonplace, everyday, mediocre, medium, middling, moderate, normal, ordinary, run-of-the-mill, standard, typical, unexceptional, usual.

averse *adj.* unwilling, disinclined.

■ disinclined, indisposed, loath, opposed, reluctant, resistant, unwilling.

aversion *n.* strong dislike or unwillingness.

■ antipathy, dislike, disinclination, distaste, hatred, loathing, reluctance, unwillingness.

avert *v.* **1** turn away. **2** ward off.

■ **1** deflect, divert, turn aside *or* away. **2** fend off, forestall, prevent, stave off, ward off.

aviary *n.* large cage or building for keeping birds.

aviation *n.* flying an aircraft.

avid *adj.* eager, greedy. □ **avidly** *adv.*, **avidity** *n.*

avocado *n.* (*pl.* **-os**) pear-shaped tropical fruit.

avocet *n.* wading bird with a long upturned bill.

avoid *v.* refrain or keep away from. □ **avoidable** *adj.*, **avoidance** *n.*

■ abstain from, refrain from; bypass, circumvent, dodge, elude, escape, evade, give a wide berth to, shun, sidestep, skirt round, steer clear of.

avow *v.* declare. □ **avowal** *n.*

avuncular *adj.* of or like a kindly uncle.

await *v.* wait for.

awake *v.* (**awoke, awoken**) wake. ● *adj.* **1** not asleep. **2** alert.

■ *v.* awaken, come to, get up, rouse oneself, wake (up). ● *adj.* **1** conscious, up, wide awake. **2** alert, on one's guard, on one's toes, vigilant, watchful.

awaken *v.* awake.

award *v.* give officially as a prize, payment, or penalty. ● *n.* thing awarded.

■ *v.* bestow on, confer on, give, grant, present with. ● *n.* gift, prize, reward, trophy.

aware *adj.* having knowledge or realization. □ **awareness** *n.*

awash *adj.* washed over by water.

away *adv.* **1** to or at a distance. **2** into non-existence. **3** persistently. ● *adj.* played on an opponent's ground.

awe *n.* reverential fear or wonder. ● *v.* fill with awe.

■ *n.* admiration, amazement, dread, fear, respect, reverence, veneration, wonder.

aweigh *adv.* (of anchor) raised just clear of the sea bottom.

awesome *adj.* causing awe.

awful *adj.* **1** extremely bad or unpleasant. **2** (*colloq.*) very great. □ **awfully** *adv.*

■ **1** abominable, abysmal, appalling, atrocious, bad, *colloq.* beastly, deplorable, dreadful, *colloq.* frightful, ghastly, gruesome, horrendous, horrible, *colloq.* lousy, nasty, obnoxious, repellent, *colloq.* shocking, terrible, unpleasant.

awhile *adv.* for a short time.

awkward *adj.* **1** difficult to use. **2** clumsy. **3** embarrassing. **4** ill at ease. □ **awkwardly** *adv.*, **awkwardness** *n.*

■ **1** cumbersome, unmanageable, unwieldy; *colloq.* fiddly. **2** blundering, bungling, clumsy, gauche, gawky, *colloq.* ham-fisted, inept, maladroit, ungainly, ungraceful, unskilful. 3 delicate, difficult, embarrassing, *colloq.* sticky, ticklish, tricky, troublesome. 4 disconcerted, embarrassed, ill at ease, self-conscious, uncomfortable, uneasy.

awning *n.* roof-like canvas shelter.

awoke, awoken *see* **awake.**

awry /əˈrī/ *adv.* & *adj.* **1** twisted to one side. **2** amiss.

axe *n.* chopping tool. ● *v.* **(axing)** abolish, dismiss.

axiom *n.* accepted general truth or principle. □ **axiomatic** *adj.*

axis *n.* (*pl.* **axes**) line through the centre of an object, round which it rotates if spinning. □ **axial** *adj.*

axle *n.* rod on which wheels turn.

ayatollah *n.* religious leader in Iran.

aye *adv.* yes. ● *n.* vote in favour of a proposal.

azalea *n.* shrub-like flowering plant.

Aztec *n.* member of a former Indian people of Mexico.

azure *adj.* & *n.* sky-blue.

Bb

baa *n.* & *v.* bleat.
babble *v.* **1** chatter indistinctly or foolishly. **2** (of a stream) murmur. ● *n.* babbling talk or sound.
babe *n.* baby.
babel *n.* confused noise.
baboon *n.* large monkey.
baby *n.* very young child or animal. ▫ **babyish** *adj.*

> ■ babe, child, infant, toddler, tot. ▫ **babyish** childish, immature, infantile, juvenile, puerile, silly.

babysit *v.* look after a child while its parents are out. ▫ **babysitter** *n.*
baccarat /bákkəraa/ *n.* gambling card game.
bachelor *n.* **1** unmarried man. **2** person with university degree.
bacillus *n.* (*pl.* **-li**) rod-like bacterium.
back *n.* **1** surface or part furthest from the front. **2** rear part of the human body from shoulders to hips. **3** corresponding part of an animal's body. **4** defensive player positioned near the goal in football etc. ● *adj.* **1** situated behind. **2** of or for past time. ● *adv.* **1** at or towards the rear. **2** in or into a previous time, position, or state. **3** in return. ● *v.* **1** move backwards. **2** help, support. **3** lay a bet on. ▫ **back down** withdraw a claim or argument. **back-pedal** *v.* reverse one's previous action or opinion. **back seat** inferior position or status. **back up** support.

> ■ *adj.* **1** end, hind, posterior, rear. ● *v.* **1** reverse. **2** aid, assist, back up, encourage, endorse, help, promote, second, side with, support, uphold; finance, fund, sponsor, subsidize, underwrite.

backache *n.* pain in one's back.
backbencher *n.* MP not entitled to sit on the front benches.
backbiting *n.* spiteful talk.
backbone *n.* column of bones down the centre of the back.
backchat *n.* verbal insolence.
backcloth *n.* painted cloth at the back of a stage or scene.
backdate *v.* declare to be valid from an earlier date.
backdrop *n.* backcloth.
backfire *v.* **1** make an explosion in an exhaust pipe. **2** produce an undesired effect.
backgammon *n.* game played on a board with draughts and dice.
background *n.* **1** conditions surrounding something. **2** person's history. **3** back part of a scene or picture.

> ■ **1** circumstances, context, setting, surroundings. **2** experience, history, qualifications, training, upbringing.

backhand *n.* backhanded stroke.
backhanded *adj.* **1** performed with the back of the hand turned forwards. **2** said with underlying sarcasm.
backhander *n.* **1** backhanded stroke. **2** (*sl.*) bribe.
backlash *n.* violent hostile reaction.
backlog *n.* arrears of work.
backpack *n.* rucksack.
backside *n.* (*colloq.*) buttocks.
backslide *v.* slip back from good behaviour into bad.
backstage *adj.* & *adv.* behind a theatre stage.
backstroke *n.* stroke used in swimming on one's back.
backtrack *v.* **1** retrace one's route. **2** reverse one's opinion.
backward *adj.* **1** directed backwards. **2** having made less than normal progress. **3** shy. ● *adv.* backwards.
backwards *adv.* **1** towards the back. **2** with the back foremost.
backwash *n.* **1** receding waves created by a ship etc. **2** reaction.
backwater *n.* **1** stagnant water joining a stream. **2** place unaffected by new ideas or progress.
backwoods *n.pl.* remote region.
bacon *n.* salted or smoked meat from a pig.
bacteriology *n.* study of bacteria. ▫ **bacteriologist** *n.*
bacterium *n.* (*pl.* **-ia**) microscopic organism. ▫ **bacterial** *adj.*
bad *adj.* (**worse, worst**) **1** of poor quality. **2** unpleasant. **3** wicked or offensive. **4** naughty. **5** harmful. **6** serious, severe. **7**

decayed, polluted. □ **bad-tempered** irritable. **badly** *adv.*

■ **1** abominable, abysmal, awful, *colloq.* chronic, deplorable, disgraceful, dreadful, inadequate, incompetent, inferior, *colloq.* lousy, poor, second-rate, *colloq.* shocking, shoddy, substandard, unsatisfactory, useless, worthless. **2** awful, *colloq.* beastly, disagreeable, dreadful, *colloq.* lousy, nasty, terrible, unpleasant. **3** corrupt, criminal, cruel, depraved, evil, immoral, malevolent, malicious, offensive, sinful, vicious, vile, villainous, wicked, wrong. **4** disobedient, mischievous, naughty, rebellious, unruly, wayward, wild. **5** dangerous, deleterious, detrimental, harmful, hurtful, injurious, noxious, unhealthy. **6** appalling, awful, dire, disastrous, distressing, dreadful, ghastly, grave, horrible, serious, severe, terrible. **7** decayed, decomposing, foul, mildewed, mouldy, off, polluted, putrid, rancid, rank, rotten, sour, stale, tainted. □ **bad-tempered** cantankerous, crabby, cross, crotchety, disagreeable, fractious, grumpy, irascible, irritable, peevish, petulant, prickly, quarrelsome, snappy, splenetic, surly, testy.

bade *see* **bid[2]**.

badge *n.* thing worn to show membership, rank, etc.

■ crest, emblem, insignia, logo, trade mark, symbol, token.

badger *n.* burrowing animal. ● *v.* pester.

badminton *n.* game like tennis, played with a shuttlecock.

baffle *v.* **1** be too difficult for. **2** frustrate. □ **bafflement** *n.*

■ **1** *colloq.* bamboozle, bewilder, confound, confuse, floor, *colloq.* flummox, mystify, perplex, puzzle, *colloq.* stump. **2** baulk, foil, frustrate, stymie, thwart.

bag *n.* **1** flexible container. **2** handbag. **3** (*pl., colloq.*) large amount. ● *v.* (**bagged**) take for oneself.

baggage *n.* luggage.

baggy *adj.* (**-ier**, **-iest**) hanging in loose folds.

bagpipes *n.pl.* wind instrument with air stored in a bag and pressed out through pipes.

bail[1] *n.* money pledged as security that an accused person will return for trial. ● *v.* **bail out 1** obtain or allow the release of (a person) on bail. **2** relieve by financial help.

bail[2] *n.* each of two crosspieces resting on the stumps in cricket.

bail[3] *v.* scoop water out of. See also **bale**.

bailey *n.* outer wall of a castle.

bailiff *n.* law officer empowered to seize goods for non-payment of fines or debts.

bailiwick *n.* area of authority.

bait *n.* an enticement, esp. placed to attract prey. ● *v.* **1** torment. **2** place bait on or in.

■ *n.* enticement, decoy, inducement, lure, temptation. ● *v.* **1** annoy, goad, harass, hound, persecute, pester, provoke, taunt, tease, torment.

baize *n.* thick green woollen cloth used for covering billiard tables.

bake *v.* cook or harden by dry heat.

baker *n.* person who bakes and sells bread.

bakery *n.* place where bread is baked for sale.

baking powder mixture used to make cakes rise.

balaclava (helmet) woollen cap covering the head and neck.

balalaika *n.* Russian guitar-like instrument with a triangular body.

balance *n.* **1** even distribution of weight or amount. **2** remainder. **3** difference between credits and debits. **4** weighing apparatus. **5** regulating apparatus of a clock. ● *v.* **1** consider by comparing. **2** be, put, or keep in a state of balance.

■ *n.* **1** equality, equilibrium, evenness, parity, poise, steadiness, symmetry. **2** difference, remainder, residue, rest, surplus. ● *v.* **1** compare, consider, counterbalance, evaluate, offset, weigh. **2** even up, keep balanced, level, steady, stabilize.

balcony *n.* **1** projecting platform with a rail or parapet. **2** upper floor of seats in a theatre etc.

bald *adj.* **1** with scalp wholly or partly hairless. **2** without details. **3** (of tyres) with the tread worn away. □ **baldly** *adv.*, **baldness** *n.*

balderdash *n.* nonsense.

balding *adj.* becoming bald.

bale *n.* **1** large bound bundle of straw etc. **2** large package of goods. ● *v.* make into a bale or bales. □ **bale out** (also **bail out**) make an emergency parachute jump from an aircraft etc.

baleful *adj.* menacing, destructive. □ **balefully** *adv.*

ball[1] *n.* **1** spherical object used in games. **2** rounded part or mass. **3** delivery of a ball by a bowler. □ **ball-bearing** *n.* **1** bearing using small steel balls. **2** one such ball. **ballpoint** *n.* pen with a tiny ball as its writing-point.

ball[2] *n.* social gathering for dancing.

ballad *n.* song telling a story.

ballast *n.* heavy material placed in a ship's hold to steady it.

ballcock *n.* device with a floating ball controlling the water level in a cistern.

ballerina *n.* female ballet dancer.

ballet *n.* performance of dancing and mime to music.

ballistics *n.pl.* study of projectiles. □ **ballistic** *adj.*

balloon *n.* bag inflated with air or lighter gas. ● *v.* swell like this.

ballot *n.* **1** method of voting by writing on a slip of paper. **2** votes recorded in this way. ● *v.* (**balloted**) (cause to) vote by ballot.

ballroom *n.* large room where dances are held.

ballyhoo *n.* **1** fuss. **2** extravagant publicity.

balm *n.* **1** soothing influence. **2** fragrant herb. **3** ointment.

balmy *adj.* (**-ier, -iest**) **1** fragrant. **2** (of air) soft and warm.

balsa *n.* **1** tropical American tree. **2** its lightweight wood.

balsam *n.* **1** soothing oil. **2** a kind of flowering plant.

baluster *n.* short stone pillar in a balustrade.

balustrade *n.* row of short pillars supporting a rail or coping.

bamboo *n.* giant tropical grass with hollow stems.

bamboozle *v.* (*colloq.*) **1** mystify. **2** trick.

ban *v.* (**banned**) forbid officially. ● *n.* order banning something.

> ■ *v.* bar, debar, disallow, forbid, outlaw, prevent, prohibit, proscribe, stop, suppress, veto. ● *n.* boycott, embargo, interdict, moratorium, prohibition, taboo, veto.

banal *adj.* commonplace, uninteresting. □ **banality** *n.*

> ■ boring, commonplace, *colloq.* corny, dull, hackneyed, humdrum, ordinary, pedestrian, platitudinous, stale, stereotyped, trite, unimaginative, unoriginal.

banana *n.* **1** curved yellow fruit. **2** tropical tree bearing this.

band *n.* **1** strip of material. **2** stripe. **3** organized group of people. **4** group of musicians. **5** range of values or wavelengths. □ **bandmaster** *n.*, **bandsman** *n.*

> ■ **1** belt, ribbon, sash. **2** bar, border, line, ring, streak, striation, strip, stripe. **3** body, clique, club, company, crew, gang, group, horde, party, team, troop. **4** ensemble, group, orchestra.

bandage *n.* strip of material for binding a wound. ● *v.* bind with this.

bandit *n.* member of a band of robbers.

> ■ brigand, buccaneer, desperado, gangster, highwayman, marauder, outlaw, pirate, robber, thief.

bandstand *n.* covered outdoor platform for musicians.

bandwagon *n.* **climb on the bandwagon** join a movement heading for success.

bandy[1] *v.* pass to and fro.

bandy[2] *adj.* (**-ier, -iest**) (of legs) curving apart at the knees.

bane *n.* cause of trouble or ruin.

bang *n.* **1** noise of or like an explosion. **2** sharp blow. ● *v.* **1** make this noise. **2** strike. **3** shut noisily. ● *adv.* **1** abruptly. **2** exactly.

> ■ *n.* **1** blast, boom, clap, crash, detonation, explosion, report, shot.

banger *n.* **1** firework that explodes noisily. **2** (*sl.*) noisy old car. **3** (*sl.*) sausage.

bangle *n.* bracelet of rigid material.

banish *v.* **1** condemn to exile. **2** dismiss from one's presence or thoughts. □ **banishment** *n.*

> ■ **1** deport, drive out, eject, exile, expel, oust, outlaw.

banisters *n.pl.* uprights and handrail of a staircase.

banjo *n.* guitar-like musical instrument with a circular body.

bank[1] *n.* **1** slope, esp. at the side of a river. **2** raised mass of earth etc. **3** row of lights, switches, etc. ● *v.* **1** build up into a bank. **2** tilt sideways in rounding a curve.

bank[2] *n.* **1** establishment for safe keeping of money. **2** storage place. ● *v.* **1** place money in a bank. **2** base one's hopes.

banknote *n.* piece of paper money.

bankrupt *adj.* unable to pay one's debts. ● *n.* bankrupt person. ● *v.* make bankrupt. □ **bankruptcy** *n.*

banner *n.* kind of flag.

> ■ ensign, flag, pennant, pennon, standard, streamer.

banns *n.pl.* announcement in church of an intended marriage.

banquet *n.* elaborate ceremonial public meal.

banquette *n.* long upholstered seat attached to a wall.

banshee *n.* spirit whose wail is said to foretell a death.

bantam *n.* small kind of fowl.

banter *n.* good-humoured joking. ● *v.* joke thus.

■ *n.* **jesting, joking, pleasantries, repartee, teasing.**

Bantu *adj.* & *n.* (*pl.* **Bantu** or **-us**) (member) of a group of African Negroid peoples.

bap *n.* large soft bread roll.

baptism *n.* religious rite of sprinkling with water as a sign of purification, usu. with name-giving. □ **baptismal** *adj.*

Baptist *n.* member of a Protestant sect believing that baptism should be by immersion.

baptistery *n.* place where baptism is performed.

baptize *v.* **1** perform baptism on. **2** name, nickname.

bar[1] *n.* **1** long piece of solid material. **2** strip. **3** barrier. **4** place containing a counter where alcohol or refreshments are served, this counter. **5** vertical line dividing music into units, this unit. **6** barristers, their profession. ● *v.* (**barred**) **1** fasten or secure with bar(s). **2** obstruct. **3** prohibit. ● *prep.* except.

■ *n.* **1 beam, pole, rail, railing, rod, shaft, stake, stick; block, cake, lump, tablet. 2 band, line, streak, strip, stripe. 3 barrier, deterrent, hindrance, impediment, obstacle, obstruction. 4 café, pub, wine bar.** ● *v.* **1 barricade, fasten, make fast, secure. 2 block, frustrate, hamper, hinder, impede, obstruct, prevent, stop. 3 ban, debar, exclude, forbid, prohibit.**

bar[2] *n.* unit of atmospheric pressure.

barb *n.* **1** backward-pointing part of an arrow. **2** wounding remark.

barbarian *n.* uncivilized person.

■ **boor,** *colloq.* **brute, hooligan, ignoramus, lout, oaf, philistine, ruffian, vandal,** *sl.* **yob; savage.**

barbaric *adj.* **1** brutal, cruel. **2** uncivilized.

■ **1 barbarous, bestial, brutal, brutish, cruel, ferocious, inhuman, ruthless, savage, vicious, violent.**

barbarity *n.* savage cruelty.

barbarous *adj.* **1** cruel. **2** uncivilized.

barbecue *n.* **1** frame for grilling food above an open fire. **2** this food. **3** open-air party where such food is served. ● *v.* cook on a barbecue.

barbed *adj.* having barbs. □ **barbed wire** wire with many sharp points.

barber *n.* men's hairdresser.

barbican *n.* **1** outer defence to a city or castle. **2** double tower over a gate or bridge.

barbiturate *n.* sedative drug.

bar code machine-readable striped code identifying a commodity, its price, etc.

bard *n.* **1** Celtic minstrel. **2** poet. □ **bardic** *adj.*

bare *adj.* **1** not clothed or covered. **2** lacking vegetation. **3** not adorned. **4** only just enough. ● *v.* reveal. □ **barely** *adv.*

■ *adj.* **1 nude, (stark) naked, unclothed, uncovered, undressed; defoliated, denuded, leafless, shorn, stripped. 2 barren, bleak, desolate, open, treeless, windswept. 3 austere, plain, simple, unadorned, undecorated, unfurnished. 4 basic, meagre, mere, minimal, scant, scanty.** ● *v.* **expose, reveal, show, uncover, unmask, unveil.**

bareback *adv.* without a saddle.

barefaced *adj.* shameless, undisguised.

bargain *n.* **1** agreement with obligations on both sides. **2** thing obtained cheaply. ● *v.* discuss the terms of an agreement. □ **bargain for** expect.

■ *n.* **1 agreement, arrangement, contract, deal, pact, settlement, understanding. 2 good buy,** *sl.* **snip,** *colloq.* **steal.** ● *v.* **barter, haggle, negotiate.** □ **bargain for be prepared for, envisage, expect, foresee.**

barge *n.* large flat-bottomed boat used on rivers and canals. ● *v.* move clumsily. □ **barge in** intrude.

baritone *n.* male voice between tenor and bass.

barium *n.* white metallic element.

bark[1] *n.* outer layer of a tree. ● *v.* scrape skin off accidentally.

bark[2] *n.* sharp harsh cry of a dog. ● *v.* **1** make this sound. **2** utter sharply.

barley *n.* **1** cereal plant. **2** its grain. □ **barley sugar** sweet made of boiled sugar. **barley water** drink made from pearl barley.

barman, barmaid *ns.* person serving in a pub etc.

barmy *adj.* (**-ier, -iest**) (*sl.*) crazy.

barn *n.* building for storing grain or hay etc.

barnacle *n.* shellfish that attaches itself to objects under water.

barometer *n.* instrument measuring atmospheric pressure. □ **barometric** *adj.*

baron *n.* member of the lowest rank of nobility. □ **baroness** *n.*, **baronial** *adj.*

baronet *n.* lowest hereditary title in Britain. ▫ **baronetcy** *n.*
baroque /bərók/ *adj.* of the ornate architectural style of the 17th–18th centuries. ● *n.* this style.
barque *n.* sailing ship.
barrack *v.* shout or jeer (at).
barracks *n.pl.* building(s) for soldiers to live in.
barracuda *n.* large voracious tropical sea fish.
barrage *n.* **1** heavy bombardment. **2** artificial barrier.
barrel *n.* **1** large round container with flat ends. **2** tube-like part esp. of a gun. ▫ **barrel organ** musical instrument with a rotating pin-studded cylinder.

■ **1** butt, cask, churn, drum, keg, tank, tub, water-butt.

barren *adj.* **1** not fertile. **2** without vegetation. ▫ **barrenness** *n.*

■ **1** infertile, sterile, unfruitful, unproductive. **2** arid, bare, desert, desolate, dry, lifeless, waste.

barricade *n.* barrier. ● *v.* block or defend with a barricade.
barrier *n.* thing that prevents or controls advance or access.

■ bar, barricade, boom, bulwark, dam, fence, hurdle, obstruction, railing, stockade, wall; handicap, hindrance, impediment, obstacle, stumbling block.

barrister *n.* lawyer representing clients in court.
barrow[1] *n.* **1** wheelbarrow. **2** cart pushed or pulled by hand.
barrow[2] *n.* prehistoric burial mound.
barter *n.* & *v.* trade by exchange of goods for other goods.
basalt /bássawlt/ *n.* dark rock of volcanic origin.
base *n.* **1** part on which a thing rests or is supported. **2** point from which something is developed. **3** headquarters. **4** substance capable of combining with an acid to form a salt. **5** each of four stations to be reached by a batter in baseball. ● *v.* **1** use as a basis (for). **2** station. ● *adj.* **1** dishonourable. **2** of inferior value.

■ *n.* **1** bottom, foot, pedestal, plinth, stand, support. **2** basis, foundation, starting point, underlying principle. **3** camp, centre, headquarters. ● *v.* **1** build, establish, found, ground. **2** locate, place, position, station. ● *adj.* **1** contemptible, corrupt, cowardly, despicable, dishonourable, disreputable, ignoble, immoral, low, mean, shabby, shameful, sordid, undignified, unworthy, vile, wicked. **2** cheap, inferior, mean, poor, second-rate, shoddy, worthless.

baseball *n.* American team game played with bat and ball, in which the batter has to hit the ball and run round a circuit.
baseless *adj.* groundless.
basement *n.* storey below ground level.
bash *v.* strike violently. ● *n.* violent blow or knock.
bashful *adj.* shy, self-conscious.

■ abashed, coy, demure, diffident, embarrassed, meek, reserved, retiring, self-conscious, self-effacing, shamefaced, sheepish, shy, timid.

basic *adj.* **1** very important. **2** simple. **3** lowest in level. **4** deeply rooted. ● *n.* (usu. *pl.*) fundamental fact or principle. ▫ **basically** *adv.*

■ *adj.* **1** central, chief, crucial, essential, fundamental, important, intrinsic, key, main, necessary, primary, principal, underlying, vital. **2** bare, crude, primitive, simple, spartan. **3** beginner's, early, elementary, first, initial, introductory, preliminary, primary, rudimentary. **4** elemental, inborn, inherent, innate, instinctive, profound, radical. ● *n.* (**basics**) elements, essentials, first principles, foundations, fundamentals, *sl.* nitty-gritty, rudiments.

basil *n.* sweet-smelling herb.
basilica *n.* oblong hall or church with an apse at one end.
basilisk *n.* **1** American lizard. **2** mythical reptile said to cause death by its glance or breath.
basin *n.* **1** deep open container for liquids. **2** washbasin. **3** sunken place. **4** area drained by a river. ▫ **basinful** *n.*
basis *n.* (*pl.* **bases**) **1** foundation or support. **2** main principle.

■ **1** base, footing, foundation, support; grounds, justification, reason. **2** cornerstone, essence, underlying principle.

bask *v.* sit or lie comfortably in warmth and light.
basket *n.* container made of interwoven cane or wire.
basketball *n.* team game in which the aim is to throw the ball through a high net.
basketwork *n.* **1** art of weaving cane etc. **2** work so produced.

Basque *n.* & *adj.* **1** (member) of a people living in the western Pyrenees. **2** (of) their language.

bas-relief *n.* sculpture or carving in low relief.

bass[1] /bass/ *n.* (*pl.* **bass**) fish of the perch family.

bass[2] /bayss/ *adj.* deep-sounding, of the lowest pitch in music. ● *n.* (*pl.* **basses**) **1** lowest male voice. **2** bass pitch. **3** double bass.

basset *n.* short-legged hound.

bassoon *n.* woodwind instrument with a deep tone.

bast *n.* **1** inner bark of the lime tree. **2** similar fibre.

bastard *n.* **1** illegitimate child. **2** (*sl.*) unpleasant or difficult person or thing. □ **bastardy** *n.*

baste[1] *v.* sew together temporarily with loose stitches.

baste[2] *v.* **1** moisten with fat during cooking. **2** thrash.

bastion *n.* **1** projecting part of a fortified place. **2** stronghold.

bat[1] *n.* **1** wooden implement for striking a ball in games. **2** batsman. ● *v.* (**batted**) perform or strike with the bat in cricket etc. □ **batsman** *n.* player batting in cricket.

bat[2] *n.* nocturnal flying animal with a mouse-like body.

batch *n.* set of people or things dealt with as a group.

bated *adj.* **with bated breath** anxiously.

bath *n.* **1** washing (of the whole body) by immersion. **2** container used for this. ● *v.* wash in a bath. □ **bathroom** *n.* room with a bath.

bathe *v.* **1** immerse in liquid. **2** swim for pleasure. ● *n.* swim.

bathos *n.* anticlimax, descent from an important thing to a trivial one.

batik *n.* method of printing designs on textiles by waxing parts not to be dyed.

batman *n.* soldier acting as an officer's personal servant.

baton *n.* short stick, esp. used by a conductor or police officer.

batrachian /bətráykiən/ *n.* amphibian that discards gills and tail when fully developed.

battalion *n.* army unit of several companies.

batten[1] *n.* bar of wood or metal, esp. holding something in place. ● *v.* fasten with batten(s).

batten[2] *v.* **batten on** thrive at another's expense.

batter[1] *v.* hit hard and often. ● *n.* beaten mixture of flour, eggs, and milk, used in cooking.

■ *v.* bash, beat, belabour, bludgeon, *sl.* clobber, cudgel, hit, pound, pummel, strike, thrash.

batter[2] *n.* player batting in baseball.

battering ram iron-headed beam formerly used in war for breaking through walls or gates.

battery *n.* **1** group of big guns. **2** artillery unit. **3** set of similar or connected units of equipment, poultry cages, etc. **4** electric cell(s) supplying current. **5** unlawful blow or touch.

battle *n.* **1** fight between armed forces. **2** contest, struggle. ● *v.* struggle.

■ *n.* **1** action, clash, combat, conflict, encounter, engagement, fight, fray. **2** competition, contest, match; campaign, crusade, fight, struggle, war. ● *v.* campaign, crusade, fight, struggle, wage war, wrestle.

battleaxe *n.* **1** medieval weapon. **2** (*colloq.*) formidable woman.

battlefield *n.* scene of battle.

battlements *n.pl.* parapet with gaps for firing from.

battleship *n.* warship of the most heavily armed kind.

batty *adj.* (**-ier, -iest**) (*sl.*) crazy.

bauble *n.* valueless ornament.

baulk *v.* **1** shirk. **2** frustrate. ● *n.* hindrance.

bauxite *n.* mineral from which aluminium is obtained.

bawdy *adj.* (**-ier, -iest**) humorously indecent. □ **bawdiness** *n.*

bawl *v.* **1** shout. **2** weep noisily. □ **bawl out** (*colloq.*) reprimand.

■ **1** bellow, roar, shout, thunder, yell. **2** blubber, cry, howl, sob, wail, weep.

bay[1] *n.* a kind of laurel.

bay[2] *n.* part of a sea or lake within a wide curve of the shore.

■ cove, creek, estuary, fiord, firth, gulf, harbour, inlet.

bay[3] *n.* recess, compartment. □ **bay window** one projecting from an outside wall.

■ alcove, booth, compartment, opening, recess.

bay[4] *n.* deep cry of a large dog or of hounds. ● *v.* make this sound. □ **at bay** forced to face attackers.

bay[5] *adj.* & *n.* reddish-brown (horse).

bayonet *n.* stabbing blade fixed to the muzzle of a rifle.

bazaar *n.* **1** oriental market. **2** sale of goods to raise funds.

bazooka *n.* portable weapon for firing anti-tank rockets.

be *v.* **1** exist, occur. **2** have a certain position, quality, or condition. ● *v.aux.* used to form tenses of other verbs.

beach *n.* sandy or pebbly shore of the sea. ● *v.* bring on shore from water. □ **beachcomber** *n.* person who salvages things on a beach. **beachhead** *n.* fortified position set up on a beach by an invading army.

■ *n.* coast, littoral, sands, seashore, shore, strand.

beacon *n.* fire or light used as a signal.

bead *n.* **1** small shaped piece of hard material pierced for threading with others on a string. **2** drop of liquid.

beading *n.* strip of trimming for wood.

beadle *n.* (formerly) minor parish official.

beady *adj.* (**-ier, -iest**) (of eyes) small and bright.

beagle *n.* small hound used for hunting hares.

beak *n.* **1** bird's horny projecting jaws. **2** any similar projection. **3** (*sl.*) magistrate.

beaker *n.* tall drinking cup.

beam *n.* **1** long piece of timber or metal used in house-building etc. **2** ray of light or other radiation. **3** bright smile. **4** ship's breadth. ● *v.* **1** send out light etc. **2** smile radiantly.

■ *n.* **1** bar, girder, joist, plank, rafter, support, timber. **2** gleam, ray, shaft. **3** grin, smile. ● *v.* **1** emit, radiate, send out, shine. **2** grin, smile.

bean *n.* **1** plant with kidney-shaped seeds in long pods. **2** seed of this or of coffee.

bear[1] *n.* large heavy animal with thick fur.

bear[2] *v.* (**bore, borne**) **1** carry. **2** support. **3** keep in thought or memory. **4** endure, tolerate. **5** be fit for. **6** produce, give birth to. **7** take (a specified direction).

■ **1** bring, carry, convey, deliver, take, transport. **2** carry, hold, support, sustain, take. **3** harbour, keep, retain. **4** cope with, endure, experience, go through, suffer, undergo, weather, withstand; abide, brook, put up with, stand, stomach, *colloq.* stick, tolerate. **5** be worthy of, deserve, merit, rate. **6** develop, generate, produce, yield; give birth to, have, spawn. **7** fork, go, swing, turn, veer.

bearable *adj.* endurable.

■ acceptable, endurable, supportable, tolerable.

beard *n.* hair on and round a man's chin. ● *v.* confront boldly.

bearing *n.* **1** posture. **2** outward behaviour. **3** relevance. **4** compass direction. **5** device reducing friction where a part turns.

■ **1** carriage, deportment, posture, stance. **2** air, aspect, attitude, behaviour, conduct, demeanour, manner, mien, presence. **3** applicability, application, connection, pertinence, relationship, relevance, significance.

bearskin *n.* guardsman's tall furry cap.

beast *n.* **1** large four-footed animal. **2** brutal person.

■ **1** animal, creature. **2** brute, fiend, monster, ogre, savage.

beastly *adj.* (**-ier, -iest**) (*colloq.*) very unpleasant.

beat *v.* (**beat, beaten**) **1** hit repeatedly. **2** mix vigorously. **3** (of the heart) pump rhythmically. **4** do better than, defeat. ● *n.* **1** recurring emphasis marking rhythm. **2** throbbing action or sound. **3** appointed course of a policeman or sentinel. □ **beat up** assault violently.

■ *v.* **1** bash, batter, belabour, *sl.* belt, bludgeon, *sl.* clobber, clout, flog, hammer, hit, *colloq.* lambaste, lash, *colloq.* lay into, pound, pummel, punch, strike, thrash, thump, thwack, *sl.* wallop, *colloq.* whack, whip. **2** blend, mix, stir, whip, whisk. **3** palpitate, pound, pulsate, pulse, throb. **4** conquer, crush, defeat, get the better of, *colloq.* lick, outdo, outstrip, overcome, overpower, overwhelm, rout, subdue, surpass, thrash, trounce, vanquish, worst. ● *n.* **1** measure, rhythm, stress, tempo. **2** palpitation, pulsation, pulse, throb. **3** circuit, course, path, route, way.

beatific *adj.* showing great happiness. □ **beatifically** *adv.*

beatify *v.* (*RC Church*) declare blessed, as first step in canonization. □ **beatification** *n.*

beatitude *n.* blessedness.

beauteous *adj.* (*poetic*) beautiful.

beautician *n.* person who gives beauty treatment.

beautiful *adj.* **1** having beauty. **2** excellent. □ **beautifully** *adv.*

■ **1** alluring, appealing, attractive, *poetic* beauteous, *Sc.* bonny, comely, *colloq.* divine, exquisite, glamorous, good-looking, *colloq.* gorgeous, handsome, lovely, pretty, *colloq.* stunning; artistic, charming, decorative, delightful, elegant, fine, graceful, picturesque, scenic, tasteful. **2** admirable, excellent, magnificent, marvellous, splendid, superb, wonderful.

beautify *v.* make beautiful. □ **beautification** *n.*

■ adorn, deck, decorate, embellish, ornament, *colloq.* titivate.

beauty *n.* **1** combination of qualities giving pleasure to the sight, mind, etc. **2** beautiful person or thing.

■ **1** attractiveness, charm, elegance, glamour, grace, loveliness, prettiness, radiance, splendour. **2** belle, *colloq.* knockout, *colloq.* stunner.

beaver *n.* small amphibious rodent. ● *v.* work hard.

becalmed *adj.* unable to move because there is no wind.

because *conj.* for the reason that. □ **because of** by reason of.

beck[1] *n.* **at the beck and call of** ready and waiting to obey.

beck[2] *n.* mountain stream.

beckon *v.* summon by a gesture.

become *v.* (**became, become**) **1** come to be, begin to be. **2** suit.

■ **1** change into, develop into, grow into, mature into, turn into. **2** flatter, look good on, suit.

bed *n.* **1** thing to sleep or rest on. **2** framework with a mattress and coverings. **3** flat base, foundation. **4** bottom of a sea or river etc. **5** layer. **6** garden plot.

bedbug *n.* bug infesting beds.

bedclothes *n.pl.* sheets, blankets, etc.

bedding *n.* beds and bedclothes.

bedevil *v.* (**bedevilled**) afflict with difficulties.

bedlam *n.* scene of uproar.

Bedouin /bédoo-in/ *n.* (*pl.* **Bedouin**) member of an Arab people living in tents in the desert.

bedpan *n.* pan for use as a lavatory by an invalid in bed.

bedraggled *adj.* limp and untidy.

■ dishevelled, drenched, messy, *colloq.* scruffy, soaked, sodden, unkempt, untidy, wet.

bedridden *adj.* permanently confined to bed through illness.

bedrock *n.* **1** solid rock beneath loose soil. **2** basic facts.

bedroom *n.* room for sleeping in.

bedsit *n.* (also **bedsitter**) room for living and sleeping in.

bedsore *n.* sore developed by lying in bed for a long time.

bedspread *n.* covering spread over a bed during the day.

bedstead *n.* framework of a bed.

bee *n.* insect that produces honey.

beech *n.* tree with smooth bark and glossy leaves.

beef *n.* **1** meat from ox, bull, or cow. **2** muscular strength. **3** (*sl.*) grumble. ● *v.* (*sl.*) grumble.

beefburger *n.* hamburger.

beefeater *n.* warder in the Tower of London.

beefy *adj.* (**-ier, -iest**) having a solid muscular body.

beehive *n.* structure in which bees live.

beeline *n.* **make a beeline for** go straight or rapidly towards.

beep *n.* & *v.* bleep. □ **beeper** *n.*

beer *n.* alcoholic drink made from malt and hops. □ **beery** *adj.*

beeswax *n.* yellow substance secreted by bees, used as polish.

beet *n.* plant with a fleshy root used as a vegetable or for making sugar.

beetle[1] *n.* insect with hard wing-covers.

beetle[2] *n.* tool for ramming or crushing things.

beetle[3] *v.* overhang, project.

beetroot *n.* beet with a dark red root used as a vegetable.

befall *v.* (**befell, befallen**) **1** happen. **2** happen to.

befit *v.* (**befitted**) be suitable for.

before *adv., prep.,* & *conj.* **1** at an earlier time (than). **2** ahead, in front of. **3** in preference to.

beforehand *adv.* in advance.

befriend *v.* show kindness towards.

beg *v.* (**begged**) **1** ask for as a gift or charity. **2** request earnestly.

■ **1** cadge, scrounge, sponge. **2** beseech, crave, entreat, implore, importune, petition, plead with, pray, request, solicit, supplicate.

beggar *n.* person who lives by begging. ● *v.* reduce to poverty.

begin *v.* (**began, begun, beginning**) **1** perform the first part of. **2** come into existence. **3** be the first to do a thing.

■ **1** commence, embark on, get going, inaugurate, initiate, set about, set in motion, set out (on), start. **2** arise, come into being, get under way, originate, start.

beginner *n.* person just beginning to learn a skill.

■ **apprentice, initiate, learner, novice, recruit, trainee.**

beginning *n.* **1** first part. **2** starting point, source or origin.

■ **1 commencement, inauguration, introduction, opening, start. 2 creation, dawn, dawning, genesis, inception, onset, origin, outset, source, start, starting point.**

begonia *n.* garden plant with bright leaves and flowers.

begrudge *v.* be unwilling to give or allow.

beguile *v.* **1** entertain pleasantly. **2** deceive.

■ **1 absorb, amuse, charm, divert, entertain. 2 cheat, deceive, delude, dupe, fool, hoodwink, mislead, trick.**

begum *n.* title of a Muslim married woman.

behalf *n.* **on behalf of** as the representative of.

behave *v.* **1** act or react in a specified way. **2** (also **behave oneself**) show good manners.

■ **1 act, conduct oneself, function, operate, perform, react.**

behaviour *n.* way of behaving.

■ **actions, attitude, bearing, conduct, demeanour, deportment, manners.**

behead *v.* cut the head off.

beheld *see* **behold.**

behind *adv.* & *prep.* **1** in or to the rear (of). **2** in arrears. **3** remaining after others' departure. ● *n.* buttocks.

behold *v.* (**beheld**) (*poetic*) **1** see. **2** observe. □ **beholder** *n.*

beholden *adj.* owing thanks.

behove *v.* be incumbent on.

beige *adj.* & *n.* light fawn (colour).

being *n.* **1** existence. **2** thing that exists and has life.

■ **1 actuality, existence, life, reality. 2 animal, creature, human being, individual, mortal, person, soul.**

belabour *v.* **1** beat. **2** attack.

belated *adj.* coming (too) late.

■ **delayed, late, overdue, tardy, unpunctual.**

belch *v.* send out wind noisily from the stomach through the mouth. ● *n.* act of belching.

beleaguer *v.* besiege.

belfry *n.* **1** bell tower. **2** space for bells in a tower.

belie *v.* contradict, fail to confirm.

belief *n.* **1** what one believes. **2** trust, confidence.

■ **1 conviction, creed, doctrine, opinion, persuasion, principle(s), tenet, view; faith, religion. 2 certainty, confidence, credence, faith, reliance, sureness, trust.**

believe *v.* **1** accept as true or as speaking truth. **2** think, suppose. □ **believe in 1** have faith in the existence of. **2** feel sure of the worth of. **believer** *n.*

■ **1 accept, be certain of, credit, find credible, put one's faith in. 2 assume, feel, gather, hold, imagine, maintain, presume, suppose, think.** □ **believe in have faith in, rely on, swear by, trust in.**

belittle *v.* disparage.

■ **criticize, decry, denigrate, deprecate, detract from, discredit, disparage, minimize, play down, run down, slight, undervalue.**

bell *n.* **1** cup-shaped metal instrument that makes a ringing sound when struck. **2** its sound.

belle *n.* beautiful woman.

belles-lettres /bel-létrə/ *n.pl.* literary studies.

bellicose *adj.* eager to fight.

belligerent *adj.* **1** aggressive. **2** engaged in a war. □ **belligerently** *adv.*, **belligerence** *n.*

■ **1 aggressive, antagonistic, argumentative, bellicose, combative, contentious, fierce, hostile, militant, pugnacious, quarrelsome, truculent. 2 martial, warring.**

bellow *v.* utter a deep loud roar. ● *n.* a bellowing sound.

■ *v.* **bawl, howl, roar, shout, thunder, trumpet, yell.**

bellows *n.pl.* apparatus for driving air into something.

belly *n.* **1** abdomen. **2** stomach.

bellyful *n.* (*colloq.*) as much as one wants or rather more.

belong *v.* have a rightful place. □ **belong to 1** be the property of **2** be a member of.

belongings *n.pl.* possessions.

■ **chattels, effects, goods, possessions, property, stuff, things.**

beloved *adj.* & *n.* dearly loved (person).

below *adv.* & *prep.* at or to a lower position or amount (than).

belt *n.* **1** strip of cloth or leather etc., worn round the waist. **2** distinct region.

● *v.* **1** put a belt round. **2** (*sl.*) hit. **3** (*sl.*) rush.

■ *n.* **1** girdle, sash. **2** area, district, quarter, region, sector, stretch, strip, tract, zone.

bemoan *v.* complain about.

bemuse *v.* bewilder.

bench *n.* **1** long seat of wood or stone. **2** long working-table. **3** judges or magistrates hearing a case.

benchmark *n.* **1** surveyor's fixed point. **2** point of reference.

bend *v.* (**bent**) **1** make or become curved. **2** turn downwards, stoop. **3** turn in a new direction. ● *n.* curve, turn.

■ *v.* **1** arch, bow, crook, curl, curve, flex. **2** crouch, duck, lean, stoop. **3** bear, incline, swerve, swing, turn, veer, wind. ● *n.* angle, arc, corner, crook, curvature, curve, turn, turning, twist.

bender *n.* (*sl.*) drinking spree.

beneath *adv.* & *prep.* **1** below, underneath. **2** not worthy of.

benediction *n.* spoken blessing.

benefactor *n.* one who gives financial or other help. □ **benefaction** *n.*, **benefactress** *n.*

■ backer, donor, fairy godmother, investor, patron, philanthropist, sponsor, supporter, underwriter.

beneficent *adj.* **1** doing good. **2** actively kind. □ **beneficence** *n.*

beneficial *adj.* helpful, useful. □ **beneficially** *adv.*

■ advantageous, constructive, favourable, gainful, good, helpful, profitable, useful, valuable, worthwhile; healthy, salutary, salubrious, wholesome.

beneficiary *n.* one who receives a benefit or legacy.

■ heir, heiress, inheritor, legatee, recipient, successor.

benefit *n.* **1** something helpful, favourable, or profitable. **2** payment made from government funds etc. ● *v.* (**benefited**, **benefiting**) **1** do good to. **2** receive benefit.

■ *n.* **1** advantage, gain, good, help, profit; asset, attraction, plus. ● *v.* **1** aid, assist, better, boost, enhance, further, help, improve, promote. **2** gain, profit.

benevolent *adj.* kindly and helpful. □ **benevolence** *n.*

■ altruistic, beneficent, benign, caring, charitable, compassionate, considerate, friendly, generous, good, helpful, humane, kind, kind-hearted, kindly, sympathetic, thoughtful, unselfish, well-disposed.

benighted *adj.* **1** in darkness. **2** ignorant.

benign *adj.* **1** mild and gentle, kindly. **2** (of a tumour) not malignant. □ **benignly** *adv.*

■ **1** amiable, benevolent, compassionate, gentle, genial, good, friendly, kind, kind-hearted, kindly, mild, well-disposed.

bent *see* **bend**. *adj.* **1** curved. **2** (*sl.*) dishonest. ● *n.* natural skill or liking. □ **bent on** seeking or determined to do.

benzene *n.* liquid obtained from coal tar, used as a solvent.

benzine *n.* liquid mixture of hydrocarbons used in dry-cleaning.

bequeath *v.* leave as a legacy.

■ hand down, leave, make over, pass on, will.

bequest *n.* legacy.

■ heritage, gift, inheritance, legacy, patrimony.

berate *v.* scold.

bereave *v.* deprive, esp. of a relative, by death. □ **bereavement** *n.*

bereft *adj.* deprived.

beret /bérray/ *n.* round flat cap with no peak.

beriberi *n.* disease caused by lack of vitamin B.

berry *n.* small round juicy fruit with no stone.

berserk *adj.* **go berserk** go into a wild destructive rage.

berth *n.* **1** sleeping place in a ship. **2** place for a ship to tie up at a wharf. ● *v.* moor at a berth. □ **give a wide berth to** keep a safe distance from.

beryl *n.* transparent green gem.

beseech *v.* (**besought**) implore.

■ appeal to, ask, beg, entreat, implore, importune, plead with, pray, supplicate.

beset *v.* (**beset**, **besetting**) **1** affect or trouble persistently. **2** hem in, surround.

■ **1** afflict, assail, attack, bedevil, beleaguer, harass, hound, *colloq.* plague, torment, trouble. **2** besiege, encircle, hem in, surround.

beside *prep.* **1** at the side of, close to. **2** compared with. □ **beside oneself** frantic with anger or worry etc. **beside the point** irrelevant.

besides *prep.* in addition to, other than. ● *adv.* also.

besiege *v.* **1** lay siege to. **2** crowd round eagerly. **3** harass with requests.

■ **1** beleaguer, blockade, lay siege to. **2** beset, encircle, hem in, surround. **3** beleaguer, beset, bombard, harass, hound, pester, *colloq.* plague.

besotted *adj.* infatuated.

besought *see* **beseech.**

bespeak *v.* (**bespoke, bespoken**) **1** engage beforehand. **2** be evidence of.

bespoke *adj.* (of clothes) made to a customer's order.

best *adj.* most excellent. ● *adv.* **1** in the best way. **2** most usefully. ● *n.* best thing. □ **best man** bridegroom's chief attendant. **best part of** most of.

■ *adj.* choicest, excellent, finest, first-class, incomparable, optimum, outstanding, peerless, pre-eminent, superb, superlative, supreme, top, unequalled, unrivalled, unsurpassed.

bestial *adj.* **1** savage. **2** of or like a beast. □ **bestiality** *n.*

■ **1** barbarous, barbaric, brutal, cruel, ferocious, fierce, inhuman, savage, vicious, violent.

bestir *v.* (**bestirred**) **bestir oneself** exert oneself.

bestow *v.* confer as a gift. □ **bestowal** *n.*

■ award, confer, give, grant, present.

bestride *v.* (**bestrode**) stand astride over.

bet *n.* sum of money etc. risked on the outcome of an unpredictable event. ● *v.* (**bet** or **betted**) **1** make a bet. **2** risk (money) as a bet. **3** (*colloq.*) predict.

■ *n. colloq.* flutter, stake, wager. ● *v.* **1** gamble, *colloq.* have a flutter, wager. **2** chance, hazard, risk, stake, venture.

beta *n.* second letter of the Greek alphabet, = b.

betake *v.* (**betook, betaken**) **betake oneself** go.

betide *v.* happen to.

betimes *adv.* (*poetic*) in good time, early.

betoken *v.* be a sign of.

betray *v.* **1** be disloyal to (a person, one's country). **2** reveal involuntarily. □ **betrayal** *n.*

■ **1** be disloyal to, denounce, double-cross, inform against *or* on, let down, rat on, sell out, *sl.* shop. **2** disclose, divulge, give away, *sl.* let on, reveal, show, tell.

betroth *v.* cause to be engaged to marry. □ **betrothal** *n.*

better *adj.* **1** more excellent. **2** partly or fully recovered from illness. ● *adv.* **1** in a better manner. **2** more usefully. ● *v.* **1** improve. **2** surpass. □ **better part** more than half. **get the better of 1** overcome. **2** outwit.

■ *adj.* **1** preferable, superior. **2** convalescent, fitter, healthier, on the mend, progressing, recovering; cured, recovered, well. ● *v.* **1** ameliorate, amend, enhance, improve, mend, polish, rectify, refine, reform. **2** beat, cap, eclipse, exceed, excel, outdo, outshine, outstrip, surpass.

betting shop bookmaker's office.

between *prep.* **1** in the space, time, or quality bounded by (two limits). **2** separating. **3** to and from. **4** connecting. **5** shared by. ● *adv.* between points or limits etc.

bevel *n.* sloping edge. ● *v.* (**bevelled**) give a sloping edge to.

beverage *n.* any drink.

bevy *n.* company, large group.

bewail *v.* wail over.

beware *v.* be on one's guard.

■ be careful, be on one's guard, look out, take care, take heed, watch out, watch one's step.

bewilder *v.* puzzle, confuse. □ **bewilderment** *n.*

■ baffle, *colloq.* bamboozle, bemuse, confound, confuse, floor, *colloq.* flummox, mystify, perplex, puzzle, *colloq.* stump.

bewitch *v.* **1** delight very much. **2** cast a spell on.

■ **1** captivate, charm, delight, enchant, enrapture, entrance, fascinate, hypnotize, mesmerize.

beyond *adv.* & *prep.* **1** at or to the further side (of). **2** outside the range of.

biannual *adj.* happening twice a year. □ **biannually** *adv.*

bias *n.* predisposition, prejudice. ● *v.* (**biased**) influence.

■ *n.* bent, inclination, leaning, partiality, predilection, predisposition, preference, prejudice, proclivity, propensity, tendency. ● *v.* affect, colour, influence, predispose, prejudice, sway.

bib *n.* covering put under a child's chin while feeding.

Bible *n.* Christian or Jewish scripture.

biblical *adj.* of or in the Bible.

bibliography *n.* list of books about a subject or by a specified author. □ **bibliographer** *n.*, **bibliographical** *adj.*

bibliophile *n.* book-lover.

bibulous *adj.* fond of drinking.

bicentenary *n.* 200th anniversary.

bicentennial *adj.* happening every 200 years. ● *n.* bicentenary.

biceps *n.* large muscle at the front of the upper arm.

bicker *v.* quarrel pettily.

bicycle *n.* two-wheeled vehicle driven by pedals. ● *v.* ride a bicycle.

bid[1] *n.* **1** offer of a price, esp. at an auction. **2** (*colloq.*) attempt. **3** statement of the number of tricks a player proposes to win in a card game. ● *v.* (**bid, bidding**) make a bid (of), offer.

■ *n.* **1** offer, price, proposal, tender. **2** attempt, effort, endeavour, go, *colloq.* stab, try. ● *v.* offer, proffer, propose, tender.

bid[2] *v.* (**bid** or **bade, bidden, bidding**) **1** command. **2** say as a greeting.

biddable *adj.* willing to obey.

bide *v.* await (one's time).

bidet /beeday/ *n.* low washbasin that one can sit astride to wash the genital and anal regions.

biennial *adj.* **1** lasting for two years. **2** happening every second year. ● *n.* plant that flowers and dies in its second year.

bier /beer/ *n.* movable stand for a coffin.

biff *v.* & *n.* (*sl.*) hit.

bifocals *n.pl.* spectacles with lenses that have two parts for distant and close focusing.

bifurcate *v.* fork. □ **bifurcation** *n.*

big *adj.* (**bigger, biggest**) **1** large in size, amount, or intensity. **2** important.

■ **1** *colloq.* almighty, astronomical, broad, capacious, colossal, commodious, considerable, elephantine, enormous, extensive, gargantuan, giant, gigantic, great, hefty, huge, *colloq.* hulking, immense, king-sized, large, mammoth, massive, mighty, monstrous, monumental, mountainous, oversized, prodigious, roomy, sizeable, spacious, substantial, *colloq.* terrific, titanic, tremendous, vast, voluminous. **2** crucial, important, major, momentous, serious, significant.

bigamy *n.* crime of going through a form of marriage while a previous marriage is still valid. □ **bigamist** *n.*, **bigamous** *adj.*

bigot *n.* intolerant adherent of a creed or view. □ **bigoted** *adj.*, **bigotry** *n.*

bike (*colloq.*) *n.* **1** bicycle. **2** motor cycle. ● *v.* ride a bicycle or motor cycle. □ **biker** *n.*

bikini *n.* (*pl.* **-is**) woman's scanty two-piece beach garment.

bilateral *adj.* **1** having two sides. **2** existing between two groups. □ **bilaterally** *adv.*

bilberry *n.* **1** small dark blue berry. **2** shrub producing this.

bile *n.* bitter yellowish liquid produced by the liver.

bilge *n.* **1** ship's bottom. **2** water collecting there. **3** (*sl.*) nonsense.

bilharzia *n.* disease caused by a tropical parasitic flatworm.

bilingual *adj.* written in or able to speak two languages.

bilious *adj.* sick, esp. from trouble with bile or liver.

bilk *v.* defraud of payment.

bill[1] *n.* **1** written statement of charges to be paid. **2** poster. **3** programme. **4** draft of a proposed law. **5** (*US*) banknote. ● *v.* **1** send a statement of charges to. **2** announce, advertise.

■ *n.* **1** account, *US* check, invoice, statement. **2** advertisement, notice, placard, poster, sign.

bill[2] *n.* bird's beak. ● *v.* **bill and coo** exchange caresses.

billabong *n.* (*Austr.*) backwater.

billet *n.* lodging for troops. ● *v.* (**billeted**) place in a billet.

billhook *n.* pruning-instrument with a concave edge.

billiards *n.* game played with cues and three balls on a table.

billion *n.* one thousand million.

billow *n.* great wave. ● *v.* **1** rise or move like waves. **2** swell out.

■ *v.* ripple, roll, surge, swell, undulate.

bin *n.* large rigid container or receptacle.

binary *adj.* of two. □ **binary digit** either of two digits (0 and 1) used in the **binary scale**, system of numbers using only these.

bind *v.* (**bound**) **1** tie or fasten tightly. **2** place under an obligation or legal agreement. **3** edge with braid etc. **4** fasten into a cover. ● *n.* (*colloq.*) nuisance.

■ *v.* **1** attach, fasten, hitch, join, make fast, rope, secure, tether, tie (up), truss; bandage, dress. **2** compel, constrain, force, obligate, oblige, require.

binding *n.* **1** book cover. **2** braid etc. used to bind an edge.

bindweed *n.* wild convolvulus.

binge *n.* (*sl.*) bout of excessive eating, drinking, etc.

bingo *n.* gambling game using cards marked with numbered squares.

binocular *adj.* using two eyes.

binoculars *n.pl.* instrument with lenses for both eyes, making distant objects seem larger.

binomial *adj.& n.* (expression or name) consisting of two terms.

biochemistry *n.* chemistry of living organisms. □ **biochemical** *adj.*, **biochemist** *n.*

biodegradable *adj.* able to be decomposed by bacteria.

biography *n.* story of a person's life. □ **biographer** *n.*, **biographical** *adj.*

biology *n.* study of the life and structure of living things. □ **biological** *adj.*, **biologist** *n.*

bionic *adj.* (of a person or faculties) operated electronically.

biopsy *n.* examination of tissue cut from a living body.

biorhythm *n.* any of the recurring cycles of activity in a person's life.

bipartite *adj.* **1** consisting of two parts. **2** involving two groups.

biped *n.* two-footed animal.

biplane *n.* aeroplane with two pairs of wings.

birch *n.* tree with smooth bark.

bird *n.* feathered animal.

birth *n.* **1** emergence of young from the mother's body. **2** parentage. □ **birth control** prevention of unwanted pregnancy.

> ■ **1** childbirth, confinement, delivery, nativity, parturition. **2** ancestry, blood, descent, extraction, family, line, lineage, origins, parentage, pedigree, stock.

birthday *n.* anniversary of the day of one's birth.

birthmark *n.* unusual coloured mark on the skin at birth.

birthright *n.* thing that is one's right through being born into a certain family or country.

biscuit *n.* small flat thin piece of pastry baked crisp.

bisect *v.* divide into two equal parts. □ **bisection** *n.*, **bisector** *n.*

bisexual *adj.* sexually attracted to members of both sexes. □ **bisexuality** *n.*

bishop *n.* **1** clergyman of high rank. **2** mitre-shaped chess piece.

bishopric *n.* diocese of a bishop.

bismuth *n.* **1** metallic element. **2** compound of this used in medicines.

bison *n.* (*pl.* **bison**) wild ox.

bistro *n.* (*pl.* **-os**) small bar or restaurant.

bit[1] *n.* **1** small piece or quantity. **2** short time or distance. **3** mouthpiece of a bridle. **4** cutting part of a tool etc.

> ■ **1** atom, chip, crumb, dollop, drop, fraction, fragment, grain, iota, jot, morsel, mouthful, part, particle, piece, pinch, portion, sample, scrap, section, segment, share, shred, slice, sliver, snippet, speck, spot, taste, titbit, trace. **2** instant, minute, moment, second, *colloq.* tick, while; inch, little.

bit[2] *n.* (in computers) binary digit.

bit[3] *see* **bite**.

bitch *n.* **1** female dog. **2** (*sl.*) spiteful woman. ● *v.* (*colloq.*) speak spitefully or sourly. □ **bitchy** *adj.*, **bitchiness** *n.*

bite *v.* (**bit, bitten**) **1** cut with the teeth. **2** penetrate. **3** grip or act effectively. ● *n.* **1** act of biting. **2** wound so made. **3** small meal.

> ■ *v.* **1** champ, chew, gnaw, munch, nibble, nip.

biting *adj.* **1** causing a smarting pain. **2** sharply critical.

bitter *adj.* **1** tasting sharp, not sweet. **2** causing or feeling mental pain or resentment. **3** piercingly cold. ● *n.* bitter beer. □ **bitterly** *adv.*, **bitterness** *n.*

> ■ *adj.* **1** acid, acrid, harsh, sharp, sour, vinegary. **2** cruel, distressing, grievous, harrowing, heartbreaking, hurtful, painful, upsetting; aggrieved, embittered, rancorous, resentful. **3** biting, cold, freezing, icy, keen, perishing, piercing, raw, wintry.

bittern *n.* a kind of marsh bird.

bitty *adj.* (**-ier, -iest**) made up of unrelated bits. □ **bittiness** *n.*

bitumen *n.* black substance made from petroleum. □ **bituminous** *adj.*

bivalve *n.* shellfish with a hinged double shell.

bivouac *n.* temporary camp without tents or other cover. ● *v.* (**bivouacked**) camp thus.

bizarre *adj.* very odd in appearance or effect.

> ■ curious, eccentric, extraordinary, fantastic, freakish, grotesque, odd, offbeat, outlandish, peculiar, strange, surreal, unconventional, unusual, weird.

blab *v.* (**blabbed**) talk indiscreetly.

black *adj.* **1** of the very darkest colour, like coal or soot. **2** dismal, gloomy. **3** hostile. **4** wicked. **5** having a black skin. ● *n.* **1** black colour or thing. **2** member of a dark-skinned race. □ **black eye** bruised eye. **black hole** region in outer space from which matter and radiation cannot escape. **blacklist** *n.* list of people who are disapproved of. *v.* put on a blacklist. **black market** illegal buying and selling. **black out** cover windows so that no light can penetrate. **black pudding** sausage of blood and suet. **black sheep** scoundrel. **in the black** with a credit balance.

■ *adj.* **1** dusky, ebony, jet-black, pitch-black, pitch-dark, raven, sable, sooty. **2** dark, dismal, gloomy, funereal, murky, overcast, sombre. **3** angry, furious, hostile, resentful, sulky, unfriendly, wrathful. **4** bad, diabolical, evil, iniquitous, nefarious, unspeakable, vile, villainous, wicked.

blackberry *n.* **1** bramble. **2** its edible dark berry.

blackbird *n.* European songbird.

blackboard *n.* board for writing on with chalk in front of a class.

blacken *v.* **1** make or become black. **2** say evil things about.

blackguard /blággaard/ *n.* scoundrel. □ **blackguardly** *adj.*

blackhead *n.* small dark lump blocking a pore in the skin.

blackleg *n.* person refusing to join a strike.

blackmail *v.* demand payment or action from (a person) by threats. ● *n.* money demanded thus. □ **blackmailer** *n.*

blackout *n.* temporary loss of consciousness or memory.

blacksmith *n.* smith who works in iron.

blackthorn *n.* thorny shrub bearing white flowers and sloes.

bladder *n.* **1** sac in which urine collects in the body. **2** inflatable bag.

blade *n.* **1** flattened cutting part of a knife or sword. **2** flat part of an oar or propeller. **3** flat narrow leaf of grass. **4** broad bone.

blame *v.* assign responsibility to. ● *n.* **1** responsibility for a fault. **2** act of blaming.

■ *v.* accuse, censure, condemn, criticize, find fault with, hold responsible, reprehend, reprimand, reproach, reprove, scold. ● *n.* **1** culpability, guilt, *sl.* rap, responsibility. **2** castigation, censure, condemnation, criticism, recrimination, reproach, reproof.

blameless *adj.* not subject to blame.

■ above suspicion, faultless, guiltless, innocent, irreproachable, unimpeachable, upright.

blameworthy *adj.* deserving blame.

blanch *v.* make or become white or pale.

blancmange /bləmónj/ *n.* flavoured jelly-like pudding.

bland *adj.* **1** mild. **2** insipid. □ **blandly** *adv.*

■ **1** gentle, mild, smooth, soothing. **2** boring, characterless, dull, flat, insipid, nondescript, tame, tasteless, unexciting, uninspiring, uninteresting, vapid, wishy-washy.

blandishments *n.pl.* flattering or coaxing words.

blank *adj.* **1** not written or printed on, unmarked. **2** without interest or expression. ● *n.* **1** blank space. **2** cartridge containing no bullet. □ **blank cheque** one with the amount left blank for the payee to fill in. **blank verse** unrhymed verse.

■ *adj.* **1** clean, clear, empty, new, unadorned, undecorated, unmarked, unused, virgin. **2** deadpan, emotionless, expressionless, impassive, poker-faced, vacant, vacuous.

blanket *n.* **1** warm covering made of woollen or similar material. **2** thick covering mass.

blare *v.* sound loudly and harshly. ● *n.* this sound.

blarney *n.* smooth talk that flatters and deceives.

blasé /blaázay/ *adj.* bored or unimpressed by things.

blaspheme *v.* utter blasphemies (about). □ **blasphemer** *n.*

blasphemy *n.* irreverent talk about sacred things. □ **blasphemous** *adj.*

■ □ **blasphemous** disrespectful, impious, irreverent, profane, sacrilegious, sinful, ungodly, wicked.

blast *n.* **1** strong gust. **2** explosion. **3** loud sound of a wind instrument, car horn, whistle, etc. **4** (*colloq.*) severe reprimand. ● *v.* **1** blow up with explosives. **2** cause to wither, destroy. **3** (*colloq.*) reprimand severely. □ **blast off** be launched by firing of rockets.

■ *n.* **1** gale, gust, wind. **2** bang, burst, crack, detonation, eruption, explosion. **3** blare, din, noise, racket. ● *v.* **1** blow up, dynamite, explode. **2** blight, dash, destroy, kill, put an end to, ruin.

blatant *adj.* flagrant, shameless. □ **blatantly** *adv.*

■ barefaced, brazen, flagrant, glaring, obvious, overt, palpable, shameless, sheer, undisguised, unmistakable.

blaze[1] *n.* **1** bright flame or fire. **2** bright light or display. **3** outburst. ● *v.* burn or shine brightly.

■ *n.* **1** conflagration, fire, flame, inferno. **3** eruption, explosion, outbreak, outburst. ● *v.* burn, flame, flare; flash, gleam, glitter, glow, shine, sparkle.

blaze[2] *n.* **1** white mark on an animal's face. **2** mark chipped in the bark of a tree to mark a route. □ **blaze a trail 1** make such marks. **2** pioneer.

blazer *n.* loose-fitting jacket, esp. in the colours or bearing the badge of a school, team, etc.

blazon *n.* coat of arms. ● *v.* **1** proclaim. **2** ornament with (a coat of arms).

bleach *v.* whiten by sunlight or chemicals. ● *n.* bleaching substance or process.

bleak *adj.* **1** bare and cold. **2** cheerless. □ **bleakness** *n.*

■ **1** bare, barren, chilly, cold, desolate, exposed, grim, inhospitable, windswept, wintry. **2** cheerless, dark, depressing, disheartening, dismal, dreary, gloomy, melancholy, miserable, sombre, uninviting, unpromising.

bleary *adj.* (**-ier**, **-iest**) (of eyes) watery and seeing indistinctly.

bleat *n.* cry of a sheep or goat. ● *v.* **1** utter this cry. **2** speak or say plaintively.

bleed *v.* (**bled**) **1** leak blood or other fluid. **2** draw blood or fluid from. **3** extort money from.

bleep *n.* short high-pitched sound. ● *v.* make this sound. □ **bleeper** *n.*

blemish *n.* flaw or defect that spoils the perfection of a thing. ● *v.* spoil with a blemish.

■ *n.* blot, blotch, defect, disfigurement, fault, flaw, imperfection, mark, scar, smudge, spot, stain. ● *v.* deface, disfigure, flaw, mar, mark, scar, spoil, stain, tarnish.

blench *v.* flinch.

blend *v.* mix into a harmonious compound. ● *n.* mixture.

■ *v.* amalgamate, combine, fuse, integrate, merge, mingle, mix, synthesize, unite. ● *n.* amalgam, amalgamation, combination, composite, compound, fusion, mix, mixture, synthesis, union.

blender *n.* machine for liquidizing food.

blenny *n.* sea fish with spiny fins.

bless *v.* **1** make sacred or holy. **2** praise (God). **3** call God's favour upon.

■ **1** consecrate, dedicate, sanctify. **2** exalt, extol, glorify, praise.

blessed *adj.* **1** holy, sacred. **2** in paradise. **3** fortunate. **4** (*colloq.*) damned.

blessing *n.* **1** act of seeking or giving (esp. divine) favour. **2** something one is glad of.

■ **1** benediction, prayer; approbation, approval, consent, permission, sanction. **2** advantage, asset, boon, godsend, help.

blight *n.* **1** disease of plants. **2** malignant influence. ● *v.* **1** affect with blight. **2** spoil.

blind *adj.* **1** without sight. **2** without adequate foresight, understanding, or information. **3** not governed by purpose. ● *v.* deprive of sight or of the power of judgement. ● *n.* **1** screen for a window. **2** pretext. □ **blind to** unwilling or unable to appreciate (a factor). **blindly** *adv.*, **blindness** *n.*

■ *adj.* **1** eyeless, sightless, visually handicapped. **2** blinkered, ignorant, imperceptive, insensitive. **3** indiscriminate, mindless, stupid, thoughtless, unreasoning, unthinking. ● *n.* **1** curtain, screen, shade, shutter(s). **2** front, pretext, smokescreen. □ **blind to** heedless of, impervious to, oblivious of *or* to, unaffected by, unaware of, unmoved by.

blindfold *n.* cloth used to cover the eyes and block the sight. ● *v.* cover the eyes of (a person) thus.

blink *v.* **1** open and shut one's eyes rapidly. **2** shine unsteadily. ● *n.* **1** act of blinking. **2** gleam.

■ *v.* **1** flutter, wink. **2** flash, flicker, gleam, glimmer, shimmer, sparkle, twinkle.

blinker *n.* leather piece fixed to a bridle to prevent a horse from seeing sideways. ● *v.* obstruct the sight or understanding of.

blip *n.* **1** quick sound or movement. **2** small image on a radar screen.

bliss *n.* perfect happiness. □ **blissful** *adj.*, **blissfully** *adv.*

■ delight, ecstasy, euphoria, felicity, glee, happiness, joy, pleasure, rapture.

blister *n.* **1** bubble-like swelling on skin. **2** raised swelling on a surface. ● *v.* **1** cause blister(s) on. **2** be affected with blister(s).

blithe *adj.* casual and carefree. □ **blithely** *adv.*

blitz *n.* violent attack, esp. from aircraft. ● *v.* attack in a blitz.

blizzard *n.* severe snowstorm.

bloat *v.* inflate, swell.

bloater *n.* salted smoked herring.

blob *n.* **1** drop of liquid. **2** round mass.
bloc *n.* group of parties or countries who combine for a purpose.
block *n.* **1** solid piece of hard substance. **2** obstruction. **3** (*sl.*) head. **4** pulley(s) mounted in a case. **5** compact mass of buildings. **6** large building divided into flats or offices. **7** large quantity treated as a unit. **8** pad of paper for drawing or writing on. ● *v.* obstruct, prevent the movement or use of. □ **block letters** plain capital letters.

■ *n.* **1** bar, brick, cake, chunk, hunk, ingot, lump, mass, piece, slab. **2** bar, barrier, blockage, impediment, obstacle, obstruction, stumbling block. ● *v.* bar, barricade, choke, clog, congest, obstruct, stop (up); hamper, hinder, impede, prevent, thwart.

blockade *n.* blocking of access to a place, to prevent entry of goods. ● *v.* set up a blockade of.
blockage *n.* obstruction.
blockhead *n.* stupid person.
bloke *n.* (*sl.*) man.
blond *adj.* & *n.* fair-haired (man).
blonde *adj.* & *n.* fair-haired (woman).
blood *n.* **1** red liquid circulating in the bodies of animals. **2** temper, courage. **3** race, descent, parentage. **4** kindred. ● *v.* **1** give a first taste of blood to (a hound). **2** initiate (a person). □ **blood-curdling** *adj.* horrifying. **blood sports** sports involving killing. **blood vessel** tubular structure conveying blood within the body.
bloodhound *n.* large keen-scented dog, used in tracking.
bloodless *adj.* without bloodshed. □ **bloodlessly** *adv.*
bloodshed *n.* killing.
bloodshot *adj.* (of eyes) red from dilated veins.
bloodstream *n.* circulating blood.
bloodsucker *n.* **1** leech. **2** person who extorts money.
bloodthirsty *adj.* eager for bloodshed.

■ brutal, cruel, ferocious, fierce, homicidal, murderous, pitiless, ruthless, sadistic, savage, vicious, violent, warlike.

bloody *adj.* (**-ier, -iest**) **1** bloodstained. **2** with much bloodshed. **3** cursed. ● *adv.* (*sl.*) extremely. ● *v.* stain with blood. □ **bloody-minded** *adj.* (*colloq.*) deliberately uncooperative.
bloom *n.* **1** flower. **2** beauty, perfection. ● *v.* **1** bear flowers. **2** flourish.

■ *n.* **1** blossom, bud, efflorescence, floret, flower. **2** beauty, perfection, prime. ● *v.* **1** blossom, bud, burgeon, come out, effloresce, flower, open. **2** blossom, flourish, prosper, thrive.

blossom *n.* flower(s), esp. of a fruit tree. ● *v.* **1** open into flowers. **2** develop and flourish.
blot *n.* **1** spot of ink etc. **2** something ugly or disgraceful. ● *v.* (**blotted**) make blot(s) on. □ **blot out** cross out thickly, obscure.

■ *n.* **1** blob, blotch, mark, smear, smudge, splodge, splotch, spot, stain. **2** blemish, defect, eyesore, fault, flaw, imperfection. ● *v.* blemish, mar, mark, smudge, spot, stain; discredit, dishonour, disgrace, spoil, sully, tarnish. □ **blot out** cross out, delete, erase, obliterate, score out; conceal, cover up, hide, obscure.

blotch *n.* large irregular mark. □ **blotchy** *adj.*
blouse *n.* shirt-like garment worn by women.
blow[1] *v.* (**blew, blown**) **1** send out a current of air or breath. **2** drive or be driven by a current of air. **3** sound (a wind instrument). **4** puff and pant. **5** break with explosives. **6** make or shape by blowing. **7** (of a fuse) melt. ● *n.* blowing. □ **blow-out** *n.* burst tyre. **blow up 1** explode. **2** inflate. **3** exaggerate. **4** enlarge (a photograph). **5** (*colloq.*) lose one's temper. **6** (*colloq.*) reprimand severely.

■ *v.* **1** breathe, exhale, puff. **2** drive, move, toss, waft, whirl; flap, flutter, stream, wave. **3** play, sound, toot. □ **blow up 1** burst, explode, go off, shatter; blast, bomb, detonate, dynamite, set off. **2** dilate, expand, inflate, pump up. **3** amplify, exaggerate, overstate. **4** enlarge, magnify. **5** explode, flare up, get angry, lose one's temper.

blow[2] *n.* **1** hard stroke with a hand, tool, or weapon. **2** shock, disaster.

■ **1** bang, bash, *colloq.* clip, clout, cuff, hit, knock, punch, rap, slap, smack, stroke, *colloq.* swipe, thump, *colloq.* whack. **2** bombshell, calamity, disappointment, disaster, misfortune, shock, surprise, upset.

blowfly *n.* bluebottle.
blowlamp *n.* portable burner for directing a very hot flame.
blowpipe *n.* tube through which air etc. is blown, e.g. to heat a flame or send out a missile.
blowy *adj.* (**-ier, -iest**) windy.
blowzy *adj.* (**-ier, -iest**) red-faced and coarse-looking.
blub *v.* (**blubbed**) (*sl.*) weep.

blubber[1] *n.* whale fat.

blubber[2] *v.* weep noisily.

bludgeon *n.* heavy club. ● *v.* **1** strike with this. **2** coerce.

blue *adj.* **1** of a colour like the cloudless sky. **2** unhappy. **3** indecent. ● *n.* **1** blue colour or thing. **2** (*pl.*) melancholy jazz melodies. **3** (**the blues**) state of depression. □ **blue-blooded** *adj.* of aristocratic descent. **out of the blue** unexpectedly. **bluish** *adj.*

■ *adj.* **1** aquamarine, azure, cerulean, cobalt, indigo, navy, sapphire, saxe-blue, sky-blue, turquoise, ultramarine. **2** dejected, depressed, despondent, dispirited, downcast, gloomy, glum, melancholy, sad, unhappy.

bluebell *n.* plant with blue bell-shaped flowers.

blueberry *n.* **1** edible blue berry. **2** shrub bearing this.

bluebottle *n.* large bluish fly.

blueprint *n.* **1** blue photographic print of building plans. **2** detailed scheme.

bluff[1] *adj.* **1** with a broad steep front. **2** abrupt, frank, hearty. ● *n.* bluff cliff etc.

bluff[2] *v.* pretend to have strength, knowledge, etc. ● *n.* bluffing.

blunder *v.* **1** move clumsily and uncertainly. **2** make a bad mistake. ● *n.* bad mistake.

■ *v.* **1** flounder, lurch, stagger, stumble. **2** *sl.* screw up, *colloq.* slip up. ● *n. colloq.* boob, error, *faux pas*, gaffe, howler, miscalculation, misjudgement, mistake, slip, *colloq.* slip-up.

blunderbuss *n.* old type of gun firing many balls at one shot.

blunt *adj.* **1** without a sharp edge or point. **2** speaking or expressed plainly. ● *v.* make blunt. □ **bluntly** *adv.*, **bluntness** *n.*

■ *adj.* **1** blunted, dull, unsharpened, worn. **2** abrupt, bluff, brusque, candid, curt, direct, downright, forthright, frank, impolite, outspoken, rude, straightforward, tactless, undiplomatic, ungracious.

blur *n.* something that appears indistinct. ● *v.* (**blurred**) **1** smear. **2** make or become indistinct.

■ *v.* **1** smear, smudge. **2** cloud, conceal, dim, hide, mask, obscure, veil.

blurb *n.* written description praising something.

blurt *v.* **blurt out** utter abruptly or tactlessly.

blush *v.* become red-faced from shame or embarrassment. ● *n.* **1** blushing. **2** pink tinge.

■ *v.* colour, flush, go red, redden.

blusher *n.* rouge.

bluster *v.* **1** talk aggressively. **2** blow in gusts. ● *n.* blustering talk. □ **blustery** *adj.*

BMX *n.* **1** bicycle racing on a dirt track. **2** bicycle for this.

boa /bṓə/ *n.* large snake that crushes its prey.

boar *n.* male pig.

board *n.* **1** flat piece of wood or other stiff material. **2** committee. **3** daily meals supplied in return for payment or services. ● *v.* **1** enter (a ship etc.). **2** provide with or receive meals and accommodation for payment. **3** cover with boards. □ **on board** on or in a ship etc.

■ *n.* **1** beam, plank, slat, timber. **2** cabinet, committee, council, directorate, panel. **3** food, meals. ● *v.* **1** embark, enter, get on, go aboard. **2** accommodate, billet, lodge, put up, quarter.

boarder *n.* **1** person who boards with someone. **2** resident pupil.

boarding house, **boarding school** one taking boarders.

boardroom *n.* room where a board of directors meets.

boast *v.* **1** speak with great pride, trying to impress people. **2** possess something to be proud of. ● *n.* **1** boastful statement. **2** thing one is proud of. □ **boastful** *adj.*

■ *v.* **1** brag, crow, show off, swagger, vaunt. □ **boastful** bragging, bumptious, *colloq.* cocky, conceited, egotistical, pompous, proud, vain.

boat *n.* vessel for travelling on water. ● *v.* travel in a boat, esp. for pleasure.

■ craft, cruiser, launch, motor boat, rowing boat, ship, skiff, speedboat, vessel, yacht.

boater *n.* flat-topped straw hat.

boathouse *n.* shed at the water's edge for boats.

boatman *n.* man who rows, sails, or rents out boats.

boatswain /bṓs'n/ *n.* ship's officer in charge of rigging, boats, etc.

bob *v.* (**bobbed**) **1** move quickly up and down. **2** cut (hair) short to hang loosely. ● *n.* **1** bobbing movement. **2** bobbed hair.

bobbin *n.* small spool holding thread or wire in a machine.

bobble *n.* small woolly ball on a hat etc.

bobsleigh *n.* sledge with two sets of runners in tandem.

bode *v.* be a sign of, promise.

bodice *n.* **1** part of a dress from shoulder to waist. **2** undergarment for this part of the body.

bodily *adj.* of the human body or physical nature. ● *adv.* **1** in person, physically. **2** as a whole.

body *n.* **1** physical structure of a person or animal. **2** corpse. **3** main part. **4** group regarded as a unit. **5** strong texture or quality. **6** separate piece of matter. **7** woman's one-piece garment covering the torso. □ **body-blow** *n.* severe blow. **bodywork** *n.* outer shell of a vehicle.

■ **2** cadaver, carcass, corpse, remains, *sl.* stiff. **3** core, essence, heart, main part, substance. **4** association, band, committee, company, corporation, council, federation, group, league, party, society. **5** firmness, fullness, richness, solidity, substance.

bodyguard *n.* escort or personal guard of an important person.

■ escort, guard, *sl.* minder, protector.

Boer /bṓər/ *n.* Afrikaner.

boffin *n.* (*colloq.*) person engaged in scientific research.

bog *n.* permanently wet spongy ground. ● *v.* **bog down** make or become stuck and unable to progress. □ **boggy** *adj.*

■ *n.* fen, marsh, mire, morass, quagmire, quicksand, slough, swamp.

bogey *n.* thing causing fear.

boggle *v.* be bewildered.

bogie *n.* undercarriage on wheels, pivoted at each end.

bogus *adj.* false.

■ counterfeit, fake, false, fraudulent, imitation, *colloq.* phoney, sham, spurious.

bohemian *adj.* socially unconventional.

boil[1] *n.* inflamed swelling producing pus.

boil[2] *v.* **1** bubble up with heat. **2** heat so that liquid does this. **3** be very angry.

boiler *n.* container in which water is heated. □ **boiler suit** protective one-piece garment.

boisterous *adj.* **1** noisily exuberant. **2** violent, rough. □ **boisterously** *adv.*

■ **1** exuberant, frisky, high-spirited, irrepressible, lively, noisy, rollicking, rowdy, *colloq.* rumbustious, unruly, wild. **2** blustery, rough, squally, stormy, tempestuous, turbulent, violent.

bold *adj.* **1** confident, courageous. **2** impudent. **3** distinct, vivid. □ **boldly** *adv.*, **boldness** *n.*

■ **1** adventurous, audacious, brave, confident, courageous, daring, dauntless, fearless, gallant, heroic, intrepid, plucky, spirited, unafraid, valiant, valorous; daredevil, foolhardy, rash, reckless. **2** brash, brazen, cheeky, forward, impertinent, impudent, insolent, pert, presumptuous, rude. **3** clear, conspicuous, distinct, prominent, pronounced, striking, strong, vivid.

bole *n.* trunk of a tree.

bolero *n.* **1** Spanish dance. **2** woman's short jacket with no fastening.

boll *n.* round seed vessel of cotton, flax, etc.

bollard *n.* short thick post.

boloney *n.* (*sl.*) nonsense.

bolster *n.* long pad placed under a pillow. ● *v.* support, prop up.

■ *v.* buttress, hold up, prop up, reinforce, shore up, strengthen, support, sustain.

bolt *n.* **1** sliding bar for fastening a door. **2** strong metal pin. **3** sliding part of a rifle-breech. **4** shaft of lightning. **5** roll of cloth. **6** arrow from a crossbow. ● *v.* **1** fasten with bolt(s). **2** run away. **3** gulp (food) hastily. □ **bolt-hole** *n.* place into which one can escape.

■ *v.* **1** fasten, latch, lock, secure. **2** dash away, *sl.* do a bunk, escape, flee, run away, rush away, *colloq.* skedaddle, take to one's heels. **3** gobble, gulp, *colloq.* scoff, wolf.

bomb *n.* case of explosive or incendiary material to be set off by impact or a timing device. ● *v.* attack with bombs.

bombard *v.* **1** attack with artillery. **2** attack with questions etc. □ **bombardment** *n.*

bombardier *n.* artillery non-commissioned officer.

bombastic *adj.* using pompous words.

■ flowery, grandiloquent, pompous, pretentious, rhetorical, turgid.

bomber *n.* **1** aircraft that carries and drops bombs. **2** person who throws or places bombs.

bombshell *n.* great shock.

bona fide /bṓnə fídi/ genuine.

bonanza *n.* sudden great wealth.

bond *n.* **1** uniting force. **2** (usu. *pl.*) thing that restrains. **3** binding agreement. **4** document issued by a government or company acknowledging that money has been lent to it and will be repaid with interest. ● *v.* unite with a bond. □ **in bond** stored in a customs warehouse until duties are paid.

■ *n.* **1** attachment, connection, link, relationship, tie, union. **2** (**bonds**) chains, cords, fetters, manacles, restraints, ropes, shackles, ties. **3** agreement, contract, covenant, pact; guarantee, oath, pledge, promise, word.

bondage *n.* slavery, captivity.

■ captivity, confinement, enslavement, servitude, slavery.

bone *n.* each of the hard parts making up the vertebrate skeleton. ● *v.* remove bones from. □ **bone china** made of clay and bone ash.

bonfire *n.* open-air fire.

bongo *n.* each of a pair of small drums played with the fingers.

bonhomie /bónnomeé/ *n.* geniality.

bonnet *n.* **1** hat with strings that tie under the chin. **2** hinged cover over the engine of a motor vehicle.

bonny *adj.* (**-ier, -iest**) **1** healthy-looking. **2** (*Sc.*) attractive.

bonsai *n.* **1** miniature tree or shrub. **2** art of growing these.

bonus *n.* extra payment or benefit.

■ commission, dividend, gratuity, handout, *colloq.* perk, reward, tip; advantage, benefit, extra, plus.

bony *adj.* (**-ier, -iest**) **1** like bones. **2** having bones with little flesh.

boo *int.* exclamation of disapproval. ● *v.* shout 'boo' (at).

boob *n.* & *v.* (*colloq.*) blunder.

booby *n.* foolish person. □ **booby prize** one given as a joke to the competitor with the lowest score. **booby trap 1** practical joke in the form of a trap. **2** hidden bomb.

book *n.* **1** written or printed work bound into a cover. **2** main division of a literary work. ● *v.* **1** reserve (a seat etc.) in advance. **2** enter in a book or list. □ **booking** *n.* reservation.

■ *n.* **1** hardback, paperback, publication, tome, volume, work. ● *v.* **1** earmark, order, reserve, save.

bookcase *n.* piece of furniture with shelves for books.

bookie *n.* (*colloq.*) bookmaker.

bookish *adj.* fond of reading.

bookkeeping *n.* systematic recording of business transactions.

booklet *n.* small thin book.

bookmaker *n.* person whose business is the taking of bets.

bookmark *n.* strip of paper etc. to mark a place in a book.

bookworm *n.* **1** grub that eats holes in books. **2** person fond of reading.

boom[1] *v.* **1** make a deep resonant sound. **2** be suddenly prosperous or successful. ● *n.* **1** booming sound. **2** period of prosperity.

■ *v.* **1** bellow, resonate, resound, reverberate, roar, rumble, thunder. **2** burgeon, flourish, grow, increase, prosper, succeed, thrive. ● *n.* **1** reverberation, roar, rumble. **2** growth, improvement, increase, upsurge, upturn.

boom[2] *n.* **1** long pole. **2** floating barrier.

boomerang *n.* Australian missile of curved wood that returns to the thrower.

boon[1] *n.* benefit.

boon[2] *adj.* **boon companion** close companion.

boor *n.* ill-mannered person. □ **boorish** *adj.*, **boorishness** *n.*

■ □ **boorish** churlish, coarse, crude, ill-mannered, loutish, uncivilized, uncouth, vulgar.

boost *v.* **1** increase the strength or reputation of. **2** push upwards. ● *n.* **1** an increase. **2** upward thrust. □ **booster** *n.*

■ *v.* **1** aid, assist, build up, enhance, help, improve, increase, promote, strengthen, support.

boot *n.* **1** sturdy shoe covering both foot and ankle. **2** luggage compartment in a car. **3** (*colloq.*) dismissal. ● *v.* kick.

bootee *n.* baby's woollen boot.

booth *n.* **1** small shelter. **2** cubicle.

bootleg *adj.* smuggled, illicit. □ **bootlegger** *n.*, **bootlegging** *n.*

booty *n.* loot.

■ contraband, haul, loot, plunder, spoils, *sl.* swag.

booze (*colloq.*) *v.* drink alcohol. ● *n.* alcoholic drink. □ **boozer** *n.*, **boozy** *adj.*

borage *n.* blue-flowered plant.

borax *n.* compound of boron used in detergents.

border *n.* **1** edge, boundary, or part near it. **2** flower bed round part of a garden. ● *v.* put or be a border to. □ **border on 1** be next to. **2** come close to being.

■ *n.* **1** edge, fringe, margin, perimeter, periphery, rim, verge; borderline, boundary, frontier; edging, frame, frieze, surround.

borderline *n.* line of demarcation.

bore[1] *see* **bear**[2].
bore[2] *v.* make (a hole), esp. with a revolving tool or by excavation. ● *n.* **1** hole bored. **2** hollow inside of a cylinder. **3** its diameter.

■ *v.* dig (out) drill, excavate, gouge (out), sink, tunnel.

bore[3] *v.* weary by dullness. ● *n.* boring person or thing. □ **boredom** *n.*, **boring** *adj.*

■ □ **boring** dreary, dry, dull, flat, humdrum, interminable, monotonous, mundane, repetitive, soporific, stodgy, tedious, tiresome, unexciting, uninspiring, uninteresting, wearisome.

bore[4] *n.* tidal wave in an estuary.
born *adj.* **1** brought forth by birth. **2** having a specified natural quality. □ **born-again** *adj.* reconverted to religion.
borne *see* **bear**[2].
boron *n.* chemical element very resistant to high temperatures.
borough *n.* town or district with rights of local government.
borrow *v.* get temporary use of (a thing or money). □ **borrower** *n.*
Borstal *n.* former name of an institution for young offenders.
borzoi *n.* Russian wolfhound.
bosom *n.* breast. □ **bosom friend** very dear friend.
boss[1] (*colloq.*) *n.* employer, manager, person in charge. ● *v.* give orders to.

■ *n.* chief, director, employer, foreman, forewoman, *colloq.* gaffer, head, leader, manager, overseer, superintendent, supervisor.

boss[2] *n.* projecting knob.
bossy *adj.* (**-ier**, **-iest**) (*colloq.*) fond of giving orders to people. □ **bossily** *adv.*, **bossiness** *n.*

■ authoritarian, dictatorial, domineering, high-handed, imperious, officious, overbearing.

botany *n.* study of plants. □ **botanical** *adj.*, **botanist** *n.*
botch *v.* spoil by poor work.
both *adj.*, *pron.*, & *adv.* the two.
bother *v.* **1** trouble, worry, annoy. **2** take trouble. **3** feel concern. ● *n.* worry, minor trouble. □ **bothersome** *adj.*

■ *v.* **1** concern, disconcert, distress, disturb, perturb, trouble, unsettle, upset, worry; *colloq.* aggravate, annoy, *sl.* bug, inconvenience, irritate; badger, harass, *colloq.* hassle, nag, pester, *colloq.* plague. ● *n.* annoyance, bind, *colloq.* hassle, headache, inconvenience, irritation, nuisance, trouble, worry; ado, commotion, disturbance, fuss, to-do.

bottle *n.* glass or plastic container for liquid. ● *v.* **1** store in bottles. **2** preserve in jars.
bottleneck *n.* **1** narrow place where traffic cannot flow freely. **2** obstruction to an even flow of work etc.
bottom *n.* **1** lowest part or place. **2** buttocks. **3** ground under a stretch of water. ● *adj.* lowest in position, rank, or degree.

■ *n.* **1** base, bed, floor, foot, foundation, underside.

bottomless *adj.* extremely deep.
botulism *n.* poisoning by bacteria in food.
bougainvillaea *n.* tropical shrub with red or purple bracts.
bough *n.* large branch coming from the trunk of a tree.
bought *see* **buy**.
boulder *n.* large rounded stone.
boulevard *n.* wide street.
bounce *v.* **1** rebound. **2** move energetically. **3** (*sl.*, of a cheque) be sent back by a bank as worthless. ● *n.* **1** bouncing movement or power. **2** (*colloq.*) liveliness. □ **bouncy** *adj.*

■ *v.* **1** rebound, ricochet. **2** bound, caper, gambol, hop, jump, leap, prance, skip, spring. ● *n.* **1** bound, hop, jump, leap, spring; elasticity, springiness. **2** animation, energy, go, life, liveliness, pep, spirit, verve, vigour, *colloq.* vim, vitality, vivacity, zest, zip.

bouncer *n.* (*sl.*) person employed to eject troublemakers.
bound[1] *v.* limit, be a boundary of. ● *n.* (usu. *pl.*) limit. □ **out of bounds** beyond the permitted area.
bound[2] *v.* leap, spring. ● *n.* bounding movement.

■ *v.* bounce, caper, gambol, hop, jump, leap, prance, romp, skip, spring, vault.

bound[3] *see* **bind**. *adj.* obstructed by a specified thing (*snow-bound*). □ **bound to** certain to.
bound[4] *adj.* going in a specified direction.
boundary *n.* **1** line that marks a limit. **2** hit to the boundary in cricket.

■ **1** border, borderline, frontier; bounds, confines, edge, fringe, limit, margin, perimeter.

bounden *adj.* **bounden duty** duty dictated by conscience.

boundless *adj.* without limits.

■ endless, immeasurable, incalculable, inexhaustible, infinite, limitless, unbounded, unending, unlimited, untold.

bountiful *adj.* **1** giving generously. **2** abundant.

bounty *n.* **1** generosity. **2** official reward. □ **bounteous** *adj.*

■ **1** beneficence, charity, generosity, largesse, liberality, munificence, philanthropy.

bouquet /bookáy/ *n.* **1** bunch of flowers. **2** perfume of wine.

■ **1** arrangement, bunch, corsage, nosegay, posy, spray. **2** aroma, fragrance, perfume, scent, smell.

bouquet garni /bookáy gaárni/ bunch of herbs for flavouring.

bourbon /búrb'n/ *n.* whisky made mainly from maize.

bourgeois /bóorzhwaá/ *adj.* urban middle-class.

bourgeoisie /bóorzhwaazée/ *n.* bourgeois society.

bout *n.* **1** period of exercise, work, or illness. **2** boxing contest.

■ **1** period, run, session, spell, stint, stretch, time; attack, fit. **2** competition, contest, encounter, fight, match, round.

boutique *n.* small shop selling fashionable clothes etc.

bovine *adj.* **1** of cattle. **2** dull and stupid.

bow[1] /bō/ *n.* **1** weapon for shooting arrows. **2** rod with horsehair stretched between its ends, for playing a violin etc. **3** knot with loops in a ribbon etc.

bow[2] /bow/ *v.* **1** incline the head or body in greeting, acknowledgement, etc. **2** submit. ● *n.* act of bowing.

■ *v.* **1** curtsy, genuflect; nod. **2** capitulate, defer, give in, give way, submit, succumb, surrender, yield.

bow[3] /bow/ *n.* **1** front end of a boat or ship. **2** oarsman nearest the bow.

bowdlerize *v.* expurgate. □ **bowdlerization** *n.*

bowel *n.* **1** (often *pl.*) intestine. **2** (*pl.*) innermost parts.

■ **1** (**bowels**) entrails, guts, *colloq.* innards, *colloq.* insides, intestines, viscera, vitals. **2** (**bowels**) centre, core, depths, heart, inside, interior.

bower *n.* leafy shelter.

bowie knife long hunting knife.

bowl[1] *n.* **1** basin. **2** hollow rounded part of a spoon etc.

bowl[2] *n.* **1** heavy ball weighted to roll in a curve. **2** (*pl.*) game played with such balls. **3** ball used in skittles. ● *v.* **1** roll (a ball etc.). **2** go rapidly. **3** send a ball to a batsman, dismiss by knocking bails off with this. **4** play bowls. □ **bowl over 1** knock down. **2** overwhelm.

bowler[1] *n.* player who bowls.

bowler[2] *n.* (in full **bowler hat**) hard felt hat with a rounded top.

bowling *n.* playing bowls, skittles, or a similar game.

box[1] *n.* **1** container or receptacle with a flat base. **2** facility at a newspaper office for holding replies to an advertisement. **3** compartment in a theatre, stable, etc. **4** small shelter. ● *v.* put into a box. □ **box office** office for booking seats at a theatre etc.

■ *n.* **1** caddy, carton, case, casket, chest, coffer, container, crate, receptacle, trunk.

box[2] *v.* fight with fists as a sport, usu. in padded gloves. ● *n.* slap. □ **boxing** *n.*

box[3] *n.* **1** small evergreen shrub. **2** its wood. □ **boxwood** *n.*

boxer *n.* **1** person who engages in the sport of boxing. **2** dog of a breed resembling a bulldog.

boy *n.* male child. □ **boyfriend** *n.* person's regular male companion or lover. **boyhood** *n.*, **boyish** *adj.*

boycott *v.* refuse to deal with or trade with. ● *n.* such a refusal.

■ *v.* avoid, ostracize, reject, shun; blacklist. ● *n.* ban, blacklist, embargo, prohibition.

bra *n.* woman's undergarment worn to support the breasts.

brace *n.* **1** device that holds things together or in position. **2** pair. **3** (*pl.*) straps supporting trousers from the shoulders. ● *v.* give support or firmness to.

bracelet *n.* ornamental band worn on the arm.

bracing *adj.* invigorating.

■ crisp, exhilarating, fresh, invigorating, refreshing, restorative, stimulating, tonic.

bracken *n.* **1** large fern growing on open land. **2** mass of such ferns.

bracket *n.* **1** projecting support. **2** any of the marks used in pairs for enclosing words or figures, (), [], {}. ● *v.* **1** enclose by brackets. **2** put together as similar.

brackish *adj.* slightly salty.

bract *n.* leaf-like part of a plant.

brag *v.* (**bragged**) boast. ● *n.* boastful statement.

braggart *n.* person who brags.

Brahman *n.* (also **Brahmin**) member of the Hindu priestly caste.

braid *n.* **1** woven ornamental trimming. **2** plait of hair. ● *v.* **1** trim with braid. **2** plait.

Braille *n.* system of representing letters etc. by raised dots which blind people read by touch.

brain *n.* **1** mass of soft grey matter in the skull, centre of the nervous system in animals. **2** (also *pl.*) mind, intelligence.

brainchild *n.* person's invention or plan.

brainstorm *n.* **1** violent mental disturbance. **2** (*US*) bright idea.

brainwash *v.* force (a person) to change their views by subjecting them to great mental pressure.

brainwave *n.* bright idea.

brainy *adj.* (**-ier, -iest**) clever.

braise *v.* cook slowly with little liquid in a closed container.

brake *n.* device for reducing speed or stopping motion. ● *v.* slow by use of this.

bramble *n.* wild thorny shrub, esp. the blackberry.

bran *n.* ground inner husks of grain, sifted from flour.

branch *n.* **1** arm-like part of a tree. **2** subdivision of a river, subject, family, etc. **3** local shop or office belonging to a large organization. ● *v.* send out or divide into branches.

> ■ *n.* **1** bough, limb. **2** department, division, offshoot, part, ramification, section, subdivision.

brand *n.* **1** goods of a particular make. **2** trade mark. **3** identifying mark made with hot metal. ● *v.* mark with a brand. □ **brand new** new, unused.

brandish *v.* wave, flourish.

brandy *n.* strong alcoholic spirit distilled from wine or fermented fruit juice.

brash *adj.* vulgarly self-assertive. □ **brashly** *adv.*, **brashness** *n.*

brass *n.* **1** yellow alloy of copper and zinc. **2** musical wind instruments made of this. ● *adj.* made of brass.

brasserie *n.* restaurant (orig. one serving beer with food).

brassière /brázziər/ *n.* bra.

brassy *adj.* (**-ier, -iest**) **1** like brass. **2** bold and vulgar.

brat *n.* (*derog.*) child.

bravado *n.* show of boldness.

brave *adj.* **1** able to face and endure danger or pain. **2** spectacular. ● *v.* face and endure bravely. □ **bravely** *adv.*

> ■ *adj.* **1** adventurous, audacious, bold, courageous, daring, dauntless, fearless, gallant, game, *colloq.* gutsy, heroic, indomitable, intrepid, mettlesome, plucky, resolute, spirited, stout, unafraid, undaunted, valiant, valorous, venturesome. ● *v.* confront, endure, face, stand up to, weather, withstand.

bravery *n.* brave conduct.

> ■ audacity, boldness, courage, daring, determination, fearlessness, firmness, fortitude, gallantry, *colloq.* grit, *colloq.* guts, heroism, intrepidity, mettle, nerve, pluck, resoluteness, spirit, valour.

bravo *int.* well done!

brawl *n.* noisy quarrel or fight. ● *v.* take part in a brawl.

> ■ *n.* affray, altercation, *colloq.* bust-up, commotion, fight, fracas, fray, mêlée, quarrel, row, rumpus, *colloq.* scrap, scuffle, squabble, tussle. ● *v.* fight, quarrel, row, *colloq.* scrap, scuffle, squabble, tussle, wrangle.

brawn *n.* **1** muscular strength. **2** pressed meat from a pig's or calf's head. □ **brawny** *adj.*

bray *n.* **1** donkey's cry. **2** similar sound. ● *v.* make a bray.

braze *v.* solder with an alloy of brass.

brazen *adj.* **1** shameless, impudent. **2** like or made of brass. ● *v.* **brazen it out** be defiantly unrepentant. □ **brazenly** *adv.*

> ■ *adj.* **1** barefaced, blatant, flagrant, shameless, unashamed; cheeky, forward, impertinent, impudent, insolent, presumptuous, rude.

brazier *n.* basket-like stand for holding burning coals.

breach *n.* **1** breaking or neglect of a rule or contract. **2** estrangement. **3** broken place, gap. ● *v.* **1** break through. **2** make a breach in.

> ■ *n.* **1** break, contravention, infringement, transgression, violation. **2** break, estrangement, rift, rupture, separation, split. **3** aperture, break, crack, fissure, gap, hole, opening, space.

bread *n.* baked dough of flour and liquid, usu. leavened by yeast. □ **breadfruit** *n.* tropical fruit with bread-like pulp. **breadwinner** *n.* person whose work supports a family.

breadline *n.* **on the breadline** living in extreme poverty.

breadth *n.* width, broadness.

break *v.* (**broke, broken**) **1** separate into pieces under a blow or strain. **2** become unusable. **3** fail to keep (a promise or law). **4** make or become discontinuous. **5** make or become weak. **6** reveal (news). **7** surpass (a record). **8** appear suddenly. **9** (of a ball) change direction after touching the ground. ● *n.* **1** act or instance of breaking. **2** sudden dash. **3** gap. **4** interval. **5** opportunity, piece of luck. **6** points scored continuously in snooker. □ **break down 1** fail, collapse. **2** give way to emotion. **3** analyse. **break even** make gains and losses that balance exactly.

■ *v.* **1** burst, *colloq.* bust, crack, crumble, fracture, fragment, shatter, smash, snap, splinter, split. **2** break down, *sl.* conk out. **3** contravene, defy, disobey, disregard, fail to observe, flout, infringe, transgress, violate. **4** cut off, discontinue, interrupt, suspend. **5** debilitate, drain, exhaust, sap, weaken, wear out, weary. **6** announce, disclose, divulge, make public, reveal, tell. **7** beat, better, cap, exceed, outdo, outstrip, surpass. ● *n.* **1** breach, breakage, burst, fracture, rift, rupture, split. **2** bolt, dart, dash, run, rush. **3** aperture, chink, crack, gap, hole, opening, slit, space. **4** breather, interlude, intermission, interval, lull, pause, respite, rest. **5** chance, opening, opportunity, piece of luck.

breakable *adj.* easily broken.

breakage *n.* breaking.

breakdown *n.* **1** mechanical failure. **2** analysis. **3** collapse of health or mental stability.

breaker *n.* heavy ocean wave.

breakfast *n.* first meal of the day. ● *v.* eat breakfast.

breakneck *adj.* dangerously fast.

breakthrough *n.* major advance in knowledge or negotiation.

breakwater *n.* barrier to break the force of waves.

bream *n.* fish of the carp family.

breast *n.* **1** upper front part of the body. **2** either of the two milk-producing organs on a woman's chest. □ **breastfeed** *v.* feed (a baby) from the breast. **breaststroke** *n.* swimming stroke performed face downwards.

breastbone *n.* bone down the upper front of the body.

breath *n.* **1** air drawn into and sent out of the lungs in breathing. **2** breathing in. **3** gentle blowing. □ **out of breath** panting after exercise. **under one's breath** in a whisper.

breathalyse *v.* test by a breathalyser.

breathalyser *n.* [P.] device measuring the alcohol in a person's breath.

breathe *v.* **1** draw (air) into the lungs and send it out again. **2** utter.

■ **1** exhale, inhale, pant, respire. **2** murmur, say, tell, utter, whisper.

breather *n.* **1** pause for rest. **2** short period in fresh air.

breathless *adj.* out of breath.

breathtaking *adj.* amazing.

bred *see* **breed**.

breech *n.* **1** buttocks. **2** back part of a gun barrel, where it opens.

breeches *n.pl.* trousers reaching to just below the knees.

breed *v.* (**bred**) **1** produce offspring. **2** raise (livestock). **3** give rise to. **4** train, bring up. ● *n.* **1** variety of animals etc. within a species. **2** sort. □ **breeder** *n.*

■ *v.* **1** procreate, reproduce, spawn. **2** farm, raise, rear. **3** cause, create, engender, foster, generate, give rise to. **4** bring up, raise, rear, train. ● *n.* **1** stock, strain. **2** kind, sort, species, type, variety.

breeding *n.* good manners resulting from training or background.

breeze *n.* light wind. □ **breezy** *adj.*

■ draught, flurry, gust, wind, *poetic* zephyr. □ **breezy** airy, blowy, draughty, fresh, gusty, windy.

breeze-blocks *n.pl.* lightweight building blocks.

brethren *n.pl.* brothers.

Breton *adj.* & *n.* (native) of Brittany.

breve *n.* **1** mark (˘) over a short vowel. **2** (in music) long note.

breviary *n.* book of prayers to be said by RC priests.

brevity *n.* briefness.

brew *v.* **1** make (beer) by boiling and fermentation. **2** make (tea) by infusion. **3** concoct, plan. **4** be forming. ● *n.* liquid or amount brewed.

■ *v.* **3** concoct, contrive, *colloq.* cook up, devise, hatch, plan, plot. **4** be imminent, be in the wind, develop, form, gather force, impend, loom. ● *n.* concoction, drink, infusion, potion.

brewer *n.* person whose trade is brewing beer.

brewery *n.* building where beer is brewed commercially.

briar *n.* = **brier**.

bribe *n.* thing offered to influence a person to act in favour of the giver. ● *v.* persuade by this. □ **bribery** *n.*

■ *n. sl.* backhander, inducement, payola, *colloq.* sweetener. ● *v.* corrupt, *sl.* nobble, *colloq.* square, suborn.

bric-à-brac *n.* odd items of ornaments, furniture, etc.

brick *n.* **1** block of baked or dried clay used to build walls. **2** rectangular block. ● *v.* block with a brick structure.

brickbat *n.* **1** missile hurled at someone. **2** criticism.

bricklayer *n.* person who builds with bricks.

bridal *adj.* of a bride or wedding.

bride *n.* woman on her wedding day or when newly married.

bridegroom *n.* man on his wedding day or when newly married.

bridesmaid *n.* girl or unmarried woman attending a bride.

bridge[1] *n.* **1** structure providing a way across something. **2** something that joins or connects different things. **3** captain's platform on a ship. **4** bony upper part of the nose. ● *v.* make or be a bridge over.

■ *n.* **1** aqueduct, causeway, flyover, footbridge, overpass, viaduct. **2** connection, link, tie. ● *v.* cross, pass over, span, traverse; connect, join, link, unite.

bridge[2] *n.* card game developed from whist.

bridgehead *n.* position held on the enemy's side of a river.

bridle *n.* harness on a horse's head. ● *v.* **1** put a bridle on. **2** restrain. **3** show offence or resentment. □ **bridle path** path suitable for horse-riding.

brief *adj.* **1** lasting only for a short time. **2** concise. **3** scanty. ● *n.* set of instructions and information, esp. to a barrister about a case. ● *v.* **1** inform or instruct in advance. **2** employ a barrister. □ **briefly** *adv.*, **briefness** *n.*

■ *adj.* **1** ephemeral, fleeting, fugitive, momentary, short, transient, transitory; cursory, hasty, quick, rapid, speedy, swift. **2** compact, concise, pithy, succinct, to the point. ● *n.* directions, instructions. ● *v.* **1** advise, apprise, *colloq.* fill in, inform, instruct, prepare, prime.

briefcase *n.* flat document case.

briefs *n.pl.* short pants or knickers.

brier *n.* thorny bush, wild rose.

brigade *n.* army unit forming part of a division.

brigadier *n.* officer commanding a brigade or of similar status.

brigand *n.* member of a band of robbers.

bright *adj.* **1** giving out or reflecting much light, shining. **2** cheerful. **3** promising. **4** clever. □ **brightly** *adv.*, **brightness** *n.*

■ **1** beaming, dazzling, gleaming, glistening, glittering, glowing, incandescent, light, luminous, radiant, resplendent, shimmering, shining, sparkling, twinkling; glossy, lustrous, polished, shiny. **2** cheerful, cheery, gay, happy, light-hearted, perky, merry, sunny. **3** auspicious, favourable, hopeful, optimistic, promising, rosy. **4** brainy, clever, gifted, intelligent, quick, sharp, smart, talented.

brighten *v.* make or become brighter.

brilliant *adj.* **1** very bright, sparkling. **2** very clever. ● *n.* cut diamond with many facets. □ **brilliantly** *adv.*, **brilliance** *n.*

■ *adj.* **1** bright, dazzling, glittering, radiant, resplendent, shining, sparkling. **2** bright, clever, gifted, intelligent, talented.

brim *n.* **1** edge of a cup or hollow. **2** projecting edge of a hat. ● *v.* (**brimmed**) be full to the brim.

brimstone *n.* (*old use*) sulphur.

brindled *adj.* brown with streaks of another colour.

brine *n.* **1** salt water. **2** sea water.

bring *v.* (**brought**) **1** convey. **2** cause to come or be present. □ **bring about** cause to happen. **bring off** do successfully. **bring out 1** show clearly. **2** publish. **bring up 1** look after and train. **2** draw attention to.

■ **1** bear, carry, convey, deliver, fetch, take, transport. **2** attract, draw, lead. □ **bring about** cause, engender, generate, give rise to, occasion, produce, work. **bring off** accomplish, achieve, carry out, do, pull off, succeed in. **bring up 1** care for, look after, nurture, raise, rear, train. **2** broach, draw attention to, introduce, mention, raise.

brink *n.* **1** edge of a steep place or of a stretch of water. **2** point just before a change.

■ **1** border, brim, edge, lip, margin, rim. **2** threshold, verge.

brinkmanship *n.* policy of pursuing a dangerous course to the brink of catastrophe.

briny *adj.* of brine or sea water.

briquette *n.* block of compressed coal dust.

brisk *adj.* lively, quick. ◻ **briskly** *adv.*

■ energetic, fast, lively, quick, rapid, snappy, speedy, spirited, sprightly, spry, vigorous; active, bustling, busy.

brisket *n.* joint of beef from the breast.

bristle *n.* **1** short stiff hair. **2** one of the stiff pieces of hair or wire in a brush. ● *v.* **1** raise bristles in anger or fear. **2** show indignation. **3** be thickly set with bristles. ◻ **bristly** *adj.*

British *adj.* of Britain or its people.

Briton *n.* British person.

brittle *adj.* easily broken.

■ breakable, crisp, delicate, fragile, friable.

broach *v.* **1** raise for discussion. **2** open and start using.

broad *adj.* **1** large across, wide. **2** in general terms. **3** full and clear. **4** (of humour) rather coarse. ◻ **broad bean** edible bean with flat seeds. **broad-minded** *adj.* having tolerant views. **broadly** *adv.*

■ **1** expansive, extensive, large, spacious, vast, wide. **2** approximate, general, generalized, rough. **4** coarse, improper, indecent, indelicate, rude, vulgar. ◻ **broad-minded** liberal, permissive, tolerant, unbiased, unprejudiced.

broadcast *v.* (**broadcast**) **1** send out by radio or television. **2** make generally known. **3** scatter (seed) etc. ● *n.* broadcast programme. ◻ **broadcaster** *n.*, **broadcasting** *n.*

■ *v.* **1** radio, relay, televise, transmit. **2** advertise, announce, circulate, disseminate, make known, make public, proclaim, promulgate, publish, report. ◻ **broadcaster** announcer, commentator, newsreader, presenter.

broaden *v.* make or become broader.

broadsheet *n.* large-sized newspaper.

broadside *n.* firing of all guns on one side of a ship.

brocade *n.* fabric woven with raised patterns.

broccoli *n.* vegetable with green or purple flower heads.

brochure /brṓshər/ *n.* booklet or leaflet giving information.

brogue *n.* **1** strong shoe with ornamental perforated bands. **2** dialectal esp. Irish accent.

broil *v.* **1** grill. **2** make or become very hot.

broiler *n.* chicken suitable for broiling.

broke *see* **break**. *adj.* (*colloq.*) **1** having no money. **2** bankrupt.

broken *see* **break**. *adj.* (of a language) badly spoken by a foreigner.

broken-hearted *adj.* overwhelmed with grief.

broker *n.* agent who buys and sells on behalf of others.

bromide *n.* chemical compound used to calm nerves.

bromine *n.* poisonous liquid element.

bronchial *adj.* of the branched tubes into which the windpipe divides.

bronchitis *n.* inflammation of the bronchial tubes.

bronco *n.* (*pl.* **-os**) wild or half-tamed horse of western N. America.

brontosaurus *n.* large plant-eating dinosaur.

bronze *n.* **1** brown alloy of copper and tin. **2** thing made of this. **3** its colour. ● *v.* make or become suntanned.

brooch /brōch/ *n.* ornamental hinged pin fastened with a clasp.

brood *n.* young produced at one hatching or birth. ● *v.* **1** sit on (eggs) and hatch them. **2** worry or ponder, esp. resentfully.

■ *n.* litter, offspring, progeny, young. ● *v.* **2** agonize, fret, worry; meditate, muse, ponder, reflect, ruminate.

broody *adj.* **1** (of a hen) wanting to brood. **2** thoughtful and depressed.

brook[1] *n.* small stream.

■ beck, *Sc.* burn, rivulet, stream, watercourse.

brook[2] *v.* tolerate, allow.

broom *n.* **1** long-handled brush for sweeping floors. **2** shrub with white, yellow, or red flowers.

broomstick *n.* broom-handle.

broth *n.* thin meat or fish soup.

brothel *n.* house where women work as prostitutes.

brother *n.* **1** son of the same parents as another person. **2** man who is a fellow member of a group, trade union, or Church. **3** monk who is not a priest. ◻ **brother-in-law** *n.* (*pl.* **brothers-in-law**) **1** brother of one's husband or wife. **2** husband of one's sister. **brotherly** *adj.*

brotherhood *n.* **1** relationship of brothers. **2** comradeship.

brought *see* **bring**.

brow *n.* **1** eyebrow. **2** forehead. **3** projecting or overhanging part.

browbeat *v.* (**-beat, -beaten**) intimidate.

■ bully, cow, frighten, hector, intimidate, persecute, terrorize, threaten, tyrannize.

brown *adj.* of the colour of dark wood. ● *v.* make or become brown. □ **browned off** (*colloq.*) bored, fed up. **brownish** *adj.*

browse *v.* **1** feed on leaves or grass. **2** read or look around casually.

bruise *n.* injury that discolours skin without breaking it. ● *v.* cause bruise(s) on.

bruiser *n.* (*colloq.*) tough brutal person.

brunch *n.* meal combining breakfast and lunch.

brunette *n.* woman with brown hair.

brunt *n.* chief stress or strain.

brush *n.* **1** implement with bristles. **2** the action of brushing. **3** skirmish. **4** undergrowth. **5** fox's tail. ● *v.* **1** use a brush on. **2** touch lightly in passing. □ **brush off 1** reject curtly. **2** snub. **brush up 1** smarten. **2** revise one's knowledge of.

> ■ *n.* **1** besom, broom. **3** altercation, clash, conflict, confrontation, dispute, encounter, fracas, scrimmage, skirmish, tussle. **4** bracken, brushwood, scrub, undergrowth. ● *v.* **1** clean, scrub, sweep; groom. **2** graze, touch. □ **brush off** dismiss, put down, rebuff, reject, snub, spurn.

brushwood *n.* **1** undergrowth. **2** cut or broken twigs.

brusque /broŏsk/ *adj.* curt and offhand. □ **brusquely** *adv.*

Brussels sprout edible bud of a kind of cabbage.

brutal *adj.* cruel, without mercy. □ **brutally** *adv.*, **brutality** *n.*

> ■ barbaric, bestial, bloodthirsty, callous, cold-blooded, cruel, ferocious, hard-hearted, heartless, inhuman, inhumane, merciless, murderous, pitiless, ruthless, sadistic, savage, severe, vicious, violent, wild.

brutalize *v.* **1** make brutal. **2** treat brutally. □ **brutalization** *n.*

brute *n.* **1** animal other than a human being. **2** brutal person. **3** (*colloq.*) unpleasant person. ● *adj.* **1** unable to reason. **2** unreasoning. □ **brutish** *adj.*

bryony *n.* climbing hedge plant.

BSE *abbr.* bovine spongiform encephalopathy (cattle disease).

Bt. *abbr.* Baronet.

bubble *n.* **1** thin ball of liquid enclosing air or gas. **2** air-filled cavity. ● *v.* rise in bubbles.

> ■ *n.* **1** (**bubbles**) effervescence, foam, froth, lather, spume, suds. ● *v.* boil, effervesce, fizz, foam, froth, seethe.

bubbly *adj.* **1** full of bubbles. **2** lively, vivacious.

> ■ **1** effervescent, fizzy, foaming, frothy, sparkling. **2** animated, *colloq.* bouncy, buoyant, cheerful, ebullient, exuberant, high-spirited, lively, vivacious.

bubonic *adj.* (of plague) characterized by swellings.

buccaneer *n.* **1** pirate. **2** adventurer. □ **buccaneering** *n.* & *adj.*

buck¹ *n.* male of deer, hare, or rabbit. ● *v.* (of a horse) jump with the back arched. □ **buck up** (*colloq.*) **1** make haste. **2** make or become more cheerful.

buck² *n.* article placed before the dealer in a game of poker. □ **pass the buck** shift the responsibility.

buck³ *n.* (*US & Austr. sl.*) dollar.

bucket *n.* open container with a handle, for carrying or holding liquid. ● *v.* pour heavily.

buckle *n.* device through which a belt or strap is threaded to secure it. ● *v.* **1** fasten with a buckle. **2** crumple under pressure. □ **buckle down to** set about doing.

buckwheat *n.* **1** cereal plant. **2** its seed.

bucolic *adj.* rustic.

bud *n.* leaf or flower not fully open. ● *v.* (**budded**) **1** put forth buds. **2** begin to develop.

Buddhism *n.* Asian religion based on the teachings of Buddha. □ **Buddhist** *adj.* & *n.*

buddleia *n.* tree or shrub with purple or yellow flowers.

buddy *n.* (*colloq.*) friend.

budge *v.* move slightly.

budgerigar *n.* a kind of Australian parakeet.

budget *n.* **1** plan of income and expenditure. **2** amount allowed. ● *v.* (**budgeted**) allow or arrange in a budget.

buff *n.* **1** fawn colour. **2** (*colloq.*) enthusiast. ● *v.* polish with soft material.

buffalo *n.* (*pl.* **buffalo** or **-oes**) wild ox.

buffer *n.* **1** thing that lessens the effect of impact. **2** (*sl.*) man.

buffet¹ /bŏŏfay/ *n.* **1** counter where food and drink are served. **2** self-service meal.

buffet² /búffit/ *n.* blow, esp. with a hand. ● *v.* (**buffeted**) deal blows to.

buffoon *n.* foolish person. □ **buffoonery** *n.*

bug *n.* **1** small unpleasant insect. **2** (*sl.*) virus, infection. **3** (*colloq.*) defect. **4** secret microphone. ● *v.* (**bugged**) **1** install a secret microphone in. **2** (*sl.*) annoy.

■ *n.* **1** creepy-crawly, insect. **2** bacterium, germ, microbe, micro-organism, virus; disease, infection. **4** tap.

bugbear *n.* thing feared or disliked.

buggy *n.* **1** light carriage. **2** small sturdy vehicle. **3** pushchair.

bugle *n.* brass instrument like a small trumpet. □ **bugler** *n.*

build *v.* (**built**) construct by putting parts or material together. ● *n.* bodily shape. □ **build up 1** establish gradually. **2** increase. **build-up** *n.* this process. **builder** *n.*

■ *v.* assemble, construct, erect, fabricate, make, put together, put up, set up. ● *n.* body, figure, physique. □ **build up 1** develop, establish, expand, extend. **2** increase, intensify, strengthen.

building *n.* house or similar structure. □ **building society** organization that accepts deposits of money and lends to people buying houses.

■ construction, edifice, pile, structure.

built *see* **build**. □ **built-in** *adj.* forming part of a structure. **built-up** *adj.* covered with buildings.

bulb *n.* **1** rounded base of the stem of certain plants. **2** thing (esp. an electric lamp) shaped like this. □ **bulbous** *adj.*

bulge *n.* rounded swelling. ● *v.* swell outwards.

■ *n.* bump, distension, excrescence, knob, lump, projection, protrusion, protuberance, swelling. ● *v.* project, protrude, stick out, swell out.

bulk *n.* **1** size, magnitude (esp. large). **2** greater part. ● *v.* increase the size or thickness of.

■ *n.* **1** amount, extent, magnitude, mass, quantity, size, volume, weight. **2** best part, body, greater part, majority, preponderance.

bulkhead *n.* partition in a ship etc.

bulky *adj.* (**-ier, -iest**) **1** large. **2** unwieldy.

■ **1** beefy, big, brawny, burly, chunky, corpulent, heavy, hefty, large. **2** awkward, cumbersome, unwieldy, voluminous.

bull[1] *n.* **1** male of ox, whale, elephant, etc. **2** bull's-eye. □ **bull's-eye** *n.* centre of a target. **bull terrier** terrier resembling a bulldog.

bull[2] *n.* pope's official edict.

bull[3] *n.* (*sl.*) absurd statement.

bulldog *n.* powerful dog with a short thick neck.

bulldozer *n.* powerful tractor with a device for clearing ground.

bullet *n.* small missile fired from a rifle or revolver.

bulletin *n.* short official statement of news.

■ announcement, communication, communiqué, dispatch, message, newsflash, report, statement.

bullfighting *n.* baiting and killing bulls as entertainment.

bullfinch *n.* songbird with a strong beak and pinkish breast.

bullion *n.* gold or silver in bulk or bars, before manufacture.

bullock *n.* castrated bull.

bully[1] *n.* person who hurts or intimidates others. ● *v.* behave as a bully towards.

■ *v.* browbeat, hector, intimidate, persecute, terrorize, threaten, torment, tyrannize, victimize.

bully[2] *v.* **bully off** put the ball into play in hockey.

bulrush *n.* a kind of tall rush.

bulwark *n.* **1** wall of earth built as a defence. **2** ship's side above the deck.

bum[1] *n.* (*sl.*) buttocks.

bum[2] *n.* (*US sl.*) beggar, loafer.

bumble *v.* move or act in a blundering way.

bumble-bee *n.* large bee.

bump *n.* **1** dull-sounding blow or collision. **2** swelling caused by this. ● *v.* **1** hit or come against with a bump. **2** travel with a jolting movement. □ **bumpy** *adj.*

■ *n.* **1** blow, collision, knock, thud, thump. **2** bulge, lump, protuberance, swelling, welt. ● *v.* **1** bang, collide with, hit, knock against, ram, run into, strike. **2** bounce, jerk, jolt, lurch.

bumper *n.* **1** horizontal bar at the front and back of a motor vehicle to lessen the effect of collision. **2** something unusually large.

bumpkin *n.* country person with awkward manners.

bumptious *adj.* conceited.

bun *n.* **1** small sweet cake. **2** coil of hair at the back of the head.

bunch *n.* cluster of things growing or fastened together. ● *v.* **1** make into bunch(es). **2** form a group.

■ *n.* batch, bundle, clump, cluster, sheaf, tuft; bouquet, nosegay, posy, spray.

bundle *n.* collection of things loosely fastened or wrapped together. ● *v.* **1** make into a bundle. **2** push hurriedly.

■ *n.* bale, bunch, collection, package, packet, parcel, sheaf. ● *v.* **1** gather together, package, pack up, tie up. **2** cram, push, ram, shove, squeeze, thrust.

bung *n.* stopper for the hole in a barrel or jar. ● *v.* (*sl.*) throw.

bungalow *n.* one-storeyed house.

bungle *v.* mismanage. ● *n.* bungled attempt. □ **bungler** *n.*

■ *v.* botch, *sl.* fluff, *colloq.* make a hash of, make a mess of, mess up, mismanage, *colloq.* muff, *sl.* screw up, spoil.

bunion *n.* swelling at the base of the big toe, with thickened skin.

bunk[1] *n.* shelf-like bed.

bunk[2] *n.* **do a bunk** (*sl.*) run away.

bunker *n.* **1** container for fuel. **2** sandy hollow on a golf course. **3** reinforced underground shelter.

bunkum *n.* nonsense, humbug.

Bunsen burner device burning mixed air and gas in a single very hot flame.

bunting[1] *n.* bird related to the finches.

bunting[2] *n.* decorative flags.

buoy /boy/ *n.* anchored floating object serving as a navigation mark. ● *v.* **buoy up 1** keep afloat. **2** sustain, hearten.

buoyant /bóyənt/ *adj.* **1** able to float. **2** cheerful, lively. □ **buoyancy** *n.*

■ **2** carefree, cheerful, cheery, ebullient, exuberant, happy, high-spirited, jaunty, light-hearted, lively, vivacious.

bur *n.* plant's seed case that clings to clothing etc.

burble *v.* **1** make a gentle murmuring sound. **2** speak lengthily.

burden *n.* **1** load, esp. a heavy one. **2** oppressive duty, obligation, etc. ● *v.* put a burden on. □ **burdensome** *adj.*

■ *n.* **1** cargo, load, weight. **2** cross, duty, imposition, millstone, obligation, onus, responsibility, trial, trouble, worry. ● *v.* encumber, load, lumber, oppress, overload, saddle, tax, trouble, weigh down. □ **burdensome** arduous, difficult, exacting, onerous, oppressive, taxing, tiring, troublesome, wearisome, worrying.

bureau /byóorō/ *n.* (*pl.* **-eaux**) **1** writing desk with drawers. **2** office, department.

bureaucracy /byoorókrəsi/ *n.* **1** government by unelected officials. **2** excessive administration.

bureaucrat *n.* government official. □ **bureaucratic** *adj.*

burgeon *v.* begin to grow rapidly.

burglar *n.* person who breaks into a building, esp. in order to steal. □ **burglary** *n.*

■ housebreaker, intruder, robber, thief.

burgle *v.* rob as a burglar.

burial *n.* burying.

burlesque *n.* mocking imitation. ● *v.* imitate mockingly.

■ *n.* caricature, imitation, mockery, parody, satire, *colloq.* spoof, take-off.

burly *adj.* (**-ier**, **-iest**) large and sturdy. □ **burliness** *n.*

■ beefy, brawny, heavy, hefty, muscular, powerful, stocky, strapping, strong, sturdy, thickset.

burn[1] *v.* (**burned** or **burnt**) **1** (cause to) be destroyed by fire. **2** blaze or glow with fire. **3** (cause to) be injured or damaged by fire, sun, etc. **4** feel a sensation (as) of heat. **5** use or be used as fuel. ● *n.* mark or sore made by burning.

■ *v.* **1** cremate, fire, ignite, incinerate, set fire to, set on fire. **2** blaze, flame, glow, smoulder. **3** char, scald, scorch, sear, singe.

burn[2] *n.* (*Sc.*) brook.

burner *n.* part that shapes the flame in a lamp or cooker etc.

burning *adj.* **1** on fire, very hot. **2** intense. **3** hotly discussed.

■ **1** ablaze, alight, blazing, fiery, flaming, on fire, smouldering. **2** ardent, fervent, fervid, fierce, impassioned, intense, passionate, vehement.

burnish *v.* polish by rubbing.

burnt *see* **burn**[1].

burp *n.* & *v.* (*colloq.*) belch.

burr *n.* **1** whirring sound. **2** rough pronunciation of 'r'. **3** country accent using this.

burrow *n.* hole dug by a fox or rabbit as a dwelling. ● *v.* **1** dig a burrow. **2** search deeply, delve.

bursar *n.* treasurer of a college etc.

bursary *n.* **1** grant, esp. a scholarship. **2** bursar's office.

burst *v.* (**burst**) **1** break violently apart, explode. **2** appear, move, speak, etc. suddenly or violently. ● *n.* **1** bursting. **2** outbreak. **3** brief violent effort, spurt.

■ *v.* **1** break apart, blow up, *colloq.* bust, explode, give way, rupture, shatter, split.

bury *v.* **1** place (a dead body) in the earth or a tomb. **2** put or hide underground, cover up. **3** involve (oneself) deeply.

■ **1** inter, lay to rest. **2** hide, secrete, *colloq.* stash away; conceal, cover up, obscure, shroud. **3** engross, immerse, occupy, plunge.

bus *n.* (*pl.* **buses**) large passenger vehicle. ● *v.* (**bussed**) travel or transport by bus.

bush[1] *n.* **1** shrub. **2** thick growth. **3** wild uncultivated land.

bush[2] *n.* **1** metal lining for a hole in which something fits. **2** electrically insulating sleeve.

bushy *adj.* (**-ier, -iest**) **1** covered with bushes. **2** growing thickly.

business *n.* **1** occupation, trade. **2** task, duty. **3** thing to be dealt with. **4** buying and selling, trade. **5** commercial establishment. □ **businessman, businesswoman** *ns.* person engaged in trade or commerce.

■ **1** calling, career, employment, field, job, line of work, occupation, profession, trade, vocation. **2** duty, function, responsibility, role, task. **3** affair, issue, matter(s) in hand, question, problem, subject; agenda. **4** commerce, industry, trade; dealings, transactions. **5** company, concern, corporation, enterprise, firm, organization, partnership, practice, venture.

businesslike *adj.* practical, systematic.

■ down-to-earth, efficient, hard-headed, level-headed, logical, methodical, orderly, practical, pragmatic, professional, sensible, systematic, well-organized.

busk *v.* perform esp. music in the street for tips. □ **busker** *n.*

bust[1] *n.* **1** sculptured head, shoulders, and chest. **2** bosom.

bust[2] *v.* (**busted** or **bust**) (*colloq.*) burst, break. □ **bust-up** *n.* (*colloq.*) quarrel. **go bust** (*colloq.*) become bankrupt.

bustle[1] *v.* move busily and energetically. ● *n.* excited activity.

■ *v.* dash, hasten, hurry, hustle, rush, scamper, scramble, scurry, scuttle. ● *n.* activity, commotion, excitement, flurry, fuss, haste, hurly-burly, hustle, stir, to-do.

bustle[2] *n.* (*old use*) padding to puff out the top of a skirt at the back.

busy *adj.* (**-ier, -iest**) **1** working, having much to do. **2** full of activity. □ **busily** *adv.*

■ **1** hard-working, industrious, *colloq.* on the go, tireless, working; engaged, occupied, tied up. **2** active, brisk, bustling, eventful, full, hectic.

busybody *n.* meddlesome person.

but *adv.* only. ● *prep.* & *conj.* **1** however. **2** except.

butane *n.* inflammable liquid used as fuel.

butch *adj.* (*sl.*) strongly masculine.

butcher *n.* **1** person who cuts up and sells meat. **2** brutal murderer. ● *v.* kill needlessly or brutally. □ **butchery** *n.*

butler *n.* chief manservant, in charge of the wine cellar.

butt[1] *n.* large cask or barrel.

butt[2] *n.* **1** thicker end of a tool or weapon. **2** short remnant, stub.

butt[3] *n.* **1** target for ridicule or teasing. **2** mound behind a target. **3** (*pl.*) shooting range.

butt[4] *v.* **1** push with the head. **2** meet or place edge to edge. □ **butt in 1** interrupt. **2** meddle.

butter *n.* fatty food substance made from cream. ● *v.* spread with butter. □ **butter up** flatter.

buttercup *n.* wild plant with yellow cup-shaped flowers.

butterfly *n.* **1** insect with four large often brightly coloured wings. **2** swimming stroke with both arms lifted at the same time.

buttermilk *n.* liquid left after butter is churned from milk.

butterscotch *n.* hard toffee-like sweet.

buttock *n.* either of the two fleshy rounded parts at the lower end of the back of the body.

button *n.* **1** disc or knob sewn to a garment as a fastener or ornament. **2** small rounded object. **3** knob etc. pressed to operate a device. ● *v.* fasten with button(s).

buttonhole *n.* **1** slit through which a button is passed to fasten clothing. **2** flower worn in the buttonhole of a lapel. ● *v.* accost and talk to.

buttress *n.* **1** support built against a wall. **2** thing that supports. ● *v.* reinforce, prop up.

■ *n.* prop, support. ● *v.* bolster, brace, prop up, reinforce, shore up, strengthen, support, sustain.

buxom *adj.* plump and healthy.

buy *v.* (**bought**) obtain in exchange for money. ● *n.* purchase. □ **buyer** *n.*

■ *v.* acquire, come by, get, obtain, pay for, procure, purchase. □ **buyer** client, consumer, customer, *colloq.* punter, purchaser, shopper.

buzz *n.* **1** vibrating humming sound. **2** rumour. **3** thrill. ● *v.* **1** make or be filled with a buzz. **2** go about busily. □ **buzzword** *n.* (*colloq.*) fashionable word.

buzzard *n.* a kind of hawk.

buzzer *n.* device that produces a buzzing sound as a signal.

by *prep.* & *adv.* **1** near, beside, in reserve. **2** along, via, past. **3** during. **4** through the agency or means of. **5** not later than. □ **by and by** before long. **by and large** on the whole. **by-election** *n.* election of an MP to replace one who has died or resigned. **by-law** *n.* regulation made by a local authority or corporation. **by oneself** alone, without help. **by-product** *n.* thing produced while making something else.

bye *n.* **1** run scored from a ball not hit by the batsman. **2** having no opponent for one round of a tournament.

bygone *adj.* belonging to the past. ● *n.pl.* bygone things.

bypass *n.* road taking traffic round a town. ● *v.* avoid.

byre *n.* cowshed.

byroad *n.* minor road.

bystander *n.* person standing near when something happens.

■ eyewitness, observer, onlooker, passer-by, spectator, witness.

byte *n.* (in computers) group of bits.

byway *n.* minor road.

byword *n.* **1** notable example. **2** familiar saying.

Byzantine *adj.* **1** of Byzantium. **2** complicated, underhand.

Cc

C *abbr.* **1** Celsius. **2** centigrade.

cab *n.* **1** taxi. **2** compartment for the driver of a train, lorry, etc.

cabaret /kábbəray/ *n.* entertainment in a nightclub etc.

cabbage *n.* vegetable with a round head of green or purple leaves.

cabby *n.* (*colloq.*) taxi driver.

caber *n.* trimmed tree trunk.

cabin *n.* **1** small hut. **2** compartment in a ship or aircraft.

■ **1** chalet, hut, lodge, shack, shanty, shelter. **2** berth, compartment, room, stateroom.

cabinet *n.* **1** cupboard with drawers or shelves. **2** (**Cabinet**) group of senior ministers in government.

cable *n.* **1** thick rope of fibre or wire. **2** set of insulated wires for carrying electricity or signals. □ **cable car** car of a **cable railway** drawn on an endless cable by a stationary engine. **cable television** transmission by cable to subscribers.

■ **1** chain, guy, hawser, mainstay, rope, wire. **2** cord, flex, lead.

cacao *n.* **1** seed from which cocoa and chocolate are made. **2** tree producing this.

cache /kash/ *n.* **1** hiding place for treasure or stores. **2** things in this. ● *v.* put into a cache.

cachet /káshay/ *n.* **1** prestige. **2** distinctive mark or feature.

cackle *n.* **1** clucking of hens. **2** chattering talk. **3** loud silly laugh. ● *v.* utter a cackle.

cacophony *n.* harsh discordant sound. □ **cacophonous** *adj.*

■ □ **cacophonous** discordant, dissonant, grating, harsh, jangling, noisy, raucous, strident.

cactus *n.* (*pl.* **-ti** or **-tuses**) fleshy plant, usu. with spines.

cadaver *n.* corpse.

cadaverous *adj.* gaunt and pale.

caddie *n.* golfer's attendant carrying clubs. ● *v.* act as caddie.

caddis-fly *n.* four-winged insect living near water.

caddy *n.* small box for tea.

cadence *n.* **1** rhythm. **2** rise and fall of the voice in speech. **3** end of a musical phrase.

■ **1** accent, beat, lilt, measure, metre, rhythm, tempo.

cadenza *n.* elaborate passage for a solo instrument or singer.

cadet *n.* young trainee in the armed forces or police.

cadge *v.* ask for as a gift, beg.

cadmium *n.* metallic element.

cadre *n.* small group, esp. of soldiers.

caecum /séékəm/ *n.* (*pl.* **-ca**) pouch between the small and large intestines.

Caesarean section delivery of a child by cutting into the mother's abdomen.

café *n.* informal restaurant.

cafeteria *n.* self-service restaurant.

caffeine *n.* stimulant found in tea and coffee.

caftan *n.* long loose robe or dress.

cage *n.* enclosure of wire or with bars, esp. for birds or animals.

cagey *adj.* (**-ier**, **-iest**) (*colloq.*) cautious and noncommittal. □ **cagily** *adv.*, **caginess** *n.*

cagoule *n.* light hooded waterproof jacket.

cahoots *n.* (*sl.*) partnership.

cairn *n.* mound of stones as a memorial or landmark. □ **cairn terrier** small shaggy short-legged terrier.

caisson *n.* watertight chamber used in underwater construction work.

cajole *v.* persuade by flattery. □ **cajolery** *n.*

■ coax, entice, inveigle, persuade, seduce, wheedle.

cake *n.* **1** baked sweet bread-like food. **2** small flattened mass. ● *v.* form into a compact mass.

calamine *n.* lotion containing zinc carbonate.

calamity *n.* disaster. □ **calamitous** *adj.*, **calamitously** *adv.*

■ cataclysm, catastrophe, disaster, misfortune, tragedy.

calcify *v.* harden by a deposit of calcium salts. □ **calcification** *n.*

calcium *n.* whitish metallic element.

calculate *v.* **1** reckon mathematically. **2** estimate. **3** plan deliberately. □ **calculation** *n.*

■ **1** add up, compute, determine, reckon, work out. **2** assess, estimate, evaluate, gauge, weigh up.

calculator *n.* electronic device for making calculations.

calculus *n.* (*pl.* **-li**) **1** method of calculating in mathematics. **2** stone formed in the body.

Caledonian *adj.* of Scotland. ● *n.* Scottish person.

calendar *n.* chart showing dates of days of the year.

calf[1] *n.* (*pl.* **calves**) young of cattle, also of elephant, whale, and seal.

calf[2] *n.* (*pl.* **calves**) fleshy part of the human leg below the knee.

calibrate *v.* **1** mark or correct the units of measurement on (a gauge). **2** find the calibre of. □ **calibration** *n.*

calibre *n.* **1** diameter of a gun, tube, or bullet. **2** level of ability.

■ **1** bore, diameter, gauge, size. **2** ability, capability, capacity, competence, merit, proficiency, quality, stature, talent.

calico *n.* a kind of cotton cloth.

caliph *n.* (formerly) Muslim ruler.

call *v.* **1** shout to attract attention. **2** summon. **3** rouse from sleep. **4** communicate (with) by telephone or radio. **5** name, describe, or address as. **6** make a brief visit. **7** utter a characteristic cry. ● *n.* **1** shout. **2** invitation. **3** need. **4** demand. **5** telephone conversation. **6** short visit. **7** bird's cry. □ **call box** telephone kiosk. **call off** cancel. **caller** *n.*

■ *v.* **1** bawl, bellow, cry (out), hail, shout, yell. **2** assemble, convene, convoke, rally, summon. **3** awake, awaken, get up, knock up, rouse, wake. **4** dial, *colloq.* phone, ring up, telephone. **5** baptize, christen, designate, dub, name, nickname. **6** drop in, pay a visit, visit. ● *n.* **1** bellow, cry, shout, yell. **2** command, invitation, request, summons. **3** cause, justification, need, occasion, reason. **4** demand, desire, market.

calligraphy *n.* (beautiful) handwriting. □ **calligraphic** *adj.*

calling *n.* vocation, profession.

calliper *n.* splint for a weak leg.

callisthenics *n.pl.* exercises to develop strength and grace.

callous *adj.* feeling no pity or sympathy. □ **callously** *adv.*, **callousness** *n.*

■ cold, cruel, hard, hard-boiled, hard-hearted, heartless, inhumane, insensitive, stony, uncaring, unfeeling, unsympathetic.

callow *adj.* immature and inexperienced.

■ green, immature, inexperienced, juvenile, naive, unsophisticated.

callus *n.* patch of hardened skin.

calm *adj.* **1** still, not windy. **2** not excited or agitated. ● *n.* calm condition. ● *v.* make calm. □ **calmly** *adv.*, **calmness** *n.*

■ *adj.* **1** peaceful, quiet, still, tranquil, windless. **2** collected, composed, controlled, cool, dispassionate, equable, impassive, imperturbable, nonchalant, relaxed, sedate, self-controlled, self-possessed, serene, stoical, *colloq.* unflappable, unperturbed, unruffled, untroubled. ● *n.* calmness, composure, equanimity, peace, peacefulness, placidity, quiet, quietness, serenity, stillness, tranquillity. ● *v.* appease, assuage, lull, mollify, pacify, placate, quiet, quieten, settle, soothe.

calorie *n.* **1** unit of heat. **2** unit of the energy-producing value of food.

calorific *adj.* heat-producing.

calumniate *v.* slander.

calumny *n.* slander.

calve *v.* give birth to a calf.

Calvinism *n.* teachings of the Protestant reformer John Calvin or his followers. □ **Calvinist** *n.*

calypso *n.* topical W. Indian song.

calyx *n.* ring of sepals covering a flower bud.

cam *n.* device changing rotary to to-and-fro motion. □ **camshaft** *n.*

camaraderie *n.* comradeship.

camber *n.* slight convex curve given to a surface esp. of a road.

cambric *n.* thin linen or cotton cloth.

camcorder *n.* combined video and sound recorder.

came *see* **come.**

camel *n.* **1** four-legged animal with one hump or two. **2** fawn colour.

camellia *n.* evergreen flowering shrub.

cameo *n.* **1** stone in a ring or brooch with coloured layers carved in a raised design. **2** small part in a play or film taken by a famous actor or actress.

camera *n.* apparatus for taking photographs or film pictures. □ **in camera** in private.

camiknickers *n.pl.* woman's undergarment combining camisole and knickers.

camisole *n.* woman's cotton bodice-like garment.

camomile *n.* aromatic herb.

camouflage *n.* disguise, concealment, by colouring or covering. ● *v.* disguise or conceal in this way.

■ *n.* concealment, cover, disguise, façade, front, mask, screen. ● *v.* cloak, conceal, cover (up), disguise, hide, mask, screen, veil.

camp[1] *n.* **1** temporary accommodation in tents. **2** place where troops are lodged or trained. **3** fortified site. ● *v.* encamp, be in a camp. □ **camp bed** portable folding bed. **camper** *n.*

camp[2] *adj.* **1** affected, exaggerated. **2** homosexual. ● *n.* camp behaviour. ● *v.* act or behave in a camp way.

campaign *n.* **1** organized course of action. **2** series of military operations. ● *v.* take part in a campaign. □ **campaigner** *n.*

■ *n.* **1** crusade, drive, effort, move, movement, plan, push, scheme. **2** manoeuvre(s), offensive, operation(s).

campanology *n.* **1** study of bells. **2** bell-ringing. □ **campanologist** *n.*

campanula *n.* plant with bell-shaped flowers.

camphor *n.* strong-smelling white substance used in medicine and moth-balls. □ **camphorated** *adj.*

campion *n.* wild plant with pink or white flowers.

campus *n.* (*pl.* **-puses**) grounds of a university or college.

can[1] *n.* container in which food etc. is sealed and preserved. ● *v.* (**canned**) preserve in a can.

can[2] *v.aux.* is or are able or allowed to.

Canadian *adj.* & *n.* (native, inhabitant) of Canada.

canal *n.* **1** artificial watercourse. **2** duct.

canalize *v.* **1** convert into a canal. **2** channel. □ **canalization** *n.*

canapé /kánnəpi/ *n.* small piece of bread etc. with savoury topping.

canary *n.* small yellow songbird.

cancan *n.* high-kicking dance.

cancel *v.* (**cancelled**) **1** declare that (something arranged) will not take place. **2** order to be discontinued, make invalid. **3** cross out. □ **cancel out** neutralize. **cancellation** *n.*

■ **1** call off, postpone, *colloq.* scrub. **2** abolish, annul, countermand, do away with, invalidate, nullify, quash, repeal, rescind, retract, revoke, withdraw. **3** cross out, delete, erase, rub out. □ **cancel out** compensate for, counterbalance, make up for, neutralize, offset.

cancer *n.* **1** malignant tumour. **2** spreading evil. □ **cancerous** *adj.*

candela *n.* unit measuring the brightness of light.

candelabrum *n.* (also **-bra**) large branched candlestick.

candid *adj.* frank. □ **candidly** *adv.*

■ blunt, direct, forthright, frank, honest, ingenuous, open, outspoken, plain, sincere, straightforward, truthful, unequivocal.

candidate *n.* person applying for a job or taking an exam. □ **candidacy** *n.*, **candidature** *n.*

■ applicant, aspirant, competitor, contender, examinee, interviewee.

candied *adj.* encrusted or preserved in sugar.

candle *n.* stick of wax enclosing a wick which is burnt to give light.

candlestick *n.* holder for a candle.

candlewick *n.* fabric with a tufted pattern.

candour *n.* frankness.

candy *n.* (*US*) sweets, a sweet.

candyfloss *n.* fluffy mass of spun sugar.

candy stripe alternate stripes of white and esp. pink.

candytuft *n.* garden plant with flowers in flat clusters.

cane *n.* **1** stem of a tall reed or grass or slender palm. **2** light walking stick.

canine /káynīn/ *adj.* of dog(s). ● *n.* (in full **canine tooth**) a pointed tooth between incisors and molars.

canister *n.* small metal container.

canker *n.* **1** disease of animals or plants. **2** influence that corrupts.

cannabis *n.* **1** hemp plant. **2** drug made from this.

cannibal *n.* person who eats human flesh. □ **cannibalism** *n.*

cannibalize *v.* use parts from (a machine) to repair another.

cannon *n.* **1** large gun. **2** hitting of two balls in one shot in billiards. ● *v.* bump heavily (into).

cannonade *n.* continuous gunfire. ● *v.* bombard with this.

cannot negative form of **can²**.

canny *adj.* (**-ier, -iest**) shrewd. ◻ **cannily** *adv.*

canoe *n.* light boat propelled by paddle(s). ● *v.* go in a canoe. ◻ **canoeist** *n.*

canon *n.* **1** member of cathedral clergy. **2** general rule or principle. **3** set of writings accepted as genuine. ◻ **canonical** *adj.*

canonize *v.* declare officially to be a saint. ◻ **canonization** *n.*

canopy *n.* covering hung up over a throne, bed, person, etc.

cant *n.* **1** insincere talk. **2** jargon.

cantaloup *n.* small ribbed melon.

cantankerous *adj.* bad-tempered. ◻ **cantankerously** *adv.*

> ■ bad-tempered, choleric, crabby, cross, crotchety, gruff, grumpy, fractious, irascible, irritable, peevish, petulant, prickly, quarrelsome, surly, testy, waspish.

cantata *n.* choral composition.

canteen *n.* **1** restaurant for employees. **2** case of cutlery.

canter *n.* gentle gallop. ● *v.* go at a canter.

cantilever *n.* projecting beam or girder supporting a structure.

canto *n.* division of a long poem.

canton *n.* division of Switzerland.

canvas *n.* **1** strong coarse cloth. **2** a painting on this.

canvass *v.* **1** ask for political support. **2** ascertain opinions of.

canyon *n.* deep gorge.

> ■ defile, gorge, pass, ravine.

cap *n.* **1** soft brimless hat, often with a peak. **2** headdress worn as part of a uniform. **3** cover or top. **4** explosive device for a toy pistol. ● *v.* (**capped**) **1** put a cap on. **2** form the top of. **3** surpass.

capable *adj.* competent, able. ◻ **capably** *adv.*, **capability** *n.*

> ■ able, accomplished, adept, clever, competent, efficient, experienced, practised, proficient, qualified, skilful, skilled, talented.

capacious *adj.* roomy.

> ■ ample, commodious, roomy, sizeable, spacious.

capacitance *n.* ability to store an electric charge.

capacitor *n.* device storing a charge of electricity.

capacity *n.* **1** amount that can be contained or produced. **2** mental power. **3** function or position. **4** ability to contain, receive, experience, or produce.

> ■ **1** content, dimensions, magnitude, proportions, size, volume. **2** ability, brains, capability, cleverness, competence, intelligence, perspicacity, wit. **3** duty, function, job, office, place, position, post, responsibility, role. **4** potential.

cape¹ *n.* a short cloak.

cape² *n.* coastal promontory.

caper¹ *v.* move friskily. ● *n.* **1** frisky movement. **2** (*sl.*) activity.

> ■ *v.* bound, cavort, frisk, frolic, gambol, hop, jump, leap, prance, skip, spring.

caper² *n.* **1** bramble-like shrub. **2** one of its pickled buds.

capercaillie *n.* (also **capercailzie**) large grouse.

capillary *n.* very fine hair-like tube or blood vessel.

capital *adj.* **1** chief, very important. **2** involving the death penalty. **3** (of a letter of the alphabet) of the kind used to begin a name or sentence. ● *n.* **1** chief city of a country etc. **2** money with which a business is started. **3** capital letter. **4** top part of a pillar.

> ■ *adj.* **1** cardinal, central, chief, foremost, leading, main, major, paramount, preeminent, primary, prime, principal. ● *n.* **2** assets, cash, finances, funds, means, money, principal, resources, savings, stocks, wealth, *colloq.* wherewithal.

capitalism *n.* system in which trade and industry are controlled by private owners.

capitalist *n.* person who has money invested in businesses.

capitalize *v.* **1** convert into or provide with capital. **2** write as or with a capital letter. ◻ **capitalize on** make advantageous use of. **capitalization** *n.*

capitulate *v.* surrender, yield. ◻ **capitulation** *n.*

> ■ acquiesce, concede, give in, relent, submit, succumb, surrender, throw in the towel, yield.

caprice /kəpreéss/ *n.* **1** whim. **2** lively piece of music.

capricious *adj.* impulsive and unpredictable. ◻ **capriciously** *adv.*, **capriciousness** *n.*

> ■ changeable, erratic, fickle, impulsive, inconsistent, inconstant, mercurial, moody, temperamental, unpredictable, unreliable, unstable, variable, volatile, wayward.

capsicum *n.* tropical plant with pungent seeds.

capsize *v.* overturn.

■ flip over, keel over, overturn, tip over, turn turtle, turn upside down.

capstan *n.* revolving post or spindle on which a cable etc. winds.

capsule *n.* **1** small soluble case containing medicine. **2** detachable compartment of a spacecraft. **3** plant's seed case.

captain *n.* **1** leader of a group or sports team. **2** person commanding a ship or civil aircraft. **3** naval officer next below rear admiral. **4** army officer next below major. ● *v.* be captain of. □ **captaincy** *n.*

caption *n.* **1** short title or heading. **2** explanation on an illustration.

captious *adj.* fond of finding fault, esp. about trivial matters.

captivate *v.* fascinate, charm. □ **captivation** *n.*

■ attract, bewitch, charm, dazzle, delight, enchant, enrapture, enthral, entrance, fascinate, hypnotize, mesmerize.

captive *adj.* taken prisoner, confined, unable to escape. ● *n.* captive person or animal. □ **captivity** *n.*

■ *adj.* caged, captured, confined, imprisoned, incarcerated, jailed, locked up. ● *n.* detainee, hostage, internee, prisoner. □ **captivity** bondage, confinement, custody, detention, imprisonment, incarceration, internment, slavery.

captor *n.* one who takes a captive.

capture *v.* **1** take prisoner. **2** take or obtain by force or skill. **3** cause (data) to be stored in a computer. ● *n.* act of capturing.

■ *v.* **1** apprehend, arrest, catch, collar, ensnare, kidnap, *sl.* nab, *sl.* nick, seize, take prisoner. **2** carry, carry off, gain, get, obtain, secure, take, win.

car *n.* **1** motor vehicle for a small number of passengers. **2** (*US*) railway carriage. **3** compartment in a cable railway, lift, etc.

■ **1** *US* automobile, *sl.* banger, motor, motor car, vehicle.

carafe /kəráf/ *n.* glass bottle for serving wine or water.

caramel *n.* **1** brown syrup made from heated sugar. **2** toffee tasting like this. □ **caramelize** *v.*

carapace *n.* upper shell of a tortoise.

carat *n.* unit of purity of gold.

caravan *n.* **1** dwelling on wheels, able to be towed by a horse or car. **2** company travelling together across desert. □ **caravanner** *n.*, **caravanning** *n.*

caraway *n.* (plant with) spicy seeds used for flavouring cakes.

carbine *n.* automatic rifle.

carbohydrate *n.* energy-producing compound in food.

carbolic *n.* a kind of disinfectant.

carbon *n.* non-metallic element occurring as diamond, graphite, and charcoal, and in all living matter. □ **carbon copy 1** copy made with carbon paper. **2** exact copy. **carbon paper** paper coated with pigment for making a copy as something is typed or written.

carbonate *n.* compound releasing carbon dioxide when mixed with acid. ● *v.* impregnate with carbon dioxide.

carboniferous *adj.* producing coal.

carborundum *n.* compound of carbon and silicon used for grinding and polishing things.

carbuncle *n.* **1** severe abscess. **2** bright red gem.

carburettor *n.* apparatus mixing air and petrol in a motor engine.

carcass *n.* dead body of an animal.

carcinogen *n.* cancer-producing substance. □ **carcinogenic** *adj.*

carcinoma *n.* cancerous tumour.

card[1] *n.* **1** piece of cardboard or thick paper. **2** this printed with a greeting or invitation. **3** postcard. **4** playing card. **5** credit card. **6** (*pl.*, *colloq.*) employee's official documents, held by his employer. □ **card-sharp** *n.* swindler at card games.

card[2] *v.* clean or comb (wool) with a wire brush.

cardboard *n.* stiff substance made by pasting together sheets of paper.

cardiac *adj.* of the heart.

cardigan *n.* knitted jacket.

cardinal *adj.* chief, most important. ● *n.* prince of the RC Church. □ **cardinal number** whole number (1, 2, 3, etc.).

cardiogram *n.* record of heart movements. □ **cardiograph** *n.* instrument producing this.

cardiology *n.* study of diseases of the heart. □ **cardiologist** *n.*

cardphone *n.* public telephone operated by a plastic machine-readable card.

care *n.* **1** serious attention and thought. **2** caution to avoid damage or loss. **3** (cause of) worry, anxiety. **4** protection. ● *v.* feel concern or interest. □ **care for 1** like, love. **2** look after.

■ *n.* **1** attention, consideration, deliberation, diligence, heed, meticulousness,

pains, prudence, punctiliousness, thought. 2 caution, circumspection, mindfulness, vigilance, watchfulness. 3 anxiety, concern, disquiet, distress, worry; problem, sorrow, trouble, woe. 4 charge, custody, guardianship, keeping, protection, responsibility, trust. □ care for 1 adore, be fond of, be in love with, cherish, dote on, like, love, treasure. 2 look after, mind, minister to, nurse, provide for, see to, support, take care of, tend.

careen *v.* tilt or keel over.

career *n.* **1** way of making one's living, profession. **2** course through life. **3** swift course. ● *v.* go swiftly or wildly.

■ *n.* **1 calling, job, line of work, livelihood, occupation, profession, trade, vocation.** ● *v.* **dash, fly, hurtle, race, rush, shoot, speed, sprint, tear, zoom.**

careerist *n.* person intent on advancement in a career.

carefree *adj.* light-hearted, free from anxieties.

■ **airy, blithe, cheerful, debonair, easy-going, happy, happy-go-lucky, insouciant, light-hearted, nonchalant, relaxed.**

careful *adj.* acting or done with care. □ **carefully** *adv.*, **carefulness** *n.*

■ **alert, cautious, chary, circumspect, guarded, on one's guard, prudent, vigilant, wary, watchful; accurate, conscientious, diligent, methodical, meticulous, orderly, organized, painstaking, precise, punctilious, scrupulous, systematic, thorough, well-organized.**

careless *adj.* not careful. □ **carelessly** *adv.*, **carelessness** *n.*

■ **absent-minded, heedless, inattentive, incautious, irresponsible, neglectful, negligent, remiss, scatterbrained, thoughtless, unguarded, unthinking; casual, cursory, perfunctory; disorganized, inaccurate, shoddy, slapdash, slipshod, sloppy, slovenly.**

carer *n.* person who looks after a sick or disabled person at home.

caress *n.* loving touch, kiss. ● *v.* give a caress to.

■ *v.* **cuddle, embrace, fondle, hug, kiss, nuzzle, pat, pet, stroke.**

caret *n.* omission mark.

caretaker *n.* person employed to look after a building.

careworn *adj.* showing signs of prolonged worry.

cargo *n.* (*pl.* **-oes**) goods carried by ship or aircraft.

■ **consignment, freight, goods, load, payload, shipment.**

Caribbean *adj.* of the W. Indies or their inhabitants.

caribou *n.* N. American reindeer.

caricature *n.* exaggerated portrayal of a person for comic effect. ● *v.* make a caricature of.

■ *n.* **burlesque, cartoon, parody, satire,** *colloq.* **spoof, take-off.** ● *v.* **burlesque, lampoon, parody, satirize, send up, take off.**

caries *n.* decay of tooth or bone.

carillon /kərílyən/ *n.* **1** set of bells sounded mechanically. **2** tune played on these.

Carmelite *n.* member of an order of white-cloaked friars or nuns.

carmine *adj.* & *n.* vivid crimson.

carnage *n.* great slaughter.

■ **bloodshed, butchery, killing, massacre, slaughter.**

carnal *adj.* **1** of the body or flesh, worldly. **2** sensual.

■ **1 animal, bodily, corporeal, physical; earthly, material, worldly. 2 erotic, lascivious, lustful, sensual, sexual.**

carnation *n.* clove-scented pink.

carnival *n.* public festivities, usu. with a procession.

carnivore *n.* carnivorous animal.

carnivorous *adj.* feeding on flesh.

carol *n.* Christmas hymn. ● *v.* (**carolled**) **1** sing carols. **2** sing joyfully.

carotid *adj.* & *n.* (artery) carrying blood to the head.

carouse *v.* drink and be merry. □ **carousal** *n.*, **carouser** *n.*

carousel *n.* **1** (*US*) merry-go-round. **2** rotating conveyor.

carp[1] *n.* freshwater fish.

carp[2] *v.* keep finding fault.

■ **cavil, complain, find fault,** *colloq.* **gripe, grumble,** *colloq.* **nit-pick, pick holes,** *colloq.* **whinge.**

carpenter *n.* person who makes or repairs wooden objects and structures. □ **carpentry** *n.*

carpet *n.* fabric for covering a floor. ● *v.* (**carpeted**) **1** cover with a carpet. **2** reprimand.

carport *n.* roofed open-sided shelter for a car.

carpus *n.* set of small bones forming the wrist joint.

carriage *n.* **1** railway passenger vehicle. **2** horse-drawn vehicle. **3** conveying of goods etc., cost of this. **4** bearing, deportment. □ **carriage clock** small portable clock with a handle on top.

> ■ **1** *US* car, coach, wagon. **2** coach, gig, landau, trap. **3** conveyance, haulage, shipping, transport, transportation. **4** bearing, demeanour, deportment, posture, stance.

carriageway *n.* that part of the road on which vehicles travel.

carrier *n.* **1** person or thing carrying something. **2** paper or plastic bag with handles.

carrion *n.* dead decaying flesh.

carrot *n.* **1** plant with edible tapering orange root. **2** this root. **3** incentive.

carry *v.* **1** transport, convey. **2** support. **3** involve, entail. **4** stock (goods for sale). **5** take (a process etc.) to a specified point. **6** win acceptance for (a motion etc.). **7** get the support of. **8** be audible at a distance. □ **carry off 1** remove by force. **2** win. **carry on 1** continue. **2** (*colloq.*) behave excitedly. **carry out** put into practice.

> ■ **1** bear, bring, *sl.* cart, convey, deliver, ferry, haul, lug, move, ship, take, transfer, transport; conduct, transmit. **2** bear, hold, support, take. **3** entail, include, involve. **4** have in stock, keep, sell, stock. □ **carry off 1** abduct, capture, kidnap, make off with, snatch. **2** gain, secure, win. **carry out** effect, execute, implement, perform, put into practice.

cart *n.* wheeled structure for carrying loads. ● *v.* (*sl.*) carry.

carte blanche /kaárt blónsh/ complete freedom to do as one thinks best.

cartel *n.* manufacturer's or producer's union to control prices.

carthorse *n.* horse of heavy build.

cartilage *n.* firm elastic tissue in skeletons of vertebrates, gristle.

cartography *n.* map-drawing. □ **cartographer** *n.*

carton *n.* cardboard or plastic container.

cartoon *n.* **1** humorous drawing. **2** film consisting of an animated sequence of drawings. **3** sketch for a painting. □ **cartoonist** *n.*

cartridge *n.* **1** case containing explosive for firearms. **2** sealed cassette. □ **cartridge paper** thick strong paper.

cartwheel *n.* sideways somersault with arms and legs extended.

carve *v.* **1** make, inscribe, or decorate by cutting. **2** cut (meat) into slices for eating.

> ■ **1** chisel, engrave, fashion, hew, inscribe, sculpt, sculpture, whittle.

cascade *n.* **1** waterfall. **2** thing falling or hanging like this. ● *v.* fall in this way.

case[1] *n.* **1** instance of a thing's occurring. **2** situation. **3** lawsuit. **4** set of facts or arguments supporting something. □ **in case** lest.

> ■ **1** example, illustration, instance, occurrence. **2** occasion, position, situation, state of affairs; eventuality. **3** action, lawsuit, suit.

case[2] *n.* **1** container or covering enclosing something. **2** item of luggage. ● *v.* **1** enclose in a case. **2** (*sl.*) examine (a building etc.) in preparation for a crime.

> ■ *n.* **1** box, carton, casket, container, crate, holder, receptacle. **2** bag, grip, holdall, suitcase, trunk.

casement *n.* window opening on vertical hinges.

cash *n.* money in the form of coins or banknotes. ● *v.* give or obtain cash for (a cheque etc.). □ **cash in (on)** get profit or advantage (from).

> ■ *n.* banknotes, *US* bills, change, coins, currency, *sl.* dough, money, notes.

cash card plastic card with magnetic code for drawing money from a machine.

cashew *n.* a kind of edible nut.

cashier[1] *n.* person employed to receive money.

cashier[2] *v.* dismiss from military service in disgrace.

cashmere *n.* **1** very fine soft wool. **2** fabric made from this.

cashpoint *n.* machine dispensing cash.

casino *n.* (*pl.* **-os**) public building or room for gambling.

cask *n.* barrel for liquids.

casket *n.* **1** small box for valuables. **2** (*US*) coffin.

cassava *n.* **1** tropical plant. **2** flour made from its roots.

casserole *n.* **1** covered dish in which meat etc. is cooked and served. **2** food cooked in this. ● *v.* cook in a casserole.

cassette *n.* small case containing a reel of film or magnetic tape.

cassock *n.* long robe worn by clergy and choristers.

cassowary *n.* large flightless bird related to the emu.

cast *v.* (**cast**) **1** throw. **2** shed. **3** direct (a glance). **4** register (one's vote). **5** select actors for a play or film, assign a role to. **6** shape (molten metal) in a mould. ● *n.* **1** throw of dice, fishing line, etc. **2** set of actors in a play etc. **3** type, quality. **4** moulded mass of solidified material. **5** slight squint. □ **cast-iron** *adj.* very strong. **cast-off** *adj.* & *n.* discarded (thing).

■ *v.* **1** *colloq.* chuck, fling, hurl, lob, pitch, *colloq.* sling, throw, toss. ● *n.* **1** pitch, shy, toss, throw. **2** company, performers, players, troupe. **3** kind, quality, sort, style, type, variety.

castanets *n.pl.* pair of shell-shaped pieces of wood clicked in the hand to accompany dancing.

castaway *n.* shipwrecked person.

caste *n.* exclusive social class, esp. in the Hindu system.

castigate *v.* punish or rebuke severely. □ **castigation** *n.*

■ berate, *colloq.* blast, chastise, chide, criticize, rebuke, reprimand, reproach, scold, upbraid; discipline, punish.

casting vote deciding vote when those on each side are equal.

castle *n.* large fortified residence.

castor *n.* **1** small swivelling wheel on a leg of furniture. **2** small container with a perforated top for sprinkling sugar etc.

castor oil laxative and lubricant vegetable oil.

castrate *v.* remove the testicles of. □ **castration** *n.*

casual *adj.* **1** chance. **2** not regular or permanent. **3** not interested or concerned. **4** careless. **5** (of clothes etc.) informal. □ **casually** *adv.*

■ **1** accidental, chance, coincidental, fortuitous, random, unexpected, unforeseen, unlooked-for, unplanned. **2** irregular, occasional, part-time, temporary. **3** blasé, blithe, dispassionate, indifferent, insouciant, lackadaisical, nonchalant, offhand, relaxed, unconcerned, unenthusiastic, uninterested. **4** careless, disorganized, haphazard, hit-or-miss, *colloq.* slap-happy, unmethodical, unsystematic.

casualty *n.* **1** person killed or injured. **2** thing lost or destroyed.

■ **1** fatality, victim; (**casualties**) dead, injured, wounded.

casuist *n.* **1** theologian who studies moral problems. **2** sophist, quibbler. □ **casuistic** *adj.*, **casuistry** *n.*

cat *n.* **1** small furry domesticated animal. **2** wild animal related to this. **3** whip with knotted lashes. □ **cat's cradle** child's game with string. **cat's-paw** *n.* person used as a tool by another.

cataclysm *n.* violent upheaval or disaster. □ **cataclysmic** *adj.*

catacomb /káttəkoōm/ *n.* underground cemetery.

catafalque *n.* decorated bier, esp. for a state funeral.

catalepsy *n.* seizure or trance with rigidity of the body. □ **cataleptic** *adj.*

catalogue *n.* systematic list of items. ● *v.* list in a catalogue.

■ *n.* directory, index, inventory, list, record, register, roll. ● *v.* index, itemize, list, make an inventory of, record, register.

catalyse *v.* subject to the action of a catalyst. □ **catalysis** *n.*

catalyst *n.* substance that aids a chemical reaction while remaining unchanged.

catamaran *n.* boat with twin hulls.

catapult *n.* device with elastic for shooting small stones. ● *v.* hurl from or as if from a catapult.

cataract *n.* **1** large waterfall. **2** opaque area clouding the eye.

catarrh *n.* inflammation of mucous membrane, esp. of the nose, with a watery discharge.

catastrophe /kətástrəfi/ *n.* sudden great disaster. □ **catastrophic** *adj.*, **catastrophically** *adv.*

■ accident, calamity, cataclysm, disaster, fiasco, misfortune, mishap, tragedy.

catcall *n.* whistle of disapproval.

catch *v.* (**caught**) **1** capture. **2** detect or surprise. **3** reach or overtake. **4** grasp and hold. **5** become infected with. **6** perceive. **7** be in time for. ● *n.* **1** act of catching. **2** amount of thing caught. **3** thing or person caught or worth catching. **4** concealed difficulty. **5** fastener. □ **catch on** (*colloq.*) **1** become popular. **2** understand what is meant. **catch out** detect in a mistake etc. **catchphrase** *n.* phrase in frequent current use, slogan. **catch-22** *n.* dilemma where the victim is bound to suffer. **catch up 1** come abreast with. **2** do arrears of work.

■ *v.* **1** apprehend, arrest, capture, collar, *sl.* cop, *sl.* nab, *sl.* nick, take prisoner; ensnare, hook, land, net, snare, trap. **2** detect, discover, find, surprise. **3** draw level with, overtake, pass, reach. **4** clasp, clutch, grab, grasp, grip, intercept, seize, snatch, take hold of. **5** contract, develop, get, pick up. **6** comprehend, follow, grasp, hear, make out, perceive, understand, take in. ● *n.* **2** harvest, haul, take, yield. **3** acquisition, conquest, find, trophy. **4** difficulty, disadvantage, drawback, hitch, problem, snag, stumbling block. **5** bolt, clasp, clip, fastener, fastening, hook, latch, lock.

catching *adj.* infectious.

catchment area 1 area from which rainfall drains into a river. **2** area from which a hospital draws patients or a school draws pupils.

catchword *n.* catchphrase.

catchy *adj.* (**-ier**, **-iest**) (of a tune) pleasant and easy to remember.

catechism *n.* series of questions and answers.

catechize *v.* put a series of questions to.

categorical *adj.* absolute, explicit. □ **categorically** *adv.*

■ absolute, complete, decided, definite, direct, downright, emphatic, explicit, express, firm, flat, point-blank, positive, unambiguous, unconditional, unequivocal, unmitigated, unqualified.

categorize *v.* place in a category. □ **categorization** *n.*

category *n.* class of things.

■ class, grade, group, kind, league, order, rank, set, sort, type, variety.

cater *v.* **1** supply food. **2** provide what is needed or wanted. □ **caterer** *n.*

caterpillar *n.* larva of butterfly or moth. □ **Caterpillar track** [P.] steel band with treads, passing round a vehicle's wheels.

caterwaul *v.* howl like a cat.

catgut *n.* gut as thread.

catharsis *n.* **1** purgation. **2** emotional release. □ **cathartic** *adj.*

cathedral *n.* principal church of a diocese.

Catherine wheel rotating firework.

catheter *n.* tube inserted into the bladder to extract urine.

cathode *n.* electrode by which current leaves a device. □ **cathode ray** beam of electrons from the cathode of a vacuum tube.

catholic *adj.* **1** universal. **2** of wide sympathies or interests. **3** of all Churches or all Christians. **4** (**Catholic**) Roman Catholic. ● *n.* (**Catholic**) Roman Catholic. □ **Catholicism** *n.*

■ *adj.* **1** general, universal, widespread. **2** broad-minded, eclectic, inclusive, liberal, tolerant, varied, wide, wide-ranging.

cation *n.* positively charged ion.

catkin *n.* hanging flower of willow, hazel, etc.

catmint *n.* strong-smelling plant attractive to cats.

catnap *n.* short nap.

catnip *n.* catmint.

Catseye *n.* [P.] reflector stud on a road.

cattery *n.* (*pl.* **-ies**) boarding place for cats.

cattle *n.pl.* large animals with horns and cloven hoofs.

catty *adj.* (**-ier**, **-iest**) spiteful. □ **cattily** *adv.*, **cattiness** *n.*

catwalk *n.* narrow strip for walking on.

caucus *n.* **1** (often *derog.*) local committee of a political party. **2** (*US*) meeting of party leaders.

caught *see* **catch**.

cauldron *n.* large deep pot for boiling things in.

cauliflower *n.* cabbage with a white flower head.

caulk *v.* stop up (a ship's seams) with waterproof material.

causal *adj.* relating to cause (and effect). □ **causality** *n.*

causation *n.* causality.

cause *n.* **1** that which produces an effect. **2** reason or motive for action. **3** lawsuit. **4** principle supported. ● *v.* be the cause of.

■ *n.* **1** basis, genesis, origin, root, source; initiator, instigator, originator. **2** grounds, justification, motive, occasion, reason. ● *v.* bring about, create, effect, engender, generate, give rise to, induce, lead to, occasion, precipitate, produce, provoke, result in, trigger off.

causeway *n.* raised road across low or wet ground.

caustic *adj.* **1** burning, corrosive. **2** sarcastic, biting. ● *n.* caustic substance. □ **caustically** *adv.*

■ **2** biting, bitter, critical, cutting, mordant, sarcastic, sardonic, scathing, sharp.

cauterize *v.* burn (tissue) to destroy infection or stop bleeding. □ **cauterization** *n.*

caution *n.* **1** avoidance of rashness. **2** warning. ● *v.* **1** warn. **2** admonish. □ **cautionary** *adj.*

■ *n.* **1** care, carefulness, circumspection, forethought, heed, prudence, vigilance, wariness, watchfulness. **2** advice, caveat, counsel, warning. ● *v.* **1** advise, counsel, forewarn, warn. **2** admonish, reprimand, *colloq.* tell off, *colloq.* tick off.

cautious *adj.* having or showing caution. □ **cautiously** *adv.*

■ alert, *colloq.* cagey, careful, chary, circumspect, discreet, guarded, heedful, prudent, vigilant, wary, watchful.

cavalcade *n.* procession.

Cavalier *n.* supporter of Charles I in the English Civil War.

cavalier *adj.* arrogant, offhand.

cavalry *n.* troops who fight on horseback.

cave *n.* natural hollow. ● *v.* explore caves. □ **cave in 1** collapse. **2** yield.

■ *n.* cavern, grotto, hole, pothole. □ **cave in 1** buckle, collapse, crumble, crumple, give (way), subside. **2** capitulate, give in *or* up *or* way, submit, surrender, yield.

caveat /kávviat/ *n.* warning.

caveman *n.* person of prehistoric times living in a cave.

cavern *n.* large cave.

cavernous *adj.* like a cavern.

caviar *n.* pickled roe of sturgeon or other large fish.

cavil *v.* (**cavilled**) raise petty objections. ● *n.* petty objection.

cavity *n.* hollow within a solid body.

■ crater, gap, hole, hollow, pit.

cavort *v.* caper excitedly.

caw *n.* harsh cry of a rook etc. ● *v.* utter a caw.

cayenne *n.* hot red pepper.

cayman *n.* S. American alligator.

CB *abbr.* citizens' band.

cc *abbr.* cubic centimetre(s).

CD *abbr.* compact disc.

cease *v.* bring or come to an end, stop. □ **ceasefire** *n.* signal to stop firing guns.

■ desist (from), discontinue, *US* quit, refrain (from), stop, suspend, terminate; come to an end, die away, end, finish.

ceaseless *adj.* not ceasing.

■ constant, continual, continuous, endless, incessant, interminable, never-ending, non-stop, perpetual, unceasing, unending, unremitting.

cedar *n.* **1** evergreen tree. **2** its hard fragrant wood.

cede *v.* surrender (territory etc.).

cedilla *n.* mark written under c (ç) pronounced as s.

ceilidh /káyli/ *n.* informal gathering for music and dancing.

ceiling *n.* **1** surface of the top of a room. **2** upper limit or level.

celandine *n.* small wild plant with yellow flowers.

celebrate *v.* mark with or engage in festivities. □ **celebration** *n.*

■ commemorate, keep, observe, solemnize. □ **celebration** commemoration, observance, solemnization; festival, gala, jamboree, party; festivities, merrymaking, revelry.

celebrated *adj.* famous.

celebrity *n.* **1** famous person. **2** fame.

■ **1** luminary, notable, personage, personality, public figure, star. **2** eminence, fame, glory, prominence, stardom, renown.

celeriac *n.* variety of celery.

celerity *n.* swiftness.

celery *n.* plant with edible crisp juicy stems.

celestial *adj.* **1** of the sky or heavenly bodies. **2** heavenly.

■ **1** planetary, stellar. **2** divine, ethereal, heavenly, immortal, spiritual, sublime.

celibate *adj.* abstaining from sexual intercourse. □ **celibacy** *n.*

cell *n.* **1** small room for a monk or prisoner. **2** compartment in a honeycomb. **3** device for producing electric current chemically. **4** microscopic unit of living matter. **5** small group as a nucleus of political activities.

cellar *n.* **1** underground room. **2** stock of wine.

cello /chéllō/ *n.* bass instrument of the violin family. □ **cellist** *n.*

Cellophane *n.* [P.] thin transparent wrapping material.

celluloid *n.* plastic made from cellulose nitrate and camphor.

cellulose *n.* substance in plant tissues used in making plastics.

Celsius *adj.* of a scale of temperature on which water freezes at 0° and boils at 100°.

Celt *n.* member of an ancient European people or their descendants. □ **Celtic** *adj.*

cement *n.* **1** substance of lime and clay setting like stone. **2** adhesive. ● *v.* **1** join with cement. **2** unite firmly.

cemetery *n.* burial ground other than a churchyard.

cenotaph *n.* tomb-like monument to people buried elsewhere.

censer *n.* container for burning incense.

censor *n.* person authorized to examine letters, books, films, etc., and remove or ban anything regarded as harmful. ● *v.* remove or ban thus. ▫ **censorship** *n.*

censorious *adj.* severely critical.

censure *n.* severe criticism and rebuke. ● *v.* criticize and rebuke severely.

> ■ *n.* castigation, condemnation, criticism, stricture. ● *v.* admonish, berate, castigate, condemn, criticize, denounce, rebuke, reproach, scold, take to task, upbraid.

census *n.* official counting of population.

cent *n.* **1** 100th part of a dollar or other currency. **2** coin worth this.

centaur *n.* mythical creature half man, half horse.

centenarian *n.* person 100 years old or more.

centenary *n.* 100th anniversary.

centennial *adj.* of a centenary. ● *n.* (*US*) centenary.

centigrade *adj.* Celsius.

centilitre *n.* 100th of a litre.

centimetre *n.* 100th of a metre.

centipede *n.* small crawling creature with many legs.

central *adj.* **1** of, at, or forming a centre. **2** most important. ▫ **central heating** heating of a building from one source. **centrally** *adv.*, **centrality** *n.*

> ■ **1** inner, medial, mid, middle. **2** basic, cardinal, chief, crucial, essential, fundamental, key, main, major, primary, principal, vital.

centralize *v.* bring under the control of a central authority. ▫ **centralization** *n.*

centre *n.* middle point or part. ● *v.* (**centred**, **centring**) place in or at a centre.

> ■ *n.* core, focus, heart, hub, kernel, middle, nub, nucleus.

centrifugal *adj.* moving away from the centre.

centrifuge *n.* machine using centrifugal force for separating substances.

centripetal *adj.* moving towards the centre.

centurion *n.* commander in the ancient Roman army.

century *n.* **1** period of 100 years. **2** 100 runs at cricket.

cephalic *adj.* of the head.

cephalopod *n.* mollusc with tentacles (e.g. an octopus).

ceramic *adj.* of pottery or a similar substance. ● *n.* (**ceramics**) art of making pottery.

cereal *n.* **1** grass plant with edible grain. **2** this grain. **3** breakfast food made from it.

cerebral *adj.* **1** of the brain. **2** intellectual. ▫ **cerebrally** *adv.*

cerebrum *n.* main part of the brain.

ceremonial *adj.* of or used in ceremonies, formal. ● *n.* **1** ceremony. **2** rules for this.

> ■ *adj.* commemorative, ritual, state; ceremonious, dignified, formal, majestic, solemn, stately.

ceremonious *adj.* full of ceremony. ▫ **ceremoniously** *adv.*

ceremony *n.* **1** formal occasion. **2** formalities.

> ■ **1** ceremonial, rite(s), ritual, service; celebration, function, pageant. **2** conventions, formalities, protocol.

cerise /səréez/ *adj.* & *n.* light red.

certain *adj.* **1** feeling sure. **2** indisputable. **3** that may be relied on to happen or be effective. **4** specific but not named. **5** some.

> ■ **1** assured, confident, convinced, positive, satisfied, sure. **2** definite, incontestable, incontrovertible, indubitable, irrefutable, undeniable, undisputed, undoubted, unquestionable. **3** bound, destined, fated, sure; dependable, infallible, reliable, unfailing. **4** unnamed, unspecified.

certainly *adv.* **1** without doubt. **2** yes.

certainty *n.* **1** being certain. **2** thing that is certain.

> ■ **1** assurance, certitude, confidence, conviction, sureness. **2** actuality, fact, reality, truth; *colloq.* cinch, foregone conclusion.

certificate *n.* official document attesting certain facts.

certify *v.* **1** declare formally. **2** officially declare insane. ▫ **certifiable** *adj.*, **certifiably** *adv.*

> ■ **1** affirm, attest (to), confirm, guarantee, swear (to), testify (to), verify, vouch for.

certitude *n.* feeling of certainty.

cerulean *adj.* sky-blue.

cervix *n.* **1** neck. **2** neck-like structure, esp. of the womb. ▫ **cervical** *adj.*

cessation *n.* ceasing.

cesspit *n.* (also **cesspool**) pit for liquid waste or sewage.

cetacean /sitáysh'n/ *adj.* & *n.* (member) of the whale family.

cf. *abbr.* compare.

CFC *abbr.* chlorofluorocarbon, gaseous compound that harms the earth's atmosphere.

chafe *v.* **1** warm by rubbing. **2** make or become sore by rubbing. **3** show irritation.

chafer *n.* large beetle.

chaff *n.* **1** corn husks separated from seed. **2** chopped hay and straw. **3** banter. ● *v.* banter, tease.

chaffinch *n.* European finch.

chafing dish heated pan for keeping food warm at the table.

chagrin *n.* annoyance and embarrassment.

chain *n.* **1** series of connected metal links. **2** (*pl.*) restraining force. **3** connected series or sequence. ● *v.* fasten with chain(s). □ **chain reaction** change causing further changes.

■ *n.* **2** **(chains)** bonds, fetters, shackles; constraints, restraints, restrictions. **3** combination, concatenation, sequence, series, set, string, succession, train; line, range, row. ● *v.* bind, confine, fasten, fetter, manacle, restrain, shackle, tie up.

chair *n.* **1** movable seat for one person. **2** (position of) chairman. **3** position of a professor. ● *v.* act as chairman of. □ **chairlift** *n.* series of chairs on a cable for carrying people up a mountain. **chairman, chairperson, chairwoman** *ns.* person who presides over a meeting or committee.

chaise longue /sháyz lóngg/ sofa with one armrest.

chalcedony *n.* type of quartz.

chalet /shállay/ *n.* **1** Swiss hut or cottage. **2** small villa. **3** small hut in a holiday camp etc.

chalice *n.* large goblet.

chalk *n.* **1** white soft limestone. **2** piece of this or similar coloured substance used for drawing. □ **chalky** *adj.*

challenge *n.* **1** call to try one's skill or strength. **2** demand to respond or identify oneself. **3** formal objection. **4** demanding task. ● *v.* **1** make a challenge to. **2** question the truth or rightness of. □ **challenger** *n.*

■ *v.* **1** dare, defy, invite. **2** contest, dispute, object to, oppose, protest against, query, question, take exception to.

chamber *n.* **1** hall used for meetings of an assembly. **2** (*old use*) room, bedroom. **3** (*pl.*) set of rooms. **4** cavity or compartment. □ **chamber music** music for performance in a room rather than a hall. **chamber pot** bedroom receptacle for urine.

chamberlain *n.* official managing a royal or noble household.

chambermaid *n.* cleaner of hotel bedrooms.

chameleon /kəmméeliən/ *n.* small lizard that changes colour according to its surroundings.

chamfer *v.* **(chamfered)** bevel the edge of.

chamois *n.* **1** /shámwaa/ small mountain antelope. **2** /shámmi/ a kind of soft leather.

champ *v.* **1** munch noisily. **2** show impatience.

champagne *n.* **1** sparkling white wine. **2** its pale straw colour.

champion *n.* **1** person or thing that defeats all others in a competition. **2** person who fights or speaks in support of another or of a cause. ● *v.* support as champion. □ **championship** *n.*

■ *n.* **1** prizewinner, title-holder, victor, winner. **2** advocate, backer, defender, patron, protector, supporter, upholder. ● *v.* back, defend, fight for, protect, stand up for, support, uphold.

chance *n.* **1** way things happen, luck. **2** unplanned occurrence. **3** likelihood. **4** opportunity. ● *adj.* happening by chance. ● *v.* **1** happen. **2** risk.

■ *n.* **1** destiny, fate, fortune, kismet, luck. **2** accident, coincidence, fluke. **3** likelihood, odds, possibility, probability, prospect. **4** occasion, opportunity, turn. ● *adj.* accidental, arbitrary, casual, coincidental, fortuitous, inadvertent, random, unexpected, unforeseen, unlooked-for, unplanned, unpremeditated. ● *v.* **1** befall, come about, happen, occur, take place. **2** hazard, risk, venture.

chancel *n.* part of a church near the altar.

chancellor *n.* **1** government minister in charge of the nation's budget. **2** state or law official of various other kinds. **3** non-resident head of a university. □ **chancellorship** *n.*

Chancery *n.* division of the High Court of Justice.

chancy *adj.* **(-ier, -iest)** risky.

chandelier *n.* hanging support for several lights.

chandler *n.* dealer in ropes, canvas, etc., for ships.

change *v.* **1** make or become different. **2** take or use another instead of, go from one to another. **3** put fresh clothes or coverings on. **4** get or give money or different currency for. ● *n.* **1** act or instance of changing. **2** money in small units or returned as balance. □ **changeable** *adj.*

■ *v.* **1** adapt, adjust, alter, amend, convert, modify, modulate, transfigure, transform, transmute; mutate, metamorphose; fluctuate, shift, oscillate, vary; affect, have an impact on, influence. **2** exchange, interchange, replace, swap, switch, substitute, transpose. ● *n.* **1** adaptation, adjustment, alteration, amendment, conversion, fluctuation, metamorphosis, modification, modulation, mutation, shift, swing, transfiguration, transformation, transition, transposition; U-turn, volte-face; exchange, interchange, replacement, substitution. **2** cash, coins, coppers, silver.

changeling *n.* child believed to be a substitute for another.

channel *n.* **1** stretch of water connecting two seas. **2** passage for liquid. **3** medium of communication. **4** band of broadcasting frequencies. ● *v.* (**channelled**) direct through a channel.

■ *n.* **1** sound, strait. **2** canal, conduit, ditch, duct, groove, gully, gutter, moat, pipe, sluice, trench, watercourse, waterway. **3** avenue, means, medium, method, path, route, way.

chant *n.* **1** melody for psalms. **2** monotonous singing. **3** rhythmic shout. ● *v.* **1** sing, esp. to a chant. **2** shout rhythmically.

chantry *n.* chapel founded for priests to sing masses for the founder's soul.

chaos *n.* great disorder. □ **chaotic** *adj.*, **chaotically** *adv.*

■ bedlam, confusion, disarray, disorder, disorganization, havoc, mayhem, pandemonium, shambles, tumult, turmoil, uproar. □ **chaotic** confused, disorderly, disorganized, haphazard, haywire, higgledy-piggledy, jumbled, topsy-turvy, untidy, upside down.

chap[1] *n.* (*colloq.*) man.

chap[2] *n.* crack in skin. ● *v.* (**chapped**) **1** cause chaps in. **2** suffer chaps.

chaparral *n.* (*US*) dense tangled brushwood.

chapatti *n.* (*pl.* **-is**) small flat cake of unleavened bread.

chapel *n.* **1** place used for Christian worship, other than a cathedral or parish church. **2** place with a separate altar within a church.

chaperon *n.* person looking after another or others in public. ● *v.* act as chaperon to.

chaplain *n.* clergyman of an institution, private chapel, ship, regiment, etc. □ **chaplaincy** *n.*

chapter *n.* **1** division of a book. **2** canons of a cathedral.

char[1] *n.* charwoman.

char[2] *v.* (**charred**) make or become black by burning.

character *n.* **1** qualities making a person or thing what he, she, or it is. **2** moral strength. **3** person in a novel or play etc. **4** letter or sign used in writing, printing, etc. **5** noticeable or eccentric person. **6** reputation.

■ **1** disposition, make-up, nature, personality, spirit, temperament; attributes, features, properties, qualities, traits; distinctiveness, flavour, stamp. **2** decency, fibre, honesty, honour, integrity, morality, rectitude, respectability. **3** part, role. **4** figure, hieroglyph, letter, mark, rune, sign, symbol.

characteristic *adj.* & *n.* (feature) forming part of the character of a person or thing. □ **characteristically** *adv.*

■ *adj.* distinctive, distinguishing, idiosyncratic, representative, symptomatic, typical. ● *n.* aspect, attribute, feature, hallmark, idiosyncrasy, peculiarity, property, quality, symptom, trait.

characterize *v.* **1** describe the character of. **2** be characteristic of. □ **characterization** *n.*

■ **1** delineate, depict, describe, identify, mark, paint, portray, present, represent.

charade /shəraad/ *n.* **1** (*pl.*) game which involves guessing words from acted clues. **2** absurd pretence.

charcoal *n.* black substance made by burning wood slowly.

charge *n.* **1** price asked for goods or services. **2** accusation. **3** rushing attack. **4** task, duty. **5** custody. **6** person or thing entrusted. **7** quantity of explosive. **8** electricity contained in a substance. ● *v.* **1** ask (an amount) as a price. **2** ask (a person) for an amount as a price. **3** accuse formally. **4** give as a task or duty. **5** rush forward in attack. **6** load or fill with explosive. **7** give an electric charge to. □ **charge card** a kind of credit card. **in**

charge in command. **take charge** take control.

■ *n.* **1** cost, fare, fee, payment, price, rate, tariff, toll. **2** accusation, allegation, imputation, indictment. **3** assault, attack, foray, incursion, onslaught, raid, sally, sortie. **4** burden, duty, obligation, responsibility, task. **5** care, control, custody, guardianship, keeping, protection, safe-keeping, trust. **6** protégé, ward. ● *v.* **1** ask, claim, demand, expect. **2** bill, invoice. **3** accuse, arraign, impeach, indict. **4** entrust, trust; burden, saddle. **5** assail, assault, attack, rush, storm.

chargé d'affaires (*pl.* **-gés**) ambassador's deputy.

chariot *n.* two-wheeled horse-drawn vehicle used in ancient times in battle and in racing.

charioteer *n.* driver of a chariot.

charisma /kərízmə/ *n.* power to inspire or attract others. □ **charismatic** *adj.*

charitable *adj.* **1** generous to those in need. **2** lenient in judging others. **3** of or belonging to charities. □ **charitably** *adv.*

■ **1** beneficent, bountiful, generous, liberal, munificent, open-handed, philanthropic, public-spirited, unselfish. **2** kind, indulgent, lenient, magnanimous, sympathetic, tolerant.

charity *n.* **1** loving kindness. **2** lenience in judging others. **3** help given voluntarily to those in need. **4** organization etc. for helping those in need.

■ **1** altruism, beneficence, benevolence, generosity, humanity, kindness, love, philanthropy, unselfishness. **2** indulgence, lenience, magnanimity, sympathy, tolerance, understanding. **3** aid, alms, assistance, help, largesse, relief, support.

charlady *n.* charwoman.

charlatan *n.* person falsely claiming to be an expert.

charlotte *n.* pudding of cooked fruit with breadcrumbs.

charm *n.* **1** attractiveness, power of arousing love or admiration. **2** act, object, or words believed to have magic power. **3** small ornament worn on a bracelet etc. ● *v.* **1** give pleasure to. **2** influence by personal charm or as if by magic. □ **charmer** *n.*

■ *n.* **1** allure, appeal, attraction, attractiveness, charisma, fascination, magnetism, pull, seductiveness. **2** amulet, mascot, talisman; incantation, spell. ● *v.* **1** captivate, delight, enrapture, enchant, entrance, please. **2** bewitch, enthral, fascinate, hold spellbound, hypnotize, mesmerize, seduce.

charming *adj.* delightful.

charnel house place containing corpses or bones.

chart *n.* **1** map for navigators. **2** table, diagram, or outline map. **3** list of recordings that are currently most popular. ● *v.* make a chart of.

charter *n.* **1** official document granting rights. **2** chartering of aircraft etc. ● *v.* **1** grant a charter to. **2** let or hire (an aircraft, ship, or vehicle). □ **chartered accountant** one qualified according to the rules of an association holding a royal charter.

chartreuse /shaartrő̋z/ *n.* fragrant green or yellow liqueur.

charwoman *n.* woman employed to clean a house etc.

chary *adj.* (**-ier**, **-iest**) cautious.

chase *v.* go quickly after in order to capture, overtake, or drive away. ● *n.* **1** chasing, pursuit. **2** hunting. **3** steeplechase.

■ *v.* follow, hound, hunt, pursue, run after, track, trail. ● *n.* **1** hunt, pursuit, search.

chasm *n.* deep cleft.

■ abyss, canyon, cleft, crater, crevasse, fissure, hole, opening, pit, ravine, rift, split.

chassis /shássi/ *n.* (*pl.* **chassis**) base frame of a vehicle.

chaste *adj.* **1** celibate. **2** virtuous. **3** simple in style, not ornate. □ **chastely** *adv.*

■ **1** celibate, pure, virginal, unsullied. **2** decent, good, irreproachable, moral, sinless, virtuous. **3** austere, plain, severe, simple, unadorned.

chasten *v.* **1** discipline by punishment. **2** subdue the pride of.

■ **1** castigate, chastise, discipline, punish. **2** humble, subdue.

chastise *v.* **1** rebuke. **2** punish, beat. □ **chastisement** *n.*

■ **1** berate, castigate, chide, rebuke, reprimand, reproach, reprove, scold, take to task, upbraid. **2** castigate, chasten, discipline, punish; beat, cane, flog, spank, thrash, whip.

chastity *n.* being chaste.

chat *n.* informal conversation. ● *v.* (**chatted**) have a chat.

■ *n.* chatter, conversation, gossip, *colloq.* natter, talk, tête-à-tête. ● *v.* chatter, gossip, *colloq.* natter, prattle, talk.

chateau *n.* (*pl.* **-eaux**) French castle or large country house.

chatelaine *n.* mistress of a large house.

chattel *n.* movable possession.

chatter *v.* **1** talk quickly and continuously about unimportant matters. **2** (of teeth) rattle together. ● *n.* chattering talk.

chatterbox *n.* talkative person.

chatty *adj.* (**-ier**, **-iest**) **1** fond of chatting. **2** resembling chat.

chauffeur *n.* person employed to drive a car.

chauvinism *n.* exaggerated patriotism. □ **male chauvinism** prejudiced belief in male superiority over women. **chauvinist** *n.*, **chauvinistic** *adj.*

cheap *adj.* **1** low in cost or value. **2** poor in quality. □ **cheaply** *adv.*, **cheapness** *n.*

■ **1** inexpensive, knock-down, low-priced, reasonable. **2** base, inferior, poor, second-rate, shoddy, tawdry, trashy, worthless.

cheapen *v.* **1** make or become cheap. **2** degrade.

cheat *v.* **1** deceive or trick. **2** gain unfair advantage. ● *n.* **1** person who cheats. **2** deception.

■ *v.* **1** *colloq.* bamboozle, beguile, *sl.* con, deceive, defraud, double-cross, dupe, *colloq.* fiddle, fleece, hoodwink, *colloq.* rip off, short-change, swindle, take in, trick. ● *n.* **1** charlatan, deceiver, fraud, impostor, quack, swindler. **2** *sl.* con, confidence trick, deception, *colloq.* fiddle, *colloq.* rip-off, ruse, swindle, trick.

check[1] *v.* **1** test or examine, inspect. **2** stop, slow the motion (of). ● *n.* **1** inspection. **2** pause. **3** restraint. **4** exposure of a chess king to capture. **5** (*US*) bill in a restaurant. **6** (*US*) cheque. □ **check in** register on arrival. **check out** register on departure or dispatch. **check-out** *n.* desk where goods are paid for in a supermarket. **checker** *n.*

■ *v.* **1** examine, go over, inspect, investigate, look into, monitor, scrutinize, test; authenticate, confirm, corroborate, validate, verify. **2** arrest, brake, control, curb, halt, slow (down), stanch, stem, stop; block, hamper, hinder, impede, obstruct; contain, hold back, repress, restrain. ● *n.* **1** examination, inspection, investigation, *colloq.* once-over, scrutiny, test. **2** break, delay, halt, hesitation, interruption, pause, stop, stoppage. **3** constraint, curb, hindrance, impediment, restraint, restriction.

check[2] *n.* pattern of squares or crossing lines. □ **checked** *adj.*

checkmate *n.* **1** situation in chess where capture of a king is inevitable. **2** complete defeat, deadlock. ● *v.* **1** put into checkmate. **2** defeat, foil.

cheek *n.* **1** side of the face below the eye. **2** impudent speech, arrogance. ● *v.* speak cheekily to. □ **cheek by jowl** close together.

cheeky *adj.* (**-ier**, **-iest**) showing cheerful lack of respect. □ **cheekily** *adv.*

■ disrespectful, forward, impertinent, impudent, irreverent, pert, presumptuous, rude, saucy.

cheep *n.* weak shrill cry like that of a young bird. ● *v.* make this cry.

cheer *n.* **1** shout of applause. **2** cheerfulness. ● *v.* **1** utter a cheer, applaud with a cheer. **2** gladden. □ **cheer up** make or become more cheerful.

■ *n.* **1** cry, hooray, hurrah, shout, whoop. **2** cheerfulness, gaiety, gladness, happiness, joy. ● *v.* **1** applaud, clap, shout, whoop, yell. **2** *colloq.* buck up, buoy up, cheer up, comfort, encourage, gladden, hearten, uplift.

cheerful *adj.* **1** in good spirits. **2** pleasantly bright. □ **cheerfully** *adv.*, **cheerfulness** *n.*

■ **1** buoyant, cheery, chirpy, exuberant, gay, glad, happy, jaunty, jolly, jovial, joyful, light-hearted, merry, optimistic, perky, sunny, *colloq.* upbeat. **2** appealing, attractive, bright, gay, pleasant.

cheerless *adj.* gloomy, dreary.

■ bleak, dark, depressing, dingy, dismal, dispiriting, drab, dreary, gloomy, grim, melancholy, miserable, sombre.

cheery *adj.* (**-ier**, **-iest**) cheerful.

cheese *n.* food made from pressed milk curds. □ **cheese-paring** *adj.* stingy.

cheeseburger *n.* hamburger with cheese on it.

cheesecake *n.* **1** open tart filled with flavoured cream cheese. **2** (*sl.*) sexually stimulating display of women.

cheesecloth *n.* thin loosely-woven cotton fabric.

cheetah *n.* a kind of leopard.

chef *n.* professional cook.

chemical *adj.* of or made by chemistry. ● *n.* substance obtained by or used in a chemical process. □ **chemically** *adv.*

chemise *n.* woman's loose-fitting undergarment or dress.

chemist *n.* **1** expert in chemistry. **2** dealer in medicinal drugs.

chemistry *n.* **1** study of substances and their reactions. **2** structure and properties of a substance.

chemotherapy *n.* treatment of disease by drugs etc.

chenille *n.* velvety fabric.

cheque *n.* **1** written order to a bank to pay out money from an account. **2** printed form for this. □ **cheque card** card guaranteeing payment of cheques.

chequer *n.* pattern of squares, esp. of alternating colours.

chequered *adj.* **1** marked with a chequer pattern. **2** having frequent changes of fortune.

cherish *v.* **1** tend lovingly. **2** be fond of. **3** cling to (hopes etc.).

> ■ **1** care for, look after, nurse, protect, take care of, tend. **2** hold dear, love, prize, treasure, value.

cheroot *n.* cigar with both ends open.

cherry *n.* **1** small soft round fruit with a stone. **2** tree bearing this. **3** deep red.

cherub *n.* **1** (*pl.* **cherubim**) angelic being. **2** (in art) chubby infant with wings. **3** angelic child. □ **cherubic** *adj.*

chervil *n.* herb with aniseed flavour.

chess *n.* game for two players using 32 **chessmen** on a chequered **chessboard** with 64 squares.

chest *n.* **1** large strong box. **2** upper front surface of the body. □ **chest of drawers** piece of furniture with drawers for clothes etc.

chesterfield *n.* sofa with a padded back, seat, and ends.

chestnut *n.* **1** tree with a hard brown nut. **2** this nut. **3** reddish-brown. **4** horse of this colour. **5** (*colloq.*) old joke or anecdote.

chevron *n.* V-shaped symbol.

chew *v.* work or grind between the teeth.

> ■ bite, champ, crunch, gnaw, masticate, munch, nibble.

chewing gum flavoured gum used for prolonged chewing.

chewy *adj.* (**-ier**, **-iest**) needing much chewing. □ **chewiness** *n.*

chiaroscuro /kiaárəskoŏrō/ *n.* **1** light and shade effects. **2** use of contrast.

chic *adj.* stylish and elegant. ● *n.* stylishness, elegance.

> ■ *adj.* elegant, fashionable, modish, sophisticated, stylish, tasteful.

chicane /shikáyn/ *n.* barriers on a motor racing course.

chicanery *n.* trickery.

chick *n.* newly hatched bird.

chicken *n.* **1** young domestic fowl. **2** its flesh as food. ● *adj.* (*colloq.*) cowardly. ● *v.* **chicken out** (*colloq.*) withdraw through cowardice. □ **chicken feed** (*colloq.*) trifling amount of money.

chickenpox *n.* disease with a rash of small red blisters.

chickpea *n.* pea with yellow seeds used as a vegetable.

chicory *n.* blue-flowered plant grown for its salad leaves.

chide *v.* (**chided** or **chid**, **chidden**) rebuke.

chief *n.* **1** leader, ruler. **2** person with the highest rank. ● *adj.* **1** highest in rank. **2** most important.

> ■ *n. colloq.* boss, captain, commander, director, *colloq.* gaffer, governor, head, leader, manager, master, overseer, president, principal, ringleader, ruler, superintendent, supervisor, supremo, *colloq.* top dog. ● *adj.* **1** first, foremost, greatest, head, highest, leading, premier, senior, supreme, top. **2** cardinal, central, essential, fundamental, key, main, major, overriding, paramount, predominant, primary, prime, principal.

chiefly *adv.* mainly.

> ■ by and large, especially, essentially, largely, mainly, mostly, on the whole, particularly, predominantly, primarily, principally.

chieftain *n.* chief of a clan or tribe.

chiffon *n.* diaphanous fabric.

chignon /sheényon/ *n.* coil of hair at the back of the head.

chihuahua /chiwaáwə/ *n.* very small smooth-haired dog.

chilblain *n.* painful swelling caused by exposure to cold.

child *n.* (*pl.* **children**) **1** young human being. **2** son or daughter. □ **childhood** *n.*, **childless** *adj.*

> ■ **1** adolescent, babe, baby, boy, *derog.* brat, chit, girl, infant, juvenile, *colloq.* kid, lad, lass, minor, stripling, teenager, toddler, tot, urchin, youngster, youth. **2** daughter, descendant, heir, son; (**children**) family, issue, offspring, progeny. □ **childhood** adolescence, boyhood, girlhood, infancy, minority, puberty, teens, youth.

childbirth *n.* process of giving birth to a child.

childish *adj.* **1** of or like a child. **2** immature, silly.

■ **1** childlike, boyish, girlish, youthful. **2** babyish, immature, infantile, juvenile, puerile, silly.

childlike *adj.* simple and innocent.

■ artless, guileless, ingenuous, innocent, naive, simple, trustful, unsophisticated, youthful.

chill *n.* **1** unpleasant coldness. **2** illness with feverish shivering. ● *adj.* chilly. ● *v.* **1** make or become chilly. **2** preserve (food or drink) by cooling.

chilli *n.* (*pl.* **-ies**) dried pod of red pepper.

chilly *adj.* (**-ier**, **-iest**) **1** rather cold. **2** unfriendly in manner.

■ **1** cold, cool, crisp, frosty, icy, *colloq.* nippy, raw, wintry. **2** aloof, cold, cool, distant, frigid, frosty, reserved, standoffish, unforthcoming, unfriendly.

chime *n.* **1** tuned set of bells. **2** series of notes from these. ● *v.* ring as a chime. □ **chime in** put in a remark.

chimera /kīmeérə/ *n.* **1** legendary monster with a lion's head, goat's body, and serpent's tail. **2** fantastic product of the imagination.

chimney *n.* (*pl.* **-eys**) structure for carrying off smoke or gases. □ **chimney pot** pipe on top of a chimney.

chimpanzee *n.* African ape.

chin *n.* front of the lower jaw.

china *n.* **1** fine earthenware, porcelain. **2** things made of this.

chinchilla *n.* **1** S. American rodent. **2** its soft grey fur.

chine *n.* **1** animal's backbone. **2** ravine in southern England.

Chinese *adj.* & *n.* (native, language) of China.

chink[1] *n.* narrow opening, slit.

■ aperture, cleft, crack, cranny, crevice, fissure, gap, opening, rift, slit, split.

chink[2] *n.* sound of glasses or coins striking together. ● *v.* make this sound.

chintz *n.* glazed cotton cloth used for furnishings.

chip *n.* **1** small piece cut or broken off something hard. **2** fried oblong strip of potato. **3** counter used in gambling. ● *v.* (**chipped**) **1** break or cut the edge or surface of. **2** shape in this way. □ **chip in** (*colloq.*) **1** interrupt. **2** contribute money.

chipboard *n.* board made of compressed wood chips.

chipmunk *n.* striped squirrel-like animal of N. America.

chipolata *n.* small sausage.

chiropody /kiróppədi/ *n.* treatment of minor ailments of the feet. □ **chiropodist** *n.*

chiropractic /kī́rōpráktik/ *n.* treatment of physical disorders by manipulation of the spinal column. □ **chiropractor** *n.*

chirp *n.* short sharp sound made by a small bird or grasshopper. ● *v.* make this sound.

chirpy *adj.* (**-ier**, **-iest**) cheerful.

chisel *n.* tool with a sharp bevelled end for shaping wood or stone etc. ● *v.* (**chiselled**) cut with this.

chit[1] *n.* young child.

chit[2] *n.* short written note.

chivalry *n.* courtesy and consideration, inclination to help weaker people. □ **chivalrous** *adj.*

■ □ **chivalrous** considerate, courteous, courtly, gallant, gentlemanly, gracious, kind, well-mannered.

chive *n.* small herb with onion-flavoured leaves.

chivvy *v.* urge to hurry.

chloride *n.* compound of chlorine and another element.

chlorinate *v.* treat or sterilize with chlorine. □ **chlorination** *n.*

chlorine *n.* poisonous gas used for bleaching and disinfecting.

chloroform *n.* liquid giving off vapour that causes unconsciousness when inhaled.

chlorophyll *n.* green colouring matter in plants.

choc *n.* (*colloq.*) chocolate. □ **choc ice** bar of ice cream coated with chocolate.

chock *n.* block or wedge for preventing something from moving. ● *v.* wedge with chock(s). □ **chock-a-block** *adj.* & *adv.* crammed, crowded together.

chocolate *n.* **1** edible substance made from cacao seeds. **2** sweet made or coated with this, drink made with this. **3** dark brown.

choice *n.* **1** act of choosing. **2** power to choose. **3** variety from which to choose. **4** person or thing chosen. ● *adj.* of especially good quality. □ **choicely** *adv.*

■ *n.* **1** election, pick, selection. **2** alternative, option. **3** assortment, diversity, range, selection, variety. ● *adj.* excellent, exceptional, fine, first-class, first-rate, prime, outstanding, splendid, superior.

choir *n.* **1** group of singers, esp. in church. **2** part of a church where these sit, chancel.

choirboy *n.* boy singer in a church choir.

choke *v.* **1** stop (a person) breathing by squeezing or blocking the windpipe. **2** be unable to breathe. **3** clog, block. ● *n.* valve controlling the flow of air into a petrol engine.

■ *v.* **1** asphyxiate, garrotte, smother, strangle, suffocate, throttle. **2** gag, retch. **3** block, clog, congest, constrict, dam, fill, obstruct, silt up.

choker *n.* close-fitting necklace.

cholera *n.* serious often fatal bacterial disease.

choleric *adj.* easily angered.

cholesterol *n.* fatty animal substance thought to cause hardening of arteries.

choose *v.* (**chose, chosen**) **1** select out of a greater number. **2** decide.

■ **1** adopt, decide on, go for, opt for, pick, plump for, select, single out. **2** decide, determine, elect, make up one's mind, resolve.

choosy *adj.* (**-ier, -iest**) (*colloq.*) careful in choosing, hard to please. □ **choosiness** *n.*

chop *v.* (**chopped**) **1** cut by a blow with an axe or knife. **2** hit with a short downward movement. ● *n.* **1** chopping stroke. **2** thick slice of meat, usu. including a rib.

chopper *n.* **1** chopping tool. **2** (*colloq.*) helicopter.

choppy *adj.* (**-ier, -iest**) **1** full of short broken waves. **2** jerky.

chopstick *n.* each of a pair of sticks used in China, Japan, etc., to lift food to the mouth.

chop suey Chinese dish of meat or fish fried with vegetables.

choral *adj.* for or sung by a chorus.

chorale *n.* choral composition using the words of a hymn.

chord[1] *n.* **1** string of a harp etc. **2** straight line joining two points on a curve.

chord[2] *n.* combination of notes sounded together.

chore *n.* routine task.

choreography *n.* composition of stage dances. □ **choreographer** *n.*, **choreographic** *adj.*

chorister *n.* member of a choir.

chortle *n.* gleeful chuckle. ● *v.* utter a chortle.

chorus *n.* **1** group of singers. **2** simultaneous utterance. **3** refrain of a song. **4** group of singing dancers in a musical comedy etc. ● *v.* say as a group.

chose, chosen *see* **choose**.

choux pastry /shoo/ light pastry for making small cakes.

chow *n.* **1** long-haired dog of a Chinese breed. **2** (*sl.*) food.

chowder *n.* stew of shellfish with bacon and onions etc.

chow mein Chinese dish of fried noodles and shredded meat etc.

christen *v.* **1** admit to the Christian Church by baptism. **2** name.

Christendom *n.* all Christians or Christian countries.

Christian *adj.* **1** of or believing in Christianity. **2** kindly, humane. ● *n.* believer in Christianity. □ **Christian name** personal name given at a christening. **Christian Science** religious system by which health and healing are sought by prayer alone.

Christianity *n.* religion based on the teachings of Christ.

Christmas *n.* festival (25 Dec.) commemorating Christ's birth. □ **Christmas tree** tree decorated at Christmas.

chromatic *adj.* of colour, in colours. □ **chromatic scale** music scale proceeding by semitones. **chromatically** *adv.*

chrome *n.* **1** chromium. **2** yellow pigment from a compound of this.

chromium *n.* metallic element that does not rust.

chromosome *n.* thread-like structure carrying genes in animal and plant cells.

chronic *adj.* **1** (of a disease) long-lasting. **2** (of a patient) having a chronic illness. **3** (*colloq.*) bad. □ **chronically** *adv.*

chronicle *n.* record of events. ● *v.* record in a chronicle. □ **chronicler** *n.*

■ *n.* account, annals, archive, diary, history, journal, narrative, record, story.

chronological *adj.* arranged in the order in which things occurred. □ **chronologically** *adv.*

chronology *n.* arrangement of events in order of occurrence.

chronometer *n.* time-measuring instrument.

chrysalis *n.* **1** form of an insect in the stage between larva and adult insect. **2** case enclosing it.

chrysanthemum *n.* garden plant flowering in autumn.

chub *n.* river fish.

chubby *adj.* (**-ier**, **-iest**) round and plump. □ **chubbiness** *n.*

■ dumpy, fat, plump, podgy, pudgy, roly-poly, stout, tubby.

chuck[1] *v.* (*colloq.*) throw carelessly or casually.

chuck[2] *n.* **1** part of a lathe holding the drill. **2** part of a drill holding the bit. **3** cut of beef from neck to ribs.

chuckle *n.* quiet laugh. ● *v.* utter a chuckle.

chug *v.* (**chugged**) make or move with a dull short repeated sound. ● *n.* this sound.

chukka *n.* period of play in a polo game.

chum *n.* (*colloq.*) close friend. □ **chummy** *adj.*

chump *n.* (*colloq.*) foolish person. □ **chump chop** chop from the thick end of a loin of mutton.

chunk *n.* **1** thick piece. **2** substantial amount.

chunky *adj.* (**-ier**, **-iest**) **1** short and thick. **2** in chunks, containing chunks. □ **chunkiness** *n.*

church *n.* **1** building for public Christian worship. **2** religious service in this. **3** (**the Church**) Christians collectively, particular denomination of these.

churchwarden *n.* parish representative, assisting with church business.

churchyard *n.* enclosed land round a church, used for burials.

churlish *adj.* ill-mannered, surly.

churn *n.* **1** machine in which milk is beaten to make butter. **2** very large milk can. ● *v.* **1** beat (milk) or make (butter) in a churn. **2** stir or swirl violently. □ **churn out** produce rapidly.

chute *n.* slide for sending things to a lower level.

chutney *n.* (*pl.* **-eys**) seasoned mixture of fruit, vinegar, spices, etc., eaten with meat or cheese.

cicada *n.* chirping insect resembling a grasshopper.

cicatrice *n.* scar.

cider *n.* fermented drink made from apples.

cigar *n.* roll of tobacco leaf for smoking.

cigarette *n.* roll of shredded tobacco in thin paper for smoking.

cinch *n.* (*colloq.*) **1** certainty. **2** easy task.

cinder *n.* piece of partly burnt coal or wood.

cine *adj.* cinematographic.

cinema *n.* **1** theatre where films are shown. **2** films as an art form or industry.

cinematography *n.* art of making films. □ **cinematographic** *adj.*

cinnamon *n.* spice made from the bark of a south-east Asian tree.

cipher *n.* **1** symbol 0 representing nought or zero. **2** numeral. **3** person of no importance. **4** secret or disguised writing.

circa *prep.* about.

circle *n.* **1** perfectly round plane figure. **2** group with similar interests. **3** curved tier of seats at a theatre etc. ● *v.* **1** move in a circle. **2** form a circle round.

■ *n.* **1** disc, ring, round. **2** clique, coterie, faction, group, set, society. ● *v.* **1** gyrate, loop, orbit, revolve, rotate, spin, spiral. **2** encircle, enclose, encompass, girdle, hem in, ring, surround.

circlet *n.* **1** small circle. **2** circular band worn as an ornament.

circuit *n.* **1** line, route, or distance round a place. **2** path of an electric current.

■ **1** lap, orbit, revolution.

circuitous *adj.* roundabout, indirect. □ **circuitously** *adv.*

■ indirect, meandering, roundabout, serpentine, tortuous, twisting, winding.

circuitry *n.* circuits.

circular *adj.* shaped like or moving round a circle. ● *n.* letter or leaflet sent to a circle of people. □ **circularity** *n.*

circulate *v.* **1** go or send round. **2** mingle among guests etc.

■ **1** course, flow, go round, move round; advertise, broadcast, disseminate, distribute, issue, make known, promulgate, publicize, send round, spread. **2** fraternize, mingle, mix, socialize.

circulation *n.* **1** movement from and back to a starting point, esp. that of blood to and from the heart. **2** transmission, distribution. **3** number of copies sold.

circumcise *v.* cut off the foreskin of. □ **circumcision** *n.*

circumference *n.* **1** boundary of a circle. **2** distance round this.

circumflex accent the accent (ˆ).

circumlocution *n.* roundabout, verbose, or evasive expression. □ **circumlocutory** *adj.*

circumnavigate *v.* sail round. □ **circumnavigation** *n.*

circumscribe *v.* **1** draw a line round. **2** restrict.

circumspect *adj.* cautious and watchful, wary. □ **circumspection** *n.*, **circumspectly** *adv.*

circumstance *n.* occurrence or fact.

■ affair, episode, event, fact, happening, incident, occasion, occurrence, situation.

circumstantial *adj.* **1** detailed. **2** consisting of facts that suggest something but do not prove it.

circumvent *v.* evade (a difficulty etc.). □ **circumvention** *n.*

circus *n.* travelling show with performing animals, acrobats, etc.

cirrhosis /sirṓsiss/ *n.* disease of the liver.

cirrus *n.* (*pl.* **cirri**) high wispy white cloud.

cistern *n.* tank for storing water.

citadel *n.* fortress overlooking a city.

cite *v.* quote or mention as an example etc. □ **citation** *n.*

■ adduce, bring up, mention, name, quote, refer to.

citizen *n.* **1** inhabitant of a city. **2** person with full rights in a country. □ **citizen's band** system of local intercommunication by radio. **citizenship** *n.*

■ **2** denizen, inhabitant, national, native, resident, subject.

citric acid acid in the juice of lemons, limes, etc.

citrus *n.* tree of a group including lemon, orange, etc.

city *n.* **1** important town. **2** town with special rights given by charter.

■ **1** borough, conurbation, metropolis, municipality, town.

civet *n.* **1** cat-like animal of central Africa. **2** musky substance obtained from its glands.

civic *adj.* of a city or citizenship.

civil *adj.* **1** of ordinary citizens, not of the armed forces or the Church. **2** polite and obliging. □ **civil engineering** designing and construction of roads, bridges, etc. **Civil List** annual allowance for the sovereign's household expenses. **civil servant** employee of the **civil service**, government departments other than the armed forces. **civil war** war between citizens of the same country. **civilly** *adv.*

■ **1** civilian, lay, non-military, secular. **2** affable, civilized, cordial, courteous, gracious, obliging, pleasant, polite, respectful, urbane, well-mannered.

civilian *n.* & *adj.* (of) person(s) not in the armed forces.

civility *n.* politeness.

civilization *n.* **1** advanced stage or system of social development. **2** peoples regarded as having achieved this. **3** making or becoming civilized.

civilize *v.* **1** improve the behaviour of. **2** bring out of a primitive stage of society.

clack *n.* short sharp sound. ● *v.* make this sound.

clad *adj.* clothed.

cladding *n.* boards or metal plates as a protective covering.

claim *v.* **1** demand as one's right. **2** assert. ● *n.* **1** demand. **2** assertion. **3** right or title.

■ *v.* **1** ask for, demand, exact, insist on, request, require. **2** affirm, allege, assert, contend, declare, insist, maintain, profess, state. ● *n.* **1** demand, petition, request, requisition. **2** affirmation, assertion, contention, declaration.

claimant *n.* person making a claim.

clairvoyance *n.* power of seeing the future. □ **clairvoyant** *n.*

clam *n.* shellfish with a hinged shell. ● *v.* (**clammed**) **clam up** (*colloq.*) refuse to talk.

clamber *v.* climb with difficulty.

clammy *adj.* (**-ier**, **-iest**) unpleasantly moist and sticky.

■ damp, dank, humid, moist, muggy, sticky, sweaty.

clamour *n.* **1** loud confused noise. **2** loud protest etc. □ **clamorous** *adj.*

■ **1** babel, commotion, din, hubbub, hullabaloo, noise, outcry, racket, uproar.

clamp *n.* **1** device for holding things tightly. **2** device for immobilizing an illegally parked car. ● *v.* **1** grip with a clamp, fix firmly. **2** immobilize an illegally parked car with a clamp. □ **clamp down on** become firmer about, put a stop to.

clan *n.* group of families with a common ancestor. □ **clannish** *adj.*

■ dynasty, family, house, line, tribe.

clandestine *adj.* secret.

clang *n.* loud ringing sound. ● *v.* make this sound.

clanger *n.* (*sl.*) blunder.

clangour *n.* clanging noise.

clank *n.* sound like metal striking metal. ● *v.* make or cause to make this sound.

clap *v.* (**clapped**) **1** strike the palms together, esp. in applause. **2** put or place quickly or vigorously. ● *n.* **1** act or sound of clapping. **2** explosive noise, esp. of thunder. □ **clapped out** (*sl.*) worn-out.

■ *v.* **1** applaud, cheer. **2** fling, place, put, slap, *colloq.* stick.

clapper *n.* tongue or striker of a bell.

clapperboard *n.* device in film-making for making a sharp clap to synchronize picture and sound at the start of a scene.

claptrap *n.* insincere talk.

claret *n.* a dry red wine.

clarify *v.* make or become clear. □ **clarification** *n.*

■ clear up, elucidate, explain, make plain, simplify, spell out.

clarinet *n.* woodwind instrument with finger-holes and keys. □ **clarinettist** *n.*

clarion *adj.* loud, rousing.

clarity *n.* clearness.

clash *n.* **1** conflict. **2** discordant sounds or colours. ● *v.* **1** make a clashing sound. **2** come into conflict, be at variance.

■ *n.* **1** altercation, argument, conflict, disagreement, dispute, fight, quarrel, squabble. ● *v.* **1** bang, clang, clank, crash, smash. **2** argue, conflict, differ, disagree, dispute, quarrel, squabble.

clasp *n.* **1** device for fastening things, with interlocking parts. **2** grasp, handshake. ● *v.* **1** fasten, join with a clasp. **2** grasp, embrace closely.

■ *n.* **1** brooch, buckle, catch, clip, fastener, fastening, hook, pin. **2** embrace, grasp, grip, handshake, hold, hug. ● *v.* **1** clip, fasten, pin, secure. **2** clutch, embrace, grasp, grip, hold, hug, seize, take hold of.

class *n.* **1** set of people or things with characteristics in common. **2** rank of society. **3** high quality. **4** set of students taught together. ● *v.* place in a class.

■ *n.* **1** category, denomination, genre, genus, group, kind, league, set, sort, species, type. **2** caste, grade, level, order, rank, stratum. ● *v.* categorize, classify, grade, group, order, rank, rate.

classic *adj.* **1** of recognized high quality. **2** typical. **3** having enduring worth. **4** simple in style. ● *n.* **1** classic author or work etc. **2** (*pl.*) study of ancient Greek and Roman literature, history, etc. □ **classicism** *n.*, **classicist** *n.*

■ *adj.* **1** excellent, exemplary, first-rate, noteworthy, outstanding, superior. **2** archetypal, standard, typical. **3** ageless, enduring, immortal, time-honoured, timeless, undying, vintage.

classical *adj.* **1** classic. **2** of the ancient Greeks and Romans. **3** traditional and standard. □ **classically** *adv.*

classify *v.* **1** arrange systematically, class. **2** designate as officially secret. □ **classifiable** *adj.*, **classification** *n.*

■ **1** arrange, catalogue, categorize, class, grade, group, order, organize, pigeon-hole, sort, systematize, tabulate.

classroom *n.* room where a class of students is taught.

classy *adj.* (*colloq.*) superior.

clatter *n.* rattling sound. ● *v.* make this sound.

clause *n.* **1** single part in a treaty, law, or contract. **2** distinct part of a sentence, with its own verb.

claustrophobia *n.* abnormal fear of being in an enclosed space. □ **claustrophobic** *adj.*

clavichord *n.* early small keyboard instrument.

claw *n.* **1** pointed nail on an animal's or bird's foot. **2** claw-like device for gripping and lifting things. ● *v.* scratch or pull with a claw or hand.

clay *n.* stiff sticky earth, used for making bricks and pottery. □ **clay pigeon** breakable disc thrown up as a target for shooting. **clayey** *adj.*

claymore *n.* Scottish two-edged broad-bladed sword.

clean *adj.* free from dirt or impurities, not soiled or used. ● *v.* make clean. □ **cleaner** *n.*, **cleanly** *adv.*

■ *adj.* decontaminated, disinfected, pure, purified, sanitary, sterile, sterilized, uncontaminated, undefiled, unpolluted, unsoiled, unstained, unsullied, untainted; laundered, scrubbed, washed; immaculate, spotless, unmarked, untouched, unused. ● *v.* cleanse, launder, mop, scour, scrub, sponge, sweep, wash, wipe.

cleanly /klénli/ *adj.* attentive to cleanness. □ **cleanliness** *n.*

cleanse /klenz/ *v.* make clean. □ **cleanser** *n.*

clear *adj.* **1** not clouded. **2** transparent. **3** convinced. **4** not confused or doubtful. **5** easily seen, heard, or understood. **6** not obstructed. ● *v.* **1** make or become clear. **2** prove innocent. **3** get past or over. **4** make as net profit. □ **clear off** (*colloq.*) go away. **clear out 1** empty. **2** remove. **3** (*colloq.*) go away. **clearly** *adv.*

■ *adj.* **1** cloudless, fair, fine, sunny. **2** crystalline, limpid, pellucid, translucent, transparent. **3** assured, certain, confident, convinced, positive, sure. **4** definite, evident, incontrovertible, indisputable, manifest, obvious, palpable, patent, plain, undeniable, unmistakable. **5** distinct, sharp, vivid, well-defined; legible, readable; coherent, comprehensible, explicit, intelligible, lucid, precise, unambiguous, understandable. **6** free, open, passable, unblocked, unobstructed. ● *v.* **1** brighten, lighten; clarify, purify; open up, unblock, unclog. **2** absolve, acquit, exonerate, vindicate. **3** jump over, leap over, vault over.

clearance *n.* **1** clearing. **2** permission. **3** space allowed for one object to pass another.

clearing *n.* space cleared of trees in a forest.

clearway *n.* road where vehicles must not stop.

cleat *n.* projecting piece for fastening ropes to.

cleavage *n.* **1** split, separation. **2** hollow between full breasts.

cleave *v.* (**cleaved**, **clove**, or **cleft**; **cloven** or **cleft**) split.

cleaver *n.* butcher's chopper.

clef *n.* symbol on a stave in music, showing the pitch of notes.

cleft *adj.* & *n.* split.

clematis *n.* climbing plant with showy flowers.

clemency *n.* **1** mildness. **2** mercy. □ **clement** *adj.*

clementine *n.* a kind of small orange.

clench *v.* close tightly.

clergy *n.* people ordained for religious duties. □ **clergyman** *n.*, **clergywoman** *n.*

cleric *n.* member of the clergy.

clerical *adj.* **1** of clerks. **2** of clergy.

■ **1** secretarial, white-collar. **2** canonical, ecclesiastical, episcopal, ministerial, priestly.

clerk *n.* person employed to do written work in an office.

clever *adj.* **1** quick to learn and understand. **2** skilful. **3** ingenious. □ **cleverly** *adv.*, **cleverness** *n.*

■ **1** able, acute, astute, brainy, bright, brilliant, canny, discerning, gifted, intellectual, intelligent, perceptive, quick, sharp, shrewd, smart. **2** adept, adroit, deft, dexterous, handy, skilful, skilled, talented. **3** cunning, imaginative, ingenious, inventive, resourceful.

cliché /kleéshay/ *n.* hackneyed phrase or idea. □ **clichéd** *adj.*

■ banality, *colloq.* chestnut, platitude, stereotype, truism.

click *n.* short sharp sound. ● *v.* **1** make or cause to make a click. **2** (*colloq.*) be a success. **3** (*colloq.*) be understood.

client *n.* **1** customer. **2** person using the services of a professional person.

clientele /kleé-ontél/ *n.* clients.

cliff *n.* steep rock face, esp. on a coast. □ **cliffhanger** *n.* story or contest full of suspense.

■ bluff, crag, escarpment, precipice, scarp.

climacteric *n.* period of life when physical powers begin to decline.

climate *n.* regular weather conditions of an area.

climax *n.* point of greatest interest or intensity.

■ acme, culmination, height, high point, peak, summit, zenith.

climb *v.* go up or over. ● *n.* ascent made by climbing. □ **climber** *n.*

■ *v.* ascend, clamber up, go up, mount, scale, shin up.

clime *n.* **1** climate. **2** region.

clinch *v.* **1** fasten securely. **2** settle conclusively. **3** (of boxers) hold on to each other. ● *n.* clinching. □ **clincher** *n.*

cling *v.* (**clung**) **1** stick. **2** hold on tightly. □ **cling film** thin polythene wrapping.

clinic *n.* **1** place or occasion for giving medical treatment. **2** private or specialized hospital.

clinical *adj.* of or used in treatment of patients. □ **clinically** *adv.*

clink *n.* thin sharp sound. ● *v.* make this sound.

clinker *n.* fused coal ash.

clip[1] *n.* device for holding things tightly or together. ● *v.* (**clipped**) fix or fasten with clip(s).

clip[2] *v.* (**clipped**) **1** cut with shears or scissors. **2** (*colloq.*) hit sharply. ● *n.* **1** act of clipping. **2** piece clipped from something. **3** (*colloq.*) sharp blow.

■ *v.* **1** crop, cut, lop, prune, shear, snip, trim.

clipper *n.* **1** fast sailing ship. **2** (*pl.*) instrument for clipping.

clipping *n.* **1** piece clipped off. **2** newspaper cutting.

clique /kleek/ *n.* small exclusive group.

clitoris *n.* small erectile part of female genitals.

cloak *n.* loose sleeveless outer garment. ● *v.* cover, conceal.

■ *n.* cape, mantle, wrap. ● *v.* conceal, cover, disguise, hide, mask, screen, shroud, veil, wrap.

cloakroom *n.* room where outer garments can be left, often containing a lavatory.

clobber (*sl.*) *n.* **1** equipment. **2** belongings. ● *v.* **1** hit hard. **2** defeat heavily.

cloche *n.* translucent cover for protecting plants.

clock *n.* instrument indicating time. □ **clock in** or **on, out** or **off** register one's time of arrival or departure. **clock up** achieve.

clockwise *adv.* & *adj.* moving in the direction of the hands of a clock.

clockwork *n.* mechanism with wheels and springs.

clod *n.* lump of earth.

clog *n.* wooden-soled shoe. ● *v.* (**clogged**) **1** cause an obstruction in. **2** become blocked.

cloister *n.* **1** covered walk along the side of a church etc. **2** life in a monastery or convent.

cloistered *adj.* secluded.

clone *n.* group of plants or organisms produced asexually from one ancestor. ● *v.* grow in this way.

close[1] /klōss/ *adj.* **1** situated at a short distance or interval. **2** dear to each other. **3** dense, compact. **4** concentrated. **5** secretive. **6** stingy. **7** stuffy or humid. **8** (of a danger etc.) narrowly avoided. ● *adv.* **1** closely. **2** in a near position. ● *n.* **1** street closed at one end. **2** grounds round a cathedral or abbey. □ **close-up** *n.* photograph etc. taken at close range. **closely** *adv.*, **closeness** *n.*

■ *adj.* **1** adjacent, near. **2** affectionate, devoted, friendly, inseparable, intimate, loving, *colloq.* pally, *colloq.* thick. **3** compact, cramped, dense, tight. **4** assiduous, careful, concentrated, minute, painstaking, precise, rigorous, searching, thorough. **5** reserved, reticent, secretive, uncommunicative. **6** mean, miserly, niggardly, parsimonious, stingy, *colloq.* tight, tight-fisted. **7** airless, fuggy, fusty, humid, muggy, oppressive, stale, stifling, stuffy, suffocating, unventilated.

close[2] /klōz/ *v.* **1** shut. **2** bring or come to an end. **3** come nearer together. ● *n.* conclusion, end.

■ *v.* **1** bolt, fasten, lock, padlock, seal, secure, shut. **2** complete, conclude, end, finish, terminate, wind up. ● *n.* completion, conclusion, culmination, end, finish, halt, termination.

closet *n.* **1** cupboard. **2** store room. ● *v.* (**closeted**) shut away in private conference or study.

closure *n.* closing, closed state.

clot *n.* **1** thickened mass of liquid. **2** (*colloq.*) stupid person. ● *v.* (**clotted**) form clot(s).

cloth *n.* **1** woven or felted material. **2** piece of this.

clothe *v.* put clothes on, provide with clothes.

■ attire, dress, garb, kit out, robe, *colloq.* tog out.

clothes *n.pl.* things worn to cover the body.

■ apparel, attire, clothing, finery, garb, garments, get-up, kit, *colloq.* togs, vestments, wardrobe.

clothier *n.* person who deals in cloth and men's clothes.

clothing *n.* clothes for the body.

cloud *n.* **1** visible mass of watery vapour floating in the sky. **2** mass of smoke or dust. ● *v.* **1** become covered with clouds or gloom. **2** make unclear.

cloudburst *n.* violent storm of rain.

cloudy *adj.* (**-ier, -iest**) **1** covered with clouds. **2** (of liquid) not transparent. □ **cloudiness** *n.*

■ **1** clouded, dark, dull, gloomy, grey, leaden, murky, overcast, starless, sunless. **2** muddy, opaque, turbid.

clout *n.* **1** blow. **2** (*colloq.*) power of effective action. ● *v.* hit.

clove[1] *n.* dried bud of a tropical tree, used as spice.

clove[2] *n.* one division of a compound bulb such as garlic.

clove[3], **cloven** *see* **cleave**[1].

clove hitch knot used to fasten a rope round a pole etc.

cloven hoof divided hoof like that of sheep, cows, etc.

clover *n.* plant with three-lobed leaves. □ **in clover** in luxury.

clown *n.* **1** comic entertainer. **2** foolish or playful person. ● *v.* perform or behave as a clown.

■ *n.* **1** comedian, comedienne, comic, jester, joker. **2** buffoon, fool, idiot.

cloy *v.* sicken by glutting with sweetness or pleasure.

club *n.* **1** heavy stick used as a weapon. **2** group who meet for social or sporting purposes, their premises. **3** organization offering benefit to subscribers. **4** stick with a wooden or metal head, used in golf. **5** playing card of the suit marked with black clover leaves. ● *v.* (**clubbed**) strike with a club. □ **club together** join in subscribing.

■ *n.* **1** bat, baton, cosh, cudgel, truncheon. **2** association, circle, federation, fellowship, group, guild, organization, society, union. ● *v.* beat, belabour, bludgeon, cudgel, thrash.

cluck *n.* throaty cry of hen. ● *v.* utter a cluck.

clue *n.* fact or idea giving a guide to the solution of a problem.

■ hint, indication, lead, pointer, suggestion, tip-off; idea, inkling.

clump *n.* cluster, esp. of trees. ● *v.* **1** form a clump. **2** tread heavily.

clumsy *adj.* (**-ier, -iest**) **1** awkward in movement or shape. **2** tactless. □ **clumsily** *adv.*, **clumsiness** *n.*

■ **1** awkward, blundering, bungling, fumbling, gangling, gawky, graceless, *colloq.* ham-fisted, *colloq.* hulking, inept, lumbering, maladroit, uncoordinated, ungainly, ungraceful; bulky, cumbersome, ponderous, unmanageable, unwieldy. **2** gauche, ill-judged, insensitive, tactless, thoughtless, undiplomatic.

clung *see* **cling**.

cluster *n.* close group of similar people or things. ● *v.* form a cluster.

■ *n.* batch, bunch, clump, collection; flock, gathering, group, huddle, knot, swarm, throng. ● *v.* assemble, bunch, collect, congregate, flock, gather, group.

clutch[1] *v.* grasp tightly. ● *n.* **1** tight grasp. **2** device for connecting and disconnecting moving parts.

■ *v.* clasp, grab, grasp, grip, seize, snatch, take hold of.

clutch[2] *n.* **1** set of eggs for hatching. **2** chickens hatched from these.

clutter *n.* things lying about untidily. ● *v.* fill with clutter.

■ *n.* confusion, disorder, jumble, litter, mess, muddle. ● *v.* litter, mess up, strew.

Co. *abbr.* **1** Company. **2** County.

c/o *abbr.* care of.

co- *pref.* joint, jointly.

coach *n.* **1** single-decker bus. **2** large horse-drawn carriage. **3** railway carriage. **4** private tutor. **5** instructor in sports. ● *v.* train, teach.

■ *n.* **3** *US* car, carriage, wagon. **4** instructor, teacher, trainer, tutor. ● *v.* drill, guide, instruct, prepare, school, teach, train, tutor.

coagulate *v.* change from liquid to semi-solid, clot. □ **coagulant** *n.*, **coagulation** *n.*

■ clot, congeal, curdle, *colloq.* jell, set, solidify, thicken.

coal *n.* hard black mineral used for burning as fuel. □ **coalfield** *n.* area yielding coal.

coalesce *v.* combine. □ **coalescence** *n.*

coalition *n.* union, esp. temporary union of political parties.

coarse *adj.* **1** rough or loose in texture or grain. **2** crude in manner, vulgar. □ **coarse fish** freshwater fish other than salmon and trout. **coarsely** *adv.*, **coarseness** *n.*

■ **1** bristly, prickly, rough, scratchy. **2** boorish, crude, loutish, rude, uncouth, vulgar.

coarsen *v.* make or become coarse.

coast *n.* seashore and land near it. ● *v.* **1** sail along a coast. **2** ride a bicycle or drive a motor vehicle without using power. □ **coastal** *adj.*

■ *n.* beach, littoral, seashore, seaside, shore, strand.

coaster *n.* **1** ship trading along a coast. **2** mat for a glass.

coastguard *n.* officer of an organization that keeps watch on the coast.

coat *n.* **1** outdoor garment with sleeves. **2** fur or hair covering an animal's body. **3** covering layer. ● *v.* cover with a layer. □ **coat of arms** design on a shield as the emblem of a family etc.

■ *n.* **1** *colloq.* mac, mackintosh, overcoat, raincoat, topcoat. **2** fleece, fur, hide, pelt, skin. **3** coating, covering, film, layer. ● *v.* cover, paint, spread.

coating *n.* covering layer.

coax *v.* **1** persuade gently. **2** manipulate carefully or slowly.

■ **1** cajole, inveigle, manipulate, persuade, talk into, wheedle.

coaxial *adj.* (of cable) containing two conductors, one surrounding but insulated from the other.

cob *n.* **1** sturdy short-legged horse. **2** large hazelnut. **3** stalk of an ear of maize. **4** small round loaf.

cobalt *n.* **1** metallic element. **2** deep-blue pigment made from it.

cobber *n.* (*Austr.* & *NZ colloq.*) friend, mate.

cobble¹ *n.* rounded stone formerly used for paving roads.

cobble² *v.* mend roughly.

cobbler *n.* shoe-mender.

cobra *n.* poisonous snake.

cobweb *n.* network spun by a spider.

cocaine *n.* drug used illegally as a stimulant.

coccyx /kóksiks/ *n.* bone at the base of the spinal column.

cochineal *n.* red colouring matter used in food.

cock *n.* **1** male bird. **2** tap or valve controlling a flow. ● *v.* **1** tilt or turn upwards. **2** set (a gun) for firing. ◻ **cock-a-hoop** *adj.* very pleased. **cock-eyed** *adj.* (*colloq.*) **1** askew. **2** absurd.

cockade *n.* rosette worn on a hat.

cockatoo *n.* crested parrot.

cocker *n.* breed of spaniel.

cockerel *n.* young male fowl.

cockle *n.* edible shellfish.

cockney *n.* native or dialect of the East End of London.

cockpit *n.* compartment for the pilot in a plane, or for the driver in a racing car.

cockroach *n.* beetle-like insect.

cockscomb *n.* cock's crest.

cocksure *adj.* very self-confident.

cocktail *n.* mixed alcoholic drink.

cocky *adj.* (**-ier, -iest**) conceited and arrogant. ◻ **cockily** *adv.*

cocoa *n.* **1** powder of crushed cacao seeds. **2** drink made from this.

coconut *n.* **1** nut of a tropical palm. **2** its edible lining.

cocoon *n.* **1** silky sheath round a chrysalis. **2** protective wrapping.

cod *n.* large edible sea fish.

coda *n.* final part of a musical composition.

coddle *v.* cherish and protect.

code *n.* **1** system of signals or symbols etc. used for secrecy, brevity, etc. **2** systematic set of laws, rules, etc.

codeine *n.* substance made from opium, used to relieve pain.

codicil *n.* appendix to a will.

codify *v.* arrange (laws etc.) into a code. ◻ **codification** *n.*

coeducation *n.* education of boys and girls in the same classes. ◻ **coeducational** *adj.*

coefficient *n.* **1** multiplier. **2** mathematical factor.

coeliac disease /seéliak/ disease causing inability to digest gluten.

coerce *v.* compel by threats or force. ◻ **coercion** *n.*, **coercive** *adj.*

> ■ blackmail, bludgeon, browbeat, bully, compel, constrain, dragoon, drive, force, pressurize, railroad.

coeval *adj.* of the same age or epoch.

coexist *v.* exist together, esp. harmoniously. ◻ **coexistence** *n.*, **coexistent** *adj.*

coffee *n.* **1** bean-like seeds of a tropical shrub, roasted and ground for making a drink. **2** this drink. **3** light brown. ◻ **coffee table** small low table.

coffer *n.* **1** large strong box for holding money and valuables. **2** (*pl.*) financial resources.

coffin *n.* box in which a corpse is placed for burial or cremation.

cog *n.* one of a series of projections on the edge of a wheel, engaging with those of another.

cogent *adj.* convincing. ◻ **cogently** *adv.*, **cogency** *n.*

> ■ compelling, convincing, effective, forceful, persuasive, potent, powerful, strong, weighty.

cogitate *v.* think deeply. ◻ **cogitation** *n.*

cognac *n.* French brandy.

cognate *adj.* akin, related. ● *n.* **1** relative. **2** cognate word.

cognition *n.* knowing, perceiving. ◻ **cognitive** *adj.*

cognizant *adj.* aware, having knowledge. ◻ **cognizance** *n.*

cohabit *v.* live together as man and wife. ◻ **cohabitation** *n.*

cohere *v.* stick together.

coherent *adj.* **1** connected logically. **2** intelligible. ◻ **coherently** *adv.*, **coherence** *n.*

> ■ **1** consistent, orderly, organized, logical, rational, reasonable. **2** articulate, clear, comprehensible, intelligible, lucid, understandable.

cohesion *n.* tendency to cohere.

cohesive *adj.* cohering.

cohort *n.* **1** tenth part of a Roman legion. **2** people banded together.

coiffure /kwaafyóor/ *n.* hairstyle.

coil *v.* wind into rings or a spiral. ● *n.* **1** something coiled. **2** one ring or turn in this.

> ■ *v.* curl, loop, snake, spiral, twine, twirl, twist, wind. ● *n.* convolution, curl, helix, loop, spiral, twist, whirl, whorl.

coin *n.* piece of metal money. ● *v.* **1** make (coins) by stamping metal. **2** invent (a word). □ **coiner** *n.*

coinage *n.* **1** coining. **2** coins, system of these. **3** invented word.

coincide *v.* **1** occur at the same time. **2** agree or be identical.

■ **2** accord, agree, correspond, match, tally.

coincidence *n.* **1** occurring together. **2** remarkable concurrence of events etc., apparently by chance. □ **coincidental** *adj.*, **coincidentally** *adv.*

coition *n.* coitus.

coitus *n.* sexual intercourse.

coke[1] *n.* solid substance left after gas and tar have been extracted from coal, used as fuel.

coke[2] *n.* (*sl.*) cocaine.

col *n.* depression in a range of mountains.

colander *n.* bowl-shaped perforated vessel for draining food.

cold *adj.* **1** at or having a low temperature. **2** not affectionate, enthusiastic, or kind. ● *n.* **1** cold conditions, weather etc. **2** illness causing catarrh and sneezing. □ **cold-blooded** *adj.* **1** having a blood temperature varying with that of the surroundings. **2** unfeeling, ruthless. **cold feet** fear. **cold-shoulder** *v.* treat with deliberate unfriendliness. **coldly** *adv.*, **coldness** *n.*

■ *adj.* **1** arctic, bitter, bleak, chill, chilly, cool, draughty, freezing, frigid, frosty, glacial, icy, keen, *colloq.* nippy, raw, wintry; chilled, frozen, shivering, shivery. **2** aloof, apathetic, chilly, cool, distant, frigid, frosty, indifferent, lukewarm, standoffish, undemonstrative, unemotional, unfriendly, unresponsive; callous, hard-hearted, heartless, stony, uncaring, unfeeling, unsympathetic.

coleslaw *n.* salad of shredded raw cabbage coated in dressing.

colic *n.* severe abdominal pain.

colitis *n.* inflammation of the colon.

collaborate *v.* work in partnership. □ **collaboration** *n.*, **collaborator** *n.*, **collaborative** *adj.*

collage /kóllaazh/ *n.* picture made by gluing pieces of paper etc. to a backing.

collapse *v.* **1** fall down suddenly. **2** lose strength suddenly. ● *n.* **1** collapsing. **2** breakdown.

■ *v.* **1** cave in, crumble, crumple, fall apart, fall down, give way, subside, tumble down. **2** faint, keel over, pass out; *colloq.* crack up.

collapsible *adj.* made so as to fold up.

collar *n.* **1** band round the neck of a garment. **2** leather band round an animal's neck. **3** band holding part of a machine. ● *v.* seize, take for oneself.

collate *v.* collect and arrange systematically. □ **collator** *n.*

collateral *adj.* **1** parallel. **2** additional but subordinate. ● *n.* additional security pledged. □ **collaterally** *adv.*

collation *n.* **1** collating. **2** light meal.

colleague fellow worker esp. in a business or profession.

collect *v.* **1** bring or come together. **2** fetch. **3** obtain specimens of, esp. as a hobby. □ **collector** *n.*

■ **1** assemble, congregate, convene, converge, gather, meet, rally; accumulate, amass, compile, garner, heap up, hoard, pile up, save, stockpile, store up. **2** bring, fetch, get, pick up.

collected *adj.* calm and controlled.

■ calm, composed, controlled, cool, imperturbable, level-headed, nonchalant, sedate, self-possessed, serene, tranquil, unperturbed, unruffled.

collection *n.* **1** act or process of collecting. **2** things collected.

■ **2** accumulation, assortment, heap, hoard, pile, stack, store; anthology, compendium, compilation, miscellany.

collective *adj.* of or denoting a group taken or working as a unit. □ **collective noun** noun (singular in form) denoting a group (e.g. *army*, *herd*). **collectively** *adv.*

colleen *n.* (*Ir.*) girl.

college *n.* **1** establishment for higher or professional education. **2** organized group of professional people. □ **collegiate** *adj.*

collide *v.* come into collision.

collie *n.* dog with a pointed muzzle and shaggy hair.

colliery *n.* coal mine.

collision *n.* violent striking of one thing against another.

■ accident, bump, crash, pile-up, smash.

collocate *v.* place (words) together. □ **collocation** *n.*

colloquial *adj.* suitable for informal speech or writing. □ **colloquially** *adv.*, **colloquialism** *n.*

collude *v.* conspire. □ **collusion** *n.*, **collusive** *adj.*

colon[1] *n.* lower part of the large intestine. □ **colonic** *adj.*
colon[2] *n.* punctuation mark (:).
colonel /kŏn'l/ *n.* army officer next below brigadier.
colonial *adj.* of a colony or colonies. ● *n.* inhabitant of a colony.
colonialism *n.* policy of acquiring or maintaining colonies.
colonize *v.* establish a colony in. □ **colonization** *n.*, **colonist** *n.*
colonnade *n.* row of columns.
colony *n.* **1** settlement or settlers in new territory, remaining subject to the parent state. **2** people of one nationality or occupation living in a particular area. **3** group of animals living close together.
coloration *n.* colouring.
colossal *adj.* immense. □ **colossally** *adv.*

> ■ enormous, gargantuan, giant, gigantic, huge, immense, mammoth, massive, monumental, prodigious, titanic, vast.

colossus *n.* (*pl.* **colossi**) immense statue.
colour *n.* **1** one, or any mixture, of the constituents into which light is separated in a rainbow etc. **2** pigment, paint. **3** (usu. *pl.*) flag of a ship or regiment. ● *v.* **1** apply colour to. **2** blush. **3** influence. □ **colour-blind** *adj.* unable to distinguish between certain colours.

> ■ *n.* **1** hue, shade, tincture, tint, tone. **2** dye, paint, pigment. **3** (**colours**) banner, flag, ensign, pennant, standard. ● *v.* **1** dye, paint, stain, tinge, tint. **2** blush, flush, go red, redden. **3** affect, bias, distort, influence, prejudice, slant, sway.

colourant *n.* colouring matter.
colourful *adj.* **1** full of colour. **2** full of interest, vivid. □ **colourfully** *adv.*

> ■ **1** bright, gay, iridescent, multicoloured, vivid. **2** graphic, interesting, picturesque, vivid.

colourless *adj.* **1** without colour. **2** lacking interest.

> ■ **1** ashen, pale, pallid, wan, washed out, white. **2** bland, boring, drab, dreary, dull, insipid, lacklustre, lifeless, tame, uninspiring, uninteresting, vapid.

colt *n.* young male horse.
coltsfoot *n.* wild plant with yellow flowers.
columbine *n.* garden flower with pointed projections on its petals.
column *n.* **1** round pillar. **2** thing shaped like this. **3** vertical division of a page, printed matter in this. **4** long narrow formation of troops, vehicles, etc.
columnist *n.* journalist contributing regularly to a newspaper.
coma *n.* deep unconsciousness.
comatose *adj.* **1** in a coma. **2** drowsy.
comb *n.* **1** toothed strip of stiff material for tidying hair. **2** fowl's fleshy crest. **3** honeycomb. ● *v.* **1** tidy with a comb. **2** search thoroughly.
combat *n.* battle, contest. ● *v.* (**combated**) counter. □ **combative** *adj.*

> ■ *n.* action, battle, conflict, contest, duel, encounter, engagement, fight, skirmish, struggle. ● *v.* counter, fight, oppose, resist, strive against, struggle against.

combatant *adj.* & *n.* (person or nation) engaged in fighting.
combination *n.* **1** act or process of combining. **2** set of things or people combined. □ **combination lock** lock controlled by a series of positions of dial(s).

> ■ amalgamation, blend, mix, mixture, synthesis, union.

combine *v.* /kəmbīn/ join into a group, set, or mixture. ● *n.* /kómbīn/ combination of people or firms acting together.

> ■ *v.* associate, club together, gang up, join forces, link, team up, unite; add together, amalgamate, blend, integrate, join, merge, mix, pool, put together, synthesize.

combustible *adj.* capable of catching fire.
combustion *n.* **1** burning. **2** process in which substances combine with oxygen and produce heat.
come *v.* (**came, come**) **1** move towards or reach a place, time, situation, etc. **2** occur. □ **come about** happen. **come across** meet or find unexpectedly. **comeback** *n.* **1** return to a former successful position. **2** retort. **come by** obtain. **come-down** *n.* fall in status. **come down with** contract (an illness). **come into** inherit. **come off** be successful. **come out 1** become visible. **2** emerge. **come round 1** recover from fainting. **2** be converted to the speaker's opinion. **come to 1** regain consciousness. **2** amount to. **come up** arise for discussion.

> ■ **1** appear, approach, arrive, draw near, *colloq.* show up, turn up. □ **come about** befall, happen, occur, take place. **come across** chance on, discover, encounter, find, happen on, hit on, light on, run into. **come by** acquire, get, obtain, procure, secure. **come to 1** awake, revive, wake up.

comedian *n.* humorous entertainer or actor.

comedienne *n.* female comedian.

comedy *n.* **1** light amusing drama. **2** humour.

comely *adj.* (**-ier**, **-iest**) good-looking. □ **comeliness** *n.*

comestibles *n.pl.* things to eat.

comet *n.* heavenly body with a luminous 'tail'.

comfort *n.* **1** state of ease and contentment. **2** relief of suffering or grief. **3** person or thing giving this. ● *v.* give comfort to. □ **comforter** *n.*

■ *n.* **1** content, contentment, ease. **2** consolation, reassurance, relief, solace, support. ● *v.* cheer (up), console, gladden, hearten, reassure, relieve, solace, soothe.

comfortable *adj.* giving or having ease and contentment. □ **comfortably** *adv.*

■ congenial, cosy, pleasant, relaxing, snug; contented, easy, relaxed, untroubled.

comic *adj.* **1** causing amusement. **2** of comedy. ● *n.* **1** comedian. **2** periodical with a series of strip cartoons. □ **comical** *adj.*, **comically** *adv.*

■ *adj.* **1** amusing, comical, diverting, droll, funny, hilarious, humorous, *sl.* priceless, witty.

comma *n.* punctuation mark (,) .

command *n.* **1** statement, given with authority, that an action must be performed. **2** holding of authority. **3** mastery. **4** forces or district under a commander. ● *v.* **1** give a command to. **2** have authority over.

■ *n.* **1** decree, dictate, direction, directive, edict, injunction, instruction, order. **2** authority, control, dominion, government, jurisdiction, leadership, management, mastery, power, sovereignty, supervision, sway. **3** grasp, knowledge, mastery. ● *v.* **1** adjure, bid, direct, instruct, order, require. **2** be in charge of, control, dominate, govern, head, hold sway over, lead, preside over, rule.

commandant *n.* officer in command of a fortress etc.

commandeer *v.* seize for use.

■ appropriate, confiscate, hijack, requisition, seize.

commander *n.* **1** person in command. **2** naval officer next below captain.

commandment *n.* divine command.

commando *n.* member of a military unit specially trained for making raids and assaults.

commemorate *v.* keep in the memory by a celebration or memorial. □ **commemoration** *n.*, **commemorative** *adj.*

■ celebrate, observe, solemnize; honour, pay homage to, pay tribute to, salute.

commence *v.* begin. □ **commencement** *n.*

■ begin, embark on, inaugurate, initiate, launch, open, start.

commend *v.* **1** praise. **2** entrust. □ **commendation** *n.*

commendable *adj.* worthy of praise. □ **commendably** *adv.*

■ admirable, creditable, deserving, laudable, meritorious, praiseworthy, worthy.

commensurable *adj.* measurable by the same standard. □ **commensurability** *n.*

commensurate *adj.* **1** of the same size. **2** proportionate.

comment *n.* **1** opinion, remark. **2** explanatory note. ● *v.* make comment(s).

■ *n.* **1** observation, opinion, remark, view. **2** annotation, explanation, footnote, note. ● *v.* observe, opine, remark, say.

commentary *n.* spoken or written description of something.

commentate *v.* act as commentator.

commentator *n.* person who writes or speaks a commentary.

commerce *n.* buying and selling, trading.

commercial *adj.* of, engaged in, or financed by commerce. □ **commercially** *adv.*

commercialize *v.* **1** make commercial. **2** make profitable. □ **commercialization** *n.*

commingle *v.* mix.

commiserate *v.* express or feel sympathy. □ **commiseration** *n.*

commission *n.* **1** committing. **2** giving of authority to perform a task. **3** task given. **4** group of people given such authority. **5** warrant conferring authority on an officer in the armed forces. **6** payment to an agent selling goods or services. ● *v.* **1** give commission to. **2** place an order for. □ **in commission** ready for service. **out of commission** not in working order.

commissionaire *n.* uniformed door attendant.

commissioner *n.* **1** member of a commission. **2** government official in charge of a district abroad.

commit *v.* (**committed**) **1** do, perform. **2** entrust, consign. **3** pledge to a course of action. □ **committal** *n.*

■ **1** carry out, do, execute, perform, perpetrate. **2** assign, consign, deliver, entrust, give, hand over, transfer. **3** pledge, promise, swear, undertake, vow.

commitment *n.* **1** committing. **2** obligation or pledge, state of being involved in this.

committee *n.* group of people appointed for a special function.

■ board, body, cabinet, commission, council, panel.

commode *n.* **1** chest of drawers. **2** chamber pot in a chair or box.

commodious *adj.* roomy.

commodity *n.* article of trade.

commodore *n.* **1** naval officer next below rear admiral. **2** president of a yacht club.

common *adj.* **1** of or affecting all. **2** occurring often. **3** ordinary. **4** of inferior quality. ● *n.* area of unfenced grassland for all to use. □ **common law** unwritten law based on custom and precedent. **common room** room shared by students or teachers for social purposes. **common sense** normal good sense in practical matters. **common time** 4 crotchets in the bar in music.

■ *adj.* **1** communal, general, joint, *colloq.* mutual, public, shared, universal. **2** customary, familiar, frequent, habitual, prevalent, usual; hackneyed, overused, stale, trite. **3** average, commonplace, conventional, everyday, mediocre, middling, normal, ordinary, run-of-the-mill, standard, stock, typical, unexceptional, workaday.

commoner *n.* person below the rank of peer.

commonly *adv.* usually, frequently.

commonplace *adj.* **1** lacking originality. **2** ordinary.

■ average, banal, hackneyed, humdrum, ordinary, pedestrian, predictable, prosaic, run-of-the-mill, standard, stock, trite, undistinguished, unremarkable.

commonwealth *n.* **1** independent state. **2** federation of states.

commotion *n.* fuss and disturbance.

■ ado, bother, din, fracas, furore, fuss, hubbub, hullabaloo, hurly-burly, *colloq.* kerfuffle, *colloq.* palaver, rumpus, stir, to-do, upheaval, uproar.

communal *adj.* shared among a group. □ **communally** *adv.*

commune[1] /kəmyóon/ *v.* communicate mentally or spiritually.

commune[2] /kómyoon/ *n.* **1** group sharing accommodation and goods. **2** district of local government in France etc.

communicant *n.* person who receives Holy Communion.

communicate *v.* **1** make known. **2** pass news etc. to and fro. **3** transmit (disease etc.). □ **communicator** *n.*, **communicable** *adj.*

■ **1** announce, broadcast, convey, disclose, divulge, impart, intimate, make known, pass on, proclaim, promulgate, reveal, spread, transfer, transmit. **2** be in communication, be in touch, commune, converse, correspond, talk.

communication *n.* **1** act of communicating. **2** letter or message. **3** means of access.

■ **1** disclosure, intimation, promulgation, spread, transmission. **2** communiqué, dispatch, epistle, letter, line, message, note.

communicative *adj.* talkative, willing to give information.

■ chatty, expansive, forthcoming, frank, informative, open, outgoing, sociable, talkative.

communion *n.* **1** fellowship. **2** social dealings. **3** branch of the Christian Church. **4** (**Holy Communion**) Eucharist.

communiqué /kəmyóonikay/ *n.* **1** official report. **2** agreed statement.

communism *n.* **1** social system based on common ownership of property etc. **2** political doctrine or movement advocating this. □ **communist** *n.*

community *n.* group of people living in one district or having common interests or origins.

commute *v.* **1** exchange for something else. **2** travel regularly by train or bus to and from one's work. □ **commuter** *n.*

compact[1] /kómpakt/ *n.* pact, contract.

compact[2] *adj.* /kəmpákt/ **1** closely or neatly packed together. **2** concise. ● *v.* /kəmpákt/ make compact. ● *n.* /kómpakt/ small flat case for face powder. □ **compact disc** small disc from which sound etc. is reproduced by laser action.

■ *adj.* **1** compressed, dense, firm, solid. **2** brief, compendious, concise, condensed, laconic, pithy, succinct, terse.

companion *n.* **1** person who accompanies or associates with another. **2** handbook, reference book. □ **companionway** *n.* staircase from a ship's deck to cabins etc. **companionship** *n.*

■ **1** associate, *colloq.* buddy, *colloq.* chum, comrade, confidant(e), crony, fellow, friend, mate, *colloq.* pal, partner; chaperon, escort. **2** guide, handbook, manual, reference book.

companionable *adj.* sociable.

company *n.* **1** being with another or others. **2** people assembled. **3** guest(s). **4** actors etc. working together. **5** commercial business. **6** subdivision of an infantry battalion. □ **keep company with** associate with habitually.

■ **1** companionship, fellowship, society. **2** assembly, audience, band, crowd, gathering, party, throng, troop; entourage, retinue, suite, train. **3** caller(s), guest(s), visitor(s). **4** cast, ensemble, performers, troupe. **5** business, concern, corporation, establishment, firm, house, partnership.

comparable *adj.* suitable to be compared, similar. □ **comparability** *n.*, **comparably** *adv.*

comparative *adj.* **1** involving comparison. **2** of the grammatical form expressing 'more'. ● *n.* comparative form of a word. □ **comparatively** *adv.*

compare *v.* **1** estimate the similarity of. **2** declare to be similar. **3** bear comparison.

■ **1** contrast, correlate, juxtapose, relate, weigh up. **2** equate, liken.

comparison *n.* **1** act or instance of comparing. **2** similarity.

■ **2** analogy, comparability, correspondence, likeness, parallel, relation, relationship, resemblance, similarity.

compartment *n.* partitioned space. □ **compartmental** *adj.*

■ alcove, bay, booth, cubby hole, cubicle, niche, pigeon-hole, section, slot, space.

compass *n.* **1** device showing the direction of the magnetic or true north. **2** range, scope. **3** (*pl.*) hinged instrument for drawing circles. ● *v.* encompass.

compassion *n.* feeling of pity. □ **compassionate** *adj.*

compatible *adj.* **1** able to coexist. **2** consistent. □ **compatibly** *adv.*, **compatibility** *n.*

■ **1** like-minded, similar, well-matched, well-suited. **2** congruent, consistent, consonant.

compatriot *n.* person from one's own country.

compel *v.* (**compelled**) **1** force. **2** arouse (a feeling) irresistibly.

■ **1** coerce, constrain, dragoon, drive, force, make, oblige, order, press-gang, pressure, pressurize, railroad, require.

compendious *adj.* giving much information concisely.

compendium *n.* (*pl.* **-dia** or **-s**) **1** summary. **2** collection of information etc.

compensate *v.* **1** make a suitable payment in return for loss or damage. **2** counterbalance. □ **compensation** *n.*, **compensatory** *adj.*

■ **1** recompense, reimburse, repay, requite; atone, make amends.

compère *n.* person who introduces performers in a variety show. ● *v.* act as compère to.

compete *v.* **1** strive. **2** take part in a contest etc.

■ **1** battle, contend, fight, strive, struggle, vie. **2** participate, take part.

competence *n.* ability, authority.

competent *adj.* **1** having ability or authority to do what is required. **2** satisfactory. **3** proficient. □ **competently** *adv.*

■ **1** capable, fit, qualified. **2** acceptable, adequate, all right, *colloq.* OK, satisfactory. **3** able, accomplished, adept, capable, experienced, expert, practised, proficient, skilled.

competition *n.* **1** friendly contest. **2** competing. **3** those who compete.

■ **1** championship, contest, event, game, match, race, tournament. **2** contention, rivalry, struggle.

competitive *adj.* involving competition. □ **competitively** *adv.*, **competitiveness** *n.*

competitor *n.* one who competes.

■ contender, contestant; adversary, opponent, rival.

compile *v.* collect and arrange into a list or book. □ **compilation** *n.*, **compiler** *n.*

complacent *adj.* self-satisfied. □ **complacency** *n.*

complain *v.* **1** express dissatisfaction. **2** say one is suffering from pain. □ **complainant** *n.*

■ **1** *sl.* beef, carp, *colloq.* gripe, *colloq.* grouch, *colloq.* grouse, grumble, moan, object, protest, whine, *colloq.* whinge.

complaint *n.* **1** cause of dissatisfaction. **2** illness.

■ **1** *sl.* beef, grievance, *colloq.* gripe, *colloq.* grouse, grumble. **2** ailment, disease, disorder, illness, malady, sickness.

complaisant *adj.* willing to please others. □ **complaisance** *n.*

complement *n.* thing that completes. ● *v.* form a complement to. □ **complementary** *adj.*

complete *adj.* **1** having all its parts. **2** finished. **3** thorough, in every way. ● *v.* **1** finish, make complete. **2** fill in (a form etc.). □ **completely** *adv.*, **completeness** *n.*, **completion** *n.*

■ *adj.* **1** entire, full, intact, total, unabridged, unbroken, uncut, undivided, unexpurgated, whole. **2** concluded, done, finished, over. **3** absolute, out and out, perfect, pure, thorough, total, unmitigated, unqualified, utter. ● *v.* **1** accomplish, achieve, clinch, conclude, finalize, finish, end, settle; crown, perfect, round off.

complex *adj.* **1** made up of many parts. **2** complicated. ● *n.* **1** complex whole. **2** set of feelings that influence behaviour. **3** set of buildings. □ **complexity** *n.*

■ *adj.* **1** composite, compound. **2** complicated, difficult, intricate, involved, knotty, labyrinthine.

complexion *n.* **1** colour and texture of the skin of the face. **2** general character of things.

compliant *adj.* complying, obedient. □ **compliance** *n.*

complicate *v.* make complicated. □ **complication** *n.*

complicated *adj.* complex and difficult.

■ Byzantine, complex, difficult, elaborate, intricate, involved, knotty, labyrinthine, tangled.

complicity *n.* involvement in wrongdoing.

compliment *n.* **1** polite expression of praise. **2** (*pl.*) formal greetings. ● *v.* congratulate.

■ *n.* **1** commendation, tribute. **2** (**compliments**) best wishes, felicitations, greetings, regards, salutations. ● *v.* commend, congratulate, eulogize, felicitate, laud, pay tribute to, praise, salute.

complimentary *adj.* **1** expressing a compliment. **2** free of charge.

■ **1** appreciative, congratulatory, eulogistic, favourable, flattering, laudatory. **2** free, gratis.

comply *v.* act in accordance (with a request etc.).

■ accede, acquiesce, agree, assent, concur, conform, consent, obey, submit, yield.

component *n.* one of the parts of which a thing is composed.

comport *v.* **comport oneself** behave.

compose *v.* **1** create in music or literature. **2** constitute, make up. **3** arrange in good order. □ **compose oneself** become calm. **composer** *n.*

■ **1** create, write. **2** constitute, form, make up. **3** arrange, dispose, lay out, order, organize. □ **compose oneself** calm down, control oneself, quieten down, settle down.

composite *adj.* & *n.* (thing) made up of parts.

composition *n.* **1** act of composing. **2** structure. **3** thing composed.

■ **1** creation, writing. **2** arrangement, construction, formation, layout, make-up, organization, structure. **3** essay, piece, work.

compositor *n.* typesetter.

compost *n.* mixture of decayed organic matter used as fertilizer.

composure *n.* calmness.

compote *n.* fruit in syrup.

compound *adj.* /kómpownd/ made up of two or more ingredients. ● *n.* /kómpownd/ compound substance. ● *v.* /kəmpównd/ **1** combine. **2** increase. **3** settle by agreement.

■ *adj.* complex, composite. ● *n.* alloy, amalgam, blend, combination, mixture, synthesis. ● *v.* **1** amalgamate, blend, combine, fuse, mix, put together, synthesize, unite. **2** add to, aggravate, augment, exacerbate, heighten, increase, intensify, worsen.

comprehend *v.* **1** understand. **2** include.

■ **1** appreciate, apprehend, fathom, grasp, realize, see, take in, understand. **2** comprise, embrace, include, take in.

comprehensible *adj.* intelligible.

comprehension *n.* understanding.

comprehensive *adj.* including much or all. ● *n.* (in full **comprehensive school**) secondary school for children of all abilities. □ **comprehensively** *adv.*

■ *adj.* broad, complete, encyclopedic, exhaustive, extensive, full, inclusive, sweeping, thorough, wide-ranging.

compress *v.* /kəmpréss/ squeeze, force into less space. ● *n.* /kómpress/ pad to stop bleeding or to relieve inflammation. □ **compression** *n.*, **compressor** *n.*

comprise *v.* **1** include. **2** consist of.

compromise *n.* settlement reached by mutual concessions. ● *v.* **1** make a settlement in this way. **2** expose to suspicion.

compulsion *n.* **1** compelling, being compelled. **2** irresistible urge.

compulsive *adj.* **1** compelling. **2** resulting or acting (as if) from compulsion. **3** irresistible. ▫ **complusively** *adv.*

■ **1** compelling, overwhelming, uncontrollable, urgent. **2** habitual, incorrigible, incurable, inveterate, obsessive. **3** compelling, fascinating, gripping, irresistible.

compulsory *adj.* required by law or rule. ▫ **compulsorily** *adv.*

compunction *n.* regret, scruple.

■ misgiving, qualm, regret, second thought, scruple.

compute *v.* **1** calculate. **2** use a computer. ▫ **computation** *n.*

computer *n.* electronic apparatus for analysing or storing data, making calculations, etc.

computerize *v.* equip with or perform or operate by computer. ▫ **computerization** *n.*

comrade *n.* companion, associate. ▫ **comradeship** *n.*

■ associate, colleague, companion, compatriot, crony, partner.

con[1] (*sl.*) *v.* (**conned**) swindle, deceive. ● *n.* confidence trick.

con[2] *v.* (**conned**) direct the steering of (a ship).

con[3] *see* **pro and con.**

concatenation *n.* combination.

concave *adj.* curved like the inner surface of a ball.

conceal *v.* hide, keep secret. ▫ **concealment** *n.*

■ camouflage, cloak, cover (up), disguise, hide, keep secret, mask; bury, secrete, tuck away; gloss over, whitewash.

concede *v.* **1** admit to be true. **2** admit defeat. **3** grant.

■ **1** accept, acknowledge, admit, allow, confess, grant, own, recognize. **2** capitulate, give in, submit, surrender, throw in the towel, yield.

conceit *n.* too much pride in oneself. ▫ **conceited** *adj.*

■ arrogance, egotism, narcissism, pride, self-admiration, self-love, vanity. ▫ **conceited** arrogant, bumptious, cocky, egotistical, narcissistic, proud, self-important, self-satisfied, smug, *colloq.* stuck-up, vain.

conceive *v.* **1** become pregnant. **2** form (a plan etc.). ▫ **conceive of** imagine. **conceivable** *adj.*

■ **2** contrive, design, devise, dream up, form, hatch, plan, *colloq.* think up. ▫ **conceive of** envisage, imagine, think of.

concentrate *v.* **1** employ all one's attention or effort. **2** bring together. **3** make less dilute. ● *n.* concentrated substance.

■ *v.* **1** apply oneself, focus, think. **2** cluster, collect, congregate, gather, group. **3** condense, distil.

concentration *n.* **1** concentrating. **2** concentrated thing. ▫ **concentration camp** camp for political prisoners in Nazi Germany.

concentric *adj.* having the same centre. ▫ **concentrically** *adv.*

concept *n.* idea, general notion.

conception *n.* **1** conceiving. **2** idea.

conceptual *adj.* of concepts.

conceptualize *v.* form a concept of. ▫ **conceptualization** *n.*

concern *v.* **1** be relevant or important to. **2** worry. ● *n.* **1** matter of interest or importance. **2** care, consideration. **3** anxiety. **4** business. ▫ **concern oneself** interest or involve oneself.

■ *v.* **1** affect, have a bearing on, interest, involve, matter to, pertain to, refer to, relate to. **2** bother, disturb, perturb, trouble, upset, worry. ● *n.* **1** affair, business, duty, problem, responsibility, task. **2** attention, care, consideration, heed, regard, thought. **3** anxiety, disquiet, distress, solicitude, uneasiness, worry. **4** business, company, establishment, firm, house, organization.

concerned *adj.* anxious.

■ anxious, bothered, distressed, perturbed, troubled, uneasy, upset, worried.

concerning *prep.* with reference to.

■ about, apropos, as regards, re, regarding, relating to, with reference to, with regard to, with respect to.

concert *n.* musical entertainment.

concerted *adj.* done in combination.

concertina *n.* portable musical instrument with bellows and keys.

concerto /kənchάirtō/ *n.* musical composition for solo instrument and orchestra.

concession *n.* **1** conceding. **2** thing conceded. **3** special privilege. **4** right granted.

conch *n.* spiral shell.

conciliate *v.* **1** soothe the hostility of. **2** reconcile. ◻ **conciliation** *n.*, **conciliatory** *adj.*

concise *adj.* brief and comprehensive. ◻ **concisely** *adv.*, **conciseness** *n.*

■ brief, compact, compendious, laconic, pithy, short, succinct, terse; abbreviated, abridged, shortened.

conclude *v.* **1** end. **2** infer. **3** settle finally.

■ **1** close, complete, end, finish, halt, stop, terminate, wind up. **2** assume, deduce, draw the conclusion, gather, infer, judge, presume, suppose, surmise, understand. **3** clinch, negotiate, pull off, settle.

conclusion *n.* **1** ending. **2** opinion reached.

■ **1** cessation, close, completion, end, finish, termination; denouement, ending, finale. **2** decision, deduction, inference, judgement, verdict.

conclusive *adj.* ending doubt, convincing. ◻ **conclusively** *adv.*

■ certain, convincing, decisive, definite, incontrovertible, indisputable, irrefutable, undeniable, unequivocal, unquestionable.

concoct *v.* **1** prepare from ingredients. **2** invent. ◻ **concoction** *n.*

concomitant *adj.* accompanying.

concord *n.* agreement, harmony.

concordance *n.* **1** agreement. **2** index of words.

concordant *adj.* being in concord.

concourse *n.* **1** crowd. **2** open area at a railway station etc.

concrete *n.* mixture of gravel and cement etc. used for building. ● *adj.* **1** existing in material form. **2** definite. ● *v.* cover with or embed in concrete.

■ *adj.* actual, definite, genuine, material, physical, real, substantial, tangible.

concretion *n.* solidified mass.

concubine *n.* woman who lives with a man as his wife.

concur *v.* (**concurred**) **1** agree in opinion. **2** happen together, coincide. ◻ **concurrence** *n.*, **concurrent** *adj.*, **concurrently** *adv.*

concuss *v.* affect with concussion.

concussion *n.* injury to the brain caused by a hard blow.

condemn *v.* **1** express strong disapproval of. **2** doom. **3** convict. **4** sentence. **5** declare unfit for use. ◻ **condemnation** *n.*

■ **1** blame, censure, criticize, decry, denounce, disparage, rebuke, reprove, revile, *colloq.* slam, *colloq.* slate, upbraid. **2** destine, doom, fate, ordain.

condense *v.* **1** make denser or briefer. **2** change from gas or vapour to liquid. ◻ **condensation** *n.*

condescend *v.* **1** consent to do something less dignified or fitting than is usual. **2** pretend to be on equal terms with (an inferior). ◻ **condescension** *n.*

■ **1** deign, demean oneself, lower oneself, stoop.

condiment *n.* seasoning for food.

condition *n.* **1** thing that must exist if something else is to exist or occur. **2** state of being. **3** ailment. **4** (*pl.*) circumstances. ● *v.* **1** bring to the desired condition. **2** have a strong effect on. **3** accustom. ◻ **conditioner** *n.*

■ *n.* **1** prerequisite, proviso, requirement, requisite, stipulation. **2** fitness, form, health, shape, state. **3** ailment, complaint, disease, disorder, illness. **4** (**conditions**) circumstances, environment, surroundings. ● *v.* **1** adapt, get ready, make ready, modify, prepare. **2** affect, determine, govern, influence, shape. **3** acclimatize, accustom, habituate, inure.

conditional *adj.* subject to specified conditions. ◻ **conditionally** *adv.*

condole *v.* express sympathy. ◻ **condolence** *n.*

condom *n.* contraceptive sheath.

condone *v.* forgive, overlook.

conducive *adj.* helping to cause or produce.

conduct *v.* /kəndúkt/ **1** lead, guide. **2** manage. **3** transmit (heat or electricity). **4** be the conductor of. ● *n.* /kóndukt/ **1** behaviour. **2** way of conducting business, war, etc.

■ *v.* **1** escort, guide, lead, show, usher. **2** administer, control, direct, handle, manage, operate, run, supervise. **3** carry, channel, convey, transmit. ● *n.* **1** actions, attitude, behaviour, demeanour, deportment, manners. **2** administration, control, direction, government, handling, management, regulation, supervision.

conduction *n.* conducting of heat or electricity. ◻ **conductive** *adj.*, **conductivity** *n.*

conductor *n.* **1** director of orchestra etc. **2** thing that conducts heat or electricity.

conduit *n.* **1** pipe or channel for liquid. **2** tube protecting wires.

cone *n.* **1** tapering object with a circular base. **2** cone-shaped thing. **3** dry fruit of pine or fir.

confection *n.* prepared dish or delicacy.

confectioner *n.* maker or seller of confectionery.

confectionery *n.* sweets, cakes, and pastries.

confederacy *n.* league of states.

confederate *adj.* joined by treaty or agreement. ● *n.* **1** member of a confederacy. **2** accomplice.

confederation *n.* union of states, people, or organizations.

confer *v.* (**conferred**) **1** grant. **2** hold a discussion. ◻ **conferment** *n.*

■ **1** award, bestow, give, grant, present. **2** consult, converse, negotiate, parley, talk.

conference *n.* meeting for discussion.

confess *v.* **1** acknowledge, admit. **2** declare one's sins to a priest.

■ **1** acknowledge, admit, concede, own (up to); disclose, divulge, reveal.

confession *n.* **1** act of confessing. **2** statement of principles.

confessional *n.* enclosed stall in a church for hearing confessions.

confessor *n.* priest who hears confessions and gives counsel.

confetti *n.* bits of coloured paper thrown at a bride and groom.

confidant *n.* (*fem.* **confidante**) person one confides in.

confide *v.* **1** tell or talk confidentially. **2** entrust.

confidence *n.* **1** firm trust. **2** feeling of certainty, trust in one's own ability. **3** thing told confidentially. ◻ **confidence trick** swindle worked by gaining a person's trust.

■ **1** belief, faith, trust. **2** certainty, certitude, conviction; aplomb, assurance, courage, nerve, panache, self-assurance, self-possession, self-reliance.

confident *adj.* feeling confidence. ◻ **confidently** *adv.*

■ assured, certain, convinced, positive, sure; bold, cocksure, cool, courageous, fearless, self-assured, self-possessed, unafraid.

confidential *adj.* **1** to be kept secret. **2** entrusted with secrets. ◻ **confidentially** *adv.*, **confidentiality** *n.*

configuration *n.* shape, outline.

confine *v.* **1** keep within limits. **2** keep shut up.

■ **1** limit, restrict. **2** cage, coop up, immure, imprison, shut in.

confinement *n.* **1** confining, being confined. **2** childbirth.

confines *n.pl.* boundaries.

confirm *v.* make firmer or definite. ◻ **confirmatory** *adj.*

■ authenticate, back up, corroborate, prove, reinforce, ratify, strengthen, substantiate, support, validate, verify.

confirmation *n.* **1** confirming. **2** thing that confirms.

confiscate *v.* take or seize by authority. ◻ **confiscation** *n.*

■ appropriate, commandeer, expropriate, impound, remove, seize, sequestrate, take away.

conflagration *n.* great fire.

conflate *v.* blend or fuse together. ◻ **conflation** *n.*

conflict *n.* /kónflikt/ **1** state of opposition. **2** fight, struggle. ● *v.* /kənflíkt/ be in disagreement.

■ *n.* **1** battle, combat, war; antagonism, disagreement, discord, friction, opposition. **2** affray, fracas, fray, skirmish, struggle; altercation, dispute, feud, quarrel, row, squabble, wrangle. ● *v.* be at odds, be at variance, be incompatible, clash, differ, disagree.

confluence *n.* place where two rivers unite. ◻ **confluent** *adj.*

conform *v.* comply with rules or general custom. ◻ **conformity** *n.*

■ comply, keep in step, obey, toe the line.

conformist *n.* person who conforms to rules or custom. ◻ **conformism** *n.*

confound *v.* **1** astonish and perplex. **2** confuse.

confront *v.* **1** be, come, or bring face to face with. **2** face boldly. ◻ **confrontation** *n.*

confuse *v.* **1** bewilder. **2** mix up. **3** make unclear. ◻ **confusion** *n.*

■ **1** baffle, bemuse, bewilder, confound, daze, discomfit, disconcert, floor, *colloq.* flummox, fluster, mystify, nonplus, perplex, puzzle, *colloq.* stump, *colloq.* throw. **2** disarrange, disorder, jumble, mess up, mix up, muddle. **3** blur, cloud, obscure. ◻ **confusion** bemusement, bewilderment, discomfiture, mystification, perplexity, puzzlement; chaos, disarray, disorder, jumble, mess, muddle, shambles, turmoil.

confute *v.* prove wrong. □ **confutation** *n.*

conga *n.* dance in which people form a long winding line.

congeal *v.* coagulate, solidify.

■ clot, coagulate, curdle, harden, set, solidify, stiffen, thicken.

congenial *adj.* pleasant, agreeable. □ **congenially** *adv.*

■ agreeable, amiable, friendly, genial, likeable, nice, pleasant, pleasing, sympathetic.

congenital *adj.* being so from birth. □ **congenitally** *adv.*

conger *n.* large sea eel.

congest *v.* make abnormally full. □ **congestion** *n.*

conglomerate *adj.* /kənglómmərət/ gathered into a mass. ● *n.* /kənglómmərət/ coherent mass. ● *v.* /kənglómmərayt/ collect into a coherent mass. □ **conglomeration** *n.*

congratulate *v.* tell (a person) that one admires his or her success. □ **congratulation** *n.*, **congratulatory** *adj.*

■ applaud, compliment, felicitate, praise.

congregate *v.* flock together.

■ assemble, cluster, collect, convene, converge, gather, mass, meet, rally, swarm, throng.

congregation *n.* people assembled at a church service.

congress *n.* **1** formal meeting of delegates for discussion. **2** (**Congress**) law-making assembly, esp. of the USA. □ **congressional** *adj.*

congruent *adj.* **1** suitable, consistent. **2** having exactly the same shape and size. □ **congruence** *n.*

conic *adj.* of a cone.

conical *adj.* cone-shaped.

conifer *n.* tree bearing cones. □ **coniferous** *adj.*

conjecture *n.* & *v.* guess.

conjugal *adj.* of marriage.

conjunction *n.* word such as 'and' or 'or' that connects others.

conjunctivitis *n.* inflammation of the membrane (**conjunctiva**) connecting eyeball and eyelid.

conjure *v.* do sleight-of-hand tricks. □ **conjuror** *n.*

conk *n.* (*sl.*) nose, head. ● *v.* (*sl.*) hit. □ **conk out** (*colloq.*) break down.

connect *v.* **1** join, be joined. **2** associate mentally. **3** (of a train etc.) arrive so that passengers are in time to catch another. □ **connection** *n.*, **connective** *adj.*, **connector** *n.*

■ **1** attach, combine, couple, fasten, fit together, fix, join, link, put together, tie, unite. **2** associate, link, relate, tie in. □ **connection** association, bond, link, linkage, relationship, tie, tie-up, union.

connive *v.* **connive at** tacitly consent to. □ **connivance** *n.*

connoisseur /kónnəsőr/ *n.* person with expert understanding.

connote *v.* imply in addition to its basic meaning. □ **connotation** *n.*

conquer *v.* overcome in war or by effort. □ **conqueror** *n.*

■ beat, crush, defeat, get the better of, *colloq.* lick, master, overpower, overthrow, rout, subjugate, thrash, triumph over, trounce, vanquish; overcome, surmount.

conquest *n.* **1** conquering. **2** thing won by conquering.

conscience *n.* person's sense of right and wrong.

conscientious *adj.* diligent, careful. □ **conscientiously** *adv.*, **conscientiousness** *n.*

■ attentive, careful, diligent, meticulous, painstaking, punctilious, rigorous, scrupulous, sedulous, thorough.

conscious *adj.* **1** with mental faculties awake. **2** intentional. **3** aware. □ **consciously** *adv.*, **consciousness** *n.*

■ **1** alert, awake. **2** calculated, deliberate, intentional, purposeful, studied, wilful.

conscript *v.* /kənskrípt/ summon for compulsory military service. ● *n.* /kónskript/ conscripted person. □ **conscription** *n.*

consecrate *v.* **1** make sacred. **2** dedicate to the service of God. □ **consecration** *n.*

consecutive *adj.* following continuously. □ **consecutively** *adv.*

consensus *n.* general agreement.

consent *v.* say one is willing to do or allow what is asked. ● *n.* agreement, permission.

■ *v.* accede, acquiesce, agree, comply, concede, concur. ● *n.* acquiescence, agreement, approval, assent, authorization, compliance, concurrence, go-ahead, *colloq.* OK, permission.

consequence *n.* **1** result. **2** importance.

■ **1** effect, outcome, repercussion, result, upshot. **2** account, import, importance, moment, note, significance.

consequent *adj.* resulting.

consequential *adj.* **1** consequent. **2** important. **3** self-important.

consequently *adv.* as a result.

conservancy *n.* commission controlling a river etc.

conservation *n.* conserving, esp. of the natural environment. ▫ **conservationist** *n.*

> ■ care, maintenance, preservation, protection, safeguarding, upkeep.

conservative *adj.* **1** opposed to change. **2** (of an estimate) purposely low. ● *n.* conservative person. ▫ **conservatism** *n.*

> ■ **1** conventional, hidebound, old-fashioned, orthodox, reactionary, traditional, unadventurous. **2** cautious, tentative.

conservatory *n.* greenhouse built on to a house.

conserve[1] /kənsérv/ *v.* keep from harm, decay, or loss.

> ■ husband, keep, preserve, save, store up; maintain, protect, safeguard, take care of.

conserve[2] /kónserv/ *n.* jam made from fresh fruit and sugar.

consider *v.* **1** think about, esp. in order to decide. **2** be of the opinion. **3** allow for.

> ■ **1** contemplate, deliberate, mull over, ponder, reflect (on), ruminate (on), study, think about. **2** believe, deem, judge, reckon, regard, think. **3** allow for, take into account *or* consideration, respect.

considerable *adj.* fairly great in amount or importance. ▫ **considerably** *adv.*

> ■ appreciable, large, respectable, sizeable, substantial; distinguished, important, notable, noteworthy.

considerate *adj.* careful not to hurt or inconvenience others. ▫ **considerately** *adv.*

> ■ caring, courteous, gracious, helpful, kind, kindly, neighbourly, obliging, polite, solicitous, sympathetic, tactful, thoughtful, unselfish.

consideration *n.* **1** careful thought. **2** being considerate. **3** fact that must be kept in mind. **4** payment given as a reward.

considering *prep.* taking into account.

consign *v.* **1** deposit, entrust. **2** send (goods etc.).

consignee *n.* person to whom goods are sent.

consignment *n.* **1** consigning. **2** batch of goods.

consist *v.* **consist of** be composed of.

consistency *n.* **1** being consistent. **2** degree of thickness or solidity.

consistent *adj.* **1** unchanging. **2** not contradictory. ▫ **consistently** *adv.*

> ■ **1** constant, dependable, predictable, reliable, steady, unchanging, undeviating. **2** compatible, consonant, harmonious, in accordance, in agreement, in harmony, of a piece.

consolation *n.* **1** consoling. **2** thing that consoles.

console[1] /kənsṓl/ *v.* comfort in time of sorrow.

> ■ cheer (up), comfort, hearten, reassure, solace, soothe.

console[2] /kónsōl/ *n.* **1** bracket supporting a shelf. **2** panel for switches, controls, etc.

consolidate *v.* **1** combine. **2** make or become secure and strong. ▫ **consolidation** *n.*

consommé /kənsómmay/ *n.* clear soup.

consonant *n.* **1** letter other than a vowel. **2** sound it represents. ● *adj.* consistent, harmonious.

consort *n.* /kónsort/ husband or wife, esp. of a monarch. ● *v.* /kənsórt/ keep company.

consortium *n.* (*pl.* **-tia**) combination of firms acting together.

conspicuous *adj.* easily seen, attracting attention. ▫ **conspicuously** *adv.*

> ■ apparent, blatant, clear, evident, flagrant, glaring, manifest, marked, noticeable, obtrusive, obvious, perceptible, prominent, pronounced, salient, striking, unmistakable, visible.

conspiracy *n.* **1** secret plan to commit a crime or do harm. **2** act of conspiring.

> ■ **1** intrigue, machination, plot, scheme.

conspirator *n.* one who conspires. ▫ **conspiratorial** *adj.*

conspire *v.* **1** plan secretly against others. **2** (of events) seem to combine.

constable *n.* policeman or policewoman of the lowest rank.

constabulary *n.* police force.

constancy *n.* **1** quality of being unchanging. **2** faithfulness.

constant *adj.* **1** happening repeatedly or all the time. **2** unchanging. **3** faithful. ● *n.* unvarying quantity. ▫ **constantly** *adv.*

> ■ *adj.* **1** ceaseless, continual, continuous, endless, incessant, non-stop, perpetual, persistent, steady, unending, uninterrup-

ted, unremitting. **2** fixed, invariable, unchanging, uniform, unvarying. **3** dependable, devoted, faithful, loyal, reliable, staunch, steadfast, true, trustworthy, trusty, unswerving.

constellation *n.* group of stars.

consternation *n.* great surprise and anxiety.

constipation *n.* difficulty in emptying the bowels.

constituency *n.* **1** group of voters who elect a representative. **2** area represented in this way.

constituent *adj.* forming part of a whole. ● *n.* **1** constituent part. **2** member of a constituency.

constitute *v.* be the parts of.

constitution *n.* **1** principles by which a state is organized. **2** bodily condition.

constitutional *adj.* in accordance with a constitution. ● *n.* walk taken for exercise.

constrain *v.* compel, oblige.

constraint *n.* **1** constraining. **2** restriction. **3** strained manner.

constrict *v.* make narrow or tight, squeeze. □ **constriction** *n.*

construct *v.* fit together, build. ● *n.* thing constructed, esp. in the mind. □ **constructor** *n.*

■ *v.* assemble, build, create, erect, fabricate, fashion, frame, make, put up, set up.

construction *n.* **1** constructing. **2** thing constructed. **3** words put together to form a phrase. **4** interpretation.

constructive *adj.* helpful, useful. □ **constructively** *adv.*

■ beneficial, helpful, positive, practical, useful, valuable, worthwhile.

construe *v.* interpret.

consul *n.* official representative of a state in a foreign city. □ **consular** *adj.*

consulate *n.* consul's position or premises.

consult *v.* seek information or advice from. □ **consultation** *n.*

■ ask, confer with, deliberate with, refer to, speak to, talk to.

consultant *n.* specialist consulted for professional advice. □ **consultancy** *n.*

consultative *adj.* **1** of or for consultation. **2** advisory.

consume *v.* **1** eat or drink. **2** use up. **3** destroy.

■ **1** devour, drink, eat, gobble, guzzle, *colloq.* scoff. **2** deplete, drain, exhaust, expend, go through, use up. **3** destroy, devastate, gut, lay waste, ravage, ruin.

consumer *n.* person who buys or uses goods or services.

consummate *v.* accomplish, complete (esp. marriage by sexual intercourse). □ **consummation** *n.*

consumption *n.* **1** consuming. **2** (*old use*) tuberculosis.

consumptive *adj.* & *n.* (person) suffering from tuberculosis.

contact *n.* **1** touching, communication. **2** electrical connection. **3** one who may be contacted for information or help. ● *v.* get in touch with. □ **contact lens** very small lens worn in the eye.

■ *v.* get in touch with, reach, ring (up), speak to, telephone, write to.

contagion *n.* spreading of disease by contact. □ **contagious** *adj.*

contain *v.* **1** have within itself. **2** include. **3** control, restrain.

■ **1** carry, hold. **2** comprise, consist of, include. **3** check, control, curb, hold back, repress, restrain, stifle, suppress.

container *n.* receptacle, esp. to transport goods.

containment *n.* prevention of hostile expansion.

contaminate *v.* pollute. □ **contamination** *n.*

■ adulterate, corrupt, debase, defile, infect, poison, pollute, soil, spoil, taint.

contemplate *v.* **1** survey with the eyes or the mind. **2** consider as a possibility, intend. □ **contemplation** *n.*

■ **1** eye, gaze at, look at, observe, scrutinize, survey, view; brood on, cogitate on, consider, meditate on, mull over, muse on, ponder on, reflect on, ruminate on, think about. **2** consider, intend, plan, propose, think about *or* of.

contemplative *adj.* **1** meditative. **2** of religious meditation.

contemporaneous *adj.* existing or occurring at the same time.

contemporary *adj.* **1** of the same period or age. **2** modern in style. ● *n.* person of the same age.

■ *adj.* **1** coeval, coexistent, concurrent. **2** fashionable, in, modern, new, stylish, *colloq.* trendy, up to date.

contempt *n.* **1** feeling of despising a person or thing. **2** the condition of being despised. **3** disrespect. □ **contemptible** *adj.*

> ■ **1** abhorrence, disdain, disgust, hatred, loathing, odium, scorn. □ **contemptible** dastardly, despicable, disgraceful, dishonourable, low, shabby, shameful, wretched.

contemptuous *adj.* showing contempt. □ **contemptuously** *adv.*

> ■ derisive, disdainful, insolent, insulting, scornful, sneering, *colloq.* snooty, supercilious.

contend *v.* **1** strive, compete. **2** assert. □ **contender** *n.*

> ■ **1** compete, contest, strive, struggle, vie. **2** affirm, allege, argue, assert, claim, maintain.

content[1] /kəntént/ *adj.* satisfied. ● *n.* being content. ● *v.* make content. □ **contented** *adj.*, **contentment** *n.*

> ■ *adj.* contented, happy, glad, gratified, pleased, satisfied. ● *v.* gladden, gratify, please, satisfy. □ **contentment** gratification, happiness, pleasure, satisfaction.

content[2] /kóntent/ *n.* what is contained in something.

contention *n.* **1** contending. **2** assertion made in argument.

contentious *adj.* **1** quarrelsome. **2** likely to cause contention.

contest *v.* /kəntést/ **1** compete for or in. **2** dispute. ● *n.* /kóntest/ **1** struggle for victory. **2** competition. □ **contestant** *n.*

> ■ *v.* **1** compete for, contend for, fight for, vie for. **2** argue against, challenge, dispute, oppose, query, question. ● *n.* **1** battle, conflict, fight, struggle. **2** championship, competition, game, match, tournament.

context *n.* **1** what precedes or follows a word or statement. **2** circumstances. □ **contextual** *adj.*

contiguous *adj.* adjacent.

continent *n.* one of the main land masses of the earth. □ **continental** *adj.*

contingency *n.* **1** something unforeseen. **2** thing that may occur.

contingent *adj.* **1** fortuitous. **2** possible but not certain. **3** conditional. ● *n.* group of troops etc. contributed to a larger group.

continual *adj.* constantly or frequently recurring. □ **continually** *adv.*

> ■ ceaseless, constant, continuous, endless, *colloq.* eternal, everlasting, incessant, non-stop, perennial, perpetual, persistent, recurrent, repeated, steady, unbroken, unceasing, unending, uninterrupted, unremitting.

continuance *n.* continuing.

continue *v.* **1** not cease. **2** remain in existence. **3** resume. □ **continuation** *n.*

> ■ **1** carry on (with), keep up, maintain, persevere (in), persist (in), proceed (with), pursue, sustain. **2** endure, go on, last, remain. **3** carry on (with), resume, return to, take up.

continuous *adj.* without interval, uninterrupted. □ **continuously** *adv.*, **continuity** *n.*

> ■ ceaseless, constant, continual, endless, incessant, interminable, non-stop, unbroken, unceasing, uninterrupted, unremitting.

continuum *n.* (*pl.* **-tinua**) continuous thing.

contort *v.* force or twist out of normal shape. □ **contortion** *n.*

contortionist *n.* performer who adopts contorted postures.

contour *n.* **1** outline. **2** line on a map showing height above sea level.

contra- *pref.* against.

contraband *n.* smuggled goods.

contraception *n.* prevention of conception, birth control.

contraceptive *adj.* & *n.* (drug or device) preventing conception.

contract *n.* /kóntrakt/ formal agreement. ● *v.* /kəntrákt/ **1** enter into an agreement. **2** catch (an illness). **3** make or become smaller or shorter. □ **contraction** *n.*, **contractor** *n.*, **contractual** *adj.*

> ■ *n.* agreement, arrangement, bargain, compact, deal, pact. ● *v.* **1** agree, covenant, promise, undertake. **2** catch, come *or* go down with, develop, get. **3** draw together, narrow, reduce, shrink.

contradict *v.* **1** deny, oppose verbally. **2** be contrary to. □ **contradiction** *n.*, **contradictory** *adj.*

contraflow *n.* flow (esp. of traffic) in a direction opposite to and alongside the usual flow.

contralto *n.* lowest female voice.

contraption *n.* strange device or machine.

contrapuntal *adj.* of or in counterpoint.

contrariwise *adv.* **1** on the other hand. **2** in the opposite way.

contrary /kóntrəri/ *adj.* **1** opposite in nature, tendency, or direction. **2** /kəntráiri/ perverse. ● *n.* the opposite.

● *adv.* in opposition. □ **on the contrary** the opposite is true.

■ *adj.* **1** conflicting, contradictory, different, opposed, opposing, opposite. **2** awkward, *colloq.* cussed, difficult, obstinate, perverse, refractory, self-willed, stubborn, uncooperative, unhelpful.

contrast *n.* /kóntraast/ difference shown by comparison. ● *v.* /kəntraást/ **1** compare to reveal contrast. **2** show contrast.

■ *n.* comparison, juxtaposition; difference, disparity, dissimilarity, distinction. ● *v.* **1** compare, juxtapose; differentiate, distinguish. **2** conflict, differ, diverge.

contravene *v.* break (a rule etc.). □ **contravention** *n.*

contretemps /káwntrətoN/ *n.* unfortunate happening.

contribute *v.* **1** give to a common fund or effort. **2** help to bring about. □ **contribution** *n.*, **contributor** *n.*, **contributory** *adj.*

■ **1** bestow, *colloq.* chip in, donate, give, grant, present, provide, supply.

contrite *adj.* **1** penitent. **2** sorry. □ **contritely** *adv.*, **contrition** *n.*

contrivance *n.* **1** contriving. **2** contrived thing, device.

contrive *v.* plan, make, or do something resourcefully.

control *n.* **1** power to give orders or restrain. **2** means of restraining or regulating. **3** standard of comparison for checking results of an experiment. ● *v.* (**controlled**) **1** have control of, regulate. **2** restrain.

■ *n.* **1** authority, charge, command, direction, guidance, jurisdiction, leadership, management, mastery, power, rule, supervision, sway; restraint, self-restraint. **2** brake, check, curb. ● *v.* **1** be in charge (of), command, conduct, direct, dominate, govern, guide, hold sway over, lead, manage, oversee, regulate, rule, run, superintend, steer, supervise. **2** check, contain, curb, hold back, keep in check, master, repress, restrain, subdue.

controversial *adj.* causing controversy. □ **controversially** *adv.*

controversy *n.* prolonged dispute.

■ argument, debate, disagreement, dispute, quarrel, wrangle.

controvert *v.* deny the truth of.

contusion *n.* bruise.

conundrum *n.* riddle, puzzle.

conurbation *n.* large urban area formed where towns have spread and merged.

convalesce *v.* regain health after illness. □ **convalescence** *n.*, **convalescent** *adj.* & *n.*

■ get better, improve, recover, recuperate, regain strength.

convection *n.* transmission of heat within a liquid or gas by movement of heated particles.

convene *v.* assemble.

convenience *n.* **1** being convenient. **2** convenient thing. **3** lavatory.

convenient *adj.* **1** serving one's comfort or interests. **2** well situated. □ **conveniently** *adv.*

■ **1** expedient, helpful, suitable, useful. **2** accessible, available, at hand, handy, nearby, well situated, within reach.

convent *n.* residence of a community of nuns.

convention *n.* **1** accepted custom. **2** assembly. **3** formal agreement.

■ **1** custom, practice, rule, tradition, usage. **2** assembly, conference, congress, convocation, gathering, meeting, symposium.

conventional *adj.* **1** depending on or according with convention. **2** bound by social conventions. **3** usual. □ **conventionally** *adv.*

■ **1** agreed, established, orthodox, traditional. **2** conservative, old-fashioned, strait-laced, *colloq.* stuffy. **3** customary, everyday, habitual, normal, ordinary, standard, usual.

converge *v.* come to or towards the same point. □ **convergence** *n.*, **convergent** *adj.*

conversant *adj.* **conversant with** having knowledge of.

conversation *n.* informal talk between people. □ **conversational** *adj.*

■ chat, dialogue, discourse, discussion, gossip, *colloq.* natter, talk, tête-à-tête.

converse[1] *v.* /kənvérss/ talk.

converse[2] /kónverss/ *adj.* opposite, contrary. ● *n.* converse idea or statement. □ **conversely** *adv.*

convert *v.* /kənvért/ **1** change from one form or use to another. **2** cause to change an attitude or belief. ● *n.* /kónvert/ person converted, esp. to a religious faith. □ **conversion** *n.*

■ *v.* **1** alter, change, modify, remodel, transform, transmute.

convertible *adj.* able to be converted. ● *n.* car with a folding or detachable roof.

convex *adj.* curved like the outer surface of a ball. ◻ **convexity** *n.*

convey *v.* **1** carry, transport, transmit. **2** communicate (meaning etc.).

■ **1** bear, bring, carry, deliver, ferry, send, ship, take, transfer, transport; conduct, transmit. **2** communicate, get *or* put across, impart, make known.

conveyance *n.* **1** conveying. **2** means of transport, vehicle.

conveyancing *n.* branch of law dealing with the transfer of property.

conveyor *n.* **1** person or thing that conveys. **2** continuous moving belt conveying objects.

convict *v.* /kənvíkt/ prove or declare guilty. ● *n.* /kónvikt/ sentenced criminal.

conviction *n.* **1** convicting. **2** firm opinion.

convince *v.* make (a person) feel certain that something is true.

convivial *adj.* sociable and lively.

convocation *n.* **1** convoking. **2** assembly convoked.

convoke *v.* summon to assemble.

convoluted *adj.* coiled, twisted.

convolution *n.* coil, twist.

convolvulus *n.* twining plant with trumpet-shaped flowers.

convoy *n.* ships or vehicles travelling together.

convulse *v.* cause violent movement or a fit of laughter in.

convulsion *n.* **1** violent involuntary movement of the body. **2** upheaval. ◻ **convulsive** *adj.*

coo *v.* make a soft murmuring sound like a dove. ● *n.* this sound.

cook *v.* **1** prepare (food) by heating. **2** undergo this process. **3** (*colloq.*) falsify (accounts etc.). ● *n.* person who cooks, esp. as a job. ◻ **cook up** (*colloq.*) concoct.

cooker *n.* stove for cooking food.

cookery *n.* art of cooking.

cookie *n.* (*US*) sweet biscuit.

cool *adj.* **1** fairly cold. **2** calm. **3** not enthusiastic. **4** unfriendly. ● *n.* **1** coolness. **2** (*sl.*) calmness. ● *v.* make or become cool. ◻ **coolly** *adv.*, **coolness** *n.*

■ *adj.* **1** chilly, cold, *colloq.* nippy. **2** calm, collected, composed, imperturbable, level-headed, phlegmatic, quiet, relaxed, self-possessed, serene, unemotional, unexcited, *colloq.* unflappable, unflustered, unruffled. **3** apathetic, half-hearted, lukewarm, unenthusiastic, uninterested. **4** aloof, cold, detached, distant, frosty, standoffish, unfriendly, unsociable, unwelcoming.

coolant *n.* fluid for cooling machinery.

coomb /ko͞om/ *n.* valley.

coop *n.* cage for poultry. ● *v.* confine, shut in.

co-op *n.* (*colloq.*) **1** cooperative society. **2** shop run by this.

cooper *n.* person who makes or repairs casks and barrels.

cooperate *v.* work or act together. ◻ **cooperation** *n.*

■ collaborate, join forces, team up, unite, work together.

cooperative *adj.* **1** willing to help. **2** based on economic cooperation. ● *n.* farm or firm etc. run on this basis.

■ *adj.* **1** accommodating, amenable, considerate, helpful, obliging, willing.

co-opt *v.* appoint to a committee by invitation of existing members, not election.

coordinate[1] /kō-órdinət/ *adj.* equal in importance. ● *n.* either of two numbers or letters used to give the position of a point on a graph or map.

coordinate[2] /kō-órdinayt/ *v.* cause to function together efficiently. ◻ **coordination** *n.*, **coordinator** *n.*

coot *n.* a kind of waterbird.

cop (*sl.*) *n.* **1** police officer. **2** capture. ● *v.* (**copped**) catch.

cope *v.* deal effectively, manage.

■ get by, make do, manage, muddle through, survive.

copier *n.* copying machine.

coping *n.* sloping top row of masonry in a wall.

copious *adj.* plentiful. ◻ **copiously** *adv.*

■ abundant, ample, bountiful, generous, lavish, liberal, luxuriant, overflowing, plentiful, profuse, unstinting, voluminous.

copper[1] *n.* **1** reddish-brown metal. **2** coin containing this. **3** its colour. ● *adj.* made of copper.

copper[2] *n.* (*sl.*) policeman.

coppice *n.* (also **copse**) group of small trees and undergrowth.

copulate *v.* have sexual intercourse. ◻ **copulation** *n.*

copy *n.* **1** thing made to look like another. **2** specimen of a book etc. ● *v.* **1** make a copy of. **2** imitate.

■ *n.* **1** duplicate, facsimile, imitation, likeness, replica, reproduction, transcript; carbon copy, photocopy; counterfeit, fake, forgery. ● *v.* **1** duplicate, replicate, reproduce, transcribe. **2** ape, echo, imitate, impersonate, mimic.

copyright *n.* sole right to publish a work. ● *v.* secure copyright for.

coquette *n.* woman who flirts. □ **coquettish** *adj.*, **coquetry** *n.*

coracle *n.* small wicker boat.

coral *n.* **1** hard red, pink, or white substance built by tiny sea creatures. **2** reddish-pink.

cor anglais /kór óngglay/ woodwind instrument like the oboe but lower in pitch.

corbel *n.* stone or wooden support projecting from a wall.

cord *n.* **1** long thin flexible material made from twisted strands. **2** piece of this. **3** corduroy.

■ **1** cable, flex, rope, string, twine.

cordial *adj.* warm and friendly. ● *n.* fruit-flavoured drink. □ **cordially** *adv.*

■ *adj.* affable, amiable, courteous, friendly, genial, hospitable, pleasant, polite, warm.

cordon *n.* **1** line or circle of police etc. **2** fruit tree pruned to grow as a single stem. ● *v.* enclose by a cordon.

cordon bleu /kórdon blő/ of the greatest excellence in cookery.

corduroy *n.* cloth with velvety ridges.

core *n.* **1** central or most important part. **2** horny central part of an apple etc., containing seeds. ● *v.* remove the core from.

■ *n.* **1** centre, crux, essence, heart, kernel, *sl.* nitty-gritty, nub, quintessence.

co-respondent *n.* person with whom the respondent in a divorce suit is said to have committed adultery.

corgi *n.* dog of a small Welsh breed with short legs.

coriander *n.* plant whose leaves and seeds are used for flavouring.

cork *n.* **1** light tough bark of a Mediterranean oak. **2** piece of this used as a float. **3** bottle stopper. ● *v.* stop up with a cork.

corkage *n.* restaurant's charge for serving wine.

corkscrew *n.* **1** tool for extracting corks from bottles. **2** spiral thing.

corm *n.* bulb-like underground stem from which buds grow.

cormorant *n.* large black seabird.

corn[1] *n.* **1** wheat, oats, or maize. **2** its grain.

corn[2] *n.* small area of horny hardened skin, esp. on the foot.

corncrake *n.* bird with a harsh cry.

cornea *n.* transparent outer covering of the eyeball.

cornelian *n.* reddish or white semiprecious stone.

corner *n.* **1** angle or area where two lines, sides, or streets meet. **2** free kick or hit from the corner of the field in football or hockey. ● *v.* **1** force into a position from which there is no escape. **2** obtain a monopoly of. **3** drive fast round a corner.

cornerstone *n.* **1** basis. **2** vital foundation.

cornet *n.* **1** brass instrument like a small trumpet. **2** cone-shaped wafer holding ice cream.

cornflour *n.* flour made from maize.

cornflower *n.* blue-flowered plant growing among corn.

cornice *n.* ornamental moulding round the top of an indoor wall.

Cornish *adj.* of Cornwall. ● *n.* Celtic language of Cornwall.

cornucopia *n.* horn-shaped container overflowing with fruit and flowers, symbol of abundance.

corny *adj.* (**-ier, -iest**) (*colloq.*) hackneyed.

corollary *n.* proposition that follows logically from another.

corona *n.* (*pl.* **-nae**) ring of light round something.

coronary *n.* **1** one of the arteries supplying blood to the heart. **2** thrombosis in this.

coronation *n.* ceremony of crowning a monarch or consort.

coroner *n.* officer holding inquests.

coronet *n.* small crown.

corporal[1] /kórprəl/ *n.* non-commissioned officer next below sergeant.

corporal[2] /kórpərəl/ *adj.* of the body. □ **corporal punishment** whipping or beating.

corporate *adj.* of or belonging to a corporation or group.

corporation *n.* group in business or elected to govern a town.

corporeal *adj.* having a body, tangible. □ **corporeally** *adv.*

corps /kor/ *n.* **1** military unit. **2** organized group of people.

corpse *n.* dead body.

■ body, cadaver, carcass, remains, *sl.* stiff.

corpulent *adj.* having a bulky body, fat. □ **corpulence** *n.*

corpus *n.* (*pl.* **corpora**) set of writings.

corpuscle *n.* blood cell.

corral *n.* (*US*) enclosure for cattle. ● *v.* (**corralled**) put or keep in a corral.

correct *adj.* **1** true, accurate. **2** in accordance with an approved way of behaving or working. ● *v.* **1** make correct. **2** mark errors in. **3** reprove. □ **correctly** *adv.*, **correctness** *n.*

■ *adj.* **1** accurate, exact, factual, faithful, precise, right, true. **2** acceptable, appropriate, decent, decorous, fitting, proper, seemly, suitable. ● *v.* **1** amend, cure, fix, put right, rectify, redress, remedy, repair. **2** grade, mark. **3** admonish, berate, rebuke, reprimand, reprove, scold.

correction *n.* **1** correcting. **2** alteration correcting something.

corrective *adj.* & *n.* (thing) correcting what is bad or harmful.

correlate *v.* compare, connect, or be connected systematically. □ **correlation** *n.*

correspond *v.* **1** be similar or equivalent. **2** exchange letters.

■ **1** accord, agree, concur, coincide, match, square, tally. **2** communicate, write.

correspondence *n.* **1** similarity. **2** (exchange of) letters.

correspondent *n.* **1** person who writes letters. **2** person employed to write or report for a newspaper or TV news station.

corridor *n.* **1** passage in a building or train. **2** strip of territory giving access to somewhere.

corroborate *v.* give support to, confirm. □ **corroboration** *n.*, **corroborative** *adj.*

■ authenticate, back up, confirm, prove, substantiate, support, validate, verify.

corrode *v.* destroy (metal etc.) gradually by chemical action. □ **corrosion** *n.*, **corrosive** *adj.*

corrugated *adj.* shaped into ridges. □ **corrugation** *n.*

corrupt *adj.* **1** dishonest, accepting bribes. **2** immoral, wicked. **3** decaying. ● *v.* **1** make corrupt. **2** spoil, taint. □ **corruption** *n.*

■ *adj.* **1** *sl.* bent, *colloq.* crooked, dishonest, dishonourable, unscrupulous, untrustworthy, venal. **2** decadent, degenerate, depraved, dissolute, evil, immoral, perverted, wicked. ● *v.* **1** bribe, suborn; deprave, pervert. **2** contaminate, defile, poison, pollute, spoil, taint.

corsage /korsaázh/ *n.* (*US*) flowers worn by a woman.

corset *n.* close-fitting undergarment worn to shape or support the body.

cortège /kortáyzh/ *n.* funeral procession.

cortex *n.* (*pl.* **-ices**) outer part of the brain.

cortisone *n.* hormone produced by adrenal glands or synthetically.

corvette *n.* small fast gunboat.

cos[1] *n.* long-leaved lettuce.

cos[2] *abbr.* cosine.

cosh *n.* weighted weapon for hitting people. ● *v.* hit with a cosh.

cosine *n.* ratio of the side adjacent to an acute angle in a right-angled triangle to the hypotenuse.

cosmetic *n.* substance for beautifying the complexion etc. ● *adj.* improving the appearance.

cosmic *adj.* of the universe. □ **cosmic rays** radiation from outer space.

cosmogony *n.* (theory of) the origin of the universe.

cosmology *n.* science or theory of the universe. □ **cosmological** *adj.*, **cosmologist** *n.*

cosmopolitan *adj.* **1** of or from all parts of the world. **2** free from national prejudices. ● *n.* cosmopolitan person.

cosmos *n.* universe.

Cossack *n.* member of a S. Russian people, famous as horsemen.

cost *v.* (**cost**) **1** have as its price. **2** involve the sacrifice or loss of. **3** (**costed**) estimate the cost of. ● *n.* what a thing costs.

■ *n.* charge, expenditure, expense, outlay, payment, price, rate, tariff.

costermonger *n.* person selling fruit etc. from a barrow.

costly *adj.* (**-ier**, **-iest**) expensive.

costume *n.* **1** style of clothes, esp. that of a historical period. **2** clothing for a specified activity.

cosy *adj.* (**-ier**, **-iest**) warm and comfortable. ● *n.* cover to keep a teapot hot. □ **cosily** *adv.*, **cosiness** *n.*

■ *adj.* comfortable, homely, relaxing, restful, snug, warm.

cot *n.* child's bed with high sides. □ **cot death** unexplained death of a sleeping baby.

coterie *n.* select group.

cotoneaster /kətṓniástər/ *n.* shrub or tree with red berries.

cottage *n.* small simple house in the country. □ **cottage cheese** mild white lumpy cheese. **cottage pie** shepherd's pie.

cotton *n.* **1** soft white substance round the seeds of a tropical plant. **2** this plant. **3** thread or fabric made from cotton. ● *v.* **cotton on** (*colloq.*) understand. □ **cotton wool** fluffy material orig. made from raw cotton.

couch *n.* long piece of furniture for lying or sitting on. ● *v.* express in a specified way.

couch grass weed with long creeping roots.

cougar /ko͞ogər/ *n.* (*US*) puma.

cough *v.* expel air etc. from lungs with a sudden sharp sound. ● *n.* **1** act or sound of coughing. **2** illness causing coughing.

could *see* **can²**.

coulomb /ko͞olom/ *n.* unit of electric charge.

council *n.* **1** (meeting of) an advisory or administrative body. **2** local administrative body of a town etc.

> ■ **1** assembly, conference, congress, convention, meeting; board, body, committee, panel.

councillor *n.* member of a council.

counsel *n.* **1** advice. **2** barrister(s). ● *v.* (**counselled**) advise, recommend. □ **counsellor** *n.*

> ■ *n.* **1** advice, guidance, judgement, opinion, recommendation. ● *v.* advise, advocate, exhort, recommend, suggest, urge.

count¹ *v.* **1** say numbers in order. **2** find the total of. **3** include or be included in a reckoning. **4** regard as. **5** be important. ● *n.* **1** counting, number reached by this. **2** point being considered. □ **count on 1** rely on. **2** expect confidently.

> ■ *v.* **2** add up, calculate, compute, reckon, total, *colloq.* tot up. **3** allow for, include, take into account. **4** consider, deem, judge, rate, reckon, regard as. **5** be important, matter, signify. □ **count on** bank on, be sure of, depend on, lean on, rely on, swear by, trust.

count² *n.* foreign nobleman corresponding to earl.

countdown *n.* counting seconds backwards to zero.

countenance *n.* **1** face. **2** expression. **3** approval. ● *v.* give approval to.

counter¹ *n.* **1** flat-topped fitment in a shop etc. over which business is conducted. **2** small disc used in board games.

counter² *adv.* in the opposite direction. ● *adj.* opposed. ● *v.* take opposing action against.

counter- *pref.* **1** rival. **2** retaliatory. **3** reversed. **4** opposite.

counteract *v.* neutralize. □ **counteraction** *n.*

> ■ cancel out, counterbalance, neutralize, nullify, offset.

counter-attack *n.* & *v.* attack in reply to an opponent's attack.

counterbalance *n.* weight or influence balancing another. ● *v.* act as a counterbalance to.

counterfeit *adj.*, *n.*, & *v.* fake.

> ■ *adj.* artificial, bogus, fake, false, forged, fraudulent, imitation, *colloq.* phoney, sham, spurious; feigned, insincere, pretended, simulated. ● *n.* fake, forgery, imitation, *colloq.* phoney. ● *v.* copy, fake, falsify, forge, imitate; feign, pretend, simulate.

counterfoil *n.* section of a cheque or receipt kept as a record.

countermand *v.* cancel.

counterpane *n.* bedspread.

counterpart *n.* person or thing corresponding to another.

counterpoint *n.* method of combining melodies.

counter-productive *adj.* having the opposite of the desired effect.

countersign *n.* password. ● *v.* add a confirming signature to.

countersink *v.* (**-sunk**) sink (a screw-head) into a shaped cavity so that the surface is level.

counter-tenor *n.* male alto.

countervail *v.* avail against.

countess *n.* **1** count's or earl's wife or widow. **2** woman with the rank of count or earl.

countless *adj.* too many to be counted.

countrified *adj.* rustic.

country *n.* **1** nation's territory, a state. **2** land of a person's birth or citizenship. **3** land consisting of fields etc. with few buildings. **4** region with regard to its associations etc. **5** national population.

> ■ **1** kingdom, nation, power, realm, state. **2** fatherland, homeland, motherland, native land. **3** countryside, green belt. **4** land, terrain, territory.

countryman, countrywoman *ns.* **1** person living in the country. **2** person of one's own country.

countryside *n.* rural district.

county *n.* **1** major administrative division of a country. **2** families of high social class long established in a county.

coup /koō/ *n.* sudden action taken to obtain power etc.

coup de grâce /koō də graáss/ finishing stroke.

coup d'état /koō daytaá/ sudden overthrow of a government by force or illegal means.

couple *n.* **1** two people or things. **2** married or engaged pair. ● *v.* **1** link together. **2** copulate.

■ *n.* **1** brace, pair. **2** duo, pair, twosome. ● *v.* **1** connect, fasten, hitch, join, link, unite, yoke.

couplet *n.* two successive rhyming lines of verse.

coupling *n.* connecting device.

coupon *n.* **1** form or ticket entitling the holder to something. **2** entry form for a football pool.

courage *n.* ability to control fear, bravery. □ **courageous** *adj.*, **courageously** *adv.*

■ boldness, bravery, daring, fearlessness, fortitude, gallantry, *colloq.* grit, *colloq.* guts, heroism, intrepidity, mettle, nerve, pluck, spirit, valour. □ **courageous** bold, brave, daring, dauntless, fearless, gallant, game, *colloq.* gutsy, heroic, intrepid, mettlesome, plucky, spirited, valiant, valorous.

courgette *n.* small vegetable marrow.

courier *n.* **1** messenger carrying documents. **2** person employed to guide and assist tourists.

course *n.* **1** direction taken or intended. **2** onward progress. **3** golf course, racecourse, etc. **4** series of lessons or treatments. **5** layer of stone etc. in a building. **6** one part of a meal. ● *v.* move or flow freely. □ **of course** without doubt.

■ *n.* **1** direction, passage, path, route, tack, track, way.

court *n.* **1** courtyard. **2** area where tennis, squash, etc., are played. **3** sovereign's establishment with attendants. **4** room or building where legal cases are heard or judged. ● *v.* **1** try to win the favour, support, or love of. **2** invite (danger etc.). □ **court martial** judicial court of military officers. **court-martial** *v.* (**-martialled**) try by court martial.

courteous *adj.* polite. □ **courteously** *adv.*, **courteousness** *n.*

■ chivalrous, civil, considerate, gentlemanly, ladylike, polite, respectful, urbane, well-mannered.

courtesan *n.* prostitute with wealthy or upper-class clients.

courtesy *n.* courteous behaviour or act.

courtier *n.* person who attends a sovereign at court.

courtly *adj.* dignified and polite.

courtship *n.* courting.

courtyard *n.* space enclosed by walls or buildings.

cousin *n.* (also **first cousin**) child of one's uncle or aunt. □ **second cousin** child of one's parent's cousin.

couture *n.* design and making of fashionable clothes.

couturier /kootyoōriay/ *n.* designer of fashionable clothes.

cove *n.* small bay.

coven *n.* assembly of witches.

covenant *n.* formal agreement, contract. ● *v.* make a covenant.

cover *v.* **1** place or be or spread over. **2** conceal or protect thus. **3** travel over (a distance). **4** be enough to pay for. **5** deal with (a subject etc.). **6** protect by insurance etc. **7** report for a newspaper etc. ● *n.* **1** thing that covers, esp. a lid, wrapper, etc. **2** screen, shelter, protection. **3** place laid at a meal. □ **cover up** conceal. **cover-up** *n.*

■ *v.* **1** coat, extend over, spread over. **2** bury, camouflage, cloak, conceal, hide, mask, shroud, veil; enclose, envelop, swaddle, wrap; protect, screen, shelter, shield. **3** travel, traverse. **4** be enough for, defray, pay for. **5** comprise, deal with, encompass, include, incorporate, take in. ● *n.* **1** cap, covering, lid, top; binding, dust cover *or* jacket, wrapper. **2** camouflage, cloak, cover-up, disguise, front, mask, pretence, screen, smokescreen; hiding place, refuge, retreat, shelter; concealment, protection.

coverage *n.* **1** process of covering. **2** area or risk etc. covered.

coverlet *n.* bedspread.

covert *n.* thick undergrowth where animals hide. ● *adj.* done secretly. □ **covertly** *adv.*

covet *v.* (**coveted**) desire (a thing belonging to another person). □ **covetous** *adj.*

covey *n.* group of partridges.

cow[1] *n.* fully grown female of cattle or other large animals.
cow[2] *v.* intimidate.
coward *n.* person who lacks courage. □ **cowardly** *adj.*

■ □ **cowardly** afraid, *colloq.* chicken, craven, faint-hearted, fearful, frightened, pusillanimous, spineless, scared, timid, timorous, unheroic, *colloq.* yellow.

cowardice *n.* lack of courage.
cowboy *n.* **1** man in charge of cattle on a ranch. **2** (*colloq.*) person with reckless methods in business.
cower *v.* crouch or shrink in fear.
cowl *n.* **1** monk's hood or hooded robe. **2** hood-shaped covering.
cowling *n.* removable metal cover on an engine.
cowrie *n.* a kind of seashell.
cowslip *n.* wild plant with small fragrant yellow flowers.
cox *n.* coxswain. ● *v.* act as cox of.
coxswain /kóks'n/ *n.* steersman.
coy *adj.* pretending to be shy or embarrassed. □ **coyly** *adv.*
coyote /koyṓti/ *n.* N. American wild dog.
coypu *n.* beaver-like water animal.
crab *n.* ten-legged shellfish. □ **crab apple** small sour apple.
crabbed *adj.* **1** bad-tempered. **2** (of handwriting) hard to read.
crabby *adj.* bad-tempered.
crack *n.* **1** sudden sharp noise. **2** sharp blow. **3** narrow opening. **4** line where a thing is broken but not separated. **5** (*colloq.*) joke. ● *adj.* (*colloq.*) first-rate. ● *v.* **1** make or cause to make the sound of a crack. **2** break without parting completely. **3** knock sharply. **4** find a solution to (a problem). **5** collapse under strain. **6** tell (a joke). **7** (of the voice) become harsh. □ **crack-brained** *adj.* (*colloq.*) crazy. **crack down on** (*colloq.*) take severe measures against. **crack up** (*colloq.*) have a physical or mental breakdown.

■ *n.* **1** bang, clap, report, snap, shot. **2** bang, blow, *colloq.* clip, clout, knock, rap, smack. **3** breach, break, chink, cleft, cranny, crevice, fissure, gap, opening, rift, rupture, slit, split. ● *v.* **2** break, fracture, rupture, split. **3** bang, bash, hit, knock, strike, *colloq.* whack. **4** decipher, figure out, solve, work out. **5** break down, cave in, collapse, give way, yield.

crackdown *n.* (*colloq.*) severe measures against something.
cracker *n.* **1** small explosive firework. **2** toy paper tube made to give an explosive crack when pulled apart. **3** thin dry biscuit.
crackers *adj.* (*sl.*) crazy.
crackle *v.* make light cracking sounds. ● *n.* these sounds.
crackling *n.* crisp skin on roast pork.
crackpot (*sl.*) *n.* eccentric person. ● *adj.* **1** crazy. **2** unworkable.
cradle *n.* **1** baby's bed usu. on rockers. **2** place where something originates. **3** supporting structure. ● *v.* hold or support gently.
craft *n.* **1** skill, technique. **2** occupation requiring this. **3** cunning, deceit. **4** (*pl.* **craft**) vessel, aircraft, or spacecraft.

■ **1** ability, craftsmanship, dexterity, expertise, flair, know-how, skill, talent, technique. **2** calling, occupation, profession, trade, vocation. **3** artfulness, craftiness, cunning, deceit, guile, trickery, wiliness. **4** boat, ship, vessel; aeroplane, aircraft, plane; rocket, spacecraft, spaceship.

craftsman *n.* person skilled in a craft. □ **craftsmanship** *n.*
crafty *adj.* (**-ier**, **-iest**) cunning, using underhand methods. □ **craftily** *adv.*, **craftiness** *n.*

■ artful, canny, clever, cunning, deceitful, designing, devious, guileful, machiavellian, scheming, shrewd, sly, underhand, wily.

crag *n.* steep or rugged rock.
craggy *adj.* (**-ier**, **-iest**) rugged.
cram *v.* (**crammed**) **1** fill to bursting. **2** force into a space. **3** study intensively for an exam. □ **crammer** *n.*

■ **1** jam, overcrowd, pack, stuff. **2** push, ram, shove, squeeze, thrust. **3** study, *colloq.* swot.

cramp *n.* painful involuntary tightening of a muscle. ● *v.* keep within too narrow limits.
crampon *n.* spiked plate worn on boots for climbing on ice.
cranberry *n.* **1** small red sharp-tasting berry. **2** shrub bearing this.
crane *n.* **1** large wading bird. **2** apparatus for lifting and moving heavy objects. ● *v.* stretch (one's neck) to see something. □ **crane-fly** *n.* long-legged flying insect.
crank[1] *n.* L-shaped part for converting to-and-fro into circular motion. ● *v.* turn with a crank. □ **crankshaft** *n.* shaft turned in this way.

crank² *n.* person with very strange ideas. ▫ **cranky** *adj.*

cranny *n.* crevice.

craps *n.pl.* (*US*) gambling game played with a pair of dice.

crash *n.* **1** sudden loud noise. **2** violent collision. **3** financial collapse. ● *v.* **1** make a crashing noise. **2** be or cause to be involved in a collision. **3** undergo financial ruin. **4** (*colloq.*) gatecrash. ● *adj.* involving intense effort to achieve something rapidly. ▫ **crash helmet** padded helmet worn to protect the head in a crash. **crash-land** *v.* land (an aircraft) in emergency, causing damage.

■ *n.* **1** bang, boom, clash, explosion, smash. **2** accident, collision, pile-up, smash. ● *v.* **1** bang, boom, clash, explode, smash. **2** bang together, collide; smash, wreck. **3** collapse, fail, fold, go bankrupt, *colloq.* go bust, *colloq.* go broke, go out of business, go under.

crass *adj.* **1** very stupid. **2** insensitive.

crate *n.* **1** packing-case made of wooden slats. **2** (*sl.*) old aircraft or car. ● *v.* pack in crate(s).

crater *n.* bowl-shaped cavity.

cravat *n.* **1** short scarf. **2** necktie.

crave *v.* long or beg for.

craven *adj.* cowardly.

craving *n.* intense longing.

craw *n.* bird's crop.

crawfish *n.* large spiny sea lobster.

crawl *v.* **1** move on hands and knees or with the body on the ground. **2** move very slowly. **3** (*colloq.*) seek favour by servile behaviour. ● *n.* **1** crawling movement or pace. **2** overarm swimming stroke. ▫ **crawler** *n.*

crayfish *n.* **1** freshwater shellfish like a small lobster. **2** crawfish.

crayon *n.* stick of coloured wax etc. ● *v.* draw or colour with crayon(s).

craze *n.* **1** temporary enthusiasm. **2** object of this.

■ enthusiasm, fad, fashion, mania, obsession, trend.

crazy *adj.* (**-ier**, **-iest**) **1** insane. **2** very foolish. **3** (*colloq.*) madly eager. ▫ **crazy paving** paving made of irregular pieces. **crazily** *adv.*, **craziness** *n.*

■ **1** *sl.* barmy, *sl.* batty, *colloq.* crack-brained, *sl.* crackers, demented, deranged, insane, lunatic, mad, *sl.* nuts, *sl.* nutty, out of one's mind, *sl.* potty, unbalanced. **2** absurd, asinine, *sl.* crackpot, daft, foolish, hare-brained, idiotic, illogical, imbecilic, impractical, inane, laughable, ludicrous, nonsensical, preposterous, ridiculous, senseless, silly, stupid, unrealistic, unreasonable, unwise.

creak *n.* harsh squeak. ● *v.* make this sound. ▫ **creaky** *adj.*

cream *n.* **1** fatty part of milk. **2** its colour, yellowish-white. **3** cream-like substance. **4** best part. ● *adj.* cream-coloured. ● *v.* **1** remove the cream from. **2** make creamy. ▫ **cream cheese** soft rich cheese. **cream cracker** crisp unsweetened biscuit. **creamy** *adj.*

crease *n.* **1** line made by crushing or pressing. **2** line marking the limit of the bowler's or batsman's position in cricket. ● *v.* **1** make a crease in. **2** develop creases.

create *v.* **1** bring into existence. **2** invest with a new rank. **3** (*sl.*) make a fuss. ▫ **creation** *n.*, **creator** *n.*

■ **1** bring into being, cause, design, devise, dream up, engender, fashion, forge, generate, give rise to, imagine, invent, make, manufacture, originate, produce, spawn, *colloq.* think up.

creative *adj.* inventive, imaginative. ▫ **creatively** *adv.*, **creativity** *n.*

■ artistic, imaginative, ingenious, inventive, original, resourceful.

creature *n.* animal, person.

crèche /kresh/ *n.* day nursery.

credence *n.* belief.

credentials *n.pl.* documents showing that a person is who or what he or she claims to be.

credible *adj.* believable. ▫ **credibly** *adv.*, **credibility** *n.*

■ believable, conceivable, feasible, likely, plausible, possible, probable.

credit *n.* **1** belief that a thing is true. **2** acknowledgement of merit. **3** good reputation. **4** system of allowing payment to be deferred. **5** sum at a person's disposal in a bank. **6** entry in an account for a sum received. **7** acknowledgement in a book or film. ● *v.* (**credited**) **1** believe. **2** enter as credit. ▫ **credit card** plastic card containing machine-readable magnetic code enabling the holder to make purchases on credit. **credit a person with** ascribe to a person.

■ *n.* **1** belief, credence, faith, trust. **2** acclaim, commendation, praise, recognition, tribute. **3** name, reputation, repute, standing, stature, status. ● *v.* **1** accept, believe, have faith in, rely on, trust.

creditable *adj.* deserving praise. □ **creditably** *adv.*

■ admirable, commendable, estimable, laudable, meritorious, praiseworthy.

creditor *n.* person to whom money is owed.

credulous *adj.* too ready to believe things. □ **credulity** *n.*

creed *n.* set of beliefs.

■ belief(s), doctrine, dogma, philosophy, principles.

creek *n.* **1** narrow inlet of water, esp. on a coast. **2** (*US*) tributary.

creep *v.* (**crept**) **1** move with the body close to the ground. **2** move timidly, slowly, or stealthily. **3** develop gradually. **4** (of a plant) grow along the ground or a wall etc. **5** (of flesh) shudder with horror etc. ● *n.* **1** creeping. **2** (*sl.*) unpleasant person. **3** (*pl.*) nervous sensation.

■ *v.* **1** crawl, slither, squirm, worm, wriggle. **2** edge, inch, sidle, slink, sneak, steal, tiptoe.

creepy *adj.* (**-ier, -iest**) causing horror or fear. □ **creepy-crawly** *n.* (*pl.* **-crawlies**) small insect.

cremate *v.* burn (a corpse) to ashes. □ **cremation** *n.*

crematorium *n.* (*pl.* **-ia**) place where corpses are cremated.

crenellated *adj.* having battlements.

Creole *n.* **1** descendant of European settlers in the W. Indies or S. America. **2** their dialect. **3** hybrid language.

creosote *n.* oily wood-preservative distilled from coal tar.

crêpe /krayp/ *n.* fabric with a wrinkled surface.

crept *see* **creep**.

crepuscular *adj.* active at twilight.

crescendo /krishéndō/ *adv.* & *n.* (*pl.* **-os**) increasing in loudness.

crescent *n.* **1** narrow curved shape tapering to a point at each end. **2** curved street of houses.

cress *n.* plant with hot-tasting leaves used in salads.

crest *n.* **1** tuft or outgrowth on a bird's or animal's head. **2** plume on a helmet. **3** top of a mountain, wave, etc. **4** design above a shield on a coat of arms.

crestfallen *adj.* dejected.

cretaceous *adj.* chalky.

cretin *n.* person who is deformed and mentally retarded as the result of a thyroid deficiency. □ **cretinous** *adj.*

crevasse *n.* deep open crack esp. in a glacier.

crevice *n.* narrow gap in a surface.

■ chink, cleft, crack, cranny, fissure, gap, opening, rift, split.

crew[1] *see* **crow**.

crew[2] *n.* group of people working together, esp. manning a ship, aircraft, etc. □ **crew-cut** *n.* closely cropped haircut.

■ band, body, company, corps, gang, group, party, team.

crib *n.* **1** model of the manger scene at Bethlehem. **2** cot. **3** (*colloq.*) translation for students' use. ● *v.* (**cribbed**) copy unfairly.

cribbage *n.* a card game.

crick *n.* painful stiffness in the neck or back.

cricket[1] *n.* outdoor game for two teams of 11 players with ball, bats, and wickets. □ **cricketer** *n.*

cricket[2] *n.* brown insect resembling a grasshopper.

crime *n.* **1** act that breaks a law. **2** illegal acts.

■ **1** misdeed, misdemeanour, offence, transgression, wrong. **2** lawlessness, wrongdoing.

criminal *n.* person guilty of a crime. ● *adj.* of or involving crime. □ **criminally** *adv.*, **criminality** *n.*

■ *n.* convict, *colloq.* crook, culprit, desperado, lawbreaker, malefactor, miscreant, offender, thug, transgressor, villain, wrongdoer. ● *adj. sl.* bent, *colloq.* crooked, dishonest, illegal, illicit, lawless, unlawful.

criminology *n.* study of crime. □ **criminologist** *n.*

crimp *v.* press into ridges.

crimson *adj.* & *n.* deep red.

cringe *v.* **1** cower. **2** behave obsequiously.

■ **1** blench, cower, flinch, quail, recoil, shrink back, wince. **2** crawl, fawn, grovel, kowtow.

crinkle *n.* & *v.* wrinkle.

crinoline *n.* hooped petticoat.

cripple *n.* lame person. ● *v.* **1** make lame. **2** weaken seriously.

■ *v.* **1** disable, handicap, incapacitate, lame, maim. **2** damage, debilitate, hamstring, impair, paralyse, weaken.

crisis *n.* (*pl.* **crises**) **1** time of acute difficulty. **2** decisive moment.

■ **1** calamity, catastrophe, disaster, emergency. **2** critical moment, turning point, watershed.

crisp *adj.* **1** brittle. **2** cold and bracing. **3** brisk and decisive. ● *n.* thin slice of potato fried crisp. □ **crisply** *adv.*, **crispness** *n.*, **crispy** *adj.*

■ *adj.* **1** brittle, crispy, crunchy, friable. **2** bracing, exhilarating, fresh, invigorating, refreshing, stimulating. **3** brisk, decisive, incisive, vigorous.

criss-cross *n.* pattern of crossing lines. ● *adj.* & *adv.* in this pattern. ● *v.* mark, form, or move in this way, intersect.

criterion *n.* (*pl.* **-ia**) standard of judgement.

critic *n.* **1** person who points out faults. **2** one skilled in criticism.

critical *adj.* **1** looking for faults. **2** of or at a crisis. **3** expressing criticism. □ **critically** *adv.*

■ **1** captious, censorious, deprecatory, derogatory, disparaging, *colloq.* nit-picking, uncomplimentary. **2** dangerous, grave, perilous, serious, severe, touch-and-go; crucial, important, momentous, vital.

criticism *n.* **1** pointing out of faults. **2** judging of merit of literary or artistic works.

■ **1** censure, condemnation, disapproval, disparagement, fault-finding. **2** analysis, appraisal, assessment, evaluation, judgement.

criticize *v.* **1** find fault with. **2** discuss the merit of literary or artistic works.

■ **1** carp at, cast aspersions on, censure, condemn, disparage, *colloq.* get at, find fault with, impugn, *sl.* knock, *colloq.* pan, pick holes in, *sl.* slam, *colloq.* slate. **2** analyse, appraise, assess, evaluate, judge, review.

critique *n.* critical essay.

croak *n.* deep hoarse cry or sound like that of a frog. ● *v.* **1** utter or speak with a croak. **2** (*sl.*) die.

crochet /krṓshay/ *n.* handiwork done with a thread and a hooked needle. ● *v.* make by crochet.

crock[1] *n.* **1** earthenware pot. **2** broken piece of this.

crock[2] *n.* (*colloq.*) old or worn-out person or vehicle.

crockery *n.* household china.

crocodile *n.* large amphibious reptile. □ **crocodile tears** pretence of sorrow.

crocus *n.* spring-flowering plant growing from a corm.

croft *n.* small rented farm in Scotland.

crofter *n.* tenant of a croft.

croissant /krwússon/ *n.* rich crescent-shaped roll.

crone *n.* withered old woman.

crony *n.* close friend or companion.

crook *n.* **1** hooked stick. **2** bent thing. **3** (*colloq.*) criminal. ● *v.* bend.

crooked *adj.* **1** not straight. **2** (*colloq.*) dishonest. □ **crookedly** *adv.*

■ **1** askew, bent, contorted, gnarled, lopsided, misshapen, twisted, warped. **2** *sl.* bent, criminal, dishonest, illegal, illicit, unlawful, wrong.

croon *v.* sing softly. □ **crooner** *n.*

crop *n.* **1** batch of plants grown for their produce. **2** harvest from this. **3** group or amount produced at one time. **4** pouch in a bird's gullet where food is broken up for digestion. **5** whip-handle. **6** very short haircut. ● *v.* (**cropped**) **1** cut or bite off. **2** produce or gather as harvest. □ **crop up** occur unexpectedly.

cropper *n.* (*sl.*) heavy fall.

croquet /krṓkay/ *n.* game played on a lawn with balls and mallets.

croquette /krəkét/ *n.* fried ball or roll of potato, meat, or fish.

cross *n.* **1** stake with a transverse bar used in crucifixion, this as an emblem of Christianity. **2** affliction. **3** hybrid animal or plant. **4** mixture of or compromise between two things. **5** mark made by drawing one line across another. **6** thing shaped like this. ● *v.* **1** go or extend across. **2** pass in different directions. **3** draw line(s) across (a cheque) so that it must be paid into a bank. **4** oppose the wishes of. **5** cause to interbreed. ● *adj.* **1** peevish, angry. **2** reaching from side to side. **3** reciprocal. □ **at cross purposes** misunderstanding each other. **cross out** obliterate by drawing lines across. **crossly** *adv.*, **crossness** *n.*

■ *n.* **1** crucifix. **2** affliction, burden, misfortune, problem, trial, tribulation, trouble, woe, worry. **3** cross-breed, hybrid, mixture, mongrel. ● *v.* **1** cross over, go across, pass over, span, traverse. **2** intersect, join, meet. ● *adj.* **1** angry, annoyed, bad-tempered, cantankerous, crabby, crotchety, fractious, grumpy, huffy, irascible, irate, irritable, irritated, peevish, pettish, *colloq.* shirty, testy, vexed. □ **cross out** blot out, cancel, delete, erase, obliterate, score out, strike out.

crossbar *n.* horizontal bar.

crossbow *n.* mechanical bow fixed across a wooden stock.

cross-breed *n.* animal produced by interbreeding. ● *v.* cause to interbreed. □ **cross-bred** *adj.*

cross-check *v.* check again by a different method.

cross-examine *v.* question, esp. in a law court. □ **cross-examination** *n.*

cross-eyed *adj.* squinting.

crossfire *n.* gunfire crossing another line of fire.

crossing *n.* **1** journey across water. **2** place where things cross. **3** place for pedestrians to cross a road.

crosspatch *n.* (*colloq.*) bad-tempered person.

cross-ply *adj.* (of a tyre) having fabric layers with cords lying crosswise.

cross-question *v.* cross-examine.

cross-reference *n.* reference to another place in the same book.

crossroads *n.* place where roads intersect.

cross-section *n.* **1** diagram showing internal structure. **2** representative sample.

crosswise *adv.* in the form of a cross.

crossword *n.* puzzle in which intersecting words have to be inserted into a diagram.

crotch *n.* place where things fork, esp. where legs join the trunk.

crotchet *n.* note in music, half a minim.

crotchety *adj.* peevish.

crouch *v.* stoop low with legs tightly bent. ● *n.* this position.

croupier /kroo͞piər/ *n.* person in charge of a gaming table.

croûton /kroo͞ton/ *n.* small piece of fried or toasted bread.

crow *n.* **1** large black bird. **2** cock's cry. ● *v.* (**crowed** or **crew**) **1** (of a cock) utter a loud cry. **2** express gleeful triumph.

crowbar *n.* bar of iron, usually with a bent end, used as a lever.

crowd *n.* large group. ● *v.* **1** come together in a crowd. **2** fill or occupy fully.

■ *n.* cluster, drove, flock, herd, horde, host, mob, multitude, pack, swarm, throng. ● *v.* **1** assemble, cluster, collect, congregate, flock, gather, herd, mass, press, swarm, throng.

crown *n.* **1** monarch's ceremonial headdress, usu. a circlet of gold. **2** (**the Crown**) supreme governing power in a monarchy. **3** top part of a head, hat, etc. ● *v.* **1** place a crown on. **2** make king or queen. **3** be a climax to. **4** (*sl.*) hit on the head.

■ *n.* **1** circlet, coronet, diadem. **2** (**the Crown**) king, monarch, queen, ruler, sovereign. ● *v.* **2** enthrone. **3** cap, complete, perfect, round off, top.

crucial *adj.* very important, decisive. □ **crucially** *adv.*

■ critical, decisive, essential, key, momentous, vital.

crucible *n.* pot in which metals are melted.

crucifix *n.* model of a cross with a figure of Christ on it.

crucifixion *n.* **1** crucifying. **2** (**the Crucifixion**) that of Christ.

cruciform *adj.* cross-shaped.

crucify *v.* **1** put to death by nailing or binding to a transverse bar. **2** cause extreme pain to.

crude *adj.* **1** in a natural or raw state. **2** rough, unpolished. **3** lacking good manners, vulgar. □ **crudely** *adv.*, **crudity** *n.*

■ **1** natural, raw, unprocessed, unrefined. **2** primitive, rough, rudimentary, simple, unfinished, unpolished. **3** boorish, coarse, uncouth, vulgar.

cruel *adj.* (**crueller**, **cruellest**) **1** feeling pleasure in another's suffering. **2** causing pain or suffering, esp. deliberately. □ **cruelly** *adv.*, **cruelty** *n.*

■ **1** callous, cold-blooded, hard, hard-hearted, harsh, heartless, merciless, pitiless, ruthless, sadistic, unkind, unmerciful. **2** barbaric, bloodthirsty, brutal, diabolical, ferocious, fiendish, inhuman, savage, vicious.

cruet *n.* set of containers for oil, vinegar, and salt at the table.

cruise *v.* **1** sail for pleasure or on patrol. **2** travel at a moderate speed. ● *n.* cruising voyage.

cruiser *n.* **1** fast warship. **2** motor boat with a cabin.

crumb *n.* small fragment of bread etc.

crumble *v.* break into small fragments. ● *n.* dish of cooked fruit with crumbly topping.

crumbly *adj.* easily crumbled.

crummy *adj.* (**-ier**, **-iest**) (*sl.*) **1** dirty, squalid. **2** inferior.

crumpet *n.* flat soft yeasty cake eaten toasted.

crumple *v.* **1** crush or become crushed into creases. **2** collapse.

■ **1** crease, crinkle, crush, rumple. **2** cave in, collapse, give way.

crunch *v.* **1** crush noisily with the teeth. **2** make this sound. ● *n.* **1** crunching sound. **2** decisive event. □ **crunchy** *adj.*

crupper *n.* strap looped under a horse's tail from the saddle.

crusade *n.* **1** campaign against an evil. **2** medieval Christian military expedition to recover the Holy Land from Muslims. ● *v.* take part in a crusade. □ **crusader** *n.*

■ *n.* **1** battle, campaign, drive, offensive, war. ● *v.* battle, campaign, fight, lobby.

crush *v.* **1** press so as to break, injure, or wrinkle. **2** pound into fragments. **3** defeat or subdue completely. ● *n.* **1** crowded mass of people. **2** (*colloq.*) infatuation.

■ *v.* **1** break, shatter, smash; mangle, mash, press, pulp, squeeze; crease, crinkle, crumple, rumple, wrinkle. **2** crumble, pound, pulverize. **3** beat, conquer, defeat, overcome, overwhelm, quash, quell, subdue, suppress, thrash, vanquish; devastate, humiliate, mortify.

crust *n.* hard outer layer, esp. of bread.

crustacean *n.* animal with a hard shell (e.g. lobster).

crusty *adj.* (**-ier**, **-iest**) **1** with a crisp crust. **2** irritable, curt.

crutch *n.* **1** support for a lame person. **2** crotch.

crux *n.* (*pl.* **cruces**) **1** vital part of a problem. **2** difficult point.

cry *n.* **1** loud wordless sound. **2** urgent appeal. **3** spell of weeping. ● *v.* **1** shed tears. **2** call loudly. **3** appeal for help.

■ *n.* **1** bellow, howl, roar, scream, screech, shout, shriek, whoop, wail, yell, yelp, yowl. **2** appeal, entreaty, plea, request, supplication. ● *v.* **1** bawl, blubber, grizzle, shed tears, snivel, sob, wail, weep, whimper. **2** bellow, call, roar, scream, shout, shriek, yell.

cryogenics *n.* branch of physics dealing with very low temperatures. □ **cryogenic** *adj.*

crypt *n.* underground room, esp. beneath a church, used usu. as a burial place.

■ catacomb, sepulchre, mausoleum, tomb, vault.

cryptic *adj.* **1** obscure in meaning. **2** secret, mysterious.

■ **1** mystifying, obscure, puzzling. **2** arcane, enigmatic, esoteric, mysterious, occult, secret.

cryptogram *n.* thing written in cipher.

crystal *adj.* **1** glass-like mineral. **2** high-quality glass. **3** symmetrical piece of a solidified substance.

crystalline *adj.* **1** like or made of crystal. **2** clear.

crystallize *v.* **1** form into crystals. **2** make or become definite in form. □ **crystallization** *n.*

cub *n.* young of certain animals.

cubby hole small compartment.

cube *n.* **1** solid object with six equal square sides. **2** product of a number multiplied by itself twice. □ **cube root** number of which a given number is the cube.

cubic *adj.* of three dimensions.

cubicle *n.* small division of a large room, screened for privacy.

cubism *n.* style of painting in which objects are shown as geometrical shapes. □ **cubist** *n.*

cuckold *n.* man whose wife commits adultery. ● *v.* make a cuckold of.

cuckoo *n.* bird with a call sounding similar to its name.

cucumber *n.* long green fleshy fruit used in salads.

cud *n.* food that cattle bring back from the stomach into the mouth and chew again.

cuddle *v.* **1** hug lovingly. **2** nestle together. ● *n.* gentle hug. □ **cuddly** *adj.* pleasant to cuddle.

■ *v.* **1** caress, embrace, hug. **2** nestle, snuggle.

cudgel *n.* short thick stick used as a weapon. ● *v.* (**cudgelled**) beat with a cudgel.

cue[1] *n.* & *v.* (**cueing**) signal to do something.

■ *n.* hint, reminder, sign, signal.

cue[2] *n.* long rod for striking balls in billiards etc. ● *v.* strike with a cue.

cuff *n.* **1** end part of a sleeve. **2** blow with the open hand. ● *v.* strike with the open hand. □ **cuff link** fastener to hold shirt cuffs together.

cuisine /kwizeen/ *n.* style of cooking.

cul-de-sac *n.* (*pl.* **culs-de-sac**) street closed at one end.

culinary *adj.* of or for cooking.

cull *v.* **1** gather, select. **2** select and kill (surplus animals). ● *n.* **1** culling. **2** animal(s) culled.

culminate *v.* reach its highest point or degree. □ **culmination** *n.*

culottes *n.pl.* women's trousers styled to resemble a skirt.

culpable *adj.* deserving blame. □ **culpably** *adv.*, **culpability** *n.*

culprit *n.* guilty person.

> ■ criminal, malefactor, miscreant, offender, wrongdoer.

cult *n.* **1** system of religious worship. **2** excessive admiration of a person or thing.

cultivate *v.* **1** prepare and use (land) for crops. **2** raise (crops). **3** further one's acquaintance with (a person). **4** improve (manners etc.). □ **cultivation** *n.*, **cultivator** *n.*

> ■ **1** farm, plough, till, work. **2** grow, produce, raise, tend. **3** develop, encourage, foster, further, nurture, promote. **4** civilize, educate, polish, refine.

culture *n.* **1** developed appreciation of the arts etc. **2** customs and civilization of a particular time or people. **3** rearing of organisms. **4** bacteria grown for study. ● *v.* grow in artificial conditions. □ **cultural** *adj.*, **culturally** *adv.*

> ■ *n.* **1** education, erudition, learning, refinement, taste. **2** civilization, customs, way of life.

cultured *adj.* educated to appreciate the arts etc.

culvert *n.* drain under a road.

cumbersome *adj.* clumsy to carry or use.

cumin *n.* plant with aromatic seed.

cummerbund *n.* sash for the waist.

cumulative *adj.* increasing by additions. □ **cumulatively** *adv.*

cumulus *n.* (*pl.* **-li**) cloud in heaped-up rounded masses.

cuneiform *n.* ancient writing done in wedge-shaped strokes.

cunning *adj.* **1** deceitful, crafty. **2** ingenious. ● *n.* craftiness, ingenuity. □ **cunningly** *adv.*

> ■ *adj.* **1** artful, crafty, deceitful, devious, guileful, machiavellian, sly, tricky, wily. **2** adroit, clever, ingenious, shrewd, skilful, subtle. ● *n.* artfulness, chicanery, craftiness, deceit, deviousness, duplicity, guile, trickery; cleverness, ingenuity, skill.

cup *n.* **1** drinking vessel usu. with a handle at the side. **2** prize. **3** wine or fruit juice with added flavourings. ● *v.* (**cupped**) make cup-shaped. □ **cupful** *n.*

cupboard *n.* recess or piece of furniture with a door, in which things may be stored.

cupidity *n.* greed for gain.

cupola *n.* small dome.

cur *n.* bad-tempered or scruffy dog.

curacy *n.* position of curate.

curare /kyooraári/ *n.* vegetable poison that induces paralysis.

curate *n.* member of the clergy who assists a parish priest.

curator *n.* person in charge of a museum or other collection.

curb *n.* means of restraint. ● *v.* restrain.

> ■ *v.* bridle, check, contain, control, hold back, repress, restrain, subdue, suppress.

curds *n.pl.* thick soft substance formed when milk turns sour.

curdle *v.* (cause to) form curds.

cure *v.* **1** restore to health. **2** rid (of a disease, trouble, etc.). **3** preserve by salting, drying, etc. ● *n.* **1** curing. **2** substance or treatment that cures disease etc.

> ■ *v.* **1** heal, make better, restore to health. **3** dry, pickle, preserve, salt, smoke. ● *n.* **2** antidote, medicine, remedy; therapy, treatment.

curette *n.* surgical scraping instrument. □ **curettage** *n.*

curfew *n.* signal or time after which people must stay indoors.

curio *n.* (*pl.* **-os**) unusual and therefore interesting object.

curiosity *n.* **1** desire to find out and know things. **2** curio.

> ■ **1** inquisitiveness, interest, *colloq.* nosiness. **2** curio, *objet d'art*, oddity, rarity.

curious *adj.* **1** eager to learn or know something. **2** strange, unusual. □ **curiously** *adv.*

> ■ **1** inquiring, inquisitive, interested, *colloq.* nosy. **2** bizarre, extraordinary, funny, odd, offbeat, outlandish, peculiar, quaint, queer, singular, strange, surprising, unusual, weird.

curl *v.* curve, esp. in a spiral shape or course. ● *n.* **1** curled thing or shape. **2** coiled lock of hair.

curler *n.* device for curling hair.

curlew *n.* wading bird with a long curved bill.

curling *n.* game like bowls played on ice.

curly *adj.* (**-ier**, **-iest**) full of curls.

curmudgeon *n.* bad-tempered person. □ **curmudgeonly** *adj.*

currant *n.* **1** dried grape used in cookery. **2** small round edible berry. **3** shrub producing this.

currency *n.* **1** money in use. **2** state of being widely known.

current *adj.* **1** belonging to the present time. **2** in general use. ● *n.* **1** body of water or air moving in one direction. **2** flow of electricity. □ **currently** *adv.*

> ■ *adj.* **1** contemporary, latest, ongoing, present, up to date. **2** accepted, common, popular, prevailing, prevalent, widespread. ● *n.* **1** flow, stream, tide.

curriculum *n.* (*pl.* **-la**) course of study. □ **curriculum vitae** brief account of one's career.

curry[1] *n.* **1** seasoning made with hot-tasting spices. **2** dish flavoured with this.

curry[2] *v.* groom (a horse) with a **curry-comb**, a pad with rubber or plastic projections. □ **curry favour** win favour by flattery.

curse *n.* **1** call for evil to come on a person or thing. **2** violent exclamation of anger. **3** thing causing evil or harm. ● *v.* **1** utter a curse (against). **2** afflict.

> ■ *n.* **1** execration, imprecation. **2** blasphemy, expletive, oath, obscenity, profanity, swear word. **3** bane, blight, evil, misfortune. ● *v.* **1** damn, execrate; blaspheme (at), swear (at). **2** afflict, burden, saddle, weigh down.

cursive *adj.* & *n.* (writing) done with joined letters.

cursor *n.* movable indicator on a VDU screen.

cursory *adj.* hasty, hurried. □ **cursorily** *adv.*

> ■ desultory, hasty, hurried, perfunctory, quick, rapid, summary, superficial.

curt *adj.* noticeably or rudely brief. □ **curtly** *adv.*, **curtness** *n.*

> ■ abrupt, blunt, brusque, laconic, rude, short, terse, ungracious.

curtail *v.* cut short, reduce. □ **curtailment** *n.*

> ■ abbreviate, abridge, cut, cut short, shorten; cut down, reduce.

curtain *n.* piece of cloth hung as a screen, esp. at a window.

curtsy *n.* movement of respect made by bending the knees. ● *v.* make a curtsy.

curvaceous *adj.* (*colloq.*) shapely.

curvature *n.* **1** curving. **2** curved form.

curve *n.* line or surface with no part straight or flat. ● *v.* form (into) a curve.

cushion *n.* **1** stuffed bag used as a pad, esp. for leaning against. **2** padded part. **3** body of air supporting a hovercraft. ● *v.* **1** protect with a pad. **2** lessen the impact of.

cushy *adj.* (**-ier**, **-iest**) (*colloq.*) pleasant and easy.

cusp *n.* pointed part where curves meet.

cuss (*colloq.*) *n.* **1** curse. **2** awkward person. ● *v.* curse.

cussed /kússid/ *adj.* (*colloq.*) awkward and stubborn.

custard *n.* sauce made with milk and eggs or flavoured cornflour.

custodian *n.* guardian, keeper.

custody *n.* **1** guardianship. **2** imprisonment.

> ■ **1** care, charge, guardianship, keeping, protection, safe-keeping. **2** confinement, detention, imprisonment, incarceration.

custom *n.* **1** usual way of behaving or acting. **2** regular dealing by customer(s). **3** (*pl.*) duty on imported goods.

> ■ **1** convention, fashion, form, habit, practice, routine, tradition, usage, way, wont. **2** business, patronage, support, trade.

customary *adj.* usual. □ **customarily** *adv.*

> ■ accepted, accustomed, common, conventional, everyday, habitual, normal, ordinary, regular, routine, traditional, usual.

customer *n.* person buying goods or services from a shop etc.

> ■ buyer, client, consumer, patron, purchaser.

cut *v.* (**cut**, **cutting**) **1** divide, wound, or penetrate with a sharp-edged instrument. **2** reduce the length of, edit. **3** reduce (prices, wages, etc.). **4** intersect. **5** ignore (a person). **6** divide (a pack of cards). **7** stop filming. ● *n.* **1** wound or mark made by a sharp edge. **2** reduction. **3** (*colloq.*) share. **4** piece cut off. **5** style of cutting hair, clothes, etc. **6** hurtful remark. □ **cut in 1** interrupt. **2** pull in too closely in front of another vehicle.

> ■ *v.* **1** carve, slice; gash, lacerate, slash, slit, wound. **2** chop off, clip, crop, lop, shear, snip, trim; mow; abbreviate, abridge, curtail, edit, précis, shorten, truncate. **3** lower, reduce, slash. **4** cross, intersect, join, meet. **5** cold-shoulder, ignore, rebuff, snub. ● *n.* **1** gash, graze, incision, laceration, nick, slash, wound. **2** curtailment,

decrease, reduction. 3 percentage, portion, share.

cute *adj.* (*colloq.*) **1** clever. **2** ingenious. **3** attractive, pretty. ▫ **cutely** *adv.*, **cuteness** *n.*

cuticle *n.* skin at the base of a nail.

cutlass *n.* short curved sword.

cutler *n.* maker of cutlery.

cutlery *n.* table knives, forks, and spoons.

cutlet *n.* **1** neck-chop. **2** mince cooked in this shape. **3** thin piece of veal.

cutthroat *adj.* merciless. ● *n.* murderer.

cutting *adj.* (of remarks) hurtful. ● *n.* **1** passage cut through high ground for a railway etc. **2** piece of a plant for replanting.

■ *adj.* biting, caustic, harsh, hurtful, malicious, sarcastic, sardonic, scathing, scornful, venomous, wounding.

cuttlefish *n.* sea creature that ejects black fluid when attacked.

cyanide *n.* a strong poison.

cybernetics *n.* science of systems of control and communication in animals and machines.

cyclamen *n.* plant with petals that turn back.

cycle *n.* **1** recurring series of events etc. **2** bicycle, tricycle. ● *v.* ride a bicycle. ▫ **cyclist** *n.*

■ *n.* **1** course, pattern, rotation, round, series, sequence.

cyclic *adj.* (also **cyclical**) happening in cycles. ▫ **cyclically** *adv.*

cyclone *n.* violent wind rotating round a central area. ▫ **cyclonic** *adj.*

cyclotron *n.* apparatus for accelerating charged particles in a spiral path.

cygnet *n.* young swan.

cylinder *n.* object with straight sides and circular ends. ▫ **cylindrical** *adj.*, **cylindrically** *adv.*

cymbal *n.* brass plate struck with another to make a ringing sound.

cynic *n.* person who believes people's motives are usually bad or selfish. ▫ **cynical** *adj.*, **cynically** *adv.*, **cynicism** *n.*

cypress *n.* evergreen tree with dark feathery leaves.

cyst *n.* sac of fluid or soft matter on or in the body.

cystic *adj.* of the bladder. ▫ **cystic fibrosis** hereditary disease usu. resulting in respiratory infections.

cystitis *n.* inflammation of the bladder.

cytology *n.* study of biological cells. ▫ **cytological** *adj.*

czar *n.* = **tsar**.

Dd

dab[1] *n.* quick light blow or pressure. ● *v.* (**dabbed**) strike or press lightly or feebly.

dab[2] *n.* a kind of small flatfish.

dabble *v.* **1** splash about gently or playfully. **2** work at something in an amateur way. □ **dabbler** *n.*

dace *n.* (*pl.* **dace**) small freshwater fish.

dachshund *n.* small dog with a long body and short legs.

dad *n.* (*colloq.*) father.

daddy *n.* (*colloq.*) father.

daddy-long-legs *n.* crane-fly.

daffodil *n.* yellow flower with a trumpet-shaped central part.

daft *adj.* silly, crazy.

■ absurd, asinine, *sl.* barmy, crazy, fatuous, foolish, hare-brained, idiotic, imbecilic, ludicrous, mad, nonsensical, *sl.* potty, ridiculous, risible, senseless, silly, stupid, unwise.

dagger *n.* short pointed two-edged weapon used for stabbing.

dahlia *n.* garden plant with bright flowers.

daily *adj.* happening or appearing on every day or every weekday. ● *adv.* once a day. ● *n.* **1** daily newspaper. **2** (*colloq.*) charwoman.

dainty *adj.* (**-ier, -iest**) **1** delicately pretty. **2** choice. **3** fastidious. □ **daintily** *adv.*, **daintiness** *n.*

■ **1** charming, delicate, elegant, exquisite, fine, graceful, pretty. **2** appetizing, choice, delectable, delicious, tasty. **3** *colloq.* choosy, fastidious, fussy, sensitive, squeamish.

daiquiri /dákkəri/ *n.* cocktail of rum and lime juice.

dairy *n.* place where milk and its products are processed or sold.

dais /dáyiss/ *n.* low platform, esp. at the end of a hall.

daisy *n.* small white flower with a yellow centre. □ **daisy wheel** printing device with radiating spokes.

dale *n.* valley.

dally *v.* **1** idle, dawdle. **2** flirt. □ **dalliance** *n.*

■ **1** dawdle, delay, *colloq.* dilly-dally, idle, linger, loiter.

Dalmatian *n.* large white dog with dark spots.

dam[1] *n.* barrier built across a river to hold back water. ● *v.* (**dammed**) **1** hold back with a dam. **2** obstruct (a flow).

dam[2] *n.* mother of an animal.

damage *n.* **1** something done that reduces the value or usefulness of the thing affected or spoils its appearance. **2** (*pl.*) money as compensation for injury. ● *v.* cause damage to.

■ *n.* **1** destruction, harm, hurt, injury, impairment, mutilation. ● *v.* deface, harm, hurt, impair, injure, mar, mutilate, sabotage, spoil, vandalize.

damask *n.* fabric woven with a pattern visible on either side.

dame *n.* **1** (*US sl.*) woman. **2** (**Dame**) title of a woman with an order of knighthood.

damn /dam/ *v.* **1** condemn to hell. **2** condemn as a failure. **3** swear at. ● *int.* & *n.* uttered curse. ● *adj.* & *adv.* damned.

damnable *adj.* hateful, annoying.

damnation *n.* eternal punishment in hell. ● *int.* exclamation of annoyance.

damp *n.* moisture. ● *adj.* slightly wet. ● *v.* **1** make damp. **2** take the force or vigour out of. **3** stop the vibration of. □ **dampness** *n.*

■ *adj.* clammy, dank, humid, moist, muggy, steamy, wet. ● *v.* **1** dampen, moisten, wet. **2** dampen, diminish, discourage, dull, lessen, moderate, subdue, reduce, temper.

dampen *v.* **1** make or become damp. **2** make less forceful or vigorous.

damper *n.* **1** plate controlling the draught in a flue. **2** depressing influence. **3** pad that damps the vibration of a piano string.

damsel *n.* (*old use*) young woman.

damson *n.* small purple plum.

dance *v.* **1** move with rhythmical steps and gestures, usu. to music. **2** move in a quick or lively way. ● *n.* **1** social gathering for dancing. **2** piece of dancing, music for this. □ **dance attendance on** follow about and help dutifully. **dancer** *n.*

dandelion *n.* wild plant with bright yellow flowers.

dandified *adj.* like a dandy.
dandle *v.* dance or nurse (a child) in one's arms.
dandruff *n.* scurf from the scalp.
dandy *n.* man who pays excessive attention to his appearance.
Dane *n.* native or inhabitant of Denmark.
danger *n.* **1** likelihood of harm or death. **2** thing causing this.

■ hazard, jeopardy, peril, risk, threat.

dangerous *adj.* causing danger, not safe. □ **dangerously** *adv.*

■ chancy, *sl.* dicey, hazardous, perilous, precarious, risky, treacherous, unsafe.

dangle *v.* **1** hang or swing loosely. **2** hold out temptingly.
Danish *adj.* & *n.* (language) of Denmark.
dank *adj.* damp and cold.
dapper *adj.* neat and smart.

■ fashionable, neat, smart, spruce, stylish, trim, well-dressed.

dapple *v.* mark with patches of colour or shade. □ **dapple-grey** *adj.* grey with darker markings.
dare *v.* **1** be bold enough (to do something). **2** challenge to do something. ● *n.* this challenge.

■ **1** hazard, risk, venture. **2** challenge, defy.

daredevil *adj.* & *n.* recklessly daring (person).
daring *adj.* bold. ● *n.* boldness.
dark *adj.* **1** with little or no light. **2** of a deep or sombre colour. **3** having dark hair or skin. **4** gloomy. **5** secret, mysterious. ● *n.* **1** absence of light. **2** time of darkness, night. □ **dark horse** competitor of whom little is known. **darkroom** *n.* darkened room for processing photographs. **darkly** *adv.*, **darkness** *n.*

■ *adj.* **1** black, pitch-black, pitch-dark, sunless, starless, unlit, unlighted, unilluminated; dim, dusky, gloomy, murky, overcast, shadowy. **2** inky, sooty; drab, dreary, dull, funereal, sombre. **3** brunette; brown, swarthy, tanned. **4** bleak, cheerless, depressing, dismal, gloomy, melancholy, mournful, pessimistic. **5** arcane, deep, incomprehensible, mysterious, obscure, secret, unfathomable.

darken *v.* make or become dark.
darling *n.* & *adj.* **1** loved or lovable (person or thing). **2** favourite.
darn *v.* mend by weaving thread across a hole. ● *n.* place darned.
dart *n.* **1** small pointed missile, esp. for throwing at the target in the game of darts. **2** darting movement. **3** tapering tuck. ● *v.* **1** run suddenly. **2** send out (a glance etc.) rapidly.
dartboard *n.* target in the game of darts.
dash *v.* **1** run rapidly, rush. **2** knock or throw forcefully against something. **3** destroy (hopes). ● *n.* **1** rapid run, rush. **2** small amount of liquid or flavouring added. **3** vigour. **4** dashboard. **5** punctuation mark (—) showing a break in the sense.

■ *v.* **1** bolt, bound, dart, flash, fly, hasten, hurry, hurtle, race, run, rush, scoot, scurry, speed, sprint, tear, whiz, zoom. **2** knock, smash, strike; cast, fling, hurl, pitch, throw, toss. **3** destroy, put paid to, ruin, spoil. ● *n.* **1** bolt, bound, dart, run, rush, sprint. **2** bit, drop, hint, pinch, sprinkling, suggestion, touch, trace. **3** élan, energy, impetuosity, liveliness, spirit, verve, vigour, vivacity.

dashboard *n.* instrument panel of a motor vehicle.
dashing *adj.* spirited, showy.
dastardly *adj.* contemptible.
data *n.pl.* **1** facts on which a decision is to be based. **2** facts to be processed by computer.
data bank large store of computerized data.
database *n.* organized store of computerized data.
date[1] *n.* **1** day, month, or year of a thing's occurrence. **2** period to which a thing belongs. **3** (*colloq.*) appointment to meet socially. **4** (*colloq.*) person to be met thus. ● *v.* **1** mark with a date. **2** assign a date to. **3** originate from a particular date. **4** become out of date. **5** (*colloq.*) make a social appointment (with). □ **out of date** old-fashioned, obsolete. **to date** until now. **up to date** fashionable, modern.

■ *n.* **2** age, day, epoch, era, season, time, year. □ **out of date** anachronistic, antiquated, behind the times, obsolete, old, old-fashioned, outdated, outmoded. **up to date** contemporary, current, fashionable, modern, *colloq.* trendy.

date[2] *n.* small brown edible fruit. □ **date palm** tree bearing this.
datum *n.* (*pl.* **data**) item of data.
daub *v.* smear roughly. ● *n.* **1** clumsily painted picture. **2** smear.
daughter *n.* female in relation to her parents. □ **daughter-in-law** *n.* (*pl.* **daughters-in-law**) son's wife.

daunt *v.* discourage, intimidate.

■ demoralize, deter, discourage, dishearten, dismay, frighten, intimidate, overawe, put off, scare, unnerve.

dauntless *adj.* brave.
dauphin *n.* title of the eldest son of former kings of France.
davit *n.* small crane on a ship.
dawdle *v.* walk slowly and idly, take one's time. □ **dawdler** *n.*

■ dally, delay, *colloq.* dilly-dally, hang about, idle, lag behind, linger, loiter, straggle, take one's time, waste time.

dawn *n.* **1** first light of day. **2** beginning. ● *v.* begin to grow light. □ **dawn on** become evident to. **dawning** *n.*

■ *n.* **1** break of day, daybreak, first light, sunrise. **2** beginning, commencement, dawning, emergence, genesis, inception, origin, start.

day *n.* **1** time while the sun is above the horizon. **2** period of 24 hours. **3** hours given to work during a day. **4** specified day. **5** time, period.
daybreak *n.* first light of day.
daydream *n.* pleasant idle thoughts. ● *v.* have daydreams.

■ *n.* dream, fancy, fantasy, pipedream, reverie. ● *v.* dream, fantasize.

daylight *n.* **1** light of day. **2** dawn.
daytime *n.* part of the day when there is natural daylight.
daze *v.* cause to feel stunned or bewildered. ● *n.* dazed state.

■ *v.* amaze, astonish, astound, bowl over, dumbfound, *colloq.* flabbergast, stagger, startle, stun, stupefy, surprise, take aback; baffle, bemuse, bewilder, confuse, nonplus, perplex, puzzle.

dazzle *v.* **1** blind temporarily with bright light. **2** impress with skill, beauty, etc.
de- *pref.* implying removal or reversal.
deacon *n.* **1** member of the clergy ranking below priest. **2** lay officer in Nonconformist churches.
dead *adj.* **1** no longer alive. **2** obsolete, extinct. **3** numb. **4** without brightness or resonance. **5** dull. **6** abrupt, complete. ● *adv.* completely. □ **dead beat** tired out. **dead end** road closed at one end. **dead heat** race in which two or more competitors finish exactly even. **dead letter** law or rule no longer observed.

■ *adj.* **1** deceased, defunct, departed, gone, late, lifeless. **2** disused, extinct, obsolete, outmoded. **3** deadened, numb, paralysed. **5** boring, dull, flat, lifeless, tedious, uninteresting. **6** abrupt, sudden; absolute, complete, downright, entire, out and out, outright, total, unqualified. ● *adv.* absolutely, completely, entirely, totally, utterly.

deaden *v.* deprive of or lose vitality, loudness, feeling, etc.

■ blunt, cushion, dampen, dull, moderate, muffle, soften; numb, paralyse.

deadline *n.* time limit.
deadlock *n.* state of unresolved conflict. ● *v.* reach this.

■ *n.* impasse, stalemate, standstill.

deadly *adj.* (**-ier**, **-iest**) **1** causing death or serious damage. **2** death-like. **3** very dreary. ● *adv.* **1** as if dead. **2** extremely. □ **deadly nightshade** plant with poisonous black berries.

■ **1** dangerous, fatal, lethal, mortal, poisonous, toxic. **2** deathly, ghastly, ghostly, livid. **3** boring, dreary, dull, tedious, tiresome.

deadpan *adj.* expressionless.
deaf *adj.* **1** wholly or partly unable to hear. **2** refusing to listen. □ **deafness** *n.*
deafen *v.* make unable to hear by a very loud noise.
deal[1] *n.* fir or pine timber.
deal[2] *v.* (**dealt**) **1** distribute. **2** cause to be received. **3** hand out (cards) to players in a card game. ● *n.* **1** business transaction. **2** large amount. **3** player's turn to deal in a card game. □ **deal in** sell. **deal with 1** take action about. **2** be about or concerned with. **3** do business with.

■ *v.* **1** allot, dispense, distribute, dole out, give out, hand out, mete out, share out. **2** administer, deliver, inflict. ● *n.* **1** agreement, arrangement, bargain, contract, transaction, understanding. □ **deal with 1** attend to, do, look after, organize, see to, sort out, take care of, take charge of. **2** analyse, consider, discuss, examine, explore, investigate, study, treat.

dealer *n.* **1** person who deals. **2** trader.
dealings *n.pl.* conduct or transactions.
dean *n.* **1** head of a cathedral chapter. **2** university official.
deanery *n.* dean's position or residence.
dear *adj.* **1** much loved. **2** expensive. ● *n.* dear person. ● *int.* exclamation of surprise or distress. □ **dearly** *adv.*

■ *adj.* **1** adored, beloved, cherished, darling, loved, precious, prized, treasured, valued. **2** costly, exorbitant, expensive, high-priced, *colloq.* steep.

dearth *n.* scarcity, lack.

death *n.* **1** process of dying, end of life. **2** ending. **3** destruction. **4** state of being dead. □ **death duty** tax levied on property after the owner's death. **death trap** very dangerous place. **death-watch beetle** beetle whose larvae bore into wood and make a ticking sound.

■ **1** decease, demise, dying, end, expiration, passing away. **2** cessation, ending, finish, termination. **3** annihilation, destruction, extinction, obliteration.

deathly *adj.* (**-ier**, **-iest**) like death.

debacle /daybaák'l/ *n.* general collapse.

debar *v.* (**debarred**) exclude.

debase *v.* lower in quality or value. □ **debasement** *n.*

■ cheapen, degrade, demean, devalue, diminish, lower, reduce.

debatable *adj.* questionable.

■ arguable, controversial, disputable, doubtful, dubious, open to question, problematic(al), questionable, uncertain.

debate *n.* formal discussion. ● *v.* **1** discuss or dispute. **2** consider.

■ *n.* argument, discussion, disputation, dispute. ● *v.* **1** argue about, discuss, dispute, wrangle over. **2** consider, deliberate, meditate on, mull over, ponder, reflect on, think over.

debauchery over-indulgence in harmful or immoral pleasures.

debilitate *v.* weaken. □ **debilitation** *n.*

debility *n.* weakness of health.

debit *n.* entry in an account for a sum owing. ● *v.* (**debited**) enter as a debit, charge. □ **direct debit** instruction allowing an organization to take regular payments from one's bank account.

debonair *adj.* having a carefree self-confident manner.

debouch *v.* come out from a narrow into an open area.

debrief *v.* question to obtain facts about a completed mission.

debris /débree/ *n.* scattered broken pieces or rubbish.

■ flotsam, fragments, litter, pieces, refuse, remains, rubbish, rubble, ruins, waste, wreckage.

debt /det/ *n.* something owed. □ **in debt** owing something.

debtor *n.* person who owes money.

debug *v.* (**debugged**) remove bugs from.

debunk *v.* (*colloq.*) show up as exaggerated or false.

debut /dáy-byōō/ *n.* first public appearance.

deca- *pref.* ten.

decade *n.* ten-year period.

decadent *adj.* in a state of moral deterioration. □ **decadence** *n.*

■ corrupt, debased, degenerate, dissipated, dissolute, immoral.

decaffeinated *adj.* with caffeine removed or reduced.

decagon *n.* geometric figure with ten sides. □ **decagonal** *adj.*

decamp *v.* go away suddenly or secretly.

decant *v.* pour off (wine etc.) leaving sediment behind.

decanter *n.* bottle for decanted wine etc.

decapitate *v.* behead. □ **decapitation** *n.*

decarbonize *v.* remove carbon deposit from (an engine). □ **decarbonization** *n.*

decathlon *n.* athletic contest involving ten events.

decay *v.* **1** rot. **2** lose quality or strength. ● *n.* **1** rotten state. **2** decline.

■ *v.* **1** decompose, go bad, go off, moulder, perish, putrefy, rot, spoil. **2** atrophy, crumble, decline, degenerate, deteriorate, disintegrate, waste away, wither. ● *n.* **1** decomposition, putrefaction, rot. **2** atrophy, decline, degeneration, deterioration, disintegration.

decease *n.* death.

deceased *adj.* dead.

deceit *n.* **1** process of deceiving. **2** trick.

■ **1** cheating, chicanery, craftiness, cunning, deceitfulness, deception, dishonesty, dissimulation, double-dealing, duplicity, fraud, guile, hypocrisy, slyness, treachery, trickery. **2** artifice, deception, manoeuvre, ploy, ruse, stratagem, subterfuge, trick, wile.

deceitful *adj.* intending to deceive. □ **deceitfully** *adv.*

■ crafty, cunning, dishonest, disingenuous, double-dealing, false, fraudulent, guileful, hypocritical, insincere, lying, scheming, sly, treacherous, two-faced, underhand, untrustworthy, wily.

deceive *v.* **1** cause to believe what is false. **2** be sexually unfaithful to. □ **deceiver** *n.*

■ **1** *colloq.* bamboozle, beguile, betray, cheat, *sl.* con, delude, double-cross, dupe, fool, fox, hoax, hoodwink, mislead, pull the wool over somone's eyes, swindle, take in, trick.

decelerate *v.* reduce the speed (of). □ **deceleration** *n.*

decennial *adj.* **1** happening every tenth year. **2** lasting ten years. □ **decennially** *adv.*

decent *adj.* **1** conforming to accepted standards of what is proper. **2** respectable. **3** kind. □ **decently** *adv.*, **decency** *n.*

■ **1** acceptable, appropriate, correct, fitting, proper, right, suitable. **2** decorous, presentable, respectable, seemly. **3** accommodating, considerate, courteous, generous, kind, nice, obliging, pleasant, thoughtful.

decentralize *v.* transfer from central to local control. □ **decentralization** *n.*

deception *n.* **1** deceit. **2** trick.

deceptive *adj.* misleading.

■ deceiving, delusive, false, illusory, misleading, unreliable; deceitful, dishonest, untruthful.

deci- *pref.* one-tenth.

decibel *n.* unit for measuring the relative loudness of sound.

decide *v.* **1** make up one's mind. **2** settle a contest or argument. □ **decide on** choose.

■ **1** determine, make up one's mind, resolve. **2** adjudicate, arbitrate, judge, resolve, settle. □ **decide on** choose, go for, opt for, pick out, plump for, select.

decided *adj.* **1** having firm opinions. **2** clear, definite. □ **decidedly** *adv.*

■ **1** decisive, determined, firm, resolute, uncompromising, unswerving, unwavering. **2** clear, definite, evident, indisputable, marked, obvious, pronounced, undeniable, unequivocal, unmistakable, unquestionable.

deciduous *adj.* shedding its leaves annually.

decimal *adj.* reckoned in tens or tenths. ● *n.* decimal fraction. □ **decimal fraction** fraction based on powers of ten, shown as figures after a dot. **decimal point** this dot.

decimalize *v.* convert into a decimal. □ **decimalization** *n.*

decimate *v.* destroy a large proportion of. □ **decimation** *n.*

decipher *v.* make out the meaning of (code, bad handwriting).

■ decode, disentangle, figure out, interpret, make out, unravel, work out.

decision *n.* **1** deciding. **2** judgement so reached. **3** resoluteness.

■ **1** arbitration, determination, resolution, settlement. **2** conclusion, judgement, resolution; decree, finding, ruling, verdict. **3** decisiveness, determination, resoluteness, resolution, resolve.

decisive *adj.* **1** conclusive. **2** quick to decide. □ **decisively** *adv.*, **decisiveness** *n.*

■ **1** conclusive, convincing, definitive, incontrovertible, irrefutable, unquestionable. **2** definite, determined, firm, forthright, resolute.

deck[1] *n.* floor or storey of a ship or bus. □ **deckchair** *n.* folding canvas chair.

deck[2] *v.* decorate, dress up.

declaim *v.* speak or say impressively. □ **declamation** *n.*, **declamatory** *adj.*

declare *v.* **1** announce openly or formally. **2** state firmly. □ **declaration** *n.*, **declaratory** *adj.*

■ **1** announce, broadcast, make known, proclaim, promulgate, pronounce. **2** affirm, assert, attest, avow, claim, profess, protest, state, swear. □ **declaration** announcement, proclamation, promulgation, pronouncement; affirmation, assertion, avowal, deposition, profession, protestation, statement, testimony.

declassify *v.* cease to classify as secret. □ **declassification** *n.*

decline *v.* **1** refuse. **2** slope downwards. **3** decrease, lose strength or vigour. ● *n.* gradual decrease or loss of strength.

■ *v.* **1** *colloq.* pass up, refuse, turn down. **2** descend, dip, slope downwards. **3** decrease, diminish, drop, dwindle, ebb, flag, lessen, subside, tail off, taper off, wane; decay, degenerate, deteriorate, fail, weaken. ● *n.* decrease, diminution, drop, fall, lessening, reduction.

declivity *n.* downward slope.

declutch *v.* disengage the clutch of a motor.

decode *v.* **1** put (a coded message) into plain language. **2** make (an electronic signal) intelligible. □ **decoder** *n.*

decompose *v.* (cause to) rot or decay. □ **decomposition** *n.*

decompress *v.* **1** release from compression. **2** reduce air pressure in. □ **decompression** *n.*

decongestant *n.* medicinal substance that relieves congestion.

decontaminate *v.* rid of contamination. □ **decontamination** *n.*

decor *n.* style of decoration used in a room.

decorate *v.* **1** make look attractive by adding objects or details. **2** paint or paper the walls of. **3** confer a medal or award on. □ **decoration** *n.*

■ **1** adorn, beautify, deck, dress, embellish, embroider, festoon, garnish, ornament, smarten up, spruce up, *colloq.* tart up, trim. **2** do up, paper, paint, redecorate, refurbish, renovate.

decorative *adj.* ornamental. □ **decoratively** *adv.*

decorator *n.* person who decorates professionally.

decorous *adj.* having or showing decorum. □ **decorously** *adv.*

■ correct, decent, demure, dignified, gentlemanly, ladylike, polite, proper, refined, respectable, seemly, well-behaved.

decorum *n.* correctness and dignity of behaviour.

■ correctness, decency, dignity, good manners, politeness, propriety, respectability, seemliness.

decoy *n.* person or animal used to lure others into danger. ● *v.* lure by a decoy.

■ *n.* bait, enticement, lure, trap. ● *v.* attract, draw, entice, inveigle, lure, seduce, tempt, trick.

decrease *v.* reduce, diminish. ● *n.* **1** decreasing. **2** amount of this.

■ *v.* curtail, cut, lessen, lower, reduce; abate, decline, diminish, drop, dwindle, ebb, fall, go down, shrink, subside, tail off, taper off, wane. ● *n.* abatement, curtailment, cut, decline, diminution, drop, dwindling, ebb, fall, lessening, reduction, shrinkage.

decree *n.* order given by a government or other authority. ● *v.* (**decreed**) order by decree.

■ *n.* command, dictate, dictum, directive, edict, injunction, law, order, ordinance, proclamation, regulation, ruling, statute. ● *v.* command, dictate, direct, ordain, order, proclaim, pronounce, rule.

decrepit *adj.* **1** made weak by age or infirmity. **2** dilapidated. □ **decrepitude** *n.*

■ **1** debilitated, doddery, feeble, frail, infirm, weak, worn-out. **2** crumbling, decaying, derelict, dilapidated, ramshackle, rickety, tumbledown.

decry *v.* disparage.

dedicate *v.* devote to a person, use, or cause. □ **dedication** *n.*

■ commit, consecrate, devote, give, pledge. □ **dedication** allegiance, commitment, devotion, faithfulness, fidelity, loyalty.

deduce *v.* infer. □ **deducible** *adj.*

■ assume, conclude, gather, infer, presume, suppose, surmise, understand.

deduct *v.* subtract.

deduction *n.* **1** deducting. **2** thing deducted. **3** deducing. **4** conclusion deduced.

deductive *adj.* based on reasoning.

deed *n.* **1** thing done, act. **2** legal document.

■ **1** accomplishment, achievement, act, action, exploit, feat.

deem *v.* consider to be.

deep *adj.* **1** going or situated far down or in. **2** (of colours) intense. **3** low-pitched. **4** heartfelt. **5** absorbed. **6** profound. □ **deeply** *adv.*

■ **1** bottomless, unfathomable, unfathomed. **2** dark, intense, rich, strong. **3** bass, booming, low, low-pitched, resonant, resounding, sonorous. **4** ardent, earnest, genuine, heartfelt, intense, profound, sincere. **5** absorbed, engrossed, immersed, intent, involved, lost, preoccupied, rapt. **6** abstruse, difficult, esoteric, heavy, profound, weighty.

deepen *v.* make or become deeper.

deer *n.* (*pl.* **deer**) hoofed animal, male of which usu. has antlers.

deerstalker *n.* cloth cap with a peak in front and at the back.

deface *v.* disfigure. □ **defacement** *n.*

■ blemish, damage, disfigure, harm, impair, injure, mar, mutilate, ruin, spoil.

de facto existing in fact.

defamatory *adj.* defaming.

defame *v.* attack the good reputation of. □ **defamation** *n.*

default *v.* fail to fulfil one's obligations or to appear. ● *n.* this failure. □ **defaulter** *n.*

defeat *v.* **1** win victory over. **2** cause to fail. ● *n.* act or process of defeating or being defeated.

■ *v.* **1** beat, be victorious over, conquer, crush, destroy, get the better of, *colloq.* lick, overcome, overpower, overthrow, overwhelm, prevail over, rout, subdue, suppress, thrash, triumph over, trounce, vanquish. **2** end, foil, frustrate, stop, thwart. ● *n.* beating, conquest, overthrow, rout, trouncing; end, frustration.

defeatism *n.* readiness to accept defeat. □ **defeatist** *n.* & *adj*

defecate *v.* discharge faeces from the body. □ **defecation** *n.*

defect *n.* /déefekt/ shortcoming, imperfection. ● *v.* /difékt/ desert one's country etc. for another. □ **defection** *n.*, **defector** *n.*

■ *n.* deficiency, failing, fault, imperfection, shortcoming, weakness, weak point; blemish, flaw, mark, spot, stain.

defective *adj.* having defect(s). □ **defectiveness** *n.*

■ broken, deficient, faulty, flawed, impaired, imperfect, incomplete, out of order.

defence *n.* **1** (means of) defending. **2** arguments against an accusation. □ **defenceless** *adj.*

■ **1** barrier, cover, guard, protection, safeguard, shelter, shield. **2** argument, excuse, explanation, justification, plea, reason, vindication.

defend *v.* **1** protect from attack. **2** speak or write in favour of. **3** represent (the defendant). □ **defender** *n.*

■ **1** guard, keep safe, preserve, protect, safeguard, screen, shelter, shield, watch over. **2** argue for, back, champion, plead for, stand by, stand up for, *colloq.* stick up for, support, uphold.

defendant *n.* person accused or sued in a lawsuit.

defensible *adj.* able to be defended.

defensive *adj.* **1** intended for defence. **2** in an attitude of defence. □ **defensively** *adv.*, **defensiveness** *n.*

defer[1] *v.* (**deferred**) postpone. □ **deferment** *n.*, **deferral** *n.*

■ adjourn, delay, postpone, put off, shelve, suspend.

defer[2] *v.* (**deferred**) yield to a person's wishes etc.

■ bow, capitulate, give way, submit, yield.

deference *n.* polite respect. □ **deferential** *adj.*

■ civility, consideration, courtesy, politeness, regard, respect.

defiance *n.* **1** bold resistance. **2** open disobedience. □ **defiant** *adj.*, **defiantly** *adv.*

■ □ **defiant** aggressive, antagonistic, belligerent, bold, pugnacious, truculent; disobedient, insubordinate, mutinous, obstinate, rebellious, recalcitrant, refractory.

deficiency *n.* **1** lack, shortage. **2** thing or amount lacking.

deficient *adj.* incomplete or insufficient.

■ defective, faulty, incomplete, lacking, imperfect, inadequate, insufficient, short, unsatisfactory, wanting.

deficit *n.* amount by which a total falls short.

defile[1] *v.* make dirty, pollute.

■ contaminate, corrupt, dirty, foul, poison, pollute, soil, stain, sully, taint.

defile[2] *n.* narrow pass or gorge.

define *v.* **1** state or explain precisely. **2** mark the boundary of.

■ **1** describe, detail, explain, specify, spell out. **2** bound, circumscribe, delineate, demarcate, outline.

definite *adj.* **1** exact. **2** clear and distinct. **3** certain, sure. □ **definitely** *adv.*

■ **1** exact, precise, specific. **2** clear, explicit, distinct, obvious, plain, unambiguous, unequivocal, well-defined. **3** certain, fixed, positive, settled, sure.

definition *n.* **1** statement of meaning. **2** clearness of outline.

definitive *adj.* **1** finally fixing or settling something. **2** most authoritative. □ **definitively** *adv.*

■ **1** conclusive, decisive, final, ultimate. **2** authoritative, reliable.

deflate *v.* (cause to) collapse through release of air. □ **deflation** *n.*

deflect *v.* turn aside. □ **deflection** *n.*, **deflector** *n.*

■ avert, divert, fend off, head off, sidetrack, turn aside; deviate, swerve, swing away, veer.

defoliate *v.* remove the leaves of. □ **defoliant** *n.*, **defoliation** *n.*

deforest *v.* clear of trees. □ **deforestation** *n.*

deform *v.* spoil the shape of. □ **deformation** *n.*

deformity *n.* abnormality of shape, esp. of a part of the body.

defraud *v.* cheat by fraud.

■ cheat, *sl.* con, dupe, fleece, *colloq.* rip off, swindle, trick.

defray *v.* provide money to pay (costs). □ **defrayal** *n.*

defrost *v.* thaw.

deft *adj.* skilful, esp. with one's hands. □ **deftly** *adv.*

■ adept, adroit, clever, dexterous, expert, handy, skilful.

defunct *adj.* **1** dead. **2** no longer existing or functioning.

defuse *v.* **1** remove the fuse from (a bomb). **2** reduce the tension in (a situation).

defy *v.* **1** refuse to obey. **2** present insuperable obstacles to. **3** challenge to do something.

■ **1** disobey, flout, go against. **2** defeat, foil, frustrate, resist, thwart, withstand. **3** challenge, dare.

degenerate *v.* /dijénnərayt/ become worse. ● *adj.* /dijénnərət/ having lost normal or good qualities. □ **degeneration** *n.*, **degeneracy** *n.*

■ *v.* become worse, decay, decline, deteriorate, *colloq.* go to pot, go to rack and ruin, retrogress, run to seed, worsen. ● *adj.* corrupt, corrupted, debased, decadent, depraved, rotten.

degrade *v.* **1** reduce to a lower rank. **2** humiliate, disgrace. □ **degradation** *n.*

■ **1** demote, downgrade. **2** abase, cheapen, debase, demean, disgrace, dishonour, humiliate, shame.

degree *n.* **1** stage in a scale or series. **2** stage in intensity or amount. **3** academic award for proficiency. **4** unit of measurement for angles or temperature.

■ **1** level, place, point, stage, step. **2** amount, extent, intensity, magnitude, measure.

dehumanize *v.* **1** remove human qualities from. **2** make impersonal. □ **dehumanization** *n.*

dehydrate *v.* **1** make dry. **2** lose moisture. □ **dehydration** *n.*

de-ice *v.* free from ice. □ **de-icer** *n.*

deify *v.* treat as a god. □ **deification** *n.*

deign *v.* condescend.

■ condescend, demean oneself, lower oneself, stoop.

deity *n.* god, goddess.

déjà vu /dáyzhaa vōō/ feeling of having experienced the present situation before.

dejected *adj.* in low spirits.

■ blue, crestfallen, depressed, despondent, disconsolate, dispirited, downcast, downhearted, glum, heavy-hearted, in the doldrums, melancholy, miserable, sad, unhappy, woebegone.

dejection *n.* lowness of spirits.

delay *v.* **1** make or be late. **2** postpone. ● *n.* delaying.

■ *v.* **1** hinder, hold up, impede, retard, set back, slow up *or* down; dally, dawdle, *colloq.* dilly-dally, hang about *or* back, hesitate, lag behind, linger, loiter, mark time, procrastinate, stall, temporize. **2** defer, postpone, put off, shelve, suspend. ● *n.* hiatus, hold-up, interlude, interruption, lull, stoppage, wait; deferment, deferral, postponement.

delectable *adj.* delightful.

delectation *n.* enjoyment.

delegate *n.* /délligət/ representative. ● *v.* /délligayt/ entrust (a task or power) to an agent.

■ *n.* agent, ambassador, emissary, envoy, go-between, plenipotentiary, representative, spokesperson. ● *v.* assign, entrust, give, hand over, transfer.

delegation *n.* **1** delegating. **2** group of representatives.

delete *v.* strike out (a word etc.). □ **deletion** *n.*

■ blot out, cancel, cross out, efface, erase, expunge, obliterate, remove, rub out, strike out.

deleterious *adj.* harmful.

deliberate *adj.* /dilíbbərət/ **1** intentional. **2** slow and careful. ● *v.* /dilíbbərayt/ **1** think carefully. **2** discuss. □ **deliberately** *adv.*

■ *adj.* **1** calculated, conscious, considered, intended, intentional, planned, preconceived, premeditated, purposeful, studied, wilful. **2** careful, cautious, methodical, orderly, painstaking, punctilious, systematic, thoughtful, thorough, unhurried. ● *v.* **1** cogitate, meditate, ponder, reflect, ruminate, think. **2** consider, debate, discuss.

deliberation *n.* **1** deliberating. **2** being deliberate.

delicacy *n.* **1** being delicate. **2** choice food.

■ **1** beauty, daintiness, grace; fragility, frailty, weakness; awkwardness, difficulty, sensitivity, trickiness; consideration, discretion, tact, thoughtfulness. **2** luxury, titbit, treat.

delicate *adj.* **1** exquisite. **2** not robust or strong. **3** requiring or using tact. □ **delicately** *adv.*

■ **1** beautiful, dainty, elegant, exquisite, fine, graceful. **2** feeble, fragile, frail, sickly, unhealthy, weak. **3** awkward, difficult, embarrassing, sensitive, *colloq.* sticky, ticklish, tricky; considerate, discreet, tactful, thoughtful.

delicatessen *n.* shop selling prepared delicacies.

delicious *adj.* delightful, esp. to taste or smell. ▫ **deliciously** *adv.*

■ appetizing, choice, delectable, delightful, luscious, savoury, *colloq.* scrumptious, tasty, *colloq.* yummy.

delight *n.* **1** great pleasure. **2** thing giving this. ● *v.* please greatly. ▫ **delight in** take great pleasure in. **delightful** *adj.*, **delightfully** *adv.*

■ *n.* **1** bliss, ecstasy, enjoyment, gratification, joy, pleasure, rapture, satisfaction. **2** joy, pleasure, treat. ● *v.* amuse, captivate, charm, enchant, entertain, entrance, gladden, gratify, please, satisfy, thrill. ▫ **delight in** enjoy, glory in, like, love, relish, revel in. **delightful** agreeable, amusing, captivating, charming, enchanting, enjoyable, entertaining, *colloq.* heavenly, *colloq.* lovely, pleasant, pleasing.

delimit *v.* determine the limits or boundaries of. ▫ **delimitation** *n.*

delineate *v.* outline. ▫ **delineation** *n.*, **delineator** *n.*

delinquent *adj.* & *n.* (person) guilty of persistent law-breaking. ▫ **delinquency** *n.*

delirium *n.* **1** disordered state of mind. **2** wild excitement. ▫ **delirious** *adj.*, **deliriously** *adv.*

■ ▫ **delirious** crazy, demented, frantic, frenzied, hysterical, incoherent, irrational, rambling, raving, unhinged; beside oneself, ecstatic, thrilled, wild.

deliver *v.* **1** distribute (letters, goods, etc.) to their destination. **2** hand over. **3** utter. **4** launch or aim (a blow etc.). **5** rescue, set free. **6** assist in the birth (of). ▫ **deliverer** *n.*, **delivery** *n.*

■ **1** bring, carry, convey, distribute, give *or* hand out, take round, transport. **2** commit, give up, hand over, relinquish, surrender, yield. **3** give, make, present, read, recite, utter. **4** administer, aim, deal, direct, inflict, launch, send, throw. **5** emancipate, liberate, release, rescue, save, set free.

deliverance *n.* rescue, freeing.

dell *n.* small wooded hollow.

delphinium *n.* tall garden plant with usu. blue flowers.

delta *n.* **1** fourth letter of the Greek alphabet, = d. **2** triangular patch of alluvial land at the mouth of a river.

delude *v.* deceive.

deluge *n.* & *v.* flood.

delusion *n.* false belief or impression. ▫ **delusive** *adj.*

■ error, illusion, misapprehension, misconception, mistake.

de luxe **1** of superior quality. **2** luxurious.

delve *v.* search deeply.

demagogue *n.* person who wins support by appealing to popular feelings and prejudices. ▫ **demagogic** *adj.*, **demagogy** *n.*

demand *n.* **1** firm or official request. **2** desire for goods or services. **3** urgent claim. ● *v.* **1** make a demand for. **2** need.

■ *n.* **1** call, command, order, request. **2** call, desire, market, need, requirement. **3** call, claim. ● *v.* **1** ask for, claim, insist on, require, requisition. **2** necessitate, need, require, want.

demanding *adj.* **1** making demands, hard to satisfy. **2** requiring great skill or effort.

■ **1** difficult, importunate, insistent, persistent, troublesome, trying. **2** arduous, challenging, difficult, exacting, hard, laborious, onerous, strenuous, taxing, tough.

demarcation *n.* marking of a boundary or limits.

demean *v.* lower the dignity of.

demeanour *n.* behaviour.

demented *adj.* mad.

dementia *n.* a type of insanity.

demerara *n.* brown raw cane sugar.

demi- *pref.* half.

demilitarize *v.* remove military forces from. ▫ **demilitarization** *n.*

demise *n.* death.

demist *v.* clear mist from (a windscreen etc.). ▫ **demister** *n.*

demobilize *v.* disband (troops) etc. ▫ **demobilization** *n.*

democracy *n.* **1** government by all the people, usu. through elected representatives. **2** country governed in this way.

democrat *n.* person favouring democracy.

democratic *adj.* of or according to democracy. ▫ **democratically** *adv.*

demography *n.* statistical study of human populations. ▫ **demographic** *adj.*

demolish *v.* **1** pull or knock down. **2** destroy. ▫ **demolition** *n.*

■ **1** destroy, dismantle, knock down, level, pull to pieces, raze, tear down, topple, wreck. **2** destroy, put an end to, put paid to, ruin, shatter, spoil, wreck.

demon *n.* **1** devil, evil spirit. **2** cruel or forceful person. □ **demonic** *adj.*, **demoniac(al)** *adj.*

demonstrable *adj.* able to be demonstrated. □ **demonstrability** *n.*, **demonstrably** *adv.*

demonstrate *v.* **1** show evidence of, prove. **2** show the working of. **3** take part in a public protest. □ **demonstration** *n.*, **demonstrator** *n.*

■ **1** display, establish, evidence, evince, exhibit, indicate, make evident, manifest, prove, show. **2** describe, explain, illustrate. **3** march, parade, protest.

demonstrative *adj.* showing one's feelings readily, affectionate. □ **demonstratively** *adv.*

■ affectionate, effusive, emotional, expansive, open, uninhibited, unreserved.

demoralize *v.* dishearten. □ **demoralization** *n.*

demote *v.* reduce to a lower rank or category. □ **demotion** *n.*

demur *v.* (**demurred**) raise objections. ● *n.* objection raised.

demure *adj.* **1** quiet, modest. **2** coy. □ **demurely** *adv.*

den *n.* **1** wild animal's lair. **2** person's small private room.

denationalize *v.* privatize. □ **denationalization** *n.*

denature *v.* **1** change the properties of. **2** make (alcohol) unfit for drinking.

denial *n.* act or instance of denying or refusing.

■ contradiction, disavowal, disclaimer, negation, refutation, repudiation; refusal, rejection.

denier /dényər/ *n.* unit of weight measuring the fineness of yarn.

denigrate *v.* disparage the reputation of. □ **denigration** *n.*

denim *n.* **1** strong twilled fabric. **2** (*pl.*) trousers made of this.

denizen *n.* inhabitant or occupant.

denominate *v.* **1** name. **2** describe as.

denomination *n.* **1** specified Church or sect. **2** class of units of measurement or money. **3** name. □ **denominational** *adj.*

■ **1** Church, faith, order, persuasion, sect. **2** class, kind, size, sort, type, value. **3** designation, name, term, title.

denominator *n.* number below the line in a vulgar fraction.

denote *v.* **1** be the sign or symbol of. **2** indicate. □ **denotation** *n.*

■ **1** betoken, represent, signify, stand for, symbolize. **2** connote, imply, indicate, mean, suggest.

denouement /daynoomon/ *n.* final outcome of a play or story.

denounce *v.* **1** speak against, condemn. **2** inform against.

■ **1** attack, castigate, censure, condemn, criticize, impugn, pillory, rail against, vilify, vituperate. **2** betray, incriminate, inform against, report.

dense *adj.* **1** closely compacted in substance, thick. **2** crowded together. **3** stupid. □ **densely** *adv.*, **denseness** *n.*

■ **1** close, compact, compressed, heavy, impenetrable, solid, thick. **2** congested, crammed, crowded, packed. **3** *colloq.* dim, obtuse, slow, stupid, *colloq.* thick.

density *n.* **1** denseness. **2** relation of weight to volume.

dent *n.* hollow left by a blow or pressure. ● *v.* make or become dented.

dental *adj.* **1** of or for teeth. **2** of dentistry.

dentifrice *n.* substance for cleaning teeth.

dentist *n.* person qualified to treat, extract, etc., teeth.

dentistry *n.* dentist's work.

denture *n.* set of artificial teeth.

denude *v.* strip of covering or property. □ **denudation** *n.*

denunciation *n.* denouncing.

deny *v.* **1** say that (a thing) is untrue or does not exist. **2** disown. **3** prevent from having.

■ **1** challenge, contradict, controvert, disclaim, dispute, gainsay. **2** disown, renounce, repudiate. **3** deprive of, forbid, refuse.

deodorant *n.* substance that removes or conceals smells.

deodorize *v.* destroy the odour of. □ **deodorization** *n.*

depart *v.* go away, leave.

■ go, go away, leave, retire, retreat, set off *or* out, withdraw.

department *n.* section of an organization. □ **department store** large shop selling many kinds of goods. **departmental** *adj.*

departure *n.* **1** departing. **2** new course of action.

depend *v.* **depend on 1** be determined by. **2** rely on. **3** be unable to do without.

■ **1** be conditional *or* dependent on, be subject to, hang on, hinge on. **2** bank on, count on, lean on, put one's faith in, reckon on, rely on, trust (in).

dependable *adj.* reliable.

dependant *n.* one who depends on another for support.

dependence *n.* depending.

dependency *n.* dependent state.

dependent *adj.* **1** depending. **2** controlled by another.

depict *v.* represent in a picture or in words. □ **depiction** *n.*

■ characterize, delineate, describe, draw, paint, picture, portray, represent, show.

depilatory *adj.* & *n.* (substance) removing hair.

deplete *v.* reduce by using quantities of. □ **depletion** *n.*

deplorable *adj.* **1** regrettable. **2** very bad. □ **deplorably** *adv.*

■ **1** despicable, disgraceful, lamentable, regrettable, reprehensible, scandalous, shameful, shocking. **2** abominable, appalling, atrocious, awful, bad, dreadful, execrable, terrible.

deplore *v.* find or call deplorable.

deploy *v.* spread out, organize for effective use. □ **deployment** *n.*

depopulate *v.* reduce the population of. □ **depopulation** *n.*

deport *v.* remove (a person) from a country. □ **deportation** *n.*

deportment *n.* behaviour, bearing.

depose *v.* remove from power.

deposit *v.* (**deposited**) **1** put or lay down. **2** entrust for safe keeping. **3** pay as a deposit. **4** leave as a layer of matter. ● *n.* **1** sum entrusted or left as guarantee. **2** something deposited. □ **depositor** *n.*

■ *v.* **1** dump, lay (down), leave, park, place, put (down), set (down). **2** bank, consign, entrust, lodge, save, set aside, *colloq.* stash, store. ● *n.* **2** accumulation, alluvium, dregs, precipitate, sediment, silt.

deposition *n.* **1** deposing. **2** depositing. **3** sworn statement.

depository *n.* storehouse.

depot /déppō/ *n.* **1** storage area, esp. for vehicles. **2** (*US*) bus or railway station.

deprave *v.* corrupt morally.

depravity *n.* moral corruption.

deprecate *v.* express disapproval of. □ **deprecation** *n.*, **deprecatory** *adj.*

depreciate *v.* make or become lower in value. □ **depreciation** *n.*

depredation *n.* plundering.

depress *v.* **1** press down. **2** reduce (trade etc.). **3** make sad. □ **depressant** *adj.* & *n.*

■ **3** demoralize, discourage, dishearten, dismay, dispirit, sadden.

depression *n.* **1** extreme dejection. **2** long period of inactivity in trading. **3** sunken place. **4** area of low atmospheric pressure. □ **depressive** *adj.*

■ **1** dejection, despair, despondency, gloom, glumness, melancholy, sadness, the blues, unhappiness. **2** economic decline, recession, slump. **3** cavity, dent, dimple, dip, hollow, indentation.

deprive *v.* prevent from having or enjoying. □ **deprivation** *n.*

■ dispossess, divest, rob, strip.

depth *n.* **1** deepness, measure of this. **2** deepest or most central part. □ **depth charge** bomb that explodes under water. **in depth** thoroughly. **out of one's depth** in water too deep to stand in.

deputation *n.* group of people sent to represent others.

depute *v.* appoint to act as one's representative.

deputize *v.* act as deputy.

deputy *n.* person appointed to act as a substitute or representative.

■ proxy, replacement, reserve, stand-in, substitute, surrogate; agent, delegate, emissary, envoy, spokesperson.

derail *v.* cause (a train) to leave the rails. □ **derailment** *n.*

derange *v.* **1** disrupt. **2** make insane. □ **derangement** *n.*

derelict *adj.* left to fall into ruin.

■ abandoned, decaying, decrepit, deserted, dilapidated, neglected, ruined, tumbledown.

dereliction *n.* **1** abandonment. **2** neglect (of duty).

deride *v.* scoff at.

■ jeer at, laugh at, make fun of, mock (at), poke fun at, ridicule, scoff at, taunt.

derision *n.* scorn, ridicule.

derisive *adj.* scornful, showing derision. □ **derisively** *adv.*

derisory *adj.* **1** showing derision. **2** deserving derision.

derivative *adj.* & *n.* derived (thing).

derive *v.* **1** obtain from a source. **2** have its origin. ◻ **derivation** *n.*

■ **1** draw, extract, gain, get, obtain, procure, secure. **2** arise, develop, emanate, originate, proceed, spring, stem.

dermatitis *n.* inflammation of the skin.

dermatology *n.* study of skin diseases. ◻ **dermatologist** *n.*

derogatory *adj.* disparaging.

derrick *n.* **1** crane. **2** framework over an oil well etc.

derv *n.* fuel for diesel engines.

dervish *n.* member of a Muslim religious order known for their whirling dance.

desalinate *v.* remove salt from. ◻ **desalination** *n.*

descant *n.* treble accompaniment to a main melody.

descend *v.* **1** come, go, or slope down. **2** stoop to unworthy behaviour. ◻ **be descended from** have as one's ancestor(s).

■ **1** climb down, come down, go down, move down; decline, drop, fall, plunge, sink, slope down. **2** condescend, lower oneself, stoop.

descendant *n.* person descended from another.

descent *n.* **1** descending. **2** downward slope. **3** lineage.

■ **1** drop, fall. **2** declivity, dip, incline, slant, slope. **3** ancestry, blood, extraction, family, heredity, lineage, origins, parentage, pedigree, stock.

describe *v.* **1** give a description of. **2** mark the outline of.

■ **1** chronicle, give an account of, narrate, outline, recount, relate, report, speak of, tell of; characterize, depict, paint, portray.

description *n.* **1** representation of a person or thing. **2** sort, kind.

■ **1** account, narrative, report, story; characterization, depiction, portrait, portrayal, representation, sketch. **2** category, character, kind, nature, sort, type, variety.

descriptive *adj.* describing.

descry *v.* catch sight of, discern.

desecrate *v.* violate the sanctity of. ◻ **desecration** *n.*, **desecrator** *n.*

desegregate *v.* abolish segregation in or of. ◻ **desegregation** *n.*

deselect *v.* decline to select or retain as a constituency candidate. ◻ **deselection** *n.*

desert[1] /dézzert/ *n.* barren uninhabited often sandy area. ● *adj.* desolate, barren.

■ *n.* waste, wilderness. ● *adj.* arid, bare, barren, desolate, empty, lonely, uncultivated, uninhabited, unpeopled, wild.

desert[2] /dizért/ *v.* **1** abandon. **2** leave one's service in the armed forces without permission. ◻ **deserter** *n.*, **desertion** *n.*

■ **1** abandon, *sl.* ditch, jilt, forsake, leave, leave in the lurch, walk out on; maroon, strand. **2** abscond, defect, run away.

deserts *n.pl.* what one deserves.

deserve *v.* be worthy of or entitled to. ◻ **deservedly** *adv.*

■ be entitled to, be worthy of, earn, justify, merit, rate, warrant.

desiccate *v.* dry out moisture from. ◻ **desiccation** *n.*

design *n.* **1** plan or sketch for a product. **2** general form or arrangement. **3** lines or shapes as decoration. **4** mental plan. ● *v.* **1** prepare a design for. **2** intend. ◻ **designedly** *adv.*, **designer** *n.*

■ *n.* **1** blueprint, draft, drawing, model, pattern, plan, sketch. **2** arrangement, composition, configuration, form, format, layout, organization, shape, style. **3** motif, pattern. **4** aim, goal, intention, object, objective, plan, purpose, target. ● *v.* **1** conceive (of), create, devise, invent, originate, plan; develop, draft, draw, fashion, form, make, shape, sketch. **2** aim, intend, mean, plan, purpose.

designate *adj.* /dézzignət/ appointed but not yet installed. ● *v.* /dézzignayt/ **1** name as. **2** specify. **3** appoint to a position. ◻ **designation** *n.*

■ *v.* **1** call, christen, describe as, dub, label, name, nickname. **2** appoint, establish, fix, pinpoint, set, specify, state, stipulate. **3** appoint, assign, choose, elect, name, nominate, select.

designing *adj.* scheming.

■ artful, calculating, crafty, cunning, devious, guileful, machiavellian, scheming, sly, wily.

desirable *adj.* arousing desire, worth desiring. ◻ **desirability** *n.*

■ alluring, attractive, captivating, seductive, sexy; covetable, enviable; advantageous, beneficial, preferable, profitable, worthwhile.

desire *n.* **1** feeling of wanting something strongly. **2** thing desired. ● *v.* feel a desire for.

■ *n.* **1** appetite, craving, hankering, hunger, itch, longing, thirst, yearning, *colloq.* yen; lasciviousness, lust, passion. ● *v.* covet, crave, hanker after, *colloq.* have a yen for, hope for, hunger for, itch for, long for, lust after, pine for, want, wish for, yearn for.

desirous *adj.* desiring.

desist *v.* cease.

desk *n.* **1** piece of furniture for reading or writing at. **2** counter. **3** section of a newspaper office.

desolate *adj.* **1** left alone. **2** ruined, uninhabited. **3** miserable. □ **desolation** *n.*, **desolated** *adj.*

■ **1** abandoned, alone, bereft, lonely, neglected, solitary. **2** bare, barren, bleak, deserted, dreary, empty, ruined, uninhabited. **3** dejected, despondent, disconsolate, forlorn, melancholy, miserable, mournful, sad, sorrowful, unhappy, woebegone, wretched.

despair *n.* complete lack of hope. ● *v.* feel despair.

■ *n.* dejection, depression, desperation, despondency, hopelessness, misery, wretchedness.

desperado *n.* (*pl.* **-oes**) reckless criminal.

desperate *adj.* **1** reckless through despair. **2** extremely serious or dangerous. □ **desperately** *adv.*, **desperation** *n.*

■ **1** foolhardy, impetuous, rash, reckless, wild; at one's wits' end, despairing, frantic. **2** acute, critical, grave, pressing, serious, urgent; dangerous, hazardous, hopeless, perilous, precarious.

despicable *adj.* contemptible. □ **despicably** *adv.*

despise *v.* regard as worthless.

■ abhor, be contemptuous of, detest, disdain, hate, loathe, look down on, scorn.

despite *prep.* in spite of.

despoil *v.* plunder. □ **despoliation** *n.*

despondent *adj.* dejected. □ **despondently** *adv.*, **despondency** *n.*

despot *n.* dictator. □ **despotic** *adj.*, **despotism** *n.*

dessert *n.* sweet course of a meal. □ **dessertspoon** *n.* medium-sized spoon for dessert.

destination *n.* place to which a person or thing is going.

destine *v.* settle the future of, set apart for a purpose.

destiny *n.* **1** fate. **2** one's future destined by fate.

■ doom, fate, fortune, kismet, lot.

destitute *adj.* without means to live. □ **destitution** *n.*

■ down-and-out, impecunious, impoverished, indigent, needy, penniless, poor, poverty-stricken.

destroy *v.* **1** pull or break down. **2** ruin. **3** kill (an animal). **4** defeat. □ **destruction** *n.*, **destructive** *adj.*

■ **1** annihilate, demolish, devastate, knock down, lay waste, obliterate, pull down, ravage, raze, ruin, tear down, vandalize, wipe out, wreak havoc on, wreck. **2** bring to an end, dash, end, finish, kill, put an end to, ruin, shatter, spoil, terminate. □ **destruction** annihilation, demolition, devastation, ruin, ruination.

destroyer *n.* **1** one who destroys. **2** fast warship.

destruct *v.* destroy deliberately. □ **destructible** *adj.*

desultory *adj.* going from one subject to another.

detach *v.* release or separate. □ **detachable** *adj.*

■ cut off, disconnect, disengage, disentangle, free, pull off, release, remove, separate, uncouple, undo, unfasten.

detached *adj.* **1** separate. **2** impartial, unemotional.

■ **1** disconnected, separate, separated, unattached. **2** disinterested, dispassionate, impartial, neutral, objective, unbiased, unemotional, unprejudiced.

detachment *n.* **1** detaching. **2** being detached. **3** military group.

detail *n.* **1** small fact or item. **2** such items collectively. **3** small military detachment. ● *v.* **1** relate in detail. **2** assign to special duty.

■ *n.* **1** aspect, component, element, fact, factor, feature, item, nicety, particular, point, respect, technicality; (**details**) minutiae, specifics. **3** detachment, group, party, squad, unit. ● *v.* **1** enumerate, itemize, list, recount, relate, specify, spell out.

detailed *adj.* containing many details.

detain *v.* **1** keep in custody. **2** cause delay to. □ **detainment** *n.*

■ **1** confine, imprison, intern, keep in custody, remand. **2** delay, hold back *or* up, impede, keep, retard, slow down.

detainee *n.* person detained in custody.

detect *v.* discover or perceive the existence or presence of. □ **detection** *n.*, **detector** *n.*

■ ascertain, discover, ferret out, find, locate, uncover, unearth; become aware of, discern, feel, identify, note, notice, observe, perceive, scent, sense, smell.

detective *n.* person, esp. a police officer, who investigates crimes.

détente /daytóNt/ *n.* easing of tension between states.

detention *n.* **1** detaining. **2** imprisonment.

deter *v.* (**deterred**) **1** discourage from action. **2** prevent.

■ **1** discourage, dissuade, frighten off, put off, scare off. **2** hinder, impede, obstruct, prevent, stop.

detergent *adj.* & *n.* cleansing (substance, esp. other than soap).

deteriorate *v.* become worse. □ **deterioration** *n.*

■ decline, degenerate, get worse, *colloq.* go to pot, worsen; decay, disintegrate, fall apart.

determination *n.* **1** firmness of purpose. **2** process of deciding.

■ **1** firmness, fortitude, *colloq.* grit, *colloq.* guts, perseverance, persistence, resoluteness, resolution, resolve, tenacity, willpower.

determine *v.* **1** establish precisely. **2** decide, settle. **3** be a decisive factor in.

■ **1** ascertain, discover, establish, find out, learn. **2** choose, decide, resolve, select, settle. **3** affect, govern, influence, shape.

determined *adj.* resolute.

deterrent *n.* thing that deters. □ **deterrence** *n.*

detest *v.* dislike intensely. □ **detestable** *adj.*, **detestation** *n.*

■ abhor, abominate, despise, execrate, hate, loathe.

dethrone *v.* remove from a throne. □ **dethronement** *n.*

detonate *v.* explode. □ **detonation** *n.*, **detonator** *n.*

detour *n.* deviation from a direct or intended course.

detract *v.* **detract from** reduce, diminish. □ **detraction** *n.*

■ devalue, diminish, lessen, reduce, take away from.

detractor *n.* person who criticizes a thing unfavourably.

detriment *n.* harm. □ **detrimental** *adj.*, **detrimentally** *adv.*

■ damage, disadvantage, harm, hurt, impairment, injury. □ **detrimental** adverse, damaging, deleterious, disadvantageous, harmful, hurtful, injurious, prejudicial, unfavourable.

deuterium *n.* heavy form of hydrogen.

Deutschmark /dóychmaark/ *n.* unit of money in Germany.

devalue *v.* reduce the value of. □ **devaluation** *n.*

devastate *v.* **1** cause great destruction to. **2** upset deeply. □ **devastation** *n.*

■ **1** annihilate, demolish, destroy, lay waste, obliterate, ravage, raze, ruin, wreck. **2** overwhelm, shatter, shock, stagger.

devastating *adj.* overwhelming.

develop *v.* (**developed**) **1** make or become larger or more mature or organized. **2** begin to exhibit or suffer from. **3** come into existence. **4** make usable or profitable, build on (land). **5** treat (a film) so as to make a picture visible. □ **developer** *n.*, **development** *n.*

■ **1** broaden, build up, extend, increase, promote, strengthen; amplify, elaborate on, enlarge on, expand on; advance, evolve, improve, mature, progress. **2** acquire, contract, get, pick up. **3** arise, begin, come about, happen, occur. □ **development** enlargement, evolution, expansion, extension, growth, increase; advance, improvement, progress.

deviant *adj.* & *n.* (person or thing) deviating from normal behaviour.

deviate *v.* turn aside. □ **deviation** *n.*

■ diverge, stray, swerve, turn aside, veer; digress, drift, get off the subject, wander.

device *n.* **1** thing made or used for a purpose. **2** scheme.

■ **1** apparatus, appliance, contraption, contrivance, gadget, implement, instrument, invention, machine, tool, utensil. **2** gambit, manoeuvre, plan, ploy, ruse, scheme, stratagem, tactic, trick.

devil *n.* **1** evil spirit. **2** (**the Devil**) supreme spirit of evil. **3** cruel or annoying person. **4** person of mischievous energy or cleverness. □ **devil's advocate** person who tests a proposition by arguing against it. **devilish** *adj.*

devilled *adj.* cooked with hot spices.

devilment *n.* mischief.

devilry *n.* **1** wickedness. **2** devilment.

devious *adj.* **1** indirect. **2** underhand. □ **deviously** *adv.*, **deviousness** *n.*

■ **1** circuitous, indirect, roundabout, serpentine, tortuous, winding, zigag. **2** artful, crafty, cunning, deceitful, deceptive, designing, dishonest, insincere, misleading, scheming, slippery, sly, treacherous, underhand, untrustworthy, wily.

devise *v.* plan, invent.

■ conceive, concoct, contrive, *colloq.* cook up, create, design, dream up, form, invent, make up, plan, scheme, *colloq.* think up, work out.

devoid *adj.* **devoid of** lacking, free from.

devolution *n.* **1** devolving. **2** delegation of power from central to local administration.

devolve *v.* pass or be passed to a deputy or successor.

devote *v.* give or use for a particular purpose.

devoted *adj.* showing devotion.

■ ardent, constant, dedicated, faithful, loving, loyal, staunch, steadfast, true, zealous.

devotee *n.* enthusiast.

devotion *n.* **1** great love or loyalty. **2** worship. **3** (*pl.*) prayers.

■ **1** ardour, fervour, love, passion, zeal; adherence, allegiance, commitment, constancy, dedication, fidelity, loyalty.

devotional *adj.* used in worship.

devour *v.* **1** eat hungrily or greedily. **2** (of fire etc.) destroy. **3** take in avidly. □ **devourer** *n.*

■ **1** bolt, consume, gobble, gulp, guzzle, *colloq.* scoff, wolf. **2** annihilate, consume, destroy, engulf, ravage. **3** absorb, drink in, take in.

devout *adj.* earnestly religious or sincere. □ **devoutly** *adv.*

■ faithful, God-fearing, pious, religious, reverent, staunch; earnest, genuine, heartfelt, sincere.

dew *n.* drops of condensed moisture on a surface.

dewclaw *n.* small claw on the inner side of a dog's leg.

dewlap *n.* fold of loose skin at the throat of cattle etc.

dexterity *n.* skill.

dexterous *adj.* (also **dextrous**) skilful. □ **dexterously** *adv.*

■ adept, adroit, clever, deft, expert, handy, skilful.

diabetes *n.* disease in which sugar and starch are not properly absorbed by the body. □ **diabetic** *adj.* & *n.*

diabolic *adj.* of the Devil.

diabolical *adj.* very cruel or wicked. □ **diabolically** *adv.*

diabolism *n.* worship of the Devil.

diaconate *n.* **1** office of deacon. **2** body of deacons. □ **diaconal** *adj.*

diadem *n.* crown.

diagnose *v.* make a diagnosis of.

diagnosis *n.* (*pl.* **-oses**) identification of a disease or condition from its symptoms. □ **diagnostic** *adj.*

diagonal *adj.* & *n.* (line) crossing from corner to corner. □ **diagonally** *adv.*

diagram *n.* drawing that shows the parts or operation of something. □ **diagrammatic** *adj.*

dial *n.* **1** face of a clock or watch. **2** similar plate or disc with a movable pointer. **3** movable disc on the front of a telephone. ● *v.* (**dialled**) select by using a dial or numbered buttons.

dialect *n.* local form of a language. □ **dialectal** *adj.*

■ idiom, language, patois, speech, tongue, vernacular.

dialectic *n.* investigation of truths in philosophy etc. by systematic reasoning. □ **dialectical** *adj.*

dialogue *n.* talk between people.

dialysis *n.* purification of blood by causing it to flow through a suitable membrane.

diamanté /diəmóntay/ *adj.* decorated with artificial jewels.

diameter *n.* **1** straight line from side to side through the centre of a circle or sphere. **2** its length.

diametrical *adj.* **1** of or along a diameter. **2** (of opposition) direct. □ **diametrically** *adv.*

diamond *n.* **1** very hard brilliant precious stone. **2** four-sided figure with equal sides and with angles that are not right angles. **3** playing card marked with such shapes. □ **diamond wedding** 60th anniversary.

diaper *n.* (*US*) baby's nappy.

diaphanous *adj.* almost transparent. □ **diaphanously** *adv.*

diaphragm /díəfram/ *n.* **1** a muscular partition between chest and abdomen. **2** contraceptive cap fitting over the cervix.

diarrhoea /díəreéə/ *n.* condition with frequent fluid faeces.

diary *n.* **1** daily record of events. **2** book for this. ▫ **diarist** *n.*

■ **1** annals, chronicle, journal, log, memoirs, record.

diatribe *n.* violent verbal attack.

dibber *n.* tool to make holes in ground for young plants.

dice *n.* (*pl.* **dice**) small cube marked on each side with 1–6 spots, used in games of chance. ● *v.* cut into small cubes. ▫ **dice with death** take great risks.

dicey *adj.* (**-ier**, **-iest**) (*sl.*) **1** risky. **2** unreliable.

dichotomy /dīkóttəmi/ *n.* division into two parts or kinds.

dicky *adj.* (**-ier**, **-iest**) (*sl.*) shaky, unsound.

dictate *v.* **1** state or order authoritatively. **2** give orders officiously. **3** say (words) aloud to be written or recorded. ● *n.* command. ▫ **dictation** *n.*

■ *v.* **1** command, decree, ordain, order, prescribe, state. ● *n.* command, decree, edict, instruction, order, requirement.

dictator *n.* **1** ruler with unrestricted authority. **2** domineering person. ▫ **dictatorship** *n.*

■ **1** autocrat, despot, tyrant.

dictatorial *adj.* of or like a dictator. ▫ **dictatorially** *adv.*

■ authoritarian, autocratic, despotic, totalitarian, tyrannical; *colloq.* bossy, dogmatic, domineering, high-handed, imperious, overbearing.

diction *n.* manner of uttering or pronouncing words.

dictionary *n.* book that lists and explains the words of a language or the topics of a subject.

dictum *n.* (*pl.* **-ta**) formal saying.

did *see* **do**.

didactic *adj.* meant or meaning to instruct. ▫ **didactically** *adv.*

die[1] *v.* (**dying**) **1** cease to be alive. **2** cease to exist or function. ▫ **be dying to** or **for** feel an intense longing to or for.

■ **1** *sl.* croak, expire, go, lose one's life, pass away, perish, *sl.* snuff it. **2** cease, come to an end, decline, diminish, disappear, dwindle, ebb, end, fade (away), fizzle out, stop, subside, vanish, wane, wilt, wither (away).

die[2] *n.* device that stamps a design or that cuts or moulds material into shape.

diehard *n.* very conservative or stubborn person.

diesel *n.* **1** diesel engine. **2** vehicle driven by this. ▫ **diesel-electric** *adj.* using an electric generator driven by a diesel engine. **diesel engine** oil-burning engine in which ignition is produced by the heat of compressed air.

diet[1] *n.* **1** usual food. **2** restricted selection of food. ● *v.* limit one's diet. ▫ **dietary** *adj.*, **dieter** *n.*

diet[2] *n.* congress, parliamentary assembly in certain countries.

dietetic *adj.* of diet and nutrition. ● *n.pl.* study of diet and nutrition.

dietitian *n.* expert in dietetics.

differ *v.* **1** be unlike. **2** disagree.

■ **1** be different, be dissimilar, contrast. **2** argue, be at odds, be at variance, clash, conflict, disagree, fall out, quarrel.

difference *n.* **1** being different. **2** way in which things differ. **3** remainder after subtraction. **4** disagreement. **5** notable change.

■ **1** discrepancy, dissimilarity, dissimilitude, diversity, inconsistency, variation. **2** contrast, disparity, distinction. **3** balance, remainder, residue, rest, surplus. **4** argument, conflict, disagreement, dispute, quarrel, tiff. **5** alteration, change, transformation.

different *adj.* **1** not the same. **2** distinct, separate. **3** unusual. ▫ **differently** *adv.*

■ **1** conflicting, contrasting, dissimilar, divergent, diverse, inconsistent, opposite, unalike, unlike. **2** disparate, distinct, separate. **3** bizarre, extraordinary, new, offbeat, original, peculiar, singular, strange, unconventional, unique, unorthodox, unusual.

differential *adj.* of, showing, or depending on a difference. ● *n.* **1** agreed difference in wage-rates. **2** arrangement of gears allowing a vehicle's wheels to revolve at different speeds when cornering.

differentiate *v.* **1** be a difference between. **2** distinguish between. **3** develop differences. ▫ **differentiation** *n.*

difficult *adj.* **1** needing much effort or skill to do, deal with, or understand. **2** troublesome. ▫ **difficulty** *n.*

■ **1** arduous, burdensome, challenging, demanding, exacting, laborious, onerous, taxing, tough; awkward, delicate, sensitive, ticklish, tricky; baffling, complex, complicated, hard, intricate, knotty, perplexing, problematic(al), puzzling. **2** *colloq.* bloody-minded, demanding, intractable, naughty, obstinate, obstreperous, obstructive, recalcitrant, refractory, stubborn, tiresome, troublesome, trying, uncooperative, unmanageable. □ **difficulty** catch, hindrance, impediment, obstacle, pitfall, problem, snag.

diffident *adj.* lacking confidence. □ **diffidently** *adv.*, **diffidence** *n.*

■ bashful, hesitant, inhibited, meek, modest, reserved, reticent, retiring, self-conscious, self-effacing, shy, tentative, timid, unassertive, unassuming.

diffract *v.* break up a beam of light into a series of coloured or dark and light bands. □ **diffraction** *n.*, **diffractive** *adj.*

diffuse *adj.* /difyóoss/ not concentrated. ● *v.* /difyóoz/ spread widely or thinly. □ **diffusion** *n.*, **diffusive** *adj.*, **diffusible** *adj.*

■ *v.* circulate, disperse, disseminate, distribute, scatter, spread.

dig *v.* (**dug, digging**) **1** break up and move soil. **2** make (a way or hole) thus. **3** poke. ● *n.* **1** excavation. **2** poke. **3** cutting remark. **4** (*pl.*) lodgings. □ **dig out** or **up** find by investigation.

■ *v.* **1** fork over, hoe, plough, till, turn over. **2** burrow, excavate, gouge (out), hollow (out), scoop (out), tunnel. **3** elbow, jab, nudge, poke, prod, stab. □ **dig out** bring to light, discover, ferret out, find, turn up, unearth.

digest *v.* /dījést/ **1** break down (food) in the body. **2** absorb into the mind. ● *n.* /díjest/ methodical summary. □ **digestible** *adj.*

digestion *n.* process or power of digesting food.

digestive *adj.* of or aiding digestion. □ **digestive biscuit** wholemeal biscuit.

digger *n.* **1** one who digs. **2** mechanical excavator.

digit *n.* **1** any numeral from 0 to 9. **2** finger or toe.

digital *adj.* of or using digits. □ **digital clock** one that shows the time as a row of figures. **digitally** *adv.*

digitalis *n.* heart stimulant prepared from foxglove leaves.

dignified *adj.* showing dignity.

■ august, courtly, distinguished, elegant, formal, grave, lofty, majestic, noble, regal, sedate, serious, sober, solemn, stately.

dignify *v.* give dignity to.

dignitary *n.* person holding high rank or position.

dignity *n.* **1** calm and serious manner. **2** high rank or position.

■ **1** decorum, grandeur, gravity, majesty, nobility, seriousness, solemnity, stateliness.

digress *v.* depart from the main subject. □ **digression** *n.*

■ deviate, drift, go off at a tangent, ramble, wander.

dike *n.* = **dyke**.

dilapidated *adj.* in disrepair.

■ crumbling, decaying, decrepit, derelict, falling down, gone to rack and ruin, gone to seed, in disrepair, in ruins, ramshackle, rickety, ruined, tumbledown.

dilapidation *n.* dilapidated state.

dilate *v.* make or become wider. □ **dilation** *n.*, **dilatation** *n.*

dilatory /díllətəri/ *adj.* delaying, not prompt.

dilemma *n.* **1** situation in which a difficult choice must be made. **2** difficult situation.

■ **1** catch-22. **2** deadlock, fix, impasse, *colloq.* pickle, plight, predicament, quandary.

dilettante /dillitánti/ *n.* person who dabbles in a subject.

diligent *adj.* working or done with care and effort. □ **diligently** *adv.*, **diligence** *n.*

■ assiduous, careful, conscientious, earnest, hard-working, industrious, meticulous, painstaking, punctilious, scrupulous, sedulous, thorough.

dill *n.* herb with spicy seeds.

dilly-dally *v.* (*colloq.*) **1** dawdle. **2** waste time by indecision.

dilute *v.* **1** reduce the strength of (fluid) by adding water etc. **2** weaken in effect. □ **dilution** *n.*

dim *adj.* (**dimmer, dimmest**) **1** lit faintly, not bright. **2** indistinct. **3** (*colloq.*) stupid. ● *v.* (**dimmed**) make or become dim. □ **dimly** *adv.*, **dimness** *n.*

■ *adj.* **1** faint, pale, weak; dark, dusky, gloomy, murky, shadowy, sombre. **2** blurred, clouded, foggy, fuzzy, hazy, ill-defined, indistinct, misty, nebulous, obscure, unclear, vague. ● *v.* cloud, darken, obscure, shade.

dime *n.* 10-cent coin of the USA.

dimension *n.* **1** measurable extent. **2** scope. □ **dimensional** *adj.*

diminish *v.* make or become smaller or less.

■ **abate, decline, decrease, dwindle, ease off, ebb, fade, lessen, shrink, subside, taper off, wane; curtail, lower, reduce.**

diminuendo *adv.* & *n.* (*pl.* **-os**) decreasing in loudness.

diminution *n.* decrease.

diminutive *adj.* tiny. ● *n.* affectionate form of a name.

dimple *n.* small dent, esp. in the skin. ● *v.* **1** show dimple(s). **2** produce dimples in.

din *n.* loud annoying noise. ● *v.* (**dinned**) force (information) into a person by constant repetition.

■ *n.* **babel, clamour, clatter, commotion, hubbub, hullabaloo, noise, racket, row, rumpus, shouting, tumult, uproar, yelling.**

dinar /deénaar/ *n.* unit of money in some Balkan and Middle Eastern countries.

dine *v.* eat dinner. □ **diner** *n.*

dinghy *n.* small open boat or inflatable rubber boat.

dingle *n.* deep dell.

dingo *n.* (*pl.* **-oes**) Australian wild dog.

dingy *adj.* (**-ier**, **-iest**) drab, dirty-looking. □ **dinginess** *n.*

■ **cheerless, dark, depressing, dim, dirty, dismal, drab, dreary, dull, frowzy, gloomy.**

dining room room in which meals are eaten.

dinky *adj.* (**-ier**, **-iest**) (*colloq.*) attractively small and neat.

dinner *n.* **1** chief meal of the day. **2** formal evening meal. □ **dinner jacket** man's usu. black jacket for evening wear.

dinosaur *n.* prehistoric reptile.

dint *n.* dent. □ **by dint of** by means of.

diocese *n.* district under the care of a bishop. □ **diocesan** *adj.*

diode *n.* semiconductor allowing the flow of current in one direction only.

dioxide *n.* oxide with two atoms of oxygen to one of a metal or other element.

dip *v.* (**dipped**) **1** plunge briefly into liquid. **2** lower, go downwards. ● *n.* **1** dipping. **2** short bathe. **3** downward slope. **4** liquid or mixture into which something is dipped. □ **dip into** read briefly from (a book).

■ *v.* **1 bathe, douse, duck, dunk, immerse, plunge, submerge. 2 decline, descend, drop, fall, go down, sag, sink, slump, subside.** ● *n.* **1 immersion; decline, drop, fall, slump. 2 bathe, swim. 3 declivity, incline, slope.**

diphtheria *n.* infectious disease with inflammation of the throat.

diphthong *n.* compound vowel sound (as *ou* in *loud*).

diploma *n.* certificate awarded on completion of a course of study.

diplomacy *n.* **1** handling of international relations. **2** tact.

■ **1 statesmanship. 2 delicacy, discretion, finesse, politeness, sensitivity, tact, tactfulness.**

diplomat *n.* **1** member of the diplomatic service. **2** tactful person.

diplomatic *adj.* **1** of or engaged in diplomacy. **2** tactful. □ **diplomatically** *adv.*

■ **2 considerate, discreet, judicious, polite, prudent, sensitive, tactful, thoughtful.**

dipper *n.* **1** diving bird. **2** ladle.

dipsomania *n.* uncontrollable craving for alcohol. □ **dipsomaniac** *n.*

diptych /díptik/ *n.* pair of pictures on two hinged panels.

dire *adj.* **1** dreadful. **2** extreme and urgent. **3** ominous.

■ **1 appalling, awful, calamitous, disastrous, dreadful, frightful, grim, horrible, terrible, unfortunate, wretched. 2 acute, desperate, drastic, extreme, grievous, serious, urgent. 3 dark, fateful, ominous, sinister, threatening.**

direct *adj.* **1** straight, not roundabout. **2** with nothing or no one between. **3** straightforward, frank. ● *adv.* by a direct route. ● *v.* **1** tell how to do something or reach a place. **2** address (a letter etc.). **3** control. **4** order. **5** cause to move in a certain direction. □ **directness** *n.*

■ *adj.* **1 shortest, straight, through, undeviating, unswerving. 3 blunt, candid, forthright, frank, honest, open, outspoken, plain, sincere, straightforward.** ● *v.* **1 advise, counsel, instruct; escort, guide, give directions, lead, show the way, usher. 3 administer, be in charge of, conduct, control, govern, handle, manage, oversee, regulate, run, steer, superintend, supervise. 4 bid, command, instruct, order, require. 5 aim, level, point, target, train, turn.**

direction *n.* **1** directing. **2** line along which a person or thing moves or faces. **3** (usu. *pl.*) instruction. □ **directional** *adj.*

■ **1 administration, control, guidance, management, regulation, supervision. 2**

bearing, course, path, route, way. 3 (directions) guidelines, instructions, orders.

directive *n.* general instruction issued by authority.

directly *adv.* **1** in a direct way. **2** very soon. ● *conj.* as soon as.

director *n.* **1** supervisor. **2** member of a board directing a business. **3** one who supervises acting and filming. □ **directorship** *n.*

directorate *n.* **1** office of director. **2** board of directors.

directory *n.* list of telephone subscribers, members, etc.

dirge *n.* song of mourning.

dirigible *n.* airship.

dirk *n.* a kind of dagger.

dirndl *n.* full gathered skirt.

dirt *n.* **1** unclean matter. **2** soil. **3** foul words. **4** scandal.

dirty *adj.* (**-ier, -iest**) **1** soiled, not clean. **2** dishonourable. **3** lewd, obscene. ● *v.* make or become dirty. □ **dirtily** *adv.*, **dirtiness** *n.*

■ *adj.* **1** dingy, dusty, filthy, foul, grimy, grubby, insanitary, marked, messy, *colloq.* mucky, muddy, polluted, smeary, soiled, sooty, sordid, squalid, stained, sullied, tainted, tarnished, unclean, unwashed. **2** dishonest, dishonourable, ignoble, low-down, mean, sordid, treacherous, unfair, unscrupulous, unsporting. ● *v.* defile, foul, mark, muddy, pollute, smear, smudge, soil, stain, sully, tarnish.

disability *n.* thing that disables.

disable *v.* **1** deprive of some ability. **2** make unfit. □ **disablement** *n.*

disabled *adj.* having a disability.

disabuse *v.* disillusion.

disadvantage *n.* something that hinders or is unhelpful. □ **disadvantaged** *adj.*, **disadvantageous** *adj.*

■ drawback, handicap, liability, set-back, weakness; detriment, disservice, harm, hurt, injury, loss.

disaffected *adj.* discontented, no longer loyal. □ **disaffection** *n.*

disagree *v.* **1** have a different opinion. **2** quarrel. **3** fail to correspond. □ **disagreement** *n.*

■ **1** be at odds, be at variance, differ, dissent. **2** argue, bicker, dispute, fall out, fight, quarrel, squabble, wrangle. **3** be different, be incompatible, conflict. □ **disagreement** conflict, controversy, dissension, dissent; altercation, argument, clash, dispute, quarrel, squabble, strife, tiff, wrangle; difference, divergence, incompatibility, disparity, dissimilarity.

disagreeable *adj.* **1** unpleasant. **2** bad-tempered. □ **disagreeably** *adv.*

■ **1** displeasing, distasteful, nasty, objectionable, offensive, repellent, unpleasant, unsavoury. **2** bad-tempered, cantankerous, crabby, cross, grumpy, impolite, irritable, quarrelsome, rude, sour, surly, uncooperative, unfriendly.

disallow *v.* refuse to sanction.

disappear *v.* pass from sight or existence. □ **disappearance** *n.*

■ become invisible, dissolve, evaporate, fade (away), melt (away), vanish.

disappoint *v.* fail to do what was desired or expected. □ **disappointment** *n.*

■ disenchant, disillusion, dissatisfy, fail, let down.

disapprobation *n.* disapproval.

disapprove *v.* **disapprove of** have or express an unfavourable opinion of. □ **disapproval** *n.*

■ censure, condemn, criticize, deplore, frown on, object to, take exception to. □ **disapproval** censure, condemnation, criticism, disapprobation, disfavour, displeasure, dissatisfaction.

disarm *v.* **1** deprive of weapon(s). **2** reduce armed forces. **3** make less hostile.

disarmament *n.* reduction of a country's forces or weapons.

disarrange *v.* put into disorder.

disarray *n.* & *v.* disorder.

disaster *n.* **1** sudden great misfortune. **2** complete failure. □ **disastrous** *adj.*, **disastrously** *adv.*

■ **1** accident, calamity, cataclysm, catastrophe, misfortune, mishap, tragedy. **2** debacle, fiasco, *sl.* flop, *colloq.* wash-out. □ **disastrous** appalling, calamitous, cataclysmic, catastrophic, devastating, dire, dreadful, horrendous, ruinous, terrible, tragic.

disavow *v.* disclaim. □ **disavowal** *n.*

disband *v.* separate, disperse.

disbelieve *v.* refuse or be unable to believe. □ **disbelief** *n.*

disburse *v.* pay out (money). □ **disbursement** *n.*

disc *n.* **1** thin circular plate or layer. **2** record bearing recorded sound. **3** = **disk**. □ **disc jockey** person who introduces and plays pop records at a disco or on the radio.

discard *v.* /diskaárd/ reject as useless or unwanted. ● *n.* /dískaard/ discarded thing.

> ■ *v.* abandon, cast off, dispose of, *sl.* ditch, get rid of, jettison, reject, scrap, throw away.

discern *v.* perceive with the mind or senses. □ **discernible** *adj.*, **discernment** *n.*

> ■ become aware of, catch sight of, descry, detect, distinguish, glimpse, make out, notice, observe, perceive, pick out, see, *colloq.* spot, spy.

discerning *adj.* perceptive, showing sensitive insight.

> ■ astute, clever, discriminating, intelligent, judicious, perceptive, perspicacious, sagacious, sage, sharp, shrewd, wise.

discharge *v.* **1** emit, pour out. **2** release. **3** dismiss. **4** pay (a debt). **5** perform (a duty etc.). ● *n.* **1** discharging. **2** substance discharged.

> ■ *v.* **1** emit, exude, leak, ooze, pour out, produce, send out. **2** free, let go, liberate, release, set free. **3** cashier, dismiss, eject, expel, fire, lay off, *colloq.* sack, throw out. **4** pay, settle, square. **5** accomplish, carry out, do, execute, fulfil, perform. ● *n.* **1** dismissal, ejection, expulsion; payment, settlement; execution, fulfilment, performance.

disciple *n.* **1** person following the teachings of another. **2** one of the original followers of Christ.

> ■ **1** adherent, admirer, devotee, follower; apprentice, learner, pupil, student.

disciplinarian *n.* person who enforces strict discipline.

disciplinary *adj.* of or for discipline.

discipline *n.* **1** control exercised over people or animals. **2** mental or moral training. **3** branch of learning. ● *v.* **1** train to be orderly. **2** punish.

> ■ *n.* **1** authority, control, order, regulation, restraint, rule. **2** drill, drilling, instruction, schooling, training. ● *v.* **1** coach, drill, indoctrinate, instruct, school, teach, train; control, curb, govern, keep in check, restrain. **2** castigate, chastise, penalize, punish, rebuke, reprimand, reprove.

disclaim *v.* disown.

> ■ deny, disown, reject, renounce, repudiate.

disclaimer *n.* statement disclaiming something.

disclose *v.* reveal. □ **disclosure** *n.*

> ■ blurt out, betray, divulge, leak, make known, reveal, tell.

disco *n.* (*pl.* **-os**) nightclub, party, etc. where people dance to recorded pop music.

discolour *v.* **1** change in colour. **2** stain. □ **discoloration** *n.*

discomfit *v.* (**discomfited**) disconcert. □ **discomfiture** *n.*

discomfort *n.* **1** being uncomfortable. **2** thing causing this.

discommode *v.* inconvenience.

disconcert *v.* disturb the composure of, fluster.

> ■ agitate, bewilder, confuse, discomfit, disturb, fluster, nonplus, perplex, perturb, puzzle, *colloq.* rattle, unsettle, upset, worry.

disconnect *v.* **1** break the connection of. **2** cut off power supply of. □ **disconnection** *n.*

disconsolate *adj.* unhappy, disappointed. □ **disconsolately** *adv.*

discontent *n.* dissatisfaction. □ **discontented** *adj.*

> ■ displeasure, dissatisfaction, unhappiness. □ **discontented** annoyed, *colloq.* browned off, disaffected, disgruntled, displeased, dissatisfied, fed up, unhappy.

discontinue *v.* **1** put an end to. **2** cease. □ **discontinuance** *n.*

discontinuous *adj.* not continuous. □ **discontinuity** *n.*

discord *n.* **1** disagreement, strife. **2** harsh sound. □ **discordance** *n.*, **discordant** *adj.*

> ■ **1** conflict, disagreement, disharmony, dissension, strife. **2** cacophony, din, dissonance, jangle. □ **discordant** conflicting, contrary, differing, disagreeing, divergent, incompatible, opposed, opposite; cacophonous, dissonant, grating, harsh, jangling, jarring, shrill, strident, tuneless, unharmonious, unmusical.

discount *n.* /dískownt/ amount deducted from the full price. ● *v.* /diskównt/ disregard.

> ■ *n.* allowance, deduction, rebate, reduction. ● *v.* dismiss, disregard, ignore, omit, overlook, pay no attention to.

discourage *v.* **1** dishearten. **2** dissuade. □ **discouragement** *n.*

> ■ **1** daunt, demoralize, depress, dishearten, dismay, unnerve. **2** deter, dissuade.

discourse *n.* /dískorss/ **1** conversation. **2** lecture. **3** treatise. ● *v.* /diskórss/ utter or write a discourse.

discourteous *adj.* lacking courtesy. ▫ **discourteously** *adv.*, **discourtesy** *n.*

discover *v.* find or find out, by effort or chance. ▫ **discovery** *n.*

■ come across, detect, dig up, ferret out, find, happen on, hit on, locate, track down, turn up, uncover, unearth; ascertain, determine, discern, find out, learn, notice, perceive, realize.

discredit *v.* (**discredited**) **1** damage the reputation of. **2** cause to be disbelieved. ● *n.* (thing causing) damage to a reputation.

■ *v.* **1** bring into disrepute, defame, disgrace, dishonour. **2** *colloq.* debunk, disprove, explode, rebut, refute. ● *n.* disgrace, disrepute, dishonour, humiliation, ignominy, shame.

discreditable *adj.* bringing discredit.

discreet *adj.* **1** circumspect, tactful. **2** unobtrusive. ▫ **discreetly** *adv.*

■ **1** careful, cautious, chary, circumspect, guarded, prudent, wary; considerate, delicate, diplomatic, judicious, tactful, thoughtful. **2** inconspicuous, low-key, unobtrusive.

discrepancy *n.* failure to tally.

discrete *adj.* separate, not continuous. ▫ **discretely** *adv.*

discretion *n.* **1** being discreet. **2** freedom to decide something.

discretionary *adj.* done or used at a person's discretion.

discriminate *v.* make a distinction (between). ▫ **discriminate against** treat unfairly. **discriminatory** *adj.*

■ differentiate, distinguish, tell apart, tell the difference.

discriminating *adj.* having good judgement.

■ astute, discerning, keen, perceptive, perspicacious, selective, shrewd.

discrimination *n.* **1** unfavourable treatment based on prejudice. **2** good taste or judgement.

■ **1** bias, bigotry, favouritism, prejudice, unfairness. **2** acumen, astuteness, discernment, insight, judgement, perception, perceptiveness, perspicacity, shrewdness, taste.

discursive *adj.* rambling, not keeping to the main subject.

discus *n.* heavy disc thrown in contests of strength.

discuss *v.* talk or write about. ▫ **discussion** *n.*

■ chat about, confer about, converse about, consider, debate, deliberate, examine, talk about, thrash out. ▫ **discussion** chat, conference, consultation, conversation, debate, deliberation, dialogue, discourse, parley.

disdain *v.* & *n.* scorn. ▫ **disdainful** *adj.*, **disdainfully** *adv.*

■ ▫ **disdainful** contemptuous, derisive, haughty, hoity-toity, scornful, snobbish, *colloq.* stuck-up, supercilious, superior.

disease *n.* **1** unhealthy condition. **2** specific illness. ▫ **diseased** *adj.*

■ **1** affliction, ailment, complaint, disorder, illness, infection, malady, sickness. **2** *sl.* bug, infection, *colloq.* virus.

disembark *v.* put or go ashore. ▫ **disembarkation** *n.*

disembodied *adj.* (of a voice) apparently not produced by anyone.

disembowel *v.* (**disembowelled**) take out the bowels of. ▫ **disembowelment** *n.*

disenchant *v.* disillusion. ▫ **disenchantment** *n.*

disenfranchise *v.* deprive of the right to vote. ▫ **disenfranchisement** *n.*

disengage *v.* detach or separate. ▫ **disengagement** *n.*

disentangle *v.* free or become free from tangles or confusion. ▫ **disentanglement** *n.*

disfavour *n.* dislike, disapproval.

disfigure *v.* spoil the appearance of. ▫ **disfigurement** *n.*

■ blemish, damage, deface, deform, distort, impair, injure, mar, mutilate, ruin, scar, spoil.

disgorge *v.* eject, pour forth.

disgrace *n.* **1** loss of reputation or respect. **2** shameful or very bad person or thing. ● *v.* bring disgrace upon. ▫ **disgraceful** *adj.*, **disgracefully** *adv.*

■ *n.* degradation, discredit, dishonour, disrepute, embarrassment, humiliation, ignominy, mortification, opprobrium, scandal, shame. ● *v.* discredit, dishonour, embarrass, humiliate, mortify, shame. ▫ **disgraceful** bad, base, contemptible, degrading, deplorable, despicable, discreditable, dishonourable, humiliating, ignominious, outrageous, scandalous, shameful, shocking.

disgruntled *adj.* discontented.

disguise *v.* conceal the identity of. ● *n.* **1** thing that disguises. **2** disguising, disguised condition.

■ *v.* camouflage, conceal, cover up, hide, mask, screen, veil. ● *n.* **1** camouflage, cover, front, guise, mask, smokescreen.

disgust *n.* strong dislike. ● *v.* cause disgust in. □ **disgusting** *adj.*

■ *n.* abhorrence, aversion, contempt, dislike, distaste, hatred, loathing, repugnance, revulsion. ● *v.* appal, repel, revolt, shock, sicken. □ **disgusting** abhorrent, distasteful, foul, loathsome, nasty, obnoxious, offensive, *colloq.* off-putting, repellent, repugnant, repulsive, sickening, vile.

dish *n.* **1** shallow bowl, esp. for food. **2** food prepared for the table. □ **dish out** (*colloq.*) distribute. **dish up** serve out food.

disharmony *n.* lack of harmony.

dishearten *v.* cause to lose hope or confidence.

■ daunt, demoralize, depress, deter, discourage, dismay.

dishevelled *adj.* ruffled and untidy. □ **dishevelment** *n.*

■ bedraggled, messy, ruffled, rumpled, *colloq.* scruffy, tangled, tousled, untidy, windswept.

dishonest *adj.* not honest. □ **dishonestly** *adv.*, **dishonesty** *n.*

■ *sl.* bent, cheating, corrupt, criminal, *colloq.* crooked, deceitful, deceptive, dishonourable, double-dealing, false, fraudulent, hypocritical, insincere, lying, mendacious, perfidious, shady, slippery, treacherous, two-faced, underhand, unprincipled, unscrupulous, untrustworthy, untruthful.

dishonour *v.* & *n.* disgrace.

dishonourable *adj.* **1** causing disgrace, ignominious. **2** unprincipled. □ **dishonourably** *adv.*

■ **1** degrading, discreditable, disgraceful, humiliating, ignominious, shameful. **2** base, contemptible, despicable, dishonest, disloyal, faithless, ignoble, low, mean, perfidious, shabby, shameless, treacherous, two-faced, unprincipled, unscrupulous, untrustworthy, unworthy.

disillusion *v.* free from pleasant but mistaken beliefs. □ **disillusionment** *n.*

disincentive *n.* thing that discourages an action or effort.

disinclination *n.* unwillingness.

disincline *v.* make unwilling.

disinfect *v.* cleanse of infection. □ **disinfection** *n.*

■ clean, cleanse, decontaminate, fumigate, purify, sanitize, sterilize.

disinfectant *n.* substance used for disinfecting things.

disinformation *n.* deliberately misleading information.

disingenuous *adj.* insincere.

disinherit *v.* reject from being one's heir. □ **disinheritance** *n.*

disintegrate *v.* **1** separate into small pieces. **2** decay. □ **disintegration** *n.*

■ **1** break up, come *or* fall apart, crumble, fall to pieces, fragment, shatter. **2** decay, decompose, rot.

disinter *v.* (**disinterred**) dig up, unearth. □ **disinterment** *n.*

disinterested *adj.* unbiased.

■ detached, dispassionate, fair, impartial, neutral, objective, unbiased, uninvolved, unprejudiced.

disjointed *adj.* lacking orderly connection.

disk *n.* flat circular device on which computer data can be stored.

dislike *n.* feeling of not liking something. ● *v.* feel dislike for.

■ *n.* antipathy, aversion, detestation, disgust, distaste, hatred, loathing, repugnance. ● *v.* despise, detest, hate, loathe.

dislocate *v.* **1** displace from its position. **2** disrupt. □ **dislocation** *n.*

dislodge *v.* move or force from an established position.

disloyal *adj.* not loyal. □ **disloyally** *adv.*, **disloyalty** *n.*

■ faithless, false, perfidious, traitorous, treacherous, two-faced, unfaithful, untrustworthy.

dismal *adj.* **1** gloomy. **2** (*colloq.*) feeble. □ **dismally** *adv.*

■ **1** bleak, cheerless, dark, depressing, dreary, dull, funereal, gloomy, grim, solemn, sombre; lugubrious, miserable, morose, mournful, sad, unhappy.

dismantle *v.* take to pieces.

dismay *n.* feeling of anxiety and discouragement. ● *v.* cause dismay to.

■ *n.* alarm, anxiety, apprehension, consternation, disappointment, discouragement, trepidation. ● *v.* alarm, disconcert, discourage, frighten, horrify, startle, take aback, unnerve.

dismember *v.* **1** remove the limbs of. **2** split into pieces. □ **dismemberment** *n.*

dismiss *v.* **1** send away from one's presence or employment. **2** reject. □ **dismissal** *n.*, **dismissive** *adj.*

■ **1** pack off, send away, send packing; discharge, fire, *colloq.* give a person the boot, *colloq.* kick out, lay off, make redundant, *colloq.* sack, send packing, throw out. **2** discount, disregard, reject.

dismount *v.* get off a thing on which one is riding.

disobedient *adj.* not obedient. □ **disobediently** *adv.*, **disobedience** *n.*

■ badly behaved, contrary, defiant, headstrong, insubordinate, intractable, mischievous, mutinous, naughty, obstinate, obstreperous, perverse, rebellious, recalcitrant, refractory, self-willed, uncontrollable, unmanageable, unruly, wayward, wilful.

disobey *v.* disregard orders.

■ break, contravene, defy, disregard, flout, ignore, infringe, transgress, violate.

disoblige *v.* fail to help or oblige. □ **disobliging** *adj.*

disorder *n.* **1** lack of order or of discipline. **2** ailment. ● *v.* throw into disorder. □ **disorderly** *adj.*

■ *n.* **1** chaos, clutter, confusion, disarray, disorganization, jumble, mess, muddle, shambles, tangle, untidiness; bedlam, commotion, disturbance, fracas, hullabaloo, pandemonium, riot, rumpus, tumult, turmoil, unrest, upheaval, uproar. **2** ailment, disease, illness, malady, sickness.

disorganize *v.* upset the arrangement of. □ **disorganization** *n.*

disorientate *v.* cause (a person) to lose his or her sense of direction. □ **disorientation** *n.*

disown *v.* **1** refuse to acknowledge. **2** reject all connection with.

disparage *v.* suggest that something is of little value or importance. □ **disparagement** *n.*

■ belittle, criticize, decry, denigrate, deprecate, minimize, run down, undervalue.

disparate *adj.* different in kind. □ **disparately** *adv.*, **disparity** *n.*

dispassionate *adj.* **1** not emotional. **2** impartial. □ **dispassionately** *adv.*

■ **1** calm, composed, cool, equable, even-tempered, level-headed, placid, self-controlled, sober, unemotional, *colloq.* unflappable. **2** detached, disinterested, impartial, neutral, objective, unbiased, unprejudiced.

dispatch *v.* **1** send off to a destination or for a purpose. **2** kill. **3** complete (a task) quickly. ● *n.* **1** dispatching. **2** promptness. **3** news report. **4** official message. □ **dispatch box** container for official documents. **dispatch rider** motorcyclist carrying messages.

dispel *v.* (**dispelled**) **1** drive away. **2** scatter. □ **dispeller** *n.*

dispensable *adj.* not essential.

dispensary *n.* place where medicines are dispensed.

dispensation *n.* **1** dispensing. **2** distributing. **3** exemption.

dispense *v.* **1** deal out. **2** prepare and give out (medicine etc.). □ **dispense with** **1** do without. **2** make unnecessary. **dispenser** *n.*

disperse *v.* go or send in different directions, scatter. □ **dispersal** *n.*, **dispersion** *n.*

■ diffuse, disseminate, distribute, spread; disband, dismiss, scatter, send away.

dispirited *adj.* dejected. □ **dispiriting** *adj.*

displace *v.* **1** shift. **2** take the place of. **3** oust. □ **displacement** *n.*

display *v.* show, arrange conspicuously. ● *n.* **1** act or instance of displaying. **2** thing(s) displayed. □ **displayer** *n.*

■ *v.* demonstrate, evince, manifest, reveal, show; arrange, dispose, exhibit, present, unveil; flash, flaunt, parade, show off. ● *n.* **1** demonstration, manifestation, revelation. **2** array, exhibit, exhibition, panoply, parade, presentation, show, spectacle.

displease *v.* irritate, annoy.

■ *colloq.* aggravate, anger, annoy, *sl.* bug, exasperate, infuriate, irk, irritate, nettle, offend, put out, *colloq.* rile, upset, vex.

displeasure *n.* disapproval.

disport *v.* **disport oneself** frolic.

disposable *adj.* **1** that can be disposed of. **2** designed to be thrown away after use. □ **disposability** *n.*

disposal *n.* disposing. □ **at one's disposal** available for one's use.

dispose *v.* **1** place, arrange. **2** make willing or ready to do something. □ **dispose of** **1** get rid of. **2** finish. **be well disposed** be friendly or favourable.

■ **1** arrange, array, distribute, organize, place, position, put, range. **2** incline, induce, influence, lead, move, persuade, tempt. □ **dispose of** **1** discard, dump, get rid of, jettison, scrap, throw away *or* out. **2** complete, conclude, finish, settle.

disposition *n.* **1** arrangement. **2** person's character. **3** tendency.

■ **1** arrangement, disposal, grouping, organization. **2** attitude, character, make-up, nature, personality, temperament. **3** inclination, leaning, predisposition, preference, proclivity, propensity, tendency.

dispossess *v.* deprive of the possession of. □ **dispossession** *n.*

disproportionate *adj.* relatively too large or too small. □ **disproportionately** *adv.*

disprove *v.* show to be wrong.

■ confute, *colloq.* debunk, destroy, demolish, discredit, explode, invalidate, negate, rebut, refute.

disputable *adj.* questionable.

disputant *n.* person engaged in a dispute.

disputation *n.* argument, debate.

disputatious *adj.* argumentative.

dispute *v.* **1** question the validity of. **2** argue, quarrel. ● *n.* **1** debate. **2** quarrel.

■ *v.* **1** challenge, contest, deny, disagree with, object to, oppose, query, question. **2** argue, debate, fight, quarrel, squabble, wrangle. ● *n.* **1** argument, controversy, debate, discussion, disputation. **2** altercation, clash, disagreement, fight, quarrel, squabble, wrangle.

disqualify *v.* make ineligible or unsuitable. □ **disqualification** *n.*

disquiet *n.* uneasiness, anxiety. ● *v.* cause disquiet to.

disregard *v.* ignore, treat as unimportant. ● *n.* lack of attention.

■ *v.* discount, dismiss, ignore, overlook, pass over, pay no attention to, take no notice of.

disrepair *n.* bad condition caused by neglect.

disreputable *adj.* not respectable. □ **disreputably** *adv.*

■ discreditable, disgraceful, dishonourable, shady, shameful, sleazy, unprincipled, unscrupulous, untrustworthy.

disrepute *n.* discredit.

disrespect *n.* lack of respect. □ **disrespectful** *adj.*

■ □ **disrespectful** cheeky, discourteous, impolite, impudent, insolent, irreverent, pert, rude, saucy, uncivil.

disrobe *v.* undress.

disrupt *v.* **1** cause to break up. **2** interrupt the continuity of. □ **disruption** *n.*, **disruptive** *adj.*

dissatisfaction *n.* lack of satisfaction or of contentment.

■ annoyance, chagrin, disappointment, discontent, displeasure, exasperation, frustration, irritation, unhappiness.

dissatisfied *adj.* not satisfied.

■ *colloq.* browned off, disaffected, disappointed, discontented, disgruntled, displeased, fed up, frustrated, unhappy.

dissect *v.* cut apart so as to examine the internal structure. □ **dissection** *n.*, **dissector** *n.*

dissemble *v.* conceal (feelings).

disseminate *v.* spread widely. □ **dissemination** *n.*

dissension *n.* disagreement that gives rise to strife.

dissent *v.* have a different opinion. ● *n.* difference in opinion. □ **dissenter** *n.*, **dissentient** *adj.* & *n.*

dissertation *n.* detailed discourse.

disservice *n.* unhelpful or harmful action.

dissident *adj.* disagreeing. ● *n.* person who disagrees, esp. with the authorities. □ **dissidence** *n.*

dissimilar *adj.* unlike. □ **dissimilarity** *n.*, **dissimilitude** *n.*

■ contrasting, different, distinct, distinguishable, diverse, unalike, unlike, unrelated.

dissimulate *v.* dissemble. □ **dissimulation** *n.*

dissipate *v.* fritter away. □ **dissipation** *n.*

dissipated *adj.* dissolute.

dissociate *v.* **1** regard as separate. **2** declare to be unconnected. □ **dissociation** *n.*

dissolute *adj.* lacking moral restraint or self-discipline.

■ corrupt, debauched, decadent, degenerate, depraved, dissipated, immoral, licentious, profligate, rakish, wanton.

dissolution *n.* dissolving of an assembly or partnership.

dissolve *v.* **1** make or become liquid or dispersed in liquid. **2** disappear gradually. **3** disperse (an assembly). **4** end (a partnership, esp. marriage).

■ **1** disintegrate, liquefy, melt. **2** disappear, fade (away), melt (away), vanish. **3** disband, dismiss, disperse, wind up.

dissonant *adj.* discordant. □ **dissonance** *n.*

dissuade *v.* persuade against a course of action. □ **dissuasion** *n.*

distaff *n.* cleft stick holding wool etc. in spinning. □ **distaff side** female branch of a family.

distance *n.* **1** length of space between two points. **2** distant part. **3** aloofness. **4** remoteness. ● *v.* separate.

■ **1** gap, interval, space, span, stretch. **3** aloofness, coolness, detachment, reserve, stiffness.

distant *adj.* **1** at a specified or considerable distance away. **2** aloof. □ **distantly** *adv.*

■ **1** far, far-away, far-off, outlying, remote. **2** aloof, chilly, cold, cool, detached, haughty, reserved, reticent, standoffish, stiff, unapproachable, unfriendly, withdrawn.

distaste *n.* dislike, disapproval.

distasteful *adj.* arousing distaste.

■ disagreeable, disgusting, displeasing, nasty, objectionable, offensive, *colloq.* off-putting, unpalatable, unpleasant, unsavoury.

distemper *n.* **1** disease of dogs. **2** paint for use on walls. ● *v.* paint with distemper.

distend *v.* swell from pressure within. □ **distension** *n.*

distil *v.* (**distilled**) **1** treat or make by distillation. **2** undergo distillation.

distillation *n.* **1** process of vaporizing and condensing a liquid so as to purify it or to extract elements. **2** something distilled.

distiller *n.* one who makes alcoholic liquor by distillation.

distillery *n.* place where alcohol is distilled.

distinct *adj.* **1** clearly perceptible. **2** different in kind. □ **distinctly** *adv.*

■ **1** apparent, clear, definite, evident, manifest, noticeable, obvious, palpable, patent, perceptible, plain, recognizable, sharp, unambiguous, unequivocal, unmistakable, visible, well-defined. **2** contrasting, different, dissimilar, unalike, unlike.

distinction *n.* **1** act or instance of distinguishing. **2** difference. **3** thing that differentiates. **4** excellence. **5** mark of honour.

■ **1** discrimination, differentiation. **2** contrast, difference, dissimilarity, disparity. **3** characteristic, feature, mark, peculiarity. **4** eminence, excellence, fame, glory, greatness, honour, importance, merit, note, prestige, renown, reputation, significance, superiority, value, worth.

distinctive *adj.* distinguishing, characteristic. □ **distinctively** *adv.*, **distinctiveness** *n.*

distinguish *v.* **1** be or see a difference between. **2** discern. **3** make notable. □ **distinguishable** *adj.*

■ **1** differentiate, separate, tell apart; set apart, single out. **2** descry, detect, discern, identify, make out, notice, perceive, pick out, recognize.

distinguished *adj.* famous for great achievements.

■ celebrated, eminent, famous, great, honoured, illustrious, notable, noted, prominent, renowned.

distort *v.* **1** pull out of shape. **2** misrepresent. □ **distortion** *n.*

■ **1** bend, contort, deform, disfigure, twist, warp. **2** falsify, misrepresent, twist.

distract *v.* draw away the attention of.

distracted *adj.* distraught.

distraction *n.* **1** distracting. **2** thing that distracts. **3** entertainment. **4** distraught state.

distraught *adj.* nearly crazy with grief or worry.

■ agitated, at one's wits' end, beside oneself, distracted, distressed, disturbed, frantic, frenetic, hysterical, overwrought, troubled, wild.

distress *n.* suffering, unhappiness. ● *v.* cause distress to. □ **in distress** in danger and needing help.

■ *n.* affliction, agony, anguish, anxiety, desolation, grief, heartache, misery, pain, sadness, sorrow, suffering, torment, trouble, unhappiness, woe, wretchedness. ● *v.* afflict, disturb, grieve, hurt, pain, perturb, sadden, torment, torture, trouble, upset, worry.

distribute *v.* **1** divide and share out. **2** scatter, place at different points. □ **distribution** *n.*

■ **1** allocate, allot, apportion, assign, *colloq.* dish out, dispense, divide up, dole out, give out, hand out, issue, partition, share (out). **2** disperse, disseminate, scatter, spread, strew.

distributor *n.* **1** one who distributes. **2** device for passing electric current to spark plugs.

district *n.* part (of a country, county, or city) with a particular feature or regarded as a unit.

■ area, locality, neighbourhood, parish, part, province, quarter, region, sector, ward, zone.

distrust *n.* lack of trust, suspicion. ● *v.* feel distrust in. □ **distrustful** *adj.*, **distrustfully** *adv.*

■ *n.* doubt, misgiving, mistrust, scepticism, suspicion, uncertainty. ● *v.* be sceptical of, doubt, have misgivings about, have qualms about, mistrust, question, suspect.

disturb *v.* **1** break the quiet, rest, or calm of. **2** disorganize. □ **disturbance** *n.*

■ **1** agitate, alarm, bother, concern, disconcert, distress, fluster, perturb, ruffle, shake, trouble, unsettle, upset, worry; disrupt, interrupt, intrude on; distract, inconvenience, put out. **2** disarrange, disorder, disorganize, jumble (up), move, muddle (up).

disturbed *adj.* mentally or emotionally unstable or abnormal.

disuse *n.* state of not being used.

disused *adj.* no longer used.

ditch *n.* long narrow trench for drainage. ● *v.* **1** make or repair ditches. **2** (*sl.*) abandon.

dither *v.* hesitate indecisively.

ditto *n.* (in lists) the same again.

ditty *n.* short simple song.

diuretic *adj.* & *n.* (substance) causing more urine to be excreted.

diurnal *adj.* of or in the day.

divan *n.* **1** couch without back or arms. **2** bed resembling this.

dive *v.* **1** plunge head first into water. **2** plunge or move quickly downwards. **3** go under water. **4** rush headlong. ● *n.* **1** diving. **2** sharp downward movement or fall. **3** (*colloq.*) disreputable place.

diver *n.* **1** one who dives. **2** person who works underwater.

diverge *v.* **1** separate and go in different directions. **2** depart from a path etc. □ **divergence** *n.*, **divergent** *adj.*

■ **1** branch, divide, fork, ramify, separate, split. **2** deviate, drift, stray, turn aside, wander.

diverse *adj.* of differing kinds.

■ assorted, heterogeneous, miscellaneous, mixed, multifarious, varied, various, varying.

diversify *v.* **1** introduce variety into. **2** vary. □ **diversification** *n.*

diversion *n.* **1** act of diverting. **2** thing that diverts attention. **3** entertainment. **4** route round a closed road.

■ **1** deviation, digression. **2** distraction, interruption. **3** amusement, entertainment, game, pastime, recreation, relaxation.

diversity *n.* variety.

divert *v.* **1** turn from a course or route. **2** entertain, amuse.

■ **1** avert, deflect, redirect, sidetrack, switch, turn aside. **2** absorb, amuse, beguile, engage, entertain, interest, occupy.

divest *v.* **divest of** strip of.

divide *v.* **1** separate into parts or from something else. **2** distribute. **3** cause to disagree. **4** find how many times one number contains another. **5** be able to be divided. ● *n.* dividing line.

■ *v.* **1** break up, cut up, partition, separate, split (up), subdivide. **2** allot, apportion, distribute, dole out, share (out). **3** alienate, estrange, part, separate, split.

dividend *n.* **1** share of profits payable. **2** number to be divided.

divider *n.* **1** thing that divides. **2** (*pl.*) measuring compasses.

divination *n.* divining.

divine *adj.* **1** of, from, or like God or a god. **2** (*colloq.*) excellent, beautiful. ● *v.* discover by intuition or magic. □ **divinely** *adv.*, **diviner** *n.*

■ *adj.* **1** celestial, godlike, heavenly, holy. ● *v.* forecast, foresee, foretell, predict, prognosticate.

divining rod dowser's stick.

divinity *n.* **1** being divine. **2** god.

divisible *adj.* able to be divided. □ **divisibility** *n.*

division *n.* **1** act or instance of dividing. **2** dividing line. **3** one of the parts into which a thing is divided. □ **divisional** *adj.*

■ **1** dividing, partition, separation, splitting. **2** border, borderline, boundary, frontier. **3** compartment, part, section, segment.

divisive *adj.* tending to cause disagreement.

divisor *n.* number by which another is to be divided.

divorce *n.* **1** legal termination of a marriage. **2** separation. ● *v.* **1** end the marriage of (a person) by divorce. **2** separate.

divorcee *n.* divorced person.

divulge *v.* reveal (information).

Diwali *n.* Hindu festival at which lamps are lit, held between September and November.

dizzy *adj.* (**-ier**, **-iest**) **1** giddy, dazed. **2** causing giddiness. □ **dizzily** *adv.*, **dizziness** *n.*

djellaba /jélləbə/ *n.* Arab cloak.

do *v.* (**did**, **done**) **1** perform, complete. **2** deal with. **3** act, behave. **4** fare. **5** be suitable. **6** suffice. ● *v.aux.* used to form present or past tense, for emphasis, or to avoid repeating a verb just used. ● *n.* (*pl.* **dos** or **do's**) entertainment, party. □ **do away with 1** abolish, get rid of. **2** kill. **do down** (*colloq.*) swindle. **do for** (*colloq.*) ruin, destroy. **do-gooder** *n.* well-meaning but unrealistic promoter of social work or reform. **do in** (*sl.*) **1** ruin, kill. **2** tire out. **do out** clean, redecorate. **do up 1** fasten. **2** repair, redecorate. **do without** manage without.

■ *v.* **1** accomplish, achieve, bring off, carry out, complete, discharge, execute, finish, fulfil, perform, pull off. **2** attend to, deal with, look after, organize, see to, sort out, take charge of. **3** act, behave, conduct oneself. **4** get on, fare, manage. **5** be acceptable, be suitable, satisfy, serve. **6** be enough, suffice. □ **do away with 1** abolish, cancel, dispense with, dispose of, eliminate, get rid of.

Dobermann pinscher dog of a large smooth-coated breed.

docile *adj.* willing to obey. □ **docilely** *adv.*, **docility** *n.*

■ biddable, complaisant, compliant, cooperative, manageable, meek, obedient, passive, submissive, tame, tractable, yielding.

dock[1] *n.* enclosed harbour for loading, unloading, and repair of ships. ● *v.* **1** bring or come into dock. **2** connect (spacecraft) in space, be joined thus.

dock[2] *n.* enclosure for the prisoner in a criminal court.

dock[3] *v.* **1** cut short. **2** reduce, take away part of.

dock[4] *n.* weed with broad leaves.

docker *n.* labourer who loads and unloads ships in a dockyard.

docket *n.* **1** document listing goods delivered. **2** voucher. ● *v.* (**docketed**) label with a docket.

dockyard *n.* area and buildings round a shipping dock.

doctor *n.* **1** qualified medical practitioner. **2** holder of a doctorate. ● *v.* **1** tamper with. **2** castrate, spay.

doctorate *n.* highest degree at a university.

doctrinaire *adj.* applying theories or principles rigidly.

doctrine *n.* principle(s) of a religious, political, or other group. □ **doctrinal** *adj.*

■ belief, conviction, creed, dogma, precept, principle, tenet.

document *n.* piece of paper giving information or evidence. ● *v.* provide or prove with documents. □ **documentation** *n.*

documentary *adj.* **1** consisting of documents. **2** giving a factual report. ● *n.* documentary film.

dodder *v.* totter, esp. from age. □ **dodderer** *n.*, **doddery** *adj.*

dodge *v.* **1** move quickly to one side so as to avoid (a thing). **2** evade. ● *n.* **1** clever trick, ingenious action. **2** quick evasive action. □ **dodger** *n.*

■ *v.* **1** bob, duck, sidestep, swerve, veer. **2** avoid, elude, escape from, evade, get away from, get out of. ● *n.* **1** device, manoeuvre, ploy, ruse, scheme, stratagem, subterfuge, trick.

dodgem *n.* small car at a funfair, driven so as to bump others in an enclosure.

dodo *n.* (*pl.* **-os**) large extinct bird.

doe *n.* female of deer, hare, or rabbit.

doff *v.* take off (one's hat).

dog *n.* **1** four-legged carnivorous wild or domesticated animal. **2** male of this or of fox or wolf. **3** (**the dogs**) greyhound racing. ● *v.* (**dogged**) follow persistently. □ **dog collar** (*colloq.*) clerical collar fastening at the back of the neck. **dog-eared** *adj.* with page-corners crumpled.

dog cart two-wheeled cart with back-to-back seats.

dogfish *n.* a kind of small shark.

dogged *adj.* determined. □ **doggedly** *adv.*

doggerel *n.* bad verse.

doggo *adv.* **lie doggo** (*sl.*) remain motionless or making no sign.

doggy *adj.* & *n.* (of) a dog. □ **doggy bag** bag for carrying away leftovers.

dogma *n.* doctrine(s) put forward by authority.

dogmatic *adj.* imposing personal opinions. □ **dogmatically** *adv.*

■ arrogant, assertive, dictatorial, domineering, high-handed, imperious, intolerant, opinionated, overbearing, peremptory.

dog rose wild hedge-rose.

dogsbody *n.* (*colloq.*) drudge.

doh *n.* name for the keynote of a scale in music, or the note C.

doily *n.* small ornamental mat.

doldrums *n.pl.* equatorial regions with little or no wind. □ **in the doldrums** in low spirits.

dole *v.* **dole out** distribute. ● *n.* (*colloq.*) unemployment benefit.

doleful *adj.* mournful. □ **dolefully** *adv.*, **dolefulness** *n.*

doll *n.* small model of a human figure, esp. as a child's toy.

dollar *n.* unit of money in the USA, Australia, etc.

dollop *n.* mass of a soft substance.

dolly *n.* **1** (*children's use*) doll. **2** movable platform for a cine camera.

dolman sleeve tapering sleeve cut in one piece with the body of a garment.

dolmen *n.* megalithic structure of a large flat stone laid on two upright ones.

dolomite *n.* a type of limestone rock. □ **dolomitic** *adj.*

dolphin *n.* sea animal like a large porpoise, with a beak-like snout.

dolt *n.* stupid person. □ **doltish** *adj.*

domain *n.* **1** area under a person's control. **2** field of activity.

dome *n.* **1** rounded roof with a circular base. **2** thing shaped like this. □ **domed** *adj.*

domestic *adj.* **1** of home or household. **2** of one's own country. **3** domesticated. ● *n.* household servant. □ **domestically** *adv.*

domesticate *v.* **1** train (an animal) to live with humans. **2** accustom to household work and home life. □ **domestication** *n.*

domesticity *n.* domestic life.

domicile *n.* place of residence. □ **domiciliary** *adj.*

dominant *adj.* dominating, prevailing. □ **dominance** *n.*

■ authoritative, commanding, dominating, influential, leading, ruling; chief, main, outstanding, predominant, pre-eminent, prevailing, primary, principal.

dominate *v.* **1** have a commanding influence over. **2** be the most influential or conspicuous person or thing. **3** tower over. □ **domination** *n.*

■ **1** command, control, direct, govern, have under one's thumb, lead, reign over, rule; be at the wheel, be in control, have the upper hand. **2** predominate, preponderate, prevail. **3** dwarf, overshadow, tower over.

domineer *v.* behave forcefully, making others obey.

dominion *n.* **1** authority to rule, control. **2** ruler's territory.

domino *n.* (*pl.* **-oes**) small oblong piece marked with pips, used in the game of **dominoes**.

don[1] *v.* (**donned**) put on.

don[2] *n.* head, fellow, or tutor of a college. □ **donnish** *adj.*

donate *v.* give as a donation.

■ contribute, give, grant, pledge, present, provide, supply.

donation *n.* gift (esp. of money) to a fund or institution.

done *see* **do**. *adj.* (*colloq.*) socially acceptable.

donkey *n.* animal of the horse family, with long ears. □ **donkey jacket** thick weatherproof jacket. **donkey's years** (*colloq.*) a very long time. **donkey work** drudgery.

donor *n.* one who gives or donates something.

doodle *v.* scribble idly. ● *n.* drawing or marks made thus.

doom *n.* terrible and inevitable fate. ● *v.* destine to a grim fate.

■ *n.* destiny, fate, fortune, kismet, lot; death, destruction, downfall, end, ruin.

doomsday *n.* Judgement Day.

door *n.* **1** hinged, sliding, or revolving barrier closing an opening. **2** doorway.

doorway *n.* opening filled by a door.

dope *n.* (*sl.*) **1** drug. **2** information. **3** stupid person. ● *v.* drug.

dopey *adj.* (**-ier, -iest**) **1** half asleep. **2** stupid.

dormant *adj.* **1** sleeping, temporarily inactive. **2** in abeyance. □ **dormancy** *n.*

■ **1** asleep, at rest, hibernating, inactive, inert, quiescent, quiet, resting, sleeping. **2** hidden, in abeyance, latent, potential, untapped.

dormer *n.* upright window under a small gable on a sloping roof.

dormitory *n.* room with several beds, esp. in a school. □ **dormitory town** commuter town or suburb.

dormouse *n.* (*pl.* **-mice**) mouse-like animal that hibernates.

dorsal *adj.* of or on the back.

dory *n.* edible sea fish.

dosage *n.* size of a dose.

dose *n.* **1** amount of medicine to be taken at one time. **2** amount of radiation received. ● *v.* give dose(s) of medicine to.

doss *v.* (*sl.*) sleep in a doss-house or on a makeshift bed etc. □ **dosser** *n.*

doss-house *n.* cheap hostel.

dossier *n.* set of documents about a person or event.

dot *n.* small round mark. ● *v.* (**dotted**) **1** mark with dot(s). **2** scatter here and there. **3** (*sl.*) hit. □ **on the dot** exactly on time.

dotage *n.* senility.

dote *v.* **dote on** feel great fondness for. □ **doting** *adj.*

■ adore, be fond of, be infatuated with, idolize, love, worship.

dotty *adj.* (**-ier, -iest**) (*colloq.*) **1** feeble-minded. **2** eccentric. **3** silly. □ **dottily** *adv.*, **dottiness** *n.*

double *adj.* **1** consisting of two things or parts. **2** twice as much or as many. **3** designed for two people or things. ● *adv.* **1** twice as much. **2** in twos. ● *n.* **1** person or thing very like another. **2** double quantity or thing. **3** (*pl.*) game with two players on each side. ● *v.* **1** make or become twice as much or as many. **2** fold in two. **3** turn back sharply. **4** act two parts. **5** have two uses. □ **at the double** running, hurrying. **double bass** lowest-pitched instrument of the violin family. **double-breasted** *adj.* (of a coat) with fronts overlapping. **double chin** chin with a roll of fat below. **double cream** thick cream. **double-cross** *v.* cheat, deceive. **double-dealing** *n.* deceit, esp. in business. *adj.* deceitful. **double-decker** *n.* bus with two decks. **double Dutch** gibberish. **double figures** numbers from 10 to 99. **double glazing** two sheets of glass in a window. **double take** delayed reaction just after one's first reaction. **double-talk** *n.* talk with deliberately ambiguous meaning. **doubly** *adv.*

double entendre /doob'l aantaandrə/ phrase with two meanings, one usu. indecent.

doublet *n.* **1** each of a pair of similar things. **2** (*old use*) man's close-fitting jacket.

doubt *n.* **1** feeling of uncertainty or disbelief. **2** being undecided. ● *v.* feel doubt about, hesitate to believe. □ **doubter** *n.*

■ *n.* **1** anxiety, apprehension, disquiet, distrust, hesitation, incredulity, misgiving, mistrust, qualm(s), reservation(s), scepticism, suspicion, uncertainty. **2** indecision, irresolution. ● *v.* be sceptical of, distrust, feel uncertain about, have misgivings about, mistrust, query, question, suspect.

doubtful *adj.* **1** feeling or causing doubt. **2** unreliable. □ **doubtfully** *adv.*

■ **1** distrustful, hesitant, incredulous, mistrustful, sceptical, suspicious, uncertain, unconvinced, undecided, unsure; ambiguous, debatable, disputable, dubious, equivocal, problematic(al), questionable, suspect. **2** unpredictable, unreliable, untrustworthy.

doubtless *adj.* certainly.

douche /doosh/ *n.* **1** jet of water applied to the body. **2** device for applying this. ● *v.* use a douche (on).

dough /dō/ *n.* **1** thick mixture of flour etc. and liquid, for baking. **2** (*sl.*) money. □ **doughy** *adj.*

doughnut /dōnut/ *n.* small cake of fried sweetened dough.

dour /door/ *adj.* stern, gloomy-looking.

douse /dowss/ *v.* **1** extinguish (a light). **2** throw water on. **3** put into water.

dove *n.* **1** bird with a thick body and short legs. **2** advocate of peaceful policies.

dovecote *n.* shelter for domesticated pigeons.

dovetail *n.* wedge-shaped joint interlocking two pieces of wood. ● *v.* combine neatly.

dowager *n.* woman holding a title or property from her dead husband.

dowdy *adj.* (**-ier, -iest**) **1** dull, not stylish. **2** dressed dowdily. □ **dowdily** *adv.*, **dowdiness** *n.*

■ drab, dull, frowzy, frumpish, frumpy, unfashionable.

dowel *n.* headless wooden or metal pin holding pieces of wood or stone together.

dowelling *n.* rods for cutting into dowels.

down[1] *n.* area of open undulating land, esp. (*pl.*) chalk uplands.

down[2] *n.* very fine soft furry feathers or short hairs.

down[3] *adv.* **1** to, in, or at a lower place or state etc. **2** to a smaller size. **3** from an earlier to a later time. **4** recorded in writing. **5** to the source or place where a thing is. **6** as (partial) payment at the time of purchase. ● *prep.* **1** downwards along or through or into. **2** at a lower part of. ● *adj.* **1** directed downwards. **2** travelling away from a central place. ● *v.* (*colloq.*) **1** knock, bring, or put down. **2** swallow. □ **down-and-out** *adj.* & *n.* destitute (person). **have a down on** (*colloq.*) be hostile to. **down-to-earth** *adj.* sens-

ible, practical. **down under** in the antipodes, esp. Australia.
downcast *adj.* **1** dejected. **2** (of eyes) looking downwards.
downfall *n.* **1** fall from prosperity or power. **2** thing causing this.
downgrade *v.* reduce to a lower grade.
downhearted *adj.* in low spirits.

> ■ **blue, dejected, depressed, despondent, dispirited, downcast, gloomy, glum, melancholy, miserable, sad, unhappy.**

downhill *adj.* & *adv.* going or sloping downwards.
downpour *n.* great fall of rain.
downright *adj.* **1** frank, straightforward. **2** utter. ● *adv.* thoroughly.

> ■ **1 blunt, candid, direct, forthright, frank, honest, open, outspoken, plain, straightforward. 2 absolute, categorical, complete, out and out, outright, sheer, total, thorough, unmitigated, utter.**

Down's syndrome abnormal congenital condition causing physical abnormalities and learning difficulties.
downstairs *adv.* & *adj.* to or on a lower floor.
downstream *adj.* & *adv.* in the direction in which a stream flows.
downtrodden *adj.* oppressed.
downward *adj.* moving or leading down. ● *adv.* downwards.
downwards *adv.* towards a lower place etc.
downy *adj.* **(-ier, -iest)** of, like, or covered with soft down.
dowry *n.* property brought by a bride to her husband.
dowse /dowz/ *v.* search for underground water or minerals by using a stick which dips when these are present. □ **dowser** *n.*
doxology *n.* hymn of praise to God.
doyen *n.* (*fem.* **doyenne**) senior member of a group.
doze *v.* sleep lightly. ● *n.* short light sleep.
dozen *n.* **1** set of twelve. **2** (*pl., colloq.*) very many.
Dr *abbr.* Doctor.
drab *adj.* dull, uninteresting.

> ■ **cheerless, colourless, depressing, dingy, dismal, dreary, dull, grey, sombre; boring, lacklustre, lifeless, tedious, uninspiring, uninteresting.**

drachm /dram/ *n.* one-eighth of an ounce or of a fluid ounce.
drachma *n.* (*pl.* **-as** or **-ae**) unit of money in Greece.
draconian *adj.* (of laws) harsh.
draft[1] *n.* **1** preliminary written version. **2** written order to a bank to pay money. **3** (*US*) conscription. ● *v.* **1** prepare a draft of. **2** (*US*) conscript.
draft[2] *n.* (*US*) draught.
drag *v.* **(dragged) 1** pull along with effort or difficulty. **2** trail on the ground. **3** (of time etc.) pass slowly or tediously. **4** search (water) with nets or hooks. ● *n.* **1** thing that slows progress. **2** (*colloq.*) draw at a cigarette. **3** (*sl.*) women's clothes worn by men. □ **dragnet** *n.* net for dragging water.

> ■ *v.* **1 draw, haul, lug, pull, tow, trail, tug.**

dragon *n.* **1** mythical reptile able to breathe fire. **2** fierce person.
dragonfly *n.* long-bodied insect with gauzy wings.
dragoon *n.* cavalryman. ● *v.* force into action.
drain *v.* **1** draw off (liquid) by channels or pipes etc. **2** flow away. **3** deprive gradually of (strength or resources). **4** drink all of. ● *n.* **1** channel or pipe carrying away water or sewage. **2** thing that drains one's strength etc.

> ■ *v.* **1 draw off, extract, pump out, remove, tap. 2 drip, ebb, flow away, seep, trickle. 3 debilitate, deplete, exhaust, sap, weaken.** ● *n.* **1 channel, conduit, ditch, drainpipe, gutter, outlet, pipe, sewer, trench, watercourse.**

drainage *n.* **1** draining. **2** system of drains. **3** what is drained off.
drake *n.* male duck.
dram *n.* **1** drachm. **2** small drink of spirits.
drama *n.* **1** play(s) for acting on the stage or broadcasting. **2** dramatic quality or series of events.
dramatic *adj.* **1** of drama. **2** exciting and unexpected, impressive. □ **dramatically** *adv.*

> ■ **1 histrionic, theatrical. 2 breathtaking, exciting, impressive, sensational, spectacular, striking, sudden, thrilling.**

dramatist *n.* writer of plays.
dramatize *v.* make into a drama. □ **dramatization** *n.*
drank *see* **drink**.
drape *v.* cover or arrange loosely.
drastic *adj.* having a strong or violent effect. □ **drastically** *adv.*
draught *n.* **1** current of air. **2** single act of drinking. **3** amount so drunk. **4** pulling. **5** depth of water needed to float a

ship. **6** (*pl.*) game played with 24 round pieces on a chessboard. □ **draught beer** beer drawn from a cask.

■ **1** breeze, current, puff, wind. **2** drink, gulp, pull, sip, swallow.

draughtsman *n.* one who draws plans or sketches.

draughty *adj.* (**-ier, -iest**) letting in sharp currents of air.

draw *v.* (**drew, drawn**) **1** pull. **2** attract. **3** take in (breath etc.). **4** take from or out. **5** produce (a picture etc.) by making marks. **6** finish a contest with scores equal. **7** make one's way, come. **8** obtain by lottery. **9** infuse. **10** promote or allow a draught of air (in). ● *n.* **1** act of drawing. **2** thing that draws custom or attention. **3** drawing of lots. **4** drawn game. □ **draw in** (of days) become shorter. **draw out 1** prolong. **2** (of days) become longer. **drawstring** *n.* string that can be pulled to tighten an opening. **draw the line at** refuse to do or tolerate. **draw up 1** halt. **2** compose (a contract etc.). **3** make (oneself) stiffly erect.

■ *v.* **1** drag, haul, lug, pull, tow, tug. **2** attract, entice, lure, pull. **3** breathe, inhale, take in. **4** extract, pull out, remove, take out, withdraw. **5** paint, sketch; delineate, depict, outline, portray, represent. **6** finish equal, tie. ● *n.* **1** pull, tug. **2** attraction, lure. **3** lottery, raffle. **4** stalemate, tie. □ **draw out 1** extend, prolong, protract, spin out, stretch out. **draw up 1** halt, pull up, stop. **2** compose, draft, prepare, put together.

drawback *n.* disadvantage.

■ catch, difficulty, disadvantage, hindrance, hitch, obstacle, problem, snag, stumbling block.

drawbridge *n.* bridge over a moat, hinged for raising.

drawer *n.* **1** person who draws. **2** one who writes a cheque. **3** horizontal sliding compartment. **4** (*pl.*) knickers, underpants.

drawing *n.* picture made with a pencil etc. □ **drawing-pin** *n.* pin for fastening paper to a surface. **drawing room** formal sitting room.

drawl *v.* speak lazily or with drawn-out vowel sounds. ● *n.* drawling manner of speaking.

drawn *see* **draw**. *adj.* looking tired and strained.

dread *n.* great fear. ● *v.* be in great fear of. ● *adj.* dreaded.

■ *n.* alarm, apprehension, awe, dismay, fear, fright, horror, terror, trepidation. ● *v.* be afraid of, fear, recoil from, shrink from.

dreadful *adj.* **1** appalling. **2** very bad. □ **dreadfully** *adv.*

■ **1** alarming, appalling, dire, frightening, frightful, ghastly, grim, gruesome, harrowing, hideous, horrible, horrifying, shocking, terrible. **2** atrocious, awful, bad, *colloq.* chronic, deplorable, disgraceful, *colloq.* frightful, *colloq.* lousy, terrible.

dream *n.* **1** series of scenes in a sleeping person's mind. **2** fantasy. **3** aspiration. ● *v.* (**dreamed** or **dreamt**) **1** have dream(s). **2** think of as a possibility. □ **dream up** imagine, invent. **dreamer** *n.*, **dreamless** *adj.*

■ *n.* **2** daydream, delusion, fantasy, hallucination, illusion, mirage, pipedream, reverie. **3** ambition, aspiration, hope, ideal, wish. ● *v.* **1** daydream, fantasize, hallucinate. □ **dream up** conceive, *colloq.* cook up, create, devise, hatch, imagine, invent, make up, *colloq.* think up.

dreamy *adj.* (**-ier, -iest**) daydreaming. □ **dreamily** *adv.*

dreary *adj.* (**-ier, -iest**) dull, gloomy. □ **drearily** *adv.*, **dreariness** *n.*

■ boring, colourless, dull, monotonous, tedious, tiresome, uninteresting, wearisome; bleak, cheerless, depressing, dismal, gloomy, miserable, sombre.

dredge *v.* **1** remove (silt) from (a river or channel). **2** sprinkle with flour etc.

dredger *n.* **1** boat that dredges. **2** container with perforated lid for sprinkling flour etc.

dregs *n.pl.* **1** sediment. **2** worst part.

■ **1** deposit, grounds, lees, remains, residue, sediment

drench *v.* wet thoroughly.

dress *n.* **1** outer clothing. **2** woman's or girl's garment with a bodice and skirt. ● *v.* **1** put clothes on. **2** clothe oneself. **3** arrange, decorate, trim. **4** put a dressing on. □ **dress circle** first gallery in a theatre. **dress rehearsal** final one, in costume. **dress shirt** shirt for wearing with evening dress.

■ *n.* **1** apparel, attire, clothes, clothing, costume, garb, garments, get-up, outfit. **2** frock, gown. ● *v.* **1** attire, clothe, garb, robe. **2** dress oneself, get dressed. **3** adorn, arrange, deck, decorate, embellish, garnish, trim. **4** bandage, bind up, swathe.

dressage /dréssazh/ *n.* management of a horse to show its obedience and deportment.

dresser[1] *n.* one who dresses a person or thing.

dresser[2] *n.* kitchen sideboard with shelves for dishes etc.

dressing *n.* **1** sauce for food. **2** fertilizer etc. spread over land. **3** bandage or ointment etc. for a wound. □ **dressing down** scolding. **dressing gown** loose robe worn when one is not fully dressed. **dressing table** table with a mirror, for use while dressing.

dressmaker *n.* person who makes women's clothes.

dressy *adj.* (**-ier**, **-iest**) smart, elegant.

drew *see* **draw**.

drey *n.* squirrel's nest.

dribble *v.* **1** have saliva flowing from the mouth. **2** flow or let flow in drops. **3** (in football etc.) move the ball forward with slight touches. ● *n.* act or flow of dribbling.

dried *adj.* (of food etc.) preserved by removal of moisture.

drift *v.* **1** be carried by a current of water or air. **2** go casually or aimlessly. ● *n.* **1** mass of snow, sand, etc. piled up by the wind. **2** general meaning of a speech etc. **3** slow movement or variation.

■ *v.* **1** coast, float, waft. **2** maunder, meander, ramble, roam, stray, wander. ● *n.* **1** bank, dune, heap, mound, pile. **2** essence, gist, import, meaning, purport, significance, substance.

drifter *n.* aimless person.

driftwood *n.* wood floating on the sea or washed ashore.

drill[1] *n.* **1** tool or machine for boring holes or sinking wells. **2** training. **3** (*colloq.*) routine procedure. ● *v.* **1** use a drill, make (a hole) with a drill. **2** train, be trained.

■ *n.* **2** discipline, exercise, instruction, practice, training. **3** custom, procedure, routine. ● *v.* **1** bore, perforate, pierce. **2** coach, discipline, exercise, instruct, school, teach, train.

drill[2] *n.* strong twilled fabric.

drily *adv.* in a dry way.

drink *v.* (**drank**, **drunk**) **1** swallow (liquid). **2** take alcoholic drink, esp. in excess. **3** pledge good wishes (to) by drinking. ● *n.* **1** (specified amount of) liquid for drinking. **2** (glass of) alcoholic liquor. □ **drink in** watch or listen to eagerly. **drinker** *n.*

■ *v.* **1** *colloq.* down, gulp, imbibe, lap (up), quaff, sip, swallow, swill. **2** *colloq.* booze, carouse, tipple. ● *n.* **1** draught, gulp, sip, swallow; beverage, potation. **2** dram, nightcap, nip, *sl.* snifter, *colloq.* tipple, tot; alcohol, *colloq.* booze, liquor, spirits.

drip *v.* (**dripped**) fall or let fall in drops. ● *n.* **1** liquid falling in drops. **2** sound of this. **3** device administering a liquid at a very slow rate, esp. intravenously. □ **drip-dry** *v.* & *adj.* (able to) dry easily without ironing. **drip-feed** *n.* & *v.* feed(-ing) by a drip.

dripping *n.* fat melted from roast meat.

drive *v.* (**drove**, **driven**) **1** send or urge onwards. **2** operate (a vehicle) and direct its course. **3** travel or convey in a private vehicle. **4** cause, compel. **5** (of wind etc.) carry along. **6** make (a bargain). ● *n.* **1** journey in a private vehicle. **2** energy, motivation. **3** organized effort. **4** track for a car, leading to a private house. **5** transmission of power to machinery. □ **drive at** intend to convey as a meaning. **drive-in** *adj.* (of a cinema etc.) able to be used without getting out of one's car.

■ *v.* **1** propel, push, send, urge; herd, shepherd. **2** control, handle, manoeuvre, operate, steer. **3** go, journey, move, ride, travel; bring, convey, give a person a lift, take. **4** compel, constrain, coerce, force, impel, make, press, pressurize. ● *n.* **1** excursion, jaunt, journey, outing, ride, run, spin, trip. **2** ambition, determination, energy, enterprise, enthusiasm, go, initiative, keenness, motivation, vigour, *colloq.* vim, zeal. **3** campaign, crusade, effort.

drivel *n.* silly talk, nonsense.

driver *n.* **1** person who drives. **2** golf club for driving from a tee.

drizzle *n.* & *v.* rain in very fine drops.

droll *adj.* amusing in an odd way. □ **drolly** *adv.*, **drollery** *n.*

dromedary *n.* camel with one hump, bred for riding.

drone *n.* **1** male bee. **2** deep humming sound. ● *v.* **1** make this sound. **2** speak monotonously.

drool *v.* **1** slaver, dribble. **2** show gushing appreciation.

droop *v.* bend or hang down limply. ● *n.* drooping attitude. □ **droopy** *adj.*

drop *n.* **1** small rounded mass of liquid. **2** very small quantity. **3** steep descent, distance of this. **4** fall. **5** (*pl.*) medicine measured by drops. ● *v.* (**dropped**) **1** shed, let fall. **2** allow to fall. **3** make or become lower. **4** omit. **5** abandon. **6** set down (a passenger etc.). **7** utter casually. □ **drop in** pay a casual visit. **drop off** fall asleep. **drop out** cease to participate. **drop-out** *n.* one who drops out from a course of study or from conventional society.

> ■ *n.* **1** bead, blob, drip, droplet, globule, tear. **2** bit, dash, pinch, spot, touch. **3** declivity, descent, fall, slope. **4** decline, decrease, fall, reduction. ● *v.* **1** dribble, drip, fall, trickle. **3** descend, dive, fall, plummet, plunge, sink. **4** eliminate, exclude, leave out, omit.

droplet *n.* small drop of liquid.

dropper *n.* device for releasing liquid in drops.

droppings *n.pl.* animal dung.

dropsy *n.* oedema. □ **dropsical** *adj.*

dross *n.* **1** scum on molten metal. **2** impurities, rubbish.

drought *n.* long spell of dry weather.

drove *see* **drive**. *n.* moving herd, flock, or crowd.

drover *n.* herder of cattle.

drown *v.* **1** kill or be killed by submersion in liquid. **2** flood, drench. **3** deaden (grief etc.) with drink. **4** overpower (sound) with greater loudness.

> ■ **2** deluge, drench, engulf, flood, immerse, inundate, overwhelm, submerge, swamp.

drowse *v.* be lightly asleep. □ **drowsy** *adj.*, **drowsily** *adv.*, **drowsiness** *n.*

drudge *n.* person who does laborious or menial work. ● *v.* do such work. □ **drudgery** *n.*

drug *n.* substance used in medicine or as a stimulant or narcotic. ● *v.* (**drugged**) add or give a drug to.

> ■ *n.* medicament, medicine; narcotic, opiate, stimulant, tranquillizer. ● *v.* anaesthetize, dope, dose, knock out, sedate.

drugstore *n.* (*US*) combined chemist's shop and café.

Druid *n.* priest of an ancient Celtic religion. □ **Druidical** *adj.*

drum *n.* **1** hollow percussion instrument covered at one or both ends with skin etc. **2** cylindrical object. **3** eardrum. ● *v.* (**drummed**) **1** tap continually. **2** din. □ **drum up** obtain by vigorous effort.

drummer *n.* player of drum(s).

drumstick *n.* **1** stick for beating a drum. **2** lower part of a cooked fowl's leg.

drunk *see* **drink**. *adj.* excited or stupefied by alcoholic drink. ● *n.* drunken person.

> ■ *adj.* fuddled, inebriated, intoxicated, *sl.* sloshed, *colloq.* sozzled, *colloq.* tiddly, *colloq.* tight, tipsy.

drunkard *n.* person who is often drunk.

drunken *adj.* **1** drunk. **2** often in this condition. □ **drunkenly** *adv.*, **drunkenness** *n.*

dry *adj.* (**drier**, **driest**) **1** without water, moisture, or rainfall. **2** uninteresting. **3** (of a sense of humour) ironic, understated. **4** not allowing the sale of alcohol. **5** (of wine) not sweet. **6** (*colloq.*) thirsty. ● *v.* **1** make or become dry. **2** preserve (food) by removing its moisture. □ **dry-clean** *v.* clean by solvent that evaporates quickly. **dry rot** decay of wood that is not ventilated. **dry run** (*colloq.*) dummy run. **dry up 1** dry washed dishes. **2** (*colloq.*) cease talking. **dryness** *n.*

> ■ *adj.* **1** arid, dehydrated, desiccated, parched, waterless. **2** boring, dreary, dull, prosaic, tedious, uninspired, uninteresting. **3** droll, ironic, subtle, understated, wry.

dryad *n.* wood nymph.

dual *adj.* composed of two parts, double. □ **dual carriageway** road with a dividing strip between traffic travelling in opposite directions. **duality** *n.*

dub[1] *v.* (**dubbed**) give a nickname to.

dub[2] *v.* (**dubbed**) replace the soundtrack of a film.

dubbin *n.* grease for softening and waterproofing leather.

dubiety *n.* feeling of doubt.

dubious *adj.* doubtful. □ **dubiously** *adv.*

ducal *adj.* of a duke.

ducat /dúkkət/ *n.* former gold coin of various European countries.

duchess *n.* **1** duke's wife or widow. **2** woman with the rank of duke.

duchy *n.* territory of a duke.

duck *n.* **1** swimming bird of various kinds. **2** female of this. **3** batsman's score of 0. **4** ducking movement. ● *v.* **1** push (a person) or dip one's head under water. **2** bob down, esp. to avoid being seen or hit. **3** dodge (a task etc.).

> ■ *v.* **1** dip, dunk, immerse, push under, submerge. **2** bend, bob down, crouch, dodge, stoop.

duckboards *n.pl.* boards forming a narrow path.

duckling *n.* young duck.

duct *n.* channel or tube conveying liquid or air. □ **ductless** *adj.*

ductile *adj.* (of metal) able to be drawn into fine strands.

dud *n.* & *adj.* (*sl.*) (thing) that is counterfeit or fails to work.

dude *n.* (*sl.*) **1** fellow. **2** dandy. □ **dude ranch** ranch used as a holiday centre.

dudgeon *n.* indignation.

due *adj.* **1** owing or payable as a debt. **2** merited, appropriate. **3** scheduled to do something or to arrive. ● *adv.* exactly. ● *n.* **1** a person's right, what is owed to him or her. **2** (*pl.*) fees. □ **be due to** be attributable to.

■ *adj.* **1** outstanding, owed, owing, payable, unpaid. **2** correct, deserved, fitting, just, merited, proper, right, rightful, suitable, well-earned; adequate, appropriate, enough, necessary, sufficient. ● *n.* **1** prerogative, privilege, right.

duel *n.* contest between two people or sides. □ **duellist** *n.*

duenna *n.* older woman acting as chaperon to girls, esp. in Spain.

duet *n.* musical composition for two performers.

duff *adj.* (*sl.*) dud.

duffer *n.* (*colloq.*) inefficient or stupid person.

duffle-coat *n.* heavy woollen coat with a hood.

dug[1] *see* **dig**.

dug[2] *n.* udder, teat.

dugong *n.* Asian sea mammal.

dugout *n.* **1** underground shelter. **2** canoe made from a hollowed tree trunk.

duke *n.* **1** nobleman of the highest hereditary rank. **2** ruler of certain small states. □ **dukedom** *n.*

dulcet *adj.* sounding sweet.

dulcimer *n.* musical instrument with strings struck by two hammers.

dull *adj.* **1** not bright. **2** stupid. **3** boring. **4** not sharp. **5** not resonant. ● *v.* make or become dull. □ **dully** *adv.*, **dullness** *n.*

■ *adj.* **1** dark, drab, dreary; cloudy, dismal, gloomy, grey, murky, overcast, sombre, sunless. **2** dense, *colloq.* dim, obtuse, slow, stupid, unintelligent. **3** boring, humdrum, monotonous, pedestrian, tedious, unimaginative, uninspiring, uninteresting. **4** blunt, blunted. **5** deadened, muffled, muted.

dullard *n.* stupid person.

duly *adv.* in a suitable way.

dumb *adj.* **1** unable to speak. **2** silent. **3** (*colloq.*) stupid. □ **dumb-bell** *n.* short bar with weighted ends, lifted to exercise muscles. **dumbly** *adv.*, **dumbness** *n.*

■ **1** mute, voiceless. **2** *colloq.* mum, quiet, silent, speechless, tongue-tied.

dumbfound *v.* astonish.

dumdum bullet soft-nosed bullet that expands on impact.

dummy *n.* **1** sham article. **2** model of the human figure, esp. as used to display clothes. **3** rubber teat for a baby to suck. ● *adj.* sham. □ **dummy run 1** trial attempt. **2** rehearsal.

dump *v.* **1** dispose of as rubbish. **2** put down carelessly. **3** sell abroad at a lower price. ● *n.* **1** rubbish heap. **2** temporary store. **3** (*colloq.*) dull place.

dumpling *n.* ball of dough cooked in stew or with fruit inside.

dumpy *adj.* (**-ier**, **-iest**) short and fat. □ **dumpiness** *n.*

dun *adj.* & *n.* greyish-brown.

dunce *n.* person slow at learning.

dune *n.* mound of drifted sand.

dung *n.* animal excrement.

dungarees *n.pl.* overalls of coarse cotton cloth.

dungeon *n.* strong underground cell for prisoners.

dunk *v.* dip into liquid.

duo *n.* (*pl.* **-os**) pair of performers.

duodecimal *adj.* reckoned in twelves or twelfths.

duodenum *n.* part of the intestine next to the stomach. □ **duodenal** *adj.*

dupe *v.* deceive, trick. ● *n.* duped person.

duple *adj.* **1** having two parts. **2** (in music) having two beats to the bar.

duplex *adj.* having two elements.

duplicate *n.* /dyóoplikət/ exact copy. ● *adj.* /dyóoplikət/ exactly like another. ● *v.* /dyóoplikayt/ **1** make or be a duplicate. **2** do twice. □ **duplication** *n.*

■ *n.* clone, copy, double, facsimile, match, replica, reproduction, twin; carbon copy, photocopy. ● *adj.* identical, matching, twin. ● *v.* **1** imitate, reproduce; copy, photocopy.

duplicity *n.* deceitfulness.

durable *adj.* likely to last. ● *n.pl.* durable goods. □ **durably** *adv.*, **durability** *n.*

■ *adj.* dependable, enduring, hard-wearing, long-lasting, reliable, stout, strong, sturdy, substantial, tough.

duration *n.* time during which a thing continues.

duress *n.* use of force or threats.

during *prep.* **1** throughout. **2** at a point in the continuance of.

dusk *n.* darker stage of twilight.

dusky *adj.* (**-ier, -iest**) **1** shadowy. **2** dark-coloured. □ **duskiness** *n.*

dust *n.* fine particles of earth or other matter. ● *v.* **1** sprinkle with dust or powder. **2** clear of dust by wiping, clean a room etc. thus. □ **dust bowl** area denuded of vegetation and reduced to desert. **dust cover, jacket** paper jacket on a book. **dusty** *adj.*

dustbin *n.* bin for household rubbish.

duster *n.* cloth for dusting things.

dustman *n.* person employed to empty dustbins.

dustpan *n.* container into which dust is brushed from a floor.

Dutch *adj.* & *n.* (language) of the Netherlands. □ **Dutch courage** that obtained by drinking alcohol. **go Dutch** share expenses on an outing. **Dutchman** *n.*, **Dutchwoman** *n.*

dutiable *adj.* on which customs or other duties must be paid.

dutiful *adj.* doing one's duty, obedient. □ **dutifully** *adv.*

■ compliant, conscientious, considerate, deferential, diligent, faithful, loyal, obedient, reliable, respectful, responsible.

duty *n.* **1** moral or legal obligation. **2** task, action to be performed. **3** tax on goods or imports. □ **on duty** at work.

■ **1** obligation, responsibility; burden, charge, onus. **2** assignment, chore, function, job, task. **3** customs, excise, levy, tariff, tax.

duvet /dōōvay/ *n.* thick soft quilt used as bedclothes.

dwarf *n.* (*pl.* **-fs**) **1** person or thing much below the usual size. **2** (in fairy tales) small being with magic powers. ● *adj.* very small. ● *v.* **1** stunt. **2** make seem small.

dwell *v.* (**dwelt**) live as an inhabitant. □ **dwell on** speak or think lengthily about. **dweller** *n.*

■ live, lodge, reside, stay. □ **dwell on** elaborate on, emphasize, harp on, labour; agonize over, brood on, fret about, worry about.

dwelling *n.* house etc. to live in.

■ abode, domicile, habitation, home, house, lodging, quarters, residence.

dwindle *v.* become less or smaller.

dye *v.* (**dyeing**) colour, esp. by dipping in liquid. ● *n.* **1** substance used for dyeing things. **2** colour given by dyeing. □ **dyer** *n.*

dying *see* **die**[1].

dyke *n.* **1** embankment to prevent flooding. **2** drainage ditch.

dynamic *adj.* **1** energetic, forceful. **2** of force producing motion. □ **dynamically** *adv.*

■ **1** active, animated, energetic, enterprising, enthusiastic, forceful, go-ahead, go-getting, lively, powerful, spirited, vigorous, vital, zealous.

dynamics *n.* branch of physics dealing with matter in motion.

dynamism *n.* energizing power.

dynamite *n.* powerful explosive made of nitroglycerine. ● *v.* fit or blow up with dynamite.

dynamo *n.* (*pl.* **-os**) small generator producing electric current.

dynasty *n.* line of hereditary rulers. □ **dynastic** *adj.*

dysentery *n.* disease causing severe diarrhoea.

dysfunction *n.* malfunction.

dyslexia *n.* condition causing difficulty in reading and spelling. □ **dyslexic** *adj.* & *n.*

dyspepsia *n.* indigestion. □ **dyspeptic** *adj.* & *n.*

dystrophy *n.* wasting of a part of the body.

Ee

E. *abbr.* **1** east. **2** eastern.

each *adj.* & *pron.* every one of two or more.

eager *adj.* full of desire, enthusiastic. □ **eagerly** *adv.*, **eagerness** *n.*

> ■ animated, ardent, avid, earnest, enthusiastic, excited, fervent, fervid, hungry, keen, passionate, zealous. □ **eagerness** animation, appetite, ardour, avidity, desire, enthusiasm, excitement, fervour, hunger, keenness, longing, passion, thirst, zeal, zest.

eagle *n.* large bird of prey.

ear[1] *n.* **1** organ of hearing. **2** external part of this. **3** ability to distinguish sounds accurately. □ **eardrum** *n.* membrane inside the ear.

ear[2] *n.* seed-bearing part of corn.

earl *n.* British nobleman ranking between marquess and viscount. □ **earldom** *n.*

early *adj.* & *adv.* (**-ier, -iest**) **1** before the usual or expected time. **2** not far on in development or in a series.

> ■ *adj.* **1** premature, untimely. **2** basic, first, initial, introductory, original. ● *adv.* **1** ahead of time, beforehand, in advance, prematurely; *poetic* betimes, in good time.

earmark *n.* distinguishing mark. ● *v.* **1** put such mark on. **2** set aside for a particular purpose.

earn *v.* **1** get or deserve for work or merit. **2** (of money) gain as interest.

> ■ **1** be paid, clear, get, gross, make, net, receive; be entitled to, be worthy of, deserve, merit, warrant, win. **2** pay, yield.

earnest *adj.* showing serious feeling or intention. □ **in earnest** seriously. **earnestly** *adv.*

> ■ assiduous, committed, conscientious, dedicated, determined, devoted, diligent, hard-working, industrious, serious, sober, solemn, thoughtful; fervent, heartfelt, profound, sincere, wholehearted.

earshot *n.* range of hearing.

earth *n.* **1** the planet we live on. **2** soil. **3** land and sea as opposed to sky. **4** fox's den. **5** connection of an electrical circuit to ground. ● *v.* connect an electrical circuit to earth. □ **run to earth** find after a long search.

> ■ *n.* **1** globe, planet, world. **2** clay, dirt, loam, sod, soil, turf.

earthen *adj.* made of earth or of baked clay. □ **earthenware** *n.* pottery made of baked clay.

earthly *adj.* **1** of the earth. **2** of human life on earth, worldly.

> ■ **1** terrestrial. **2** human, material, mortal, physical, secular, temporal, worldly.

earthquake *n.* violent movement of part of the earth's crust.

earthwork *n.* bank built of earth.

earthworm *n.* worm living in the soil.

earthy *adj.* (**-ier, -iest**) **1** like earth or soil. **2** coarse, crude.

earwig *n.* small insect with pincers at the end of its body.

ease *n.* freedom from pain, worry, or effort. ● *v.* **1** relieve from pain, anxiety, etc. **2** make easier. **3** become less burdensome or severe. **4** move gently or gradually.

> ■ *n.* calmness, comfort, composure, contentment, peace, peace and quiet, relaxation, relief, repose, rest, serenity, tranquillity; easiness, effortlessness, facility; aplomb, insouciance, naturalness, nonchalance. ● *v.* **1** calm, comfort, pacify, quieten, relieve, soothe; allay, alleviate, appease, assuage, mollify. **2** aid, assist, expedite, facilitate, help. **3** abate, decrease, drop, diminish, lessen, *colloq.* let up, moderate, subside. **4** edge, guide, inch, manoeuvre.

easel *n.* frame to support a painting or blackboard etc.

east *n.* **1** point on the horizon where the sun rises. **2** direction in which this lies. **3** eastern part. ● *adj.* **1** in the east. **2** (of wind) from the east. ● *adv.* towards the east.

Easter *n.* festival commemorating Christ's resurrection. □ **Easter egg** chocolate egg given at Easter.

easterly *adj.* towards or blowing from the east.

eastern *adj.* of or in the east.

easternmost *adj.* furthest east.

eastward *adj.* towards the east. ◻ **eastwards** *adv.*

easy *adj.* (**-ier**, **-iest**) **1** not difficult. **2** free from pain, trouble, or anxiety. **3** easy-going. ● *adv.* in an easy way. ◻ **easy chair** large comfortable chair. **easily** *adv.*, **easiness** *n.*

■ *adj.* **1** basic, effortless, elementary, rudimentary, simple, straightforward, uncomplicated, undemanding. **2** carefree, comfortable, *colloq.* cushy, peaceful, relaxing, restful, serene, tranquil, untroubled. **3** agreeable, affable, amiable, easygoing, friendly, genial, natural, pleasant.

easygoing *adj.* relaxed in manner, tolerant.

■ accommodating, agreeable, affable, amenable, amiable, carefree, casual, cheerful, even-tempered, flexible, friendly, genial, happy-go-lucky, natural, placid, pleasant, relaxed; indulgent, lenient, permissive, tolerant.

eat *v.* (**ate**, **eaten**) **1** chew and swallow (food). **2** have a meal. ◻ **eat away (at)** destroy gradually.

■ **1** bolt, consume, devour, gobble, gulp, guzzle, munch, nibble, *colloq.* scoff, *colloq.* tuck into, wolf. **2** breakfast, dine, have a bite, have a meal, lunch. ◻ **eat away (at)** consume, corrode, destroy, erode, wear away.

eatables *n.pl.* food.

eau-de-Cologne /ṓdəkəlṓn/ *n.* a delicate perfume.

eaves *n.pl.* overhanging edge of a roof.

eavesdrop *v.* (**-dropped**) listen secretly to a private conversation. ◻ **eavesdropper** *n.*

ebb *n.* **1** outflow of the tide, away from the land. **2** decline. ● *v.* **1** flow away. **2** decline.

■ *v.* **1** drain away, flow back, go down, recede, subside. **2** decline, decrease, diminish, dwindle, flag, lessen, wane.

ebony *n.* hard black wood of a tropical tree. ● *adj.* black as ebony.

ebullient *adj.* full of high spirits. ◻ **ebulliently** *adv.*, **ebullience** *n.*

EC *abbr.* **1** European Community. **2** European Commission.

eccentric *adj.* **1** unconventional. **2** not concentric. **3** (of an orbit or wheel) not circular. ● *n.* eccentric person. ◻ **eccentrically** *adv.*, **eccentricity** *n.*

■ *adj.* **1** aberrant, abnormal, bizarre, curious, odd, offbeat, outlandish, out of the ordinary, peculiar, quaint, queer, quirky, singular, strange, unconventional, unusual, weird.

ecclesiastical *adj.* of the Church or clergy.

echelon /éshəlon/ *n.* **1** staggered formation of troops etc. **2** level of rank or authority.

echo *n.* (*pl.* **-oes**) **1** repetition of sound by reflection of sound waves. **2** close imitation. ● *v.* (**echoed**, **echoing**) **1** repeat by an echo. **2** imitate.

■ *v.* **1** resound, reverberate, ring. **2** ape, copy, emulate, imitate, mimic, repeat.

éclair *n.* finger-shaped cake with cream filling.

eclectic *adj.* choosing or accepting from various sources.

eclipse *n.* **1** blocking of light from one heavenly body by another. **2** loss of brilliance or power etc. ● *v.* **1** cause an eclipse of. **2** outshine.

■ *v.* **1** block, blot out, conceal, cover, hide, obscure, shroud, veil. **2** outshine, outstrip, overshadow, surpass, top.

eclogue *n.* short pastoral poem.

ecology *n.* **1** (study of) relationships of living things to their environment. **2** protection of the natural environment. ◻ **ecological** *adj.*, **ecologically** *adv.*, **ecologist** *n.*

economic *adj.* **1** of economics. **2** enough to give a good return for money or effort outlaid.

economical *adj.* thrifty, avoiding waste. ◻ **economically** *adv.*

■ careful, frugal, provident, prudent, sparing, thrifty; cheap, inexpensive, reasonable.

economics *n.* **1** the science of the production and use of goods or services. **2** (as *pl.*) the financial aspects of something.

economist *n.* expert in economics.

economize *v.* use or spend less.

■ cut back, retrench, save, scrimp, skimp, spend less.

economy *n.* **1** being economical. **2** community's system of using its resources to produce wealth. **3** state of a country's prosperity.

■ **1** frugality, husbandry, prudence, thrift, thriftiness.

ecstasy *n.* intense delight. ◻ **ecstatic** *adj.*, **ecstatically** *adv.*

■ bliss, delight, elation, euphoria, exultation, happiness, joy, rapture. ◻ **ecstatic**

delighted, elated, enraptured, euphoric, exhilarated, exultant, happy, joyful, overjoyed, rapturous, thrilled.

ecu /ékyoo/ *abbr.* European currency unit.

ecumenical *adj.* **1** of the whole Christian Church. **2** seeking worldwide Christian unity.

eczema *n.* skin disease causing scaly itching patches.

eddy *n.* swirling patch of water or air etc. ● *v.* swirl in eddies.

edelweiss /áyd'lvīss/ *n.* alpine plant with woolly white bracts.

edge *n.* **1** outer limit of a surface or area, narrow surface of a thin object. **2** sharpness. **3** sharpened side of a blade. ● *v.* **1** border. **2** move gradually.

■ *n.* **1** boundary, fringe, margin, perimeter, periphery; border, brim, brink, lip, rim, side, verge. **2** acuteness, keenness, sharpness. ● *v.* **1** border, fringe, trim. **2** crawl, creep, inch, sidle, steal, worm.

edgeways *adv.* (also **edgewise**) with the edge forwards or outwards.

edging *n.* thing placed round an edge to define or decorate it.

edgy *adj.* (**-ier, -iest**) tense and irritable.

edible *adj.* suitable for eating.

■ eatable, palatable, wholesome.

edict /eedikt/ *n.* order proclaimed by authority.

edifice *n.* large building.

edify *v.* be an uplifting influence on the mind of. □ **edification** *n.*

edit *v.* (**edited**) **1** prepare for publication. **2** prepare (a film) by arranging sections in sequence.

edition *n.* **1** form in which something is published. **2** number of objects issued at one time.

editor *n.* **1** person responsible for the contents of a newspaper etc. or a section of this. **2** one who edits.

editorial *adj.* of an editor. ● *n.* newspaper article giving the editor's comments.

educate *v.* **1** train the mind and abilities of. **2** provide such training for. □ **education** *n.*, **educational** *adj.*

■ bring up, civilize, edify, enlighten, inform, instruct, school, teach, train, tutor. □ **education** instruction, schooling, teaching, training, tuition, upbringing.

Edwardian *adj.* of the reign of Edward VII (1901–10).

EEC *abbr.* European Economic Community.

eel *n.* snake-like fish.

eerie *adj.* (**-ier, -iest**) mysterious and frightening. □ **eerily** *adv.*

■ creepy, frightening, ghostly, mysterious, *colloq.* scary, *colloq.* spooky, strange, uncanny, unearthly, weird.

efface *v.* **1** rub out, obliterate. **2** make inconspicuous. □ **effacement** *n.*

effect *n.* **1** result of an action etc. **2** efficacy. **3** impression. **4** (*pl.*) property. **5** state of being operative. ● *v.* cause to occur.

■ *n.* **1** conclusion, consequence, outcome, repercussion, result, upshot. **2** effectiveness, efficacy, impact, influence, force, power. **3** feeling, impression, sense. **4** (**effects**) belongings, chattels, possessions, property, things. ● *v.* accomplish, achieve, bring about, carry out, cause, create, execute, make, produce, secure.

effective *adj.* **1** producing the intended result. **2** striking. **3** operative. □ **effectively** *adv.*, **effectiveness** *n.*

■ **1** effectual, efficacious, efficient, functional, serviceable, useful. **2** impressive, outstanding, powerful, striking. **3** functioning, in operation, operational, operative.

effectual *adj.* answering its purpose. □ **effectually** *adv.*

effeminate *adj.* (of a man) feminine in appearance or manner. □ **effeminacy** *n.*

effervesce *v.* give off bubbles. □ **effervescence** *n.*, **effervescent** *adj.*

■ bubble, fizz, foam, froth. □ **effervescent** bubbling, bubbly, carbonated, fizzy, foaming, frothing, frothy, sparkling.

effete *adj.* having lost its vitality.

efficacious *adj.* producing the desired result. □ **efficacy** *n.*

efficient *adj.* **1** producing results with little waste of effort. **2** capable, competent. □ **efficiently** *adv.*, **efficiency** *n.*

■ **1** effective, effectual, efficacious. **2** businesslike, capable, competent, organized, professional, proficient, skilled, systematic, well-organized.

effigy *n.* model of person.

effloresce *v.* flower. □ **efflorescence** *n.*

effluent *n.* outflow, sewage.

effluvium *n.* (*pl.* **-ia**) outflow, esp. unpleasant or harmful.

effort *n.* **1** strenuous use of energy. **2** attempt. □ **effortless** *adj.*

■ **1** exertion, pains, strain, striving, struggle, toil, trouble, work. **2** attempt, endeavour, essay, shot, try, venture.

effrontery *n.* bold insolence.
effusion *n.* outpouring.
effusive *adj.* expressing emotion in an unrestrained way. □ **effusively** *adv.*, **effusiveness** *n.*
e.g. *abbr.* (Latin *exempli gratia*) for example.
egalitarian *adj.* & *n.* (person) holding the principle of equal rights for all. □ **egalitarianism** *n.*
egg[1] *n.* **1** hard-shelled oval body produced by the female of birds, esp. that of the domestic hen. **2** ovum. □ **eggshell** *n.*
egg[2] *v.* **egg on** urge on.
eggplant *n.* aubergine.
ego *n.* **1** self. **2** self-esteem.
egocentric *adj.* self-centred.
egoism *n.* self-centredness. □ **egoist** *n.*, **egoistic** *adj.*
egotism *n.* conceit, selfishness. □ **egotist** *n.*, **egotistic(al)** *adj.*

■ conceit, narcissism, pride, self-admiration, self-importance, selfishness, self-love, vanity.

egregious /igreéjəss/ *adj.* **1** shocking. **2** (*old use*) remarkable.
egress *n.* **1** departure. **2** way out.
egret *n.* a kind of heron.
Egyptian *adj.* & *n.* (native) of Egypt.
Egyptology *n.* study of Egyptian antiquities. □ **Egyptologist** *n.*
eider *n.* northern species of duck.
eiderdown *n.* quilt stuffed with soft material.
eight *adj.* & *n.* one more than seven. □ **eighth** *adj.* & *n.*
eighteen *adj.* & *n.* one more than seventeen. □ **eighteenth** *adj.* & *n.*
eighty *adj.* & *n.* ten times eight. □ **eightieth** *adj.* & *n.*
either *adj.* & *pron.* **1** one or other of two. **2** each of two. ● *adv.* & *conj.* as one alternative.
ejaculate *v.* **1** utter suddenly. **2** eject (semen). □ **ejaculation** *n.*
eject *v.* **1** drive out forcefully. **2** emit. □ **ejection** *n.*, **ejector** *n.*

■ **1** drive out, evict, expel, *colloq.* kick out, oust, remove, send packing, throw out; banish, deport, exile, send away; discharge, dismiss, fire, *colloq.* sack. **2** disgorge, emit, exude, send out, spew, spout.

eke *v.* **eke out 1** supplement. **2** make (a living) laboriously.
elaborate *adj.* /ilábbərət/ **1** carefully worked out. **2** with many parts or details. ● *v.* /ilábbərayt/ add detail to. □ **elaborately** *adv.*, **elaboration** *n.*

■ *adj.* **1** careful, detailed, exhaustive, meticulous, painstaking, thorough. **2** baroque, decorative, fancy, fussy, ornamental, ornate, rococo, showy; complex, complicated, intricate, involved. ● *v.* add to, amplify, develop, embellish, embroider, enlarge on, expand, fill out.

élan /aylóɴ/ *n.* vivacity, vigour.
eland /éelənd/ *n.* large African antelope.
elapse *v.* (of time) pass away.
elastic *adj.* **1** able to go back to its original length or shape after being stretched or squeezed. **2** adaptable. ● *n.* cord or material made elastic by interweaving strands of rubber etc. □ **elasticity** *n.*

■ *adj.* **1** bendable, flexible, pliable, pliant, resilient, springy, stretchable, stretchy. **2** adaptable, adjustable, flexible.

elate *v.* cause to feel very pleased or proud. □ **elation** *n.*
elbow *n.* **1** joint between the forearm and upper arm. **2** part of a sleeve covering this. **3** sharp bend. ● *v.* thrust with one's elbow. □ **elbow grease** (*joc.*) vigorous polishing. **elbow room** enough space to move or work in.
elder[1] *adj.* older. ● *n.* **1** older person. **2** official in certain Churches.
elder[2] *n.* tree with dark berries.
elderberry *n.* fruit of the elder tree.
elderly *adj.* rather old.
eldest *adj.* first-born or oldest surviving (son, daughter, etc.).
elect *v.* **1** choose by vote. **2** choose. ● *adj.* chosen.

■ *v.* **1** appoint, choose, designate, name, nominate, pick, select, vote for. **2** adopt, choose, decide on, go for, opt for, pick, select.

election *n.* **1** process of electing. **2** occasion for this.

■ **1** choice, nomination, selection. **2** ballot, plebiscite, poll, referendum, vote.

electioneer *v.* busy oneself in an election campaign.
elective *adj.* **1** chosen by election. **2** entitled to elect. **3** optional.
elector *n.* person entitled to vote in an election. □ **electoral** *adj.*
electorate *n.* body of electors.

electric *adj.* of, producing, or worked by electricity.

electrical *adj.* of or worked by electricity. □ **electrically** *adv.*

electrician *n.* person who installs or maintains electrical equipment.

electricity *n.* **1** form of energy occurring in certain particles. **2** supply of electric current.

electrify *v.* **1** charge with electricity. **2** convert to electric power. □ **electrification** *n.*

electrocardiogram *n.* record of the electric current generated by heartbeats.

electrocute *v.* kill by electricity. □ **electrocution** *n.*

electrode *n.* solid conductor through which electricity enters or leaves a vacuum tube etc.

electroencephalogram *n.* record of the electrical activity of the brain.

electrolyte *n.* solution that conducts electric current.

electromagnet *n.* magnet consisting of a metal core magnetized by a current-carrying coil round it.

electromagnetism *n.* magnetic forces produced by electricity. □ **electromagnetic** *adj.*

electron *n.* particle with a negative electric charge. □ **electron microscope** very powerful one using a focused beam of electrons instead of light.

electronic *adj.* **1** produced or worked by a flow of electrons. **2** of electronics. □ **electronically** *adv.*

electronics *n.* science concerned with the movement of electrons in a vacuum, gas, semiconductor, etc.

elegant *adj.* tasteful and dignified. □ **elegantly** *adv.*, **elegance** *n.*

■ dignified, exquisite, fine, graceful, refined, tasteful; debonair, polished, sophisticated, suave, urbane; artistic, beautiful, chic, fashionable, smart, stylish, *colloq.* swish.

elegy *n.* sorrowful or serious poem. □ **elegiac** *adj.*

element *n.* **1** component part. **2** suitable or satisfying environment. **3** (*pl.*) basic principles. **4** substance that cannot be broken down into other substances. **5** trace. **6** wire that gives out heat in an electrical appliance. **7** (*pl.*) atmospheric forces. □ **elemental** *adj.*

■ **1** component, constituent, detail, factor, feature, ingredient, part, piece, unit. **2** domain, environment, habitat, medium, sphere, surroundings, territory. **3** (**elements**) basics, essentials, first principles, foundations, fundamentals, rudiments.

elementary *adj.* dealing with the simplest facts of a subject.

■ basic, fundamental, initial, introductory, primary, rudimentary; easy, simple, straightforward, uncomplicated.

elephant *n.* very large animal with a trunk and ivory tusks.

elephantine *adj.* **1** of or like elephants. **2** very large, clumsy.

elevate *v.* raise to a higher position or level.

■ lift, raise; advance, exalt, promote, upgrade.

elevation *n.* **1** elevating. **2** altitude. **3** hill. **4** drawing showing one side of a structure.

elevator *n.* **1** thing that hoists something. **2** (*US*) lift.

eleven *adj.* & *n.* one more than ten. □ **eleventh** *adj.* & *n.*

elevenses *n.* mid-morning snack.

elf *n.* (*pl.* **elves**) imaginary small being with magic powers. □ **elfin** *adj.*

elicit *v.* draw out.

■ bring out, draw out, evoke, extract, get, wrest, wring.

eligible *adj.* qualified to be chosen or allowed something. □ **eligibility** *n.*

■ appropriate, entitled, fit, qualified, suitable, worthy.

eliminate *v.* **1** get rid of. **2** exclude. □ **elimination** *n.*, **eliminator** *n.*

■ **1** dispense with, dispose of, do away with, eradicate, get rid of, remove, root out, stamp out. **2** drop, exclude, leave out, omit, rule out.

elite /ayleet/ *n.* group regarded as superior and favoured.

elitism /ayleetiz'm/ *n.* favouring of or dominance by a selected group. □ **elitist** *n.*

elixir *n.* fragrant liquid used as medicine or flavouring.

Elizabethan *adj.* of Elizabeth I's reign (1558–1603).

elk *n.* large deer.

ellipse *n.* regular oval.

ellipsis *n.* (*pl.* **-pses**) omission of words.

elliptical *adj.* **1** shaped like an ellipse. **2** having omissions. □ **elliptically** *adv.*

elm *n.* **1** tree with rough serrated leaves. **2** its wood.

elocution *n.* art of speaking.

elongate *v.* lengthen.

elope *v.* run away secretly with a lover. □ **elopement** *n.*

eloquence *n.* fluent speaking. □ **eloquent** *adj.*, **eloquently** *adv.*

■ □ **eloquent** articulate, expressive, fluent, lucid.

else *adv.* **1** besides. **2** otherwise.

elsewhere *adv.* somewhere else.

elucidate *v.* throw light on, explain. □ **elucidation** *n.*

elude *v.* **1** escape skilfully from. **2** avoid. **3** escape the memory or understanding of. □ **elusion** *n.*

■ **1** avoid, dodge, escape, evade, get away from, give a person the slip. **2** avoid, duck, sidestep.

elusive *adj.* difficult to find, catch, or remember.

elver *n.* young eel.

emaciated *adj.* thin from illness or starvation. □ **emaciation** *n.*

■ bony, cadaverous, gaunt, haggard, scraggy, scrawny, skeletal, skinny, starved, thin, underfed, undernourished, wizened.

emanate *v.* issue, originate from a source. □ **emanation** *n.*

emancipate *v.* liberate, free from restraint. □ **emancipation** *n.*

■ deliver, free, liberate, release, set free.

emasculate *v.* deprive of force, weaken. □ **emasculation** *n.*

embalm *v.* preserve (a corpse) by using spices or chemicals.

embankment *n.* bank constructed to confine water or carry a road or railway.

embargo *n.* (*pl.* **-oes**) order forbidding commerce or other activity.

embark *v.* board a ship. □ **embark on** begin an undertaking. **embarkation** *n.*

embarrass *v.* cause to feel awkward or ashamed. □ **embarrassment** *n.*

■ abash, discomfit, disconcert; disgrace, humiliate, mortify, shame. □ **embarrassment** awkwardness, discomfort, self-consciousness; chagrin, mortification.

embassy *n.* **1** ambassador and staff. **2** their headquarters.

embed *v.* (**embedded**) fix firmly in a surrounding mass.

embellish *v.* **1** ornament. **2** improve (a story) with invented details. □ **embellishment** *n.*

■ **1** adorn, beautify, deck, decorate, dress (up), embroider, ornament, trim. **2** elaborate, embroider, exaggerate.

embers *n.pl.* small pieces of live coal or wood in a dying fire.

embezzle *v.* take (money etc.) fraudulently for one's own use. □ **embezzler** *n.*

embitter *v.* rouse bitter feelings in. □ **embitterment** *n.*

emblem *n.* symbol, design used as a badge etc.

■ badge, crest, insignia, mark, seal, sign, symbol, token; logo, trade mark.

emblematic *adj.* serving as an emblem. □ **emblematically** *adv.*

embody *v.* **1** express (principles or ideas) in visible form. **2** incorporate. □ **embodiment** *n.*

■ **1** epitomize, exemplify, express, manifest, personify, represent, typify. **2** comprise, embrace, encompass, include, incorporate.

embolden *v.* encourage.

embolism *n.* obstruction of an artery by a blood clot etc.

emboss *v.* **1** decorate by a raised design. **2** mould in relief.

embrace *v.* **1** hold closely and lovingly, hold each other thus. **2** accept, adopt. **3** include. ● *n.* act of embracing, hug.

■ *v.* **1** clasp, cuddle, enfold, grasp, hold, hug. **2** accept, adopt, espouse, welcome. **3** comprise, embody, encompass, include, incorporate. ● *n.* clasp, cuddle, hug, squeeze.

embrocation *n.* liquid for rubbing on the body to relieve muscular pain.

embroider *v.* **1** ornament with needlework. **2** embellish (a story). □ **embroidery** *n.*

embroil *v.* involve in an argument or quarrel etc.

embryo *n.* (*pl.* **-os**) animal developing in a womb or egg. □ **embryonic** *adj.*

embryology *n.* study of embryos.

emend *v.* alter to remove errors. □ **emendation** *n.*

emerald *n.* **1** bright green precious stone. **2** its colour.

emerge *v.* **1** come up or out into view. **2** become known. □ **emergence** *n.*, **emergent** *adj.*

■ **1** appear, arise, come into view, come out, surface. **2** become known, be revealed, come to light, transpire, turn out.

emergency *n.* serious situation needing prompt attention.

■ crisis, danger, difficulty, exigency, predicament.

emery *n.* coarse abrasive. □ **emery board** emery-coated cardboard strip for filing the nails.

emetic *n.* medicine used to cause vomiting.

emigrate *v.* leave one country and go to settle in another. □ **emigration** *n.*, **emigrant** *n.*

eminence *n.* **1** state of being eminent. **2** rising ground.

eminent *adj.* famous, distinguished. □ **eminently** *adv.*

■ celebrated, distinguished, esteemed, exalted, famous, great, honoured, illustrious, important, notable, noteworthy, outstanding, pre-eminent, prominent, renowned, respected, well-known.

emir /emeer/ *n.* Muslim ruler. □ **emirate** *n.* his territory.

emissary *n.* person sent to conduct negotiations.

emit *v.* (**emitted**) **1** send out (light, heat, fumes, etc.). **2** utter. □ **emission** *n.*, **emitter** *n.*

■ **1** discharge, eject, expel, exude, give off *or* out, radiate, send out.

emollient *adj.* softening, soothing. ● *n.* emollient substance.

emolument *n.* **1** fee. **2** salary.

emotion *n.* strong instinctive feeling.

■ feeling, passion, sensation, sentiment.

emotional *adj.* **1** of, expressing, or arousing great emotion. **2** liable to excessive emotion. □ **emotionally** *adv.*, **emotionalism** *n.*

■ **1** ardent, fervent, fervid, heartfelt, heated, impassioned, passionate; emotive, moving, pathetic, poignant, touching. **2** excitable, highly-strung, temperamental, volatile.

emotive *adj.* rousing emotion.

empathize *v.* show empathy.

empathy *n.* ability to identify oneself mentally with, and so understand, a person or thing.

emperor *n.* ruler of an empire.

emphasis *n.* (*pl.* **-ases**) **1** special importance. **2** vigour of expression etc. **3** stress on word(s).

■ **1** importance, priority, prominence, significance, stress, weight.

emphasize *v.* lay emphasis on.

■ accent, accentuate, draw attention to, feature, highlight, point up, spotlight, stress, underline.

emphatic *adj.* using or showing emphasis. □ **emphatically** *adv.*

■ assertive, categorical, decided, definite, explicit, firm, forceful, insistent, strong, uncompromising, unequivocal, vigorous.

emphysema *n.* disease of the lungs, causing breathlessness.

empire *n.* **1** group of countries ruled by a supreme authority. **2** large organization controlled by one person or group.

empirical *adj.* based on observation or experiment, not on theory. □ **empirically** *adv.*, **empiricism** *n.*, **empiricist** *n.*

emplacement *n.* place or platform for a gun or guns.

employ *v.* **1** use the services of. **2** make use of. □ **employer** *n.*

■ **1** engage, hire, sign, recruit, take on. **2** make use of, use, utilize. □ **employer** *colloq.* boss, chief, *colloq.* gaffer, head, manager, owner, proprietor.

employee *n.* person employed by another in return for wages.

■ hand, member of staff, worker; (**employees**) staff, workforce.

employment *n.* **1** act of employing or state of being employed. **2** person's trade or profession.

■ **1** engagement, hire, hiring, recruitment; application, use, utilization. **2** business, craft, job, occupation, profession, trade, vocation, work.

empower *v.* authorize, enable.

empress *n.* **1** woman emperor. **2** wife of an emperor.

empty *adj.* (**-ier, -iest**) **1** containing nothing. **2** without occupant(s). **3** meaningless, insincere. ● *v.* make or become empty. □ **emptiness** *n.*

■ *adj.* **1** unfilled, void; blank, clean, new, unused; drained, emptied; unfurnished. **2** deserted, uninhabited, unoccupied, vacant; bare, barren, desolate, waste. **3** hollow, hypocritical, insincere, meaningless, pointless, shallow, valueless, worthless. ● *v.* clear (out), evacuate, vacate; drain; discharge, unload, void.

emu *n.* large Australian bird resembling an ostrich.

emulate *v.* try to do as well as. □ **emulation** *n.*, **emulator** *n.*

emulsify *v.* convert or be converted into emulsion. □ **emulsification** *n.*, **emulsifier** *n.*

emulsion *n.* **1** creamy liquid. **2** light-sensitive coating on photographic film.

enable *v.* give the authority or means to do something.

> ■ allow, authorize, empower, entitle, license, permit, qualify.

enact *v.* **1** make into a law. **2** perform (a play etc.). □ **enactment** *n.*

enamel *n.* **1** glass-like coating for metal or pottery. **2** glossy paint. **3** hard outer covering of teeth. ● *v.* (**enamelled**) coat with enamel.

enamoured *adj.* fond.

en bloc /ON blók/ all together.

encamp *v.* settle in a camp.

encampment *n.* camp.

encapsulate *v.* **1** enclose (as) in a capsule. **2** summarize.

encase *v.* enclose in a case.

encephalitis *n.* inflammation of the brain.

enchant *v.* bewitch. □ **enchantment** *n.*, **enchanter** *n.*, **enchantress** *n.*

> ■ bewitch, captivate, cast a spell on, charm, delight, enthral, entrance, fascinate, hold spellbound.

encircle *v.* surround. □ **encirclement** *n.*

enclave *n.* small territory wholly within the boundaries of another.

enclose *v.* **1** shut in on all sides. **2** include with other contents.

> ■ **1** bound, circle, confine, encircle, encompass, envelop, fence in, hedge in, hem in, immure, pen, ring, shut in, surround, wall in.

enclosure *n.* **1** enclosing. **2** enclosed area. **3** thing enclosed.

encompass *v.* **1** encircle. **2** include.

encore *n.* a (call for) repetition of a performance. ● *int.* this call.

encounter *v.* **1** meet by chance. **2** be faced with. ● *n.* **1** chance meeting. **2** meeting in conflict.

> ■ *v.* **1** come across, meet, run into. **2** be faced with, experience, face, meet (with). ● *n.* **2** altercation, battle, brush, clash, conflict, confrontation, dispute, fight, *colloq.* scrap, skirmish, struggle, tussle.

encourage *v.* **1** give hope, confidence, or stimulus to. **2** urge. □ **encouragement** *n.*

> ■ **1** buoy up, cheer (up), embolden, fortify, hearten, inspire, pep up, reassure, support; advance, aid, assist, boost, foster, help, promote, stimulate. **2** egg on, exhort, incite, spur on, urge. □ **encouragement** help, inspiration, moral support, reassurance, stimulation, support; exhortation, incitement.

encroach *v.* **encroach on** intrude on someone's territory or rights. □ **encroachment** *n.*

> ■ impinge on, infringe on, intrude on, invade, trespass on.

encrust *v.* cover with a crust of hard material. □ **encrustation** *n.*

encumber *v.* be a burden to, hamper. □ **encumbrance** *n.*

encyclical *n.* pope's letter for circulation to churches.

encyclopedia *n.* book of information on many subjects. □ **encyclopedic** *adj.*

end *n.* **1** extreme limit, furthest point or part. **2** final part. **3** destruction, death. **4** objective. ● *v.* bring or come to an end. □ **end up** reach a certain place or condition eventually. **make ends meet** live within one's income.

> ■ *n.* **1** boundary, edge, extremity, limit, tip. **2** cessation, completion, conclusion, culmination, finish, termination; close, coda, denouement, finale. **3** annihilation, death, destruction, extinction, ruin. **4** aim, aspiration, design, goal, intention, object, objective, plan, purpose; destination. ● *v.* bring to an end, conclude, discontinue, halt, put paid to, stop, terminate, wind up; cease, come to an end, die (away), fizzle out.

endanger *v.* cause danger to.

> ■ expose to risk, imperil, jeopardize, put in jeopardy, threaten.

endear *v.* cause to be loved.

endearment *n.* word(s) expressing love.

endeavour *v.* & *n.* attempt.

endemic *adj.* commonly found in a specified area or people.

ending *n.* final part.

endive *n.* **1** curly-leaved plant used in salads. **2** (*US*) chicory.

endless *adj.* **1** infinite. **2** continual. □ **endlessly** *adv.*

> ■ **1** boundless, immeasurable, infinite, unbounded, unlimited; eternal, everlasting, perpetual. **2** ceaseless, constant, continual, continuous, everlasting, incessant,

interminable, never-ending, non-stop, perpetual, unceasing, unending, unremitting.

endocrine gland gland secreting hormones into the blood.

endorse *v.* **1** declare approval of. **2** sign the back of (a cheque). **3** note an offence on (a driving licence etc.). ◻ **endorsement** *n.*

■ **1** agree to, approve, assent to, authorize, back, give the go-ahead to, *colloq.* OK, rubber-stamp, sanction.

endow *v.* provide with a permanent income. ◻ **endowment** *n.*

endurance *n.* power of enduring.

■ durability, *colloq.* grit, fortitude, patience, perseverance, persistence, resilience, stamina, staying power, strength, tenacity.

endure *v.* **1** experience and survive (pain or hardship). **2** tolerate. **3** last. ◻ **endurable** *adj.*

■ **1** brave, face, survive, take, undergo, weather, withstand. **2** abide, bear, cope with, put up with, stand, stomach, suffer, tolerate. **3** carry on, continue, last, persist, remain, stay, survive.

enema *n.* liquid injected into the rectum.

enemy *n.* one who is hostile to and seeks to harm another.

■ adversary, antagonist, foe, opponent, rival.

energetic *adj.* full of energy. ◻ **energetically** *adv.*

■ active, animated, brisk, bubbly, dynamic, enthusiastic, go-ahead, go-getting, indefatigable, lively, spirited, sprightly, spry, tireless, unflagging, untiring, vibrant, vigorous, vivacious, zestful.

energize *v.* **1** give energy to. **2** cause electricity to flow into.

energy *n.* **1** vigour. **2** capacity for activity. **3** ability of matter or radiation to do work. **4** oil etc. as fuel.

■ **1** animation, *colloq.* bounce, drive, dynamism, élan, enthusiasm, go, gusto, life, liveliness, pep, spirit, verve, *colloq.* vim, vigour, vitality, vivacity, zeal, zest, zip. **2** force, forcefulness, might, power, strength.

enervate *v.* cause to lose vitality. ◻ **enervation** *n.*

enfant terrible /óNfoN tereébla/ person whose behaviour is embarrassing or irresponsible.

enfeeble *v.* make feeble. ◻ **enfeeblement** *n.*

enfold *v.* **1** wrap up. **2** clasp.

enforce *v.* compel obedience to. ◻ **enforcement** *n.*

enfranchise *v.* give the right to vote. ◻ **enfranchisement** *n.*

engage *v.* **1** employ (a person). **2** occupy the attention of. **3** reserve. **4** begin a battle with. **5** interlock.

■ **1** employ, hire, recruit, sign, take on. **2** absorb, attract, capture, catch, draw, hold, occupy. **3** book, reserve, take. **4** clash (with), encounter, fight (against), join in battle (with), wage war (against).

engaged *adj.* **1** having promised to marry a specified person. **2** occupied. **3** in use.

engagement *n.* **1** act of engaging something. **2** appointment. **3** promise to marry a specified person. **4** battle.

■ **2** appointment, assignation, *colloq.* date, meeting, rendezvous. **3** betrothal. **4** battle, combat, conflict, encounter, fight, fray, skirmish, war.

engaging *adj.* attractive.

engender *v.* give rise to.

engine *n.* **1** machine using fuel and supplying power. **2** railway locomotive.

engineer *n.* **1** person skilled in engineering. **2** one in charge of machines and engines. ● *v.* contrive, bring about.

engineering *n.* application of science for the design and building of machines etc.

English *adj.* & *n.* (language) of England. ◻ **Englishman** *n.*, **Englishwoman** *n.*

engrave *v.* **1** cut (a design) into a hard surface. **2** ornament thus. ◻ **engraver** *n.*

■ **1** carve, chisel, cut, etch, inscribe.

engraving *n.* print made from an engraved metal plate.

engross *v.* occupy fully by absorbing the attention.

engulf *v.* swamp, overwhelm.

enhance *v.* intensify, improve. ◻ **enhancement** *n.*

■ add to, augment, boost, heighten, improve, increase, intensify, raise, strengthen.

enigma *n.* puzzling person or thing. ◻ **enigmatic** *adj.*

■ conundrum, mystery, problem, puzzle, riddle.

enjoy *v.* **1** get pleasure from. **2** have the use or benefit of. ◻ **enjoyable** *adj.*, **enjoyment** *n.*

■ **1** appreciate, *colloq.* be into, be keen on, be partial to, delight in, like, love, relish, revel in, savour, take pleasure in. **2** benefit from, have, make use of, possess, use, utilize. □ **enjoyable** agreeable, amusing, delightful, entertaining, pleasant, pleasing, pleasurable, satisfying.

enlarge *v.* make or become larger. □ **enlarge upon** say more about. **enlargement** *n.*

■ add to, augment, broaden, develop, elongate, expand, extend, increase, lengthen, magnify, stretch, supplement, widen; dilate, distend, inflate, swell. □ **enlarge upon** amplify, go into detail about, elaborate on, embellish, embroider, expand on, expatiate on.

enlighten *v.* **1** inform. **2** free from ignorance. □ **enlightenment** *n.*

■ **1** advise, apprise, inform, make aware. **2** civilize, edify, educate, instruct, teach.

enlist *v.* **1** enrol for military service. **2** get the support of. □ **enlistment** *n.*

enliven *v.* make more lively. □ **enlivenment** *n.*

■ animate, arouse, energize, galvanize, inspire, invigorate, kindle, pep up, refresh, rouse, stimulate, stir (up), vivify.

en masse /ᴏɴ máss/ all together.

enmesh *v.* entangle.

enmity *n.* hostility, hatred.

ennoble *v.* make noble. □ **ennoblement** *n.*

ennui /onwée/ *n.* boredom.

enormity *n.* (act of) great wickedness.

enormous *adj.* very large.

■ colossal, elephantine, gargantuan, giant, gigantic, huge, immense, mammoth, massive, monstrous, mountainous, prodigious, stupendous, titanic, vast.

enough *adj.*, *adv.*, & *n.* as much or as many as necessary.

■ *adj.* adequate, ample, sufficient.

enquire *v.* ask. □ **enquiry** *n.*

enrage *v.* make furious.

■ anger, drive berserk, exasperate, incense, inflame, infuriate, madden.

enrapture *v.* delight intensely.

enrich *v.* make richer. □ **enrichment** *n.*

enrol *v.* (**enrolled**) admit as or become a member. □ **enrolment** *n.*

en route /ᴏɴ ro͞ot/ on the way.

ensconce *v.* establish securely or comfortably.

ensemble /onsómb'l/ *n.* **1** thing viewed as a whole. **2** set of performers. **3** outfit.

enshrine *v.* set in a shrine.

ensign *n.* military or naval flag.

enslave *v.* make slave(s) of. □ **enslavement** *n.*

ensnare *v.* trap.

ensue *v.* happen afterwards or as a result.

en suite /ᴏɴ swéet/ forming a unit.

ensure *v.* make certain or safe.

entail *v.* necessitate or involve.

■ demand, involve, mean, necessitate, require.

entangle *v.* **1** tangle. **2** entwine and trap. □ **entanglement** *n.*

entente /ontónt/ *n.* friendly understanding between countries.

enter *v.* **1** go or come in or into. **2** penetrate. **3** put on a list or into a record etc. **4** register as a competitor.

■ **2** go through, penetrate, perforate, pierce, puncture; infiltrate, invade. **3** jot down, list, log, note, put down, record, register, write down.

enteritis *n.* inflammation of the intestines.

enterprise *n.* **1** bold undertaking. **2** initiative. **3** business activity.

■ **1** adventure, endeavour, project, scheme, undertaking, venture. **2** ambition, courage, daring, determination, drive, dynamism, energy, initiative, resourcefulness. **3** business, company, concern, corporation, firm, organization.

enterprising *adj.* full of initiative.

■ adventurous, ambitious, bold, courageous, daring, determined, dynamic, eager, energetic, enthusiastic, go-ahead, go-getting, imaginative, ingenious, innovative, inventive, keen, *colloq.* pushy, resourceful, spirited, vigorous, zealous.

entertain *v.* **1** amuse, occupy pleasantly. **2** receive with hospitality. **3** consider (an idea etc.). □ **entertainer** *n.*

■ **1** amuse, delight, make laugh, please, tickle; absorb, beguile, divert, engage, interest, occupy. **2** accommodate, feed, fête, receive, regale, treat. **3** consider, contemplate.

entertainment *n.* **1** entertaining, being entertained. **2** thing that entertains, performance.

■ **1** amusement, distraction, diversion; enjoyment, fun, pleasure, recreation. **2** extravaganza, performance, presentation, show, spectacle.

enthral *v.* (**enthralled**) hold spellbound. □ **enthralment** *n.*

enthrone *v.* place on a throne. □ **enthronement** *n.*

enthuse *v.* fill with or show enthusiasm.

enthusiasm *n.* **1** eager liking or interest. **2** object of this. □ **enthusiastic** *adj.*, **enthusiastically** *adv.*

■ **1** appetite, ardour, devotion, eagerness, fervour, gusto, interest, keenness, liking, love, passion, predilection, relish, zeal, zest. **2** craze, fad, mania, passion; hobby, pastime. □ **enthusiastic** animated, ardent, avid, committed, dedicated, devoted, eager, ebullient, exuberant, fervent, hearty, impassioned, keen, passionate, vehement, vigorous, warm, wholehearted, zealous; *colloq.* crazy, excited, mad, wild.

enthusiast *n.* person who is full of enthusiasm for something.

entice *v.* attract by offering something pleasant. □ **enticement** *n.*

■ allure, attract, coax, inveigle, lure, seduce, tempt, wheedle.

entire *adj.* complete. □ **entirely** *adv.*

■ complete, full, total, whole, unabridged, uncut; intact, perfect, unbroken, undamaged.

entirety *n.* **in its entirety** as a whole.

entitle *v.* **1** give a title to (a book etc.). **2** give (a person) a right or claim. □ **entitlement** *n.*

■ **1** call, name. **2** allow, authorize, empower, enable, license, permit, qualify.

entity *n.* a separate thing.

entomology *n.* study of insects. □ **entomological** *adj.*, **entomologist** *n.*

entourage /óntooraázh/ *n.* people accompanying an important person.

entrails *n.pl.* intestines.

entrance[1] /éntrənss/ *n.* **1** act or instance of entering. **2** door or passage by which one enters. **3** right of admission, fee for this.

■ **1** appearance, arrival, entry. **2** access, door, doorway, entry, gate, way in. **3** admission, admittance, ingress, right of entry.

entrance[2] /intraánss/ *v.* fill with intense delight.

entreat *v.* request earnestly or emotionally. □ **entreaty** *n.*

■ □ **entreaty** appeal, call, cry, petition, plea, request, supplication.

entrench *v.* establish firmly. □ **entrenchment** *n.*

entrepreneur *n.* person who organizes a commercial undertaking. □ **entrepreneurial** *adj.*

entrust *v.* give as a responsibility, place in a person's care.

entry *n.* **1** act or instance of entering. **2** entrance. **3** item entered in a list etc.

entwine *v.* twine round.

enumerate *v.* mention (items) one by one. □ **enumeration** *n.*

enunciate *v.* **1** pronounce. **2** state clearly. □ **enunciation** *n.*

envelop *v.* (**enveloped**) wrap, cover on all sides. □ **envelopment** *n.*

■ clothe, cover, encase, enclose, enfold, sheathe, swaddle, swathe, wrap; bury, cloak, conceal, hide, mask, shroud, veil.

envelope *n.* folded gummed cover for a letter.

enviable *adj.* desirable enough to arouse envy. □ **enviably** *adv.*

envious *adj.* full of envy. □ **enviously** *adv.*

■ begrudging, bitter, covetous, grudging, jaundiced, jealous, resentful.

environment *n.* **1** surroundings. **2** natural world. □ **environmental** *adj.*, **environmentally** *adv.*

■ **1** ambience, conditions, environs, surroundings; element, habitat, medium, milieu; background, context, setting, situation.

environmentalist *adj.* & *n.* (person) seeking to protect the natural environment.

environs *n.pl.* surrounding districts, esp. of a town.

envisage *v.* imagine, foresee.

envoy *n.* messenger, esp. to a foreign government.

envy *n.* **1** discontent aroused by another's possessions or success. **2** object of this. ● *v.* feel envy of.

■ *n.* **1** bitterness, ill will, jealousy, resentment; craving, hankering, longing. ● *v.* begrudge, be jealous of, grudge, resent; covet, crave, hanker after, long for.

enzyme *n.* protein formed in living cells (or produced synthetically) and assisting chemical processes.

epaulette *n.* ornamental shoulder-piece.

ephemeral *adj.* lasting only a short time. □ **ephemerally** *adv.*

■ brief, fleeting, fugitive, momentary, passing, short, transient, transitory.

epic *n.* long poem, story, or film about heroic deeds or history. ● *adj.* of or like an epic.

epicentre *n.* point where an earthquake reaches the earth's surface.

epicure *n.* person with refined tastes in food and drink. □ **epicurean** *adj.* & *n.*

epidemic *n.* outbreak of a disease etc. spreading through a community.

epidermis *n.* outer layer of the skin.

epidural *n.* anaesthetic injected close to the spinal cord.

epiglottis *n.* cartilage that covers the larynx in swallowing.

epigram *n.* short witty saying. □ **epigrammatic** *adj.*

epilepsy *n.* disorder of the nervous system, causing fits. □ **epileptic** *adj.* & *n.*

epilogue *n.* short concluding section.

episcopal *adj.* of or governed by bishop(s).

episcopalian *adj.* & *n.* (member) of an episcopal church.

episode *n.* **1** event forming one part of a sequence. **2** one part of a serial. □ **episodic** *adj.*

■ **1** event, happening, incident, occasion, occurrence. **2** chapter, instalment, part.

epistle *n.* letter. □ **epistolary** *adj.*

epitaph *n.* words in memory of a dead person, esp. on a tomb.

epithet *n.* descriptive word(s).

epitome /ipíttəmi/ *n.* person or thing embodying a quality etc.

■ archetype, embodiment, incarnation, personification, quintessence.

epitomize *v.* be an epitome of.

epoch /éepok/ *n.* particular period.

eponymous *adj.* after whom something is named.

equable *adj.* moderate, even-tempered. □ **equably** *adv.*

equal *adj.* **1** same in size, amount, value, etc. **2** evenly matched. **3** having the same rights or status. ● *n.* person or thing equal to another. ● *v.* (**equalled**) **1** be equal to. **2** do something equal to. □ **equally** *adv.*, **equality** *n.*

■ *adj.* **1** commensurate, equivalent, identical, indistinguishable, interchangeable, level, on a par, similar, the same. **2** balanced, matched, proportionate, symmetrical; level pegging, neck and neck. ● *n.* counterpart, equivalent, fellow, match, peer. ● *v.* **2** be on a par with, compare with, match, parallel, resemble, rival. □ **equality** equivalence, identity, parity, sameness, similarity.

equalize *v.* **1** make or become equal. **2** equal an opponent's score. □ **equalization** *n.*

equalizer *n.* equalizing goal etc.

equanimity *n.* composure.

equate *v.* consider to be equal or equivalent.

equation *n.* mathematical statement that two expressions are equal.

equator *n.* imaginary line round the earth at an equal distance from the North and South Poles. □ **equatorial** *adj.*

equestrian *adj.* **1** of horse-riding. **2** on horseback.

equidistant *adj.* at an equal distance.

equilateral *adj.* having all sides equal.

equilibrium *n.* state of balance.

equine /ékwīn/ *adj.* of or like a horse.

equinox *n.* time of year when night and day are of equal length. □ **equinoctial** *adj.*

equip *v.* (**equipped**) supply with what is needed.

■ arm, fix up, furnish, kit (out), provide, rig (out), stock, supply.

equipment *n.* **1** necessary tools, clothing, etc. for a purpose. **2** process of equipping or being equipped.

■ **1** accoutrements, apparatus, *sl.* clobber, gear, kit, materials, outfit, paraphernalia, stuff, tackle, things, tools, trappings.

equipoise *n.* equilibrium.

equitable *adj.* fair and just. □ **equitably** *adv.*

equitation *n.* horse-riding.

equity *n.* **1** fairness, impartiality. **2** (*pl.*) stocks and shares not bearing fixed interest.

equivalent *adj.* equal in importance, value, or meaning etc. ● *n.* equivalent thing. □ **equivalence** *n.*

■ *adj.* akin, analogous, commensurate, comparable, corresponding, equal, identical, interchangeable, parallel, similar, the same, tantamount.

equivocal *adj.* **1** ambiguous. **2** questionable. □ **equivocally** *adv.*

■ **1** ambiguous, evasive, indefinite, misleading, vague. **2** doubtful, dubious, questionable, suspect, suspicious.

equivocate *v.* use words ambiguously. □ **equivocation** *n.*

■ be evasive, evade the issue, hedge, prevaricate, quibble.

era *n.* period of history.

■ age, day(s), epoch, period, time.

eradicate *v.* wipe out, destroy. □ **eradication** *n.*

■ destroy, eliminate, expunge, extirpate, get rid of, obliterate, remove, root out, stamp out, uproot, wipe out.

erase *v.* rub out, obliterate. ▫ **eraser** *n.*, **erasure** *n.*

■ cancel, cross out, delete, efface, expunge, obliterate, rub out, strike out.

erect *adj.* **1** upright. **2** (of a part of the body) rigid, esp. from sexual excitement. ● *v.* set up, build. ▫ **erection** *n.*

■ *adj.* **1** perpendicular, standing, straight, upright, upstanding, vertical. ● *v.* assemble, build, construct, put together, put up, set up; pitch.

ergonomics *n.* study of work and its environment in order to improve efficiency. ▫ **ergonomic** *adj.*, **ergonomically** *adv.*

ermine *n.* **1** stoat. **2** its white winter fur.

erode *v.* wear away gradually. ▫ **erosion** *n.*, **erosive** *adj.*

■ corrode, eat away (at), gnaw away, grind down, wash away, wear away, whittle away.

erogenous *adj.* (of a part of the body) sexually sensitive.

erotic *adj.* of or arousing sexual desire. ▫ **erotically** *adv.*, **eroticism** *n.*

err *v.* (**erred**) **1** be mistaken or incorrect. **2** sin.

■ **1** be in the wrong, be mistaken, be wrong, blunder, make a mistake, miscalculate, *colloq.* slip up. **2** do wrong, sin, transgress.

errand *n.* **1** short journey to take or fetch something. **2** its purpose.

errant *adj.* misbehaving.

erratic *adj.* inconsistent or uncertain in conduct, movement, etc. ▫ **erratically** *adv.*

■ capricious, changeable, fickle, inconsistent, unpredictable, unreliable, wayward; aimless, haphazard, meandering, wandering; fitful, irregular, spasmodic, sporadic, uneven, variable.

erroneous *adj.* incorrect. ▫ **erroneously** *adv.*

error *n.* **1** mistake. **2** condition of being wrong. **3** amount of inaccuracy.

■ **1** blunder, *colloq.* boob, *sl.* clanger, fault, flaw, gaffe, *colloq.* howler, inaccuracy, mistake, oversight, slip, *colloq.* slip-up; misprint; misapprehension, miscalculation, misconception, misunderstanding. **2** misconduct, sin, transgression, wrongdoing.

erstwhile *adj.* former.

erudite *adj.* learned. ▫ **erudition** *n.*

erupt *v.* **1** break out or through. **2** eject lava. ▫ **eruption** *n.*

■ **1** break out, burst forth, explode, flare up. **2** gush, shoot out, spew, spout.

escalate *v.* increase in intensity or extent. ▫ **escalation** *n.*

escalator *n.* moving staircase.

escalope *n.* slice of boneless meat, esp. veal.

escapade *n.* piece of reckless or mischievous conduct.

escape *v.* **1** get free. **2** leak out of its container. **3** avoid. **4** be forgotten or unnoticed by. ● *n.* **1** act or means of escaping. **2** leakage.

■ *v.* **1** abscond, bolt, decamp, *sl.* do a bunk, flee, fly, get away, get free, run away, *colloq.* skedaddle, slip away, take to one's heels. **2** drain, leak, ooze, pour, seep. **3** avoid, dodge, elude, evade. ● *n.* **1** flight, flit, getaway.

escapee *n.* one who escapes.

escapism *n.* escape from the realities of life. ▫ **escapist** *n.* & *adj.*

escapologist *n.* person who entertains by escaping from confinement.

escarpment *n.* steep slope at the edge of a plateau etc.

eschew *v.* abstain from.

escort *n.* /éskort/ **1** person(s) or vehicle(s) accompanying another as a protection or honour. **2** person accompanying a person of the opposite sex socially. **3** person acting as a guide on a journey etc. ● *v.* /iskórt/ act as escort to.

■ *n.* **1** bodyguard, chaperon, guard, guardian, protector; entourage, retinue, suite, train. **2** companion, *colloq.* date, partner. **3** guide, leader. ● *v.* chaperon, guard, protect, watch over; accompany, attend, conduct, shepherd, take, usher.

escudo *n.* (*pl.* **-os**) unit of money in Portugal.

Eskimo *n.* (*pl.* **Eskimo** or **-os**) member or language of a people living in Arctic regions. ● *adj.* of Eskimos or their language. (The Eskimos of N. America prefer the name *Inuit.*)

esoteric *adj.* intended only for people with special knowledge.

espadrille *n.* canvas shoe with a sole of plaited fibre.

espalier *n.* **1** trellis. **2** shrub or tree trained on this.

esparto *n.* a kind of grass used in making paper.

especial *adj.* special, notable.

especially *adv.* **1** in particular. **2** more than in other cases.

espionage *n.* spying.

esplanade *n.* promenade.

espouse *v.* **1** support (a cause). **2** marry. □ **espousal** *n.*

espresso *n.* (*pl.* **-os**) coffee made by forcing steam through powdered coffee beans.

esprit de corps /espréé də kór/ loyalty uniting a group.

espy *v.* catch sight of.

Esq. *abbr.* Esquire, courtesy title placed after a man's surname.

essay *n.* /éssay/ short piece of writing. ● *v.* /esáy/ attempt.

essence *n.* **1** indispensable quality or element. **2** concentrated extract.

> ■ **1** core, crux, heart, pith, quintessence, soul, substance. **2** concentrate, distillation, extract.

essential *adj.* **1** absolutely necessary. **2** fundamental. ● *n.* essential thing. □ **essentially** *adv.*

> ■ **1** crucial, indispensable, key, necessary, requisite, vital. **2** basic, fundamental, inherent, innate, intrinsic, quintessential.

establish *v.* **1** set up. **2** settle. **3** prove. **4** cause to be accepted.

> ■ **1** begin, create, form, found, inaugurate, institute, organize, set up, start. **2** ensconce, entrench, install, settle. **3** authenticate, certify, confirm, corroborate, demonstrate, determine, prove, show, substantiate, validate, verify.

establishment *n.* **1** establishing. **2** firm or institution. **3** premises or personnel of this. **4** (**the Establishment**) people established in authority.

> ■ **1** creation, formation, foundation, inauguration, institution, introduction, setting up. **2** business, company, concern, enterprise, firm, institution, organization.

estate *n.* **1** landed property. **2** property left at one's death. **3** residential or industrial district planned as a unit. □ **estate car** car that can carry passengers and goods in one compartment.

> ■ **2** assets, belongings, capital, effects, fortune, possessions, property, wealth.

esteem *v.* think highly of. ● *n.* favourable opinion, respect.

> ■ *v.* admire, appreciate, honour, prize, regard highly, respect, revere, treasure, value. ● *n.* admiration, favour, (high) opinion, (high) regard, respect, reverence, veneration.

estimable *adj.* worthy of esteem.

estimate *n.* /éstimət/ judgement of a thing's approximate value, amount, cost, etc. ● *v.* /éstimayt/ form an estimate of. □ **estimation** *n.*

> ■ *n.* approximation, calculation, estimation, guess; appraisal, assessment, evaluation. ● *v.* appraise, assess, calculate, consider, evaluate, gauge, guess, judge, reckon, *colloq.* size up, surmise, work out.

estrange *v.* make hostile or indifferent. □ **estrangement** *n.*

estuary *n.* mouth of a large river, affected by tides.

etc. *abbr.* = **et cetera** and other things of the same kind.

etch *v.* engrave with acids. □ **etcher** *n.*, **etching** *n.*

eternal *adj.* **1** existing always. **2** unchanging. **3** (*colloq.*) constant. □ **eternally** *adv.*

> ■ **1** endless, everlasting, immortal, perpetual, undying, unending. **2** enduring, lasting, immutable, unchangeable, unchanging.

eternity *n.* **1** infinite time. **2** endless period of life after death. □ **eternity ring** jewelled finger ring symbolizing eternal love.

ether *n.* **1** upper air. **2** liquid used as an anaesthetic and solvent.

ethereal *adj.* **1** light and delicate. **2** heavenly. □ **ethereally** *adv.*

ethic *n.* **1** moral principle. **2** (*pl.*) moral philosophy.

ethical *adj.* **1** morally correct, honourable. **2** relating to morals or ethics. □ **ethically** *adv.*

> ■ **1** correct, decent, fair, good, honest, honourable, just, moral, noble, principled, righteous, upright, virtuous.

ethnic *adj.* of a group sharing a common origin, culture, or language. □ **ethnic cleansing** mass expulsion or killing of people from opposing ethnic groups. **ethnically** *adv.*, **ethnicity** *n.*

ethnology *n.* comparative study of human races. □ **ethnological** *adj.*, **ethnologist** *n.*

ethos /eéthoss/ *n.* characteristic spirit and beliefs.

etiolate /eétiōlayt/ *v.* make pale through lack of light. □ **etiolation** *n.*

etiquette *n.* rules of correct behaviour.

etymology *n.* account of a word's origin and development. ▫ **etymological** *adj.*

eucalyptus *n.* (*pl.* **-tuses** or **-ti**) evergreen tree with leaves that yield a strong-smelling oil.

Eucharist *n.* **1** Christian sacrament in which bread and wine are consumed. **2** this bread and wine. ▫ **Eucharistic** *adj.*

eugenics *n.* science of improving the human race by control of inherited characteristics.

eulogy *n.* piece of spoken or written praise. ▫ **eulogistic** *adj.*, **eulogize** *v.*

eunuch *n.* castrated man.

euphemism *n.* mild word(s) substituted for improper or blunt one(s). ▫ **euphemistic** *adj.*, **euphemistically** *adv.*

euphonium *n.* tenor tuba.

euphony *n.* pleasantness of sounds, esp. in words. ▫ **euphonious** *adj.*

euphoria *n.* feeling of happiness. ▫ **euphoric** *adj.*

Eurasian *adj.* **1** of Europe and Asia. **2** of mixed European and Asian parentage. ● *n.* Eurasian person.

eureka *int.* I have found it! (announcing a discovery etc.).

Euro- *pref.* European.

European *adj.* of Europe or its people. ● *n.* European person.

Eustachian tube /yoōstáysh'n/ passage between the ear and the throat.

euthanasia *n.* bringing about an easy death, esp. to end suffering.

evacuate *v.* **1** remove from a dangerous place. **2** empty or leave (a place). ▫ **evacuation** *n.*

> ■ **2** abandon, desert, leave, move out of, pull out of, quit, vacate, withdraw from.

evacuee *n.* evacuated person.

evade *v.* avoid, escape from.

> ■ avoid, circumvent, dodge, duck, elude, escape from, get away from, get out of, shirk, sidestep.

evaluate *v.* **1** find out or state the value of. **2** assess. ▫ **evaluation** *n.*

> ■ **1** calculate, estimate, gauge, judge, reckon. **2** appraise, assess, value.

evangelical *adj.* of or preaching the gospel. ▫ **evangelicalism** *n.*

evangelist *n.* **1** author of one of the Gospels. **2** person who preaches the gospel. ▫ **evangelism** *n.*, **evangelistic** *adj.*

evaporate *v.* **1** turn into vapour. **2** (cause to) disappear. ▫ **evaporation** *n.*

> ■ **1** vaporize. **2** disappear, dissolve, fade (away), melt away, vanish.

evasion *n.* **1** evading. **2** evasive answer or excuse.

evasive *adj.* seeking to evade something. ▫ **evasively** *adv.*, **evasiveness** *n.*

> ■ ambiguous, *colloq.* cagey, equivocal, indirect, misleading, oblique, prevaricating.

eve *n.* evening, day, or time just before a special event.

even *adj.* **1** level, smooth. **2** uniform. **3** calm. **4** equal. **5** exactly divisible by two. ● *v.* make or become even. ● *adv.* (used for emphasis or in comparing things). ▫ **evenly** *adv.*

> ■ *adj.* **1** flat, flush, level, plane, smooth, straight. **2** consistent, constant, measured, regular, rhythmical, steady, unbroken, uniform, unvarying. **3** calm, cool, equable, even-tempered, imperturbable, placid, sedate, self-possessed, serene, tranquil. **4** equal, identical, the same.

evening *n.* latter part of the day, before nightfall.

event *n.* **1** something that happens, esp. something important. **2** item in a sports programme, or the programme as a whole.

> ■ **1** affair, circumstance, episode, experience, happening, incident, occurrence. **2** championship, competition, contest, game, match, tournament.

eventful *adj.* full of incidents.

eventual *adj.* coming at last, ultimate. ▫ **eventually** *adv.*

eventuality *n.* possible event.

ever *adv.* **1** always. **2** at any time.

evergreen *adj.* having green leaves throughout the year. ● *n.* evergreen tree or shrub.

everlasting *adj.* lasting for ever or for a very long time.

> ■ endless, eternal, immortal, perpetual, timeless, undying; ceaseless, constant, continual, continuous, incessant, interminable, never-ending, unceasing.

evermore *adv.* for ever, always.

every *adj.* **1** each one without exception. **2** each in a series. **3** all possible.

everybody *pron.* every person.

everyday *adj.* **1** worn or used on ordinary days. **2** ordinary.

everyone *pron.* everybody.

everything *pron.* **1** all things. **2** all that is important.

everywhere *adv.* in every place.

evict *v.* expel (a tenant) by legal process. □ **eviction** *n.*, **evictor** *n.*

evidence *n.* **1** anything that gives reason for believing something. **2** statements made in a law court to support a case. ● *v.* be evidence of. □ **be in evidence** be conspicuous. **evidential** *adj.*

■ *n.* **1** data, documentation, facts, grounds, indication, proof, sign. **2** affidavit, attestation, deposition, statement, testimony. ● *v.* attest, demonstrate, display, evince, exhibit, manifest, prove, show.

evident *adj.* obvious to the eye or mind. □ **evidently** *adv.*

■ apparent, clear, discernible, manifest, noticeable, obvious, palpable, patent, perceptible, plain, unmistakable, visible.

evil *adj.* **1** morally bad. **2** harmful. **3** very unpleasant. ● *n.* evil thing, sin, harm. □ **evilly** *adv.*, **evildoer** *n.*

■ *adj.* **1** bad, base, corrupt, depraved, diabolical, heinous, immoral, iniquitous, malevolent, nefarious, perverted, sinful, vicious, vile, villainous, wicked, wrong. **2** harmful, hurtful, injurious, malignant, pernicious. **3** disagreeable, disgusting, foul, nasty, offensive, repulsive, unpleasant, vile. ● *n.* depravity, immorality, iniquity, sin, turpitude, vice, viciousness, villainy, wickedness, wrongdoing.

evince *v.* show, indicate.

eviscerate *v.* disembowel. □ **evisceration** *n.*

evoke *v.* inspire (feelings etc.). □ **evocation** *n.*, **evocative** *adj.*

■ arouse, awaken, draw forth, elicit, excite, inspire, provoke, stir up.

evolution *n.* **1** process of evolving. **2** development of species from earlier forms. □ **evolutionary** *adj.*

evolve *v.* develop gradually.

■ develop, grow, mature, progress, unfold.

ewe *n.* female sheep.

ewer *n.* pitcher, water jug.

ex- *pref.* former.

exacerbate /igzássərbayt/ *v.* **1** make worse. **2** irritate. □ **exacerbation** *n.*

exact *adj.* accurate, correct in all details. ● *v.* insist on and obtain. □ **exaction** *n.*, **exactness** *n.*

■ *adj.* accurate, correct, faithful, faultless, identical, perfect, precise, true. ● *v.* claim, compel, demand, extort, extract, get, insist on, require.

exacting *adj.* making great demands, requiring great effort.

exactly *adv.* **1** in an exact manner. **2** quite so, as you say.

exaggerate *v.* make seem greater or larger than it really is. □ **exaggeration** *n.*, **exaggerator** *n.*

■ blow up, magnify, overemphasize, overstate, overstress; amplify, embellish, embroider.

exalt *v.* **1** raise in rank. **2** praise highly. **3** make more excellent or sublime. □ **exaltation** *n.*

exam *n.* examination.

examination *n.* **1** examining. **2** test of knowledge or ability.

■ **1** analysis, appraisal, assessment, inspection, investigation, *colloq.* once-over, probe, scrutiny, study, survey; cross-examination, grilling, interrogation, questioning.

examine *v.* **1** inquire into. **2** look at closely. **3** question formally. □ **examiner** *n.*

■ **1** analyse, check, explore, inquire into, look into, investigate, probe, research, sift (through), *colloq.* vet. **2** go over, inspect, peruse, pore over, scan, scrutinize, study. **3** catechize, cross-examine, cross-question, grill, interrogate, pump, question, quiz.

examinee *n.* person being tested in an examination.

example *n.* **1** thing characteristic of its kind or illustrating a general rule. **2** person or thing worthy of imitation. □ **make an example of** punish as a warning to others.

■ **1** case, illustration, instance, occurrence, sample, specimen. **2** exemplar, model, pattern.

exasperate *v.* annoy greatly. □ **exasperation** *n.*

■ *colloq.* aggravate, anger, annoy, *sl.* bug, drive mad, enrage, gall, infuriate, irk, irritate, madden, nettle, *colloq.* rile, vex.

excavate *v.* **1** make (a hole) by digging, dig out. **2** reveal by digging. □ **excavation** *n.*, **excavator** *n.*

■ **1** dig (out), gouge (out), hollow (out), scoop (out). **2** disinter, exhume, reveal, uncover, unearth.

exceed *v.* **1** be greater than. **2** go beyond the limit of.

■ **1** beat, better, excel, outdo, outstrip, overtake, pass, surpass, top, transcend.

exceedingly *adv.* very.

excel *v.* (**excelled**) **1** be or do better than. **2** be very good at something.

■ **1** beat, be superior to, eclipse, outclass, outdo, outshine, surpass, top.

excellent *adj.* extremely good. ▫ **excellently** *adv.*, **excellence** *n.*

■ admirable, *colloq.* divine, exceptional, *colloq.* fabulous, *colloq.* fantastic, fine, first-class, first-rate, *colloq.* great, magnificent, marvellous, outstanding, *colloq.* smashing, splendid, sterling, *colloq.* super, superb, superlative, *colloq.* terrific, *colloq.* tremendous, wonderful.

except *prep.* not including. ● *v.* exclude from a statement etc.

excepting *prep.* except.

exception *n.* **1** excepting. **2** thing that does not follow the general rule. ▫ **take exception** object.

exceptionable *adj.* offensive.

exceptional *adj.* **1** unusual. **2** very good. ▫ **exceptionally** *adv.*

■ **1** anomalous, atypical, extraordinary, odd, out of the ordinary, peculiar, rare, singular, special, strange, surprising, uncommon, unexpected, unheard-of, unprecedented, unusual. **2** excellent, first-rate, outstanding, magnificent, marvellous, splendid, superb, wonderful.

excerpt *n.* extract from a book, film, etc.

■ citation, extract, passage, quotation, selection.

excess *n.* **1** exceeding of due limits. **2** amount by which one quantity etc. exceeds another. **3** intemperance in eating and drinking. ● *adj.* exceeding a limit.

■ *n.* **1** glut, over-abundance, overflow, plethora, superfluity, surfeit, surplus. **2** balance, difference. **3** extravagance, immoderation, intemperance, over-indulgence.

excessive *adj.* too much. ▫ **excessively** *adv.*

■ disproportionate, exorbitant, extravagant, extreme, immoderate, inordinate, outrageous, superfluous, undue, unjustifiable, unreasonable.

exchange *v.* give or receive in place of another thing. ● *n.* **1** exchanging. **2** price at which one currency is exchanged for another. **3** place where merchants, brokers, or dealers assemble to do business. **4** centre where telephone lines are connected. ▫ **exchangeable** *adj.*

■ *v.* barter, change, interchange, substitute, swap, switch, trade. ● *n.* **1** swap, trade, transfer.

excise¹ /éksīz/ *n.* duty or tax on certain goods and licences.

excise² /iksī́z/ *v.* cut out or away. ▫ **excision** *n.*

excitable *adj.* easily excited. ▫ **excitably** *adv.*, **excitability** *n.*

■ emotional, fiery, highly-strung, mercurial, nervous, quick-tempered, temperamental, volatile.

excitation *n.* **1** exciting, arousing. **2** stimulation.

excite *v.* **1** rouse the emotions or feelings of. **2** provoke an (action etc.). ▫ **excitement** *n.*

■ **1** animate, arouse, fire, inflame, inspire, intoxicate, kindle, rouse, stimulate; thrill, titillate, *colloq.* turn on. **2** cause, give rise to, instigate, occasion, provoke, stir up.

exclaim *v.* cry out suddenly.

■ bellow, call, cry out, shout, shriek, yell.

exclamation *n.* **1** exclaiming. **2** word(s) exclaimed. ▫ **exclamation mark** punctuation mark (!) placed after an exclamation. **exclamatory** *adj.*

exclude *v.* **1** keep out from a place or group or privilege etc. **2** omit, ignore as irrelevant. **3** make impossible. ▫ **exclusion** *n.*

■ **1** ban, bar, debar, forbid, keep out, prohibit, proscribe, shut out, veto. **2** eliminate, except, leave out, omit, rule out.

exclusive *adj.* **1** excluding others. **2** catering only for the wealthy. **3** not obtainable elsewhere. ▫ **exclusive of** not including. **exclusively** *adv.*, **exclusiveness** *n.*

excommunicate *v.* cut off from a Church or its sacraments. ▫ **excommunication** *n.*

excoriate *v.* **1** strip skin from. **2** censure severely. ▫ **excoriation** *n.*

excrement *n.* faeces.

excrescence *n.* outgrowth on an animal or plant.

excreta *n.pl.* matter (esp. faeces) excreted from the body.

excrete *v.* expel (waste matter) from the body or tissues. ▫ **excretion** *n.*, **excretory** *adj.*

excruciating *adj.* intensely painful.

excursion *n.* short journey to and from a place, made for pleasure.

■ expedition, jaunt, journey, outing, tour, trip.

excuse *v.* /ikskyo͞oz/ **1** try to lessen the blame attaching to (a person, act, etc.). **2** overlook (a fault). **3** exempt. ● *n.* /ikskyo͞oss/ reason put forward to justify a fault etc. □ **excusable** *adj.*

■ *v.* **1** condone, extenuate, explain, justify, mitigate, palliate, vindicate, warrant. **2** disregard, forgive, ignore, overlook, pardon, pass over. ● *n.* apology, defence, explanation, justification, plea, pretext, reason, vindication.

ex-directory *adj.* deliberately not listed in a telephone directory.

execrable *adj.* abominable.

execrate *v.* **1** express loathing for. **2** curse. □ **execration** *n.*

execute *v.* **1** carry out. **2** put (a condemned person) to death. □ **execution** *n.*

■ **1** accomplish, bring off, carry out, discharge, do, effect, implement, perform, pull off.

executioner *n.* one who executes condemned person(s).

executive *n.* person or group with managerial powers, or with authority to put government decisions into effect. ● *adj.* having such power or authority.

executor *n.* person appointed to carry out the terms of one's will.

exemplar *n.* **1** model. **2** typical instance.

exemplary *adj.* **1** fit to be imitated, outstandingly good. **2** serving as a warning to others.

■ **1** model, perfect; admirable, commendable, excellent, outstanding, meritorious, praiseworthy. **2** cautionary, warning.

exemplify *v.* serve as an example of. □ **exemplification** *n.*

■ demonstrate, depict, embody, epitomize, illustrate, personify, represent, show, typify.

exempt *adj.* free from a customary obligation or payment etc. ● *v.* make exempt. □ **exemption** *n.*

exercise *n.* **1** use of one's powers or rights. **2** activity, esp. designed to train the body or mind. ● *v.* **1** use (powers etc.). **2** (cause to) take exercise. □ **exercise book** book for writing in.

■ *n.* **1** application, employment, use, utilization. **2** activity, exertion, movement, practice, training; **(exercises)** aerobics, callisthenics, gymnastics. ● *v.* **1** apply, employ, put to use, use, utilize, wield. **2** drill, keep fit, train, warm up, work out.

exert *v.* use. □ **exert oneself** make an effort. **exertion** *n.*

■ deploy, employ, exercise, put to use, use, utilize, wield. □ **exert oneself** apply oneself, do one's best, make an effort, push oneself, strive, try. **exertion** action, diligence, effort, endeavour, industry, labour, strain, struggle, toil, trouble, work.

exfoliate *v.* come off in scales or layers. □ **exfoliation** *n.*

ex gratia /eks gráyshə/ done or given as a concession, without legal obligation.

exhale *v.* **1** breathe out. **2** give off in vapour. □ **exhalation** *n.*

exhaust *v.* **1** use up completely. **2** tire out. ● *n.* **1** waste gases from an engine etc. **2** device through which they are expelled. □ **exhaustible** *adj.*, **exhaustion** *n.*

■ *v.* **1** consume, deplete, expend, go through, finish, fritter away, squander, spend, use up. **2** debilitate, drain, enervate, fatigue, *sl.* knacker, prostrate, sap, tire (out), wear out, weary. □ **exhaustion** debilitation, enervation, fatigue, lassitude, tiredness, weariness.

exhaustive *adj.* thorough, comprehensive. □ **exhaustively** *adv.*

■ complete, comprehensive, extensive, far-reaching, sweeping, thorough, wide-ranging.

exhibit *v.* **1** display, present for the public to see. **2** manifest. ● *n.* thing exhibited. □ **exhibitor** *n.*

■ *v.* **1** display, offer, present, show. **2** demonstrate, display, evidence, evince, manifest, reveal, show.

exhibition *n.* **1** public display. **2** act or instance of exhibiting.

■ demonstration, display, exposition, presentation, show.

exhibitionism *n.* tendency to behave in a way designed to attract attention. □ **exhibitionist** *n.*

exhilarate *v.* make joyful or lively. □ **exhilaration** *n.*

■ □ **exhilaration** animation, elation, exuberance, gaiety, glee, happiness, joy, joyfulness.

exhort *v.* urge or advise earnestly. □ **exhortation** *n.*, **exhortative** *adj.*

■ advise, counsel, encourage, press, recommend, urge.

exhume *v.* dig up (a buried corpse). ◻ **exhumation** *n.*

exigency *n.* (also **exigence**) **1** urgent need. **2** emergency.

exigent *adj.* **1** urgent. **2** requiring much, exacting.

exiguous *adj.* very small.

exile *n.* **1** banishment or long absence from one's country or home. **2** exiled person. ● *v.* send into exile.

■ *n.* **1** banishment, deportation, expatriation, expulsion. **2** expatriate, refugee. ● *v.* banish, deport, drive out, eject, expatriate, expel.

exist *v.* **1** have being. **2** be found. **3** maintain life. ◻ **existence** *n.*, **existent** *adj.*

■ **1** be extant, live. **2** be found, be present, occur. **3** get by, keep going, stay alive, subsist, survive.

existentialism *n.* philosophical theory emphasizing that individuals are free to choose their actions. ◻ **existentialist** *n.*

exit *n.* **1** departure from a stage or place. **2** way out. ● *v.* make one's exit.

■ *n.* **1** departure, leave-taking, retreat, withdrawal. **2** door, egress, gate, way out.

exodus *n.* departure of many people.

ex officio /éks əfíshiō/ because of one's official position.

exonerate *v.* declare or show to be blameless. ◻ **exoneration** *n.*

exorbitant *adj.* (of a price or demand) much too great.

■ disproportionate, excessive, extortionate, extravagant, immoderate, inordinate, outrageous, unjustifiable, unreasonable.

exorcize *v.* **1** drive out (an evil spirit) by prayer. **2** free (a person or place) of an evil spirit. ◻ **exorcism** *n.*, **exorcist** *n.*

exotic *adj.* **1** brought from abroad. **2** strange, unusual. ◻ **exotically** *adv.*

■ **1** alien, foreign, imported. **2** different, extraordinary, odd, outlandish, out of the ordinary, peculiar, remarkable, singular, strange, unfamiliar, unusual.

expand *v.* **1** increase in size or importance. **2** become more genial. **3** spread out flat. ◻ **expand on** give a fuller account of. **expandable** *adj.*, **expansion** *n.*

■ **1** dilate, distend, enlarge, extend, increase, inflate, spread (out), stretch, swell; augment, broaden, develop, widen. ◻ **expand on** amplify, develop, elaborate on, embellish, embroider, enlarge on.

expanse *n.* wide area or extent.

■ area, extent, range, space, spread, stretch, tract.

expansive *adj.* **1** able to expand. **2** genial, communicative.

expatiate /ikspáyshiayt/ *v.* speak or write at length.

expatriate *adj.* living abroad. ● *n.* expatriate person.

expect *v.* **1** believe that (a person or thing) will come or (a thing) will happen. **2** be confident of receiving. **3** think, suppose.

■ **1** anticipate, await, look forward to, wait for; bargain for, contemplate, envisage, foresee. **2** count on, demand, hope for, rely on, require, want. **3** assume, believe, conjecture, guess, imagine, presume, suppose, surmise, think.

expectant *adj.* filled with expectation. ◻ **expectant mother** pregnant woman. **expectantly** *adv.*, **expectancy** *n.*

expectation *n.* **1** expecting. **2** thing expected. **3** probability.

expectorant *n.* medicine for causing a person to expectorate.

expectorate *v.* cough and spit phlegm. ◻ **expectoration** *n.*

expedient *adj.* advantageous rather than right or just. ● *n.* means of achieving an end. ◻ **expediency** *n.*

■ *adj.* advantageous, advisable, beneficial, desirable, helpful, opportune, practical, prudent, useful. ● *n.* contrivance, device, manoeuvre, means, measure, method, resort, stratagem.

expedite *v.* help or hurry the progress of.

expedition *n.* **1** journey for a purpose. **2** people and equipment for this. ◻ **expeditionary** *adj.*

■ **1** excursion, exploration, journey, pilgrimage, tour, trek, trip, voyage.

expeditious *adj.* speedy and efficient. ◻ **expeditiously** *adv.*

expel *v.* (**expelled**) **1** deprive of membership. **2** compel to leave.

■ **1** ban, bar, debar, exclude. **2** dismiss, drive out, eject, evict, force out, *colloq.* kick out, oust, push out, throw out, *colloq.* turf out, turn out; banish, deport, exile.

expend *v.* **1** spend. **2** use up.

expendable *adj.* **1** able to be expended. **2** not worth saving.

expenditure *n.* **1** expending of money etc. **2** amount expended.

expense *n.* **1** cost. **2** (*pl.*) amount spent doing a job, reimbursement of this.

expensive *adj.* costing or charging more than average. □ **expensively** *adv.*

■ costly, dear, over-priced, *colloq.* steep.

experience *n.* **1** personal observation or contact. **2** knowledge or skill gained by this. **3** event that affects one. ● *v.* **1** have experience of. **2** feel.

■ *n.* **1** involvement, observation, participation, practice. **2** expertise, know-how, knowledge, judgement, skill, wisdom. **3** adventure, episode, event, happening, incident, occurrence; ordeal, trial. ● *v.* **1** encounter, endure, face, go through, live through, meet, suffer, undergo. **2** be aware of, feel, sense, taste.

experienced *adj.* having had much experience.

experiment *n.* & *v.* test to find out or prove something. □ **experimentation** *n.*

■ *n.* investigation, test, trial, try-out.

experimental *adj.* **1** of or used in experiments. **2** still being tested. □ **experimentally** *adv.*

expert *adj.* well-informed or skilful in a subject. ● *n.* expert person. □ **expertly** *adv.*

■ *adj.* accomplished, adept, adroit, dexterous, experienced, knowledgeable, masterly, practised, proficient, qualified, skilful, skilled, trained. ● *n.* ace, authority, connoisseur, master, professional, pundit, specialist, virtuoso, wizard.

expertise *n.* expert knowledge or skill.

■ adroitness, dexterity, judgement, know-how, knowledge, mastery, proficiency, skill.

expiate *v.* make amends for. □ **expiation** *n.*, **expiatory** *adj.*

expire *v.* **1** cease to be valid. **2** die. **3** exhale. □ **expiration** *n.*

■ **1** cease, come to an end, finish, run out, terminate. **2** decease, die, pass away, perish.

expiry *n.* termination of validity.

explain *v.* **1** make clear, show the meaning of. **2** account for. □ **explanation** *n.*, **explanatory** *adj.*

■ **1** clarify, clear up, define, disentangle, elucidate, expound, interpret, make clear, make plain, simplify, spell out, unravel. **2** account for, excuse, give reasons for, justify, rationalize. □ **explanation** account, definition, description, exposition, interpretation; excuse, justification, rationalization.

expletive *n.* violent exclamation, oath.

explicable *adj.* explainable.

explicit *adj.* stated plainly. □ **explicitly** *adv.*, **explicitness** *n.*

■ categorical, clear, definite, distinct, express, overt, plain, positive, precise, specific, unambiguous, unequivocal, unmistakable.

explode *v.* **1** (cause to) expand and break with a loud noise. **2** show sudden violent emotion. **3** discredit. **4** increase suddenly. □ **explosion** *n.*

■ **1** blow up, burst, erupt, go off, fly apart, shatter; blast, detonate, set off. **2** *colloq.* blow up, flare up, lose one's temper. **3** *colloq.* debunk, discredit, disprove, refute. □ **explosion** bang, blast, boom, clap, crack, eruption, report; detonation; outbreak, outburst, paroxysm, spasm.

exploit *n.* /éksployt/ notable deed. ● *v.* /iksplóyt/ **1** make good use of. **2** use selfishly. □ **exploitation** *n.*, **exploiter** *n.*

■ *n.* accomplishment, achievement, attainment, deed, feat. ● *v.* **1** capitalize on, cash in on, make use of, profit from, take advantage of, use, utilize. **2** abuse, manipulate, misuse, take advantage of.

explore *v.* **1** travel into (a country etc.) in order to learn about it. **2** inquire into. **3** examine (a part of the body). □ **exploration** *n.*, **exploratory** *adj.*, **explorer** *n.*

■ **1** reconnoitre, survey, tour, travel through, traverse. **2** analyse, examine, inquire into, inspect, investigate, look into, probe, research, study.

explosive *adj.* & *n.* (substance) able or liable to explode.

exponent *n.* one who favours a specified theory etc.

export *v.* /ekspórt/ send (goods etc.) to another country for sale. ● *n.* /éksport/ **1** exporting. **2** thing exported. □ **exportation** *n.*, **exporter** *n.*

expose *v.* **1** leave uncovered or unprotected. **2** disclose, make public. **3** allow light to reach (film etc.). □ **expose to** subject to a risk etc. **exposure** *n.*

■ **2** bring to light, disclose, make known, make public, reveal, uncover, unmask, unveil. □ **expose to** lay open to, make vulnerable to, put at risk of, subject to.

exposé /ekspṓzay/ *n.* **1** statement of facts. **2** disclosure.

exposition *n.* **1** expounding. **2** explanation. **3** large exhibition.

expostulate *v.* protest, remonstrate. □ **expostulation** *n.*

expound *v.* explain in detail.

express *adj.* **1** definitely stated. **2** travelling rapidly, designed for high speed. ● *adv.* at high speed. ● *n.* fast train or bus making few or no stops. ● *v.* **1** make (feelings, thoughts, etc.) known by words, gestures, etc. **2** represent by symbols. **3** squeeze out. □ **express oneself** say what one thinks, feels, or means. **expressible** *adj.*

■ *adj.* **1** clear, definite, explicit, plain, specific, unambiguous, unmistakable. ● *v.* **1** air, articulate, communicate, enunciate, give vent to, put into words, state, utter, verbalize, voice; phrase, put, word; betoken, convey, demonstrate, denote, evidence, evince, indicate, intimate, make known, manifest, reveal, show. **2** denote, represent, signify, symbolize. **3** extract, squeeze out, wring out.

expression *n.* **1** act or instance of expressing. **2** word or phrase. **3** look or manner that expresses feeling.

■ **1** articulation, utterance, voicing; indication, manifestation, sign, token. **2** idiom, phrase, saying, term. **3** air, appearance, aspect, countenance, face, look, mien.

expressionism *n.* style of art seeking to express feelings rather than represent objects realistically. □ **expressionist** *n.*

expressive *adj.* **1** expressing something. **2** full of expression. □ **expressively** *adv.*

expressly *adv.* **1** explicitly. **2** for a particular purpose.

expropriate *v.* **1** seize (property). **2** dispossess. □ **expropriation** *n.*

expulsion *n.* **1** expelling. **2** being expelled. □ **expulsive** *adj.*

expunge *v.* wipe out.

expurgate *v.* remove (objectionable matter) from (a book etc.). □ **expurgation** *n.*, **expurgator** *n.*, **expurgatory** *adj.*

exquisite *adj.* **1** having exceptional beauty. **2** acute, keenly felt. □ **exquisitely** *adv.*

■ **1** beautiful, delicate, fine, *colloq.* heavenly, *colloq.* gorgeous, lovely, perfect, *colloq.* stunning. **2** acute, agonizing, excruciating, intense, keen, sharp.

extant *adj.* still existing.

extemporize *v.* speak, perform, or produce without preparation. □ **extemporary** *adj.*, **extemporization** *n.*

extend *v.* **1** make longer. **2** stretch out. **3** reach. **4** offer. □ **extendible** *adj.*, **extensible** *adj.*

■ **1** add to, broaden, elongate, enlarge, expand, increase, lengthen, widen; prolong, protract. **2** give, hold out, offer, present, proffer, reach out, stretch out. **3** range, reach, stretch.

extension *n.* **1** extending. **2** extent. **3** additional part or period. **4** subsidiary telephone, its number.

extensive *adj.* large in area or scope. □ **extensively** *adv.*

■ big, considerable, enormous, great, huge, immense, large, sizeable, spacious, substantial, vast; broad, catholic, comprehensive, far-reaching, sweeping, wide, widespread.

extent *n.* **1** space over which a thing extends. **2** width of application.

■ **1** amplitude, area, dimensions, expanse, length, magnitude, size, space, span. **2** breadth, compass, degree, range, scope.

extenuate *v.* make (an offence) seem less great by providing a partial excuse. □ **extenuation** *n.*

exterior *adj.* on or coming from the outside. ● *n.* exterior aspect or surface.

■ *adj.* external, outer, outside, outward, superficial, surface. ● *n.* façade, face, front, outside, shell, surface.

exterminate *v.* destroy utterly. □ **extermination** *n.*, **exterminator** *n.*

■ annihilate, destroy, eliminate, eradicate, extirpate, get rid of, liquidate, obliterate, wipe out.

external *adj.* of or on the outside. □ **externally** *adv.*

■ exterior, outer, outside, outward, superfical; apparent, perceptible, visible.

extinct *adj.* **1** no longer existing or burning. **2** (of a volcano) no longer active.

extinction *n.* **1** extinguishing. **2** making or becoming extinct.

extinguish *v.* **1** put out (a flame or light). **2** end the existence of.

■ **1** blow out, douse, put out, quench, snuff out; switch off, turn off. **2** annihilate, destroy, eliminate, end, eradicate, exterminate, kill, obliterate, wipe out.

extinguisher *n.* device for discharging liquid chemicals or foam to extinguish a fire.

extirpate *v.* root out, destroy. □ **extirpation** *n.*

extol *v.* (**extolled**) praise enthusiastically.
extort *v.* obtain by force or threats. □ **extortion** *n.*
extortionate *adj.* exorbitant.
extra *adj.* additional, more than is usual or expected. ● *adv.* **1** more than usually. **2** in addition. ● *n.* **1** extra thing. **2** person employed as one of a crowd in a film.

■ *adj.* added, additional, auxiliary, further, spare, supplementary.

extra- *pref.* outside, beyond.
extract *v.* /ikstrákt/ **1** take out or obtain by force or effort. **2** obtain by suction, pressure, distillation, etc. **3** derive (pleasure etc.). ● *n.* /ékstrakt/ **1** substance extracted from another. **2** passage from a book, play, film, or music. □ **extractor** *n.*

■ *v.* **1** draw out, pull out, remove, take out, withdraw; extort, obtain, wrest, wring. ● *n.* **1** concentrate, distillation, essence. **2** citation, excerpt, passage, quotation, selection.

extraction *n.* **1** extracting. **2** lineage.
extradite *v.* hand over (an accused person) for trial in the country where a crime was committed. □ **extradition** *n.*
extramarital *adj.* (of sexual relationships) outside marriage.
extramural *adj.* additional to ordinary teaching or studies.
extraneous *adj.* **1** of external origin. **2** not relevant.
extraordinary *adj.* unusual, remarkable. □ **extraordinarily** *adv.*

■ abnormal, exceptional, rare, remarkable, singular, special, uncommon, unheard-of, unprecedented, unusual; bizarre, curious, odd, peculiar, strange, uncanny; amazing, astonishing, astounding, incredible, marvellous, miraculous, notable, noteworthy, outstanding, phenomenal, sensational, unbelievable.

extrapolate *v.* estimate on the basis of available data. □ **extrapolation** *n.*
extrasensory *adj.* achieved by some means other than the known senses.
extraterrestrial *adj.* of or from outside the earth or its atmosphere.
extravagant *adj.* **1** spending excessively. **2** going beyond what is reasonable. □ **extravagantly** *adv.*, **extravagance** *n.*

■ **1** improvident, lavish, prodigal, profligate, spendthrift, wasteful. **2** excessive, immoderate, inordinate, unjustifiable, unreasonable.

extravaganza *n.* lavish spectacular display or entertainment.
extreme *adj.* **1** very great or intense. **2** severe, not moderate. **3** at the end(s), outermost. ● *n.* **1** either of two things as different or as remote as possible. **2** highest degree. **3** thing at either end of anything. □ **extremely** *adv.*

■ *adj.* **1** considerable, enormous, great, immense, tremendous. **2** draconian, drastic, excessive, harsh, immoderate, severe, stiff, stringent, unreasonable. **3** endmost, farthest, furthermost, outermost, utmost. ● *n.* **2** apex, extremity, height, peak, pinnacle, summit, zenith; depth, nadir.

extremist *n.* person holding extreme views. □ **extremism** *n.*
extremity *n.* **1** extreme point. **2** extreme degree of need or danger. **3** (*pl.*) hands and feet.
extricate *v.* free from an entanglement or difficulty. □ **extrication** *n.*, **extricable** *adj.*
extrinsic *adj.* **1** not intrinsic. **2** extraneous. □ **extrinsically** *adv.*
extrovert *n.* lively sociable person. □ **extroversion** *n.*
extrude *v.* thrust or squeeze out. □ **extrusion** *n.*, **extrusive** *adj.*
exuberant *adj.* **1** high-spirited. **2** growing profusely. □ **exuberantly** *adv.*, **exuberance** *n.*

■ **1** animated, boisterous, bubbly, buoyant, cheerful, ebullient, effusive, energetic, enthusiastic, exhilarated, happy, high-spirited, joyful, lively, spirited, sprightly, spry, vigorous, vivacious. **2** abundant, copious, lush, luxuriant, plentiful, profuse, prolific.

exude *v.* **1** ooze. **2** give off like sweat or a smell. □ **exudation** *n.*
exult *v.* rejoice greatly. □ **exultant** *adj.*, **exultation** *n.*

■ □ **exultant** delighted, ecstatic, elated, gleeful, joyful, jubilant, overjoyed, rejoicing, triumphant.

eye *n.* **1** organ of sight. **2** iris of this. **3** region round it. **4** power of seeing. **5** thing like an eye, spot, hole. ● *v.* (**eyed**, **eyeing**) look at, watch. □ **eye-opener** *n.* thing that brings enlightenment or great surprise. **eye-shade** *n.* device to protect the eyes from strong light. **eye-shadow** *n.* cosmetic applied to the skin round the eyes. **eye-tooth** *n.* canine tooth in the upper jaw.

eyeball *n.* whole of the eye within the eyelids.

eyebrow *n.* fringe of hair on the ridge above the eye socket.

eyelash *n.* one of the hairs fringing the eyelids.

eyelet *n.* **1** small hole. **2** ring strengthening this.

eyelid *n.* upper or lower fold of skin closing to cover the eye.

eyepiece *n.* lens(es) to which the eye is applied in a telescope or microscope etc.

eyesight *n.* **1** ability to see. **2** range of vision.

eyesore *n.* ugly object.

eyewitness *n.* person who actually saw something happen.

eyrie /íri/ *n.* **1** eagle's nest. **2** house etc. perched high up.

Ff

F *abbr.* Fahrenheit.

fable *n.* fictional tale, often legendary or moral. □ **fabled** *adj.*

fabric *n.* **1** cloth or knitted material. **2** walls etc. of a building.

> ■ **1** cloth, material, textile.

fabricate *v.* **1** construct, manufacture. **2** invent (a story etc.). □ **fabrication** *n.*, **fabricator** *n.*

fabulous *adj.* **1** incredibly great. **2** legendary. **3** (*colloq.*) excellent. □ **fabulously** *adv.*

> ■ **1** amazing, astonishing, astounding, extraordinary, inconceivable, incredible, phenomenal, unbelievable. **2** fabled, fictional, fictitious, imaginary, legendary, mythical.

façade /fəsaád/ *n.* **1** front of a building. **2** outward appearance.

face *n.* **1** front of the head. **2** expression shown by its features. **3** front or main side. **4** outward aspect. **5** dial of a clock. **6** coalface. ● *v.* **1** have or turn the face towards. **2** meet firmly. **3** put a facing on. □ **face flannel** cloth for washing one's face. **facelift** *n.* **1** operation for tightening the skin of the face. **2** improvement in appearance. **face up to** accept bravely.

> ■ *n.* **1** countenance, features, *sl.* mug, physiognomy, visage. **2** appearance, aspect, expression, look. **3** exterior, façade, front, outside, surface. **4** façade, front, mask, veneer. ● *v.* **1** be opposite, front onto, look towards, overlook. **2** brave, confront, cope with, deal with, encounter, meet.

faceless *adj.* **1** without identity. **2** purposely not identifiable.

facet *n.* **1** one of many sides of a cut stone or jewel. **2** one aspect.

facetious *adj.* intended or intending to be amusing. □ **facetiously** *adv.*, **facetiousness** *n.*

facia /fáyshə/ *n.* **1** dashboard. **2** nameplate over a shop front.

facial *adj.* of the face. ● *n.* beauty treatment for the face.

facile /fássīl/ *adj.* **1** easily achieved. **2** superficial.

facilitate *v.* make easy or easier. □ **facilitation** *n.*

facility *n.* **1** absence of difficulty. **2** means for doing something.

> ■ **1** ease, effortlessness, smoothness. **2** amenity, convenience, resource.

facing *n.* **1** outer covering. **2** material at the edge of a garment for strength or contrast.

facsimile *n.* a reproduction of a document etc.

fact *n.* **1** reality. **2** (*pl.*) evidence. **3** thing known to have happened or to be true.

> ■ **1** actuality, certainty, reality, truth. **2** (**facts**) data, details, evidence, information *colloq.* low-down, particulars.

faction *n.* small united group within a larger one.

factitious *adj.* **1** made for a special purpose. **2** artificial.

factor *n.* **1** circumstance that contributes towards a result. **2** number by which a given number can be divided exactly.

> ■ **1** aspect, circumstance, consideration element, fact, influence, ingredient.

factory *n.* building(s) in which goods are manufactured.

factotum *n.* servant or assistant doing all kinds of work.

factual *adj.* based on or containing facts. □ **factually** *adv.*

> ■ accurate, actual, authentic, bona fide, faithful, genuine, objective, real, realistic, true, unbiased, unvarnished, verifiable.

faculty *n.* **1** any of the powers of the body or mind. **2** university department.

fad *n.* craze, whim.

faddy *adj.* having petty likes and dislikes, esp. about food.

fade *v.* **1** (cause to) lose colour, freshness or vigour. **2** disappear gradually.

> ■ **1** blanch, discolour, (grow) dim *or* pale, droop, wilt, wither; decline, deteriorate, die away, diminish, dwindle, ebb, flag, languish, wane. **2** disappear, dissolve, melt away, vanish.

faeces /féesseez/ *n.pl.* waste matter discharged from the bowels. □ **faecal** *adj.*

fag (*colloq.*) *v.* (**fagged**) exhaust. ● *n.* **1** tedious task. **2** cigarette.

faggot *n.* **1** tied bundle of sticks or twigs. **2** ball of chopped seasoned liver, baked or fried.

Fahrenheit *adj.* of a temperature scale on which water freezes at 32° and boils at 212°.

faience /fíonss/ *n.* painted glazed earthenware.

fail *v.* **1** be unsuccessful. **2** become weak, cease functioning. **3** disappoint. **4** become bankrupt. **5** neglect or be unable. **6** declare to be unsuccessful. ● *n.* failure.

■ *v.* **1** be unsuccessful, come to grief, fall through, *sl.* flop, founder, go wrong, miscarry, misfire. **2** deteriorate, diminish, disappear, dwindle, ebb, fade, give out, wane, weaken. **3** disappoint, let down. **4** crash, fold, go bankrupt, *colloq.* go bust, *colloq.* go broke, go under. **5** neglect, omit.

failing *n.* weakness or fault. ● *prep.* in default of.

■ *n.* blemish, defect, fault, flaw, imperfection, shortcoming, weakness, weak spot.

failure *n.* **1** lack of success. **2** person or thing that fails.

■ **1** collapse, defeat, disappointment. **2** disaster, fiasco, *sl.* flop, *colloq.* wash-out.

faint *adj.* **1** indistinct, not intense. **2** about to faint. **3** slight. ● *v.* collapse unconscious. ● *n.* act or state of fainting. □ **faint-hearted** *adj.* timid. **faintly** *adv.*, **faintness** *n.*

■ *adj.* **1** blurred, dim, feeble, hazy, ill-defined, indistinct, pale, subdued, weak; hushed, low, muffled, muted, quiet, soft, stifled. **2** dizzy, giddy, light-headed, unsteady, weak. **3** remote, slight. ● *v.* black out, collapse, *colloq.* flake out, keel over, pass out.

fair¹ *n.* **1** funfair. **2** gathering for a sale of goods, often with entertainments. **3** trade exhibition. □ **fairground** *n.* open space where a funfair is held.

fair² *adj.* **1** just, unbiased. **2** light in colour, having light-coloured hair. **3** of moderate quality or amount. **4** (of weather) fine and dry, (of wind) favourable. ● *adv.* fairly.

■ *adj.* **1** disinterested, equitable, honest, honourable, impartial, just, lawful, legitimate, proper, right, unbiased, unprejudiced. **2** light, pale; blond(e), fair-haired, flaxen-haired. **3** acceptable, adequate, average, mediocre, middling, *colloq.* OK, passable, satisfactory, tolerable. **4** bright, clear, cloudless, dry, fine, pleasant, sunny.

fairing *n.* streamlining structure.

fairy *n.* imaginary small being with magical powers. □ **fairy godmother** benefactress. **fairy lights** strings of small coloured lights used as decorations. **fairy story, tale 1** tale about fairies or magic. **2** falsehood.

fairyland *n.* **1** world of fairies. **2** very beautiful place.

fait accompli /fáyt əkómplee/ thing already done and not reversible.

faith *n.* **1** complete trust. **2** (system of) religious belief. **3** loyalty. □ **faith healing** cure etc. dependent on faith.

■ **1** belief, certainty, certitude, confidence, conviction, credence, trust. **2** denomination, sect; belief, creed, religion. **3** allegiance, dedication, devotion, faithfulness, fidelity, loyalty.

faithful *adj.* **1** loyal, trustworthy. **2** true, accurate. □ **faithfully** *adv.*, **faithfulness** *n.*

■ **1** constant, dedicated, dependable, devoted, loyal, reliable, staunch, steadfast, trusted, trustworthy. **2** accurate, exact, perfect, precise, true.

faithless *adj.* disloyal.

fake *n.* a person or thing that is not genuine. ● *adj.* faked. ● *v.* **1** make an imitation of. **2** pretend.

■ *n.* charlatan, cheat, fraud, impostor, *colloq.* phoney, quack; copy, counterfeit, forgery, imitation, sham. ● *adj.* artificial, bogus, counterfeit, false, forged, fraudulent, imitation, mock, *colloq.* phoney, sham, simulated, spurious. ● *v.* **1** counterfeit, fabricate, forge. **2** affect, feign, pretend, simulate.

fakir /fáykeer/ *n.* Muslim or Hindu religious mendicant or ascetic.

falcon *n.* a kind of small hawk.

falconry *n.* breeding and training of hawks. □ **falconer** *n.*

fall *v.* (**fell, fallen**) **1** come or go down freely. **2** lose balance and come suddenly to the ground. **3** decrease. **4** be captured or conquered. **5** die in battle. **6** pass into a specified state. **7** occur. **8** lose power or status. **9** (of the face) show dismay. ● *n.* **1** act or an instance of falling. **2** amount of this. **3** (*US*) autumn. **4** (*pl.*) waterfall. □ **fall back on** have recourse to. **fall for 1** fall in love with. **2** be deceived by. **fall out 1** quarrel. **2** happen. **fallout** *n.* airborne radioactive debris. **fall short** be inadequate. **fall through** (of a plan) fail to be achieved.

■ *v.* **1** come down, descend, dive, drop down, nosedive, plummet, plunge; cascade. **2** collapse, keel over, overbalance, stumble, topple (over), trip (over), tumble. **3** decline, decrease, diminish, drop, dwindle, sink, slump, subside. **4** be defeated, be captured, be overthrown; capitulate, succumb, surrender, yield. **5** die, perish. **6** become, grow, get. **7** happen, occur, take place. ● *n.* **1** descent, dive, drop, nosedive, plunge, tumble; collapse, decline, decrease, diminution, slump; capture, conquest, defeat, overthrow; capitulation, submission, surrender. □ **fall out 1** clash, disagree, dispute, fight, quarrel, squabble, wrangle.

fallacy *n.* false belief or reasoning. □ **fallacious** *adj.*

■ delusion, error, miscalculation, misconception, misjudgement, mistake.

fallible *adj.* liable to make mistakes. □ **fallibility** *n.*

Fallopian tube either of the two tubes from the ovary to the womb.

fallow *adj.* (of land) left unplanted. ● *n.* such land.

fallow deer reddish-brown deer with white spots.

false *adj.* **1** incorrect. **2** deceitful, unfaithful. **3** not genuine, sham. □ **falsely** *adv.*, **falseness** *n.*

■ **1** erroneous, fallacious, fictitious, flawed, imprecise, inaccurate, incorrect, invalid, misleading, mistaken, untrue, wrong. **2** deceitful, dishonest, disloyal, double-dealing, faithless, lying, treacherous, two-faced, unfaithful, untrustworthy. **3** affected, artificial, bogus, counterfeit, fake, forged, imitation, insincere, mock, *colloq.* phoney, sham, simulated, spurious, synthetic.

falsehood *n.* lie(s).

falsetto *n.* (*pl.* **-os**) unusually high male voice.

falsify *v.* **1** alter fraudulently. **2** misrepresent. □ **falsification** *n.*

■ **1** alter, *colloq.* cook, doctor. **2** distort, misrepresent, twist.

falsity *n.* **1** falseness. **2** falsehood.

falter *v.* **1** stumble, go unsteadily. **2** lose courage. **3** speak hesitantly.

fame *n.* **1** condition of being famous. **2** reputation.

■ celebrity, eminence, illustriousness, prominence, renown, reputation, repute, stardom; notoriety.

familial *adj.* of a family.

familiar *adj.* **1** well-known. **2** too informal. □ **familiar with** knowing a thing well. **familiarly** *adv.*, **familiarity** *n.*

■ **1** common, commonplace, customary, everyday, habitual, traditional, usual, well-known. **2** disrespectful, forward, impertinent, impudent, insolent, over-friendly, presumptuous. □ **familiar with** acquainted with, at home with, aware of, conversant with, knowledgeable about, versed in, well-informed about.

familiarize *v.* make familiar. □ **familiarization** *n.*

family *n.* **1** set of relations, esp. parents and children. **2** a person's children. **3** lineage. **4** group of related plants, animals, or things.

■ **1** flesh and blood, folk, kindred, kinsfolk, kith and kin, relations, relatives. **2** brood, children, *colloq.* kids, offspring, progeny. **3** ancestors, forebears, forefathers; ancestry, descent, dynasty, genealogy, house, line, lineage, parentage, pedigree, stock.

famine *n.* extreme scarcity (esp. of food) in a region.

famished *adj.* extremely hungry.

famous *adj.* known to very many people. □ **famously** *adv.*

■ celebrated, distinguished, eminent, illustrious, *colloq.* legendary, notable, noted, prominent, renowned, well-known; notorious.

fan[1] *n.* hand-held or mechanical device to create a current of air. ● *v.* (**fanned**) **1** cool with a fan. **2** spread from a central point. □ **fan belt** belt driving a fan that cools a car engine.

fan[2] *n.* enthusiastic admirer or supporter. □ **fan mail** letters from fans.

■ admirer, *colloq.* buff, devotee, enthusiast, fanatic, fiend, follower, lover, supporter.

fanatic *n.* person filled with excessive enthusiasm for something. □ **fanatical** *adj.*, **fanatically** *adv.*, **fanaticism** *n.*

■ □ **fanatical** excessive, extreme, fervent, fervid, maniacal, obsessive, passionate, rabid, zealous.

fanciful *adj.* **1** imaginative. **2** imaginary. □ **fancifully** *adv.*

fancy *n.* **1** inclination. **2** unfounded idea. **3** imagination. ● *adj.* ornamental, elaborate. ● *v.* **1** imagine. **2** suppose. **3** (*colloq.*) desire, find attractive. □ **fancy dress** costume representing an animal, historical character, etc., worn for a party.

■ *n.* **1** fondness, inclination, liking, partiality, penchant, predilection, taste. **2** caprice, notion, vagary, whim. **3** creativity, imagination, inventiveness. ● *adj.* decorated, decorative, elaborate, embellished, embroidered, intricate, ornamental, ornate, rococo. ● *v.* **1** envisage, imagine, picture, visualize. **2** conjecture, guess, imagine, presume, suppose, surmise, think.

fandango *n.* (*pl.* **-oes**) lively Spanish dance.

fanfare *n.* short showy or ceremonious sounding of trumpets.

fang *n.* **1** long sharp tooth. **2** snake's tooth that injects venom.

fanlight *n.* small window above a door or larger window.

fantasia *n.* imaginative musical or other composition.

fantasize *v.* daydream.

fantastic *adj.* **1** absurdly fanciful. **2** (*colloq.*) excellent. □ **fantastically** *adv.*

■ **1** absurd, extravagant, fanciful, illusory, imaginary, irrational, strange, whimsical, wild.

fantasy *n.* **1** imagination. **2** thing(s) imagined. **3** fanciful design.

■ **1** fancy, imagination, inventiveness. **2** chimera, delusion, hallucination, illusion, mirage; daydream, dream, pipedream.

far *adv.* at or to or by a great distance. ● *adj.* distant, remote. □ **far-away, far-off** *adjs.* remote. **Far East** countries of east and south-east Asia. **far-fetched** *adj.* very unlikely.

farad *n.* unit of capacitance.

farce *n.* **1** light comedy. **2** absurd and useless proceedings, pretence. □ **farcical** *adj.*

fare *n.* **1** price charged for a passenger to travel. **2** passenger paying this. **3** food provided. ● *v.* get on or be treated.

farewell *int.* & *n.* goodbye.

farinaceous *adj.* starchy.

farm *n.* unit of land used for raising crops or livestock. ● *v.* **1** grow crops, raise livestock. **2** use (land) for this. □ **farmer** *n.*

farmhouse *n.* farmer's house.

farmyard *n.* enclosed area round farm buildings.

farrago /fəraágō/ *n.* (*pl.* **-os**) hotchpotch.

farrier *n.* smith who shoes horses.

farrow *v.* give birth to piglets. ● *n.* litter of piglets.

farther *adv.* & *adj.* at or to a greater distance, more remote.

farthest *adv.* & *adj.* at or to the greatest distance, most remote.

fascinate *v.* **1** capture the interest of. **2** attract irresistibly. □ **fascination** *n.*

■ bewitch, captivate, charm, enchant, enthral, entrance, hold spellbound, intrigue, mesmerize, rivet, transfix.

fascism /fáshiz'm/ *n.* system of extreme right-wing dictatorship. □ **fascist** *n.*

fashion *n.* **1** manner of doing something. **2** style popular at a given time. ● *v.* shape, make.

■ *n.* **1** manner, method, mode, style, way. **2** craze, custom, fad, style, trend, vogue. ● *v.* build, carve, construct, create, form, frame, make, shape.

fashionable *adj.* **1** of or conforming to current fashion. **2** used by stylish people. □ **fashionably** *adv.*

■ chic, in, in fashion, in vogue, *sl.* snazzy, stylish, *colloq.* swish, *colloq.* trendy, up to date.

fast[1] *adj.* **1** moving or done quickly. **2** firmly fixed or attached. **3** capable of or intended for high speed. **4** showing a time ahead of the correct one. ● *adv.* **1** quickly. **2** firmly, tightly.

■ *adj.* **1** brisk, expeditious, *colloq.* nippy, quick, rapid, speedy, swift; hasty, hurried, precipitate. **2** attached, bound, fastened, fixed, secured, tied; firm, lasting, secure, settled, unshakeable, unwavering.

fast[2] *v.* go without food. ● *n.* act or period of fasting.

fasten *v.* make or become fixed or secure.

■ affix, anchor, attach, bind, bolt, clasp, close, connect, do up, join, link, lock, pin, secure, stick, tether, tie.

fastener *n.* (also **fastening**) device that fastens something.

fastidious *adj.* **1** choosing only what is good. **2** easily disgusted.

■ **1** *colloq.* choosy, dainty, finicky, fussy, nice, particular, *colloq.* pernickety, selective. **2** queasy, squeamish.

fastness *n.* stronghold, fortress.

fat *n.* white or yellow substance found in animal bodies and certain seeds. ● *adj.* (**fatter, fattest**) **1** excessively plump. **2** containing much fat. **3** thick, substantial. □ **fatness** *n.*, **fatty** *adj.*

■ *adj.* **1** bulky, chubby, corpulent, dumpy, flabby, fleshy, heavy, obese, overweight,

plump, podgy, portly, pot-bellied, pudgy, roly-poly, rotund, stout, tubby.

fatal *adj.* causing or ending in death or disaster. □ **fatally** *adv.*

■ **deadly, final, lethal, mortal, terminal.**

fatalism *n.* belief that all events are determined by fate and therefore inevitable. □ **fatalist** *n.*

fatality *n.* death caused by accident or in war etc.

fate *n.* **1** power thought to control all events. **2** person's destiny.

■ **1 chance, destiny, fortune, kismet, luck. 2 destiny, lot.**

fated *adj.* **1** destined by fate. **2** doomed.

■ **1 destined, ineluctable, predestined, preordained, unavoidable.**

fateful *adj.* bringing great usu. unpleasant events.

father *n.* **1** male parent or ancestor. **2** founder, originator. **3** title of certain priests. ● *v.* **1** be the father of. **2** originate. □ **father-in-law** *n.* (*pl.* **fathers-in-law**) father of one's wife or husband. **fatherhood** *n.*, **fatherless** *adj.*, **fatherly** *adj.*

■ *n.* **2 author, creator, designer, founder, inventor, originator.**

fatherland *n.* native country.

fathom *n.* measure (1.82 m) of the depth of water. ● *v.* understand. □ **fathomable** *adj.*

fatigue *n.* **1** tiredness. **2** weakness in metal etc., caused by stress. **3** soldier's non-military task. ● *v.* cause fatigue to.

■ *n.* **1 enervation, exhaustion, lassitude, lethargy, listlessness, tiredness, weakness, weariness.**

fatstock *n.* livestock fattened for slaughter as food.

fatten *v.* make or become fat.

fatuous *adj.* foolish, silly. □ **fatuously** *adv.*, **fatuousness** *n.*

faucet *n.* tap.

fault *n.* **1** defect, imperfection. **2** error, offence. **3** responsibility for something wrong. **4** break in layers of rock. ● *v.* find fault(s) in. □ **at fault** responsible for a mistake etc. **faultless** *adj.*, **faulty** *adj.*

■ *n.* **1 blemish, defect, deficiency, failing, flaw, imperfection, shortcoming, weakness. 2 error, mistake, oversight, slip, *colloq.* slip-up; misdeed, misdemeanour, offence, sin, transgression. 3 blame, culpability, liability, responsibility.** □ **faultless exemplary, flawless, immaculate, impeccable, irreproachable, perfect. faulty broken, damaged, defective, flawed, impaired, imperfect, out of order, unsound.**

faun *n.* Latin rural deity with a goat's legs and horns.

fauna *n.pl.* animals of an area or period.

faux pas /fō paā/ (*pl.* ***faux pas*** /paās/) embarrassing blunder.

favour *n.* **1** liking, approval. **2** kindly or helpful act. **3** favouritism. ● *v.* **1** regard or treat with favour. **2** facilitate.

■ *n.* **1 approbation, approval, goodwill, liking. 2 courtesy, good turn, kindness. 3 bias, favouritism, partiality, partisanship.** ● *v.* **1 have a liking for, incline to, like, opt for, side with, support; advocate, back, endorse, prefer, recommend. 2 advance, aid, assist, benefit, encourage, facilitate, help, promote.**

favourable *adj.* **1** approving. **2** promising. **3** pleasing, satisfactory. □ **favourably** *adv.*

■ **1 approving, complimentary, encouraging, enthusiastic, good, laudatory, positive, sympathetic. 2 auspicious, hopeful, promising, propitious. 3 good, pleasing, satisfactory.**

favourite *adj.* liked above others. ● *n.* **1** favoured person or thing. **2** competitor expected to win.

■ *adj.* **beloved, best-liked, chosen, favoured, pet, preferred.** ● *n.* **1 beloved, darling, idol, pet.**

favouritism *n.* unfair favouring of one at the expense of others.

fawn[1] *n.* **1** a deer in its first year. **2** light yellowish-brown. ● *adj.* fawn-coloured.

fawn[2] *v.* **1** (of a dog) show affection. **2** try to win favour by obsequiousness.

fax *n.* **1** facsimile transmission by electronic scanning. **2** document produced thus. ● *v.* transmit by this process.

fear *n.* unpleasant sensation caused by nearness of danger or pain. ● *v.* **1** feel fear of. **2** be afraid.

■ *n.* **alarm, apprehension, consternation, dismay, dread, fright, horror, panic, terror, trepidation.**

fearful *adj.* **1** feeling fear. **2** terrible, extremely unpleasant. □ **fearfully** *adv.*

■ **1 afraid, alarmed, anxious, apprehensive, edgy, frightened, jumpy, nervous, panicky, panic-stricken, scared, terrified. 2 appalling, atrocious, awful, disgusting, dreadful, frightful, ghastly, gruesome, horrendous, horrible, horrific, horrifying,**

loathsome, monstrous, repugnant, repulsive, revolting, terrible.

fearless *adj.* feeling no fear. □ **fearlessly** *adv.*

■ **bold, brave, courageous, daring, dauntless, gallant, game, *colloq.* gutsy, heroic, intrepid, plucky, resolute, spirited, unafraid, undaunted, valiant, valorous.**

fearsome *adj.* frightening.

feasible *adj.* **1** able to be done. **2** plausible. □ **feasibly** *adv.*, **feasibility** *n.*

■ **1 attainable, possible, practicable, practical, viable, workable. 2 believable, credible, likely, plausible, reasonable.**

feast *n.* **1** large elaborate meal. **2** sensual or mental pleasure. **3** joyful festival. ● *v.* **1** eat heartily. **2** give a feast to.

■ *n.* **1 banquet, *colloq.* spread. 2 delight, gratification, pleasure, treat.** ● *v.* **1 dine, gorge (oneself). 2 entertain, feed, regale.**

feat *n.* remarkable achievement.

■ **accomplishment, achievement, act, action, attainment, deed, exploit, performance, *tour de force*.**

feather *n.* each of the structures with a central shaft and fringe of fine strands, growing from a bird's skin. ● *v.* **1** cover or fit with feathers. **2** turn (an oar-blade etc.) to pass through the air edgeways. □ **feather-bed** *v.* make things financially easy for. **feather one's nest** enrich oneself. **feathery** *adj.*

featherweight *n.* very lightweight thing or person.

feature *n.* **1** characteristic or distinctive part. **2** (usu. *pl.*) part of the face. **3** prominent article in a newspaper etc. **4** full-length cinema film. ● *v.* **1** give prominence to. **2** be a feature of or in.

■ *n.* **1 aspect, attribute, characteristic, facet, hallmark, idiosyncrasy, mark, peculiarity, property, quality, trait. 2 (features) countenance, face, *sl.* mug, physiognomy, visage. 3 article, column, piece.** ● *v.* **1 call attention to, emphasize, highlight, spotlight, stress. 2 act, perform, star.**

febrile /feébrīl/ *adj.* of fever.

feckless *adj.* incompetent and irresponsible. □ **fecklessness** *n.*

fecund *adj.* fertile. □ **fecundity** *n.*

fed *see* **feed**. *adj.* **fed up** discontented, bored.

federal *adj.* of a system in which states unite under a central authority but are independent in internal affairs. □ **federalism** *n.*, **federalist** *n.*, **federally** *adv.*

federate *v.* /féddərayt/ unite on a federal basis. ● *adj.* /féddərət/ united thus. □ **federative** *adj.*

federation *n.* **1** federating. **2** federal group.

fee *n.* payment for professional advice or services.

feeble *adj.* **1** weak. **2** ineffective. □ **feebly** *adv.*, **feebleness** *n.*

■ **1 ailing, debilitated, decrepit, delicate, enfeebled, fragile, frail, infirm, puny, sickly, weak. 2 flimsy, insubstantial, lame, poor, thin, unconvincing, weak; impotent, ineffective, ineffectual, namby-pamby, spineless, *colloq.* wet, wishy-washy.**

feed *v.* (**fed**) **1** give food to. **2** (of animals) take food. **3** supply (material) to a machine etc. ● *n.* **1** meal. **2** food for animals.

■ *v.* **1 nourish, nurture; breast-feed, suckle; cater for, provide for, support, sustain. 2 browse, eat, graze, pasture.**

feedback *n.* **1** return of part of a system's output to its source. **2** return of information about a product etc. to its supplier.

feeder *n.* **1** one that feeds. **2** baby's feeding bottle. **3** feeding apparatus in a machine. **4** road or railway line linking outlying areas to a central system.

feel *v.* (**felt**) **1** explore or perceive by touch. **2** be conscious of (being). **3** experience. **4** have an impression, think. **5** give a sensation. ● *n.* **1** sense of touch. **2** act of feeling. **3** sensation produced by a thing touched. □ **feel like** be in the mood for.

■ *v.* **1 finger, handle, manipulate, touch; caress, fondle, *colloq.* paw, stroke. 2 be aware of, be conscious of, detect, discern, notice, perceive, sense. 3 bear, endure, experience, go through, undergo. 4 believe, consider, have a feeling, have a hunch, get the impression, judge, think. 5 appear, seem, strike one as.**

feeler *n.* **1** long slender organ of touch in certain animals. **2** tentative suggestion.

feeling *n.* **1** power to feel things. **2** mental or physical awareness. **3** (*pl.*) emotional susceptibilities. **4** opinion or notion, esp. a vague one. **5** readiness to feel sympathy or compassion. **6** impression.

■ **1 sensation, sense of touch, sensitivity. 2 awareness, consciousness, perception, sensation, sense. 3 (feelings) emotions, susceptibilities, sympathies. 4 hunch, idea, impression, inkling, instinct, intuition, no-**

tion, suspicion; premonition, presentiment. **5** empathy, sensibility, sensitivity, sympathy, understanding. **6** air, ambience, atmosphere, aura, impression.

feet *see* **foot**.

feign /fayn/ *v.* pretend.

feint /faynt/ *n.* sham attack made to divert attention. ● *v.* make a feint. ● *adj.* (of ruled lines) faint.

feldspar *n.* white or red mineral containing silicates.

felicitate *v.* congratulate. □ **felicitation** *n.*

felicitous *adj.* well-chosen, apt.

felicity *n.* **1** happiness. **2** pleasing manner or style.

feline *adj.* of cats, cat-like. ● *n.* animal of the cat family.

fell[1] *n.* stretch of moor or hilly land, esp. in northern England.

fell[2] *v.* strike or cut down.

■ cut down, floor, knock down, prostrate, strike down.

fell[3] *see* **fall**.

fellow *n.* **1** associate, comrade. **2** (*colloq.*) man, boy. **3** thing like another. **4** member of a learned society or governing body of a college. □ **fellow feeling** sympathy.

fellowship *n.* **1** friendly association with others. **2** society, membership of this. **3** position of a college fellow.

felt[1] *n.* cloth made by matting and pressing fibres. ● *v.* **1** make or become matted. **2** cover with felt.

felt[2] *see* **feel**.

female *adj.* **1** of the sex that can bear offspring or produce eggs. **2** (of plants) fruit-bearing. **3** (of a socket etc.) hollow. ● *n.* female animal or plant.

feminine *adj.* **1** of, like, or traditionally considered suitable for women. **2** of the grammatical form suitable for names of females. ● *n.* feminine word. □ **femininity** *n.*

feminism *n.* belief in the principle that women should have the same rights and opportunities as men. □ **feminist** *n.* & *adj.*

femur *n.* thigh-bone. □ **femoral** *adj.*

fen *n.* low-lying marshy land.

fence *n.* barrier round the boundary of a field or garden etc. ● *v.* **1** surround with a fence. **2** engage in the sport of fencing. □ **fencer** *n.*

■ *n.* barricade, barrier, hedge, palisade, railing, rampart, stockade, wall. ● *v.* **1** bound, circumscribe, encircle, enclose, hedge, surround.

fencing *n.* **1** fences, their material. **2** sport of fighting with foils.

fend *v.* **1 fend for** look after (esp. oneself). **2 fend off** ward off.

■ **2** deflect, fight off, hold at bay, keep away, parry, repel, stave off, ward off.

fender *n.* **1** low frame bordering a fireplace. **2** pad hung over a moored vessel's side to protect against bumping. **3** (*US*) mudgard or bumper of a vehicle.

feral *adj.* wild.

ferment *v.* **1** undergo or subject to fermentation. **2** stir up, excite.

■ **1** boil, bubble, effervesce, foam, froth, seethe. **2** excite, foment, incite, inflame, instigate, provoke, rouse, stir up.

fermentation *n.* breakdown of a substance by yeasts, bacteria, etc.

fern *n.* flowerless plant with feathery green leaves.

ferocious *adj.* fierce, savage. □ **ferociously** *adv.*, **ferocity** *n.*

■ barbaric, bestial, bloodthirsty, brutal, cruel, fierce, inhuman, merciless, murderous, pitiless, savage, vicious, violent, wild.

ferret *n.* small animal of the weasel family. ● *v.* (**ferreted**) search, rummage. □ **ferret out** discover by searching.

ferroconcrete *n.* reinforced concrete.

ferrous *adj.* containing iron.

ferrule *n.* metal ring or cap on the end of a stick or tube.

ferry *v.* **1** convey in a boat across water. **2** transport. ● *n.* **1** boat used for ferrying. **2** place where it operates. **3** service it provides.

fertile *adj.* **1** able to produce vegetation, fruit, or young. **2** capable of growth. **3** inventive. □ **fertility** *n.*

■ **1** fecund, fruitful, productive, prolific, rich.

fertilize *v.* **1** make fertile. **2** introduce pollen or sperm into. □ **fertilization** *n.*

fertilizer *n.* material added to soil to make it more fertile.

fervent *adj.* showing fervour. □ **fervently** *adv.*, **fervency** *n.*

■ ardent, burning, eager, emotional, enthusiastic, fanatical, fervid, fiery, impassioned, intense, keen, passionate, spirited, vehement, zealous.

fervid *adj.* fervent. □ **fervidly** *adv.*

fervour *n.* intensity of feeling.

■ **ardour, eagerness, enthusiasm, fervency, gusto, intensity, passion, spirit, vehemence, warmth, zeal, zest.**

fester *v.* **1** make or become septic. **2** cause continuing resentment. **3** rot.

■ **1 suppurate. 2 rankle, smoulder. 3 decay, decompose, putrefy, rot.**

festival *n.* **1** day or period of celebration. **2** series of cultural events in a town etc.

■ **1 anniversary, carnival, feast, fête, fiesta, gala, jubilee.**

festive *adj.* **1** of or suitable for a festival. **2** joyful.

■ **2 cheerful, cheery, convivial, gay, happy, jolly, jovial, joyful, joyous, light-hearted, merry, mirthful.**

festivity *n.* **1** gaiety. **2** festive proceedings.

■ **1 gaiety, glee, jollity, joyfulness, jubilation, merriment, merrymaking, mirth, rejoicing, revelry.**

festoon *n.* hanging chain of flowers or ribbons etc. ● *v.* decorate with hanging ornaments.

fetch *v.* **1** go for and bring back. **2** be sold for (a price).

■ **1 bring (back), get, go for, retrieve. 2 earn, make, sell for.**

fête /fayt/ *n.* **1** festival. **2** outdoor entertainment or sale, esp. in aid of charity. ● *v.* entertain in celebration of an achievement.

fetid *adj.* stinking.

fetish *n.* **1** object worshipped as having magical powers. **2** thing given excessive respect.

fetter *n.* & *v.* shackle.

feud *n.* prolonged hostility. ● *v.* conduct a feud.

■ ***n.*** **argument, conflict, dispute, falling out, quarrel, vendetta.**

feudal *adj.* of or like the feudal system. □ **feudal system** medieval system of holding land by giving one's services to the owner. **feudalism** *n.*, **feudalistic** *adj.*

fever *n.* **1** abnormally high body temperature. **2** disease causing it. **3** nervous excitement. □ **fevered** *adj.*

feverish *adj.* **1** having symptoms of fever. **2** excited, restless.

■ **1 febrile, fevered, flushed, hot. 2 excited, frantic, frenetic, frenzied, hectic, restless.**

few *adj.* & *n.* not many. □ **a few** some. **quite a few** (*colloq.*) a fairly large number.

fey *adj.* having a strange dreamy charm, whimsical.

fez *n.* (*pl.* **fezzes**) Muslim man's high flat-topped red cap.

fiancé (*fem.* **fiancée**) *n.* person one is engaged to marry.

fiasco *n.* (*pl.* **-os**) ludicrous failure.

fib *n.* unimportant lie. ● *v.* (**fibbed**) tell a fib. □ **fibber** *n.*

fibre *n.* **1** thread-like strand. **2** substance formed of fibres. **3** fibrous matter in food. **4** strength of character. □ **fibre optics** transmission of information by infra-red signals along thin glass fibres. **fibrous** *adj.*

fibreglass *n.* material made of or containing glass fibres.

fibroid *adj.* consisting of fibrous tissue. ● *n.* benign fibroid tumour.

fibrositis *n.* rheumatic pain in fibrous tissue.

fibula *n.* (*pl.* **-lae**) bone on the outer side of the shin.

fiche /feesh/ *n.* microfiche.

fickle *adj.* often changing, not loyal. □ **fickleness** *n.*

■ **capricious, changeable, disloyal, erratic, faithless, inconstant, mutable, unfaithful, unpredictable, unreliable, unstable, unsteady.**

fiction *n.* **1** invented story. **2** non-factual literature, esp. novels. □ **fictional** *adj.*

fictitious *adj.* **1** imaginary. **2** not genuine.

■ **1 apocryphal, fictional, imaginary, imagined, invented, made-up, unreal, untrue. 2 bogus, false, *colloq.* phoney, spurious.**

fiddle *n.* **1** violin. **2** (*colloq.*) swindle. ● *v.* **1** fidget with something. **2** (*colloq.*) cheat, falsify. **3** play the violin. □ **fiddler** *n.*

fiddlesticks *n.* nonsense.

fiddly *adj.* (**-ier**, **-iest**) (*colloq.*) awkward to do or use.

fidelity *n.* **1** faithfulness, loyalty. **2** accuracy.

fidget *v.* (**fidgeted**) **1** move restlessly. **2** make or be uneasy. ● *n.* **1** one who fidgets. **2** (*pl.*) restless mood. □ **fidgety** *adj.*

■ ***v.*** **1 jig about, squirm, wiggle, wriggle.** □ **fidgety *colloq.* jittery, jumpy, nervous, nervy, restive, restless, uneasy.**

fiduciary *adj.* held or given etc. in trust. ● *n.* trustee.

fief *n.* land held under the feudal system.

field *n.* **1** piece of open ground, esp. for pasture or cultivation. **2** sports ground. **3** all competitors in a race or contest. **4** sphere of action or interest. **5** area rich in a natural product. ● *v.* **1** be a fielder, stop and return (a ball). **2** put (a team) into a contest. □ **field day** day of much activity. **field events** athletic contests other than races. **field glasses** binoculars. **Field Marshal** army officer of the highest rank.

■ *n.* **1** *poetic* lea, meadow, paddock, pasture. **2** ground, pitch, playing field. **3** competition, competitors, contestants, participants, players. **4** area, domain, province, realm, speciality, sphere, subject, territory.

fielder *n.* **1** person who fields a ball. **2** member of the side not batting.

fieldwork *n.* practical work done by surveyors, social workers, etc. □ **fieldworker** *n.*

fiend /feend/ *n.* **1** evil spirit. **2** devotee. **3** wicked, mischievous, or annoying person. □ **fiendish** *adj.*

fierce *adj.* **1** violent in manner or action. **2** eager, intense. □ **fiercely** *adv.*, **fierceness** *n.*

■ **1** aggressive, barbaric, barbarous, bestial, bloodthirsty, brutal, brutish, cruel, dangerous, ferocious, homicidal, inhuman, murderous, savage, vicious, violent, wild. **2** ardent, eager, fiery, furious, intense, vehement.

fiery *adj.* (**-ier**, **-iest**) **1** consisting of or like fire. **2** spirited.

■ **1** blazing, burning, flaming; gleaming, glowing, incandescent; hot, red-hot, white-hot. **2** eager, excitable, excited, fierce, lively, passionate, spirited.

fiesta *n.* festival in Spanish-speaking countries.

fife *n.* small shrill flute.

fifteen *adj.* & *n.* one more than fourteen. □ **fifteenth** *adj.* & *n.*

fifth *adj.* & *n.* next after fourth. □ **fifthly** *adv.*

fifty *adj.* & *n.* five times ten. □ **fifty-fifty** *adj.* & *adv.* half-and-half, equally. **fiftieth** *adj.* & *n.*

fig *n.* **1** soft fruit with many seeds. **2** tree bearing this.

fight *v.* (**fought**) **1** struggle against, esp. in physical combat or war. **2** argue, quarrel. **3** strive to overcome. **4** strive to achieve something. ● *n.* **1** combat, struggle. **2** battle. **3** argument. **4** boxing match.

■ *v.* **1** battle, brawl, clash, conflict, contend, engage, grapple, *colloq.* scrap, scuffle, skirmish, spar, tussle, wage war, wrestle. **2** argue, bicker, disagree, dispute, fall out, quarrel, row, squabble, wrangle. **3** confront, defy, make a stand against, oppose, resist, struggle against. **4** campaign, strive, struggle. ● *n.* **1** brawl, brush, clash, combat, contest, fracas, fray, mêlée, *colloq.* scrap, scrimmage, scuffle, skirmish, struggle, tussle. **2** battle, conflict, encounter, engagement. **3** altercation, argument, *colloq.* bust-up, disagreement, dispute, quarrel, row, squabble, tiff, wrangle.

fighter *n.* **1** one who fights. **2** aircraft designed for attacking others.

figment *n.* thing that does not exist except in the imagination.

figurative *adj.* metaphorical. □ **figuratively** *adv.*

figure *n.* **1** external form, bodily shape. **2** representation of a person or animal. **3** numerical symbol or number. **4** diagram. **5** value, amount of money. **6** (*pl.*) arithmetic. **7** geometric shape. ● *v.* **1** appear or be mentioned. **2** represent in a diagram etc. **3** calculate. □ **figurehead** *n.* **1** carved image at the prow of a ship. **2** leader with only nominal power. **figure of speech** word(s) used for effect and not literally. **figure out** work out by arithmetic or logic.

■ *n.* **1** body, build, form, physique, shape. **2** bust, effigy, image, representation, sculpture, statue. **3** cipher, digit, number, numeral, symbol. **4** diagram, drawing, illustration, picture, plate, sketch.

figured *adj.* with a woven pattern.

figurine *n.* statuette.

filament *n.* **1** strand. **2** wire giving off light in an electric lamp.

filbert *n.* nut of a cultivated hazel.

filch *v.* pilfer, steal.

file[1] *n.* tool with a rough surface for smoothing things. ● *v.* shape or smooth with a file.

file[2] *n.* **1** cover or box etc. for holding documents. **2** its contents. **3** line of people or things one behind another. **4** set of data in a computer. ● *v.* **1** place in a file or among records. **2** march in a file.

■ *n.* **1** case, folder, portfolio. **2** documents, dossier, papers. **3** column, line, queue, rank, row.

filial *adj.* of or due from a son or daughter. □ **filially** *adv.*

filigree *n.* lace-like work in metal.

filings *n.pl.* particles filed off.

fill *v.* **1** make or become full. **2** block. **3** occupy completely. **4** spread over or through. **5** appoint to (a vacant post). ● *n.* **1** enough to fill a thing. **2** enough to satisfy a person's appetite or desire. □ **fill in 1** complete. **2** act as substitute. **3** (*colloq.*) inform more fully. **fill out 1** enlarge. **2** become enlarged or plumper. **fill up** fill completely.

■ *v.* **1** fill up, top up. **2** block, close, plug, stop (up). **3** cram, crowd (into), occupy, pack, squeeze into, stuff. **4** permeate, pervade, suffuse.

filler *n.* thing or material used to fill a gap or increase bulk.

fillet *n.* piece of boneless meat or fish. ● *v.* (**filleted**) remove bones from.

filling *n.* substance used to fill a cavity etc. □ **filling station** place selling petrol to motorists.

filly *n.* young female horse.

film *n.* **1** thin layer. **2** motion picture. **3** sheet or rolled strip of light-sensitive material for taking photographs. ● *v.* **1** make a film of. **2** cover or become covered with a thin layer.

■ *n.* **1** coat, coating, covering, layer, membrane, skin. **2** motion picture, *US colloq.* movie, picture; video.

filmy *adj.* (**-ier**, **-iest**) thin and almost transparent.

filter *n.* **1** device or substance for holding back impurities in liquid or gas passing through it. **2** screen for absorbing or modifying light or electrical or sound waves. **3** arrangement for filtering traffic. ● *v.* **1** pass through a filter. **2** pass gradually in or out. **3** (of traffic) be allowed to pass while other traffic is held up.

■ *v.* **1** filtrate, leach, percolate, sieve, sift, strain; clarify, purify, refine.

filth *n.* **1** disgusting dirt. **2** obscenity. □ **filthy** *adj.*

■ **1** dirt, filthiness, grime, *colloq.* muck, slime; dung, excrement, ordure; garbage, refuse, rubbish. □ **filthy** dirty, grimy, grubby, *colloq.* mucky, muddy, polluted, slimy, soiled, sooty, sordid, squalid, stained, unclean, unwashed.

filtrate *n.* filtered liquid. ● *v.* filter. □ **filtration** *n.*

fin *n.* **1** thin projection from a fish's body, used for propelling and steering itself. **2** similar projection to improve the stability of aircraft etc.

final *adj.* **1** at the end, coming last. **2** conclusive. ● *n.* **1** last heat or game. **2** last edition of a day's newspaper. **3** (*pl.*) final exams. □ **finally** *adv.*, **finality** *n.*

■ *adj.* **1** closing, concluding, eventual, finishing, last, terminal, terminating, ultimate. **2** conclusive, decisive, definitive, irrevocable, unalterable.

finale /finaáli/ *n.* final section of a drama or musical composition.

finalist *n.* competitor in a final.

finalize *v.* **1** bring to an end. **2** put in final form.

■ clinch, complete, conclude, put the finishing touches to, settle.

finance *n.* **1** management of money. **2** (*pl.*) money resources. ● *v.* provide money for. □ **financial** *adj.*, **financially** *adv.*

■ *n.* **2** (**finances**) assets, capital, cash, funds, money, resources, wealth, *colloq.* wherewithal. ● *v.* back, fund, invest in, pay for, subsidize, underwrite.

financier *n.* person engaged in financing businesses.

finch *n.* a kind of small bird.

find *v.* (**found**) **1** discover by effort or chance. **2** succeed in obtaining. **3** discover by experience. **4** consider to be. **5** (of a jury etc.) decide and declare. ● *n.* **1** discovery. **2** thing found. □ **find out** detect, discover. **finder** *n.*

■ *v.* **1** chance on, come across, dig out, dig up, discover, encounter, ferret out, happen on, hit on, light on, locate, trace, track down, turn up, uncover, unearth. **2** achieve, acquire, gain, get, obtain, procure, secure, win. **3** discover, note, notice, observe, perceive, realize, see. **4** consider, deem, judge, think.

fine[1] *n.* money paid as a penalty. ● *v.* punish by a fine.

fine[2] *adj.* **1** of high quality or merit. **2** bright, free from rain. **3** thin, delicate. **4** subtle. **5** in small particles. ● *adv.* finely. □ **finely** *adv.*, **fineness** *n.*

■ *adj.* **1** choice, first-class, first-rate, prime, select, superior, supreme; admirable, commendable, excellent, exceptional, exquisite, good, great, magnificent, marvellous, meritorious, outstanding, splendid. **2** bright, clear, cloudless, dry, fair, nice, pleasant, sunny. **3** delicate, diaphanous, filmy, flimsy, fragile, gauzy, thin, translucent. **4** nice, precise, subtle. **5** powdery.

finery *n.* showy clothes etc.

finesse *n.* **1** delicate manipulation. **2** tact.

finger *n.* **1** each of the five parts extending from each hand. **2** any of these other than the thumb. **3** finger-like object. **4** measure (about 20 mm) of alcohol in a glass. ● *v.* touch or feel with the fingers. □ **finger-stall** *n.* sheath to cover an injured finger.

fingerprint *n.* impression of ridges on the pad of a finger.

finish *v.* **1** bring to an end, complete. **2** consume or use all of. **3** come to an end, cease. **4** reach the end, esp. of a race. ● *n.* **1** last stage. **2** point where a race etc. ends. **3** completed state. □ **finish off** **1** (*colloq.*) kill, overcome completely. **2** consume the whole of.

■ *v.* **1** accomplish, achieve, bring to an end, carry out, clinch, complete, fulfil, round off. **2** consume, devour, drain, drink (up), eat (up), finish off, get *or* go through, polish off, use (up). **3** cease, close, come to an end, conclude, end, halt, stop, terminate, wind up; culminate. ● *n.* **1** close, completion, conclusion, culmination, end, ending, finale, termination, winding-up.

finite *adj.* limited.

■ bounded, countable, delimited, limited, measurable, restricted.

fiord /fyord/ *n.* narrow inlet of the sea esp. in Norway.

fir *n.* evergreen cone-bearing tree.

fire *n.* **1** state of combustion. **2** flame. **3** destructive burning. **4** burning fuel. **5** electric or gas heater. **6** firing of guns. **7** angry or excited feeling. ● *v.* **1** send a bullet or shell from (a gun). **2** detonate. **3** discharge (a missile). **4** dismiss from a job. **5** set fire to. **6** catch fire. **7** excite. **8** bake (pottery etc.). □ **fire brigade** organized group of people employed to extinguish fires. **fire engine** vehicle with equipment for putting out fires. **fire escape** special staircase or apparatus for escape from a burning building.

■ *n.* **3** blaze, conflagration, flames, holocaust, inferno. **6** barrage, bombardment, cannonade, firing, fusillade, gunfire, salvo, shelling, volley. **7** animation, ardour, energy, enthusiasm, excitement, fervour, intensity, passion, spirit, vigour. ● *v.* **1** open fire, shoot. **2** detonate, explode, let off, set off. **3** catapult, discharge, launch, propel. **4** discharge, dismiss, *colloq.* give a person the boot *or* the sack, give a person notice, lay off, make redundant, *colloq.* sack, throw out. **5** burn, ignite, kindle, set alight, set fire to, set on fire. **7** animate, awaken, excite, inflame, inspire, motivate, rouse, stimulate, stir (up).

firearm *n.* gun, pistol, etc.

firebreak *n.* open space as an obstacle to the spread of fire.

firedamp *n.* explosive mixture of methane and air in mines.

firefly *n.* phosphorescent beetle.

fireman *n.* member of a fire brigade.

fireplace *n.* recess with a chimney for a domestic fire.

fireside *n.* space round a fireplace.

firework *n.* device containing chemicals that burn or explode spectacularly.

firing squad group detailed to shoot a condemned person.

firm[1] *n.* business company.

■ business, company, concern, corporation, establishment, organization, partnership.

firm[2] *adj.* **1** not yielding when pressed or pushed. **2** fixed, steady. **3** resolute. **4** securely established. ● *v.* make or become firm. □ **firmness** *n.*

■ *adj.* **1** compact, compressed, dense, hard, inflexible, rigid, set, solid, solidified, stiff, unyielding. **2** anchored, fast, fixed, immovable, moored, secure, stable, steady, tight. **3** adamant, decided, determined, dogged, resolute, resolved, stubborn, unshakeable, unwavering. **4** abiding, constant, devoted, enduring, faithful, longstanding, staunch, steadfast.

firmament *n.* sky with its clouds and stars.

first *adj.* coming before all others in time or order or importance. ● *n.* **1** first thing or occurrence. **2** first day of a month. ● *adv.* before all others or another. □ **at first** at the beginning. **first aid** treatment given for an injury etc. before a doctor arrives. **first-class** *adj.* & *adv.* **1** of the best quality. **2** in the best category of accommodation. **first cousin** (*see* **cousin**). **first name** personal name. **first-rate** *adj.* excellent.

■ *adj.* earliest, oldest, original; basic, elementary, fundamental, initial, introductory, opening, preliminary, rudimentary; chief, foremost, head, key, leading, main, paramount, pre-eminent, premier, primary, prime, principal.

firstly *adv.* first.

firth *n.* estuary, inlet.

fiscal *adj.* of public revenue.

fish *n.* (*pl.* usu. **fish**) **1** cold-blooded vertebrate living in water. **2** its flesh as food. ● *v.* **1** try to catch fish (from). **2** search by reaching into something.

fishery *n.* **1** place where fish are caught or reared. **2** business of fishing.

fishmeal *n.* dried ground fish used as a fertilizer.

fishmonger *n.* shopkeeper who sells fish.

fishy *adj.* (**-ier, -iest**) **1** like fish. **2** causing disbelief or suspicion.

fissile *adj.* **1** tending to split. **2** capable of undergoing nuclear fission.

fission *n.* splitting (esp. of an atomic nucleus, with release of energy).

fissure *n.* cleft.

fist *n.* hand when tightly closed.

fisticuffs *n.* fighting with fists.

fistula *n.* abnormal or artificial passage in the body.

fit[1] *n.* **1** sudden attack of illness or its symptoms, or of convulsions or loss of consciousness. **2** sudden short bout of activity, feeling, etc.

■ **1** attack, bout, convulsion, paroxysm, seizure, spasm. **2** bout, burst, outbreak, outburst, spell.

fit[2] *adj.* (**fitter, fittest**) **1** suitable. **2** right and proper. **3** in good health. ● *v.* (**fitted**) **1** be the right shape and size for. **2** put into place. **3** make or be suitable or competent. ● *n.* way a thing fits. ▫ **fitly** *adv.*, **fitness** *n.*

■ *adj.* **1** adapted, appropriate, apt, fitting, suitable, suited. **2** becoming, correct, proper, right. **3** hale, healthy, in good health, *colloq.* in the pink, robust, strong, well. ● *v.* **2** connect, insert, join, position, put in place, put together. **3** adapt, adjust, alter, change, modify; prepare, prime, qualify, train.

fitful *adj.* spasmodic, intermittent. ▫ **fitfully** *adv.*

fitment *n.* piece of fixed furniture.

fitter *n.* **1** person who supervises the fitting of clothes. **2** mechanic.

fitting *adj.* right and proper.

fittings *n.pl.* fixtures, fitments.

five *adj.* & *n.* one more than four.

fiver *n.* (*colloq.*) five-pound note.

fix *v.* **1** put something firmly in place. **2** establish, specify. **3** repair. **4** direct (eyes etc.) steadily. **5** (*colloq.*) arrange fraudulently. ● *n.* **1** awkward situation. **2** position determined by taking bearings. ▫ **fix up 1** organize. **2** provide. **fixer** *n.*

■ *v.* **1** affix, anchor, attach, fasten, make fast, secure, stick, pin. **2** agree on, arrange, arrive at, conclude, decide, establish, name, organize, set, settle, specify. **3** adjust, correct, cure, emend, mend, patch up, put right, rectify, remedy, repair. **5** *colloq.* fiddle, rig. ● *n.* **1** catch-22, difficulty, dilemma, *colloq.* hole, *colloq.* jam, mess, *colloq.* pickle, plight, predicament, quandary.

fixated *adj.* having an obsession.

fixation *n.* **1** fixing. **2** obsession.

fixative *n.* & *adj.* (substance) for keeping things in position, or preventing fading or evaporation.

fixedly *adv.* intently.

fixity *n.* fixed state, stability, permanence.

fixture *n.* **1** thing fixed in position. **2** firmly established person or thing.

fizz *v.* hiss or splutter, esp. when gas escapes in bubbles from a liquid. ● *n.* **1** effervescence. **2** fizzing drink. ▫ **fizziness** *n.*, **fizzy** *adj.*

■ *v.* bubble, effervesce, fizzle, froth, hiss, sizzle, splutter, sputter.

fizzle *v.* fizz feebly. ▫ **fizzle out** end feebly or unsuccessfully.

flab *n.* (*colloq.*) flabbiness, fat.

flabbergast *v.* (*colloq.*) astound.

flabby *adj.* (**-ier, -iest**) fat and limp, not firm. ▫ **flabbiness** *n.*

flaccid *adj.* loose or wrinkled, not firm. ▫ **flaccidity** *n.*

flag *n.* **1** piece of cloth attached by one edge to a staff or rope, used as a signal or symbol. **2** similarly shaped device. **3** flagstone. ● *v.* (**flagged**) **1** droop. **2** lose vigour. **3** mark or signal (as) with a flag. ▫ **flag day** day on which small emblems are sold for a charity.

■ *n.* **1** banner, colours, ensign, pennant, pennon, standard, streamer. ● *v.* **1** droop, sag, wilt. **2** grow tired, tire, weaken; decline, die away, diminish, dwindle, ebb, fade, fail, languish, subside, wane.

flagellate *v.* whip, flog. ▫ **flagellant** *n.*, **flagellation** *n.*

flageolet /flájəlét/ *n.* small flute.

flagged *adj.* paved with flagstones.

flagon *n.* large bottle or other vessel for wine, cider, etc.

flagrant /fláygrənt/ *adj.* (of an offence or offender) very bad and obvious. ▫ **flagrantly** *adv.*

flagship *n.* **1** admiral's ship. **2** principal shop, product, etc.

flagstone *n.* large paving-stone.

flail *n.* implement formerly used for threshing grain. ● *v.* thrash or swing about wildly.

flair *n.* natural ability.

■ **ability, aptitude, bent, genius, gift, instinct, knack, skill, talent.**

flak *n.* **1** anti-aircraft shells. **2** barrage of criticism.

flake *n.* small thin piece. ● *v.* come off in flakes. □ **flake out** (*colloq.*) faint, fall asleep from exhaustion. **flaky** *adj.*

■ *n.* **bit, chip, fragment, particle, piece, scale, shaving, sliver, wafer.**

flamboyant *adj.* showy in appearance or manner. □ **flamboyantly** *adv.*, **flamboyance** *n.*

■ **colourful, elaborate, extravagant, flashy, gaudy, ornate, ostentatious, showy, theatrical.**

flame *n.* bright tongue-shaped portion of gas burning visibly. ● *v.* **1** burn with flames. **2** become bright red. □ **old flame** (*colloq.*) former sweetheart.

flamenco *n.* (*pl.* **-os**) Spanish style of singing and dancing.

flamingo *n.* (*pl.* **-os**) wading bird with long legs and pink feathers.

flammable *adj.* able to be set on fire. □ **flammability** *n.*

flan *n.* open pastry or sponge case with filling.

flange *n.* projecting rim.

flank *n.* side, esp. of the body between ribs and hip. ● *v.* place or be at the side of.

flannel *n.* **1** woollen fabric. **2** face flannel. **3** (*pl.*) trousers of flannel. **4** (*sl.*) nonsense, flattery. ● *v.* (**flannelled**) (*sl.*) flatter.

flannelette *n.* napped cotton fabric like flannel.

flap *v.* (**flapped**) **1** move up and down with a sharp sound. **2** (*colloq.*) show agitation. ● *n.* **1** act or sound of flapping. **2** hanging or hinged piece. **3** (*colloq.*) agitation.

flare *v.* **1** blaze with bright unsteady flame. **2** burst into activity or anger. **3** widen outwards. ● *n.* **1** sudden blaze. **2** device producing flame as a signal or illumination. **3** flared shape.

■ *v.* **1 blaze, burn, flame, flash, flicker, gleam, glitter, shine, sparkle. 2 blow up, erupt, explode. 3 broaden, spread out, swell, widen.**

flash *v.* **1** (cause to) emit a sudden bright light. **2** move rapidly. **3** show suddenly or ostentatiously. **4** come suddenly into sight or mind. ● *n.* **1** sudden burst of flame or light. **2** very brief time. **3** sudden show of wit or feeling. **4** device producing a brief bright light in photography. ● *adj.* (*colloq.*) flashy. □ **flash flood** sudden destructive flood.

■ *v.* **1 beam, blaze, flare, flicker, glare, gleam, glimmer, glint, glitter, shimmer, shine, sparkle, twinkle. 2 dart, dash, fly, hasten, hurry, race, run, rush, shoot, speed, sprint, streak, tear, whiz, zoom. 3 display, flaunt, flourish, show off.** ● *n.* **1 blaze, flare, flickering, gleam, glimmer, glint, glitter, shimmer, sparkle, twinkle. 2 instant, minute, moment, split second.**

flashback *n.* change of scene in a story or film to an earlier period.

flashing *n.* strip of metal covering a joint in a roof etc.

flashlight *n.* electric torch.

flashpoint *n.* temperature at which a vapour ignites.

flashy *adj.* (**-ier, -iest**) showy, gaudy. □ **flashily** *adv.*

flask *n.* **1** narrow-necked bottle. **2** vacuum flask.

flat *adj.* (**flatter, flattest**) **1** horizontal, level. **2** absolute. **3** monotonous. **4** dejected. **5** having lost effervescence or power to generate electric current. **6** below the correct pitch in music. ● *adv.* **1** lying at full length. **2** completely, exactly. ● *n.* **1** flat surface, level ground. **2** set of rooms on one floor, used as a residence. **3** (sign indicating) music note lowered by a semitone. □ **flatfish** *n.* fish with a flattened body, swimming on its side. **flat out 1** at top speed. **2** with maximum effort.

■ *adj.* **1 even, horizontal, level, plane, smooth, unbroken. 2 absolute, categorical, definite, direct, downright, firm, outright, plain, unambiguous, unequivocal, unqualified. 3 boring, dead, dull, featureless, insipid, lacklustre, lifeless, monotonous, prosaic, tedious, unexciting, uninteresting. 4 dejected, depressed, dispirited, listless, low.** ● *adv.* **1 outstretched, prone, prostrate, recumbent, spread-eagled, supine.**

flatlet *n.* small flat.

flatten *v.* make or become flat.

flatter *v.* **1** compliment insincerely. **2** enhance the appearance of. □ **flatterer** *n.*, **flattery** *n.*

■ **1** butter up, compliment, curry favour with, fawn on, *sl.* flannel, overpraise, *colloq.* suck up to, toady to. **2** become, set off, suit.

flatulent *adj.* causing or suffering from formation of gas in the digestive tract. □ **flatulence** *n.*

flaunt *v.* display proudly or ostentatiously.

flautist *n.* flute player.

flavour *n.* **1** distinctive taste. **2** characteristic quality. ● *v.* give flavour to.

■ *n.* **1** piquancy, savour, tang, taste. **2** air, atmosphere, character, feeling, quality, spirit, stamp, style. ● *v.* season, spice.

flavouring *n.* substance used to give flavour to food.

flaw *n.* imperfection. ● *v.* spoil with a flaw. □ **flawless** *adj.*

■ *n.* blemish, blot, defect, error, failing, fault, imperfection, mistake, shortcoming, weakness. □ **flawless** clean, faultless, immaculate, impeccable, perfect, pristine, pure, spotless, undamaged, unsoiled, unsullied, untarnished.

flax *n.* **1** blue-flowered plant. **2** textile fibre from its stem.

flaxen *adj.* **1** made of flax. **2** (of hair) pale yellow.

flay *v.* **1** strip off the skin or hide of. **2** criticize severely.

flea *n.* small jumping blood-sucking insect. □ **flea market** market for second-hand goods.

fleck *n.* **1** small patch of colour or light. **2** speck. ● *v.* mark with flecks.

fled *see* **flee**.

fledged *adj.* (of a young bird) able to fly.

fledgeling *n.* young bird.

flee *v.* (**fled**) run away (from).

■ abscond, bolt, dash away, decamp, *sl.* do a bunk, escape, fly, make a getaway, make off, run away, rush away, *colloq.* scram, *colloq.* skedaddle, take flight, take to one's heels, vanish.

fleece *n.* sheep's woolly coat. ● *v.* rob by trickery. □ **fleecy** *adj.*

fleet[1] *n.* **1** navy. **2** ships sailing together. **3** vehicles or aircraft under one command or ownership.

■ **2** armada, convoy, flotilla, squadron.

fleet[2] *adj.* moving swiftly, nimble.

fleeting *adj.* transitory, brief.

■ brief, ephemeral, fugitive, impermanent, momentary, passing, short, transient, transitory.

flesh *n.* **1** soft substance of animal bodies. **2** body as opposed to mind or soul. **3** pulpy part of fruits and vegetables. □ **flesh and blood 1** human nature. **2** one's relatives. **flesh wound** superficial wound.

fleshy *adj.* (**-ier**, **-iest**) **1** of or like flesh. **2** having much flesh.

flew *see* **fly**[2].

flex[1] *v.* **1** bend. **2** move (a muscle) to bend a joint. □ **flexion** *n.*

flex[2] *n.* flexible insulated wire for carrying electric current.

flexible *adj.* **1** able to bend easily. **2** adaptable. □ **flexibly** *adv.*, **flexibility** *n.*

■ **1** bendable, elastic, lithe, plastic, pliable, pliant, resilient, springy, stretchable, supple, whippy, willowy, yielding. **2** accommodating, adaptable, amenable, compliant, cooperative, easygoing.

flexitime *n.* system of flexible working hours.

flick *n.* quick light blow. ● *v.* move or strike with a flick. □ **flick knife** knife with a blade that springs out. **flick through** glance through by turning over (pages etc.) rapidly.

flicker *v.* **1** burn or shine unsteadily. **2** occur briefly. **3** quiver. ● *n.* **1** flickering light or movement. **2** brief occurrence.

flier *n.* = **flyer**.

flight[1] *n.* **1** flying. **2** movement or path of a thing through the air. **3** journey by air. **4** birds or aircraft flying together. **5** series of stairs. **6** feathers etc. on a dart or arrow. □ **flight deck 1** cockpit of a large aircraft. **2** deck of an aircraft carrier. **flight recorder** electronic device in an aircraft recording details of its flight.

flight[2] *n.* hasty retreat.

■ departure, escape, exit, exodus, getaway, retreat.

flightless *adj.* unable to fly.

flighty *adj.* (**-ier**, **-iest**) frivolous.

flimsy *adj.* (**-ier**, **-iest**) **1** light and thin. **2** fragile. **3** unconvincing. □ **flimsily** *adv.*, **flimsiness** *n.*

■ **1** delicate, diaphanous, filmy, fine, light, sheer, thin. **2** breakable, fragile, frail, gimcrack, makeshift, ramshackle, rickety, shaky. **3** feeble, implausible, inadequate, lame, poor, unbelievable, unconvincing, unsatisfactory, weak.

flinch *v.* draw back in fear, wince.
■ blench, cower, cringe, draw back, quail, recoil, shrink back, shy away, start, wince.

fling *v.* (**flung**) **1** throw violently or hurriedly. **2** rush angrily or violently. ● *n.* spell of indulgence in pleasure.
■ *v.* **1** cast, *colloq.* chuck, *colloq.* heave, hurl, launch, lob, pitch, *colloq.* sling, throw, toss.

flint *n.* **1** very hard stone. **2** piece of hard alloy producing sparks when struck.

flip *v.* (**flipped**) **1** flick. **2** toss with a sharp movement. ● *n.* action of flipping. ● *adj.* (*colloq.*) glib, flippant.

flippant *adj.* not showing proper seriousness. □ **flippantly** *adv.*, **flippancy** *n.*
■ cheeky, disrespectful, facetious, *colloq.* flip, frivolous, impertinent, impudent, irreverent, pert, saucy.

flipper *n.* **1** sea animal's limb used in swimming. **2** large flat rubber attachment to the foot for underwater swimming.

flirt *v.* **1** behave in a frivolously amorous way. **2** toy (with an idea etc.). ● *n.* person who flirts. □ **flirtation** *n.*, **flirtatious** *adj.*
■ *v.* **1** dally, philander. **2** play, toy, trifle. ● *n.* coquette, philanderer, tease.

flit *v.* (**flitted**) **1** fly or move lightly and quickly. **2** decamp stealthily. ● *n.* act of flitting.

flitter *v.* flit about.

float *v.* **1** rest or drift on the surface of liquid. **2** start (a company or scheme). **3** have or allow (currency) to have a variable rate of exchange. ● *n.* **1** thing designed to float on liquid. **2** money for minor expenditure or giving change.

flocculent *adj.* like tufts of wool.

flock[1] *n.* **1** number of animals or birds together. **2** large number of people, congregation. ● *v.* gather or go in a flock.
■ *n.* **1** drove, gaggle, herd, pack, pride, school, shoal, skein, swarm. **2** assembly, band, body, bunch, cluster, company, congregation, crowd, gang, gathering, group, horde, host, multitude, throng, troop. ● *v.* assemble, collect, congregate, crowd, gather, herd, mass, meet, swarm, throng.

flock[2] *n.* shredded wool, cotton, etc. used as stuffing.

floe *n.* sheet of floating ice.

flog *v.* (**flogged**) **1** beat severely. **2** (*sl.*) sell. □ **flogging** *n.*
■ **1** beat, cane, flagellate, flay, lash, scourge, thrash, whip.

flood *n.* **1** overflow of water on a place usually dry. **2** great outpouring. **3** inflow of the tide. **4** (**the Flood**) the flood described in the Old Testament. ● *v.* **1** cover or fill with a flood, overflow. **2** come in great quantities.
■ *n.* **1** deluge, inundation, overflow, overflowing. **2** flow, outpouring, rush, spate, stream, surge, tide, torrent. ● *v.* **1** deluge, drown, engulf, fill, inundate, overflow, submerge, swamp. **2** flow, gush, pour, surge, swarm.

floodlight *n.* lamp producing a broad bright beam. ● *v.* (**floodlit**) illuminate with this.

floor *n.* **1** lower surface of a room. **2** right to speak in an assembly. **3** storey. ● *v.* **1** provide with a floor. **2** knock down. **3** baffle. □ **floor show** cabaret.

flooring *n.* material for a floor.

flop *v.* (**flopped**) **1** hang or fall heavily and loosely. **2** (*sl.*) be a failure. ● *n.* **1** flopping movement or sound. **2** (*sl.*) failure.

floppy *adj.* (**-ier, -iest**) tending to flop. □ **floppy disk** flexible disk for storing computer data.

flora *n.* plants of an area or period.

floral *adj.* of flowers.

floret *n.* each of the small flowers of a composite flower.

florid *adj.* **1** ornate. **2** ruddy.

florist *n.* person who sells or grows flowers as a business.

flotation *n.* floating, esp. of a commercial venture.

flotilla *n.* **1** small fleet. **2** fleet of small ships.

flotsam *n.* floating wreckage. □ **flotsam and jetsam** odds and ends.

flounce[1] *v.* go in an impatient annoyed manner. ● *n.* flouncing movement.

flounce[2] *n.* deep frill attached by its upper edge. □ **flounced** *adj.*

flounder[1] *n.* small flatfish.

flounder[2] *v.* **1** move clumsily, as in mud. **2** become confused when trying to do something.
■ **1** blunder, fumble, grope, lurch, stagger, struggle, stumble.

flour *n.* fine powder made from grain, used in cooking. ● *v.* cover with flour. □ **floury** *adj.*

flourish *v.* **1** grow vigorously. **2** prosper. **3** wave dramatically. ● *n.* **1** dramatic gesture. **2** ornamental curve. **3** fanfare.

■ *v.* **1** bloom, blossom, burgeon, flower, thrive. **2** be successful, boom, do well, prosper, succeed, thrive. **3** brandish, flaunt, shake, swing, twirl, wave, wield.

flout *v.* disobey openly.

flow *v.* **1** glide along as a stream. **2** gush out. **3** proceed evenly. **4** hang loosely. ● *n.* **1** flowing movement or liquid. **2** amount flowing. **3** inflow of the tide. □ **flow chart** diagram showing a sequence of processes.

■ *v.* **1** course, glide, go, run, stream, swirl, trickle. **2** cascade, flood, gush, spew, spout, spurt, squirt, stream, well. ● *n.* **1** current, drift, movement, stream. **2** gush, outflow, outpouring, rush, surge.

flower *n.* **1** part of a plant where fruit or seed develops. **2** plant grown for this. **3** best part. ● *v.* produce flowers.

■ *n.* **1** bloom, blossom, floret. ● *v.* bloom, blossom, burgeon, come out, effloresce, open, unfold.

flowered *adj.* ornamented with a design of flowers.

flowery *adj.* **1** full of flowers. **2** full of ornamental phrases.

flown *see* **fly²**.

flu *n.* (*colloq.*) influenza.

fluctuate *v.* vary, esp. irregularly. □ **fluctuation** *n.*

■ alternate, change, oscillate, see-saw, shift, swing, vacillate, vary, waver.

flue *n.* **1** smoke-duct in a chimney. **2** channel for conveying heat.

fluent *adj.* speaking or spoken smoothly and readily. □ **fluently** *adv.*, **fluency** *n.*

■ articulate, eloquent, glib, voluble; effortless, flowing, smooth.

fluff *n.* soft mass of fibres or down. ● *v.* **1** shake into a soft mass. **2** (*sl.*) bungle. □ **fluffy** *adj.*

fluid *adj.* **1** consisting of particles that move freely among themselves. **2** not stable. ● *n.* fluid substance. □ **fluidity** *n.*

■ *adj.* **1** aqueous, flowing, liquefied, liquid, molten, runny. **2** changeable, fluctuating, shifting, uncertain, unstable, variable.

fluke¹ *n.* success due to luck.

fluke² *n.* **1** barbed arm of an anchor etc. **2** lobe of a whale's tail.

flummox *v.* (*colloq.*) baffle.

flung *see* **fling**.

fluoresce *v.* be or become fluorescent.

fluorescent *adj.* taking in radiations and sending them out as light. □ **fluorescence** *n.*

fluoridate *v.* add fluoride to (a water supply). □ **fluoridation** *n.*

fluoride *n.* compound of fluorine with metal.

fluorine *n.* pungent corrosive gas.

flurry *n.* **1** short rush of wind, rain, or snow. **2** commotion. **3** nervous agitation. ● *v.* fluster.

flush¹ *v.* **1** become red in the face. **2** cleanse or dispose of with a flow of water. ● *n.* **1** blush. **2** rush of emotion. **3** rush of water. ● *adj.* **1** level, in the same plane. **2** (*colloq.*) having plenty of money.

■ *v.* **1** blush, colour, glow, go red, redden.

flush² *v.* drive out from cover.

fluster *v.* make nervous or confused. ● *n.* flustered state.

■ *v.* agitate, bewilder, bother, confuse, discomfit, disconcert, disturb, perplex, perturb, put off, puzzle, *colloq.* rattle, ruffle, *colloq.* throw, unsettle, upset.

flute *n.* **1** wind instrument, pipe with a mouth-hole at the side. **2** ornamental groove.

flutter *v.* **1** move wings hurriedly. **2** wave or flap quickly. **3** (of the heart) beat irregularly. ● *n.* **1** fluttering movement or beat. **2** nervous excitement. **3** (*sl.*) small bet.

fluvial *adj.* of or found in rivers.

flux *n.* **1** flow. **2** continuous succession of changes. **3** substance mixed with metal etc. to assist fusion.

fly¹ *n.* two-winged insect. □ **fly-blown** *adj.* tainted by flies' eggs.

fly² *v.* (**flew, flown**) **1** move through the air on wings or in an aircraft. **2** control the flight of. **3** go quickly. **4** flee. **5** display (a flag). ● *n.* **1** flying. **2** (*pl.*) fastening down the front of trousers. □ **fly-post** *v.* display (posters etc.) in unauthorized places. **fly-tip** *v.* dump (waste) illegally.

■ *v.* **1** soar, take to the air, take wing, wing. **2** control, operate, pilot. **3** dart, dash, hasten, hurry, race, run, scoot, shoot, speed, sprint, tear, whiz, zoom. **4** abscond, bolt, decamp, *sl.* do a bunk, escape, flee, make a getaway, make off, run away, rush away, *colloq.* skedaddle, take flight, take to one's heels.

flyer *n.* **1** one that flies. **2** airman. **3** fast animal or vehicle.

flying *adj.* able to fly. □ **flying buttress** one based on separate structure, usu.

forming an arch. **flying colours** great credit. **flying fox** fruit-eating bat. **flying saucer** unidentified object reported as seen in the sky.

flyleaf *n.* blank leaf at the beginning or end of a book.

flyover *n.* bridge carrying one road or railway over another.

flywheel *n.* heavy wheel revolving on a shaft to regulate machinery.

foal *n.* young of the horse. ● *v.* give birth to a foal.

foam *n.* **1** collection of small bubbles. **2** spongy rubber or plastic. ● *v.* form foam. □ **foamy** *adj.*

■ *n.* **1** bubbles, effervescence, fizz, froth, lather, spume, suds.

fob *v.* (**fobbed**) **fob off 1** palm off. **2** get (a person) to accept something inferior.

focal *adj.* of or at a focus.

fo'c's'le /fóks'l/ *n.* forecastle.

focus *n.* (*pl.* **-cuses** or **-ci**) **1** point where rays meet. **2** distance at which an object is most clearly seen. **3** adjustment on a lens to produce a clear image. **4** centre of activity or interest. ● *v.* (**focused**) **1** adjust the focus of. **2** bring into focus. **3** concentrate.

fodder *n.* food for animals.

foe *n.* enemy.

foetus /féetəss/ *n.* (*pl.* **-tuses**) developed embryo in a womb or egg. □ **foetal** *adj.*

fog *n.* thick mist. ● *v.* (**fogged**) cover or become covered with fog or condensed vapour. □ **foghorn** *n.* horn warning ships in fog. **foggy** *adj.*

fogey *n.* (also **fogy**) (*pl.* **-eys** or **-ies**) old-fashioned person.

foible *n.* harmless peculiarity in a person's character.

foil[1] *n.* **1** paper-thin sheet of metal. **2** person or thing emphasizing another's qualities by contrast.

foil[2] *v.* thwart, frustrate.

■ baffle, baulk, check, defeat, frustrate, hamper, hinder, impede, outwit, *sl.* scupper, stymie, thwart.

foil[3] *n.* long thin sword with a button on the point.

foist *v.* impose (an unwelcome person or thing).

fold[1] *v.* **1** bend so that one part lies on another. **2** clasp, embrace. **3** cease to function. ● *n.* **1** folded part. **2** line or hollow made by folding.

■ *v.* **1** bend, crease, crimp, gather, pleat. **2** clasp, embrace, enclose, enfold, envelop, hold, hug, wrap. **3** close down, go bankrupt, *colloq.* go broke, go out of business, go under, fail. ● *n.* **1** crease, pleat; crinkle, pucker, wrinkle.

fold[2] *n.* enclosure for sheep. ● *v.* enclose (sheep) in a fold.

folder *n.* **1** folding cover for loose papers. **2** leaflet.

foliage *n.* leaves.

foliate *v.* split into thin layers. □ **foliation** *n.*

folk *n.* **1** people. **2** (*pl.*) one's relatives. □ **folk dance, song** dance or song in the traditional style of a country. **folk-tale** *n.* popular or traditional story.

■ **1** citizenry, people, population; clan, ethnic group, race, tribe. **2** (**folks**) family, parents, kin, kinsfolk, kith and kin, relations, relatives.

folklore *n.* traditional beliefs and tales of a community.

folksy *adj.* informal and friendly.

follicle *n.* small cavity, esp. for a hair-root. □ **follicular** *adj.*

follow *v.* **1** go or come after. **2** go along (a road etc.). **3** accept the ideas of. **4** take an interest in the progress of. **5** grasp the meaning of. **6** be a natural consequence of. □ **follow suit** follow a person's example. **follow up** investigate further. **follow-up** *n.* **follower** *n.*

■ **1** come *or* go after, walk behind; chase, dog, pursue, shadow, stalk, *colloq.* tail, track, trail; replace, succeed, supersede, supplant, take the place of. **3** abide by, accept, adhere to, be guided by, comply with, conform to, heed, obey, observe. **5** appreciate, comprehend, fathom, *colloq.* get, grasp, see, take in, understand. **6** develop, ensue, result.

following *n.* group of supporters. ● *adj.* now to be mentioned. ● *prep.* as a sequel to.

folly *n.* **1** foolishness. **2** foolish act. **3** ornamental building.

■ **1** absurdity, foolishness, idiocy, insanity, lunacy, madness, silliness, stupidity.

foment *v.* stir up (trouble). □ **fomentation** *n.*

■ excite, ferment, incite, instigate, kindle, provoke, stimulate, stir up, whip up.

fond *adj.* **1** affectionate, doting. **2** foolishly optimistic. □ **fondly** *adv.*, **fondness** *n.*

■ **1** adoring, affectionate, caring, devoted, doting, loving, tender, warm. **2** credulous, foolish, naive.

fondant *n.* soft sugary sweet.

fondle *v.* handle lovingly.

■ caress, cuddle, pat, pet, stroke.

fondue *n.* dish of flavoured melted cheese.

font *n.* basin in a church, holding water for baptism.

fontanelle *n.* soft spot where the bones of an infant's skull have not yet grown together.

food *n.* substance (esp. solid) taken into the body of an animal or plant to maintain its life.

■ nourishment, nutriment, sustenance; comestibles, eatables, foodstuffs, *sl.* grub, provisions, refreshments.

foodstuff *n.* substance used as food.

fool *n.* **1** foolish person. **2** creamy fruit-flavoured pudding. ● *v.* **1** joke, tease. **2** trick. □ **fool about** play about idly. **make a fool of** make (a person) look foolish, trick.

■ *n.* **1** ass, blockhead, booby, buffoon, *colloq.* chump, *colloq.* clot, dolt, *sl.* dope, *colloq.* duffer, halfwit, idiot, imbecile, jackass, ninny, *colloq.* nitwit, *sl.* sap, simpleton, *sl.* twerp, *sl.* twit, *sl.* wally. ● *v.* **1** banter, jest, joke, *colloq.* kid, tease. **2** *colloq.* bamboozle, *sl.* con, deceive, delude, dupe, hoax, hoodwink, make a fool of, pull the wool over someone's eyes, take in, trick.

foolery *n.* foolish acts.

foolhardy *adj.* taking foolish risks.

foolish *adj.* lacking good sense or judgement. □ **foolishly** *adv.*, **foolishness** *n.*

■ absurd, asinine, *sl.* barmy, *sl.* batty, crazy, daft, fatuous, foolhardy, hare-brained, idiotic, ill-considered, imbecilic, imprudent, inane, incautious, laughable, ludicrous, mad, misguided, nonsensical, rash, reckless, ridiculous, senseless, short-sighted, silly, stupid, thoughtless, unintelligent, unwise, witless.

foolproof *adj.* simple and easy to use, unable to go wrong.

foot *n.* (*pl.* **feet**) **1** part of the leg below the ankle. **2** lower part or end. **3** measure of length, = 12 inches (30.48 cm). **4** unit of rhythm in verse. ● *v.* **1** walk. **2** be the one to pay (a bill). □ **foot-and-mouth disease** contagious virus disease of cattle.

football *n.* **1** large round or elliptical inflated ball. **2** game played with this. □ **football pool** form of gambling on the results of football matches. **footballer** *n.*

footfall *n.* sound of footsteps.

foothills *n.pl.* low hills near the bottom of a mountain or range.

foothold *n.* **1** place just wide enough for one's foot. **2** small but secure position gained.

footing *n.* **1** foothold. **2** balance. **3** status, conditions.

footlights *n.pl.* row of lights along the front of a stage floor.

footling *adj.* (*sl.*) trivial.

footloose *adj.* independent, without responsibilities.

footman *n.* manservant, usu. in livery.

footnote *n.* note printed at the bottom of a page.

footpath *n.* path for pedestrians.

footprint *n.* impression left by a foot or shoe.

footsore *adj.* with feet sore from walking.

footstep *n.* **1** step. **2** sound of this.

footwork *n.* manner of moving or using the feet in sports etc.

fop *n.* dandy.

for *prep.* **1** in place of. **2** as the price or penalty of. **3** in defence or favour of. **4** with a view to. **5** in the direction of. **6** intended to be received or used by. **7** because of. **8** during. ● *conj.* because.

forage *v.* **1** go searching. **2** rummage. ● *n.* **1** foraging. **2** food for horses and cattle.

foray *n.* sudden attack, raid. ● *v.* make a foray.

forbade *see* **forbid**.

forbear *v.* (**forbore**, **forborne**) refrain (from).

forbearance *n.* patience, tolerance.

forbearing *adj.* patient, tolerant.

forbid *v.* (**forbade**, **forbidden**) **1** refuse to allow. **2** order not to.

■ **1** ban, bar, disallow, outlaw, prohibit, proscribe, rule out, veto.

forbidding *adj.* having an uninviting appearance, stern.

■ grim, hostile, menacing, ominous, sinister, stern, threatening, unfriendly, uninviting.

force *n.* **1** strength, power. **2** intense effort. **3** compulsion. **4** effectiveness. **5** group of troops or police. **6** organized or available group. **7** influence tending to cause movement. ● *v.* **1** use force upon, esp. in order to get or do something. **2** break open by force. **3** propel. **4** strain to the utmost, overstrain. **5** produce by effort. **6** impose.

■ *n.* **1** dynamism, energy, impact, intensity, might, muscle, potency, power, strength, vigour, violence. **2** effort, exertion, strain. **3** coercion, compulsion, constraint, pressure. **4** cogency, effectiveness, persuasiveness, power, validity, weight. **5** army, corps, detachment, division, squad, squadron, regiment, unit. ● *v.* **1** coerce, compel, constrain, dragoon, drive, impel, make, oblige, press-gang, pressurize, railroad. **2** break open, jemmy, prise open, wrench open. **3** drive, propel, push, thrust. **6** foist, impose, inflict.

forceful *adj.* **1** powerful and vigorous. **2** cogent. □ **forcefully** *adv.*, **forcefulness** *n.*

■ **1** dynamic, energetic, mighty, potent, powerful, strong, vigorous. **2** cogent, compelling, convincing, effective, persuasive.

forceps *n.* (*pl.* **forceps**) pincers used in surgery etc.

forcible *adj.* done by force. □ **forcibly** *adv.*

ford *n.* shallow place where a stream may be crossed. ● *v.* cross thus.

fore *adj.* & *adv.* in, at, or towards the front. ● *n.* fore part. □ **to the fore** in front, conspicuous.

forearm[1] *n.* arm from the elbow downwards.

forearm[2] *v.* arm or prepare in advance against possible danger.

forebears *n.pl.* ancestors.

foreboding *n.* feeling that trouble is coming.

■ intimation, premonition, presentiment; anxiety, apprehension, dread, fear, misgiving.

forecast *v.* (**forecast**) tell in advance (what is likely to happen). ● *n.* statement that does this. □ **forecaster** *n.*

■ *v.* foresee, foretell, predict, presage, prognosticate, prophesy. ● *n.* prediction, prognosis, prognostication, prophecy.

forecastle /fóks'l/ *n.* forward part of certain ships.

foreclose *v.* repossess property when a loan is not duly repaid. □ **foreclosure** *n.*

forecourt *n.* enclosed space in front of a building.

forefathers *n.pl.* ancestors.

forefinger *n.* finger next to the thumb.

forefoot *n.* animal's front foot.

forefront *n.* the very front.

foregoing *adj.* preceding.

foregone *adj.* **foregone conclusion** predictable result.

foreground *n.* part of a scene etc. that is nearest to the observer.

forehand *n.* stroke played with the palm of the hand turned forwards. ● *adj.* of or made with this stroke. □ **forehanded** *adj.*

forehead *n.* part of the face above the eyes.

foreign *adj.* **1** of, from, or dealing with a country that is not one's own. **2** not belonging naturally.

■ **1** alien, exotic, imported, overseas; international.

foreigner *n.* person born in or coming from another country.

foreknowledge *n.* knowledge of a thing before it occurs.

foreleg *n.* animal's front leg.

forelock *n.* lock of hair just above the forehead.

foreman *n.* **1** worker superintending others. **2** president and spokesperson of a jury.

foremost *adj.* **1** most important. **2** most advanced in position or rank. ● *adv.* in the foremost position.

■ *adj.* chief, first, leading, main, paramount, pre-eminent, primary, prime, principal, supreme, top.

forename *n.* first name.

forensic *adj.* of or used in law courts. □ **forensic medicine** medical knowledge used in police investigations etc.

foreplay *n.* stimulation preceding sexual intercourse.

forerunner *n.* person or thing that comes in advance of another which it foreshadows.

■ ancestor, antecedent, herald, precursor, predecessor.

foresee *v.* (**foresaw**, **foreseen**) be aware of or realize beforehand. □ **foreseeable** *adj.*

■ anticipate, envisage, forecast, foretell, predict, prophesy.

foreshadow *v.* be an advance sign of (a future event etc.).

foreshore *n.* shore that the tide flows over.

foreshorten *v.* show or portray with apparent shortening giving an effect of distance.

foresight *n.* ability to foresee and prepare for future needs.

foreskin *n.* loose skin at the end of the penis.

forest *n.* trees and undergrowth covering a large area.

forestall *v.* prevent or foil by taking action first.

forestry *n.* science of planting and caring for forests.

foretaste *n.* experience in advance of what is to come.

foretell *v.* (**foretold**) forecast.

forethought *n.* careful thought and planning for the future.

forewarn *v.* warn beforehand.

foreword *n.* introductory remarks in a book.

forfeit *n.* thing that has to be paid or given up as a penalty. ● *v.* give or lose as a penalty. ● *adj.* forfeited. □ **forfeiture** *n.*

■ *n.* fee, fine, penalty. ● *v.* concede, give up, lose, relinquish, surrender.

forgather *v.* assemble.

forgave *see* **forgive.**

forge[1] *v.* advance by effort.

forge[2] *n.* **1** blacksmith's workshop. **2** furnace where metal is heated. ● *v.* **1** shape (metal) by heating and hammering. **2** make a fraudulent copy of. □ **forger** *n.*

■ *v.* **1** construct, fashion, hammer out, make, manufacture, shape. **2** copy, counterfeit, fake, falsify.

forgery *n.* **1** forging. **2** thing forged.

■ **2** copy, counterfeit, fake, imitation, *colloq.* phoney, sham.

forget *v.* (**forgot, forgotten**) cease to remember or think about. □ **forget-me-not** *n.* plant with small blue flowers.

forgetful *adj.* apt to forget. □ **forgetfully** *adv.*, **forgetfulness** *n.*

forgive *v.* (**forgave, forgiven**) cease to feel angy or resentful towards, pardon. □ **forgivable** *adj.*, **forgiveness** *n.*

■ absolve, acquit, clear, exonerate, let off, reprieve; condone, disregard, ignore, overlook, pass over; excuse, pardon, make allowances for.

forgo *v.* (**forwent, forgone**) give up, go without.

■ abandon, abstain from, deny oneself, do without, eschew, give up, go without, *colloq.* pass up, refrain from, sacrifice.

fork *n.* **1** pronged instrument or tool. **2** thing or part divided like this. **3** each of its divisions. ● *v.* **1** lift or dig with a fork. **2** separate into two branches. **3** follow one of these branches. □ **fork-lift truck** truck with a forked device for lifting and carrying loads.

forlorn *adj.* left alone and unhappy. □ **forlorn hope** the only faint hope left. **forlornly** *adv.*

■ abandoned, alone, bereft, dejected, deserted, desolate, forsaken, friendless, lonely, melancholy, miserable, sad, solitary, sorrowful, unhappy, woebegone, woeful, wretched.

form *n.* **1** shape, arrangement of parts. **2** visible aspect. **3** kind or variety. **4** customary or correct behaviour. **5** condition of health and training. **6** way in which a thing exists. **7** school class. **8** document with blank spaces for details. **9** bench. ● *v.* **1** shape, produce. **2** bring into existence. **3** constitute. **4** take shape, be formed. **5** develop.

■ *n.* **1** appearance, arrangement, composition, configuration, construction, shape, structure. **2** body, build, figure, physique, shape. **3** brand, category, class, genre, kind, make, sort, type, variety. **4** code, convention, custom, etiquette, manners, practice, procedure, protocol, routine, rule, tradition, way. **5** condition, health, shape, state. ● *v.* **1** construct, fabricate, fashion, forge, make, manufacture, model, mould, produce, shape, turn. **2** conceive, concoct, contrive, create, design, devise, dream up, *colloq.* think up. **3** constitute, make up. **4** appear, arise, develop, emerge, grow, materialize, take shape. **5** acquire, contract, develop.

formal *adj.* **1** in accordance with rules, convention, or ceremony. **2** regular in design. □ **formally** *adv.*

■ **1** conventional, customary, correct, established, official, prescribed, proper, set, standard; ceremonial, ceremonious, dignified, solemn, stately.

formaldehyde *n.* colourless gas used in solution as a preservative and disinfectant.

formality *n.* **1** being formal. **2** formal act, esp. one required by rules.

formalize *v.* make formal or official. □ **formalization** *n.*

format *n.* **1** shape and size of a book etc. **2** style of arrangement. ● *v.* (**formatted**) arrange in a format.

formation *n.* **1** forming. **2** thing formed. **3** particular arrangement.

formative *adj.* **1** forming. **2** of formation.

former *adj.* **1** of an earlier period. **2** mentioned first of two.

■ **1** bygone, old, past; earlier, erstwhile, ex-, previous, prior, recent, sometime.

formerly *adv.* in former times.

formidable *adj.* **1** inspiring fear, awe, or respect. **2** difficult to do. □ **formidably** *adv.*

■ **1** alarming, dreadful, fearsome, frightening, frightful, intimidating, terrible, terrifying; awesome, imposing, impressive, redoubtable. **2** arduous, challenging, daunting, difficult, onerous, tough.

formula *n.* (*pl.* **-ae** or **-as**) **1** symbols showing chemical constituents or a mathematical statement. **2** fixed series of words for use on social or ceremonial occasions. **3** list of ingredients. **4** classification of a racing car. □ **formulaic** *adj.*

formulate *v.* express systematically. □ **formulation** *n.*

fornicate *v.* have extramarital sexual intercourse. □ **fornication** *n.*, **fornicator** *n.*

forsake *v.* (**forsook**, **forsaken**) withdraw one's help or companionship etc. from.

■ abandon, desert, *sl.* ditch, drop, jilt, leave, leave in the lurch, reject.

fort *n.* fortified place or building.

forth *adv.* **1** out. **2** onwards. □ **back and forth** to and fro.

forthcoming *adj.* **1** about to occur or appear. **2** communicative.

forthright *adj.* frank, outspoken.

■ blunt, candid, direct, frank, honest, open, outspoken, plain, straightforward, unequivocal.

forthwith *adv.* immediately.

fortification *n.* **1** fortifying. **2** defensive wall or building etc.

fortify *v.* **1** provide with fortifications. **2** strengthen.

■ **1** defend, guard, protect, safeguard, strengthen. **2** buoy up, cheer, embolden, encourage, hearten, inspire, invigorate, strengthen, sustain.

fortitude *n.* courage in bearing pain or trouble.

■ bravery, courage, determination, endurance, *colloq.* grit, *colloq.* guts, nerve, pluck, resilience, resoluteness, stoicism, strength, valour, will-power.

fortnight *n.* period of two weeks.

fortnightly *adj.* & *adv.* (happening or appearing) once a fortnight.

fortress *n.* fortified building or town.

fortuitous *adj.* happening by chance. □ **fortuitously** *adv.*

fortunate *adj.* lucky, auspicious. □ **fortunately** *adv.*

■ blessed, lucky; advantageous, auspicious, favourable, fortuitous, happy, opportune, promising, propitious, providential, timely.

fortune *n.* **1** chance as a power in humankind's affairs. **2** destiny. **3** prosperity, wealth. **4** (*colloq.*) large sum of money. □ **fortune-teller** *n.* person claiming to foretell one's destiny.

■ **1** accident, chance, destiny, fate, luck. **2** destiny, fate, kismet, lot. **3** affluence, prosperity, riches, wealth.

forty *adj.* & *n.* four times ten. □ **fortieth** *adj.* & *n.*

forum *n.* place or meeting where a public discussion is held.

forward *adj.* **1** directed towards the front or its line of motion. **2** having made more than normal progress. **3** presumptuous. ● *n.* attacking player in football or hockey. ● *adv.* **1** forwards. **2** towards the future. **3** in advance, ahead. ● *v.* **1** send on (a letter, goods) to a final destination. **2** help to advance, promote.

■ *adj.* **1** advancing, onward, progressing. **2** advanced, precocious, well-developed. **3** bold, brash, brazen, cheeky, familiar, impertinent, impudent, insolent, pert, presumptuous, saucy. ● *adv.* **1** ahead, forwards, on, onward, onwards, towards the front. ● *v.* **1** dispatch, mail, post, send on. **2** accelerate, advance, aid, assist, benefit, expedite, further, help the progress of, promote, speed up.

forwards *adv.* **1** towards the front. **2** with forward motion. **3** so as to make progress. **4** with the front foremost.

fossil *n.* hardened remains or traces of a prehistoric animal or plant. ● *adj.* **1** of or like a fossil. **2** (of fuel) extracted from the ground.

fossilize *v.* turn or be turned into a fossil. □ **fossilization** *n.*

foster *v.* **1** promote the growth of. **2** rear (a child that is not one's own). □ **foster-child** *n.* child reared thus. **foster home** home in which a foster-child is reared. **foster-parent** *n.* person who fosters a child.

■ **1** advance, aid, assist, encourage, forward, further, help, promote, stimulate. **2** bring up, look after, raise, rear, take care of.

fought *see* **fight**.

foul *adj.* **1** causing disgust. **2** unfair, against the rules of a game. ● *adv.* unfairly. ● *n.* action that breaks rules. ● *v.* **1** make or become foul. **2** (cause to) become entangled or blocked. **3** commit a foul against. ▫ **foul-mouthed** *adj.* using foul language. **foully** *adv.*, **foulness** *n.*

■ *adj.* **1** disgusting, *colloq.* horrible, horrid, loathsome, nasty, obnoxious, odious, offensive, repellent, repulsive, revolting, unpleasant, vile; fetid, putrid, rancid, rank, rotten, stinking; contaminated, defiled, dirty, filthy, grimy, polluted, soiled, sordid, squalid, unclean. **2** dirty, dishonest, underhand, unfair, unscrupulous, unsporting, unsportsmanlike.

found[1] *see* **find**.

found[2] *v.* **1** establish (an institution etc.). **2** base.

■ **1** create, establish, inaugurate, initiate, institute, launch, organize, originate, set up, start. **2** base, build, construct, ground.

found[3] *v.* **1** melt or mould (metal or glass). **2** make (an object) in this way.

foundation *n.* **1** base, first layer. **2** (*pl.*) basic principles. **3** act or instance of founding. **4** institution or fund founded.

■ **1** base, bottom, substructure. **2** (**foundations**) basics, elements, essentials, fundamentals, groundwork, principles, rudiments. **3** creation, establishment, formation, founding, inauguration, initiation, institution, origination, setting up. **4** establishment, institute, institution.

founder *v.* **1** stumble or fall. **2** (of a ship) sink. **3** fail completely. ● *n.* person who has founded an institution etc.

■ **1** collapse, fall, lurch, stumble, topple, trip. **2** go down *or* under, sink. **3** come to grief, go wrong, fail, fall through, *sl.* flop, miscarry, misfire.

foundling *n.* deserted child of unknown parents.

foundry *n.* workshop where metal or glass founding is done.

fount *n.* **1** fountain, source. **2** one size and style of printing type.

fountain *n.* **1** spring or jet of water. **2** structure provided for this. **3** source. ▫ **fountainhead** *n.* source. **fountain pen** pen that can be filled with ink.

four *adj.* & *n.* one more than three. ▫ **four-poster** *n.* bed with four posts that support a canopy. **four-wheel drive** motive power acting on all four wheels of a vehicle.

fourfold *adj.* & *adv.* four times as much or as many.

foursome *n.* group of four people.

fourteen *adj.* & *n.* one more than thirteen. ▫ **fourteenth** *adj.* & *n.*

fourth *adj.* next after the third. ● *n.* **1** fourth thing, class, etc. **2** quarter. ▫ **fourthly** *adv.*

fowl *n.* kind of bird kept to supply eggs and flesh for food.

fox *n.* **1** wild animal of the dog family with a bushy tail. **2** its fur. **3** crafty person. ● *v.* deceive or puzzle by acting craftily.

foxglove *n.* tall plant with purple or white flowers.

foxhole *n.* small trench as a military shelter.

foxtrot *n.* **1** dance with slow and quick steps. **2** music for this.

foyer /fóyay/ *n.* entrance hall of a theatre, cinema, or hotel.

fracas /frákkaa/ *n.* (*pl.* **fracas**) noisy quarrel or disturbance.

fraction *n.* **1** number that is not a whole number. **2** small part or amount. ▫ **fractional** *adj.*, **fractionally** *adv.*

fractious *adj.* irritable, peevish.

fracture *n.* break, esp. of bone. ● *v.* break.

fragile *adj.* **1** easily broken. **2** not strong. ▫ **fragility** *n.*

■ **1** breakable, brittle, delicate, flimsy, frail, insubstantial, rickety. **2** delicate, feeble, frail, weak.

fragment *n.* /frágmənt/ **1** piece broken off something. **2** isolated part. ● *v.* /fragmént/ break into fragments. ▫ **fragmentation** *n.*

■ *n.* bit, chip, crumb, part, particle, piece, scrap, shard, shred, sliver, snippet, splinter.

fragmentary *adj.* consisting of fragments.

fragrant *adj.* having a pleasant smell. ▫ **fragrance** *n.*

■ aromatic, balmy, perfumed, redolent, scented, sweet-smelling.

frail *adj.* **1** fragile. **2** physically weak. ▫ **frailty** *n.*

■ **1** breakable, brittle, delicate, flimsy, fragile, insubstantial. **2** ailing, delicate, feeble, ill, infirm, sickly, unwell, weak.

frame *n.* **1** rigid structure supporting other parts. **2** case or border enclosing a picture or pane of glass etc. **3** single exposure on cine film. ● *v.* **1** put or form a frame round. **2** construct. **3** (*sl.*) arrange false evidence against. □ **frame of mind** temporary state of mind.

■ *n.* **1** framework, shell, skeleton, structure; bodywork, chassis. **2** border, edge, edging, frieze, mount, surround. ● *v.* **1** box in, encase, enclose, surround. **2** assemble, build, construct, fabricate, fashion, make, put up.

framework *n.* supporting frame.

franc *n.* unit of money in France, Belgium, and Switzerland.

franchise *n.* **1** right to vote in public elections. **2** authorization to sell a company's goods or services in a certain area. ● *v.* grant a franchise to.

Franco- *pref.* French.

frank[1] *adj.* showing one's thoughts and feelings openly. □ **frankly** *adv.*, **frankness** *n.*

■ artless, blunt, candid, direct, downright, forthright, honest, open, outspoken, plain, sincere, straightforward, truthful.

frank[2] *v.* mark (a letter etc.) to show that postage has been paid.

frankfurter *n.* smoked sausage.

frankincense *n.* sweet-smelling gum burnt as incense.

frantic *adj.* wildly excited or agitated. □ **frantically** *adv.*

■ delirious, excited, feverish, frenetic, frenzied, hysterical, wild, worked up; agitated, anxious, at one's wits' end, beside oneself, desperate, distraught, *colloq.* fraught, overwrought, panicky, panic-stricken.

fraternal *adj.* of a brother or brothers. □ **fraternally** *adv.*

fraternity *n.* brotherhood.

fraternize *v.* associate with others in a friendly way. □ **fraternization** *n.*

fratricide *n.* killing or killer of own brother or sister. □ **fratricidal** *adj.*

fraud *n.* **1** criminal deception. **2** dishonest trick. **3** person carrying this out.

■ **1** cheating, chicanery, deceit, deception, dishonesty, double-dealing, duplicity, sharp practice, swindling, trickery. **2** hoax, *colloq.* rip-off, ruse, swindle, trick. **3** charlatan, cheat, deceiver, fake, impostor, swindler.

fraudulent *adj.* of, involving, or guilty of fraud. □ **fraudulence** *n.*, **fraudulently** *adv.*

■ bogus, counterfeit, fake, false, falsified, forged, *colloq.* phoney, spurious; cheating, deceitful, deceptive, dishonest, double-dealing, duplicitous.

fraught *adj.* (*colloq.*) causing or suffering anxiety. □ **fraught with** filled with, involving.

fray[1] *n.* fight, conflict.

fray[2] *v.* **1** make or become worn so that there are loose threads. **2** strain (nerves or temper).

frazzle *n.* exhausted state.

freak *n.* abnormal person or thing. ● *v.* **freak out** (*colloq.*) (cause to) hallucinate or become wildly excited. □ **freakish** *adj.*, **freaky** *adj.*

freckle *n.* light brown spot on the skin. ● *v.* spot or become spotted with freckles. □ **freckled** *adj.*

free *adj.* (**freer, freest**) **1** not in the power of another, not a slave. **2** unrestricted, not confined. **3** without charge. **4** not occupied, not in use. **5** lavish. ● *v.* (**freed**) **1** make free. **2** disentangle. □ **free fall** unrestricted fall under the force of gravity. **free from** without, not subject to. **free hand** right of taking what action one chooses. **freehand** *adj.* (of drawing) done without ruler or compasses etc. **free house** pub not controlled by one brewery. **freelance** *adj.* & *n.* (person) selling services to various employers. **free-range** *adj.* **1** (of hens) allowed to range freely in search of food. **2** (of eggs) from such hens. **freewheel** *v.* ride a bicycle without pedalling.

■ *adj.* **1** autonomous, democratic, independent, self-governing, self-ruling, sovereign. **2** at large, at liberty, loose, out, unconfined, unconstrained, unencumbered, unfettered, unimpeded, unrestricted, untrammelled; emancipated, liberated, released. **3** complimentary, free of charge, gratis. **4** available, empty, not in use, unoccupied, vacant. **5** bountiful, generous, lavish, liberal, munificent, open-handed, unstinting. ● *v.* **1** emancipate, let go, let loose, liberate, loose, release, set free; unchain, unfetter, unleash, unloose, untie. **2** disengage, disentangle, extricate, release.

freedom *n.* **1** being free. **2** right or power to do as one pleases. **3** unrestricted use. **4** honorary citizenship.

■ **1** autonomy, independence, liberty, self-determination, self-government, sovereignty; emancipation, liberation. **2** ability, authority, discretion, free hand, latitude, leeway, licence, permission, power, privilege, right, scope.

freehold *n.* holding of land or a house etc. in absolute ownership. □ **freeholder** *n.*

Freemason *n.* member of a fraternity for mutual help, with elaborate secret rituals. □ **Freemasonry** *n.*

freesia *n.* fragrant flower.

freeze *v.* (**froze, frozen**) **1** change from liquid to solid by extreme cold. **2** be so cold that water turns to ice. **3** chill or be chilled by extreme cold or fear. **4** preserve by refrigeration. **5** make (assets) unable to be realized. **6** hold (prices or wages) at a fixed level. **7** stop, stand very still. ● *n.* **1** period of freezing weather. **2** freezing of prices etc. □ **freeze-dry** *v.* freeze and dry by evaporation of ice in a vacuum.

freezer *n.* refrigerated container for preserving and storing food.

freight *n.* **1** cargo. **2** transport of goods. ● *v.* **1** load with freight. **2** transport as freight.

freighter *n.* ship or aircraft carrying mainly freight.

French *adj.* & *n.* (language) of France. □ **French horn** brass wind instrument with a coiled tube. **French-polish** *v.* polish (wood) with shellac polish. **French window** one reaching to the ground, used also as a door. **Frenchman** *n.*, **Frenchwoman** *n.*

frenetic *adj.* in a state of frenzy. □ **frenetically** *adv.*

frenzy *n.* violent excitement or agitation. □ **frenzied** *adj.*

frequency *n.* **1** frequent occurrence. **2** rate of recurrence (of a vibration etc.).

frequent¹ /fréekwənt/ *adj.* happening or appearing often. □ **frequently** *adv.*

■ common, constant, continual, customary, familiar, habitual, persistent, recurrent, regular, repeated, usual.

frequent² /frikwént/ *v.* go frequently to, be often in (a place).

fresco *n.* (*pl.* **-os**) picture painted on a wall or ceiling before the plaster is dry.

fresh *adj.* **1** new, not stale or faded. **2** different. **3** additional. **4** refreshing, invigorating. **5** pure. **6** not preserved by tinning or freezing etc. **7** not salty. □ **freshly** *adv.*, **freshness** *n.*

■ **1** brand new, new, newly-made. **2** alternative, different, innovative, new, novel, original, unconventional, unfamiliar, unusual, up to date. **3** additional, extra, further, new, supplementary. **4** bracing, breezy, cool, crisp, exhilarating, invigorating, refreshing. **5** clean, pure, uncontaminated, unpolluted.

freshen *v.* make or become fresh.

fresher *n.* first-year university student.

fret¹ *v.* (**fretted**) be worried or distressed. □ **fretful** *adj.*

fret² *n.* each of the ridges on the finger-board of a guitar etc.

fretsaw *n.* very narrow saw used for fretwork.

fretwork *n.* woodwork cut in decorative patterns.

friable *adj.* easily crumbled. □ **friability** *n.*

friar *n.* member of certain religious orders of men.

friary *n.* monastery of friars.

fricassee *n.* dish of pieces of meat served in a thick sauce. ● *v.* make a fricassee of.

friction *n.* **1** rubbing of one thing against another. **2** resistance so encountered. **3** conflict of people who disagree. □ **frictional** *adj.*

■ **1** abrasion, chafing, grating, rubbing, scraping. **3** animosity, antagonism, bad feeling, bickering, conflict, discord, dispute, dissension, dissent, hostility, ill will, quarrelling, strife, wrangling.

fridge *n.* (*colloq.*) refrigerator.

fried *see* **fry¹**.

friend *n.* **1** person (other than a relative or lover) with whom one is on terms of mutual affection. **2** helper, sympathizer. □ **friendship** *n.*

■ **1** *colloq.* buddy, *colloq.* chum, companion, confidant(e), crony, intimate, mate, *colloq.* pal. **2** backer, benefactor, helper, patron, supporter, sympathizer. □ **friendship** closeness, fellowship, harmony, intimacy, rapport.

friendly *adj.* (**-ier, -iest**) **1** kind and pleasant. **2** on good terms. □ **friendliness** *n.*

■ **1** affable, affectionate, agreeable, amiable, amicable, approachable, benevolent, benign, civil, companionable, congenial, convivial, cordial, easygoing, genial, good-natured, gracious, helpful, hospitable, kind, kind-hearted, kindly, likeable, neighbourly, outgoing, pleasant, sociable, sympathetic, warm, welcoming, well-disposed. **2** *colloq.* chummy, close, intimate, on good terms, *colloq.* pally, *colloq.* thick.

frieze *n.* band of decoration round the top of a wall.

frigate *n.* small fast naval ship.

fright *n.* **1** sudden great fear. **2** instance of this. **3** ridiculous-looking person or thing.

■ **1** alarm, dread, fear, horror, panic, terror, trepidation. **2** scare, shock, start.

frighten *v.* **1** cause fright to. **2** drive or compel by fright.

■ **1** alarm, daunt, horrify, intimidate, petrify, *colloq.* put the wind up, scare, shock, startle, terrify.

frightful *adj.* **1** causing horror. **2** ugly. **3** (*colloq.*) extremely great or bad. □ **frightfully** *adv.*

■ **1** appalling, awful, dreadful, fearsome, ghastly, grisly, gruesome, hideous, horrendous, horrible, horrid, horrific, horrifying, loathsome, shocking, terrible, vile.

frigid *adj.* **1** intensely cold. **2** very cold in manner. **3** unresponsive sexually. □ **frigidity** *n.*

frill *n.* **1** gathered or pleated strip of trimming attached at one edge. **2** unnecessary extra. □ **frilled** *adj.*, **frilly** *adj.*

fringe *n.* **1** ornamental edging of hanging threads or cords. **2** front hair cut short to hang over the forehead. **3** edge of an area or group etc. ● *v.* edge. □ **fringe benefit** one provided by an employer in addition to wages.

frippery *n.* showy finery.

frisk *v.* **1** leap or skip playfully. **2** (*sl.*) search (a person) for concealed weapons etc. ● *n.* playful leap or skip.

frisky *adj.* (**-ier**, **-iest**) lively, playful. □ **friskily** *adv.*

fritter[1] *n.* fried batter-coated slice of fruit or meat etc.

fritter[2] *v.* waste little by little on trivial things.

frivolous *adj.* **1** trivial. **2** not serious, silly. □ **frivolously** *adv.*, **frivolity** *n.*

■ **1** *sl.* footling, inconsequential, insignificant, minor, petty, trifling, trivial, unimportant. **2** flighty, *colloq.* flip, flippant, foolish, giddy, shallow, silly, superficial.

frizz *v.* curl into a wiry mass. □ **frizzy** *adj.*, **frizziness** *n.*

frizzle *v.* fry crisp.

fro *see* **to and fro.**

frock *n.* woman's or girl's dress. □ **frock-coat** *n.* man's long-skirted coat.

frog *n.* small amphibian with long web-footed hind legs. □ **frog in one's throat** hoarseness.

frogman *n.* swimmer with a rubber suit and oxygen supply for use under water.

frogmarch *v.* hustle (a person) forcibly, holding the arms.

frolic *v.* (**frolicked**) play about in a lively way. ● *n.* such play.

■ *v.* caper, cavort, frisk, gambol, leap, play, prance, romp, skip.

from *prep.* **1** having as the starting point, source, or cause. **2** as separated, distinguished, or unlike. □ **from time to time** at intervals of time.

frond *n.* leaf-like part of a fern or palm tree etc.

front *n.* **1** side or part normally nearer or towards the spectator or line of motion. **2** battle line. **3** outward appearance. **4** cover for secret activities. **5** promenade of a seaside resort. **6** boundary between warm and cold air-masses. ● *adj.* of or at the front. ● *v.* **1** face, have the front towards. **2** (*sl.*) serve as a cover for secret activities. □ **front runner** leading contestant. **in front** at the front.

■ *n.* **1** façade, face, fore, forefront, frontage. **2** van, vanguard. **3** appearance, aspect, façade, face, look, show. ● *adj.* first, foremost, leading. ● *v.* **1** face, look out on, lie opposite, overlook.

frontage *n.* **1** front of a building. **2** land bordering this.

frontal *adj.* of or on the front.

frontier *n.* boundary between countries.

frontispiece *n.* illustration opposite the title-page of a book.

frost *n.* **1** freezing weather condition. **2** white frozen dew or vapour. ● *v.* **1** injure with frost. **2** cover (as) with frost. **3** make (glass) opaque by roughening its surface. □ **frostbite** *n.* injury to body tissue from freezing. **frostbitten** *adj.*

frosting *n.* sugar icing.

frosty *adj.* (**-ier**, **-iest**) **1** covered with frost. **2** unfriendly.

froth *n.* & *v.* foam. □ **frothy** *adj.*

frown *v.* wrinkle one's brow in thought or disapproval. ● *n.* frowning movement or look. □ **frown on** disapprove of.

■ *v.* & *n.* glare, glower, grimace, scowl.

frowzy *adj.* (**-ier**, **-iest**) **1** fusty. **2** dingy. □ **frowziness** *n.*

froze, **frozen** *see* **freeze.**

frugal *adj.* **1** careful and economical. **2** scanty, costing little. □ **frugally** *adv.*, **frugality** *n.*

■ **1** abstemious, careful, economical, provident, prudent, sparing, thrifty. **2** meagre, paltry, poor, scanty, skimpy, small.

fruit *n.* **1** seed-containing part of a plant. **2** this used as food. **3** (usu. *pl.*) product of labour. ● *v.* produce or allow to produce fruit. □ **fruit machine** coin-operated gambling machine.

fruiterer *n.* shopkeeper selling fruit.

fruitful *adj.* **1** producing much fruit. **2** successful. □ **fruitfully** *adv.*

■ **1** fecund, fertile, productive, rich. **2** beneficial, productive, profitable, rewarding, successful, useful, worthwhile.

fruition /froo-ish'n/ *n.* realization of aims or hopes.

fruitless *adj.* producing little or no result. □ **fruitlessly** *adv.*

■ abortive, futile, ineffective, pointless, profitless, unprofitable, unrewarding, unsuccessful, useless, vain.

fruity *adj.* (**-ier**, **-iest**) like fruit in smell or taste. □ **fruitiness** *n.*

frump *n.* dowdy woman. □ **frumpish** *adj.*, **frumpy** *adj.*

frustrate *v.* prevent from achieving something or from being achieved. □ **frustration** *n.*

■ baulk, block, check, defeat, foil, hamper, hinder, impede, prevent, *sl.* scupper, stop, stymie, thwart.

frustrated *adj.* discontented, not satisfied.

fry[1] *v.* (**fried**) cook or be cooked in very hot fat.

fry[2] *n.* (*pl.* **fry**) young fish. □ **small fry** people of little importance.

ft *abbr.* foot or feet (as a measure).

fuchsia /fyóoshə/ *n.* ornamental shrub with drooping flowers.

fuddle *v.* stupefy, esp. with drink.

fuddy-duddy *adj.* & *n.* (*sl.*) (person who is) old-fashioned.

fudge *n.* soft sweet made of milk, sugar, and butter. ● *v.* make or do clumsily or dishonestly.

fuel *n.* **1** material burnt as a source of energy. **2** thing that increases anger etc. ● *v.* (**fuelled**) supply with fuel.

fug *n.* stuffy atmosphere in a room etc. □ **fuggy** *adj.*

fugitive *n.* person who is fleeing or escaping. ● *adj.* **1** fleeing. **2** transient.

■ *n.* deserter, escapee, refugee, runaway. ● *adj.* **1** escaped, fleeing, runaway. **2** brief, ephemeral, fleeting, momentary, passing, short, transient, transitory.

fulcrum *n.* (*pl.* **-cra**) point of support on which a lever pivots.

fulfil *v.* (**fulfilled**) **1** accomplish, carry out (a task). **2** satisfy, do what is required by (a contract etc.). □ **fulfil oneself** realize one's potential. **fulfilment** *n.*

■ **1** accomplish, achieve, carry out, complete, discharge, do, effect, execute, perform, realize. **2** answer, comply with, meet, satisfy.

full *adj.* **1** holding or having as much as is possible. **2** copious. **3** complete. **4** plump. **5** (of tone) deep and mellow. **6** (of clothes) made with plenty of material. ● *adv.* **1** completely. **2** exactly. □ **full-blooded** *adj.* vigorous, hearty. **full-blown** *adj.* fully developed. **full moon** moon with the whole disc illuminated. **full-scale** *adj.* of actual size, not reduced. **full stop 1** dot used as a punctuation mark at the end of a sentence or abbreviation. **2** complete stop. **fullness** *n.*

■ *adj.* **1** brimming, chock-a-block, crammed, crowded, filled, jam-packed, packed, stuffed; replete, sated, satiated. **2** abundant, ample, copious, extensive, plentiful. **3** complete, comprehensive, exhaustive, detailed, thorough, unabridged. **4** buxom, plump, rounded, shapely, voluptuous. **5** deep, mellow, resonant, rich, sonorous.

fully *adv.* completely. □ **fully-fledged** *adj.* mature.

fulminate *v.* protest loudly and bitterly. □ **fulmination** *n.*

fulsome *adj.* **1** excessive. **2** insincere.

fumble *v.* **1** touch or handle (a thing) awkwardly. **2** grope about.

fume *n.* pungent smoke or vapour. ● *v.* **1** emit fumes. **2** seethe with anger. **3** subject to fumes.

fumigate *v.* disinfect by fumes. □ **fumigation** *n.*, **fumigator** *n.*

fun *n.* **1** light-hearted amusement. **2** source of this. □ **funfair** *n.* group of stalls, amusements, sideshows, etc. **make fun of** tease, ridicule.

■ **1** amusement, enjoyment, gaiety, glee, jollity, joy, merriment, mirth, pleasure. □ **make fun of** deride, jeer at, laugh at, mock, poke fun at, *sl.* rag, *colloq.* rib, ridicule, send up, taunt, tease.

function *n.* **1** special activity or purpose of a person or thing. **2** important ceremony. **3** (in mathematics) quantity whose value depends on varying values of others. ● *v.* **1** fulfil a function. **2** be in action.

■ *n.* **1** activity, business, capacity, duty, job, occupation, office, place, position, responsibility, role, task; purpose, use. **2** ceremony, *colloq.* do, occasion, party, reception. ● *v.* **1** act, serve. **2** be in working order, go, operate, run, work.

functional *adj.* **1** of or serving a function. **2** practical, not decorative. **3** able to function. ◻ **functionally** *adv.*

■ **2** practical, serviceable, useful, utilitarian. **3** functioning, going, operating, operational, running.

functionary *n.* official.

fund *n.* **1** sum of money for a special purpose. **2** stock, supply. **3** (*pl.*) money resources. ● *v.* provide with money.

■ *n.* **1** collection, kitty; endowment, nest egg. **2** cache, hoard, mine, pool, reserve, stock, store, supply. **3** (**funds**) assets, capital, cash, investments, means, money, resources, savings, wealth, *colloq.* wherewithal. ● *v.* back, finance, pay for, subsidize, support.

fundamental *adj.* **1** basic. **2** essential. ● *n.* fundamental fact or principle. ◻ **fundamentally** *adv.*

■ *adj.* basic, central, chief, crucial, essential, important, inherent, instrinsic, main, necessary, primary, prime, principal, quintessential, underlying, vital.

fundamentalism *n.* strict adherence to traditional religious beliefs. ◻ **fundamentalist** *n.*

funeral *n.* ceremonial burial or cremation of the dead.

funerary *adj.* of or used for a burial or funeral.

funereal *adj.* **1** suitable for a funeral. **2** dismal, dark.

fungicide *n.* substance that kills fungus. ◻ **fungicidal** *adj.*

fungus *n.* (*pl.* **-gi**) plant without green colouring matter (e.g. mushroom, mould). ◻ **fungal** *adj.*, **fungous** *adj.*

funnel *n.* **1** tube with a wide top for pouring liquid into small openings. **2** chimney on a steam engine or ship. ● *v.* (**funnelled**) move through a narrowing space.

funny *adj.* (**-ier, -iest**) **1** causing amusement. **2** puzzling, odd. ◻ **funny bone** part of the elbow where a very sensitive nerve passes. **funnily** *adv.*

■ **1** amusing, comic, comical, diverting, droll, entertaining, hilarious, humorous, witty. **2** bizarre, curious, mysterious, mystifying, odd, peculiar, perplexing, puzzling, queer, surprising, weird.

fur *n.* **1** short fine hair of certain animals. **2** skin with this used for clothing. **3** coating, incrustation. ● *v.* (**furred**) cover or become covered with fur.

furbish *v.* **1** clean up. **2** renovate.

furious *adj.* **1** full of anger. **2** violent, intense. ◻ **furiously** *adv.*

■ **1** angry, beside oneself, cross, enraged, fuming, incensed, infuriated, irate, *colloq.* livid, *colloq.* mad, seething, wild. **2** fierce, intense, savage, violent, wild.

furl *v.* roll up and fasten.

furlong *n.* one-eighth of a mile.

furnace *n.* enclosed fireplace for intense heating or smelting.

furnish *v.* **1** equip with furniture. **2** provide, supply.

furnishings *n.pl.* furniture and fitments etc.

furniture *n.* movable articles (e.g. chairs, beds) for use in a room.

furore /fyoorórі/ *n.* uproar of enthusiastic admiration or fury.

furrier *n.* person who deals in furs or fur clothes.

furrow *n.* **1** trench made by a plough. **2** groove. ● *v.* make furrows in.

furry *adj.* (**-ier, -iest**) **1** like fur. **2** covered with fur. ◻ **furriness** *n.*

further *adv.* & *adj.* **1** more distant. **2** to a greater extent. **3** additional(ly). ● *v.* help the progress of. ◻ **further education** that provided for people above school age. **furtherance** *n.*

■ *adv.* **3** additionally, also, besides, furthermore, in addition, moreover, too. ● *adj.* **3** additional, auxiliary, extra, fresh, more, new, supplementary. ● *v.* advance, aid, assist, expedite, facilitate, forward, foster, help, promote.

furthermore *adv.* moreover.

furthest *adj.* most distant. ● *adv.* at or to the greatest distance.

furtive *adj.* sly, stealthy. ◻ **furtively** *adv.*, **furtiveness** *n.*

■ clandestine, covert, crafty, secret, secretive, sly, stealthy, surreptitious, underhand.

fury *n.* **1** wild anger. **2** violence.

■ **1** anger, ire, rage, wrath. **2** ferocity, fierceness, savagery, violence.

furze *n.* gorse.

fuse[1] *v.* **1** melt with intense heat. **2** blend by melting. **3** fit with a fuse. **4** stop functioning through melting of a fuse. ● *n.* strip of wire placed in an electric circuit to melt and interrupt the current when the circuit is overloaded.

fuse[2] *n.* length of easily burnt material for igniting a bomb or explosive. ● *v.* fit a fuse to.

fuselage *n.* body of an aeroplane.

fusible *adj.* able to be fused. □ **fusibility** *n.*

fusion *n.* **1** fusing. **2** union of atomic nuclei, with release of energy.

fuss *n.* **1** excited commotion, nervous activity. **2** excessive concern about a trivial thing. **3** vigorous protest. ● *v.* **1** make a fuss. **2** agitate.

■ *n.* **1** ado, ballyhoo, bother, bustle, commotion, excitement, flap, flurry, fluster, hubbub, *colloq.* kerfuffle, *colloq.* palaver, stir, to-do. **3** complaint, objection, protest, *colloq.* stink.

fussy *adj.* (**-ier, -iest**) **1** often fussing. **2** fastidious. **3** with much unnecessary detail or decoration. □ **fussily** *adv.*

■ **2** *colloq.* choosy, faddy, fastidious, finicky, particular, *colloq.* pernickety. **3** elaborate, fancy, ornate, over-decorated.

fusty *adj.* (**-ier, -iest**) smelling stale and stuffy. □ **fustiness** *n.*

futile *adj.* producing no result. □ **futilely** *adv.*, **futility** *n.*

■ abortive, fruitless, profitless, unprofitable, unsuccessful, useless, vain.

futon /fo͞oton/ *n.* light orig. Japanese kind of mattress.

future *adj.* belonging to the time after the present. ● *n.* **1** future time, events, or condition. **2** prospect of success etc. □ **in future** from now on.

futuristic *adj.* suitable for the future, not traditional.

fuzz *n.* fluff, fluffy or frizzy thing.

fuzzy *adj.* (**-ier, -iest**) **1** like or covered with fuzz. **2** blurred, indistinct. □ **fuzzily** *adv.*

Gg

g *abbr.* gram(s).
gabardine *n.* strong twilled fabric.
gabble *v.* talk quickly and indistinctly. ● *n.* gabbled talk.
gable *n.* triangular part of a wall, between sloping roofs. ◻ **gabled** *adj.*
gad *v.* (**gadded**) **gad about** go about idly in search of pleasure. ◻ **gadabout** *n.* idle pleasure-seeker.
gadfly *n.* fly that bites cattle.
gadget *n.* small mechanical device or tool. ◻ **gadgetry** *n.*

■ apparatus, appliance, contraption, device, implement, invention, machine, tool, utensil.

Gaelic /gáylik/ *n.* Celtic language of Scots or Irish.
gaff *n.* stick with a hook for landing large fish. ● *v.* seize with a gaff.
gaffe *n.* blunder.
gaffer *n.* **1** elderly man. **2** (*colloq.*) boss, foreman.
gag *n.* **1** thing put in or over a person's mouth to silence them. **2** device to hold the mouth open. **3** joke. ● *v.* (**gagged**) **1** put a gag on. **2** deprive of freedom of speech. **3** retch.
gaggle *n.* **1** flock (of geese). **2** disorderly group.
gaiety *n.* **1** cheerfulness. **2** merrymaking. **3** bright appearance.

■ **1** cheerfulness, glee, happiness, high spirits, jollity, joy, joyfulness, lightheartedness, merriment, mirth. **2** celebration, festivity, fun, jollification, merrymaking, revelry.

gaily *adv.* with gaiety.
gain *v.* **1** obtain, secure. **2** acquire more of something. **3** earn. **4** reach. **5** (of a clock) become fast. ● *n.* increase, profit. ◻ **gain on** get nearer to in pursuit.

■ *v.* **1** achieve, acquire, attain, come by, get, obtain, secure, pick up, procure, reap, win. **2** gather, pick up; put on. **3** clear, earn, get, make. **4** arrive at, come to, get to, reach.

gainful *adj.* profitable. ◻ **gainfully** *adv.*

■ advantageous, lucrative, profitable, productive, remunerative.

gainsay *v.* (**gainsaid**) deny, contradict.
gait *n.* manner of walking or running.
gaiter *n.* cloth or leather covering for the lower part of the leg.
gala /gaálə/ *n.* **1** festive occasion. **2** fête.
galaxy *n.* system of stars, esp. (**the Galaxy**) the one of which the solar system is a part. ◻ **galactic** *adj.*
gale *n.* **1** very strong wind. **2** noisy outburst.
gall[1] *n.* **1** impudence. **2** bitterness of feeling. **3** bile. ◻ **gall bladder** organ storing bile.

■ **1** cheek, effrontery, impertinence, impudence, insolence, temerity. **2** acrimony, asperity, bitterness, rancour, resentment.

gall[2] *n.* sore made by rubbing. ● *v.* **1** rub and make sore. **2** vex.

■ *v.* **1** chafe, rub, scrape, scratch. **2** *colloq.* aggravate, annoy, *sl.* bug, exasperate, irk, irritate, nettle, *colloq.* rile, vex.

gall[3] *n.* abnormal growth on a plant, esp. on an oak tree.
gallant *adj.* **1** brave. **2** chivalrous, attentive to women. ◻ **gallantly** *adv.*, **gallantry** *n.*

■ **1** bold, brave, courageous, daring, dauntless, fearless, heroic, intrepid, plucky, unafraid, valorous. **2** attentive, chivalrous, considerate, courteous, courtly, gentlemanly, gracious, polite.

galleon *n.* large Spanish sailing ship in the 15th–17th centuries.
gallery *n.* **1** balcony in a hall or theatre etc. **2** long room or passage, esp. used for special purpose. **3** room or building for showing works of art.
galley *n.* (*pl.* **-eys**) **1** ancient ship, esp. propelled by oars. **2** kitchen in a ship or aircraft. **3** (also **galley proof**) printer's proof in a long narrow form.
Gallic *adj.* **1** of ancient Gaul. **2** French.
gallon *n.* measure for liquids, = 4 quarts (4.546 litres).
gallop *n.* **1** horse's fastest pace. **2** ride at this. ● *v.* (**galloped**) **1** go or ride at a gallop. **2** progress rapidly.
gallows *n.* framework with a noose for hanging criminals.
gallstone *n.* small hard mass formed in the gall bladder.
galore *adv.* in plenty.

galosh *n.* rubber overshoe.

galvanize *v.* **1** stimulate into activity. **2** coat with zinc.

gambit *n.* opening move.

gamble *v.* **1** play games of chance for money. **2** bet (a sum of money). **3** risk in hope of gain. ● *n.* **1** gambling. **2** risky undertaking. □ **gambler** *n.*

■ *v.* **1** bet, game, *colloq.* have a flutter. **2** bet, hazard, risk, stake, venture, wager.

gambol *v.* (**gambolled**) jump about in play. ● *n.* gambolling movement.

game[1] *n.* **1** play or sport, esp. with rules. **2** section of this as a scoring unit. **3** scheme. **4** wild animals hunted for sport or food. **5** their flesh as food. ● *v.* gamble for money stakes. ● *adj.* **1** brave. **2** willing. □ **gamely** *adv.*

■ *n.* **1** recreation, sport; competition, contest. **2** bout, heat, match, round. **3** design, plan, plot, ploy, scheme, stratagem, strategy. ● *adj.* **1** adventurous, bold, brave, courageous, daring, *colloq.* gutsy, plucky, spirited. **2** eager, enthusiastic, prepared, ready, willing.

game[2] *adj.* lame.

gamekeeper *n.* person employed to protect and breed game.

gamesmanship *n.* art of winning games by psychological means.

gamete *n.* cell able to unite with another in sexual reproduction.

gamine *n.* girl with mischievous charm.

gamma *n.* third letter of the Greek alphabet, = g.

gammon *n.* cured or smoked ham.

gammy *adj.* (*sl.*) = **game**[2].

gamut *n.* whole range or scope.

gamy *adj.* smelling or tasting like high game. □ **gaminess** *n.*

gander *n.* male goose.

gang *n.* group of people working or going about together. ● *v.* **gang up** combine in a gang.

■ *n.* band, company, crew, crowd, group, party, team, troop; circle, clique, coterie, set.

gangling *adj.* tall and awkward.

ganglion *n.* (*pl.* **-ia**) **1** group of nerve cells. **2** cyst on a tendon.

gangplank *n.* plank placed for walking into or out of a boat.

gangrene *n.* decay of body tissue. □ **gangrenous** *adj.*

gangster *n.* member of a gang of violent criminals.

gangway *n.* **1** passage, esp. between rows of seats. **2** passageway on a ship. **3** bridge from a ship to land.

gannet *n.* large seabird.

gantry *n.* overhead bridge-like framework supporting railway signals or a travelling crane etc.

gaol *n.* = **jail**.

gap *n.* **1** opening, space, interval. **2** wide difference. **3** deficiency.

■ **1** aperture, breach, cavity, chink, crack, cranny, crevice, hole, opening, rift, space; break, hiatus, interlude, intermission, interruption, interval, lull, pause, recess, respite. **2** difference, discrepancy, disparity, divergence, inconsistency.

gape *v.* **1** open the mouth wide. **2** stare in surprise. **3** be wide open. ● *n.* **1** yawn. **2** stare.

garage *n.* **1** building for storing motor vehicle(s). **2** commercial establishment for refuelling or repairing motor vehicles. ● *v.* put or keep in a garage.

garb *n.* clothing. ● *v.* clothe.

garbage *n.* domestic rubbish.

garble *v.* distort or confuse (a message or story etc.).

■ confuse, distort, jumble, mix up, scramble, twist.

garden *n.* **1** piece of cultivated ground, esp. attached to a house. **2** (*pl.*) ornamental public grounds. ● *v.* tend a garden. □ **gardener** *n.*

gardenia *n.* tree or shrub with fragrant flowers.

gargantuan *adj.* gigantic.

gargle *v.* wash the inside of the throat with liquid held there by the breath. ● *n.* liquid for this.

gargoyle *n.* grotesque carved face or figure on a building.

garish *adj.* gaudy. □ **garishly** *adv.*, **garishness** *n.*

garland *n.* wreath of flowers etc. as a decoration. ● *v.* adorn with garland(s).

garlic *n.* onion-like plant. □ **garlicky** *adj.*

garment *n.* article of clothing.

garner *v.* store up, collect.

garnet *n.* red semiprecious stone.

garnish *v.* decorate (esp. food). ● *n.* thing used for garnishing.

garret *n.* attic, esp. a poor one.

garrison *n.* **1** troops stationed in a town or fort. **2** building they occupy. ● *v.* occupy thus.

garrotte *n.* wire or metal collar used to strangle a victim. ● *v.* strangle with this.

garrulous *adj.* talkative. □ **garrulously** *adv.*, **garrulousness** *n.*

garter *n.* band worn round the leg to keep a stocking up.

gas *n.* (*pl.* **gases**) **1** substance with particles that can move freely. **2** such a substance used as a fuel or anaesthetic. **3** (*colloq.*) empty talk. **4** (*US*) petrol. ● *v.* (**gassed**) **1** kill or overcome by poisonous gas. **2** (*colloq.*) talk lengthily. □ **gas chamber** room filled with poisonous gas to kill people. **gas mask** device worn over face as a protection against poisonous gas. **gassy** *adj.*

gaseous *adj.* of or like a gas.

gash *n.* long deep cut. ● *v.* make a gash in.

gasify *v.* convert or be converted into gas. □ **gasification** *n.*

gasket *n.* piece of rubber etc. sealing a joint between metal surfaces.

gasoline *n.* (*US*) petrol.

gasp *v.* **1** breathe quickly and noisily. **2** breathe in sharply as in surprise or pain. ● *n.* breath drawn in thus.

■ *v.* **1** blow, heave, pant, puff, wheeze.

gastric *adj.* of the stomach.

gastropod *n.* mollusc, such as a snail, that moves by means of a ventral organ.

gate *n.* **1** hinged movable barrier in a wall or fence etc. **2** gateway. **3** number of spectators paying to attend a sporting event, amount of money taken.

gateau /gáttō/ *n.* (*pl.* **-eaux**) large rich cream cake.

gatecrash *v.* go to (a party) uninvited. □ **gatecrasher** *n.*

gateway *n.* **1** opening or structure framing a gate. **2** entrance.

gather *v.* **1** bring or come together. **2** collect. **3** understand, conclude. **4** develop a higher degree of. **5** draw together in folds.

■ **1** assemble, cluster, collect, come together, congregate, convene, flock, forgather, group, mass, meet, throng; bring together, marshal, rally, round up. **2** accumulate, amass, collect, garner, heap (up), pile (up), stockpile. **3** assume, conclude, deduce, infer, surmise, understand. **4** gain, increase, pick up.

gathering *n.* people assembled.

■ assembly, conference, congregation, convention, *colloq.* get-together, meeting, rally.

gauche /gōsh/ *adj.* socially awkward. □ **gaucherie** *n.*

gaucho /gówchō/ *n.* (*pl.* **-os**) S. American cowboy.

gaudy *adj.* (**-ier, -iest**) showy or bright in a tasteless way. □ **gaudily** *adv.*, **gaudiness** *n.*

■ garish, *colloq.* flash, flashy, loud, ostentatious, showy, tasteless, tawdry, vulgar.

gauge /gayj/ *n.* **1** standard measure to which thing must conform. **2** device for measuring things. **3** distance between pairs of rails or wheels. ● *v.* **1** measure **2** estimate.

■ *n.* **1** benchmark, measure, rule, pattern, yardstick. ● *v.* **1** calculate, compute, determine, measure. **2** assess, estimate, evaluate, judge, rate.

gaunt *adj.* lean and haggard. □ **gauntness** *n.*

■ bony, cadaverous, emaciated, haggard, lean, scraggy, scrawny, skeletal, thin.

gauntlet[1] *n.* **1** glove with a long wide cuff. **2** this cuff.

gauntlet[2] *n.* **run the gauntlet** be exposed to criticism or risk.

gauze *n.* **1** thin transparent fabric. **2** fine wire mesh. □ **gauzy** *adj.*

gave *see* **give**.

gavel *n.* mallet used by an auctioneer or chairman etc. to call for attention or order.

gawky *adj.* (**-ier, -iest**) awkward and ungainly. □ **gawkiness** *n.*

gawp *v.* (*colloq.*) stare stupidly.

gay *adj.* **1** happy and full of fun. **2** brightly coloured. **3** homosexual. ● *n.* homosexual person. □ **gayness** *n.*

■ *adj.* **1** bubbly, buoyant, carefree, cheerful, cheery, chirpy, ebullient, exuberant, gleeful, happy, high-spirited, jolly, jovial, joyful, light-hearted, lively, merry, vivacious. **2** bright, colourful, vivid.

gaze *v.* look long and steadily. ● *n.* long steady look.

■ *v.* gape, *colloq.* gawp, goggle, look fixedly, stare.

gazebo /gəzeébō/ *n.* (*pl.* **-os**) summer house with a wide view.

gazelle *n.* small antelope.

gazette *n.* **1** newspaper. **2** official journal containing public notices.

gazetteer *n.* index of places, rivers, mountains, etc.

gazump *v.* disappoint (an intended house buyer) by raising the price agreed.

GB *abbr.* Great Britain.

gear *n.* **1** equipment. **2** set of toothed wheels working together in machinery.

● *v.* **1** provide with gear(s). **2** adapt (to a purpose). ▫ **in gear** with gear mechanism engaged.

■ *n.* **1** accoutrements, apparatus, *sl.* clobber, equipment, kit, materials, paraphernalia, stuff, tackle, things, tools, trappings.

gearbox *n.* case enclosing gear mechanism.

gecko *n.* (*pl.* **-os**) tropical lizard.

geese *see* **goose.**

Geiger counter /gígər/ device for measuring radioactivity.

geisha /gáyshə/ *n.* Japanese woman trained to entertain men.

gel *n.* jelly-like substance.

gelatine *n.* clear substance made by boiling bones. ▫ **gelatinous** *adj.*

geld *v.* castrate, spay.

gelding *n.* gelded horse.

gelignite *n.* explosive containing nitroglycerine.

gem *n.* **1** precious stone. **2** thing of great beauty or excellence.

gender *n.* **1** one's sex. **2** grammatical classification corresponding roughly to sex.

gene *n.* one of the factors controlling heredity.

genealogy *n.* **1** list of ancestors. **2** study of pedigrees. ▫ **genealogical** *adj.*, **genealogist** *n.*

genera *see* **genus.**

general *adj.* **1** including or affecting all or most parts, things, or people. **2** prevalent, usual. **3** not detailed or specific. **4** (in titles) chief. ● *n.* army officer next below field marshal. ▫ **general election** election of parliamentary representatives from the whole country. **general practitioner** community doctor treating cases of all kinds. **in general 1** usually. **2** for the most part. **generally** *adv.*

■ *adj.* **1** comprehensive, extensive, global, universal, worldwide; communal, popular, public, shared. **2** common, familiar, habitual, normal, ordinary, prevailing, prevalent, regular, usual, widespread. **3** approximate, generalized, imprecise, indefinite, inexact, loose, rough, vague.

generality *n.* **1** being general. **2** general statement.

generalize *v.* **1** draw a general conclusion. **2** speak in general terms. ▫ **generalization** *n.*

generate *v.* bring into existence, produce.

■ breed, bring about, cause, create, engender, give rise to, produce.

generation *n.* **1** generating. **2** single stage in descent or pedigree. **3** all people born at about the same time. **4** period of about 30 years.

generator *n.* machine converting mechanical energy into electricity.

generic *adj.* of a whole genus or group. ▫ **generically** *adv.*

generous *adj.* **1** giving or given freely. **2** magnanimous. **3** plentiful. ▫ **generously** *adv.*, **generosity** *n.*

■ **1** bountiful, charitable, free, lavish, liberal, munificent, open-handed, unstinting. **2** benevolent, forgiving, humane, magnanimous, noble, philanthropic, public-spirited, selfless, unselfish. **3** abundant, ample, bountiful, copious, lavish, overflowing, plentiful; big, considerable, large, sizeable, substantial. ▫ **generosity** bounty, largesse, liberality, munificence.

genesis *n.* origin.

genetic *adj.* of genes or genetics. ▫ **genetically** *adv.*

genetics *n.* science of heredity.

genial *adj.* kindly and cheerful. ▫ **genially** *adv.*, **geniality** *n.*

■ agreeable, affable, amiable, cheerful, congenial, convivial, cordial, easygoing, friendly, good-humoured, good-natured, hospitable, kindly, likeable, nice, pleasant, sociable, sympathetic.

genie *n.* (*pl.* **genii**) magical spirit or goblin.

genital *adj.* of animal reproduction or sex organs. ● *n.pl.* external sex organs.

genius *n.* (*pl.* **-uses**) **1** exceptionally great natural ability. **2** person having this.

■ **1** ability, aptitude, capability, flair, gift, knack, talent; brains, brilliance, intellect, intelligence. **2** adept, expert, maestro, mastermind, prodigy, virtuoso, wizard, *colloq.* whiz-kid.

genocide *n.* deliberate extermination of a race of people.

genre /zhóɴrə/ *n.* kind, esp. of art or literature.

genteel *adj.* affectedly polite and refined. ▫ **genteelly** *adv.*

gentian /jénsh'n/ *n.* alpine plant with usu. deep-blue flowers.

gentile *n.* non-Jewish person.

gentility *n.* good manners and elegance.

gentle *adj.* **1** mild or kind. **2** moderate, not severe. ● *v.* coax. □ **gently** *adv.*, **gentleness** *n.*

■ *adj.* **1** benign, humane, kind, kindly, lenient, mellow, merciful, mild, peaceful, placid, serene, sweet-tempered, tender, tranquil. **2** light, mild, moderate, soft.

gentleman *n.* **1** man, esp. of good social position. **2** well-mannered man. □ **gentlemanly** *adj.*

gentrify *v.* alter (an area) to conform to middle-class tastes. □ **gentrification** *n.*

gentry *n.pl.* **1** people next below nobility. **2** (*derog.*) people.

genuflect *v.* bend the knee and lower the body, esp. in worship. □ **genuflection** *n.*

genuine *adj.* really what it is said to be. □ **genuinely** *adv.*, **genuineness** *n.*

■ authentic, bona fide, legitimate, real, true, veritable.

genus *n.* (*pl.* **genera**) **1** group of animals or plants, usu. containing several species. **2** kind.

geocentric *adj.* **1** having the earth as a centre. **2** as viewed from the earth's centre.

geode *n.* **1** cavity lined with crystals. **2** rock containing this.

geography *n.* **1** study of earth's physical features, climate, etc. **2** features of a place. □ **geographical** *adj.*, **geographically** *adv.*, **geographer** *n.*

geology *n.* **1** study of earth's crust. **2** features of earth's crust. □ **geological** *adj.*, **geologist** *n.*

geometry *n.* branch of mathematics dealing with lines, angles, surfaces, and solids. □ **geometric(al)** *adj.*, **geometrician** *n.*

Georgian *adj.* of the time of the Georges, kings of England, esp. 1714–1830.

geranium *n.* garden plant with red, pink, or white flowers.

gerbil *n.* rodent with long hind legs.

geriatrics *n.* branch of medicine dealing with the diseases and care of old people. □ **geriatric** *adj.* & *n.*

germ *n.* **1** micro-organism, esp. one capable of causing disease. **2** portion of an organism capable of developing into a new organism. **3** basis from which a thing may develop.

■ **1** bacterium, *sl.* bug, microbe, micro-organism, virus. **3** basis, beginning, origin, root, seed, source, start.

German *adj.* & *n.* (native, language) of Germany. □ **German measles** disease with symptoms like mild measles. **German shepherd dog** dog of a large strong smooth-haired breed.

germane *adj.* relevant.

germinate *v.* begin or cause to grow. □ **germination** *n.*

gerontology *n.* study of ageing.

gerrymander *v.* arrange boundaries of (a constituency etc.) so as to gain unfair electoral advantage.

gerund *n.* English verbal noun ending in *-ing*.

Gestapo *n.* Nazi secret police.

gestation *n.* **1** carrying in the womb between conception and birth. **2** period of this.

gesticulate *v.* make expressive movements with the hands and arms. □ **gesticulation** *n.*

gesture *n.* expressive movement or action. ● *v.* make a gesture.

■ *n.* gesticulation, motion, movement, sign, signal. ● *v.* gesticulate, indicate, motion, signal, wave.

get *v.* (**got, getting**) **1** come into possession of. **2** earn. **3** win. **4** fetch. **5** capture. **6** contract (an illness). **7** bring or come into a certain state. **8** persuade. **9** (*colloq.*) understand. **10** (*colloq.*) annoy. **11** prepare (a meal). □ **get across** manage to communicate. **get along** live harmoniously, be on good terms. **get away** escape. **get at 1** reach. **2** (*colloq.*) imply. **3** (*colloq.*) nag. **get by** manage to survive. **get off** be acquitted. **get on 1** manage. **2** make progress. **3** be on harmonious terms. **4** advance in age. **get out of** evade. **get-out** *n.* means of evading something. **get over 1** recover from. **2** surmount. **3** communicate. **get round 1** successfully coax or cajole. **2** evade (a law or rule). **get through 1** pass (an exam etc.). **2** use up (resources). **3** make contact by telephone. **get-together** *n.* (*colloq.*) social gathering. **get up 1** rise, esp. from bed. **2** prepare, organize. **get-up** *n.* outfit.

■ **1** acquire, be given, come by, obtain, pick up, procure, receive, secure; buy, purchase. **2** be paid, clear, earn, gross, make, net. **3** achieve, attain, find, gain, win. **4** collect, fetch, go for, retrieve. **5** apprehend, arrest, capture, grab, seize, take. **6** be afflicted with, catch, come down with, contract, develop, fall ill with, have, suffer from. **7** become, come to be, grow, turn. **8** cajole, coax, convince, induce, persuade, prevail on, talk into, wheedle into. **9** appreciate, comprehend, follow, grasp, see, understand. □ **get across** communicate, convey, get over, impart, make clear, put over. **get by** cope, exist, make do, make ends meet, manage, pull through, scrape by, struggle along, survive. **get on 1** cope, fare, manage. **2** advance, be successful, do well, make progress, progress, succeed. **3** be friendly, be on good terms, get along, hit it off. **get out of** avoid, dodge, escape, evade, shirk, sidestep.

getaway *n.* escape after a crime.

gewgaw *n.* gaudy ornament etc.

geyser /geézər/ *n.* **1** spring that spouts hot water or steam. **2** a kind of water heater.

ghastly *adj.* (**-ier**, **-iest**) **1** causing horror. **2** pale and ill-looking. **3** (*colloq.*) very bad.

■ **1** appalling, awful, dreadful, grim, grisly, gruesome, frightful, hideous, horrendous, horrible, horrid, horrifying, shocking, terrible, terrifying. **2** ashen, livid, pale, pallid, wan.

gherkin *n.* small cucumber used for pickling.

ghetto *n.* (*pl.* **-os**) slum area occupied by a particular group. □ **ghetto-blaster** *n.* large portable cassette player.

ghost *n.* dead person's spirit. ● *v.* write as a ghost writer. □ **ghost writer** person who writes a book etc. for another to pass off as his or her own. **ghostly** *adj.*, **ghostliness** *n.*

■ *n.* apparition, phantom, spectre, spirit, *colloq.* spook, wraith.

ghoul /gōōl/ *n.* **1** person who enjoys gruesome things. **2** (in Muslim stories) spirit preying on corpses. □ **ghoulish** *adj.*

giant *n.* **1** (in fairy tales) a being of superhuman size. **2** abnormally large person, animal, or thing. **3** person of outstanding ability. ● *adj.* very large.

gibber *v.* make meaningless sounds, esp. in shock or terror.

gibberish *n.* unintelligible talk, nonsense.

gibbet *n.* gallows.

gibbon *n.* long-armed ape.

gibe /jīb/ *n.* & *v.* jeer.

giblets *n.pl.* edible organs from a bird.

giddy *adj.* (**-ier**, **-iest**) **1** having or causing the feeling that everything is spinning round. **2** excitable, flighty. □ **giddily** *adv.*, **giddiness** *n.*

■ **1** dizzy, faint, light-headed, unsteady. **2** capricious, excitable, impulsive, volatile; flighty, foolish, frivolous, irresponsible, scatterbrained, silly.

gift *n.* **1** thing given or received without payment. **2** natural ability. **3** easy task. ● *v.* bestow.

■ *n.* **1** benefaction, bequest, donation, handout, offering, present. **2** ability, aptitude, capability, capacity, flair, genius, instinct, knack, skill, talent.

gifted *adj.* talented.

■ able, accomplished, bright, brilliant, capable, clever, expert, skilful, skilled, talented.

gig[1] *n.* light two-wheeled horse-drawn carriage.

gig[2] (*colloq.*) *n.* engagement to play music. ● *v.* (**gigged**) perform a gig.

giga- *pref.* multiplied by 10^9 (as in *gigametre*).

gigantic *adj.* very large.

■ colossal, enormous, gargantuan, giant, huge, immense, mammoth, massive, monumental, titanic, vast.

giggle *v.* give small bursts of half-suppressed laughter. ● *n.* this laughter.

■ *v.* & *n.* cackle, chortle, chuckle, laugh, snicker, snigger, titter.

gigolo /zhíggəlō/ *n.* (*pl.* **-os**) man paid by a woman to be her escort or lover.

gild *v.* (**gilded**) cover with a thin layer of gold or gold paint.

gill[1] /gil/ *n.* **1** (usu. *pl.*) respiratory organ of a fish etc. **2** each of the vertical plates on the underside of a mushroom cap.

gill[2] /jil/ *n.* one-quarter of a pint.

gilt *adj.* gilded, gold-coloured. ● *n.* **1** substance used in gilding. **2** gilt-edged investment. □ **gilt-edged** *adj.* (of an investment etc.) very safe.

gimbals *n.pl.* contrivance of rings to keep instruments horizontal in a moving ship etc.

gimcrack /jímcrak/ *adj.* cheap and flimsy.

gimlet *n.* small tool with a screw-like tip for boring holes.

gimmick *n.* trick or device to attract attention. □ **gimmicky** *adj.*

gin *n.* alcoholic spirit flavoured with juniper berries.

ginger *n.* **1** hot-tasting root of a tropical plant. **2** this plant. **3** reddish-yellow. ● *adj.* ginger-coloured. □ **ginger ale, beer**

ginger-flavoured fizzy drinks. **ginger group** group urging a more active policy. **gingery** *adj.*

gingerbread *n.* ginger-flavoured cake or biscuit.

gingerly *adj.* & *adv.* cautious(ly).

gingham *n.* cotton fabric with a checked or striped pattern.

ginseng *n.* plant with a fragrant root used in medicine.

gipsy *n.* = **gypsy**.

giraffe *n.* long-necked African animal.

girder *n.* metal beam supporting part of a building or bridge.

girdle *n.* **1** cord worn round the waist. **2** elastic corset. **3** connected ring of bones in the body. ● *v.* surround.

girl *n.* **1** female child. **2** young woman. **3** female assistant or employee. **4** man's girlfriend. □ **girlfriend** *n.* female friend, esp. man's usual companion. **girlhood** *n.*, **girlish** *adj.*

giro /jírō/ *n.* (*pl.* **-os**) **1** banking system by which payment can be made by transferring credit from one account to another. **2** cheque or payment made by this.

girth *n.* **1** distance round something. **2** band under a horse's belly, holding a saddle in place.

gist /jist/ *n.* essential points or general sense of a speech etc.

■ core, heart, essence, nub, substance; drift, import, meaning, purport, sense.

give *v.* (**gave, given**) **1** cause to receive or have. **2** provide with. **3** make or perform (an action etc.). **4** yield as a product or result. **5** yield to pressure, collapse. ● *n.* springiness, elasticity. □ **give and take** willingness to make reciprocal concessions. **give away 1** give as a gift. **2** reveal (a secret etc.) unintentionally. **give-away** *n.* (*colloq.*) unintentional disclosure. **give in** acknowledge defeat. **give off** emit. **give out 1** announce. **2** distribute. **3** become exhausted or used up. **give over 1** devote. **2** (*colloq.*) cease. **give up 1** cease. **2** abandon hope or an attempt. **3** part with. **give way 1** yield. **2** collapse. **3** allow other traffic to go first. **giver** *n.*

■ *v.* **1** award, bestow, confer, contribute, donate, grant, hand over, make over, present; bequeath, leave, will; commit, consign, entrust; communicate, convey, deliver, impart, pass on, send. **2** furnish with, provide with, supply with. **3** emit, utter. **4** afford, make, produce, yield. **5** break, buckle, collapse, come *or* fall apart, give way. □ **give away 2** disclose, divulge, leak, reveal; betray, grass on, inform on. **give in** admit defeat, capitulate, concede, give up, submit, surrender, throw in the towel, yield. **give out 1** announce, broadcast, declare, make known, make public. **2** allocate, allot, deal out, *colloq.* dish out, distribute, dole out, hand out. **give up 1** abandon, cease, desist from, *US* quit, stop. **2** admit defeat, capitulate, give in, submit, surrender. **give way 1** back down, capitulate, give in, submit, surrender, yield. **2** buckle, cave in, collapse, crumple, subside.

given *see* **give**. *adj.* **1** specified. **2** having a tendency.

gizzard *n.* bird's second stomach, in which food is ground.

glacé /glássay/ *adj.* (of fruit) preserved in sugar.

glacial *adj.* **1** icy. **2** of or from glaciers. □ **glacially** *adv.*

glaciated *adj.* covered with or affected by a glacier. □ **glaciation** *n.*

glacier *n.* mass or river of ice moving very slowly.

glad *adj.* **1** pleased. **2** cheerful. **3** giving joy. □ **gladly** *adv.*, **gladness** *n.*

■ **1** delighted, elated, gratified, happy, overjoyed, pleased, thrilled; eager, keen, ready, willing. **2** cheerful, cheery, gay, joyful, joyous, merry. **3** cheering, gratifying, pleasing, welcome.

gladden *v.* make glad.

glade *n.* open space in a forest.

gladiator *n.* man trained to fight at public shows in ancient Rome. □ **gladiatorial** *adj.*

gladiolus *n.* (*pl.* **-li**) garden plant with spikes of flowers.

glamour *n.* alluring or exciting beauty or charm. □ **glamorize** *v.*, **glamorous** *adj.*

■ allure, attraction, beauty, charisma, charm, fascination, excitement, romance. □ **glamorous** alluring, attractive, beautiful, fascinating, romantic.

glance *v.* **1** look briefly. **2** strike and bounce off. ● *n.* brief look.

■ *v.* **1** peek, peep. **2** bounce, rebound, ricochet.

gland *n.* organ that secretes substances to be used or expelled by the body. □ **glandular** *adj.*

glare *v.* **1** shine with a harsh dazzling light. **2** stare angrily. ● *n.* glaring light or stare.

glaring *adj.* conspicuous.

■ blatant, conspicuous, flagrant, obvious, manifest, obtrusive, patent, prominent.

glass *n.* **1** hard brittle usu. transparent substance. **2** things made of this. **3** mirror. **4** glass drinking vessel. **5** barometer. **6** (*pl.*) spectacles, binoculars. ● *v.* fit or cover with glass.

glasshouse *n.* **1** greenhouse. **2** (*sl.*) military prison.

glassy *adj.* **1** like glass. **2** (of eyes etc.) dull, expressionless.

glaze *v.* **1** fit or cover with glass. **2** coat with a glossy surface. **3** become glassy. ● *n.* shiny surface or coating.

■ *n.* coat, coating; gloss, lustre, patina, polish, sheen, shine.

glazier *n.* person whose trade is to fit glass in windows etc.

gleam *n.* **1** faint or brief light. **2** brief show of a quality. ● *v.* shine faintly, send out gleams.

■ *v.* flicker, glimmer, glint, glisten, glitter, glow, shimmer, shine, sparkle, twinkle.

glean *v.* **1** pick up (grain left by harvesters). **2** gather scraps of. □ **gleaner** *n.*, **gleanings** *n.pl.*

glee *n.* lively or triumphant joy. □ **gleeful** *adj.*, **gleefully** *adv.*

■ delight, elation, exultation, gaiety, joy, joyfulness, jubilation, merriment, mirth. □ **gleeful** delighted, elated, exultant, joyful, jubilant, merry, mirthful, triumphant.

glen *n.* narrow valley.

glib *adj.* speaking or spoken fluently but insincerely.

■ *colloq.* flip, fluent, insincere, slick, smooth.

glide *v.* **1** move smoothly. **2** fly in a glider or aircraft without engine power. ● *n.* gliding movement.

■ *v.* **1** coast, float, flow, sail, skim, slide, slip, stream.

glider *n.* aeroplane with no engine.

glimmer *n.* faint gleam. ● *v.* gleam faintly.

glimpse *n.* brief view. ● *v.* have a brief view of.

glint *n.* very brief flash of light. ● *v.* send out a glint.

glisten *v.* shine like something wet.

■ gleam, glint, glitter, shimmer, shine, sparkle, twinkle.

glitter *v.* & *n.* sparkle.

gloaming *n.* evening twilight.

gloat *v.* be full of greedy or malicious delight.

global *adj.* covering or affecting the whole world. □ **global warming** increase in temperature of the earth's atmosphere. **globally** *adv.*

■ international, universal, world, worldwide.

globe *n.* **1** ball-shaped object, esp. with a map of the earth on it. **2** the world. **3** hollow round glass object. □ **globetrotting** *n.* travelling widely as a tourist.

globular *adj.* shaped like a globe.

globule *n.* small rounded drop.

glockenspiel *n.* musical instrument of tuned steel bars or tubes struck by hammers.

gloom *n.* **1** semi-darkness. **2** melancholy, depression.

■ **1** dark, darkness, dimness, dusk, murk, obscurity, shade, shadow. **2** dejection, depression, despair, despondency, melancholy, misery, sadness, sorrow, unhappiness.

gloomy *adj.* (**-ier**, **-iest**) **1** dark. **2** depressed. **3** depressing.

■ **1** dark, dim, dusky, murky, shadowy, shady. **2** dejected, depressed, disconsolate, dispirited, downcast, downhearted, forlorn, glum, lugubrious, melancholy, miserable, morose, sad, sorrowful, unhappy. **3** bleak, cheerless, depressing, dismal, dreary.

glorify *v.* **1** praise highly. **2** make seem grander than it is. □ **glorification** *n.*

■ **1** acclaim, applaud, commend, eulogize, extol, honour, laud, pay tribute to, praise.

glorious *adj.* **1** having or bringing glory. **2** (*colloq.*) excellent. □ **gloriously** *adv.*

■ **1** celebrated, distinguished, eminent, famed, famous, honoured, illustrious, renowned; admirable, excellent, impressive, magnificent, marvellous, outstanding, spectacular, splendid, superb.

glory *n.* **1** fame, honour, and praise. **2** thing bringing this. **3** magnificence. ● *v.* rejoice. □ **glory in** take pride in.

■ *n.* **1** celebrity, distinction, eminence, fame, honour, kudos, prestige, renown. **3** grandeur, magnificence, majesty, pomp, splendour. □ **glory in** delight in, exult in, rejoice in, relish, revel in, take pride in.

gloss *n.* shine on a smooth surface. ● *v.* make glossy. □ **gloss over** cover up (a mistake etc.).

■ *n.* gleam, lustre, patina, sheen, shine. ● *v.* buff, burnish, polish, shine. □ **gloss over** camouflage, conceal, cover up, disguise, hide, whitewash.

glossary *n.* list of technical or special words, with definitions.

glossy *adj.* (**-ier, -iest**) shiny.

■ gleaming, lustrous, shining, shiny, sleek.

glottis *n.* opening at the upper end of the windpipe between the vocal cords. □ **glottal** *adj.*

glove *n.* covering for the hand, usu. with separate divisions for fingers and thumb.

glow *v.* **1** send out light and heat without flame. **2** have a warm or flushed look, colour, or feeling. ● *n.* glowing state, look, or feeling. □ **glow-worm** *n.* beetle that can give out a greenish light.

■ *v.* **1** gleam, glimmer, shine. **2** blush, colour, flush, go red, redden. ● *n.* brightness, incandescence, radiance; blush, flush, rosiness, ruddiness; ardour, enthusiasm, fervour, passion, warmth.

glower /glowr/ *v.* scowl.

glucose *n.* form of sugar found in fruit juice.

glue *n.* sticky substance used for joining things together. ● *v.* (**gluing**) **1** fasten with glue. **2** attach closely. □ **gluey** *adj.*

glum *adj.* (**glummer, glummest**) sad and gloomy.

■ crestfallen, dejected, depressed, despondent, disconsolate, dispirited, downcast, downhearted, gloomy, low, melancholy, miserable, morose, sad, sorrowful, unhappy, woebegone.

glut *v.* (**glutted**) **1** supply with more than is needed. **2** overload with food. ● *n.* excessive supply.

■ *v.* **1** deluge, flood, inundate, overload, oversupply, saturate, swamp. **2** overfeed, sate, satiate, surfeit. ● *n.* excess, overabundance, oversupply, superfluity, surfeit, surplus.

gluten *n.* sticky protein substance found in cereals.

glutinous *adj.* glue-like, sticky.

glutton *n.* **1** greedy person. **2** one who is eager for something. □ **gluttonous** *adj.*, **gluttony** *n.*

■ **1** gourmand, *colloq.* hog, *colloq.* pig. □ **gluttonous** greedy, *colloq.* gutsy, insatiable, *colloq.* piggy, voracious. **gluttony** greed, greediness, insatiability, voracity.

glycerine *n.* thick sweet liquid used in medicines etc.

gnarled /naarld/ *adj.* knobbly, twisted.

■ contorted, crooked, distorted, knobbly, knotty, misshapen, twisted.

gnash /nash/ *v.* grind (one's teeth).

gnat /nat/ *n.* small biting fly.

gnaw /naw/ *v.* bite persistently.

■ bite, champ, chew, crunch, masticate, munch.

gnome /nōm/ *n.* dwarf, goblin.

gnomic /nṓmik/ *adj.* sententious.

gnu /noō/ *n.* ox-like antelope.

go *v.* (**went, gone**) **1** walk, travel, proceed. **2** depart. **3** extend. **4** function. **5** (of time) pass. **6** belong in a specified place. **7** become. **8** match. **9** be spent or used up. **10** collapse, fail, die. **11** make (a specified movement or sound). **12** be sold. ● *n.* (*pl.* **goes**) **1** energy. **2** turn, try. □ **go-ahead** *n.* signal to proceed. *adj.* enterprising. **go back on** fail to keep (a promise). **go-between** *n.* one who acts as messenger or negotiator. **go down with** become ill with. **go for 1** go to fetch. **2** choose. **3** (*sl.*) attack. **go-getter** *n.* pushily enterprising person. **go-getting** *adj.* aggressively ambitious. **go-kart** *n.* miniature racing car. **go off 1** explode. **2** decay. **go out** be extinguished. **go over** inspect the details of. **go round 1** spin, revolve. **2** be enough for everyone. **go slow** work at a deliberately slow pace as a protest. **go-slow** *n.* this procedure. **go through 1** undergo. **2** use up. **go under 1** succumb. **2** fail. **go up 1** rise in price. **2** explode, burn rapidly. **go with 1** accompany. **2** match, harmonize with. **on the go** (*colloq.*) in constant motion, active.

■ *v.* **1** advance, betake oneself, journey, move, proceed, progress, travel, walk, wend one's way. **2** depart, leave, make off, move away, retire, retreat, withdraw; set off *or* out. **3** extend, lead, stretch. **4** be operational, function, operate, run, work. **5** elapse, pass, slip by *or* away. **7** become, get, grow, turn. **8** be compatible, be suited, complement each other, harmonize, match, suit each other. ● *n.* **1** animation, drive, dynamism, energy, enthusiasm, initiative, pep, verve, vigour, *colloq.* vim, vitality, vivacity, zest. **2** attempt, bid, shot, *colloq.* stab, try; chance, opportunity, turn. □ **go-ahead** *n.* approval, authorization, *colloq.*

green light, *colloq.* OK, permission. *adj.* ambitious, dynamic, enterprising, go-getting, keen, *colloq.* pushy, resourceful. **go back on** break, fail to keep, renege on, repudiate, retract. **go-between** intermediary, mediator, messenger, middleman, negotiator, representative. **go off 1** blow up, burst, erupt, explode. **2** decay, decompose, go bad, go stale, moulder, rot, spoil. **go over** examine, inspect, investigate, look at, read, review, scrutinize, study. **go through 1** bear, endure, experience, live through, put up with, stand, suffer, undergo, withstand. **2** consume, exhaust, spend, use up. **go under 2** collapse, fail, fold, go bankrupt, *colloq.* go bust, go out of business.

goad *n.* **1** pointed stick for driving cattle. **2** stimulus to activity. ● *v.* stimulate by annoying.

■ *n.* **2** impetus, incitement, instigation, motivation, provocation, spur, stimulus.

goal *n.* **1** structure or area into which players try to send the ball in certain games. **2** point scored thus. **3** objective. □ **goalpost** *n.* either of the posts marking the limit of a goal.

■ **3** aim, ambition, aspiration, design, end, intention, object, objective, purpose, target.

goalie *n.* (*colloq.*) goalkeeper.

goalkeeper *n.* player whose task is to keep the ball out of the goal.

goat *n.* small horned animal.

gobble *v.* **1** eat quickly and greedily. **2** make a throaty sound like a turkeycock.

■ **1** bolt, gulp, guzzle, *colloq.* scoff, wolf.

gobbledegook *n.* (*colloq.*) pompous language used by officials.

goblet *n.* drinking glass with a stem and foot.

goblin *n.* mischievous ugly elf.

god *n.* **1** superhuman being worshipped as having power over nature and human affairs. **2** (**God**) creator and ruler of the universe in Christian, Jewish, and Muslim teaching. **3** person or thing that is greatly admired or adored. □ **God-fearing** *adj.* sincerely religious. **godforsaken** *adj.* wretched, dismal.

godchild *n.* child in relation to its godparent(s).

god-daughter *n.* female godchild.

goddess *n.* female god.

godfather *n.* male godparent.

godhead *n.* **1** divine nature. **2** deity.

godmother *n.* female godparent.

godparent *n.* person who promises at a child's baptism to see that it is brought up as a Christian.

godsend *n.* piece of unexpected good fortune.

godson *n.* male godchild.

goggle *v.* stare with wide-open eyes.

goggles *n.pl.* spectacles for protecting the eyes.

goitre /góytər/ *n.* enlarged thyroid gland.

gold *n.* **1** yellow metal of high value. **2** coins or articles made of this. **3** its colour. ● *adj.* made of, coloured, or shining like gold. □ **goldfield** *n.* area where gold is found. **gold rush** rush to a newly discovered goldfield.

golden *adj.* **1** made of gold. **2** like gold in colour. **3** precious, excellent. □ **golden handshake** generous cash payment to a person dismissed or forced to retire. **golden jubilee, wedding** 50th anniversary.

goldfinch *n.* songbird with a band of yellow across each wing.

goldfish *n.* (*pl.* **goldfish**) reddish carp kept in a bowl or pond.

goldsmith *n.* person whose trade is making articles in gold.

golf *n.* game in which a ball is struck with clubs into a series of holes. ● *v.* play golf. □ **golf course** land on which golf is played. **golfer** *n.*

golliwog *n.* black-faced soft doll with fuzzy hair.

gonad *n.* animal organ producing gametes.

gondola *n.* Venetian canal-boat.

gondolier *n.* man who propels a gondola by means of a pole.

gone *see* **go.**

gong *n.* **1** metal plate that resounds when struck. **2** (*sl.*) medal.

gonorrhoea *n.* a venereal disease.

goo *n.* (*colloq.*) sticky wet substance. □ **gooey** *adj.*

good *adj.* (**better, best**) **1** having the right or desirable qualities, satisfactory. **2** proper, expedient. **3** morally correct, kindly. **4** well-behaved. **5** enjoyable. **6** beneficial. **7** efficient. **8** thorough. **9** considerable, full. **10** valid. ● *n.* **1** that which is beneficial or morally good. **2** (*pl.*) movable property, articles of trade, things to be carried by road or rail. □ **as good as** practically, almost. **good-for-nothing** *adj.* & *n.* worthless (person). **Good Friday** Friday before Easter, commemorating the Crucifixion. **good-looking** *adj.* attractive. **good name** good reputation. **good-natured** *adj.* kind, friendly.

■ *adj.* **1** acceptable, adequate, commendable, fine, *colloq.* OK, satisfactory. **2** appropriate, correct, expedient, fitting, proper, right, suitable. **3** ethical, honourable, moral, noble, respectable, righteous, upright, virtuous, worthy; benevolent, charitable, considerate, humane, kind, kind-hearted, kindly, nice. **4** manageable, obedient, well-behaved, well-mannered. **5** agreeable, amusing, enjoyable, entertaining, pleasant, pleasing, pleasurable. **6** beneficial, healthy, salutary, wholesome. **7** accomplished, adept, capable, competent, proficient, skilful, skilled; efficient, reliable, safe, sound, trustworthy. **8** careful, methodical, meticulous, painstaking, scrupulous, systematic, throrough. **9** considerable, large, respectable, sizeable, substantial. **10** authentic, believable, credible, genuine, legitimate, sound, valid. ● *n.* **1** advantage, benefit, gain, profit, use. **2** (**goods**) belongings, effects, possessions, property, things; commodities, stocks, wares; cargo, freight. □ **good-looking** attractive, beautiful, *Sc.* bonny, comely, handsome, personable, pretty, *colloq.* stunning. **good-natured** amiable, charitable, friendly, generous, genial, helpful, kind, kind-hearted, kindly, likeable, nice, pleasant.

goodbye *int.* & *n.* expression used when parting.

goodness *n.* **1** quality of being good. **2** good part of something.

goodwill *n.* **1** friendly feeling. **2** established popularity of a business, treated as a saleable asset.

goody *n.* (*colloq.*) something good or attractive, esp. to eat. □ **goody-goody** *adj.* & *n.* smugly virtuous (person).

goose *n.* (*pl.* **geese**) **1** web-footed bird larger than a duck. **2** female of this. □ **goose-flesh**, **-pimples** *ns.* bristling skin caused by cold or fear. **goose-step** *n.* way of marching without bending the knees.

gooseberry *n.* **1** thorny shrub. **2** its edible (usu. green) berry.

gore[1] *n.* clotted blood from a wound.

gore[2] *v.* pierce with a horn or tusk.

gore[3] *n.* triangular or tapering section of a skirt or sail.

gorge *n.* narrow steep-sided valley. ● *v.* **1** devour greedily. **2** satiate.

■ *n.* canyon, defile, pass, ravine. ● *v.* **1** bolt, devour, gobble, guzzle, wolf. **2** fill, glut, satiate, surfeit.

gorgeous *adj.* **1** richly coloured, magnificent. **2** (*colloq.*) beautiful. □ **gorgeously** *adv.*

■ **1** brilliant, exquisite, magnificent, resplendent, rich, splendid, sumptuous. **2** attractive, beautiful, good-looking, lovely, *colloq.* stunning.

gorgon *n.* terrifying woman.

gorilla *n.* large powerful ape.

gorse *n.* wild evergreen thorny shrub with yellow flowers.

gory *adj.* (**-ier, -iest**) **1** covered with blood. **2** involving bloodshed.

gosling *n.* young goose.

gospel *n.* **1** teachings of Christ. **2** (**Gospel**) any of the first four books of the New Testament. **3** thing regarded as definitely true.

gossamer *n.* **1** fine filmy piece of cobweb. **2** flimsy delicate material.

gossip *n.* **1** casual talk, esp. about other people's affairs. **2** person fond of gossiping. ● *v.* (**gossiped**) engage in gossip.

■ *n.* **1** chat, conversation, small talk, talk; hearsay, rumour, scandal, tittle-tattle. ● *v.* chat, chatter, *colloq.* natter, prattle, talk; spread rumours, tell tales, tittle-tattle.

got *see* **get**.

Gothic *adj.* **1** of an architectural style of the 12th–16th centuries, with pointed arches. **2** (of a novel etc.) in a horrific style popular in the 18th–19th centuries.

gouge *n.* chisel with a concave blade. ● *v.* **1** cut out with a gouge. **2** scoop or force out.

goulash *n.* stew of meat and vegetables, seasoned with paprika.

gourd *n.* **1** fleshy fruit of a climbing plant. **2** container made from its dried rind.

gourmand *n.* glutton.

gourmet /goórmay/ *n.* connoisseur of good food and drink.

gout *n.* disease causing inflammation of the joints. □ **gouty** *adj.*

govern *v.* **1** rule with authority. **2** keep under control. **3** influence, determine. □ **governable** *adj.*, **governor** *n.*

■ **1** be in charge of, command, control, direct, hold sway over, manage, rule, run, preside over, steer. **2** bridle, check, control, curb, repress, restrain, subdue, suppress. 3 affect, determine, influence.

governance *n.* governing, control.

government *n.* **1** governing. **2** group or organization governing a country. □ **governmental** *adj.*

■ **1** administration, command, control, direction, governance, management, rule. **2** administration, leadership, regime.

gown *n.* **1** loose flowing garment. **2** woman's long dress. **3** official robe.

GP *abbr.* general practitioner.

grab *v.* (**grabbed**) **1** grasp suddenly. **2** take greedily. ● *n.* **1** sudden clutch or attempt to seize. **2** mechanical device for gripping things.

■ *v.* **1** clutch, grasp, grip, seize, take hold of. **2** appropriate, *sl.* nab, seize, snap up, snatch.

grace *n.* **1** attractiveness and elegance, esp. of manner or movement. **2** courteous good will. **3** mercy. **4** short prayer of thanks for a meal. ● *v.* **1** add grace to, enhance. **2** confer honour or dignity on.

■ *n.* **1** attractiveness, ease, elegance, finesse, gracefulness, poise, polish, refinement, suavity, tastefulness; agility, suppleness. **2** courtesy, decency, manners, politeness, tact. **3** clemency, compassion, forgiveness, leniency, mercifulness, mercy. ● *v.* **1** adorn, beautify, decorate, embellish, enhance. **2** dignify, distinguish, honour.

graceful *adj.* having or showing grace. □ **gracefully** *adv.*

■ agile, lissom, lithe, nimble, supple; elegant, polished, refined.

graceless *adj.* lacking grace or charm.

gracious *adj.* kind and pleasant towards inferiors. □ **graciously** *adv.*, **graciousness** *n.*

■ affable, agreeable, beneficent, benevolent, charitable, civil, considerate, cordial, courteous, friendly, good-natured, kind, kindly, obliging, pleasant, polite.

gradation *n.* **1** stage in a process of gradual change. **2** this process.

grade *n.* **1** level of rank, quality, or value. **2** mark given to a student for his or her standard of work. ● *v.* **1** arrange in grades. **2** assign a grade to. □ **make the grade** be successful.

■ *n.* **1** class, degree, echelon, level, position, rank, stage, standing, station, status. **2** mark, score. ● *v.* **1** categorize, class, classify, group, order, organize, sort. **2** assess, evaluate, mark, rate.

gradient *n.* slope, amount of this.

gradual *adj.* taking place by degrees. □ **gradually** *adv.*

■ progressive, slow, steady, step by step.

graduate *n.* /grádyooət/ person who holds a university degree. ● *v.* /grádyooayt/ **1** take a university degree. **2** divide into graded sections. **3** mark into regular divisions. □ **graduation** *n.*

graffiti *n.pl.* words or drawings scribbled on a wall etc.

graft *n.* **1** shoot fixed into a cut in a tree to form a new growth. **2** living tissue transplanted surgically. **3** (*sl.*) hard work. ● *v.* **1** put a graft in or on. **2** join inseparably. **3** (*sl.*) work hard.

grain *n.* **1** small hard seed(s) of a food plant such as wheat or rice. **2** these plants. **3** small hard particle. **4** unit of weight (0.065 g). **5** texture or pattern made by fibres or particles. □ **grainy** *adj.*

gram *n.* one-thousandth of a kilogram.

grammar *n.* use of words in their correct forms and relationships.

grammatical *adj.* according to the rules of grammar. □ **grammatically** *adv.*

grampus *n.* dolphin-like sea animal.

gran *n.* (*colloq.*) grandmother.

granary *n.* storehouse for grain.

grand *adj.* **1** splendid, imposing. **2** of chief importance. **3** (*colloq.*) very good. ● *n.* grand piano. □ **grand piano** large full-toned piano with horizontal strings. **grandly** *adv.*, **grandness** *n.*

■ *adj.* **1** dignified, fine, imposing, impressive, lofty, luxurious, magnificent, majestic, monumental, noble, palatial, splendid, stately, sumptuous. **2** chief, head, leading, main, principal.

grandad *n.* (*colloq.*) grandfather.

grandchild *n.* child of one's son or daughter.

granddaughter *n.* female grandchild.

grandeur *n.* **1** splendour. **2** nobility of character.

■ **1** magnificence, majesty, pomp, splendour. **2** dignity, eminence, greatness, nobility.

grandfather *n.* male grandparent. □ **grandfather clock** one in a tall wooden case.

grandiloquent *adj.* using pompous language. □ **grandiloquence** *n.*

grandiose *adj.* **1** imposing. **2** planned on a large scale. □ **grandiosely** *adv.*, **grandiosity** *n.*

> ■ **1** imposing, impressive, magnificent, majestic, splendid. **2** ambitious, extravagant, flamboyant, ostentatious, pretentious.

grandma *n.* (*colloq.*) grandmother.

grandmother *n.* female grandparent.

grandpa *n.* (*colloq.*) grandfather.

grandparent *n.* parent of one's father or mother.

grandson *n.* male grandchild.

grandstand *n.* main stand for spectators at sports ground.

grange *n.* country house with farm buildings that belong to it.

granite *n.* hard grey stone.

granny *n.* (*colloq.*) grandmother. □ **granny flat** self-contained accommodation in one's house for a relative.

grant *v.* **1** consent to fulfil. **2** give or allow as a privilege. **3** admit to be true. ● *n.* money given for a particular purpose, esp. from public funds. □ **take for granted** assume to be true or sure to happen or continue.

> ■ *v.* **1** accede to, agree to, assent to, concede to, consent to. **2** allow, award, give, permit. **3** acknowledge, admit, allow, concede, confess. ● *n.* allocation, allowance, award; subsidy, subvention; bursary, scholarship.

granular *adj.* like grains.

granulate *v.* **1** form into grains. **2** roughen the surface of. □ **granulation** *n.*

granule *n.* small grain.

grape *n.* green or purple berry used for making wine. □ **grapevine** *n.* **1** vine bearing grapes. **2** way news spreads unofficially.

grapefruit *n.* large round yellow citrus fruit.

graph *n.* diagram showing the relationship between quantities.

graphic *adj.* **1** of drawing, painting, or engraving. **2** giving a vivid description. ● *n.pl.* diagrams etc. used in calculation and design. □ **graphically** *adv.*

> ■ *adj.* **2** clear, colourful, detailed, lifelike, picturesque, realistic, striking, vivid.

graphite *n.* a form of carbon.

graphology *n.* study of handwriting. □ **graphologist** *n.*

grapnel *n.* **1** small anchor with several hooks. **2** hooked device for dragging a river bed.

grapple *v.* fight at close quarters, struggle. □ **grapple with** try to manage or overcome. **grappling-iron** *n.* grapnel.

grasp *v.* **1** seize and hold. **2** understand. ● *n.* **1** firm hold or grip. **2** understanding.

> ■ *v.* **1** clasp, clutch, grab, grip, hold, seize, snatch, take hold of. **2** comprehend, fathom, follow, *colloq.* get, *colloq.* get the hang of, see, take in, understand.

grasping *adj.* greedy, avaricious.

> ■ acquisitive, avaricious, greedy, rapacious.

grass *n.* **1** wild plant with green blades eaten by animals. **2** species of this (e.g. a cereal plant). **3** ground covered with grass. ● *v.* cover with grass. □ **grass roots 1** fundamental level or source. **2** rank-and-file members. **grass widow** wife whose husband is absent temporarily. **grassy** *adj.*

grasshopper *n.* jumping insect that makes a chirping noise.

grassland *n.* wide grass-covered area with few trees.

grate[1] *n.* **1** metal frame keeping fuel in a fireplace. **2** hearth.

grate[2] *v.* **1** shred finely by rubbing against a jagged surface. **2** make a harsh noise by rubbing. □ **grate on** have an irritating effect on.

> ■ **1** grind, shred. **2** rasp, scrape, scratch. □ **grate on** annoy, get on one's nerves, irk, irritate, jar on.

grateful *adj.* feeling that one values a kindness or benefit received. □ **gratefully** *adv.*

> ■ appreciative, thankful; beholden, indebted, obliged.

grater *n.* device for grating food.

gratify *v.* **1** please. **2** satisfy (wishes). □ **gratification** *n.*

> ■ **1** delight, gladden, please. **2** fulfil, indulge, pander to, satisfy.

grating *n.* screen of spaced bars placed across an opening.

gratis *adj.* & *adv.* free of charge.

gratitude *n.* being grateful.

gratuitous *adj.* **1** given or done free. **2** uncalled-for. □ **gratuitously** *adv.*

> ■ **2** needless, uncalled-for, unjustified, unprovoked, unsolicited, unwarranted.

gratuity *n.* money given in recognition of services rendered.

grave[1] *n.* hole dug to bury a corpse.

grave[2] *adj.* **1** serious, causing great anxiety. **2** solemn. □ **grave accent** /graav/ the accent (`). **gravely** *adv.*

■ **1** important, pressing, serious, urgent, weighty; acute, critical, dangerous, perilous, severe, threatening. **2** earnest, serious, sober, solemn, sombre, unsmiling.

gravel *n.* coarse sand with small stones. □ **gravelly** *adj.*

gravestone *n.* stone placed over a grave.

graveyard *n.* burial ground.

gravitate *v.* move or be attracted towards something.

gravitation *n.* **1** gravitating. **2** force of gravity. □ **gravitational** *adj.*

gravity *n.* **1** seriousness, importance. **2** solemnity. **3** force that attracts bodies towards the centre of the earth.

■ **1** importance, magnitude, seriousness, significance; acuteness, severity, urgency. **2** dignity, soberness, solemnity, sombreness.

gravy *n.* **1** juice from cooked meat. **2** sauce made from this.

gray *adj.* & *n.* = **grey**.

graze[1] *v.* **1** feed on growing grass. **2** pasture animals in (a field).

graze[2] *v.* **1** injure by scraping the skin. **2** touch or scrape lightly in passing. ● *n.* abrasion.

■ *v.* **1** bark, scrape, scratch, skin. **2** brush, shave, touch. ● *n.* abrasion, scrape, scratch.

grease *n.* fatty or oily matter, esp. as a lubricant. ● *v.* put grease on. □ **grease-paint** *n.* make-up used by actors. **greasy** *adj.*

great *adj.* **1** much above average in size, amount, or intensity. **2** of remarkable ability etc. **3** important, distinguished. **4** (*colloq.*) very good. □ **greatness** *n.*

■ **1** big, colossal, enormous, extensive, gigantic, huge, immense, large, massive, prodigious, tremendous, vast; considerable, extreme, intense, marked, pronounced. **2** accomplished, brilliant, excellent, exceptional, gifted, outstanding, skilled, talented. **3** celebrated, distinguished, eminent, famous, illustrious, important, notable, prominent, renowned, well-known.

great- *pref.* (of a family relationship) one generation removed in ancestry or descent.

greatly *adv.* very much.

grebe *n.* a diving bird.

Grecian *adj.* Greek.

greed *n.* excessive desire, esp. for food or wealth. □ **greedy** *adj.*, **greedily** *adv.*, **greediness** *n.*

■ acquisitiveness, avarice, cupidity, greediness; gluttony, voracity. □ **greedy** acquisitive, avaricious, covetous, grasping, materialistic; gluttonous, *colloq.* gutsy, insatiable, ravenous, voracious.

Greek *adj.* & *n.* (native, language) of Greece.

green *adj.* **1** of the colour of growing grass. **2** unripe. **3** inexperienced, easily deceived. **4** concerned with protecting the environment. ● *n.* **1** green colour or thing. **2** piece of grassy public land. **3** (*pl.*) green vegetables. □ **green belt** area of open land round a town. **green fingers** skill in growing plants. **green light** signal or (*colloq.*) permission to proceed. **Green Paper** government report of proposals being considered. **green-room** *n.* room in a theatre for the use of actors when off stage. **greenish** *adj.*

■ *adj.* **2** unready, unripe, unripened. **3** callow, credulous, gullible, immature, inexperienced, innocent, naive, unsophisticated.

greenery *n.* green foliage or plants.

greenfinch *n.* finch with green and yellow feathers.

greenfly *n.* (*pl.* **-fly**) small green insect that sucks juices from plants.

greengage *n.* round plum with a greenish skin.

greengrocer *n.* shopkeeper selling vegetables and fruit.

greenhorn *n.* novice.

greenhouse *n.* glass building for rearing plants. □ **greenhouse effect** trapping of the sun's radiation by pollution in the atmosphere. **greenhouse gas** gas contributing to this.

greet *v.* **1** address politely on meeting or arrival. **2** receive or acknowledge in a particular way. **3** become apparent to (the eye or ear).

■ **1,2** hail, meet, receive, salute, welcome; acknowledge.

gregarious *adj.* **1** fond of company. **2** living in flocks etc. □ **gregariousness** *n.*

■ **1** companionable, convivial, friendly, outgoing, sociable.

gremlin *n.* (*colloq.*) mischievous spirit blamed for faults or problems.

grenade *n.* small bomb thrown by hand or fired from a rifle.

grenadine *n.* flavouring syrup made from pomegranates etc.

grew *see* **grow**.

grey *adj.* of the colour between black and white. ● *n.* grey colour or thing. ● *v.* make or become grey. □ **greyish** *adj.*

greyhound *n.* slender dog noted for its swiftness.

grid *n.* **1** grating. **2** system of numbered squares for map references. **3** network of lines, power cables, etc. **4** gridiron.

gridiron *n.* **1** framework of metal bars for cooking on. **2** field for American football, marked with parallel lines.

grief *n.* deep sorrow. □ **come to grief 1** meet with disaster. **2** fail.

■ anguish, desolation, distress, heartache, misery, pain, sadness, sorrow, unhappiness.

grievance *n.* cause for complaint.

■ *sl.* beef, complaint, *colloq.* gripe, *colloq.* grouse, objection.

grieve *v.* (cause to) feel grief.

■ distress, hurt, sadden, upset, wound; be in mourning, cry, mourn, sorrow, weep.

grievous *adj.* **1** causing grief. **2** serious. □ **grievously** *adv.*

griffin *n.* mythological creature with an eagle's head and wings and a lion's body.

griffon *n.* **1** small terrier-like dog. **2** a kind of vulture. **3** griffin.

grill *n.* **1** metal grid, grating. **2** device on a cooker for radiating heat downwards. **3** grilled food. ● *v.* **1** cook under a grill or on a gridiron. **2** question closely and severely.

grille *n.* grating, esp. in a door or window.

grim *adj.* (**grimmer**, **grimmest**) **1** stern, severe. **2** ghastly. **3** cheerless, uninviting. □ **grimly** *adv.*, **grimness** *n.*

■ **1** dour, fierce, forbidding, implacable, relentless, severe, sombre, stern, stony, unrelenting, unsmiling, unyielding. **2** appalling, awful, dire, dreadful, ghastly, grisly, gruesome, hideous, horrendous, horrible, horrid, shocking, terrible. **3** bleak, cheerless, depressing, dismal, dreary, gloomy, uninviting.

grimace *n.* contortion of the face in pain or disgust, or to amuse. ● *v.* make a grimace.

grime *n.* ingrained dirt. ● *v.* blacken with grime. □ **grimy** *adj.*

grin *v.* (**grinned**) smile broadly, showing the teeth. ● *n.* broad smile.

grind *v.* (**ground**) **1** crush into grains or powder. **2** sharpen or smooth by friction. **3** rub harshly together. ● *n.* **1** grinding process. **2** hard monotonous work. □ **grind down** oppress by cruelty. **grinder** *n.*

■ *v.* **1** crush, granulate, mince, pound, powder, pulverize. **2** sharpen, whet; file, polish, smooth. **3** gnash, grate. □ **grind down** crush, oppress, persecute, subdue, suppress.

grindstone *n.* thick revolving disc for sharpening or grinding.

grip *v.* (**gripped**) **1** take or keep firm hold of. **2** hold the attention of. ● *n.* **1** firm grasp or hold. **2** way of or thing for gripping. **3** understanding. **4** travelling bag.

■ *v.* **1** clasp, clutch, grab, grasp, hold, take hold of. **2** absorb, captivate, enthral, entrance, fascinate, hold spellbound, hypnotize, mesmerize, rivet. ● *n.* **1** clasp, grasp, hold. **2** foothold, purchase, toe-hold; handle. **3** apprehension, awareness, comprehension, grasp, perception, understanding.

gripe *v.* (*colloq.*) grumble. ● *n.* **1** colic pain. **2** (*colloq.*) grievance.

grisly *adj.* (**-ier**, **-iest**) causing fear, horror, or disgust.

gristle *n.* tough tissue of animal bodies, esp. in meat. □ **gristly** *adj.*

grit *n.* **1** particles of stone or sand. **2** (*colloq.*) courage and endurance. ● *v.* (**gritted**) **1** make a grating sound. **2** clench. **3** spread grit on. □ **gritty** *adj.*

grizzle *v.* & *n.* whimper, whine.

grizzled *adj.* grey-haired.

grizzly bear large brown bear of N. America.

groan *v.* make a long deep sound in pain, grief, or disapproval. ● *n.* sound made by groaning.

■ *v.* cry out, moan, sigh, wail, whimper.

grocer *n.* shopkeeper selling foods and household stores.

grocery *n.* grocer's shop or goods.

grog *n.* drink of spirits mixed with water.

groggy *adj.* (**-ier**, **-iest**) weak and unsteady, esp. after illness. □ **groggily** *adv.*, **grogginess** *n.*

groin *n.* **1** groove where each thigh joins the trunk. **2** curved edge where two vaults meet.

grommet *n.* **1** insulating washer. **2** tube placed through the eardrum.

groom *n.* **1** person employed to look after horses. **2** bridegroom. ● *v.* **1** clean and brush (an animal). **2** make neat and trim. **3** prepare (a person) for a career or position.

■ *v.* **1** brush, clean, curry. **2** neaten up, smarten up, spruce up, tidy up, *colloq.* titivate. **3** coach, drill, make ready, prepare, prime, school, train.

groove *n.* long narrow channel. ● *v.* make groove(s) in.

grope *v.* feel about as one does in the dark.

gross *adj.* **1** thick, large-bodied. **2** vulgar. **3** outrageous. **4** total, without deductions. ● *n.* (*pl.* **gross**) twelve dozen. ● *v.* produce or earn as total profit. □ **grossly** *adv.*

■ *adj.* **1** bloated, bulky, corpulent, fat, fleshy, heavy, large, massive, obese, overfed, overweight. **2** coarse, crude, indecent, indelicate, obscene, offensive, rude, vulgar. **3** blatant, flagrant, glaring, monstrous, outrageous, shocking. **4** entire, inclusive, overall, total, whole.

grotesque *adj.* very odd or ugly. ● *n.* **1** comically distorted figure. **2** design using fantastic forms. □ **grotesquely** *adv.*

■ *adj.* bizarre, freakish, hideous, misshapen, monstrous, odd, outlandish, peculiar, strange, ugly, unnatural.

grotto *n.* (*pl.* **-oes**) picturesque cave.

grouch *v.* & *n.* (*colloq.*) grumble.

ground[1] *n.* **1** solid surface of earth. **2** soil. **3** (*pl.*) reason for a belief or action. **4** (*pl.*) enclosed land attached to a house etc. **5** area designated for special use. **6** (*pl.*) dregs. ● *v.* **1** prevent (an aircraft or pilot) from flying. **2** base. **3** give basic training to. □ **ground-rent** *n.* rent paid for land leased for building. **ground swell** slow heavy waves.

■ *n.* **1** earth, land, terra firma. **2** clay, dirt, earth, loam, sod, soil. **3** (**grounds**) basis, cause, excuse, foundation, justification, motive, reason. **4** (**grounds**) estate, lands; campus. **5** field, pitch, playing field. **5** (**grounds**) dregs, lees, sediment. ● *v.* **2** base, establish, found. **3** coach, inform, instruct, prepare, teach, train.

ground[2] *see* **grind**.

grounding *n.* basic training.

groundless *adj.* without foundation.

groundnut *n.* peanut.

groundsheet *n.* waterproof sheet for spreading on the ground.

groundsman *n.* person employed to look after a sports ground.

groundwork *n.* preliminary or basic work.

group *n.* number of people or things near, belonging, classed, or working together. ● *v.* form or gather into group(s).

■ *n.* assortment, batch, category, class, collection, series, set; alliance, assembly, association, band, body, circle, clique, club, cluster, company, congregation, corps, coterie, crew, crowd, faction, flock, gang, gathering, league, party, society, team, throng, troop, union. ● *v.* categorize, class, classify, grade, rank, sort; assemble, bring together, collect, gather.

grouse[1] *n.* a kind of game bird.

grouse[2] *v.* & *n.* (*colloq.*) grumble.

grout *n.* thin fluid mortar. ● *v.* fill with grout.

grove *n.* group of trees.

grovel *v.* (**grovelled**) **1** behave obsequiously. **2** crawl face downwards.

■ **1** abase oneself, crawl, cringe, fawn, kowtow, toady.

grow *v.* (**grew, grown**) **1** increase in size or amount. **2** develop or exist as a living plant. **3** become. **4** produce by cultivation. □ **grow up** become adult or mature. **grown-up** *adj.* & *n.* adult. **grower** *n.*

■ **1** broaden, develop, enlarge, evolve, expand, extend, lengthen, multiply, spread, stretch, swell, widen; flourish, prosper, thrive. **2** bud, burgeon, germinate, shoot up, spring up, sprout. **3** become, get. **4** cultivate, farm, produce, propagate, raise.

growl *v.* make a low threatening sound. ● *n.* this sound.

grown *see* **grow**.

growth *n.* **1** process of growing. **2** thing that grows or has grown. **3** tumour. □ **growth industry** one developing faster than others.

■ **1** broadening, development, enlargement, evolution, expansion, extension, increase, proliferation, spread.

groyne *n.* wall built out into the sea to prevent erosion.

grub *n.* **1** larva of certain insects. **2** (*sl.*) food. ● *v.* (**grubbed**) **1** dig the surface of soil. **2** dig up by the roots. **3** rummage.

grubby *adj.* (**-ier, -iest**) dirty.

grudge *n.* feeling of resentment or ill will. ● *v.* begrudge, resent.

■ *v.* begrudge, be jealous of, envy, mind, resent.

gruel *n.* thin oatmeal porridge.

gruelling *adj.* very tiring.
■ arduous, demanding, draining, exhausting, hard, laborious, strenuous, taxing, tiring, tough.

gruesome *adj.* filling one with horror or disgust.
■ abhorrent, appalling, awful, disgusting, dreadful, fearful, frightful, ghastly, grim, grisly, hideous, horrendous, horrible, horrid, horrific, loathsome, repellent, repugnant, repulsive, revolting, shocking, terrible.

gruff *adj.* **1** (of the voice) low and hoarse. **2** surly. □ **gruffly** *adv.*
■ **1** deep, guttural, hoarse, husky, low, throaty. **2** bad-tempered, blunt, brusque, cantankerous, churlish, crabby, crotchety, crusty, curt, grumpy, irascible, short-tempered, surly, terse.

grumble *v.* **1** complain peevishly. **2** rumble. ● *n.* **1** complaint. **2** rumble. □ **grumbler** *n.*
■ *v.* **1** carp, complain, *colloq.* gripe, *colloq.* grouch, *colloq.* grouse, make a fuss, moan, *colloq.* whinge.

grumpy *adj.* (**-ier, -iest**) bad-tempered. □ **grumpily** *adv.*
■ bad-tempered, cantankerous, crabby, cross, crotchety, crusty, disagreeable, gruff, fractious, irritable, peevish, snappy, splenetic, surly, testy.

grunt *n.* gruff snorting sound made by a pig. ● *v.* make this or a similar sound.

gryphon *n.* = **griffin**.

G-string *n.* narrow strip of cloth covering the genitals, attached to a string round the waist.

guano /gwaánō/ *n.* dung of seabirds, used as manure.

guarantee *n.* **1** formal promise to do something or that a thing is of specified quality and durability. **2** thing offered as security. **3** guarantor. ● *v.* **1** give or serve as a guarantee for. **2** promise.
■ *n.* **1** assurance, oath, pledge, promise, word; warrant, warranty. **2** bond, pledge, security, surety. ● *v.* **1** answer for, attest to, certify, vouch for. **2** give one's word, pledge, promise, swear.

guarantor *n.* giver of a guarantee.

guard *v.* **1** watch over and protect or supervise. **2** restrain. ● *n.* **1** state of watchfulness. **2** person(s) guarding something. **3** protecting part or device. **4** railway official in charge of a train. **5** defensive attitude in boxing, cricket, etc. □ **guard against** take precautions against. **on** or **off one's guard** prepared (or unprepared) for some surprise or difficulty.
■ *v.* **1** defend, keep safe, look after, mind, patrol, protect, shield, supervise, watch over. **2** contain, control, curb, keep in check, restrain. ● *n.* **1** lookout, vigil, watch. **2** bodyguard, custodian, guardian, *sl.* minder, protector; sentinel, sentry, watchman; jailer, warder. **3** defence, protection, safeguard, shield.

guarded *adj.* cautious, discreet.
■ *colloq.* cagey, careful, cautious, chary, circumspect, discreet, noncommittal, reticent, wary.

guardian *n.* **1** one who guards or protects. **2** person undertaking legal responsibility for an orphan. □ **guardianship** *n.*
■ **1** custodian, defender, keeper, preserver, protector.

guardsman *n.* soldier acting as guard.

guava /gwaávə/ *n.* orange-coloured fruit of a tropical tree.

gudgeon[1] *n.* small freshwater fish.

gudgeon[2] *n.* **1** a kind of pivot. **2** socket for a rudder. **3** metal pin.

guerrilla *n.* member of a small fighting force, taking independent irregular action.

guess *v.* **1** estimate without calculation or measurement. **2** think likely. ● *n.* opinion formed by guessing. □ **guesser** *n.*
■ *v.* **1** estimate, make a guess (at). **2** assume, believe, conjecture, deduce, fancy, feel, infer, reckon, suppose, surmise, suspect, think. ● *n.* conjecture, estimate, speculation, supposition, surmise.

guesswork *n.* guessing.

guest *n.* **1** person entertained at another's house or table etc., or lodging at a hotel. **2** visiting performer. □ **guest house** superior boarding house.

guffaw *n.* coarse noisy laugh. ● *v.* utter a guffaw.

guidance *n.* **1** guiding. **2** advising or advice on problems.
■ **1** conduct, control, direction, leadership. **2** advice, briefing, counsel, counselling, information, instruction, teaching.

guide *n.* **1** person who shows others the way, esp. one employed to point out interesting sights to travellers. **2** adviser. **3** book of information. **4** thing directing actions or movements. ● *v.* **1** act as guide to. **2** lead, direct.

■ *n.* **1** conductor, director, escort, leader; courier. **2** adviser, counsellor, guru, mentor, teacher. **3** companion, guidebook, handbook, manual. **4** example, model, pattern, standard; beacon, indicator, landmark, light, marker, sign, signal, signpost. ● *v.* conduct, direct, lead, manoeuvre, pilot, shepherd, show, steer, usher; advise, counsel, instruct, teach, train.

guidebook *n.* book of information about a place.

guild *n.* **1** society for mutual aid or with a common purpose. **2** medieval association of craftsmen.

guilder *n.* unit of money of the Netherlands.

guile *n.* treacherous cunning, craftiness. □ **guileful** *adj.*, **guileless** *adj.*

guillotine *n.* **1** machine for beheading criminals. **2** machine for cutting paper or metal. **3** fixing of times for voting in Parliament, to prevent a lengthy debate. ● *v.* use a guillotine on.

guilt *n.* **1** fact of having committed an offence. **2** responsibility for an offence. **3** feeling that one is to blame. □ **guiltless** *adj.*

■ **1** criminality, misconduct, sinfulness, wrongdoing. **2** blame, culpability, responsibility. **3** contrition, penitence, regret, remorse, repentance, self-reproach, shame.

guilty *adj.* (**-ier**, **-iest**) **1** having done wrong. **2** feeling or showing guilt. □ **guiltily** *adv.*

■ **1** at fault, blameworthy, culpable, responsible. **2** ashamed, conscience-stricken, contrite, penitent, remorseful, repentant, rueful, shamefaced, sheepish, sorry.

guinea *n.* **1** former British coin worth 21 shillings (£1.05). **2** this amount. □ **guinea pig 1** rodent kept as a pet or for biological experiments. **2** person used in an experiment.

guise *n.* **1** false outward manner or appearance. **2** pretence.

guitar *n.* a kind of stringed musical instrument. □ **guitarist** *n.*

gulf *n.* **1** large area of sea partly surrounded by land. **2** deep hollow. **3** wide difference in opinion.

gull *n.* seabird with long wings.

gullet *n.* passage by which food goes from mouth to stomach.

gullible *adj.* easily deceived. □ **gullibility** *n.*

■ credulous, green, innocent, naive, unsuspecting.

gully *n.* narrow channel cut by water or carrying rainwater from a building.

gulp *v.* **1** swallow (food etc.) hastily or greedily. **2** make a gulping movement. ● *n.* **1** act of gulping. **2** large mouthful.

■ *v.* **1** bolt, devour, gobble, guzzle, *colloq.* scoff, swallow, wolf; quaff, swill.

gum¹ *n.* firm flesh in which teeth are rooted.

gum² *n.* **1** sticky substance exuded by certain trees. **2** adhesive. **3** chewing gum. **4** gumdrop. ● *v.* (**gummed**) smear or stick together with gum. □ **gum tree** tree that exudes gum, esp. eucalyptus. **gummy** *adj.*

gumboil *n.* abscess on the gum.

gumboot *n.* rubber boot.

gumdrop *n.* hard gelatine sweet.

gumption *n.* (*colloq.*) resourcefulness, common sense.

gun *n.* **1** weapon that sends shells or bullets from a metal tube. **2** device operating similarly. ● *v.* (**gunned**) shoot with a gun.

gunfire *n.* firing of guns.

gunman *n.* man armed with a gun.

gunner *n.* **1** artillery soldier. **2** naval officer in charge of a battery of guns.

gunnery *n.* construction and operating of large guns.

gunny *n.* **1** coarse sackcloth. **2** sack made of this.

gunpowder *n.* explosive of saltpetre, sulphur, and charcoal.

gunrunning *n.* smuggling of firearms. □ **gunrunner** *n.*

gunshot *n.* shot fired from a gun.

gunsmith *n.* maker and repairer of small firearms.

gunwale /gúnn'l/ *n.* upper edge of a small ship's or boat's side.

guppy *n.* very small brightly coloured tropical fish.

gurgle *n.* low bubbling sound. ● *v.* make or utter with this sound.

guru *n.* (*pl.* **-us**) **1** Hindu spiritual teacher. **2** revered teacher.

gush *v.* **1** flow or pour suddenly or in great quantities. **2** talk effusively. ● *n.* **1** sudden or great outflow. **2** effusiveness.

■ *v.* **1** cascade, flood, flow, pour, rush, spout, spurt, stream, surge. **2** be effusive, enthuse. ● *n.* 1 cascade, flood, flow, jet, rush, spout, spurt, stream, torrent.

gusset *n.* piece of cloth inserted to strengthen or enlarge a garment etc. □ **gusseted** *adj.*

gust *n.* sudden rush of wind, rain, smoke, or sound. ● *v.* blow in gusts. □ **gusty** *adj.*

gustatory *adj.* of the sense of taste.

gusto *n.* zest.

■ delight, eagerness, enjoyment, enthusiasm, pleasure, relish, verve, vigour, zest.

gut *n.* **1** intestine. **2** thread made from animal intestines. **3** (*pl.*) abdominal organs. **4** (*pl., colloq.*) courage and determination. ● *v.* (**gutted**) **1** remove guts from (fish). **2** remove or destroy internal fittings or parts of.

■ *n.* **3** (**guts**) bowels, entrails, *colloq.* innards, *colloq.* insides, intestines, viscera, vitals. **4** (**guts**) boldness, bravery, courage, daring, fearlessness, *colloq.* grit, mettle, nerve, pluck, spirit, valour. ● *v.* **1** disembowel, eviscerate. **2** destroy, devastate, ravage; empty, loot, pillage, plunder, ransack, strip.

gutsy *adj.* (*colloq.*) **1** courageous. **2** greedy.

gutta-percha *n.* rubbery substance made from latex.

gutter *n.* **1** trough round a roof, or channel at a roadside, for carrying away rainwater. **2** (**the gutter**) slum environment. ● *v.* (of a candle) burn unsteadily.

guttural *adj.* throaty, harsh-sounding. □ **gutturally** *adv.*

guy[1] *n.* **1** effigy of Guy Fawkes burnt on 5 Nov. **2** (*colloq.*) man. ● *v.* ridicule.

guy[2] *n.* rope or chain used to keep a thing steady or secured.

guzzle *v.* eat or drink greedily.

■ bolt, devour, gobble, gulp, *colloq.* scoff, wolf.

gybe *v.* **1** (of a sail or boom) swing across. **2** (of a boat) change course thus.

gym *n.* (*colloq.*) **1** gymnasium. **2** gymnastics.

gymkhana *n.* horse-riding competition.

gymnasium *n.* room equipped for physical training and gymnastics.

gymnast *n.* expert in gymnastics.

gymnastics *n.pl.* exercises to develop the muscles or demonstrate agility. □ **gymnastic** *adj.*

gynaecology /gīnikólləji/ *n.* study of the physiological functions and diseases of women. □ **gynaecological** *adj.*, **gynaecologist** *n.*

gypsophila *n.* garden plant with many small white flowers.

gypsum *n.* chalk-like substance.

gypsy *n.* member of a nomadic people of Europe.

gyrate *v.* move in circles or spirals, revolve. □ **gyration** *n.*

■ circle, revolve, rotate, spin, swirl, swivel twirl, wheel, whirl.

gyratory *adj.* gyrating, following a circular or spiral path.

gyro *n.* (*pl.* **-os**) (*colloq.*) gyroscope.

gyroscope *n.* rotating device used to keep navigation instruments steady.

Hh

ha *int.* exclamation of triumph.

habeas corpus order requiring a person to be brought into court.

habit *n.* **1** usual behaviour or practice. **2** tendency to act in a particular way. **3** monk's or nun's long dress. **4** a woman's riding-dress.

■ **1** convention, custom, practice, routine, rule, usage, wont. **2** idiosyncrasy, mannerism, quirk, peculiarity, tendency, trick.

habitable *adj.* suitable for living in.

habitat *n.* animal's or plant's natural environment.

■ domain, element, environment, surroundings, territory.

habitation *n.* place to live in.

habitual *adj.* **1** done or doing something constantly, esp. as a habit. **2** usual. □ **habitually** *adv.*

■ **1** constant, continual, perpetual, persistent; chronic, complusive, inveterate. **2** accustomed, common, conventional, customary, established, everyday, familiar, normal, regular, routine, set, settled, standard, traditional, usual.

habituate *v.* accustom. □ **habituation** *n.*

hacienda *n.* ranch or large estate in S. America.

hack[1] *n.* **1** horse for ordinary riding. **2** person doing routine work, esp. as a writer. ● *v.* ride on horseback at an ordinary pace.

hack[2] *v.* cut, chop, or hit roughly. ● *n.* blow given thus.

hacker *n.* (*colloq.*) computer enthusiast, esp. one gaining unauthorized access to files.

hacking *adj.* (of a cough) dry and frequent.

hackles *n.pl.* feathers or hairs on some birds or animals, raised in anger.

hackneyed *adj.* (of sayings) over-used and therefore lacking impact.

hacksaw *n.* saw for metal.

haddock *n.* (*pl.* **haddock**) edible sea fish like a small cod.

haematology /hee-/ *n.* study of blood. □ **haematologist** *n.*

haemoglobin /hee-/ *n.* red oxygen-carrying substance in blood.

haemophilia /hee-/ *n.* failure of the blood to clot causing excessive bleeding. □ **haemophiliac** *n.*

haemorrhage /hém-/ *n.* profuse bleeding. ● *v.* bleed profusely.

haemorrhoids /hém-/ *n.pl.* varicose veins at or near the anus.

haft *n.* handle of a knife etc.

hag *n.* ugly old woman.

haggard *adj.* looking ugly from exhaustion. □ **haggardness** *n.*

■ careworn, drawn, exhausted, gaunt, run down, worn.

haggis *n.* Scottish dish made from sheep's offal.

haggle *v.* argue about price or terms when settling a bargain.

ha-ha *n.* sunk fence.

hail[1] *v.* **1** greet. **2** acclaim. **3** signal to and summon.

■ **1** accost, address, call, greet, salute. **2** acclaim, applaud, extol, honour, laud, praise.

hail[2] *n.* **1** pellets of frozen rain falling in a shower. **2** shower of blows, questions, etc. ● *v.* pour down as or like hail. □ **hailstone** *n.*, **hailstorm** *n.*

■ *n.* **2** barrage, onslaught, shower, storm, torrent, volley. ● *v.* pelt, pour, rain, shower, volley.

hair *n.* **1** fine thread-like strand growing from the skin. **2** mass of these, esp. on the head. □ **hair-raising** *adj.* terrifying. **hair-trigger** *n.* trigger operated by the slightest pressure.

haircut *n.* **1** shortening of hair by cutting it. **2** style of this.

hairdo *n.* (*pl.* **-dos**) arrangement of the hair.

hairdresser *n.* person who cuts and arranges hair. □ **hairdressing** *n.*

hairgrip *n.* springy hairpin.

hairline *n.* **1** edge of the hair on the forehead etc. **2** very narrow crack or line.

hairpin *n.* U-shaped pin for keeping hair in place. □ **hairpin bend** sharp U-shaped bend in a road.

hairy *adj.* (**-ier**, **-iest**) **1** covered with hair. **2** (*sl.*) unpleasant, difficult. □ **hairiness** *n.*

■ **1** bearded, bristly, hirsute, shaggy, unshaven, whiskered, whiskery.

Haitian *adj.* & *n.* (native) of Haiti.

hajji *n.* Muslim who has been to Mecca on pilgrimage.

hake *n.* (*pl.* **hake**) edible sea fish of the cod family.

halal *n.* meat from an animal killed according to Muslim law.

halcyon *adj.* (of a period) happy and peaceful.

hale *adj.* strong and healthy.

half *n.* (*pl.* **halves**) **1** each of two equal parts. **2** this amount. **3** (*colloq.*) half-back, half-pint, etc. ● *adj.* amounting to a half. ● *adv.* to the extent of a half, partly. □ **half a dozen** six. **half and half** half one thing and half another. **half-back** *n.* player between forwards and full back(s). **half-brother, -sister** *ns.* one having only one parent in common. **half-hearted** *adj.* not very enthusiastic. **half-life** *n.* time after which radioactivity etc. is half its original level. **at half-mast** (of a flag) lowered in mourning. **half nelson** wrestling hold. **half-term** *n.* short holiday halfway through a school term. **half-timbered** *adj.* built with a timber frame with brick or plaster filling. **half-time** *n.* interval between two halves of a game. **halfway** *adj.* & *adv.* at a point equidistant between two others. **halfwit** *n.* halfwitted person. **halfwitted** *adj.* stupid.

halfpenny /háypni/ *n.* (*pl.* **-pennies** for single coins, **-pence** for a sum of money) coin worth half a penny.

halibut *n.* (*pl.* **halibut**) large edible flatfish.

halitosis *n.* breath that smells unpleasant.

hall *n.* **1** large room or building for meetings, concerts, etc. **2** space inside the front entrance of a house etc. **3** large country house.

■ **1** auditorium, lecture hall, theatre. **2** corridor, entrance hall, lobby, passage, passageway, vestibule.

hallmark *n.* **1** official mark on precious metals to indicate their standard. **2** distinguishing characteristic.

hallo *int.* & *n.* = **hello**.

Hallowe'en *n.* 31 Oct., eve of All Saints' Day.

hallucinate *v.* experience hallucinations.

hallucination *n.* illusion of seeing or hearing something not actually present. □ **hallucinatory** *adj.*

■ apparition, chimera, illusion, mirage, vision.

hallucinogenic *adj.* causing hallucinations.

halo *n.* (*pl.* **-oes**) circle of light esp. round the head of a sacred figure.

halogen *n.* any of a group of certain non-metallic elements.

halt *n.* & *v.* stop.

■ *n.* cessation, close, end, interruption, standstill, stop, stoppage, termination. ● *v.* bring to an end, discontinue, end, finish, put an end to, terminate; come to a standstill, draw up, pull up, stop; cease, come to an end, finish.

halter *n.* strap round the head of a horse for leading or fastening it.

halting *adj.* slow and hesitant.

halve *v.* **1** divide or share equally between two. **2** reduce by half.

halyard *n.* rope for raising or lowering a sail or flag.

ham *n.* **1** upper part of a pig's leg cured for food. **2** (*colloq.*) poor actor or performer. **3** (*colloq.*) amateur radio operator. ● *v.* (**hammed**) (*colloq.*) overact. □ **ham-fisted** *adj.* (*colloq.*) clumsy.

hamburger *n.* flat round cake of minced beef.

hamlet *n.* small village.

hammer *n.* **1** tool with a head for hitting things or driving nails in. **2** metal ball attached to a wire for throwing as an athletic contest. ● *v.* **1** hit or beat with a hammer. **2** strike loudly.

hammock *n.* hanging bed of canvas or netting.

hamper[1] *n.* large basket, usu. with a hinged lid and containing food.

hamper[2] *v.* **1** obstruct the movement of. **2** hinder.

■ **1** block, encumber, impede, inhibit, obstruct, prevent. **2** baulk, check, delay, frustrate, handicap, hinder, hold up, restrict, retard, slow down, thwart.

hamster *n.* small rodent with cheek-pouches.

hamstring *n.* tendon at the back of a knee or hock. ● *v.* (**hamstrung**) **1** cripple by cutting hamstring(s). **2** cripple the activity of.

hand *n.* **1** end part of the arm, below the wrist. **2** (often *pl.*) control. **3** influence, or help in doing something. **4** manual worker. **5** style of handwriting. **6** pointer on a dial etc. **7** (*colloq.*) round of applause. **8** (right or left) side. **9** round of a card game, player's cards. ● *v.* give or pass. □ **at hand** close by. **handout** *n.* thing distributed free of charge. **hands down** easily. **in hand** receiving attention. **lend a hand** help. **on hand** available. **out of hand** out of control. **to hand** within reach.

■ *n.* **2** (**hands**) care, charge, control, custody, guardianship, hold, jurisdiction, keeping, possession, power. **3** agency, influence, involvement. **4** employee, labourer, operative, worker. **5** handwriting, writing. **6** indicator, needle, pointer. **7** clap, ovation, round of applause. ● *v.* deliver, give, pass, present. □ **at hand** accessible, at one's disposal, available, close (by), convenient, handy, near, nearby, on hand, *colloq.* on tap, to hand, within reach. **hands down** comfortably, easily, effortlessly, without difficulty.

handbag *n.* bag to hold a purse and small personal articles.

handbill *n.* printed notice circulated by hand.

handbook *n.* small book giving useful facts.

handcuff *n.* metal ring linked to another, for securing a prisoner's wrists. ● *v.* put handcuffs on.

handful *n.* **1** quantity that fills the hand. **2** a few. **3** (*colloq.*) difficult person or task.

handicap *n.* **1** disadvantage imposed on a superior competitor to equalize chances. **2** race etc. in which handicaps are imposed. **3** physical or mental disability. **4** thing that makes progress difficult. ● *v.* (**handicapped**) impose or be a handicap on.

■ *n.* **4** bar, barrier, disadvantage, encumbrance, hindrance, impediment, limitation, obstacle, stumbling block. ● *v.* encumber, hamper, hinder, hold back, impede, limit, put at a disadvantage, restrict.

handkerchief *n.* (*pl.* **-fs**) small square of cloth for wiping the nose etc.

handle *n.* part by which a thing is to be held, carried, or controlled. ● *v.* **1** touch or move with the hands. **2** deal with, manage. **3** deal in.

■ *v.* **1** feel, finger, *colloq.* paw, touch. **2** attend to, deal with, do, look after, manage, organize, see to, sort out, take charge of; control, cope with. **3** deal in, sell, stock, trade in.

handlebar *n.* steering bar of a bicycle etc.

handler *n.* person in charge of a trained dog etc.

handrail *n.* rail beside stairs etc.

handshake *n.* act of shaking hands as a greeting etc.

handsome *adj.* **1** good-looking. **2** generous. **3** considerable.

■ **1** attractive, good-looking, personable. **2** generous, lavish, liberal, munificent, princely. **3** big, considerable, large, sizeable, substantial.

handstand *n.* balancing on one's hands with feet in the air.

handwriting *n.* **1** writing by hand. **2** style of this.

handy *adj.* (**-ier**, **-iest**) **1** convenient. **2** clever with one's hands. □ **handily** *adv.*

■ **1** accessible, at hand, available, close by, convenient, nearby, to hand, within reach; helpful, serviceable, useful. **2** adept, adroit, clever, deft, dexterous, skilful, skilled, proficient.

handyman *n.* person who does odd jobs.

hang *v.* (**hung**) **1** support or be supported from above with the lower end free. **2** let droop. **3** remain or be hung. **4** (**hanged**) kill or be killed by suspension from a rope round the neck. ● *n.* way a thing hangs. □ **get the hang of** (*colloq.*) get the knack of, understand. **hang about** loiter. **hang back** hesitate. **hang-glider** *n.* frame used in hang-gliding. **hang-gliding** *n.* sport of gliding in an airborne frame in which the pilot is suspended. **hang on 1** hold tightly. **2** depend on. **3** (*sl.*) wait. **hang-up** *n.* (*sl.*) emotional problem or inhibition.

■ *v.* **1** attach, fasten, put up, suspend; be suspended, dangle, sway, swing. **2** droop, drop, let fall. **4** execute, string up. □ **hang about** dally, dawdle, *colloq.* dilly-dally, hover, idle, linger, loaf about, loiter, wait.

hangar *n.* shed for aircraft.

hangdog *adj.* shamefaced.

hanger *n.* **1** loop or hook by which a thing is hung. **2** shaped piece of wood etc. to hang a garment on.

hangings *n.pl.* draperies hung on walls.

hangman *n.* person who hangs people condemned to death.

hangnail *n.* torn skin at the root of a fingernail.

hangover *n.* after-effects of an excess of alcohol.

hank *n.* coil or length of thread.

hanker *v.* **hanker after** long for. ◻ **hankering** *n.*

■ crave, desire, hunger for, long for, thirst for, want, yearn for.

hanky *n.* (*colloq.*) handkerchief.

Hanukkah *n.* Jewish festival of lights, beginning in December.

haphazard *adj.* done or chosen at random. ◻ **haphazardly** *adv.*

■ aimless, arbitrary, chaotic, hit-or-miss, indiscriminate, random, unmethodical, unplanned, unsystematic.

hapless *adj.* unlucky.

happen *v.* occur. ◻ **happen on** discover by chance. **happen to** be the fate or experience of.

■ come about, occur, take place. ◻ **happen on** come across, discover, encounter, find, hit on, light on, meet by chance. **happen to** become of, befall, betide.

happy *adj.* (**-ier**, **-iest**) **1** contented, pleased. **2** fortunate. ◻ **happy-go-lucky** *adj.* taking events cheerfully. **happily** *adv.*, **happiness** *n.*

■ **1** buoyant, carefree, cheerful, cheery, chirpy, content, contented, delighted, ecstatic, elated, euphoric, exhilarated, gay, glad, gleeful, high-spirited, in high spirits, jovial, joyful, jubilant, merry, overjoyed, pleased, sunny, thrilled. **2** auspicious, favourable, fortunate, lucky, propitious. ◻ **happy-go-lucky** blithe, carefree, easygoing, insouciant, light-hearted, relaxed. **happiness** bliss, cheerfulness, delight, elation, euphoria, exhilaration, felicity, gaiety, glee, high spirits, joy, joyfulness, jubilation, pleasure.

harangue *n.* lengthy earnest speech. ● *v.* make a harangue to.

■ *n.* address, declamation, diatribe, exhortation, lecture, oration, sermon, speech, tirade.

harass *v.* **1** worry or annoy continually. **2** make repeated attacks on. ◻ **harassment** *n.*

■ **1** annoy, badger, bait, bother, harry, *colloq.* hassle, hound, nag, pester, *colloq.* plague, torment, trouble, worry.

harbour *n.* place of shelter for ships. ● *v.* **1** shelter. **2** keep in one's mind.

hard *adj.* **1** firm, not easily cut. **2** difficult. **3** not easy to bear. **4** harsh. **5** strenuous. **6** (of drugs) strong and addictive. **7** (of currency) not likely to drop suddenly in value. **8** (of drinks) strongly alcoholic. **9** (of water) containing mineral salts that prevent soap from lathering freely. ● *adv.* **1** strenuously, intensively, copiously. **2** with difficulty. ◻ **hard-boiled** *adj.* **1** (of eggs) boiled until yolk and white are set. **2** callous. **hard copy** material produced in printed form by a computer. **hard-headed** *adj.* shrewd and practical. **hard-hearted** *adj.* unfeeling. **hard of hearing** slightly deaf. **hard sell** aggressive salesmanship. **hard shoulder** extra strip of road beside a motorway, for use in an emergency. **hard up** short of money. **hardness** *n.*

■ *adj.* **1** dense, firm, impenetrable, rigid, set, solid, solidified, stiff, tough, unmalleable, unyielding; flinty, rocky, steely, stony. **2** baffling, complex, complicated, difficult, intricate, knotty, perplexing, problematic(al), puzzling. **3** calamitous, distressing, grievous, insupportable, intolerable, painful, unbearable, unendurable, unpleasant. **4** callous, cold, cruel, hard-boiled, hard-hearted, harsh, heartless, insensitive, ruthless, severe, stern, stony, strict, unfeeling, unkind, unsympathetic. **5** arduous, demanding, exacting, exhausting, gruelling, herculean, laborious, onerous, strenuous, taxing, tiring, tough. ◻ **hard-headed** astute, businesslike, level-headed, practical, pragmatic, realistic, sensible, sharp, shrewd. **hard up** *colloq.* broke, impecunious, impoverished, penniless, poor, short of money, *sl.* skint.

hardbitten *adj.* tough and tenacious.

hardboard *n.* stiff board made of compressed wood pulp.

harden *v.* become or make hard or hardy.

■ cake, coagulate, congeal, set, solidify, stiffen; strengthen, toughen.

hardly *adv.* **1** only with difficulty. **2** scarcely.

hardship *n.* harsh circumstance.

■ adversity, affliction, deprivation, difficulty, misfortune, privation, trouble.

hardware *n.* **1** tools and household implements sold by a shop. **2** weapons. **3** machinery used in a computer system.

hardwood *n.* hard heavy wood of deciduous trees.

hardy *adj.* (**-ier**, **-iest**) capable of enduring cold or harsh conditions. ◻ **hardiness** *n.*

■ fit, hale, healthy, robust, stalwart, strong, sturdy, tough.

hare *n.* field animal like a large rabbit. ● *v.* run rapidly. ◻ **hare-brained** *adj.* wild, rash.

harem /haáreem/ *n.* **1** women of a Muslim household. **2** their apartments.

haricot bean /hárrikō/ white dried seed of a kind of bean.

hark *v.* listen. ◻ **hark back** return to an earlier subject.

harlequin *adj.* in varied colours.

harm *n.* damage, injury. ● *v.* cause harm to. ◻ **harmful** *adj.*, **harmless** *adj.*

■ *n.* damage, hurt, injury, mischief, misfortune. ● *v.* abuse, damage, deface, hurt, ill-treat, impair, injure, maltreat, mar, wound. ◻ **harmful** bad, damaging, dangerous, deleterious, destructive, detrimental, hurtful, injurious, noxious, pernicious, poisonous, toxic. **harmless** benign, innocuous, mild, safe.

harmonic *adj.* full of harmony.

harmonica *n.* mouth-organ.

harmonious *adj.* **1** sweet-sounding. **2** forming a pleasing or consistent whole. **3** free from ill feeling. ◻ **harmoniously** *adv.*

■ **1** dulcet, euphonious, mellifluous, melodious, musical, sweet-sounding, tuneful. **2** compatible, complementary, concordant, consonant. **3** amicable, cordial, friendly, peaceable, peaceful.

harmonium *n.* musical instrument like a small organ.

harmonize *v.* **1** make or be harmonious. **2** add notes to form chords. ◻ **harmonization** *n.*

harmony *n.* **1** combination of musical notes to form chords, esp. with a pleasing effect. **2** agreement between people or things.

■ **1** euphony, melodiousness, tunefulness. **2** accord, agreement, compatibility, concord, consistency; friendship, rapport.

harness *n.* **1** straps and fittings by which a horse is controlled. **2** similar fastenings. ● *v.* **1** put harness on, attach by this. **2** control and use.

harp *n.* musical instrument with strings in a triangular frame. ● *v.* **harp on** dwell on tediously. ◻ **harpist** *n.*

harpoon *n.* spear-like missile with a rope attached. ● *v.* spear with a harpoon.

harpsichord *n.* piano-like instrument.

harpy *n.* grasping unscrupulous woman.

harridan *n.* bad-tempered old woman.

harrow *n.* heavy frame with metal spikes or discs for breaking up clods. ● *v.* **1** draw a harrow over. **2** distress greatly.

harry *v.* harass.

harsh *adj.* **1** unpleasantly rough to the senses. **2** severe, cruel. ◻ **harshly** *adv.*, **harshness** *n.*

■ **1** bristly, coarse, rough, scratchy; cacophonous, discordant, dissonant, grating, guttural, hoarse, jangling, jarring, rasping, raucous, shrill, strident; acrid, bitter, sour. **2** cruel, draconian, hard, heartless, inhuman, merciless, pitiless, ruthless, severe, stern, stringent; austere, bleak, comfortless, grim, spartan, stark.

hart *n.* adult male deer.

hartebeest *n.* large African antelope.

harum-scarum *adj.* & *n.* wild and reckless (person).

harvest *n.* **1** gathering of crop(s). **2** season for this. **3** season's yield of a natural product. ● *v.* **1** gather a crop. **2** receive as the consequence of actions. ◻ **harvester** *n.*

■ *v.* **1** collect, garner, gather, glean, pick, reap. **2** earn, gain, get, make, obtain.

hash *n.* **1** dish of chopped re-cooked meat. **2** jumble. ● *v.* make into hash. ◻ **make a hash of** (*colloq.*) bungle.

hashish *n.* narcotic drug obtained from hemp.

hasp *n.* clasp fitting over a staple, secured by a pin or padlock.

hassle *n.* & *v.* (*colloq.*) **1** quarrel. **2** trouble, inconvenience.

hassock *n.* thick firm cushion for kneeling on in church.

haste *n.* hurry. ◻ **make haste** hurry.

hasten *v.* hurry.

■ dash, hurry, fly, make haste, race, run, rush, scamper, scurry, scuttle, speed, sprint.

hasty *adj.* (**-ier**, **-iest**) **1** hurried. **2** acting or done too quickly. ◻ **hastily** *adv.*, **hastiness** *n.*

■ **1** brisk, fast, hurried, quick, rapid, speedy, swift. **2** heedless, impetuous, impulsive, incautious, precipitate, rash, reckless, unthinking; careless, cursory, perfunctory, slapdash.

hat *n.* covering for the head, worn out of doors. ◻ **hat trick** three successes in a row, esp. in sports.

hatch[1] *n.* **1** opening in a door, floor, ship's deck, etc. **2** its cover.

hatch[2] *v.* **1** emerge or produce (young) from an egg. **2** devise (a plot). ● *n.* brood hatched.

■ *v.* **2** conceive, concoct, contrive, *colloq.* cook up, devise, dream up, form, invent, plan, *colloq.* think up.

hatch[3] *v.* mark with close parallel lines. □ **hatching** *n.* these marks.

hatchback *n.* a car with a back door that opens upwards.

hatchery *n.* place for hatching eggs.

hatchet *n.* small axe. □ **bury the hatchet** cease quarrelling and become friendly.

hatchway *n.* = **hatch**[1].

hate *n.* hatred. ● *v.* dislike greatly.

■ *v.* abhor, abominate, despise, detest, dislike, execrate, loathe.

hateful *adj.* arousing hatred.

■ abhorrent, abominable, contemptible, despicable, detestable, disgusting, execrable, foul, loathsome, odious, repugnant, repulsive, revolting.

hatred *n.* violent dislike.

■ abhorrence, antagonism, antipathy, aversion, contempt, detestation, disgust, dislike, hate, hostility, ill will, loathing, odium, repugnance, revulsion.

haughty *adj.* (**-ier**, **-iest**) proud, superior. □ **haughtily** *adv.*, **haughtiness** *n.*

■ aloof, arrogant, conceited, contemptuous, disdainful, hoity-toity, lofty, pretentious, proud, scornful, self-important, snobbish, *colloq.* snooty, *colloq.* stuck-up, supercilious, superior, *colloq.* uppity.

haul *v.* **1** pull or drag forcibly. **2** transport by truck etc. ● *n.* **1** process of hauling. **2** amount gained by effort, booty.

■ *v.* **1** drag, draw, heave, lug, pull, tow, tug. **2** carry, cart, convey, move, transport.

haulage *n.* transport of goods.

haulier *n.* person or firm that transports goods by road.

haunch *n.* **1** fleshy part of the buttock and thigh. **2** leg and loin of meat.

haunt *v.* **1** linger in the mind of. **2** (esp. of a ghost) repeatedly visit (a person or place). ● *n.* place frequented by a particular person.

■ *v.* **1** beset, obsess, preoccupy, prey on, torment, trouble.

haute couture /ōt kootyŏŏr/ high fashion.

have *v.* **1** possess. **2** contain. **3** experience, undergo. **4** give birth to. **5** cause to be or do or be done. **6** allow. **7** receive, accept. **8** (*colloq.*) cheat, deceive. ● *v.aux.* used with past participle to form past tenses. □ **have it out** settle a problem by frank discussion. **have up** bring (a person) to trial. **haves and have-nots** people with and without wealth or privilege.

■ *v.* **1** hold, keep, own, possess. **2** consist of, contain, include. **3** endure, experience, feel, suffer, undergo. **4** bear, deliver, give birth to. **6** allow, brook, permit, put up with, *colloq.* stand for, tolerate. **7** accept, acquire, get, obtain, receive, take.

haven *n.* **1** refuge. **2** harbour.

■ **1** asylum, refuge, retreat, sanctuary, shelter. **2** anchorage, harbour, port.

haversack *n.* strong bag carried on the back or shoulder.

havoc *n.* great destruction or disorder.

■ damage, destruction, devastation, ruin; chaos, confusion, disorder, disruption, turmoil.

haw *n.* hawthorn berry.

hawk[1] *n.* **1** bird of prey. **2** person who favours an aggressive policy. □ **hawk-eyed** *adj.* having very keen sight.

hawk[2] *v.* clear one's throat of phlegm noisily.

hawser *n.* heavy rope or cable for mooring or towing a ship.

hawthorn *n.* thorny tree or shrub with small red berries.

hay *n.* grass mown and dried for fodder. □ **hay fever** irritation of nose, throat, etc. caused by pollen or dust.

haymaking *n.* mowing grass and spreading it to dry.

haystack *n.* regular pile of hay firmly packed for storing.

haywire *adj.* badly disorganized.

hazard *n.* **1** risk, danger. **2** source of this. ● *v.* risk. □ **hazardous** *adj.*

■ *n.* danger, jeopardy, peril, pitfall, risk, threat. ● *v.* endanger, imperil, jeopardize, put in jeopardy, risk. □ **hazardous** dangerous, *sl.* dicey, perilous, precarious, risky, uncertain, unpredictable, unreliable, unsafe.

haze *n.* thin mist.

hazel *n.* **1** bush with small edible nuts. **2** light brown. □ **hazelnut** *n.*

hazy *adj.* (**-ier**, **-iest**) **1** misty. **2** vague, confused. ▫ **hazily** *adv.*, **haziness** *n.*

■ **1** foggy, misty. **2** dim, faint, fuzzy, indistinct, nebulous, vague; confused, muddled, uncertain, unclear.

H-bomb *n.* hydrogen bomb.

he *pron.* male previously mentioned. ● *n.* male animal.

head *n.* **1** part of the body containing the eyes, nose, mouth, and brain. **2** intellect. **3** thing like the head in form or position. **4** top or leading part or position. **5** person in charge, esp. of a school. **6** (*colloq.*) headache. **7** individual person or animal. **8** foam on beer etc. **9** body of water or steam confined for exerting pressure. **10** (*pl.*) side of a coin showing a head, turned upwards after being tossed. ● *adj.* chief, principal. ● *v.* **1** be at the head or top of. **2** direct one's course. **3** strike (a ball) with one's head. ▫ **head-hunt** *v.* seek to recruit (senior staff) from another firm. **head off** force to turn by getting in front. **head-on** *adj.* & *adv.* with head or front foremost. **head wind** wind blowing from directly in front.

■ *n.* **1** *sl.* block, *sl.* conk, *sl.* nut, skull. **2** brain(s), intellect, intelligence, *sl.* loaf, mind, wit. **4** apex, brow, crest, crown, peak, summit, tip, top; fore, forefront, front, van, vanguard. **5** *colloq.* boss, chairman, chairwoman, chief, director, governor, leader, manager, president, supervisor; headmaster, headmistress, principal. ● *adj.* chief, first, foremost, leading, main, pre-eminent, prime, principal, senior. ● *v.* **1** be in charge of, be in control of, command, direct, lead, supervise. **2** aim, face, go, move, point, proceed, steer, turn.

headache *n.* **1** continuous pain in the head. **2** worrying problem.

headdress *n.* ornamental covering worn on the head.

header *n.* **1** dive with the head first. **2** heading of the ball in football.

headgear *n.* hat or headdress.

heading *n.* word(s) at the top of written matter as a title.

headlamp *n.* headlight.

headland *n.* promontory.

headlight *n.* powerful light on the front of a vehicle etc.

headline *n.* **1** heading in a newspaper. **2** (*pl.*) summary of broadcast news.

headlong *adj.* & *adv.* **1** falling or plunging with the head first. **2** in a hasty and rash way.

headmaster, **headmistress** *ns.* teacher in charge of a school.

headphone *n.* receiver held over the ear(s) by a band over the head.

headquarters *n.pl.* organization's administrative centre.

headstone *n.* stone set up at the head of a grave.

headstrong *adj.* self-willed and obstinate.

■ defiant, disobedient, intractable, intransigent, obstinate, pigheaded, recalcitrant, refractory, self-willed, stubborn, wilful.

headway *n.* progress.

heady *adj.* (**-ier**, **-iest**) likely to cause intoxication.

heal *v.* **1** make or become healthy after injury or illness. **2** put right (differences). ▫ **healer** *n.*

■ **1** cure; be on the mend, get better, improve, mend. **2** mend, patch up, put right, reconcile, rectify, remedy, repair, settle.

health *n.* **1** state of being well and free from illness. **2** condition of the body.

■ **1** fitness, healthiness, robustness, strength, vigour, well-being. **2** condition, constitution, form, shape.

healthful *adj.* health-giving.

healthy *adj.* (**-ier**, **-iest**) having, showing, or producing good health. ▫ **healthily** *adv.*, **healthiness** *n.*

■ blooming, fit, flourishing, hale, in good health, *colloq.* in the pink, robust, strong, thriving, vigorous, well; healthful, nourishing, nutritious, salubrious, wholesome.

heap *n.* **1** a number of things or particles lying one on top of another. **2** (usu. *pl.*, *colloq.*) plenty. ● *v.* **1** pile or become piled in a heap. **2** load copiously.

■ *n.* **1** accumulation, collection, mass, mound, mountain, pile, stack. ● *v.* **1** accumulate, amass, collect, gather, pile up, stack. **2** burden, load, lumber, overwhelm, weigh down.

hear *v.* (**heard**) **1** perceive (sounds) with the ear. **2** pay attention to. **3** receive information. ▫ **hear! hear!** I agree. **hearer** *n.*

■ **1** catch, perceive. **2** attend to, heed, listen to, mark, note, pay attention to, take notice of. **3** be informed, be told, discover, find out, get wind of, learn.

hearing *n.* **1** ability to hear. **2** opportunity to be heard. **3** trial of a lawsuit. ▫ **hearing aid** small sound-amplifier worn by a partially deaf person.

hearsay *n.* rumour, gossip.

hearse *n.* vehicle for carrying the coffin at a funeral.

heart *n.* **1** muscular organ that keeps blood circulating. **2** centre of a person's emotions or inmost thoughts. **3** courage or enthusiasm. **4** central part. **5** figure representing a heart. **6** playing card of the suit marked with these. □ **break the heart of** cause overwhelming grief to. **by heart** memorized thoroughly. **heart attack** sudden failure of the heart to function normally. **heart-searching** *n.* examination of one's own feelings and motives. **heart-to-heart** *adj.* frank and personal. **heart-warming** *adj.* emotionally moving and encouraging.

■ **3** bravery, boldness, courage, *colloq.* guts, mettle, nerve, pluck; enthusiasm, spirit, verve. **4** centre, core, hub, middle, nucleus; crux, essence, kernel, nub, pith, quintessence, substance. □ **heart-to-heart** candid, frank, intimate, personal. **heart-warming** cheering, encouraging, moving, poignant, touching, uplifting.

heartache *n.* mental anguish.

heartbeat *n.* pulsation of the heart.

heartbreak *n.* overwhelming grief.

heartbroken *adj.* broken-hearted.

■ broken-hearted, desolate, devastated, grief-stricken, inconsolable, miserable, sad, sorrowful.

heartburn *n.* burning sensation in the chest from indigestion.

hearten *v.* encourage.

heartfelt *adj.* felt deeply, sincere.

■ ardent, deep, devout, earnest, genuine, honest, profound, sincere, unfeigned, wholehearted.

hearth *n.* **1** floor of a fireplace. **2** fireside.

heartless *adj.* not feeling pity or sympathy. □ **heartlessly** *adv.*

■ brutal, callous, cold, cold-blooded, cruel, hard, hard-hearted, inhuman, inhumane, insensitive, merciless, pitiless, ruthless, unfeeling, unkind, unsympathetic.

heart-throb *n.* (*colloq.*) object of romantic affection.

hearty *adj.* (**-ier**, **-iest**) **1** vigorous. **2** enthusiastic. **3** (of meals) large. **4** warm, friendly. □ **heartily** *adv.*, **heartiness** *n.*

■ **1** energetic, full-blooded, hale, healthy, robust, strong, vigorous. **2** eager, enthusiastic, spirited, vigorous. **3** ample, large, satisfying, sizeable, substantial. **4** affectionate, friendly, genial, warm.

heat *n.* **1** condition or sensation of being hot. **2** form of energy produced by movement of molecules. **3** passion, anger. **4** preliminary contest. ● *v.* make or become hot. □ **heatstroke** *n.* illness caused by overexposure to sun.

■ *n.* **1** warmness, warmth. **3** ardour, eagerness, fervour, intensity, passion, vehemence, warmth, zeal; anger, fury. **4** game, round.

heated *adj.* (of a person or discussion) angry. □ **heatedly** *adv.*

heater *n.* device supplying heat.

heath *n.* flat uncultivated land with low shrubs.

heathen *n.* person who does not believe in an established religion. ● *adj.* of or relating to heathens.

heather *n.* evergreen plant with purple, pink, or white flowers.

heave *v.* **1** lift or haul with great effort. **2** utter (a sigh). **3** (*colloq.*) throw. **4** rise and fall like waves. **5** pant, retch. ● *n.* act of heaving.

■ *v.* **1** hoist, lift, raise; drag, draw, haul, pull, tug. **2** breathe, utter.

heaven *n.* **1** abode of God. **2** place or state of bliss. **3** (**the heavens**) the sky as seen from the earth.

heavenly *adj.* **1** of heaven, divine. **2** of or in the heavens. **3** (*colloq.*) very pleasing. □ **heavenly bodies** sun, moon, stars, etc.

■ **1** angelic, celestial, divine, ethereal, holy, immortal, seraphic, spiritual. **2** celestial, planetary, stellar.

heavy *adj.* (**-ier**, **-iest**) **1** having great weight, force, or intensity. **2** dense. **3** hard to digest or understand. **4** dull, tedious. **5** serious. □ **heavy-hearted** *adj.* sad. **heavy industry** that producing metal or heavy machines etc. **heavily** *adv.*, **heaviness** *n.*

■ **1** hefty, massive, ponderous, weighty; bulky, burly, corpulent, fat, large, overweight, portly, stout, tubby; forceful, intense, severe, torrential, violent. **2** dense, solid, thick. **3** indigestible, stodgy; abstruse, complex, deep, difficult, esoteric, profound. **4** boring, dull, tedious, uninteresting, wearisome. **5** grave, important, serious, weighty.

heavyweight *adj.* having great weight or influence. ● *n.* heavyweight person.

Hebrew *n.* & *adj.* **1** (member) of a Semitic people in ancient Palestine. **2** (of) their language or a modern form of this. □ **Hebraic** *adj.*

heckle *v.* interrupt (a public speaker) with aggressive questions and abuse. □ **heckler** *n.*

hectare *n.* unit of area, 10,000 sq. metres (about 2½ acres).

hectic *adj.* with feverish activity. □ **hectically** *adv.*

■ busy, chaotic, confused, excited, feverish, frantic, frenetic, turbulent, wild.

hectogram *n.* 100 grams.

hector *v.* intimidate by bullying.

hedge *n.* **1** fence of bushes or shrubs. **2** barrier. ● *v.* **1** surround with a hedge. **2** make or trim hedges. **3** avoid giving a direct answer or commitment.

hedgehog *n.* small animal covered in stiff spines.

hedgerow *n.* bushes etc. forming a hedge.

hedonist *n.* person who believes pleasure is the chief good. □ **hedonism** *n.*, **hedonistic** *adj.*

heed *v.* pay attention to. ● *n.* careful attention. □ **heedful** *adj.*, **heedless** *adj.*

■ *v.* attend to, be mindful of, listen to, mark, mind, note, pay attention to, take notice of. ● *n.* attention, consideration, notice, thought. □ **heedful** alert, attentive, careful, cautious, mindful, observant, on the lookout, vigilant, watchful. **heedless** careless, inattentive, incautious, thoughtless, unmindful, unobservant, unthinking.

heel[1] *n.* **1** back part of the human foot. **2** part of a stocking or shoe covering or supporting this. **3** (*sl.*) dishonourable man. ● *v.* make or repair the heel(s) of. □ **down at heel** shabby. **take to one's heels** run away.

heel[2] *v.* tilt (a ship) or become tilted to one side. ● *n.* this tilt.

hefty *adj.* (**-ier**, **-iest**) large and heavy.

■ beefy, big, brawny, burly, muscular, powerful, robust, strapping; bulky, heavy, large, massive, substantial, weighty.

hegemony /hijémmәni/ *n.* leadership, esp. by one country.

Hegira /héjirә/ *n.* Muhammad's flight from Mecca (AD 622), from which the Muslim era is reckoned.

heifer /héffәr/ *n.* young cow.

height *n.* **1** measurement from base to top. **2** distance above ground or sea level. **3** highest degree of something.

■ **2** altitude, elevation. **3** acme, climax, culmination, high point, peak, pinnacle, summit, zenith.

heighten *v.* make or become higher or more intense.

■ elevate, lift, raise; add to, amplify, augment, deepen, enhance, increase, intensify, strengthen.

heinous /háynәss/ *adj.* very wicked.

heir /air/ *n.* person entitled to inherit property or a rank etc.

heiress /áiriss/ *n.* female heir, esp. to great wealth.

heirloom /áirlōōm/ *n.* possession handed down in a family for several generations.

held *see* **hold[1]**.

helical *adj.* like a helix.

helicopter *n.* aircraft with blades that revolve horizontally.

heliport *n.* helicopter station.

helium *n.* light colourless gas that does not burn.

helix *n.* (*pl.* **-ices**) spiral.

hell *n.* **1** place of punishment for the wicked after death. **2** place or state of supreme misery. □ **hell-bent** *adj.* recklessly determined.

Hellenistic *adj.* of Greece in the 4th–1st centuries BC.

hello *int.* & *n.* exclamation used in greeting or to call attention.

helm *n.* tiller or wheel by which a ship's rudder is controlled.

helmet *n.* protective head-covering.

helmsman *n.* person controlling a ship's helm.

help *v.* **1** provide with the means to what is sought or needed. **2** be useful (to). **3** improve (a situation). **4** serve with food. ● *n.* **1** act of helping. **2** person or thing that helps. □ **helper** *n.*

■ *v.* **1** aid, assist, lend a hand; encourage, fortify, succour, support; back, finance, fund, sponsor, subsidize. **2** be of use (to), be useful (to), serve. **3** alleviate, ease, improve, mitigate, relieve, remedy; expedite, facilitate, further. ● *n.* aid, assistance, backing, succour, support. □ **helper** accessory, accomplice, aide, assistant, helper, partner, right-hand man *or* woman, supporter.

helpful *adj.* giving help, useful. □ **helpfully** *adv.*, **helpfulness** *n.*

■ accommodating, cooperative, kind, neighbourly, obliging, supportive; beneficial, constructive, practical, useful, valuable.

helping *n.* portion of food served.

helpless *adj.* **1** unable to manage without help. **2** powerless. ▫ **helplessly** *adv.*, **helplessness** *n.*

■ **1** feeble, incapable, infirm, weak. **2** defenceless, exposed, powerless, unprotected, vulnerable.

helpline *n.* telephone service providing help with problems.

helter-skelter *adv.* in disorderly haste. ● *n.* spiral slide at a funfair.

hem *n.* edge (of cloth) turned under and sewn or fixed down. ● *v.* (**hemmed**) sew thus. ▫ **hem in** surround and restrict.

hemisphere *n.* **1** half a sphere. **2** half the earth. ▫ **hemispherical** *adj.*

hemlock *n.* poisonous plant.

hemp *n.* **1** plant with coarse fibres used in making rope etc. **2** narcotic drug made from it.

hempen *adj.* made of hemp.

hen *n.* female bird, esp. of the domestic fowl. ▫ **hen-party** *n.* (*colloq.*) party for women only.

hence *adv.* **1** from this time. **2** for this reason.

henceforth *adv.* (also **henceforward**) from this time on.

henchman *n.* trusty supporter.

henna *n.* **1** reddish dye used esp. on the hair. **2** tropical plant from which it is made. ▫ **hennaed** *adj.*

henpecked *adj.* (of a man) nagged by his wife.

henry *n.* unit of inductance.

hepatic *adj.* of the liver.

hepatitis *n.* inflammation of the liver.

heptagon *n.* geometric figure with seven sides. ▫ **heptagonal** *adj.*

heptathlon *n.* athletic contest involving seven events.

her *pron.* objective case of *she.* ● *adj.* belonging to her.

herald *n.* person or thing heralding something. ● *v.* proclaim the approach of.

heraldic *adj.* of heraldry.

heraldry *n.* study of armorial bearings.

herb *n.* plant used in making medicines or flavourings.

herbaceous *adj.* soft-stemmed. ▫ **herbaceous border** border containing esp. perennial plants.

herbal *adj.* of herbs. ● *n.* book about herbs.

herbalist *n.* dealer in medicinal herbs.

herbicide *n.* substance used to destroy plants. ▫ **herbicidal** *adj.*

herbivore *n.* plant-eating animal. ▫ **herbivorous** *adj.*

herculean *adj.* needing or showing great strength or effort.

herd *n.* **1** group of animals feeding or staying together. **2** mob. ● *v.* **1** gather, stay, or drive as a group. **2** tend (a herd). ▫ **herdsman** *n.*

■ *n.* **1** drove, flock, pack. **2** crowd, group, horde, mob, multitude, swarm, throng. ● *v.* **1** assemble, collect, congregate, flock, gather, huddle; drive, round up. **2** look after, take care of, tend.

here *adv.* **1** in, at, or to this place. **2** at this point. ● *n.* this place.

hereabouts *adv.* near here.

hereafter *adv.* from now on. ● *n.* **1** the future. **2** the next world.

hereby *adv.* by this act.

hereditary *adj.* **1** inherited. **2** holding a position by inheritance.

heredity *n.* inheritance of characteristics from parents.

herein *adv.* in this place or book etc.

heresy *n.* **1** opinion contrary to accepted beliefs. **2** holding of this.

heretic *n.* person who holds a heresy. ▫ **heretical** *adj.*

hereto *adv.* to this

herewith *adv.* with this.

heritage *n.* **1** thing(s) inherited. **2** nation's historic buildings etc.

■ **1** bequest, birthright, inheritance, legacy, patrimony.

hermaphrodite *n.* creature with male and female sexual organs.

hermetic *adj.* with an airtight seal. ▫ **hermetically** *adv.*

hermit *n.* person living in solitude.

hermitage *n.* hermit's dwelling.

hernia *n.* protrusion of part of an organ through the wall of the cavity containing it.

hero *n.* (*pl.* **-oes**) **1** man admired for his brave deeds. **2** chief male character in a story etc.

heroic *adj.* very brave. ● *n.pl.* over-dramatic behaviour. ▫ **heroically** *adv.*

■ *adj.* bold, brave, courageous, daring, dauntless, fearless, gallant, intrepid, manly, spirited, valiant, valorous.

heroin *n.* powerful drug prepared from morphine.

heroine *n.* female hero.

heroism *n.* heroic conduct.

heron *n.* long-legged wading bird.

herpes /hérpeez/ *n.* virus disease causing blisters.

herring *n.* edible N. Atlantic fish. □ **herringbone** *n.* zigzag pattern or arrangement.

hers *poss.pron.* belonging to her.

herself *pron.* emphatic and reflexive form of *she* and *her*.

hertz *n.* (*pl.* **hertz**) unit of frequency of electromagnetic waves.

hesitant *adj.* hesitating. □ **hesitantly** *adv.*, **hesitancy** *n.*

■ ambivalent, cautious, doubtful, hesitating, in two minds, irresolute, tentative, undecided, unsure.

hesitate *v.* **1** pause doubtfully. **2** be reluctant. □ **hesitation** *n.*

■ **1** be in two minds, be uncertain *or* undecided, dither, hang back, pause, shilly-shally, vacillate, waver. **2** be disinclined, be reluctant, have misgivings *or* qualms, scruple.

hessian *n.* strong coarse cloth of hemp or jute.

heterodox *adj.* not orthodox.

heterogeneous *adj.* made up of people or things of various sorts. □ **heterogeneity** *n.*

heterosexual *adj.* & *n.* (person) sexually attracted to people of the opposite sex. □ **heterosexuality** *n.*

hew *v.* (**hewn**) **1** chop or cut with an axe etc. **2** cut into shape.

hexagon *n.* geometric figure with six sides. □ **hexagonal** *adj.*

hey *int.* exclamation of surprise or interest, or to attract attention.

heyday *n.* time of greatest success.

hi *int.* exclamation calling attention or greeting.

hiatus *n.* (*pl.* **-tuses**) break or gap in a sequence or series.

hibernate *v.* spend the winter in sleep-like state. □ **hibernation** *n.*

Hibernian *adj.* & *n.* (native) of Ireland.

hibiscus *n.* shrub or tree with trumpet-shaped flowers.

hiccup *n.* cough-like stopping of breath. ● *v.* (**hiccuped**) make this sound.

hide[1] *v.* (**hid, hidden**) **1** put or keep out of sight. **2** keep secret. **3** conceal oneself. □ **hideout** *n.* (*colloq.*) hiding place.

■ **1** bury, conceal, cover up, put out of sight, secrete, *colloq.* stash away. **2** camouflage, conceal, cover up, disguise, keep secret, mask, suppress. **3** go into hiding, *sl.* lie doggo, lie low, lurk.

hide[2] *n.* animal's skin.

hidebound *adj.* rigidly conventional.

■ conservative, conventional, narrow-minded, reactionary, set in one's ways, strait-laced.

hideous *adj.* very ugly. □ **hideously** *adv.*, **hideousness** *n.*

■ disgusting, ghastly, grisly, grotesque, gruesome, monstrous, repellent, repulsive, revolting, ugly, unsightly.

hiding *n.* (*colloq.*) thrashing.

hierarchy *n.* system with grades of status. □ **hierarchical** *adj.*

hieroglyph *n.* pictorial symbol used in ancient Egyptian and other writing. □ **hieroglyphic** *adj.*, **hieroglyphics** *n.pl.*

hi-fi *adj.* & *n.* (*colloq.*) high fidelity, (equipment) reproducing sound with little or no distortion.

higgledy-piggledy *adj.* & *adv.* in complete confusion.

high *adj.* **1** extending far or a specified distance upwards. **2** far above ground or sea level. **3** ranking above others. **4** greater than normal. **5** (of sound or a voice) not deep or low. **6** (of meat) slightly decomposed. **7** (*sl.*) intoxicated, under the influence of a drug. ● *n.* **1** high level. **2** area of high pressure. ● *adv.* in, at, or to a high level. □ **higher education** education at university etc. **high-handed** *adj.* using authority arrogantly. **high-rise** *adj.* with many storeys. **high road** main road. **high sea(s)** sea outside a country's territorial waters. **high season** busiest season. **high-spirited** *adj.* lively, cheerful. **high spirits** liveliness, cheerfulness. **high street** principal shopping street. **high tea** early evening meal with tea and cooked food. **high-tech** *adj.* involving advanced technology and electronics. **high-water mark** level reached by the tide at its highest level.

■ *adj.* **1** elevated, lofty, tall, towering. **3** chief, distinguished, eminent, foremost, important, leading, principal, superior. **4** excessive, exorbitant, extreme, *colloq.* steep, stiff. **5** high-pitched, penetrating, piercing, piping, shrill, squeaky, treble. **6** gamy. ● *n.* **1** height, peak, record, summit. □ **high-handed** arrogant, autocratic, *colloq.* bossy, dictatorial, domineering, imperious, overbearing, peremptory. **high-spirited** animated, bubbly, buoyant, cheerful, ebullient, exuberant, lively, spirited, vivacious.

highbrow *adj.* very intellectual, cultured. ● *n.* highbrow person.

■ *adj.* cultured, erudite, intellectual, learned, scholarly.

highlands *n.pl.* mountainous region. □ **highland** *adj.*, **highlander** *n.*

highlight *n.* **1** bright area in a picture. **2** best feature. ● *v.* emphasize.

■ *v.* accent, accentuate, draw attention to, emphasize, feature, focus attention on, point up, spotlight, stress, underline.

highly *adv.* **1** in a high degree, extremely. **2** very favourably. □ **highly-strung** *adj.* sensitive, nervous.

■ **1** exceptionally, extraordinarily, extremely, immensely, tremendously. **2** admiringly, appreciatively, favourably, warmly, well. □ **highly-strung** emotional, excitable, jumpy, nervous, nervy, sensitive, temperamental, touchy, volatile.

highway *n.* **1** public road. **2** main route.

highwayman *n.* person (usu. on horseback) who robbed travellers in former times.

hijack *v.* seize control illegally of (a vehicle or aircraft in transit). ● *n.* hijacking. □ **hijacker** *n.*

hike *n.* long walk. ● *v.* go for a hike. □ **hiker** *n.*

hilarious *adj.* **1** extremely funny. **2** boisterous and merry. □ **hilariously** *adv.*, **hilarity** *n.*

■ **1** comical, funny, *sl.* priceless. **2** boisterous, jolly, merry, rollicking, uproarious.

hill *n.* **1** raised part of earth's surface, lower than a mountain. **2** slope in a road etc. **3** mound. □ **hill-billy** *n.* (*US*) rustic person.

■ **1** elevation, hillock, hummock, knoll, mound. **2** gradient, incline, rise, slope. **3** heap, mound, pile, stack.

hillock *n.* small hill, mound.

hilt *n.* handle of a sword or dagger. □ **to the hilt** completely.

him *pron.* objective case of *he*.

Himalayan *adj.* of the Himalaya Mountains.

himself *pron.* emphatic and reflexive form of *he* and *him*.

hind¹ *n.* female deer.

hind² *adj.* situated at the back.

hinder *v.* delay progress of.

■ baulk, delay, foil, forestall, frustrate, hamper, handicap, hold up, impede, obstruct, prevent, set back, slow down, thwart.

Hindi *n.* group of languages of northern India.

hindmost *adj.* furthest behind.

hindrance *n.* **1** thing that hinders. **2** hindering, being hindered.

■ **1** catch, drawback, hitch, impediment, obstacle, obstruction, snag, stumbling block.

hindsight *n.* wisdom about an event after it has occurred.

Hindu *n.* person whose religion is Hinduism. ● *adj.* of Hindus or Hinduism.

Hinduism *n.* principal religion and philosophy of India.

Hindustani *n.* language of much of northern India and Pakistan.

hinge *n.* movable joint on which a door or lid turns. ● *v.* attach or be attached by hinge(s). □ **hinge on** depend on.

hint *n.* **1** slight or indirect indication or suggestion. **2** piece of practical information. **3** faint trace. ● *v.* make a hint.

■ *n.* **1** allusion, clue, implication, indication, inkling, innuendo, insinuation, intimation, suggestion. **2** piece of advice, suggestion, tip. **3** shade, suggestion, touch, trace. ● *v.* imply, indicate, insinuate, intimate, suggest.

hinterland *n.* district behind that lying along a coast etc.

hip¹ *n.* projection of the pelvis on each side of body.

hip² *n.* fruit of wild rose.

hippopotamus *n.* (*pl.* **-muses**) large African river animal with a thick skin.

hire *v.* engage or grant temporary use of, for payment. ● *n.* hiring. □ **hire purchase** system of purchase by paying in instalments. **hirer** *n.*

■ *v.* charter, lease, rent; employ, engage, take on. ● *n.* lease, rental.

hireling *n.* (*derog.*) hired helper.

hirsute /húrsyoot/ *adj.* hairy.

his *adj.* & *poss.pron.* belonging to him.

Hispanic *adj.* & *n.* (native) of Spain or a Spanish-speaking country.

hiss *n.* sound like 's'. ● *v.* **1** make this sound. **2** utter with a hiss. **3** express disapproval in this way.

histamine *n.* substance present in the body associated with allergic reactions.

histology *n.* study of organic tissues. □ **histological** *adj.*

historian *n.* expert in or writer of history.

historic *adj.* famous in history.

■ celebrated, famous, great, important, memorable, momentous, notable, noteworthy, significant, unforgettable.

historical *adj.* of or concerned with history. □ **historically** *adv.*

■ authentic, documented, factual, recorded, true, verifiable.

history *n.* **1** continuous record of (esp. public) events. **2** study of past events. **3** series of events, facts, etc. connected with a person, thing, or place.

■ **1** account, annals, chronicle, description, narrative, record, story. **3** antecedents, background, life, past.

histrionic *adj.* **1** of acting. **2** theatrical in manner. ● *n.pl.* **1** theatricals. **2** theatrical behaviour.

hit *v.* (**hit, hitting**) **1** strike with a blow or missile. **2** come forcefully against. **3** affect badly. **4** reach. **5** encounter. ● *n.* **1** blow, stroke. **2** shot that hits its target. **3** success. □ **hit it off** get on well together. **hit on** find, esp. by chance. **hit-or-miss** *adj.* aimed or done carelessly. **hitter** *n.*

■ *v.* **1** bash, batter, beat, *sl.* belt, *sl.* biff, box, buffet, *colloq.* clip, *sl.* clobber, clout, cuff, pound, pummel, slap, smack, *colloq.* sock, spank, strike, swat, *colloq.* swipe, thrash, thump, thwack, *sl.* wallop, *colloq.* whack. **2** bang into, collide with, crash into, knock against, run into, smash into, strike. **3** affect, have an impact on, leave a mark on. **4** arrive at, come *or* get to, reach. **5** be faced with, encounter, experience, meet (with). □ **hit on** come across, discover, find, happen on, light on.

hitch *v.* **1** fasten with a loop or hook. **2** move (a thing) with a slight jerk. **3** hitchhike, obtain (a lift) in this way. ● *n.* **1** snag. **2** slight jerk. **3** noose or knot of various kinds.

■ *v.* **1** attach, connect, couple, fasten, hook, join, link, tether, tie. ● *n.* **1** catch, difficulty, hindrance, impediment, obstacle, problem, snag, stumbling block.

hitchhike *v.* travel by seeking free lifts in passing vehicles. □ **hitchhiker** *n.*

hi-tech *adj.* high-tech.

hither *adv.* to or towards this place.

hitherto *adv.* until this time.

HIV *abbr.* human immunodeficiency virus (causing Aids).

hive *n.* structure in which bees live. ● *v.* **hive off** separate from a larger group.

hives *n.pl.* skin eruption, esp. nettle-rash.

hoard *v.* save and put away. ● *n.* things hoarded. □ **hoarder** *n.*

■ *v.* accumulate, amass, collect, lay up, put by, reserve, save, *colloq.* stash away, stockpile, store (up). ● *n.* accumulation, cache, collection, reserve, stock, stockpile, store, supply.

hoarding *n.* fence of boards, often bearing advertisements.

hoar-frost *n.* white frost.

hoarse *adj.* **1** (of a voice) sounding rough as if from a dry throat. **2** having such a voice. □ **hoarsely** *adv.*, **hoarseness** *n.*

■ croaking, gruff, guttural, husky, rough, throaty.

hoary *adj.* (**-ier, -iest**) **1** grey with age. **2** (of a joke etc.) old.

hoax *v.* deceive jokingly. ● *n.* joking deception. □ **hoaxer** *n.*

■ *v.* *colloq.* bamboozle, deceive, dupe, fool, hoodwink, pull the wool over someone's eyes, swindle, trick. ● *n.* cheat, *sl.* con, joke, practical joke, trick.

hob *n.* top of a cooker, with hotplates.

hobble *v.* **1** walk lamely. **2** fasten the legs of (a horse) to limit its movement. ● *n.* **1** hobbling walk. **2** rope etc. used to hobble a horse.

hobby *n.* thing done often and for pleasure in one's spare time.

■ interest, leisure activity, pastime, pursuit, recreation, sideline.

hobby horse **1** stick with a horse's head, as a toy. **2** favourite topic.

hobgoblin *n.* mischievous imp.

hobnail *n.* heavy-headed nail for boot-soles. □ **hobnailed** *adj.*

hobnob *v.* (**-nobbed**) spend time together in a friendly way.

■ consort, fraternize, keep company, mingle, mix, socialize.

hock[1] *n.* middle joint of an animal's hind leg.

hock[2] *n.* German white wine.

hockey *n.* **1** field game played with curved sticks and a small hard ball. **2** ice hockey.

hocus-pocus *n.* trickery.

hod *n.* **1** trough on a pole for carrying mortar or bricks. **2** portable container for coal.

hoe *n.* tool for loosening soil or scraping up weeds. ● *v.* (**hoeing**) dig or scrape with a hoe.

hog *n.* **1** castrated male pig reared for meat. **2** (*colloq.*) greedy person. ● *v.*

(**hogged**) (*colloq.*) **1** take greedily. **2** hoard selfishly.

hoick *v.* (*colloq.*) lift or jerk.

hoi polloi ordinary people.

hoist *v.* raise or haul up. ● *n.* apparatus for hoisting things.

■ *v.* elevate, haul up, heave up, lift, raise, winch. ● *n.* crane, davit, lift, winch.

hoity-toity *adj.* haughty.

hokum *n.* (*sl.*) nonsense.

hold[1] *v.* (**held**) **1** keep in one's arms or hands etc. **2** keep in a particular position. **3** contain. **4** possess (property etc.). **5** bear the weight of. **6** detain. **7** continue. **8** occupy, engross. **9** cause to take place. **10** believe. ● *n.* **1** act, manner, or means of holding. **2** means of exerting influence. □ **hold back** impede the progress of. **hold dear** regard with affection. **hold out 1** offer. **2** last. **3** continue to make a demand. **hold up 1** support. **2** hinder. **3** stop and rob by use of threats or force. **hold-up** *n.* **1** delay. **2** robbery. **hold with** (*colloq.*) approve of. **holder** *n.*

■ *v.* **1** clasp, clench, clutch, grasp, grip, hang on to; cradle, embrace, enfold, hug. **2** keep, maintain, sustain. **3** accommodate, carry, contain, take. **4** have, keep, own, possess, retain. **5** bear, carry, support, sustain, take. **6** confine, detain, keep in custody, shut up, restrain. **7** carry on, continue, go on, keep up, last, persist. **8** absorb, engage, engross, involve, occupy. **9** assemble, call, convene, convoke. **10** believe, consider, deem, judge, maintain, think. ● *n.* **1** clasp, clutch, grasp, grip; foothold, purchase. **2** ascendancy, authority, *colloq.* clout, control, dominance, influence, leverage, mastery, power, sway. □ **hold back** check, control, curb, hinder, impede, inhibit, restrain. **hold out 1** extend, offer, present, proffer, reach out, stretch out. **2** carry on, continue, last, persevere, persist. **hold up 1** bolster, buttress, prop up, shore up, support. **2** delay, hinder, impede, obstruct, set back, slow down.

hold[2] *n.* storage cavity below a ship's deck.

holdall *n.* large soft travel bag.

holding *n.* **1** something held or owned. **2** land held by an owner or tenant.

hole *n.* **1** hollow place. **2** burrow. **3** aperture. **4** (*colloq.*) wretched place. **5** (*colloq.*) awkward situation. ● *v.* make hole(s) in.

■ *n.* **1** cave, cavity, crater, dent, depression, dip, hollow, indentation, niche, nook, pit, pocket, pothole, recess. **2** burrow, den, lair, sett, tunnel. **3** aperture, breach, fissure, gap, opening, orifice, perforation, puncture, rip, slit, slot, tear, vent.

holey *adj.* full of holes.

holiday *n.* day(s) of recreation. ● *v.* spend a holiday. □ **holiday-maker** *n.* person on holiday.

■ *n.* leave, recess, *colloq.* vac, *US* vacation. □ **holiday-maker** sightseer, tourist, traveller, tripper, visitor.

holiness *n.* being holy.

holistic *adj.* (of treatment) involving the mind, body, social factors, etc.

hollow *adj.* **1** empty within, not solid. **2** sunken. **3** echoing as if in something hollow. **4** worthless. **5** insincere. ● *n.* **1** cavity. **2** sunken place. **3** valley. ● *v.* make hollow.

■ *adj.* **1** empty, void, unfilled. **2** cavernous, concave, indented, recessed, sunken. **3** echoing, low, rumbling. **4** empty, fruitless, futile, meaningless, pointless, profitless, useless, vain, valueless, worthless. **5** empty, false, hypocritical, insincere. ● *n.* **1,2** cavity, concavity, crater, dent, depression, dip, hole, indentation, pit, trough. **3** coomb, dale, dell, dingle, glen, valley. ● *v.* dig, excavate, furrow, gouge.

holly *n.* evergreen shrub with prickly leaves and red berries.

hollyhock *n.* plant with large flowers on a tall stem.

holocaust *n.* large-scale destruction, esp. by fire.

hologram *n.* three-dimensional photographic image.

holograph[1] *v.* record as a hologram. □ **holography** *n.*

holograph[2] *adj.* & *n.* (document) written wholly in the handwriting of the author.

holster *n.* leather case holding a pistol or revolver.

holy *adj.* (**-ier, -iest**) **1** belonging or devoted to God. **2** consecrated. **3** morally and spiritually excellent. □ **holy of holies** most sacred place.

■ **1** divine, sacred. **2** blessed, consecrated, sacred, sanctified. **3** devout, God-fearing, pious, religious, reverent, saintly, virtuous.

homage *n.* things said or done as a mark of respect or loyalty.

home *n.* **1** place where one lives. **2** dwelling house. **3** native land. **4** institution where those needing care may live. ● *adj.* **1** of one's home or country. **2** played on one's own ground. ● *adv.* **1** at or to one's home. **2** to the point aimed at. ● *v.* make its way home. ☐ **home in on** be guided to a destination. **at home 1** in one's own home. **2** at ease. **3** well-informed. **home truth** unpleasant truth about oneself.

■ *n.* **1** abode, domicile, habitation, house, lodging, quarters, residence. **2** *US* apartment, digs, dwelling, flat, house, lodgings. **3** country, fatherland, homeland, motherland, native land. **4** almshouse, hospice, institution, nursing home. ☐ **home in on** aim at *or* for, head for, make a beeline for, target, zero in on.

homeland *n.* native land.

homeless *adj.* lacking a home. ☐ **homelessness** *n.*

homely *adj.* (**-ier, -iest**) **1** simple and informal. **2** (*US*) plain, not beautiful. ☐ **homeliness** *n.*

■ **1** modest, ordinary, plain, simple, unassuming, unpretentious, unsophisticated; comfortable, cosy, folksy, friendly, informal, relaxed, snug, welcoming. **2** plain, ugly, unattractive.

homesick *adj.* longing for home.

homeward *adj.* & *adv.* going towards home. ☐ **homewards** *adv.*

homework *n.* work set for a pupil to do away from school.

homicide *n.* killing of one person by another. ☐ **homicidal** *adj.*

homily *n.* moralizing lecture. ☐ **homiletic** *adj.*

hominid *adj.* & *n.* (member) of the family including humans and their fossil ancestors.

homoeopathy /hṓmióppəthi/ *n.* treatment of a disease by very small doses of a substance that would produce the same symptoms in a healthy person. ☐ **homoeopathic** *adj.*

homogeneous *adj.* of the same kind, uniform. ☐ **homogeneously** *adv.*, **homogeneity** *n.*

homogenize *v.* treat (milk) so that cream does not separate and rise to the top.

homonym *n.* word with the same spelling as another.

homophobia *n.* hatred or fear of homosexuals.

homophone *n.* word with the same sound as another.

homosexual *adj.* & *n.* (person) sexually attracted to people of the same sex. ☐ **homosexuality** *n.*

hone *v.* sharpen on a whetstone.

honest *adj.* **1** truthful. **2** trustworthy. **3** fairly earned. ☐ **honestly** *adv.*, **honesty** *n.*

■ **1** candid, direct, forthright, frank, *colloq.* on the level, open, sincere, straight, straightforward, truthful, veracious. **2** decent, dependable, ethical, fair, good, honourable, just, law-abiding, loyal, moral, principled, reliable, trustworthy, upright, virtuous. **3** above board, bona fide, legitimate. ☐ **honesty** candour, directness, forthrightness, frankness, sincerity, truthfulness, veracity; equity, fairness, goodness, honour, integrity, probity, rectitude, trustworthiness, virtue, virtuousness.

honey *n.* (*pl.* **-eys**) **1** sweet substance made by bees from nectar. **2** darling. ☐ **honey bee** common bee living in a hive.

honeycomb *n.* **1** bees' wax structure for holding their honey and eggs. **2** pattern of six-sided sections.

honeydew melon melon with pale skin and sweet green flesh.

honeyed *adj.* sweet, sweet-sounding.

honeymoon *n.* **1** holiday spent together by a newly married couple. **2** initial period of goodwill. ● *v.* spend a honeymoon.

honeysuckle *n.* climbing shrub with fragrant flowers.

honk *n.* **1** sound of a car horn. **2** cry of a wild goose. ● *v.* make this noise.

honorary *adj.* **1** given as an honour. **2** unpaid.

honour *n.* **1** great respect or public regard. **2** mark of this, privilege. **3** adherence to what is right. **4** reputation, good name. ● *v.* **1** respect highly. **2** confer honour on. **3** pay (a cheque) or fulfil (a promise etc.).

■ *n.* **1** distinction, esteem, glory, kudos, prestige, regard, renown, respect, reverence, veneration. **2** award, reward, tribute; distinction, pleasure, privilege. **3** decency, fairness, goodness, honesty, justice, integrity, morality, probity, rectitude, righteousness, virtue, virtuousness. ● *v.* **1** admire, esteem, respect, revere, venerate. **2** acclaim, applaud, eulogize, glorify, laud, pay tribute to, praise. **3** pay, redeem; carry out, discharge, fulfil.

honourable *adj.* deserving, possessing, or showing honour. ☐ **honourably** *adv.*

■ decent, ethical, fair, good, honest, incorruptible, just, moral, noble, principled, respectable, trustworthy, upright, virtuous, worthy.

hood[1] *n.* **1** covering for the head and neck, esp. forming part of a garment. **2** hood-like thing or cover. **3** folding roof over a car. □ **hooded** *adj.*

hood[2] *n.* (*US*) gangster, gunman.

hoodlum *n.* hooligan.

hoodoo *n.* (*US*) **1** bad luck. **2** thing causing this.

hoodwink *v.* deceive.

■ cheat, *sl.* con, deceive, double-cross, dupe, hoax, mislead, pull the wool over someone's eyes, take in, trick.

hoof *n.* (*pl.* **hoofs** or **hooves**) horny part of a horse's foot.

hook *n.* **1** bent or curved device for catching hold or hanging things on. **2** short blow made with the elbow bent. ● *v.* **1** grasp, catch, or fasten with hook(s). **2** (in sports) send (a ball) in a curving or deviating path. □ **hook-up** *n.* interconnection. **off the hook** freed from a difficulty.

hookah *n.* tobacco pipe with a long tube passing through water.

hooked *adj.* hook-shaped. □ **hooked on** (*sl.*) addicted to.

hookworm *n.* parasitic worm with hooklike mouthparts.

hooligan *n.* young ruffian. □ **hooliganism** *n.*

■ delinquent, hoodlum, lout, ruffian, tearaway, thug, tough, vandal, *sl.* yob.

hoop *n.* **1** circular band of metal or wood. **2** metal arch used in croquet.

hoopla *n.* game in which rings are thrown to encircle a prize.

hoopoe *n.* bird with a crest and striped plumage.

hooray *int.* & *n.* = **hurrah.**

hoot *n.* **1** owl's cry. **2** sound of a hooter. **3** cry of laughter or disapproval. **4** cause of laughter. ● *v.* (cause to) make a hoot.

hooter *n.* thing that hoots, esp. a siren or car's horn.

Hoover *n.* [P.] a kind of vacuum cleaner. ● *v.* (**hoover**) clean with a vacuum cleaner.

hop[1] *v.* (**hopped**) **1** (of a person) jump on one foot. **2** (of an animal) jump from both or all feet. **3** (*colloq.*) make a quick short trip. ● *n.* **1** hopping movement. **2** informal dance. **3** short journey.

hop[2] *n.* **1** climbing plant cultivated for its cones. **2** (*pl.*) these cones, used to give a bitter flavour to beer.

hope *n.* **1** feeling of expectation and desire. **2** person or thing giving cause for this. **3** what one hopes for. ● *v.* feel hope.

■ *n.* **1** anticipation, expectancy, expectation. **3** ambition, aspiration, dream, wish.

hopeful *adj.* **1** feeling or inspiring hope. **2** promising. □ **hopefully** *adv.*

■ **1** confident, full of hope, optimistic, sanguine. **2** auspicious, bright, encouraging, favourable, promising, propitious, rosy.

hopeless *adj.* **1** feeling no hope. **2** inadequate, incompetent. **3** not likely to improve or succeed. □ **hopelessly** *adv.*, **hopelessness** *n.*

■ **1** dejected, desolate, despairing, despondent, downcast, inconsolable, in despair, melancholy, miserable, wretched. **2** bad, inadequate, incompetent, inept, poor. **3** beyond hope, desperate, irremediable, irreparable, irretrievable; futile, impossible, impracticable, pointless, useless, vain, worthless.

hopper *n.* **1** one who hops. **2** container with an opening at its base through which its contents can be discharged.

hopscotch *n.* game involving hopping over marked squares.

horde *n.* large group or crowd.

■ crowd, flock, gang, herd, host, mob, multitude, swarm, throng.

horizon *n.* **1** line at which earth and sky appear to meet. **2** limit of knowledge or interests.

horizontal *adj.* parallel to the horizon, going straight across. □ **horizontally** *adv.*

hormone *n.* secretion (or synthetic substance) that stimulates an organ or growth. □ **hormonal** *adj.*

horn *n.* **1** hard pointed growth on the heads of certain animals. **2** substance of which this is made. **3** similar projection. **4** wind instrument with a trumpet-shaped end. **5** device for sounding a warning signal. □ **horn-rimmed** *adj.* with frames of material like horn or tortoiseshell.

hornblende *n.* dark mineral constituent of granite etc.

hornet *n.* a kind of large wasp.

hornpipe *n.* lively solo dance associated esp. with sailors.

horny *adj.* (**-ier, -iest**) **1** of or like horn. **2** hardened and calloused. □ **horniness** *n.*

horology *n.* art of making clocks etc. □ **horologist** *n.*

horoscope *n.* forecast of events based on the positions of stars.

horrendous *adj.* horrifying □ **horrendously** *adv.*

horrible *adj.* **1** causing horror. **2** (*colloq.*) unpleasant. □ **horribly** *adv.*

■ **1** appalling, awful, dreadful, fearful, frightful, ghastly, grim, grisly, gruesome, hideous, horrendous, horrid, horrific, horrifying, repulsive, revolting, shocking, terrible. **2** awful, *colloq.* beastly, disagreeable, disgusting, *colloq.* frightful, horrid, nasty, objectionable, obnoxious, offensive, *colloq.* off-putting, repellent, unpleasant.

horrid *adj.* horrible.

horrific *adj.* horrifying. □ **horrifically** *adv.*

horrify *v.* arouse horror in, shock.

■ appal, disgust, dismay, frighten, outrage, repel, revolt, scandalize, scare, shock, startle, terrify.

horror *n.* **1** extreme fear, dread. **2** intense dislike or dismay. **3** person or thing causing horror.

■ **1** alarm, dread, fear, fright, panic, perturbation, terror, trepidation. **2** abhorrence, antipathy, aversion, detestation, dislike, distaste, hatred, hostility, loathing, odium, repugnance, revulsion.

hors d'oeuvre /or dőrvrə/ food served as an appetizer.

horse *n.* **1** four-legged animal with a mane and tail. **2** padded structure for vaulting over in a gymnasium. ● *v.* (*colloq.*) fool, play. □ **horse chestnut 1** brown shiny nut. **2** tree bearing this. **horse sense** (*colloq.*) common sense.

horseback *n.* **on horseback** riding on a horse.

horsebox *n.* closed vehicle for transporting a horse.

horseman, **horsewoman** *ns.* rider on horseback. □ **horsemanship** *n.*

horseplay *n.* boisterous play.

horsepower *n.* unit for measuring the power of an engine.

horseradish *n.* plant with a hot-tasting root used to make sauce.

horseshoe *n.* **1** U-shaped strip of metal nailed to a horse's hoof. **2** thing shaped like this.

horsy *adj.* **1** of or like a horse. **2** interested in horses.

horticulture *n.* art of garden cultivation. □ **horticultural** *adj.*, **horticulturist** *n.*

hose *n.* **1** (also **hosepipe**) flexible tube for conveying water. **2** stockings and socks. ● *v.* water or spray with a hosepipe.

hosiery *n.* stockings, socks, etc.

hospice *n.* hospital or home for the terminally ill.

hospitable *adj.* giving hospitality. □ **hospitably** *adv.*

■ convivial, genial, friendly, kind, neighbourly, sociable, warm, welcoming.

hospital *n.* institution for treatment of sick or injured people.

■ clinic, infirmary, sanatorium.

hospitality *n.* friendly and generous entertainment of guests.

hospitalize *v.* send or admit to a hospital. □ **hospitalization** *n.*

host¹ *n.* large number of people or things.

■ army, crowd, herd, horde, legion, mass, mob, multitude, pack, swarm, throng.

host² *n.* **1** person who entertains guest(s). **2** organism on which another lives as a parasite. ● *v.* act as host to.

hostage *n.* person held as security that the holder's demands will be satisfied.

hostel *n.* lodging house for students, nurses, etc.

hostess *n.* woman who entertains guests.

hostile *adj.* **1** of an enemy. **2** unfriendly, opposed.

■ **1** aggressive, belligerent, combative, militant, opposing, warring. **2** inhospitable, inimical, malevolent, unfriendly, unsympathetic, unwelcoming; antagonistic, averse, opposed.

hostility *n.* **1** being hostile, enmity. **2** (*pl.*) acts of warfare.

■ **1** animosity, animus, antagonism, antipathy, aversion, enmity, hatred, ill will, malevolence, opposition, unfriendliness. **2** (**hostilities**) action, combat, fighting, war, warfare.

hot *adj.* (**hotter**, **hottest**) **1** at or having a high temperature. **2** producing a burning sensation to the taste. **3** excited. **4** eager. ● *v.* (**hotted**) **hot up** (*colloq.*) make or become hot or exciting. □ **hot air** (*colloq.*) excited or boastful talk. **hot dog** hot sausage in a bread roll. **hot line** direct line for speedy communication. **hot-tempered** *adj.* impulsively angry. **in hot water** in trouble.

■ *adj.* **1** burning, fiery, red-hot, scorching, sultry, sweltering, torrid, white-hot. **2** peppery, piquant, sharp, spicy. **3** ardent, enthusiastic, excited, fervent, fervid, impassioned, intense, passionate, vehement.

4 anxious, avid, eager, keen. □ **hot-tempered** excitable, irascible, irritable, quick-tempered, stormy, temperamental, volatile.

hotbed *n.* place encouraging vice, intrigue, etc.

hotchpotch *n.* jumble.

hotel *n.* building providing meals and rooms for travellers.

hotelier *n.* hotel-keeper.

hotfoot *adv.* in eager haste.

■ hastily, helter-skelter, hurriedly, pell-mell, rapidly, swiftly.

hothead *n.* impetuous person.

hotheaded *adj.* impetuous.

■ foolhardy, hasty, impetuous, impulsive, madcap, precipitate, rash, reckless, thoughtless, wild.

hothouse *n.* heated greenhouse.

hotplate *n.* heated surface on a cooker.

hotpot *n.* stew of meat and vegetables.

houmous *n.* = **hummus**.

hound *n.* dog used in hunting. ● *v.* pursue, harass.

■ *v.* chase, hunt, pursue; annoy, badger, harass, harry, *colloq.* hassle, nag, persecute, pester, *colloq.* plague.

hour *n.* **1** one twenty-fourth part of a day and night. **2** point of time. **3** occasion. **4** (*pl.*) period for daily work. □ **hourly** *adj.* & *adv.*

hourglass *n.* glass containing sand that takes one hour to trickle from upper to lower section through a narrow opening.

houri /hoori/ *n.* (*pl.* **-is**) beautiful young woman of the Muslim paradise.

house *n.* /howss/ **1** building for people (usu. one family) to live in, or for a specific purpose. **2** household. **3** legislative assembly. **4** business firm. **5** family, dynasty. **6** theatre audience or performance. ● *v.* /howz/ **1** provide accommodation or storage space for. **2** encase. □ **house arrest** detention in one's own home. **house-proud** *adj.* attentive to the appearance of one's home. **house-trained** *adj.* trained to be clean in the house. **house-warming** *n.* party celebrating a move to a new home.

■ *n.* **1** abode, domicile, dwelling, habitation, home, lodging(s), residence; building, edifice, structure. **2** household, ménage. **3** legislature, parliament. **4** business, company, concern, corporation, enterprise, establishment, firm, organization. **5** clan, dynasty, family, line, lineage. ● *v.* **1** accommodate, harbour, lodge, put up, quarter, shelter. **2** contain, cover, encase, enclose.

houseboat *n.* barge-like boat equipped for living in.

housebound *adj.* confined to one's house through illness etc.

housebreaker *n.* burglar. □ **housebreaking** *n.*

housecoat *n.* woman's garment for informal wear in the house.

household *n.* occupants of a house. □ **household word** familiar saying or name.

householder *n.* person owning or renting a house or flat.

housekeeper *n.* person employed to look after a household.

housekeeping *n.* **1** management of household affairs. **2** money to be used for this.

housemaster, housemistress *ns.* teacher in charge of a school boarding house.

housewife *n.* woman managing a household. □ **housewifely** *adj.*

housework *n.* cleaning and cooking etc. done in the home.

housing *n.* **1** accommodation. **2** rigid case enclosing machinery.

hovel *n.* small miserable dwelling.

hover *v.* **1** (of a bird etc.) remain in one place in the air. **2** linger, wait close at hand. □ **hover fly** wasp-like insect that hovers.

hovercraft *n.* (*pl.* **-craft**) vehicle supported by air thrust downwards from its engines.

how *adv.* **1** by what means, in what way. **2** to what extent or amount etc. **3** in what condition.

howdah *n.* seat, usu. with a canopy, on an elephant's back.

however *adv.* **1** in whatever way, to whatever extent. **2** nevertheless.

howitzer *n.* short gun firing shells at high elevation.

howl *n.* long loud wailing cry or sound. ● *v.* **1** make or utter with a howl. **2** weep loudly.

■ *n.* cry, scream, ululation, wail, yelp, yowl. ● *v.* **1** bay, bellow, cry, scream, ululate, wail, yelp, yowl. **2** bawl, cry, wail, weep.

howler *n.* **1** animal etc. that howls. **2** (*colloq.*) stupid mistake.

hoyden *n.* boisterous girl.

h.p. *abbr.* **1** hire purchase. **2** horsepower.

hub *n.* **1** central part of a wheel. **2** centre of activity, interest, etc. □ **hubcap** *n.* cover for the hub of a car wheel.

■ **2** centre, core, focal point, focus, heart, nucleus.

hubbub *n.* confused noise of voices.

hubris /hyo͞obriss/ *n.* arrogant pride.

huckleberry *n.* **1** N. American shrub. **2** its fruit.

huddle *v.* crowd into a small place. ● *n.* close mass.

■ *v.* cluster, crowd, flock, gather, squeeze. ● *n.* bunch, clump, cluster, crowd, group, throng.

hue[1] *n.* colour.

■ colour, shade, tincture, tinge, tint, tone.

hue[2] *n.* **hue and cry** outcry.

huff *n.* fit of annoyance. ● *v.* blow. □ **huffy** *adj.*, **huffily** *adv.*

hug *v.* (**hugged**) **1** squeeze tightly in one's arms. **2** keep close to. ● *n.* hugging movement.

■ *v.* **1** clasp, cuddle, embrace, squeeze.

huge *adj.* extremely large. □ **hugely** *adv.*, **hugeness** *n.*

■ colossal, enormous, gargantuan, giant, gigantic, huge, immense, mammoth, massive, monumental, prodigious, titanic, tremendous, vast.

hula *n.* Polynesian women's dance. □ **hula hoop** large hoop for spinning round the body.

hulk *n.* **1** body of an old ship. **2** large clumsy-looking person or thing.

hulking *adj.* (*colloq.*) large and clumsy.

hull[1] *n.* framework of a ship.

hull[2] *n.* **1** pod of a pea or bean. **2** cluster of leaves on a strawberry. ● *v.* remove the hull of.

hullabaloo *n.* uproar.

■ clamour, commotion, din, disorder, disturbance, fracas, furore, hubbub, pandemonium, racket, rumpus, tumult, uproar.

hullo *int.* = **hello**.

hum *v.* (**hummed**) **1** make a low continuous sound like a bee. **2** sing with closed lips. **3** (*colloq.*) be in state of activity. ● *n.* humming sound.

■ *v.* **1** buzz, drone, murmur, purr, whirr.

human *adj.* **1** of humankind. **2** having the weaknesses or strengths of humankind. ● *n.* human being. □ **human being** man, woman, or child.

humane *adj.* kind-hearted, merciful. □ **humanely** *adv.*

■ benevolent, caring, charitable, compassionate, considerate, forbearing, forgiving, generous, good, good-natured, kind, kind-hearted, kindly, lenient, merciful, sympathetic, tender, warm.

humanism *n.* non-religious philosophy based on liberal human values. □ **humanist** *n.*, **humanistic** *adj.*

humanitarian *adj.* promoting human welfare and reduction of suffering. □ **humanitarianism** *n.*

humanity *n.* **1** human nature or qualities. **2** kindness. **3** human race. **4** (*pl.*) arts subjects.

humanize *v.* make human or humane. □ **humanization** *n.*

humankind *n.* human beings in general.

■ human beings, humanity, man, mankind, the human race.

humble *adj.* **1** having or showing a modest estimate of one's own importance. **2** of low rank. **3** not large or expensive. ● *v.* lower the rank or self-importance of. □ **humbly** *adv.*

■ *adj.* **1** modest, self-effacing, unassuming; deferential, meek, obsequious, servile, submissive, subservient. **2** inferior, insignificant, low, lowly, mean, undistinguished, unimportant. ● *v.* demote, downgrade; chasten, humiliate, subdue.

humbug *n.* **1** misleading behaviour or talk to win support or sympathy. **2** person behaving thus. **3** peppermint-flavoured boiled sweet. ● *v.* (**humbugged**) delude.

humdrum *adj.* dull, commonplace.

■ boring, commonplace, dull, mundane, ordinary, routine, run-of-the-mill, tedious, tiresome, uneventful, unexciting, uninteresting, wearisome.

humerus *n.* (*pl.* **-ri**) bone of the upper arm. □ **humeral** *adj.*

humid *adj.* (of air) damp. □ **humidity** *n.*

■ clammy, close, damp, moist, muggy, oppressive, steamy, sticky, sultry.

humidify *v.* keep (air) moist in a room etc. □ **humidifier** *n.*

humiliate *v.* cause to feel disgraced. □ **humiliation** *n.*

■ abase, chasten, crush, degrade, demean, disgrace, embarrass, humble, mortify, shame. □ **humiliation** degradation, disgrace, dishonour, embarrassment, ignominy, indignity, mortification, shame.

humility *n.* humble condition or attitude of mind.

■ **lowliness, meekness, modesty, self-effacement, servility, submissiveness, subservience.**

hummock *n.* hillock.

hummus /hŏŏmməss/ *n.* (also **houmous**) dip of chickpeas, sesame oil, lemon juice, and garlic.

humour *n.* **1** quality of being amusing. **2** state of mind. ● *v.* keep (a person) contented by doing as he or she wishes. □ **sense of humour** ability to perceive and enjoy humour. **humorous** *adj.*, **humorously** *adv.*

■ *n.* **1 comedy, drollery, hilarity, jocularity, wit, wittiness. 2 disposition, mood, state of mind, spirits, temper.** ● *v.* **appease, gratify, indulge, mollify, pander to, placate, please.** □ **humorous amusing, comic(al), droll, entertaining, funny, hilarious, witty.**

hump *n.* **1** rounded projecting part. **2** curved deformity of the spine. ● *v.* **1** form into a hump. **2** hoist and carry. □ **humpback bridge** small steeply arched bridge.

■ *n.* **1 bulge, bump, knob, lump, node, projection, protrusion, protuberance, swelling; hillock, hummock, mound.** ● *v.* **1 arch, bend, crook, curve, hunch.**

humus *n.* soil-fertilizing substance formed by decay of dead leaves and plants etc.

hunch *v.* bend into a hump. ● *n.* **1** intuitive feeling. **2** hump.

■ *n.* **1 feeling, impression, intuition, premonition, presentiment, suspicion.**

hundred *n.* ten times ten. □ **hundredth** *adj.* & *n.*

hundredfold *adj.* & *adv.* 100 times as much or as many.

hundredweight *n.* measure of weight, 112 lb (50.80 kg).

hung *see* **hang**. *adj.* (of a council, parliament, etc.) with no party having a clear majority. □ **hung-over** *adj.* (*colloq.*) having a hangover.

Hungarian *adj.* & *n.* (native, language) of Hungary.

hunger *n.* **1** pain or discomfort felt when one has not eaten for some time. **2** strong desire. ● *v.* feel hunger. □ **hunger for** have a strong desire for. **hunger strike** refusal of food as a protest.

■ *n.* **2 appetite, craving, desire, hankering, itch, longing, thirst, yearning,** *colloq.* **yen.** □ **hunger for crave, desire, hanker after, long for, thirst for, want, wish for, yearn for.**

hungry *adj.* (**-ier, -iest**) feeling hunger. □ **hungrily** *adv.*

■ **famished,** *colloq.* **peckish, ravenous,** *colloq.* **starving; avid, desirous, eager, greedy, longing, thirsty, yearning.**

hunk *n.* large piece cut off.

hunt *v.* **1** pursue (wild animals) for food or sport. **2** (of animals) pursue prey. ● *n.* **1** process of hunting. **2** hunting group. □ **hunt for** seek, search for.

■ *v.* **chase, pursue, stalk, track, trail.** ● *n.* **1 chase, pursuit, search, quest.** □ **hunt for go in search of, look for, search for, seek.**

hunter *n.* **1** one who hunts. **2** horse used for hunting.

hurdle *n.* **1** portable frame with bars, used as a temporary fence. **2** frame to be jumped over in a race. **3** obstacle, difficulty. □ **hurdler** *n.*

■ **3 barrier, complication, difficulty, hindrance, impediment, obstacle, obstruction, problem, snag, stumbling block.**

hurl *v.* throw violently. ● *n.* violent throw.

■ *v.* **cast,** *colloq.* **chuck, fling,** *colloq.* **heave, lob, pitch,** *colloq.* **sling, throw, toss.**

hurly-burly *n.* boisterous activity.

hurrah *int.* & *n.* (also **hurray**) exclamation of joy or approval.

hurricane *n.* storm with violent wind. □ **hurricane lamp** lamp with the flame protected from the wind.

■ **cyclone, storm, tornado, typhoon, whirlwind.**

hurried *adj.* done with great haste. □ **hurriedly** *adv.*

hurry *v.* **1** act or move with eagerness or too quickly. **2** cause to do this. ● *n.* hurrying.

■ *v.* **1 dash, hasten, fly, make haste, race, run, rush, scurry, scuttle, shoot, speed, tear, zoom. 2 accelerate, expedite, quicken, speed up.**

hurt *v.* (**hurt**) **1** cause pain, harm, or distress (to). **2** feel pain. ● *n.* injury, harm. □ **hurtful** *adj.*

■ *v.* **1 injure, wound; damage, harm, impair, mar, spoil; afflict, distress, grieve, pain, upset. 2 ache, be painful** *or* **sore, smart, sting, throb.** ● *n.* **damage, harm, injury; agony, anguish, distress, pain, suffering.** □ **hurtful cruel, cutting, malicious, mean, nasty, spiteful, unkind, wounding;**

damaging, deleterious, detrimental, harmful, injurious.

hurtle *v.* move or hurl rapidly.

husband *n.* married man in relation to his wife. ● *v.* use economically, try to save.

husbandry *n.* **1** farming. **2** management of resources.

hush *v.* make or become silent. ● *n.* silence. □ **hush-hush** *adj.* secret.

■ *v.* quieten, *colloq.* shush, silence; fall silent, *colloq.* shut up. ● *n.* peace, quiet, silence, stillness, tranquillity.

husk *n.* dry outer covering of certain seeds and fruits. ● *v.* remove the husk from.

husky[1] *adj.* **(-ier, -iest) 1** hoarse. **2** burly. □ **huskily** *adv.*

husky[2] *n.* Arctic sledge-dog.

hustle *v.* **1** push roughly. **2** hurry. ● *n.* hustling.

■ *v.* **1** elbow, jostle, push, shove, thrust. **2** hasten, hurry, rush, scurry, scuttle, sprint.

hut *n.* small simple or roughly made house or shelter.

hutch *n.* box-like pen for rabbits.

hyacinth *n.* plant with fragrant bell-shaped flowers.

hybrid *n.* **1** offspring of two different species or varieties. **2** thing made by combining different elements. ● *adj.* produced in this way. □ **hybridism** *n.*

■ *n.* **1** cross, cross-breed, mongrel. **2** blend, composite, compound, mix, mixture.

hybridize *v.* **1** cross-breed. **2** produce hybrids. **3** interbreed. □ **hybridization** *n.*

hydrangea *n.* shrub with pink, blue, or white flowers in clusters.

hydrant *n.* outlet for drawing water from a main.

hydrate *n.* chemical compound of water with another substance.

hydraulic *adj.* operated by pressure of fluid conveyed in pipes. ● *n.* **(hydraulics)** science of hydraulic operations.

hydrocarbon *n.* compound of hydrogen and carbon.

hydrochloric acid solution of hydrogen chloride in water.

hydrodynamics *n.* science of forces acting or or exerted by liquids. □ **hydrodynamic** *adj.*

hydroelectric *adj.* using water-power to produce electricity.

hydrofoil *n.* **1** boat with a structure that raises its hull out of the water when the boat is in motion. **2** this structure.

hydrogen *n.* odourless gas, the lightest element. □ **hydrogen bomb** powerful bomb releasing energy by fusion of hydrogen nuclei.

hydrolysis *n.* decomposition by chemical reaction with water. □ **hydrolytic** *adj.*

hydrometer *n.* device measuring the density of liquids.

hydrophobia *n.* **1** abnormal fear of water. **2** rabies.

hydroponics *n.* art of growing plants in water impregnated with chemicals.

hydrostatic *adj.* of the pressure and other characteristics of liquid at rest.

hydrotherapy *n.* use of water to treat diseases etc.

hydrous *adj.* containing water.

hyena *n.* wolf-like animal with a howl that sounds like laughter.

hygiene *n.* cleanliness as a means of preventing disease. □ **hygienic** *adj.*, **hygienically** *adv.*, **hygienist** *n.*

■ □ **hygienic** aseptic, clean, disinfected, sanitary, sterile.

hymen *n.* membrane partly closing the opening of the vagina of a virgin girl or woman.

hymn *n.* song of praise to God or a sacred being.

hyper- *pref.* excessively.

hyperactive *adj.* abnormally active. □ **hyperactivity** *n.*

hypermarket *n.* very large supermarket.

hypersonic *adj.* of speeds more than five times that of sound.

hypertension *n.* **1** abnormally high blood pressure. **2** extreme tension.

hyphen *n.* the sign - used to join words together or divide a word into parts. ● *v.* hyphenate.

hyphenate *v.* join or divide with a hyphen. □ **hyphenation** *n.*

hypnosis *n.* sleep-like condition produced in a person who then obeys suggestions.

hypnotic *adj.* of or producing hypnosis. □ **hypnotically** *adv.*

hypnotism *n.* hypnosis.

hypnotize *v.* **1** produce hypnosis in. **2** fascinate. □ **hypnotist** *n.*

■ **1** mesmerize. **2** bewitch, captivate, entrance, fascinate, hold spellbound, mesmerize.

hypochondria *n.* state of constantly imagining that one is ill. □ **hypochondriac** *n.* & *adj.*

hypocrisy *n.* **1** falsely pretending to be virtuous. **2** insincerity.

■ **deceit, deceitfulness, deception, duplicity, falseness, falsity, insincerity, sanctimoniousness.**

hypocrite *n.* person guilty of hypocrisy. □ **hypocritical** *adj.*, **hypocritically** *adv.*

■ □ **hypocritical deceitful, dishonest, disingenuous, dissembling, duplicitous, insincere, sanctimonious, two-faced.**

hypodermic *adj.* (of a drug, syringe, etc.) introduced under the skin. ● *n.* hypodermic syringe.

hypotenuse *n.* longest side of a right-angled triangle.

hypothermia *n.* abnormally low body temperature.

hypothesis *n.* (*pl.* **-theses**) supposition put forward as a basis for reasoning or investigation.

■ **assumption, premiss, proposition, supposition, theory.**

hypothetical *adj.* supposed but not necessarily true. □ **hypothetically** *adv.*

■ **notional, putative, speculative, supposed, theoretical, unproven.**

hysterectomy *n.* surgical removal of the womb.

hysteria *n.* wild uncontrollable emotion. □ **hysterical** *adj.*, **hysterically** *adv.*

hysterics *n.pl.* hysterical outburst.

Hz *abbr.* hertz.

Ii

I *pron.* person speaking or writing and referring to himself or herself.

iambic *adj.* & *n.* (verse) using iambuses, metrical feet of one long and one short syllable.

iatrogenic *adj.* (of disease) caused unintentionally by medical treatment.

Iberian *adj.* of the peninsula comprising Spain and Portugal.

ibex *n.* (*pl.* **ibex** or **ibexes**) mountain goat with curving horns.

ibis *n.* wading bird found in warm climates.

ice *n.* **1** frozen water. **2** portion of ice cream. ● *v.* **1** become frozen. **2** make very cold. **3** decorate with icing. □ **ice cream** sweet creamy frozen food. **ice hockey** game like hockey played on ice by skaters. **ice lolly** water ice or ice cream on a stick.

iceberg *n.* mass of ice floating in the sea.

Icelandic *adj.* & *n.* (language) of Iceland.

ichthyology /ikthiólləji/ *n.* study of fish. □ **ichthyologist** *n.*

icicle *n.* hanging ice formed when dripping water freezes.

icing *n.* mixture of powdered sugar etc. used to decorate food.

icon *n.* sacred painting, mosaic, etc.

iconoclast *n.* person who attacks cherished beliefs. □ **iconoclasm** *n.*, **iconoclastic** *adj.*

icy *adj.* (**-ier**, **-iest**) **1** very cold. **2** covered with ice. **3** very unfriendly. □ **icily** *adv.*, **iciness** *n.*

■ **1** arctic, bitter, chilling, chilly, cold, freezing, frigid, frosty, glacial, raw. **2** frosty, frozen, slippery. **3** aloof, chilly, cold, cool, frigid, unfriendly, unwelcoming.

idea *n.* **1** plan etc. formed by mental effort. **2** opinion. **3** mental impression, vague belief.

■ **1** brainwave, inspiration; concept, conception, design, notion, plan, scheme, stratagem, thought. **2** belief, conviction, opinion, theory, view. **3** feeling, hunch, impression, inkling, intimation, notion, suspicion.

ideal *adj.* satisfying one's idea of what is perfect. ● *n.* person or thing regarded as perfect or as a standard to aim at. □ **ideally** *adv.*

■ *adj.* excellent, exemplary, faultless, flawless, model, perfect.

idealist *n.* person with high ideals. □ **idealism** *n.*, **idealistic** *adj.*

idealize *v.* regard or represent as perfect. □ **idealization** *n.*

identical *adj.* **1** the same. **2** exactly alike. □ **identically** *adv.*

■ **1** selfsame, (very) same. **2** alike, corresponding, duplicate, equal, equivalent, indistinguishable, interchangeable, like, matching, twin.

identify *v.* establish the identity of. □ **identify with 1** closely associate with. **2** associate (oneself) with in feeling or interest. **identifiable** *adj.*, **identification** *n.*

■ discern, distinguish, pick out, recognize; diagnose, establish, find out. □ **identify with 2** empathize with, relate to, sympathize with.

identity *n.* **1** who or what a person or thing is. **2** sameness.

■ **1** distinctiveness, individuality, personality, uniqueness. **2** congruence, correspondence, likeness, sameness.

ideology *n.* ideas that form the basis of a political or economic theory. □ **ideological** *adj.*

■ beliefs, convictions, doctrine, ideas, philosophy, principles, teachings, tenets.

idiocy *n.* **1** state of being an idiot. **2** extreme foolishness.

idiom *n.* **1** phrase etc. established by usage and not immediately comprehensible from the words used. **2** form of expression peculiar to a language. □ **idiomatic** *adj.*

■ **1** expression, phrase, saying. **2** dialect, jargon, parlance, patois, phraseology.

idiosyncrasy *n.* person's own characteristic way of behaving. □ **idiosyncratic** *adj.*

■ characteristic, foible, habit, mannerism, peculiarity, quirk.

idiot *n.* very stupid person. □ **idiotic** *adj.*, **idiotically** *adv.*

■ ass, blockhead, *colloq.* clot, dolt, *sl.* dope, *colloq.* duffer, dullard, ignoramus, imbecile, ninny, *colloq.* nitwit, *sl.* twit, *sl.* wally.

idle *adj.* **1** not employed or in use. **2** lazy. **3** aimless. ● *v.* **1** be idle. **2** (of an engine) run slowly in neutral gear. □ **idle away** pass (time) aimlessly. **idly** *adv.*, **idleness** *n.*, **idler** *n.*

■ *adj.* **1** inactive, not in use, not working, stationary; jobless, out of work, redundant, unemployed. **2** indolent, lazy, shiftless, slothful. **3** aimless, casual, offhand, purposeless. ● *v.* **1** hang about, laze around, loaf about, lounge about. **2** tick over. □ **idle away** fritter away, squander, waste, while away.

idol *n.* **1** image worshipped as a god. **2** idolized person or thing.

■ **1** fetish, icon, totem. **2** hero, heroine, star.

idolatry *n.* worship of idols. □ **idolater** *n.*, **idolatrous** *adj.*

idolize *v.* love or admire excessively. □ **idolization** *n.*

■ admire, adore, deify, lionize, love, revere, venerate, worship.

idyll /íddil/ *n.* **1** peaceful or romantic scene or incident. **2** description of this, usu. in verse. □ **idyllic** *adj.*, **idyllically** *adv.*

i.e. *abbr.* (Latin *id est*) that is.

if *conj.* **1** on condition that. **2** supposing that. **3** whether. ● *n.* condition, supposition.

igloo *n.* Eskimo snow house.

igneous *adj.* (of rock) formed by volcanic action.

ignite *v.* **1** set fire to. **2** catch fire.

■ **1** light, set alight, set fire to, set on fire. **2** burst into flames, catch fire, kindle.

ignition *n.* **1** igniting. **2** mechanism producing a spark to ignite the fuel in an engine.

ignoble *adj.* not noble in character, aims, or purpose.

ignominy *n.* disgrace, humiliation. □ **ignominious** *adj.*, **ignominiously** *adv.*

■ discredit, disgrace, dishonour, humiliation, infamy, mortification, obloquy, opprobrium, shame.

ignoramus *n.* (*pl.* **-muses**) ignorant person.

ignorant *adj.* **1** lacking knowledge. **2** rude through lack of respect for good manners. □ **ignorantly** *adv.*, **ignorance** *n.*

■ **1** benighted, illiterate, uneducated, unenlightened, uninformed, unlettered, unschooled. **2** bad-mannered, boorish, ill-mannered, rude, uncouth.

ignore *v.* take no notice of.

■ cold-shoulder, cut, snub; discount, disregard, leave out, omit, overlook, pass over, pay no attention to, take no notice of.

iguana *n.* tropical tree lizard.

il- *pref. see* **in-**.

ileum *n.* part of the small intestine.

ill *adj.* **1** unwell. **2** bad. **3** harmful. **4** hostile, unkind. ● *adv.* badly. ● *n.* evil, harm, injury. □ **ill-advised** *adj.* unwise. **ill at ease** uncomfortable, embarrassed. **ill-gotten** *adj.* gained by evil or unlawful means. **ill-mannered** *adj.* having bad manners. **ill-treat** *v.* treat badly or cruelly. **ill will** hostility, unkind feeling.

■ *adj.* **1** ailing, in bad health, indisposed, infirm, not well, off colour, poorly, sick, sickly, under the weather, unhealthy, unwell. **2** bad, inauspicious, unfavourable, unfortunate, unlucky, unpromising. **3** adverse, damaging, dangerous, deleterious, detrimental, harmful, hurtful, injurious, noxious, pernicious. **4** antagonistic, cruel, hostile, malevolent, malicious, unfriendly, unkind. ● *adv.* adversely, badly, critically, unfavourably. ● *n.* evil, injustice, wrong; damage, harm, hurt, injury, mischief, misfortune. □ **ill-advised** foolhardy, foolish, impolitic, imprudent, incautious, misguided, rash, reckless, short-sighted, unwise. **ill-mannered** discourteous, ignorant, impertinent, impolite, impudent, insolent, rude, uncivil, uncouth, ungentlemanly, ungracious, unladylike. **ill-treat** abuse, harm, hurt, injure, knock about, maltreat, misuse. **ill will** acrimony, animosity, animus, antipathy, bad feeling, dislike, enmity, hate, hatred, hostility, loathing, malevolence, malice, rancour, resentment, unfriendliness, venom.

illegal *adj.* against the law. □ **illegally** *adv.*, **illegality** *n.*

■ actionable, banned, criminal, forbidden, illegitimate, illicit, outlawed, prohibited, unauthorized, unlawful, unlicensed.

illegible *adj.* not legible. □ **illegibly** *adv.*, **illegibility** *n.*

■ crabbed, indecipherable, scrawled, scribbled, unintelligible, unreadable.

illegitimate *adj.* **1** born of parents not married to each other. **2** against the law, illegal. □ **illegitimately** *adv.*, **illegitimacy** *n.*

illicit *adj.* unlawful, illegal. □ **illicitly** *adv.*

illiterate *adj.* **1** unable to read and write. **2** uneducated. □ **illiteracy** *n.*

illness *n.* **1** state of being ill. **2** particular form of ill health.

■ **1** bad health, ill health, infirmity, sickness. **2** ailment, *sl.* bug, complaint, condition, disorder, indisposition, infection, infirmity, sickness, *colloq.* virus.

illogical *adj.* not logical. □ **illogically** *adv.*, **illogicality** *n.*

illuminate *v.* **1** light up. **2** throw light on (a subject). **3** decorate with lights. □ **illumination** *n.*

■ **1** brighten, light (up), lighten, illumine, irradiate, shed *or* throw light on. **2** clarify, elucidate, explain, shed *or* throw light on.

illumine *v.* **1** light up. **2** enlighten.

illusion *n.* **1** false belief. **2** thing wrongly supposed to exist. □ **illusive** *adj.*, **illusory** *adj.*

■ **1** delusion, fallacy, misapprehension, misconception, mistake, mistaken impression. **2** chimera, fantasy, figment of the imagination, hallucination, mirage, vision. □ **illusory** deceptive, delusive, fallacious, false, fanciful, illusive, imaginary, imagined, misleading, unreal, untrue.

illusionist *n.* conjuror.

illustrate *v.* **1** supply (a book etc.) with drawings or pictures. **2** make clear by example(s) or picture(s) etc. **3** serve as an example of. □ **illustrative** *adj.*, **illustrator** *n.*

■ **2** clarify, elucidate, explain, illuminate, shed *or* throw light on. **3** demonstrate, epitomize, exemplify, represent, typify.

illustration *n.* **1** drawing etc. in a book. **2** explanatory example. **3** act or instance of illustrating.

■ **1** diagram, drawing, figure, picture, plate, sketch. **2** case, example, exemplification, instance. **3** depiction, representation; clarification, elucidation, explication.

illustrious *adj.* distinguished.

■ acclaimed, celebrated, distinguished, eminent, esteemed, famed, famous, great, *colloq.* legendary, notable, noted, renowned, well-known.

im- *pref. see* **in-**.

image *n.* **1** representation of an object. **2** reputation. **3** appearance of a thing as seen in a mirror or through a lens. **4** mental picture. ● *v.* picture.

■ *n.* **1** effigy, figure, icon, likeness, portrait, representation, sculpture, statue. **2** character, persona, reputation. **3** reflection. **4** concept, conception, idea, impression, mental picture, vision.

imaginary *adj.* existing only in the imagination, not real.

■ fabulous, fanciful, fictional, fictitious, illusory, imagined, invented, made-up, unreal.

imagination *n.* **1** process of imagining. **2** ability to imagine or to plan creatively. □ **imaginative** *adj.*, **imaginatively** *adv.*

■ **1** conception, visualization. **2** creativity, ingenuity, invention, inventiveness. □ **imaginative** creative, ingenious, innovative, inspired, inventive.

imagine *v.* **1** form a mental image of. **2** think, suppose.

■ **1** conceive of, envisage, picture, think of, visualize. **2** assume, be of the opinion, believe, expect, guess, presume, reckon, suppose, surmise, suspect.

imago /imáygō/ *n.* (*pl.* **-gines** or **-os**) insect in its fully developed adult stage.

imam *n.* Muslim spiritual leader.

imbalance *n.* lack of balance.

imbecile *n.* extremely stupid person. ● *adj.* idiotic. □ **imbecilic** *adj.*, **imbecility** *n.*

imbibe *v.* **1** drink. **2** absorb (ideas).

imbroglio /imbrṓliō/ *n.* (*pl.* **-os**) confused situation.

imbue *v.* fill with feelings, qualities, or emotions.

imitate *v.* **1** try to act or be like. **2** copy. □ **imitable** *adj.*, **imitator** *n.*

■ **1** copy, echo, emulate. **2** ape, copy, impersonate, mimic, parody, take off.

imitation *n.* **1** act or instance of imitating. **2** copy. ● *adj.* counterfeit, fake.

■ *n.* **1** emulation, impersonation, mimicry, parody. **2** copy, replica, reproduction; counterfeit, fake, forgery. ● *adj.* artificial, counterfeit, fake, mock, *colloq.* phoney, sham, simulated, synthetic.

imitative *adj.* imitating.

immaculate *adj.* **1** spotlessly clean. **2** free from blemish or fault. □ **immaculately** *adv.*

immanent *adj.* inherent. □ **immanence** *n.*

immaterial *adj.* **1** having no physical substance. **2** of no importance.

immature *adj.* not mature. □ **immaturity** *n.*

■ young, youthful; babyish, callow, childish, green, inexperienced, infantile, juvenile, naive, puerile.

immeasurable *adj.* not measurable, immense. □ **immeasurably** *adv.*, **immeasurability** *n.*

immediate *adj.* **1** with no delay. **2** nearest. □ **immediately** *adv.*, **immediacy** *n.*

■ **1** instant, instantaneous, on the spot, prompt, rapid, speedy, swift. **2** adjacent, closest, nearest, next, proximate.

immemorial *adj.* existing from before what can be remembered.

immense *adj.* extremely great. □ **immensely** *adv.*, **immensity** *n.*

■ colossal, enormous, extensive, gargantuan, giant, gigantic, great, huge, mammoth, massive, monumental, prodigious, titanic, tremendous, vast.

immerse *v.* **1** put completely into liquid. **2** involve deeply.

■ **1** bathe, dip, douse, duck, dunk, plunge, sink, submerge. **2** absorb, bury, engage, engross, involve, occupy.

immersion *n.* immersing. □ **immersion heater** electric heater placed in the liquid to be heated.

immigrate *v.* come into a foreign country as a permanent resident. □ **immigrant** *adj.* & *n.*, **immigration** *n.*

imminent *adj.* about to occur. □ **imminence** *n.*, **imminently** *adv.*

■ approaching, drawing near, forthcoming, impending, looming.

immobile *adj.* **1** immovable. **2** not moving. □ **immobility** *n.*

immobilize *v.* make or keep immobile. □ **immobilization** *n.*

immoderate *adj.* excessive. □ **immoderately** *adv.*

immolate *v.* kill as a sacrifice. □ **immolation** *n.*

immoral *adj.* morally wrong. □ **immorally** *adv.*, **immorality** *n.*

■ bad, corrupt, decadent, degenerate, depraved, dishonest, dissipated, dissolute, evil, iniquitous, nefarious, sinful, unprincipled, unscrupulous, wicked.

immortal *adj.* **1** living or lasting for ever. **2** divine. **3** famous for all time. ● *n.* immortal being. □ **immortality** *n.*

■ *adj.* **1** endless, enduring, eternal, everlasting, incorruptible, indestructible, lasting, perpetual, undying. **2** celestial, divine, heavenly. **3** celebrated, classic, famous, *colloq.* legendary, timeless.

immortalize *v.* make immortal.

immovable *adj.* **1** unable to be moved. **2** unyielding. □ **immovably** *adv.*, **immovability** *n.*

■ **1** anchored, fast, fixed, immobile, riveted, rooted, set, unmovable; immutable, unalterable, unchangeable. **2** adamant, determined, dogged, firm, inflexible, obdurate, resolute, steadfast, stubborn, unbending, uncompromising, unshakeable, unswerving, unwavering, unyielding.

immune *adj.* having immunity. □ **immune from** or **to** free or exempt from.

immunity *n.* **1** ability to resist infection. **2** special exemption.

immunize *v.* make immune to infection □ **immunization** *n.*

immunodeficiency *n.* reduction in normal resistance to infection.

immure *v.* imprison, shut in.

immutable *adj.* unchangeable. □ **immutability** *n.*

imp *n.* **1** small devil. **2** mischievous child.

impact *n.* /ímpakt/ **1** collision, force of this. **2** strong effect. ● *v.* /impákt/ press or wedge firmly. □ **impaction** *n.*

■ *n.* **1** bang, bump, collision, crash, smash; brunt, force, weight, thrust. **2** effect, impression, influence.

impair *v.* damage, weaken. □ **impairment** *n.*

■ damage, debilitate, harm, hurt, injure, mar, ruin, spoil, weaken.

impala *n.* (*pl.* **impala**) small antelope.

impale *v.* fix or pierce with a pointed object. □ **impalement** *n.*

impalpable *adj.* intangible.

impart *v.* **1** give. **2** make (information etc.) known.

impartial *adj.* not favouring one more than another. □ **impartially** *adv.*, **impartiality** *n.*

■ detached, disinterested, dispassionate, equitable, fair, just, neutral, objective, unbiased, unprejudiced.

impassable *adj.* impossible to travel on or over.

impasse /ámpass/ *n.* deadlock.

impassioned *adj.* passionate.

impassive *adj.* not feeling or showing emotion. □ **impassively** *adv.*

impatient *adj.* **1** feeling or showing lack of patience. **2** restlessly eager. □ **impatiently** *adv.*, **impatience** *n.*

■ **1** abrupt, brusque, curt, irritable, quick-tempered, short-tempered, snappy. **2** edgy, fidgety, nervous, restive, restless; agog, anxious, eager, keen.

impeach *v.* accuse of a serious crime against the state and bring for trial. □ **impeachment** *n.*

impeccable *adj.* faultless. □ **impeccably** *adv.*, **impeccability** *n.*

impecunious *adj.* having little or no money.

impedance *n.* resistance of an electric circuit to the flow of current.

impede *v.* hinder.

■ bar, baulk, block, check, delay, frustrate, hamper, handicap, hinder, hold up, obstruct, retard, slow down, stymie, thwart.

impediment *n.* **1** hindrance, obstruction. **2** defect in speech, esp. a lisp or stammer.

impel *v.* (**impelled**) **1** drive, force. **2** propel.

impending *adj.* imminent.

impenetrable *adj.* unable to be penetrated. □ **impenetrably** *adv.*, **impenetrability** *n.*

imperative *adj.* **1** essential. **2** (of a verb) expressing a command. ● *n.* **1** command. **2** essential thing.

■ *adj.* **1** compulsory, essential, indispensable, mandatory, necessary, obligatory, required, vital.

imperceptible *adj.* too slight to be noticed. □ **imperceptibly** *adv.*

■ inaudible, indiscernible, indistinguishable, invisible, undetectable, unnoticeable.

imperfect *adj.* **1** not perfect. **2** (of a tense) implying action going on but not completed. □ **imperfectly** *adv.*, **imperfection** *n.*

■ **1** damaged, defective, faulty, flawed; deficient, inadequate, incomplete, patchy, rudimentary, sketchy, unfinished, unpolished. □ **imperfection** blemish, defect, failing, fault, flaw, shortcoming.

imperial *adj.* **1** of an empire. **2** majestic. **3** (of measures) belonging to the British official non-metric system. □ **imperially** *adv.*

imperialism *n.* policy of having or extending an empire. □ **imperialist** *n.*, **imperialistic** *adj.*

imperil *v.* (**imperilled**) endanger.

imperious *adj.* domineering. □ **imperiously** *adv.*

■ *colloq.* bossy, dictatorial, domineering, high-handed, magisterial, overbearing, peremptory.

impersonal *adj.* not showing or influenced by personal feeling. □ **impersonally** *adv.*, **impersonality** *n.*

■ cold, cool, formal, starchy, stiff, stilted, unfriendly; detached, disinterested, dispassionate, impartial, neutral, objective, unbiased, unprejudiced.

impersonate *v.* pretend to be (another person). □ **impersonation** *n.*, **impersonator** *n.*

impertinent *adj.* not showing proper respect. □ **impertinently** *adv.*, **impertinence** *n.*

■ brazen, cheeky, disrespectful, forward, impolite, impudent, insolent, pert, rude, saucy, uncivil.

imperturbable *adj.* not excitable, calm. □ **imperturbably** *adv.*, **imperturbability** *n.*

impervious *adj.* **impervious to** not able to be penetrated or influenced by.

impetigo /impitī́gō/ *n.* contagious skin disease.

impetuous *adj.* acting or done on impulse. □ **impetuously** *adv.*, **impetuosity** *n.*

■ abrupt, hasty, hotheaded, impromptu, impulsive, precipitate, quick, rash, reckless, spontaneous, sudden, unplanned, unpremeditated, unthinking.

impetus *n.* moving or driving force.

■ energy, force, momentum; goad, impulse, incentive, inducement, motivation, spur, stimulation, stimulus.

impinge *v.* **1** make an impact. **2** encroach. □ **impingement** *n.*

impious *adj.* not reverent, wicked. □ **impiously** *adv.*

■ blasphemous, irreligious, irreverent, profane, sacrilegious, sinful, ungodly, unholy, wicked.

implacable *adj.* relentless. □ **implacably** *adv.*

■ hard, inexorable, inflexible, merciless, pitiless, relentless, ruthless, unforgiving, unrelenting, unyielding.

implant *v.* /impla͡ant/ **1** plant, insert. **2** insert (tissue) in a living thing. ● *n.* /ímplaant/ implanted tissue. □ **implantation** *n.*

implement *n.* tool. ● *v.* put into effect. ◻ **implementation** *n.*

■ *n.* appliance, device, gadget, instrument, tool utensil. ● *v.* accomplish, achieve, bring about, carry out, effect, execute, fulfil, perform, put into practice, realize.

implicate *v.* show or cause to be involved in a crime etc.

■ associate, connect, embroil, entangle, include, incriminate, involve.

implication *n.* **1** thing implied. **2** implicating.

■ **1** hint, inference, innuendo, insinuation, intimation, suggestion.

implicit *adj.* **1** implied. **2** absolute. ◻ **implicitly** *adv.*

■ **1** implied, indirect, tacit, undeclared, unspoken. **2** absolute, complete, entire, perfect, total, unquestioning, unreserved, utter.

implode *v.* (cause to) burst inwards. ◻ **implosion** *n.*

implore *v.* request earnestly.

imply *v.* **1** suggest without stating directly. **2** mean.

■ **1** hint (at), insinuate, intimate, make out, suggest. **2** connote, denote, indicate, mean, signify; entail, involve, necessitate.

impolitic *adj.* not advisable.

imponderable *adj.* not able to be estimated. ● *n.* imponderable thing. ◻ **imponderably** *adv.*, **imponderability** *n.*

import *v.* /impórt/ **1** bring in from abroad or from an outside source. **2** imply. ● *n.* /ímport/ **1** importing. **2** thing imported. **3** meaning. **4** importance. ◻ **importation** *n.*, **importer** *n.*

important *adj.* **1** having a great effect. **2** having great authority or influence. ◻ **importance** *n.*

■ **1** consequential, grave, momentous, pressing, serious, significant, urgent, weighty. **2** distinguished, eminent, foremost, high-ranking, influential, leading, notable, noted, noteworthy, outstanding, powerful, prominent, respected, worthy. ◻ **importance** consequence, gravity, import, momentousness, seriousness, significance, weight; distinction, eminence, influence, note, prominence, standing, status, worth.

importunate *adj.* making persistent requests. ◻ **importunity** *n.*

importune *v.* solicit.

impose *v.* **1** inflict (a tax etc.). **2** enforce compliance with. ◻ **impose on** take advantage of.

imposing *adj.* impressive.

imposition *n.* **1** act of imposing. **2** unfair demand or burden.

impossible *adj.* **1** not possible. **2** unendurable. ◻ **impossibly** *adv.*, **impossibility** *n.*

■ **1** impracticable, out of the question, unfeasible, unworkable. **2** insupportable, intolerable, unbearable, unendurable.

impostor *n.* person who assumes a false identity.

imposture *n.* fraudulent deception.

impotent *adj.* **1** powerless. **2** (of a male) incapable of sexual intercourse. ◻ **impotently** *adv.*, **impotence** *n.*

impound *v.* **1** take (property) into legal custody. **2** confiscate.

impoverish *v.* **1** cause to become poor. **2** exhaust the strength or fertility of. ◻ **impoverishment** *n.*

imprecation *n.* spoken curse.

impregnable *adj.* safe against attack. ◻ **impregnability** *n.*

■ impenetrable, invincible, inviolable, invulnerable, safe, secure, unassailable, unconquerable.

impregnate *v.* **1** make pregnant. **2** saturate. ◻ **impregnation** *n.*

impresario *n.* (*pl.* **-os**) organizer of public entertainments.

impress *v.* **1** affect or influence deeply. **2** arouse admiration or respect in. **3** press a mark into.

■ **1** affect, influence, inspire, move, strike, sway, touch.

impression *n.* **1** effect produced on the mind. **2** uncertain idea. **3** imitation done for entertainment. **4** impressed mark. **5** reprint.

■ **1** awareness, consciousness, feeling, sensation, sense; effect, impact, influence. **2** feeling, hunch, idea, notion, suspicion. **3** imitation, impersonation, parody, take-off. **4** brand, mark, stamp. **5** edition, reprint.

impressionable *adj.* easily influenced.

■ persuadable, receptive, suggestible, susceptible.

impressionism *n.* style of painting etc. giving a general impression without detail. ◻ **impressionist** *n.*, **impressionistic** *adj.*

mpressive *adj.* arousing respect or admiration. ▫ **impressively** *adv.*

▪ awe-inspiring, awesome, breathtaking, formidable, imposing, magnificent, majestic, redoubtable, splendid, striking.

mprint *n.* /ímprint/ **1** mark made by pressing on a surface. **2** publisher's name etc. on a title-page. ● *v.* /imprínt/ impress or stamp a mark etc. on.

mprison *v.* **1** put into prison. **2** confine. ▫ **imprisonment** *n.*

▪ **1** detain, incarcerate, jail, lock up, remand. **2** confine, coop up, immure, shut in *or* up. ▫ **imprisonment** confinement, custody, detention, incarceration.

mprobable *adj.* not likely to be true or to happen. ▫ **improbably** *adv.*, **improbability** *n.*

▪ doubtful, dubious, far-fetched, implausible, incredible, unbelievable, unconvincing, unlikely.

mpromptu *adj.* & *adv.* without preparation or rehearsal.

▪ *adj.* ad lib, extemporary, improvised, spontaneous, unprepared, unrehearsed.

mproper *adj.* **1** indecent, unseemly. **2** not conforming to social conventions. ▫ **improperly** *adv.*, **impropriety** *n.*

▪ **1** immodest, indecent, indecorous, indelicate, unbecoming, ungentlemanly, unladylike, unseemly. **2** inappropriate, unacceptable, unfitting, unsuitable.

mprove *v.* make or become better. ▫ **improvement** *n.*

▪ ameliorate, better, enhance, perfect, polish, put right, rectify, refine; modernize, overhaul, refurbish, renovate, repair, revamp, touch up; amend, correct, edit, emend; be on the mend, get better, look up, make progress, pick up, rally, recover, recuperate. ▫ **improvement** amelioration, amendment, correction, enhancement, rectification, reform; modernization, overhaul, refurbishment; rally, recovery, upswing, upturn.

mprovident *adj.* not providing for future needs. ▫ **improvidently** *adv.*, **improvidence** *n.*

▪ careless, heedless, imprudent, incautious, injudicious, short-sighted; extravagant, prodigal, spendthrift, thriftless, uneconomical, wasteful.

mprovise *v.* **1** compose impromptu. **2** provide from whatever materials are at hand. ▫ **improvisation** *n.*

imprudent *adj.* unwise. ▫ **imprudently** *adv.*, **imprudence** *n.*

▪ foolhardy, foolish, hasty, heedless, ill-advised, impolitic, incautious, injudicious, irresponsible, misguided, precipitate, rash, reckless, short-sighted, thoughtless, unwise, wild.

impudent *adj.* impertinent. ▫ **impudently** *adv.*, **impudence** *n.*

▪ cheeky, disrespectful, forward, impertinent, impolite, insolent, pert, rude, saucy.

impugn /impyóon/ *v.* express doubts about the truth or honesty of.

impulse *n.* **1** impetus. **2** stimulating force in a nerve. **3** sudden urge to do something.

▪ **1** impetus, incentive, motivation, spur, stimulation, stimulus. **3** caprice, desire, fancy, instinct, urge, whim.

impulsive *adj.* acting or done on impulse. ▫ **impulsively** *adv.*, **impulsiveness** *n.*

▪ impetuous, involuntary, snap, spontaneous, spur-of-the-moment, unplanned, unpremeditated; hasty, hotheaded, madcap, precipitate, quick, rash, reckless.

impunity *n.* freedom from punishment or injury.

impure *adj.* not pure.

▪ adulterated, contaminated, dirty, polluted, tainted, unclean; immoral, sinful, unchaste, wanton.

impurity *n.* **1** being impure. **2** impure thing or part.

impute *v.* attribute (a fault etc.). ▫ **imputation** *n.*

in *prep.* **1** having as a position or state within (limits of space, time, surroundings, etc.). **2** having as a state or manner. **3** into, towards. ● *adv.* **1** in or to a position bounded by limits. **2** inside. **3** in fashion, season, or office. ● *adj.* **1** internal. **2** living etc. inside. **3** fashionable. ▫ **in for 1** about to experience. **2** competing in. **ins and outs** details. **in so far as** to the extent that.

in- *pref.* (**il-** before *l*; **im-** before *b*, *m*, *p*; **ir-** before *r*) **1** not. **2** without, lacking.

in. *abbr.* inch(es).

inability *n.* being unable.

inaction *n.* lack of action.

inactive *adj.* not active. ▫ **inactivity** *n.*

▪ dormant, immobile, inert, motionless, passive, quiescent, stagnant, static, stationary, still, unmoving; idle, inoperative, unoccupied.

inadequate *adj.* **1** not adequate. **2** not sufficiently able. □ **inadequately** *adv.*, **inadequacy** *n.*

■ **1** deficient, insufficient, meagre, scanty, skimpy, sparse. **2** incapable, incompetent, ineffective, ineffectual, inept, not up to scratch, *colloq.* pathetic, useless.

inadmissible *adj.* not allowable.

inadvertent *adj.* unintentional.

■ accidental, chance, unconscious, unintended, unintentional, unplanned, unpremeditated, unthinking.

inalienable *adj.* not able to be given or taken away.

inane *adj.* silly, lacking sense. □ **inanely** *adv.*, **inanity** *n.*

inanimate *adj.* **1** lacking animal life. **2** showing no signs of life.

inappropriate *adj.* unsuitable.

■ inapt, inapposite, infelicitous, inopportune, out of keeping, unsuitable, unsuited, untimely.

inarticulate *adj.* **1** unable to express oneself clearly. **2** (of speech) not clear or well-expressed.

■ **1** incoherent, speechless, tongue-tied. **2** disjointed, garbled, faltering, incoherent, indistinct, muddled, muffled, mumbled, muttered, rambling, unclear, unintelligible.

inasmuch *adv.* **inasmuch as** seeing that, because.

inattentive *adj.* not paying attention. □ **inattentiveness** *n.*

inaugural *adj.* of an inauguration.

inaugurate *v.* **1** admit to office ceremonially. **2** open (a building etc.) formally. **3** begin, introduce. □ **inauguration** *n.*

■ **1** enthrone, induct, install, invest, ordain. **3** begin, initiate, introduce, launch, start.

inborn *adj.* existing in a person or animal from birth, natural.

■ congenital, hereditary, inbred, inherent, inherited, innate, native, natural.

inbred *adj.* **1** produced by inbreeding. **2** inborn.

inbreeding *n.* breeding from closely related individuals.

Inc. *abbr.* (*US*) Incorporated.

incalculable *adj.* unable to be calculated. □ **incalculably** *adv.*

incandescent *adj.* glowing with heat. □ **incandescence** *n.*

incantation *n.* words or sounds uttered as a magic spell.

incapable *adj.* **1** not capable. **2** helpless. □ **incapability** *n.*

incapacitate *v.* **1** disable. **2** make ine igible. □ **incapacitation** *n.*

incapacity *n.* inability, lack of sufficien strength or power.

incarcerate *v.* imprison. □ **incarcera tion** *n.*

incarnate *adj.* embodied, esp. in huma form.

incarnation *n.* **1** embodiment, esp. i human form. **2** (**the Incarnation**) that c God as Christ.

incautious *adj.* rash. □ **incautiously** *adv*

incendiary *adj.* designed to cause fire ● *n.* **1** incendiary bomb. **2** arsonis □ **incendiarism** *n.*

incense¹ /insenss/ *n.* **1** substance burn to produce fragrant smoke, esp. in rel gious ceremonies. **2** this smoke.

incense² /insénss/ *v.* make angry.

■ anger, drive berserk, enrage, exasperate inflame, infuriate, madden, outrage.

incentive *n.* thing that encourages a action or effort.

■ carrot, encouragement, impetus, i ducement, lure, motivation, spur, stimulus

inception *n.* beginning.

incessant *adj.* not ceasing. □ **incess antly** *adv.*

■ ceaseless, constant, continual, co tinuous, endless, *colloq.* eternal, everlas ing, interminable, never-ending, non-stop perpetual, persistent, relentless, unbroken unceasing, unending, uninterrupted, ur remitting.

incest *n.* sexual intercourse betwee very closely related people. □ **incestuou** *adj.*

inch *n.* measure of length (= 2.54 cm). ● move gradually.

incidence *n.* **1** rate at which a thin occurs. **2** falling of a ray, line, etc. on surface.

incident *n.* event, esp. one causin trouble.

■ *colloq.* affair, circumstance, episode experience, event, happening, occasion occurrence; clash, confrontation, contre temps, disturbance, fight, fracas, quarre row, skirmish, to-do.

incidental *adj.* **1** occurring in conne tion with something more important. casual.

■ **1** peripheral, secondary, subordinate subsidiary, supplementary. **2** accidenta casual, chance, fortuitous, unexpected unforeseen, unlooked-for.

ncidentally *adv.* **1** in an incidental way. **2** by the way.

ncinerate *v.* burn to ashes. ▫ **incineration** *n.*, **incinerator** *n.*

ncipient *adj.* beginning to exist.

ncise *v.* **1** make a cut in. **2** engrave. ▫ **incision** *n.*

ncisive *adj.* **1** mentally sharp. **2** clear and effective. ▫ **incisively** *adv.*, **incisiveness** *n.*

■ **1** acute, astute, canny, clever, intelligent, keen, penetrating, perceptive, percipient, perspicacious, sharp, shrewd. **2** clear, concise, effective, pithy, succinct, terse, to the point.

ncisor *n.* any of the front teeth.

ncite *v.* **1** urge on. **2** stir up. ▫ **incitement** *n.*

■ **1** drive on, egg on, encourage, goad, prod, rouse, spur, urge. **2** foment, instigate, provoke, stir up, whip up.

ncivility *n.* rudeness.

nclination *n.* **1** slope. **2** bending. **3** tendency. **4** liking, preference.

■ **1** angle, gradient, incline, list, slant, slope, tilt. **2** bending, bow, bowing, nod, nodding. **3** bent, bias, disposition, leaning, predisposition, proclivity, propensity, tendency. **4** affection, fondness, liking, love, partiality, penchant, predilection, soft spot, weakness.

ncline *v.* /inklīn/ **1** slope. **2** bend. **3** (cause to) have a certain tendency, influence. ● *n.* /inklīn/ slope.

■ *v.* **1** angle, bank, lean, list, slant, slope, tilt. **2** bend, bow, nod, stoop. **3** cause, convince, dispose, induce, influence, persuade, predispose; gravitate, swing, tend. ● *n.* ascent, descent, dip, gradient, hill, ramp, slant, slope.

nclude *v.* **1** have or treat as part of a whole. **2** put into a specified category. ▫ **inclusion** *n.*

■ **1** comprehend, comprise, contain, embody, embrace, encompass, incorporate, take in; allow for, count, number, take into account. **2** catalogue, categorize, classify, group.

nclusive *adj.* **1** including what is mentioned. **2** including everything. ▫ **inclusively** *adv.*, **inclusiveness** *n.*

ncognito *adj.* & *adv.* with one's identity kept secret. ● *n.* (*pl.* **-os**) pretended identity.

incoherent *adj.* rambling in speech or reasoning. ▫ **incoherently** *adv.*, **incoherence** *n.*

■ confused, delirious, inarticulate, incomprehensible, rambling, raving, unintelligible; disconnected, disjointed, disordered, disorganized, garbled, illogical, jumbled, mixed-up, muddled.

incombustible *adj.* not able to be burnt. ▫ **incombustibility** *n.*

income *n.* money received during a period as wages, interest, etc.

■ pay, remuneration, salary; earnings, proceeds, profits, returns, revenue, takings, turnover.

incoming *adj.* coming in.

incommunicado *adj.* not allowed or not wishing to communicate with others.

incomparable *adj.* beyond comparison, without an equal.

■ inimitable, matchless, peerless, perfect, unequalled, unique, unmatched, unparalleled, unrivalled, unsurpassable, unsurpassed.

incomprehensible *adj.* not able to be understood. ▫ **incomprehension** *n.*

■ illegible, inarticulate, incoherent, indecipherable, unintelligible; abstruse, arcane, baffling, cryptic, dark, deep, enigmatic, mysterious, mystifying, obscure, perplexing, puzzling, recondite, unaccountable, unfathomable.

inconceivable *adj.* unable to be imagined.

inconclusive *adj.* not fully convincing. ▫ **inconclusively** *adv.*

incongruous *adj.* unsuitable, not harmonious. ▫ **incongruously** *adv.*, **incongruity** *n.*

■ discordant, inappropriate, incompatible, inconsistent, inharmonious, out of keeping, out of place, unsuitable.

inconsequential *adj.* **1** unimportant. **2** not following logically. ▫ **inconsequentially** *adv.*

inconsiderable *adj.* negligible.

inconsolable *adj.* not able to be consoled. ▫ **inconsolably** *adv.*

inconstant *adj.* **1** fickle. **2** variable. ▫ **inconstancy** *n.*

■ **1** capricious, changeable, fickle, flighty, mercurial, moody, temperamental, volatile. **2** erratic, fluctuating, inconsistent, irregular, unsettled, unstable, unsteady, variable, wavering.

incontestable *adj.* indisputable. □ **incontestably** *adv.*

incontinent *adj.* **1** unable to control one's bowels or bladder. **2** lacking self-restraint. □ **incontinence** *n.*

incontrovertible *adj.* indisputable. □ **incontrovertibly** *adv.*

inconvenience *n.* **1** lack of convenience. **2** thing causing this. ● *v.* cause inconvenience to.

> ■ *n.* **1** awkwardness, discomfort, disruption, disturbance, trouble. **2** annoyance, *colloq.* bind, bore, bother, burden, *colloq.* hassle, hindrance, irritation, nuisance. ● *v.* bother, discommode, disturb, *colloq.* hassle, put out, trouble.

inconvenient *adj.* not convenient, slightly troublesome. □ **inconveniently** *adv.*

> ■ annoying, awkward, bothersome, ill-timed, inappropriate, inexpedient, inopportune, irritating, troublesome, untimely.

incorporate *v.* **1** include as a part. **2** form into a corporation. □ **incorporation** *n.*

incorrigible *adj.* not able to be reformed. □ **incorrigibly** *adv.*

> ■ habitual, hardened, incurable, inveterate, irredeemable.

incorruptible *adj.* **1** not liable to decay. **2** not corruptible morally. □ **incorruptibility** *n.*

increase *v.* /inkreéss/ make or become greater. ● *n.* /ínkreess/ **1** increasing. **2** amount by which a thing increases.

> ■ *v.* add to, amplify, augment, boost, broaden, build up, develop, enlarge, expand, extend, heighten, intensify, lengthen, lift, maximize, raise, step up, strengthen, widen; dilate, distend, inflate, swell; accumulate, burgeon, escalate, flourish, grow, mount, multiply, mushroom, pile up, proliferate, snowball, spread. ● *n.* **1** addition, amplification, augmentation, development, enlargement, escalation, expansion, extension, multiplication, proliferation, spread. **2** gain, growth, increment, jump, rise.

increasingly *adv.* more and more.

incredible *adj.* unbelievable. □ **incredibly** *adv.*, **incredibility** *n.*

> ■ far-fetched, implausible, improbable, inconceivable, unbelievable, unimaginable, unlikely; amazing, astonishing, astounding, extraordinary, marvellous, miraculous, wonderful.

incredulous *adj.* unbelieving, showin disbelief. □ **incredulously** *adv.*, **incredu lity** *n.*

> ■ disbelieving, distrustful, doubtful, dubi ous, mistrustful, sceptical, suspicious unconvinced.

increment *n.* increase, added amount □ **incremental** *adj.*

incriminate *v.* indicate as guilty. □ **in crimination** *n.*, **incriminatory** *adj.*

incrustation *n.* **1** encrusting. **2** crust o deposit formed on a surface.

incubate *v.* **1** hatch (eggs) by warmth. **2** cause (bacteria etc.) to develop. □ **in cubation** *n.*

incubator *n.* apparatus providin warmth for hatching eggs, rearing pre mature babies, or developing bacteria.

inculcate *v.* implant (a habit etc.) by constant urging. □ **inculcation** *n.*

incumbent *adj.* forming an obligation o duty. ● *n.* holder of an office, esp. a rector or a vicar.

incur *v.* (**incurred**) become subject t (something unpleasant) as a result o one's own behaviour.

incursion *n.* brief invasion, raid.

> ■ attack, foray, invasion, raid, sally, sortie.

indebted /indéttid/ *adj.* owing a debt.

> ■ beholden, grateful, obligated, obliged thankful.

indecent *adj.* **1** offending agains standards of decency. **2** unseemly. □ **in decently** *adv.*, **indecency** *n.*

> ■ **1** blue, coarse, crude, dirty, indelicate lewd, naughty, obscene, offensive, ribald risqué, rude, salacious, smutty, suggest ive, vulgar. **2** improper, inappropriate, in decorous, unbecoming, unseemly, un suitable.

indecipherable *adj.* unable to be rea or deciphered.

indecision *n.* inability to decide some thing, hesitation.

indecorous *adj.* improper.

indeed *adv.* in truth, really.

indefatigable *adj.* untiring. □ **indefat igably** *adv.*

> ■ assiduous, indomitable, industrious tenacious, tireless, unflagging, untiring.

indefensible *adj.* unable to be defende or justified.

> ■ inexcusable, reprehensible, unforgivable unjustifiable, unjustified, unpardonable.

indefinable *adj.* unable to be defined o described clearly. □ **indefinably** *adv.*

indefinite *adj.* not clearly stated or fixed, vague. ◻ **indefinite article** the word 'a' or 'an'.

■ ambiguous, equivocal, uncertain, unclear, unsure, vague; indeterminate, undecided, undefined, unfixed, unsettled, unspecified; blurred, dim, fuzzy, hazy, indistinct, indistinguishable, obscure, unrecognizable.

indefinitely *adv.* **1** in an indefinite way. **2** for an unlimited period.

indelible *adj.* **1** (of a mark) unable to be removed. **2** permanent. ◻ **indelibly** *adv.*

indelicate *adj.* **1** slightly indecent. **2** tactless. ◻ **indelicately** *adv.*, **indelicacy** *n.*

indemnify *v.* provide indemnity to. ◻ **indemnification** *n.*

indemnity *n.* **1** compensation for loss or damage. **2** legal exemption from penalties incurred by one's own actions.

indent *v.* **1** start inwards from a margin. **2** place an official order (for goods etc.). ◻ **indentation** *n.*

indenture *n.* written contract, esp. of apprenticeship. ● *v.* bind by this.

independent *adj.* not dependent on or not controlled by another person or thing. ◻ **independently** *adv.*, **independence** *n.*

■ autonomous, free, self-governing, sovereign; footloose, individualistic, self-reliant, self-sufficient; disinterested, impartial, neutral, unbiased, unprejudiced. ◻ **independence** autonomy, freedom, liberty, self-determination, self-government; individualism, self-sufficiency.

indescribable *adj.* unable to be described. ◻ **indescribably** *adv.*

indestructible *adj.* unable to be destroyed. ◻ **indestructibly** *adv.*

■ enduring, eternal, everlasting, immortal, undying; durable, shatter-proof, tough, unbreakable.

indeterminable *adj.* impossible to discover or decide.

indeterminate *adj.* not fixed in extent or character.

index *n.* (*pl.* **indexes** or **indices**) **1** list (usu. alphabetical) of names, subjects, etc., with references. **2** figure showing the current level of prices etc. compared with a previous level. ● *v.* **1** make an index to. **2** enter in an index. **3** adjust (wages etc.) according to a price index. ◻ **index finger** forefinger. **indexation** *n.*

■ *n.* **1** catalogue, concordance, directory, inventory, list, register, table of contents.

Indian *adj.* of India or Indians. ● *n.* **1** native of India. **2** any of the original inhabitants of the American continent or their descendants. ◻ **Indian ink** a black pigment. **Indian summer** dry sunny weather in autumn.

indiarubber *n.* rubber for rubbing out pencil or ink marks.

indicate *v.* **1** point out. **2** be a sign of. **3** state briefly. ◻ **indication** *n.*, **indicative** *adj.*

■ **1** direct attention to, display, identify, point out, point to, show. **2** bespeak, betoken, denote, evidence, evince, imply, manifest, reveal, signify, suggest. **3** express, make known, state. ◻ **indication** clue, hint, inkling, intimation, manifestation, mark, omen, sign, signal, symptom, token, warning.

indicator *n.* **1** thing that indicates. **2** pointer. **3** device on a vehicle showing the direction in which it is about to turn.

indict /indīt/ *v.* accuse formally. ◻ **indictment** *n.*

■ accuse, arraign, charge, impeach, summons.

indifferent *adj.* **1** showing no interest. **2** neither good nor bad. **3** not very good. ◻ **indifferently** *adv.*, **indifference** *n.*

■ **1** aloof, apathetic, blasé, casual, cool, detached, impassive, lackadaisical, lukewarm, uncaring, unconcerned, uninterested, unmoved, unsympathetic. **2** average, commonplace, fair, mediocre, middling, ordinary, undistinguished, uninspired. **3** all right, *colloq.* OK, passable, tolerable.

indigenous *adj.* native.

indigent *adj.* needy, poor. ◻ **indigence** *n.*

indigestible *adj.* difficult or impossible to digest.

indigestion *n.* pain caused by difficulty in digesting food.

indignant *adj.* feeling or showing indignation. ◻ **indignantly** *adv.*

■ annoyed, disgruntled, exasperated, huffy, in a huff, irked, irritated, piqued, *colloq.* peeved, sore, *colloq.* uptight.

indignation *n.* anger aroused by a supposed injustice.

■ anger, annoyance, displeasure, exasperation, irritation, pique, resentment, vexation.

indignity *n.* unworthy treatment, humiliation.

indigo *n.* deep blue dye or colour.

indiscernible *adj.* unable to be discerned. □ **indiscernibly** *adv.*

indiscreet *adj.* **1** revealing secrets. **2** not judicious. □ **indiscreetly** *adv.*, **indiscretion** *n.*

■ **1** garrulous, talkative; irresponsible, untrustworthy. **2** ill-advised, ill-judged, impolitic, imprudent, incautious, injudicious, insensitive, tactless, thoughtless, undiplomatic, unguarded, unwise.

indiscriminate *adj.* **1** not discriminating. **2** haphazard. □ **indiscriminately** *adv.*

■ **1** undiscriminating, unselective. **2** aimless, arbitrary, haphazard, hit-or-miss, random, unmethodical, unsystematic, wholesale.

indispensable *adj.* essential.

■ crucial, essential, imperative, key, necessary, requisite, vital.

indisposed *adj.* **1** slightly ill. **2** unwilling. □ **indisposition** *n.*

indisputable *adj.* undeniable. □ **indisputably** *adv.*

■ certain, incontestable, incontrovertible, indubitable, irrefutable, undeniable, unquestionable.

indissoluble *adj.* firm and lasting, not able to be destroyed.

individual *adj.* **1** single, separate. **2** characteristic of one particular person or thing. ● *n.* **1** one person, animal, or plant considered separately. **2** (*colloq.*) person. □ **individually** *adv.*, **individuality** *n.*

■ *adj.* **1** distinct, particular, separate, single, specific. **2** characteristic, distinctive, idiosyncratic, peculiar, personal, special, unique. ● *n.* **2** child, human being, man, mortal, person, soul, woman.

individualist *n.* person who is independent in thought etc. □ **individualistic** *adj.*, **individualism** *n.*

indoctrinate *v.* fill (a person's mind) with particular ideas or doctrines. □ **indoctrination** *n.*

indolent *adj.* lazy. □ **indolently** *adv.*, **indolence** *n.*

indomitable *adj.* **1** unconquerable. **2** unyielding. □ **indomitably** *adv.*

■ **1** invincible, unassailable, unconquerable. **2** brave, courageous, dauntless, determined, indefatigable, persistent, resolute, staunch, steadfast, tireless, undaunted, unflagging, untiring, unyielding.

indoor *adj.* situated, used, or done inside a building. □ **indoors** *adv.*

indubitable *adj.* that cannot be doubted. □ **indubitably** *adv.*

induce *v.* **1** persuade. **2** cause. **3** bring on (labour) artificially.

■ **1** cajole, coax, convince, get, influence, inveigle, lead, persuade, press, prevail on, push, talk into, wheedle. **2** bring about, cause, create, engender, give rise to, lead to.

inducement *n.* **1** inducing. **2** incentive.

induct *v.* install (a clergyman) ceremonially into a benefice.

inductance *n.* amount of induction of electric current.

induction *n.* **1** inducting. **2** inducing. **3** reasoning (from observed examples) that a general law exists. **4** production of an electric or magnetic state by proximity of an electrified or magnetic object. **5** drawing of a fuel mixture into the cylinder(s) of an engine. □ **inductive** *adj.*

indulge *v.* **1** allow (a person) to have or do what he or she wishes. **2** gratify. □ **indulge in** allow oneself to enjoy the pleasure of. **indulgence** *n.*

■ **1** mollycoddle, pamper, spoil, treat. **2** cater to, gratify, humour, minister to, pander to, satisfy. □ **indulge in** luxuriate in, succumb to, treat oneself to, wallow in, yield to.

indulgent *adj.* lenient, willing to overlook faults. □ **indulgently** *adv.*

■ easygoing, forbearing, forgiving, kind, lax, lenient, liberal, patient, permissive, soft, tolerant, understanding.

industrial *adj.* of, for, or full of industries. □ **industrially** *adv.*

industrialist *n.* owner or manager of an industrial business. □ **industrialism** *n.*

industrialized *adj.* full of highly developed industries.

industrious *adj.* hard-working. □ **industriously** *adv.*

■ assiduous, conscientious, diligent, dogged, hard-working, indefatigable, painstaking, sedulous, tireless, zealous.

industry *n.* **1** manufacture or production of goods. **2** business activity. **3** being industrious.

inebriated *adj.* drunken. □ **inebriation** *n.*

inedible *adj.* not edible.

ineducable *adj.* incapable of being educated.

ineffable *adj.* too great to be described. □ **ineffably** *adv.*

ineluctable *adj.* inescapable, unavoidable.

inept *adj.* **1** unskilful. **2** unsuitable. **3** absurd. □ **ineptly** *adv.*, **ineptitude** *n.*

■ **1** amateurish, awkward, bumbling, bungling, clumsy, *colloq.* ham-fisted, incompetent, inefficient, inexpert, maladroit, unskilled, unskilful. **2** inapt, inappropriate, out of keeping, out of place, unsuitable.

inequality *n.* lack of equality.

■ difference, discrepancy, disparity, disproportion, dissimilarity, imbalance, incongruity, inconsistency; bias, partiality, prejudice, unfairness, injustice.

inequitable *adj.* unfair, unjust. □ **inequitably** *adv.*

ineradicable *adj.* not able to be eradicated. □ **ineradicably** *adv.*

inert *adj.* **1** without the power of moving. **2** sluggish, slow. **3** chemically inactive. □ **inertly** *adv.*, **inertness** *n.*

■ **1** immobile, inanimate, lifeless, motionless, quiescent, static, stationary, still. **2** idle, inactive, indolent, languid, languorous, lazy, leaden, listless, passive, slothful, slow, sluggish, torpid.

inertia *n.* **1** being inert. **2** property by which matter continues in its state of rest or line of motion.

■ **1** immobility, inactivity, quiescence, stasis, stillness; idleness, indolence, lassitude, laziness, listlessness, sloth, sluggishness, torpor.

inescapable *adj.* unavoidable. □ **inescapably** *adv.*

inessential *adj.* not necessary. ● *n.* inessential thing.

■ *adj.* dispensable, expendable, needless, superfluous, unnecessary, unneeded.

inestimable *adj.* too great to be estimated. □ **inestimably** *adv.*

inevitable *adj.* unavoidable, sure to happen or appear. □ **inevitably** *adv.*, **inevitability** *n.*

■ destined, fated, ineluctable, inescapable, unavoidable; assured, certain, guaranteed, sure.

inexact *adj.* not exact. □ **inexactly** *adv.*, **inexactitude** *n.*

■ approximate, estimated, general, imprecise, rough, vague; erroneous, false, inaccurate, incorrect, mistaken, wrong.

inexhaustible *adj.* available in unlimited quantity.

■ boundless, endless, infinite, limitless, unbounded, unlimited, unrestricted.

inexorable *adj.* relentless. □ **inexorably** *adv.*, **inexorability** *n.*

inexperience *n.* lack of experience. □ **inexperienced** *adj.*

■ □ **inexperienced** callow, green, immature, innocent, naive, unsophisticated, unworldly; amateurish, inexpert, unseasoned, unskilled, untrained, unversed.

inexpert *adj.* not expert, unskilful. □ **inexpertly** *adv.*

inexplicable *adj.* unable to be explained. □ **inexplicably** *adv.*

■ baffling, incomprehensible, mysterious, mystifying, perplexing, puzzling, unaccountable, unexplainable.

inextricable *adj.* unable to be extricated or disentangled. □ **inextricably** *adv.*

infallible *adj.* **1** incapable of being wrong. **2** never failing. □ **infallibly** *adv.*, **infallibility** *n.*

■ certain, dependable, faultless, flawless, foolproof, guaranteed, perfect, reliable, sure, unerring, unfailing.

infamous /ínfəməss/ *adj.* having a bad reputation. □ **infamously** *adv.*, **infamy** *n.*

■ disreputable, ill-famed, notorious.

infancy *n.* **1** early childhood. **2** early stage of development.

infant *n.* child during the earliest stage of its life.

infanticide *n.* killing or killer of an infant soon after its birth. □ **infanticidal** *adj.*

infantile *adj.* **1** of infants or infancy. **2** very childish.

■ **2** babyish, childish, immature, juvenile, puerile.

infantry *n.* troops who fight on foot.

infatuated *adj.* filled with intense love. □ **infatuation** *n.*

infect *v.* **1** affect or contaminate with a disease or its germs. **2** affect with one's feeling.

■ **1** blight, contaminate, poison, pollute, taint.

infection *n.* **1** infecting, being infected. **2** disease or condition so caused.

■ **1** contamination, pollution, tainting. **2** *sl.* bug, disease, disorder, sickness, *colloq.* virus.

infectious *adj.* **1** (of disease) able to spread by air or water. **2** infecting others.

■ **1** catching, communicable, contagious, transmissible.

infer *v.* (**inferred**) reach (an opinion) from facts or reasoning. □ **inference** *n.*

■ conclude, deduce, draw the conclusion, gather, surmise, understand.

inferior *adj.* low or lower in rank, importance, quality, or ability. ● *n.* person inferior to another, esp. in rank. □ **inferiority** *n.*

■ *adj.* lesser, lower, junior, minor, lowly, subordinate; bad, cheap, defective, faulty, gimcrack, low-quality, mediocre, poor, second-class, second-rate, shoddy, slipshod, substandard, tawdry, trashy. ● *n.* junior, subordinate, underling; *colloq.* dogsbody, drudge, menial.

infernal *adj.* **1** of hell. **2** (*colloq.*) tiresome. □ **infernally** *adv.*

inferno *n.* (*pl.* **-os**) **1** hell. **2** intensely hot place. **3** raging fire.

infest *v.* overrun in large numbers. □ **infestation** *n.*

■ invade, overrun, pervade, swarm over, take over.

infidel *n.* **1** person with no religious faith. **2** opponent of a religion, esp. Christianity.

infidelity *n.* unfaithfulness.

■ adultery, deceit, deception, disloyalty, faithlessness, falseness, treachery, unfaithfulness.

infighting *n.* hidden conflict within an organization.

infiltrate *v.* enter gradually and unperceived. □ **infiltration** *n.*, **infiltrator** *n.*

infinite /infinit/ *adj.* **1** having no limit. **2** too great or too many to be measured. □ **infinitely** *adv.*

■ bottomless, boundless, countless, immeasurable, incalculable, inestimable, inexaustible, innumerable, limitless, numberless, unbounded, unlimited, untold.

infinitesimal *adj.* extremely small. □ **infinitesimally** *adv.*

infinitive *n.* form of a verb not indicating tense, subject, etc. (e.g. *to go*).

infinity *n.* infinite number, extent, or time.

infirm *adj.* physically weak.

■ ailing, debilitated, decrepit, doddery, enfeebled, feeble, frail, ill, sick, unwell, weak.

infirmary *n.* hospital.

infirmity *n.* **1** being infirm. **2** particular physical weakness.

■ **1** debilitation, debility, decrepitude, feebleness, frailty, weakness. **2** affliction, ailment, complaint, condition, disease, disorder, illness, indisposition, malady, sickness.

inflame *v.* **1** arouse strong feeling in. **2** cause inflammation in.

■ **1** arouse, excite, fire, incite, kindle, provoke, stimulate, stir up, whip up; anger, enrage, exasperate, incense, infuriate, madden.

inflammable *adj.* able to be set on fire. □ **inflammability** *n.*

inflammation *n.* redness and heat in a part of the body.

inflammatory *adj.* arousing strong feeling or anger.

inflatable *adj.* able to be inflated.

inflate *v.* **1** fill with air or gas so as to swell. **2** increase artificially.

■ **1** blow up, pump up; dilate, distend, swell.

inflation *n.* **1** inflating. **2** general rise in prices and fall in the purchasing power of money. □ **inflationary** *adj.*

inflect *v.* **1** change the pitch of (a voice) in speaking. **2** change the ending or form of (a word) grammatically. □ **inflection** *n.*

inflexible *adj.* **1** not flexible. **2** unyielding. □ **inflexibly** *adv.*, **inflexibility** *n.*

■ **1** firm, hard, inelastic, rigid, stiff, unbendable, unmalleable. **2** adamant, determined, firm, immovable, obdurate, obstinate, resolute, rigid, stiff-necked, stubborn, unbending, uncompromising, unyielding.

inflict *v.* cause (a blow, penalty, etc.) to be suffered. □ **infliction** *n.*

■ administer, deal, deliver, impose, levy, wreak.

influence *n.* **1** effect a person or thing has on another. **2** moral power, ascendancy. **3** person or thing with this. ● *v.* exert influence on.

■ *n.* **1** bearing, effect, impact, impression. **2** ascendancy, authority, *colloq.* clout, control, leverage, mastery, power, pull, sway, weight. ● *v.* affect, alter, change, have an effect on, modify, sway; drive, force, impel, induce, motivate, move, persuade; manipulate, pressurize, pull strings with.

influential *adj.* having great influence. □ **influentially** *adv.*

■ authoritative, controlling, dominant, important, leading, powerful, strong; persuasive, significant, telling.

influenza *n.* virus disease causing fever, muscular pain, and catarrh.

influx *n.* inward flow.

inform *v.* tell. □ **inform against** or **on** give incriminating evidence about a person to police etc. **informer** *n.*

■ advise, apprise, brief, communicate to, enlighten, fill in, let know, notify, tell, tip off. □ **inform on** betray, denounce, incriminate, rat on, *sl.* shop, *sl.* squeal on.

informal *adj.* not formal, without formality or ceremony. □ **informally** *adv.*, **informality** *n.*

informant *n.* giver of information.

information *n.* facts told, heard, or discovered.

■ data, details, facts, intelligence, knowledge, *colloq.* low-down, news, tidings, word.

informative *adj.* giving information. □ **informatively** *adv.*

■ edifying, educational, enlightening, helpful, illuminating, instructive; chatty, communicative, forthcoming.

infra-red *adj.* of or using radiation with a wavelength longer than that of visible light rays.

infrastructure *n.* subordinate parts forming the basis of an enterprise.

infringe *v.* **1** break (a rule or agreement). **2** encroach, trespass. □ **infringement** *n.*

■ **1** break, contravene, disobey, flout, violate.

infuriate *v.* make very angry.

infuse *v.* **1** fill (with a quality). **2** soak to bring out flavour.

infusion *n.* **1** infusing. **2** liquid made by this.

ingenious *adj.* **1** clever at inventing things. **2** cleverly contrived. □ **ingeniously** *adv.*, **ingenuity** *n.*

■ adept, adroit, astute, brilliant, clever, creative, cunning, deft, dexterous, imaginative, inventive, original, resourceful, shrewd, skilful, skilled.

ingenuous *adj.* without artfulness. □ **ingenuously** *adv.*

■ artless, childlike, guileless, innocent, naive, trustful, unsophisticated, unsuspecting.

ingest *v.* take in as food.

ingot *n.* oblong lump of cast metal.

ingrained *adj.* deeply fixed in a surface or character.

ingratiate *v.* bring (oneself) into a person's favour, esp. to gain advantage. □ **ingratiation** *n.*

ingratitude *n.* lack of gratitude.

ingredient *n.* one element in a mixture or combination.

■ component, constituent, element, factor, part.

ingress *n.* **1** going in. **2** right of entry.

ingrowing *adj.* growing abnormally into the flesh.

inhabit *v.* live in, occupy. □ **inhabitable** *adj.*, **inhabitant** *n.*

■ dwell in, live in, lodge in, occupy, reside in; people, populate. □ **inhabitant** householder, inmate, occupant, occupier, resident, tenant; citzen, denizen, local, native.

inhalant *n.* medicinal substance to be inhaled.

inhale *v.* **1** breathe in. **2** draw tobacco smoke into the lungs.

inhaler *n.* device producing a medicinal vapour to be inhaled.

inherent *adj.* existing in a thing as a permanent quality. □ **inherently** *adv.*

■ congenital, inborn, ingrained, inherited, innate, native, natural; essential, fundamental, intrinsic.

inherit *v.* **1** receive from a predecessor, esp. someone who has died. **2** derive from parents etc. □ **inheritance** *n.*

■ □ **inheritance** bequest, birthright, heritage, legacy, patrimony.

inhibit *v.* **1** restrain, prevent. **2** cause inhibitions in.

■ **1** check, deter, discourage, frustrate, hamper, hinder, hold back, impede, interfere with, obstruct, prevent, restrain, stop.

inhibition *n.* **1** inhibiting. **2** inability to act naturally or spontaneously.

■ **1** prevention, repression, suppression. **2** embarrassment, reticence, self-consciousness, shyness; *sl.* hang-up.

inhospitable *adj.* **1** not hospitable. **2** giving no shelter.

inhuman *adj.* brutal, cruel. □ **inhumanly** *adv.*, **inhumanity** *n.*

■ barbaric, barbarous, bestial, bloodthirsty, brutal, cruel, ferocious, harsh, heartless, inhumane, merciless, pitiless, ruthless, savage, vicious.

inhumane *adj.* callous. □ **inhumanely** *adv.*, **inhumanity** *n.*

■ brutal, callous, cold, cold-blooded, cruel, hard, hard-hearted, harsh, heartless, inconsiderate, inhuman, merciless, pitiless, remorseless, ruthless, uncaring, unforgiving, unkind, unsympathetic.

inimical *adj.* hostile. □ **inimically** *adv.*

inimitable *adj.* impossible to imitate. □ **inimitably** *adv.*

iniquity *n.* **1** great injustice. **2** wickedness. □ **iniquitous** *adj.*

initial *n.* first letter of a word or name. ● *v.* (**initialled**) mark or sign with initials. ● *adj.* of the beginning. □ **initially** *adv.*

■ *adj.* early, first, inaugural, introductory, opening, original.

initiate *v.* /iníshiayt/ **1** begin, originate. **2** admit into membership. **3** give instruction to. ● *n.* /iníshiət/ initiated person. □ **initiation** *n.*, **initiator** *n.*, **initiatory** *adj.*

■ *v.* **1** begin, instigate, institute, introduce, launch, originate, pioneer, set up, start. **2** accept, admit, enrol, install, introduce. **3** coach, drill, instruct, school, teach, train, tutor.

initiative *n.* **1** ability to initiate. **2** first step in a process.

■ **1** drive, dynamism, energy, enterprise, go, *colloq.* gumption, resourcefulness.

inject *v.* force (liquid) into the body with a syringe. □ **injection** *n.*

■ □ **injection** inoculation, *colloq.* jab, shot, vaccination.

injudicious *adj.* unwise. □ **injudiciously** *adv.*

■ foolhardy, foolish, ill-advised, impolitic, improvident, imprudent, incautious, misguided, rash, reckless, short-sighted, unwise.

injunction *n.* court order.

injure *v.* **1** damage physically. **2** impair. **3** do wrong to.

■ **1** break, bruise, burn, cut, damage, disfigure, fracture, gash, harm, hurt, lacerate, maim, wound. **2** damage, impair, harm, mar, ruin, spoil, undermine, weaken. **3** abuse, ill-treat, maltreat, wrong.

injurious *adj.* hurtful, wrongful.

■ adverse, bad, damaging, dangerous, deleterious, destructive, detrimental, harmful, hurtful; unfair, unjust, wrongful.

injury *n.* **1** damage, harm. **2** form of this. **3** wrong or unjust act.

■ **1** damage, disfigurement, harm, hurt, impairment. **2** abrasion, break, bruise, contusion, cut, fracture, gash, laceration, wound. **3** disservice, injustice, ill, mischief, wrong.

injustice *n.* **1** lack of justice. **2** unjust action or treatment.

■ **1** inequality, iniquity, unfairness, unjustness, wrong; bias, discrimination, favouritism, partiality, partisanship, prejudice. **2** bad turn, disservice, injury, wrong.

ink *n.* coloured liquid used in writing, printing, etc. ● *v.* apply ink to. □ **inky** *adj.*

inkling *n.* slight suspicion, hint.

■ clue, hint, indication, intimation, suggestion, suspicion.

inlaid *see* **inlay**.

inland *adj.* & *adv.* in or towards the interior of a country.

in-laws *n.pl.* one's relatives by marriage.

inlay *v.* /inláy/ (**inlaid**) set (one thing in another) so that the surfaces are flush. ● *n.* /ínlay/ inlaid material or design.

inlet *n.* **1** strip of water extending inland. **2** way of admission.

inmate *n.* inhabitant, esp. of an institution.

inmost *adj.* furthest inward.

inn *n.* pub, esp. one in the country offering accommodation.

innards *n.pl.* (*colloq.*) **1** entrails. **2** inner parts.

innate *adj.* inborn. □ **innately** *adv.*

inner *adj.* interior, internal. □ **inner city** central densely populated urban area.

innermost *adj.* furthest inward.

innocent *adj.* **1** not guilty, free of evil. **2** naive, foolishly trustful. □ **innocently** *adv.*, **innocence** *n.*

■ **1** blameless, guiltless, irreproachable, not guilty, unimpeachable; chaste, good, moral, pure, righteous, sinless, virtuous. **2** artless, childlike, guileless, gullible, inexperienced, ingenuous, naive, simple, trustful, unsophisticated, unsuspecting, unworldly.

innocuous *adj.* harmless. □ **innocuously** *adv.*

innovate *v.* introduce something new. □ **innovation** *n.*, **innovative** *adj.*, **innovator** *n.*

innuendo *n.* (*pl.* **-oes**) insinuation.

innumerable *adj.* too many to be counted.

■ countless, incalculable, infinite, many, numberless, *sl.* umpteen, unnumbered, untold.

innumerate *adj.* not knowing basic mathematics. □ **innumeracy** *n.*

inoculate *v.* protect (against disease) with vaccines or serums. □ **inoculation** *n.*

inoperable *adj.* unable to be cured by surgical operation.

inoperative *adj.* not functioning.

inopportune *adj.* happening at an unsuitable time.

■ badly-timed, inappropriate, inconvenient, unfavourable, unfortunate, unseasonable, unsuitable, untimely, untoward.

inordinate *adj.* excessive. □ **inordinately** *adv.*

inorganic *adj.* of mineral origin, not organic. □ **inorganically** *adv.*

in-patient *n.* patient staying in a hospital during treatment.

input *n.* what is put in. ● *v.* (**input** or **inputted**) **1** put in. **2** supply (data etc.) to a computer.

inquest *n.* judicial investigation, esp. of a sudden death.

inquire *v.* seek information formally. □ **inquire into** make an investigation into. **inquirer** *n.*

■ □ **inquire into** examine, explore, investigate, look into, probe, research, scrutinize, study.

inquiry *n.* investigation.

■ examination, exploration, inquest, investigation, probe, scrutiny, study, survey.

inquisition *n.* detailed or relentless questioning. □ **inquisitor** *n.*, **inquisitorial** *adj.*

inquisitive *adj.* **1** eagerly seeking knowledge. **2** prying. □ **inquisitively** *adv.*, **inquisitiveness** *n.*

■ **1** curious, inquiring, interested. **2** intrusive, meddlesome, *colloq.* nosy, prying.

inroad *n.* incursion.

insalubrious *adj.* unhealthy.

insane *adj.* **1** mad. **2** very foolish. □ **insanely** *adv.*, **insanity** *n.*

■ **1** *sl.* barmy, *sl.* batty, certifiable, *sl.* crackers, crazy, demented, deranged, lunatic, mad, *sl.* nuts, *sl.* nutty, out of one's mind, *sl.* potty, psychotic, unbalanced. **2** absurd, asinine, crazy, daft, foolish, idiotic, imbecilic, inane, ludicrous, nonsensical, ridiculous, senseless, silly, stupid. □ **insanity** dementia, lunacy, madness, psychosis; folly, foolishness, idiocy, senselessness, stupidity.

insanitary *adj.* not clean, not hygienic.

■ dirty, filthy, insalubrious, sordid, squalid, unclean, unhealthy, unhygienic.

insatiable *adj.* unable to be satisfied. □ **insatiably** *adv.*, **insatiability** *n.*

inscribe *v.* write or engrave.

inscription *n.* words inscribed.

inscrutable *adj.* baffling, impossible to interpret. □ **inscrutably** *adv.*, **inscrutability** *n.*

■ baffling, enigmatic, mysterious, mystifying, perplexing, puzzling, unfathomable.

insect *n.* small creature with six legs, no backbone, and a segmented body.

insecticide *n.* substance for killing insects.

insectivorous *adj.* insect-eating.

inseminate *v.* insert semen into. □ **insemination** *n.*

insensible *adj.* **1** unconscious. **2** unaware. **3** callous. **4** imperceptible. □ **insensibly** *adv.*

■ **1** comatose, out, senseless, unconscious. **2** heedless, oblivious, unaware, unconscious, unmindful.

insensitive *adj.* not sensitive.

■ callous, inconsiderate, tactless, thick-skinned, thoughtless, uncaring, unfeeling, unsympathetic.

inseparable *adj.* unable to be separated or kept apart. □ **inseparably** *adv.*, **inseparability** *n.*

insert *v.* /insért/ put into or between or among. ● *n.* /insert/ thing inserted. □ **insertion** *n.*

■ *v.* implant, inset, introduce, place in, pop in, put in, *colloq.* stick in; interject, interpose.

inset *v.* /insét/ (**inset, insetting**) **1** place in. **2** decorate with an inset. ● *n.* /inset/ thing set into a larger thing.

inshore *adj.* & *adv.* near or nearer to the shore.

inside *n.* **1** inner side, surface, or part. **2** (*pl., colloq.*) stomach and bowels. ● *adj.* of or from the inside. ● *adv.* on, in, or to the inside. ● *prep.* **1** on or to the inside of. **2** within. ◻ **inside out 1** with the inner side outwards. **2** thoroughly.

■ *n.* **1** centre, core, heart, interior, middle. **2** (**insides**) bowels, entrails, guts, *colloq.* innards, intestines, viscera, vitals. ● *adj.* inner, interior, internal; indoor.

insidious *adj.* proceeding inconspicuously but harmfully. ◻ **insidiously** *adv.*, **insidiousness** *n.*

insight *n.* perception and understanding of a thing's nature.

■ acumen, discernment, perception, perceptiveness, percipience, perspicacity, sensitivity, sharpness, shrewdness, understanding.

insignia *n.pl.* badges or marks of office etc.

insignificant *adj.* unimportant. ◻ **insignificantly** *adv.*, **insignificance** *n.*

■ *sl.* footling, inconsequential, little, minor, petty, trifling, trivial, unimportant.

insinuate *v.* **1** insert gradually or craftily. **2** hint artfully. ◻ **insinuation** *n.*, **insinuator** *n.*

■ **1** insert, introduce, slip. **2** hint, imply, indicate, intimate, suggest. ◻ **insinuation** allusion, hint, implication, innuendo, intimation, suggestion.

insipid *adj.* dull, lifeless, flavourless. ◻ **insipidity** *n.*

■ boring, characterless, colourless, dull, flat, lacklustre, lifeless, tedious, unexciting, uninspiring, uninteresting, vapid; bland, flavourless, tasteless.

insist *v.* declare or demand emphatically.

■ assert, avow, declare, emphasize, hold, maintain, stress, swear; demand, require.

insistent *adj.* **1** tending to insist. **2** demanding attention. ◻ **insistently** *adv.*, **insistence** *n.*

■ **1** determined, dogged, emphatic, firm, positive, resolute, tenacious, unrelenting, unyielding. **2** intrusive, obtrusive; demanding, importunate, nagging, persistent.

in situ /in sityōō/ in its original place.

insolent *adj.* impertinently insulting. ◻ **insolently** *adv.*, **insolence** *n.*

■ brazen, cheeky, contemptuous, disrespectful, impertinent, impudent, insulting, offensive, pert, presumptuous, rude.

insoluble *adj.* **1** unable to be dissolved. **2** unable to be solved.

insolvent *adj.* unable to pay one's debts. ◻ **insolvency** *n.*

■ bankrupt, *colloq.* broke, *colloq.* gone bust, failed.

insomnia *n.* inability to sleep.

insomniac *n.* sufferer from insomnia.

insouciant *adj.* carefree, unconcerned. ◻ **insouciantly** *adv.*, **insouciance** *n.*

inspect *v.* examine critically or officially. ◻ **inspection** *n.*

■ check, examine, go over, investigate, look into, peruse, pore over, scan, scrutinize, study, survey, *colloq.* vet. ◻ **inspection** check, examination, investigation, *colloq.* once-over, scrutiny, study, survey, *colloq.* vetting.

inspector *n.* **1** person who inspects. **2** police officer above sergeant.

inspiration *n.* **1** creative force or influence. **2** sudden brilliant idea. ◻ **inspirational** *adj.*

■ **1** creativity, imagination, originality, spark; encouragement, impetus, spur, stimulation, stimulus. **2** *US* brainstorm, brainwave, bright idea.

inspire *v.* **1** stimulate to activity. **2** animate. **3** instil (a feeling or idea) into.

■ **1,2** animate, arouse, encourage, energize, enliven, excite, fortify, galvanize, invigorate, motivate, move, rouse, stimulate, stir.

install *v.* **1** place (a person) into office ceremonially. **2** set in position and ready for use. **3** establish. ◻ **installation** *n.*

■ **1** enthrone, inaugurate, induct, initiate, ordain. **2** fit, put in. **3** ensconce, establish, place, position, seat, settle, station.

instalment *n.* one of the parts in which a thing is presented or a debt paid over a period of time.

instance *n.* **1** example. **2** particular case. ● *v.* mention as an instance.

■ *n.* **1** example, exemplar, exemplification, illustration. **2** case, circumstance, event, occasion, occurrence, situation. ● *v.* adduce, allude to, cite, quote.

instant *adj.* **1** immediate. **2** (of food) quickly and easily prepared. ● *n.* exact moment. ◻ **instantly** *adv.*

■ *adj.* **1** immediate, instantaneous, on the spot, prompt, quick, speedy, sudden, swift, unhesitating. ● *n.* minute, moment, point, split second.

nstantaneous *adj.* occurring or done instantly. □ **instantaneously** *adv.*

nstead *adv.* as an alternative.

nstep *n.* **1** middle part of the foot. **2** part of a shoe etc. covering this.

nstigate *v.* **1** incite. **2** initiate. □ **instigation** *n.*, **instigator** *n.*

■ **1** encourage, foment, incite, provoke, stir up, whip up. **2** begin, get under way, initiate, set in motion, start.

nstil *v.* (**instilled**) implant (ideas etc.) gradually. □ **instillation** *n.*

nstinct *n.* **1** inborn pattern of behaviour. **2** innate impulse. **3** unconscious skill, intuition.

■ **3** aptitude, bent, flair, gift, knack, skill, talent; feeling, intuition, sixth sense.

nstinctive *adj.* **1** of or prompted by instinct. **2** apparently unconscious or automatic. □ **instinctively** *adv.*

■ **1** inborn, inbred, inherent, innate, intuitive, native, natural. **2** automatic, involuntary, knee-jerk, mechanical, spontaneous, subconscious, unconscious.

nstitute *n.* **1** organization for promotion of a specified activity. **2** its premises. ● *v.* **1** establish. **2** initiate. □ **institutor** *n.*

■ *n.* **1** association, company, establishment, foundation, institution, organization, society. ● *v.* **1** create, establish, form, found, inaugurate, launch, set up. **2** begin, get under way, initiate, instigate, set in motion, start.

nstitution *n.* **1** process of instituting. **2** organization founded esp. for educational or social purposes. **3** established rule or custom. □ **institutional** *adj.*

■ **1** creation, establishment, formation, foundation, inauguration, initiation, instigation, launch, start. **2** establishment, foundation, institute, organization; academy, college, school; clinic, hospital, (nursing) home, sanatorium. **3** convention, custom, habit, practice, routine, rule, tradition.

nstitutionalize *v.* accustom to living in an institution.

nstruct *v.* **1** teach (a person) a subject or skill. **2** give instructions to. □ **instructor** *n.*

■ **1** coach, drill, educate, ground, school, teach, train, tutor. **2** bid, command, direct, order, require, tell.

instruction *n.* **1** process of teaching. **2** (*pl.*) statements telling a person what to do.

■ **1** coaching, education, grounding, schooling, teaching, training. **2** (**instructions**) advice, commands, dictates, directions, directives, guidelines, orders, recommendations.

instructive *adj.* giving instruction, enlightening.

■ edifying, educational, enlightening, helpful, illuminating, informative, revealing.

instrument *n.* **1** implement, esp. for delicate work. **2** measuring device of an engine or vehicle. **3** device for producing musical sounds.

■ **1** appliance, device, gadget, implement, tool, utensil.

instrumental *adj.* **1** serving as a means. **2** performed on musical instruments.

instrumentalist *n.* player of a musical instrument.

insubordinate *adj.* disobedient, rebellious. □ **insubordination** *n.*

■ defiant, disobedient, insurgent, intractable, mutinous, obstreperous, rebellious, recalcitrant, refractory, uncooperative, ungovernable, unruly.

insubstantial *adj.* **1** not real. **2** weak, flimsy. □ **insubstantiality** *n.*

■ **1** fanciful, illusive, illusory, imaginary, unreal. **2** fragile, frail, rickety, shaky; feeble, flimsy, tenuous, unconvincing, weak; diaphanous, filmy, thin.

insufferable *adj.* intolerable. □ **insufferably** *adv.*

insular *adj.* **1** of an island. **2** narrow-minded. □ **insularity** *n.*

insulate *v.* **1** cover with a substance that prevents the passage of electricity, sound, or heat. **2** isolate from influences. □ **insulation** *n.*, **insulator** *n.*

■ **1** cover, lag, wrap. **2** keep apart, isolate, segregate, separate; cushion, shelter.

insulin *n.* hormone controlling the body's absorption of sugar.

insult *v.* /insúlt/ speak or act so as to offend someone. ● *n.* /ínsult/ insulting remark or action. □ **insulting** *adj.*

■ *v.* abuse, be rude to, calumniate, defame, malign, put down, revile, *sl.* slag (off), slander, vilify; affront, hurt, injure, offend, slight. ● *n.* affront, rebuff, slander, slight, slur, snub.

insuperable *adj.* unable to be overcome. □ **insuperably** *adv.*, **insuperability** *n.*

■ insurmountable, invincible, overwhelming, unconquerable.

insupportable *adj.* intolerable.

insurance *n.* **1** contract to provide compensation for loss, damage, or death. **2** sum payable as a premium or in compensation. **3** safeguard against loss or failure.

insure *v.* protect by insurance.

insurgent *adj.* in revolt, rebellious. ● *n.* rebel. □ **insurgency** *n.*

insurmountable *adj.* insuperable.

insurrection *n.* rebellion. □ **insurrectionist** *n.*

intact *adj.* undamaged, complete.

■ complete, entire, flawless, perfect, unbroken, undamaged, unharmed, unscathed, whole.

intake *n.* **1** taking things in. **2** place or amount of this.

integral *adj.* forming or necessary to form a whole.

■ complete, entire, intact, integrated, unified, whole; basic, essential, fundamental, indispensable, intrinsic, necessary.

integrate *v.* **1** combine (parts) into a whole. **2** bring or come into full membership of a community. □ **integration** *n.*

integrity *n.* honesty.

■ decency, goodness, honesty, honour, morality, principle, probity, rectitude, righteousness, trustworthiness, virtue.

intellect *n.* faculty of reasoning and acquiring knowledge.

■ brains, intelligence, *sl.* loaf, mind, reason, understanding.

intellectual *adj.* **1** of or using the intellect. **2** having a strong intellect. ● *n.* intellectual person. □ **intellectually** *adv.*

■ *adj.* **1** cerebral, mental. **2** academic, bookish, brainy, clever, erudite, highbrow, learned, scholarly. ● *n.* academic, highbrow, professor, scholar, thinker.

intelligence *n.* **1** the intellect. **2** quickness of understanding. **3** information, esp. that of military value. **4** people collecting this.

■ **1** brains, intellect, *sl.* loaf, mind, reason, understanding. **2** acumen, acuteness, astuteness, brightness, cleverness, common sense, insight, *colloq.* nous, perception, percipience, perspicacity, sharpness, shrewdness, smartness, wit.

intelligent *adj.* mentally able, cleve
□ **intelligently** *adv.*

■ able, astute, brainy, bright, brillia canny, clever, discerning, penetratin perceptive, quick, sharp, shrewd, smart.

intelligible *adj.* able to be understoo □ **intelligibly** *adv.*, **intelligibility** *n.*

■ articulate, audible, clear, coherer comprehensible, legible, lucid, plain, tional, readable, understandable.

intend *v.* have in mind as what o wishes to do or achieve.

■ aim, contemplate, design, have in min mean, plan, propose, purpose.

intense *adj.* **1** strong in quality or d gree. **2** feeling, or apt to feel, stror emotion. □ **intensely** *adv.*, **intensity** *n.*

■ **1** acute, ardent, burning, exquisite, e treme, fervent, fierce, great, heartfelt, kee overpowering, powerful, profound, sever sharp, strong, vehement, violent. **2** em tional, impassioned, passionate; highl strung, moody, sensitive, temperamenta touchy.

intensify *v.* make or become more i tense. □ **intensification** *n.*

■ add to, amplify, augment, deepe double, enhance, escalate, heighten, i crease, redouble, reinforce, step u strengthen; aggravate, exacerbat worsen.

intensive *adj.* employing much effor □ **intensively** *adv.*, **intensiveness** *n.*

■ assiduous, concentrated, exhaustiv meticulous, painstaking, rigorous, th ough.

intent *n.* intention. ● *adj.* with conce trated attention. □ **intent on** determine to. **intently** *adv.*, **intentness** *n.*

■ *adj.* absorbed, concentrating, e grossed, involved, rapt.

intention *n.* purpose, aim.

■ aim, ambition, design, goal, end, inte object, objective, purpose, target.

intentional *adj.* done on purpose. □ **i tentionally** *adv.*

■ calculated, conscious, deliberate, i tended, meant, planned, preconceive premeditated, purposeful, wilful.

inter *v.* (**interred**) bury.

inter- *prep.* between, among.

interact *v.* have an effect upon eac other. □ **interaction** *n.*, **interactive** *adj.*

interbreed *v.* (**interbred**) breed wit each other, cross-breed.

intercede *v.* intervene on someone's behalf.

intercept *v.* stop or catch in transit. □ **interception** *n.*, **interceptor** *n.*

intercession *n.* interceding.

interchange *v.* /íntərcháynj/ **1** cause to change places. **2** alternate. ● *n.* /íntərchaynj/ **1** process of interchanging. **2** road junction where streams of traffic do not cross on the same level. □ **interchangeable** *adj.*

intercom *n.* (*colloq.*) communication system operating by telephone or radio.

interconnect *v.* connect with each other. □ **interconnection** *n.*

intercontinental *adj.* between continents.

intercourse *n.* **1** dealings between people or countries. **2** sexual intercourse, copulation.

interdict *n.* formal prohibition.

interest *n.* **1** feeling of curiosity or concern. **2** hobby etc. in which one is concerned. **3** advantage. **4** legal share. **5** money paid for use of money borrowed. ● *v.* arouse the interest of.

■ *n.* **1** attention, attentiveness, concern, curiosity, eagerness, enthusiasm, fascination, inquisitiveness. **2** amusement, hobby, pastime, pursuit. **3** advantage, avail, benefit, gain, good, profit, use. **4** claim, entitlement, share, stake, title. ● *v.* absorb, appeal to, attract, captivate, engage, engross, fascinate, grip, intrigue, occupy, rivet; amuse, divert, entertain.

interested *adj.* **1** feeling interest. **2** having an interest, not impartial.

■ **1** absorbed, curious, gripped, engrossed, fascinated, intrigued, riveted. **2** biased, concerned, involved, partial, partisan, prejudiced.

interesting *adj.* arousing interest.

■ absorbing, compelling, engrossing, entertaining, exciting, fascinating, gripping, intriguing, stimulating.

interface *n.* place where interaction occurs.

interfere *v.* take part in others' affairs, esp. without right or invitation. □ **interfere with 1** obstruct, hinder. **2** assult sexually.

■ barge in, butt in, intercede, interpose, intervene, intrude, meddle. □ **interfere with 1** block, get in the way of, hamper, handicap, hinder, hold back, impede, inhibit, obstruct, slow down.

interference *n.* **1** interfering. **2** disturbance of radio signals.

interferon *n.* protein preventing the development of a virus.

interim *n.* intervening period. ● *adj.* temporary.

■ *n.* interval, meantime, meanwhile. ● *adj.* makeshift, provisional, temporary.

interior *adj.* inner. ● *n.* interior part.

interject *v.* put in (a remark) when someone is speaking.

interjection *n.* exclamation.

interlace *v.* weave or lace together.

interlink *v.* link together.

interlock *v.* fit into each other. ● *n.* fine machine-knitted fabric.

interloper *n.* intruder.

interlude *n.* **1** interval. **2** thing happening or performed in this.

■ **1** break, hiatus, intermission, interval, lull, pause, respite, stop.

intermarry *v.* (of families, races, etc.) become connected by marriage. □ **intermarriage** *n.*

intermediary *n.* mediator. ● *adj.* **1** acting as intermediary. **2** intermediate.

■ *n.* agent, arbitrator, go-between, mediator, middleman, negotiator, peacemaker, representative.

intermediate *adj.* coming between in time, place, or order.

■ halfway, intermediary, median, mid, middle, midway, transitional.

interment *n.* burial.

intermezzo /intərmétsō/ *n.* (*pl.* **-os**) short piece of music.

interminable *adj.* very long and boring. □ **interminably** *adv.*

intermission *n.* interval, pause.

intermittent *adj.* occurring at intervals. □ **intermittently** *adv.*

■ discontinuous, fitful, irregular, occasional, on and off, periodic, spasmodic, sporadic.

intern *v.* compel (esp. an enemy alien) to live in a special area.

internal *adj.* **1** of or in the inside. **2** of a country's domestic affairs. □ **internal-combustion engine** engine producing motive power from fuel exploded within a cylinder. **internally** *adv.*

international *adj.* between countries. ● *n.* **1** sports contest between players representing different countries. **2** one of these players. ▫ **internationally** *adv.*

■ *adj.* cosmopolitan, global, intercontinental, universal, world, worldwide.

internecine /intərnéessIn/ *adj.* mutually destructive.

internee *n.* interned person.

internment *n.* interning.

interplay *n.* interaction.

interpolate *v.* **1** interject. **2** insert. ▫ **interpolation** *n.*

interpose *v.* **1** insert. **2** intervene. ▫ **interposition** *n.*

interpret *v.* **1** explain the meaning of. **2** act as interpreter. ▫ **interpretation** *n.*

■ **1** clarify, elucidate, explain, illuminate, make clear, shed *or* throw light on, simplify; decipher, decode. **2** translate. ▫ **interpretation** clarification, elucidation, explanation, illumination, simplification, translation.

interpreter *n.* person who orally translates speech between people speaking different languages.

interregnum *n.* period between the rule of two successive rulers.

interrogate *v.* question closely or formally. ▫ **interrogation** *n.*, **interrogator** *n.*

■ catechize, cross-examine, cross-question, examine, grill, pump, question, quiz. ▫ **interrogation** cross-examination, debriefing, examination, grilling, questioning, quizzing.

interrogative *adj.* forming or having the form of a question.

interrupt *v.* **1** break the continuity of. **2** break the flow of (speech etc.) by a remark. ▫ **interruption** *n.*

■ **1** adjourn, bring to a halt, discontinue, halt, stop, suspend. **2** barge in on, butt in on, cut in on, disrupt, disturb, intrude on; chime in, *colloq.* chip in. ▫ **interruption** disruption, disturbance, intrusion; adjournment, break, discontinuance, suspension.

intersect *v.* divide or cross by passing or lying across. ▫ **intersection** *n.*

intersperse *v.* insert here and there.

interval *n.* **1** intervening time or space. **2** pause or break. **3** difference in musical pitch.

■ **1** interim, meantime, meanwhile, period, time; gap, space. **2** break, interlude, intermission, pause.

intervene *v.* **1** enter a situation to change its course or resolve it. **2** occur between events. ▫ **intervention** *n.*

■ **1** intercede, mediate, step in; interfere, intrude, meddle.

interview *n.* formal meeting with a person to assess his or her merits or to obtain information. ● *v.* hold an interview with. ▫ **interviewer** *n.*, **interviewee** *n.*

■ *n.* appraisal, assessment, evaluation, examination, grilling; audience, conversation, discussion, press-conference, talk.

interweave *v.* (**-wove**, **-woven**) weave together.

intestate *adj.* not having made a valid will. ▫ **intestacy** *n.*

intestine *n.* section of the alimentary canal between stomach and anus. ▫ **intestinal** *adj.*

intimate[1] /íntimət/ *adj.* **1** closely acquainted or familiar. **2** private and personal. **3** having a sexual relationship (esp. outside marriage). ● *n.* intimate friend. ▫ **intimately** *adv.*, **intimacy** *n.*

■ *adj.* **1** attached, close, devoted, friendly, inseparable, *colloq.* pally, *colloq.* thick; cherished, dear. **2** confidential, personal, private, secret. ● *n.* bosom friend, *colloq.* chum, confidant(e), crony, friend, mate, *colloq.* pal.

intimate[2] /íntimayt/ *v.* make known, esp. by hinting. ▫ **intimation** *n.*

■ hint, imply, indicate, insinuate, suggest. ▫ **intimation** hint, implication, indication, insinuation, suggestion.

intimidate *v.* influence by frightening. ▫ **intimidation** *n.*

■ browbeat, bully, cow, daunt, frighten, *colloq.* lean on, menace, overawe, pressure, pressurize, railroad, scare, terrify, terrorize, threaten.

into *prep.* **1** to the inside of, to a point within. **2** to a particular state or occupation. **3** dividing (a number) mathematically. **4** (*colloq.*) interested in.

intolerable *adj.* too bad to be endured. ▫ **intolerably** *adv.*

■ insufferable, insupportable, unbearable, unendurable.

intonation *n.* **1** pitch of the voice in speaking. **2** act of intoning.

■ **1** accent, cadence, inflection, modulation, pitch, tonality, tone.

intone *v.* chant, esp. on one note.

intoxicate *v.* **1** make drunk. **2** make greatly excited. ☐ **intoxication** *n.*

■ **1** fuddle, inebriate, stupefy. **2** arouse, elate, enrapture, excite, exhilarate, inflame, thrill.

intra- *pref.* within.

intractable *adj.* hard to deal with or control. ☐ **intractability** *n.*

■ disobedient, headstrong, insubordinate, obdurate, obstinate, recalcitrant, refractory, uncontrollable, ungovernable, unmanageable, unruly, wild.

intransigent *adj.* stubborn. ☐ **intransigence** *n.*

intra-uterine *adj.* within the uterus.

intravenous *adj.* into a vein. ☐ **intravenously** *adv.*

intrepid *adj.* fearless, brave. ☐ **intrepidly** *adv.*, **intrepidity** *n.*

intricate *adj.* very complicated. ☐ **intricately** *adv.*, **intricacy** *n.*

■ Byzantine, complex, complicated, difficult, involved, knotty, labyrinthine; detailed, elaborate, fancy, ornate, rococo.

intrigue *v.* **1** plot secretly. **2** rouse the interest of. ● *n.* **1** underhand plot or plotting. **2** secret love affair.

■ *v.* **1** conspire, plot, scheme. **2** attract, captivate, charm, enthral, fascinate, interest. ● *n.* **1** conspiracy, machination, plot, scheme.

intrinsic *adj.* belonging to the basic nature of. ☐ **intrinsically** *adv.*

■ basic, fundamental, essential, inborn, inherent, innate, native, natural.

introduce *v.* **1** make (a person) known to another. **2** present to an audience. **3** bring into use. **4** insert.

■ **1** make acquainted, present. **2** announce, present. **3** establish, inaugurate, initiate, instigate, institute, launch, set in motion. **4** add, insert, interpolate, interpose.

introduction *n.* **1** introducing. **2** introductory section or treatise.

■ **1** presentation; inauguration, initiation, institution. **2** foreword, preamble, preface, prelude, prologue.

introductory *adj.* preliminary.

introspection *n.* examination of one's own thoughts and feelings. ☐ **introspective** *adj.*

introvert *n.* introspective and shy person. ☐ **introverted** *adj.*

intrude *v.* come or join in without being invited or wanted. ☐ **intruder** *n.*, **intrusion** *n.*, **intrusive** *adj.*

■ encroach, impinge, infringe, invade, trespass; barge in, butt in, interfere, meddle, *colloq.* muscle in. ☐ **intruder** burglar, housebreaker, robber, thief; gatecrasher, interloper, invader, prowler, trespasser.

intuition *n.* (power of) knowing without learning or reasoning. ☐ **intuitive** *adj.*, **intuitively** *adv.*

■ instinct, sixth sense; feeling, foreboding, hunch, inkling, premonition, presentiment.

Inuit /ínyoo-it/ *n.* (*pl.* **Inuit** or **-s**) N. American Eskimo.

inundate *v.* flood. ☐ **inundation** *n.*

inure *v.* accustom, esp. to something unpleasant.

invade *v.* **1** enter (territory) with hostile intent. **2** swarm into. **3** encroach on. **4** (of disease etc.) attack. ☐ **invader** *n.*

■ **1** occupy, overrun, storm, take over. **2** infest, permeate, pervade, swarm into. **3** encroach on, infringe on, intrude on, trespass on.

invalid[1] /ínvəleed/ *n.* person suffering from ill health.

invalid[2] /inváIlid/ *adj.* not valid.

■ null and void, void, worthless; baseless, false, groundless, illogical, incorrect, irrational, spurious, unreasonable.

invalidate *v.* make no longer valid. ☐ **invalidation** *n.*

■ annul, cancel, nullify, quash, repeal, rescind, revoke; discredit, disprove, negate, rebut.

invaluable *adj.* having value too great to be measured.

invariable *adj.* not variable, always the same. ☐ **invariably** *adv.*

■ changeless, constant, enduring, eternal, fixed, immutable, permanent, unalterable, unchangeable, unchanging, unvarying; even, level, regular, stable, steady, uniform.

invasion *n.* hostile or harmful intrusion.

■ encroachment, infiltration, intrusion, occupation; attack, foray, incursion, inroad, offensive, onslaught, raid.

invective *n.* abusive language.

inveigle *v.* entice.

invent *v.* **1** make or design (something new). **2** make up (a lie, a story). □ **inventor** *n.*

■ **1** coin, conceive, contrive, create, design, devise, dream up, imagine, improvise, *colloq.* think up. **2** concoct, *colloq.* cook up, fabricate, make up.

invention *n.* **1** process of inventing. **2** thing invented. **3** inventiveness.

■ **1** conception, creation, design, origination. **2** brainchild, contraption, contrivance, device, gadget; fib, lie, story, tale, untruth. **3** creativity, imagination, imaginativeness, ingenuity, inspiration, inventiveness, originality, resourcefulness.

inventive *adj.* able to invent things. □ **inventiveness** *n.*

■ clever, creative, imaginative, ingenious, innovative, original, resourceful.

inventory *n.* detailed list of goods or furniture.

inverse *adj.* inverted. □ **inversely** *adv.*

invert *v.* **1** turn upside down. **2** reverse the position, order, or relationship of. □ **inverted commas** quotation marks. **inversion** *n.*

invertebrate *adj.* & *n.* (animal) having no backbone.

invest *v.* **1** use (money, time, etc.) to earn interest or bring profit. **2** confer rank or power upon. **3** endow with a quality. □ **investment** *n.*, **investor** *n.*

■ **1** contribute, devote, donate, expend, give, lay out, sink, spend.

investigate *v.* **1** study carefully. **2** inquire into. □ **investigation** *n.*, **investigator** *n.*, **investigative** *adj.*

■ analyse, examine, explore, go over, inquire into, inspect, look into, probe, research, scrutinize, study, survey. □ **investigation** analysis, examination, exploration, inquiry, inspection, review, scrutiny, study, survey.

inveterate *adj.* **1** habitual. **2** firmly established.

invidious *adj.* liable to cause resentment. □ **invidiously** *adv.*

invigilate *v.* supervise examinees. □ **invigilator** *n.*

invigorate *v.* fill with vigour, give strength or courage to.

■ animate, brace, energize, enliven, fortify, hearten, pep up, perk up, refresh, rejuvenate, restore, revive, strengthen, vivify.

invincible *adj.* unconquerable. □ **invincibly** *adv.*, **invincibility** *n.*

■ impenetrable, impregnable, indestructible, indomitable, invulnerable, unassailable, unbeatable, unconquerable.

invisible *adj.* not able to be seen. □ **invisibly** *adv.*, **invisibility** *n.*

■ imperceptible, indiscernible, indistinguishable, undetectable, unnoticeable; concealed, hidden, out of sight.

invite *v.* **1** ask (a person) politely to come or to do something. **2** ask for. **3** attract. □ **invitation** *n.*

■ **1** ask, bid, summon. **2** appeal for, ask for, call for, request, solicit. **3** allure, attract, tempt.

inviting *adj.* attractive and tempting. □ **invitingly** *adv.*

■ alluring, appealing, attractive, captivating, engaging, fascinating, intriguing, seductive, tempting.

in vitro in a test-tube or other laboratory environment.

invoice *n.* bill for goods or services. ● *v.* send an invoice to.

invoke *v.* **1** call for the help or protection of. **2** summon (a spirit).

involuntary *adj.* done without intention. □ **involuntarily** *adv.*

■ automatic, impulsive, instinctive, knee-jerk, mechanical, spontaneous, unconscious, unintentional, unpremeditated, unthinking, unwitting.

involve *v.* **1** engross, occupy fully. **2** have as a consequence. **3** include or affect. **4** implicate. □ **involvement** *n.*

■ **1** absorb, engage, engross, immerse, interest, occupy. **2** entail, imply, mean, necessitate, presuppose. **3** comprehend, comprise, cover, embody, embrace, encompass, include, incorporate; affect, apply to, be relevant to, concern, have a bearing on. **4** connect, embroil, entangle, implicate.

involved *adj.* **1** complicated. **2** concerned.

■ **1** complex, complicated, difficult, elaborate, intricate, knotty, labyrinthine, tangled. **2** affected, concerned, interested.

invulnerable *adj.* not vulnerable. □ **invulnerability** *n.*

■ impenetrable, impregnable, invincible, safe, secure, unassailable, unconquerable.

inward *adj.* **1** situated on or going towards the inside. **2** in the mind or spirit.

● *adv.* inwards. ◻ **inwardly** *adv.*, **inwards** *adv.*

odine *n.* chemical used in solution as an antiseptic.

on *n.* electrically charged particle. ◻ **ionic** *adj.*

onize *v.* convert or be converted into ions. ◻ **ionization** *n.*

onosphere *n.* ionized region of the atmosphere. ◻ **ionospheric** *adj.*

ota *n.* **1** ninth letter of the Greek alphabet, = i. **2** very small amount.

OU *n.* signed paper given as a receipt for money borrowed.

- *see* **in-**.

ascible *adj.* hot-tempered, irritable. ◻ **irascibly** *adv.*, **irascibility** *n.*

ate *adj.* angry. ◻ **irately** *adv.*

e *n.* anger.

idescent *adj.* **1** coloured like a rainbow. **2** shimmering. ◻ **iridescence** *n.*

is *n.* **1** coloured part of the eyeball, round the pupil. **2** plant with showy flowers and sword-shaped leaves.

ish *adj.* & *n.* (language) of Ireland. ◻ **Irishman** *n.*, **Irishwoman** *n.*

k *v.* annoy, be tiresome to.

■ *colloq.* aggravate, annoy, *sl.* bug, exasperate, gall, get on a person's nerves, grate on, infuriate, irritate, madden, nettle, pique, *colloq.* rile, vex.

rksome *adj.* tiresome.

■ annoying, bothersome, exasperating, infuriating, irritating, maddening, tiresome, troublesome, trying, vexatious.

ron *n.* **1** hard grey metal. **2** tool etc. made of this. **3** implement with a flat base heated for smoothing cloth or clothes. **4** (*pl.*) fetters. ● *adj.* **1** made of iron. **2** strong as iron. ● *v.* smooth (clothes etc.) with an iron. ◻ **ironing board** narrow folding table for ironing clothes on.

ronic *adj.* (also **ironical**) using irony. ◻ **ironically** *adv.*

ronmonger *n.* shopkeeper selling tools and household implements.

ronstone *n.* **1** hard iron ore. **2** a kind of hard white pottery.

rony *n.* **1** expression of meaning by use of words normally conveying the opposite. **2** apparent perversity of fate or circumstances.

rradiate *v.* **1** throw light or other radiation on. **2** treat (food) by radiation. ◻ **irradiation** *n.*

rrecoverable *adj.* unable to be recovered. ◻ **irrecoverably** *adv.*

irrefutable *adj.* unable to be refuted. ◻ **irrefutably** *adv.*

irregular *adj.* **1** not regular. **2** contrary to rules or custom. ◻ **irregularly** *adv.*, **irregularity** *n.*

■ **1** asymmetrical, lopsided; bumpy, craggy, pitted, rocky, rough, rugged, uneven; erratic, fitful, fluctuating, inconsistent, intermittent, on and off, spasmodic, sporadic, unsteady, variable. **2** improper, incorrect, unauthorized; abnormal, anomalous, exceptional, odd, peculiar, strange, uncommon, unconventional, unorthodox, unusual.

irrelevant *adj.* not relevant. ◻ **irrelevance** *n.*

■ beside the point, immaterial, inapplicable, inapposite, inappropriate, unconnected, unrelated.

irreparable *adj.* unable to be repaired. ◻ **irreparably** *adv.*

irreplaceable *adj.* unable to be replaced.

irrepressible *adj.* unable to be repressed. ◻ **irrepressibly** *adv.*

■ boisterous, bubbly, buoyant, ebullient, exuberant, headstrong, lively, spirited, vivacious.

irreproachable *adj.* blameless, faultless. ◻ **irreproachably** *adv.*

irresistible *adj.* too strong or delightful to be resisted. ◻ **irresistibly** *adv.*, **irresistibility** *n.*

irresolute *adj.* unable to make up one's mind. ◻ **irresolutely** *adv.*, **irresolution** *n.*

■ doubtful, dubious, hesitant, indecisive, in two minds, uncertain, undecided, unsure, wavering.

irrespective *adj.* **irrespective of** not taking (a thing) into account.

irresponsible *adj.* not showing a proper sense of responsibility. ◻ **irresponsibly** *adv.*, **irresponsibility** *n.*

■ careless, feckless, flighty, giddy, heedless, reckless, thoughtless, unreliable, unthinking, untrustworthy.

irreverent *adj.* **1** not reverent. **2** not respectful. ◻ **irreverently** *adv.*, **irreverence** *n.*

irreversible *adj.* not reversible, unable to be altered or revoked. ◻ **irreversibly** *adv.*

irrevocable *adj.* unalterable. ◻ **irrevocably** *adv.*

■ decided, fixed, immutable, irreversible, settled, unalterable, unchangeable.

irrigate *v.* supply (land) with water by streams, pipes, etc. □ **irrigation** *n.*

irritable *adj.* easily annoyed. □ **irritably** *adv.*, **irritability** *n.*

■ bad-tempered, cantankerous, crabby, cross, crotchety, crusty, disagreeable, fractious, grumpy, huffy, irascible, peevish, pettish, petulant, prickly, quarrelsome, querulous, short-tempered, snappy, splenetic, testy, touchy, waspish.

irritant *adj.* & *n.* (thing) causing irritation.

irritate *v.* **1** annoy. **2** cause discomfort in. □ **irritation** *n.*

■ **1** *colloq.* aggravate, annoy, *sl.* bug, exasperate, gall, *colloq.* get, get on a person's nerves, grate on, infuriate, irk, nettle, pique, provoke, *colloq.* rile, vex. **2** chafe, hurt, sting.

Islam *n.* **1** Muslim religion. **2** Muslim world. □ **Islamic** *adj.*

island *n.* piece of land surrounded by water.

islander *n.* inhabitant of an island.

isle *n.* island.

islet *n.* small island.

isobar *n.* line on a map, connecting places with the same atmospheric pressure. □ **isobaric** *adj.*

isolate *v.* **1** place apart or alone. **2** separate (esp. an infectious patient from others). □ **isolation** *n.*

■ insulate, keep apart, maroon, seclude, segregate, separate, sequester; quarantine.

isolationism *n.* policy of holding aloof from other countries or groups. □ **isolationist** *n.*

isosceles /Isóssileez/ *adj.* (of a triangle) having two sides equal.

isotherm *n.* line on a map, connecting places with the same temperature.

isotope *n.* one of two or more forms of a chemical element differing in their atomic weight. □ **isotopic** *adj.*

issue *n.* **1** outflow. **2** issuing, quantity issued. **3** one edition (e.g. of a magazine). **4** important topic. **5** offspring. ● *v.* **1** flow out. **2** supply for use. **3** send out. **4** publish.

■ *n.* **2** circulation, dissemination, distribution, issuing, promulgation, publication. **3** copy, edition, impression, number. **4** affair, business, matter, point, question, subject, topic. **5** children, descendants, heirs, offspring, progeny. ● *v.* **1** emanate, escape, flow, pour, stream. **2** furnish, provide, supply. **3** broadcast, circulate, deliver, disseminate, distribute, hand out, make public, release, send out.

isthmus /isməss/ *n.* (*pl.* **-muses**) narrow strip of land connecting two larger masses of land.

it *pron.* **1** thing mentioned or being discussed. **2** impersonal subject of a verb.

Italian *adj.* & *n.* (native, language) of Italy.

italic *adj.* (of type) sloping like *this*. ● *n.pl.* italic type.

italicize *v.* print in italics.

itch *n.* **1** tickling sensation in the skin causing a desire to scratch. **2** restless desire. ● *v.* feel an itch. □ **itchy** *adj.*

item *n.* **1** single thing in a list or collection. **2** single piece of news.

■ **1** article, object, thing; component, detail, element, entry, particular. **2** article, feature, piece, report, story.

itemize *v.* list, state the individual items of. □ **itemization** *n.*

itinerant *adj.* travelling from place to place.

■ migrant, migratory, nomadic, peripatetic, roaming, roving, travelling, wandering.

itinerary *n.* route, list of places to be visited on a journey.

its *poss.pron.* of it.

it's it is, it has.

itself *pron.* emphatic and reflexive form of *it*.

ivory *n.* **1** hard creamy-white substance forming tusks of elephant etc. **2** object made of this. **3** its colour. ● *adj.* creamy white. □ **ivory tower** seclusion from the harsh realities of life.

ivy *n.* climbing evergreen shrub.

Jj

ab *v.* (**jabbed**) poke roughly. ● *n.* **1** rough poke. **2** (*colloq.*) injection.

abber *v.* talk rapidly, often unintelligibly. ● *n.* jabbering talk.

ack *n.* **1** portable device for raising heavy weights off the ground. **2** playing card next below queen. **3** ship's small flag showing nationality. **4** electrical connection with a single plug. **5** small ball aimed at in bowls. **6** male donkey. ● *v.* **jack up** raise with a jack.

ackal *n.* dog-like wild animal.

ackass *n.* **1** male ass. **2** stupid person.

ackboot *n.* large high boot.

ackdaw *n.* bird of the crow family.

acket *n.* **1** short coat. **2** outer covering.

ackknife *n.* large folding knife. ● *v.* (of an articulated vehicle) fold against itself in an accident.

ackpot *n.* large prize of money that has accumulated until won. □ **hit the jackpot** (*colloq.*) have a sudden success.

Jacobean *adj.* of the reign of James I of England (1603–25).

Jacuzzi *n.* [P.] large bath with underwater jets of water.

ade *n.* **1** hard green, blue, or white stone. **2** its green colour.

aded *adj.* tired and bored.

agged *adj.* having sharp projections.

> ■ craggy, crenellated, irregular, ragged, rough, serrated, spiky, uneven.

aguar *n.* large flesh-eating animal of the cat family.

ail *n.* prison. ● *v.* put into jail.

> ■ *n.* gaol, lock-up, prison, reformatory. ● *v.* detain, imprison, incarcerate, remand, send to prison.

ailer *n.* person in charge of a jail or its prisoners.

am¹ *n.* thick sweet substance made by boiling fruit with sugar.

am² *v.* (**jammed**) **1** squeeze or wedge into a space. **2** become wedged. **3** crowd or block (an area). **4** make (a broadcast) unintelligible by causing interference. ● *n.* **1** crowded mass of people, vehicles, etc. **2** stoppage caused by jamming. **3** (*colloq.*) difficult situation. □ **jam-packed** *adj.* (*colloq.*) very full.

> ■ *v.* **1** cram, crowd, force, pack, push, ram, shove, squash, squeeze, stuff, thrust, wedge. **3** block, clog, congest, obstruct, stop (up). ● *n.* **1** crowd, crush, horde, mass, mob, multitude, pack, swarm, throng. **2** blockage, congestion, obstruction.

jamb *n.* side post of a door or window.

jamboree *n.* **1** large party. **2** rally.

jangle *n.* harsh metallic sound. ● *v.* (cause to) make this sound.

janitor *n.* caretaker of a building.

japan *n.* hard usu. black varnish. ● *v.* (**japanned**) coat with this.

Japanese *adj.* & *n.* (native, language) of Japan.

jar¹ *n.* cylindrical glass or earthenware container.

> ■ crock, ewer, flagon, jug, pitcher, urn.

jar² *v.* (**jarred**) jolt. ● *n.* jolt. □ **jar on** have a harsh or disagreeable effect on.

> ■ *v.* bounce, bump, jerk, jog, joggle, jolt, judder, knock, rock, shake. □ **jar on** annoy, grate on, irritate.

jargon *n.* words or expressions developed for use within a particular group of people.

> ■ cant, dialect, idiom, language, parlance, patois, slang, speech.

jasmine *n.* shrub with white or yellow flowers.

jasper *n.* a kind of quartz.

jaundice *n.* yellowing of the skin caused by liver disease etc.

jaundiced *adj.* **1** affected by jaundice. **2** envious, resentful.

jaunt *n.* short pleasure trip. ● *v.* make a jaunt.

jaunty *adj.* (**-ier**, **-iest**) cheerful, self-confident. □ **jauntily** *adv.*, **jauntiness** *n.*

> ■ buoyant, cheerful, cheery, chirpy, high-spirited, jolly, jovial, lively, merry, perky, pert, self-confident, spirited, sprightly, vivacious.

javelin *n.* light spear.

jaw *n.* **1** bone(s) forming the framework of the mouth. **2** (*pl.*) gripping-parts. **3**

(*colloq.*) lengthy talk. ● *v.* (*colloq.*) talk lengthily.

jay *n.* bird of the crow family.

jaywalking *n.* crossing a road carelessly. □ **jaywalker** *n.*

jazz *n.* type of music with strong rhythm and much syncopation.

jealous *adj.* **1** suspicious or resentful of rivalry in love. **2** envious. **3** taking watchful care. □ **jealously** *adv.*, **jealousy** *n.*

> ■ **1** distrustful, mistrustful, suspicious. **2** covetous, envious, grudging, resentful. **3** careful, mindful, protective, vigilant, watchful.

jeans *n.pl.* denim trousers.

Jeep *n.* [P.] small sturdy motor vehicle with four-wheel drive.

jeer *v.* laugh or shout rudely or scornfully (at). ● *n.* jeering.

> ■ *v.* barrack, boo, deride, gibe (at), heckle, laugh (at), mock (at), ridicule, scoff (at), taunt.

Jehovah *n.* name of God in the Old Testament.

jell *v.* (*colloq.*) **1** set as a jelly. **2** take definite form.

jelly *n.* **1** soft solid food made of liquid set with gelatine. **2** substance of similar consistency. **3** jam made of strained fruit juice.

jellyfish *n.* sea animal with a jelly-like body.

jemmy *n.* burglar's short crowbar. ● *v.* open with this.

jenny *n.* female donkey.

jeopardize /jéppərdīz/ *v.* endanger.

> ■ endanger, imperil, put at risk, put in jeopardy, risk, threaten.

jeopardy /jéppərdi/ *n.* danger.

jerboa *n.* rat-like desert animal with long hind legs.

jerk *n.* sudden sharp movement or pull. ● *v.* move, pull, or stop with jerk(s). □ **jerky** *adj.*, **jerkily** *adv.*, **jerkiness** *n.*

> ■ *n.* bump, jolt, lurch; pull, tug, tweak, wrench, *colloq.* yank. ● *v.* jiggle, jolt, lurch, start, twitch; pull, tug, tweak, wrench, *colloq.* yank.

jerkin *n.* sleeveless jacket.

jerrycan *n.* five-gallon can for petrol or water.

jersey *n.* (*pl.* **-eys**) **1** knitted woollen pullover. **2** machine-knitted fabric.

jest *n.* & *v.* joke.

jester *n.* **1** person who makes jokes. **2** entertainer at a medieval court.

jet[1] *n.* **1** hard black mineral. **2** gloss black. □ **jet-black** *adj.*

jet[2] *n.* **1** stream of water, gas, or flam from a small opening. **2** burner on a ga cooker. **3** engine or aircraft using je propulsion. ● *v.* (**jetted**) travel by je □ **jet lag** delayed tiredness etc. after long flight. **jet-propelled** *adj.* using **je propulsion**, propulsion by engines tha send out a high-speed jet of gases at th back.

jetsam *n.* goods jettisoned by a ship an washed ashore.

jettison *v.* **1** throw overboard, eject. discard.

jetty *n.* **1** breakwater. **2** landing-stage.

Jew *n.* person of Hebrew descent o whose religion is Judaism. □ **Jewish** *ad*

jewel *n.* **1** precious stone cut or set as a ornament. **2** person or thing that highly valued. □ **jewelled** *adj.*

> ■ **1** brilliant, gem, precious stone, stone. gem, marvel, treasure.

jeweller *n.* person who makes or deals i jewels or jewellery.

jewellery *n.* jewels or similar ornamen to be worn.

Jewry *n.* the Jewish people.

jib *n.* **1** triangular sail set forward from mast. **2** projecting arm of a crane. ● (**jibbed**) refuse to proceed. □ **jib at** objec to.

jig *n.* **1** lively dance. **2** device that hold work and guides tools working on it. template. ● *v.* (**jigged**) move quickly u and down.

jiggery-pokery *n.* (*colloq.*) trickery.

jiggle *v.* rock or jerk lightly.

jigsaw *n.* machine fretsaw. □ **jigsa puzzle** picture cut into pieces which ar then shuffled and reassembled fc amusement.

jilt *v.* abandon (a lover).

> ■ abandon, desert, *sl.* ditch, drop, leave (i the lurch), walk out on.

jingle *v.* (cause to) make a ringing o clinking sound. ● *n.* **1** this sound. simple rhyme.

> ■ *n.* **1** chink, clink, jangle, ring, tinkle. ditty, rhyme, song, tune.

jingoism *n.* excessive patriotism an contempt for other countries. □ **jingois** *n.*, **jingoistic** *adj.*

jinx *n.* (*colloq.*) influence causing ba luck.

jitters *n.pl.* (*colloq.*) nervousness. □ **jitter** *adj.*

jive *n.* **1** fast lively jazz. **2** dance to this. ● *v.* dance to this music.

job *n.* **1** piece of work. **2** paid position of employment. **3** function or responsibility. **4** (*colloq.*) difficult task. □ **good** or **bad job** fortunate or unfortunate state of affairs. **job lot** miscellaneous articles sold together.

■ **1** assignment, chore, piece of work, project, task, undertaking. **2** appointment, employment, occupation, position, post, situation; career, profession, trade. **3** duty, function, responsibility, role.

jobbing *adj.* doing single pieces of work for payment.

jobless *adj.* out of work.

jockey *n.* (*pl.* **-eys**) person who rides in horse races. ● *v.* manoeuvre to gain advantage.

jocose *adj.* jocular.

jocular *adj.* **1** fond of joking. **2** humorous. □ **jocularly** *adv.*, **jocularity** *n.*

jocund *adj.* merry, cheerful.

jodhpurs *n.pl.* riding-breeches fitting closely below the knee.

jog *v.* (**jogged**) **1** nudge. **2** stimulate. **3** run at a slow regular pace. ● *n.* **1** slow run. **2** nudge. □ **jogger** *n.*

■ *v.* **1** elbow, knock, nudge, poke, prod, push. **2** prompt, refresh, stimulate, stir. **3** lope, run, trot.

joggle *v.* shake slightly. ● *n.* slight shake.

jogtrot *n.* slow regular trot.

joie de vivre /zhwaá də veévrə/ exuberant enjoyment of life.

join *v.* **1** unite. **2** come into the company of. **3** become a member of. **4** take one's place in. ● *n.* place where things join. □ **join forces** combine efforts. **join in** take part. **join up** enlist for military service.

■ *v.* **1** attach, combine, connect, couple, fasten, fuse, link, marry, merge, splice, tie, unify, unite, wed, yoke. **2** accompany, go with. **3** become a member of, enlist in, enrol in, enter. **4** ally oneself with, associate oneself with, team up with. ● *n.* connection, intersection, joint, junction, juncture. □ **join forces** club together, collaborate, cooperate, team up.

joiner *n.* maker of wooden doors, windows, etc. □ **joinery** *n.*

joint *adj.* shared by two or more people. ● *n.* **1** join. **2** structure where parts or bones fit together. **3** large piece of meat. ● *v.* **1** connect by joint(s). **2** divide into joints. □ **out of joint 1** dislocated. **2** in disorder. **jointly** *adv.*

■ *adj.* collaborative, collective, combined, common, communal, *colloq.* mutual, shared.

jointure *n.* estate settled on a widow for her lifetime.

joist *n.* one of the beams supporting a floor or ceiling.

jojoba /hōhṓbə/ *n.* plant producing seeds containing oil used in cosmetics.

joke *n.* **1** thing said or done to cause laughter. **2** ridiculous person or thing. ● *v.* make jokes.

■ *n.* **1** *colloq.* crack, gag, jest, pun, quip, *colloq.* wisecrack, witticism; hoax, jape, lark, practical joke, prank, trick. ● *v.* banter, jest; fool about, *colloq.* kid, tease.

joker *n.* **1** person who jokes. **2** extra playing card in a pack.

jollification *n.* merrymaking.

jollity *n.* **1** being jolly. **2** merrymaking.

jolly *adj.* (**-ier, -iest**) **1** cheerful, merry. **2** very pleasant. ● *adv.* (*colloq.*) very. ● *v.* **jolly along** keep in good humour.

■ *adj.* **1** buoyant, cheerful, cheery, chirpy, happy, gay, glad, gleeful, high-spirited, jocular, jovial, joyful, merry, sunny.

jolt *v.* **1** shake or dislodge with a jerk. **2** move jerkily. **3** shock. ● *n.* **1** jolting movement. **2** shock.

■ *v.* **1** bump, jerk, jog, joggle, knock, rock, shake. **2** bounce, bump, jerk, judder, lurch. **3** disturb, perturb, shake, shock, startle, unnerve, unsettle, upset. ● *n.* **1** bang, bump, jerk, judder, knock, lurch. **2** blow, bombshell, shock, start, surprise.

jonquil *n.* a kind of narcissus.

joss-stick *n.* thin stick that burns with a smell of incense.

jostle *v.* push roughly.

■ elbow, hustle, push, shoulder, shove, thrust.

jot *n.* very small amount. ● *v.* (**jotted**) write down briefly.

jotter *n.* notepad, notebook.

joule *n.* unit of energy.

journal *n.* **1** daily record of events. **2** newspaper, periodical.

■ **1** chronicle, diary, log, record. **2** daily, gazette, magazine, monthly, newspaper, paper, periodical, quarterly, *derog.* rag, weekly.

journalese *n.* style of language used in inferior journalism.

journalist *n.* person employed in writing for a newspaper or magazine. ▫ **journalism** *n.*

■ columnist, correspondent, hack, reporter.

journey *n.* (*pl.* **-eys**) act of going from one place to another. ● *v.* make a journey.

■ *n.* crossing, excursion, expedition, jaunt, odyssey, outing, passage, pilgrimage, tour, trek, trip, voyage. ● *v.* go, make a trip, tour, travel, voyage.

joust *v.* & *n.* fight on horseback with lances.

jovial *adj.* merry, cheerful. ▫ **jovially** *adv.*, **joviality** *n.*

jowl *n.* **1** jaw, cheek. **2** dewlap, loose skin on the throat.

joy *n.* **1** deep feeling of pleasure. **2** thing causing delight.

■ **1** delight, ecstasy, elation, euphoria, exhilaration, exultation, gratification, happiness, jubilation, pleasure, rapture. **2** delight, pleasure, treat.

joyful *adj.* full of joy. ▫ **joyfully** *adv.*, **joyfulness** *n.*

■ cheerful, ecstatic, elated, euphoric, exhilarated, exultant, gay, glad, gleeful, happy, joyous, jubilant, merry, overjoyed, pleased, rapturous, thrilled.

joyous *adj.* joyful. ▫ **joyously** *adv.*

joyride *n.* (*colloq.*) ride for pleasure, esp. in a stolen car. ▫ **joyriding** *n.*

joystick *n.* **1** aircraft's control lever. **2** device for moving a cursor on a VDU screen.

jubilant *adj.* rejoicing, joyful. ▫ **jubilantly** *adv.*, **jubilation** *n.*

jubilee *n.* special anniversary.

Judaic *adj.* Jewish.

Judaism *n.* religion of the Jewish people.

judder *v.* shake noisily or violently. ● *n.* this movement.

judge *n.* **1** public officer appointed to hear and try legal cases. **2** person appointed to decide a dispute etc. **3** person able to give an authoritative opinion. ● *v.* **1** try a legal case. **2** act as judge of. **3** form an opinion about. **4** consider.

■ *n.* **1** *sl.* beak, justice, magistrate. **2** ajudicator, arbiter, arbitrator, assessor, examiner, mediator, referee, umpire. **3** authority, connoisseur, expert, pundit, specialist. ● *v.* **1** adjudge, adjudicate, try. **2** adjudicate, arbitrate, referee, umpire. **3** appraise, assess, estimate, evaluate, gauge, rate, weigh (up). **4** believe, consider, deem, hold, regard, think.

judgement *n.* (in law contexts **judgment**) **1** discernment, good sense. **2** opinion or estimate. **3** judge's decision. ▫ **Judgement Day** day on which God will judge humankind. **judgemental** *adj.*

■ **1** acumen, discernment, discrimination, good sense, insight, intelligence, judiciousness, perception, perceptiveness, sagacity, understanding, wisdom, wit. **2** belief, conviction, feeling, opinion, view; appraisal, assessment, estimate, estimation, evaluation. **3** decision, finding, ruling, verdict.

judicial *adj.* **1** of the administration of justice. **2** of a judge or judgement. ▫ **judicially** *adv.*

judiciary *n.* the whole body of judges in a country.

judicious *adj.* judging wisely, showing good sense. ▫ **judiciously** *adv.*, **judiciousness** *n.*

■ astute, canny, careful, diplomatic, discerning, discreet, intelligent, logical, perceptive, percipient, perspicacious, prudent, sagacious, sensible, shrewd, sound, tactful, wise.

judo *n.* Japanese system of unarmed combat. ▫ **judoist** *n.*

jug *n.* vessel with a handle and a shaped lip, for holding and pouring liquids. ▫ **jugful** *n.*

juggernaut *n.* **1** large heavy lorry. **2** overwhelming force or object.

juggle *v.* **1** toss and catch objects skilfully for entertainment. **2** manipulate skilfully. ▫ **juggler** *n.*

jugular vein either of the two large veins in the neck.

juice *n.* **1** fluid content of fruits, vegetables, or meat. **2** fluid secreted by an organ of the body. ▫ **juicy** *adj.*

ju-jitsu *n.* Japanese system of unarmed combat.

jukebox *n.* coin-operated record player.

julep *n.* drink of spirits and water flavoured esp. with mint.

jumble *v.* mix in a confused way. ● *n.* **1** jumbled articles. **2** items for a jumble sale. ▫ **jumble sale** sale of second-hand articles, esp. to raise money for charity.

■ *v.* confuse, disarrange, disorder, disorganize, mess up, mix (up), muddle, scramble, shuffle. ● *n.* **1** clutter, confusion, disarray, disorder, hotchpotch, mêlée, mess, muddle, tangle.

jumbo *n.* (*pl.* **-os**) very large thing. ▫ **jumbo jet** large jet aircraft.

jump *v.* **1** move up off the ground etc. by muscular movement of the legs. **2** move suddenly. **3** rise suddenly. **4** pass over by jumping. **5** pounce on. **6** leave (rails or track) accidentally. ● *n.* **1** jumping movement. **2** sudden rise or change. **3** gap in a series. **4** obstacle to be jumped. □ **jump-lead** *n.* cable for conveying current from one battery to another. **jump suit** one-piece garment for the whole body. **jump the gun** start prematurely. **jump the queue** take unfair precedence.

■ *v.* **1,2** bound, hop, leap, skip, spring, vault. **3** escalate, go up, increase, rise, rocket, shoot up. **4** clear, leap-frog, leap (over), vault (over). **5** attack, mug, pounce on, take unawares. ● *n.* **1** bound, hop, leap, skip, spring, vault. **2** escalation, increase, rise, surge, upsurge. **3** break, hiatus, interval, space. **4** fence, hurdle, obstacle.

jumper *n.* knitted pullover.

jumpy *adj.* (**-ier, -iest**) nervous.

■ agitated, anxious, edgy, fidgety, fretful, ill at ease, *colloq.* jittery, keyed up, nervous, nervy, overwrought, restless, tense, uneasy, *colloq.* uptight.

junction *n.* **1** join. **2** place where roads or railway lines unite.

■ **1** connection, join, union. **2** crossroads, interchange, intersection.

juncture *n.* point of time, convergence of events.

jungle *n.* **1** tropical forest with tangled vegetation. **2** scene of ruthless struggle.

junior *adj.* **1** younger in age. **2** lower in rank or authority. **3** for younger children. ● *n.* junior person.

juniper *n.* evergreen shrub with dark berries.

junk[1] *n.* useless or discarded articles, rubbish. □ **junk food** food with low nutritional value.

junk[2] *n.* flat-bottomed ship with sails, used in China seas.

junket *n.* sweet custard-like food made of milk and rennet.

junkie *n.* (*sl.*) drug addict.

junta *n.* group taking power after a *coup d'état.*

jurisdiction *n.* authority to administer justice or exercise power.

■ authority, control, dominion, influence, power, rule, sovereignty.

jurist *n.* person skilled in law.

juror *n.* member of a jury.

jury *n.* group of people sworn to give a verdict on a case in a court of law.

just *adj.* **1** fair to all concerned. **2** right in amount etc., deserved. ● *adv.* **1** exactly. **2** by only a short amount etc. **3** only a moment ago. **4** merely. **5** positively. □ **just now** a very short time ago. **justly** *adv.*

■ *adj.* **1** disinterested, dispassionate, equitable, ethical, fair, impartial, honest, honourable, moral, objective, principled, unbiased, unprejudiced, upright. **2** appropriate, apt, correct, deserved, due, fitting, justified, legitimate, merited, proper, reasonable, rightful, suitable. ● *adv.* **1** exactly, precisely. **2** barely, scarcely. **3** just now, not long ago, recently. **4** but, merely, only. **5** absolutely, altogether, positively, thoroughly, utterly.

justice *n.* **1** just treatment, fairness. **2** legal proceedings. **3** judge.

■ **1** equity, fairness, impartiality, objectivity, right, rightfulness.

justifiable *adj.* able to be justified. □ **justifiably** *adv.*

■ defensible, excusable, just, legitimate, reasonable, sound, tenable, valid, well-founded.

justify *v.* **1** show to be right or reasonable. **2** be sufficient reason for. **3** adjust (a line of type) to fill a space neatly. □ **justification** *n.*

■ **1** defend, legitimize, substantiate, validate, vindicate, warrant. **2** account for, excuse, explain.

jut *v.* (**jutted**) project.

jute *n.* fibre from the bark of certain tropical plants.

juvenile *adj.* **1** youthful. **2** childish. **3** for young people. ● *n.* young person. □ **juvenility** *n.*

■ *adj.* **1** adolescent, boyish, girlish, teenage(d), under age, young, youthful. **2** callow, childish, immature, infantile, naive, puerile, unsophisticated. ● *n.* adolescent, boy, child, girl, minor, teenager, youngster, youth.

juxtapose *v.* put (things) side by side. □ **juxtaposition** *n.*

Kk

kale *n.* cabbage with curly leaves.

kaleidoscope *n.* toy tube containing mirrors and coloured fragments reflected to produce changing patterns. □ **kaleidoscopic** *adj.*

kamikaze /kámmikaázi/ *n.* (in the Second World War) Japanese explosive-laden aircraft deliberately crashed on its target.

kangaroo *n.* Australian marsupial with strong hind legs for jumping. □ **kangaroo court** illegal court held by strikers etc.

kaolin *n.* fine white clay used in porcelain and medicine.

kapok *n.* fluffy fibre used for padding things.

karate /kəraáti/ *n.* Japanese system of unarmed combat.

karma *n.* (in Buddhism & Hinduism) person's actions as affecting his or her next reincarnation.

kayak /kíak/ *n.* small covered canoe.

kc/s *abbr.* kilocycle(s) per second.

kebabs *n.pl.* small pieces of meat cooked on a skewer.

kedge *n.* small anchor. ● *v.* move by hauling on a kedge.

kedgeree *n.* cooked dish of rice and fish or eggs.

keel *n.* timber or steel structure along the base of a ship. ● *v.* **1** overturn. **2** become tilted.

keen *adj.* **1** eager, ardent. **2** penetrating. **3** sharp. **4** very cold. **5** intense. □ **keen on** liking greatly. **keenly** *adv.*, **keenness** *n.*

> ■ **1** ardent, avid, dedicated, devoted, eager, enthusiastic, fervent, fervid, passionate, zealous. **2** acute, astute, clever, discerning, discriminating, intelligent, penetrating, perceptive, percipient, perspicacious, sharp. **3** sharp, sharpened. **4** biting, bitter, chilling, chilly, icy, piercing. **5** acute, extreme, excruciating, fierce, intense; deep, heartfelt, profound, strong. □ **keen on** devoted to, enamoured of, enthusiastic about, fond of, interested in.

keep *v.* **(kept) 1** retain possession of, have charge of. **2** remain or cause to remain in a specified state or position. **3** detain. **4** observe or respect (a law, secret, etc.). **5** provide with food and othe necessities. **6** own and look after (ani mals). **7** manage (a shop etc.). **8** con tinue doing something. **9** stock (goods fo sale). **10** remain in good condition. **1** put aside for a future time. ● *n.* **1** pe son's food and other necessities. strongly fortified structure in a castle □ **keep house** look after a house o household. **keep up 1** progress at th same pace as others. **2** continue. maintain.

> ■ *v.* **1** conserve, have, hold, maintain preserve, retain, save; accumulate, amass hoard, put by; look after, mind, protect safeguard, take care of, take charge of tend. **2** remain, stay. **3** delay, detain, hol up, slow up. **4** abide by, comply with, fol low, obey, observe, stick to; fulfil, honou respect; celebrate, commemorate. **5** feed maintain, nourish, nurture, provide fo support, sustain. **6** look after, own, raise rear. **7** be responsible for, manage, run. carry on, continue, go on, persevere in persist in. **9** carry, sell, stock. **10** last, sta fresh.

keeper *n.* person who keeps or look after something, custodian.

> ■ caretaker, curator, custodian, guardian warden; guard, jailer, warder.

keeping *n.* custody, charge. □ **in** or **ou of keeping with** suited or unsuited to.

> ■ care, charge, custody, guardianship protection, safe-keeping.

keepsake *n.* memento, esp. of a person.

keg *n.* small barrel. □ **keg beer** beer in a pressurized metal keg.

kelp *n.* large brown seaweed.

kelvin *n.* degree of the **Kelvin scale** o temperature which has zero at absolute zero (−273.15°C).

kendo *n.* Japanese sport of fencing with bamboo swords.

kennel *n.* **1** shelter for a dog. **2** (*pl.*) boarding place for dogs.

kept *see* **keep**.

kerb *n.* stone edging to a pavement.

kerfuffle *n.* (*colloq.*) fuss, commotion.

kermes *n.* insect used in making a red dye.

kernel *n.* **1** seed within a husk, nut, or fruit stone. **2** central or important part.
■ **1** nut, pip, seed, stone. **2** centre, core, essence, heart, nub, nucleus, pith.

kerosene *n.* paraffin oil.

kestrel *n.* a kind of small falcon.

ketch *n.* two-masted sailing boat.

ketchup *n.* thick sauce made from tomatoes and vinegar.

kettle *n.* container with a spout and handle, for boiling water in.

kettledrum *n.* large bowl-shaped drum.

Kevlar *n.* [P.] synthetic fibre used to reinforce rubber etc.

key *n.* **1** piece of metal shaped for moving the bolt of a lock, tightening a spring, etc. **2** thing giving access or control or insight. **3** system of related notes in music. **4** lever for a finger to press on a piano, typewriter, etc. ● *adj.* important, essential. □ **keyed up** tense, excited.
■ *adj.* central, chief, crucial, essential, important, leading, main, necessary, principal, vital.

keyboard *n.* set of keys on a piano, typewriter, or computer. ● *v.* enter (data) by using a keyboard. □ **keyboarder** *n.*

keyhole *n.* hole by which a key is put into a lock.

keynote *n.* **1** note on which a key in music is based. **2** prevailing tone.

keypad *n.* small keyboard or set of buttons for operating an electronic device, telephone, etc.

keyring *n.* ring on which keys are threaded.

keystone *n.* central stone of an arch.

keyword *n.* key to a cipher etc.

kg *abbr.* kilogram(s).

khaki *adj.* & *n.* dull brownish-yellow, colour of military uniforms.

kHz *abbr.* kilohertz.

kibbutz *n.* (*pl.* **-im**) communal settlement in Israel.

kick *v.* **1** strike or propel with the foot. **2** (of a gun) recoil when fired. ● *n.* **1** kicking action or blow. **2** (*colloq.*) thrill, interest. □ **kick-off** *n.* start of a football game. **kick out** (*colloq.*) expel or dismiss forcibly. **kick-start** *n.* lever pressed with the foot to start a motor cycle. **kick up** (*colloq.*) create (a fuss or noise).

kid *n.* **1** young goat. **2** (*colloq.*) child. ● *v.* (**kidded**) (*colloq.*) hoax, tease.

kidnap *v.* (**kidnapped**) carry off (a person) illegally in order to obtain a ransom. □ **kidnapper** *n.*
■ abduct, capture, carry off, make off with, seize, snatch.

kidney *n.* (*pl.* **-eys**) either of a pair of organs that remove waste products from the blood and secrete urine. □ **kidney bean** kidney-shaped bean.

kill *v.* **1** cause the death of. **2** put an end to. **3** spend (time) unprofitably when waiting. ● *n.* **1** killing. **2** animal(s) killed by a hunter. □ **killer** *n.*
■ *v.* **1** assassinate, butcher, dispatch, do away with, *sl.* do in, execute, *colloq.* finish off, liquidate, massacre, murder, put to death, slaughter, slay; destroy, put down. **2** bring to an end, dash, destroy, end, eradicate, extinguish, put an end to, ruin, scotch, shatter. **3** fritter away, idle away, pass, spend, squander, waste, while away.

killjoy *n.* person who spoils the enjoyment of others.

kiln *n.* oven for hardening or drying things (e.g. pottery, hops).

kilo *n.* (*pl.* **-os**) kilogram.

kilo- *pref.* one thousand.

kilocycle *n.* kilohertz.

kilogram *n.* unit of weight or mass in the metric system (2.205 lb).

kilohertz *n.* 1000 hertz.

kilometre *n.* 1000 metres (0.62 mile).

kilovolt *n.* 1000 volts.

kilowatt *n.* 1000 watts.

kilt *n.* knee-length pleated skirt of tartan wool, esp. as part of Highland man's dress.

kimono *n.* (*pl.* **-os**) **1** loose Japanese robe worn with a sash. **2** dressing gown resembling this.

kin *n.* person's relatives.
■ family, folks, kindred, kinsfolk, kinsmen, kinswomen, kith and kin, relations, relatives.

kind[1] *n.* class of similar things. □ **in kind** (of payment) in goods etc. instead of money.
■ brand, category, class, genre, genus, make, sort, species, type, variety.

kind[2] *adj.* gentle and considerate towards others. □ **kind-hearted** *adj.*, **kindness** *n.*
■ accommodating, altruistic, benevolent, benign, caring, charitable, compassionate, considerate, decent, forbearing, forgiving, generous, gentle, good-natured, gracious, helpful, humane, indulgent, kind-hearted, kindly, lenient, magnanimous, merciful,

neighbourly, obliging, sympathetic, thoughtful, tolerant, understanding, unselfish, warm-hearted.

kindergarten *n.* school for very young children.

kindle *v.* **1** set on fire. **2** arouse, stimulate. **3** become kindled.

■ **1** fire, ignite, light, set alight, set fire to, set on fire. **2** arouse, excite, foment, incite, inflame, instigate, rouse, stimulate, stir (up), whip up. **3** burst into flames, catch fire, ignite.

kindling *n.* small pieces of wood for lighting fires.

kindly *adj.* (**-ier**, **-iest**) kind. ● *adv.* **1** in a kind way. **2** please.

kindred *n.* kin. ● *adj.* **1** related. **2** of similar kind.

kinetic *adj.* of movement.

king *n.* **1** male ruler of a country by right of birth. **2** man or thing regarded as supreme. **3** chess piece to be protected. **4** playing card next above queen. □ **king-size(d)** *adj.* extra large. **kingly** *adj.*, **kingship** *n.*

kingdom *n.* **1** country ruled by a king or queen. **2** division of the natural world.

kingfisher *n.* small bird that dives to catch fish.

kingpin *n.* indispensable person or thing.

kink *n.* **1** short twist in thread or wire etc. **2** mental peculiarity. ● *v.* form or cause to form kink(s).

■ *n.* **1** bend, coil, corkscrew, curl, knot, tangle, twist. **2** eccentricity, foible, idiosyncrasy, quirk.

kinky *adj.* **1** having kinks. **2** eccentric. **3** (*colloq.*) deviant.

kinsfolk *n.pl.* kin. □ **kinsman** *n.*, **kinswoman** *n.*

kiosk *n.* booth where newspapers or refreshments are sold, or containing a public telephone.

kipper *n.* smoked herring.

kirk *n.* (*Sc.*) church.

kirsch /keersh/ *n.* colourless liqueur made from wild cherries.

kismet *n.* destiny, fate.

kiss *n.* & *v.* touch or caress with the lips.

kissogram *n.* novelty greetings message delivered with a kiss.

kit *n.* **1** set of clothing, tools, etc. **2** set of parts to be assembled. ● *v.* (**kitted**) equip with kit.

■ *n.* **1** clothes, outfit, *colloq.* rig-out, uniform; accoutrements, apparatus, equipment, gear, implements, instruments, paraphernalia, stuff, supplies, tackle, things, tools, trappings, utensils.

kitbag *n.* bag for holding kit.

kitchen *n.* room where meals are prepared. □ **kitchen garden** vegetable garden.

kitchenette *n.* small kitchen.

kite *n.* **1** large bird of prey. **2** light framework flown on a long string in the wind.

kith *n.* **kith and kin** relatives.

kitsch /kich/ *n.* art perceived as being of poor quality, esp. when garish or sentimental.

kitten *n.* young of cat, rabbit, or ferret. □ **kittenish** *adj.*

kitty *n.* communal fund.

kiwi *n.* (*pl.* **-is**) flightless New Zealand bird.

kleptomania *n.* compulsive desire to steal. □ **kleptomaniac** *n.*

km *abbr.* kilometre(s).

knack *n.* ability to do something skilfully.

■ ability, aptitude, bent, flair, genius, gift, instinct, skill, talent.

knacker *n.* person who buys and slaughters usu. old horses etc. ● *v.* (*sl.*) exhaust.

knapsack *n.* bag worn strapped on the back.

knave *n.* **1** rogue. **2** jack in playing cards.

knead *v.* **1** press and stretch (dough) with the hands. **2** massage with similar movements.

knee *n.* **1** joint between the thigh and the lower part of the leg. **2** part of a garment covering this. ● *v.* (**kneed**) touch or strike with the knee. □ **knee-jerk** *adj.* (of a reaction) automatic and predictable. **knees-up** *n.* (*colloq.*) lively party.

kneecap *n.* small bone over the front of the knee. ● *v.* (**kneecapped**) shoot in the knee as a punishment.

kneel *v.* (**knelt**) lower one's body to rest on the knees.

knell *n.* sound of a bell tolled after a death or at a funeral.

knelt *see* **kneel**.

knew *see* **know**.

knickerbockers *n.pl.* loose breeches gathered in at the knee.

knickers *n.pl.* woman's or girl's undergarment for the lower body.

knick-knack *n.* small ornament.

■ bauble, gewgaw, ornament, trinket.

knife *n.* (*pl.* **knives**) cutting instrument with a sharp blade and a handle. ● *v.* cut or stab with a knife.

knight *n.* **1** man given a rank below baronet, with the title 'Sir'. **2** chess piece shaped like horse's head. ● *v.* confer a knighthood on.

knighthood *n.* rank of knight.

knit *v.* (**knitted** or **knit**) **1** form (yarn) into fabric of interlocking loops. **2** make in this way. **3** grow together so as to unite. □ **knitter** *n.*, **knitting** *n.*

knob *n.* **1** rounded projecting part, esp. as a handle. **2** small lump. □ **knobbly** *adj.*

■ **1** dial, handle, switch. **2** boss, bump, knot, lump, node, protrusion, protuberance, stud.

knock *v.* **1** strike with an audible sharp blow. **2** strike a door to gain admittance. **3** drive or make by knocking. **4** (*sl.*) criticize insultingly. ● *n.* **1** act or sound of knocking. **2** sharp blow. □ **knock about 1** treat roughly. **2** wander casually. **knock-down** *adj.* (of price) very low. **knock-kneed** *adj.* having an abnormal inward curvature of the legs at the knees (**knock knees**). **knock off 1** (*colloq.*) cease work. **2** (*colloq.*) complete quickly. **3** (*sl.*) steal. **knock-on effect** secondary or cumulative effect. **knockout** *n.* **1** knocking a person out. **2** (*colloq.*) outstanding or irresistible person or thing. **knock out 1** make unconscious. **2** eliminate. **knock up 1** rouse by knocking at a door. **2** make or arrange hastily. **knock-up** *n.* practice or casual game at tennis etc.

■ *v.* **1** bang, bash, hammer, hit, pound, rap, *old use* smite, strike, tap, thump. ● *n.* **1** rap, tap, thump. **2** blow, *colloq.* clip, clout, cuff, rap, smack.

knocker *n.* hinged flap for rapping on a door.

knoll *n.* hillock, mound.

knot *n.* **1** intertwining of one or more pieces of thread or rope etc. as a fastening. **2** tangle. **3** hard mass esp. where a branch joins a tree trunk. **4** round spot in timber. **5** cluster. **6** unit of speed used by ships and aircraft, = one nautical mile per hour. ● *v.* (**knotted**) **1** tie or fasten with a knot. **2** entangle.

■ *n.* **2** snarl, tangle. **3** knob, lump, node, nodule, protuberance. **5** bunch, clump, cluster, collection, crowd, gathering, group, throng. ● *v.* **1** attach, bind, fasten, lash, secure, tie. **2** entangle, entwine, snarl, twist.

knotty *adj.* (**-ier, -iest**) **1** full of knots. **2** puzzling, difficult.

■ **1** gnarled, knotted, lumpy, nodular; ravelled, tangled, twisted. **2** baffling, complex, complicated, difficult, perplexing, puzzling.

know *v.* (**knew, known**) **1** have in one's mind or memory. **2** feel certain. **3** recognize, be familiar with. **4** understand. □ **in the know** (*colloq.*) having inside information. **know-how** *n.* practical knowledge or skill.

■ **1** be familiar with, comprehend, grasp, understand. **3** recall, recognize, recollect, remember; be acquainted with, be a friend of, be friendly with. **4** be aware of, realize, understand. □ **know-how** ability, capability, experience, expertise, knowledge, proficiency, skill.

knowing *adj.* **1** aware. **2** cunning. □ **knowingly** *adv.*

■ **1** astute, aware, clever, intelligent, knowledgeable, perceptive, sagacious, shrewd, wise. **2** artful, canny, crafty, cunning, sly, wily.

knowledge *n.* **1** knowing about things. **2** information. **3** all a person knows. **4** all that is known.

■ **1** awareness, cognition, comprehension, consciousness, familiarity, grasp, insight, know-how, understanding; experience, expertise, proficiency. **2** data, facts, information, intelligence.

knowledgeable *adj.* **1** intelligent. **2** well-informed.

■ **1** erudite, intelligent, learned, sagacious, well-educated, well-read, wise. **2** aware, enlightened, *colloq.* in the know, up to date, well-informed.

knuckle *n.* **1** finger-joint. **2** animal's leg-joint as meat. ● *v.* **knuckle under** yield, submit.

knuckleduster *n.* metal device worn over the knuckles to increase the effect of a blow.

koala *n.* (in full **koala bear**) Australian tree-climbing animal with thick grey fur.

kohl *n.* black powder used as eye make-up.

kohlrabi *n.* cabbage with a turnip-like edible stem.

kookaburra *n.* Australian giant kingfisher with a harsh cry.

kopeck *n.* Russian coin, one-hundredth of a rouble.

Koran *n.* Islamic sacred book.

kosher *adj.* conforming to Jewish dietary laws.

kowtow *v.* behave with exaggerated respect.

k.p.h. *abbr.* kilometres per hour.

krill *n.* tiny plankton crustaceans that are food for whales etc.

kudos *n.* honour and glory.

kudu *n.* African antelope.

kummel /ko͝omm'l/ *n.* liqueur flavoured with caraway seeds.

kumquat *n.* tiny variety of orange.

kung fu Chinese form of unarmed combat similar to karate.

Kurd *n.* member of a people of SW Asia. □ **Kurdish** *adj.*

kV *abbr.* kilovolt(s).

kW *abbr.* kilowatt(s).

l *abbr.* litre(s).

lab *n.* (*colloq.*) laboratory.

label *n.* note fixed on or beside an object to show its nature, destination, etc. ● *v.* (**labelled**) **1** fix a label to. **2** describe as.

■ *n.* docket, mark, marker, sticker, tab, tag, ticket. ● *v.* **1** docket, mark, tag, ticket. **2** brand, call, classify as, characterize as, describe as, designate, dub, name.

labial *adj.* of the lips.

laboratory *n.* room or building equipped for scientific work.

laborious *adj.* needing or showing much effort. □ **laboriously** *adv.*

■ arduous, demanding, difficult, exhausting, gruelling, hard, herculean, onerous, strenuous, taxing, tiring; forced, laboured, ponderous, strained.

labour *n.* **1** work, exertion. **2** workers. **3** contractions of the womb at childbirth. ● *v.* **1** work hard. **2** emphasize lengthily.

■ *n.* **1** donkey work, drudgery, effort, exertion, *sl.* graft, grind, industry, slog, toil, work. **2** employees, labourers, workers, workforce. ● *v.* **1** beaver (away), drudge, *sl.* graft, *colloq.* plug (away), slave (away), slog (away), strive, struggle, sweat, toil, work. **2** dwell on, harp on, overdo, overemphasize.

laboured *adj.* showing signs of great effort, not spontaneous.

labourer *n.* person employed to do unskilled work.

Labrador *n.* dog of the retriever breed with a black or golden coat.

laburnum *n.* tree with hanging clusters of yellow flowers.

labyrinth *n.* maze. □ **labyrinthine** *adj.*

lace *n.* **1** ornamental openwork fabric or trimming. **2** cord etc. threaded through holes or hooks to pull opposite edges together. ● *v.* **1** fasten with lace(s). **2** intertwine. **3** add a dash of spirits to (drink).

■ *n.* **1** filigree, net, netting, openwork. **2** cord, shoelace, shoestring, string, thong. ● *v.* **1** do up, fasten, knot, tie (up). **2** intertwine, string, thread, weave. **3** flavour, fortify, *colloq.* spike, strengthen.

lacerate *v.* **1** tear (flesh). **2** wound (feelings). □ **laceration** *n.*

■ **1** cut, gash, mangle, rip, slash, tear, wound.

lachrymose *adj.* tearful.

lack *n.* state or fact of not having something. ● *v.* be without.

■ *n.* absence, dearth, deficiency, insufficiency, need, paucity, scarcity, shortage, want. ● *v.* be deficient in, be short of, be without, need, want.

lackadaisical *adj.* lacking vigour, unenthusiastic.

■ apathetic, casual, cool, half-hearted, indifferent, indolent, languid, languorous, lazy, lethargic, listless, lukewarm, sluggish, unenthusiastic, uninterested.

lackey *n.* (*pl.* **-eys**) **1** footman, servant. **2** servile follower.

lacking *adj.* **1** undesirably absent. **2** without.

lacklustre *adj.* lacking brightness or enthusiasm.

■ bland, boring, colourless, dreary, dull, flat, insipid, prosaic, tedious, unimaginative, uninspired, uninteresting, vapid, wishy-washy.

laconic *adj.* terse. □ **laconically** *adv.*

■ brief, concise, pithy, short, succinct, terse, to the point.

lacquer *n.* hard glossy varnish. ● *v.* coat with lacquer.

lactation *n.* **1** suckling. **2** secretion of milk.

lacy *adj.* (**-ier, -iest**) of or like lace.

lad *n.* boy, young fellow.

ladder *n.* **1** set of crossbars between uprights, used as a means of climbing. **2** vertical ladder-like flaw where stitches become undone in a stocking etc. ● *v.* cause or develop a ladder (in).

laden *adj.* loaded.

ladle *n.* deep long-handled spoon for transferring liquids. ● *v.* transfer with a ladle.

lady *n.* **1** woman, esp. of good social position. **2** well-mannered woman. **3** (**Lady**) title of wives, widows, or daughters of certain noblemen.

ladybird *n.* small flying beetle, usu. red with black spots.

ladylike *adj.* polite and appropriate to a lady.

ladyship *n.* title used of or to a woman with rank of *Lady*.

lag[1] *v.* (**lagged**) go too slow, not keep up. ● *n.* lagging, delay.

■ *v.* dally, dawdle, *colloq.* dilly-dally, hang back, linger, straggle, trail.

lag[2] *v.* (**lagged**) encase in material that prevents loss of heat.

lager *n.* light beer. □ **lager lout** (*colloq.*) youth who behaves badly after drinking too much.

laggard *n.* person who lags behind.

lagging *n.* material used to lag a boiler etc.

lagoon *n.* **1** salt-water lake beside a sea. **2** freshwater lake beside a river or larger lake.

laid *see* **lay**[2].

lain *see* **lie**[2].

lair *n.* wild animal's resting-place.

■ burrow, covert, den, nest, sett, tunnel.

laird *n.* (*Sc.*) landowner.

laissez-faire /léssayfáir/ *n.* policy of non-interference.

laity *n.* laymen.

lake[1] *n.* large body of water surrounded by land.

lake[2] *n.* reddish pigment.

lam *v.* (**lammed**) (*sl.*) hit hard.

lama *n.* Buddhist priest in Tibet and Mongolia.

lamb *n.* **1** young sheep. **2** its flesh as food. **3** gentle or endearing person. ● *v.* give birth to a lamb.

lambaste *v.* (*colloq.*) **1** thrash. **2** reprimand severely.

lame *adj.* **1** unable to walk normally. **2** weak, unconvincing. ● *v.* make lame. □ **lame duck** helpless person etc. **lamely** *adv.*, **lameness** *n.*

■ *adj.* **1** crippled, disabled, game, *sl.* gammy, handicapped, incapacitated. **2** feeble, flimsy, inadequate, poor, thin, unconvincing, unsatisfactory, weak.

lamé /laámay/ *n.* fabric with gold or silver thread interwoven.

lament *n.* **1** passionate expression of grief. **2** song or poem expressing grief. ● *v.* feel or express grief or regret. □ **lamentation** *n.*

■ *n.* **1** lamentation, moan, moaning, wail, wailing. **2** dirge, elegy, requiem. ● *v.* grieve, moan, mourn, sorrow, wail, weep; bemoan, regret, rue.

lamentable *adj.* regrettable, deplorable. □ **lamentably** *adv.*

■ awful, deplorable, dreadful, pitiful, regrettable, sad, terrible, unfortunate, wretched.

laminate *n.* laminated material.

laminated *adj.* made of layers joined one upon another.

lamp *n.* device for giving light.

lamplight *n.* light from a lamp.

lampoon *n.* piece of writing that attacks a person by ridiculing him. ● *v.* ridicule in a lampoon.

■ *n.* burlesque, caricature, parody, satire, take-off. ● *v.* burlesque, caricature, make fun of, parody, ridicule, satirize, send up, take off.

lamppost *n.* tall post of a street lamp.

lamprey *n.* (*pl.* **-eys**) small eel-like water animal.

lampshade *n.* shade placed over a lamp to screen its light.

lance *n.* long spear. ● *v.* prick or cut open with a lancet. □ **lance-corporal** *n.* army rank below corporal.

lanceolate *adj.* tapering to each end like a spearhead.

lancet *n.* **1** surgeon's pointed two-edged knife. **2** tall narrow pointed arch or window.

land *n.* **1** part of earth's surface not covered by water. **2** expanse of this. **3** ground, soil. **4** country, state. **5** (*pl.*) estates. ● *v.* **1** set or go ashore. **2** come or bring to the ground. **3** bring to or reach a place or situation. **4** deal (a person) a blow. **5** bring (a fish) to land. **6** obtain (a prize, appointment, etc.). □ **landlocked** *adj.* surrounded by land.

■ *n.* **1** earth, ground, terra firma. **2** country, terrain. **3** dirt, earth, loam, sod, soil, turf. **4** country, fatherland, homeland, motherland, nation, native land, state, territory. **5** (**lands**) estates, grounds, property. ● *v.* **1** berth, dock, disembark, go ashore. **2** alight, arrive, come to rest, settle, touch down. **4** administer, deal, give. **5** catch, hook, net. **6** gain, get, obtain, procure, secure, win.

landau /lándaw/ *n.* a kind of horse-drawn carriage.

landed *adj.* **1** owning land. **2** consisting of land.

landfall *n.* approach to land after a journey by sea or air.

landfill *n.* **1** waste material etc. used in landscaping or reclaiming ground. **2** use of this.

landing *n.* **1** coming or bringing ashore or to ground. **2** place for this. **3** level area at the top of a flight of stairs. □ **landing-stage** *n.* platform for landing from a boat.

■ **1** arrival, disembarkation, touchdown. **2** dock, jetty, landing-stage, pier, quay, wharf.

landlord, landlady *ns.* **1** person who lets land or a house or room to a tenant. **2** person who keeps an inn or boarding house.

■ **1** lessor, owner, proprietor. **2** host, hotelier, innkeeper, manager, proprietor, publican.

landlubber *n.* person unfamiliar with the sea and ships.

landmark *n.* **1** conspicuous feature of a landscape. **2** event marking a stage in a thing's history.

landscape *n.* **1** scenery of a land area. **2** picture of this. ● *v.* lay out (an area) attractively with natural-looking features.

landslide *n.* **1** landslip. **2** overwhelming majority of votes.

landslip *n.* sliding down of a mass of land on a slope.

landward *adj.* & *adv.* towards the land. □ **landwards** *adv.*

lane *n.* **1** narrow road, track, or passage. **2** strip of road for a single line of traffic. **3** track to which ships or aircraft etc. must keep.

■ **1** alley, footpath, passage, passageway, path, road, track.

language *n.* **1** words and their use. **2** system of this used by a nation or group.

■ **1** diction, speech, talk, words. **2** dialect, idiom, jargon, parlance, patois, slang, tongue, vernacular.

languid *adj.* lacking vigour or vitality. □ **languidly** *adv.*

■ apathetic, half-hearted, indolent, inert, lackadaisical, languorous, lazy, lethargic, listless, sluggish, torpid, unenthusiastic.

languish *v.* lose or lack vitality. □ **languish under** live under (miserable conditions).

■ decline, droop, fade, flag, waste away, wilt.

languor *n.* **1** state of being languid. **2** tender mood or effect. □ **languorous** *adj.*

lank *adj.* **1** tall and lean. **2** straight and limp. □ **lanky** *adj.*

■ **1** lanky, lean, scrawny, skinny, spindly, tall, thin. **2** limp, straggling, straight.

lanolin *n.* fat extracted from sheep's wool, used in ointments.

lantern *n.* transparent case for holding and shielding a light.

lanyard *n.* **1** short rope for securing things on a ship. **2** cord for hanging a whistle etc. round the neck or shoulder.

lap[1] *n.* **1** flat area over the thighs of a seated person. **2** single circuit. **3** section of a journey. ● *v.* (**lapped**) **1** wrap round. **2** be lap(s) ahead of (a competitor). □ **lapdog** *n.* small pampered dog.

lap[2] *v.* (**lapped**) **1** take up (liquid) by movements of tongue. **2** flow (against) with ripples.

lapel *n.* flap folded back at the front of a coat etc.

lapidary *adj.* of stones.

lapis lazuli blue semiprecious stone.

Lapp *n.* native or language of Lapland.

lapse *v.* **1** pass gradually into a worse state or condition. **2** become void or no longer valid. ● *n.* **1** slight error. **2** lapsing. **3** passage of time.

■ *v.* **1** decline, degenerate, deteriorate, sink. **2** expire, run out, terminate. ● *n.* **1** blunder, error, mistake, omission, oversight, slip, *colloq.* slip-up. **2** decline, degeneration, deterioration. **3** break, gap, interval, period, space.

laptop *n.* portable microcomputer.

lapwing *n.* kind of plover.

larch *n.* deciduous tree of the pine family.

lard *n.* white greasy substance prepared from pig-fat. ● *v.* put strips of fat bacon in or on (meat) before cooking.

larder *n.* storeroom for food.

large *adj.* of great size or extent. □ **at large 1** free to roam about. **2** in general. **largeness** *n.*

■ ample, big, broad, capacious, considerable, enormous, extensive, great, huge, immense, massive, roomy, sizeable, spacious, substantial, vast, voluminous; brawny, burly, heavy, hefty, stocky, strapping, thickset.

largely *adv.* to a great extent.

largesse *n.* (also **largess**) **1** money or gifts generously given. **2** generosity.

lariat *n.* lasso.

lark[1] *n.* small brown bird, skylark.

lark² *n.* **1** light-hearted adventurous action. **2** amusing incident. **3** type of activity. ● *v.* **lark about** play light-heartedly.

larkspur *n.* plant with spur-shaped blue or pink flowers.

larva *n.* (*pl.* **-vae**) insect in the first stage after coming out of the egg. □ **larval** *adj.*

laryngitis *n.* inflammation of the larynx.

larynx *n.* part of the throat containing the vocal cords. □ **laryngeal** *adj.*

lasagne /ləsányə/ *n.pl.* pasta in wide ribbon-like strips.

lascivious *adj.* lustful. □ **lasciviously** *adv.*, **lasciviousness** *n.*

laser *n.* device emitting an intense narrow beam of light.

lash *v.* **1** move in a whip-like movement. **2** beat with a whip. **3** strike violently. **4** fasten with a cord etc. ● *n.* **1** flexible part of a whip. **2** stroke with this. **3** eyelash. □ **lash out 1** attack with blows or words. **2** spend lavishly.

> ■ *v.* **2** beat, *sl.* belt, birch, cane, flog, scourge, thrash, whip. **3** batter, buffet, beat, dash, pound, strike. **4** attach, bind, fasten, hitch, rope, secure, strap, tether, tie.

lashings *n.pl.* (*colloq.*) a lot.

lass *n.* (also **lassie**) (*Sc.* & *N. Engl.*) girl, young woman.

lassitude *n.* tiredness, languor.

lasso *n.* (*pl.* **-oes**) rope with a noose for catching cattle. ● *v.* (**lassoed, lassoing**) catch with a lasso.

last¹ *n.* foot-shaped block used in making and repairing shoes.

last² *adj.* & *adv.* **1** coming after all others. **2** most recent(ly). ● *n.* last person or thing. □ **at (long) last** after much delay. **last post** military bugle-call sounded at sunset or military funerals. **last straw** slight addition to difficulties, making them unbearable. **last word 1** final statement in a dispute. **2** latest fashion.

> ■ *adj.* **1** closing, concluding, final, terminal, ultimate; hindmost, rearmost.

last³ *v.* **1** continue, endure. **2** suffice for a period of time. □ **lasting** *adj.*

> ■ **1** carry on, continue, go on, endure, hold out, persist, remain, survive; keep, stay fresh. **2** do, serve, suffice.

lastly *adv.* finally.

latch *n.* **1** bar lifted from its catch by a lever, used to fasten a gate etc. **2** spring-lock that catches when a door is closed. ● *v.* fasten with a latch.

latchkey *n.* key of an outer door.

late *adj.* & *adv.* **1** after the proper or usual time. **2** far on in a day or night or period. **3** recent. **4** no longer living or holding a position. □ **of late** lately. **lateness** *n.*

> ■ *adj.* **1** belated, delayed, overdue, tardy, unpunctual. **4** dead, deceased, departed; erstwhile, ex-, former, past, sometime.

lately *adv.* recently.

latent *adj.* existing but not developed or visible. □ **latency** *n.*

> ■ concealed, dormant, hidden, potential, quiescent, undeveloped, unrevealed.

lateral *adj.* of, at, to, or from the side(s). □ **laterally** *adv.*

latex *n.* **1** milky fluid from certain plants, esp. the rubber tree. **2** similar synthetic substance.

lath *n.* (*pl.* **laths**) narrow thin strip of wood, e.g. in trellis.

lathe *n.* machine for holding and turning pieces of wood or metal etc. while they are worked.

lather *n.* **1** froth from soap and water. **2** frothy sweat. ● *v.* cover with or form lather.

Latin *n.* language of the ancient Romans. ● *adj.* **1** of or in Latin. **2** speaking a language based on Latin. □ **Latin America** parts of Central and S. America where Spanish or Portuguese is the main language.

latitude *n.* **1** distance of a place from the equator, measured in degrees. **2** region. **3** freedom from restrictions.

> ■ **3** freedom, leeway, liberty, license, scope.

latrine *n.* lavatory in a camp or barracks.

latter *adj.* **1** mentioned after another. **2** nearer to the end. **3** recent. □ **latter-day** *adj.* modern, recent.

latterly *adv.* **1** recently. **2** nowadays.

lattice *n.* framework of crossed strips.

laud *v.* praise.

laudable *adj.* praiseworthy.

> ■ admirable, commendable, creditable, estimable, meritorious, praiseworthy.

laudanum *n.* opium prepared for use as a sedative.

laudatory *adj.* praising.

laugh *v.* make sounds and movements of the face that express amusement or scorn. ● *n.* **1** act or manner of laughing. **2** (*colloq.*) amusing incident. □ **laugh at** make fun of, ridicule. **laugh off** make light of (something embarrassing or humiliating). **laughing-stock** *n.* person or thing that is ridiculed.

■ *v. colloq.* be in stitches, cackle, chortle, chuckle, giggle, guffaw, hoot, snicker, snigger, titter. ● *n.* 1 cackle, chortle, chuckle, giggle, guffaw, hoot, snicker, snigger, titter. 2 joke, hoot, lark, *sl.* scream. □ **laugh at** deride, jeer (at), make fun of, mock, *colloq.* rib, ridicule, scoff at, taunt, tease.

laughable *adj.* ridiculous.

■ absurd, farcical, ludicrous, nonsensical, preposterous, ridiculous, risible.

laughter *n.* act or sound of laughing.

launch[1] *v.* **1** put or go into action. **2** cause (a ship) to slide into the water. **3** send forth (a weapon, rocket, etc.). ● *n.* process of launching something.

■ *v.* 1 begin, embark on, establish, inaugurate, initiate, institute, introduce, set up, start. 2 float, set afloat. 3 fire, hurl, project, propel, shoot, throw.

launch[2] *n.* large motor boat.

launder *v.* **1** wash and iron (clothes etc.). **2** (*colloq.*) transfer (funds) to conceal their origin.

launderette *n.* establishment fitted with washing machines to be used for a fee.

laundry *n.* **1** place where clothes etc. are laundered. **2** batch of clothes etc. sent to or from this.

laurel *n.* **1** evergreen shrub with smooth glossy leaves. **2** (*pl.*) victories or honours gained.

lava *n.* flowing or hardened molten rock from a volcano.

lavatory *n.* **1** fixture into which urine and faeces are discharged for disposal. **2** room equipped with this.

lavender *n.* **1** shrub with fragrant purple flowers. **2** light purple. □ **lavender-water** *n.* perfume made from lavender.

laver *n.* edible seaweed.

lavish *adj.* **1** generous. **2** abundant, plentiful. ● *v.* bestow lavishly. □ **lavishly** *adv.*

■ *adj.* 1 bountiful, free, generous, liberal, munificent, unstinting. 2 abundant, copious, plentiful, prolific, profuse. ● *v.* bestow, pour, rain, shower.

law *n.* **1** rule(s) established by authority or custom. **2** their influence or operation. **3** statement of what always happens in certain circumstances. □ **law-abiding** *adj.* obeying the law.

■ 1 act, decree, directive, edict, injunction, order, ordinance, precept, regulation, rule, statute. 2 equity, justice. 3 axiom, principle, theorem, theory. □ **law-abiding** decent, good, honest, honourable, obedient, orderly, peaceable, principled, upright, virtuous.

lawful *adj.* permitted or recognized by law. □ **lawfully** *adv.*, **lawfulness** *n.*

■ allowable, authorized, constitutional, legal, legitimate, permissible, proper, rightful, valid.

lawless *adj.* disregarding the law. □ **lawlessness** *n.*

lawn[1] *n.* fine woven cotton fabric.

lawn[2] *n.* area of closely cut grass. □ **lawnmower** *n.* machine for cutting the grass of lawns. **lawn tennis** (*see* **tennis**).

lawsuit *n.* process of bringing a problem or claim etc. before a court of law for settlement.

lawyer *n.* person trained and qualified in legal matters.

lax *adj.* slack, not strict or severe. □ **laxity** *n.*

■ careless, casual, hit-or-miss, lackadaisical, neglectful, negligent, remiss, slipshod, sloppy; easygoing, indulgent, lenient, liberal, permissive.

laxative *adj.* & *n.* (medicine) stimulating the bowels to empty.

lay[1] *adj.* **1** not ordained into the clergy. **2** non-professional.

lay[2] *v.* (**laid**) **1** place on a surface. **2** arrange ready for use. **3** cause to be in a certain position or state. **4** (of a hen) produce (an egg or eggs). ● *n.* way a thing lies. □ **lay about one** hit out on all sides. **lay bare** expose, reveal. **lay into** (*colloq.*) **1** thrash. **2** scold harshly. **lay off** **1** discharge (workers) temporarily. **2** (*colloq.*) cease. **lay-off** *n.* temporary discharge. **lay on** provide. **lay out** **1** arrange. **2** prepare (a body) for burial. **3** spend (money) for a purpose. **4** knock unconscious. **lay to rest** bury in a grave. **lay up** **1** store. **2** cause (a person) to be ill. **lay waste** devastate (an area).

■ *v.* 1 deposit, park, place, put (down), set (down). 2 arrange, prepare, set. 3 fit, put in place. □ **lay bare** disclose, divulge, expose, make known, reveal, show, uncover, unmask, unveil. **lay off** 1 discharge, dismiss, fire, make redundant, *colloq.* sack. **lay on** furnish, give, provide, supply. **lay out** 1 arrange, array, display, spread out. 2 expend, invest, spend. 3 floor, knock out, prostrate. **lay up** 1 hoard, put by, save, set aside, stockpile, store. **lay waste** destroy, devastate, ravage, wreak havoc on, wreck.

lay[3] *see* **lie**[2].

layabout *n.* loafer, one who lazily avoids working for a living.

lay-by *n.* place beside a road where vehicles may stop.

layer *n.* **1** one thickness of material laid over a surface. **2** attached shoot fastened down to take root. **3** hen that lays eggs. ● *v.* **1** arrange in layers. **2** propagate (a plant) by layers.

■ *n.* **1** coat, coating, covering, film, sheet, skin, thickness.

layette *n.* clothes etc. for a newborn baby.

lay figure artist's jointed model of the human body.

layout *n.* arrangement of parts etc. according to a plan.

layperson *n.* non-professional person.

laze *v.* spend time idly. ● *n.* act or period of lazing.

■ *v.* do nothing, idle, loaf, lounge, loll.

lazy *adj.* (**-ier**, **-iest**) **1** unwilling to work, doing little work. **2** showing little energy. □ **lazybones** *n.* (*colloq.*) lazy person. **lazily** *adv.*, **laziness** *n.*

■ **1** idle, shiftless, slothful. **2** inactive, indolent, lackadaisical, languid, languorous, lethargic, listless, sluggish, torpid.

lb *abbr.* pound(s) weight.

lea *n.* (*poetic*) piece of meadow etc.

leach *v.* **1** percolate (liquid) through soil etc. **2** remove (soluble matter) or be removed in this way.

lead[1] /leed/ *v.* (**led**) **1** guide. **2** influence into an action, opinion, or state. **3** be a route or means of access. **4** pass (one's life). **5** be in first place or ahead. **6** be in charge of. ● *n.* **1** guidance. **2** clue. **3** leading place, amount by which one competitor is in front. **4** wire conveying electric current. **5** strap or cord for leading an animal. **6** chief role in a play etc. □ **lead to** have as a consequence, result in. **lead up to 1** serve as introduction to or preparation for. **2** direct conversation towards. **leading question** one worded to prompt the desired answer.

■ *v.* **1** accompany, conduct, convey, escort, guide, pilot, shepherd, show, steer, take, usher. **2** bring, cause, dispose, induce, influence, move, persuade, prompt. **3** extend, go, run. **4** have, live, pass, spend. **5** be ahead, be at the front, head. **6** be in charge of, command, control, direct, govern, head, manage, oversee, preside over, superintend, supervise. ● *n.* **1** direction, guidance, leadership; example, model, pattern. **2** clue, hint, pointer, suggestion, tip-off. **3** first *or* leading place, front, pole position, van, vanguard; gap, margin. **4** cable, cord, flex, wire. **5** chain, cord, leash, rope, tether. □ **lead to** bring about, cause, create, engender, give rise to, produce, result in.

lead[2] /led/ *n.* **1** heavy grey metal. **2** graphite in a pencil. **3** lump of lead used for sounding depths. **4** (*pl.*) strips of lead.

leaden *adj.* **1** made of lead. **2** heavy, slow. **3** lead-coloured.

leader *n.* **1** one that leads. **2** newspaper article giving editorial opinions. □ **leadership** *n.*

■ **1** chief, commander, director, governor, head, manager, ruler, supremo; captain, skipper; guide, escort; pioneer, trail-blazer, trend-setter.

leaf *n.* **1** flat (usu. green) organ growing from the stem, branch, or root of a plant. **2** single thickness of paper as a page of a book. **3** very thin sheet of metal. **4** hinged flap or extra section of a table. ● *v.* **leaf through** turn over the leaves of (a book). □ **leaf-mould** *n.* soil consisting of decayed leaves. **leafy** *adj.*

leaflet *n.* **1** small leaf of a plant. **2** printed sheet of paper giving information. ● *v.* (**leafleted**) distribute leaflets to.

■ *n.* **2** booklet, brochure, circular, handbill, pamphlet.

league *n.* **1** union of people or countries. **2** class of contestants. **3** association of sports clubs which compete against each other. □ **in league** allied, conspiring.

■ **1** alliance, association, body, coalition, confederacy, confederation, federation, fellowship, group, guild, organization, society, union. **2** category, class, grade, level, rank. □ **in league** allied, collaborating, conspiring, in alliance, *sl.* in cahoots, in collusion.

leak *n.* **1** hole through which liquid or gas makes its way wrongly. **2** liquid etc. passing through this. **3** process of leaking. **4** similar escape of an electric charge. **5** disclosure of secret information. ● *v.* **1** escape or let out from a container. **2** disclose. □ **leak out** become known. **leakage** *n.*, **leaky** *adj.*

■ *n.* **1** aperture, chink, crack, crevice, fissure, hole, opening, perforation, puncture. **2,3** discharge, escape, leakage, outflow, seepage, spillage. **5** disclosure, revelation.

● *v.* 1 dribble, escape, flow, issue, pour, ooze, seep, stream, trickle. 2 disclose, divulge, impart, give away, make known, reveal, tell.

lean[1] *adj.* **1** without much flesh. **2** (of meat) with little or no fat. **3** scanty. ● *n.* lean part of meat. □ **leanness** *n.*

■ *adj.* 1 bony, lanky, rangy, scraggy, scrawny, skinny, slender, slim, thin. 3 inadequate, insufficient, meagre, poor, scant, scanty, sparse.

lean[2] *v.* (**leaned**, **leant**) **1** put or be in a sloping position. **2** rest for support against. □ **lean on 1** depend on for help. **2** (*colloq.*) influence by intimidating. **lean-to** *n.* shed etc. against the side of a building. **lean towards** have a tendency towards.

■ 1 bend, incline, slant, slope, tilt, tip. 2 lay, place, prop, put, rest; be propped *or* supported, lie. □ **lean on 1** be dependent on, count on, depend on, rely on.

leaning *n.* inclination, preference.

■ bent, bias, disposition, inclination, liking, partiality, penchant, predilection, predisposition, preference, prejudice, proclivity, propensity, tendency.

leap *v.* (**leaped**, **leapt**) jump vigorously. ● *n.* vigorous jump. □ **leap year** year with an extra day (29 Feb.).

■ *n.* & *v.* bound, hop, jump, skip, spring.

leap-frog *n.* game in which each player vaults over another who is bending down. ● *v.* (**-frogged**) **1** perform this vault (over). **2** overtake alternately.

learn *v.* (**learned** or **learnt**) **1** gain knowledge of or skill in. **2** become aware of. □ **learner** *n.*

■ 1 become proficient in, *colloq.* get the hang of, master, pick up; study, take lessons in. 2 ascertain, discover, find out, gather, hear, realize, see.

learned /lérnid/ *adj.* having or showing great learning.

■ academic, cultured, erudite, highbrow, intellectual, knowledgeable, scholarly, well-educated, well-read.

learning *n.* knowledge obtained by study.

■ education, erudition, knowledge, lore, scholarship, wisdom.

lease *n.* contract allowing the use of land or a building for a specified time. ● *v.* allow, obtain, or hold by lease. □ **leasehold** *n.*, **leaseholder** *n.*

leash *n.* dog's lead.

least *adj.* **1** smallest in amount or degree. **2** lowest in importance. ● *n.* least amount etc. ● *adv.* in the least degree.

leather *n.* **1** material made by treating animal skins. **2** piece of soft leather for polishing with. ● *v.* thrash. □ **leather-jacket** *n.* crane-fly grub with tough skin.

leathery *adj.* **1** like leather. **2** tough.

leave *v.* (**left**) **1** go away (from). **2** cease to live at, belong to, work for, etc. **3** deposit or entrust with. **4** abandon. **5** bequeath. **6** let remain. ● *n.* **1** permission. **2** official permission to be absent from duty, period for which this lasts. □ **on leave** absent in this way. **leave out** not insert or include.

■ *v.* 1 be on one's way, depart, *sl.* do a bunk, go, go away, move away, *sl.* push off, set off; decamp, pull out, retire, retreat, withdraw; abandon, evacuate, quit, vacate. 2 give up, *US* quit. 3 consign, deposit, entrust, place, put. 4 abandon, desert, *sl.* ditch, forsake, jilt, leave in the lurch, walk out on. 5 bequeath, hand down, will. ● *n.* 1 authorization, consent, dispensation, permission, sanction.

leaven *n.* **1** raising agent. **2** enlivening influence. ● *v.* **1** add leaven to. **2** enliven.

lecher *n.* lecherous man.

lecherous *adj.* lustful. □ **lechery** *n.*

lectern *n.* stand with a sloping top from which a bible etc. is read.

lecture *n.* **1** speech giving information about a subject. **2** lengthy reproof or warning. ● *v.* **1** give lecture(s). **2** reprove at length. □ **lecturer** *n.*

■ *n.* 1 address, discourse, harangue, speech, talk. 2 dressing down, scolding, talking-to, *colloq.* telling-off, *colloq.* ticking-off. ● *v.* 1 discourse, deliver a speech, speak, talk. 2 *colloq.* bawl out, berate, carpet, castigate, rebuke, remonstrate with, reprimand, reprove, scold, *colloq.* tell off, *colloq.* tick off, upbraid.

led *see* **lead**[1].

ledge *n.* narrow horizontal shelf or projection.

■ overhang, projection, ridge, shelf, sill.

ledger *n.* book in which accounts are kept.

lee *n.* **1** sheltered side. **2** shelter in this.

leech *n.* small blood-sucking worm.

leek *n.* plant related to the onion, with a long cylindrical white bulb.

leer *v.* look slyly, maliciously, or lustfully. ● *n.* leering look.

lees *n.pl.* sediment in wine.

leeward *adj.* & *n.* (on) the side away from the wind.

leeway *n.* degree of freedom of action.

■ **freedom, latitude, room, scope, space.**

left¹ *see* **leave.**

left² *adj.* & *adv.* of, on, or to the side or region opposite right. ● *n.* **1** left side or region. **2** left hand or foot. **3** people supporting a more extreme form of socialism than others in their group. □ **left-handed** *adj.* using the left hand.

leftovers *n.pl.* things remaining when the rest is finished.

■ **leavings, remainders, remains, remnants, scraps.**

leg *n.* **1** each of the limbs on which a person, animal, etc., stands or moves. **2** part of a garment covering a person's leg. **3** projecting support of piece of furniture. **4** one section of a journey or contest.

legacy *n.* thing left to someone in a will, or handed down by a predecessor.

■ **bequest, heritage, inheritance, patrimony.**

legal *adj.* **1** of or based on law. **2** authorized or required by law. □ **legally** *adv.*, **legality** *n.*

■ **1 forensic, judicial. 2 above board, acceptable, admissible, authorized, lawful, legitimate, permissible, rightful, valid; constitutional, statutory.**

legalize *v.* make legal. □ **legalization** *n.*

■ **allow, authorize, legitimize, license, permit, sanction.**

legate *n.* envoy.

legatee *n.* recipient of a legacy.

legation *n.* **1** diplomatic minister and staff. **2** their headquarters.

legend *n.* **1** story handed down from the past. **2** such stories collectively. **3** inscription, esp. on a coin or medal.

■ **1 epic, fable, folk-tale, myth, romance, saga. 2 folklore, myth, tradition.**

legendary *adj.* **1** of or described in legend. **2** (*colloq.*) famous.

■ **1 epic, fabled, mythical, traditional; fabulous, fictional, fictitious, imaginary, fairy-tale.**

legerdemain /léjərdəmáyn/ *n.* sleight of hand.

legible *adj.* clear enough to be deciphered, readable. □ **legibly** *adv.*, **legibility** *n.*

■ **clear, decipherable, easy to read, intelligible, readable.**

legion *n.* **1** division of the ancient Roman army. **2** organized group. **3** multitude.

legionnaire *n.* member of a legion. □ **legionnaires' disease** form of bacterial pneumonia.

legislate *v.* make laws. □ **legislator** *n.*

legislation *n.* **1** legislating. **2** law(s) made.

legislative *adj.* making laws.

legislature *n.* country's legislative assembly.

legitimate *adj.* **1** in accordance with a law or rule. **2** justifiable. **3** born of parents married to each other. □ **legitimately** *adv.*, **legitimacy** *n.*

■ **1 allowable, authorized, lawful, legal, permissible, rightful; authentic, bona fide, genuine, proper, real, true. 2 justifiable, logical, rational, reasonable, valid.**

legitimize *v.* make legitimate. □ **legitimization** *n.*

legume *n.* **1** leguminous plant. **2** pod of this.

leguminous *adj.* of the family of plants bearing seeds in pods.

leisure *n.* time free from work. □ **at one's leisure** when one has time.

leisured *adj.* having plenty of leisure.

leisurely *adj.* & *adv.* without hurry.

■ *adj.* **easy, easygoing, gentle, relaxed, slow, unhurried.**

lemming *n.* mouse-like Arctic rodent (said to rush headlong into the sea and drown during migration).

lemon *n.* **1** oval fruit with acid juice. **2** tree bearing it. **3** pale yellow colour. **4** (*colloq.*) useless person or thing. □ **lemony** *adj.*

lemonade *n.* lemon-flavoured soft drink.

lemon sole a kind of plaice.

lemur *n.* nocturnal monkey-like animal of Madagascar.

lend *v.* (**lent**) **1** give or allow to use temporarily. **2** provide (money) temporarily in return for payment of interest. **3** contribute as a help or effect. □ **lend itself to** be suitable for. **lender** *n.*

■ **1 loan. 2 advance, loan. 3 add, bestow, confer, contribute, furnish, give, impart.**

length *n.* **1** measurement or extent from end to end. **2** extent of time. **3** piece (of cloth etc.). ▫ **at length** after or taking a long time.

▪ **1** distance, extent, size, span. **2** duration, period, stretch, time. ▫ **at length** at last, eventually, finally, in the end, ultimately; endlessly, interminably.

engthen *v.* make or become longer.

▪ elongate, extend; draw out, pad out, prolong, protract, spin out, stretch out.

engthways *adv.* in the direction of a thing's length. ▫ **lengthwise** *adv.* & *adj.*

engthy *adj.* (**-ier**, **-iest**) **1** very long. **2** long and boring. ▫ **lengthily** *adv.*

▪ **1** extended, long, prolonged. **2** boring, dull, endless, interminable, long-winded, protracted, tedious.

enient *adj.* merciful, not severe. ▫ **leniently** *adv.*, **lenience** *n.*

▪ charitable, compassionate, easygoing, forbearing, forgiving, generous, humane, indulgent, kind, kind-hearted, magnanimous, merciful, permissive, tolerant.

ens *n.* piece of glass or similar substance shaped for use in an optical instrument.

.ent *n.* Christian period of fasting and repentance before Easter.

ent *see* **lend**.

entil *n.* a kind of bean.

eonine *adj.* of or like a lion.

eopard /léppərd/ *n.* large flesh-eating animal of the cat family, with a dark-spotted yellowish or a black coat.

eotard *n.* close-fitting garment worn by gymnasts, dancers, etc.

eper *n.* person with leprosy.

eprechaun *n.* (in Irish folklore) elf resembling a little old man.

eprosy *n.* contagious disease of the skin and nerves. ▫ **leprous** *adj.*

esbian *adj.* & *n.* homosexual (woman). ▫ **lesbianism** *n.*

esion *n.* harmful change in the tissue of an organ of the body.

ess *adj.* **1** not so much of. **2** smaller in amount or degree. ● *adv.* to a smaller extent. ● *n.* smaller amount. ● *prep.* minus.

essee *n.* person holding property by lease.

essen *v.* make or become less.

▪ allay, alleviate, assuage, deaden, dull, ease, relieve, soothe; abate, decrease, diminish, ebb, *colloq.* let up, moderate, subside, wane.

esser *adj.* not so great as the other.

esson *n.* **1** amount of teaching given at one time. **2** thing to be learnt. **3** experience by which one can learn. **4** passage from the Bible read aloud in church.

lessor *n.* person who lets property on lease.

lest *conj.* for fear that.

let *v.* (**let**, **letting**) **1** allow, enable, or cause to. **2** allow the use of (rooms or land) in return for payment. ● *v.aux.* used in requests, commands, assumptions, or challenges. ● *n.* letting of property etc. ▫ **let alone 1** refrain from interfering with or doing. **2** not to mention. **let down 1** let out air from (a tyre etc.). **2** fail to support, disappoint. **let-down** *n.* disappointment. **let in** allow to enter. **let off 1** fire or explode (a weapon etc.). **2** not punish or compel. **let on** (*sl.*) reveal a secret. **let out 1** allow to go out. **2** make looser. **let up** (*colloq.*) relax, become less intense.

▪ *v.* **1** allow, authorize, give permission, permit, sanction. **2** hire (out), lease, rent (out), sublet. ▫ **let down 2** disappoint, disenchant, disillusion, dissatisfy, fail. **let off 1** discharge, fire; detonate, explode, set off. **2** exempt from, excuse (from), forgive, pardon, reprieve, spare. **let up** relax, slow down, take a break; abate, decrease, diminish, ease, lessen, moderate, subside.

lethal *adj.* causing death.

▪ deadly, fatal, mortal.

lethargy *n.* extreme lack of energy or vitality. ▫ **lethargic** *adj.*, **lethargically** *adv.*

▪ apathy, fatigue, indolence, inertia, languidness, languor, lassitude, laziness, sloth, sluggishness, torpor, weariness.

letter *n.* **1** symbol representing a speech sound. **2** written message, usu. sent by post. **3** (*pl.*) literature. ● *v.* inscribe letters (on). ▫ **letter box 1** slit in a door, with a movable flap, through which letters are delivered. **2** postbox.

▪ *n.* **1** character, rune, sign, symbol. **2** dispatch, epistle, line, message, note.

letterhead *n.* printed heading on stationery.

lettuce *n.* plant with broad crisp leaves used in salads.

leucocyte *n.* white blood cell.

leukaemia /lookeémiə/ *n.* disease in which leucocytes multiply uncontrollably.

levee *n.* (*US*) **1** embankment against floods. **2** quay.

level *adj.* **1** horizontal. **2** without projections or hollows. **3** having equality with something else. **4** steady, uniform. ● *n.* **1** horizontal line or plane. **2** device for testing this. **3** measured height or value etc. **4** relative position. **5** level surface or area. ● *v.* (**levelled**) **1** make or become level. **2** knock down (a building). **3** aim (a gun etc.). □ **level crossing** place where a road and railway cross at the same level. **level-headed** *adj.* sensible, calm. **level pegging** equality in score. **on the level** (*colloq.*) honest(ly).

■ *adj.* **1** flat, horizontal. **2** even, flat, flush, plane, smooth, straight, unbroken, uninterrupted. **3** even, flush, parallel; equal, equivalent, level-pegging, neck and neck, on a par, the same. **4** consistent, even, regular, stable, steady, unchanging, uniform, unvarying. ● *v.* **1** even (out *or* up), flatten (out), iron (out), smooth (out); equalize, make equal. **2** demolish, destroy, knock down, raze. **3** aim, direct, focus, point, train, turn. □ **level-headed** calm, collected, composed, cool, equable, imperturbable, rational, reasonable, sensible, *colloq.* unflappable.

lever *n.* **1** bar pivoted on a fixed point to lift something. **2** pivoted handle used to operate machinery. **3** means of power or influence. ● *v.* **1** use a lever. **2** lift by this.

leverage *n.* **1** action or power of a lever. **2** power, influence.

leveret /lévvərit/ *n.* young hare.

leviathan *n.* thing of enormous size and power.

levitate *v.* (cause to) rise and float in the air. □ **levitation** *n.*

levity *n.* humorous attitude.

levy *v.* impose (payment) or collect (an army etc.) by authority or force. ● *n.* **1** levying. **2** payment or (*pl.*) troops levied.

■ *v.* charge, demand, exact, impose, inflict; conscript, mobilize, raise.

lewd *adj.* **1** treating sexual matters vulgarly. **2** lascivious. □ **lewdly** *adv.*, **lewdness** *n.*

lexical *adj.* of words. □ **lexically** *adv.*

lexicography *n.* process of compiling a dictionary. □ **lexicographer** *n.*

lexicon *n.* dictionary.

liability *n.* **1** being liable. **2** troublesome person or thing. **3** (*pl.*) debts.

■ **1** accountability, responsibility; susceptibility, vulnerability. **2** burden, disadvantage, drawback, encumbrance, hindrance, impediment, millstone.

liable *adj.* **1** held responsible by lav legally obliged to pay a tax or penal etc. **2** likely to do or suffer something.

■ **1** accountable, answerable, responsibl **2** apt, disposed, inclined, likely, pron open, subject, susceptible, vulnerable.

liaise *v.* act as liaison.

liaison *n.* **1** communication ar cooperation. **2** illicit sexual relationship

■ **1** communication, connection, contac cooperation, link, relationship. **2** affair, i trigue, relationship.

liana *n.* climbing plant of tropical forests

liar *n.* person who tells lies.

libation *n.* drink-offering to a god.

libel *n.* **1** published false statement tha damages a person's reputation. **2** act c publishing it. **3** false defamatory stat ment. ● *v.* (**libelled**) **1** publish a lib against. **2** accuse falsely and maliciousl □ **libellous** *adj.*

■ *n.* **3** calumny, defamation, denigratio obloquy, slander, vilification. ● *v.* **2** abus calumniate, cast aspersions on, defam denigrate, malign, slander, traduce, vilify.

liberal *adj.* **1** given or giving freely. tolerant. □ **liberally** *adv.*, **liberality** *n.*

■ **1** abundant, ample, copious, plentifu profuse; bountiful, generous, free, lavis magnanimous, munificent, open-hande unsparing, unstinting. **2** broad-minde indulgent, lax, lenient, permissive, tolerar unprejudiced.

liberalize *v.* make less stric □ **liberalization** *n.*, **liberalizer** *n.*

liberate *v.* set free. □ **liberation** *n.*, **liberator** *n.*

■ deliver, emancipate, free, let go, (le loose, release, rescue, set free.

libertine *n.* man who lives an irr sponsible immoral life.

liberty *n.* freedom. □ **take liberties** b have with undue freedom or familiarity.

■ autonomy, freedom, independence, sel determination, sovereignty; emancipatio liberation; authority, latitude, licence, pe mission, power, privilege, right, scope.

librarian *n.* person in charge of or a sisting in a library.

library *n.* **1** collection of books (or re ords, films, etc.) for consulting or bo rowing. **2** room or building containin these.

lice *see* **louse**.

icence *n.* **1** official permit to own or do something. **2** permission. **3** disregard of rules etc.

■ **1** certificate, pass, permit, warrant. **2** authorization, authority, dispensation, entitlement, leave, liberty, permission, right.

icense *v.* grant a licence to or for.

■ allow, authorize, empower, enable, entitle, permit, sanction.

icensee *n.* holder of a licence.

icentiate *n.* holder of a certificate of competence in a profession.

icentious *adj.* sexually immoral. □ **licentiousness** *n.*

■ debauched, decadent, depraved, dissipated, dissolute, immoral, lascivious, lecherous, lustful, wanton.

ichen /líkən/ *n.* dry-looking plant that grows on rocks etc.

ick *v.* **1** pass the tongue over. **2** (of waves or flame) touch lightly. **3** (*colloq.*) defeat. ● *n.* **1** act of licking. **2** slight application (of paint etc.). **3** (*colloq.*) fast pace.

id *n.* **1** hinged or removable cover for a box, pot, etc. **2** eyelid.

ie¹ *n.* statement the speaker knows to be untrue. ● *v.* (**lied, lying**) tell lie(s).

■ *n.* falsehood, fib, untruth, *sl.* whopper. ● *v.* commit perjury, fib, perjure oneself, tell lies.

ie² *v.* (**lay, lain, lying**) **1** have or put one's body in a flat or resting position. **2** be at rest on something. **3** be in a specified state. **4** be situated. ● *n.* way a thing lies. □ **lie in** lie in bed late in the morning. **lie-in** *n.* **lie low** conceal oneself or one's intentions.

■ *v.* **1** be recumbent *or* prone *or* prostrate *or* supine, prostrate oneself, recline, sprawl, stretch out. **2** be supported, lean, rest. □ **lie low** go into hiding, hide, keep out of sight, *sl.* lie doggo.

ieu /lyoo/ *n.* **in lieu** instead.

ieutenant /lefténnənt/ *n.* **1** army officer next below captain. **2** naval officer next below lieutenant commander. **3** rank just below a specified officer. **4** chief assistant.

ife *n.* (*pl.* **lives**) **1** animals' and plants' ability to function and grow, state of being alive. **2** period during which life lasts. **3** living things. **4** way of living. **5** liveliness. **6** biography. □ **life cycle** series of forms into which a living thing changes. **life-jacket** *n.* jacket of buoyant material to keep a person afloat. **life-size(d)** *adj.* of the same size as a real person. **life-support** *adj.* (of equipment etc.) enabling the body to function in a hostile environment or in cases of physical failure.

■ **1** being, existence, sentience. **2** lifetime. **3** fauna, flora, living things, people. **4** existence, lifestyle, way of life. **5** animation, *colloq.* bounce, energy, enthusiasm, exuberance, go, liveliness, pep, spirit, verve, vigour, *colloq.* vim, vitality, vivacity, zest. **6** autobiography, biography, memoir.

lifebelt *n.* belt of buoyant material to keep a person afloat.

lifeboat *n.* **1** boat for rescuing people in danger on the sea. **2** ship's boat for emergency use.

lifebuoy *n.* buoyant device to keep a person afloat.

lifeguard *n.* expert swimmer employed to rescue bathers who are in danger.

lifeless *adj.* **1** without life. **2** unconscious. **3** lacking liveliness.

■ **1** dead, deceased; arid, barren, desert, desolate, empty, sterile, uninhabited, waste; inanimate. **2** comatose, insensible, out, unconscious. **3** boring, colourless, dull, flat, lacklustre, tedious, uninspiring, vapid, wooden.

lifelike *n.* closely resembling a real person or thing.

lifeline *n.* **1** rope used in rescue. **2** vital means of communication.

lifelong *adj.* for all one's life.

lifestyle *n.* way of life of a person or group.

lifetime *n.* duration of a person's life.

lift *v.* **1** raise. **2** rise. **3** remove (restrictions). **4** steal, plagiarize. ● *n.* **1** lifting. **2** apparatus for transporting people or goods from one level to another. **3** free ride in a motor vehicle. **4** feeling of elation. □ **lift-off** *n.* vertical take-off of a spacecraft etc.

■ *v.* **1** heave (up), hoist (up), pick up; boost, buoy up, elevate, raise, uplift. **2** disappear, disperse, dissipate, rise, vanish. **3** cancel, discontinue, end, put an end to, remove, rescind. **4** filch, *sl.* knock off, *sl.* nick, pilfer, *sl.* pinch, pocket, purloin, steal, *colloq.* swipe, thieve, *sl.* whip; copy, pirate, plagiarize.

ligament *n.* tough flexible tissue holding bones together.

ligature *n.* thing that ties something, esp. in surgical operations. ● *v.* tie with a ligature.

light[1] *n.* **1** a kind of radiation that stimulates sight. **2** brightness. **3** source of light, electric lamp. **4** flame or spark. **5** enlightenment. **6** aspect, way a thing appears to the mind. ● *adj.* **1** full of light, not in darkness. **2** pale. ● *v.* (**lit** or **lighted**) **1** set burning, begin to burn. **2** provide with light. □ **bring** or **come to light** reveal or be revealed. **light-pen** *n.* **1** light-emitting device for reading bar codes. **2** (also **light-gun**) device for passing information to a computer screen. **light up 1** put lights on at dusk. **2** brighten. **3** make or become animated. **4** begin to smoke a cigarette etc. **light year** distance light travels in one year.

■ *n.* **2** brightness, brilliance, gleam, glow, illumination, incandescence, luminosity, phosphorescence, radiance; daylight, lamplight, moonlight, sunlight. **3** beacon, candle, flashlight, lamp, lantern, taper, torch; headlamp, headlight. **4** flame, spark. **5** elucidation, enlightenment, illumination. **6** aspect, perspective. ● *adj.* **1** bright, illuminated, sunny, well-lit. **2** pale, pastel-coloured. ● *v.* **1** fire, ignite, set alight, set fire to. **2** brighten, cast light on, illuminate, illumine, irradiate, lighten.

light[2] *adj.* **1** not heavy, easy to lift, carry, or do. **2** of less than average weight, force, or intensity. **3** not profound or serious. **4** cheerful. **5** (of food, meals, etc.) easy to digest. ● *adv.* lightly, with little load. ● *v.* (**lit** or **lighted**) **light on** find accidentally. □ **light-fingered** *adj.* apt to steal. **light-headed** *adj.* **1** feeling faint. **2** delirious. **light-hearted** *adj.* cheerful. **light industry** that producing small or light articles. **make light of** treat as unimportant. **lightly** *adv.*, **lightness** *n.*

■ *adj.* **1** lightweight, portable; bearable, *colloq.* cushy, easy, manageable, undemanding. **2** delicate, faint, gentle, mild, slight, soft. **3** frivolous, lightweight, inconsequential, trivial. **4** cheerful, cheery, gay, glad, happy, merry, sunny. **5** digestible; modest, simple, small. ● *v.* chance on, come across, discover, encounter, find, happen on, hit on. □ **light-hearted** airy, carefree, cheerful, cheery, gay, glad, happy, in good spirits, jolly, joyful, merry, sunny, *colloq.* upbeat.

lighten[1] *v.* **1** shed light on. **2** make or become brighter.

■ brighten, cast *or* shed light on, illuminate, illumine, light up.

lighten[2] *v.* make or become less heavy.

■ alleviate, ease, lessen, reduce.

lighter[1] *n.* device for lighting cigarett[es] and cigars.

lighter[2] *n.* flat-bottomed boat for u[n]loading ships.

lighthouse *n.* tower with a beacon lig[ht] to warn or guide ships.

lighting *n.* **1** means of providing light. [2] the light itself.

lightning *n.* flash of bright light pr[o]duced from cloud by natural electricit[y]. ● *adj.* very quick.

lights *n.pl.* lungs of certain animals, us[ed] as animal food.

lightship *n.* moored ship with a ligh[t] serving as a lighthouse.

lightweight *adj.* not having great weig[ht] or influence. ● *n.* lightweight person.

lignite *n.* brown coal of woody texture.

like[1] *adj.* **1** having the qualities or a[p]pearance of. **2** characteristic of. ● *pre[p.]* in the manner of, to the same degree a[s]. ● *conj.* **1** (*colloq.*) as. **2** (*US*) as if. ● *ad[v.]* (*colloq.*) likely. ● *n.* person or thing lik[e] another. □ **like-minded** *adj.* with simila[r] tastes or opinions.

■ *adj.* **1** akin (to), analogous (to), com[]parable (to *or* with), equivalent (to), of [a] piece (with), similar (to). **2** characterist[ic] of, typical of. ● *n.* counterpart, equa[l], equivalent, match, peer, twin.

like[2] *v.* **1** find pleasant or satisfactory. [2] be fond of. **3** choose to have, prefe[r]. ● *n.pl.* things one likes or prefers.

■ *v.* **1** be keen on, be partial to, delight i[n], have a liking for, have a weakness for, e[n]joy, love. **2** be attracted to, be fond of, b[e] keen on, have a soft spot for.

likeable *adj.* pleasant, easy to like.

■ agreeable, amiable, charming, endea[r]ing, friendly, genial, good-natured, nic[e], pleasant, pleasing, sympathetic, winnin[g], winsome.

likelihood *n.* probability.

■ chance, likeliness, possibility, probab[il]ity, prospect.

likely *adj.* (**-ier**, **-iest**) **1** such as ma[y] reasonably be expected to occur or b[e] true. **2** seeming to be suitable or have [a] chance of success. ● *adv.* probabl[y]. □ **likeliness** *n.*

■ *adj.* **1** possible, probable; apt, dispose[d], liable, prone; believable, credible, feasibl[e], plausible, reasonable. **2** appropriate, fit, fi[t]ting, promising, suitable.

liken *v.* point out the likeness of (one thing to another).

■ compare, correlate, equate, parallel.

likeness *n.* **1** being like. **2** portrait, representation.

■ **1** correspondence, resemblance, similarity, similitude. **2** copy, drawing, image, model, painting, picture, portrait, portrayal, representation, reproduction, sketch.

likewise *adv.* **1** also. **2** in the same way.

liking *n.* condition of being fond of a person or thing.

■ affection, fancy, fondness, inclination, love, partiality, penchant, predilection, preference, proclivity, propensity, soft spot, taste, weakness.

lilac *n.* **1** shrub with fragrant purple or white flowers. **2** pale purple. ● *adj.* pale purple.

lilt *n.* **1** light pleasant rhythm. **2** song with this. □ **lilting** *adj.*

lily *n.* plant growing from a bulb, with large flowers.

limb *n.* **1** projecting part of a person's or animal's body, used in movement or in grasping things. **2** large branch of a tree.

limber *v.* **limber up** exercise in preparation for athletic activity.

limbo[1] *n.* intermediate inactive or neglected state.

limbo[2] *n.* (*pl.* **-os**) W. Indian dance in which the dancer bends back to pass under a bar.

lime[1] *n.* white substance used in making cement etc.

lime[2] *n.* **1** round yellowish-green fruit like a lemon. **2** its colour.

lime[3] *n.* tree with heart-shaped leaves.

limelight *n.* great publicity.

limerick *n.* a type of humorous poem with five lines.

limestone *n.* a kind of rock from which lime is obtained.

limit *n.* **1** point beyond which something does not continue. **2** greatest amount allowed. ● *v.* set or serve as a limit, keep within limits. □ **limitation** *n.*

■ *n.* **1** border, boundary, edge, frontier, perimeter, periphery; end, extent, limitation. ● *v.* check, curb, restrain, restrict; confine, delimit.

limousine *n.* large luxurious car.

limp[1] *v.* walk or proceed lamely. ● *n.* limping walk.

limp[2] *adj.* **1** not stiff or firm. **2** wilting. □ **limply** *adv.*, **limpness** *n.*

■ **1** drooping, droopy, flaccid, floppy, loose, slack. **2** enervated, fatigued, spent, tired, weary, wilting, worn-out.

limpet *n.* small shellfish that sticks tightly to rocks.

limpid *adj.* (of liquids) clear. □ **limpidly** *adv.*, **limpidity** *n.*

linchpin *n.* **1** pin passed through the end of an axle to secure a wheel. **2** person or thing vital to something.

linctus *n.* soothing cough mixture.

linden *n.* lime tree.

line[1] *n.* **1** long narrow mark. **2** furrow, wrinkle. **3** outline. **4** boundary. **5** row of people or things. **6** brief letter. **7** series, several generations of a family. **8** course or manner of procedure, conduct, etc. **9** type of activity, business, or goods. **10** piece of cord for a particular purpose. **11** row of words. **12** (*pl.*) words of an actor's part. **13** service of ships, buses, or aircraft. **14** railway track. **15** electrical or telephone cable, connection by this. **16** each of a set of military fieldworks. **17** (**the Line**) the equator. ● *v.* **1** mark with lines. **2** form a line along.

■ *n.* **1** diagonal, slash, stroke. **2** crease, crinkle, furrow, groove, wrinkle. **3** contour, outline, profile, silhouette. **4** border, borderline, boundary, edge, frontier, limit, perimeter. **5** column, file, procession, queue, rank, row, train; band, bar, belt, strip, stripe. **6** card, letter, note, postcard. **7** series, succession; ancestry, descent, family, house, lineage, parentage, stock. **8** approach, course (of action), path, policy, procedure, strategy, tack, tactic, way. **9** area, department, field, province, speciality; business, job, profession, work; brand, kind, make, type, variety. **10** cable, cord, rope, string, wire. ● *v.* **1** crease, furrow, wrinkle. **2** border, edge, fringe.

line[2] *v.* cover the inside surface of. □ **line one's pockets** make money, esp. in underhand ways.

lineage *n.* line of ancestors or descendants.

■ ancestry, descent, extraction, family, genealogy, line, parentage, pedigree, stock.

lineal *adj.* of or in a line.

linear *adj.* **1** of a line, of length. **2** arranged in a line.

linen *n.* **1** cloth made of flax. **2** articles (e.g. sheets, tablecloths) formerly made of this.

liner[1] *n.* passenger ship or aircraft of a regular line.

liner[2] *n.* removable lining.

linesman *n.* **1** umpire's assistant at the boundary line. **2** workman who maintains railway, electrical, or telephone lines.

ling[1] *n.* a kind of heather.

ling[2] *n.* sea fish of north Europe.

linger *v.* **1** be slow or reluctant to depart. **2** dawdle.

> ■ **1** hang about, *sl.* hang on, hover, loiter, remain, stay behind, wait. **2** dally, dawdle, *colloq.* dilly-dally, pause.

lingerie /lánzhəri/ *n.* women's underwear.

lingua franca language used among people whose native languages are different.

lingual *adj.* **1** of the tongue. **2** of speech or languages.

linguist *n.* person who is skilled in languages or linguistics.

linguistic *adj.* of language. □ **linguistically** *adv.*

linguistics *n.* study of language.

liniment *n.* embrocation.

lining *n.* material used to cover a surface.

link *n.* **1** each ring of a chain. **2** person or thing connecting others. **3** means of connection. ● *v.* **1** connect. **2** intertwine (hands etc.). □ **linkage** *n.*

> ■ *n.* **3** association, connection, linkage, relation, relationship. ● *v.* **1** attach, connect, fasten, join, tie, unite. **2** clasp, intertwine.

linnet *n.* a kind of finch.

lino *n.* linoleum.

linocut *n.* **1** design cut in relief on a block of linoleum. **2** print made from this.

linoleum *n.* a kind of smooth covering for floors.

linseed *n.* seed of flax.

lint *n.* **1** soft fabric for dressing wounds. **2** fluff.

lintel *n.* horizontal timber or stone over a doorway etc.

lion *n.* large flesh-eating animal of the cat family. □ **lion's share** largest part.

lionize *v.* treat as a celebrity.

lip *n.* **1** either of the fleshy edges of the opening of the mouth. **2** edge of a container or opening. **3** slight projection shaped for pouring from. □ **lip-read** *v.* understand what is said from movements of a speaker's lips.

lipsalve *n.* ointment for the lips.

lipstick *n.* cosmetic for colouring the lips.

liquefy *v.* make or become liquid. □ **liquefaction** *n.*

liqueur /likyŏŏr/ *n.* strong sweet alcoholic spirit.

liquid *n.* flowing substance like water or oil. ● *adj.* **1** in the form of liquid. **2** (of assets) easy to convert into cash. □ **liquidity** *n.*

> ■ *n.* fluid, juice, liquor, sap, solution. ● *adj.* **1** aqueous, fluid, liquefied, molten, runny, watery.

liquidate *v.* **1** pay (a debt). **2** close down (a business) and divide its assets between creditors. **3** get rid of, esp. by killing. □ **liquidation** *n.*, **liquidator** *n.*

liquidize *v.* reduce to liquid.

liquidizer *n.* machine for making purées etc.

liquor *n.* **1** alcoholic drink. **2** juice from cooked food.

liquorice *n.* **1** black substance used in medicine and as a sweet. **2** plant from whose root it is made.

lira *n.* (*pl.* **lire**) unit of money in Italy and Turkey.

lisp *n.* speech defect in which *s* and *z* are pronounced like *th*. ● *v.* speak or utter with a lisp.

lissom *adj.* lithe.

list[1] *n.* written or printed series of names, items, figures, etc. ● *v.* **1** make a list of. **2** enter in a list.

> ■ *n.* catalogue, directory, index, inventory, register, roll, roster, rota, schedule. ● *v.* **1** catalogue, index, itemize. **2** enter, log, note, record, register.

list[2] *v.* (of a ship) lean over to one side. ● *n.* listing position.

> ■ *v.* careen, heel, incline, keel, lean (over), lurch, slant, tilt, tip.

listen *v.* **1** make an effort to hear. **2** pay attention. □ **listen in 1** eavesdrop. **2** listen to a broadcast. **listen to** take notice of, be persuaded by advice etc. **listener** *n.*

> ■ **1** hark. **2** attend, pay attention. □ **listen to** be mindful of, heed, mark, mind, note, take notice of.

listless *adj.* without energy or enthusiasm. □ **listlessly** *adv.*, **listlessness** *n.*

> ■ apathetic, drained, enervated, exhausted, lackadaisical, languid, lethargic

lifeless, sluggish, tired, torpid, unenthusiastic, weary.

lit *see* **light**[1], **light**[2].

litany *n.* a set form of prayer.

literacy *n.* being literate.

literal *adj.* taking the primary meaning of a word or words, not a metaphorical or exaggerated one. □ **literally** *adv.*

literary *adj.* of literature.

literate *adj.* able to read and write.

literature *n.* writings, esp. great novels, poetry, and plays.

lithe *adj.* supple, agile.

■ agile, athletic, flexible, graceful, lissom, supple, willowy.

litho *adj.* & *n.* lithographic (process).

lithograph *n.* picture printed by lithography.

lithography *n.* printing from a design on a smooth surface. □ **lithographic** *adj.*

litigant *adj.* & *n.* (person) involved in or initiating a lawsuit.

litigate *v.* **1** carry on a lawsuit. **2** contest in law.

litigious *adj.* fond of litigation.

litmus *n.* substance turned red by acids and blue by alkalis. □ **litmus paper** paper stained with this.

litre *n.* metric unit of capacity (about 1¾ pints).

litter *n.* **1** rubbish left lying about. **2** young animals born at one birth. **3** material used as bedding for animals or to absorb their excrement. ● *v.* **1** scatter as litter. **2** make untidy by litter. **3** give birth to (a litter).

■ *n.* **1** garbage, junk, refuse, rubbish, trash. **2** brood. ● *v.* **1** scatter, spread, strew, throw. **2** clutter (up), mess up.

little *adj.* **1** small in size or amount etc. **2** relatively unimportant. ● *n.* small amount, time, or distance. ● *adv.* **1** to a small extent. **2** not at all.

■ *adj.* **1** diminutive, mini-, miniature, minuscule, petite, short, slight, small, *colloq.* teeny, tiny, *Sc.* wee. **2** inconsequential, insignificant, minor, petty, negligible, trifling, trivial, unimportant. ● *n.* bit, dash, drop, hint, modicum, piece, pinch, scrap, spot, taste, trace, touch; minute, moment, second, while. ● *adv.* **1** barely, hardly, scarcely.

littoral *adj.* & *n.* (region) of or by the shore.

liturgy *n.* set form of public worship. □ **liturgical** *adj.*

live[1] /līv/ *adj.* **1** alive. **2** burning. **3** unexploded. **4** charged with electricity. **5** (of broadcasts) transmitted while actually happening. □ **live wire** energetic forceful person.

■ **1** alive, animate, living. **2** burning, flaming, glowing, hot, smouldering.

live[2] /liv/ *v.* **1** have life, remain alive. **2** have one's home. **3** pass, spend. **4** conduct (one's life) in a certain way. **5** enjoy life fully. □ **live down** live until (scandal etc.) is forgotten. **live on** keep oneself alive on. **live through** survive. **live with** **1** share a home with. **2** tolerate.

■ **1** be alive, exist; endure, keep going, last, stay alive, subsist, survive, sustain oneself. **2** dwell, lodge, reside, stay. **3** have, lead, pass, spend. □ **live through** endure, experience, go through, survive, weather, withstand.

livelihood *n.* **1** means of living. **2** job, income.

lively *adj.* (**-ier, -iest**) full of energy or action. □ **liveliness** *n.*

■ active, animated, *colloq.* bouncy, bubbly, buoyant, chirpy, ebullient, energetic, exuberant, frisky, full of life, high-spirited, irrepressible, perky, playful, skittish, spirited, sprightly, spry, vibrant, vivacious; brisk, fast, quick, rapid, snappy, swift; busy, bustling, crowded, hectic.

liven *v.* make or become lively.

liver *n.* **1** large organ in the abdomen, secreting bile. **2** animal's liver as food.

liveried *adj.* wearing livery.

livery *n.* distinctive uniform worn by male servants.

livestock *n.* farm animals.

livid *adj.* **1** bluish-grey. **2** (*colloq.*) furiously angry.

living *adj.* **1** having life, not dead. **2** currently in use. ● *n.* **1** manner of life. **2** livelihood. □ **living room** room for general daytime use.

lizard *n.* reptile with four legs and a long tail.

llama *n.* S. American animal related to the camel.

load *n.* **1** thing or quantity carried. **2** burden of responsibility or worry. **3** (*pl.*, *colloq.*) plenty. **4** amount of electric current supplied. ● *v.* **1** put a load in or on. **2** receive a load. **3** weight. **4** put ammunition into (a gun) or film into (a camera). **5** put (data etc.) into (a computer).

■ *n.* **1** cargo, consignment, freight, shipment. **2** burden, cross, millstone, onus, responsibility, weight, worry. ● *v.* **1** cram, fill, pack, stuff; heap, pile, stack. **3** burden, encumber, overload, lumber, saddle, weigh down, weight.

loaf[1] *n.* (*pl.* **loaves**) **1** mass of baked bread. **2** (*sl.*) head.

loaf[2] *v.* spend time idly.

loam *n.* rich soil. □ **loamy** *adj.*

loan *n.* **1** lending. **2** thing lent, esp. money. ● *v.* grant a loan of.

loath *adj.* unwilling.

■ disinclined, reluctant, unwilling.

loathe *v.* hate, detest. □ **loathing** *n.*, **loathsome** *adj.*

■ abhor, abominate, despise, detest, execrate, hate, shudder at. □ **loathing** abhorrence, antipathy, detestation, disgust, hatred, odium, repugnance, revulsion. **loathsome** abhorrent, abominable, contemptible, despicable, detestable, disgusting, execrable, hateful, nasty, odious, repellent, repugnant, repulsive, revolting, sickening, vile.

lob *v.* (**lobbed**) send or strike (a ball) slowly in a high arc. ● *n.* lobbed ball.

lobar *adj.* of a lobe, esp. of the lung.

lobby *n.* **1** porch, entrance hall, anteroom. **2** group of people seeking to influence legislation. ● *v.* seek to persuade (an MP etc.) to support one's cause.

lobbyist *n.* person who lobbies an MP etc.

lobe *n.* **1** rounded part or projection. **2** lower soft part of the ear.

lobelia *n.* garden plant.

lobster *n.* **1** shellfish with large claws. **2** its flesh as food.

local *adj.* of or affecting a particular place or small area. ● *n.* **1** inhabitant of a particular district. **2** (*colloq.*) nearby pub. □ **local government** administration of a district by representatives elected locally. **locally** *adv.*

■ *adj.* endemic, indigenous, native; municipal, regional, provincial; nearby, neighbouring. ● *n.* **1** denizen, native, national, resident, townsman, townswoman.

locality *n.* **1** thing's position. **2** district or neighbourhood.

■ **1** location, place, position, whereabouts. **2** area, district, environs, neighbourhood, region, vicinity.

localize *v.* **1** confine within an area. **2** decentralize.

locate *v.* **1** discover the position of. **2** situate in a particular location.

■ **1** ascertain, determine, detect, discover, find out, identify, pinpoint, track down. **2** place, position, put, site, situate, station.

location *n.* **1** place where a thing is situated. **2** act or process of locating. □ **on location** (of filming) in a suitable environment, not in a film studio.

■ **1** locality, place, position, setting, site, situation, spot, venue.

loch *n.* (*Sc.*) lake, arm of the sea.

lock[1] *n.* **1** device (opened by a key) for fastening a door or lid etc. **2** gated section of a canal where the water level can be changed. **3** secure hold in wrestling. **4** interlocked or jammed state. ● *v.* **1** fasten with a lock. **2** make or become rigidly fixed. □ **lock in** or **up** shut into a locked place. **lock out** shut out by locking a door. **lock-up** *n.* **1** premises that can be locked. **2** place where prisoners can be kept temporarily. **lockable** *adj.*

■ *n.* **1** combination lock, latch, padlock, mortise lock. ● *v.* **1** fasten, latch, padlock, secure. **2** jam, seize up, stick. □ **lock up** confine, detain, imprison, incarcerate, jail, shut in *or* up.

lock[2] *n.* **1** portion of hair that hangs together. **2** (*pl.*) hair.

■ **1** curl, ringlet, strand, tress.

locker *n.* cupboard where things can be stowed securely.

locket *n.* small ornamental case worn on a chain round the neck.

lockjaw *n.* tetanus.

lockout *n.* employer's procedure of locking out employees during a dispute.

locksmith *n.* maker and mender of locks.

locomotion *n.* ability to move from place to place.

locomotive *n.* self-propelled engine for moving railway trains. ● *adj.* of or effecting locomotion.

locum *n.* temporary stand-in for a doctor, clergyman, etc.

locus *n.* (*pl.* **-ci**) **1** thing's exact place. **2** line or curve etc. formed by certain points or by the movement of a point or line.

locust *n.* a kind of grasshopper that devours vegetation.

lode *n.* vein of metal ore.

lodestar *n.* star (esp. pole star) used as a guide in navigation.

lodestone *n.* oxide of iron used as a magnet.

lodge *n.* **1** cabin for use by hunters, skiers, etc. **2** gatekeeper's house. **3** porter's room at the entrance to a building. **4** members or meeting place of a branch of certain societies. **5** beaver's or otter's lair. ● *v.* **1** provide with sleeping quarters or temporary accommodation. **2** live as a lodger. **3** deposit. **4** be or become embedded.

■ *n.* **1** cabin, chalet, cottage. ● *v.* **1** accommodate, billet, house, put up. **2** board, dwell, live, reside, stay. **3** deposit, store, *colloq.* stash.

lodger *n.* person paying for accommodation in another's house.

lodging *n.* **1** temporary accommodation. **2** (*pl.*) room(s) rented for living in.

■ **1** accommodation, shelter. **2** (**lodgings**) accommodation, digs, quarters, rooms.

loft *n.* **1** space under a roof. **2** gallery in a church. ● *v.* send (a ball) in a high arc.

lofty *adj.* (**-ier**, **-iest**) **1** very tall. **2** noble. **3** haughty. □ **loftily** *adv.*, **loftiness** *n.*

■ **1** high, tall, towering. **2** exalted, grand, noble, sublime. **3** aloof, arrogant, contemptuous, condescending, disdainful, haughty, hoity-toity, patronizing, scornful, snobbish, *colloq.* snooty, supercilious, superior, *colloq.* uppity.

log[1] *n.* **1** piece cut from a trunk or branch of a tree. **2** device for gauging a ship's speed. **3** logbook, entry in this. ● *v.* (**logged**) enter (facts) in a logbook. □ **logbook** *n.* book in which details of a voyage or journey are recorded. **log on** or **off** begin or finish operations at a computer terminal.

log[2] *n.* logarithm.

loganberry *n.* large dark red fruit resembling a blackberry.

logarithm *n.* one of a series of numbers set out in tables, used to simplify calculations.

logic *n.* **1** science or method of reasoning. **2** correct reasoning.

logical *adj.* **1** of or according to logic. **2** reasonable. **3** reasoning correctly. □ **logically** *adv.*, **logicality** *n.*

■ **1** rational, reasoned. **2** defensible, justifiable, legitimate, reasonable, right, sensible, sound, valid. **3** intelligent, judicious, rational, sensible, wise.

logician *n.* person skilled in logic.

logistics *n.* organization of supplies and services. □ **logistical** *adj.*

logo *n.* (*pl.* **-os**) design used as an emblem.

loin *n.* side and back of the body between ribs and hip bone.

loincloth *n.* cloth worn round the loins.

loiter *v.* linger, stand about idly. □ **loiterer** *n.*

loll *v.* **1** stand, sit, or rest lazily. **2** hang loosely.

lollipop *n.* large usu. flat boiled sweet on a small stick. □ **lollipop lady, man** official using a circular sign on a stick to halt traffic for children to cross a road.

lollop *v.* (**lolloped**) (*colloq.*) **1** move in clumsy bounds. **2** flop.

lolly *n.* **1** (*colloq.*) lollipop, ice lolly. **2** (*sl.*) money.

lone *adj.* solitary.

lonely *adj.* **1** solitary. **2** sad because lacking companions. **3** not much frequented. □ **loneliness** *n.*

■ **1** lone, single, sole, solitary. **2** alone, companionless, forlorn, friendless, lonesome. **3** desolate, isolated, out of the way, remote, solitary, unfrequented.

loner *n.* person who prefers not to associate with others.

lonesome *adj.* lonely.

long[1] *adj.* of great or specified length. ● *adv.* **1** for a long time. **2** throughout a specified time. □ **as** or **so long as** provided that. **long-distance** *adj.* travelling or operated between distant places. **long face** dismal expression. **long johns** (*colloq.*) underpants with long legs. **long-life** *adj.* (of milk etc.) treated to prolong its shelf-life. **long-lived** *adj.* living or lasting for a long time. **long-range** *adj.* **1** having a relatively long range. **2** relating to a long period of future time. **long shot** wild guess or venture. **long-sighted** *adj.* able to see clearly only what is at a distance. **long-standing** *adj.* having existed for a long time. **long-suffering** *adj.* bearing provocation patiently. **long-term** *adj.* of or for a long period. **long ton** (*see* **ton**). **long wave** radio wave of frequency less than 300 kHz. **long-winded** *adj.* talking or writing at tedious length.

■ *adj.* drawn-out, extended, interminable, lengthy, prolonged, protracted; in length, lengthwise. □ **long-suffering** forbearing, patient, resigned, stoical, uncomplaining. **long-winded** lengthy, rambling, verbose, wordy.

long[2] *v.* **long for** feel a longing for.

■ crave, desire, hanker after, hunger for, itch for, thirst for, want, wish for, yearn for.

longevity /lonjévviti/ *n.* long life.

longhand *n.* ordinary writing, not shorthand or typing etc.

longhorn *n.* one of a breed of cattle with long horns.

longing *n.* intense wish.

> ■ craving, desire, hankering, hunger, itch, thirst, wish, yearning, *colloq.* yen.

longitude *n.* distance east or west (measured in degrees on a map) from the Greenwich meridian.

longitudinal *adj.* **1** of longitude. **2** of length, lengthwise. □ **longitudinally** *adv.*

loo *n.* (*colloq.*) lavatory.

loofah *n.* dried pod of a gourd, used as a rough sponge.

look *v.* **1** direct one's sight. **2** make a search. **3** face. **4** seem. ● *n.* **1** act of looking. **2** inspection, search. **3** appearance. □ **look after 1** take care of. **2** attend to. **look at 1** turn one's eyes in a particular direction. **2** consider. **look down on** despise. **look for** seek. **look forward to** await eagerly. **look into** investigate. **lookout** *n.* **1** watch. **2** watcher(s). **3** observation post. **4** prospect. **5** person's own concern. **look out** be vigilant. **look up 1** search for information about. **2** improve in prospects. **3** go to visit. **look up to** admire and respect.

> ■ *v.* **1** gaze, glance, peek, peep, peer, stare. **2** hunt, search. **3** face, point. **4** appear, seem. ● *n.* **1** gaze, glance, glimpse, peek, peep, squint, stare. **2** check, examination, inspection, once-over; hunt, search. **3** air, appearance, aspect, bearing, countenance, demeanour, expression, mien. □ **look after 1** care for, guard, mind, nurse, tend, take care of. **2** attend to, deal with, do, handle, see to, sort out, take charge of. **look at 1** contemplate, eye, examine, inspect, observe, regard, scan, scrutinize, study, survey, view, watch. **look for** hunt for, search for, seek, try to find. **look into** examine, explore, go over, investigate, inquire into, research, scrutinize, study. **look up to** admire, esteem, idolize, respect, revere, venerate, worship.

looker-on *n.* (*pl.* **lookers-on**) mere spectator.

loom[1] *n.* apparatus for weaving.

loom[2] *v.* **1** appear, esp. close at hand or threateningly. **2** be ominously close.

> ■ **1** appear, become visible, emerge, materialize, take shape.

loop *n.* **1** curve that is U-shaped or crosses itself. **2** thing shaped like this, esp. length of cord or wire etc. fastened at the crossing. ● *v.* **1** form into loop(s). **2** fasten or join with loop(s). **3** enclose in a loop. □ **loop the loop** fly in a vertical circle.

loophole *n.* means of evading a rule or contract.

loose *adj.* **1** not tight. **2** not fastened, held, or fixed. **3** inexact. ● *adv.* loosely. ● *v.* **1** release. **2** untie, loosen. □ **at a loose end** without a definite occupation. **loose-leaf** *adj.* with each page removable. **loosely** *adv.*

> ■ *adj.* **1** baggy. **2** flowing, unbound, untied; at large, free, unconfined; unattached, unconnected, unfastened. **3** broad, general, imprecise, inexact, rough. ● *v.* **1** let go, release, set free. **2** ease, loosen, slacken, undo, unfasten, untie.

loosen *v.* make or become loose or looser.

> ■ ease, loose, release, slacken, undo, unfasten, unlace, untie.

loot *n.* goods taken from an enemy or by theft. ● *v.* **1** take loot (from). **2** take as loot. □ **looter** *n.*

> ■ *n.* booty, haul, plunder, spoils, *sl.* swag. ● *v.* pillage, plunder, raid, ransack, rob, sack.

lop *v.* (**lopped**) **1** cut branches or twigs of. **2** cut off.

lope *v.* run with a long bounding stride. ● *n.* this stride.

lop-eared *adj.* with drooping ears.

lopsided *adj.* with one side lower, smaller, or heavier.

> ■ askew, asymmetrical, awry, *colloq.* cockeyed, crooked, unbalanced, uneven, unsymmetrical.

loquacious *adj.* talkative. □ **loquaciously** *adv.*, **loquacity** *n.*

> ■ chatty, garrulous, talkative, voluble.

lord *n.* **1** master, ruler. **2** nobleman. **3** (**Lord**) title of certain peers or high officials. **4** (**the Lord**) God. **5** (**Our Lord**) Christ.

lordship *n.* title used of or to a man with the rank of *Lord.*

lore *n.* body of traditions and knowledge.

> ■ beliefs, folklore, mythology, myths, tradition(s); knowledge, learning.

lorgnette /lornyét/ *n.* eyeglasses or opera-glasses held to the eyes on a long handle.

lorry *n.* large motor vehicle for transporting heavy loads.

lose *v.* (**lost**) **1** cease to have or maintain. **2** become unable to find. **3** forfeit (right to something). **4** waste (time or an opportunity). **5** get rid of. **6** be defeated in a contest etc. **7** suffer loss (of). **8** cause the loss of. □ **loser** *n.*

■ **1,2** mislay, misplace. **3** forfeit, give up. **4** miss, squander, waste. **5** elude, escape from, evade, get rid of, give a person the slip.

loss *n.* **1** act or instance of losing. **2** person or thing or amount lost. **3** disadvantage caused by losing something. □ **be at a loss** not know what to do or say. **loss-leader** *n.* article sold at a loss to attract customers.

■ **1** forfeiture, mislaying, misplacing, privation. **2** casualty, fatality; depletion, diminution, reduction; debit, deficit. **3** damage, detriment, disadvantage, harm, injury.

lost *see* **lose**. *adj.* strayed or separated from its owner.

lot *n.* **1** each of a set of objects drawn at random to decide something. **2** item being sold at auction. **3** person's share or destiny. **4** piece of land. **5** number of people or things of the same kind. **6** (*pl.*) large number or amount, much. **7** (**the lot**) the total quantity. □ **bad lot** person of bad character.

■ **3** *colloq.* cut, part, portion, quota, ration, share; destiny, fate, fortune. **4** piece of land, plot. **5** batch, collection, consignment, group, set. **6** (**lots**) *colloq.* heaps, *colloq.* loads, *colloq.* oodles, *colloq.* piles, plenty, quantities, *colloq.* stacks.

lotion *n.* medicinal or cosmetic liquid applied to the skin.

■ balm, cream, embrocation, emollient, liniment, moisturizer, salve, unguent.

lottery *n.* **1** system of raising money by selling numbered tickets and giving prizes to holders of numbers drawn at random. **2** thing where the outcome is governed by luck.

lotus *n.* (*pl.* **-uses**) **1** tropical water lily. **2** mythical fruit.

loud *adj.* **1** producing much noise. **2** gaudy. ● *adv.* loudly. □ **loudly** *adv.*, **loudness** *n.*

■ *adj.* **1** blaring, booming, deafening, noisy, raucous, stentorian, thunderous. **2** flashy, garish, gaudy, showy, tasteless, tawdry.

loudspeaker *n.* apparatus that converts electrical impulses into audible sound.

lough /lok/ *n.* (*Ir.*) = **loch**.

lounge *v.* **1** loll. **2** sit or stand about idly. ● *n.* **1** sitting room. **2** waiting-room at an airport etc. □ **lounge suit** man's ordinary suit for day wear. **lounger** *n.*

■ *v.* **1** laze, lie, loll, recline, sprawl. **2** idle, hang about, loaf, loiter. ● *n.* drawing room, living room, salon, sitting room.

louse *n.* **1** (*pl.* **lice**) small parasitic insect. **2** (*pl.* **louses**) contemptible person.

lousy *adj.* (**-ier, -iest**) **1** infested with lice. **2** (*colloq.*) very bad.

lout *n.* clumsy ill-mannered young man. □ **loutish** *adj.*

louvre *n.* each of a set of overlapping slats arranged to admit air but exclude light or rain. □ **louvred** *adj.*

lovable *adj.* easy to love.

■ adorable, appealing, charming, *colloq.* cute, darling, dear, enchanting, endearing, engaging, likeable, *colloq.* sweet, taking, winning, winsome.

lovage *n.* herb used for flavouring.

love *n.* **1** warm liking or affection. **2** sexual passion. **3** loved person. **4** (in games) no score, nil. ● *v.* **1** feel love for. **2** like greatly. □ **in love** feeling (esp. sexual) love for another person. **love affair** romantic or sexual relationship between two people. **lovebird** *n.* kind of parakeet. **make love** have sexual intercourse.

■ *n.* **1** adoration, affection, devotion, fondness, tenderness, warmth; liking, partiality, relish, taste. **2** ardour, lust, passion. **3** beloved, darling, dear, lover, sweet, sweetheart. ● *v.* **1** adore, be devoted to, be in love with, be infatuated with, care for, cherish, dote on, *colloq.* have a crush on, hold dear, idolize, treasure, worship. **2** be fond of, be partial to, delight in, have a soft spot for, have a weakness for, like, relish. □ **love affair** affair, intrigue, liaison, relationship, romance.

lovelorn *adj.* pining with love.

lovely *adj.* (**-ier, -iest**) **1** beautiful, attractive. **2** (*colloq.*) delightful. □ **loveliness** *n.*

■ **1** attractive, beautiful, comely, exquisite, good-looking, *colloq.* gorgeous, pretty, *colloq.* stunning. **2** agreeable, delightful, enjoyable, nice, pleasant, pleasurable, *colloq.* super.

lover *n.* **1** person in love with another or having an illicit love affair. **2** one who likes or enjoys something.

loving *adj.* feeling or showing love. □ **lovingly** *adv.*

■ adoring, affectionate, caring, devoted, doting, fond, tender; amorous, ardent, passionate.

low[1] *n.* deep sound made by cattle. ● *v.* make this sound.

low[2] *adj.* **1** not high, not extending or lying far up. **2** ranking below others. **3** ignoble, vulgar. **4** less than normal in amount or intensity. **5** not loud or shrill. **6** lacking vigour, depressed. ● *n.* **1** low level. **2** area of low pressure. ● *adv.* in, at, or to a low level. □ **low-down** *adj.* dishonourable. *n.* (*colloq.*) relevant information. **low-key** *adj.* restrained, not intense or emotional. **low season** season that is least busy.

■ *adj.* **1** little, short, small, stubby, stumpy. **2** humble, inferior, lowly, menial. **3** base, contemptible, despicable, dishonourable, ignoble, mean; coarse, crude, rude, vulgar. **4** inadequate, insufficient, limited, short, scant, scanty, sparse. **5** hushed, indistinct, muffled, muted, quiet, soft, subdued. **6** dejected, depressed, despondent, dispirited, downcast, downhearted, heavy-hearted, gloomy, glum, miserable, sad, unhappy.

lowbrow *adj.* not intellectual or cultured. ● *n.* lowbrow person.

lower *v.* **1** let or haul down. **2** make or become lower. □ **lower case** letters that are not capitals.

■ **1** drop, haul down, let down, put down. **2** cut, decrease, diminish, lessen, mark down, reduce, slash.

lowlands *n.pl.* low-lying land. □ **lowland** *adj.*, **lowlander** *n.*

lowly *adj.* (**-ier**, **-iest**) of humble rank or condition. □ **lowliness** *n.*

loyal *adj.* firm in one's allegiance. □ **loyally** *adv.*, **loyalty** *n.*

■ constant, dependable, devoted, faithful, reliable, staunch, steadfast, steady, true, trusted, trustworthy, trusty, unswerving, unwavering.

loyalist *n.* person who is loyal, esp. while others revolt.

lozenge *n.* **1** four-sided diamond-shaped figure. **2** small tablet to be dissolved in the mouth.

Ltd. *abbr.* Limited.

lubricant *n.* lubricating substance.

lubricate *v.* oil or grease (machinery etc.). □ **lubrication** *n.*

lubricious *adj.* **1** slippery. **2** lewd.

lucid *adj.* **1** clearly expressed. **2** sane. □ **lucidly** *adv.*, **lucidity** *n.*

■ **1** clear, coherent, comprehensible, eloquent, intelligible, understandable. **2** rational, sane.

luck *n.* **1** chance regarded as the bringer of good or bad fortune. **2** success due to chance.

■ **1** chance, destiny, fate, fortune. **2** good fortune, serendipity.

luckless *adj.* unlucky.

■ hapless, unfortunate, unlucky.

lucky *adj.* (**-ier**, **-iest**) having, bringing, or resulting from good luck. □ **lucky dip** tub containing articles which people may choose at random. **luckily** *adv.*

■ blessed, favoured, fortunate; fortuitous, happy, opportune, timely.

lucrative *adj.* producing much money, profitable.

lucre /lóokər/ *n.* (*derog.*) money.

ludicrous *adj.* ridiculous. □ **ludicrously** *adv.*, **ludicrousness** *n.*

■ absurd, asinine, crazy, farcical, foolish, laughable, nonsensical, preposterous, ridiculous, risible, silly, stupid.

ludo *n.* simple game played with counters on a special board.

lug[1] *v.* (**lugged**) drag or carry with great effort.

lug[2] *n.* **1** ear-like projection. **2** (*colloq.*) ear.

luggage *n.* suitcases, bags, etc. for a traveller's belongings.

■ baggage, bags, belongings, cases, paraphernalia, suitcases, things.

lugubrious *adj.* dismal, mournful. □ **lugubriously** *adv.*

■ dismal, forlorn, gloomy, melancholy, miserable, mournful, sad, woebegone, woeful.

lukewarm *adj.* **1** only slightly warm. **2** not enthusiastic.

■ **1** tepid. **2** apathetic, cool, half-hearted, indifferent, lackadaisical, unenthusiastic, uninterested.

lull *v.* **1** soothe or send to sleep. **2** calm. **3** become quiet. ● *n.* period of quiet or inactivity.

■ *v.* **2** allay, assuage, calm, ease, quiet, quieten, pacify, soothe. **3** abate, decrease, diminish, lessen, quieten, *colloq.* let up, moderate, subside. ● *n.* break, hiatus, intermission, interval, pause, respite; calm, calmness, hush, quiet, quietness, silence, stillness.

lullaby *n.* soothing song sung to send a child to sleep.

lumbago *n.* rheumatic pain in muscles of the lower back.

lumbar *adj.* of or in the lower back.

lumber *n.* **1** useless or cumbersome articles. **2** timber sawn into planks. ● *v.* **1** encumber. **2** move heavily and clumsily.

■ *n.* **1** clutter, junk, odds and ends, white elephants. **2** planks, timber, wood. ● *v.* **1** burden, encumber, load, overload, saddle. **2** clump, plod, stump, trudge.

lumberjack *n.* (*US*) workman cutting or conveying lumber.

luminary *n.* **1** natural light-giving body, esp. the sun or moon. **2** eminent person.

luminescent *adj.* emitting light without heat. □ **luminescence** *n.*

luminous *adj.* emitting light, glowing in the dark. □ **luminously** *adv.*, **luminosity** *n.*

■ bright, brilliant, gleaming, glistening, radiant, shimmering, shining; fluorescent, glowing, incandescent, luminescent, phosphorescent.

lump *n.* **1** hard or compact mass. **2** swelling. ● *v.* put or consider together. □ **lump sum** money paid as a single amount.

■ *n.* **1** cake, chunk, clod, clot, nugget, pat, piece, wedge, *colloq.* wodge. **2** bulge, bump, excrescence, growth, knob, nodule, protrusion, protuberance, swelling, tumescence, tumour.

lumpectomy *n.* surgical removal of a lump from the breast.

lumpy *adj.* (**-ier, -iest**) **1** full of lumps. **2** covered in lumps. □ **lumpiness** *n.*

lunacy *n.* **1** insanity. **2** great folly.

■ **1** dementia, derangement, insanity, madness. **2** craziness, folly, foolishness, idiocy, insanity, madness, senselessness, stupidity.

lunar *adj.* of the moon. □ **lunar month** period between new moons (29½ days), four weeks.

lunatic *n.* **1** insane person. **2** wildly foolish person. ● *adj.* **1** insane. **2** very foolish.

lunation *n.* lunar month

lunch *n.* midday meal. ● *v.* eat lunch.

luncheon *n.* lunch. □ **luncheon meat** tinned cured meat ready for serving. **luncheon voucher** voucher given to an employee, exchangeable for food.

lung *n.* either of the pair of breathing-organs in the chest.

lunge *n.* **1** sudden forward movement of the body. **2** thrust. ● *v.* make this movement.

lupin *n.* garden plant with tall spikes of flowers.

lurch[1] *n.* **leave in the lurch** leave (a person) in difficulties.

lurch[2] *v.* & *n.* (make) an unsteady swaying movement, stagger.

■ *v.* reel, stagger, stumble, sway; heel, list, pitch, roll.

lure *v.* entice. ● *n.* **1** enticement. **2** bait or decoy to attract wild animals.

■ *v.* attract, coax, draw, entice, inveigle, seduce, tempt. ● *n.* **1** enticement, inducement, temptation. **2** bait, decoy.

lurid *adj.* **1** in glaring colours. **2** sensational or shocking. □ **luridly** *adv.*

■ **1** garish, gaudy, loud. **2** graphic, melodramatic, sensational, shocking, startling, vivid; ghastly, gory, grisly, gruesome.

lurk *v.* **1** wait furtively or keeping out of sight. **2** be latent.

luscious *adj.* **1** delicious. **2** voluptuously attractive.

■ **1** appetizing, delectable, delicious, *colloq.* scrumptious, succulent, tasty, *colloq.* yummy.

lush *adj.* **1** (of grass etc.) luxuriant. **2** luxurious.

lust *n.* **1** intense sexual desire. **2** any intense desire. ● *v.* **lust after** or **for** feel lust for. □ **lustful** *adj.*, **lustfully** *adv.*

■ *n.* **1** desire, passion; lasciviousness, lechery, licentiousness. **2** craving, desire, greed, hunger, longing, thirst, yearning. ● *v.* covet, crave, desire, hanker after, hunger for, thirst for, yearn for. □ **lustful** amorous, passionate; lascivious, lecherous, lewd, licentious, salacious.

lustre *n.* **1** soft brightness of a surface. **2** glory. **3** metallic glaze on pottery. □ **lustrous** *adj.*

lusty *adj.* (**-ier, -iest**) strong and vigorous. □ **lustily** *adv.*

■ full-blooded, hale, healthy, hearty, robust, strong, vigorous.

lute *n.* guitar-like instrument. □ **lutenist** *n.*

luxuriant *adj.* growing profusely. □ **luxuriantly** *adv.*, **luxuriance** *n.*

■ abundant, copious, lush, profuse, rich, thick.

luxuriate *v.* **luxuriate in** feel great enjoyment in something.

■ bask in, delight in, indulge in, relish, revel in, wallow in.

luxurious *adj.* supplied with luxuries, very comfortable. □ **luxuriously** *adv.*

■ de luxe, grand, magnificent, opulent, palatial, plush, *colloq.* posh, splendid, sumptuous.

luxury *n.* **1** choice and costly surroundings, food, etc. **2** self-indulgence. **3** thing that is enjoyable but not essential.

■ **1** grandeur, magnificence, opulence, splendour, sumptuousness. **2** hedonism, indulgence, self-indulgence, voluptuousness. **3** extra, frill, treat.

lying *see* **lie**[1], **lie**[2].

lymph *n.* colourless fluid from body tissue. □ **lymphatic** *adj.*

lynch *v.* execute or punish violently by a mob, without trial.

lynx *n.* wild animal of the cat family with keen sight.

lyre *n.* ancient musical instrument with strings in a U-shaped frame. □ **lyre-bird** *n.* Australian bird with a lyre-shaped tail.

lyric *adj.* of poetry that expresses the poet's emotions. ● *n.* **1** lyric poem. **2** (*pl.*) words of a song.

lyrical *adj.* **1** resembling or using language suitable for lyric poetry. **2** (*colloq.*) highly enthusiastic. □ **lyrically** *adv.*

lyricist *n.* writer of lyrics.

Mm

m *abbr.* **1** metre(s). **2** mile(s). **3** million(s).
ma *n.* (*colloq.*) mother.
ma'am *n.* madam.
mac *n.* (*colloq.*) mackintosh.
macabre *adj.* gruesome.

■ appalling, awful, dreadful, fearsome, frightening, frightful, ghastly, gory, gruesome, grotesque, grisly, horrific, terrible.

macadam *n.* layers of broken stone used in road-making.
macadamize *v.* pave with macadam.
macaroni *n.* tube-shaped pasta.
macaroon *n.* biscuit or small cake made with ground almonds.
macaw *n.* American parrot.
mace[1] *n.* ceremonial staff carried or placed before an official.
mace[2] *n.* spice made from the dried outer covering of nutmeg.
machete /məchétti/ *n.* broad heavy knife.
machiavellian *adj.* elaborately cunning or deceitful.
machination *n.* (usu. *pl.*) intrigue, plot.
machine *n.* **1** apparatus for applying mechanical power. **2** thing operated by this. **3** controlling system of an organization etc. ● *v.* produce or work on with a machine. □ **machine-gun** *n.* gun that can fire continuously. **machine-readable** *adj.* in a form that a computer can process.

■ *n.* **1** apparatus, appliance, contraption, device, gadget, mechanism. **2** engine, motor; car, vehicle; automaton, robot.

machinery *n.* **1** machines. **2** mechanism.
machinist *n.* person who works machinery.
machismo *n.* **1** manly courage. **2** show of this.
macho *adj.* ostentatiously manly.
mackerel *n.* (*pl.* **mackerel**) edible sea fish.
mackintosh *n.* **1** cloth waterproofed with rubber. **2** raincoat.
macramé /məkraámi/ *n.* **1** art of knotting cord in patterns. **2** items made in this way.
macrocosm *n.* **1** the universe. **2** any great whole.
mad *adj.* (**madder, maddest**) **1** not sane. **2** extremely foolish. **3** wildly enthusiastic. **4** frenzied. **5** (*colloq.*) very annoyed. □ **madly** *adv.*, **madness** *n.*

■ **1** *sl.* barmy, *sl.* batty, certifiable, *sl.* crackers, crazy, demented, deranged, insane, lunatic, *colloq.* mental, *sl.* nuts, *sl.* nutty, out of one's mind, *sl.* potty, psychotic, unbalanced, unhinged. **2** absurd, asinine, *sl.* barmy, *colloq.* crack-brained, crazy, daft, foolhardy, foolish, hare-brained, idiotic, ludicrous, madcap, nonsensical, reckless, ridiculous, senseless, silly, stupid, unwise. **3** ardent, *colloq.* crazy, enthusiastic, fanatical, fervent, fervid, keen, passionate, wild. **4** frantic, frenzied, frenetic, uncontrolled, unrestrained, wild. **5** angry, annoyed, cross, enraged, exasperated, furious, fuming, irate, incensed, irritated, *colloq.* livid. □ **madness** dementia, derangement, lunacy, mania, mental illness, psychosis; absurdity, craziness, folly, foolishness, idiocy, insanity, senselessness, stupidity.

madam *n.* polite form of address to a woman.
madcap *adj.* & *n.* wildly impulsive (person).
madden *v.* make mad or angry.

■ *colloq.* aggravate, anger, annoy, *sl.* bug, enrage, exasperate, gall, incense, inflame, infuriate, irk, irritate, nettle, *colloq.* rile, vex.

made *see* **make.**
Madeira *n.* fortified wine from Madeira. □ **Madeira cake** rich plain cake.
madonna *n.* picture or statue of the Virgin Mary.
madrigal *n.* song for several voices.
maelstrom /máylstrəm/ *n.* great whirlpool.
maestro /mīstrō/ *n.* (*pl.* **-ri**) **1** great conductor or composer of music. **2** master of any art.
magazine *n.* **1** illustrated periodical. **2** store for arms or explosives. **3** chamber holding cartridges in a gun, slides in a projector, etc.

■ **1** journal, monthly, periodical, publication, quarterly, weekly. **2** armoury, arsenal.

magenta *adj.* & *n.* purplish-red.
maggot *n.* larva, esp. of the bluebottle.

magic *n.* **1** supposed art of controlling things by supernatural power. **2** conjuring tricks. ● *adj.* using or used in magic. □ **magical** *adj.*, **magically** *adv.*

■ *n.* **1** necromancy, sorcery, voodoo, witchcraft, wizardry. **2** conjuring, legerdemain, sleight of hand.

magician *n.* **1** person skilled in magic. **2** conjuror.

■ **1** enchanter, enchantress, necromancer, sorcerer, sorceress, witch, wizard. **2** conjuror, illusionist.

magisterial *adj.* **1** of a magistrate. **2** imperious.

magistrate *n.* official or citizen with authority to hold preliminary hearings and judge minor cases. □ **magistracy** *n.*

magnanimous *adj.* noble and generous in conduct, not petty. □ **magnanimity** *n.*

■ beneficent, benevolent, charitable, generous, forgiving, good, humane, kind, merciful, noble, philanthropic, unselfish.

magnate *n.* wealthy influential person, esp. in business.

■ *colloq.* mogul, tycoon.

magnesia *n.* compound of magnesium used in medicine.

magnesium *n.* silvery metallic element.

magnet *n.* **1** piece of iron or steel that can attract iron and point north when suspended. **2** thing exerting powerful attraction.

magnetic *adj.* **1** having the properties of a magnet. **2** produced or acting by magnetism. □ **magnetic tape** strip of plastic with magnetic particles, used in recording, computers, etc. **magnetically** *adv.*

magnetism *n.* **1** properties and effects of magnetic substances. **2** great charm and attraction.

■ **2** allure, appeal, attraction, attractiveness, charisma, charm, pull, seductiveness.

magnetize *v.* **1** make magnetic. **2** attract. □ **magnetization** *n.*

magneto /magneétō/ *n.* (*pl.* **-os**) small electric generator using magnets.

magnification *n.* magnifying.

magnificent *adj.* **1** splendid in appearance etc. **2** excellent in quality. □ **magnificently** *adv.*, **magnificence** *n.*

■ **1** glorious, gorgeous, grand, imposing, impressive, majestic, regal, resplendent, splendid, stately, striking; luxurious, opulent, palatial, sumptuous. **2** excellent, fine, marvellous, outstanding, splendid, superb, *colloq.* terrific, wonderful.

magnify *v.* **1** make (an object) appear larger by use of a lens. **2** exaggerate. □ **magnifier** *n.*

magnitude *n.* **1** largeness, size. **2** importance.

■ **1** extent, greatness, immensity, largeness, size. **2** consequence, importance, significance.

magnolia *n.* tree with large wax-like white or pink flowers.

magnum *n.* wine bottle twice the standard size.

magpie *n.* kind of crow with black and white plumage.

Magyar *adj.* & *n.* (member, language) of a people now predominant in Hungary.

maharajah *n.* former title of certain Indian princes. □ **maharanee** *n.* maharajah's wife or widow.

maharishi *n.* great Hindu sage.

mahatma *n.* (in India etc.) title of a man regarded with reverence.

mah-jong *n.* Chinese game played with 136 or 144 pieces.

mahogany *n.* **1** hard reddish-brown wood. **2** its colour.

maid *n.* woman servant.

maiden *n.* (*old use*) young unmarried woman, virgin. ● *adj.* **1** unmarried. **2** first. □ **maiden name** woman's surname before marriage. **maidenly** *adj.*

maidenhair *n.* fern with very thin stalks and delicate foliage.

maidservant *n.* female servant.

mail[1] *n.* = **post**[3]. ● *v.* send by post. □ **mail order** purchase of goods by post. **mailshot** *n.* material sent to potential customers in an advertising campaign.

mail[2] *n.* body-armour made of metal rings or chains.

maim *v.* wound or injure so that a part of the body is useless.

■ cripple, disable, hamstring, injure, incapacitate, lame, mutilate, wound.

main *adj.* principal, most important, greatest in size or extent. ● *n.* main pipe or channel conveying water, gas, or (*pl.*) electricity. □ **mainly** *adv.*

■ *adj.* basic, cardinal, chief, dominant, first, foremost, fundamental, key, leading, major, paramount, predominant, preeminent, primary, prime, principal.

mainframe *n.* large computer.

mainland *n.* country or continent without its adjacent islands.

mainmast *n.* principal mast.

mainsail *n.* lowest sail or sail set on the after part of the mainmast.

mainspring *n.* **1** chief spring of a watch or clock. **2** chief motivating force.

mainstay *n.* **1** cable securing the mainmast. **2** chief support.

mainstream *n.* dominant trend of opinion or style etc.

maintain *v.* **1** cause to continue, keep in existence. **2** keep in repair. **3** bear the expenses of. **4** assert.

■ **1** carry on, continue, keep up, perpetuate, persevere in, persist in, preserve, sustain. **2** keep in good condition, keep up, look after, service, take care of. **3** keep, provide for, support. **4** assert, avow, claim, declare, hold, insist, profess, state.

maintenance *n.* **1** process of maintaining something. **2** provision of means to support life, allowance of money for this.

maisonette *n.* **1** small house. **2** flat on more than one floor.

maize *n.* **1** tall cereal plant. **2** its grain.

majestic *adj.* stately and dignified, imposing. □ **majestically** *adv.*

■ dignified, glorious, grand, imperial, imposing, kingly, magnificent, noble, princely, queenly, regal, royal, splendid, stately.

majesty *n.* **1** impressive stateliness. **2** sovereign power. **3** title of a king or queen.

■ **1** dignity, glory, grandeur, magnificence, nobility, pomp, splendour, stateliness.

major *adj.* **1** greater or relatively great. **2** very important. ● *n.* army officer next below lieutenant colonel. ● *v.* (*US*) specialize (in a subject) at college. □ **major-general** *n.* army officer next below lieutenant general.

■ *adj.* **1** bigger, greater, larger, main. **2** crucial, foremost, important, leading, paramount, primary, prime, principal, significant, vital.

majority *n.* **1** greatest part of a group or class. **2** number by which votes for one party etc. exceed those for the next or for all combined. **3** age when a person legally becomes adult.

■ **1** best *or* better part, bulk, lion's share, preponderance.

make *v.* (**made**) **1** form, prepare, produce. **2** cause to exist, be, or become. **3** succeed in arriving at or achieving. **4** gain, acquire. **5** amount to. **6** calculate or estimate. **7** compel. **8** execute or perform (an action, speech, etc.). ● *n.* brand of goods. □ **make believe** pretend. **make-believe** *n.* pretence. **make do** manage with something not fully satisfactory. **make for 1** try to reach. **2** tend to bring about. **make good 1** become successful. **2** repair or pay compensation for. **make it** (*colloq.*) **1** arrive in time. **2** be successful. **make much of 1** treat as important. **2** give flattering attention to. **make off** go away hastily. **make off with** carry away, steal. **make out 1** write out (a list etc.). **2** manage to see or understand. **3** imply, suggest. **make over 1** transfer the ownership of. **2** refashion (a garment etc.). **make up 1** form, constitute. **2** invent (a story). **3** become reconciled after (a quarrel). **4** complete (an amount). **5** apply cosmetics (to). **make up for** compensate for. **make-up** *n.* **1** cosmetics applied to the skin. **2** way a thing is made. **3** person's character. **make up one's mind** decide.

■ *v.* **1** assemble, build, construct, create, erect, fabricate, fashion, forge, form, frame, manufacture, produce, shape; cook, get, prepare; compose, draft, draw up. **2** bring about, cause, engender, generate, give rise to, occasion, produce; appoint, designate, elect, name, nominate. **3** arrive at, reach, get to; achieve, accomplish, attain. **4** acquire, clear, gain, get, gross, net, obtain, reap, win; fetch, sell for. **5** add up to, amount to, come to, total. **6** calculate, estimate, gauge, reckon, think. **7** cause, compel, force, impel, induce, oblige, persuade, press, pressure, pressurize, prevail on. **8** execute, perform; give, deliver, present. ● *n.* brand, kind, sort, type, variety. □ **make believe** daydream, dream, fantasize, pretend. **make do** cope, get by, manage, muddle through, scrape by, survive. **make off** abscond, *colloq.* clear off, dash away, decamp, flee, fly, make a getaway, run away, *colloq.* scram, *colloq.* skedaddle, take to one's heels. **make out 2** descry, detect, discern, distinguish, espy, perceive, pick out, see; comprehend, fathom, figure out, follow, grasp, understand. **3** imply, insinuate, intimate, suggest. **make up 1** compose, comprise, consitute, form. **2** concoct, dream up, fabricate, invent, manufacture. **3** be reconciled, bury the hatchet. **make up for** atone for, compensate for, make amends *or* reparation for, make good. **make-up 1** cosmetics, greasepaint. **2** arrangement, composition, configuration, construction, format. **3** character, disposition, personality, temper, temperament.

maker *n.* **1** one who makes something. **2** manufacturer.

■ **1** author, creator, designer, father, inventor, originator.

makeshift *adj.* & *n.* (thing) used as an improvised substitute.

■ *adj.* improvised, provisional, stopgap, temporary.

makeweight *n.* something added to make up for a deficiency.

maladjusted *adj.* not happily adapted to one's circumstances. □ **maladjustment** *n.*

maladminister *v.* manage badly or improperly.

maladroit *adj.* bungling, clumsy.

malady *n.* illness, disease.

malaise *n.* feeling of illness or uneasiness.

malapropism *n.* comical confusion of words.

malaria *n.* disease causing a recurring fever. □ **malarial** *adj.*

Malay *adj.* & *n.* (member, language) of a people of Malaysia and Indonesia.

malcontent *n.* discontented person.

male *adj.* **1** of the sex that can fertilize egg cells produced by a female. **2** (of a plant) producing pollen, not seeds. **3** (of a screw etc.) for insertion into a corresponding hollow part. ● *n.* male person, animal, or plant.

malefactor /mállifaktər/ *n.* wrongdoer.

malevolent *adj.* wishing evil to others. □ **malevolently** *adv.*, **malevolence** *n.*

■ baleful, evil, hostile, malign, malicious, spiteful, venomous, vicious, vindictive.

malformation *n.* faulty formation. □ **malformed** *adj.*

malfunction *n.* faulty functioning. ● *v.* function faultily.

malice *n.* desire to harm others. □ **malicious** *adj.*, **maliciously** *adv.*

■ animosity, hatred, hostility, ill will, malevolence, rancour, spite, spitefulness, venom, vindictiveness. □ **malicious** *colloq.* bitchy, bitter, hurtful, malevolent, malign, malignant, mean, nasty, spiteful, unkind, venomous, vicious, vindictive, vitriolic.

malign /məlín/ *adj.* **1** harmful. **2** showing malice. ● *v.* slander.

■ *v.* blacken the reputation of, calumniate, defame, denigrate, slander, smear, traduce, vilify.

malignant *adj.* **1** showing great ill will. **2** (of a tumour) growing harmfully and uncontrollably. □ **malignancy** *n.*

malinger *v.* pretend illness to avoid work. □ **malingerer** *n.*

mall *n.* **1** sheltered walk or promenade. **2** shopping precinct.

mallard *n.* wild duck, male of which has a glossy green head.

malleable *adj.* **1** able to be hammered into shape. **2** easy to influence. □ **malleability** *n.*

■ **1** ductile, plastic. **2** adaptable, biddable, compliant, impressionable, tractable.

mallet *n.* **1** hammer, usu. of wood. **2** instrument for striking the ball in croquet or polo.

malmsey /maámzi/ *n.* strong sweet wine.

malnutrition *n.* insufficient nutrition.

malodorous *adj.* stinking.

malpractice *n.* **1** wrongdoing. **2** improper professional conduct.

malt *n.* **1** barley or other grain prepared for brewing or distilling. **2** (*colloq.*) whisky made with this. □ **malted milk** drink made from dried milk and malt.

maltreat *v.* ill-treat. □ **maltreatment** *n.*

mamba *n.* poisonous S. African snake.

mammal *n.* member of the class of animals that suckle their young. □ **mammalian** *adj.*

mammary *adj.* of the breasts.

mammoth *n.* large extinct elephant. ● *adj.* huge.

man *n.* (*pl.* **men**) **1** adult male person. **2** human being, person. **3** humankind. **4** male servant or employee. **5** (usu. *pl.*) soldier, sailor, etc. **6** piece in chess, draughts, etc. ● *v.* (**manned**) supply with people to guard or operate something. □ **man-hour** *n.* one hour's work by one person. **manhunt** *n.* organized search for a person, esp. a criminal. **man-made** *adj.* synthetic.

■ *n.* **1** *sl.* bloke, *colloq.* chap, *colloq.* fellow, gentleman, *colloq.* guy, male. **2** human being, individual, mortal, person, soul. **3** human beings, humanity, humankind, mankind, the human race. **4** batman, footman, manservant, valet; employee, hand, worker.

manacle *n.* & *v.* handcuff.

manage *v.* **1** have control of. **2** contrive. **3** deal with (a person) tactfully. **4** succeed with limited resources. **5** operate (a tool etc.) effectively. □ **manageable** *adj.*

■ **1** administer, be in charge of, conduct, direct, govern, handle, head, organize, oversee, preside over, run, supervise. **2** contrive, succeed. **3** control, cope with, deal with, handle. **4** cope, get by, make do, make ends meet, muddle through, scrape by, survive. **5** handle, use, wield.

management *n.* **1** managing. **2** people managing a business.

■ **1** administration, conduct, control, direction, handling, operation, running. **2** *colloq.* bosses, directorate, directors, executives.

manager *n.* person in charge of a business etc. □ **managerial** *adj.*

■ administrator, *colloq.* boss, director, employer, head, overseer, superintendent, supervisor.

manageress *n.* woman manager, esp. of a shop, hotel, etc.

manatee *n.* large tropical aquatic mammal.

Mandarin *n.* northern variety of the Chinese language.

mandarin *n.* **1** senior influential official. **2** small orange.

mandate *n.* & *v.* (give) authority to perform certain tasks.

mandatory *adj.* compulsory.

■ compulsory, essential, necessary, obligatory, required, requisite.

mandible *n.* jaw or jaw-like part.

mandolin *n.* guitar-like musical instrument.

mandrake *n.* poisonous plant with a large yellow fruit.

mandrill *n.* large baboon.

mane *n.* long hair on a horse's or lion's neck.

manganese *n.* hard brittle grey metal or its black oxide.

mange *n.* skin disease affecting hairy animals.

mangel-wurzel *n.* large beet used as cattle food.

manger *n.* open trough for horses or cattle to feed from.

mangle *v.* damage by cutting or crushing roughly, mutilate.

■ crush, damage, disfigure, lacerate, maim, mutilate.

mango *n.* (*pl.* **-oes**) **1** tropical fruit with juicy flesh. **2** tree bearing it.

mangrove *n.* tropical tree growing in shore-mud and swamps.

mangy *adj.* **1** having mange. **2** squalid.

manhandle *v.* **1** move by human effort alone. **2** treat roughly.

manhole *n.* opening through which a person can enter a drain etc. to inspect it.

manhood *n.* **1** state of being a man. **2** manly qualities.

mania *n.* **1** violent madness. **2** extreme enthusiasm, obsession.

■ **1** dementia, derangement, insanity, lunacy, madness. **2** craze, fad, fixation, obsession, passion.

maniac *n.* person with a mania.

■ lunatic, psychopath; enthusiast, fan, fanatic, fiend.

maniacal *adj.* of or like a mania or maniac.

manic *adj.* of or affected by mania.

manicure *n.* cosmetic treatment of fingernails. ● *v.* apply such treatment to. □ **manicurist** *n.*

manifest *adj.* clear and unmistakable. ● *v.* show clearly, give signs of. ● *n.* list of cargo or passengers carried by a ship or aircraft. □ **manifestation** *n.*

■ *adj.* apparent, blatant, clear, conspicuous, discernible, evident, obvious, palpable, patent, perceptible, plain, unambiguous, unmistakable, visible. ● *v.* demonstrate, display, evince, exhibit, indicate, reveal, show. □ **manifestation** demonstration, display, exhibition, indication, show, sign.

manifesto *n.* (*pl.* **-os**) public declaration of policy.

manifold *adj.* of many kinds. ● *n.* (in a machine) pipe or chamber with several openings.

manikin *n.* little man, dwarf.

manila *n.* brown paper used for wrapping and for envelopes.

manipulate *v.* handle or manage skilfully or cunningly. □ **manipulation** *n.*, **manipulator** *n.*

■ control, handle, operate, use; exploit, influence, manoeuvre; *colloq.* cook, doctor, falsify, *colloq.* fiddle, juggle, rig, tamper with.

mankind *n.* humankind.

manly *adj.* **1** brave, strong. **2** considered suitable for a man. □ **manliness** *n.*

mannequin /mánnikin/ *n.* **1** fashion model. **2** dummy for display of clothes.

manner *n.* **1** way a thing is done or happens. **2** person's way of behaving towards others. **3** kind, sort. **4** (*pl.*) polite behaviour.

■ **1** fashion, method, mode, procedure, style, technique, way. **2** air, attitude, bearing, behaviour, conduct, demeanour. **3** class, kind, sort, type, variety. **4** (**manners**) decorum, etiquette, (good) form, politeness.

mannered *adj.* **1** having manners of a certain kind. **2** stilted.

mannerism *n.* distinctive personal habit or way of doing something.

■ characteristic, foible, habit, idiosyncrasy, peculiarity, quirk, trick.

manoeuvre *n.* **1** planned movement of a vehicle, troops, etc. **2** skilful or crafty proceeding. ● *v.* **1** perform manoeuvre(s). **2** move skilfully. **3** manipulate by scheming. □ **manoeuvrable** *adj.*

■ *n.* **1** deployment, movement. **2** dodge, gambit, move, plan, plot, ploy, scheme, stratagem, tactic, trick. ● *v.* **1** guide, navigate, steer. **2** contrive, engineer, manipulate, *colloq.* wangle.

manor *n.* large country house, usu. with lands. □ **manorial** *adj.*

manpower *n.* number of people available for work or service.

manservant *n.* (*pl.* **menservants**) male servant.

mansion *n.* large stately house.

manslaughter *n.* act of killing a person unlawfully but not intentionally.

mantelpiece *n.* shelf above a fireplace.

mantilla *n.* Spanish lace veil worn over a woman's hair and shoulders.

mantis *n.* grasshopper-like insect.

mantle *n.* **1** loose cloak. **2** covering.

manual *adj.* **1** of the hands. **2** done or operated by the hand(s). ● *n.* handbook. □ **manually** *adv.*

manufacture *v.* **1** make or produce (goods) on a large scale by machinery. **2** invent. ● *n.* process of manufacturing. □ **manufacturer** *n.*

■ *v.* **1** assemble, construct, fabricate, make, produce. **2** concoct, fabricate, invent, make up. ● *n.* assembly, building, construction, fabrication, production.

manure *n.* substance, esp. dung, used as a fertilizer. ● *v.* apply manure to.

manuscript *n.* book or document written by hand or typed, not printed.

Manx *adj.* & *n.* (language) of the Isle of Man.

many *adj.* numerous, great in number. ● *n.* many people or things.

■ *adj.* a lot of, countless, innumerable, lots of, multitudinous, numerous, *sl.* umpteen. ● *n.* a lot, crowds, droves, *colloq.* heaps, hordes, *colloq.* loads, lots, masses, multitudes, *colloq.* oodles, plenty, quantities, *colloq.* stacks.

Maori /mówrī/ *n.* & *adj.* (*pl.* **Maori** or **-is**) (member, language) of the indigenous race of New Zealand.

map *n.* representation of earth's surface or a part of it. ● *v.* (**mapped**) make a map of. □ **map out** plan in detail.

maple *n.* tree with broad leaves.

mar *v.* (**marred**) damage, spoil.

■ blemish, damage, disfigure, impair, spoil, stain, tarnish.

maraca *n.* club-like gourd containing beads etc., shaken as a musical instrument.

marathon *n.* **1** long-distance foot race. **2** long test of endurance.

marauding *adj.* & *n.* pillaging. □ **marauder** *n.*

marble *n.* **1** a kind of limestone that can be polished. **2** piece of sculpture in this. **3** small ball of glass or clay used in children's games. ● *v.* give a veined or mottled appearance to.

marcasite *n.* crystals of iron pyrites, used in jewellery.

march *v.* **1** walk in a regular rhythm or an organized column. **2** walk purposefully. **3** cause to march or walk. **4** take part in a protest march. ● *n.* **1** act or instance of marching. **2** procession as a protest or demonstration. **3** progress. **4** music suitable for marching to. □ **marcher** *n.*

■ *v.* **1** file, parade. **2** stalk, stride, walk. ● *n.* **1** parade, procession, walk. **2** demonstration, parade, rally. **3** advance, passage, progress.

marchioness *n.* **1** wife or widow of a marquess. **2** woman with the rank of marquess.

mare *n.* female of the horse or a related animal.

margarine *n.* substance made from animal or vegetable fat and used like butter.

marge *n.* (*colloq.*) margarine.

margin *n.* **1** edge or border of a surface. **2** blank space round the edges of a page.

3 amount by which a thing exceeds, falls short, etc.

> **■ 1** border, brink, edge, lip, perimeter, rim, side, verge.

marginal *adj.* **1** of or in a margin. **2** insignificant. □ **marginally** *adv.*

> **■ 2** insignificant, minimal, negligible, slight, small.

marginalize *v.* make or treat as insignificant. □ **marginalization** *n.*

marguerite *n.* large daisy.

marigold *n.* plant with golden or bright yellow flowers.

marijuana /márrihwaánə/ *n.* dried hemp, smoked as a drug.

marimba *n.* a kind of xylophone.

marina *n.* harbour for yachts and pleasure boats.

marinade *n.* flavoured liquid in which meat or fish is steeped before cooking. ● *v.* steep in a marinade.

marine *adj.* **1** of the sea. **2** of shipping. ● *n.* **1** soldier trained to serve on land or sea. **2** a country's shipping.

mariner *n.* sailor, seaman.

marionette *n.* puppet worked by strings.

marital *adj.* of marriage.

maritime *adj.* **1** living or found near the sea. **2** of seafaring.

marjoram *n.* fragrant herb.

mark[1] *n.* **1** thing that visibly breaks the uniformity of a surface. **2** distinguishing feature of a person or animal. **3** thing indicating the presence of a quality or feeling etc. **4** symbol. **5** point given for merit. **6** lasting effect. **7** object, target, goal. **8** line or object serving to indicate a position. **9** numbered design of a piece of equipment etc. ● *v.* **1** make a mark on. **2** characterize. **3** assign marks of merit to. **4** pay attention to. **6** keep close to (an opponent in football etc.). □ **mark time** move the feet as if marching but without advancing.

> **■** *n.* **1** blemish, blot, blotch, bruise, scar, smear, smudge, splodge, splotch, spot, stain, streak; dent, nick, scratch. **2** blaze, marking, stripe; birthmark. **3** indication, sign, symbol, token. **4** brand, emblem, hallmark, logo, seal, symbol, trademark, watermark. **5** grade, rating, score. **6** effect, impact, impression. **7** aim, end, goal, object, objective, purpose, target. **8** indicator, landmark, marker, milestone, signpost. ● *v.* **1** blemish, smear, smudge, splodge, splotch, spot, stain, streak; nick, scratch. **2** characterize, distinguish. **3** appraise, assess, correct, evaluate, grade. **4** attend to, heed, mind, note, pay attention to, take notice of.

mark[2] *n.* Deutschmark.

marked *adj.* clearly noticeable. □ **markedly** *adv.*

> **■** conspicuous, decided, distinct, noticeable, perceptible, pronounced, unmistakable.

marker *n.* person or object that marks something.

market *n.* **1** gathering or place for the sale of provisions, livestock, etc. **2** demand (for a commodity). ● *v.* **1** sell in a market. **2** offer for sale. □ **market garden** one in which vegetables are grown for market. **on the market** offered for sale.

marking *n.* **1** mark(s). **2** colouring of feathers, fur, etc.

marksman *n.* person who is a skilled shot. □ **marksmanship** *n.*

marl *n.* soil composed of clay and lime, used as a fertilizer.

marmalade *n.* preserve made from citrus fruit, esp. oranges.

marmoset *n.* small bushy-tailed monkey of tropical America.

marmot *n.* small burrowing animal of the squirrel family.

maroon[1] *n.* **1** brownish-red colour. **2** explosive device used as a warning signal. ● *adj.* brownish-red.

maroon[2] *v.* **1** put and leave (a person) ashore in a desolate place. **2** leave stranded.

marquee *n.* large tent used for a party or exhibition etc.

marquess *n.* nobleman ranking between duke and earl.

marquetry *n.* inlaid work in wood etc.

marquis *n.* foreign nobleman ranking between duke and count.

marram *n.* type of grass that grows esp. in sand.

marriage *n.* **1** legal union of a man and woman. **2** act or ceremony of marrying.

> **■ 1** matrimony, wedlock. **2** nuptials, wedding.

marriageable *adj.* suitable or old enough for marriage.

marrow *n.* **1** soft fatty substance in the cavities of bones. **2** large gourd used as a vegetable.

marry *v.* **1** take, join, or give in marriage. **2** unite (things).

> **■ 1** espouse, get married to, wed. **2** amalgamate, combine, couple, join, link, unify, unite.

marsh *n.* low-lying watery ground. ◻ **marshy** *adj.*

■ bog, fen, morass, quagmire, slough, swamp.

marshal *n.* **1** high-ranking officer. **2** official controlling an event or ceremony. ● *v.* (**marshalled**) **1** arrange in proper order. **2** assemble. **3** usher.

marshmallow *n.* soft sweet made from sugar, egg white, and gelatine.

marsupial *n.* mammal that usu. carries its young in a pouch.

mart *n.* market.

marten *n.* weasel-like animal with thick soft fur.

martial *adj.* **1** of war. **2** warlike. ◻ **martial law** military government suspending ordinary law.

■ **1** military, soldierly. **2** aggressive, bellicose, belligerent, militant, pugnacious, warlike.

martin *n.* bird of the swallow family.

martinet *n.* person who demands strict obedience.

martyr *n.* person who undergoes death or suffering for his or her beliefs. ● *v.* kill or torment as a martyr. ◻ **martyrdom** *n.*

marvel *n.* wonderful thing. ● *v.* (**marvelled**) feel wonder.

■ *n.* miracle, phenomenon, sensation, wonder. ● *v.* be amazed *or* awed, wonder.

marvellous *adj.* wonderful. ◻ **marvellously** *adv.*

■ amazing, astounding, breathtaking, extraordinary, incredible, magnificent, miraculous, outstanding, phenomenal, remarkable, sensational, stupendous, superb, wonderful; excellent, *colloq.* fabulous, *colloq.* fantastic, glorious, *colloq.* great, *colloq.* heavenly, *colloq.* smashing, splendid, *colloq.* super, *colloq.* terrific, *colloq.* tremendous.

Marxism *n.* socialist theories of Karl Marx. ◻ **Marxist** *adj.* & *n.*

marzipan *n.* edible paste made from ground almonds.

mascara *n.* cosmetic for darkening the eyelashes.

mascot *n.* thing believed to bring good luck to its owner.

masculine *adj.* **1** of, like, or traditionally considered suitable for men. **2** of the grammatical form suitable for the names of males. ● *n.* masculine word. ◻ **masculinity** *n.*

mash *n.* **1** soft mixture of grain or bran. **2** mashed potatoes. ● *v.* beat into a soft mass.

mask *n.* **1** covering worn over the face as a disguise or protection. **2** disguise. ● *v.* **1** cover with a mask. **2** conceal.

■ *n.* **2** camouflage, cover-up, disguise, guise, pretence, semblance, show. ● *v.* **2** camouflage, cloak, conceal, cover (up), disguise, hide, screen, shroud, veil.

masochism *n.* pleasure in suffering pain. ◻ **masochist** *n.*, **masochistic** *adj.*

mason *n.* person who builds or works with stone.

masonry *n.* stonework.

masquerade *n.* false show or pretence. ● *v.* pretend to be what one is not.

mass[1] *n.* **1** celebration (esp. in the RC Church) of the Eucharist. **2** form of liturgy used in this.

mass[2] *n.* **1** quantity of matter without a regular shape. **2** large quantity, heap, or expanse. **3** quantity of matter a body contains. **4** (**the masses**) ordinary people. ● *v.* gather or assemble into a mass. ◻ **mass-produce** *v.* manufacture in large quantities by a standardized process.

■ *n.* **1** block, chunk, concretion, lump, piece. **2** crowd, herd, horde, host, multitude, swarm, throng; abundance, accumulation, assortment, bunch, collection, conglomeration, heap, mound, mountain, pile, profusion, quantity, *colloq.* stack. ● *v.* accumulate, amass, collect, gather, pile up; assemble, congregate, convene, flock together, group; marshal, mobilize, rally.

massacre *n.* great slaughter. ● *v.* slaughter in large numbers.

■ *n.* annihilation, butchery, killing, slaughter. ● *v.* annihilate, butcher, exterminate, kill, murder, slaughter, slay.

massage *n.* rubbing and kneading the body to reduce pain or stiffness. ● *v.* treat in this way.

masseur *n.* (*fem.* **masseuse**) person who practises massage professionally.

massive *adj.* **1** large and heavy or solid. **2** huge. ◻ **massively** *adv.*

■ big, bulky, colossal, enormous, gargantuan, gigantic, huge, immense, large, mammoth, mighty, monstrous, monumental, mountainous, titanic, vast.

mast[1] *n.* tall pole, esp. supporting a ship's sails.

mast[2] *n.* fruit of beech, oak, etc., used as food for pigs.

mastectomy *n.* surgical removal of a breast.

master *n.* **1** man who has control of people or things. **2** male teacher. **3** person with very great skill, great artist. **4** thing from which a series of copies is made. **6** (**Master**) title of a boy not old enough to be called *Mr.* ● *adj.* **1** superior. **2** principal. **3** controlling others. ● *v.* **1** bring under control. **2** acquire knowledge or skill in. □ **master-key** *n.* key that opens a number of different locks. **master-stroke** *n.* very skilful act of policy.

■ *n.* **1** *colloq.* boss, captain, chief, commander, governor, head, leader, owner, ruler, skipper. **2** instructor, schoolmaster, teacher, tutor. **3** adept, authority, expert, genius, maestro, mastermind, virtuoso, wizard. ● *adj.* **1** excellent, exceptional, expert, outstanding, superior, supreme. **2** chief, main, principal. ● *v.* **1** check, conquer, control, curb, defeat, overcome, quell, repress, subdue, subjugate, suppress. **2** become proficient in, *colloq.* get the hang of, grasp, learn.

masterful *adj.* domineering. □ **masterfully** *adv.*

masterly *adj.* very skilful.

mastermind *n.* **1** person of outstanding mental ability. **2** one directing an enterprise. ● *v.* plan and direct.

masterpiece *n.* outstanding piece of work.

mastery *n.* **1** control, supremacy. **2** thorough knowledge or skill.

■ **1** ascendancy, command, control, dominance, supremacy, superiority, the upper hand. **2** command, comprehension, grasp, knowledge, understanding.

mastic *n.* **1** gum or resin from certain trees. **2** a kind of cement.

masticate *v.* chew. □ **mastication** *n.*

mastiff *n.* large strong dog.

mastodon *n.* extinct animal resembling an elephant.

mastoid *n.* part of a bone behind the ear.

masturbate *v.* stimulate the genitals (of) manually. □ **masturbation** *n.*

mat *n.* piece of material placed on a floor or other surface as an ornament or to protect it. ● *v.* (**matted**) make or become tangled into a thick mass.

matador *n.* bullfighter.

match[1] *n.* short stick tipped with material that catches fire when rubbed on a rough surface.

match[2] *n.* **1** contest in a game or sport. **2** person or thing exactly like or corresponding or equal to another. **3** matrimonial alliance. ● *v.* **1** equal in ability or achievement. **2** be alike. **3** find a match for. □ **match against** or **with** place in competition with.

■ *n.* **1** bout, competition, contest, game, tournament. **2** equal, equivalent, like, peer; copy, counterpart, double, duplicate, twin. **3** marriage, partnership, union. ● *v.* **1** be in the same league as, be on a par with, compare with, equal, measure up to, rival, touch. **2** agree, correspond, tally.

matchmaking *n.* scheming to arrange marriages. □ **matchmaker** *n.*

matchstick *n.* stick of a match.

matchwood *n.* **1** wood that splinters easily. **2** wood broken into splinters.

mate[1] *n.* **1** companion or fellow worker. **2** male or female of mated animals. **3** merchant ship's officer. ● *v.* come or bring (animals) together to breed.

■ *n.* **1** *colloq.* buddy, *colloq.* chum, companion, crony, friend, *colloq.* pal; associate, colleague, comrade, fellow worker.

mate[2] *n.* checkmate.

material *n.* **1** that from which something is or can be made. **2** cloth, fabric. ● *adj.* **1** of matter. **2** not spiritual. **3** significant. □ **materially** *adv.*

■ *n.* **1** matter, stuff, substance. **2** cloth, fabric, textile. ● *adj.* **1** concrete, palpable, physical, real, solid, tangible. **2** earthly, mundane, secular, temporal, worldly. **3** consequential, important, relevant, significant.

materialism *n.* **1** belief that only the material world exists. **2** excessive concern with material possessions. □ **materialist** *n.*, **materialistic** *adj.*

materialize *v.* **1** appear, become visible. **2** become a fact, happen. □ **materialization** *n.*

■ **1** appear, become visible, emerge, take shape. **2** come about, happen, occur, take place.

maternal *adj.* **1** of a mother. **2** motherly. **3** related through one's mother. □ **maternally** *adv.*

■ **2** affectionate, caring, gentle, kind, loving, motherly, nurturing, protective, tender, warm.

maternity *n.* motherhood. ● *adj.* of or for women in pregnancy and childbirth.

mathematician *n.* person skilled in mathematics.

mathematics *n.* science of numbers, quantities, and measurements. ▫ **mathematical** *adj.*
maths *n.* mathematics.
matinée *n.* afternoon performance. ▫ **matinée coat** baby's jacket.
matriarch *n.* female head of a family or tribe. ▫ **matriarchal** *adj.*
matriarchy *n.* social organization in which a female is head of the family.
matricide *n.* killing or killer of own mother. ▫ **matricidal** *adj.*
matriculate *v.* admit or be admitted to a university. ▫ **matriculation** *n.*
matrimony *n.* marriage. ▫ **matrimonial** *adj.*
matrix *n.* (*pl.* **matrices**) **1** mould in which a thing is cast or shaped. **2** rectangular array of mathematical quantities.
matron *n.* **1** married woman. **2** woman in charge of domestic affairs or nursing in a school etc.
matronly *adj.* like or characteristic of a staid or dignified married woman.
matt *adj.* dull, not shiny.
matter *n.* **1** that which occupies space in the visible world. **2** specified substance, material, or things. **3** business etc. being discussed. **4** pus. ● *v.* be of importance. ▫ **what is the matter?** what is amiss?

> ■ *n.* **2** material, stuff, substance. **3** affair, business, concern, issue, question, subject, topic. ● *v.* be important, be of consequence, count, signify.

mattress *n.* fabric case filled with padding or springy material, used on or as a bed.
maturation *n.* maturing.
mature *adj.* **1** fully grown or developed. **2** (of a bill etc.) due for payment. ● *v.* make or become mature. ▫ **maturity** *n.*

> ■ *adj.* **1** adult, developed, experienced, fully-fledged, fully-grown, grown-up, of age; mellow, ready, ripe, ripened. ● *v.* age, grow up, mellow, ripen.

maudlin *adj.* sentimental in a silly or tearful way.

> ■ mawkish, mushy, sentimental, *colloq.* soppy.

maul *v.* treat roughly, injure by tearing flesh.

> ■ maltreat, manhandle, mistreat, molest; claw, lacerate, mutilate, savage.

maunder *v.* **1** talk in a dreamy or rambling way. **2** move idly.
mausoleum *n.* magnificent tomb.
mauve /mōv/ *adj.* & *n.* pale purple.

maverick *n.* unorthodox or undisciplined person.
mawkish *adj.* sentimental in a sickl[y] way.
maxim *n.* sentence giving a general trut[h] or rule of conduct.

> ■ aphorism, axiom, motto, proverb, saw[,] saying.

maximize *v.* increase to a maximum. ▫ **maximization** *n.*
maximum *adj.* & *n.* (*pl.* **-ima**) greates[t] (amount) possible. ▫ **maximal** *adj.* **maximally** *adv.*

> ■ *adj.* greatest, highest, most, top, topmost, utmost, uttermost. ● *n.* limit, peak[,] top, utmost, uttermost.

may[1] *v.aux.* (**might**) used to express [a] wish, possibility, or permission.
may[2] *n.* hawthorn blossom.
maybe *adv.* perhaps.

> ■ conceivably, perhaps, possibly.

mayday *n.* international radio signal o[f] distress.
May Day 1 May, esp. as a festival.
mayfly *n.* insect with long hair-like tails[,] living in spring.
mayhem *n.* violent action.
mayonnaise *n.* creamy sauce made wit[h] eggs and oil.
mayor *n.* head of the municipal corporation of a city or borough. ▫ **mayoral** *adj.* **mayoralty** *n.*
mayoress *n.* **1** female mayor. **2** mayor's wife, or other woman with her ceremonial duties.
maypole *n.* tall pole for dancing roun[d] on May Day.
maze *n.* complex and baffling network o[f] paths, lines, etc.
me *pron.* objective case of *I*.
mead *n.* alcoholic drink of fermente[d] honey and water.
meadow *n.* field of grass.

> ■ field, *poetic* lea, paddock.

meagre *adj.* scant in amount.

> ■ inadequate, insufficient, *sl.* measly[,] *colloq.* pathetic, poor, scant, scanty[,] skimpy, sparse.

meal[1] *n.* **1** occasion when food is eaten. **2** the food itself.

> ■ banquet, feast, *formal* repast, *colloq.* spread; bite, snack.

meal[2] *n.* coarsely ground grain.
mealy *adj.* of or like meal. ▫ **mealy mouthed** *adj.* afraid to speak plainly.

mean[1] *adj.* **1** miserly, not generous. **2** ignoble. **3** of low degree or poor quality. **4** malicious, unkind. **5** (*US*) vicious. □ **meanly** *adv.*, **meanness** *n.*

■ **1** close, miserly, near, niggardly, parsimonious, penny-pinching, stingy, *colloq.* tight, tight-fisted, ungenerous. **2** base, dishonourable, ignoble, low, small-minded. **3** humble, inferior, lowly; miserable, poor, *colloq.* scruffy, seedy, shabby, squalid, wretched. **4** cruel, malicious, nasty, spiteful, unkind.

mean[2] *adj.* & *n.* **1** (thing) midway between two extremes. **2** average.

mean[3] *v.* (**meant**) **1** have as one's purpose. **2** have as an equivalent in another language, signify. **3** entail, involve. **4** portend, be likely to result in. **5** be of specified importance.

■ **1** aim, contemplate, design, intend, plan, propose, purpose. **2** connote, denote, indicate, represent, signify. **3** entail, imply, involve. **4** augur, foreshadow, portend, presage, promise.

meander *v.* **1** follow a winding course. **2** wander in a leisurely way. ● *n.* winding course.

■ *v.* **1** snake, turn, twist, wind. **2** amble, drift, ramble, stroll, wander.

meaning *n.* **1** what is meant. **2** significance. ● *adj.* expressive.

■ *n.* **1** connotation, denotation, drift, gist, import, purport, sense, signification, substance. **2** consequence, importance, significance, value, worth.

meaningful *adj.* **1** full of meaning. **2** significant.

■ **1** expressive, pointed, pregnant, suggestive. **2** consequential, important, serious, significant.

meaningless *adj.* having no meaning or significance.

■ empty, hollow, inconsequential, insignificant, pointless, purposeless, senseless, trivial, unimportant, valueless, worthless.

means *n.pl* (often treated as *sing.*) **1** that by which a result is brought about. **2** resources. □ **by all means** certainly. **by no means** not at all. **means test** official inquiry to establish need before giving financial help from public funds.

■ **1** agency, manner, medium, method, mode, process, technique, vehicle, way. **2** capital, cash, funds, money, resources, *colloq.* wherewithal.

meant *see* **mean**[3].

meantime *adv.* meanwhile.

meanwhile *adv.* **1** in the intervening period. **2** at the same time.

measles *n.* infectious disease producing red spots on the body.

measly *adj.* (*sl.*) meagre.

measure *n.* **1** size or quantity found by measuring. **2** unit, device, or system used in measuring. **3** rhythm. **4** (usu. *pl.*) action taken for a purpose. **5** (proposed) law. ● *v.* **1** find the size etc. of by comparison with a known standard. **2** be of a certain size. □ **measure out** distribute in measured quantities. **measure up to** reach the standard required by. **measurable** *adj.*, **measurably** *adv.*

■ *n.* **1** amount, amplitude, bulk, dimension(s), extent, magnitude, measurement(s), quantity, proportions, scope, size, weight; breadth, depth, height, length, width; capacity, volume. **2** gauge, rule, ruler, tape-measure; scale, system. **3** beat, cadence, metre, rhythm. **4** (**measures**) action, expedients, procedures, steps, tactics. **5** act, bill, statute. ● *v.* **1** ascertain, assess, calculate, calibrate, compute, determine, evaluate, gauge, size up, weigh. □ **measure out** allocate, allot, apportion, dispense, distribute, deal out, dole out, give out, hand out, mete out, share out. **measure up to** be equal to, equal, fulfil, match, meet.

measured *adj.* **1** rhythmical. **2** carefully considered.

measurement *n.* **1** measuring. **2** size etc. found by measuring.

■ **1** assessment, calculation, computation, evaluation, gauging, mensuration. **2** amount, amplitude, dimension, extent, magnitude, measure, size; breadth, depth, height, length, width; capacity, volume; acreage, yardage.

meat *n.* animal flesh as food.

meaty *adj.* (**-ier, -iest**) **1** like meat. **2** full of meat. **3** full of subject-matter. □ **meatiness** *n.*

mechanic *n.* person skilled in using or repairing machinery.

mechanical *adj.* **1** of or worked by machinery. **2** done without conscious thought. □ **mechanically** *adv.*

■ **1** automated, robotic. **2** automatic, instinctive, involuntary, knee-jerk, unconscious, unthinking.

mechanics *n.* **1** study of motion and force. **2** science of machinery. **3** (as *pl.*) way a thing works.

mechanism *n.* **1** structure or parts of a machine. **2** system of parts working together.

mechanize *v.* equip with machinery. □ **mechanization** *n.*

medal *n.* coin-like piece of metal commemorating an event or awarded for an achievement.

medallion *n.* **1** large medal. **2** circular ornamental design.

medallist *n.* winner of a medal.

meddle *v.* interfere in people's affairs.

■ butt in, interfere, intervene, intrude, pry.

meddlesome *adj.* often meddling.

media *see* **medium**. *n.pl.* **(the media)** newspapers and broadcasting as conveying information to the public.

mediaeval *adj.* = **medieval**.

medial *adj.* situated in the middle. □ **medially** *adv.*

median *adj.* in or passing through the middle. ● *n.* median point or line.

mediate *v.* **1** act as peacemaker between disputants. **2** bring about (a settlement) in this way. □ **mediation** *n.*, **mediator** *n.*

■ **1** arbitrate, intercede, intervene, liaise, negotiate. □ **mediator** arbitrator, arbiter, go-between, intermediary, middleman, negotiator, peacemaker.

medical *adj.* of the science of medicine. ● *n.* (*colloq.*) medical examination. □ **medically** *adv.*

medicament *n.* any medicine, ointment, etc.

medicate *v.* treat with a medicinal substance.

medication *n.* **1** medicinal drug. **2** treatment using drugs.

medicinal *adj.* having healing properties. □ **medicinally** *adv.*

■ healing, health-giving, restorative, therapeutic.

medicine *n.* **1** science of the prevention and cure of disease. **2** substance used to treat disease. □ **medicine man** witch-doctor.

■ **2** drug, medicament, medication, remedy.

medieval *adj.* of the Middle Ages.

mediocre *adj.* **1** of medium quality. **2** second-rate. □ **mediocrity** *n.*

■ **1** average, commonplace, fair, indifferent, middling, ordinary, pedestrian, run-of-the-mill, undistinguished, unremarkable. **2** inferior, poor, second-rate.

meditate *v.* think deeply. □ **meditation** *n.*, **meditative** *adj.*

■ cogitate, muse, ponder, reflect, ruminate, think.

medium *n.* (*pl.* **-dia** or **-s**) **1** middle size, quality, etc. **2** substance or surroundings in which a thing exists. **3** agency, means. ● *adj.* **1** intermediate. **2** average.

■ *n.* **2** ambience, atmosphere, environment, milieu. **3** agency, avenue, channel, means, method, mode, vehicle. ● *adj.* **1** intermediate, median, mid, middle. **2** average, middling, standard.

medley *n.* (*pl.* **-eys**) assortment.

■ assortment, collection, hotchpotch, miscellany, mixture, pot-pourri.

medulla *n.* **1** spinal or bone marrow. **2** hindmost segment of the brain. □ **medullary** *adj.*

meek *adj.* quiet and obedient, not protesting. □ **meekly** *adv.*, **meekness** *n.*

■ acquiescent, compliant, deferential, docile, humble, manageable, obedient, quiet, submissive, timid, tractable, unassuming.

meerschaum /méershəm/ *n.* **1** tobacco pipe with a white clay bowl. **2** this clay.

meet *v.* **(met) 1** come into contact (with). **2** come together. **3** make the acquaintance of. **4** experience. **5** satisfy (needs etc.). **6** be present at the arrival of. ● *n.* assembly for a hunt etc.

■ *v.* **1** come across, encounter, happen on, run into. **2** converge, intersect, join, link up, touch; assemble, collect, congregate, convene, gather. **3** be introduced to, make the acquaintance of. **4** encounter, endure, experience, go through, undergo. **5** comply with, fulfil, measure up to, satisfy. **6** greet, receive, welcome.

meeting *n.* **1** coming together. **2** an assembly for discussion.

■ **1** appointment, assignation, *colloq.* date, encounter, engagement, rendezvous; confluence, conjunction, convergence, intersection, junction. **2** assembly, conference, convention, convocation, gathering, *colloq.* get-together, session.

mega- *pref.* **1** large. **2** one million (as in *megavolts, megawatts*).

megabyte *n.* one million bytes.

megalith *n.* large stone, esp. as a prehistoric monument. □ **megalithic** *adj.*

megalomania *n.* excessive self-esteem, esp. as a form of insanity. □ **megalomaniac** *adj.* & *n.*

megaphone *n.* funnel-shaped device for amplifying the voice.

megaton *n.* unit of explosive power equal to one million tons of TNT.

melancholy *n.* mental depression, sadness. ● *adj.* sad, gloomy.

■ *n.* dejection, depression, despondency, gloom, misery, sadness, sorrow, unhappiness. ● *adj.* blue, cheerless, crestfallen, dejected, depressed, despondent, downcast, downhearted, gloomy, glum, forlorn, heavy-hearted, in low spirits, in the doldrums, lugubrious, miserable, morose, mournful, sad, sorrowful, unhappy, woebegone, woeful.

melanin *n.* dark pigment in the skin, hair, etc.

mêlée /méllay/ *n.* **1** confused fight, scuffle. **2** muddle.

mellifluous *adj.* sweet-sounding.

mellow *adj.* **1** (of fruit) ripe and sweet. **2** (of sound or colour) soft and rich. **3** (of people) having become kindly, e.g. with age. ● *v.* make or become mellow.

■ *adj.* **1** mature, ripe, sweet. **2** deep, full, resonant, rich, sonorous; muted, soft, subtle. **3** easygoing, genial, good-natured, friendly, kind, pleasant, warm. ● *v.* age, mature, ripen.

melodic *adj.* **1** of melody **2** melodious. □ **melodically** *adv.*

melodious *adj.* full of melody. □ **melodiously** *adv.*

■ dulcet, euphonious, harmonious, melodic, musical, silvery, sweet-sounding, tuneful.

melodrama *n.* sensational drama. □ **melodramatic** *adj.*, **melodramatically** *adv.*

■ □ **melodramatic** exaggerated, overdone, over-dramatic, sensational, theatrical.

melody *n.* **1** (piece of) sweet music. **2** main part in a piece of harmonized music.

■ **1** air, song, strain, tune; euphony, harmony, melodiousness, tunefulness.

melon *n.* large sweet fruit.

melt *v.* **1** make into or become liquid, esp. by heat. **2** soften through pity or love. **3** disappear unobtrusively.

■ **1** dissolve, liquefy, thaw. **2** disarm, mellow, soften, thaw. **3** disappear, disperse, evaporate, fade (away), vanish.

meltdown *n.* melting of an overheated reactor core.

member *n.* person or thing belonging to a particular group or society. □ **membership** *n.*

membrane *n.* thin flexible skin-like tissue. □ **membranous** *adj.*

memento *n.* (*pl.* **-oes**) souvenir.

■ keepsake, reminder, souvenir, token.

memo *n.* (*pl.* **-os**) (*colloq.*) memorandum.

memoir /mémwaar/ *n.* historical account written from personal knowledge.

■ account, biography, chronicle, journal, life, record.

memorable *adj.* **1** worth remembering. **2** easy to remember. □ **memorably** *adv.*

■ **1** great, historic, important, momentous, notable, noteworthy, remarkable, significant. **2** catchy, unforgettable.

memorandum *n.* **1** (*pl.* **-da**) note written as a reminder. **2** (*pl.* **-dums**) written message from one colleague to another.

memorial *n.* object or custom etc. established in memory of an event or person(s). ● *adj.* serving as a memorial.

■ *n.* cairn, cenotaph, monument, plaque, statue.

memorize *v.* learn (a thing) so as to know it from memory.

memory *n.* **1** ability to remember things. **2** thing(s) remembered. **3** computer store for data etc.

■ **2** recollection, remembrance, reminiscence.

men *see* **man.**

menace *n.* **1** threat. **2** annoying or troublesome person or thing. ● *v.* threaten. □ **menacingly** *adv.*

■ *n.* **1** danger, hazard, peril, risk, threat. **2** annoyance, nuisance, pest, troublemaker. ● *v.* browbeat, bully, cow, intimidate, threaten, terrorize.

ménage /maynaázh/ *n.* household.

menagerie *n.* small zoo.

mend *v.* **1** repair. **2** make or become better. ● *n.* repaired place. □ **on the mend** recovering after illness.

■ *v.* **1** darn, fix, patch (up), repair. **2** correct, improve, put *or* set right, rectify, remedy; convalesce, get better, heal, recover, recuperate. □ **on the mend** convalescent, convalescing, improving, recovering, recuperating.

mendacious *adj.* untruthful. □ **mendacity** *n.*

mendicant *adj.* & *n.* (person) living by begging.

menfolk *n.* **1** men in general. **2** men of one's family.

menhir /ménheer/ *n.* tall stone set up in prehistoric times.

menial *adj.* lowly, degrading. ● *n.* person who does menial tasks. □ **menially** *adv.*

■ *adj.* **degrading, demeaning, humble, lowly, servile.** ● *n.* **drudge,** *colloq.* **dogsbody, lackey, minion, slave, underling.**

meningitis *n.* inflammation of the membranes covering the brain and spinal cord.

meniscus *n.* **1** curved surface of a liquid. **2** lens convex on one side and concave on the other.

menopause *n.* **1** ceasing of menstruation. **2** time in a woman's life when this occurs. □ **menopausal** *adj.*

menorah *n.* seven-armed candelabrum used in Jewish worship.

menstrual *adj.* of or in menstruation.

menstruate *v.* experience a monthly discharge of blood from the womb. □ **menstruation** *n.*

mensuration *n.* **1** measuring. **2** mathematical rules for this.

mental *adj.* **1** of, in, or performed by the mind. **2** (*colloq.*) mad. □ **mentally** *adv.*

mentality *n.* characteristic attitude of mind.

■ **attitude, character, disposition, frame of mind, make-up, outlook.**

menthol *n.* camphor-like substance.

mentholated *adj.* impregnated with menthol.

mention *v.* **1** speak or write about briefly. **2** refer to by name. ● *n.* act of mentioning.

■ *v.* **1 allude to, bring up, broach, refer to, speak about, touch on, write about. 2 cite, name, quote.** ● *n.* **allusion, citation, reference.**

mentor *n.* trusted adviser.

menu *n.* (*pl.* **-us**) **1** list of dishes to be served. **2** list of options displayed on a computer screen.

mercantile *adj.* trading, of trade or merchants.

mercenary *adj.* primarily concerned with money or reward. ● *n.* professional soldier hired by a foreign country.

■ *adj.* **acquisitive, avaricious, covetous, grasping, greedy.**

merchandise *n.* goods bought and sold or for sale. ● *v.* **1** trade. **2** promote sales of (goods).

■ *n.* **commodities, goods, products, stock, wares.** ● *v.* **1 market, retail, sell, trade. 2 advertise, promote, publicize.**

merchant *n.* **1** wholesale trader. **2** (*US & Sc.*) retail trader. □ **merchant bank** one dealing in commercial loans and the financing of businesses. **merchant navy** shipping employed in commerce. **merchant ship** ship carrying merchandise.

merchantable *adj.* saleable.

merchantman *n.* merchant ship.

merciful *adj.* showing mercy. □ **mercifulness** *n.*

■ **charitable, clement, compassionate, forbearing, forgiving, humane, kind, kind-hearted, kindly, magnanimous, lenient.**

mercifully *adv.* **1** in a merciful way. **2** (*colloq.*) fortunately.

merciless *adj.* showing no mercy. □ **mercilessly** *adv.*

■ **barbaric, barbarous, brutal, callous, cold, cruel, hard, hard-hearted, harsh, heartless, inhuman, inhumane, implacable, pitiless, ruthless, savage, unforgiving, unmerciful.**

mercurial *adj.* **1** volatile. **2** of or containing mercury.

■ **1 capricious, changeable, inconstant, temperamental, unpredictable, volatile.**

mercury *n.* heavy silvery usu. liquid metal. □ **mercuric** *adj.*

mercy *n.* **1** kindness shown to an offender or enemy etc. who is in one's power. **2** merciful act. □ **at the mercy of** wholly in the power of or subject to.

■ **1 charity, clemency, compassion, forbearance, forgiveness, generosity, humanity, kindness, leniency, magnanimity, pity, quarter.**

mere[1] *adj.* no more or no better than what is specified. □ **merely** *adv.*

mere[2] *n.* (*poetic*) lake.

merge *v.* **1** combine into a whole. **2** blend gradually.

■ **1 amalgamate, coalesce, combine, consolidate, join, unite. 2 blend, mingle, mix.**

merger *n.* combining of commercial companies etc. into one.

■ **alliance, amalgamation, coalition, combination, merging, union.**

meridian *n.* any of the circles round the earth that pass through both the North and South Poles.

meringue /məráng/ *n.* **1** baked mixture of sugar and egg white. **2** small cake of this.

merino *n.* (*pl.* **-os**) **1** a kind of sheep with fine soft wool. **2** soft woollen fabric.

merit *n.* **1** feature or quality that deserves praise. **2** worthiness. ● *v.* (**merited**) deserve.

■ *n.* **1** advantage, asset, good point. **2** excellence, value, worth, worthiness. ● *v.* be entitled to, be worthy of, deserve, earn, rate, warrant.

meritocracy *n.* government by people selected for merit.

meritorious *adj.* deserving praise.

■ admirable, commendable, creditable, estimable, excellent, exemplary, laudable, outstanding, praiseworthy.

merlin *n.* a kind of falcon.

mermaid *n.* imaginary half-human sea creature with a fish's tail instead of legs.

merry *adj.* (**-ier, -iest**) cheerful and lively, joyous. □ **merry-go-round** *n.* roundabout at a funfair. **merrymaking** *n.* revelry. **merrily** *adv.*, **merriment** *n.*

■ bubbly, buoyant, carefree, cheerful, cheery, chirpy, gay, gleeful, happy, in good spirits, jolly, joyful, light-hearted, lively, vivacious. □ **merriment** cheer, cheerfulness, gaiety, glee, high spirits, joyfulness, mirth.

mesh *n.* **1** network fabric. **2** space between threads in net, sieve, etc. ● *v.* (of a toothed wheel) engage with another.

mesmerize *v.* **1** hypnotize. **2** fascinate.

■ **2** bewitch, captivate, enthral, fascinate, grip, hold spellbound.

mesolithic *adj.* of the period between palaeolithic and neolithic.

mess *n.* **1** dirty or untidy condition. **2** difficult or confused situation, trouble. **3** something split etc. **4** (in the armed forces) group who eat together, their dining room. ● *v.* **mess up 1** make untidy or dirty. **2** muddle, bungle. □ **make a mess of** bungle. **mess about 1** potter. **2** fool about.

■ *n.* **1** chaos, clutter, confusion, disarray, disorder, disorganization, hotchpotch, jumble, muddle, shambles, tangle, untidiness. **2** difficulty, fix, *colloq.* hole, *colloq.* jam, *colloq.* pickle, plight, predicament, quandary, trouble. ● *v.* **1** clutter up, disarrange, disarray, disorder; dishevel, ruffle, rumple, tousle. **2** botch, bungle, *colloq.* make a hash of, make a mess of, *colloq.* muff, ruin, *sl.* screw up, spoil.

message *n.* **1** spoken or written communication. **2** moral or social teaching.

■ **1** commmunication, communiqué, dispatch; letter, *colloq.* memo, memorandum, note.

messenger *n.* bearer of a message.

■ courier, emissary, envoy, go-between, intermediary.

Messiah *n.* **1** deliverer expected by Jews. **2** Christ as this. □ **Messianic** *adj.*

Messrs *see* **Mr.**

messy *adj.* (**-ier, -iest**) untidy or dirty. □ **messily** *adv.*

■ chaotic, cluttered, disordered, disorderly, in a muddle, in disarray, topsy-turvy, untidy; dishevelled, rumpled, *colloq.* scruffy, unkempt; dirty, grubby, *colloq.* mucky.

met *see* **meet.**

metabolism *n.* process by which nutrition takes place. □ **metabolic** *adj.*, **metabolically** *adv.*

metabolize *v.* process (food) in metabolism.

metal *n.* **1** any of a class of mineral substances such as gold, silver, iron, etc., or an alloy of these. **2** road-metal. ● *v.* make or mend (a road) with road-metal.

metallic *adj.* of or like metal.

metallurgy *n.* science of extracting and working metals.

metamorphose *v.* change by metamorphosis.

metamorphosis *n.* (*pl.* **-phoses**) change of form or character. □ **metamorphic** *adj.*

metaphor *n.* application of a word or phrase to something that it does not apply to literally (e.g. the *evening* of one's life, *food* for thought). □ **metaphorical** *adj.*, **metaphorically** *adv.*

metaphysics *n.* branch of philosophy dealing with the nature of existence and of knowledge. □ **metaphysical** *adj.*

mete *v.* **mete out** deal out.

meteor *n.* small mass of matter from outer space.

meteoric *adj.* **1** of meteors. **2** swift and brilliant.

meteorite *n.* meteor fallen to earth.

meteorology *n.* study of atmospheric conditions in order to forecast weather. □ **meteorological** *adj.*, **meteorologist** *n.*

meter[1] *n.* device measuring and indicating the quantity supplied, distance travelled, time elapsed, etc. ● *v.* measure by a meter.

meter[2] *n.* (*US*) = **metre.**

methane *n.* colourless inflammable gas.
method *n.* **1** procedure or way of doing something. **2** orderliness.

> ■ **1** approach, manner, means, mode, procedure, process, system, technique, way. **2** neatness, order, orderliness, organization, regularity, system.

methodical *adj.* orderly, systematic. □ **methodically** *adv.*

> ■ businesslike, careful, deliberate, ordered, orderly, organized, painstaking, systematic.

meths *n.* (*colloq.*) methylated spirit.
methylated spirit form of alcohol used as a solvent and for heating.
meticulous *adj.* very careful and exact. □ **meticulously** *adv.*

> ■ careful, exact, painstaking, particular, precise, punctilious, scrupulous, thorough.

metre *n.* **1** metric unit of length (about 39.4 inches). **2** rhythm in poetry.
metric *adj.* **1** of or using the metric system. **2** of poetic metre. □ **metric system** decimal system of weights and measures, using the metre, litre, and gram as units.
metrical *adj.* of or composed in rhythmic metre, not prose.
metronome *n.* device used to indicate tempo while practising music.
metropolis *n.* chief city of country or region.
metropolitan *adj.* of a metropolis.
mettle *n.* courage, strength of character.
mettlesome *adj.* spirited, brave.
mew *n.* cat's characteristic cry. ● *v.* make this sound.
mews *n.* set of stables converted into dwellings etc.
mezzanine *n.* extra storey set between two others.
mezzotint *n.* method of engraving.
mg *abbr.* milligram(s).
MHz *abbr.* megahertz.
miaow *n.* & *v.* = **mew**.
miasma *n.* unpleasant or unwholesome air.
mica /mī́kə/ *n.* transparent mineral with a layered structure.
mice *see* **mouse**.
micro- *pref.* **1** extremely small. **2** one-millionth part of (as in *microgram*).
microbe *n.* micro-organism.
microbiology *n.* study of micro-organisms.
microchip *n.* tiny piece of a semiconductor holding a complex electronic circuit.
microcomputer *n.* computer in which the central processor is contained on microchip(s).
microcosm *n.* community or complex resembling something else but on a very small scale.
microfiche /mī́krōfeesh/ *n.* (*pl.* **-fiche**) small sheet of microfilm.
microfilm *n.* length of film bearing miniature photographs of document(s). ● *v.* photograph on this.
microlight *n.* a kind of motorized hang-glider.
micrometer *n.* gauge for small-scale measurement.
micron *n.* one-millionth of a metre.
micro-organism *n.* organism invisible to the naked eye.
microphone *n.* instrument for amplifying or broadcasting sound.
microprocessor *n.* data processor contained on microchip(s).
microscope *n.* instrument with lenses that magnify very small things and make them visible.
microscopic *adj.* **1** of a microscope. **2** very small. **3** visible only with a microscope. □ **microscopically** *adv.*, **microscopy** *n.*
microsurgery *n.* surgery using a microscope.
microwave *n.* **1** electromagnetic wave of length between about 50 cm and 1 mm. **2** oven using such waves to cook or heat food quickly.
mid *adj.* middle.
midday *n.* noon.
middle *adj.* occurring at an equal distance from extremes or outer limits. ● *n.* middle point, position, area, etc. □ **middle age** part of life between youth and old age. **Middle Ages** period of European history from *c.* 1000–1453. **middle class** class of society between upper and working classes. **Middle East** area from Egypt to Iran inclusive.

> ■ *adj.* central, halfway, medial, median, mid. ● *n.* centre, halfway point, heart, midst.

middleman *n.* **1** trader handling a commodity between producer and consumer. **2** intermediary.
middling *adj.* moderately good.

> ■ adequate, average, fair, indifferent, mediocre, moderate, *colloq.* OK, ordinary, passable, run-of-the-mill, tolerable, unremarkable.

midge *n.* small biting insect.

midget *n.* extremely small person or thing. ● *adj.* extremely small.

Midlands *n.pl.* inland counties of central England. □ **midland** *adj.*

midnight *n.* 12 o'clock at night.

midriff *n.* front part of the body just above the waist.

midshipman *n.* naval rank just below sub-lieutenant.

midst *n.* middle. □ **in the midst of** among.

midway *adv.* halfway.

midwife *n.* person trained to assist at childbirth.

mien /meen/ *n.* person's manner or bearing.

might[1] *n.* great strength or power.

■ energy, force, muscle, potency, power, strength.

might[2] *see* **may**[1]. *v.aux.* used to request permission or (like *may*) to express possibility.

mighty *adj.* (**-ier**, **-iest**) **1** very strong or powerful. **2** very great. □ **mightily** *adv.*

■ **1** potent, powerful, strong; brawny, burly, hefty, muscular, robust, strapping, sturdy. **2** big, colossal, enormous, gigantic, great, huge, large, massive, monumental, tremendous.

migraine /méegrayn/ *n.* severe form of headache.

migrant *adj.* & *n.* migrating (person or animal).

migrate *v.* **1** leave one place and settle in another. **2** (of birds etc.) go from one place to another at each season. □ **migration** *n.*, **migratory** *adj.*

mihrab *n.* niche or slab in a mosque, showing the direction of Mecca.

mild *adj.* **1** gentle. **2** moderate in intensity, not harsh or drastic. **3** not strongly flavoured. □ **mildly** *adv.*, **mildness** *n.*

■ **1** affable, amiable, benign, easygoing, gentle, good-natured, inoffensive, kind, kindly, peaceable, peaceful, placid, serene, tranquil. **2** gentle, light, moderate, soft. **3** bland, insipid.

mildew *n.* tiny fungi forming a coating on things exposed to damp. □ **mildewed** *adj.*

mile *n.* **1** measure of length, 1760 yds (about 1.609 km). **2** (*colloq.*) great distance. □ **nautical mile** unit used in navigation, 2025 yds (1.852 km).

mileage *n.* distance in miles.

milestone *n.* **1** stone showing the distance to a certain place. **2** significant event or stage reached.

milieu /milyő/ *n.* (*pl.* **-eus**) environment, surroundings.

■ background, environment, setting, surroundings.

militant *adj.* & *n.* (person) prepared to take aggressive action. □ **militancy** *n.*

■ *adj.* aggressive, belligerent, combative, pugnacious. ● *n.* activist, extremist, fighter.

militarism *n.* reliance on military attitudes. □ **militaristic** *adj.*

military *adj.* of soldiers or the army or all armed forces.

militate *v.* serve as a strong influence.

militia /milíshə/ *n.* a military force, esp. of trained civilians available in an emergency.

milk *n.* **1** white fluid secreted by female mammals as food for their young. **2** cow's milk as food for human beings. **3** milk-like liquid. ● *v.* **1** draw milk from. **2** exploit. □ **milk teeth** first (temporary) teeth in young mammals.

milkman *n.* person who delivers milk to customers.

milky *adj.* **1** of or like milk. **2** containing much milk. □ **Milky Way** broad luminous band of stars.

mill *n.* **1** machinery for grinding or processing specified material. **2** building containing this. ● *v.* **1** process in a mill. **2** produce grooves in (metal). **3** move in a confused mass. □ **miller** *n.*

millennium *n.* (*pl.* **-ums**) **1** period of 1000 years. **2** future period of great happiness for everyone.

millepede *n.* small crawling creature with many legs.

millet *n.* **1** cereal plant. **2** its seeds.

milli- *pref.* one-thousandth part of (as in *milligram*, *millilitre*, *millimetre*).

milliner *n.* maker or seller of women's hats. □ **millinery** *n.*

million *n.* one thousand thousand. □ **millionth** *adj.* & *n.*

millionaire *n.* person who possesses a million pounds.

millstone *n.* **1** circular stone for grinding corn. **2** great burden.

milometer *n.* instrument measuring the distance in miles travelled by a vehicle.

milt *n.* sperm of male fish.

mime *n.* acting with gestures without words. ● *v.* act with mime.

mimic *v.* (**mimicked**) imitate, esp. playfully or for entertainment. ● *n.* person who is clever at mimicking others. □ **mimicry** *n.*

■ *v.* ape, caricature, copy, imitate, impersonate, parody, send up, take off. ● *n.* imitator, impersonator, impressionist.

mimosa *n.* tropical shrub with small ball-shaped flowers.

minaret *n.* tall slender tower on or beside a mosque.

mince *v.* **1** cut into small pieces in a mincer. **2** walk or speak with affected refinement. ● *n.* minced meat. □ **mince pie** pie containing mincemeat.

mincemeat *n.* mixture of dried fruit, sugar, etc., used in pies.

mincer *n.* machine with revolving blades for cutting food into very small pieces.

mind *n.* **1** ability to be aware of things and to think and reason. **2** person's attention or ability to remember. **3** normal condition of one's mental faculties. ● *v.* **1** have charge of. **2** object to. **3** remember and be careful (about). **4** heed. □ **be in two minds** be undecided. **be of one mind** agree. **have (it) in mind** intend. **out of one's mind** mad. **to my mind** in my opinion.

■ *n.* **1** brain(s), intellect, intelligence, reason, sense, understanding, wit. **2** attention, attentiveness, concentration; memory. ● *v.* **1** care for, have charge of, look after, take care of, tend, watch. **2** be annoyed by, be offended by, be upset by, object to, resent, take offence at. **4** attend to, heed, listen to, mark, note, pay attention to, take note of. □ **be in two minds** be undecided, dither, hesitate, shilly-shally, vacillate, waver. **have (it) in mind** aim, intend, mean, plan, propose, purpose.

minded *adj.* having inclinations or interests of a certain kind.

minder *n.* **1** person employed to look after a person or thing. **2** (*sl.*) bodyguard.

mindful *adj.* taking thought or care (of something).

mindless *adj.* without intelligence. □ **mindlessness** *n.*

■ asinine, fatuous, foolish, idiotic, senseless, silly, stupid, thoughtless, witless.

mine[1] *adj.* & *poss.pron.* belonging to me.

mine[2] *n.* **1** excavation for extracting metal or coal etc. **2** abundant source (of information etc.). **3** explosive device laid in or on the ground or in water. ● *v.* **1** dig for minerals, extract in this way. **2** lay explosive mines under or in.

■ *n.* **1** colliery, excavation, quarry, pit. **2** fund, repository, reserve, source, store. ● *v.* **1** dig, excavate, quarry.

minefield *n.* area where explosive mines have been laid.

miner *n.* worker in a mine.

mineral *n.* **1** inorganic natural substance. **2** fizzy soft drink. ● *adj.* of or containing minerals. □ **mineral water** water naturally containing dissolved mineral salts or gases.

mineralogy *n.* study of minerals. □ **mineralogist** *n.*

minesweeper *n.* ship for clearing away mines laid in the sea.

mingle *v.* **1** blend together. **2** mix socially.

■ **1** amalgamate, blend, combine, intermingle, merge, mix, unite. **2** circulate, fraternize, hobnob, mix, socialize.

mini- *pref.* miniature.

miniature *adj.* very small, on a small scale. ● *n.* small-scale portrait, copy, or model.

■ *adj.* diminutive, little, microscopic, miniminuscule, minute, small, tiny, *colloq.* wee.

miniaturize *v.* make miniature, produce in a very small version. □ **miniaturization** *n.*

minibus *n.* small bus-like vehicle with seats for only a few people.

minim *n.* **1** note in music, lasting half as long as a semibreve. **2** one-sixtieth of a fluid drachm.

minimal *adj.* very small, least possible □ **minimally** *adv.*

minimize *v.* **1** reduce to a minimum. **2** represent as small or unimportant. □ **minimization** *n.*

■ **1** curtail, cut, decrease, diminish, lessen, reduce. **2** belittle, deprecate, make light of, play down, underestimate, undervalue.

minimum *adj.* & *n.* (*pl.* **-ima**) smallest (amount) possible.

minion *n.* (*derog.*) assistant.

minister *n.* **1** head of a government department. **2** clergyman. **3** senior diplomatic representative. ● *v.* **minister to** attend to the needs of. □ **ministerial** *adj.*

■ *n.* **2** chaplain, clergyman, clergywoman cleric, padre, *colloq.* parson, pastor, priest rector, vicar. **3** ambassador, chargé d'affaires, consul, diplomat, emissary plenipotentiary. ● *v.* attend to, care for look after, see to, take care of.

ministry *n.* **1** government department headed by a minister. **2** period of government under one leader. **3** work of a clergyman.

mink *n.* **1** small stoat-like animal. **2** its fur. **3** coat made of this.

minnow *n.* small freshwater carp.

minor *adj.* lesser or comparatively small in size or importance. ● *n.* person not yet legally of adult age.

■ *adj.* lesser, secondary, smaller, subordinate, subsidiary; inconsequential, inconsiderable, insignificant, negligible, slight, trifling, trivial, unimportant. ● *n.* adolescent, boy, child, girl, juvenile, teenager, youngster, youth.

minority *n.* **1** smallest part of a group or class. **2** small group differing from others. **3** age when a person is not yet legally adult.

minstrel *n.* medieval singer and musician.

mint[1] *n.* place authorized to make a country's coins. ● *v.* make (coins). □ **in mint condition** new-looking.

mint[2] *n.* **1** fragrant herb. **2** peppermint. **3** sweet flavoured with this. □ **minty** *adj.*

minuet /mínyoo-ét/ *n.* slow stately dance.

minus *prep.* **1** reduced by subtraction of. **2** below zero. **3** (*colloq.*) without. ● *adj.* **1** less than zero. **2** less than the amount indicated.

minuscule *adj.* extremely small.

minute[1] /mínnit/ *n.* **1** one-sixtieth of an hour or degree. **2** moment of time. **3** (*pl.*) official summary of an assembly's proceedings. ● *v.* record in the minutes.

■ *n.* **2** flash, instant, moment, split second, *colloq.* tick.

minute[2] /mīnyōōt/ *adj.* **1** extremely small. **2** very precise. □ **minutely** *adv.*, **minuteness** *n.*

■ **1** diminutive, infinitesimal, little, microscopic, mini-, miniature, minuscule, small, *colloq.* teeny, tiny, *colloq.* wee. **2** accurate, detailed, exact, meticulous, painstaking, precise.

minutiae /mīnyōōshi-ee/ *n.pl.* very small details.

miracle *n.* **1** wonderful event attributed to a supernatural agency. **2** remarkable event or thing. □ **miraculous** *adj.*, **miraculously** *adv.*

■ marvel, phenomenon, prodigy, wonder. □ **miraculous** amazing, astonishing, astounding, breathtaking, extraordinary, incredible, magical, marvellous, phenomenal, remarkable, stupendous, wonderful.

mirage *n.* optical illusion caused by atmospheric conditions.

■ hallucination, illusion, vision.

mire *n.* **1** swampy ground, bog. **2** mud or sticky dirt. □ **miry** *adj.*

mirror *n.* glass coated so that reflections can be seen in it. ● *v.* reflect in a mirror.

mirth *n.* merriment, laughter.

■ cheerfulness, gaiety, glee, fun, high spirits, jollity, joviality, joyousness, laughter, merriment.

mis- *pref.* badly, wrongly.

misadventure *n.* piece of bad luck.

misanthrope *n.* (also **misanthropist**) person who hates humankind. □ **misanthropy** *n.*, **misanthropic** *adj.*

misapply *v.* apply (esp. funds) wrongly. □ **misapplication** *n.*

misapprehend *v.* misunderstand. □ **misapprehension** *n.*

misappropriate *v.* take dishonestly. □ **misappropriation** *n.*

■ embezzle, filch, *sl.* nick, pilfer, *sl.* pinch, pocket, purloin, steal.

misbehave *v.* behave badly. □ **misbehaviour** *n.*

miscalculate *v.* calculate incorrectly. □ **miscalculation** *n.*

miscarriage *n.* abortion occurring naturally.

miscarry *v.* **1** have a miscarriage. **2** go wrong, be unsuccessful.

■ **2** be unsuccessful, come to grief, *colloq.* come unstuck, fall through, founder, go amiss, go wrong, misfire.

miscellaneous *adj.* assorted.

■ assorted, diverse, heterogeneous, mixed, motley, multifarious, sundry, varied, various.

miscellany *n.* collection of assorted items.

■ assortment, diversity, hotchpotch, jumble, medley, mixture, pot-pourri, variety.

mischance *n.* misfortune.

mischief *n.* **1** children's annoying but not malicious conduct. **2** playful malice. **3** harm, damage.

■ **1,2** devilment, devilry, misbehaviour, mischievousness, naughtiness, playfulness. **3** damage, detriment, harm, hurt, injury, trouble.

mischievous *adj.* full of mischief. ▫ **mischievously** *adv.*, **mischievousness** *n.*

■ devilish, disobedient, naughty, playful, rascally, roguish, wicked.

misconception *n.* wrong interpretation.

■ error, misapprehension, misconstruction, misinterpretation, misjudgement, mistake, misunderstanding, wrong idea.

misconduct *n.* **1** bad behaviour. **2** mismanagement.

■ **1** bad behaviour, disobedience, mischief, misdeeds, misdemeanours, naughtiness.

misconstrue *v.* misinterpret. ▫ **misconstruction** *n.*

miscreant *n.* wrongdoer.

■ blackguard, criminal, *colloq.* crook, knave, lawbreaker, malefactor, offender, reprobate, rogue, ruffian, scoundrel, villain, wretch, wrongdoer.

misdeed *n.* wrongful act.

■ crime, misdemeanour, offence, sin, transgression, wrong.

misdemeanour *n.* misdeed.

miser *n.* person who hoards money and spends as little as possible. ▫ **miserly** *adj.*, **miserliness** *n.*

■ hoarder, niggard, skinflint. ▫ **miserly** cheese-paring, close, mean, niggardly, parsimonious, penny-pinching, stingy, *colloq.* tight, tight-fisted.

miserable *adj.* **1** full of misery. **2** wretchedly poor in quality or surroundings. ▫ **miserably** *adv.*

■ **1** blue, broken-hearted, dejected, depressed, desolate, despondent, disconsolate, downcast, downhearted, forlorn, gloomy, glum, heartbroken, melancholy, mournful, sad, sorrowful, tearful, unhappy, upset, woebegone, woeful, wretched. **2** mean, poor, shabby, sordid, sorry, squalid, wretched.

misery *n.* **1** great unhappiness or discomfort. **2** thing causing this. **3** (*colloq.*) disagreeable person.

■ **1** anguish, depression, desolation, despair, despondency, distress, gloom, heartache, melancholy, sadness, sorrow, unhappiness, woe, wretchedness. **2** adversity, affliction, hardship, misfortune, suffering, trial, tribulation. **3** killjoy, moaner, pessimist, *colloq.* sourpuss, spoilsport.

misfire *v.* **1** (of a gun or engine) fail to fire correctly. **2** go wrong.

misfit *n.* person not well suited to his or her environment.

misfortune *n.* **1** bad luck. **2** unfortunate event.

■ **1** bad luck, ill luck. **2** accident, calamity, catastrophe, disaster, misadventure, mischance, mishap, reverse, tragedy.

misgiving *n.* slight feeling of doubt, fear, or mistrust.

■ anxiety, apprehension, concern, disquiet, doubt, mistrust, suspicion, uncertainty, unease; qualm, reservation, scruple.

misguided *adj.* mistaken in one's opinions or actions.

■ foolish, ill-advised, impolitic, imprudent, injudicious, misled, mistaken, unwise, wrong.

mishap *n.* unlucky accident.

■ accident, calamity, disaster, misadventure, mischance, misfortune, reverse.

misinform *v.* give wrong information to, mislead. ▫ **misinformation** *n.*

misinterpret *v.* interpret incorrectly. ▫ **misinterpretation** *n.*

■ misapprehend, misconstrue, misjudge, misread, mistake, misunderstand.

misjudge *v.* form a wrong opinion or estimate of. ▫ **misjudgement** *n.*

mislay *v.* (**mislaid**) lose temporarily.

mislead *v.* (**misled**) **1** cause to form a wrong impression. **2** deceive.

■ deceive, delude, fool, hoodwink, lead astray, misinform, pull the wool over someone's eyes, take in, trick.

mismanage *v.* manage badly or wrongly. ▫ **mismanagement** *n.*

■ botch, bungle, *sl.* fluff, *colloq.* make a hash of, make a mess of, mess up, *colloq.* muff.

misnomer *n.* wrongly applied name or description.

misogynist *n.* person who hates women. ▫ **misogyny** *n.*

misplace *v.* **1** put in a wrong place, mislay. **2** place (confidence etc.) unwisely.

misprint *n.* error in printing.

misquote *v.* quote incorrectly. ▫ **misquotation** *n.*

misread *v.* (**-read**) read or interpret incorrectly.

misrepresent *v.* represent wrongly. ▫ **misrepresentation** *n.*

misrule *n.* bad government.

Miss *n.* (*pl.* **Misses**) title of a girl or unmarried woman.

miss *v.* **1** fail to hit, catch, see, hear, understand, etc. **2** notice or regret the absence or loss of. ● *n.* failure to hit or attain what is aimed at.

■ *v.* **2** feel nostalgic for, long for, pine for, yearn for.

misshapen *adj.* badly shaped.

■ contorted, crooked, distorted, gnarled, grotesque, malformed, twisted, warped.

missile *n.* object thrown or fired at a target.

missing *adj.* **1** not present. **2** not in its place, lost.

mission *n.* **1** task assigned to a person or group. **2** this group. **3** missionaries' headquarters.

■ 1 assignment, commission, duty, errand, job, task. **2** commission, delegation, deputation, group.

missionary *n.* person sent to spread religious faith.

misspell *v.* (**misspelt**) spell incorrectly. □ **misspelling** *n.*

mist *n.* **1** water vapour near the ground or clouding a window etc. **2** thing resembling this. ● *v.* cover or become covered with mist.

■ *n.* **1** cloud, fog, haze, vapour.

mistake *n.* **1** incorrect idea or opinion. **2** thing done incorrectly. ● *v.* (**mistook, mistaken**) **1** misunderstand. **2** choose or identify wrongly.

■ *n.* **1** misapprehension, miscalculation, misconception, misjudgement. **2** *colloq.* boob, blunder, *sl.* clanger, error, *faux pas*, gaffe, howler, oversight, slip, *colloq.* slip-up. ● *v.* **1** misconstrue, misinterpret, misjudge, misunderstand.

mistle thrush large thrush.

mistletoe *n.* plant with white berries, growing on trees.

mistral *n.* cold north or north-west wind in southern France.

mistress *n.* **1** woman who has control of people or things. **2** female teacher. **3** man's illicit female lover.

mistrust *v.* feel no trust in. ● *n.* lack of trust. □ **mistrustful** *adj.*

■ *v.* be suspicious *or* doubtful of, distrust, doubt, have reservations about, question, suspect. ● *n.* distrust, doubt, scepticism, suspicion, wariness.

misty *adj.* (**-ier, -iest**) **1** full of mist. **2** indistinct. □ **mistily** *adv.*

■ 1 cloudy, foggy, hazy. **2** blurred, dim, fuzzy, indistinct, unclear, vague.

misunderstand *v.* (**-stood**) fail to understand correctly. □ **misunderstanding** *n.*

■ misapprehend, misconstrue, misinterpret, misjudge, misread.

misuse *v.* /míssyo͞oz/ **1** use wrongly. **2** treat badly. ● *n.* /míssyo͞os/ wrong use.

■ *v.* **1** abuse, misapply. **2** abuse, harm, hurt, ill-treat, knock about, maltreat, mistreat. ● *n.* abuse, misapplication.

mite *n.* **1** very small spider-like animal. **2** small creature, esp. a child. **3** small amount.

mitigate *v.* make seem less serious or severe. □ **mitigation** *n.*

■ allay, alleviate, assuage, decrease, ease, lessen, lighten, moderate, palliate, reduce, relieve, soften, temper, tone down.

mitre *n.* **1** pointed headdress of bishops and abbots. **2** join with tapered ends that form a right angle. ● *v.* join in this way.

mitt *n.* mitten.

mitten *n.* glove with no partitions between the fingers, or leaving the fingertips bare.

mix *v.* **1** combine (different things). **2** prepare by doing this. **3** be compatible. **4** come or be together socially. ● *n.* mixture. □ **mix up 1** mix thoroughly. **2** confuse. **mixer** *n.*

■ *v.* **1** alloy, amalgamate, blend, combine, intermingle, merge, mingle, mix, unite. **4** associate, consort, fraternize, hobnob, keep company, socialize.

mixed *adj.* **1** composed of various elements. **2** of or for both sexes. □ **mixed-up** *adj.* **1** muddled. **2** (*colloq.*) not well-adjusted emotionally.

■ 1 assorted, diverse, diversified, heterogeneous, miscellaneous, motley, sundry, varied, various.

mixture *n.* **1** thing made by mixing. **2** process of mixing things.

■ 1 alloy, amalgam, amalgamation, blend, compound, mix; assortment, collection, hotchpotch, jumble, medley, miscellany, pot-pourri, variety.

ml *abbr.* millilitre(s).

mm *abbr.* millimetre(s).

mnemonic /nimónnik/ *adj.* & *n.* (verse etc.) aiding the memory.

moan *n.* **1** low mournful sound. **2** grumble. ● *v.* **1** make or utter with a moan. **2** grumble. □ **moaner** *n.*

■ *n.* **1** groan, lament, lamentation, sigh, sough, wail. **2** *sl.* beef, complaint, grievance, *colloq.* gripe, *colloq.* grouse, grumble. ● *v.* **1** groan, lament, sigh, sough, wail. **2** *sl.* beef, complain, *colloq.* gripe, *colloq.* grouse, grumble, whine, *colloq.* whinge.

moat *n.* deep wide usu. water-filled ditch round a castle.

mob *n.* **1** large disorderly crowd. **2** (*sl.*) gang. ● *v.* (**mobbed**) crowd round in great numbers.

■ *n.* **1** crowd, herd, horde, gaggle, multitude, rabble, swarm, throng. ● *v.* besiege, crowd around, surround, swarm around.

mobile *adj.* able to move or be moved easily. ● *n.* artistic hanging structure whose parts move in currents of air. □ **mobility** *n.*

mobilize *v.* assemble (troops etc.) for active service. □ **mobilization** *n.*, **mobilizer** *n.*

■ assemble, marshal, organize, prepare, rally, ready.

moccasin *n.* soft flat-soled leather shoe.

mock *v.* **1** ridicule, scoff at. **2** mimic contemptuously. ● *adj.* imitation. □ **mock-up** *n.* model for testing or study.

■ *v.* **1** deride, gibe (at), jeer (at), laugh at, make fun of, poke fun at, *colloq.* rib, ridicule, scoff at, taunt, tease. **2** ape, burlesque, caricature, imitate, mimic, parody, satirize, send up, take off. ● *adj.* artificial, fake, false, imitation, *colloq.* phoney, simulated, synthetic.

mockery *n.* **1** derision. **2** absurd or unsatisfactory imitation.

■ **1** derision, ridicule, scorn, taunting. **2** charade, farce, parody, travesty.

mode *n.* **1** way a thing is done. **2** current fashion.

■ **1** approach, manner, method, procedure, system, technique, way. **2** fashion, style, trend, vogue.

model *n.* **1** three-dimensional reproduction, usu. on a smaller scale. **2** exemplary person or thing. **3** particular design or style. **4** person employed to pose for an artist or display clothes in a shop etc. by wearing them. ● *adj.* exemplary. ● *v.* (**modelled**) **1** make a model of. **2** work as artist's or fashion model, display (clothes) in this way.

■ *n.* **1** miniature, mock-up, replica, reproduction; dummy, effigy. **2** archetype, epitome, exemplar, ideal, paragon, pattern, prototype, standard, type. **3** design, kind, style, type, version. **4** sitter, subject; mannequin. ● *adj.* archetypal, exemplar[y], ideal, inimitable, perfect. ● *v.* **1** fashio[n], form, make, mould, sculpt, sculptur[e], shape. **2** pose, sit; display, show.

modem *n.* device for sending and receiving computer data through a telephone line.

moderate *adj.* /móddərət/ **1** avoidin[g] extremes, temperate in conduct etc. [**2**] fairly large or good. ● *n.* /móddərə[t]/ holder of moderate views. ● [*v.*] /móddərayt/ make or become moderat[e]. □ **moderately** *adv.*

■ *adj.* **1** calm, controlled, cool, mil[d], reasonable, restrained, sensible, tempe[r]ate. **2** acceptable, average, fair, mediu[m], mediocre, middling, modest, unexceptional. ● *v.* cushion, decrease, diminis[h], lessen, mitigate, reduce, relieve, softe[n], temper, tone down; abate, ease, *colloq.* le[t] up, subside.

moderation *n.* moderating. □ **in moderation** in moderate amounts.

modern *adj.* **1** of present or recent time[s]. **2** in current style. □ **modernity** *n.*

■ **1** contemporary, current, today's. [**2**] fashionable, in, in fashion, in vogue, new[,] newfangled, *colloq.* trendy, up to date.

modernist *n.* person who favours modern ideas or methods. □ **modernism** *n.*

modernize *v.* make modern, adapt t[o] modern ways. □ **modernization** *n*[.], **modernizer** *n.*

■ do up, redecorate, redesign, refurbis[h], renovate, update.

modest *adj.* **1** not vain or boastful. [**2**] moderate in size etc. **3** not showy. [**4**] showing regard for conventional decencies. □ **modestly** *adv.*, **modesty** *n.*

■ **1** humble, self-effacing, unassuming. [**2**] adequate, moderate, unexceptional. [**3**] homely, humble, ordinary, plain, simpl[e,] unpretentious. **4** decent, decorous, demure, proper, seemly.

modicum *n.* small amount.

modify *v.* **1** make less severe. **2** mak[e] partial changes in. □ **modification** *n.*

■ **1** decrease, diminish, lessen, moderat[e], reduce, soften, temper, tone down. **2** a[d]apt, adjust, alter, amend, change, revise.

modish /mṓdish/ *adj.* fashionable.

modulate *v.* **1** regulate, moderate. **2** var[y] in tone or pitch. □ **modulation** *n.*

■ **1** adjust, moderate, modify, regulat[e,] temper. **2** lower, soften, tone down, tur[n] down.

module *n.* **1** standardized part or independent unit. **2** unit of training or education. □ **modular** *adj.*

mogul *n.* (*colloq.*) important or influential person.

mohair *n.* **1** fine silky hair of the angora goat. **2** yarn or fabric made from this.

moiety /móyəti/ *n.* half.

moist *adj.* slightly wet.

■ clammy, damp, humid, muggy, wet.

moisten *v.* make or become moist.

■ damp, dampen, water, wet.

moisture *n.* water or other liquid diffused through a substance or as vapour or condensed on a surface.

■ condensation, damp, dampness, liquid, water, wetness.

moisturize *v.* make (skin) less dry. □ **moisturizer** *n.*

molar *n.* back tooth with a broad top, used in chewing.

molasses *n.* **1** syrup from raw sugar. **2** (*US*) treacle.

mole[1] *n.* small dark spot on human skin.

mole[2] *n.* **1** small burrowing animal with dark fur. **2** spy established within an organization.

molecule *n.* very small unit (usu. a group of atoms) of a substance. □ **molecular** *adj.*

molehill *n.* mound of earth thrown up by a mole.

molest *v.* attack or interfere with, esp. sexually. □ **molestation** *n.*

mollify *v.* soothe the anger of. □ **mollification** *n.*

■ appease, calm, pacify, placate, soothe.

mollusc *n.* animal with a soft body and often a hard shell.

mollycoddle *v.* pamper.

molten *adj.* liquefied by heat.

molybdenum *n.* hard metal used in steel.

moment *n.* **1** point or brief portion of time. **2** importance.

■ **1** juncture, point; flash, instant, minute, split second, *colloq.* tick. **2** consequence, import, importance, seriousness, significance, weight.

momentary *adj.* lasting only a moment. □ **momentarily** *adv.*

■ brief, ephemeral, fleeting, fugitive, impermanent, passing, short, transient, transitory.

momentous *adj.* of great importance. □ **momentously** *adv.*

■ consequential, critical, crucial, historic, important, serious, significant, weighty.

momentum *n.* impetus gained by a moving body.

■ drive, energy, force, impetus, power, push, thrust.

monarch *n.* ruler with the title of king, queen, emperor, or empress. □ **monarchic** *adj.*, **monarchical** *adj.*

■ emperor, empress, king, potentate, queen, ruler, sovereign.

monarchist *n.* supporter of monarchy. □ **monarchism** *n.*

monarchy *n.* **1** government headed by a monarch. **2** country governed thus.

monastery *n.* residence of a community of monks.

monastic *adj.* of monks or monasteries. □ **monasticism** *n.*

monetarism *n.* control of the money supply as a method of curbing inflation. □ **monetarist** *n.*

monetary *adj.* of money or currency.

money *n.* **1** coins and banknotes. **2** (*pl.* **-eys**) sums of money. **3** wealth. □ **money-spinner** *n.* profitable thing.

■ **1** banknotes, *US* bills, cash, change, coins, currency, *sl.* dough, legal tender, *sl.* lolly, lucre, notes, *colloq.* wherewithal. **3** assets, funds, means, resources, riches, wealth.

moneyed *adj.* wealthy.

■ affluent, prosperous, rich, wealthy, *colloq.* well-heeled, well-off, well-to-do.

Mongol *n.* & *adj.* (native) of Mongolia.

mongoose *n.* (*pl.* **-gooses**) small flesh-eating tropical mammal.

mongrel *n.* animal (esp. a dog) of mixed breed. ● *adj.* of mixed origin or character.

monitor *n.* **1** device used to observe or test the operation of something. **2** pupil with special duties in a school. ● *v.* **1** act as monitor of. **2** maintain regular surveillance over.

■ *v.* examine, *colloq.* keep tabs on, observe, oversee, study, supervise, survey, watch.

monk *n.* member of a male religious community. □ **monkish** *adj.*

monkey *n.* (*pl.* **-eys**) **1** any of various primates, esp. a long-tailed one. **2** mischievous person. ● *v.* (**monkeyed**) tamper mischievously. □ **monkey-nut** *n.*

peanut. **monkey-puzzle** *n.* evergreen tree with sharp stiff leaves. **monkey wrench** wrench with an adjustable jaw.

mono *adj.* & *n.* (*pl.* **-os**) monophonic (sound or recording).

mono- *pref.* one, alone, single.

monochrome *adj.* done in only one colour or in black and white.

monocle *n.* single eyeglass.

monocular *adj.* with or for one eye.

monogamy *n.* system of being married to only one person at a time. □ **monogamous** *adj.*

monogram *n.* letters (esp. a person's initials) combined in a design. □ **monogrammed** *adj.*

monolith *n.* **1** single upright block of stone. **2** massive organization etc. □ **monolithic** *adj.*

monologue *n.* long speech.

monomania *n.* obsession with one idea or interest. □ **monomaniac** *n.*

monophonic *adj.* using only one transmission channel for reproduction of sound.

monoplane *n.* aeroplane with only one set of wings.

monopolize *v.* **1** have a monopoly of. **2** not allow others to share in. □ **monopolization** *n.*

■ control, corner, dominate, take control of, take over.

monopoly *n.* sole possession or control of something, esp. of trade in a specified commodity. □ **monopolist** *n.*

monorail *n.* railway in which the track is a single rail.

monosodium glutamate substance added to food to enhance its flavour.

monosyllable *n.* word of one syllable. □ **monosyllabic** *adj.*

monotheism *n.* doctrine that there is only one God. □ **monotheist** *n.*, **monotheistic** *adj.*

monotone *n.* level unchanging tone of voice.

monotonous *adj.* lacking in variety. □ **monotonously** *adv.*, **monotony** *n.*

■ boring, colourless, dreary, dull, repetitious, repetitive, soporific, tedious, unchanging, uneventful, unexciting, uninteresting, unvaried, unvarying, wearisome.

monsoon *n.* **1** seasonal wind in S. Asia. **2** rainy season accompanying this.

monster *n.* **1** imaginary creature, usually large and frightening. **2** very cruel or wicked person. **3** misshapen animal or plant. **4** large ugly animal or thing.

■ **1** dragon, giant, ogre, troll. **2** beast, brute, demon, fiend, ogre. **3** freak, mutant, mutation. **4** eyesore, horror, monstrosity.

monstrosity *n.* monstrous thing.

monstrous *adj.* **1** like a monster. **2** huge. **3** outrageous, atrocious.

■ **1** disgusting, dreadful, frightful, ghastly, grotesque, gruesome, hideous, horrible, nightmarish, repellent, repulsive, revolting, ugly. **2** colossal, enormous, gargantuan, giant, gigantic, huge, immense, mammoth, massive, titanic, tremendous, vast. **3** appalling, atrocious, barbaric, barbarous, disgraceful, dreadful, evil, foul, heinous, outrageous, scandalous, shameful, shocking, terrible, vile, wicked.

montage *n.* **1** making of a composite picture from pieces of others. **2** this picture. **3** joining of disconnected shots in a cinema film.

month *n.* **1** each of the twelve portions into which the year is divided. **2** period of 28 days.

monthly *adj.* & *adv.* (produced or occurring) once a month. ● *n.* monthly periodical.

monument *n.* thing that commemorates, esp. a structure, building, or memorial stone.

■ cairn, memorial, obelisk, shrine, statue, cenotaph, mausoleum, sepulchre, tomb, gravestone, headstone, tombstone.

monumental *adj.* **1** extremely great, stupendous. **2** massive. **3** of or serving as a monument.

■ **1** amazing, awesome, awe-inspiring, impressive, magnificent, marvellous, outstanding, prodigious, remarkable, spectacular, splendid, stupendous, wonderful. **2** colossal, enormous, gigantic, huge, immense, massive, tremendous, vast. **3** commemorative, memorial.

moo *n.* cow's low deep cry. ● *v.* make this sound.

mooch *v.* (*colloq.*) walk slowly and aimlessly.

mood *n.* **1** temporary state of mind or spirits. **2** fit of bad temper or depression.

■ **1** disposition, frame of mind, humour, state of mind, temper.

moody *adj.* (**-ier**, **-iest**) **1** given to changes of mood. **2** bad-tempered, sullen.

■ **1** capricious, changeable, erratic, mercurial, temperamental, unpredictable, volatile. **2** bad-tempered, cantankerous

crabby, crotchety, irritable, peevish, petulant, snappy, testy, touchy; dour, gloomy, glum, lugubrious, morose, saturnine, sulky, sullen.

moon *n.* **1** earth's satellite, made visible by light it reflects from the sun. **2** natural satellite of any planet. ● *v.* behave dreamily.

moonlight *n.* light from the moon. ● *v.* (*colloq.*) have two paid jobs, one by day and the other in the evening.

moonlit *adj.* lit by the moon.

moonstone *n.* pearly semiprecious stone.

Moor *n.* member of a Muslim people of north-west Africa. □ **Moorish** *adj.*

moor[1] *n.* stretch of open uncultivated land with low shrubs.

moor[2] *v.* secure (a boat etc.) to a fixed object.

moorhen *n.* small waterbird.

moorings *n.pl.* cables or place for mooring a boat.

moose *n.* (*pl.* **moose**) elk of N. America.

mop *n.* **1** pad or bundle of yarn on a stick, used for cleaning things. **2** thick mass of hair. ● *v.* (**mopped**) clean with a mop. □ **mop up** wipe up with a mop or cloth etc.

mope *v.* be unhappy and listless.

moped *n.* motorized bicycle.

moraine *n.* mass of stones etc. deposited by a glacier.

moral *adj.* **1** concerned with right and wrong conduct. **2** virtuous. ● *n.* **1** moral lesson or principle. **2** (*pl.*) person's standards of behaviour. □ **moral support** encouragement. **moral victory** a triumph though without concrete gain. **morally** *adv.*

■ *adj.* **1 ethical. 2 decent, ethical, fair, good, honest, honourable, incorruptible, principled, proper, righteous, upright, virtuous.** ● *n.* **1 lesson, message, point, teaching. 2 (morals) ethics, morality, principles, standards, values.**

morale /mərraal/ *n.* state of a person's or group's spirits and confidence.

moralist *n.* person who expresses or teaches moral principles.

morality *n.* **1** moral principles or rules. **2** goodness or rightness.

■ **1 ethics, morals, principles, standards. 2 decency, goodness, honesty, honour, integrity, probity, rectitude, right, virtue.**

moralize *v.* talk or write about the morality of something, esp. in a self-righteous way.

■ **pontificate, preach, sermonize.**

morass *n.* **1** complex entanglement. **2** marsh, bog.

■ **1 confusion, entanglement, muddle, tangle. 2 bog, fen, marsh, mire, quagmire, slough, swamp.**

moratorium *n.* (*pl.* **-ums**) temporary agreed ban on an activity.

morbid *adj.* (of mind, ideas, etc.) having or showing an interest in gloomy or unpleasant things. □ **morbidly** *adv.*, **morbidity** *n.*

■ **ghoulish, gruesome, macabre, sick, unhealthy, unwholesome.**

mordant *adj.* (of wit etc.) caustic.

■ **biting, caustic, cutting, sarcastic, scathing, stinging, trenchant, vitriolic.**

more *adj.* greater in quantity or intensity etc. ● *n.* greater amount or number. ● *adv.* **1** to a greater extent. **2** again. □ **more or less** approximately.

■ *adj.* **added, additional, extra, fresh, further, supplementary.**

moreover *adv.* besides.

■ **also, besides, further, furthermore, in addition, what is more.**

morgue /morg/ *n.* mortuary.

moribund *adj.* in a dying state.

Mormon *n.* member of a Christian sect founded in the USA.

morning *n.* part of the day before noon or the midday meal.

morocco *n.* goatskin leather.

moron *n.* **1** adult with a mental age of 8–12. **2** (*colloq.*) stupid person. □ **moronic** *adj.*

morose *adj.* gloomy, sullen. □ **morosely** *adv.*, **moroseness** *n.*

■ **dejected, depressed, gloomy, glum, lugubrious, melancholy, miserable, moody, sad, saturnine, sombre, sulky, sullen, unhappy.**

morphia *n.* morphine.

morphine *n.* drug made from opium, used to relieve pain.

morphology *n.* study of forms of animals and plants or of words. □ **morphological** *adj.*

morris dance English folk dance by people in costume.

Morse code code of signals using short and long sounds or flashes of light.

morsel *n.* **1** mouthful. **2** small piece (esp. of food).

■ **1** bite, mouthful, nibble, spoonful, taste. **2** bit, crumb, grain, fragment, particle, piece, scrap, shred, sliver.

mortal *adj.* **1** subject to death. **2** fatal. **3** implacable. ● *n.* mortal being. ▫ **mortally** *adv.*

■ *adj.* **1** earthly, human, temporal, worldly; ephemeral, transient, transitory. **2** deadly, fatal, lethal, terminal. **3** bitter, implacable, relentless, unrelenting. ● *n.* creature, human (being), individual, man, person, soul, woman.

mortality *n.* **1** being mortal. **2** loss of life on a large scale. **3** death rate.

mortar *n.* **1** mixture of lime or cement with sand and water for joining bricks or stones. **2** bowl in which substances are pounded with a pestle. **3** short cannon.

mortarboard *n.* stiff square cap worn as part of academic dress.

mortgage /mórgij/ *n.* **1** loan for purchase of property, in which the property itself is pledged as security. **2** agreement effecting this. ● *v.* pledge (property) as security in this way.

mortgagee /mórgijeé/ *n.* borrower in a mortgage.

mortgager /mórgijər/ *n.* (also **mortgagor**) lender in a mortgage.

mortify *v.* **1** humiliate greatly. **2** (of flesh) become gangrenous. ▫ **mortification** *n.*

■ **1** abash, chasten, crush, embarrass, humble, humiliate, shame.

mortise *n.* hole in one part of a framework shaped to receive the end of another part. ▫ **mortise lock** lock set in (not on) a door.

mortuary *n.* place where dead bodies may be kept temporarily.

mosaic *n.* pattern or picture made with small pieces of glass or stone of different colours.

Moslem *adj.* & *n.* = **Muslim**.

mosque *n.* Muslim place of worship.

mosquito *n.* (*pl.* **-oes**) small biting insect.

moss *n.* small flowerless plant forming a dense growth in moist places. ▫ **mossy** *adj.*

most *adj.* greatest in quantity or intensity etc. ● *n.* greatest amount or number. ● *adv.* **1** to the greatest extent. **2** very. ▫ **at most** not more than. **for the most part 1** mainly. **2** usually.

mostly *adv.* for the most part.

motel *n.* roadside hotel for motorists.

moth *n.* **1** nocturnal insect like a butterfly. **2** similar insect whose larvae feed on cloth or fur. ▫ **moth-eaten** *adj.* damaged by moths.

mothball *n.* small ball of pungent substance for keeping moths away from clothes.

mother *n.* **1** female parent. **2** title of the female head of a religious community. ● *v.* look after in a motherly way. ▫ **mother-in-law** *n.* (*pl.* **mothers-in-law**) mother of one's wife or husband. **mother-of-pearl** *n.* pearly substance lining shells of oysters and mussels etc. **mother tongue** one's native language. **motherhood** *n.*, **motherless** *adj.*

motherland *n.* native country.

motherly *adj.* showing a mother's kindness.

■ affectionate, caring, devoted, fond, gentle, kind, loving, maternal, nurturing, protective, tender, warm.

motif *n.* recurring design, feature, or melody.

motion *n.* **1** moving. **2** movement. **3** formal proposal put to a meeting for discussion. **4** emptying of the bowels, faeces. ● *v.* make a gesture directing (a person) to do something. ▫ **motion picture** cinema film.

■ *n.* **1** action, movement, moving, progress, transit, travelling. **2** gesticulation, gesture, movement, sign, signal. **3** proposal, proposition, suggestion. ● *v.* gesticulate, gesture, sign, signal, wave.

motionless *adj.* not moving.

■ immobile, static, stationary, still, stock-still, unmoving.

motivate *v.* **1** supply a motive to. **2** cause to feel active interest. ▫ **motivation** *n.*

■ **1** cause, drive, encourage, incite, induce, influence, inspire, lead, move, persuade, prompt, spur, stimulate. **2** excite, galvanize, inspire, rouse, stir.

motive *n.* that which induces a person to act in a certain way. ● *adj.* producing movement or action.

■ *n.* incentive, inducement, motivation, stimulus; ground(s), justification, rationale, reason.

motley *adj.* **1** multicoloured **2** assorted. ● *n.* (*old use*) jester's particoloured costume.

motor *n.* **1** machine supplying motive power. **2** car. ● *adj.* **1** producing motion. **2** driven by a motor. ▫ **motor bike**

(*colloq.*) motor cycle. **motor cycle** motor-driven two-wheeled vehicle. **motor cyclist** rider of a motor cycle. **motor vehicle** vehicle with a motor engine, for use on ordinary roads.

motorcade *n.* procession or parade of motor vehicles.

motorist *n.* driver of a car.

motorize *v.* equip with motor(s) or motor vehicles.

motorway *n.* road designed for fast long-distance traffic.

mottled *adj.* patterned with irregular patches of colour.

■ **blotched, blotchy, brindled, dappled, flecked, freckled, marbled, piebald, pied, speckled, spotted, stippled, streaked.**

motto *n.* (*pl.* **-oes**) **1** short sentence or phrase expressing an ideal or rule of conduct. **2** maxim, riddle, etc., inside a paper cracker.

■ **1 aphorism, axiom, maxim, proverb, saw, saying.**

mould[1] *n.* **1** hollow container into which a liquid is poured to set in a desired shape. **2** pudding etc. made in this. ● *v.* **1** shape. **2** influence the development of.

■ *n.* **1 cast, die, matrix.** ● *v.* **1 fashion, form, make, model, sculpt, sculpture, shape, work. 2 control, direct, form, guide, influence, shape.**

mould[2] *n.* furry growth of tiny fungi on a damp substance. □ **mouldy** *adj.*

mould[3] *n.* soft fine earth rich in organic matter.

moulder *v.* decay and rot away.

moult *v.* shed feathers, hair, or skin before new growth. ● *n.* process of moulting.

mound *n.* **1** mass of piled-up earth, stones, etc. **2** heap, pile. **3** small hill.

■ **1 barrow. 2 heap, mountain, pile, stack. 3 hill, hillock, hummock, knoll, rise.**

mount[1] *n.* mountain, hill.

mount[2] *v.* **1** go up. **2** get on a horse etc. **3** increase. **4** fix on or in support(s) or setting. **5** organize, arrange. ● *n.* **1** thing on which something is fixed. **2** horse for riding.

■ *v.* **1 ascend, clamber up, climb (up), go up, make one's way up, scale. 3 escalate, grow, increase, intensify, rise; accumulate, build up, multiply, pile up. 4 frame. 5 arrange, coordinate, organize, prepare, stage.**

mountain *n.* **1** mass of land rising to a great height. **2** large heap or pile. □ **mountain ash** rowan tree. **mountain bike** strong bicycle suitable for riding on rough hilly ground.

■ **1 elevation, eminence, mount, peak, tor. 2 heap, mass, mound, pile, stack.**

mountaineer *n.* person who climbs mountains. ● *v.* climb mountains as a recreation.

mountainous *adj.* **1** full of mountains. **2** huge.

■ **1 alpine, craggy, hilly. 2 colossal, enormous, gigantic, high, huge, immense, massive, mighty, towering.**

mourn *v.* feel or express sorrow or regret about (a dead person or lost thing). □ **mourner** *n.*

■ **grieve, lament, sorrow, weep; regret, rue.**

mournful *adj.* sorrowful. □ **mournfully** *adv.*

■ **desolate, despondent, disconsolate, dispirited, doleful, downcast, downhearted, forlorn, heavy-hearted, lugubrious, melancholy, miserable, sad, sorrowful, woebegone, woeful.**

mourning *n.* **1** expression of sorrow for a dead person, esp. by wearing dark clothes. **2** such clothes.

mouse *n.* (*pl.* **mice**) **1** small rodent with a long tail. **2** quiet timid person. **3** small rolling device for moving the cursor on a VDU screen.

moussaka *n.* Greek dish of minced meat and aubergine.

mousse *n.* **1** frothy creamy dish. **2** substance of similar texture.

moustache /məstaásh/ *n.* hair on the upper lip.

mousy *adj.* **1** dull greyish-brown. **2** quiet and timid.

mouth *n.* /mowth/ **1** opening in the head through which food is taken in and sounds uttered. **2** opening of a bag, cave, cannon, etc. **3** place where a river enters the sea. ● *v.* /mowth/ form (words) soundlessly with the lips. □ **mouth-organ** *n.* small instrument played by blowing and sucking.

mouthful *n.* **1** quantity of food etc. that fills the mouth. **2** small quantity.

mouthpiece *n.* **1** part of an instrument placed between or near the lips. **2** spokesperson.

mouthwash *n.* liquid for cleansing the mouth.

move *v.* **1** (cause to) be in motion, or change place or position. **2** progress. **3** change one's residence. **4** provoke an emotion or reaction in. **5** take action. **6** propose for discussion. ● *n.* **1** act of moving. **2** calculated action. **3** moving of a piece in chess etc. □ **movable** *adj.*, **mover** *n.*

■ *v.* **1** budge, make a move, stir; carry, shift, transfer, transport. **2** advance, go, pass, proceed, progress, travel, walk. **3** emigrate, leave, relocate. **4** affect, have an effect on, have an impact on, hit, impress, make an impression on, touch; dispose, incline, influence, inspire, lead, motivate, persuade, prompt, provoke, rouse. **5** act, take action, take the initiative. **6** advance, propose, put forward, submit, suggest. ● *n.* **1** action, motion, movement; relocation, shift, transfer. **2** act, gambit, manoeuvre, ploy, ruse, stratagem, tactic; initiative, step.

movement *n.* **1** act or instance of moving or being moved. **2** moving parts. **3** group with a common cause. **4** campaign undertaken by such a group. **5** section of a long piece of music.

■ **1** flow, migration, move, relocation, shift, transfer; action, activity, stir, stirring; gesticulation, gesture, motion, sign, signal. **2** machinery, mechanism, works. **3** faction, group, lobby, party. **4** campaign, crusade, drive.

movie *n.* (*US colloq.*) cinema film.

moving *adj.* arousing pity or sympathy. □ **movingly** *adv.*

■ affecting, emotional, emotive, pathetic, poignant, touching.

mow *v.* (**mown**) cut (grass etc.) with a scythe or a machine. □ **mow down** kill or destroy by a moving force. **mower** *n.*

m.p.h. *abbr.* miles per hour.

Mr *n.* (*pl.* **Messrs**) title prefixed to a man's name.

Mrs *n.* (*pl.* **Mrs**) title prefixed to a married woman's name.

Ms *n.* title prefixed to a married or unmarried woman's name.

much *adj.* & *n.* (existing in) great quantity. ● *adv.* **1** in a great degree. **2** to a great extent.

mucilage *n.* **1** sticky substance obtained from plants. **2** adhesive gum.

muck *n.* **1** farmyard manure. **2** (*colloq.*) dirt. □ **mucky** *adj.*

muckraking *n.* seeking and exposing scandal.

mucous *adj.* like or covered with mucus.

mucus *n.* slimy substance coating the inner surface of hollow organs of the body.

mud *n.* wet soft earth.

■ dirt, mire, *colloq.* muck, silt, sludge.

muddle *v.* **1** bring into disorder. **2** bewilder. ● *n.* muddled condition or things. □ **muddle along** progress in a haphazard way. **muddle through** succeed by perseverance rather than skill.

■ *v.* **1** confuse, disarrange, disorganize, jumble, mess up, mix up, scramble. **2** bemuse, bewilder, confuse, confound, mystify, perplex, puzzle. ● *n.* clutter, confusion, disarray, disorder, hotchpotch, jumble, mess, tangle.

muddy *adj.* (**-ier**, **-iest**) **1** like mud. **2** covered in or full of mud. **3** (of liquid etc.) not clear. ● *v.* make muddy.

■ *adj.* **2** dirty, filthy, grubby, miry, *colloq.* mucky, muddied, soiled; boggy, marshy, swampy. **3** clouded, cloudy, opaque, turbid.

mudguard *n.* curved cover above a wheel as a protection against spray.

muesli *n.* food of mixed crushed cereals, dried fruit, nuts, etc.

muezzin /moo-ézzin/ *n.* man who proclaims the hours of prayer for Muslims.

muff[1] *n.* tube-shaped usu. furry covering for the hands.

muff[2] *v.* (*colloq.*) bungle.

muffin *n.* light round cake eaten toasted and buttered.

muffle *v.* wrap for warmth or protection, or to deaden sound.

■ cloak, cover, envelop, swaddle, swathe, wrap; deaden, dull, silence, quieten, smother, stifle, suppress.

muffler *n.* scarf worn for warmth.

mufti *n.* civilian clothes.

mug *n.* **1** large drinking vessel with a handle, for use without a saucer. **2** (*sl.*) face. **3** (*sl.*) person who is easily outwitted. ● *v.* (**mugged**) attack and rob, esp. in a public place. □ **mugger** *n.*

muggy *adj.* (**-ier**, **-iest**) (of weather) oppressively damp and warm. □ **mugginess** *n.*

■ close, damp, humid, oppressive, steamy, sticky, sultry.

mulberry *n.* **1** edible purple or white fruit. **2** tree bearing this. **3** dull purplish-red.

mulch *n.* mixture of wet straw, leaves, etc., spread on ground to protect plants or retain moisture. ● *v.* cover with mulch.

mulct *v.* take money from (a person) by a fine, taxation, etc.

mule[1] *n.* animal that is the offspring of a horse and a donkey.

mule[2] *n.* backless slipper.

mull[1] *v.* heat (wine etc.) with sugar and spices, as a drink.

mull[2] *v.* **mull over** think over.

■ cogitate on, consider, contemplate, meditate on, muse on, ponder, reflect on, ruminate on, think about *or* over, weigh.

mullah *n.* Muslim learned in Islamic law.

mullet *n.* small edible sea fish.

mulligatawny *n.* highly seasoned soup.

mullion *n.* upright bar between the sections of a tall window.

multi- *pref.* many.

multicultural *adj.* of or involving several cultural or ethnic groups. □ **multiculturalism** *n.*

multifarious *adj.* very varied.

multinational *adj.* & *n.* (business company) operating in several countries.

multiple *adj.* having or affecting many parts. ● *n.* quantity exactly divisible by another.

multiplex *adj.* having many elements.

multiplication *n.* multiplying.

multiplicity *n.* great variety.

■ abundance, diversity, profusion, range, variety.

multiply *v.* **1** add a quantity to itself a specified number of times. **2** increase in number.

■ **2** grow, increase, mushroom, pile up, proliferate, snowball, spread.

multiracial *adj.* of or involving people of several races.

multitude *n.* great number of things or people.

■ crowd, flock, horde, host, legion, mass, myriad, swarm, throng.

multitudinous *adj.* very numerous.

mum[1] *adj.* (*colloq.*) silent.

mum[2] *n.* (*colloq.*) mother.

mumble *v.* speak or utter indistinctly. ● *n.* indistinct speech.

mumbo-jumbo *n.* **1** meaningless ritual. **2** deliberately obscure language.

mummify *v.* preserve (a corpse) by embalming as in ancient Egypt. □ **mummification** *n.*

mummy[1] *n.* corpse embalmed and wrapped for burial; esp. in ancient Egypt.

mummy[2] *n.* (*colloq.*) mother.

mumps *n.* virus disease with painful swellings in the neck.

munch *v.* chew vigorously.

■ champ, chew, crunch, masticate, scrunch.

mundane *adj.* **1** dull, routine. **2** worldly.

■ **1** boring, commonplace, dull, everyday, humdrum, ordinary, pedestrian, prosaic, routine, run-of-the-mill, tedious, unexciting, uninteresting. **2** earthly, secular, temporal, worldly.

municipal *adj.* of a town or city.

municipality *n.* self-governing town or district.

■ borough, city, district, metropolis, town.

munificent *adj.* splendidly generous. □ **munificence** *n.*

munitions *n.pl.* weapons, ammunition, etc., used in war.

mural *adj.* of or on a wall. ● *n.* a painting made on a wall.

murder *n.* intentional unlawful killing. ● *v.* kill intentionally and unlawfully. □ **murderer** *n.*

■ *n.* assassination, butchery, genocide, homicide, infanticide, killing, massacre, matricide, parricide, patricide, regicide, slaughter. ● *v.* assassinate, butcher, *sl.* do in, exterminate, kill, liquidate, massacre, put to death, slaughter, slay.

murderous *adj.* involving or capable of murder.

■ deadly, fatal, lethal, mortal; barbarous, bloodthirsty, brutal, cruel, ferocious, fierce, inhuman, homicidal, savage, vicious.

murk *n.* darkness, gloom. □ **murky** *adj.*

■ □ **murky** clouded, cloudy, dark, dim, dismal, dreary, funereal, gloomy, overcast, shadowy.

murmur *n.* **1** low continuous sound. **2** softly spoken words. ● *v.* **1** make a murmur. **2** speak or utter softly.

■ *n.* **1** hum, humming, murmuring, rumble, whirr. **2** mumble, whisper. ● *v.* **1** hum, rumble, whirr. **2** mumble, mutter, whisper.

murrain *n.* infectious disease of cattle.

muscle /múss'l/ *n.* **1** strip of fibrous tissue able to contract and so move a part of the body. **2** muscular power. **3** strength. ● *v.* **muscle in** (*colloq.*) force one's way.

muscular *adj.* **1** of muscles. **2** having well-developed muscles. □ **muscularity** *n.*

■ **2** athletic, beefy, brawny, burly, husky, powerful, robust, sinewy, strapping, strong, sturdy.

muse *v.* ponder.

■ brood, cogitate, consider, deliberate, meditate, ponder, reflect, ruminate, think.

museum *n.* place where objects of historical interest are collected and displayed.

mush *n.* soft pulp.

mushroom *n.* edible fungus with a stem and a domed cap. ● *v.* appear or develop rapidly.

mushy *adj.* **1** as or like mush. **2** feebly sentimental.

music *n.* **1** arrangement of sounds of one or more voices or instruments. **2** written form of this.

musical *adj.* **1** of or involving music. **2** fond of or skilled in music. **3** sweet-sounding. ● *n.* play with songs and dancing. □ **musically** *adv.*

■ **3** dulcet, euphonious, mellifluous, melodic, melodious, sweet-sounding, tuneful.

musician *n.* person skilled in music.

musicology *n.* study of the history and forms of music. □ **musicologist** *n.*

musk *n.* substance secreted by certain animals or produced synthetically, used in perfumes. □ **musky** *adj.*

musket *n.* long-barrelled gun formerly used by infantry.

Muslim *adj.* of or believing in Muhammad's teaching. ● *n.* believer in this faith.

muslin *n.* thin cotton cloth.

musquash *n.* **1** N. American water animal. **2** its fur.

mussel *n.* edible bivalve mollusc.

must *v.aux.* used to express necessity or obligation, certainty, or insistence. ● *n.* (*colloq.*) thing that must be done or visited etc.

mustang *n.* wild horse of Mexico and California.

mustard *n.* **1** sharp-tasting yellow condiment made from the seeds of a plant. **2** this plant.

musty *adj.* (**-ier**, **-iest**) mouldy, stale. □ **mustiness** *n.*

■ damp, mildewed, mouldy; airless, fusty, stale.

mutable *adj.* liable to change. □ **mutability** *n.*

mutant *adj.* & *n.* (living thing) differing from its parents as a result of genetic change.

mutate *v.* change in form.

mutation *n.* **1** change in form. **2** mutant.

■ **1** alteration, change, metamorphosis, modification, transformation, transmutation. **2** freak, monstrosity, mutant.

mute *adj.* **1** silent. **2** dumb. ● *n.* **1** dumb person. **2** device muffling the sound of a musical instrument. ● *v.* deaden or muffle the sound of. □ **mutely** *adv.*

■ *adj.* **1** *colloq.* mum, quiet, silent, speechless, tongue-tied; tacit, undeclared, unspoken. **2** dumb, voiceless.

mutilate *v.* injure or disfigure by cutting off a part. □ **mutilation** *n.*

■ cripple, damage, disable, disfigure, maim, mangle.

mutineer *n.* person who mutinies.

mutinous *adj.* rebellious, ready to mutiny. □ **mutinously** *adv.*

■ insurgent, rebellious, revolutionary, seditious, subversive; defiant, disobedient, insubordinate, recalcitrant, refractory, uncontrollable, ungovernable, unruly.

mutiny *n.* rebellion against authority, esp. by members of the armed forces. ● *v.* engage in mutiny.

■ *n.* insurgency, insurrection, rebellion, revolt, uprising. ● *v.* rebel, revolt.

mutter *v.* **1** speak or utter in a low unclear tone. **2** utter subdued grumbles. ● *n.* muttering.

■ *v.* **1** mumble, murmur, whisper. **2** complain, grumble, *colloq.* grouch, *colloq.* gripe, moan.

mutton *n.* flesh of sheep as food.

mutual *adj.* **1** felt or done by each to the other. **2** (*colloq.*) common to two or more. □ **mutually** *adv.*, **mutuality** *n.*

■ **1** reciprocal, reciprocated. **2** common, communal, joint, shared.

muzzle *n.* **1** projecting nose and jaws of certain animals. **2** open end of a firearm. **3** strap etc. over an animal's head to prevent it from biting or feeding. ● *v.* **1** put a muzzle on. **2** prevent from expressing opinions freely.

muzzy *adj.* dazed, confused.

my *adj.* belonging to me.

mycology *n.* study of fungi.

myna *n.* bird of the starling family that can mimic sounds.

myopia /mīṓpiə/ *n.* short sight. □ **myopic** /-óppik/ *adj.*

myriad *n.* vast number.

myrrh /mur/ *n.* gum resin used in perfumes, incense, etc.

myrtle *n.* evergreen shrub.

myself *pron.* emphatic and reflexive form of *I* and *me*.

mysterious *adj.* full of mystery, puzzling. □ **mysteriously** *adv.*

■ baffling, bewildering, confusing, curious, enigmatic, inexplicable, incomprehensible, inscrutable, mystifying, perplexing, puzzling, strange, uncanny, unfathomable, weird; abstruse, arcane, dark, occult, recondite, secret.

mystery *n.* **1** a matter that remains unexplained. **2** quality of being unexplained or obscure. **3** story dealing with a puzzling crime.

■ **1** conundrum, enigma, puzzle, riddle. **2** ambiguity, inscrutability, obscurity, secrecy. **3** detective story, thriller, *colloq.* whodunit.

mystic *adj.* **1** having a hidden or symbolic meaning, esp. in religion. **2** inspiring a sense of mystery and awe. ● *n.* person who seeks to obtain union with God by spiritual contemplation. □ **mystical** *adj.*, **mystically** *adv.*, **mysticism** *n.*

mystify *v.* cause to feel puzzled. □ **mystification** *n.*

■ baffle, *colloq.* bamboozle, bewilder, confound, confuse, *colloq.* flummox, perplex, puzzle, *colloq.* stump.

mystique *n.* aura of mystery or mystical power.

myth *n.* **1** traditional tale(s) containing beliefs about ancient times or natural events. **2** imaginary person or thing. □ **mythical** *adj.*

■ **1** fable, legend, story, folk-tale. □ **mythical** fabled, fabulous, fairy-tale, legendary, mythological.

mythology *n.* **1** myths. **2** study of myths. □ **mythological** *adj.*

■ **1** folklore, legend, lore, myth(s), tradition.

myxomatosis *n.* fatal virus disease of rabbits.

Nn

N. *abbr.* **1** north. **2** northern.

nab *v.* (**nabbed**) (*sl.*) **1** catch in wrong-doing, arrest. **2** seize.

nadir *n.* lowest point.

naevus /néevəss/ *n.* (*pl.* **-vi**) red birth-mark.

nag¹ *n.* (*colloq.*) horse.

nag² *v.* (**nagged**) **1** criticize or scold persistently. **2** (of pain) be felt persistently.

> ■ **1** berate, carp at, criticize, *colloq.* get at, find fault with, *colloq.* keep on at, pick on, scold.

naiad /níad/ *n.* water nymph.

nail *n.* **1** layer of horny substance over the outer tip of a finger or toe. **2** claw. **3** small metal spike. ● *v.* **1** fasten with nail(s). **2** catch, arrest.

naive /naa-éev/ *adj.* showing lack of experience or judgement. □ **naively** *adv.*, **naivety** *n.*

> ■ artless, childlike, guileless, ingenuous, innocent, trustful, unsophisticated, unworldly; callow, credulous, green, gullible, immature, inexperienced.

naked *adj.* **1** without clothes on. **2** without coverings. □ **naked eye** the eye unassisted by a telescope or microscope etc. **nakedness** *n.*

> ■ **1** bare, nude, unclothed, undressed. **2** uncovered, unprotected, unsheathed.

namby-pamby *adj.* & *n.* feeble or unmanly (person).

name *n.* **1** word(s) by which a person, place, or thing is known or indicated. **2** reputation. ● *v.* **1** give as a name. **2** nominate. **3** mention, specify. □ **in name only** as a mere formality.

> ■ *n.* **1** denomination, designation, term, title; alias, nickname, *nom de plume*, pen-name, pseudonym. ● *v.* **1** call, dub, label; baptize, christen. **2** appoint, choose, designate, elect, nominate, select. **3** cite, identify, mention, specify.

namely *adv.* that is to say.

namesake *n.* person or thing with the same name as another.

nanny *n.* child's nurse. □ **nanny goat** female goat.

nano- *pref.* one thousand millionth.

nap¹ *n.* short sleep, esp. during the day. ● *v.* (**napped**) have a nap. □ **catch a person napping** catch him or her unawares.

> ■ *n.* catnap, doze, rest, siesta, sleep, snooze.

nap² *n.* short raised fibres on the surface of cloth or leather.

napalm /náypaam/ *n.* jelly-like petrol substance used in incendiary bombs.

nape *n.* back part of neck.

naphtha /náfthə/ *n.* inflammable oil.

naphthalene *n.* pungent white substance obtained from coal tar.

napkin *n.* **1** piece of cloth or paper used to protect clothes or for wiping one's lips at meals. **2** nappy.

nappy *n.* piece of absorbent material worn by a baby to absorb or retain its excreta.

narcissism *n.* abnormal self-admiration. □ **narcissistic** *adj.*

> ■ conceit, egotism, self-admiration, self-love, vanity.

narcissus *n.* (*pl.* **-cissi**) flower of the group including the daffodil.

narcotic *adj.* & *n.* (drug) causing sleep or drowsiness.

narrate *v.* tell (a story), give an account of. □ **narration** *n.*, **narrator** *n.*

> ■ chronicle, describe, detail, recount, relate, report, tell.

narrative *n.* spoken or written account of something. ● *adj.* in this form.

> ■ *n.* account, chronicle, description, history, report, story, tale.

narrow *adj.* **1** small across, not wide. **2** of limited scope. **3** with little margin. ● *v.* make or become narrower. □ **narrow-minded** *adj.* restricted in one's views, intolerant. **narrowly** *adv.*, **narrowness** *n.*

> ■ *adj.* **1** attenuated, slender, slim, tapering, thin; confined, constricted, cramped. **2** circumscribed, limited, restricted. **3** close, lucky, near. ● *v.* constrict, limit, reduce, restrict; contract, decrease, diminish, lessen. □ **narrow-minded** bigoted, conservative, hidebound, insular, intolerant, parochial, prejudiced, provincial, small-minded, strait-laced, *colloq.* stuffy.

narwhal *n.* Arctic whale with a spirally grooved tusk.

nasal *adj.* **1** of the nose. **2** sounding as if breath came out through the nose. □ **nasally** *adv.*

nascent *adj.* just coming into existence. □ **nascence** *n.*

nasty *adj.* (**-ier**, **-iest**) **1** unpleasant. **2** unkind. **3** difficult. □ **nastily** *adv.*, **nastiness** *n.*

■ **1** disagreeable, disgusting, distasteful, foul, *colloq.* horrible, horrid, loathsome, objectionable, obnoxious, odious, offensive, repellent, repugnant, repulsive, revolting, unpleasant, unsavoury, vile. **2** cruel, malicious, mean, spiteful, unkind. **3** difficult, problematic(al), tricky.

natal *adj.* of or from one's birth.

nation *n.* people of mainly common descent and history usu. inhabiting a particular country under one government.

■ country, land, realm, state.

national *adj.* of or common to a nation. ● *n.* citizen of a particular country. □ **nationally** *adv.*

nationalism *n.* **1** patriotic feeling. **2** policy of national independence. □ **nationalist** *n.*, **nationalistic** *adj.*

nationality *n.* condition of belonging to a particular nation.

nationalize *v.* convert from private to state ownership. □ **nationalization** *n.*

native *adj.* **1** natural, inborn. **2** belonging to a place by birth. ● *n.* **1** person born in a specified place. **2** local inhabitant.

■ *adj.* **1** inborn, inherent, innate, instinctive, natural. **2** aboriginal, indigenous. ● *n.* **2** citizen, inhabitant, local, national, resident.

nativity *n.* **1** birth. **2** (**the Nativity**) that of Christ.

natter *v.* & *n.* (*colloq.*) chat.

natural *adj.* **1** of or produced by nature. **2** normal. **3** innate. **4** not seeming artificial or affected. ● *n.* person or thing that seems naturally suited for something. □ **natural history** study of animals and plants. **naturalness** *n.*

■ *adj.* **2** common, customary, everyday, habitual, normal, regular, routine, standard, typical, usual. **3** inborn, inherent, innate, instinctive, native. **4** artless, easy, genuine, guileless, ingenuous, spontaneous, straightforward, unaffected, unsophisticated, unstudied.

naturalism *n.* realism in art and literature. □ **naturalistic** *adj.*

naturalist *n.* expert in natural history.

naturalize *v.* **1** admit (a foreigner) to full citizenship of a country. **2** introduce (a plant etc.) into a region. **3** make look natural. □ **naturalization** *n.*

naturally *adv.* **1** in a natural manner. **2** as might be expected, of course.

nature *n.* **1** the world with all its features and living things. **2** physical power producing these. **3** kind, sort. **4** thing's or person's essential qualities or character.

■ **3** category, class, description, kind, sort, type, variety. **4** character, disposition, make-up, personality, temperament; attributes, essence, features, properties, qualities.

naturist *n.* nudist. □ **naturism** *n.*

naughty *adj.* (**-ier**, **-iest**) **1** behaving badly, disobedient. **2** slightly indecent. □ **naughtily** *adv.*, **naughtiness** *n.*

■ **1** devilish, disobedient, mischievous, obstreperous, rascally, refractory, roguish, uncooperative, unmanageable, ungovernable, unruly, wayward, wicked.

nausea *n.* feeling of sickness.

nauseate *v.* affect with nausea.

nauseous *adj.* affected with or causing nausea.

nautical *adj.* of sailors or seamanship.

nautilus *n.* (*pl.* **-luses**) mollusc with a spiral shell.

naval *adj.* of a navy.

nave *n.* main part of a church.

navel *n.* small hollow in the centre of the abdomen.

navigable *adj.* **1** suitable for ships. **2** able to be steered and sailed. □ **navigability** *n.*

navigate *v.* **1** sail in or through (a sea or river etc.). **2** direct the course of (a ship or vehicle etc.). □ **navigation** *n.*, **navigator** *n.*

■ **1** cross, cruise, sail (across). **2** direct, guide, manoeuvre, pilot, steer.

navvy *n.* labourer making roads etc. where digging is necessary.

navy *n.* **1** a country's warships. **2** crew of these. **3** navy blue. □ **navy blue** very dark blue.

NB *abbr.* (Latin *nota bene*) note well.

NE *abbr.* **1** north-east. **2** north-eastern.

neap tide tide when there is least rise and fall of water.

near *adv.* **1** at, to, or within a short distance or interval. **2** nearly. ● *prep.* near to. ● *adj.* **1** close to, in place or time. **2**

closely related. **3** with little margin. **4** of the left side of a horse, vehicle, or road. **5** stingy. ● *v.* draw near. ◻ **nearness** *n.*

■ *adv.* **1** close (by), nearby, in the vicinity, not far away, within reach. ● *adj.* **1** adjacent, nearby, neighbouring; approaching, close, forthcoming, imminent, impending, looming. **2** close, intimate. **3** close, lucky, narrow. **5** close, mean, miserly, niggardly, parsimonious, penny-pinching, stingy, *colloq.* tight, tight-fisted.

nearby *adj.* & *adv.* near in position.

nearly *adv.* **1** closely. **2** almost.

■ **2** about, all but, almost, approximately, around, as good as, more or less, practically, virtually, well-nigh.

neat *adj.* **1** orderly in appearance or workmanship. **2** undiluted. ◻ **neatly** *adv.*, **neatness** *n.*

■ **1** orderly, organized, shipshape, smart, spruce, straight, tidy, trim, uncluttered. **2** pure, straight, unadulterated, undiluted.

neaten *v.* make neat.

■ put in order, smarten up, spruce up, straighten, tidy.

nebula *n.* (*pl.* **-ae**) a cloud of gas or dust in space. ◻ **nebular** *adj.*

nebulous *adj.* indistinct. ◻ **nebulously** *adv.*, **nebulosity** *n.*

■ dim, faint, foggy, fuzzy, hazy, indistinct, obscure, shadowy, unclear, vague.

necessarily *adv.* as a necessary result, inevitably.

necessary *adj.* **1** essential in order to achieve something. **2** happening or existing by necessity.

■ **1** compulsory, essential, imperative, indispensable, needed, needful, obligatory, required, requisite, vital. **2** inescapable, inevitable, unavoidable.

necessitate *v.* **1** make necessary. **2** involve as a condition or result.

necessitous *adj.* needy.

necessity *n.* **1** indispensable thing. **2** state of being indispensable. **3** pressure of circumstances. **4** poverty.

■ **1** essential, fundamental, need, prerequisite, requirement, requisite. **3** exigency, need, urgency. **4** destitution, hardship, indigence, need, penury, poverty, want.

neck *n.* **1** narrow part connecting the head to the body. **2** part of a garment round this. **3** narrow part of a bottle, cavity, etc. ◻ **neck and neck** running level in a race.

necklace *n.* piece of jewellery etc. worn round the neck.

neckline *n.* outline formed by the edge of a garment at the neck.

necromancy *n.* art of predicting things by communicating with the dead. ◻ **necromancer** *n.*

necrosis *n.* death of bone or tissue. ◻ **necrotic** *adj.*

nectar *n.* **1** sweet fluid from plants, collected by bees. **2** any delicious drink.

nectarine *n.* smooth-skinned variety of peach.

née /nay/ *adj.* (before a married woman's maiden name) born.

need *n.* **1** requirement. **2** circumstances requiring action. **3** crisis, emergency. **4** poverty. ● *v.* be in need of, require. ◻ **need to** be obliged to.

■ *n.* **1** demand, necessity, requirement, want. **3** crisis, difficulty, distress, emergency, exigency, trouble. **4** destitution, indigence, penury, poverty. ● *v.* demand, require, want; be without, lack.

needful *adj.* necessary.

needle *n.* **1** small thin pointed piece of steel used in sewing. **2** thing shaped like this. **3** pointer of a compass or gauge. ● *v.* annoy, provoke.

needless *adj.* unnecessary. ◻ **needlessly** *adv.*

■ gratuitous, pointless, superfluous, uncalled-for, unnecessary, unwarranted.

needlework *n.* sewing or embroidery.

needy *adj.* (**-ier**, **-iest**) very poor.

■ destitute, impecunious, impoverished, indigent, necessitous, on the breadline, penniless, penurious, poor, poverty-stricken.

nefarious *adj.* wicked.

negate *v.* nullify, disprove. ◻ **negation** *n.*

negative *adj.* **1** expressing denial, refusal, or prohibition. **2** lacking positive attributes. **3** (of a quantity) less than zero. **4** (of a battery terminal) through which electric current leaves. ● *n.* **1** negative statement or word. **2** negative quality or quantity. **3** photograph with lights and shades or colours reversed, from which positive pictures can be obtained. ◻ **negatively** *adv.*

■ *adj.* **1** anti-, contradictory, contrary, opposing. **2** apathetic, defeatist, pessimistic, unenthusiastic, uninterested, unresponsive.

neglect *v.* **1** fail to care for or do. **2** forget the need to. **3** pay no attention to.

● *n.* neglecting, being neglected. □ **neglectful** *adj.*

■ *v.* **2** fail, forget, omit. **3** disregard, ignore, overlook, pay no attention to. ● *n.* dereliction, laxity, negligence.

negligee /néglizhay/ *n.* woman's light dressing gown.

negligence *n.* lack of proper care or attention. □ **negligent** *adj.*, **negligently** *adv.*

■ □ **negligent** careless, heedless, inattentive, neglectful, remiss, thoughtless, unmindful, unthinking.

negligible *adj.* too small to be worth taking into account.

■ inconsequential, insignificant, minor, petty, slight, small, trifling, trivial, unimportant.

negotiate *v.* **1** hold a discussion so as to reach agreement. **2** arrange by such discussion. **3** get past (an obstacle) successfully. □ **negotiation** *n.*, **negotiator** *n.*

■ **1** bargain, haggle; debate, parley, speak, talk. **2** arrange, bring about, engineer, get, obtain, organize, pull off, work out.

Negress *n.* female Negro.

Negro *n.* (*pl.* **-oes**) member of the black-skinned race that originated in Africa.

neigh *n.* horse's long high-pitched cry. ● *v.* make this cry.

neighbour *n.* person living next door or nearby.

neighbourhood *n.* district, vicinity. □ **neighbourhood watch** systematic vigilance by residents to deter crime in their area.

■ area, district, environs, locality, quarter, region, vicinity.

neighbouring *adj.* living or situated nearby.

neighbourly *adj.* kind and friendly. □ **neighbourliness** *n.*

■ affable, agreeable, amiable, considerate, cordial, friendly, helpful, kind, kindly, obliging, sociable, well-disposed.

neither *adj., pron.,* & *adv.* not either.

nemesis /némmisiss/ *n.* inevitable retribution.

neo- *pref.* new.

neolithic *adj.* of the later part of the Stone Age.

neologism *n.* new word.

neon *n.* inert gas used in illuminated signs.

nephew *n.* one's brother's or sister's son.

nephritis *n.* inflammation of the kidneys.

nepotism *n.* favouritism shown to relatives in appointing them to jobs. □ **nepotistic** *adj.*

nerve *n.* **1** fibre carrying impulses of sensation or movement between the brain and a part of the body. **2** courage. **3** (*colloq.*) impudence. **4** (*pl.*) nervousness, effect of mental stress. ● *v.* give courage to. □ **get on a person's nerves** irritate a person.

■ *n.* **2** boldness, bravery, courage, daring, fearlessness, fortitude, *colloq.* grit, *colloq.* guts, intrepidity, mettle, pluck, spirit, valour. **3** cheek, effrontery, gall, impertinence, impudence, insolence, presumption, temerity. **4** (**nerves**) anxiety, *colloq.* the jitters, nervousness, tension, worry.

nervous *adj.* **1** of the nerves. **2** easily alarmed. **3** slightly afraid, anxious. □ **nervously** *adv.*, **nervousness** *n.*

■ **2** edgy, excitable, jumpy, highly-strung, nervy, *colloq.* uptight. **3** agitated, anxious, apprehensive, edgy, fearful, fidgety, frightened, *colloq.* in a tizzy, *colloq.* jittery, jumpy, keyed up, on tenterhooks, restless, scared, troubled, uneasy, worried.

nervy *adj.* nervous.

nest *n.* **1** structure or place in which a bird lays eggs and shelters its young. **2** breeding place, lair. **3** snug place, shelter. **4** set of articles (esp. tables) designed to fit inside each other. ● *v.* make or have a nest. □ **nest egg** money saved for future use.

nestle *v.* **1** settle oneself comfortably. **2** lie sheltered.

nestling *n.* bird too young to leave the nest.

net[1] *n.* **1** openwork material of thread, cord, or wire etc. **2** piece of this used for a particular purpose. ● *v.* (**netted**) **1** place nets in or on. **2** catch in a net.

■ *n.* **1** mesh, netting, openwork, web, webbing. **2** dragnet, fishing net, seine, trawl. ● *v.* **2** capture, catch, ensnare, snare, trap.

net[2] *adj.* **1** remaining after all deductions. **2** (of weight) not including wrappings etc. ● *v.* (**netted**) obtain or yield as net profit.

netball *n.* game in which a ball has to be thrown into a high net.

nether *adj.* lower.

netting *n.* netted fabric.

nettle *n.* wild plant with leaves that sting when touched. ● *v.* irritate, provoke.

□ **nettle-rash** *n.* skin eruption like nettle stings.

network *n.* **1** arrangement with intersecting lines. **2** complex system. **3** group of interconnected people or things.

neural *adj.* of nerves.

neuralgia *n.* sharp pain along a nerve. □ **neuralgic** *adj.*

neuritis *n.* inflammation of nerve(s).

neurology *n.* study of nerve systems. □ **neurological** *adj.*, **neurologist** *n.*

neurosis *n.* (*pl.* **-oses**) mental disorder producing depression or abnormal behaviour.

neurotic *adj.* **1** of or caused by a neurosis. **2** subject to abnormal anxieties or obsessive behaviour. ● *n.* neurotic person. □ **neurotically** *adv.*

neuter *adj.* **1** neither masculine nor feminine. **2** without male or female parts. ● *v.* castrate, spay.

neutral *adj.* **1** not supporting either side in a conflict. **2** without distinctive or positive characteristics. ● *n.* **1** neutral person, country, or colour. **2** neutral gear. □ **neutral gear** position of gear mechanism in which the engine is disconnected from driven parts. **neutrally** *adv.*, **neutrality** *n.*

> ■ *adj.* **1** non-belligerent, non-combatant; disinterested, impartial, unbiased, unprejudiced.

neutralize *v.* make ineffective. □ **neutralization** *n.*

> ■ cancel out, counteract, counterbalance, invalidate, negate, offset, undo.

neutrino *n.* (*pl.* **-os**) particle with zero electric charge and (probably) zero mass.

neutron *n.* nuclear particle with no electric charge. □ **neutron bomb** nuclear bomb that kills people but does little damage to buildings etc.

never *adv.* **1** at no time, on no occasion. **2** not. **3** (*colloq.*) surely not. □ **never mind** do not worry.

nevermore *adv.* at no future time.

nevertheless *adv.* in spite of that.

> ■ anyhow, anyway, despite that, however, in spite of that, notwithstanding, still, yet.

new *adj.* **1** not existing before, recently made, discovered, experienced, etc. **2** not worn or used. **3** different from the previous one. **4** additional. **5** unfamiliar, unaccustomed. **6** modern. ● *adv.* newly, recently. □ **new moon** moon seen as a crescent. **New Testament** part of the Bible concerned with Christ and hi teaching. **New Year's Day** 1 Jan.

> ■ **1** brand-new, fresh, recent; different innovative, novel, original. **2** in mint condi tion, pristine, unused, unworn, virgin. additional, extra, fresh, further, more, sup plementary. **5** different, strange, unaccus tomed, unfamiliar, unusual. **6** avant-garde contemporary, latest, modern, newfangled present-day, *colloq.* trendy, up to date.

newcomer *n.* one who has arrived re cently.

> ■ alien, colonist, immigrant, outsider, set tler, stranger.

newel *n.* **1** top or bottom post of the handrail of a stair. **2** central pillar of a winding stair.

newfangled *adj.* objectionably new in method or style.

newly *adv.* recently, freshly. □ **newlywed** *adj.* & *n.* recently married (person).

news *n.* **1** new or interesting information about recent events. **2** (**the news**) broadcast report of this. □ **newsy** *adj.*

> ■ **1** facts, information, intelligence, word gossip, hearsay, *colloq.* low-down, rumour talk.

newsagent *n.* shopkeeper who sells newspapers.

newscaster *n.* newsreader.

newsflash *n.* brief news item.

newsletter *n.* informal printed report containing news of interest to members of a club etc.

newspaper *n.* **1** printed daily or weekly publication containing news reports. **2** sheets of paper forming this.

> ■ **1** broadsheet, daily, gazette, journal paper, publication, *derog.* rag, tabloid weekly.

newsprint *n.* type of paper on which newspapers are printed.

newsreader *n.* person who reads broadcast news reports.

newsworthy *adj.* worth reporting as news.

newt *n.* small lizard-like amphibious creature.

next *adj.* nearest in position or time etc ● *adv.* **1** in the next place or degree. **2** on the next occasion. ● *n.* next person or thing. □ **next door** in the next house or room. **next of kin** one's closest relative.

> ■ *adj.* adjacent, bordering, closest, contiguous, nearest, neighbouring; following, subsequent, succeeding.

nexus *n.* (*pl.* **-uses**) connected group or series.

nib *n.* metal point of a pen.

nibble *v.* take small quick or gentle bites (at). ● *n.* **1** small quick bite. **2** snack. □ **nibbler** *n.*

nice *adj.* **1** pleasant, satisfactory. **2** fine, subtle. **3** fastidious. □ **nicely** *adv.*, **niceness** *n.*

■ **1** agreeable, affable, amiable, amicable, charming, cordial, courteous, friendly, genial, good-natured, kind, kindly, likeable, pleasant, polite; delightful, enjoyable, good, *colloq.* lovely, pleasing, pleasurable, satisfactory. **2** fine, precise, subtle. **3** dainty, fastidious, finicky, fussy, particular, *colloq.* pernickety.

nicety *n.* **1** precision. **2** detail. □ **to a nicety** exactly.

■ **1** accuracy, exactness, precision. **2** detail, nuance, subtlety.

niche *n.* **1** shallow recess esp. in a wall. **2** suitable position in life or employment.

■ **1** alcove, cubby hole, nook, recess. **2** place, position, slot.

nick *n.* small cut or notch. ● *v.* **1** make a nick in. **2** (*sl.*) steal. **3** (*sl.*) arrest. □ **in good nick** (*colloq.*) in good condition. **in the nick of time** only just in time.

nickel *n.* **1** hard silvery-white metal. **2** (*US*) 5-cent piece.

nickname *n.* name given humorously to a person or thing. ● *v.* give as a nickname.

nicotine *n.* poisonous substance found in tobacco.

niece *n.* one's brother's or sister's daughter.

niggardly *adj.* stingy. □ **niggard** *n.*

niggle *v.* fuss over details.

nigh *adv.* & *prep.* near.

night *n.* **1** dark hours between sunset and sunrise. **2** nightfall. □ **night-life** *n.* entertainment available at night. **night school** instruction provided in the evening.

nightcap *n.* (alcoholic) drink taken just before going to bed.

nightclub *n.* club providing entertainment etc. at night.

nightdress *n.* woman's or child's loose garment for sleeping in.

nightfall *n.* onset of night.

nightgown *n.* nightdress.

nightie *n.* (*colloq.*) nightdress.

nightingale *n.* small thrush, male of which sings melodiously.

nightjar *n.* night-flying bird with a harsh cry.

nightly *adj.* & *adv.* (happening) at night or every night.

nightmare *n.* **1** frightening dream. **2** (*colloq.*) unpleasant experience. □ **nightmarish** *adj.*

nightshade *n.* plant with poisonous berries.

nightshirt *n.* man's or boy's long shirt for sleeping in.

nihilism *n.* rejection of all religious and moral principles. □ **nihilist** *n.*, **nihilistic** *adj.*

nil *n.* nothing.

nimble *adj.* able to move quickly. □ **nimbly** *adv.*

■ agile, fleet, graceful, lissom, lithe, lively, *colloq.* nippy, quick, sprightly, spry, swift.

nincompoop *n.* foolish person.

nine *adj.* & *n.* one more than eight. □ **ninth** *adj.* & *n.*

ninepins *n.* game of skittles played with nine objects.

nineteen *adj.* & *n.* one more than eighteen. □ **nineteenth** *adj.* & *n.*

ninety *adj.* & *n.* nine times ten. □ **ninetieth** *adj.* & *n.*

ninny *n.* foolish person.

nip[1] *v.* (**nipped**) **1** pinch or squeeze sharply. **2** bite quickly with the front teeth. **3** (*sl.*) go quickly. ● *n.* **1** sharp pinch, squeeze, or bite. **2** biting coldness.

nip[2] *n.* small drink of spirits.

■ dram, finger, *colloq.* shot, *sl.* snifter, tot.

nipple *n.* **1** small projection at the centre of a breast. **2** similar protuberance. **3** teat of a feeding bottle.

nippy *adj.* (**-ier**, **-iest**) (*colloq.*) **1** nimble, quick. **2** bitingly cold.

nirvana *n.* (in Buddhism and Hinduism) state of perfect bliss achieved by the soul.

nit *n.* egg of a louse or similar parasite. □ **nit-picking** *n.* & *adj.* (*colloq.*) petty fault-finding.

nitrate *n.* substance formed from nitric acid, esp. used as a fertilizer.

nitric acid corrosive acid containing nitrogen.

nitrogen *n.* gas forming about four-fifths of the atmosphere.

nitroglycerine *n.* explosive yellow liquid.

nitrous oxide gas used as an anaesthetic.

nitty-gritty *n.* (*sl.*) basic facts or realities of a matter.

nitwit *n.* (*colloq.*) stupid person.

no *adj.* **1** not any. **2** not a. ● *adv.* **1** (used as a denial or refusal of something). **2** not at all. ● *n.* (*pl.* **noes**) negative reply, vote against a proposal. □ **no-go area** area to which entry is forbidden or restricted. **no man's land** area not controlled by anyone, esp. between opposing armies. **no one** no person, nobody.

No. *abbr.* number.

nobble *v.* (*sl.*) **1** seize. **2** tamper with or influence dishonestly.

nobility *n.* **1** nobleness of character or of rank. **2** titled people.

■ **1** dignity, excellence, grandeur, greatness, honesty, integrity, magnanimity, probity, rectitude. **2** aristocracy, peerage, *colloq.* upper crust.

noble *adj.* **1** aristocratic. **2** of excellent character, not mean or petty. **3** impressive in appearance. ● *n.* member of the nobility. □ **nobly** *adv.*, **nobleness** *n.*

■ *adj.* **1** aristocratic, blue-blooded, titled, upper class, *colloq.* upper crust. **2** decent, generous, good, honest, honourable, magnanimous, moral, principled, righteous, upright, virtuous. **3** grand, imposing, impressive, magnificent, majestic, splendid, stately. ● *n.* aristocrat, lady, lord, nobleman, noblewoman, peer, peeress.

nobleman, noblewoman *ns.* member of the nobility.

nobody *pron.* no person. ● *n.* person of no importance.

nocturnal *adj.* of, happening in, or active in the night. □ **nocturnally** *adv.*

nod *v.* (**nodded**) **1** incline the head, indicate (agreement or casual greeting) thus. **2** let the head droop, be drowsy. **3** (of flowers etc.) bend and sway. ● *n.* nodding movement, esp. in agreement or greeting.

node *n.* **1** knob-like swelling. **2** point on a stem where a leaf or bud grows out. □ **nodal** *adj.*

nodule *n.* small rounded lump, small node. □ **nodular** *adj.*

noise *n.* sound, esp. loud, harsh, or undesired.

■ babel, cacophony, caterwauling, clamour, clangour, commotion, din, hubbub, hullabaloo, racket, row, rumpus, sound, thunder, thundering, tumult, uproar.

noisy *adj.* (**-ier, -iest**) making much noise. □ **noisily** *adv.*

■ blaring, booming, cacophonous, clamorous, deafening, discordant, dissonant, loud, piercing, raucous, shrill, thunderou[s], boisterous, obstreperous, riotous, rowd[y], uproarious.

nomad *n.* **1** member of a tribe that roam[s] seeking pasture for its animals. **2** wa[n]derer. □ **nomadic** *adj.*

nom de plume writer's pseudonym.

nomenclature *n.* system of names, e.[g.] in a science.

nominal *adj.* **1** in name only. **2** (of a fe[e]) very small. □ **nominal value** face valu[e] of a coin etc. **nominally** *adv.*

■ **1** in name only, self-styled, titular. **2** i[n]considerable, insignificant, minimal, mino[r], minuscule, small, tiny, trifling, trivial, toke[n].

nominate *v.* **1** propose as a candidate. [2] appoint to an office. **3** appoint a place o[r] date. □ **nomination** *n.*, **nominator** *n.*

■ **1** propose, put forward, recommen[d], submit, suggest. **2** appoint, choose, de[s]ignate, name, select. **3** choose, name, s[e]lect.

nominee *n.* person nominated.

non- *pref.* not.

nonagenarian *n.* person in his or he[r] nineties.

nonchalant *adj.* calm and casua[l]. □ **nonchalantly** *adv.*, **nonchalance** *n.*

■ blasé, calm, casual, collected, com[]posed, easygoing, happy-go-lucky, impe[r]turbable, insouciant, offhand, phlegmati[c], relaxed, unconcerned, *colloq.* unflappabl[e], unperturbed.

noncommittal *adj.* not revealing one['s] opinion.

■ careful, *colloq.* cagey, cautious, ci[r]cumspect, guarded, wary.

nonconformist *n.* **1** person not co[n]forming to established practices. [2] (**Nonconformist**) member of a Protesta[nt] sect not conforming to Anglican pra[c]tices.

■ **1** dissenter, dissident, iconoclast, ind[i]vidualist, maverick, rebel, renegade.

nondescript *adj.* lacking distinctiv[e] characteristics.

■ bland, characterless, colourless, dra[b], insipid, ordinary, unexceptional, uninte[r]esting, unremarkable.

none *pron.* **1** not any. **2** no person(s). ● *adv.* not at all.

nonentity *n.* person of no importance.

non-event *n.* event that was expected t[o] be important but proves disappointing.

non-existent *adj.* not existing.

■ fanciful, fictional, fictitious, illusory, imaginary, imagined, mythical, unreal.

nonplussed *adj.* completely perplexed.

■ astonished, astounded, baffled, bemused, bewildered, confounded, dumbfounded, *colloq.* flummoxed, mystified, perplexed, puzzled, taken aback.

nonsense *n.* **1** absurd or meaningless words or ideas. **2** foolish talk or behaviour. □ **nonsensical** *adj.*

■ **1** balderdash, *sl.* bilge, *sl.* boloney, bunkum, double Dutch, drivel, gibberish, *colloq.* gobbledegook, mumbo-jumbo, *colloq.* piffle, *sl.* poppycock, *sl.* rot, rubbish, *sl.* tripe, twaddle. **2** antics, buffoonery, clowning, foolishness, silliness. □ **nonsensical** absurd, asinine, crazy, foolish, idiotic, laughable, ludicrous, mad, preposterous, ridiculous, senseless, silly, stupid.

non sequitur conclusion that does not follow from the evidence given.

non-starter *n.* **1** horse entered for a race but not running in it. **2** person or idea etc. not worth considering for a purpose.

non-stop *adj.* & *adv.* **1** without ceasing. **2** (of a train etc.) not stopping at intermediate places.

■ *adj.* **1** ceaseless, constant, continual, endless, *colloq.* eternal, incessant, never-ending, persistent, unceasing, unending, unremitting. **2** direct, through.

noodles *n.pl.* pasta in narrow strips, used in soups etc.

nook *n.* **1** secluded place. **2** recess.

noon *n.* twelve o'clock in the day.

noose *n.* loop of rope etc. with a knot that tightens when pulled.

nor *conj.* & *adv.* and not.

norm *n.* standard or pattern that is typical.

■ custom, pattern, rule, standard.

normal *adj.* **1** conforming to what is standard or usual. **2** free from mental or emotional disorders. □ **normally** *adv.*, **normality** *n.*

■ **1** average, common, conventional, customary, natural, ordinary, regular, routine, run-of-the-mill, standard, typical, usual. **2** rational, sane, well-adjusted.

north *n.* **1** point or direction to the left of person facing east. **2** northern part. ● *adj.* **1** in the north. **2** (of wind) from the north. ● *adv.* towards the north. □ **north-east** *n.* point or direction midway between north and east. **north-easterly** *adj.* & *n.*, **north-eastern** *adj.* **north-west** *n.* point or direction midway between north and west. **north-westerly** *adj.* & *n.*, **north-western** *adj.*

northerly *adj.* towards or blowing from the north.

northern *adj.* of or in the north.

northerner *n.* native of the north.

northernmost *adj.* furthest north.

northward *adj.* towards the north. □ **northwards** *adv.*

Norwegian *adj.* & *n.* (native, language) of Norway.

Nos. *abbr.* numbers.

nose *n.* **1** organ at the front of the head, used in breathing and smelling. **2** sense of smell. **3** open end of a tube. **4** front end or projecting part. ● *v.* **1** detect or search by use of the sense of smell. **2** push one's nose against or into. **3** push one's way cautiously ahead.

nosebag *n.* bag of fodder for hanging on a horse's head.

nosedive *n.* steep downward plunge, esp. of an aeroplane. ● *v.* make this plunge.

nostalgia *n.* sentimental memory of or longing for things of the past. □ **nostalgic** *adj.*

■ □ **nostalgic** homesick, maudlin, sentimental, wistful.

nostril *n.* either of the two external openings in the nose.

nosy *adj.* (**-ier**, **-iest**) (*colloq.*) inquisitive.

not *adv.* expressing a negative or denial or refusal.

notable *adj.* **1** worthy of notice, remarkable. **2** eminent. ● *n.* eminent person. □ **notably** *adv.*

■ *adj.* **1** conspicuous, impressive, memorable, noteworthy, outstanding, pre-eminent, remarkable, singular, striking, unforgettable, unusual. **2** celebrated, distinguished, eminent, famed, famous, great, illustrious, important, prominent, well-known. ● *n.* celebrity, dignitary, luminary, personage, worthy.

notation *n.* system of signs or symbols representing numbers, quantities, musical notes, etc.

notch *n.* V-shaped cut or indentation. ● *v.* make notch(es) in. □ **notch up** score, achieve.

note *n.* **1** brief record written down to aid memory. **2** short or informal letter or message. **3** short written comment. **4** banknote. **5** eminence. **6** notice, attention. **7** musical tone of definite pitch. **8**

symbol representing the pitch and duration of a musical sound. **9** each of the keys on a piano etc. ● *v.* **1** notice, pay attention to. **2** write down.

■ *n.* **1** *colloq.* memo, memorandum, record, reminder. **2** card, letter, line, *colloq.* memo, memorandum, message, postcard. **3** annotation, comment, observation, remark. **4** banknote, *US* bill. **5** consequence, distinction, eminence, importance, prestige, renown, reputation, repute. **6** attention, heed, regard, notice, thought. ● *v.* **1** mark, notice, observe, pay attention to, perceive, register, see. **2** jot down, put down, record, register, write down.

notebook *n.* book with blank pages on which to write notes.

notecase *n.* wallet for banknotes.

noted *adj.* famous, well-known.

■ acclaimed, celebrated, distinguished, eminent, famed, famous, illustrious, prominent, renowned, well-known.

notelet *n.* small folded card etc. for a short informal letter.

notepad *n.* notebook.

notepaper *n.* paper for writing letters on.

noteworthy *adj.* worthy of notice, remarkable.

■ exceptional, extraordinary, important, impressive, memorable, notable, outstanding, remarkable, signal, significant, singular.

nothing *n.* **1** no thing, not anything. **2** no amount, nought. **3** non-existence. **4** person or thing of no importance. ● *adv.* not at all. □ **nothingness** *n.*

notice *n.* **1** attention, observation. **2** intimation, warning. **3** formal announcement of the termination of an agreement or employment. **4** written or printed information displayed. **5** review in a newspaper. ● *v.* perceive, observe. □ **take notice** show interest. **take notice of** pay attention to.

■ *n.* **1** attention, awareness, cognizance, consciousness, observation, perception. **2** intimation, notification, warning. **4** advertisment, bill, placard, poster, sign. **5** review, *colloq.* write-up. ● *v.* descry, detect, discern, make out, observe, perceive, see, *colloq.* spot.

noticeable *adj.* easily seen or noticed. □ **noticeably** *adv.*

■ apparent, clear, conspicuous, detectable, discernible, distinct, distinguishable, evident, manifest, observable, obvious, palpable, perceptible, recognizable, visible

notifiable *adj.* that must be notified.

notify *v.* **1** inform. **2** report, make known. □ **notification** *n.*

■ **1** advise, apprise, inform, tell, warn. announce, declare, make known, report.

notion *n.* **1** concept, idea. **2** vague belief or understanding. **3** whim.

■ **1** concept, conception, idea, opinion, thought, view. **2** feeling, impression, inkling, suspicion. **3** caprice, fancy, vagary, whim.

notional *adj.* hypothetical, imaginary. □ **notionally** *adv.*

notorious *adj.* well-known, esp. unfavourably. □ **notoriously** *adv.*, **notoriety** *n.*

■ disreputable, ill-famed, infamous.

notwithstanding *prep.* in spite of. ● *adv.* nevertheless.

nougat /nōōgaa/ *n.* chewy sweet.

nought *n.* **1** the figure 0. **2** nothing.

noun *n.* word used as the name of a person, place, or thing.

nourish *v.* **1** keep alive and well by food. **2** cherish (a feeling).

nourishment *n.* food.

■ food, *sl.* grub, nutriment, nutrition, sustenance.

nous /nowss/ *n.* (*colloq.*) common sense.

nova *n.* (*pl.* **-ae**) star that suddenly becomes much brighter for a short time.

novel *n.* book-length story. ● *adj.* of a new kind.

■ *adj.* different, fresh, innovative, new, original, unconventional, unfamiliar, unusual.

novelette *n.* short (esp. romantic) novel.

novelist *n.* writer of novels.

novelty *n.* **1** novel quality or thing. **2** small toy etc.

■ **1** freshness, newness, originality. **2** bauble, gewgaw, knick-knack, toy, trinket.

novice *n.* **1** inexperienced person. **2** probationary member of a religious order.

■ **1** apprentice, beginner, greenhorn, learner, probationer, trainee.

now *adv.* **1** at the present time. **2** immediately. **3** (with no reference to time) I wonder or am telling you. ● *conj.* as a consequence of or simultaneously with the fact that. ● *n.* the present time

□ **now and again, now and then** occasionally.

■ *adv.* **1** at present, at the present time, at this moment; nowadays, today. **2** at once, immediately, instantly, right away, straight away, without delay.

nowadays *adv.* in present times.

nowhere *adv.* not anywhere.

noxious *adj.* unpleasant and harmful.

■ dangerous, foul, harmful, injurious, nasty, pernicious, poisonous, toxic, unhealthy.

nozzle *n.* vent or spout of a hosepipe etc.

nuance /nyōō-onss/ *n.* shade of meaning.

nub *n.* **1** central point or core of a matter or problem. **2** small lump.

■ **1** centre, core, crux, essence, gist, heart, kernel, pith, substance.

nubile *adj.* (of a woman) marriageable or sexually attractive. □ **nubility** *n.*

nuclear *adj.* **1** of a nucleus. **2** of the nuclei of atoms. **3** using energy released or absorbed during reactions in these.

nucleus *n.* (*pl.* **-lei**) **1** central part or thing round which others are collected. **2** central portion of an atom, seed, or cell.

nude *adj.* naked. ● *n.* nude figure in a picture etc. □ **nudity** *n.*

■ *adj.* bare, (stark) naked, unclothed, undressed.

nudge *v.* **1** poke gently with the elbow to attract attention. **2** push slightly or gradually. ● *n.* this movement.

■ *v.* **1** elbow, jog, poke, prod, push.

nudist *n.* person who advocates or practises going unclothed. □ **nudism** *n.*

nugget *n.* rough lump of gold or platinum found in the earth.

nuisance *n.* annoying person or thing.

■ bore, menace, pest; annoyance, bother, *colloq.* hassle, headache, inconvenience, irritation, trial.

null *adj.* (esp. **null and void**) having no legal force. □ **nullity** *n.*

nullify *v.* **1** make null. **2** neutralize the effect of. □ **nullification** *n.*

■ **1** annul, cancel, invalidate, repeal, rescind, revoke. **2** counteract, counterbalance, make ineffective, negate, neutralize, offset.

numb *adj.* deprived of power to feel. ● *v.* make numb. □ **numbness** *n.*

■ *adj.* anaesthetized, dead, deadened, numbed, insensible.

number *n.* **1** symbol or word indicating how many. **2** total. **3** quantity. **4** person or thing having a place in a series, esp. single issue of magazine, item in a programme, etc. ● *v.* **1** include. **2** amount to. **3** assign a number or numbers to. □ **number one** (*colloq.*) oneself. **number plate** plate on a motor vehicle, bearing its registration number.

■ *n.* **1** digit, figure, integer, numeral. **2** aggregate, sum, total. **3** amount, quantity. **4** copy, edition, issue.

numberless *adj.* innumerable.

numeral *n.* written symbol of a number.

numerate *adj.* having a good basic understanding of mathematics. □ **numeracy** *n.*

numerator *n.* number above the line in a vulgar fraction.

numerical *adj.* of number(s). □ **numerically** *adv.*

numerous *adj.* great in number.

■ a lot of, countless, innumerable, lots of, many, myriad.

nun *n.* member of a female religious community.

nunnery *n.* residence of a community of nuns.

nuptial *adj.* of marriage or a wedding. ● *n.pl.* wedding ceremony.

nurse *n.* **1** person trained to look after sick or injured people. **2** person employed to take charge of young children. ● *v.* **1** work as a nurse, act as nurse (to). **2** feed at the breast or udder. **3** hold carefully. **4** give special care to. **5** harbour. □ **nursing home** privately run hospital or home for invalids.

■ *v.* **1** care for, look after, take care of, tend. **2** breastfeed, feed, suckle. **4** cherish, foster, nurture.

nursery *n.* **1** room(s) for young children. **2** place where plants are reared, esp. for sale. □ **nursery rhyme** traditional verse for children. **nursery school** school for children below normal school age. **nursery slopes** slopes suitable for beginners at skiing.

nurseryman *n.* person growing plants etc. at a nursery.

nurture *v.* **1** bring up. **2** nourish. **3** foster. ● *n.* nurturing.

■ *v.* **1** bring up, care for, look after, raise, rear. **2** feed, nourish. **3** cherish, foster, nurse.

nut *n.* **1** fruit with a hard shell round an edible kernel. **2** this kernel. **3** small threaded metal ring for use with a bolt. **4** (*sl.*) head.

nutcase *n.* (*sl.*) crazy person.
nuthatch *n.* small climbing bird.
nutmeg *n.* hard fragrant tropical seed ground or grated as spice.
nutria *n.* fur of the coypu.
nutrient *adj.* & *n.* nourishing (substance).
nutriment *n.* nourishing substance.
nutrition *n.* food, nourishment. □ **nutritional** *adj.*
nutritious *adj.* nourishing.

■ **healthful, healthy, nourishing, wholesome.**

nuts *adj.* (*sl.*) crazy.
nutshell *n.* hard shell of a nut. □ **in a nutshell** expressed very briefly.
nutty *adj.* (**-ier, -iest**) **1** full of nuts. **2** tasting like nuts. **3** (*sl.*) crazy.
nuzzle *v.* press or rub gently with the nose.
NW *abbr.* **1** north-west. **2** north-western.
nylon *n.* **1** very light strong synthetic fibre. **2** fabric made of this.
nymph *n.* **1** mythological semi-divine maiden. **2** young insect.
nymphomania *n.* excessive sexual desire in a woman. □ **nymphomaniac** *n.*
NZ *abbr.* New Zealand.

Oo

oaf *n.* (*pl.* **oafs**) awkward lout.

oak *n.* **1** deciduous forest tree bearing acorns. **2** its hard wood. □ **oak-apple** *n.* = **gall**[3].

oar *n.* **1** pole with a flat blade used to propel a boat by its leverage against water. **2** rower.

oasis *n.* (*pl.* **oases**) fertile spot in a desert, with a spring or well.

oath *n.* **1** solemn promise. **2** swear word.

■ **1** pledge, promise, vow, word (of honour). **2** curse, expletive, imprecation, profanity, swear word.

oatmeal *n.* **1** ground oats. **2** greyish-fawn colour.

oats *n.* **1** hardy cereal plant. **2** its grain.

obdurate *adj.* stubborn. □ **obdurately** *adv.*, **obduracy** *n.*

obedient *adj.* doing what one is told to do. □ **obediently** *adv.*, **obedience** *n.*

■ amenable, biddable, compliant, docile, dutiful, law-abiding, meek, submissive, tractable.

obeisance *n.* bow or curtsy.

obelisk *n.* tall pillar set up as a monument.

obese *adj.* very fat. □ **obesity** *n.*

■ corpulent, fat, fleshy, gross, heavy, overweight.

obey *v.* do what is commanded (by).

■ abide by, comply (with), conform (to), defer (to), follow, knuckle under, observe, respect, submit (to), toe the line.

obfuscate *v.* **1** darken. **2** confuse, bewilder. □ **obfuscation** *n.*

obituary *n.* printed statement of person's death (esp. in a newspaper).

object *n.* /óbjikt/ **1** something solid that can be seen or touched. **2** person or thing to which an action or feeling is directed. **3** purpose, intention. **4** noun etc. acted upon by a verb or preposition. ● *v.* /əbjékt/ **1** express opposition, disapproval, or reluctance. **2** protest. □ **no object** not a limiting factor. **object lesson** practical illustration of a principle. **objector** *n.*

■ *n.* **1** article, entity, item, thing. **2** butt, destination, focus, quarry, target. **3** aim, ambition, end, goal, intent, intention, objective, purpose. ● *v.* complain, demur, disapprove, draw the line, protest, remonstrate, take exception.

objection *n.* **1** expression or feeling of disapproval or opposition. **2** reason for objecting.

objectionable *adj.* unpleasant. □ **objectionably** *adv.*

■ disagreeable, distasteful, *colloq.* horrible, horrid, loathsome, nasty, obnoxious, odious, offensive, repellent, unpleasant.

objective *adj.* **1** not influenced by personal feelings or opinions. **2** of the form of a word used when it is the object of a verb or preposition. ● *n.* object, purpose. □ **objectively** *adv.*, **objectivity** *n.*

■ *adj.* **1** detached, disinterested, dispassionate, equitable, fair, impartial, neutral, unbiased, unprejudiced. ● *n.* aim, ambition, aspiration, design, end, goal, intent, intention, object, purpose, target.

objet d'art /óbzhay daár/ (*pl.* ***objets d'art***) small artistic object.

oblation *n.* offering made to God.

obligate *v.* oblige.

obligation *n.* **1** compelling power of a law, duty, etc. **2** burdensome task, duty.

■ **1** compulsion, constraint. **2** burden, duty, charge, responsibility, task.

obligatory *adj.* compulsory.

■ compulsory, essential, mandatory, necessary, required, requisite.

oblige *v.* **1** compel. **2** help or gratify by a small service.

■ **1** compel, constrain, force, make, obligate, require. **2** gratify, indulge, humour, please.

obliged *adj.* indebted.

obliging *adj.* polite and helpful. □ **obligingly** *adv.*

■ accommodating, agreeable, amenable, considerate, cooperative, courteous, friendly, helpful, kind, kindly, neighbourly, polite, pleasant, willing.

oblique *adj.* **1** slanting. **2** indirect. □ **obliquely** *adv.*, **obliqueness** *n.*

■ **1** diagonal, slanted, slanting, sloping. **2** circuitous, circumlocutory, evasive, indirect, roundabout.

obliterate *v.* blot out, destroy. □ **obliteration** *n.*

■ blot out, delete, erase, rub out, strike out; annihilate, destroy, eradicate, expunge, extirpate, wipe out.

oblivion *n.* **1** state of being forgotten. **2** state of being oblivious.

oblivious *adj.* unaware. □ **obliviously** *adv.*, **obliviousness** *n.*

■ heedless, insensible, unaware, unconscious, unmindful.

oblong *adj.* & *n.* (having) rectangular shape with length greater than breadth.

obloquy *n.* **1** verbal abuse. **2** disgrace.

obnoxious *adj.* very unpleasant. □ **obnoxiously** *adv.*, **obnoxiousness** *n.*

■ abhorrent, disgusting, hateful, *colloq.* horrible, horrid, loathsome, nasty, objectionable, odious, repellent, repulsive, repugnant, revolting, sickening, vile.

oboe *n.* woodwind instrument of treble pitch. □ **oboist** *n.*

obscene *adj.* indecent in a repulsive or offensive way. □ **obscenely** *adv.*, **obscenity** *n.*

■ blue, coarse, dirty, filthy, gross, improper, indecent, lewd, offensive, pornographic, rude, scabrous, smutty, vulgar.

obscure *adj.* **1** not easily understood. **2** not famous. **3** indistinct. ● *v.* make obscure, conceal. □ **obscurely** *adv.*, **obscurity** *n.*

■ *adj.* **1** abstruse, arcane, baffling, confusing, cryptic, enigmatic, esoteric, incomprehensible, mysterious, mystifying, perplexing, puzzling, recondite, unclear. **2** insignificant, minor, undistinguished, unknown, unimportant. **3** dim, faint, fuzzy, hazy, indistinct, nebulous, shadowy, vague. ● *v.* cloud, darken, dim, obfuscate, shade, shadow; cloak, conceal, cover, disguise, hide, mask, screen, shroud, veil.

obsequies *n.pl.* funeral rites.

obsequious *adj.* excessively respectful. □ **obsequiously** *adv.*, **obsequiousness** *n.*

■ cringing, fawning, flattering, grovelling, ingratiating, mealy-mouthed, servile, slavish, *colloq.* smarmy, sycophantic, toadying, unctuous.

observance *n.* keeping of a law, custom, or festival.

observant *adj.* quick at noticing. □ **observantly** *adv.*

■ alert, attentive, perceptive, sharp, shrewd, vigilant, watchful, *colloq.* wide awake.

observation *n.* **1** observing, being observed. **2** remark. □ **observational** *adj.*

■ **1** attention, notice; examination, inspection, scrutiny, surveillance. **2** comment, remark, statement, utterance.

observatory *n.* building for observation of stars or weather.

observe *v.* **1** perceive, watch carefully. **2** keep (a law etc.). **3** celebrate (a festival). **4** remark. □ **observer** *n.*

■ **1** detect, discern, notice, perceive, see; contemplate, examine, eye, inspect, look at, monitor, regard, scrutinize, study, view, watch. **2** abide by, comply with, follow, keep, obey, respect. **3** celebrate, commemorate, keep, solemnize. **4** comment, declare, mention, note, remark, say, state.

obsess *v.* occupy the thoughts of continually. □ **obsessive** *adj.*

obsession *n.* **1** state of being obsessed. **2** persistent idea. □ **obsessional** *adj.*

obsolescent *adj.* becoming obsolete. □ **obsolescence** *n.*

obsolete *adj.* antiquated, no longer used.

■ antediluvian, antiquated, archaic, dated, dead, old, old-fashioned, outdated, outmoded, out of date.

obstacle *n.* thing that obstructs progress.

■ bar, barrier, catch, hindrance, hitch, hurdle, impediment, obstruction, snag, stumbling block.

obstetrics *n.* branch of medicine and surgery dealing with childbirth. □ **obstetric** *adj.*, **obstetrician** *n.*

obstinate *adj.* not easily persuaded or influenced. □ **obstinately** *adv.*, **obstinacy** *n.*

■ adamant, *colloq.* bloody-minded, dogged, headstrong, inflexible, intractable, intransigent, obdurate, pigheaded, recalcitrant, self-willed, stiff-necked, stubborn, tenacious, unbending, uncooperative, unyielding, wilful.

obstreperous *adj.* unruly, noisy.

■ boisterous, disorderly, irrepressible, riotous, rowdy, uncontrollable, unruly, wild; clamorous, loud, noisy, raucous.

obstruct *v.* **1** block. **2** hinder movement or progress. ◻ **obstructive** *adj.*
■ **1** block (up), clog, stop (up). **2** baulk, block, check, hamper, handicap, hinder, hold up, impede, inhibit, interfere with, retard, slow down, thwart.

obstruction *n.* **1** act or instance of obstructing. **2** obstacle or blockage.
■ **2** bar, barrier, hindrance, impediment, hurdle, obstacle; blockage, bottleneck.

obtain *v.* **1** get, come into possession of. **2** be customary.
■ **1** acquire, come by, gain, get, pick up, procure, secure; buy, purchase. **2** apply, be customary, be relevant, exist, prevail.

obtrude *v.* force (ideas or oneself) upon others. ◻ **obtrusion** *n.*

obtrusive *adj.* obtruding oneself, unpleasantly noticeable. ◻ **obtrusively** *adv.*, **obtrusiveness** *n.*

obtuse *adj.* **1** slow at understanding. **2** of blunt shape. **3** (of an angle) more than 90° but less than 180°. ◻ **obtusely** *adv.*, **obtuseness** *n.*
■ **1** dense, *colloq.* dim, dull, *colloq.* dumb, slow, stupid, *colloq.* thick, unintelligent.

obverse *n.* side of a coin bearing a head or the principal design.

obviate *v.* make unnecessary.

obvious *adj.* easy to perceive or understand. ◻ **obviously** *adv.*, **obviousness** *n.*
■ apparent, clear, conspicuous, distinct, evident, manifest, noticeable, open, overt, patent, perceptible, plain, prominent, self-evident, unmistakable, visible.

ocarina *n.* egg-shaped wind instrument.

occasion *n.* **1** time at which a particular event takes place. **2** special event. **3** opportunity. **4** need or cause. ● *v.* cause.
■ *n.* **2** *colloq.* affair, event, function, happening, occurrence. **3** chance, opening, opportunity, time. **4** cause, grounds, justification, need, reason. ● *v.* bring about, cause, effect, elicit, engender, evoke, generate, give rise to, prompt, provoke, result in.

occasional *adj.* **1** happening sometimes but not frequently. **2** for a special occasion. ◻ **occasionally** *adv.*
■ **1** infrequent, intermittent, irregular, odd, periodic, sporadic.

Occident /óksid'nt/ *n.* **(the Occident)** the West, the western world. ◻ **occidental** *adj.*

occlude *v.* stop up, obstruct.

occlusion *n.* upward movement of a mass of warm air caused by a cold front overtaking it.

occult *adj.* **1** secret. **2** supernatural.
■ **1** arcane, dark, esoteric, obscure, recondite, secret. **2** magic, mysterious, mystical, preternatural, supernatural.

occupant *n.* person occupying a place or dwelling. ◻ **occupancy** *n.*
■ householder, inhabitant, lessee, occupier, owner, resident, tenant.

occupation *n.* **1** occupying or being occupied. **2** pastime. **3** employment.
■ **1** occupancy, possession, tenancy, tenure; conquest, invasion, seizure, takeover. **2** activity, pastime, pursuit. **3** career, craft, employment, job, position, profession, situation, trade, work.

occupational *adj.* of or caused by one's employment. ◻ **occupational therapy** creative activities designed to assist recovery from certain illnesses.

occupy *v.* **1** dwell in. **2** take possession of (a place) by force. **3** fill (a space or time or position). **4** keep busy. ◻ **occupier** *n.*
■ **1** dwell in, inhabit, live in, reside in. **2** capture, conquer, garrison, invade, overrun, seize, take over, take possession of. **3** cover, extend over; fill, take up, use (up). **4** absorb, engage, engross, hold, involve, preoccupy.

occur *v.* **(occurred) 1** come into being as an event or process. **2** exist. ◻ **occur to** come into the mind of.
■ **1** arise, befall, come about, crop up, happen, materialize, take place.

occurrence *n.* **1** act or instance of occurring. **2** incident, event.
■ **2** *colloq.* affair, event, happening, incident, phenomenon.

ocean *n.* sea surrounding the continents of the earth. ◻ **oceanic** *adj.*

oceanography *n.* study of the ocean.

ocelot *n.* **1** leopard-like animal of Central and S. America. **2** its fur.

ochre *n.* **1** type of clay used as pigment. **2** pale brownish-yellow.

o'clock *adv.* by the clock.

octagon *n.* geometric figure with eight sides. ◻ **octagonal** *adj.*

octahedron *n.* solid with eight sides. ◻ **octahedral** *adj.*

octane *n.* hydrocarbon occurring in petrol.

octave *n.* **1** note six whole tones above or below a given note. **2** interval or notes between these.

octet *n.* **1** group of eight voices or instruments. **2** music for these.

octogenarian *n.* person in his or her eighties.

octopus *n.* (*pl.* **-puses**) sea animal with eight tentacles.

ocular *adj.* of, for, or by the eyes.

oculist *n.* specialist in the treatment of the eye.

odd *adj.* **1** unusual. **2** occasional. **3** (of a number) not divisible by 2. **4** not part of a set. **5** exceeding a round number or amount. □ **oddly** *adv.*, **oddness** *n.*

■ **1** abnormal, anomalous, atypical, bizarre, curious, eccentric, exceptional, extraordinary, funny, out of the ordinary, outlandish, peculiar, quaint, queer, strange, uncharacteristic, uncommon, unconventional, unexpected, unusual, weird. **2** irregular, occasional, random, sporadic.

oddity *n.* **1** strangeness. **2** unusual person or thing.

oddment *n.* thing left over, isolated article.

■ (**oddments**) bits, fragments, leftovers, odds and ends, pieces, remnants, scraps, snippets.

odds *n.pl.* **1** probability. **2** ratio between amounts staked by parties to a bet. □ **at odds with** in conflict with. **odds and ends** oddments. **odds-on** *adj.* with success more likely than failure.

ode *n.* type of poem addressed to a person or celebrating an event.

odious *adj.* hateful. □ **odiously** *adv.*, **odiousness** *n.*

■ abhorrent, abominable, foul, hateful, *colloq.* horrible, horrid, loathsome, nasty, obnoxious, repellent, repugnant, repulsive, revolting, unpleasant, vile.

odium *n.* widespread hatred or disgust.

odoriferous *adj.* diffusing (usu. pleasant) odours.

odour *n.* smell. □ **odorous** *adj.*

■ aroma, bouquet, fragrance, perfume, redolence, scent, smell; *sl.* pong, stench, stink.

odyssey *n.* (*pl.* **-eys**) long adventurous journey.

oedema /ideéma/ *n.* excess fluid in tissues, causing swelling.

oesophagus /eesóffəgəss/ *n.* gullet.

of *prep.* **1** belonging to. **2** from. **3** composed or made from. **4** concerning. **5** for, involving.

off *adv.* **1** away. **2** out of position, disconnected. **3** not operating, cancelled. **4** (of food) beginning to decay. ● *prep.* **1** away from. **2** below the normal standard of. ● *adj.* of the right-hand side of a horse, vehicle, or road. □ **off chance** remote possibility. **off colour** not in the best of health. **off-licence** *n.* **1** licence to sell alcohol for consumption away from the premises. **2** shop with this. **off-putting** *adj.* (*colloq.*) **1** disconcerting. **2** unpleasant. **off-white** *adj.* not quite pure white.

offal *n.* edible organs from an animal carcass.

offbeat *adj.* unconventional.

■ bizarre, eccentric, odd, outlandish, peculiar, queer, strange, unconventional, unusual, weird.

offence *n.* **1** illegal act, transgression. **2** feeling of annoyance or resentment. □ **take offence** feel upset or hurt.

■ **1** crime, fault, misdeed, misdemeanour, sin, transgression, wrong. **2** annoyance, resentment, umbrage.

offend *v.* **1** cause offence to, upset. **2** displease. **3** do wrong. □ **offender** *n.*

■ **1** affront, hurt, insult, upset, wound. **2** annoy, displease, gall, irritate, nettle, pique, *colloq.* rile. **3** sin, transgress. □ **offender** criminal, *colloq.* crook, culprit, lawbreaker, malefactor, miscreant, sinner, transgressor, wrongdoer.

offensive *adj.* **1** causing offence, insulting. **2** disgusting. **3** used in attacking. ● *n.* aggressive action. □ **offensively** *adv.*, **offensiveness** *n.*

■ *adj.* **1** discourteous, disrespectful, impolite, impudent, insolent, insulting, rude, uncivil. **2** disgusting, distasteful, foul, *colloq.* horrible, horrid, nasty, obnoxious, repellent, repugnant, repulsive, revolting, sickening, unsavoury, vile. ● *n.* assault, attack, charge, onslaught, strike.

offer *v.* (**offered**) **1** present for acceptance or refusal, or for consideration or use. **2** express readiness, show intention. **3** make available for sale. ● *n.* **1** expression of willingness to do, give, or pay something. **2** amount offered.

■ *v.* **1** proffer, tender; advance, propose, put forward, submit, suggest. **2** volunteer. ● *n.* **1** bid, proposal, proposition, tender.

offering *n.* gift, contribution.

offhand *adj.* casual or curt in manner. ● *adv.* in an offhand way. □ **offhanded** *adj.*

■ *adj.* blasé, careless, casual, cavalier, insouciant, nonchalant, offhanded, unceremonious, unconcerned; abrupt, brusque, curt, perfunctory.

office *n.* **1** room or building used for clerical and similar work. **2** position of authority or trust.

officer *n.* **1** official. **2** person holding authority on a ship or in the armed services. **3** policeman or policewoman.

official *adj.* **1** of office or officials. **2** authorized. ● *n.* person holding office. □ **officially** *adv.*

■ *adj.* **1** bureaucratic; ceremonial, ceremonious, formal. **2** authorized, certified, documented, endorsed, lawful, legal, legitimate, licensed, proper, recognized, sanctioned, valid. ● *n.* appointee, bureaucrat, commissioner, functionary, officer.

officiate *v.* act in an official capacity, be in charge.

officious *adj.* asserting one's authority. □ **officiously** *adv.*

■ *colloq.* bossy, dictatorial, domineering, imperious, overbearing, *colloq.* pushy, self-important.

offload *v.* unload.

offset *v.* (**-set**, **-setting**) counterbalance, compensate for.

■ balance (out), cancel out, compensate for, counteract, counterbalance, make up for, neutralize, nullify.

offshoot *n.* **1** side shoot. **2** subsidiary product.

offside *adj.* & *adv.* in a position where one may not legally play the ball (in football etc.).

offspring *n.* (*pl.* **-spring**) **1** person's child, children, or descendant(s). **2** animal's young.

■ **1** child(ren), descendant(s), heir(s), progeny, issue, successor(s). **2** brood, young.

often *adv.* **1** many times, at short intervals. **2** in many instances.

■ commonly, frequently, habitually, ordinarily, regularly, repeatedly, usually.

ogle *v.* eye flirtatiously.

ogre *n.* **1** man-eating giant in fairy tales. **2** terrifying person.

oh *int.* exclamation of delight or pain, or used for emphasis.

ohm *n.* unit of electrical resistance.

oil *n.* **1** thick slippery liquid that will not dissolve in water. **2** petroleum, a form of this. **3** oil paint. ● *v.* lubricate or treat with oil. □ **oil paint** paint made by mixing pigment in oil. **oil painting** picture painted in this. **oily** *adj.*

oilfield *n.* area where oil is found in the ground.

oilskin *n.* **1** cloth waterproofed by treatment with oil etc. **2** (*pl.*) waterproof clothing made of this.

ointment *n.* healing or cosmetic preparation for the skin.

■ balm, cream, embrocation, emollient, lotion, salve, unguent.

OK *adj.* & *adv.* (also **okay**) (*colloq.*) all right.

okapi *n.* (*pl.* **-is**) giraffe-like animal of Central Africa.

okra *n.* African plant with seed pods used as food.

old *adj.* **1** having lived, existed, or been known etc. for a long time. **2** of specified age. **3** shabby from age or wear. **4** former. **5** not recent or modern. □ **old age** later part of life. **old-fashioned** *adj.* no longer fashionable. **Old Testament** part of the Bible dealing with pre-Christian times. **old wives' tale** traditional but foolish belief.

■ **1** aged, ageing, ancient, elderly, getting on, hoary, senescent; long-standing, time-honoured, well-established; bygone, former. **3** crumbling, dilapidated, ramshackle, tumbledown; ragged, shabby, tattered, tatty, threadbare, worn. **4** erstwhile, ex-, former, previous, prior, recent, sometime. **5** ancient, antediluvian, antiquated, antique, dated, obsolete, old-fashioned, outdated, outmoded, out of date.

oleaginous *adj.* **1** producing oil. **2** oily.

oleander *n.* flowering shrub of Mediterranean regions.

olfactory *adj.* concerned with smelling.

oligarch *n.* member of an oligarchy.

oligarchy *n.* **1** government by a small group. **2** country governed in this way. □ **oligarchic** *adj.*

olive *n.* **1** small oval fruit from which an oil (**olive oil**) is obtained. **2** tree bearing this. **3** greenish colour. ● *adj.* **1** of this colour. **2** (of the complexion) yellowish-brown. □ **olive branch** gesture of peace.

ombudsman *n.* official appointed to investigate complaints against public authorities.

omega *n.* last letter of the Greek alphabet, = o.

omelette *n.* dish of beaten eggs cooked in a frying pan.

omen *n.* event regarded as a prophetic sign.

■ augury, portent, presage, sign.

ominous *adj.* seeming as if trouble is imminent. □ **ominously** *adv.*, **ominousness** *n.*

■ inauspicious, menacing, sinister, threatening, unfavourable, unpromising, unpropitious.

omit *v.* (**omitted**) **1** leave out, not include. **2** neglect (to do something). □ **omission** *n.*

■ **1** disregard, drop, exclude, leave out, pass over, *colloq.* skip. **2** fail, forget, neglect, overlook.

omnibus *n.* **1** bus. **2** volume containing several novels etc.

omnipotent *adj.* having unlimited power. □ **omnipotence** *n.*

omnipresent *adj.* present everywhere. □ **omnipresence** *n.*

omniscient *adj.* knowing everything. □ **omniscience** *n.*

omnivorous *adj.* feeding on all kinds of food.

on *prep.* **1** supported by, attached to, covering. **2** close to. **3** towards. **4** (of time) exactly at, during. **5** in the state or process of. **6** concerning. **7** added to. ● *adv.* **1** so as to be on or covering something. **2** further forward, towards something. **3** with continued movement or action. **4** operating, taking place. □ **be** or **keep on at** (*colloq.*) nag. **on and off** from time to time.

once *adv.* **1** on one occasion or for one time only. **2** formerly. ● *conj.* as soon as. ● *n.* one time or occasion. □ **at once 1** immediately. **2** simultaneously. **(every) once in a while** occasionally. **once-over** *n.* (*colloq.*) rapid inspection. **once upon a time** at some vague time in the past.

oncology *n.* study of tumours.

oncoming *adj.* approaching.

one *adj.* single, individual, forming a unity. ● *n.* **1** smallest whole number. **2** single thing or person. ● *pron.* **1** person. **2** any person (esp. used by a speaker or writer of himself as representing people in general). □ **one another** each other. **one day** at some unspecified date. **one-sided** *adj.* unfair, prejudiced. **one-way street** street where traffic is permitted to move in one direction only.

onerous *adj.* needing much effort, burdensome.

■ arduous, burdensome, demanding, difficult, exacting, gruelling, hard, laborious, strenuous, taxing, tough, trying.

oneself *pron.* emphatic and reflexive form of *one*.

ongoing *adj.* continuing, in progress.

onion *n.* vegetable with a bulb that has a strong taste and smell.

onlooker *n.* spectator.

■ bystander, eyewitness, looker-on, observer, passer-by, spectator, viewer, watcher, witness.

only *adj.* existing alone of its or their kind, sole. ● *adv.* **1** without anything or anyone else. **2** no longer ago than. ● *conj.* but then. □ **only too** extremely.

■ *adj.* lone, single, sole, solitary. ● *adv.* **1** exclusively, just, solely; merely, purely, simply.

onomatopoeia /ónnəmattəpéeə/ *n.* formation of words that imitate the sound of what they stand for. □ **onomatopoeic** *adj.*

onset *n.* **1** beginning. **2** attack.

■ **1** beginning, commencement, inception, start. **2** assault, attack, charge, onslaught, raid, sally, sortie, strike.

onslaught *n.* fierce attack.

onus *n.* duty or responsibility.

■ burden, charge, duty, load, obligation, responsibility, weight.

onward *adv.* & *adj.* **1** with an advancing motion. **2** further on. □ **onwards** *adv.*

onyx *n.* stone like marble.

oodles *n.pl.* (*colloq.*) very great amount.

oolite /ṓəlīt/ *n.* type of limestone.

ooze *v.* **1** trickle or flow out slowly. **2** exude. ● *n.* wet mud. □ **oozy** *adj.*

■ *v.* **1** drain, drip, flow, leak, trickle, seep. **2** exude, secrete.

opacity *n.* being opaque.

opal *n.* iridescent quartz-like stone. □ **opaline** *adj.*

opalescent *adj.* iridescent like an opal. □ **opalescence** *n.*

opaque *adj.* **1** not clear. **2** unintelligible. □ **opaqueness** *n.*

■ **1** clouded, cloudy, dark, muddy, murky, turbid. **2** baffling, enigmatic, incomprehensible, mystifying, obscure, perplexing, puzzling, unclear, unfathomable, unintelligible.

open *adj.* **1** not closed, locked, or blocked. **2** not covered, confined, or sealed. **3** unfolded. **4** frank. **5** undisguised. **6** not yet decided. ● *v.* **1** make or become open or more open. **2** begin, establish. □ **in the open air** not in a house or building etc. **open-ended** *adj.* with no fixed limit. **open-handed** *adj.* giving generously. **open house** hospitality to all comers. **open letter** one addressed to a person by name but printed in a newspaper. **open-plan** *adj.* without partition walls or fences. **open secret** one known to so many people that it is no longer secret. **open to 1** willing to receive. **2** likely to suffer from or be affected by. **openness** *n.*

■ *adj.* ● *v.* **1** ajar, unbolted, unclosed, unfastened, unlatched, unlocked; clear, passable, unblocked, unobstructed. **2** bare, exposed, unprotected; unenclosed, unfenced; uncovered, unsealed, unwrapped. **3** spread (out), unfolded, unfurled, unrolled. **4** candid, communicative, direct, forthright, frank, honest, outspoken, straightforward. **5** blatant, evident, manifest, obvious, patent, unconcealed, undisguised. **6** pending, undecided, unresolved, unsettled. ● *v.* **1** unbolt, unfasten, unlatch, unlock; undo, untie, unwrap; spread (out), unfold, unfurl, unroll. **2** begin, commence, get under way, inaugurate, initiate, launch, start; establish, set up. □ **open-handed** bountiful, free, generous, lavish, liberal, munificent, philanthropic.

opencast *adj.* (of mining) on the surface of the ground.

opener *n.* device for opening tins or bottles etc.

opening *n.* **1** gap, place where a thing opens. **2** beginning. **3** opportunity.

■ **1** aperture, breach, chink, cleft, cranny, crevice, fissure, gap, hole, orifice, slot, vent. **2** beginning, commencement, inauguration, initiation, launch, outset, start. **3** break, chance, occasion, opportunity.

openly *adv.* publicly, frankly.

openwork *n.* pattern with intervening spaces in metal, leather, lace, etc.

opera *n.* play(s) in which words are sung to music. □ **opera-glasses** *n.pl.* small binoculars.

operable *adj.* **1** able to be operated. **2** suitable for treatment by surgery.

operate *v.* **1** be in action. **2** control (a machine etc.). **3** produce an effect. **4** perform an operation.

■ **1** function, go, run, perform, work. **2** control, handle, manage, manipulate, use, work.

operatic *adj.* of or like opera.

operation *n.* **1** action, working. **2** military manoeuvre. **3** a surgical treatment.

■ **1** action, function, functioning, performance, running, working; control, handling, manipulation. **2** action, campaign, exercise, manoeuvre.

operational *adj.* **1** of or used in operations. **2** able or ready to function.

■ **2** functional, functioning, going, in use, in working order, operative, running, working.

operative *adj.* **1** working, functioning. **2** of surgical operations. ● *n.* worker, esp. in a factory.

operator *n.* person operating a machine, esp. connecting lines in a telephone exchange.

operetta *n.* short or light opera.

ophidian *adj.* & *n.* (member) of the snake family.

ophthalmic *adj.* of or for the eyes.

ophthalmology *n.* study of the eye. □ **ophthalmologist** *n.*

ophthalmoscope *n.* instrument for examining the eye.

opiate *n.* sedative containing opium.

opine *v.* express or hold as an opinion.

opinion *n.* **1** belief or judgement held without actual proof. **2** what one thinks on a particular point. □ **be of the opinion that** believe or think that.

■ belief, conviction, feeling, judgement, idea, impression, notion, point of view, sentiment, theory, thought, view, viewpoint.

opinionated *adj.* holding strong opinions obstinately.

■ doctrinaire, dogmatic, inflexible, obdurate, obstinate, pigheaded, stubborn.

opium *n.* narcotic drug made from the juice of certain poppies.

opossum *n.* small furry marsupial.

opponent *n.* one who opposes another.

■ adversary, antagonist, enemy, foe, rival.

opportune *adj.* **1** (of time) favourable. **2** well-timed.

■ **1** advantageous, auspicious, favourable, fortunate, good, happy, propitious. **2** ap-

■ propriate, convenient, seasonable, suitable, timely, well-timed.

opportunist *n.* person who grasps opportunities. □ **opportunism** *n.*, **opportunistic** *adj.*

opportunity *n.* circumstances suitable for a particular purpose.

■ break, chance, occasion, opening, time.

oppose *v.* **1** argue or fight against. **2** place opposite. **3** place in opposition or contrast.

■ **1** attack, combat, contest, counter, defy, fight (against), resist, stand up to, take a stand against; argue against, challenge, dispute, object to, protest against.

opposite *adj.* **1** facing, on the further side. **2** as different as possible from. ● *n.* opposite thing or person. ● *adv.* & *prep.* in an opposite position or direction (to).

■ *adj.* **2** antithetical, conflicting, contradictory, contrary, contrasting, different, differing, opposing. ● *n.* antithesis, contrary, converse, reverse.

opposition *n.* **1** antagonism, resistance. **2** placing or being placed opposite. **3** people opposing something.

■ **1** antagonism, antipathy, hostility, objection, resistance. **3** adversaries, competition, competitors, enemies, opponents, rivals.

oppress *v.* **1** govern or treat harshly. **2** weigh down with cares. □ **oppression** *n.*, **oppressor** *n.*

■ **1** abuse, grind down, maltreat, persecute, ride roughshod over, tyrannize. **2** afflict, burden, overload, trouble, weigh down. □ **oppression** abuse, cruelty, maltreatment, persecution, torture, tyranny.

oppressive *adj.* **1** harsh or cruel. **2** (of weather) sultry and tiring. □ **oppressively** *adv.*, **oppressiveness** *n.*

■ **1** brutal, cruel, despotic, harsh, repressive, severe, tyrannical, unjust. **2** close, muggy, stifling, stuffy, suffocating, sultry.

opprobrious *adj.* abusive.

opprobrium *n.* great disgrace from shameful conduct.

opt *v.* **1 opt for** make a choice. **2 opt out** choose not to participate.

■ **1** choose, decide on, go for, pick, plump for, select.

optic *adj.* of the eye or sight.

optical *adj.* **1** of or aiding sight. **2** visual. □ **optically** *adv.*

optician *n.* maker or seller of spectacles.

optics *n.* study of sight and of light as its medium.

optimism *n.* tendency to take a hopeful view of things. □ **optimist** *n.*, **optimistic** *adj.*, **optimistically** *adv.*

■ □ **optimistic** bright, buoyant, cheerful, confident, hopeful, sanguine, *colloq.* upbeat.

optimum *adj.* & *n.* best or most favourable (conditions etc.).

option *n.* **1** thing that is or may be chosen. **2** freedom to choose. **3** right to buy or sell a thing within a limited time.

optional *adj.* not compulsory. □ **optionally** *adv.*

■ discretionary, elective, non-compulsory, voluntary.

opulent *adj.* **1** wealthy. **2** luxurious. □ **opulently** *adv.*, **opulence** *n.*

■ **1** affluent, moneyed, prosperous, rich, wealthy, *colloq.* well-heeled, well-off, well-to-do. **2** luxurious, magnificent, palatial, plush, plushy, splendid, sumptuous.

or *conj.* **1** as an alternative. **2** because if not. **3** also known as.

oracle *n.* person or thing giving wise guidance. □ **oracular** *adj.*

■ prophet, prophetess, seer, sibyl, soothsayer.

oral *adj.* **1** spoken not written. **2** of the mouth, taken by mouth. ● *n.* spoken exam. □ **orally** *adv.*

orange *n.* **1** round juicy citrus fruit with reddish-yellow peel. **2** this colour. ● *adj.* reddish-yellow.

orangeade *n.* orange-flavoured soft drink.

orang-utan *n.* large ape of Borneo and Sumatra.

oration *n.* long speech, esp. of a ceremonial kind.

orator *n.* **1** maker of a formal speech. **2** skilful speaker.

oratorio *n.* (*pl.* **-os**) musical composition usu. with a biblical theme.

oratory *n.* **1** art of public speaking. **2** eloquent speech. □ **oratorical** *adj.*

orb *n.* sphere, globe.

orbit *n.* **1** curved path of a planet, satellite, or spacecraft round another. **2** sphere of influence. ● *v.* (**orbited**) move in an orbit (round).

■ *n.* **1** circuit, course, path, revolution, track. ● *v.* circle, go round, revolve round.

orbital *adj.* **1** of orbits. **2** (of a road) round the outside of a city.

orchard *n.* piece of land planted with fruit trees.

orchestra *n.* large group of people playing various musical instruments. □ **orchestral** *adj.*

orchestrate *v.* **1** compose or arrange (music) for an orchestra. **2** coordinate. □ **orchestration** *n.*

orchid *n.* showy often irregularly shaped flower.

ordain *v.* **1** appoint ceremonially to the Christian ministry. **2** destine. **3** decree authoritatively.

ordeal *n.* difficult experience.

■ affliction, hardship, *colloq.* nightmare, trial, tribulation, trouble.

order *n.* **1** way things are placed in relation to each other. **2** condition in which every part etc. is in its right place. **3** state of obedience to law, authority, etc. **4** authoritative direction or instruction. **5** request to supply goods etc., things supplied. **6** rank. **7** kind, quality. **8** group of plants or animals classified as similar. **9** group of people living under religious rules. ● *v.* **1** arrange in order. **2** command. **3** give an order for (goods etc.). □ **in order to** or **that** with the purpose of or intention that. **out of order 1** not working properly. **2** not in proper sequence.

■ *n.* **1** arrangement, classification, codification, disposition, grouping, form, sequence, shape, structure, system, systematization. **2** harmony, neatness, orderliness, organization, pattern, regularity, symmetry, system, tidiness. **3** calm, peace, peacefulness, quiet, serenity, tranquillity. **4** command, decree, dictate, direction, directive, edict, instruction, ordinance; injunction, warrant, writ. **5** instruction, request, requisition. **6** class, degree, grade, level, position, rank, station, status. **7** kind, nature, quality, sort, type, variety. ● *v.* **1** arrange, categorize, classify, codify, lay out, put in order, regulate, sort (out), systematize. **2** command, direct, instruct, require, tell. **3** request, requisition; book, reserve. □ **out of order 1** broken, faulty, inoperative, not in working order, not working.

orderly *adj.* **1** methodically arranged, tidy. **2** not unruly. ● *n.* **1** soldier assisting an officer. **2** hospital attendant. □ **orderliness** *n.*

■ *adj.* **1** methodical, neat, organized, shipshape, tidy, systematic, well-organized. **2** disciplined, law-abiding, peaceable, peaceful, well-behaved.

ordinal *n.* (in full **ordinal number**) number defining position in a series (*first*, *second*, etc.).

ordinance *n.* decree.

ordinary *adj.* usual, not exceptional. □ **out of the ordinary** unusual. **ordinarily** *adv.*

■ accustomed, common, customary, everyday, expected, familiar, habitual, normal, regular, routine, standard, traditional, typical, usual; boring, commonplace, humdrum, mediocre, pedestrian, prosaic, run-of-the-mill, undistinguished, unexceptional, uninspired, unremarkable; modest, plain, simple, unpretentious, workaday.

ordination *n.* ordaining.

ordnance *n.* military materials. □ **Ordnance Survey** official survey of the UK producing detailed maps.

ordure *n.* dung.

ore *n.* solid rock or mineral from which metal is obtained.

organ *n.* **1** musical instrument with pipes supplied with wind by bellows and sounded by keys. **2** distinct part with a specific function in an animal or plant body. **3** medium of communication, esp. a newspaper.

organic *adj.* **1** of bodily organ(s). **2** of or formed from living things. **3** organized as a system. **4** using no artificial fertilizers or pesticides. □ **organically** *adv.*

organism *n.* a living being, individual animal or plant.

organist *n.* person who plays the organ.

organization *n.* **1** organizing. **2** organized system or group of people. □ **organizational** *adj.*

■ **1** coordination, organizing, structuring; categorization, classification, codification. **2** arrangement, composition, configuration, constitution, design, form, order, pattern, shape, structure, system; body, coalition, confederation, consortium, federation, group, institution, league, society, syndicate; business, company, concern, corporation, firm.

organize *v.* **1** arrange systematically. **2** initiate. **3** make arrangements for. □ **organizer** *n.*

■ **1** arrange, catalogue, categorize, classify, codify, order, sort (out), systematize. **2** establish, form, found, initiate, institute, set up, start. 3 deal with, do, handle, look after, see to, take care of.

orgasm *n.* climax of sexual excitement.

orgy *n.* **1** wild revelry. **2** unrestrained activity. □ **orgiastic** *adj.*

Orient *n.* (**the Orient**) the East, the eastern world.

orient *v.* place or determine the position of (a thing) with regard to points of the compass. □ **orient oneself 1** get one's bearings. **2** become accustomed to a new situation. **orientation** *n.*

oriental *adj.* of the Orient. ● *n.* (**Oriental**) native of the Orient.

orientate *v.* orient.

orienteering *n.* sport of finding one's way across country by map and compass.

orifice *n.* opening of a cavity etc.

origin *n.* **1** point, source, or cause from which a thing begins its existence. **2** (often *pl.*) ancestry.

■ **1** base, basis, provenance, root, source; beginning, cradle, dawn, dawning, genesis, inception, start. **2** (**origins**) ancestry, descent, extraction, genealogy, lineage, pedigree, parentage.

original *adj.* **1** existing from the first, earliest. **2** being the first form of something. **3** inventive, creative. ● *n.* first form, thing from which another is copied. □ **originally** *adv.*, **originality** *n.*

■ *adj.* **1** earliest, first, initial, primary; aboriginal, indigenous, native. **2** archetypal, prototypical. **3** creative, fresh, imaginative, ingenious, innovative, inventive, novel, unusual. ● *n.* archetype, master, model, pattern, prototype, source.

originate *v.* bring or come into being. □ **origination** *n.*, **originator** *n.*

■ create, design, establish, found, inaugurate, initiate, institute, introduce, invent, launch, pioneer, set up, start; arise, begin, derive, develop, grow, proceed, spring, stem.

oriole *n.* bird with black and yellow plumage.

ormolu *n.* **1** gold-coloured alloy of copper. **2** things made of this.

ornament *n.* **1** decorative object or detail. **2** decoration. ● *v.* decorate with ornament(s), beautify. □ **ornamentation** *n.*

■ *n.* **1** bauble, gewgaw, knick-knack, trinket. **2** adornment, decoration, embellishment, ornamentation, trimming. ● *v.* adorn, beautify, deck, decorate, embellish, embroider, garnish, trim.

ornamental *adj.* serving as an ornament. □ **ornamentally** *adv.*

ornate *adj.* elaborately ornamented. □ **ornately** *adv.*

■ baroque, elaborate, fancy, florid, fussy, ostentatious, rococo, showy.

ornithology *n.* study of birds. □ **ornithological** *adj.*, **ornithologist** *n.*

orphan *n.* child whose parents are dead. ● *v.* make (a child) an orphan.

orphanage *n.* institution where orphans are cared for.

orrisroot *n.* fragrant iris root.

orthodontics *n.* correction of irregularities in teeth. □ **orthodontic** *adj.*, **orthodontist** *n.*

orthodox *adj.* of or holding conventional or currently accepted beliefs, esp. in religion. □ **Orthodox Church** Eastern or Greek Church. **orthodoxy** *n.*

■ conservative, conventional, established, prevailing, traditional.

orthopaedics /orthəpeédiks/ *n.* surgical correction of deformities in bones or muscles. □ **orthopaedic** *adj.*, **orthopaedist** *n.*

oryx *n.* large African antelope.

oscillate *v.* **1** move to and fro. **2** vary. □ **oscillation** *n.*

oscilloscope *n.* device for recording oscillations.

osier *n.* **1** willow with flexible twigs. **2** twig of this.

osmosis *n.* diffusion of fluid through a porous partition into another fluid. □ **osmotic** *adj.*

osprey *n.* (*pl.* **-eys**) large bird preying on fish in inland waters.

osseous *adj.* like bone, bony.

ossify *v.* **1** turn into bone, harden. **2** make or become rigid and unprogressive. □ **ossification** *n.*

ostensible *adj.* apparent but not necessarily real. □ **ostensibly** *adv.*

■ alleged, apparent, outward, pretended, professed, purported, supposed.

ostentation *n.* showy display intended to impress people. □ **ostentatious** *adj.*

■ □ **ostentatious** elaborate, extravagant, flamboyant, *colloq.* flash, flashy, gaudy, loud, showy.

osteopath *n.* practitioner who treats certain diseases and abnormalities by manipulating bones and muscles. □ **osteopathic** *adj.*, **osteopathy** *n.*

ostracize *v.* refuse to associate with. □ **ostracism** *n.*

> ■ avoid, boycott, cold-shoulder, cut, shun, snub.

ostrich *n.* large swift-running African bird, unable to fly.

other *adj.* **1** alternative, additional. **2** being the remaining one of a set of two or more. **3** not the same. ● *n.* & *pron.* the other person or thing. ● *adv.* otherwise. □ **the other day** or **week** a few days or weeks ago.

otherwise *adv.* **1** in a different way. **2** in other respects. **3** in different circumstances.

otter *n.* fish-eating water animal with thick brown fur.

ottoman *n.* storage box with a padded top.

ought *v.aux.* expressing duty, rightness, advisability, or strong probability.

ounce *n.* unit of weight, one-sixteenth of a pound (about 28 grams).

our *adj.*, **ours** *poss.pron.* belonging to us.

ourselves *pron.* emphatic and reflexive form of *we* and *us*.

oust *v.* drive out, eject.

> ■ depose, dismiss, drive out, eject, expel, force out, *colloq.* kick out, push out, remove, throw out, *colloq.* turf out.

out *adv.* **1** away from or not in a place, not at home. **2** not burning. **3** in error. **4** not possible. **5** unconscious. **6** into the open, so as to be heard or seen. **7** (of a secret) revealed. **8** (of a flower) blooming. ● *prep.* out of. ● *n.* way of escape. □ **be out to** be intending to. **out and out** thorough. **out of 1** from within or among. **2** without a supply of. **out of doors** in the open air. **out of the way 1** no longer an obstacle. **2** remote.

> ■ *adv.* **1** absent, away, elsewhere. **2** doused, extinguished, quenched, unlit. **4** impossible, impracticable, out of the question, unfeasible, unworkable. **5** comatose, insensible, knocked out, unconscious. □ **out and out** absolute, complete, downright, outright, perfect, sheer, thorough, total, unmitigated, unqualified, utter.

out- *pref.* more than, so as to exceed.

outboard *adj.* (of a motor) attached to the outside of a boat.

outbreak *n.* breaking out of anger, war, disease, etc.

outbuilding *n.* outhouse.

outburst *n.* explosion of feeling.

> ■ eruption, explosion, fit, outbreak, paroxysm, spasm.

outcast *n.* person driven out of a group or by society.

outclass *v.* surpass in quality.

outcome *n.* result of an event.

> ■ consequence, effect, result, upshot.

outcrop *n.* part of an underlying layer of rock that projects on the surface of the ground.

outcry *n.* **1** loud cry. **2** strong protest

> ■ clamour, commotion, hue and cry, hullabaloo, protest, protestation, uproar.

outdistance *v.* leave (a competitor) behind completely.

outdo *v.* (**-did**, **-done**) be or do better than.

> ■ beat, cap, eclipse, exceed, excel, outclass, outshine, outstrip, overshadow, surpass, top, transcend.

outdoor *adj.* of or for use in the open air. □ **outdoors** *adv.*

outer *adj.* **1** further from the centre or inside. **2** exterior, external.

outermost *adv.* furthest outward.

outface *v.* disconcert by staring or by a confident manner.

outfit *n.* set of clothes or equipment.

> ■ costume, ensemble, get-up, *colloq.* rig-out, suit, turn-out; clothes, clothing, dress, garb, *colloq.* togs; apparatus, equipment, gear, kit, paraphernalia.

outfitter *n.* supplier of equipment or men's clothing.

outflank *v.* get round the flank of (an enemy).

outflow *n.* **1** outward flow. **2** what flows out.

outgoing *adj.* **1** retiring from office. **2** sociable.

> ■ **1** departing, ex-, former, past, retiring. **2** approachable, communicative, expansive, friendly, genial, gregarious, sociable, unreserved.

outgoings *n.pl.* expenditure.

outgrow *v.* (**-grew**, **-grown**) **1** grow faster than. **2** grow too large for. **3** leave behind (childish habit etc.).

outgrowth *n.* something which grows out of another thing.

outhouse *n.* shed, barn, etc.

outing *n.* pleasure trip.

> ■ excursion, expedition, jaunt, spin, tour, trip.

outlandish *adj.* looking or sounding strange or foreign.

■ bizarre, curious, eccentric, exotic, extraordinary, odd, offbeat, peculiar, queer, strange, unfamiliar, unusual, weird.

outlast *v.* last longer than.

outlaw *n.* **1** (*old use*) person deprived of the law's protection. **2** fugitive from the law. ● *v.* **1** make (a person) an outlaw. **2** declare illegal.

■ *n.* **2** bandit, brigand, criminal, desperado, fugitive, pirate, robber. ● *v.* **2** ban, bar, disallow, forbid, prohibit, proscribe.

outlay *n.* expenditure.

■ cost, disbursement, expenditure, expense, spending.

outlet *n.* **1** way out. **2** means for giving vent to energies or feelings. **3** distributor for goods.

outline *n.* **1** line(s) showing a thing's shape or boundary. **2** summary. ● *v.* **1** draw or describe in outline. **2** mark the outline of.

■ *n.* **1** contour, profile, silhouette. **2** abstract, digest, précis, résumé, summary, synopsis. ● *v.* define, delineate, draft, sketch, trace.

outlook *n.* **1** view, prospect. **2** mental attitude. **3** future prospect.

■ **1** aspect, prospect, view, vista. **2** attitude, opinion, perspective, point of view, standpoint, view, viewpoint. **3** forecast, prospect.

outlying *adj.* remote.

■ distant, far-away, far-off, outermost, out of the way, remote.

outmoded *adj.* no longer fashionable or accepted.

outnumber *v.* exceed in number.

outpace *v.* go faster than.

out-patient *n.* person visiting a hospital for treatment but not remaining resident there.

outpost *n.* outlying settlement or detachment of troops.

output *n.* amount of electrical power etc. produced. ● *v.* (**-put** or **-putted**) (of a computer) supply (results etc.).

outrage *n.* **1** act that shocks public opinion. **2** (act of) great violence or cruelty. ● *v.* shock and anger greatly.

■ *n.* **1** disgrace, scandal. **2** atrocity, barbarism, brutality, cruelty, enormity, evil, savagery, violation, violence. ● *v.* anger, appal, disgust, enrage, horrify, incense, infuriate, madden, scandalize, shock.

outrageous *adj.* greatly exceeding what is moderate or reasonable. □ **outrageously** *adv.*

■ excessive, exorbitant, extortionate, immoderate, inordinate, unjustifiable, unreasonable; appalling, disgraceful, dreadful, egregious, monstrous, preposterous, scandalous, shameful, shocking.

outrider *n.* mounted attendant or motor cyclist riding as guard.

outrigger *n.* **1** stabilizing strip of wood fixed outside and parallel to a canoe. **2** canoe with this.

outright *adv.* **1** completely. **2** not gradually. **3** frankly. ● *adj.* thorough, complete.

■ *adv.* **1** absolutely, altogether, completely, entirely, totally, utterly, wholly. **2** at once, immediately, instantaneously, instantly, straight away. **3** candidly, directly, frankly, honestly, openly, plainly. ● *adj.* absolute, complete, downright, out and out, thorough, total, unmitigated, unqualified, utter.

outrun *v.* (**-ran**, **-run**, **-running**) run faster or further than.

outset *n.* beginning.

outshine *v.* surpass in ability.

outside *n.* outer side, surface, or part. ● *adj.* of or from the outside. ● *adv.* on, at, or to the outside. ● *prep.* **1** on, at, or to the outside of. **2** other than.

■ *n.* case, exterior, façade, face, facing, front, shell, skin, surface. ● *adj.* exterior, external, outer; outdoor. ● *adv.* out, outdoors, out of doors.

outsider *n.* **1** non-member of a group. **2** horse etc. thought to have no chance in a contest.

■ **1** alien, foreigner, newcomer, stranger; gatecrasher, interloper, intruder, trespasser.

outsize *adj.* much larger than average.

outskirts *n.pl.* outer districts.

■ borders, edge, fringes, outlying districts, periphery, suburbs.

outsmart *v.* outwit.

outspoken *adj.* very frank.

■ blunt, candid, direct, explicit, forthright, frank, honest, open, plain, straightforward.

outstanding *adj.* **1** exceptionally good. **2** not yet paid or settled. □ **outstandingly** *adv.*

■ **1** excellent, exceptional, first-class, first-rate, magnificent, marvellous, memorable, notable, noteworthy, remarkable, splendid, superb, superior, unforgettable, wonderful; celebrated, distinguished, eminent, famous, important, prominent, renowned. **2** due, owed, owing, payable, unpaid, unsettled.

outstrip *v.* (**-stripped**) **1** run faster or further than. **2** surpass.

■ **1** outdistance, outpace, outrun, overtake. **2** beat, better, cap, exceed, outclass, outdo, outshine, overshadow, surpass, top.

outvote *v.* defeat by a majority of votes.

outward *adj.* on or to the outside. ● *adv.* outwards. □ **outwardly** *adv.*, **outwards** *adv.*

■ *adj.* exterior, external, outer, outside; apparent, discernible, evident, manifest, observable, obvious, perceptible, visible.

outweigh *v.* be of greater weight or importance than.

outwit *v.* (**-witted**) defeat by one's craftiness.

■ deceive, dupe, fool, get the better of, hoodwink, outsmart, trick.

ouzel /o͞oz'l/ *n.* small bird of the thrush family.

ouzo /o͞ozō/ *n.* Greek aniseed-flavoured spirit.

ova *see* **ovum**.

oval *n.* & *adj.* (of) rounded symmetrical shape longer than it is broad.

ovary *n.* **1** organ producing egg cells. **2** that part of a pistil from which fruit is formed. □ **ovarian** *adj.*

ovation *n.* enthusiastic applause.

oven *n.* enclosed chamber in which to cook food.

over *prep.* **1** in or to a position higher than. **2** throughout, during. **3** more than. **4** above and across. ● *adv.* **1** outwards and downwards from the brink or an upright position etc. **2** from one side or end etc. to the other. **3** across a space or distance. **4** besides. **5** with repetition. **6** at an end.

over- *pref.* **1** above. **2** excessively.

overall *n.* **1** protective outer garment. **2** (*pl.*) one-piece garment of this kind covering the body and legs. ● *adj.* total, taking all aspects into account. ● *adv.* taken as a whole.

overarm *adj.* & *adv.* with the arm brought forward and down from above shoulder level.

overawe *v.* overcome with awe.

overbalance *v.* **1** lose balance and fall. **2** cause to do this.

overbearing *adj.* domineering.

■ authoritarian, autocratic, *colloq.* bossy, bullying, dictatorial, domineering, high-handed, imperious, officious, peremptory, tyrannical.

overboard *adv.* from a ship into the water. □ **go overboard** (*colloq.*) show extreme enthusiasm.

overcast *adj.* covered with cloud.

■ clouded, cloudy, dark, dreary, dull, gloomy, grey, leaden, moonless, murky, sombre, starless, sunless.

overcharge *v.* charge too much.

overcoat *n.* warm outdoor coat.

overcome *v.* **1** win a victory over. **2** succeed in subduing or dealing with. **3** make helpless.

■ **1** beat, conquer, defeat, *colloq.* lick, overpower, overthrow, overwhelm, prevail over, subdue, triumph over, trounce, vanquish. **2** conquer, get over, get the better of, master, surmount.

overdo *v.* (**-did, -done**) **1** do too much. **2** cook too long.

overdose *n.* too large a dose. ● *v.* **1** give an overdose to. **2** take an overdose.

overdraft *n.* **1** overdrawing of a bank account. **2** amount of this.

overdraw *v.* (**-drew, -drawn**) draw more from (a bank account) than the amount credited.

overdrive *n.* mechanism providing an extra gear above top gear.

overdue *adj.* not paid or arrived etc. by the due time.

■ outstanding, owed, owing, unpaid; belated, late, tardy, unpunctual.

overestimate *v.* form too high an estimate of.

overflow *v.* **1** flow over the edge or limits (of). **2** be very abundant. ● *n.* **1** what overflows. **2** outlet for excess liquid.

overgrown *adj.* **1** grown too large. **2** covered with weeds etc.

overhang *v.* project or hang over. ● *n.* fact or amount of overhanging.

overhaul *v.* **1** examine and repair. **2** overtake. ● *n.* examination and repair.

■ *v.* **1** mend, recondition, renovate, repair, service.

overhead *adj.* & *adv.* **1** above the level of one's head. **2** in the sky.

overheads *n.pl.* expenses involved in running a business etc.

overhear *v.* (**-heard**) hear accidentally or without the speaker's knowledge.

overjoyed *adj.* filled with great joy.

■ delighted, ecstatic, elated, euphoric, happy, joyful, jubilant, rapturous, thrilled.

overland *adj.* & *adv.* (travelling) by land. □ **overlander** *n.*

overlap *v.* (**-lapped**) **1** extend beyond the edge of. **2** coincide partially. ● *n.* **1** overlapping. **2** part or amount that overlaps.

overleaf *adv.* on the other side of a leaf of a book etc.

overload *v.* put too great a load on or in. ● *n.* load that is too great.

overlook *v.* **1** fail to observe or consider. **2** condone (an offence etc.). **3** have a view over.

■ **1** discount, disregard, ignore, miss, neglect, omit, pass over. **2** condone, excuse, forgive, make allowances for, pardon.

overman *v.* (**-manned**) provide with too many people as workmen or crew.

overnight *adv.* & *adj.* during a night.

overpass *n.* road crossing another by means of a bridge.

overpower *v.* overcome by greater strength or numbers.

■ beat, conquer, crush, defeat, master, overcome, overwhelm, quash, quell, rout, subjugate, triumph over, trounce, vanquish.

overpowering *adj.* (of heat or feelings) extremely intense.

■ compelling, intense, irresistible, overwhelming, powerful, strong.

overrate *v.* have too high an opinion of.

overreach *v.* **overreach oneself** fail through being too ambitious.

override *v.* (**-rode, -ridden**) **1** overrule. **2** prevail over. **3** intervene and make ineffective.

overrule *v.* set aside (a decision etc.) by using one's authority.

overrun *v.* (**-ran, -run, -running**) **1** spread over. **2** conquer (territory) by force. **3** exceed (a limit).

■ **1** infest, spread over, swarm over. **2** conquer, defeat, invade, occupy, overwhelm, storm, take over, take possession of.

overseas *adj.* & *adv.* across or beyond the sea, abroad.

oversee *v.* (**-saw, -seen**) superintend. □ **overseer** *n.*

■ be in charge of, control, direct, manage, run, superintend, supervise.

oversew *v.* (**-sewn**) sew (edges) together so that each stitch lies over the edges.

overshadow *v.* **1** cast a shadow over. **2** cause to seem unimportant in comparison.

■ **2** dominate, dwarf, eclipse, outshine, upstage.

overshoot *v.* (**-shot**) pass beyond (a target or limit etc.).

oversight *n.* **1** supervision. **2** unintentional omission or mistake.

■ **1** control, direction, management, superintendence, supervision, surveillance. **2** *colloq.* boob, error, fault, lapse, mistake, omission, slip, *colloq.* slip-up.

overspill *n.* **1** what spills over. **2** surplus population leaving one area for another.

overstay *v.* **overstay one's welcome** stay so long that one is no longer welcome.

overstep *v.* (**-stepped**) go beyond (a limit).

overt *adj.* done or shown openly. □ **overtly** *adv.*

■ apparent, clear, conspicuous, evident, manifest, observable, obvious, open, patent, plain, unconcealed, undisguised, visible.

overtake *v.* (**-took, -taken**) **1** pass (a moving person or thing). **2** (of misfortune etc.) come suddenly upon.

■ **1** outstrip, outdistance, overhaul, pass.

overthrow *v.* (**-threw, -thrown**) cause the downfall of. ● *n.* downfall, defeat.

■ *v.* conquer, defeat, depose, oust, overcome, overpower, overwhelm, rout, unseat. ● *n.* collapse, conquest, defeat, destruction, downfall, fall, ousting, suppression.

overtime *adv.* in addition to regular working hours. ● *n.* **1** time worked in this way. **2** payment for this.

overtone *n.* additional quality or implication.

■ hint, implication, indication, intimation, suggestion, undertone.

overture *n.* **1** orchestral composition forming a prelude to a performance. **2** (*pl.*) initial approach or proposal.

overturn *v.* **1** (cause to) turn over. **2** abolish.

■ **1** capsize, keel over, turn over, turn turtle, turn upside down; knock over, upend, upset. **2** abolish, annul, cancel, invalidate, nullify, override, overrule, rescind, repeal, reverse, revoke.

overview *n.* general survey.

overwhelm *v.* **1** bury beneath a huge mass. **2** overcome completely. **3** make helpless with emotion. □ **overwhelming** *adj.*

■ **1** bury, deluge, engulf, flood, inundate, submerge, swamp. **2** conquer, crush, defeat, destroy, overcome, overpower, quash, subdue, trounce, vanquish. **3** astonish, astound, bowl over, dumbfound, overcome, shock, stagger, stun, take aback.

overwrought *adj.* in a state of nervous agitation.

■ agitated, distraught, edgy, frantic, *colloq.* in a state, *colloq.* in a tizzy, jumpy, keyed up, nervous, tense, *colloq.* uptight, worked up.

oviduct *n.* tube through which ova pass from the ovary.

oviparous *adj.* egg-laying.

ovoid *adj.* egg-shaped, oval.

ovulate *v.* produce or discharge an egg cell from an ovary. □ **ovulation** *n.*

ovule *n.* germ cell of a plant.

ovum *n.* (*pl.* **ova**) egg cell, reproductive cell produced by a female.

owe *v.* **1** be under an obligation to pay or repay. **2** be indebted to a person or thing for.

owing *adj.* owed and not yet paid. □ **owing to 1** caused by. **2** because of.

■ due, outstanding, owed, payable, unpaid.

owl *n.* bird of prey usu. flying at night. □ **owlish** *adj.*

own *adj.* belonging to oneself or itself. ● *v.* **1** have as one's own. **2** acknowledge as true or belonging to one. □ **of one's own** belonging to oneself. **on one's own 1** alone. **2** independently. **own up** confess. **owner** *n.*, **ownership** *n.*

■ *adj.* individual, particular, personal, private. ● *v.* **1** be in possession of, have, hold, possess. **2** accept, acknowledge, admit, concede, confess, grant, recognize. □ **owner** holder, keeper, possessor, proprietor.

ox *n.* (*pl.* **oxen**) **1** animal of or related to the kind kept as domestic cattle. **2** fully grown bullock.

oxidation *n.* process of combining with oxygen.

oxide *n.* compound of oxygen and one other element.

oxidize *v.* **1** combine with oxygen. **2** coat with an oxide. **3** make or become rusty. □ **oxidization** *n.*

oxyacetylene *adj.* using a mixture of oxygen and acetylene, esp. in cutting or welding metals.

oxygen *n.* colourless gas existing in air.

oyster *n.* edible shellfish.

oz. *abbr.* ounce(s).

ozone *n.* **1** form of oxygen. **2** protective layer of this in the stratosphere.

Pp

pa *n.* (*colloq.*) father.

pace *n.* **1** single step in walking or running. **2** rate of progress. ● *v.* **1** walk steadily or to and fro. **2** measure by pacing. **3** set the pace for.

■ *n.* **1** footstep, step, stride. **2** rate, speed, velocity. ● *v.* **1** march, stride, walk.

pacemaker *n.* **1** person who sets the pace for another. **2** device regulating heart contractions.

pachyderm /pákkiderm/ *n.* large thick-skinned mammal such as the elephant. □ **pachydermatous** *adj.*

pacific *adj.* making or loving peace. □ **pacifically** *adv.*

pacifist *n.* person totally opposed to war. □ **pacifism** *n.*

pacify *v.* **1** calm and soothe. **2** establish peace in. □ **pacification** *n.*

■ **1** appease, calm (down), conciliate, mollify, placate, quiet, quieten, soothe.

pack *n.* **1** collection of things wrapped or tied for carrying or selling. **2** set of playing cards. **3** group of hounds or wolves. ● *v.* **1** put into or fill a container. **2** press or crowd together, fill (a space) thus. **3** cover or protect with something pressed tightly. □ **pack off** send away. **send packing** dismiss abruptly.

■ *n.* **1** bale, bundle, package, packet, parcel. ● *v.* **1** fill; bundle, cram, jam, ram, squeeze, stuff. **2** cram, crowd, jam, squash, squeeze. **3** package, wrap (up).

package *n.* **1** parcel. **2** box etc. in which goods are packed. **3** package deal. ● *v.* put together in a package. □ **package deal** set of proposals offered or accepted as a whole. **package holiday** one with a fixed inclusive price.

packet *n.* **1** small package. **2** (*colloq.*) large sum of money.

pact *n.* agreement, treaty.

■ agreement, arrangement, bargain, compact, contract, covenant, deal, entente, settlement, treaty, understanding.

pad *n.* **1** piece of padding. **2** set of sheets of paper fastened together at one edge. **3** soft fleshy part under an animal's paw. **4** flat surface for use by helicopters or for launching rockets. ● *v.* **(padded) 1** put padding on or into. **2** walk softly or steadily. □ **pad out** fill out (a book, speech, etc.).

■ *n.* **1** bolster, cushion, pillow; wad. **2** jotter, notebook, notepad. ● *v.* **1** cushion, fill, stuff, wad. □ **pad out** amplify, augment, expand, fill out, lengthen, protract, spin out.

padding *n.* soft material used to protect against jarring, add bulk, absorb fluid, etc.

paddle[1] *n.* short oar with broad blade(s). ● *v.* **1** propel by use of paddle(s). **2** row gently.

paddle[2] *v.* walk with bare feet in shallow water for pleasure.

paddock *n.* **1** small field where horses are kept. **2** enclosure for horses at a racecourse.

padlock *n.* detachable lock with a U-shaped bar secured through the object fastened. ● *v.* fasten with a padlock.

padre /paádri/ *n.* chaplain in the army etc.

paean /peéən/ *n.* song of triumph.

paediatrics /peédiátriks/ *n.* study of children's diseases. □ **paediatric** *adj.*, **paediatrician** *n.*

paella /pī-éllə/ *n.* Spanish dish of rice, saffron, seafood, etc.

pagan *adj.* & *n.* heathen.

page[1] *n.* **1** sheet of paper in a book etc. **2** one side of this.

page[2] *n.* boy attendant of a bride. ● *v.* summon by an announcement, messenger, or pager.

pageant *n.* public show or procession, esp. with people in costume. □ **pageantry** *n.*

■ display, extravaganza, parade, procession, show, spectacle, tableau, tattoo. □ **pageantry** magnificence, pomp, showiness, splendour.

pager *n.* bleeping device for summoning the wearer.

pagoda *n.* temple or sacred tower in China, India, etc.

paid *see* **pay**. **put paid to** end (hopes or prospects etc.).

pail *n.* bucket.

pain *n.* **1** bodily suffering caused by injury, illness, etc. **2** mental suffering. **3** (*pl.*) careful effort. ● *v.* cause pain to.

■ *n.* **1** ache, aching, cramp, discomfort, hurt, pang, smarting, soreness, tenderness, twinge. **2** affliction, agony, anguish, distress, grief, heartache, misery, suffering, torment, torture, woe. **3** (**pains**) effort, exertion, labour, trouble. ● *v.* distress, grieve, hurt, sadden, trouble, wound.

painful *adj.* **1** causing or suffering pain. **2** laborious. □ **painfully** *adv.*

■ **1** aching, achy, agonizing, excruciating, raw, sensitive, smarting, sore, stinging, tender, throbbing; distressing, grievous, harrowing, heartbreaking, traumatic, upsetting. **2** arduous, demanding, exacting, laborious, onerous.

painless *adj.* not causing pain. □ **painlessly** *adv.*

painstaking *adj.* very careful.

■ assiduous, careful, conscientious, diligent, meticulous, scrupulous, sedulous, thorough.

paint *n.* **1** colouring matter for applying in liquid form to a surface. **2** (*pl.*) tubes or cakes of paint. ● *v.* **1** coat with paint. **2** portray by using paint(s) or in words. **3** apply (liquid) to.

painter[1] *n.* person who paints as artist or decorator.

painter[2] *n.* rope attached to a boat's bow for tying it up.

painting *n.* painted picture.

■ fresco, landscape, mural, oil painting, picture, portrait, still life, seascape, water colour.

pair *n.* **1** set of two things or people, couple. **2** article consisting of two parts. **3** other member of a pair. ● *v.* arrange or be arranged in pair(s).

■ *n.* **1** brace; couple, duo, set of two, twosome. ● *v.* join, match (up), put together, team up, twin.

pal *n.* (*colloq.*) friend.

palace *n.* **1** official residence of a sovereign, archbishop, or bishop. **2** splendid mansion.

■ **2** castle, chateau, mansion, stately home.

palaeography /pálliógrəfi/ *n.* study of ancient writing and inscriptions. □ **palaeographer** *n.*

palaeolithic /pállió-/ *adj.* of the early part of the Stone Age.

palaeontology /pálli-/ *n.* study of life in the geological past. □ **palaeontologist** *n.*

palatable *adj.* pleasant to the taste or mind.

■ appetizing, savoury, tasty; acceptable, agreeable, pleasant, pleasing, satisfactory.

palate *n.* **1** roof of the mouth. **2** sense of taste.

palatial /pəláysh'l/ *adj.* of or like a palace, splendid.

■ de luxe, grand, luxurious, magnificent, opulent, *colloq.* posh, plush, splendid, sumptuous.

palaver *n.* (*colloq.*) fuss.

pale[1] *adj.* **1** (of face) having less colour than normal. **2** (of colour or light) faint. ● *v.* turn pale. □ **palely** *adv.*, **paleness** *n.*

■ *adj.* **1** anaemic, ashen, ashy, colourless, pallid, pasty, peaky, wan, washed out, white. **2** faded, light, pastel; dim, faint, weak. ● *v.* blanch, fade, whiten.

pale[2] *n.* **beyond the pale** outside the bounds of acceptable behaviour.

Palestinian *adj.* & *n.* (native) of Palestine.

palette *n.* board on which an artist mixes colours. □ **palette-knife** *n.* blade with a handle, used for spreading paint or for smoothing soft substances in cookery.

paling *n.* railing(s).

pall /pawl/ *n.* **1** cloth spread over a coffin. **2** heavy dark covering. ● *v.* become uninteresting.

pallbearer *n.* person helping to carry or walking beside the coffin at a funeral.

pallet[1] *n.* **1** straw mattress. **2** makeshift bed.

pallet[2] *n.* tray or platform for goods being lifted or stored.

palliasse *n.* straw mattress.

palliate *v.* **1** alleviate. **2** partially excuse. □ **palliative** *adj.*

pallid *adj.* pale, esp. from illness. □ **pallidness** *n.*, **pallor** *n.*

■ anaemic, ashen, ashy, ghastly, pale, pasty, peaky, wan, washed out, white.

pally *adj.* (*colloq.*) friendly.

palm *n.* **1** inner surface of the hand. **2** part of a glove covering this. **3** tree of warm and tropical climates, with large leaves and no branches. ● *v.* conceal in one's hand. □ **palm off** get (a thing) accepted fraudulently.

palmist *n.* person who tells people's fortunes or characters from lines in their palms. □ **palmistry** *n.*

palomino *n.* (*pl.* **-os**) golden or cream-coloured horse.

palpable *adj.* **1** able to be touched or felt. **2** obvious. □ **palpably** *adv.*, **palpability** *n.*

■ **1** solid, tangible, touchable. **2** apparent, blatant, clear, evident, manifest, obvious, patent, plain, unmistakable.

palpate *v.* examine medically by touch. □ **palpation** *n.*

palpitate *v.* **1** throb rapidly. **2** quiver with fear or excitement. □ **palpitation** *n.*

■ **1** flutter, pound, pulsate, pulse, throb. **2** quake, quaver, quiver, shake, tremble.

palsy *n.* paralysis, esp. with involuntary tremors. □ **palsied** *adj.*

paltry *adj.* (**-ier, -iest**) worthless. □ **paltriness** *n.*

■ base, contemptible, despicable, low, mean, miserable, sorry, worthless, wretched.

pampas *n.* vast grassy plains in S. America. □ **pampas-grass** *n.* large ornamental grass.

pamper *v.* treat very indulgently.

■ coddle, indulge, mollycoddle, spoil.

pamphlet *n.* small unbound booklet. □ **pamphleteer** *n.*

pan[1] *n.* **1** metal or earthenware vessel with a flat base, used in cooking. **2** similar vessel. ● *v.* (**panned**) **1** wash (gravel) in a pan in searching for gold. **2** (*colloq.*) criticize severely. □ **pan out** turn out (well). **panful** *n.*

pan[2] *v.* (**panned**) turn horizontally in filming.

pan- *pref.* all-, whole.

panacea /pánnəseéə/ *n.* remedy for all diseases or troubles.

panache *n.* confident stylish manner.

panama *n.* straw hat.

panatella *n.* thin cigar.

pancake *n.* thin round cake of fried batter.

panchromatic *adj.* sensitive to all visible colours.

pancreas *n.* gland near the stomach, discharging insulin into the blood. □ **pancreatic** *adj.*

panda *n.* **1** bear-like black and white animal native to China and Tibet. **2** racoon-like animal of India.

pandemic *adj.* (of a disease) widespread.

pandemonium *n.* uproar.

■ bedlam, chaos, confusion, disorder, havoc, tumult, turmoil, uproar.

pander *v.* **pander to** gratify by satisfying a taste or weakness.

■ cater to, fulfil, gratify, humour, indulge, satisfy.

pane *n.* sheet of glass in a window or door.

panegyric *n.* piece of written or spoken praise.

panel *n.* **1** distinct usu. rectangular section. **2** strip of board etc. forming this. **3** group assembled to discuss or decide something. **4** list of jurors, jury. ● *v.* (**panelled**) cover or decorate with panels.

panelling *n.* **1** series of wooden panels in a wall. **2** wood used for making panels.

panellist *n.* member of a panel.

pang *n.* sudden sharp pain.

pangolin *n.* scaly anteater.

panic *n.* sudden strong fear. ● *v.* (**panicked**) affect or be affected with panic. □ **panic-stricken** *adj.*, **panicky** *adj.*

■ *n.* alarm, anxiety, apprehension, consternation, dismay, dread, fear, fright, horror, terror, trepidation. ● *v.* alarm, frighten, scare, terrify, unnerve. □ **panic-stricken** afraid, agitated, alarmed, beside oneself, fearful, frightened, horrified, *colloq.* in a flap, *colloq.* in a tizzy, nervous, panicky, petrified, scared, terrified, unnerved.

pannier *n.* **1** large basket carried by a donkey etc. **2** bag fitted on a motor cycle or bicycle.

panoply *n.* splendid array.

panorama *n.* view of a wide area or set of events. □ **panoramic** *adj.*

■ □ **panoramic** comprehensive, extensive, far-reaching, overall, sweeping, wide, wide-ranging.

pansy *n.* garden flower of violet family with broad petals.

pant *v.* **1** breathe with short quick breaths. **2** utter breathlessly.

■ **1** blow, gasp, heave, huff, puff, wheeze.

pantechnicon *n.* large van for transporting furniture.

pantheism *n.* doctrine that God is in everything. □ **pantheist** *n.*, **pantheistic** *adj.*

panther *n.* leopard.

panties *n.pl.* (*colloq.*) short knickers.

pantile *n.* curved roof tile.

pantograph *n.* device for copying a plan etc. on any scale.

pantomime *n.* Christmas play based on a fairy tale.

pantry *n.* **1** room for storing china, glass, etc. **2** larder.

pants *n.pl.* **1** (*US*) trousers. **2** underpants. **3** knickers.

pap *n.* **1** soft food suitable for infants or invalids. **2** pulp.

papacy *n.* position or authority of the pope.

papal *adj.* of the pope or papacy.

papaya *n.* = **pawpaw**.

paper *n.* **1** substance manufactured in thin sheets from wood fibre, rags, etc., used for writing on, wrapping things, etc. **2** newspaper. **3** set of exam questions. **4** document. **5** dissertation. ● *v.* cover (walls etc.) with wallpaper.

> ■ *n.* **2** broadsheet, daily, gazette, journal, newspaper, organ, periodical, publication, *derog.* rag, tabloid, weekly. **4** certificate, deed, document, form. **5** article, dissertation, essay, report, study, thesis, treatise.

paperback *adj.* & *n.* (book) bound in flexible paper binding.

paperweight *n.* small heavy object to hold loose papers down.

papier mâché /pápyay máshay/ moulded paper pulp used for making small objects.

papoose *n.* young N. American Indian child.

paprika *n.* red pepper.

papyrus *n.* **1** reed-like water plant from which the ancient Egyptians made a kind of paper. **2** this paper. **3** (*pl.* **-ri**) manuscript written on this.

par *n.* **1** average or normal amount or condition etc. **2** equal footing.

parable *n.* story told to illustrate a moral or spiritual truth.

paracetamol *n.* drug that relieves pain and reduces fever.

parachute *n.* device used to slow the descent of a person or object dropping from a great height. ● *v.* descend or drop by parachute. □ **parachutist** *n.*

parade *n.* **1** formal assembly of troops. **2** procession. **3** ostentatious display. **4** public square or promenade. ● *v.* **1** assemble for parade. **2** march or walk with display. **3** make a display of.

> ■ *n.* **2** cavalcade, march, procession. **3** display, exhibition, show, spectacle. **3** esplanade, mall, promenade, walk, way. ● *v.* **2** file, march, walk. **3** display, exhibit, flaunt, show off.

paradigm /párrədīm/ *n.* example, model.

paradise *n.* **1** heaven. **2** delightful place or state.

> ■ **2** heaven, Utopia; bliss, delight, ecstasy, happiness, joy, rapture.

paradox *n.* statement that seems self-contradictory but contains a truth. □ **paradoxical** *adj.*, **paradoxically** *adv.*

paraffin *n.* oil from petroleum or shale, used as fuel.

paragon *n.* apparently perfect person or thing.

> ■ archetype, epitome, exemplar, ideal, model, pattern, quintessence.

paragraph *n.* one or more sentences on a single theme, beginning on a new line. ● *v.* arrange in paragraphs.

parakeet *n.* small parrot.

parallax *n.* apparent difference in an object's position when viewed from different points.

parallel *adj.* **1** (of lines or planes) going continuously at the same distance from each other. **2** similar, corresponding. ● *n.* **1** person or thing analogous to another. **2** line of latitude. **3** analogy. ● *v.* (**paralleled**) **1** be parallel to. **2** compare. □ **parallelism** *n.*

> ■ *adj.* **2** analogous, comparable, corresponding, equivalent, like, matching, similar. ● *n.* **1** analogue, counterpart, equal, equivalent, match. **3** analogy, correlation, correspondence, equivalence, likeness, similarity. ● *v.* **1** be analogous to, be comparable with, be similar to, correspond to match, resemble. **2** compare, equate, juxtapose, liken.

parallelogram *n.* four-sided geometric figure with its opposite sides parallel to each other.

paralyse *v.* **1** affect with paralysis. **2** render powerless.

> ■ **2** cripple, disable, incapacitate; bring to a standstill, halt, immobilize.

paralysis *n.* loss of power of movement. □ **paralytic** *adj.* & *n.*

paramedic *n.* skilled person working in support of medical staff.

parameter *n.* variable quantity or quality that restricts what it characterizes.

paramilitary *adj.* organized like a military force.

paramount *adj.* most important.

> ■ cardinal, chief, foremost, main, preeminent, primary, prime, supreme, uppermost.

paranoia *n.* **1** mental disorder in which a person has delusions of grandeur or persecution. **2** abnormal tendency to mistrust others. ▫ **paranoid** *adj.* & *n.*

parapet *n.* low wall along the edge of a balcony or bridge.

paraphernalia *n.* numerous belongings or pieces of equipment.

■ accessories, accoutrements, apparatus, belongings, *sl.* clobber, effects, equipment, gear, kit, possessions, property, stuff, tackle, things, trappings.

paraphrase *v.* express in other words. ● *n.* rewording in this way.

paraplegia *n.* paralysis of the legs and part or all of the trunk. ▫ **paraplegic** *adj.* & *n.*

parapsychology *n.* study of mental perceptions that seem outside normal abilities.

paraquat *n.* extremely poisonous weedkiller.

parasite *n.* **1** animal or plant living on or in another. **2** person living off another or others and giving no useful return. ▫ **parasitic** *adj.*

parasol *n.* light umbrella used to give shade from the sun.

parboil *v.* partly cook by boiling.

parcel *n.* **1** thing(s) wrapped for carrying or post. **2** piece of land. ● *v.* (**parcelled**) **1** wrap as a parcel. **2** divide into portions.

parch *v.* make or become hot and dry.

parched *adj.* **1** hot and dry. **2** (*colloq.*) thirsty.

■ **1** arid, dehydrated, desiccated, dried out *or* up, dry, scorched.

parchment *n.* **1** writing material made from animal skins. **2** paper resembling this.

pardon *n.* forgiveness. ● *v.* (**pardoned**) forgive. ▫ **pardonable** *adj.*, **pardonably** *adv.*

■ *n.* absolution, amnesty, exoneration, forgiveness, reprieve. ● *v.* absolve, condone, excuse, exonerate, forgive, let off, overlook, reprieve.

pare *v.* **1** trim the edges of. **2** peel. **3** reduce little by little.

parent *n.* **1** father or mother. **2** source from which other things are derived. ▫ **parental** *adj.*, **parenthood** *n.*

parentage *n.* ancestry.

■ ancestry, birth, descent, extraction, family, line, lineage, origins, pedigree, stock.

parenthesis *n.* (*pl.* **-theses**) **1** word, phrase, or sentence inserted into a passage. **2** brackets (like these) placed round this. ▫ **parenthetic** *adj.*

pariah /pərīə/ *n.* outcast.

parietal bone each of a pair of bones forming part of the skull.

paring *n.* piece pared off.

parish *n.* **1** area with its own church and clergyman. **2** local government district.

parishioner *n.* inhabitant of a parish.

Parisian *adj.* & *n.* (native) of Paris.

parity *n.* equality.

park *n.* **1** public garden or recreation ground. **2** enclosed land of a country house. **3** parking area. ● *v.* place and leave (esp. a vehicle) temporarily.

parka *n.* thick jacket with a hood.

parlance *n.* phraseology.

parley *n.* (*pl.* **-eys**) discussion, esp. between enemies, to settle a dispute. ● *v.* (**parleyed**) hold a parley.

parliament *n.* assembly that makes a country's laws. ▫ **parliamentary** *adj.*

■ congress, diet, house, legislative assembly, legislature.

Parmesan *n.* hard Italian cheese.

parochial *adj.* **1** of a church parish. **2** interested in a limited area only. ▫ **parochially** *adv.*, **parochialism** *n.*

■ **2** insular, limited, narrow-minded, provincial, restricted, small-minded.

parody *n.* comic or grotesque imitation. ● *v.* make a parody of.

■ *n.* burlesque, caricature, lampoon, mockery, satire, *colloq.* spoof, take-off. ● *v.* burlesque, caricature, lampoon, mimic, mock, satirize, send up, take off.

parole *n.* release of a prisoner before the end of his or her sentence on promise of good behaviour. ● *v.* release in this way.

paroxysm *n.* fit (of pain, rage, coughing etc.).

■ attack, convulsion, eruption, explosion, fit, outbreak, outburst, spasm.

parquet /paːrki/ *n.* flooring of wooden blocks arranged in a pattern.

parricide *n.* killing or killer of own parent. ▫ **parricidal** *adj.*

parrot *n.* **1** tropical bird with a short hooked bill. **2** unintelligent imitator. ● *v.* (**parroted**) repeat mechanically.

parry *v.* **1** ward off (a blow). **2** evade (a question) skilfully.

■ **1** avert, deflect, fend off, stave off, turn aside, ward off. **2** avoid, circumvent, dodge, elude, evade, sidestep.

parsec *n.* unit of distance used in astronomy, about 3.25 light years.

parsimonious *adj.* stingy, very sparing. □ **parsimoniously** *adv.*, **parsimony** *n.*

■ cheese-paring, close, mean, miserly, near, niggardly, penny-pinching, stingy, *colloq.* tight, tight-fisted.

parsley *n.* herb with crinkled green leaves.

parsnip *n.* vegetable with a large yellowish tapering root.

parson *n.* (*colloq.*) clergyman. □ **parson's nose** fatty lump on the rump of a cooked fowl.

parsonage *n.* rectory, vicarage.

part *n.* **1** some but not all. **2** distinct portion. **3** component. **4** portion allotted. **5** character assigned to an actor in a play etc. **6** (usu. *pl.*) region, district. ● *adv.* partly. ● *v.* separate, divide. □ **in good part** without taking offence. **part of speech** word's grammatical class (noun, verb, adjective, etc.). **part-time** *adj.* for or during only part of the working week. **part with** give up possession of.

■ *n.* **2** bit, division, piece, portion, section, segment; chapter, episode, instalment. **3** component, constituent, element, ingredient, unit. **4** allotment, percentage, quota, ration, share. **5** character, role. **6** area, district, neighbourhood, quarter, region. ● *v.* divide, separate, split (up). □ **part with** forgo, give up, relinquish, sacrifice, surrender.

partake *v.* (**-took, -taken**) **1** participate. **2** take a portion, esp. of food. □ **partaker** *n.*

partial *adj.* **1** favouring one side or person, biased. **2** not complete or total. □ **be partial to** have a strong liking for. **partially** *adv.*

■ **1** biased, discriminatory, one-sided, partisan, prejudiced, unfair. **2** fragmentary, incomplete. □ **be partial to** be fond of, be keen on, enjoy, have a soft spot for, have a weakness for, like, love.

partiality *n.* **1** bias, favouritism. **2** strong liking.

■ **1** bias, favouritism, partisanship, prejudice. **2** fondness, inclination, liking, love, penchant, predilection, preference, soft spot, taste, weakness.

participate *v.* have a share, take part in something. □ **participation** *n.*, **participant** *n.*

participle *n.* word formed from a verb, as a **past participle** (e.g. *burnt, frightened*), **present participle** (e.g. *burning, frightening*). □ **participial** *adj.*

particle *n.* **1** very small portion of matter. **2** minor part of speech.

■ **1** atom, bit, crumb, fragment, iota, jot, molecule, morsel, piece, scrap, shred, sliver, speck, spot.

particoloured *adj.* coloured partly in one colour, partly in another.

particular *adj.* **1** relating to one person or thing and not others. **2** fastidious. ● *n.* **1** detail. **2** (*pl.*) points of information. □ **in particular** specifically. **particularly** *adv.*

■ *adj.* **1** certain, distinct, individual, single, specific. **2** *colloq.* choosy, fastidious, finicky, fussy, nice, *colloq.* pernickety. ● *n.* **1** detail, element, item, specific. **2** (**particulars**) details, facts, information, *colloq.* low-down.

parting *n.* **1** leave-taking. **2** line from which hair is combed in different directions.

partisan *n.* **1** strong supporter. **2** guerrilla. ● *adj.* **1** of partisans. **2** biased. □ **partisanship** *n.*

■ *n.* **1** champion, devotee, enthusiast, fan, fanatic, follower, supporter. ● *adj.* **2** biased, bigoted, one-sided, partial, prejudiced.

partition *n.* **1** division into parts. **2** structure dividing a room or space, thin wall. ● *v.* divide into parts or by a partition.

partly *adv.* partially.

partner *n.* **1** person sharing with another or others in an activity. **2** each of a pair. **3** husband or wife or member of an unmarried couple. ● *v.* **1** be the partner of. **2** put together as partners.

■ *n.* **1** accomplice, ally, associate, colleague, collaborator, comrade, confederate. **3** consort, husband, spouse, wife; boyfriend, girlfriend.

partnership *n.* **1** being a partner or partners. **2** joint business.

■ **1** alliance, association, collaboration, cooperation, union. **2** business, company, firm, practice.

partridge *n.* plump brown game bird.

parturition *n.* **1** process of giving birth to young. **2** childbirth.

party *n.* **1** social gathering. **2** group travelling or working as a unit. **3** group with common aims, esp. in politics. **4** one side in an agreement or dispute. □ **party line** set policy of a political party. **party wall** wall common to two buildings or rooms.

■ **1** ball, celebration, do, function, gathering, *colloq.* get-together, jamboree, reception. **2** band, body company, crew, gang, group, squad, team, troop. **3** faction, movement, side. **4** defendant, disputant, litigant, plaintiff.

paschal /pásk'l/ *adj.* **1** of the Passover. **2** of Easter.

pass *v.* (**passed**) **1** move onward or past. **2** go or send to another person or place. **3** elapse. **4** happen. **5** occupy (time). **6** allow (a bill in Parliament) to proceed. **7** examine and declare satisfactory. **8** achieve the required standard in a test. **9** go beyond. **10** (in a game) refuse one's turn. **11** discharge from the body as excreta. ● *n.* **1** passing, movement made with the hands or thing held. **2** permit to enter or leave. **3** gap in mountains, allowing passage to the other side. □ **make a pass at** (*colloq.*) make sexual advances to. **pass away** die. **pass on** transmit to the next person in a series. **pass out** become unconscious. **pass over** disregard. **pass up** (*colloq.*) refuse to accept.

■ *v.* **1** go, move, proceed, progress, travel; get ahead of, get past, overtake. **3** elapse, go by, slip by *or* away. **4** befall, come about, happen, occur, take place. **5** employ, fill, kill, occupy, spend, use, while away. **6** accept, agree to, allow, authorize, endorse, sanction. **8** be successful (in), get through, succeed (in). **9** exceed, go beyond, outdo, surpass, transcend. ● *n.* **2** permit, ticket. **3** canyon, defile, gap, gorge, passage, ravine. □ **pass away** *sl.* croak, die, expire, perish, *sl.* snuff it. **pass out** black out, collapse, faint, keel over. **pass over** disregard, ignore, overlook, pay no attention to, *colloq.* skip. **pass up** decline, refuse, reject, turn down.

passable *adj.* **1** able to be traversed. **2** just satisfactory.

■ **1** clear, navigable, open. **2** acceptable, adequate, all right, average, fair, *colloq.* OK, satisfactory, tolerable.

passage *n.* **1** passing. **2** right to pass or be a passenger. **3** way through, esp. with a wall on each side. **4** journey by sea. **5** section of a literary or musical work. **6** tube-like structure in the body. □ **passageway** *n.*

■ **1** course, passing, progress, transit. **3** corridor, hall, passageway; road, route, thoroughfare, way. **4** crossing, journey, voyage, trip. **5** citation, excerpt, extract, part, quotation, section, selection; canto, stanza, verse.

passbook *n.* book recording a customer's deposits and withdrawals from a bank etc.

passenger *n.* **1** person (other than the driver, pilot, or crew) travelling in a vehicle, train, ship, or aircraft. **2** ineffective member of a team.

passer-by *n.* (*pl.* **passers-by**) person who happens to be going past.

passion *n.* **1** strong emotion. **2** sexual love. **3** (object arousing) great enthusiasm. **4** (**the Passion**) sufferings of Christ in his last days.

■ **1** ardour, emotion, fervency, fervour, fire, heat, intensity, vehemence, warmth, zeal. **2** desire, infatuation, love, lust. **3** craze, fad, mania, obsession; eagerness, enthusiasm, fascination, keenness.

passionate *adj.* full of passion, intense. □ **passionately** *adv.*

■ ardent, avid, burning, eager, emotional, enthusiastic, excited, fervent, fervid, fiery, impassioned, intense, vehement, zealous; amorous, lustful.

passive *adj.* **1** acted upon and not active. **2** not resisting, submissive. **3** showing no interest, initiative, or forceful qualities. □ **passively** *adv.*, **passivity** *n.*

■ **2** complaisant, compliant, docile, malleable, meek, submissive, tame, tractable, unresisting, yielding. **3** apathetic, impassive, inactive, inert, lifeless, listless, phlegmatic, quiescent, unresponsive.

Passover *n.* Jewish festival commemorating the escape of Jews from slavery in Egypt.

passport *n.* official document for use by a person travelling abroad.

password *n.* secret word(s), knowledge of which distinguishes friend from enemy.

past *adj.* belonging to the time before the present, gone by. ● *n.* **1** (**the past**) past time or events. **2** person's past life. ● *prep.* & *adv.* beyond. □ **past master** expert.

■ *adj.* last, recent; erstwhile, ex-, former, previous, prior; done, finished, over. ● *n.* **1** (**the past**) days gone by, days of yore, former times, yesterday. **2** background, career, history, life.

pasta *n.* **1** dried dough produced in various shapes. **2** cooked dish made with this.

paste *n.* **1** moist mixture. **2** adhesive. **3** edible doughy substance. **4** glass-like substance used in imitation gems. ● *v.* **1** fasten or coat with paste. **2** (*sl.*) thrash.

pasteboard *n.* cardboard.

pastel *n.* **1** chalk-like crayon. **2** drawing made with this. **3** light delicate shade of colour.

pasteurize *v.* sterilize by heating. □ **pasteurization** *n.*

pastille *n.* **1** small flavoured sweet for sucking. **2** lozenge.

pastime *n.* something done to pass time pleasantly.

> ■ amusement, entertainment, hobby, interest, leisure activity, recreation, sport.

pastor *n.* clergyman in charge of a church or congregation.

pastoral *adj.* **1** of country life. **2** of or appropriate to a pastor.

> ■ **1** bucolic, country, rural, rustic. **2** clerical, ecclesiastical, ministerial, priestly.

pastry *n.* **1** dough made of flour, fat, and water. **2** (item of) food made with this.

pasturage *n.* pasture land.

pasture *n.* grassy land suitable for grazing cattle. ● *v.* put (animals) to pasture, graze.

pasty[1] /pásti/ *n.* pastry with sweet or savoury filling, baked without a dish.

pasty[2] /páysti/ *adj.* (**-ier**, **-iest**) **1** of or like paste. **2** pallid.

pat *v.* (**patted**) tap gently with an open hand. ● *n.* **1** patting movement. **2** small mass of a soft substance. ● *adv.* & *adj.* known and ready.

patch *n.* **1** piece put on, esp. in mending. **2** distinct area or period. **3** piece of ground. ● *v.* **1** put patch(es) on. **2** piece (things) together. □ **not a patch on** (*colloq.*) not nearly as good as. **patch up** **1** repair. **2** settle (a quarrel etc.).

> ■ *n.* **2** area, region, section; period, spell, time. **3** lot, parcel, piece, plot. □ **patch up** **1** darn, fix, mend, repair. **2** heal, put right, resolve, settle.

patchwork *n.* **1** needlework in which small pieces of cloth are joined decoratively. **2** thing made of assorted pieces.

patchy *adj.* (**-ier**, **-iest**) **1** existing in patches. **2** uneven in quality. □ **patchily** *adv.*, **patchiness** *n.*

pâté /páttay/ *n.* paste of meat etc.

patella *n.* (*pl.* **-ae**) kneecap.

patent *adj.* **1** obvious. **2** patented. ● *v.* obtain or hold a patent for. ● *n.* **1** official right to be the sole maker or user of an invention or process. **2** invention etc. protected by this. □ **patent leather** leather with glossy varnished surface. **patently** *adv.*

> ■ *adj.* **1** apparent, clear, evident, manifest, obvious, plain, unmistakable.

patentee *n.* holder of a patent.

paternal *adj.* **1** of a father. **2** fatherly. **3** related through one's father. □ **paternally** *adv.*

> ■ **2** fatherly, fond, indulgent, kindly, loving, solicitous.

paternalism *n.* policy of making kindly provision for people's needs but giving them no responsibility. □ **paternalistic** *adj.*

paternity *n.* fatherhood.

path *n.* **1** way by which people pass on foot. **2** line along which a person or thing moves. **3** course of action.

> ■ **1** bridle path, footpath, track, trail, walk. **2** circuit, course, orbit, route, track, trajectory. **3** approach, avenue, course of action, method, procedure, strategy.

pathetic *adj.* **1** arousing pity or sadness. **2** (*colloq.*) miserably inadequate. □ **pathetically** *adv.*

> ■ **1** emotional, emotive, heartbreaking, moving, piteous, pitiable, pitiful, poignant, sad, touching.

pathogenic *adj.* causing disease.

pathology *n.* study of disease. □ **pathological** *adj.*, **pathologist** *n.*

pathos *n.* pathetic quality.

patience *n.* **1** calm endurance. **2** card game for one player.

> ■ **1** calmness, composure, endurance, equanimity, forbearance, fortitude, imperturbability, self-control, serenity, stoicism, tolerance.

patient *adj.* showing patience. ● *n.* person treated by a doctor or dentist etc. □ **patiently** *adv.*

> ■ *adj.* calm, composed, even-tempered, forbearing, long-suffering, serene, stoical, tolerant, uncomplaining.

patina *n.* sheen on a surface produced by age or use.

patio *n.* (*pl.* **-os**) paved courtyard.

patois /pátwaa/ *n.* dialect.

patriarch *n.* **1** male head of a family or tribe. **2** bishop of high rank in certain Churches. □ **patriarchal** *adj.*, **patriarchate** *n.*

patriarchy *n.* social organization in which a male is head of the family.

patricide *n.* killing or killer of own father. □ **patricidal** *adj.*

patrimony *n.* heritage.

patriot *n.* person devoted to and ready to defend his or her country. □ **patriotic** *adj.*, **patriotism** *n.*

patrol *v.* (**patrolled**) walk or travel regularly through (an area or building) to see that all is well. ● *n.* **1** patrolling. **2** person(s) patrolling.

■ *v.* defend, guard, keep guard over, police, protect. ● *n.* **1** beat, rounds. **2** guard, sentinel, sentry, watchman.

patron /páytrən/ *n.* **1** person giving influential or financial support to a cause. **2** regular customer. □ **patron saint** saint regarded as a protector.

■ **1** backer, benefactor, champion, friend, sponsor, supporter. **2** client, customer, *colloq.* regular.

patronage *n.* **1** patron's support. **2** patronizing behaviour.

patronize *v.* **1** treat in a condescending way. **2** be a regular customer of. **3** act as patron to.

■ **1** condescend to, talk down to. **2** be a customer of, frequent, shop at. **3** assist, back, help, finance, fund, sponsor, support.

patronymic *n.* name derived from that of a father or ancestor.

patter[1] *n.* sound of quick light taps or steps. ● *v.* (of rain etc.) make this sound.

patter[2] *n.* rapid glib speech.

pattern *n.* **1** decorative design. **2** model, design, or instructions showing how a thing is to be made. **3** sample of cloth etc. **4** excellent example. **5** regular manner in which things occur. □ **patterned** *adj.*

■ **1** design, motif. **2** blueprint, design, model, plan, stencil, template. **3** sample, specimen, swatch. **4** archetype, exemplar, ideal, model, paradigm, paragon, standard. **5** consistency, order, regularity, sequence, system.

patty *n.* small pie or pasty.

paucity *n.* smallness of quantity.

paunch *n.* protruding stomach.

pauper *n.* very poor person.

pause *n.* temporary stop. ● *v.* make a pause.

■ *n.* break, breather, delay, gap, hesitation, hiatus, interlude, intermission, interruption, interval, lull, respite, rest, stop. ● *v.* delay, halt, hesitate, rest, stop, take a break, wait.

pave *v.* cover (a street etc.) with a hard durable surface.

pavement *n.* paved path at the side of a road.

pavilion *n.* **1** building on a sports ground for players or spectators. **2** ornamental building.

pavlova *n.* meringue cake containing cream and fruit.

paw *n.* foot of an animal that has claws. ● *v.* **1** strike with a paw. **2** scrape (the ground) with a hoof. **3** (*colloq.*) touch with the hands.

pawl *n.* lever with a catch that engages with the notches of a ratchet.

pawn[1] *n.* **1** chessman of the smallest size and value. **2** person whose actions are controlled by others.

pawn[2] *v.* deposit with a pawnbroker as security for money borrowed.

pawnbroker *n.* person licensed to lend money on the security of personal property deposited.

pawnshop *n.* pawnbroker's premises.

pawpaw *n.* **1** fruit of a palm-like tropical tree. **2** this tree.

pay *v.* (**paid**) **1** give (money) in return for goods or services. **2** give what is owed. **3** give (attention, respect, etc.). **4** be profitable or worthwhile. **5** reward or punish. ● *n.* wages. □ **pay for 1** hand over money for. **2** bear the cost of. **3** suffer or be punished for. **pay off 1** pay (a debt) in full. **2** discharge (an employee). **3** yield good results. **pay-off** *n.* **1** (*sl.*) payment. **2** reward, retribution. **3** climax.

■ *v.* **1** disburse, expend, lay out, spend, *colloq.* stump up; remunerate, settle up with. **2** discharge, honour, pay off, recompense, refund, reimburse, repay, settle, square. **3** bestow, give, pass on. **4** avail, benefit, help, profit; be advantageous *or* profitable *or* worthwhile, pay off. **5** avenge oneself on, get revenge on, punish. ● *n.* emolument, income, remuneration, salary, stipend, wage(s).

payable *adj.* which must or may be paid.

■ due, outstanding, owed, owing, unpaid.

payee *n.* person to whom money is paid or is to be paid.

payload *n.* aircraft's or rocket's total load.

payment *n.* **1** act or instance of paying. **2** money etc. paid.

■ **1** compensation, remuneration, settlement. **2** instalment, premium, remittance.

payola *n.* bribery offered for dishonest use of influence to promote a commercial product.

payroll *n.* list of a firm's employees receiving regular pay.

pea *n.* **1** plant bearing seeds in pods. **2** its round seed used as a vegetable. □ **pea-green** *adj.* & *n.* bright green.

peace *n.* **1** calm, quiet. **2** freedom from or cessation of war.

■ **1** calm, calmness, peacefulness, quiet, repose, serenity, stillness, tranquillity. **2** accord, concord, harmony; armistice, ceasefire, truce.

peaceable *adj.* **1** fond of peace, not quarrelsome. **2** peaceful.

■ **1** amicable, cooperative, easygoing, friendly, genial, good-natured, inoffensive, mild, pacific, peace-loving.

peaceful *adj.* characterized by or not infringing peace. □ **peacefully** *adv.*, **peacefulness** *n.*

■ calm, gentle, peaceable, placid, quiet, restful, serene, tranquil; law-abiding, orderly, well-behaved.

peacemaker *n.* person who brings about peace.

■ arbitrator, conciliator, intermediary, mediator.

peach *n.* **1** round juicy fruit with a rough stone. **2** tree bearing this. **3** its yellowish-pink colour.

peacock *n.* male bird with splendid plumage and a long fan-like tail.

peafowl *n.* peacock or peahen.

peahen *n.* female peafowl.

peak *n.* **1** pointed top, esp. of a mountain. **2** projecting part of the edge of a cap. **3** point of highest value or intensity etc. □ **peaked** *adj.*

■ **1** crest, pinnacle, summit, top. **2** brim, visor. **3** acme, apex, climax, culmination, height, high point, top, zenith.

peaky *adj.* (**-ier**, **-iest**) looking drawn and sickly.

peal *n.* **1** sound of ringing bell(s). **2** set of bells with different notes. **3** loud burst of thunder or laughter. ● *v.* sound in a peal.

■ *n.* **1** chime, ring, ringing, tinkling, toll, tolling. **2** carillon, chime. **3** clap, crash, roar, rumble. ● *v.* chime, ring, toll; crash, roar, rumble, thunder.

peanut *n.* **1** plant bearing underground pods with two edible seeds. **2** this seed. **3** (*pl.*) very trivial sum of money.

pear *n.* **1** rounded fruit tapering towards the stalk. **2** tree bearing this.

pearl *n.* round usu. white gem formed inside the shell of certain oysters. □ **pearly** *adj.*

peasant *n.* person working on the land.

peasantry *n.* peasants collectively.

peat *n.* decomposed vegetable matter from bogs etc., used in horticulture or as fuel. □ **peaty** *adj.*

pebble *n.* small smooth round stone. □ **pebbly** *adj.*

pecan *n.* **1** smooth pinkish-brown nut. **2** tree bearing this.

peck *v.* **1** strike, nip, or pick up with the beak. **2** kiss hastily. ● *n.* pecking movement.

peckish *adj.* (*colloq.*) hungry.

pectin *n.* substance found in fruits that makes jam set.

pectoral *adj.* of, in, or on the chest or breast. ● *n.* pectoral fin or muscle.

peculiar *adj.* **1** strange, eccentric. **2** distinctive, special. □ **peculiar to** belonging exclusively to one person, place, or thing. **peculiarly** *adv.*

■ **1** abnormal, anomalous, bizarre, curious, different, eccentric, extraordinary, funny, odd, offbeat, outlandish, out of the ordinary, queer, strange, unconventional, unusual, weird. **2** characteristic, distinctive, distinguishing, idiosyncratic, individual, special, typical, unique.

peculiarity *n.* **1** idiosyncrasy. **2** characteristic. **3** being peculiar.

■ **1** eccentricity, idiosyncrasy, kink, oddity, quirk. **2** attribute, characteristic, feature, property, quality, trait.

pecuniary *adj.* of or in money.

pedagogue *n.* (*derog.*) person who teaches pedantically.

pedal *n.* lever operated by the foot. ● *v.* (**pedalled**) **1** work the pedal(s) of. **2** operate by pedals.

pedant *n.* person who insists on strict adherence to literal meaning or formal rules. □ **pedantic** *adj.*, **pedantry** *n.*

peddle *v.* sell (goods) as a pedlar.

pedestal *n.* base supporting a column or statue etc.

pedestrian *n.* person walking, esp. in a street. ● *adj.* **1** of or for pedestrians. **2** dull.

■ *adj.* **2** banal, boring, commonplace, dull, humdrum, mundane, ordinary, prosaic, run-of-the-mill, tedious, unimaginative, uninspired, uninteresting.

pedicure *n.* care or treatment of the feet and toenails.

pedigree *n.* line or list of (esp. distinguished) ancestors. ● *adj.* (of an animal) of recorded and pure breeding.

■ *n.* ancestry, birth, blood, descent, extraction, family, genealogy, lineage, parentage, stock.

pediment *n.* triangular part crowning the front of a building.

pedlar *n.* person who sells small articles from door to door.

pedometer *n.* device for estimating the distance travelled on foot.

peduncle *n.* stalk of a flower etc.

pee (*colloq.*) *v.* urinate. ● *n.* urine.

peek *v.* & *n.* peep, glance.

peel *n.* skin of certain fruits and vegetables etc. ● *v.* **1** remove the peel of. **2** strip off. **3** come off in strips or layers, lose skin or bark etc. thus. □ **peelings** *n.pl.*

peep *v.* **1** look furtively or through a narrow opening. **2** come slowly into view. ● *n.* brief or surreptitious look. □ **peep-hole** *n.* small hole to peep through. **peeping Tom** furtive voyeur.

■ *v.* **1** glance, look quickly, peek, squint.

peer¹ *v.* look searchingly or with difficulty or effort.

peer² *n.* **1** duke, marquess, earl, viscount, or baron. **2** one who is the equal of another in rank or merit etc.

peerage *n.* **1** peers as a group. **2** rank of peer or peeress.

peeress *n.* **1** a female peer. **2** a peer's wife.

peerless *adj.* unequalled.

■ incomparable, inimitable, matchless, superb, superlative, supreme, unequalled, unique, unparalleled, unrivalled, unsurpassed.

peeved *adj.* (*colloq.*) annoyed.

peevish *adj.* irritable. □ **peevishly** *adv.*, **peevishness** *n.*

■ bad-tempered, cantankerous, crabby, crotchety, fractious, grumpy, irascible, irritable, pettish, petulant, prickly, querulous, snappy, testy, touchy, waspish.

peewit *n.* lapwing.

peg *n.* **1** wooden or metal pin or stake. **2** clip for holding clothes on a washing-line. ● *v.* (**pegged**) **1** fix or mark by means of peg(s). **2** keep (wages or prices) at a fixed level. □ **off the peg** (of clothes) ready-made.

pejorative *adj.* derogatory.

peke *n.* Pekingese dog.

Pekingese *n.* dog of a breed with shor legs, flat face, and silky hair.

pelargonium *n.* plant with show flowers.

pelican *n.* waterbird with a pouch in it long bill for storing fish. □ **pelica crossing** pedestrian crossing with light operated by pedestrians.

pellagra *n.* deficiency disease causin cracking of the skin.

pellet *n.* **1** small round mass of a sub stance. **2** small shot.

pell-mell *adv.* **1** headlong, recklessly. in disorder.

pellucid *adj.* very clear.

pelmet *n.* ornamental strip above window etc.

pelt¹ *n.* an animal skin.

pelt² *v.* **1** throw missiles at. **2** run fas □ **at full pelt** as fast as possible.

■ **1** bombard, pepper, shower. **2** caree dash, hare, hurtle, race, run, rush, speec sprint, tear, whiz, zoom.

pelvis *n.* framework of bones round th body below the waist. □ **pelvic** *adj.*

pen¹ *n.* small fenced enclosure, esp. fo animals. ● *v.* (**penned**) shut in or as if i a pen.

■ *n.* coop, *US* corral, enclosure, folc hutch, pound, sty. ● *v.* confine, coop up corral, enclose, shut in *or* up.

pen² *n.* device with a metal point fo writing with ink. ● *v.* (**penned**) write (letter etc.). □ **penfriend** *n.* friend wit whom a person corresponds withou meeting. **pen-name** *n.* author's pseud onym.

penal *adj.* of or involving punishment.

penalize *v.* **1** inflict a penalty on. **2** pu at a disadvantage.

penalty *n.* punishment for breaking law or rule.

■ punishment, sentence; fine, forfeit.

penance *n.* act performed as an expres sion of penitence.

■ atonement, reparation.

pence *see* **penny**.

penchant /póNshoN/ *n.* liking.

■ fondness, liking, love, partiality, pred lection, preference, proclivity, soft spo taste, weakness.

pencil *n.* instrument containing graphite used for drawing or writing. ● *v.* (**pen cilled**) write, draw, or mark with a pen cil.

pendant *n.* ornament hung from a chain round the neck.

pendent *adj.* hanging.

pending *adj.* waiting to be decided or settled. ● *prep.* **1** during. **2** until.

pendulous *adj.* hanging loosely.

pendulum *n.* **1** weight hung from a cord and swinging freely. **2** rod with a weighted end that regulates a clock's movement.

penetrate *v.* **1** make a way into or through, pierce. **2** see into or through. □ **penetrable** *adj.*, **penetrability** *n.*, **penetration** *n.*

■ **1** enter, go into, make one's way through; bore into, perforate, pierce, prick, puncture.

penetrating *adj.* **1** showing great insight. **2** (of sound) piercing.

■ **1** acute, astute, clever, discerning, incisive, intelligent, keen, perceptive, percipient, perspicacious, sharp, shrewd. **2** loud, piercing, shrill.

penguin *n.* flightless seabird of Antarctic regions.

penicillin *n.* antibiotic obtained from mould fungi.

peninsula *n.* piece of land almost surrounded by water. □ **peninsular** *adj.*

penis *n.* organ by which a male mammal copulates and urinates.

penitent *adj.* feeling or showing regret that one has done wrong. ● *n.* penitent person. □ **penitently** *adv.*, **penitence** *n.*

■ *adj.* apologetic, ashamed, conscience-stricken, contrite, regretful, remorseful, repentant, rueful, sorry.

penitential *adj.* of penitence or penance.

pennant *n.* long tapering flag.

penniless *adj.* destitute.

pennon *n.* flag, esp. a long triangular or forked one.

penny *n.* (*pl.* **pennies** for separate coins, **pence** for a sum of money) **1** British bronze coin worth one-hundredth of £1. **2** former coin worth one-twelfth of a shilling. □ **penny-pinching** *adj.* niggardly.

pension *n.* income paid by the government, an ex-employer, or a private fund to a person who is retired, disabled, etc. ● *v.* pay a pension to. □ **pension off** dismiss with a pension.

pensionable *adj.* entitled or (of a job) entitling one to a pension.

pensioner *n.* person who receives a pension.

pensive *adj.* deep in thought. □ **pensively** *adv.*, **pensiveness** *n.*

■ contemplative, meditative, reflective, ruminative, thoughtful.

pentagon *n.* geometric figure with five sides. □ **pentagonal** *adj.*

pentagram *n.* five-pointed star.

pentathlon *n.* athletic contest involving five events.

Pentecost *n.* **1** Jewish harvest festival, 50 days after second day of Passover. **2** Whit Sunday.

penthouse *n.* flat on the roof of a tall building.

penultimate *adj.* last but one.

penumbra *n.* (*pl.* **-ae**) area of partial shadow.

penury *n.* poverty. □ **penurious** *adj.*

■ destitution, indigence, need, pennilessness, poverty. □ **penurious** destitute, impecunious, impoverished, indigent, necessitous, needy, penniless, poor, poverty-stricken.

peony *n.* plant with large round red, pink, or white flowers.

people *n.pl.* **1** persons in general. **2** (**the people**) citizens of a country. **3** (as *sing.*) persons composing a race or nation. **4** parents or other relatives. ● *v.* fill with people, populate.

■ *n.* **1** human beings, individuals, persons. **2** (**the people**) the *hoi polloi*, the masses, the populace, the (general) public. **3** folk, nation, race, tribe. **4** family, kin, kinsfolk, kith and kin, parents, relations, relatives.

pep *n.* vigour. ● *v.* (**pepped**) **pep up** fill with vigour. □ **pep talk** talk urging great effort.

pepper *n.* **1** hot-tasting seasoning powder made from the dried berries of certain plants. **2** capsicum. ● *v.* **1** sprinkle with pepper. **2** pelt. **3** sprinkle. □ **peppery** *adj.*

peppercorn *n.* dried black berry from which pepper is made. □ **peppercorn rent** very low rent.

peppermint *n.* **1** a kind of mint with strong fragrant oil. **2** this oil. **3** sweet flavoured with this.

pepsin *n.* enzyme in gastric juice.

peptic *adj.* of digestion.

per *prep.* **1** for each. **2** in accordance with. **3** by means of. □ **per annum** for each year. **per capita** for each person. **per cent** in or for every hundred.

perambulate *v.* walk through or round (an area). □ **perambulation** *n.*

perceive *v.* **1** become aware of by one of the senses. **2** apprehend, understand.

■ **1** catch sight of, descry, detect, discern, distinguish, espy, glimpse, make out, notice, observe, see, *colloq.* spot. **2** appreciate, apprehend, comprehend, grasp, realize, recognize, see, sense, understand.

percentage *n.* **1** rate or proportion per hundred. **2** proportion, part.

perceptible *adj.* able to be perceived. □ **perceptibly** *adv.*, **perceptibility** *n.*

■ apparent, clear, detectable, discernible, distinguishable, evident, manifest, noticeable, observable, obvious, palpable, patent, perceivable, recognizable, unmistakable, visible.

perception *n.* perceiving, ability to perceive.

■ apprehension, consciousness, grasp, realization, recognition, understanding; feeling, insight, intuition, sense, sensitivity.

perceptive *adj.* showing insight and understanding. □ **perceptively** *adv.*, **perceptiveness** *n.*

■ acute, astute, discerning, discriminating, intelligent, penetrating, percipient, perspicacious, quick, sensitive, sharp, shrewd.

perch[1] *n.* **1** bird's resting place, rod etc. for this. **2** high seat. ● *v.* rest or place on or as if on a perch.

perch[2] *n.* (*pl.* **perch**) edible freshwater fish with spiny fins.

percipient *adj.* perceptive. □ **percipience** *n.*

percolate *v.* **1** filter, esp. through small holes. **2** prepare in a percolator. □ **percolation** *n.*

percolator *n.* coffee pot in which boiling water is circulated repeatedly through ground coffee held in a perforated drum.

percussion *n.* playing of a musical instrument by striking it with a stick etc. □ **percussive** *adj.*

perdition *n.* eternal damnation.

peregrine *n.* a kind of falcon.

peremptory *adj.* imperious. □ **peremptorily** *adv.*

■ *colloq.* bossy, domineering, high-handed, imperious, overbearing.

perennial *adj.* **1** lasting a long time. **2** constantly recurring. **3** (of plants) living for several years. ● *n.* perennial plant. □ **perennially** *adv.*

perfect *adj.* /pérfikt/ **1** complete, entir[e]. **2** faultless. **3** exact. ● *v.* /pərfékt/ mak[e] perfect. □ **perfectly** *adv.*, **perfection** *n.*

■ *adj.* **1** complete, entire, intact, undam[aged], whole; absolute, out and out, ou[t]right, thorough, total, unmitigated, utter. exemplary, excellent, faultless, flawless, ideal, incomparable, inimitable, matchless, peerless, superb, superlative, suprem[e], wonderful. **3** accurate, correct, exac[t], faithful, precise. ● *v.* improve, polish, r[e]fine.

perfectionism *n.* uncompromisin[g] pursuit of perfection. □ **perfectionist** *n.*

perfidious *adj.* treacherous, disloya[l]. □ **perfidy** *n.*

perforate *v.* make hole(s) through. □ **perforation** *n.*

■ bore through, penetrate, pierce, prick, puncture.

perform *v.* **1** carry into effect. **2** g[o] through (a play, ceremony, etc.). **3** func[tion]. **4** act, sing, esp. in public. □ **per[former]** *n.*, **performance** *n.*

■ **1** accomplish, achieve, bring off, carr[y] out, complete, discharge, do, effect, ex[ecute], fulfil. **3** function, go, operate, run, work.

perfume *n.* **1** sweet smell. **2** fragrant li[quid] for applying to the body. ● *v.* give sweet smell to.

■ *n.* **1** aroma, bouquet, fragrance, scent, smell. **2** eau-de-Cologne, scent, toile[t] water.

perfumery *n.* perfumes.

perfunctory *adj.* done or doing thing[s] without much care or interest. □ **per[functorily]** *adv.*

■ brief, careless, casual, cursory, desu[l]tory, hasty, hurried, offhand, quick, rapi[d], sketchy, superficial.

pergola *n.* arch of trellis-work wit[h] climbing plants trained over it.

perhaps *adv.* it may be, possibly.

pericardium *n.* membranous sac en[closing] the heart.

perigee *n.* point nearest to the earth i[n] the moon's orbit.

peril *n.* serious danger.

■ danger, jeopardy, risk, threat.

perilous *adj.* full of risk, dangerous. □ **perilously** *adv.*

■ dangerous, hazardous, risky, unsafe.

perimeter *n.* **1** outer edge of an area. [2] length of this.

period *n.* **1** length or portion of time. **2** occurrence of menstruation. **3** sentence. **4** full stop in punctuation. ● *adj.* characteristic of a past age.

■ *n.* **1** duration, interval, patch, spell, stretch, term, time, while; age, epoch, era.

periodic *adj.* happening at esp. regular intervals. □ **periodicity** *n.*

■ cyclical, periodical, recurrent, regular, repeated.

periodical *adj.* periodic. ● *n.* magazine etc. published at regular intervals. □ **periodically** *adv.*

■ *n.* journal, magazine, monthly, newspaper, organ, paper, publication, quarterly, weekly.

peripatetic *adj.* going from place to place.

peripheral *adj.* **1** of or on the periphery. **2** of minor but not central importance to something.

■ **1** external, outer, outside. **2** incidental, inessential, marginal, minor, secondary, unimportant.

periphery *n.* **1** boundary, edge. **2** outer or surrounding area.

■ **1** border, boundary, edge, fringe, margin, perimeter.

periscope *n.* tube with mirror(s) by which a person in a submarine etc. can see things above.

perish *v.* **1** suffer destruction, die. **2** rot.

■ **1** be killed, die, expire, lose one's life. **2** decay, decompose, deteriorate, rot.

perishable *adj.* liable to decay or go bad in a short time.

perishing *adj.* (*colloq.*) very cold.

peritoneum *n.* membrane lining the abdominal cavity.

peritonitis *n.* inflammation of the peritoneum.

periwinkle[1] *n.* trailing plant with blue or white flowers.

periwinkle[2] *n.* winkle.

perjure *v.* **perjure oneself** lie under oath.

perjury *n.* deliberate giving of false evidence while under oath.

perk[1] *v.* **perk up 1** cheer up. **2** brighten or smarten up.

perk[2] *n.* (*colloq.*) perquisite.

perky *adj.* (**-ier, -iest**) lively and cheerful.

■ animated, *colloq.* bouncy, bright, bubbly, buoyant, cheerful, chirpy, energetic, jaunty, lively, spirited, vivacious.

perm[1] *n.* permanent artificial wave in the hair. ● *v.* give a perm to.

perm[2] *n.* permutation. ● *v.* make a permutation of.

permafrost *n.* permanently frozen subsoil in polar regions.

permanent *adj.* lasting indefinitely. □ **permanently** *adv.*, **permanence** *n.*

■ abiding, enduring, everlasting, immutable, indestructible, invariable, lasting, long-lasting, perennial, perpetual, unchangeable, unchanging, undying, unending.

permeable *adj.* able to be permeated by fluids etc. □ **permeability** *n.*

permeate *v.* pass or flow into every part of. □ **permeation** *n.*

■ pervade, saturate, seep through, soak through, spread through.

permissible *adj.* allowable.

■ acceptable, allowable, allowed, lawful, legal, legitimate, *colloq.* OK, permitted, sanctioned.

permission *n.* consent or authorization to do something.

■ approbation, approval, assent, authorization, consent, dispensation, leave, licence, sanction.

permissive *adj.* tolerant, liberal. □ **permissiveness** *n.*

■ broad-minded, easygoing, indulgent, lax, lenient, liberal, tolerant.

permit *v.* /pərmít/ (**permitted**) **1** give consent to, authorize. **2** make possible. ● *n.* /pérmit/ written permission, esp. for entry to a place.

■ *v.* **1** accept, agree to, allow, authorize, consent to, let, license, sanction, tolerate. **2** allow, enable, entitle. ● *n.* licence, pass, warrant.

permutation *n.* variation in the order of a set of things.

pernicious *adj.* harmful.

pernickety *adj.* (*colloq.*) fastidious.

peroration *n.* concluding part of a speech.

peroxide *n.* compound of hydrogen used to bleach hair. ● *v.* bleach with this.

perpendicular *adj.* **1** at an angle of 90° to a line or surface. **2** upright, vertical. ● *n.* perpendicular line or direction.

perpetrate *v.* commit (a crime), be guilty of (a blunder). □ **perpetration** *n.*, **perpetrator** *n.*

perpetual *adj.* **1** lasting. **2** not ceasing. □ **perpetually** *adv.*

■ **1** enduring, eternal, everlasting, lasting, never-ending, perennial, permanent, undying, unending. **2** ceaseless, constant, continual, continuous, incessant, non-stop, persistent, unceasing, uninterrupted, unremitting.

perpetuate *v.* preserve from being forgotten or from going out of use. □ **perpetuation** *n.*

perpetuity *n.* **in perpetuity** for ever.

perplex *v.* bewilder, puzzle. □ **perplexity** *n.*

■ baffle, *colloq.* bamboozle, bemuse, bewilder, confound, confuse, disconcert, floor, *colloq.* flummox, mystify, puzzle, *colloq.* stump, *colloq.* throw.

perquisite *n.* profit or privilege given in addition to wages.

perry *n.* drink resembling cider, made from fermented pears.

persecute *v.* **1** treat with hostility because of race or religion. **2** harass. □ **persecution** *n.*, **persecutor** *n.*

■ **1** abuse, ill-treat, mistreat, oppress, torment, torture, victimize, tyrannize. **2** annoy, badger, bother, harass, harry, *colloq.* hassle, hound, pester, *colloq.* plague.

persevere *v.* continue in spite of difficulties. □ **perseverance** *n.*

■ carry on, continue, keep going, persist, *colloq.* soldier on.

persimmon *n.* **1** edible orange plum-like fruit. **2** tree bearing this.

persist *v.* **1** continue firmly or obstinately. **2** continue to exist.

■ **1** be persistent, carry on, keep going, persevere, *colloq.* soldier on. **2** continue, endure, keep on, last, remain, survive.

persistent *adj.* **1** continuing obstinately, persisting. **2** constantly repeated. □ **persistence** *n.*, **persistently** *adv.*

■ **1** determined, dogged, indefatigable, insistent, obstinate, persevering, pertinacious, resolute, stubborn, tenacious, unflagging. **2** constant, continual, continuous, incessant, interminable, non-stop, perpetual, unceasing, unending, unremitting.

person *n.* **1** individual human being. **2** one's body. **3** (in grammar) one of the three classes of personal pronouns and verb forms, referring to the person(s) speaking, spoken to, or spoken of. □ **in person** physically present.

■ **1** creature, human (being), individual, soul.

persona *n.* (*pl.* **-ae**) personality as perceived by others.

personable *adj.* attractive in appearance or manner.

personage *n.* person, esp. an important one.

personal *adj.* **1** of one's own. **2** of or involving a person's private life. **3** referring to a person. **4** done in person. □ **personally** *adv.*

■ **1** individual, particular, peculiar, unique. **2** confidential, intimate, private, secret.

personality *n.* **1** person's distinctive character. **2** well-known person.

■ **1** character, disposition, make-up, nature, persona, temperament. **2** celebrity, luminary, personage, star.

personalize *v.* identify as belonging to a particular person. □ **personalization** *n.*

personify *v.* **1** represent in human form or as having human characteristics. **2** embody in one's behaviour. □ **personification** *n.*

■ **2** embody, epitomize, exemplify, represent, symbolize, typify.

personnel *n.* employees, staff.

perspective *n.* **1** art of drawing so as to give an effect of solidity and relative position. **2** point of view. □ **in perspective** **1** according to the rules of perspective. **2** not distorting a thing's relative importance.

■ **2** angle, outlook, point of view, standpoint, view, viewpoint.

perspicacious *adj.* showing great insight. □ **perspicaciously** *adv.*, **perspicacity** *n.*

perspire *v.* sweat. □ **perspiration** *n.*

persuade *v.* cause (a person) to believe or do something by reasoning. □ **persuader** *n.*

■ coax, convince, get round, induce, inveigle, prevail on, prompt, sway, talk into, wheedle into, win over.

persuasion *n.* **1** persuading. **2** persuasiveness. **3** religious belief or sect.

persuasive *adj.* able or trying to persuade people. □ **persuasively** *adv.*, **persuasiveness** *n.*

■ cogent, convincing, compelling, effective, forceful, influential, weighty.

pert *adj.* **1** cheeky. **2** lively. □ **pertly** *adv.*, **pertness** *n.*

pertain *v.* **pertain to** **1** be relevant to. **2** belong to as a part.

■ **1** affect, apply to, be relevant to, concern, have a bearing on, refer to, relate to.

pertinacious *adj.* persistent and determined. □ **pertinaciously** *adv.*, **pertinacity** *n.*

pertinent *adj.* relevant. □ **pertinently** *adv.*, **pertinence** *n.*

■ applicable, appropriate, apposite, fitting, germane, relevant, suitable, to the point.

perturb *v.* disturb greatly, make uneasy. □ **perturbation** *n.*

■ agitate, disconcert, distress, disturb, fluster, make anxious *or* uneasy, ruffle, trouble, unsettle, upset, worry.

peruse /pəro͞oz/ *v.* read carefully. □ **perusal** *n.*

■ examine, inspect, pore over, read, scan, scrutinize, study.

pervade *v.* spread throughout (a thing). □ **pervasive** *adj.*

■ fill, permeate, spread through, suffuse.

perverse *adj.* obstinately doing something different from what is reasonable or required. □ **perversely** *adv.*, **perversity** *n.*

■ awkward, contrary, *colloq.* cussed, difficult, obstinate, refractory, self-willed, stubborn, uncooperative, wayward, wilful.

pervert *v.* /pərvért/ **1** turn (a thing) aside from its proper use. **2** lead astray, corrupt. ● *n.* /pérvert/ perverted person. □ **perversion** *n.*

■ **1** distort, falsify, misapply, misconstrue, misrepresent, twist. **2** corrupt, deprave, lead astray, warp.

pervious *adj.* **1** permeable. **2** penetrable.

peseta *n.* unit of money in Spain.

peso *n.* (*pl.* **-os**) unit of money in several S. American countries.

pessimism *n.* tendency to take a gloomy view of things. □ **pessimist** *n.*, **pessimistic** *adj.*

■ cynicism, defeatism, despair, despondency, gloom, hopelessness. □ **pessimistic** cynical, despairing, despondent, gloomy, hopeless, negative.

pest *n.* **1** troublesome person or thing. **2** insect or animal harmful to plants, stored food, etc.

■ **1** annoyance, bother, irritant, menace, nuisance, trial.

pester *v.* annoy continually, esp. with requests or questions.

■ annoy, badger, bother, harass, harry, *colloq.* hassle, nag, persecute, *colloq.* plague, torment, worry.

pesticide *n.* substance used to destroy harmful insects etc.

pestilence *n.* deadly epidemic disease. □ **pestilential** *adj.*

pestle /péss'l/ *n.* instrument for pounding things to powder.

pet *n.* **1** tame animal treated with affection. **2** darling, favourite. ● *adj.* **1** kept as a pet. **2** favourite. ● *v.* (**petted**) **1** treat as a pet. **2** fondle. □ **pet name** name used affectionately.

■ *adj.* **1** domesticated, tame. **2** cherished, favoured, favourite, preferred, prized, special, treasured. ● *v.* **1** indulge, mollycoddle, pamper, spoil. **2** caress, fondle, pat, stroke.

petal *n.* one of the coloured outer parts of a flower head.

petite *adj.* of small dainty build.

petition *n.* **1** request, supplication. **2** formal written request, esp. one signed by many people. ● *v.* make a petition to.

■ *n.* **1** appeal, application, entreaty, plea, request, supplication. ● *v.* appeal to, ask, beg, beseech, entreat, plead with, solicit, supplicate.

petrel *n.* a kind of seabird.

petrify *v.* **1** turn or be turned into stone. **2** paralyse with fear or astonishment. □ **petrifaction** *n.*

■ **2** frighten, horrify, numb, paralyse, scare, terrify.

petrochemical *n.* substance obtained from petroleum or gas.

petrol *n.* inflammable liquid made from petroleum used as fuel in motor vehicles etc.

petroleum *n.* mineral oil found underground, refined for use as fuel or in drycleaning etc.

petticoat *n.* dress-length undergarment worn hanging from the shoulders or waist beneath a dress or skirt.

pettifogging *adj.* **1** trivial. **2** quibbling about petty details.

pettish *adj.* peevish, irritable.

petty *adj.* (**-ier**, **-iest**) **1** unimportant. **2** minor, on a small scale. **3** small-minded. □ **petty cash** money kept by an office etc. for small payments. **pettily** *adv.*, **pettiness** *n.*

■ **1** inconsequential, inessential, insignificant, minor, negligible, pettifogging, small, trifling, trivial, unimportant. **3** mean, small-minded, ungenerous.

petulant *adj.* peevish. □ **petulantly** *adv.*, **petulance** *n.*

■ bad-tempered, crabby, cross, crotchety, fractious, irritable, peevish, pettish, querulous, snappy, testy, touchy, waspish.

petunia *n.* garden plant with funnel-shaped flowers.

pew *n.* **1** long bench-like seat in a church. **2** (*colloq.*) seat.

pewter *n.* grey alloy of tin with lead or other metal.

phalanx *n.* compact mass esp. of people.

phallus *n.* (image of) the penis. □ **phallic** *adj.*

phantom *n.* ghost.

Pharaoh /fáirō/ *n.* title of the kings of ancient Egypt.

pharmaceutical *adj.* of or engaged in pharmacy.

pharmacist *n.* person skilled in pharmacy.

pharmacology *n.* study of the action of drugs. □ **pharmacological** *adj.*, **pharmacologist** *n.*

pharmacopoeia /faamakəpeeə/ *n.* list or stock of drugs.

pharmacy *n.* **1** preparation and dispensing of medicinal drugs. **2** pharmacist's shop, dispensary.

pharynx *n.* cavity behind the nose and throat. □ **pharyngeal** *adj.*

phase *n.* stage of change or development. ● *v.* carry out in stages. □ **phase in** or **out** bring gradually into or out of use.

pheasant *n.* game bird with bright feathers.

phenomenal *adj.* extraordinary, remarkable.

■ amazing, astonishing, exceptional, extraordinary, fabulous, marvellous, miraculous, outstanding, prodigious, remarkable, sensational, singular, staggering, uncommon, unparalleled, unprecedented, wonderful.

phenomenon *n.* (*pl.* **-ena**) **1** fact, occurrence, or change perceived by the senses or the mind. **2** remarkable person or thing.

■ **1** experience, event, fact, happening, occurrence. **2** marvel, miracle, prodigy, sensation, wonder.

phial *n.* small bottle.

philander *v.* (of a man) flirt. □ **philanderer** *n.*

philanthropy *n.* love of humankind, esp. shown in benevolent acts. □ **philanthropist** *n.*, **philanthropic** *adj.*

■ □ **philanthropic** altruistic, beneficent, benevolent, charitable, generous, humanitarian, kind, magnanimous, public-spirited, unselfish.

philately *n.* stamp-collecting. □ **philatelic** *adj.*, **philatelist** *n.*

philistine *adj.* & *n.* uncultured (person).

philology *n.* study of languages. □ **philologist** *n.*, **philological** *adj.*

philosopher *n.* **1** person skilled in philosophy. **2** philosophical person.

philosophical *adj.* **1** of philosophy. **2** bearing misfortune calmly. □ **philosophically** *adv.*

■ **2** calm, cool, even-tempered, equable, imperturbable, patient, placid, serene, stoical, tranquil, unperturbed.

philosophize *v.* **1** theorize. **2** moralize.

philosophy *n.* **1** system or study of the basic truths and principles of the universe, life, and morals, and of human understanding of these. **2** person's principles.

■ **2** beliefs, convictions, ideas, ideology, principles, tenets, values.

philtre *n.* magic potion.

phlegm /flem/ *n.* bronchial mucus ejected by coughing.

phlegmatic *adj.* **1** not easily excited or agitated. **2** sluggish, apathetic. □ **phlegmatically** *adv.*

■ **1** calm, collected, composed, cool, impassive, imperturbable, placid, serene, tranquil, *colloq.* unflappable. **2** apathetic, indifferent, lethargic, listless, sluggish, stolid, unresponsive.

phlox *n.* plant bearing a cluster of red, purple, or white flowers.

phobia *n.* abnormal fear or great dislike. □ **phobic** *adj.* & *n.*

■ abhorrence, aversion, detestation, dread, fear, hatred, horror, loathing, terror.

phoenix /feeniks/ *n.* mythical Arabian bird said to burn itself and rise again from its ashes.

phone *n.* & *v.* (*colloq.*) telephone. □ **phone book** telephone directory. **phone-in** *n.* broadcast programme in which listeners participate by telephone.

phonetic *adj.* **1** of or representing speech sounds. **2** (of spelling) corresponding to pronunciation. □ **phonetically** *adv.*

phonetics *n.* study or representation of speech sounds. □ **phonetician** *n.*

phoney (*colloq.*) *adj.* (**-ier, -iest**) sham. ● *n.* phoney person or thing.

phosphate *n.* fertilizer containing phosphorus.

phosphorescent *adj.* luminous. □ **phosphorescence** *n.*

phosphorus *n.* **1** non-metallic chemical element. **2** wax-like form of this appearing luminous in the dark.

photo *n.* (*pl.* **-os**) photograph. □ **photo finish** close finish where the winner is decided by a photograph.

photocopy *n.* photographed copy of a document. ● *v.* make a photocopy of. □ **photocopier** *n.*

photoelectric cell electronic device emitting an electric current when light falls on it.

photogenic *adj.* looking attractive in photographs.

photograph *n.* picture formed by the chemical action of light or other radiation on sensitive material. ● *v.* take a photograph of. □ **photographer** *n.*, **photography** *n.*, **photographic** *adj.*

photosynthesis *n.* process by which green plants use sunlight to convert carbon dioxide and water into complex substances.

phrase *n.* **1** group of words forming a unit. **2** unit in a melody. ● *v.* **1** express in words. **2** divide (music) into phrases. □ **phrasal** *adj.*

phraseology *n.* the way something is worded.

phylum *n.* (*pl.* **phyla**) major division of the plant or animal kingdom.

physical *adj.* **1** of the body. **2** of matter or the laws of nature. **3** of physics. □ **physical geography** study of earth's natural features. **physically** *adv.*

> ■ **1** bodily, corporal. **2** actual, concrete, corporeal, material, palpable, real, solid, tangible; carnal, earthly, worldly.

physician *n.* doctor, esp. one specializing in medicine as distinct from surgery.

physics *n.* study of the properties and interactions of matter and energy. □ **physicist** *n.*

physiognomy /fìzziónnəmi/ *n.* features of a person's face.

physiology *n.* study of the bodily functions of living organisms. □ **physiological** *adj.*, **physiologist** *n.*

physiotherapy *n.* treatment of an injury etc. by massage and exercises. □ **physiotherapist** *n.*

physique *n.* bodily structure and muscular development.

> ■ body, build, figure, form, frame, shape.

pi *n.* Greek letter π used as a symbol for the ratio of a circle's circumference to its diameter (about 3.14).

pianist *n.* person who plays the piano.

piano *n.* (*pl.* **-os**) keyboard instrument with metal strings struck by hammers.

pianoforte *n.* piano.

piazza /piátsə/ *n.* public square or market place.

picador *n.* mounted bullfighter with a lance.

picaresque *adj.* (of fiction) dealing with the adventures of rogues.

piccalilli *n.* pickle of chopped vegetables and hot spices.

piccolo *n.* (*pl.* **-os**) small flute.

pick¹ *n.* **1** pickaxe. **2** plectrum.

pick² *v.* **1** select. **2** use a pointed instrument or the fingers or beak etc. to make (a hole) in or remove bits from (a thing). **3** detach (flower or fruit) from the plant bearing it. ● *n.* **1** picking. **2** selection. **3** best part. □ **pick a lock** open it with a tool other than a key. **pick a quarrel** provoke one deliberately. **pick holes in** find fault with. **pick off 1** pluck off. **2** shoot or destroy one by one. **pick on 1** nag, find fault with. **2** select. **pick out 1** select. **2** discern. **pick up 1** lift or take up. **2** call for and take away. **3** acquire by chance or without effort. **4** succeed in seeing or hearing by use of apparatus. **5** recover health, improve. **6** gather (speed). **7** become acquainted with casually. **pick-up** *n.* **1** small open motor truck. **2** stylus-holder in a record player.

> ■ *v.* **1** choose, decide on, elect, go for, opt for, pick on *or* out, plump for, select, single out. **3** collect, gather, harvest, pluck. ● *n.* **2** choice, option, selection. **3** best part, cream, flower. □ **pick on 1** carp at, criticize, *colloq.* get at, find fault with, nag. **pick out 2** discern, distinguish, identify, make out, perceive, recognize, see, *colloq.* spot. **pick up 3** acquire, come by, gain, get, find, obtain. **5** be on the mend, get better, improve, make progress, rally, recover.

pickaxe *n.* tool with sharp-pointed iron crossbar for breaking up ground etc.

picket *n.* **1** person(s) stationed at a workplace to dissuade others from entering during a strike. **2** party of sentries. **3** pointed stake set in the ground.

● *v.* (**picketed**) **1** form a picket on (a workplace). **2** enclose with stakes.

pickings *n.pl.* odd gains or perquisites.

pickle *n.* **1** vegetables preserved in vinegar or brine. **2** this liquid. **3** (*colloq.*) plight, mess. ● *v.* preserve in pickle.

pickpocket *n.* thief who steals from people's pockets.

picnic *n.* informal outdoor meal. ● *v.* (**picnicked**) take part in a picnic. □ **picnicker** *n.*

pictograph *n.* pictorial symbol used as a form of writing.

pictorial *adj.* **1** of, in, or like a picture or pictures. **2** illustrated. □ **pictorially** *adv.*

picture *n.* **1** representation of person(s) or object(s) etc. made by painting, drawing, or photography etc. **2** beautiful thing. **3** mental image. **4** description. **5** cinema film. ● *v.* **1** depict. **2** imagine.

■ *n.* **1** drawing, illustration, painting, portrait, print, representation, sketch; photo, photograph, shot, snap, snapshot. **3** idea, image, impression, notion. **4** description, portrait, portrayal, representation. **5** film, motion picture, *US colloq.* movie. ● *v.* **1** depict, draw, illustrate, paint, portray, represent. **2** envisage, fancy, imagine, visualize.

picturesque *adj.* **1** forming a pleasant scene. **2** (of words or description) very expressive.

■ **1** beautiful, charming, delightful, idyllic, lovely, pleasing, scenic. **2** colourful, graphic, vivid.

pidgin *n.* simplified language, esp. used between speakers of different languages.

pie *n.* baked dish of meat, fish, or fruit covered with pastry or other crust. □ **pie chart** diagram representing quantities as sectors of a circle.

piebald *adj.* with irregular patches of white and black.

piece *n.* **1** part, portion. **2** thing regarded as a unit. **3** musical, literary, or artistic composition. **4** small object used in board games. ● *v.* make by putting pieces together. □ **of a piece 1** of the same kind. **2** consistent. **piece-work** *n.* work paid according to the quantity done.

■ *n.* **1** part, portion, section, segment; bit, block, chip, chunk, crumb, fragment, lump, particle, scrap, shard, shred, slice, sliver, wedge. **2** component, constituent, element, unit. **3** article, composition, essay, poem, story, work.

piecemeal *adj.* & *adv.* done piece by piece, part at a time.

pied *adj.* particoloured.

pied-à-terre /pyáydaatáir/ *n.* (*pl.* ***pieds-à-terre***) small dwelling for occasional use.

pier *n.* **1** structure built out into the sea. **2** pillar supporting an arch or bridge.

pierce *v.* **1** go into or through like a sharp-pointed instrument. **2** make (a hole) in.

■ **1** impale, lance, penetrate, puncture, skewer, spear, transfix. **2** bore into, drill into, perforate.

piercing *adj.* **1** (of cold or wind etc.) penetrating sharply. **2** (of sound) shrilly audible.

■ **1** biting, bitter, chilling, cold, freezing, icy, keen. **2** harsh, high-pitched, loud, shrill, penetrating, strident.

piety *n.* piousness.

■ devoutness, holiness, piousness, sanctity; sanctimoniousness, self-righteousness.

piffle *n.* (*colloq.*) nonsense.

pig *n.* **1** animal with short legs, cloven hooves, and a blunt snout. **2** (*colloq.*) greedy or unpleasant person. □ **pig-iron** *n.* crude iron from a smelting-furnace.

pigeon *n.* **1** bird of the dove family. **2** (*colloq.*) person's business or responsibility.

pigeon-hole *n.* small compartment in a desk or cabinet. ● *v.* **1** put away for future consideration or indefinitely. **2** classify.

piggery *n.* **1** pig farm. **2** pigsty.

piggy *adj.* like a pig. □ **piggyback** *adv.* & *n.* (ride) on a person's back or on top of a larger object. **piggy bank** money box shaped like a pig.

pigheaded *adj.* obstinate.

piglet *n.* young pig.

pigment *n.* colouring matter.

pigsty *n.* pen for pigs.

pigtail *n.* long hair worn in a plait at the back of the head.

pike *n.* **1** spear with long wooden shaft. **2** (*pl.* **pike**) large voracious freshwater fish.

pilaff *n.* = **pilau**.

pilaster *n.* rectangular usu. ornamental column.

pilau *n.* oriental dish of rice with meat, spices, etc.

pilchard *n.* small sea fish related to the herring.

pile[1] *n.* **1** a number of things lying one upon another. **2** (*colloq.*) large amount. **3** large imposing building. ● *v.* heap, stack, load. □ **pile up** accumulate. **pile-up** *n.* collision of several vehicles.

■ *n.* **1** accumulation, collection, heap, mass, mound, mountain, stack. ● *v.* heap (up), load (up), stack (up). □ **pile up** accumulate, amass, collect, hoard, stockpile. **pile-up** accident, collision, crash, smash.

pile[2] *n.* heavy beam driven vertically into ground as a support for a building or bridge.

pile[3] *n.* cut or uncut loops on the surface of fabric.

pile[4] *n.* haemorrhoid.

pilfer *v.* steal (small items or in small quantities). □ **pilferer** *n.*

■ filch, lift, misappropriate, *sl.* nick, *sl.* pinch, purloin, *colloq.* snaffle, steal, take, thieve.

pilgrim *n.* person who travels to a sacred place as an act of religious devotion. □ **pilgrimage** *n.*

pill *n.* **1** small ball or piece of medicinal substance for swallowing whole. **2** (**the pill**) contraceptive pill.

■ **1** capsule, lozenge, pastille, tablet.

pillage *n.* & *v.* plunder.

pillar *n.* vertical structure used as a support or ornament.

■ column, pilaster, pile, pole, post, stanchion, support, upright.

pillion *n.* saddle for a passenger behind the driver of a motor cycle. □ **ride pillion** ride on this.

pillory *n.* wooden frame in which offenders were locked and exposed to public abuse. ● *v.* ridicule publicly.

pillow *n.* cushion used (esp. in bed) for supporting the head. ● *v.* rest on or as if on a pillow.

pilot *n.* **1** person who operates an aircraft's flying-controls. **2** person qualified to steer ships into or out of a harbour. **3** guide. ● *v.* (**piloted**) **1** act as pilot of. **2** guide. □ **pilot-light** *n.* **1** small burning jet of gas which lights a larger burner. **2** electric indicator light.

■ *n.* **2** helmsman, navigator, steersman. **3** conductor, guide, leader. ● *v.* **1** fly; drive, navigate, sail, steer. **2** conduct, direct, guide, lead, shepherd, steer.

pimento *n.* (*pl.* **-os**) **1** allspice. **2** sweet pepper.

pimp *n.* man who solicits clients for a prostitute or brothel.

pimple *n.* small inflamed spot on the skin. □ **pimply** *adj.*

pin *n.* **1** short pointed piece of metal usu. with a round broadened head, used for fastening things together. **2** peg or stake of wood or metal. ● *v.* (**pinned**) **1** fasten with pin(s). **2** transfix. **3** hold down and make unable to move. **4** attach, fix. □ **pin down 1** establish clearly. **2** bind by a promise. **pins and needles** tingling sensation. **pin-up** *n.* (*colloq.*) picture of an attractive or famous person.

■ *n.* **2** bolt, dowel, nail, peg, spike, tack. ● *v.* **1** clip, fasten, secure, staple, tack. **2** impale, skewer, spike, transfix. **3** hold, pinion. **4** affix, attach, fix, stick.

pinafore *n.* apron. □ **pinafore dress** sleeveless dress worn over a blouse or jumper.

pincers *n.* **1** tool with pivoted jaws for gripping and pulling things. **2** claw-like part of a lobster etc.

pinch *v.* **1** squeeze between two surfaces, esp. between finger and thumb. **2** (*sl.*) steal. ● *n.* **1** pinching. **2** stress of circumstances. **3** small amount. □ **at a pinch** if really necessary.

pine[1] *n.* **1** evergreen tree with needle-shaped leaves. **2** its wood.

pine[2] *v.* **1** waste away with grief etc. **2** feel an intense longing.

pineapple *n.* **1** large juicy tropical fruit. **2** plant bearing this.

ping *n.* short sharp ringing sound. ● *v.* make this sound.

ping-pong *n.* table tennis.

pinion[1] *n.* bird's wing. ● *v.* restrain by holding or binding the arms or legs.

pinion[2] *n.* small cogwheel.

pink[1] *adj.* pale red. ● *n.* **1** pink colour. **2** garden plant with fragrant flowers. □ **in the pink** (*colloq.*) in very good health. **pinkish** *adj.*

pink[2] *v.* **1** pierce slightly. **2** cut a zigzag edge on (fabric).

pink[3] *v.* (of an engine) make slight explosive sounds when running imperfectly.

pinnacle *n.* **1** pointed ornament on a roof. **2** peak. **3** highest point.

■ **2** crest, crown, peak, summit, tip, top. **3** acme, apex, climax, culmination, height, highest point, peak, top, zenith.

pinpoint *v.* locate precisely.

pinstripe *n.* very narrow stripe in cloth fabric. □ **pinstriped** *adj.*

pint *n.* measure for liquids, one-eighth of a gallon (0.568 litre).

pioneer *n.* person who is one of the first to explore a new region or subject. ● *v.* **1** initiate, originate. **2** act as a pioneer.

pious *adj.* **1** devout in religion. **2** sanctimonious. □ **piously** *adv.*, **piousness** *n.*

■ **1** devout, faithful, God-fearing, good, holy, religious, reverent, saintly, virtuous. **2** goody-goody, hypocritical, sanctimonious, self-righteous.

pip[1] *n.* small seed in fruit.

pip[2] *n.* star showing rank on an army officer's uniform.

pip[3] *v.* (**pipped**) (*colloq.*) defeat by a small margin.

pip[4] *n.* short high-pitched sound.

pipe *n.* **1** tube through which something can flow. **2** wind instrument. **3** (*pl.*) bagpipes. **4** narrow tube with a bowl at one end for smoking tobacco. ● *v.* **1** convey through pipe(s). **2** play (music) on pipe(s). **3** utter in a shrill voice. □ **pipe down** (*colloq.*) be quiet. **pipedream** *n.* fanciful hope or scheme.

pipeline *n.* **1** long pipe for conveying oil etc. across country. **2** channel of supply or information. □ **in the pipeline** on the way, in preparation.

piper *n.* player of pipe(s).

pipette *n.* slender tube for transferring or measuring small amounts of liquid.

pipit *n.* small bird resembling a lark.

pippin *n.* a kind of apple.

piquant *adj.* pleasantly sharp in taste or smell. □ **piquancy** *n.*

■ pungent, savoury, sharp, spicy, tangy, tasty.

pique /peek/ *v.* hurt the pride of. ● *n.* feeling of hurt pride.

piquet /peékay/ *n.* card game for two players.

piranha /piraánə/ *n.* fierce S. American freshwater fish.

pirate *n.* **1** person on a ship who robs another ship at sea or raids a coast. **2** one who infringes copyright or business rights, or broadcasts without due authorization. ● *v.* reproduce (a book etc.) without due authorization. □ **piratical** *adj.*, **piracy** *n.*

pirouette *n.* & *v.* spin on the toe in dancing.

pistachio /pistaáshiō/ *n.* (*pl.* **-os**) a kind of nut.

piste *n.* ski run.

pistil *n.* seed-producing part of a flower.

pistol *n.* small gun.

piston *n.* sliding disc or cylinder inside a tube, esp. as part of an engine or pump.

pit *n.* **1** hole in the ground. **2** coal mine. **3** sunken area. **4** place where racing cars are refuelled etc. during a race. ● *v.* (**pitted**) **1** make pits or depressions in. **2** match or set in competition.

■ *n.* **1** cavity, chasm, crater, hole, trough. **2** coal mine, excavation, mine, quarry, working. **3** dent, depression, hollow, indentation.

pitch[1] *n.* dark tarry substance. □ **pitch-black**, **-dark** *adjs.*

pitch[2] *v.* **1** throw. **2** erect (a tent or camp). **3** express in a particular style or at a particular level. **4** fall heavily. **5** (of a ship) plunge forward and back alternately. ● *n.* **1** process of pitching. **2** steepness. **3** intensity. **4** degree of highness or lowness of a music note or voice. **5** place where a street trader or performer is stationed. **6** playing field. □ **pitched battle** one fought from prepared positions. **pitch in** (*colloq.*) set to work vigorously.

■ *v.* **1** cast, *colloq.* chuck, fling, hurl, launch, lob, *colloq.* sling, throw, toss. **2** erect, put up. **4** fall (headlong), plummet, plunge, topple, tumble. **5** heel, list, lurch, rock, roll, toss. ● *n.* **2** angle, inclination, slant, slope, steepness, tilt. **3** degree, height, intensity, level. **4** intonation, modulation, timbre, tone. **6** field, ground, playing field.

pitchblende *n.* mineral ore (uranium oxide) yielding radium.

pitcher[1] *n.* baseball player who delivers the ball to the batter.

pitcher[2] *n.* large jug.

pitchfork *n.* long-handled fork for lifting and tossing hay.

piteous *adj.* deserving or arousing pity. □ **piteously** *adv.*

■ distressing, heartbreaking, moving, pathetic, pitiable, pitiful, plaintive, poignant, sad.

pitfall *n.* unsuspected danger or difficulty.

■ danger, hazard, peril; catch, difficulty, snag.

pith *n.* **1** spongy tissue in stems or fruits. **2** essential part.

■ **2** core, crux, essence, gist, heart, kernel, nub, substance.

pithy *adj.* (**-ier**, **-iest**) brief and full of meaning. □ **pithily** *adv.*

■ **2** brief, compact, concise, laconic, succinct, terse.

pitiful *adj.* deserving or arousing pity or contempt. □ **pitifully** *adv.*

■ distressing, heartbreaking, moving, pathetic, piteous, pitiable, plaintive, poignant, sad, wretched; contemptible, miserable, *colloq.* pathetic, poor, sorry.

pitta *n.* a kind of flat bread.

pittance *n.* very small allowance of money.

pituitary gland gland at the base of the brain which influences bodily growth and functions.

pity *n.* **1** feeling of sorrow for another's suffering. **2** cause for regret. ● *v.* feel pity for. □ **take pity on** pity and try to help.

■ *n.* **1** compassion, sorrow, sympathy. **2** shame. ● *v.* feel sorry for, have compassion for, sympathize with, take pity on.

pivot *n.* central point or shaft on which a thing turns or swings. ● *v.* (**pivoted**) turn on a pivot. □ **pivotal** *adj.*

pixie *n.* a kind of fairy.

pizza *n.* layer of dough baked with a savoury topping.

pizzicato *adv.* by plucking the strings of a violin etc. instead of using the bow.

placard *n.* poster or similar notice. ● *v.* put up placards on.

placate *v.* conciliate. □ **placation** *n.*, **placatory** *adj.*

■ appease, calm (down), conciliate, mollify, pacify, soothe.

place *n.* **1** particular part of space or of an area etc. **2** particular town, district, building, etc. **3** position. **4** duty appropriate to one's rank. ● *v.* **1** put into a particular place, arrange. **2** locate. **3** identify. **4** put or give (an order for goods etc.). □ **be placed** (in a race) be among the first three. **out of place 1** in the wrong place. **2** unsuitable.

■ *n.* **1** area, location, point, position, scene, setting, site, spot. **2** area, district, locality, neighbourhood, quarter, region; city, hamlet, town, village. **3** grade, rank, standing, station, status; job, position, post, situation; niche, slot. **4** concern, duty, function, job, responsibility, role, task. ● *v.* **1** arrange, deposit, dispose, dump, lay, position, park, put, rest, set (out), *colloq.* stick. **2** locate, position, site, situate. **3** identify, recognize, remember.

placebo /pləséebō/ *n.* (*pl.* **-os**) harmless substance given as medicine, esp. to humour a patient.

placement *n.* placing.

placenta *n.* (*pl.* **-ae**) organ in the womb that nourishes the foetus. □ **placental** *adj.*

placid *adj.* calm, not easily upset. □ **placidly** *adv.*, **placidity** *n.*

■ calm, collected, composed, cool, easy-going, equable, even-tempered, gentle, imperturbable, mild, peaceful, phlegmatic, tranquil, *colloq.* unflappable.

plagiarize /pláyjərīz/ *v.* take and use (another's writings etc.) as one's own. □ **plagiarism** *n.*, **plagiarist** *n.*

plague *n.* **1** deadly contagious disease. **2** infestation. ● *v.* (*colloq.*) annoy, pester.

■ *n.* **1** epidemic, pestilence. ● *v.* annoy, badger, bother, harass, harry, *colloq.* hassle, hound, nag, persecute, pester, torment, worry.

plaice *n.* (*pl.* **plaice**) marine flatfish.

plaid /plad/ *n.* **1** long piece of woollen cloth worn as part of Highland costume. **2** tartan pattern.

plain *adj.* **1** unmistakable, easy to see, hear, or understand. **2** not elaborate. **3** candid. **4** ordinary, without affectation. **5** not good-looking. ● *adv.* plainly. ● *n.* large area of level country. □ **plain clothes** civilian clothes, not a uniform. **plain sailing** easy course of action. **plainly** *adv.*, **plainness** *n.*

■ *adj.* **1** apparent, clear, evident, manifest, obvious, patent, transparent, unmistakable; comprehensible, intelligible, lucid, simple, uncomplicated, understandable. **2** austere, bare, basic, simple, spartan, stark, unadorned, undecorated, unembellished, unostentatious. **3** blunt, candid, direct, forthright, frank, honest, open, outspoken, straightforward. **4** homely, modest, ordinary, simple, unassuming, unpretentious, unsophisticated, workaday. **5** *US* homely, ugly, unattractive, unprepossessing. ● *n.* grassland, pampas, prairie, savannah, steppe, tundra.

plaintiff *n.* person that brings an action in a court of law.

plaintive *adj.* sounding sad. □ **plaintively** *adv.*

■ doleful, melancholy, mournful, pathetic, piteous, pitiful, plangent, sad, sorrowful, woeful.

plait /plat/ *v.* weave (three or more strands) into one rope-like length. ● *n.* something plaited.

plan *n.* **1** diagram showing the relative position of parts of a building or town etc. **2** method thought out in advance. **3**

intention. ● *v.* (**planned**) **1** intend. **2** make a plan (of). □ **planner** *n.*

■ *n.* **1** blueprint, chart, design, diagram, drawing, layout, map. **2** method, procedure, scheme, strategy, system, tactic. **3** aim, ambition, design, end, intent, intention, purpose. ● *v.* **1** aim, contemplate, design, intend, mean, propose, purpose. **2** design, lay out, map out.

plane[1] *n.* tall spreading tree with broad leaves.

plane[2] *n.* **1** level surface. **2** level of thought or development. **3** aeroplane. ● *adj.* level.

■ *n.* **2** level, stratum. **3** aeroplane, aircraft, airliner, jet. ● *adj.* even, flat, horizontal, level.

plane[3] *n.* tool for smoothing wood by paring shavings from it. ● *v.* smooth or pare with this.

planet *n.* large body in space that revolves round the sun. □ **planetary** *adj.*

planetarium *n.* room with a domed ceiling on which lights are projected to show the positions of the stars and planets.

plangent *adj.* resonant, loud and mournful.

plank *n.* long flat piece of timber.

plankton *n.* minute life-forms floating in the sea, rivers, etc.

plant *n.* **1** living organism with neither the power of movement nor special organs of digestion. **2** factory. **3** its machinery. ● *v.* **1** place in soil for growing. **2** place in position.

plantain[1] *n.* herb whose seed is used as birdseed.

plantain[2] *n.* **1** tropical banana-like fruit. **2** tree bearing this.

plantation *n.* **1** area planted with trees or cultivated plants. **2** estate on which cotton, tobacco, or tea etc. is cultivated.

plaque *n.* **1** commemorative plate fixed on a wall. **2** film forming on teeth and gums.

plasma *n.* **1** colourless fluid part of blood. **2** a kind of gas. □ **plasmic** *adj.*

plaster *n.* **1** mixture of lime, sand, and water etc. used for coating walls. **2** sticking plaster. ● *v.* **1** cover with plaster. **2** coat, daub. □ **plaster of Paris** white paste made from gypsum. **plasterer** *n.*

plastic *adj.* **1** able to be moulded. **2** made of plastic. ● *n.* synthetic substance moulded to a permanent shape. □ **plastic surgery** repair or restoration of lost or damaged etc. tissue. **plasticity** *n.*

■ *adj.* **1** ductile, flexible, malleable, pliable, pliant, soft.

Plasticine *n.* [P.] plastic substance used for modelling.

plate *n.* **1** almost flat usu. circular utensil for holding food. **2** articles of gold, silver, or other metal. **3** flat thin sheet of metal, glass, or other material. **4** illustration on special paper in a book. **5** (*colloq.*) denture. ● *v.* cover or coat with metal. □ **plate glass** thick glass for windows etc. **plateful** *n.*

plateau *n.* (*pl.* **-eaux**) **1** area of level high ground. **2** steady state following an increase.

platform *n.* raised level surface or area, esp. from which a speaker addresses an audience.

■ dais, podium, rostrum, stage.

platinum *n.* silver-white metal that does not tarnish. □ **platinum blonde** woman with very light blonde hair.

platitude *n.* commonplace remark. □ **platitudinous** *adj.*

■ banality, *colloq.* chestnut, cliché, truism. □ **platitudinous** banal, commonplace, *colloq.* corny, hackneyed, stale, trite, unimaginative, unoriginal.

platonic *adj.* involving affection but not sexual love.

platoon *n.* subdivision of a military company.

platter *n.* large plate for food.

platypus *n.* (*pl.* **-puses**) Australian aquatic egg-laying mammal with a duck-like beak.

plaudits *n.pl.* applause, expression of approval.

plausible *adj.* **1** seeming probable. **2** persuasive but deceptive. □ **plausibly** *adv.*, **plausibility** *n.*

■ **1** believable, conceivable, convincing, credible, feasible, likely, possible, probable, reasonable, tenable.

play *v.* **1** occupy oneself in (a game) or in other recreational activity. **2** compete against in a game. **3** move (a piece, a ball, etc.) in a game. **4** act the part of. **5** perform on (a musical instrument). **6** cause (a radio, recording, etc.) to produce sound. **7** move lightly, allow (light or water) to fall on something. ● *n.* **1** playing. **2** activity, operation. **3** literary work for stage or broadcast performance. **4** freedom of movement, space for this.

□ **play at** perform in a half-hearted way. **play down** minimize the importance of. **play for time** seek to gain time by delaying. **play-pen** *n.* portable enclosure for a young child to play in. **play safe** not take risks. **play the game** behave honourably. **play up** cause trouble, be unruly. **player** *n.*

■ *v.* **1** amuse oneself, enjoy oneself, frolic, have fun, sport. **2** compete against, contend with, pit oneself against, take on. **4** act, perform, take the role *or* part of. ● *n.* **1** amusement, fun, horseplay, merrymaking, recreation, skylarking, sport. **2** action, activity, exercise, operation, working. **3** comedy, drama, farce, tragedy. **4** flexibility, freedom, give, leeway, margin, movement, room, space. □ **play down** belittle, diminish, make light of, minimize. **play for time** delay, hesitate, procrastinate, stall, temporize.

playboy *n.* pleasure-loving usu. rich man.

playful *adj.* **1** full of fun. **2** done in fun, not serious. □ **playfully** *adv.*, **playfulness** *n.*

■ **1** frisky, fun-loving, high-spirited, lively, mischievous, skittish, sportive. **2** humorous, jocular, teasing, tongue-in-cheek.

playgroup *n.* group of preschool children who play regularly together under supervision.

playhouse *n.* theatre.

playing card one of a pack or set of (usu. 52) pieces of pasteboard used in card games.

playing field field used for outdoor games.

playmate *n.* child's companion in play.

playwright *n.* writer of plays.

plea *n.* **1** defendant's answer to a charge in a law court. **2** appeal, entreaty. **3** excuse.

■ **2** appeal, cry, entreaty, petition, request, supplication. **3** excuse, explanation, justification, reason.

plead *v.* **1** give as one's plea. **2** put forward (a case) in a law court. **3** put forward as an excuse. □ **plead with** make an earnest appeal to.

■ □ **plead with** appeal to, beg, beseech, entreat, implore, petition, request, supplicate.

pleasant *adj.* **1** pleasing. **2** having an agreeable manner. □ **pleasantly** *adv.*

■ **1** agreeable, delightful, enjoyable, entertaining, *colloq.* lovely, nice, pleasing, pleasurable, satisfying. **2** affable, agreeable, amiable, charming, congenial, engaging, friendly, genial, good-natured, likeable, nice, warm, winning.

pleasantry *n.* friendly or humorous remark.

please *v.* **1** give pleasure to. **2** think fit. ● *adv.* polite word of request. □ **please oneself** do as one chooses.

■ *v.* **1** amuse, appeal to, content, delight, divert, entertain, gladden, gratify, humour, satisfy. **2** choose, desire, like, think fit, want, wish.

pleased *adj.* feeling or showing pleasure or satisfaction.

■ contented, delighted, glad, gratified, happy, overjoyed, satisfied, thrilled.

pleasurable *adj.* causing pleasure, pleasant. □ **pleasurably** *adv.*

pleasure *n.* **1** feeling of satisfaction or joy. **2** source of this.

■ **1** contentment, delight, enjoyment, gratification, happiness, joy, satisfaction. **2** delight, joy, treat.

pleat *n.* flat fold of cloth. ● *v.* make a pleat or pleats in.

plebiscite /plébbisit/ *n.* referendum.

plectrum *n.* small piece of metal, bone, or ivory for plucking the strings of a musical instrument.

pledge *n.* **1** thing deposited as a guarantee (e.g. that a debt will be paid). **2** token of something. **3** solemn promise. ● *v.* **1** deposit as a pledge. **2** promise solemnly.

■ *n.* **1** bond, collateral, guarantee, security, surety. **2** mark, sign, symbol, token. **3** bond, guarantee, oath, promise, undertaking, vow, word (of honour). ● *v.* **1** deposit, pawn. **2** promise, swear, undertake, vow.

plenary /pléenəri/ *adj.* **1** entire. **2** attended by all members.

plenipotentiary *adj.* & *n.* (envoy) with full powers to take action.

plenitude *n.* **1** abundance. **2** completeness.

plentiful *adj.* existing in large amounts. □ **plentifully** *adv.*

■ abundant, ample, bountiful, copious, generous, lavish, liberal, plenteous, profuse, rich.

plenty *n.* enough and more. ● *adv.* (*colloq.*) quite, fully. □ **plenteous** *adj.*

■ *n.* a lot, heaps, *colloq.* loads, lots, masses, *colloq.* oodles, *colloq.* piles, quantities, *colloq.* stacks.

plethora *n.* over-abundance.

pleurisy *n.* inflammation of the membrane round the lungs.

pliable *adj.* flexible. □ **pliability** *n.*

■ bendable, ductile, elastic, flexible, malleable, plastic, pliant, whippy.

pliant *adj.* pliable. □ **pliancy** *n.*

pliers *n.pl.* pincers with flat surfaces for gripping things.

plight *n.* predicament.

plimsoll *n.* canvas sports shoe.

Plimsoll line mark on a ship's side showing the legal water level when loaded.

plinth *n.* slab forming the base of a column or statue etc.

plod *v.* (**plodded**) **1** walk doggedly, trudge. **2** work slowly but steadily. □ **plodder** *n.*

■ **1** clump, lumber, stomp, tramp, trudge. **2** labour, *colloq.* plug, slave, slog, toil, work.

plonk *n.* (*colloq.*) cheap or inferior wine.

plop *n.* & *v.* (**plopped**) sound like something small dropping into water with no splash.

plot *n.* **1** small piece of land. **2** story in a play, novel, or film. **3** conspiracy, secret plan. ● *v.* (**plotted**) **1** make a map or chart of, mark on this. **2** plan secretly. □ **plotter** *n.*

■ *n.* **1** lot, parcel, patch, piece. **2** scenario, story. **3** conspiracy, intrigue, machination, scheme, stratagem. ● *v.* **1** chart, draw, map (out), plan; indicate, mark. **2** conspire, intrigue, scheme.

plough /plow/ *n.* implement for cutting furrows in soil and turning it up. ● *v.* **1** cut or turn up (soil etc.) with a plough. **2** make one's way laboriously. □ **ploughman** *n.*

plover *n.* a wading bird.

ploy *n.* cunning manoeuvre.

■ dodge, manoeuvre, ruse, stratagem, tactic, trick.

pluck *v.* **1** pick or pull out or away. **2** strip (a bird) of its feathers. ● *n.* **1** plucking movement. **2** courage.

■ *v.* **1** pick; grab, snatch, tear, *colloq.* yank. ● *n.* **2** boldness, bravery, courage, *colloq.* grit, *colloq.* guts, intrepidity, mettle, nerve, spirit.

plucky *adj.* (**-ier, -iest**) brave.

■ bold, brave, courageous, fearless, game, *colloq.* gutsy, intrepid, mettlesome, spirited, valiant, valorous.

plug *n.* **1** thing fitting into and stopping or filling a hole or cavity. **2** device of this kind for making an electrical connection. ● *v.* (**plugged**) **1** put a plug into. **2** (*colloq.*) work diligently. **3** (*colloq.*) advertise by frequent recommendation. □ **plug in** connect electrically by putting a plug into a socket.

plum *n.* **1** fruit with sweet pulp round a pointed stone. **2** tree bearing this. **3** reddish-purple. **4** something desirable, the best.

plumage *n.* bird's feathers.

plumb /plum/ *n.* lead weight hung on a cord (**plumb line**), used for testing depths or verticality. ● *adv.* **1** exactly. **2** (*US*) completely. ● *v.* **1** measure or test with a plumb line. **2** reach (depths). **3** get to the bottom of. **4** work or fit (things) as a plumber.

plumber /plúmmər/ *n.* person who fits and repairs plumbing.

plumbing /plúmming/ *n.* system of water pipes and drainage pipes etc. in a building.

plume *n.* **1** feather, esp. as an ornament. **2** thing(s) resembling this. □ **plumed** *adj.*

plummet *v.* (**plummeted**) fall steeply, plunge.

plump¹ *adj.* having a full rounded shape. ● *v.* make or become plump. □ **plumpness** *n.*

■ *adj.* buxom, chubby, fat, podgy, pudgy, roly-poly, rotund, stout, tubby.

plump² *v.* plunge abruptly. □ **plump for** choose, decide on.

plunder *v.* rob. ● *n.* **1** plundering. **2** goods etc. acquired by this.

■ *v.* despoil, loot, pillage, ransack, rifle (through), rob, sack. ● *n.* **1** depredation, looting, pillage, ransacking, robbery. **2** booty, loot, spoils, *sl.* swag.

plunge *v.* **1** thrust or go forcefully into something. **2** go down suddenly. ● *n.* plunging, dive.

■ *v.* **1** drive, push, stick, thrust. **2** descend, drop, fall (headlong), nosedive, pitch, plummet. ● *n.* descent, dive, drop, fall, nosedive.

plunger *n.* thing that works with a plunging movement.

plural *n.* form of a noun or verb used in referring to more than one person or thing. ● *adj.* **1** of this form. **2** of more than one. □ **plurality** *n.*

plus *prep.* **1** with the addition of. **2** above zero. ● *adj.* **1** more than zero. **2** more than the amount indicated. ● *n.* **1** the sign +. **2** advantage.

plush *n.* cloth with a long soft nap. ● *adj.* **1** made of plush. **2** luxurious.

■ *adj.* **2** de luxe, lavish, lush, luxurious, opulent, palatial, *colloq.* posh, plushy, sumptuous.

plushy *adj.* (**-ier**, **-iest**) luxurious.

plutocrat *n.* wealthy and influential person. □ **plutocracy** *n.*, **plutocratic** *adj.*

plutonium *n.* radioactive metallic element.

pluvial *adj.* of or caused by rain.

ply[1] *n.* **1** thickness or layer of wood or cloth etc. **2** plywood.

ply[2] *v.* **1** use or wield (a tool etc.). **2** work (at a trade). **3** keep offering or supplying. **4** (of a vehicle etc.) go to and fro.

plywood *n.* board made by gluing layers with the grain crosswise.

p.m. *abbr.* (Latin *post meridiem*) after noon.

pneumatic /nyoomáttik/ *adj.* filled with or operated by compressed air.

pneumonia /nyoomṓniə/ *n.* inflammation of the lungs.

poach *v.* **1** cook (an egg without its shell) in or over boiling water. **2** simmer in a small amount of liquid. **3** take (game or fish) illegally. **4** trespass, encroach. □ **poacher** *n.*

pocket *n.* **1** small bag-like part in or on a garment. **2** one's resources of money. **3** pouch-like compartment. **4** isolated group or area. ● *adj.* suitable for carrying in one's pocket. ● *v.* **1** put into one's pocket. **2** appropriate. □ **in** or **out of pocket** having made a profit or loss. **pocket money 1** money for small personal expenses. **2** money allowed to children. **pocketful** *n.*

pock-marked *adj.* marked by scars or pits.

pod *n.* long narrow seed case.

podgy *adj.* (**-ier**, **-iest**) short and fat. □ **podginess** *n.*

podium *n.* (*pl.* **-ia**) **1** projecting base. **2** rostrum.

poem *n.* literary composition in verse.

■ lyric, ode, rhyme, sonnet, verse.

poet *n.* writer of poems.

poetic *adj.* (also **poetical**) of or like poetry. □ **poetically** *adv.*

poetry *n.* **1** poems. **2** poet's work. **3** beautiful and graceful quality.

pogo stick stilt-like toy with a spring, for jumping about on.

pogrom *n.* organized massacre.

poignant *adj.* arousing sympathy, deeply moving. □ **poignantly** *adv.*, **poignancy** *n.*

■ affecting, emotional, emotive, moving, pathetic, piteous, pitiable, pitiful, sad, touching, upsetting.

poinsettia *n.* plant with large scarlet bracts.

point *n.* **1** tapered or sharp end, tip. **2** promontory. **3** particular place, moment, or stage. **4** unit of measurement or scoring. **5** item, detail. **6** characteristic. **7** sense, purpose. **8** advantage or value. **9** significant or essential thing. **10** electrical socket. **11** movable rail for directing a train from one line to another. ● *v.* **1** aim, direct (a finger or weapon etc.). **2** have a certain direction. **3** fill in (joints of brickwork) with mortar or cement. □ **on the point of** on the verge of (an action). **point-blank** *adj.* **1** aimed or fired at very close range. **2** (of a remark) direct. *adv.* in a point-blank manner. **point-duty** *n.* that of a policeman stationed to regulate traffic. **point of view** way of looking at a matter. **point out** draw attention to. **point to** indicate. **point up** emphasize. **to the point** relevant(ly).

■ *n.* **1** prong, spike, tine; apex, end, tip. **2** cape, headland, peninsula, promontory. **3** location, place, situation, site, spot; instant, juncture, moment, stage, time. **5** aspect, detail, element, facet, item, matter, particular, respect. **6** attribute, characteristic, feature, property, quality, trait. **7** aim, end, goal, intention, object, objective, purpose; drift, gist, meaning, sense, significance. **8** advantage, benefit, use, value. **9** crux, essence, heart, *sl.* nitty-gritty, nub, substance. ● *v.* **1** aim, direct, level, train. **2** face, look. □ **point of view** angle, outlook, perspective, standpoint, view, viewpoint. **point up** accent, accentuate, emphasize, highlight, spotlight, stress, underline. **to the point** applicable, apposite, germane, pertinent, relevant.

pointed *adj.* **1** tapering to a point. **2** (of a remark or manner) emphasized, clearly aimed at a person or thing. □ **pointedly** *adv.*

pointer *n.* **1** thing that points to something. **2** dog that points towards game which it scents.

pointless *adj.* having no purpose or meaning. □ **pointlessly** *adv.*, **pointlessness** *n.*

■ empty, fruitless, futile, hollow, meaningless, purposeless, senseless, useless, vain, worthless.

poise *n.* **1** balance. **2** dignified self-assured manner.

■ **1** balance, equilibrium, equipoise. **2** aplomb, calmness, composure, control, *sl.* cool, dignity, equanimity, self-assurance, self-possession, serenity.

poison *n.* substance that can destroy life or harm health. ● *v.* **1** give poison to. **2** put poison on or in. **3** corrupt, fill with prejudice. □ **poison-pen letter** malicious unsigned letter. **poisoner** *n.*, **poisonous** *adj.*

■ *n.* toxin, venom. ● *v.* **2** adulterate, contaminate, defile, pollute, taint. **3** corrupt, deprave, pervert, warp. □ **poisonous** deadly, lethal, noxious, toxic, venomous; evil, harmful, malevolent, malign, malignant, pernicious.

poke *v.* **1** thrust with the end of a finger or stick etc. **2** thrust forward. ● *n.* poking movement. □ **poke about** search casually, pry. **poke fun at** ridicule.

■ *v.* **1** dig, jab, prod, push, stab, stick, thrust. □ **poke fun at** chaff, deride, gibe at, jeer at, laugh at, make fun of, mock, *sl.* rag, *colloq.* rib, ridicule, taunt, tease.

poker¹ *n.* stiff metal rod for stirring up a fire.

poker² *n.* gambling card game. □ **poker-face** *n.* one that does not reveal thoughts or feelings.

poky *adj.* (**-ier**, **-iest**) small and cramped. □ **pokiness** *n.*

polar *adj.* **1** of or near the North or South Pole. **2** of a pole of a magnet. □ **polar bear** white bear of Arctic regions.

polarize *v.* **1** confine similar vibrations of (light waves) to one direction. **2** give magnetic poles to. **3** set at opposite extremes of opinion. □ **polarization** *n.*

Pole *n.* Polish person.

pole¹ *n.* long rod or post. ● *v.* push along by using a pole. □ **pole position** most favourable starting position in a motor race.

■ *n.* mast, post, rod, shaft, spar, staff, stanchion, stick, upright.

pole² *n.* **1** north (**North Pole**) or south (**South Pole**) end of earth's axis. **2** one of the opposite ends of a magnet or terminals of an electric cell or battery. □ **pole star** star near the North Pole in the sky.

polecat *n.* **1** small animal of the weasel family. **2** (*US*) skunk.

polemic *n.* verbal attack on a belief or opinion. □ **polemical** *adj.*

police *n.* civil force responsible for keeping public order. ● *v.* control or provide with police. □ **police state** country where political police supervise and control citizens' activities. **policeman** *n.*, **policewoman** *n.*

policy¹ *n.* course or general plan of action.

■ approach, course, line, method, plan, procedure, programme, scheme, strategy, system, tactic.

policy² *n.* insurance contract.

polio *n.* (*colloq.*) poliomyelitis.

poliomyelitis *n.* infectious disease causing temporary or permanent paralysis.

Polish *adj.* & *n.* (language) of Poland.

polish *v.* **1** make smooth and glossy by rubbing. **2** refine, perfect. ● *n.* **1** glossiness. **2** polishing, substance used for this. **3** elegance. □ **polish off** finish off. **polisher** *n.*

■ *v.* **1** buff, burnish, gloss, shine, wax. **2** improve, perfect, refine. ● *n.* **1** gloss, glossiness, lustre, sheen, shine. **3** elegance, refinement, sophistication, urbanity. □ **polish off** consume, devour, eat up, finish off, gobble (up), wolf (down).

polished *adj.* (of manner or performance) elegant, perfected.

■ civilized, cultivated, elegant, polite, refined, sophisticated, suave, urbane; accomplished, faultless, flawless, impeccable, masterly, perfect.

polite *adj.* **1** having good manners, courteous. **2** refined. □ **politely** *adv.*, **politeness** *n.*

■ **1** civil, courteous, diplomatic, respectful, tactful, well-mannered. **2** civilized, elegant, genteel, refined.

political *adj.* **1** of or involving politics. **2** of the way a country is governed. □ **politically** *adv.*

politician *n.* person engaged in politics.

politics *n.* **1** science and art of government. **2** political affairs or life. **3** (*pl.*) political principles.

polka *n.* lively dance for couples. □ **polka dots** round evenly spaced dots on fabric.

poll *n.* **1** votes at an election. **2** place for this. **3** estimate of public opinion made by questioning people. ● *v.* **1** receive as votes. **2** cut off the horns of (cattle) or the top of (a tree etc.). □ **poll tax** tax levied on every adult.

pollard *v.* poll (a tree) to produce a close head of young branches. ● *n.* **1** pollarded tree. **2** hornless animal.

pollen *n.* fertilizing powder from the anthers of flowers.

pollinate *v.* fertilize with pollen. □ **pollination** *n.*

pollute *v.* make dirty or impure. □ **pollution** *n.*, **pollutant** *n.*

> ■ contaminate, defile, dirty, foul, poison, soil, sully, taint.

polo *n.* game like hockey played by teams on horseback. □ **polo-neck** *n.* high turned-over collar.

polonaise *n.* slow processional dance.

poltergeist *n.* spirit that throws things about noisily.

polyandry *n.* system of having more than one husband at a time.

polyanthus *n.* (*pl.* **-thuses** or **-thus**) cultivated primrose.

polychrome *adj.* multicoloured. □ **polychromatic** *adj.*

polyester *n.* synthetic resin or fibre.

polygamy *n.* system of having more than one wife at a time. □ **polygamist** *n.*, **polygamous** *adj.*

polyglot *adj.* & *n.* (person) knowing or using several languages.

polygon *n.* geometric figure with many sides. □ **polygonal** *adj.*

polyhedron *n.* (*pl.* **-dra**) solid with many sides. □ **polyhedral** *adj.*

polymath *n.* person with knowledge of many subjects.

polymer *n.* compound whose molecule is formed from a large number of simple molecules.

polymerize *v.* combine into a polymer. □ **polymerization** *n.*

polyp *n.* **1** simple organism with a tube-shaped body. **2** abnormal growth projecting from mucous membrane.

polyphony *n.* combination of melodies. □ **polyphonal** *adj.*

polystyrene *n.* a kind of plastic.

polytheism *n.* belief in or worship of more than one god. □ **polytheist** *n.*, **polytheistic** *adj.*

polythene *n.* tough light plastic.

polyunsaturated *adj.* (of fat) not associated with the formation of cholesterol in the blood.

polyurethane *n.* synthetic resin or plastic.

pomander *n.* ball of mixed sweet-smelling substances.

pomegranate *n.* **1** tropical fruit with many seeds. **2** tree bearing this.

Pomeranian *n.* dog of a small silky-haired breed.

pommel /púmm'l/ *n.* **1** knob on the hilt of a sword. **2** upward projection on a saddle.

pomp *n.* splendid display, splendour.

> ■ glory, grandeur, magnificence, majesty, pageantry, splendour.

pom-pom *n.* decorative tuft or ball.

pompous *adj.* self-important, affectedly grand. □ **pompously** *adv.*, **pomposity** *n.*

> ■ arrogant, conceited, haughty, hoity-toity, pretentious, proud, self-important, *colloq.* stuck-up, *colloq.* uppity, vain; bombastic, flowery, grandiloquent, turgid.

poncho *n.* (*pl.* **-os**) type of cloak made like a blanket with a hole for the head.

pond *n.* small area of still water.

ponder *v.* **1** be deep in thought. **2** think over.

> ■ brood (over), cogitate, consider, contemplate, deliberate (over), meditate (on), mull over, muse (over), reflect (on), ruminate (on), think (over).

ponderous *adj.* **1** heavy, unwieldy. **2** dull, laboured. □ **ponderously** *adv.*

> ■ **1** cumbersome, heavy, hefty, unwieldy, weighty. **2** dull, laboured, strained, stodgy, tedious, turgid.

pong *n.* & *v.* (*sl.*) stink.

pontiff *n.* **1** bishop. **2** pope.

pontificate *v.* speak in a pompously dogmatic way.

pontoon[1] *n.* **1** flat-bottomed boat. **2** floating platform. □ **pontoon bridge** temporary bridge supported on pontoons.

pontoon[2] *n.* a kind of card game.

pony *n.* horse of any small breed. □ **pony-tail** *n.* long hair drawn back and tied to hang down.

poodle *n.* dog with thick curly hair.

pooh *int.* exclamation of contempt. □ **pooh-pooh** *v.* express contempt for.

pool[1] *n.* **1** small area of still water. **2** puddle. **3** swimming pool.

pool[2] *n.* **1** shared fund or supply. **2** game resembling snooker. **3** (*pl.*) football pools. ● *v.* put into a common fund or supply.

poop *n.* **1** ship's stern. **2** raised deck at the stern.

poor *adj.* **1** having little money or means. **2** not abundant. **3** not very good. **4** pitiful.

■ **1** *colloq.* broke, destitute, hard-up, impecunious, impoverished, indigent, needy, penniless, penurious, poverty-stricken, *sl.* skint. **2** inadequate, insufficient, meagre, *sl.* measly, scant, scanty, skimpy, sparse. **3** bad, incompetent, inferior, *colloq.* lousy, mediocre, second-rate, shoddy, slipshod, substandard, unacceptable, unprofessional, unsatisfactory. **4** hapless, luckless, pitiable, pitiful, unfortunate, unlucky, wretched.

poorly *adv.* in a poor way, badly. ● *adj.* unwell.

pop¹ *n.* **1** small explosive sound. **2** fizzy drink. ● *v.* (**popped**) **1** make or cause to make a pop. **2** put, come, or go quickly.

pop² *adj.* in a popular modern style. ● *n.* pop record or music.

popcorn *n.* maize heated to burst and form puffy balls.

pope *n.* head of the RC Church.

poplar *n.* tall slender tree.

poplin *n.* plain woven fabric.

poppadam *n.* large thin crisp savoury Indian biscuit.

poppy *n.* plant with showy flowers and milky juice.

poppycock *n.* (*sl.*) nonsense.

populace *n.* the general public.

popular *adj.* **1** liked, enjoyed, or used etc. by many people. **2** of, for, or prevalent among the general public. □ **popularly** *adv.*, **popularity** *n.*

■ **1** favoured, favourite, well-liked; fashionable, in, in fashion, in vogue. **2** common, current, general, prevailing, prevalent, widespread.

popularize *v.* **1** make generally liked. **2** present in a popular non-technical form.

populate *v.* **1** form the population of. **2** fill with a population.

■ **1** dwell in, inhabit, live in, occupy, reside in. **2** colonize, people, settle.

population *n.* inhabitants of an area.

populous *adj.* thickly populated.

porcelain *n.* fine china.

porch *n.* roofed shelter over the entrance of a building.

porcine *adj.* of or like a pig.

porcupine *n.* animal covered with protective spines.

pore¹ *n.* tiny opening in a surface through which fluids may pass.

pore² *v.* **pore over** study closely.

■ examine, go over, inspect, peruse, read, scrutinize, study.

pork *n.* unsalted pig-meat.

porn *n.* (*colloq.*) pornography.

pornography *n.* writings or pictures intended to stimulate erotic feelings by portraying sexual activity. □ **pornographer** *n.*, **pornographic** *adj.*

porous *adj.* permeable by fluid or air. □ **porosity** *n.*

porphyry *n.* rock containing mineral crystals.

porpoise *n.* small whale with a blunt rounded snout.

porridge *n.* food made by boiling oatmeal etc. to a thick paste.

port¹ *n.* **1** harbour. **2** town with this.

port² *n.* **1** opening in a ship's side. **2** porthole.

port³ *n.* left-hand side of a ship or aircraft.

port⁴ *n.* strong sweet wine.

portable *adj.* able to be carried. □ **portability** *n.*

portal *n.* door or entrance, esp. an imposing one.

portcullis *n.* vertical grating that slides down in grooves to block the gateway to a castle.

portend *v.* foreshadow.

■ augur, bode, foreshadow, indicate, presage.

portent *n.* omen, significant sign. □ **portentous** *adj.*

■ augury, indication, omen, presage, sign, warning.

porter¹ *n.* doorkeeper of a large building.

porter² *n.* person employed to carry luggage or goods.

portfolio *n.* (*pl.* **-os**) **1** case for loose sheets of paper. **2** set of investments.

porthole *n.* window in the side of a ship or aircraft.

portico *n.* (*pl.* **-oes**) columns supporting a roof to form a porch or similar structure.

portion *n.* **1** part, share. **2** amount of food for one person. ● *v.* divide into portions.

■ *n.* **1** allocation, allowance, *colloq.* cut, part, percentage, quota, ration, share. **2** helping, piece, serving, slice.

portly *adj.* (**-ier**, **-iest**) corpulent, stout. □ **portliness** *n.*

portrait *n.* **1** picture of a person or animal. **2** description.

■ **1** drawing, image, likeness, painting, picture, representation, sketch. **2** characterization, description, portrayal, profile.

portray *v.* **1** make a picture of. **2** describe. **3** represent in a play etc. □ **portrayal** *n.*

■ **1** delineate, depict, draw, paint, picture, represent, show. **2** characterize, depict, describe, paint, represent. **3** act, play, represent, take the part *or* role of.

Portuguese *adj.* & *n.* (native, language) of Portugal.

pose *v.* **1** put into or take a particular attitude. **2** put forward, present (a problem etc.). ● *n.* **1** attitude in which someone is posed. **2** pretence. □ **pose as** pretend to be.

■ *v.* **1** arrange, place, position; model, sit. **2** ask, present, put forward, raise, submit. ● *n.* **1** attitude, position, posture, stance. **2** act, affectation, façade, pretence, show. □ **pose as** be disguised as, imitate, impersonate, masquerade as, pretend to be.

poser *n.* **1** puzzling problem. **2** poseur.

poseur *n.* person who behaves affectedly.

posh *adj.* (*colloq.*) very smart, luxurious.

position *n.* **1** place occupied by or intended for a person or thing. **2** posture. **3** situation. **4** status. **5** job. ● *v.* place. □ **positional** *adj.*

■ *n.* **1** locality, location, place, setting, site, situation, spot, whereabouts. **2** attitude, pose, posture, stance. **3** circumstances, condition, situation, state. **4** grade, place, rank, standing, station, status. **5** appointment, job, place, post, situation. ● *v.* arrange, dispose, lay, locate, place, pose, put, set, settle, site, situate.

positive *adj.* **1** definite, explicit. **2** convinced. **3** constructive, favourable. **4** (of a quantity) greater than zero. **5** (of a battery terminal) through which electric current enters. **6** (of a photograph) with lights, shades, or colours as in the subject, not reversed. ● *n.* positive quality, quantity, or photograph. □ **positively** *adv.*, **positiveness** *n.*

■ *adj.* **1** categorical, certain, clear, conclusive, convincing, decisive, definite, explicit, firm, incontestable, indisputable, undeniable, unequivocal, unquestionable. **2** certain, confident, convinced, satisfied, sure. **3** constructive, helpful, productive, useful, valuable; complimentary, enthusiastic, favourable, good, supportive.

positron *n.* particle with a positive electric charge.

posse /póssi/ *n.* **1** group of law-enforcers. **2** strong force or company.

possess *v.* **1** own. **2** have as a quality etc. **3** dominate the mind of. □ **possessor** *n.*

■ **1** be in possession of, be the owner of, enjoy, have, own. **3** control, dominate, haunt, obsess, preoccupy.

possession *n.* **1** possessing. **2** thing possessed. □ **take possession of** become the possessor of.

■ **1** ownership, proprietorship, tenure. **2** (**possessions**) belongings, chattels, effects, goods, property, things.

possessive *adj.* **1** desiring to possess things. **2** jealous and domineering. □ **possessiveness** *n.*

possible *adj.* capable of existing, happening, being done, etc. □ **possibly** *adv.*, **possibility** *n.*

■ believable, conceivable, credible, feasible, imaginable, likely, plausible, tenable; achievable, attainable, practicable, viable.

possum *n.* (*colloq.*) opossum. □ **play possum** pretend to be unaware.

post[1] *n.* piece of timber, metal, etc. set upright to support or mark something. ● *v.* display (a notice etc.), announce thus.

■ *n.* column, picket, pile, pillar, pole, shaft, stake, stanchion, strut, upright.

post[2] *n.* **1** place of duty. **2** outpost of soldiers. **3** trading station. **4** job. ● *v.* place, station.

■ *n.* **4** appointment, job, place, position, situation.

post[3] *n.* **1** official conveyance of letters etc. **2** the letters etc. conveyed. ● *v.* send by post. □ **keep me posted** keep me informed. **postbox** *n.* box into which letters are put for sending by post. **post-haste** *adv.* with great haste. **post office** building or room where postal business is carried on.

post- *pref.* after.

postage *n.* charge for sending something by post.

postal *adj.* of or by post.

postcard *n.* card for sending messages by post without an envelope.

postcode *n.* group of letters and figures in a postal address to assist sorting.

poster *n.* large picture or notice announcing or advertising something.

poste restante /pōst restónt/ post office department where letters are kept until called for.

posterior *adj.* situated behind or at the back. ● *n.* buttocks.

posterity *n.* future generations.

postern *n.* small back or side entrance to a fortress etc.

posthumous *adj.* **1** (of a child) born after its father's death. **2** published or awarded after a person's death. □ **posthumously** *adv.*

postman *n.* person who delivers or collects letters etc.

postmark *n.* official mark stamped on something sent by post, giving place and date of marking. ● *v.* mark with this.

postmaster, **postmistress** *ns.* person in charge of certain post offices.

post-mortem *adj.* & *n.* (examination) made after death.

postnatal *adj.* after childbirth.

postpone *v.* keep (an event etc.) from occurring until a later time. □ **postponement** *n.*

■ adjourn, defer, delay, put off, shelve, suspend.

postprandial *adj.* after lunch or dinner.

postscript *n.* additional paragraph at the end of a letter etc.

postulate *v.* assume to be true, esp. as a basis for reasoning. □ **postulation** *n.*

posture *n.* **1** attitude of the body. **2** way in which a person stands, walks, etc. ● *v.* assume a posture, esp. for effect.

■ ***n.*** **1 attitude, pose, position, stance. 2 bearing, carriage, deportment.**

posy *n.* small bunch of flowers.

pot *n.* vessel for holding liquids or solids, or for cooking in. ● *v.* (**potted**) **1** put into a pot. **2** send (a ball in billiards etc.) into a pocket. □ **go to pot** (*colloq.*) deteriorate. **pot-belly** *n.* protuberant belly. **pot-boiler** *n.* literary or artistic work produced merely to make a living. **potluck** *n.* whatever is available for a meal. **pot-roast** *n.* piece of meat cooked slowly in a covered dish. *v.* cook in this way. **pot-shot** *n.* shot aimed casually.

potable *adj.* drinkable.

potash *n.* potassium carbonate.

potassium *n.* soft silvery-white metallic element.

potation *n.* **1** drinking. **2** a drink.

potato *n.* (*pl.* **-oes**) **1** plant with starchy tubers that are used as food. **2** one of these tubers.

potent *adj.* **1** powerful, strong. **2** cogen □ **potency** *n.*

■ 1 forceful, influential, mighty, powerfu strong, vigorous. 2 compelling, cogen convincing, effective, forceful, persuasive.

potentate *n.* monarch, ruler.

potential *adj.* & *n.* (ability etc.) capabl of being developed or used. □ **potentiall** *adv.*, **potentiality** *n.*

■ ***adj.*** **budding, developing, dorman future, latent, possible, undevelope unrealized. ●** ***n.*** **capability, capacity, poss bility, promise.**

pothole *n.* **1** hole formed undergroun by the action of water. **2** hole in a roa surface.

potholing *n.* caving. □ **potholer** *n.*

potion *n.* liquid for drinking as a med cine or drug.

■ brew, concoction, draught, elixir, philtr potation, tonic.

pot-pourri /pōpoori/ *n.* **1** scented mi ture of dried petals and spices. **2** medley

potsherd *n.* broken piece of earther ware.

potted *see* **pot**. *adj.* **1** preserved in a po **2** abridged.

potter¹ *n.* maker of pottery.

potter² *v.* work on trivial tasks in leisurely way.

pottery *n.* **1** vessels and other objec made of baked clay. **2** potter's work c workshop.

potty¹ *adj.* (**-ier**, **-iest**) (*sl.*) **1** trivial. crazy.

potty² *n.* (*colloq.*) child's chamber pot.

pouch *n.* small bag or bag-like formatior

pouffe *n.* padded stool.

poulterer *n.* dealer in poultry.

poultice *n.* moist usu. hot dressing a plied to inflammation. ● *v.* put a poultic on.

poultry *n.* domestic fowls.

pounce *v.* swoop down and grasp or a tack. ● *n.* pouncing movement.

pound¹ *n.* **1** unit of weight equal to 16 o (454 g). **2** unit of money in the UK an certain other countries.

pound² *n.* enclosure where stray an mals, or vehicles officially removed, a kept until claimed.

pound³ *v.* **1** beat or crush with heav strokes. **2** walk or run heavily. **3** (of th heart) beat heavily.

■ 1 beat, crush, grind, mash, pulveriz batter, belabour, bludgeon, ***sl.*** **clobbe cudgel, pummel, thump,** ***sl.*** **wallop.**

clump, stomp, tramp, trudge. **3** palpitate, pulsate, pulse, throb.

poundage *n.* charge or commission per £ or per pound weight.

pour *v.* **1** flow, cause to flow. **2** rain heavily. **3** send out freely.

■ **1** cascade, course, flood, flow, gush, run, spout, spurt, stream. **2** bucket, rain, teem.

pout *v.* **1** push out one's lips. **2** (of lips) be pushed out. ● *n.* pouting expression.

poverty *n.* **1** state of being poor. **2** scarcity, lack. **3** inferiority. □ **poverty-stricken** very poor.

■ **1** destitution, indigence, necessity, need, pennilessness, penury, want. **2** dearth, deficiency, insufficiency, lack, meagreness, paucity, scarcity, shortage, want. **3** inferiority, meanness, poorness.

powder *n.* **1** mass of fine dry particles. **2** medicine or cosmetic in this form. **3** gunpowder. ● *v.* **1** cover with powder. **2** reduce to powder. □ **powdery** *adj.*

power *n.* **1** ability to do something. **2** vigour, strength. **3** control, influence, authority. **4** influential person or country etc. **5** product of a number multiplied by itself a given number of times. **6** mechanical or electrical energy. **7** electricity supply. ● *v.* supply with (esp. motive) power. □ **power station** building where electricity is generated for distribution.

■ *n.* **1** ability, capability, capacity, potential. **2** brawn, dynamism, energy, force, forcefulness, might, muscle, potency, strength, vigour. **3** ascendancy, authority, command, control, dominance, dominion, mastery, rule, sovereignty; *colloq.* clout, influence, pull, sway, weight.

powerful *adj.* having great power or influence. □ **powerfully** *adv.*

■ dynamic, mighty, muscular, potent, robust, strapping, strong, sturdy, vigorous; authoritative, cogent, compelling, effective, forceful, persuasive, weighty; important, impressive, influential.

powerless *adj.* without power to take action, wholly unable.

■ defenceless, feeble, helpless, impotent, incapable, incapacitated, ineffective, ineffectual, unable, weak.

practicable *adj.* able to be done. □ **practicability** *n.*

■ achievable, feasible, possible, viable, workable.

practical *adj.* **1** involving activity as distinct from study or theory. **2** suitable for use. **3** good at doing and making things. **4** sensible and realistic. **5** virtual. □ **practical joke** trick played on a person. **practicality** *n.*

■ **1** empirical. **2** functional, serviceable, usable, useful, utilitarian. **4** businesslike, down-to-earth, efficient, hard-headed, pragmatic, realistic, sensible.

practically *adv.* **1** in a practical way. **2** virtually, almost.

practice *n.* **1** action as opposed to theory. **2** custom. **3** repeated exercise to improve skill. **4** doctor's or lawyer's business. □ **put into practice** apply (an idea, method, etc.).

■ **1** action, enactment, execution, operation; actuality, effect, fact, reality. **2** convention, custom, habit, routine, tradition, usage, way. **3** drill, exercise, rehearsal, training.

practise *v.* **1** do something repeatedly or habitually. **2** (of a doctor or lawyer) perform professional work.

practised *adj.* expert.

■ able, accomplished, adept, capable, experienced, expert, masterly, proficient, qualified, skilful, skilled, talented.

practitioner *n.* professional worker, esp. in medicine.

pragmatic *adj.* treating things from a practical point of view. □ **pragmatically** *adv.*, **pragmatism** *n.*, **pragmatist** *n.*

prairie *n.* large treeless area of grassland, esp. in N. America.

praise *v.* **1** express approval or admiration of. **2** honour (God) in words. ● *n.* act or instance of praising.

■ *v.* **1** acclaim, applaud, commend, compliment, congratulate, eulogize, extol, honour, laud, pay tribute to. **2** exalt, glorify, worship. ● *n.* acclaim, approbation, approval, applause, commendation, compliments, congratulations, glory, plaudits, tribute; exaltation, glorification, worship.

praiseworthy *adj.* deserving praise. □ **praiseworthiness** *n.*

■ admirable, commendable, creditable, deserving, exemplary, laudable, meritorious, worthy.

pram *n.* four-wheeled carriage for a baby.

prance *v.* move springily.

■ bound, caper, cavort, frisk, jump, leap, skip, spring.

prank *n.* piece of mischief.

prankster *n.* person playing pranks.

prattle *v.* chatter in a childish or inconsequential way. ● *n.* childish chatter.

prawn *n.* edible shellfish like a large shrimp.

pray *v.* **1** say prayers. **2** entreat.

> ■ **2** appeal to, ask, beg, beseech, entreat, implore, importune, petition, plead with, solicit, supplicate.

prayer *n.* **1** solemn request or thanksgiving to God. **2** set form of words used in this. **3** act of praying. **4** entreaty.

> ■ **4** appeal, entreaty, petition, plea, request, supplication.

pre- *pref.* **1** before. **2** beforehand.

preach *v.* **1** deliver a sermon. **2** expound (the Gospel etc.). **3** advocate. **4** give moral advice. □ **preacher** *n.*

> ■ **2** expound, proclaim, teach. **3** advocate, recommend, urge. **4** moralize, pontificate, sermonize.

preamble *n.* preliminary statement, introductory section.

prearrange *v.* arrange beforehand. □ **prearrangement** *n.*

precarious *adj.* unsafe, not secure. □ **precariously** *adv.*, **precariousness** *n.*

> ■ chancy, dangerous, *sl.* dicey, hazardous, insecure, perilous, risky, shaky, treacherous, uncertain, unpredictable, unreliable, unsafe, unsure.

precaution *n.* something done in advance to avoid a risk. □ **precautionary** *adj.*

precede *v.* come or go before in time or order etc.

precedence *n.* priority.

precedent *n.* previous case serving as an example to be followed.

precept *n.* rule for action or conduct.

> ■ command, dictate, direction, directive, guideline, instruction, law, principle, rule.

precession *n.* change by which equinoxes occur earlier in each sidereal year.

precinct *n.* **1** enclosed area, esp. round a cathedral. **2** area closed to traffic in a town. **3** (*pl.*) environs.

precious *adj.* **1** of great value. **2** beloved. **3** affectedly refined. □ **precious stone** small valuable piece of mineral.

> ■ **1** costly, expensive, invaluable, priceless, valuable. **2** adored, beloved, cherished, dear, dearest, loved, prized, treasured. **3** affected, pretentious, twee.

precipice *n.* very steep or vertical face of a cliff or rock.

precipitate *v.* /prisíppitayt/ **1** throw headlong. **2** cause to happen suddenly or soon. **3** cause (a substance) to be deposited. **4** condense (vapour). ● *n.* /prisíppitət/ **1** substance deposited from a solution. **2** moisture condensed from vapour. ● *adj.* /prisíppitət/ hasty, rash. □ **precipitately** *adv.*, **precipitation** *n.*

> ■ *v.* **1** cast, catapult, fling, hurl, launch, propel, throw. **2** accelerate, advance, bring about, expedite, facilitate, further, hasten, quicken, speed (up), trigger. ● *adj.* abrupt, fast, hasty, headlong, hurried, rapid, speedy, sudden, swift; foolhardy, harebrained, hotheaded, impetuous, impulsive, incautious, injudicious, rash, reckless.

precipitous *adj.* very steep.

précis /práysee/ *n.* (*pl.* **précis**) summary. ● *v.* make a précis of.

precise *adj.* **1** exact. **2** correct and clearly stated. **3** punctilious, scrupulous. □ **precisely** *adv.*, **precision** *n.*

> ■ **1** exact, particular, very. **2** accurate, correct, error-free, exact, faithful, perfect, true, unerring. **3** careful, conscientious, meticulous, particular, punctilious, scrupulous,

preclude *v.* **1** prevent. **2** make impossible.

> ■ **1** bar, debar, exclude, impede, inhibit, obstruct, prevent, prohibit, stop. **2** forestall, obviate, remove, rule out.

precocious *adj.* having developed earlier than is usual. □ **precociously** *adv.*

precognition *n.* foreknowledge, esp. supernatural.

preconceived *adj.* (of an idea) formed beforehand. □ **preconception** *n.*

precondition *n.* condition to be fulfilled beforehand.

precursor *n.* forerunner.

predatory /préddətəri/ *adj.* preying on others. □ **predator** *n.* predatory animal.

predecease *v.* die earlier than (another person).

predecessor *n.* **1** former holder of an office or position. **2** ancestor.

> ■ **1** antecedent, forerunner, precursor. **2** ancestor, antecedent, forebear, forefather.

predicament *n.* difficult situation.

> ■ difficulty, dilemma, fix, *colloq.* hole, *colloq.* jam, mess, *colloq.* pickle, plight, quandary, scrape.

predicate *n.* the part of a sentence that says something about the subject (e.g. 'is short' in *life is short*). □ **predicative** *adj.*

predict *v.* foretell. ▫ **prediction** *n.*, **predictive** *adj.*, **predictor** *n.*

▪ augur, forecast, foresee, foretell, presage, prognosticate, prophesy.

predictable *adj.* able to be predicted.

predilection *n.* special liking.

predispose *v.* **1** influence in advance. **2** render liable or inclined. ▫ **predisposition** *n.*

predominate *v.* **1** be most numerous or powerful. **2** exert control. ▫ **predominant** *adj.*, **predominantly** *adv.*, **predominance** *n.*

▪ **1** be prevalent, dominate, preponderate, prevail. **2** control, dominate, have the upper hand, rule. ▫ **predominant** dominant, preponderant, preponderating, prevailing, prevalent; chief, controlling, leading, main, ruling, supreme.

pre-eminent *adj.* **1** excelling others, outstanding. **2** leading, principal. ▫ **pre-eminently** *adv.*, **pre-eminence** *n.*

▪ **1** distinguished, eminent, excellent, important, inimitable, matchless, outstanding, peerless, superb, supreme, unequalled, unsurpassed. **2** chief, first, foremost, leading, main, paramount, primary, principal.

pre-empt *v.* **1** forestall. **2** take (a thing) before anyone else can do so. ▫ **pre-emption** *n.*, **pre-emptive** *adj.*

preen *v.* smooth (feathers) with the beak. ▫ **preen oneself** **1** groom oneself. **2** show self-satisfaction.

prefabricate *v.* manufacture in sections for assembly on a site. ▫ **prefabrication** *n.*

preface *n.* introductory statement. ● *v.* **1** introduce with a preface. **2** lead up to (an event).

▪ *n.* foreword, introduction, preamble, prologue. ● *v.* **1** begin, introduce, open, prefix, start.

prefect *n.* **1** senior pupil authorized to maintain discipline in a school. **2** administrative official in certain countries. ▫ **prefecture** *n.*

prefer *v.* (**preferred**) **1** choose as more desirable, like better. **2** put forward (an accusation).

preferable *adj.* more desirable. ▫ **preferably** *adv.*

preference *n.* **1** preferring or being preferred. **2** thing preferred. **3** prior right. **4** favouring.

▪ **1** inclination, leaning, liking, partiality, predilection, proclivity. **2** favourite. **4** favour, favouritism, preferential treatment, priority.

preferential *adj.* of or involving preference. ▫ **preferentially** *adv.*

▪ advantageous, better, favourable, privileged, special, superior.

preferment *n.* promotion.

prefigure *v.* foreshadow.

prefix *n.* (*pl.* **-ixes**) word or syllable placed in front of a word to change its meaning. ● *v.* add as a prefix or introduction.

pregnant *adj.* **1** having a child or young developing in the womb. **2** full of meaning. ▫ **pregnancy** *n.*

prehensile *adj.* able to grasp things. ▫ **prehensility** *n.*

prehistoric *adj.* of the period before written records were made. ▫ **prehistorically** *adv.*

▪ ancient, antediluvian, earliest, early, primal, primeval, primitive, primordial.

prejudge *v.* form a judgement on before knowing all the facts. ▫ **prejudgement** *n.*

prejudice *n.* **1** unreasoning opinion or dislike. **2** harm to rights. ● *v.* **1** cause to have a prejudice. **2** cause harm to. ▫ **prejudiced** *adj.*

▪ *n.* **1** bias, bigotry, chauvinism, discrimination, favouritism, partisanship, racialism, racism, sexism. **2** damage, detriment, harm, injury. ● *v.* **1** bias, colour, influence, poison, sway. **2** damage, harm, injure, mar, ruin, spoil. ▫ **prejudiced** biased, bigoted, intolerant, one-sided, partial, partisan, unfair.

prejudicial *adj.* harmful to rights or interests. ▫ **prejudicially** *adv.*

▪ damaging, deleterious, detrimental, disadvantageous, harmful, hurtful, injurious, unfavourable.

prelate *n.* clergyman of high rank. ▫ **prelacy** *n.*

preliminary *adj.* & *n.* (action or event etc.) preceding and preparing for a main action or event.

▪ *adj.* initial, introductory, opening, preceding, preparatory. ● *n.* beginning, introduction, opening, prelude, preparation.

prelude *n.* **1** action or event leading up to another. **2** introductory part of a poem etc.

▪ **2** introduction, overture, preamble, preface, prologue.

premarital *adj.* before marriage.

premature *adj.* coming or done before the usual or proper time. ▫ **prematurely** *adv.*

▪ early, ill-timed, too soon, unseasonable, untimely.

premedication *n.* medication in preparation for an operation.

premeditated *adj.* planned beforehand. ▫ **premeditation** *n.*

▪ calculated, conscious, deliberate, intended, intentional, planned, preconceived, wilful.

premenstrual *adj.* of the time immediately before menstruation.

premier *adj.* first in importance, order, or time. ● *n.* prime minister. ▫ **premiership** *n.*

première /prémmiair/ *n.* first public performance.

premises *n.pl.* house or other building and its grounds.

premiss *n.* statement on which reasoning is based.

▪ assumption, hypothesis, postulate, presupposition, proposition, supposition, thesis.

premium *n.* **1** amount or instalment paid for an insurance policy. **2** extra sum of money. ▫ **at a premium 1** above the nominal or usual price. **2** scarce and in demand.

premonition *n.* presentiment. ▫ **premonitory** *adj.*

▪ feeling, foreboding, hunch, intuition, presage, presentiment, suspicion.

prenatal *adj.* **1** before birth. **2** before childbirth. ▫ **prenatally** *adv.*

preoccupied *adj.* mentally engrossed.

▪ absorbed, engrossed, immersed, rapt, wrapped up.

preoccupy *v.* dominate the mind of. ▫ **preoccupation** *n.*

preparation *n.* **1** preparing or being prepared. **2** (often *pl.*) thing done to make ready. **3** substance prepared for use.

▪ **1** groundwork, organization, planning; education, instruction, teaching, training, tuition. **2** (**preparations**) arrangements, measures, plans. **3** compound, concoction, mixture, product.

preparatory *adj.* preparing for something. ● *adv.* in a preparatory way.

prepare *v.* make or get ready. ▫ **be prepared** be disposed or willing.

▪ arrange, get *or* make ready, make provisions, organize, put in order; lay, set; cook, get, make; brief, coach, groom, prime, train; adapt, equip, fit (out), modify.

preponderate *v.* be greater in number or intensity etc. ▫ **preponderant** *adj.* **preponderance** *n.*

preposition *n.* word used with a noun or pronoun to show position, time, or means (e.g. *at* home, *by* train). ▫ **prepositional** *adj.*

prepossessing *adj.* attractive.

preposterous *adj.* absurd, outrageous. ▫ **preposterously** *adv.*

▪ absurd, farcical, foolish, laughable, ludicrous, nonsensical, outrageous, ridiculous, risible, senseless, stupid.

prepuce *n.* foreskin.

prerequisite *adj.* & *n.* (thing) required before something can happen.

▪ *n.* condition, essential, *colloq.* must, necessity, precondition, proviso, requirement, requisite, stipulation.

prerogative *n.* right or privilege belonging to a person or group.

presage *n.* **1** omen. **2** presentiment. ● *v.* **1** portend. **2** foresee.

Presbyterian *adj.* & *n.* (member) of a Church governed by elders of equal rank. ▫ **Presbyterianism** *n.*

prescribe *v.* **1** advise the use of (a medicine etc.). **2** dictate as a course of action or rule to be followed.

▪ **2** decree, dictate, ordain, order, require, stipulate.

prescription *n.* **1** prescribing. **2** doctor's written instructions for the preparation and use of a medicine.

prescriptive *adj.* prescribing.

presence *n.* **1** being present. **2** person's bearing. **3** person or thing that is or seems present. ▫ **presence of mind** ability to act sensibly in a crisis.

▪ **1** attendance, company, fellowship, society; existence. **2** air, bearing, demeanour, manner. **3** ghost, spectre, spirit, wraith.

present[1] /prézz'nt/ *adj.* **1** being in the place in question. **2** existing or being dealt with now. ● *n.* present time, time now passing. ▫ **at present** now. **for the present** temporarily.

▪ *adj.* **1** here, in attendance, there. **2** contemporary, current, existing, existent.

present² *n.* /prézz'nt/ gift. ● *v.* /prizént/ **1** give as a gift or award. **2** introduce. **3** bring to the public. □ **presenter** *n.*

■ *n.* donation, gift, offering. ● *v.* **1** award, bestow, confer, give, grant, hand over. **2** introduce, make known. **3** offer, proffer, put forward, submit, tender; display, exhibit, mount, put on, stage.

presentable *adj.* fit to be presented, of good appearance. □ **presentably** *adv.*

presentation *n.* **1** presenting or being presented. **2** thing presented. **3** exhibition or theatrical performance.

■ **1** bestowal, donation, giving. **2** award, donation, gift, offering, present. **3** performance, production, show, showing, staging.

presentiment *n.* vague expectation, foreboding.

■ feeling, foreboding, hunch, intuition, premonition, presage.

presently *adv.* **1** soon. **2** (*Sc.* & *US*) now.

preservation *n.* preserving.

preservative *adj.* preserving. ● *n.* substance that preserves perishable food.

preserve *v.* **1** keep safe, unchanged, or in existence. **2** treat (food) to prevent decay. ● *n.* **1** interests etc. regarded as one person's domain. **2** (also *pl.*) jam.

■ *v.* **1** defend, guard, keep safe, protect, safeguard, shelter, shield; conserve, keep up, maintain, perpetuate, retain, sustain, uphold. **2** cure, freeze, pickle, salt, smoke. ● *n.* **1** domain, field, province, realm, sphere. **2** conserve, jam, jelly.

preside *v.* be president or chairman. □ **preside over** be in a position of authority over.

■ □ **preside over** administrate, be in charge of, control, direct, handle, lead, manage, oversee, run, supervise.

president *n.* **1** head of an institution or club. **2** head of a republic. □ **presidency** *n.*, **presidential** *adj.*

press¹ *v.* **1** apply weight or force against. **2** squeeze. **3** flatten, smooth (esp. clothes). **4** urge. **5** throng closely. ● *n.* **1** process of pressing. **2** instrument for pressing something. **3** (**the press**) newspapers and periodicals, people involved in producing these. □ **be pressed for** have barely enough of. **press conference** interview before a number of reporters. **press cutting** report etc. cut from a newspaper. **press-stud** *n.* small fastener with two parts that engage when pressed together. **press-up** *n.* exercise of pressing on the hands to raise the trunk while prone.

■ *v.* **1** depress, push down. **2** compress, crush, squeeze. **3** flatten, iron, smooth. **4** beg, entreat, exhort, implore, pressure, pressurize, urge. **5** crowd, flock, gather, mill, swarm, throng. ● *n.* **3** (**the press**) the media, the papers; journalists, reporters.

press² *v.* bring into use as a makeshift. □ **press-gang** *v.* force into service.

pressing *adj.* urgent.

■ critical, important, serious, urgent, vital.

pressure *n.* **1** exertion of force against a thing. **2** this force. **3** compelling or oppressive influence. ● *v.* pressurize (a person). □ **pressure cooker** pan for cooking things quickly by steam under pressure. **pressure group** group seeking to exert influence by concerted action.

■ *n.* **1,2** compression, force, power, strength, tension. **3** coercion, compulsion, constraint, force, inducement, influence, persuasion. ● *v.* coerce, compel, constrain, dragoon, force, press, pressurize.

pressurize *v.* **1** try to compel into an action. **2** maintain a constant atmospheric pressure in (a compartment).

prestige *n.* respect resulting from good reputation or achievements.

■ cachet, distinction, eminence, esteem, kudos, reputation, respect, standing, stature, status.

prestigious *adj.* having or giving prestige.

■ acclaimed, celebrated, distinguished, eminent, esteemed, estimable, illustrious, influential, notable, noteworthy, pre-eminent, renowned, reputable, respected.

presumably *adv.* it may be presumed.

presume *v.* **1** suppose to be true. **2** be presumptuous. □ **presumption** *n.*

■ **1** assume, imagine, infer, presuppose, suppose, surmise, take for granted. **2** dare, have the effrontery, venture.

presumptuous *adj.* unduly confident, arrogant. □ **presumptuously** *adv.*

■ arrogant, audacious, bold, brazen, cheeky, forward, impertinent, impudent, insolent, over-confident, presuming.

presuppose *v.* **1** assume beforehand. **2** assume the prior existence of. □ **presupposition** *n.*

pretence *n.* **1** pretending. **2** false show of intentions or motives, pretext. **3** claim (e.g. to merit or knowledge).

■ **1** fabrication, fiction, invention, make-believe, pretending. **2** blind, cloak, cover, façade, front, mask, semblance, veneer; affectation, masquerade, pose, show; excuse, pretext.

pretend *v.* **1** imagine to oneself in play. **2** claim or assert falsely. □ **pretender** *n.*

■ **1** fantasize, make believe. **2** affect, counterfeit, fake, feign, simulate.

pretension *n.* **1** asserting of a claim. **2** pretentiousness.

pretentious *adj.* **1** claiming great merit or importance. **2** ostentatious. □ **pretentiously** *adv.*, **pretentiousness** *n.*

■ **1** haughty, hoity-toity, lofty, pompous, self-important, snobbish, *colloq.* stuck-up, *colloq.* uppity. **2** grandiose, ostentatious, showy; bombastic, flowery, grandiloquent, turgid.

preternatural *adj.* beyond what is natural. □ **preternaturally** *adv.*

pretext *n.* reason put forward to conceal one's true reason.

■ blind, excuse, pretence.

pretty *adj.* (**-ier**, **-iest**) attractive in a delicate way. ● *adv.* fairly, moderately. □ **prettily** *adv.*, **prettiness** *n.*

■ *adj.* appealing, attractive, *Sc.* bonny, comely, *colloq.* cute, good-looking, lovely.

pretzel *n.* salted biscuit.

prevail *v.* **1** be victorious. **2** be the more usual or predominant. □ **prevail on** persuade.

■ **1** gain victory, succeed, triumph, win. **2** be prevalent *or* widespread, dominate, predominate, preponderate, reign. □ **prevail on** convince, get, induce, persuade, sway, talk into, win over.

prevalent *adj.* existing generally, widespread. □ **prevalence** *n.*

■ common, current, general, pervasive, popular, predominant, prevailing, widespread.

prevaricate *v.* speak evasively. □ **prevarication** *n.*

■ be evasive, equivocate, evade the issue, hedge, quibble.

prevent *v.* stop or hinder, make impossible. □ **prevention** *n.*, **preventable** *adj.*

■ baulk, block, curb, foil, forestall, frustrate, halt, hamper, hinder, impede, inhibit, obstruct, preclude, prohibit, stop, thwart.

preventative *adj.* & *n.* preventive.

preventive *adj.* & *n.* (thing) preventing something.

previous *adj.* coming before in time or order. □ **previously** *adv.*

■ earlier, erstwhile, former, past, prior, sometime; aforesaid, antecedent, foregoing, preceding.

prey *n.* **1** animal hunted or killed by another for food. **2** victim. ● *v.* **prey on 1** seek or take as prey. **2** cause worry to. □ **bird of prey** one that kills and eats animals.

■ *n.* **1** quarry. **2** quarry, target, victim. ● *v.* **1** eat, feed on, live off. **2** burden, haunt, oppress, torment, trouble, weigh on, worry.

price *n.* **1** amount of money for which a thing is bought or sold. **2** what must be given or done etc. to achieve something. ● *v.* fix, find, or estimate the price of.

■ *n.* **1** amount, charge, cost, expense, fee, outlay; value, worth. **2** cost, sacrifice. ● *v.* cost, evaluate, rate, value.

priceless *adj.* **1** invaluable. **2** (*colloq.*) very amusing or absurd.

■ **1** beyond price, costly, expensive, invaluable, precious, valuable.

prick *v.* **1** pierce slightly. **2** feel a pricking sensation. ● *n.* **1** act of pricking. **2** sensation of being pricked. □ **prick up one's ears 1** (of a dog etc.) erect the ears. **2** listen alertly.

■ *v.* **1** penetrate, perforate, pierce, puncture. **2** prickle, smart, sting, tingle.

prickle *n.* **1** small thorn or spine. **2** pricking sensation. ● *v.* feel or cause a pricking sensation.

■ *n.* **1** barb, bristle, needle, spike, spine, thorn. **2** itchiness, prick, sting, tingle, tingling. ● *v.* itch, smart, sting, tingle; prick.

prickly *adj.* (**-ier**, **-iest**) **1** having prickles. **2** irritable, touchy.

■ **1** barbed, bristly, spiky, spiny, thistly, thorny. **2** cantankerous, fractious, irascible, irritable, peevish, pettish, petulant, short-tempered, snappy, testy, touchy, waspish.

pride *n.* **1** feeling of pleasure or satisfaction about one's actions, qualities, or possessions. **2** source of this. **3** sense of dignity. **4** unduly high opinion of oneself. **5** group (of lions). ● *v.* **pride oneself on** be proud of. □ **pride of place** most prominent position. **take pride in** be proud of.

■ *n.* **2** boast, darling, delight, gem, jewel, joy, treasure. **3** dignity, self-esteem, self-respect. **4** arrogance, conceit, egotism, self-admiration, self-love, self-importance, self-satisfaction, smugness, vainglory, vanity. ● *v.* be proud of, glory in, preen oneself on, revel in, take pride in.

priest *n.* **1** member of the clergy. **2** official of a non-Christian religion. ▫ **priesthood** *n.*, **priestly** *adj.*

■ **1** clergyman, clergywoman, cleric, divine, ecclesiastic, minister, padre, reverend, vicar.

priestess *n.* female priest of a non-Christian religion.

prig *n.* self-righteous person. ▫ **priggish** *adj.*, **priggishness** *n.*

■ ▫ **priggish** goody-goody, prim, prissy, prudish, puritanical, self-righteous, strait-laced.

prim *adj.* (**primmer**, **primmest**) **1** formal and precise. **2** prudish. ▫ **primly** *adv.*, **primness** *n.*

■ **1** formal, precise, proper, punctilious, starchy, stiff. **2** priggish, prissy, prudish, puritanical, strait-laced, *colloq.* stuffy.

prima *adj.* **prima ballerina** chief ballerina. **prima donna** chief female singer in opera.

primal *adj.* **1** primitive, primeval. **2** fundamental.

primary *adj.* first in time, order, or importance. ● *n.* primary thing. ▫ **primary colour** one not made by mixing others, i.e. (for light) red, green, violet, or (for paint) red, blue, yellow. **primary school** one for children below the age of 11. **primarily** *adv.*

■ *adj.* earliest, initial, original; basic, cardinal, central, chief, first, foremost, fundamental, leading, main, major, paramount, pre-eminent, prime, principal.

primate *n.* **1** archbishop. **2** member of the highly developed order of animals that includes humans, apes, and monkeys.

prime[1] *adj.* **1** chief. **2** first-rate. **3** fundamental. ● *n.* state of greatest perfection. ▫ **prime minister** leader of a government. **prime number** number that can be divided exactly only by itself and one.

■ *adj.* **1** chief, foremost, leading, main, major, paramount, primary, principal. **2** choice, excellent, exceptional, finest, first-class, first-rate, outstanding, select, superior, *colloq.* tiptop. **3** basic, essential, fundamental, original.

prime[2] *v.* **1** prepare for use or action. **2** provide with information in preparation for something.

■ **1** get ready, prepare. **2** educate, instruct, teach, train, tutor; apprise, brief, inform.

primer *n.* **1** substance used to prime a surface for painting. **2** elementary textbook.

primeval *adj.* of the earliest times of the world.

primitive *adj.* **1** at an early stage of civilization. **2** simple, crude.

■ **1** ancient, antediluvian, early, prehistoric, primal, primeval, primordial. **2** basic, crude, rough, rude, rudimentary, simple.

primogeniture *n.* system by which an eldest son inherits all his parents' property.

primordial *adj.* primeval.

primrose *n.* **1** pale yellow spring flower. **2** its colour.

primula *n.* perennial plant of a kind that includes the primrose.

prince *n.* **1** male member of a royal family. **2** sovereign's son or grandson. **3** the greatest or best.

princely *adj.* **1** like a prince. **2** splendid, generous.

princess *n.* **1** female member of a royal family. **2** sovereign's daughter or granddaughter. **3** prince's wife.

principal *adj.* first in rank or importance. ● *n.* **1** head of certain schools or colleges. **2** person with highest authority or playing the leading part. **3** capital sum as distinct from interest or income.

■ *adj.* chief, first, foremost, key, leading, main, major, paramount, pre-eminent, primary, prime. ● *n.* **1** head, headmaster, headmistress. **2** *colloq.* boss, chief, director, head, manager, manageress, president, ruler; lead, star. **3** capital, funds, resources.

principality *n.* country ruled by a prince.

principally *adv.* mainly.

■ chiefly, especially, for the most part, largely, mainly, mostly, on the whole, particularly, predominantly, primarily.

principle *n.* **1** general truth used as a basis of reasoning or action. **2** (often *pl.*) personal code of conduct. **3** scientific law shown or used in the working of something. ▫ **in principle** as regards the main elements. **on principle** because of one's moral beliefs.

■ **1** axiom, canon, doctrine, fundamental, law, precept, rule, tenet, truth. **2** conscience, honesty, honour, integrity, morality, probity; **(principles)** beliefs, ethics, morals, philosophy, values.

principled *adj.* based on or having (esp. praiseworthy) principles of behaviour.

■ ethical, honest, honourable, just, moral, noble, righteous, upright, virtuous.

print *v.* **1** press (a mark) on a surface, impress (a surface etc.) in this way. **2** produce by applying inked type to paper. **3** write with unjoined letters. **4** produce a positive picture from (a photographic negative). ● *n.* **1** mark left by pressing. **2** printed lettering or words. **3** printed design, picture, or fabric. □ **printed circuit** electric circuit with lines of conducting material printed on a flat sheet.

■ *v.* **1** impress, imprint, stamp. ● *n.* **1** fingerprint, footprint, impression, imprint, indentation, mark. **2** lettering, text, type. **3** design, pattern; illustration, lithograph, picture.

printer *n.* person who prints books or newspapers etc.

printout *n.* computer output in printed form.

prior[1] *adj.* coming before in time, order, or importance.

■ earlier, foregoing, former, previous.

prior[2] *n.* monk who is head of a religious community, or one ranking next below an abbot.

prioress *n.* female prior.

prioritize *v.* treat as a priority.

priority *n.* **1** right to be first. **2** thing that should be treated as most important.

■ **1** precedence, preference; importance, pre-eminence, seniority, superiority.

priory *n.* monastery or nunnery governed by a prior or prioress.

prise *v.* force out or open by leverage.

prism *n.* **1** solid geometric shape with ends that are equal and parallel. **2** transparent object of this shape with refracting surfaces.

prismatic *adj.* **1** of or like a prism. **2** (of colours) rainbow-like.

prison *n.* place of captivity, esp. a building to which people are consigned while awaiting trial or for punishment.

■ gaol, *sl.* glasshouse, jail, lock-up, reformatory.

prisoner *n.* **1** person kept in prison. **2** (in full **prisoner of war**) person captured in war. **3** person in confinement.

■ **1** convict, inmate, internee. **2** captive, hostage.

prissy *adj.* **(-ier, -iest)** prim. □ **prissily** *adv.*, **prissiness** *n.*

pristine *adj.* in its original and unspoilt condition.

privacy *n.* being private.

■ isolation, seclusion, solitude.

private *adj.* **1** belonging to a person or group, not public. **2** confidential. **3** secluded. **4** not provided by the state. ● *n.* soldier of the lowest rank. □ **in private** privately. **privately** *adv.*

■ *adj.* **1** individual, particular, personal. **2** clandestine, confidential, hush-hush, intimate, off the record, personal, secret. **3** isolated, quiet, remote, secluded, sequestered.

privation *n.* lack of comforts or necessities.

■ deprivation, destitution, hardship, necessity, need, penury, poverty, want.

privatize *v.* transfer from state to private ownership. □ **privatization** *n.*

privet *n.* bushy evergreen shrub used for hedges.

privilege *n.* **1** right, advantage, or immunity belonging to a person etc. **2** special benefit or honour. □ **privileged** *adj.*

■ **1** advantage, prerogative, right; dispensation, exemption, freedom, immunity. **2** honour, pleasure.

prize[1] *n.* **1** award for victory or superiority. **2** thing that can be won. ● *adj.* **1** winning a prize. **2** excellent. ● *v.* value highly.

■ *n.* **1** award, cup, medal, trophy; reward. ● *adj.* **2** choice, excellent, first-class, first-rate, select. ● *v.* cherish, hold dear, set store by, treasure, value.

prize[2] *v.* = **prise**.

pro[1] *n.* (*pl.* **-os**) (*colloq.*) professional.

pro[2] *n.* **pros and cons** arguments for and against something.

pro- *pref.* in favour of.

probable *adj.* likely to happen or be true. □ **probably** *adv.*, **probability** *n.*

■ believable, conceivable, credible, feasible, likely, plausible, possible, tenable.

probate *n.* **1** official process of proving that a will is valid. **2** certified copy of a will.

probation *n.* **1** testing of behaviour or abilities. **2** system whereby certain offenders are supervised by an official (**probation officer**) instead of being imprisoned. ▫ **probationary** *adj.*

probationer *n.* person undergoing a probationary period.

probe *n.* **1** blunt surgical instrument for exploring a wound. **2** unmanned exploratory spacecraft. **3** investigation. ● *v.* **1** explore with a probe. **2** investigate.

▪ *n.* **3** examination, exploration, inquiry, investigation, scrutiny, study. ● *v.* **2** examine, explore, inquire into, investigate, look into, research, scrutinize, study.

probity *n.* honesty.

problem *n.* **1** something difficult to deal with or understand. **2** thing to be solved or dealt with.

▪ **1** complication, difficulty, headache, trouble. **2** conundrum, enigma, riddle, poser, puzzle.

problematic *adj.* (also **problematical**) **1** difficult. **2** questionable.

▪ **1** awkward, complicated, delicate, difficult, knotty, ticklish, tricky. **2** debatable, disputable, doubtful, questionable.

proboscis *n.* **1** long flexible snout. **2** insect's elongated mouthpart used for sucking things.

procedure *n.* **1** way of conducting business or performing a task. **2** series of actions. ▫ **procedural** *adj.*

▪ approach, course of action, method, plan of action, policy, process, strategy, system; *colloq.* drill, practice, routine.

proceed *v.* **1** go forward or onward. **2** continue. **3** start a lawsuit. **4** originate.

▪ **1** advance, go, make one's way, progress. **2** carry on, continue, go on. **4** arise, come, derive, develop, issue, originate, result, spring, start, stem.

proceedings *n.pl.* **1** what takes place. **2** published report of a conference etc. **3** lawsuit.

proceeds *n.pl.* profit from a sale or performance etc.

▪ gate, income, profit(s), return(s), take, takings.

process *n.* **1** series of operations used in making something. **2** procedure. **3** series of changes or events. ● *v.* subject to a process.

▪ *n.* **2** approach, course of action, method, procedure, system. **3** course, progress.

procession *n.* number of people, vehicles, or boats etc. going along in an orderly line.

▪ cavalcade, cortège, line, march, motorcade, parade.

processor *n.* machine that processes things.

proclaim *v.* announce publicly. ▫ **proclamation** *n.*

▪ advertise, announce, blazon, broadcast, declare, make known, promulgate, pronounce, trumpet. ▫ **proclamation** announcement, declaration, promulgation, pronouncement.

proclivity *n.* tendency.

procrastinate *v.* postpone action. ▫ **procrastination** *n.*

▪ be dilatory, delay, *colloq.* dilly-dally, play for time, stall, temporize.

procreate *v.* produce (offspring). ▫ **procreation** *n.*

procure *v.* **1** obtain by care or effort, acquire. **2** act as procurer. ▫ **procurement** *n.*

▪ **1** acquire, come by, get, land, obtain, pick up, secure; buy, purchase.

procurer *n.* person who obtains women for prostitution.

prod *v.* (**prodded**) **1** poke. **2** stimulate to action. ● *n.* **1** prodding action. **2** stimulus. **3** instrument for prodding things.

▪ *v.* **1** dig, elbow, jab, nudge, poke. **2** move, motivate, prompt, provoke, rouse, spur, stimulate, stir, urge. ● *n.* **1** dig, jab, nudge, poke. **2** goad, spur, stimulus.

prodigal *adj.* wasteful, extravagant. ▫ **prodigally** *adv.*, **prodigality** *n.*

▪ extravagant, improvident, profligate, spendthrift, wasteful.

prodigious *adj.* **1** amazing. **2** enormous. ▫ **prodigiously** *adv.*

▪ **1** amazing, astonishing, astounding, extraordinary, incredible, marvellous, phenomenal, remarkable, sensational, staggering, startling, wonderful. **2** colossal, enormous, huge, large, massive, tremendous, vast.

prodigy *n.* **1** person with exceptional qualities or abilities. **2** wonderful thing.

▪ **1** genius, wizard, *colloq.* whiz-kid. **2** marvel, miracle, phenomenon, sensation, wonder.

produce *v.* /prədyo͞oss/ **1** bring forward for inspection. **2** bring (a performance etc.) before the public. **3** bring into existence, cause. **4** manufacture. ● *n.* /pródyooss/ amount or thing(s) produced. ▫ **producer** *n.*, **production** *n.*

■ *v.* **1** bring forward, offer, present, show. **2** present, put on, stage. **3** bring about, cause, generate, give rise to, initiate, occasion, result in. **4** fabricate, make, manufacture. ● *n.* commodities, goods, products, staples.

product *n.* **1** thing produced. **2** number obtained by multiplying.

■ **1** artefact, commodity; **(products)** goods, merchandise, produce, wares.

productive *adj.* **1** producing things, esp. in large quantities. **2** useful.

■ **1** fecund, fertile, fruitful, prolific, rich. **2** constructive, helpful, useful, valuable, worthwhile.

productivity *n.* efficiency in industrial production.

profane *adj.* **1** secular. **2** irreverent, blasphemous. ● *v.* treat irreverently. ▫ **profanely** *adv.*, **profanity** *n.*, **profanation** *n.*

■ *adj.* **1** lay, non-religious, secular. **2** blasphemous, disrespectful, impious, irreverent, sacrilegious.

profess *v.* **1** claim (a quality etc.). **2** declare. **3** affirm faith in (a religion).

■ **1** claim, pretend. **2** affirm, assert, declare, maintain, state.

professed *adj.* **1** alleged. **2** self-acknowledged. ▫ **professedly** *adv.*

profession *n.* **1** occupation requiring advanced learning. **2** people engaged in this. **3** declaration.

■ **1** calling, employment, field, line (of work), occupation, trade, vocation. **3** affirmation, announcement, assertion, avowal, declaration, statement.

professional *adj.* **1** belonging to a profession. **2** skilful, worthy of a professional. **3** doing something for payment, not as a pastime. ● *n.* professional worker or player. ▫ **professionalism** *n.*, **professionally** *adv.*

■ *adj.* **2** experienced, expert, masterly, practised, proficient, skilful, skilled, trained; businesslike, efficient, thorough. ● *n.* adept, expert, master, *colloq.* pro, specialist.

professor *n.* university teacher of the highest rank. ▫ **professorial** *adj.*

proffer *v.* & *n.* offer.

proficient *adj.* expert, skilled. ▫ **proficiently** *adv.*, **proficiency** *n.*

■ able, accomplished, adept, capable, competent, experienced, expert, practised, professional, skilful, skilled, trained.

profile *n.* **1** side view, esp. of the face. **2** short account of a person's character or career.

profit *n.* **1** advantage, benefit. **2** money gained. ● *v.* **(profited)** be beneficial to. ▫ **profit from** obtain an advantage or benefit.

■ *n.* **1** advantage, avail, benefit, good, interest, use, usefulness, value. **2** gain, proceeds, return. ● *v.* aid, avail, be advantageous *or* beneficial to, benefit, help, serve. ▫ **profit from** capitalize on, cash in on, exploit, make use of, take advantage of, utilize.

profitable *adj.* bringing profit. ▫ **profitably** *adv.*, **profitability** *n.*

■ gainful, lucrative, remunerative, rewarding, well-paid; advantageous, beneficial, helpful, productive, useful, valuable, worthwhile.

profiteer *v.* make or seek excessive profits. ● *n.* person who makes excessive profits.

profiterole *n.* small hollow cake of choux pastry with filling.

profligate *adj.* **1** wasteful, extravagant. **2** dissolute. ● *n.* profligate person. ▫ **profligacy** *n.*

■ *adj.* **1** extravagant, improvident, prodigal, spendthrift, wasteful. **2** corrupt, debauched, degenerate, depraved, dissipated, dissolute, immoral, unprincipled, wanton.

profound *adj.* **1** intense. **2** showing or needing great insight. ▫ **profoundly** *adv.*, **profundity** *n.*

■ **1** acute, deep, extreme, heartfelt, intense, keen, overpowering, overwhelming. **2** erudite, intellectual, learned, sagacious, scholarly; abstruse, arcane, esoteric, recondite.

profuse *adj.* **1** lavish. **2** plentiful. ▫ **profusely** *adv.*, **profusion** *n.*

■ **1** bountiful, generous, lavish, liberal, unsparing, unstinting. **2** abundant, ample, copious, luxuriant, plenteous, plentiful, prolific, rich.

progenitor *n.* ancestor.

progeny *n.* offspring.

progesterone *n.* sex hormone that maintains pregnancy.

prognosis *n.* (*pl.* **-oses**) forecast, esp. of the course of a disease. ▫ **prognostic** *adj.*

prognosticate *v.* forecast. ▫ **prognostication** *n.*

program *n.* **1** (*US*) = **programme**. **2** series of coded instructions for a computer. ● *v.* (**programmed**) instruct (a computer) by means of this. ▫ **programmer** *n.*

programme *n.* **1** plan of action. **2** list of items in an entertainment. **3** broadcast performance.

■ **1** agenda, plan of action, schedule. **3** broadcast, production, show.

progress *n.* /prógress/ **1** forward movement. **2** development. ● *v.* /prəgréss/ **1** make progress. **2** develop. ▫ **in progress** taking place. **progression** *n.*

■ *n.* **1** advancement, headway, progression. **2** advance, development, evolution, expansion, growth, improvement. ● *v.* **1** advance, forge ahead, make headway, make one's way, proceed. **2** advance, develop, evolve, expand, get better, grow, improve.

progressive *adj.* **1** favouring progress or reform. **2** (of a disease) gradually increasing in its effect. ▫ **progressively** *adv.*

prohibit *v.* **1** forbid. **2** prevent. ▫ **prohibition** *n.*

■ **1** ban, bar, debar, disallow, forbid, outlaw, proscribe, veto. **2** block, hamper, hinder, impede, inhibit, obstruct, preclude, prevent, rule out, stop.

prohibitive *adj.* **1** prohibiting. **2** (of prices etc.) extremely high.

project *v.* /prəjékt/ **1** extend outwards. **2** cast, throw. **3** estimate. **4** plan. ● *n.* /prójekt/ **1** plan, undertaking. **2** task involving research.

■ *v.* **1** beetle (out), bulge (out), jut (out), overhang, protrude, stick out. **2** cast, *colloq.* chuck, fling, hurl, launch, lob, propel, throw, toss. **3** calculate, estimate, forecast, predict. **4** contemplate, plan, propose. ● *v.* **1** idea, plan, proposal, scheme; assignment, enterprise, undertaking, venture.

projectile *n.* missile.

projection *n.* **1** process of projecting something. **2** thing projecting from a surface. **3** estimate of future situations based on a study of present ones.

■ **2** bulge, ledge, outcrop, overhang, protrusion, protuberance, spur. **3** estimate, forecast, prediction.

projectionist *n.* person who operates a projector.

projector *n.* apparatus for projecting images on to a screen.

proletariat *n.* working-class people. ▫ **proletarian** *adj.* & *n.*

proliferate *v.* reproduce rapidly, multiply. ▫ **proliferation** *n.*

■ burgeon, grow, increase, multiply, mushroom, snowball.

prolific *adj.* **1** abundantly productive. **2** copious.

■ **1** fecund, fertile, fruitful, productive. **2** abundant, copious, plenteous, plentiful, profuse.

prologue *n.* introduction to a poem or play etc.

prolong *v.* lengthen in extent or duration. ▫ **prolongation** *n.*

■ draw out, extend, lengthen, protract, spin out, stretch out.

prolonged *adj.* continuing for a long time.

prom *n.* (*colloq.*) **1** promenade. **2** promenade concert.

promenade *n.* paved public walk (esp. along a sea front). ▫ **promenade concert** one where part of the audience is not seated and can move about.

prominent *adj.* **1** projecting. **2** conspicuous. **3** well-known. ▫ **prominence** *n.*

■ **1** jutting, projecting, protruding, protuberant. **2** conspicuous, discernible, evident, noticeable, obvious, pronounced, recognizable, striking. **3** acclaimed, celebrated, distinguished, eminent, famed, famous, illustrious, notable, noted, renowned, well-known.

promiscuous *adj.* **1** having sexual relations with many people. **2** indiscriminate. ▫ **promiscuously** *adv.*, **promiscuity** *n.*

promise *n.* **1** declaration that one will give or do a certain thing. **2** indication of future results. ● *v.* **1** make (a person) a promise, esp. to do or give (a thing). **2** seem likely, produce expectation of.

■ *n.* **1** assurance, bond, guarantee, oath, pledge, undertaking, vow, word (of honour). **2** ability, aptitude, capability, potential. ● *v.* **1** give one's word, guarantee, pledge, swear, undertake, vow. **2** augur, bespeak, betoken, foretell, indicate, presage.

promising *adj.* likely to turn out well or produce good results.

■ **auspicious, bright, encouraging, favourable, hopeful, optimistic, propitious, rosy.**

promontory *n.* high land jutting out into the sea or a lake.

promote *v.* **1** raise to a higher rank or office. **2** help the progress of. **3** publicize in order to sell. □ **promotion** *n.*, **promotional** *adj.*, **promoter** *n.*

■ **1 elevate, raise, upgrade. 2 advance, aid, assist, boost, encourage, forward, foster, further, help, support. 3 advertise,** *colloq.* **plug, publicize.**

prompt *adj.* **1** done without delay. **2** ready, willing. ● *adv.* punctually. ● *v.* **1** incite. **2** supply (an actor or speaker) with forgotten words or with a suggestion. □ **promptly** *adv.*, **promptness** *n.*

■ *adj.* **1 fast, immediate, instantaneous, punctual, quick, rapid, speedy, swift, unhesitating. 2 eager, keen, quick, ready, willing.** ● *v.* **1 egg on, encourage, incite, induce, make, motivate, move, prod, provoke, spur, urge.**

prompter *n.* person stationed off stage to prompt actors.

promulgate *v.* make known to the public. □ **promulgation** *n.*

prone *adj.* **1** lying face downwards. **2** likely to do or suffer something.

■ **1 prostrate, reclining, recumbent. 2 apt, disposed, given, inclined, liable, predisposed; subject, susceptible.**

prong *n.* each of the pointed parts of a fork. □ **pronged** *adj.*

pronoun *n.* word used as a substitute for a noun (e.g. *I, me, who, which*). □ **pronominal** *adj.*

pronounce *v.* **1** utter distinctly or in a certain way. **2** declare. □ **pronunciation** *n.*

■ **1 articulate, enunciate, say, utter, vocalize, voice. 2 announce, assert, declare, proclaim.**

pronounced *adj.* noticeable.

■ **clear, conspicuous, decided, definite, distinct, marked, noticeable, plain, prominent, recognizable, striking, unmistakable.**

pronouncement *n.* declaration.

proof *n.* **1** evidence that something is true or exists. **2** demonstration or act of proving. **3** copy of printed matter for correction. ● *adj.* able to resist penetration or damage. ● *v.* make (fabric) proof against something (e.g. water).

■ *n.* **1 data, documentation, evidence, facts. 2 authentication, confirmation, corroboration, substantiation, validation, verification.**

proofread *v.* read and correct (printed proofs). □ **proofreader** *n.*

prop[1] *n.* support to prevent something from falling, sagging, or failing. ● *v.* (**propped**) support (as) with a prop. □ **prop up** prevent from falling.

■ *n.* **brace, buttress, mainstay, pier, post, support, upright.** □ **prop up bolster, brace, buttress, hold up, shore up, support.**

prop[2] *n.* (*colloq.*) a stage property.

prop[3] *n.* (*colloq.*) propeller.

propaganda *n.* publicity intended to persuade or convince people.

propagate *v.* **1** breed or reproduce from parent stock. **2** spread (news etc.). **3** transmit. □ **propagation** *n.*

■ **1 breed, multiply, procreate, reproduce. 2 broadcast, circulate, disseminate, make known, promulgate, publicize, spread.**

propane *n.* hydrocarbon fuel gas.

propel *v.* (**propelled**) push forward or onward.

■ **drive, impel, move, push** *or* **thrust forward.**

propellant *n.* thing that propels something. □ **propellent** *adj.*

propeller *n.* revolving device with blades, for propelling a ship or aircraft.

propensity *n.* **1** tendency. **2** inclination.

proper *adj.* **1** suitable. **2** correct. **3** conforming to social conventions. **4** (*colloq.*) thorough. □ **proper name** or **noun** name of an individual person or thing.

■ **1 apposite, apt, appropriate, fit, fitting, right, suitable. 2 accepted, accurate, correct, established, orthodox, precise, right, true. 3 correct, decent, decorous, genteel, gentlemanly, ladylike, respectable, seemly. 4 absolute, complete, perfect, out and out, thorough, utter.**

property *n.* **1** thing(s) owned. **2** real estate, land. **3** movable object used in a play etc. **4** quality, characteristic.

■ **1 assets, belongings, effects, possessions, things. 4 attribute, characteristic, feature, hallmark, quality, trait.**

prophecy *n.* **1** power of prophesying. **2** statement prophesying something.

> ■ **1** augury, divination, fortune-telling, second sight, soothsaying. **2** forecast, prediction, prognostication.

prophesy *v.* foretell.

> ■ augur, forecast, foresee, foretell, predict, presage, prognosticate.

prophet, **prophetess** *ns.* **1** person who predicts. **2** teacher or interpreter of divine will.

> ■ **1** clairvoyant, fortune-teller, oracle, seer, sibyl, soothsayer.

prophetic *adj.* prophesying.

prophylactic *adj.* & *n.* (medicine or action etc.) preventing disease or misfortune. □ **prophylaxis** *n.*

propinquity *n.* nearness.

propitiate /prəpíshiayt/ *v.* appease. □ **propitiation** *n.*, **propitiatory** *adj.*

propitious /prəpíshəss/ *adj.* auspicious, favourable.

> ■ advantageous, auspicious, bright, favourable, fortunate, happy, lucky, promising, providential, rosy.

proponent *n.* person putting forward a proposal.

proportion *n.* **1** fraction or share of a whole. **2** ratio. **3** correct or pleasing relation between things or parts of a thing. **4** (*pl.*) dimensions. □ **proportional** *adj.*, **proportionally** *adv.*

> ■ **1** *colloq.* cut, division, part, percentage, portion, quota, ration, share. **2** ratio, relationship. **3** arrangement, balance, harmony, symmetry. **4** (**proportions**) dimensions, extent, magnitude, measurements, size.

proportionate *adj.* in due proportion. □ **proportionately** *adv.*

proposal *n.* **1** proposing of something. **2** thing proposed. **3** offer of marriage.

> ■ **1** bid, motion, offer, overture, proposition, recommendation, suggestion, tender.

propose *v.* **1** put forward for consideration. **2** intend. **3** nominate. **4** make a proposal of marriage. □ **proposer** *n.*

> ■ **1** advance, propound, put forward, recommend, submit, suggest, table. **2** aim, have in mind, intend, mean, plan, purpose. **3** nominate, put forward, recommend, suggest.

proposition *n.* **1** statement. **2** proposal, scheme proposed. **3** (*colloq.*) undertaking. ● *v.* (*colloq.*) put a proposal to.

propound *v.* put forward for consideration.

proprietary *adj.* **1** made and sold by a particular firm, usu. under a patent. **2** of an owner or ownership.

proprietor *n.* owner of a business. □ **proprietorial** *adj.*

propriety *n.* correctness of behaviour.

> ■ correctness, courtesy, decency, decorum, gentility, politeness, refinement, respectability, seemliness.

propulsion *n.* process of propelling or being propelled.

pro rata proportional(ly).

prosaic *adj.* ordinary, unimaginative. □ **prosaically** *adv.*

> ■ banal, boring, commonplace, dry, dull, flat, humdrum, lifeless, monotonous, mundane, ordinary, pedestrian, run-of-the-mill, tedious, unimaginative, uninspired, uninspiring.

proscribe *v.* forbid by law.

prose *n.* written or spoken language not in verse form.

prosecute *v.* **1** take legal proceedings against (a person) for a crime. **2** carry on, conduct. □ **prosecution** *n.*, **prosecutor** *n.*

> ■ **1** arraign, bring to trial, charge, indict, sue. **2** carry on, conduct, perform, practise, pursue.

proselyte *n.* recent convert to a religion.

proselytize *v.* seek to convert.

prospect *n.* /próspekt/ **1** (often *pl.*) expectation, esp. of success in a career etc. **2** view. ● *v.* /prəspékt/ explore in search of something. □ **prospector** *n.*

> ■ *n.* **1** chance, expectation, hope, likelihood, possibility, probability. **2** panorama, scene, sight, view, vista.

prospective *adj.* expected to be or to occur, future or possible.

prospectus *n.* document giving details of a school, business, etc.

prosper *v.* be successful, thrive.

> ■ do well, flourish, grow, make good, succeed, thrive.

prosperous *adj.* financially successful. □ **prosperity** *n.*

> ■ affluent, moneyed, rich, wealthy, *colloq.* well-heeled, well-off, well-to-do.

prostate gland gland round the neck of the bladder in males.

prosthesis *n.* (*pl.* **-theses**) artificial limb or similar appliance. □ **prosthetic** *adj.*

prostitute *n.* woman who engages in sexual intercourse for payment. ● *v.* **1** make a prostitute of. **2** put (talent etc.) to an unworthy use. ▫ **prostitution** *n.*

prostrate *adj.* /próstrayt/ **1** face downwards. **2** lying horizontally. **3** exhausted, overcome. ● *v.* /prostráyt/ cause to be prostrate. ▫ **prostration** *n.*

■ *adj.* **2** prone, recumbent, stretched out. **3** drained, exhausted, fatigued, tired, wearied, worn-out; crushed, helpless, impotent, overcome, overpowered, overwhelmed, powerless. ● *v.* fell, floor, knock down; exhaust, fatigue, *sl.* knacker, tire (out), wear out, weary; crush, overcome, overpower, overwhelm.

protagonist *n.* **1** chief person in a drama, story, etc. **2** supporter.

protean *adj.* **1** variable. **2** versatile.

protect *v.* keep from harm or injury. ▫ **protector** *n.*

■ conserve, defend, guard, keep (safe), preserve, safeguard, shelter, shield.

protection *n.* **1** protecting, being protected. **2** defence.

■ **1** care, charge, custody, guardianship, safe-keeping. **2** defence, safeguard, screen, shelter, shield; safety, security.

protectionism *n.* policy of protecting home industries by tariffs etc. ▫ **protectionist** *n.*

protective *adj.* protecting, giving protection. ▫ **protectively** *adv.*

protectorate *n.* country that is under the official protection and partial control of a stronger one.

protégé /próttizhay/ *n.* (*fem.* **protégée**) person under the protection or patronage of another.

protein *n.* organic compound forming an essential part of humans' and animals' food.

protest *n.* /prótest/ statement or action indicating disapproval. ● *v.* /prətést/ **1** express disapproval. **2** declare firmly.

■ *n.* complaint, demur, expostulation, *colloq.* gripe, *colloq.* grouse, grumble, objection, outcry, protestation, remonstrance. ● *v.* **1** complain, demur, expostulate, *colloq.* gripe, *colloq.* grouse, grumble, *colloq.* kick up, make a fuss, object, remonstrate. **2** affirm, assert, declare, maintain, profess.

Protestant *n.* member of one of the western Churches that are separated from the RC Church. ▫ **Protestantism** *n.*

protestation *n.* firm declaration.

protocol *n.* **1** etiquette applying to rank or status. **2** draft of a treaty.

proton *n.* particle of matter with a positive electric charge.

protoplasm *n.* contents of a living cell.

prototype *n.* original example from which others are developed. ▫ **prototypical** *adj.*

protozoon /prṓtəzṓ-on/ *n.* (*pl.* **-zoa**) one-celled microscopic animal. ▫ **protozoan** *adj.* & *n.*

protract *v.* prolong in duration. ▫ **protraction** *n.*

protractor *n.* instrument for measuring angles.

protrude *v.* project, stick out. ▫ **protrusion** *n.*, **protrusive** *adj.*

■ bulge, extend, jut (out), project, stick out. ▫ **protrusion** bulge, bump, excrescence, knob, lump, outgrowth, projection, protuberance, swelling, tumescence.

protuberant *adj.* bulging out. ▫ **protuberance** *n.*

■ bulbous, bulging, jutting, projecting, prominent, protruding, protrusive.

proud *adj.* **1** feeling greatly honoured or pleased. **2** haughty, arrogant. **3** imposing. **4** slightly projecting. ▫ **proudly** *adv.*

■ **1** contented, delighted, elated, gratified, honoured, pleased, satisfied. **2** arrogant, boastful, conceited, haughty, hoity-toity, self-important, self-satisfied, smug, *colloq.* snooty, *colloq.* stuck-up, supercilious, superior, vain, vainglorious. **3** grand, imposing, impressive, magnificent, majestic, splendid, stately.

prove *v.* **1** give or be proof of. **2** be found to be.

■ **1** authenticate, back up, confirm, corroborate, demonstrate, establish, show, substantiate, support, validate, verify.

proven *adj.* proved.

provenance *n.* place of origin.

proverb *n.* short well-known saying.

■ aphorism, maxim, saw, saying.

proverbial *adj.* **1** of or mentioned in a proverb. **2** well-known.

provide *v.* **1** supply, make available. **2** make preparations. ▫ **provide for** supply with the necessities of life. **provider** *n.*

■ **1** equip, furnish, lay on, provision, supply; afford, give, offer, present. **2** cater, get ready, make provision(s), prepare. ▫ **provide for** care for, keep, look after, support, take care of.

provided *conj.* on condition (that).

providence *n.* **1** being provident. **2** God's or nature's protection.

provident *adj.* showing wise forethought for future needs.

■ canny, judicious, prudent, sagacious, sage, shrewd, wise; economical, frugal, thrifty.

providential *adj.* happening very luckily. □ **providentially** *adv.*

■ fortunate, happy, lucky, opportune, timely.

providing *conj.* = **provided.**

province *n.* **1** administrative division of a country. **2** area of learning or responsibility. **3** (*pl.*) all parts of a country outside its capital city.

■ **1** area, county, district, region, state, territory, zone. **2** area, concern, domain, field, *colloq.* pigeon, preserve, responsibility, sphere.

provincial *adj.* **1** of a province or provinces. **2** having limited interests and narrow-minded views. ● *n.* inhabitant of province(s).

■ *adj.* **1** local, regional. **2** insular, limited, narrow-minded, parochial, small-minded, unsophisticated.

provision *n.* **1** process of providing things. **2** stipulation in a treaty or contract etc. **3** (*pl.*) supply of food and drink.

■ **1** providing, supply, supplying. **2** condition, proviso, requirement, stipulation. **3** (**provisions**) comestibles, eatables, food, foodstuffs, stores, supplies, viands.

provisional *adj.* arranged temporarily. □ **provisionally** *adv.*

■ interim, stopgap, temporary.

proviso /prəvízō/ *n.* (*pl.* **-os**) stipulation. □ **provisory** *adj.*

provoke *v.* **1** annoy, irritate. **2** rouse to action. **3** produce as a reaction. □ **provocation** *n.*, **provocative** *adj.*

■ **1** anger, annoy, exasperate, gall, get on one's nerves, incense, infuriate, irk, irritate, madden, needle, nettle, pique, *colloq.* rile, vex. **2** drive, induce, motivate, move, prompt, rouse, spur (on), stimulate, stir. **3** cause, engender, give rise to, lead to, occasion, produce; excite, foment, incite, instigate, kindle.

prow *n.* projecting front part of a ship or boat.

prowess *n.* great ability or daring.

■ ability, adroitness, aptitude, capability, dexterity, expertise, know-how, proficiency, skill, skilfulness; boldness, bravery, courage, daring, fearlessness, gallantry, intrepidity, mettle, pluck, valour.

prowl *v.* go about stealthily. ● *n.* prowling. □ **prowler** *n.*

■ *v.* creep, lurk, skulk, slink, sneak, steal.

proximate *adj.* nearest.

proximity *n.* nearness.

proxy *n.* **1** person authorized to represent or act for another. **2** use of such a person.

prude *n.* person who is extremely correct and proper, one who is easily shocked. □ **prudery** *n.*

prudent *adj.* acting with or showing care and foresight. □ **prudently** *adv.*, **prudence** *n.*

■ careful, canny, cautious, circumspect, sagacious, sensible, shrewd, wise; diplomatic, discreet, tactful.

prudish *adj.* showing prudery. □ **prudishly** *adv.*, **prudishness** *n.*

■ goody-goody, priggish, prim, prissy, puritanical, strait-laced, *colloq.* stuffy.

prune[1] *n.* dried plum.

prune[2] *v.* **1** trim by cutting away dead or unwanted parts. **2** reduce.

prurient *adj.* having or encouraging excessive interest in sexual matters. □ **prurience** *n.*

pry *v.* inquire or peer impertinently (often furtively).

■ be nosy, interfere, intrude, meddle, poke about, *colloq.* snoop.

psalm /saam/ *n.* sacred song.

psalmist /saám-/ *n.* writer of psalms.

psalter /sáwl-/ *n.* copy of the Book of Psalms in the Old Testament.

psephology /sef-/ *n.* study of voting etc. □ **psephologist** *n.*

pseudo- /syōōdō/ *pref.* false.

pseudonym /syōō-/ *n.* fictitious name, esp. used by an author.

psoriasis /səríəsiss/ *n.* skin disease causing scaly red patches.

psyche /síki/ *n.* **1** soul, self. **2** mind.

psychiatry /sī-/ *n.* study and treatment of mental illness. □ **psychiatric** *adj.*, **psychiatrist** *n.*

psychic /síkik/ *adj.* **1** of the soul or mind. **2** of or having apparently supernatural powers. ● *n.* person with such powers.

psychoanalyse /sí-/ *v.* treat by psychoanalysis. □ **psychoanalyst** *n.*

psychoanalysis /sí-/ *n.* method of examining and treating mental conditions by investigating the interaction of conscious and unconscious elements.

psychology /sī-/ *n.* **1** study of the mind. **2** mental characteristics. □ **psychological** *adj.*, **psychologist** *n.*

psychopath /sí-/ *n.* person suffering from a severe mental disorder sometimes resulting in antisocial or violent behaviour. □ **psychopathic** *adj.*

psychosis /sī-/ *n.* (*pl.* **-oses**) severe mental disorder involving a person's whole personality. □ **psychotic** *adj.* & *n.*

psychosomatic /sí-/ *adj.* (of illness) caused or aggravated by mental stress.

psychotherapy /sí-/ *n.* treatment of mental disorders by the use of psychological methods. □ **psychotherapist** *n.*

ptarmigan /taármigən/ *n.* bird of the grouse family.

pterodactyl /térrədáktil/ *n.* extinct reptile with wings.

ptomaine /tṓmayn/ *n.* (toxic) compound found in putrefying matter.

pub *n.* (*colloq.*) public house.

puberty *n.* stage in life at which a person's reproductive organs become able to function.

pubic *adj.* of the abdomen at the lower front part of the pelvis.

public *adj.* **1** of or for people in general. **2** done or existing openly. ● *n.* members of a community in general. □ **public house** place selling alcoholic drinks for consumption on the premises. **public school 1** secondary school for fee-paying pupils. **2** (in Scotland, USA, etc.) school run by public authorities. **public-spirited** *adj.* ready to do things for the community. **publicly** *adv.*

■ *adj.* **1** collective, common, communal, general, national, popular, universal. **2** conspicuous, known, manifest, obvious, open, overt, visible. ● *n.* citizens, community, nation, people, populace, population; *hoi polloi*, masses, proletariat, rank and file.

publican *n.* keeper of a public house.

publication *n.* **1** publishing. **2** published book, newspaper, etc.

■ **2** book, booklet, brochure, journal, leaflet, magazine, newsletter, newspaper, pamphlet, paper, periodical.

publicity *n.* **1** public attention directed upon a person or thing. **2** process of attracting this.

publicize *v.* bring to the attention of the public. □ **publicist** *n.*

■ advertise, plug, promote; air, announce, broadcast, circulate, make public, promulgate.

publish *v.* **1** issue copies of (a book etc.) to the public. **2** make generally known. □ **publisher** *n.*

■ **1** bring out, issue. **2** advertise, announce, broadcast, circulate, give out, make known *or* public, proclaim, promulgate, publicize, release, spread.

puce *adj.* & *n.* brownish-purple.

puck *n.* hard rubber disc used in ice hockey.

pucker *v.* & *n.* wrinkle.

■ *v.* crease, crinkle, furrow, purse, ruck, screw up, wrinkle.

pudding *n.* **1** baked, boiled, or steamed dish made of or enclosed in a mixture of flour and other ingredients. **2** sweet course of a meal. **3** a kind of sausage.

puddle *n.* small pool of rainwater or of liquid on a surface.

pudenda *n.pl.* genitals.

puerile *adj.* childish. □ **puerility** *n.*

■ babyish, childish, immature, infantile, juvenile.

puerperal *adj.* of or resulting from childbirth.

puff *n.* **1** short light blowing of breath, wind, smoke, etc. **2** soft pad for applying powder to the skin. **3** piece of advertising. ● *v.* **1** send (air etc.) or come out in puffs. **2** breathe hard, pant. **3** swell. □ **puffball** *n.* globular fungus. **puff pastry** very light flaky pastry.

■ *n.* **1** blast, breath, gust, waft, whiff; *colloq.* drag, draw, pull. ● *v.* **1** blow (out), breathe (out), exhale. **2** blow, gasp, heave, huff, pant, wheeze. **3** balloon, distend, expand, inflate, swell.

puffin *n.* seabird with a short striped bill.

puffy *adj.* (**-ier**, **-iest**) puffed out, swollen. □ **puffiness** *n.*

pug *n.* dog of a small breed with a flat nose and wrinkled face.

pugilist /pyóojilist/ *n.* professional boxer. □ **pugilism** *n.*

pugnacious *adj.* eager to fight, aggressive. □ **pugnaciously** *adv.*, **pugnacity** *n.*

■ aggressive, antagonistic, argumentative, bellicose, belligerent, combative, contentious, disputatious, hostile, quarrelsome, unfriendly.

puke *v.* & *n.* (*sl.*) vomit.

pull *v.* **1** exert force upon (a thing) so as to move it towards the source of the force. **2** remove by pulling. **3** exert a pulling or driving force. **4** attract. ● *n.* **1** act of pulling. **2** force exerted by this. **3** means of exerting influence. **4** deep drink. **5** draw at a pipe etc. □ **pull in** move towards the side of the road or into a stopping place. **pull off** succeed in doing or achieving. **pull out 1** withdraw. **2** move away from the side of a road or a stopping place. **pull through** come or bring successfully through an illness or difficulty. **pull up** stop.

■ *v.* **1** jerk, pluck, tug, tweak, wrench, *colloq.* yank; drag, draw, haul, heave, lug, tow, trail. **2** extract, remove, take out. **4** attract, bring in, draw. ● *n.* **1** jerk, tug, tweak, wrench, *colloq.* yank. **2** appeal, attractiveness, magnetism, seductiveness. **3** authority, *colloq.* clout, influence, leverage, weight. **4** draught, drink, gulp, swallow. **5** *sl.* drag, draw, puff. □ **pull off** accomplish, bring off, carry out, do, manage, perform, succeed in. **pull through** get better, improve, recover, survive.

pullet *n.* young hen.

pulley *n.* (*pl.* **-eys**) wheel over which a rope etc. passes, used in lifting things.

pullover *n.* knitted garment put on over the head.

pulmonary *adj.* of the lungs.

pulp *n.* fleshy part of fruit etc. ● *v.* reduce to pulp. □ **pulpy** *adj.*

pulpit *n.* raised enclosed platform from which a preacher speaks.

pulsar *n.* cosmic source of pulses of radiation.

pulsate *v.* expand and contract rhythmically. □ **pulsation** *n.*

■ beat, palpitate, pound, pulse, throb.

pulse[1] *n.* **1** rhythmical throbbing of arteries as blood is propelled along them, esp. as felt in the wrists or temples etc. **2** single beat, throb, or vibration. ● *v.* pulsate.

pulse[2] *n.* edible seed of beans, peas, lentils, etc.

pulverize *v.* **1** crush into powder. **2** become powder. **3** defeat thoroughly. □ **pulverization** *n.*

puma *n.* large brown American animal of the cat family.

pumice *n.* (in full **pumice-stone**) solidified lava used for scouring or polishing.

pummel *v.* (**pummelled**) strike repeatedly esp. with the fists.

pump *n.* machine for moving liquid, gas, or air. ● *v.* **1** use a pump. **2** cause (air, gas, water, etc.) to move (as) with a pump. **3** move vigorously up and down. **4** question persistently.

pumpkin *n.* large round orange-coloured fruit of a vine.

pun *n.* humorous use of a word to suggest another that sounds the same. □ **punning** *adj.* & *n.*

punch[1] *v.* **1** strike with the fist. **2** cut (a hole etc.) with a device. ● *n.* **1** blow with the fist. **2** device for cutting holes or impressing a design. □ **punch-drunk** *adj.* stupefied by repeated blows. **punchline** *n.* words giving the climax of a joke.

punch[2] *n.* drink made of wine or spirits mixed with fruit juices etc. □ **punch-bowl** *n.*

punctilious *adj.* **1** very careful about details. **2** conscientious. □ **punctiliously** *adv.*, **punctiliousness** *n.*

punctual *adj.* arriving or doing things at the appointed time. □ **punctually** *adv.*, **punctuality** *n.*

■ □ **punctually** in good time, on the dot, on time, promptly, sharp.

punctuate *v.* **1** insert the appropriate marks in written material to separate sentences etc. **2** interrupt at intervals. □ **punctuation** *n.*

puncture *n.* small hole made by something sharp, esp. in a tyre. ● *v.* **1** make a puncture in. **2** suffer a puncture.

pundit *n.* learned expert.

pungent *adj.* having a strong sharp taste or smell. □ **pungently** *adv.*, **pungency** *n.*

■ hot, peppery, piquant, sharp, spicy, strong, tangy.

punish *v.* **1** cause (an offender) to suffer for his or her offence. **2** inflict a penalty for. **3** treat roughly. □ **punishment** *n.*

■ **1** castigate, chasten, chastise, discipline, penalize. **3** abuse, damage, harm, maltreat, mistreat.

punitive *adj.* inflicting or intended to inflict punishment.

punk *n.* **1** (*sl.*) worthless person. **2** (devotee of) punk rock. □ **punk rock** deliberately outrageous type of rock music.

punnet *n.* small basket or similar container for fruit etc.

punt[1] *n.* shallow flat-bottomed boat with broad square ends. ● *v.* **1** propel (a punt) by thrusting with a pole against the bottom of a river. **2** travel in a punt.

punt[2] *v.* kick (a dropped football) before it touches the ground. ● *n.* this kick.

punter *n.* **1** person who gambles. **2** (*colloq.*) customer.

puny *adj.* (**-ier**, **-iest**) **1** undersized. **2** feeble. ▫ **puniness** *n.*

■ **1** diminutive, little, minute, small, tiny, undersized. **2** feeble, frail, sickly, weak.

pup *n.* **1** young dog. **2** young wolf, rat, or seal. ● *v.* (**pupped**) give birth to pup(s).

pupa *n.* (*pl.* **-ae**) chrysalis.

pupil *n.* **1** person who is taught by another. **2** opening in the centre of the iris of the eye.

puppet *n.* **1** a kind of doll made to move as an entertainment. **2** person controlled by another. ▫ **puppetry** *n.*

■ **1** marionette. **2** cat's-paw, pawn, stooge, tool.

puppy *n.* young dog.

purchase *v.* buy. ● *n.* **1** buying. **2** thing bought. **3** firm hold to pull or raise something, leverage. ▫ **purchaser** *n.*

■ *v.* acquire, buy, get, obtain, pay for, procure. ● *n.* **1** acquisition, buying, purchasing. **2** acquisition, buy. **3** foothold, footing, grasp, grip, hold, leverage, toehold.

purdah *n.* screening of Muslim or Hindu women from strangers.

pure *adj.* **1** not mixed with any other substances. **2** mere, nothing but. **3** innocent. **4** chaste, not morally corrupt. **5** (of mathematics or sciences) dealing with theory, not with practical applications.

■ **1** unadulterated, unalloyed, uncontaminated, unmixed, unpolluted; solid, sterling. **2** absolute, complete, downright, mere, nothing but, out and out, outright, sheer, unmitigated, utter. **3** blameless, guiltless, innocent. **4** chaste, undefiled, unsullied, virgin, virginal; ethical, good, guiltless, honest, honourable, incorruptible, moral, righteous, sinless, virtuous.

purée *n.* pulped fruit or vegetables etc. ● *v.* make into purée.

purely *adv.* **1** in a pure way. **2** entirely. **3** only.

purgative *adj.* strongly laxative. ● *n.* purgative substance.

purgatory *n.* place or state of suffering, esp. in which souls undergo purification. ▫ **purgatorial** *adj.*

purge *v.* **1** clear the bowels of by a purgative. **2** rid of undesirable people or things. ● *n.* process of purging.

purify *v.* make pure. ▫ **purification** *n.*, **purifier** *n.*

■ clarify, clean, cleanse, decontaminate, disinfect, refine.

purist *n.* stickler for correctness. ▫ **purism** *n.*

puritan *n.* person who is strict in morals and regards certain pleasures as sinful. ▫ **puritanical** *adj.*

■ ▫ **puritanical** ascetic, austere, narrow-minded, priggish, prim, prissy, prudish, strait-laced, *colloq.* stuffy.

purity *n.* pure state or condition.

■ cleanliness, cleanness, pureness; chastity, virtuousness, virginity; honesty, innocence, integrity, rectitude, sinlessness, virtue.

purl *n.* a kind of knitting stitch. ● *v.* produce this stitch (in).

purloin *v.* steal.

purple *adj.* & *n.* (of) a colour made by mixing red and blue. ▫ **purplish** *adj.*

purport *n.* /púrport/ meaning. ● *v.* /pərpórt/ **1** have as its apparent meaning. **2** be intended to seem. ▫ **purportedly** *adv.*

purpose *n.* **1** intended result of effort. **2** intention to act, determination. **3** reason for which something is done or made. ● *v.* intend. ▫ **on purpose** by intention. **purpose-built** *adj.* built for a particular purpose.

■ *n.* **1** aim, end, goal, intent, intention, object, objective, plan, target. **2** determination, doggedness, drive, firmness, perseverance, purposefulness, resoluteness, resolution, resolve, single-mindedness, tenacity. **3** function, use. ● *v.* aim, design, have in mind, mean, intend, plan, propose. ▫ **on purpose** deliberately, intentionally, knowingly, purposely, purposefully, wilfully.

purposeful *adj.* having or showing purpose, intentional. ▫ **purposefully** *adv.*

■ deliberate, intended, intentional, planned, wilful; determined, dogged, resolute, resolved, steadfast, strong-willed, tenacious, unfaltering, unwavering.

purposely *adv.* on purpose.

purr *n.* **1** low vibrant sound that a cat makes when pleased. **2** similar sound. ● *v.* make this sound.

purse *n.* **1** small pouch for carrying money. **2** (*US*) handbag. **3** money, funds. ● *v.* pucker (one's lips).

purser *n.* ship's officer in charge of accounts.

pursuance *n.* performance (of duties etc.).

pursue *v.* **1** follow or chase. **2** continue. **3** proceed along. **4** engage in. □ **pursuer** *n.*

■ **1** chase, dog, go *or* run after, hunt, shadow, stalk, *colloq.* tail, track, trail. **2** carry on with, continue (with), persevere in, persist in, proceed with. **3** follow, keep to. **4** engage in, practise, prosecute.

pursuit *n.* **1** act or instance of pursuing. **2** activity to which one gives time or effort.

■ **1** chase, chasing, hunt, stalking, tracking, trailing. **2** activity, hobby, interest, occupation, pastime.

purulent *adj.* of or containing pus. □ **purulence** *n.*

purvey *v.* supply (articles of food) as a trader. □ **purveyor** *n.*

pus *n.* thick yellowish matter produced from infected tissue.

push *v.* **1** move away by exerting force. **2** press. **3** make one's way by pushing. **4** make demands on the abilities or tolerance of. **5** urge or impel. **6** (*colloq.*) sell (drugs) illegally. ● *n.* **1** act or force of pushing. **2** vigorous effort. □ **push off** (*sl.*) go away. **pusher** *n.*

■ *v.* **1** drive, propel, ram, shove, thrust. **2** depress, press. **3** elbow, force, jostle, shoulder, shove, thrust. **5** encourage, induce, motivate, move, persuade, press, pressurize, prod, prompt, spur, stimulate, urge; coerce, compel, dragoon, drive, force, impel.

pushchair *n.* folding chair on wheels, in which a child can be pushed along.

pushy *adj.* (**-ier**, **-iest**) (*colloq.*) self-assertive, determined to get on. □ **pushiness** *n.*

pusillanimous *adj.* cowardly. □ **pusillanimity** *n.*

puss *n.* cat.

pussy *n.* cat. □ **pussy willow** willow with furry catkins.

pussyfoot *v.* **1** move stealthily. **2** act cautiously.

pustule *n.* pimple, blister. □ **pustular** *adj.*

put *v.* (**put**, **putting**) **1** cause to be in a certain place, position, state, or relationship. **2** express, phrase. **3** throw (the shot or weight) as an athletic exercise. ● *n.* throw of the shot or weight. □ **put across** or **over** make understood. **put by** save for future use. **put down 1** suppress. **2** snub, disparage. **3** have (an animal) killed. **4** record in writing. **put forward** suggest, present for discussion. **put off 1** postpone. **2** dissuade. **3** disconcert, offend. **put out 1** disconcert or annoy. **2** inconvenience. **3** extinguish. **4** dislocate. **put up 1** construct, build. **2** raise the price of. **3** provide (money etc.). **4** give temporary accommodation to. **put-up job** scheme concocted fraudulently. **put upon** (*colloq.*) unfairly burdened. **put up with** endure, tolerate.

■ *v.* **1** deposit, dump, lay, park, place, position, rest, set, *colloq.* shove, situate, station, *colloq.* stick; assign, attach, attribute, impute, pin. **2** couch, express, phrase, say, word. □ **put across** communicate, convey, explain, get across, make clear, make understood. **put by** hoard, lay up, *colloq.* salt away, save, set aside, stockpile, store. **put down 1** crush, overthrow, quash, quell, stamp out, suppress. **2** ignore, slight, snub; belittle, criticize, deprecate, disparage, insult. **4** jot down, note (down), record, register, write down. **put forward** advance, move, offer, present, proffer, propose, propound, recommend, submit, suggest. **put off 1** defer, delay, postpone, shelve. **2** discourage, dissuade, deter. **3** disconcert, disturb, fluster, perturb, *colloq.* throw, upset; disgust, offend, repel, revolt. **put out 1** discomfit, disconcert, dismay, *colloq.* throw; annoy, gall, irritate, exasperate, irk, offend, vex. **2** bother, discommode, disturb, impose on, inconvenience, trouble. **3** blow out, douse, extinguish, quench, snuff out; switch off, turn off. **put up 1** build, construct, erect, set up. **2** increase, raise. **3** contribute, donate, give, furnish, provide, supply. **4** accommodate, billet, lodge, house, quarter. **put up with** accept, bear, brook, endure, live with, stand (for), *colloq.* stick, stomach, swallow, suffer, tolerate.

putative *adj.* reputed, supposed.

putrefy *v.* rot. □ **putrefaction** *n.*

■ decay, decompose, go bad, go off, moulder, rot.

putrescent *adj.* rotting. □ **putrescence** *n.*

putrid *adj.* **1** rotten. **2** foul-smelling.

■ **1** bad, decayed, decaying, decomposed, decomposing, mouldy, putrescent, rotten, rotting. **2** fetid, foul-smelling, rank, stinking.

putt *v.* strike (a golf ball) gently to make it roll along the ground. ● *n.* this stroke. □ **putter** *n.* club used for this.

putty *n.* soft paste that sets hard, used for fixing glass in frames, filling up holes, etc.

puzzle *n.* **1** difficult question or problem. **2** problem or toy designed to test ingenuity etc. ● *v.* disconcert mentally, confound. □ **puzzle over** think deeply about. **puzzlement** *n.*

■ *n.* **1** enigma, mystery, problem. **2** conundrum, poser, riddle. ● *v.* baffle, bemuse, bewilder, confound, confuse, *colloq.* flummox, mystify, nonplus, perplex, *colloq.* stump. □ **puzzle over** consider, contemplate, meditate on, mull over, ponder, reflect on, think about.

pygmy *n.* **1** member of a dwarf people of equatorial Africa. **2** very small person or thing.

pyjamas *n.pl.* loose jacket and trousers esp. for sleeping in.

pylon *n.* lattice-work tower used for carrying electricity cables.

pyorrhoea /pīrée̍ə/ *n.* discharge of pus, esp. from tooth-sockets.

pyramid *n.* structure with triangular sloping sides that meet at the top. □ **pyramidal** *adj.*

pyre *n.* pile of wood etc. for burning a dead body.

pyrethrum *n.* **1** a kind of chrysanthemum. **2** insecticide made from its dried flowers.

pyrites /pīríteez/ *n.* mineral sulphide of (copper and) iron.

pyromaniac *n.* person with an uncontrollable impulse to set things on fire.

pyrotechnics *n.pl.* firework display □ **pyrotechnic** *adj.*

Pyrrhic victory /pírrik/ one gained at too great a cost.

python *n.* large snake that crushes its prey.

Qq

quack[1] *n.* duck's harsh cry. ● *v.* make this sound.

quack[2] *n.* person who falsely claims to have medical skill.

quad *n.* (*colloq.*) **1** quadrangle. **2** quadruplet.

quadrangle *n.* four-sided courtyard enclosed by buildings.

quadrant *n.* **1** one-quarter of a circle or of its circumference. **2** graduated instrument for taking angular measurements.

quadraphonic *adj.* & *n.* (sound reproduction) using four transmission channels.

quadratic *adj.* & *n.* (equation) involving the second and no higher power of an unknown quantity or variable.

quadrilateral *n.* geometric figure with four sides.

quadrille *n.* square dance.

quadruped *n.* four-footed animal.

quadruple *adj.* **1** having four parts or members. **2** four times as much as. ● *v.* increase by four times its amount.

quadruplet *n.* one of four children born at one birth.

quaff *v.* drink in large draughts.

quagmire *n.* bog, marsh.

quail[1] *n.* bird related to the partridge.

quail[2] *v.* flinch, show fear.

> ■ blench, cower, cringe, flinch, recoil, shrink back, shy away.

quaint *adj.* odd in a pleasing way. □ **quaintly** *adv.*, **quaintness** *n.*

> ■ charming, curious, *colloq.* cute, droll, fanciful, *colloq.* sweet, twee, unusual, whimsical.

quake *v.* shake or tremble, esp. with fear.

> ■ quaver, quiver, shake, shiver, shudder, tremble.

qualification *n.* **1** qualifying. **2** thing that qualifies.

qualify *v.* **1** make or become competent, eligible, or legally entitled to do something. **2** limit the meaning of. □ **qualifier** *n.*

> ■ **1** equip, fit, make eligible, prepare; enable, entitle, permit. **2** limit, modify, restrict.

qualitative *adj.* of or concerned with quality.

quality *n.* **1** (degree of) excellence. **2** characteristic, something that is special in a person or thing.

> ■ **1** calibre, grade, order, standard; distinction, excellence, merit, pre-eminence, superiority. **2** attribute, characteristic, feature, point, property, trait.

qualm /kwaam/ *n.* misgiving, pang of conscience.

> ■ compunction, doubt, misgiving, regret, reservation, scruple, second thought.

quandary *n.* state of perplexity, difficult situation.

> ■ difficulty, dilemma, fix, *colloq.* pickle, plight, predicament.

quango *n.* (*pl.* **-os**) administrative group with members appointed by the government.

quantify *v.* express as a quantity. □ **quantifiable** *adj.*

quantitative *adj.* of or concerned with quantity.

quantity *n.* **1** amount or number of things. **2** ability to be measured. **3** (*pl.*) large amounts. □ **quantity surveyor** person who measures and prices building work.

> ■ **1** amount, extent, measure, volume, weight; number, sum, total.

quantum *n.* **quantum theory** theory of physics based on the assumption that energy exists in indivisible units.

quarantine *n.* isolation imposed on those who have been exposed to an infection which they could spread. ● *v.* put into quarantine.

quark *n.* component of elementary particles.

quarrel *n.* angry disagreement. ● *v.* (**quarrelled**) engage in a quarrel.

> ■ *n.* altercation, argument, *colloq.* bust-up, disagreement, dispute, fight, row, *colloq.* scrap, squabble, tiff, wrangle. ● *v.* argue, bicker, disagree, dispute, fall out, fight, row, *colloq.* scrap, spar, squabble, wrangle.

quarrelsome *adj.* liable to quarrel.

■ **antagonistic, argumentative, belligerent, contentious, disagreeable, disputatious, irascible, irritable, pugnacious, truculent.**

quarry¹ *n.* **1** intended prey or victim. **2** object of pursuit.

■ **prey, victim; object, prize, target.**

quarry² *n.* open excavation from which stone etc. is obtained. ● *v.* obtain from a quarry.

quart *n.* quarter of a gallon, two pints (1.137 litres).

quarter *n.* **1** one of four equal parts. **2** this amount. **3** (*US* & *Canada*) (coin worth) 25 cents. **4** fourth part of a year. **5** point of time 15 minutes before or after every hour. **6** direction, district. **7** mercy towards an opponent. **8** (*pl.*) lodgings, accommodation. ● *v.* **1** divide into quarters. **2** put into lodgings. □ **quarter-final** *n.* contest preceding a semifinal.

■ *n.* **6 direction; area, district, locality, neighbourhood, region, sector, territory, zone. 7 clemency, compassion, forgiveness, mercifulness, mercy, pity. 8 (quarters) accommodation, digs, dwelling, habitation, lodging(s), residence, rooms; barracks, billet.** ● *v.* **2 accommodate, billet, board, house, lodge, put up.**

quarterdeck *n.* part of a ship's upper deck nearest the stern.

quarterly *adj.* & *adv.* (produced or occurring) once in every quarter of a year. ● *n.* quarterly periodical.

quartermaster *n.* **1** regimental officer in charge of stores etc. **2** naval petty officer in charge of steering and signals.

quartet *n.* **1** group of four instruments or voices. **2** music for these.

quartz *n.* a kind of hard mineral. □ **quartz clock** one operated by electric vibrations of a quartz crystal.

quasar *n.* star-like object that is the source of intense electromagnetic radiation.

quash *v.* **1** annul. **2** suppress.

■ **1 annul, cancel, declare null and void, invalidate, nullify, overthrow, overturn, rescind, revoke, void. 2 crush, overcome, overwhelm, put an end to, put down, quell, squash, stamp out, subdue, suppress.**

quasi- *pref.* seeming to be but not really so.

quatrain *n.* stanza of four lines.

quaver *v.* **1** tremble, vibrate. **2** speak or utter in a trembling voice. ● *n.* **1** trembling sound. **2** note in music, half a crotchet.

■ *v.* **1 quiver, shake, shiver, shudder, tremble, vibrate.**

quay /kee/ *n.* landing place built for ships to load or unload alongside. □ **quayside** *n.*

queasy *adj.* **1** nauseous. **2** squeamish. □ **queasiness** *n.*

■ **1 bilious, ill, nauseous, off colour, queer, sick. 2 fastidious, squeamish.**

queen *n.* **1** female ruler of a country by right of birth. **2** king's wife. **3** woman or thing regarded as supreme in some way. **4** piece in chess. **5** playing card bearing a picture of a queen. **6** fertile female of bee or ant etc. □ **queen mother** dowager queen who is the reigning sovereign's mother. **queenly** *adj.*

queer *adj.* **1** strange, odd, eccentric. **2** slightly ill or faint. **3** (*derog.*) homosexual. ● *n.* (*derog.*) homosexual. ● *v.* spoil.

■ *adj.* **1 abnormal, anomalous, atypical, bizarre, eccentric, extraordinary, funny, kinky, odd, offbeat, outlandish, peculiar, quaint, strange, uncanny, unconventional, unorthodox, unusual, weird. 2 groggy, ill, off colour, poorly, queasy, sick, under the weather, unwell; dizzy, faint, light-headed.**

quell *v.* suppress.

■ **crush, overcome, overpower, put down, quash, squash, stamp out, subdue, suppress.**

quench *v.* **1** extinguish (a fire or flame). **2** satisfy (one's thirst) by drinking something.

■ **1 douse, extinguish, put out, snuff out. 2 satisfy, slake.**

quern *n.* hand-mill for grinding corn or pepper.

querulous *adj.* complaining peevishly □ **querulously** *adv.*

query *n.* **1** question. **2** question mark. ● *v.* ask a question or express doubt about.

■ *n.* **1 enquiry, question.** ● *v.* **ask, enquire, challenge, contest, dispute, doubt, question.**

quest *n.* seeking, search.

question *n.* **1** sentence requesting information. **2** matter for discussion or solution. **3** doubt. ● *v.* ask or raise question(s) about. □ **in question** being discussed or disputed. **no question of** no possibility of. **out of the question** completely impracticable. **question mark** punctuation mark (?) placed after a question.

■ *n.* **1** enquiry, query. **2** concern, issue, matter, point; difficulty, problem. ● *v.* ask, catechize, examine, grill, interrogate, pump, quiz; challenge, contest, dispute, doubt, query. □ **out of the question** absurd, impossible, impracticable, inconceivable, preposterous, ridiculous, unthinkable.

questionable *adj.* open to doubt.

■ ambiguous, debatable, disputable, doubtful, dubious, problematic(al), uncertain.

questionnaire *n.* list of questions seeking information.

queue /kyōō/ *n.* line of people waiting for something. ● *v.* **(queuing)** wait in a queue.

quibble *n.* **1** petty objection. **2** evasion. ● *v.* **1** make petty objections. **2** be evasive.

■ *v.* **1** cavil, *colloq.* nit-pick. **2** be evasive, equivocate, evade the issue, hedge, prevaricate.

quiche /keesh/ *n.* savoury flan.

quick *adj.* **1** taking only a short time. **2** able to learn or think quickly. **3** (of temper) easily roused. ● *n.* sensitive flesh below the nails. □ **quickly** *adv.*

■ **1** expeditious, fast, fleet, *colloq.* nippy, rapid, speedy, swift; immediate, instantaneous, prompt; brief, cursory, hasty, hurried, perfunctory, precipitate, sudden, summary. **2** able, astute, bright, clever, intelligent, perceptive, perspicacious, sharp, shrewd, smart. **3** excitable, impatient, irritable, short, testy, touchy.

quicken *v.* make or become quicker or livelier.

■ accelerate, expedite, hasten, hurry, rush, speed up; animate, arouse, enliven, excite, galvanize, invigorate, kindle, rouse, stimulate, vivify.

quicklime *n.* = **lime**[1].

quicksand *n.* area of loose wet deep sand into which heavy objects will sink.

quicksilver *n.* mercury.

quid *n.* (*pl.* **quid**) (*sl.*) £1.

quid pro quo thing given in return.

quiescent *adj.* inactive, quiet. □ **quiescence** *n.*

quiet *adj.* **1** with little or no sound, silent. **2** free from disturbance or vigorous activity. **3** gentle, tranquil. ● *n.* quietness. ● *v.* quieten. □ **on the quiet** unobtrusively, secretly. **quietly** *adv.*, **quietness** *n.*

■ *adj.* **1** hushed, noiseless, quiescent, silent, soundless; reserved, taciturn, uncommunicative. **2** calm, motionless, still, unmoving; peaceful, secluded, sleepy. **3** calm, gentle, peaceable, peaceful, placid, serene, tranquil. ● *n.* calm, calmness, hush, peace, quietness, serenity, silence, still, stillness, tranquillity.

quieten *v.* make or become quiet.

■ hush, quiet, *colloq.* shush, *colloq.* shut up, silence; calm, lull, pacify, settle, soothe, tranquillize.

quiff *n.* upright tuft of hair.

quill *n.* **1** large wing or tail feather. **2** pen made from this. **3** each of a porcupine's spines.

quilt *n.* padded bed-covering. ● *v.* line with padding and fix with lines of stitching.

quin *n.* (*colloq.*) quintuplet.

quince *n.* **1** hard yellowish fruit. **2** tree bearing this.

quinine *n.* bitter-tasting drug.

quinsy *n.* abscess on a tonsil.

quintessence *n.* **1** essence. **2** perfect example of a quality. □ **quintessential** *adj.*, **quintessentially** *adv.*

■ **1** core, essence, heart. **2** embodiment, epitome, exemplar, ideal, incarnation, model, personification.

quintet *n.* **1** group of five instruments or voices. **2** music for these.

quintuple *adj.* **1** having five parts or members. **2** five times as much as. ● *v.* increase by five times its amount.

quintuplet *n.* one of five children born at one birth.

quip *n.* witty or sarcastic remark. ● *v.* **(quipped)** utter as a quip.

quire *n.* twenty-five or twenty-four sheets of writing paper.

quirk *n.* **1** a peculiarity of behaviour. **2** trick of fate.

■ **1** characteristic, eccentricity, foible, idiosyncrasy, kink, oddity, peculiarity, vagary.

quisling *n.* traitor who collaborates with occupying forces.

quit *v.* **(quitted) 1** leave. **2** abandon. **3** (*US*) cease.

■ **1** depart from, go away from, flee, leave, move out of, vacate. **2** abandon, forsake, give up, relinquish. **3** cease, desist from, discontinue, refrain from, stop.

quite *adv.* **1** completely. **2** somewhat. **3** (as an answer) I agree.

■ **1** absolutely, altogether, completely, entirely, fully, perfectly, thoroughly, totally, utterly, wholly. **2** fairly, moderately, pretty, rather, relatively, somewhat.

quits *adj.* on even terms after retaliation or repayment.

quiver[1] *n.* case for holding arrows.

quiver[2] *v.* shake or vibrate with a slight rapid motion. ● *n.* quivering movement or sound.

■ *v.* quaver, shake, shiver, shudder, tremble, vibrate, wobble.

quixotic *adj.* romantically chivalrous. □ **quixotically** *adv.*

quiz *n.* (*pl.* **quizzes**) series of questions testing knowledge, esp. as an entertainment. ● *v.* (**quizzed**) interrogate.

■ *v.* ask, examine, grill, interrogate, pump, question.

quizzical *adj.* done in a questioning way, esp. humorously. □ **quizzically** *adv.*

quoit /koyt/ *n.* ring thrown to encircle a peg in the game of **quoits**.

quorate *adj.* having a quorum present.

quorum *n.* minimum number of people that must be present to constitute a valid meeting.

quota *n.* **1** fixed share. **2** number of goods, people, etc. permitted or stipulated.

■ **1** allocation, allotment, allowance, *colloq.* cut, part, percentage, portion, ration, share.

quotable *adj.* worth quoting.

quotation *n.* **1** quoting. **2** passage or price quoted. □ **quotation marks** punctuation marks (‘ ’ or “ ”) enclosing words quoted.

quote *v.* **1** repeat words from a book or speech. **2** mention in support of a statement. **3** state the price of, estimate.

■ **2** allude to, cite, instance, mention, refer to.

quotidian *adj.* daily.

quotient /kwṓsh'nt/ *n.* result of a division sum.

Rr

rabbi *n.* (*pl.* **-is**) religious leader of a Jewish congregation.

rabbinical *adj.* of rabbis or Jewish doctrines or law.

rabbit *n.* burrowing animal with long ears and a short furry tail.

rabble *n.* disorderly crowd.

■ crowd, herd, horde, mob, swarm, throng.

rabid *adj.* **1** furious, fanatical. **2** affected with rabies.

rabies *n.* contagious fatal virus disease esp. of dogs.

race¹ *n.* **1** contest of speed. **2** (*pl.*) series of races for horses or dogs. ● *v.* **1** compete in a race (with). **2** move or operate at full or excessive speed. □ **racer** *n.*

■ *n.* **1** competition, contest. ● *v.* **1** compete (against), contend. **2** dash, fly, hare, hasten, hurry, run, rush, shoot, speed, sprint, tear, whiz, zip, zoom.

race² *n.* **1** large group of people with common ancestry and inherited physical characteristics. **2** genus, species, breed, or variety of animals or plants.

■ **1** ethnic group, folk, nation, people. **2** breed, genus, species, strain, variety.

racecourse *n.* ground where horse races are run.

racetrack *n.* **1** racecourse. **2** track for motor racing.

raceme *n.* flower cluster with flowers attached by short stalks along a central stem.

racial *adj.* of or based on race. □ **racially** *adv.*

racialism *n.* racism. □ **racialist** *adj.* & *n.*

racism *n.* **1** belief in the superiority of a particular race. **2** antagonism towards other races. □ **racist** *adj.* & *n.*

rack¹ *n.* **1** framework for keeping or placing things on. **2** bar with teeth that engage with those of a wheel. **3** instrument of torture on which people were tied and stretched. ● *v.* inflict great torment on. □ **rack one's brains** think hard about a problem.

rack² *n.* **rack and ruin** destruction.

racket¹ *n.* stringed bat used in tennis and similar games.

racket² *n.* **1** din, noisy fuss. **2** (*sl.*) fraudulent business or scheme.

■ **1** clamour, commotion, din, disturbance, fuss, hubbub, hue and cry, hullabaloo, noise, outcry, row, rumpus, to-do, uproar.

racketeer *n.* person who operates a fraudulent business etc. □ **racketeering** *n.*

raconteur *n.* person who is good at telling anecdotes.

racoon *n.* small arboreal mammal of N. America.

racy *adj.* (**-ier, -iest**) spirited and vigorous in style. □ **racily** *adv.*

radar *n.* system for detecting objects by means of radio waves.

radial *adj.* **1** of rays or radii. **2** having spokes or lines etc. that radiate from a central point. **3** (in full **radial-ply**) (of a tyre) having the fabric layers parallel and the tread strengthened. ● *n.* radial-ply tyre. □ **radially** *adv.*

radiant *adj.* **1** emitting rays of light or heat. **2** emitted in rays. **3** looking very bright and happy. □ **radiantly** *adv.*, **radiance** *n.*

■ **1** beaming, blazing, bright, gleaming, glittering, glowing, incandescent, luminous, shimmering, shining, sparkling. **3** ecstatic, elated, happy, joyful, overjoyed, rapturous.

radiate *v.* **1** spread outwards from a central point. **2** send or be sent out in rays.

■ **1** fan out, spread out. **2** emanate, emit, give off, send out.

radiation *n.* **1** process of radiating. **2** sending out of rays and atomic particles characteristic of radioactive substances. **3** these rays and particles.

radiator *n.* **1** apparatus that radiates heat, esp. a metal case through which steam or hot water circulates. **2** engine-cooling apparatus.

radical *adj.* **1** fundamental. **2** drastic, thorough. **3** holding extremist views. ● *n.* person desiring radical reforms or holding radical views. □ **radically** *adv.*

■ **1** basic, elemental, essential, fundamental, inherent. **2** complete, comprehensive, drastic, far-reaching, sweeping, thorough. **3** extreme, extremist, militant, revolutionary.

radicle *n.* embryo root.

radio *n.* (*pl.* **-os**) **1** process of sending and receiving messages etc. by electromagnetic waves. **2** transmitter or receiver for this. **3** sound broadcasting, station for this. ● *v.* transmit or communicate by radio.

radioactive *adj.* emitting radiation caused by decay of atomic nuclei. □ **radioactivity** *n.*

radiography *n.* production of X-ray photographs. □ **radiographer** *n.*

radiology *n.* study of X-rays and similar radiation. □ **radiological** *adj.*, **radiologist** *n.*

radiotherapy *n.* treatment of disease by X-rays or similar radiation. □ **radiotherapist** *n.*

radish *n.* plant with a crisp hot-tasting root that is eaten raw.

radium *n.* radioactive metal obtained from pitchblende.

radius *n.* (*pl.* **-dii**) **1** straight line from the centre to the circumference of a circle or sphere. **2** its length. **3** distance from a centre. **4** thicker long bone of the forearm.

raffia *n.* strips of fibre from the leaves of a kind of palm tree.

raffish *adj.* looking vulgarly flashy or rakish. □ **raffishness** *n.*

raffle *n.* lottery with an object as the prize. ● *v.* offer as the prize in a raffle.

raft *n.* flat floating structure of timber etc., used as a boat.

rafter *n.* one of the sloping beams forming the framework of a roof.

rag¹ *n.* **1** torn or worn piece of cloth. **2** (*derog.*) newspaper. **3** (*pl.*) old and torn clothes.

rag² *v.* (**ragged**) (*sl.*) tease. ● *n.* students' carnival in aid of charity.

ragamuffin *n.* person in ragged dirty clothes.

rage *n.* violent anger. ● *v.* **1** show violent anger. **2** (of a storm or battle) continue furiously. □ **all the rage** very popular.

■ *n.* anger, fury, ire, wrath. ● *v.* **1** boil, fulminate, fume, rail, rant, rave, seethe, storm.

ragged *adj.* **1** torn, frayed. **2** wearing torn clothes. **3** jagged.

■ **1** frayed, ripped, tattered, tatty, threadbare, torn, worn. **2** *colloq.* scruffy, shabby, unkempt. **3** craggy, irregular, jagged, rough, serrated, uneven.

raglan *n.* type of sleeve joined to a garment by sloping seams.

ragout /ragoo/ *n.* stew of meat and vegetables.

raid *n.* **1** brief attack to destroy or seize something. **2** surprise visit by police etc to arrest suspected people or seize illicit goods. ● *v.* make a raid on. □ **raider** *n.*

■ *n.* **1** attack, charge, foray, incursion, invasion, onset, onslaught, sally, sortie. ● *v* attack, invade, rush, storm; loot, pillage plunder, ransack, sack.

rail¹ *n.* **1** horizontal bar. **2** any of the lines of metal bars on which trains or trams run. **3** railway(s).

rail² *v.* utter angry reproaches.

railing *n.* fence of rails supported on upright metal bars.

railroad *n.* (*US*) railway. ● *v.* force into hasty action.

railway *n.* **1** set of rails on which trains run. **2** system of transport using these.

rain *n.* **1** atmospheric moisture falling as drops. **2** a fall of this. **3** shower of things. ● *v.* send down or fall as or like rain.

■ *v.* bucket, drizzle, pour (down), spit teem; lavish, shower.

rainbow *n.* arch of colours formed in rain or spray by the sun's rays.

raincoat *n.* rain-resistant coat.

raindrop *n.* single drop of rain.

rainfall *n.* total amount of rain falling in a given time.

rainforest *n.* dense wet tropical forest.

rainwater *n.* water that has fallen as rain.

rainy *adj.* (**-ier**, **-iest**) in or on which much rain falls.

raise *v.* **1** bring to or towards a higher level or an upright position. **2** increase the amount, value, or strength of. **3** cause to be heard or considered. **4** breed grow. **5** bring up (a child). **6** collect manage to obtain. ● *n.* (*US*) rise in salary. □ **raising agent** substance that makes bread etc. swell in cooking.

■ *v.* **1** elevate, haul up, hoist, lift (up), pull up; up-end. **2** increase, put up; amplify augment, heighten, intensify, step up boost, buoy up, lift, uplift. **3** bring up broach, introduce, mention, put forward suggest. **4** breed, rear; cultivate, farm grow, propagate. **5** bring up, nurture, rear educate. **6** collect, gather together, levy.

raisin *n.* dried grape.

raison d'être /ráysoɴ détrə/ reason for or purpose of a thing's existence.

rake[1] *n.* **1** tool with prongs for gathering of leaves etc. or smoothing loose soil. **2** implement used similarly. ● *v.* **1** gather or smooth with a rake. **2** search. **3** direct (gunfire etc.) along. □ **rake-off** *n.* (*colloq.*) share of profits. **rake up** revive (unwelcome) memory of.

rake[2] *n.* backward slope of an object. ● *v.* set at a sloping angle.

rake[3] *n.* dissolute man. □ **rakish** *adj.* **1** like a rake. **2** jaunty.

rally *v.* **1** bring or come (back) together for a united effort. **2** revive, recover strength. ● *n.* **1** act of rallying, recovery. **2** series of strokes in tennis etc. **3** mass meeting. **4** driving competition over public roads.

■ *v.* **1** assemble, come *or* get together, convene, congregate, group; bring *or* call together, gather, marshal, mobilize, round up, summon; reassemble, regroup. **2** get better, improve, pick up, recuperate, revive. ● *n.* **1** improvement, recovery, recuperation, revival. **3** assembly, convention, gathering, meeting.

ram *n.* **1** uncastrated male sheep. **2** striking or plunging device. ● *v.* (**rammed**) strike or push heavily, crash against.

■ *v.* bump, butt, collide with, crash against *or* into, hit, run into, strike; cram, force, jam, pack, push, squeeze, stuff, thrust, wedge.

Ramadan *n.* ninth month of the Muslim year, when Muslims fast during daylight hours.

ramble *n.* walk taken for pleasure. ● *v.* **1** take a ramble, wander. **2** talk or write disconnectedly. □ **rambler** *n.*

■ *n.* amble, constitutional, saunter, stroll, walk. ● *v.* **1** amble, meander, perambulate, range, rove, saunter, stroll, walk, wander. **2** digress, maunder, wander.

ramify *v.* **1** form branches or subdivisions. **2** become complex. □ **ramification** *n.*

ramp *n.* slope joining two levels.

rampage *v.* /rampáyj/ behave or race about violently. ● *n.* /rámpayj/ violent behaviour. □ **on the rampage** rampaging.

rampant *adj.* flourishing excessively, unrestrained.

■ exuberant, flourishing, luxuriant, rank, profuse; rife, uncontrollable, unrestrained, widespread.

rampart *n.* broad-topped defensive wall or bank of earth.

ramshackle *adj.* tumbledown, rickety.

■ crumbling, decrepit, derelict, dilapidated, rickety, ruined, shaky, tumbledown.

ran *see* **run**.

ranch *n.* **1** cattle-breeding establishment in N. America. **2** farm where certain other animals are bred. ● *v.* farm on a ranch. □ **rancher** *n.*

rancid *adj.* smelling or tasting like stale fat. □ **rancidity** *n.*

rancour *n.* bitter feeling or ill will. □ **rancorous** *adj.*

■ acrimony, animosity, animus, antipathy, bitterness, enmity, hate, hatred, hostility, ill will, malevolence, malice, resentfulness, resentment, spite, spitefulness, venom, vindictiveness.

rand *n.* unit of money in S. African countries.

random *adj.* done or made etc. at random. □ **at random** without a particular aim or purpose.

■ accidental, arbitrary, casual, chance, fortuitous, haphazard, hit-or-miss, indiscriminate, unplanned, unsystematic.

randy *adj.* (**-ier, -iest**) lustful.

rang *see* **ring**[2].

range *n.* **1** series representing variety or choice. **2** limits between which something operates or varies. **3** distance a thing can travel or be effective. **4** line or row, esp. of mountains, hills, etc. **5** large open area for grazing or hunting. **6** place with targets for shooting practice. ● *v.* **1** arrange in row(s) etc. **2** extend. **3** vary between limits. **4** go about a place. □ **rangefinder** *n.* device for calculating the distance to a target etc.

■ *n.* **1** assortment, choice, kind, selection, sort, variety. **2** ambit, area, compass, distance, extent, gamut, limit, reach, scope, span, sphere. **4** chain, line, row, series, string. ● *v.* **1** align, arrange, array, dispose, line up, order, rank. **2** extend, go, reach, stretch. **3** fluctuate, vary. **4** roam, rove, travel over, traverse, wander.

rangy *adj.* (**-ier, -iest**) tall and slim.

rank[1] *n.* **1** line of people or things. **2** place in a scale of quality or value etc. **3** high social position. **4** (*pl.*) ordinary soldiers, not officers. ● *v.* **1** arrange in a rank. **2** assign a rank to. **3** have a certain rank. □ **the rank and file** the ordinary people of an organization.

■ *n.* **1** column, file, line, row, string, tier. **2** grade, level, place, position, standing, status. ● *v.* **1** align, arrange, array, dispose, line up, order, range. **2** categorize, class, classify, grade, rate.

rank[2] *adj.* **1** growing too thickly and coarsely. **2** full of weeds. **3** foul-smelling. **4** complete and utter, unmistakable.

■ **1** abundant, dense, exuberant, flourishing, prolific, profuse. **3** fetid, foul-smelling, putrid, rancid, reeking, smelly, stinking. **4** blatant, complete, downright, flagrant, gross, out and out, sheer, unmistakable, unmitigated, utter.

rankle *v.* cause lasting resentment.

ransack *v.* **1** search thoroughly. **2** pillage.

■ **1** go through, rifle *or* rummage through, scour, search. **2** despoil, loot, pillage, plunder, raid, rob, sack.

ransom *n.* price demanded or paid for the release of a captive. ● *v.* demand or pay ransom for.

rant *v.* make a violent speech.

rap *n.* **1** quick sharp blow. **2** knocking sound. **3** monologue recited to music. **4** rock music with recited words. **5** (*sl.*) blame, punishment. ● *v.* (**rapped**) **1** strike sharply. **2** knock. **3** (*sl.*) reprimand.

rapacious *adj.* grasping, violently greedy. □ **rapacity** *n.*

■ acquisitive, avaricious, covetous, grasping, greedy.

rape[1] *v.* have sexual intercourse with (esp. a woman) without consent. ● *n.* this act or crime.

rape[2] *n.* plant grown as fodder and for its seeds, which yield oil.

rapid *adj.* quick, swift. ● *n.pl.* swift current where a river bed slopes steeply. □ **rapidly** *adv.*, **rapidity** *n.*

■ *adj.* brisk, expeditious, express, fast, fleet, prompt, quick, speedy, swift; hasty, hurried, precipitate, sudden.

rapier *n.* thin light sword.

rapist *n.* person who commits rape.

rapport /rapór/ *n.* harmonious understanding relationship.

■ affinity, empathy, sympathy, understanding.

rapt *adj.* very intent and absorbed, enraptured. □ **raptly** *adv.*

■ absorbed, captivated, engrossed, enraptured, enthralled, fascinated, intent, spellbound.

rapture *n.* intense delight. □ **rapturous** *adj.*

■ bliss, delight, ecstasy, elation, euphoria, joy, joyfulness.

rare[1] *adj.* **1** very uncommon. **2** exceptionally good. **3** of low density. □ **rarely** *adv.*, **rareness** *n.*

■ **1** atypical, exceptional, extraordinary, out of the ordinary, uncommon, unfamiliar, unusual; infrequent, scarce. **2** excellent, exquisite, incomparable, matchless, outstanding, peerless, superlative, unparalleled.

rare[2] *adj.* (of meat) underdone.

rarebit *n. see* **Welsh rabbit**.

rarefied *adj.* (of air) of low density, thin. □ **rarefaction** *n.*

rarity *n.* **1** rareness. **2** rare thing.

rascal *n.* dishonest or mischievous person. □ **rascally** *adj.*

■ devil, imp, knave, mischief-maker, rogue, scallywag, scamp, scoundrel, wretch.

rash[1] *n.* eruption of spots or patches on the skin.

rash[2] *adj.* acting or done without due consideration of the risks. □ **rashly** *adv.*, **rashness** *n.*

■ foolhardy, hare-brained, hasty, hotheaded, ill-advised, ill-considered, impetuous, imprudent, impulsive, incautious, injudicious, madcap, precipitate, reckless, thoughtless, unthinking, unwise, wild.

rasher *n.* slice of bacon or ham.

rasp *n.* **1** coarse file. **2** grating sound. ● *v.* **1** scrape with a rasp. **2** utter with or make a grating sound.

raspberry *n.* **1** edible red berry. **2** plant bearing this. **3** (*sl.*) vulgar sound of disapproval.

rat *n.* **1** rodent like a large mouse. **2** scoundrel, treacherous deserter. ● *v.* (**ratted**) **rat on** desert or betray. □ **rat race** fiercely competitive struggle for success.

ratchet *n.* bar or wheel with notches in which a pawl engages to prevent backward movement.

rate *n.* **1** numerical proportion between two sets of things, esp. as the basis of calculating amount or value. **2** charge, cost, or value. **3** rapidity. ● *v.* **1** estimate the worth or value of. **2** consider, regard as. **3** deserve. □ **at any rate 1** no matter what happens. **2** at least.

■ *n.* **1 proportion, scale. 2 charge, cost, fee, price, tariff. 3 gait, pace, speed, tempo, velocity.** ● *v.* **1 appraise, assess, calculate, estimate, evaluate, gauge, judge; class, grade, place, rank. 2 consider, count, deem, reckon, regard as. 3 be entitled to, be worthy of, deserve, merit.**

rather *adv.* **1** slightly. **2** more exactly. **3** by preference. **4** emphatically yes.

■ **1 fairly, moderately, pretty, quite, slightly, somewhat. 3 preferably, sooner.**

ratify *v.* confirm or accept (an agreement etc.) formally. □ **ratification** *n.*

■ **agree to, accept, approve, back, confirm, endorse, sanction, support, uphold; sign.**

rating *n.* **1** level at which a thing is rated. **2** non-commissioned sailor.

ratio *n.* (*pl.* **-os**) relationship between two amounts, reckoned as the number of times one contains the other.

ratiocinate *v.* reason logically. □ **ratiocination** *n.*

ration *n.* fixed allowance of food etc. ● *v.* limit to a ration.

■ *n.* **allocation, allowance, lot, measure, percentage, portion, quota, share.** ● *v.* **confine, control, limit, restrict.**

rational *adj.* **1** able to reason. **2** sane. **3** based on reasoning. □ **rationally** *adv.*, **rationality** *n.*

■ **1 discriminating, enlightened, intelligent, level-headed, reasonable, sensible, wise. 2 lucid, normal, of sound mind, sane, well-balanced. 3 logical, reasoned.**

rationale /rashənaal/ *n.* **1** fundamental reason. **2** logical basis.

rationalism *n.* treating reason as the basis of belief and knowledge. □ **rationalist** *n.*, **rationalistic** *adj.*

rationalize *v.* **1** invent a rational explanation for. **2** make more efficient by reorganizing. □ **rationalization** *n.*

■ **1 account for, excuse, explain, justify. 2 reorganize, streamline.**

rattan *n.* palm with jointed stems.

rattle *v.* **1** (cause to) make a rapid series of short hard sounds. **2** (*colloq.*) make nervous. ● *n.* **1** rattling sound. **2** device for making this. □ **rattle off** utter rapidly.

■ *v.* **1 clank, clatter, clink, jangle, shake. 2 discomfit, disconcert, disturb, fluster, perturb, put off, shake,** *colloq.* **throw, unnerve.**

rattlesnake *n.* poisonous American snake with a rattling tail.

raucous *adj.* loud and harsh. □ **raucously** *adv.*

■ **discordant, dissonant, grating, harsh, jarring, loud, noisy, rasping, shrill, strident.**

raunchy *adj.* (**-ier, -iest**) sexually provocative.

ravage *v.* do great damage to. ● *n.pl.* damages.

■ *v.* **destroy, devastate, lay waste, ruin, wreak havoc on, wreck.**

rave *v.* talk wildly or furiously. □ **rave about** or **over** be rapturously enthusiastic about.

ravel *v.* (**ravelled**) tangle.

raven *n.* black bird with a hoarse cry. ● *adj.* (of hair) glossy black.

ravenous *adj.* very hungry. □ **ravenously** *adv.*

■ **famished, hungry, starved, starving.**

ravine /rəveen/ *n.* deep narrow gorge.

raving *adj.* completely (mad).

ravioli *n.* small square pasta cases with a savoury filling.

ravish *v.* **1** rape. **2** enrapture.

raw *adj.* **1** not cooked. **2** not yet processed. **3** stripped of skin, sensitive because of this. **4** (of weather) damp and chilly. □ **raw deal** unfair treatment.

■ **2 crude, natural, unprocessed, unrefined, untreated. 3 exposed, open; painful, sensitive, sore, tender. 4 bitter, chill, chilly, cold, damp, freezing, icy,** *colloq.* **nippy.**

rawhide *n.* untanned leather.

ray[1] *n.* **1** single line or narrow beam of light. **2** radiating line.

■ **1 beam, gleam, shaft, streak.**

ray[2] *n.* large marine flatfish.

rayon *n.* synthetic fibre or fabric, made from cellulose.

raze *v.* tear down (a building).

razor *n.* sharp-edged instrument for shaving.

razzmatazz *n.* **1** excitement. **2** extravagant publicity etc.

RC *abbr.* Roman Catholic.

re *prep.* concerning.

re- *pref.* **1** again. **2** back again.

reach *v.* **1** extend, go as far as. **2** arrive at. **3** establish communication with. **4** achieve, attain. ● *n.* **1** distance over which a person or thing can reach. **2** section of a river. □ **reach for** or **out**

stretch out a hand in order to touch or take something.

■ *v.* **1** extend, go, run, stretch. **2** arrive at, come to, gain, get as far as, get to, hit, make. **3** contact, get in touch with. **4** accomplish, achieve, attain, make. ● *n.* **1** ambit, compass, extent, range, scope. □ **reach out** extend, hold out, outstretch, stretch out.

react *v.* undergo a reaction. □ **reactive** *adj.*

reaction *n.* **1** response to a stimulus or act or situation etc. **2** chemical change produced by substances acting upon each other. **3** occurrence of one condition after a period of the opposite.

reactionary *adj.* & *n.* (person) opposed to progress and reform.

reactor *n.* apparatus for the production of nuclear energy.

read *v.* (**read**) **1** understand the meaning of (written or printed words or symbols). **2** reproduce (such words etc.) mentally or vocally. **3** study or discover by reading. **4** interpret mentally. **5** have a certain wording. **6** (of an instrument) indicate as a measurement. ● *n.* (*colloq.*) session of reading.

■ *v.* **2** peruse, pore over, study; recite. **4** construe, interpret, understand.

readable *adj.* **1** pleasant to read. **2** legible. □ **readably** *adv.*

■ **1** absorbing, enjoyable, entertaining, interesting, stimulating. **2** comprehensible, decipherable, intelligible, legible.

reader *n.* **1** person who reads. **2** senior lecturer at a university. **3** device producing a readable image from a microfilm etc.

readership *n.* readers of a newspaper etc.

readily *adv.* **1** willingly. **2** easily.

■ **1** eagerly, gladly, happily, willingly. **2** easily, effortlessly.

readiness *n.* being ready.

ready *adj.* (**-ier**, **-iest**) **1** fit or available for action or use. **2** willing, inclined. **3** quick. ● *adv.* beforehand. □ **ready to** about to.

■ *adj.* **1** equipped, fit, prepared, primed, set. **2** eager, game, glad, happy, keen, pleased, willing; apt, disposed, given, inclined, likely, prone. **3** prompt, quick, rapid, speedy, swift; bright, clever, intelligent, keen, perceptive, sharp. □ **ready to** about to, in danger of, on the point of, on the verge of.

reagent *n.* substance used to produce a chemical reaction.

real *adj.* **1** existing as a thing or occurring as a fact. **2** genuine. □ **real estate** immovable assets, i.e. buildings, land.

■ **1** actual, existent, unimaginary; concrete, material, palpable, physical, tangible. **2** authentic, bona fide, genuine, legitimate, proper, true, verifiable, veritable; earnest, heartfelt, honest, sincere, unaffected, unfeigned.

realism *n.* representing or viewing things as they are in reality. □ **realist** *n.*

realistic *adj.* **1** showing realism. **2** practical. □ **realistically** *adv.*

■ **1** graphic, lifelike, naturalistic, vivid, factual. **2** businesslike, down-to-earth, hard-headed, practical, pragmatic, rational, reasonable, sensible.

reality *n.* **1** quality of being real. something real and not imaginary.

realize *v.* **1** be or become aware of. understand clearly. **3** fulfil (a hope or plan). □ **realization** *n.*

■ **1** appreciate, be aware of, be conscious of, know. **2** *colloq.* catch on to, comprehend, *colloq.* cotton on to, grasp, perceive, recognize, see, *colloq.* twig, understand. accomplish, achieve, bring about, fulfil.

really *adv.* **1** in fact. **2** thoroughly. **3** assure you, I protest.

realm *n.* **1** kingdom. **2** field of activity or interest.

■ **1** domain, empire, kingdom, principality. **2** area, domain, field, province, sphere.

ream *n.* **1** 500 sheets of paper. **2** (*pl.*) large quantity of writing.

reap *v.* **1** cut (grain etc.) as harvest. receive as the consequence of actions. □ **reaper** *n.*

■ **1** collect, garner, gather, harvest. **2** acquire, gain, get, obtain, receive, secure.

rear¹ *n.* back part. ● *adj.* situated at the rear. □ **bring up the rear** be last. **rear admiral** naval officer next below vice admiral.

■ *n.* back, end, hind part, tail; stern.

rear² *v.* **1** bring up (children). **2** breed and look after (animals). **3** cultivate (crops). **4** (of a horse etc.) raise itself on its hind legs.

■ **1** bring up, nurture, raise; educate. breed, care for, keep, look after, raise. cultivate, farm, grow, propagate, raise.

rearguard *n.* troops protecting an army's rear.

rearm *v.* arm again. ▫ **rearmament** *n.*

rearrange *v.* arrange in a different way. ▫ **rearrangement** *n.*

rearward *adj., adv.,* & *n.* (towards or at) the rear. ▫ **rearwards** *adv.*

reason *n.* **1** motive, cause, justification. **2** ability to think and draw conclusions. **3** sanity. **4** good sense or judgement. ● *v.* use one's ability to think and draw conclusions. ▫ **reason with** try to persuade by argument.

■ *n.* **1** cause, motivation, motive; defence, excuse, explanation, ground(s), justification. **2** intellect, intelligence, mind. **3** mind, sanity, senses. **4** common sense, judgement, perspicacity, sense, wisdom, wit. ▫ **reason with** argue with, persuade, plead with, prevail on.

reasonable *adj.* **1** ready to use or listen to reason. **2** in accordance with reason, logical. **3** moderate, not expensive. ▫ **reasonably** *adv.*

■ **1** intelligent, judicious, logical, rational, sane, sensible, wise. **2** justifiable, logical, reasoned, sensible, tenable, valid, well-thought-out. **3** affordable, inexpensive, low, moderate, modest.

reassure *v.* restore confidence to. ▫ **reassurance** *n.*

■ buoy up, cheer up, comfort, encourage, hearten, put at ease.

rebate *n.* partial refund.

rebel *n.* /rébb'l/ person who fights against or refuses allegiance to established government or conventions. ● *v.* /ribél/ (**rebelled**) act as a rebel. ▫ **rebellion** *n.*, **rebellious** *adj.*

■ *n.* insurgent, insurrectionist, mutineer, revolutionary; dissident, nonconformist, recusant. ● *v.* mutiny, revolt, rise (up). ▫ **rebellion** insurgency, insurrection, mutiny, revolt, revolution, uprising. **rebellious** defiant, insubordinate, insurgent, mutinous, revolutionary, seditious.

rebound *v.* /ribównd/ spring back after impact. ● *n.* /reébownd/ act of rebounding. ▫ **on the rebound** while still reacting to a disappointment etc.

rebuff *v.* & *n.* snub.

■ *v.* brush off, cut, ignore, put down, reject, slight, snub, spurn.

rebuild *v.* (**rebuilt**) build again after destruction.

rebuke *v.* reprove. ● *n.* reproof.

■ *v.* admonish, *colloq.* bawl out, berate, carpet, castigate, censure, chastise, chide, *sl.* rap, reprehend, reprimand, reproach, reprove, scold, take to task, *colloq.* tell off, *colloq.* tick off, upbraid. ● *n.* admonition, castigation, dressing down, lecture, reprimand, reproof, scolding.

rebus *n.* representation of a word by pictures etc. suggesting its parts.

rebut *v.* (**rebutted**) disprove. ▫ **rebuttal** *n.*

recalcitrant *adj.* obstinately disobedient. ▫ **recalcitrance** *n.*

■ contrary, defiant, disobedient, headstrong, intractable, obstinate, perverse, refractory, unmanageable, unruly, wayward, wilful.

recall *v.* **1** summon to return. **2** remember. ● *n.* recalling, being recalled.

recant *v.* withdraw and reject (one's former statement or belief). ▫ **recantation** *n.*

recap (*colloq.*) *v.* (**recapped**) recapitulate. ● *n.* recapitulation.

recapitulate *v.* state again briefly. ▫ **recapitulation** *n.*

recapture *v.* **1** capture again. **2** experience again. ● *n.* recapturing.

recede *v.* **1** go back. **2** become more distant.

■ **1** ebb, go *or* move back, retreat, subside.

receipt /riseét/ *n.* **1** act of receiving. **2** written acknowledgement that something has been received or money paid.

receive *v.* **1** acquire, accept, or take in. **2** experience, be treated with. **3** greet on arrival.

■ **1** accept, acquire, be given, come by, gain, get, obtain, take. **2** be subjected to, endure, experience, meet with, suffer, undergo. **3** greet, meet, welcome.

receiver *n.* **1** person or thing that receives something. **2** one who deals in stolen goods. **3** official who handles the affairs of a bankrupt person or firm. **4** apparatus that receives electrical signals and converts them into sound or a picture. **5** earpiece of a telephone.

recent *adj.* happening or established in a time shortly before the present. ▫ **recently** *adv.*

■ current, fresh, modern, new, up to date.

receptacle *n.* thing for holding what is put into it.

■ container, holder, repository; bag, box, can, case, casket, chest, tin, vessel.

reception *n.* **1** act, process, or way of receiving. **2** assembly held to receive guests. **3** place where clients etc. are received on arrival.

receptionist *n.* person employed to receive and direct clients etc.

receptive *adj.* quick to receive ideas. □ **receptiveness** *n.*, **receptivity** *n.*

■ astute, bright, intelligent, perceptive, quick, sharp; amenable, impressionable, open, responsive, sensitive.

recess *n.* **1** space set back from the line of a wall or room etc. **2** temporary cessation from work. ● *v.* make a recess in or of.

■ *n.* **1** alcove, bay, hollow, niche, nook. **2** break, breather, interlude, intermission, interval, pause, respite, rest; holiday, vacation.

recession *n.* **1** receding from a point or level. **2** temporary decline in economic activity.

■ **2** depression, economic decline, slump.

recessive *adj.* tending to recede.

recidivist *n.* person who persistently relapses into crime. □ **recidivism** *n.*

recipe /réssipi/ *n.* **1** directions for preparing a dish. **2** way of achieving something.

recipient *n.* person who receives something.

reciprocal *adj.* mutual. □ **reciprocally** *adv.*, **reciprocity** *n.*

reciprocate *v.* **1** return (affection etc.). **2** give in return. **3** (of a machine part) move backwards and forwards. □ **reciprocation** *n.*

recital *n.* **1** reciting. **2** long account of events. **3** musical entertainment.

■ **1** description, narration, reading, recitation, recounting, relation. **2** account, narrative, report, story. **3** concert, performance.

recitation *n.* **1** reciting. **2** thing recited.

recite *v.* **1** repeat aloud from memory. **2** state (facts) in order.

■ **1** declaim, quote, read aloud *or* out, repeat. **2** detail, enumerate, itemize, list, rattle off, reel off.

reckless *adj.* wildly impulsive. □ **recklessly** *adv.*, **recklessness** *n.*

■ careless, daredevil, foolhardy, foolish, hare-brained, hasty, headlong, heedless, hotheaded, ill-advised, ill-considered, imprudent, impulsive, incautious, injudicious, irresponsible, madcap, precipitate, rash, thoughtless, unwise, wild.

reckon *v.* **1** count up. **2** have as one's opinion. **3** rely. □ **reckon with** take into account.

■ **1** add up, calculate, compute, count up, total, *colloq.* tot up. **2** assume, believe, be of the opinion, imagine, presume, suppose, think; consider, deem, judge, rate, regard. **3** bank, count, depend, rely. □ **reckon with** allow for, anticipate, consider, expect, foresee, take into account *or* consideration.

reclaim *v.* **1** take action to recover possession of. **2** make (waste land) usable. □ **reclamation** *n.*

recline *v.* lean (one's body), lie down.

■ lean back, lie (down), loll, lounge, sprawl, stretch out.

recluse *n.* person who avoids social life.

recognition *n.* recognizing.

recognizance *n.* **1** pledge made to a law court or magistrate. **2** surety for this.

recognize *v.* **1** identify as already known. **2** acknowledge as genuine, valid, or worthy. **3** realize or discover the nature of. □ **recognizable** *adj.*

■ **1** identify, know (again), place, recall, recollect, remember. **2** acknowledge, accept, admit, allow, concede, grant, own. **3** appreciate, be aware *or* conscious of, discover, perceive, realize, see, understand.

recoil *v.* **1** spring back. **2** shrink back in fear or disgust. ● *n.* act of recoiling.

■ *v.* **1** jump *or* spring back, shy away, start. **2** blench, flinch, quail, shrink back.

recollect *v.* remember, call to mind. □ **recollection** *n.*

recommend *v.* **1** advise. **2** suggest as suitable for employment or use etc., speak favourably of. **3** (of qualities etc.) make desirable. □ **recommendation** *n.*

■ **1** advise, advocate, counsel, exhort, suggest, urge. **2** nominate, propose, put forward, suggest; commend, endorse, vouch for. □ **recommendation** advice, advocacy, counsel; proposal, suggestion; commendation, endorsement.

recompense *v.* repay, compensate. ● *n.* repayment.

reconcile *v.* **1** make friendly after an estrangement. **2** settle (quarrel) etc. **3** make compatible. □ **reconcile oneself to** accept something unwelcome, unpleasant, etc. **reconciliation** *n.*

■ **1** bring together, reunite, unite. **2** mend, patch up, put right, resolve, settle, sort out. □ **reconcile oneself to** accept, adjust to, come to terms with, get used to, habituate oneself to, resign oneself to.

recondite *adj.* obscure, dealing with an obscure subject.

■ abstruse, arcane, cryptic, dark, deep, esoteric, incomprehensible, obscure, profound, unfathomable.

recondition *v.* overhaul, repair.

reconnaissance *n.* preliminary survey, esp. exploration of an area for military purposes.

reconnoitre *v.* (**reconnoitring**) make a reconnaissance (of).

■ *sl.* case, examine, explore, inspect, investigate, scout (out), scrutinize, survey.

reconsider *v.* consider again, esp. for a possible change of decision. □ **reconsideration** *n.*

reconstitute *v.* **1** reconstruct. **2** restore to its original form. □ **reconstitution** *n.*

reconstruct *v.* construct or enact again. □ **reconstruction** *n.*

record *v.* /rikórd/ **1** set down in writing or other permanent form. **2** preserve (sound) on a disc or magnetic tape for later reproduction. **3** (of a measuring instrument) indicate, register. ● *n.* /rékkord/ **1** information set down in writing etc. **2** document bearing this. **3** disc bearing recorded sound. **4** facts known about a person's past. **5** best performance or most remarkable event etc. of its kind. ● *adj.* /rékkord/ best or most extreme hitherto recorded. □ **off the record** not for publication.

■ *v.* **1** catalogue, chronicle, enter, log, note *or* put *or* set *or* write down, put in writing, register. **3** display, indicate, read, register, show. ● *n.* **2** annals, archive(s), chronicle, diary, document, file, journal, log, minutes, report. **3** album, compact disc, disc, single. **4** background, career, history, life, past. □ **off the record** confidential, private, secret, unofficial.

recorder *n.* **1** person or thing that records. **2** a kind of flute.

recount *v.* narrate, tell in detail.

■ describe, narrate, recount, report, tell.

re-count *v.* count again. ● *n.* second or subsequent counting.

recoup *v.* **1** reimburse or compensate for a loss. **2** recover or regain (a loss).

recourse *n.* source of help to which one may turn. □ **have recourse to** turn to for help.

recover *v.* **1** regain possession or control of. **2** return to health. □ **recovery** *n.*

■ **1** get *or* win back, recapture, reclaim, recoup, regain, repossess, retake, retrieve; rescue, salvage, save. **2** be on the mend, convalesce, get better *or* well, improve, pick up, rally, recuperate, revive. □ **recovery** recapture, reclamation, repossession, retrieval; rescue, salvage; convalescence, improvement, recuperation.

recreation *n.* **1** pastime. **2** relaxation. □ **recreational** *adj.*

■ **1** hobby, interest, pastime. **2** amusement, diversion, enjoyment, entertainment, fun, play, sport.

recriminate *v.* make angry accusations in retaliation. □ **recrimination** *n.*, **recriminatory** *adj.*

recruit *n.* **1** newly enlisted member of the armed forces. **2** new member of a society or organization. ● *v.* **1** form by enlisting recruits. **2** enlist as a recruit. □ **recruitment** *n.*

■ *n.* **1** conscript. **2** apprentice, initiate, trainee. ● *v.* **1** form, mobilize, raise. **2** engage, enlist, enrol, take on.

rectal *adj.* of the rectum.

rectangle *n.* geometric figure with four straight sides and four right angles. □ **rectangular** *adj.*

rectify *v.* **1** put right. **2** purify, refine. **3** convert to direct current. □ **rectification** *n.*

■ **1** ameliorate, amend, better, correct, emend, fix, improve, mend, put *or* set right, remedy, repair, right.

rectilinear *adj.* bounded by straight lines.

rectitude *n.* correctness of behaviour or procedure.

■ correctness, decency, honesty, honour, integrity, morality, principle, probity, propriety, respectability, righteousness, virtue.

rector *n.* **1** clergyman in charge of a parish. **2** head of certain colleges and universities.

rectory *n.* house of a rector.

rectum *n.* last section of the intestine, between colon and anus.

recumbent *adj.* lying down.

■ lying down, prone, prostrate, reclining, supine.

recuperate *v.* recover (health, strength, or losses). □ **recuperation** *n.*, **recuperative** *adj.*

recur *v.* (**recurred**) happen again or repeatedly.

recurrent *adj.* recurring. ▫ **recurrence** *n.*

■ frequent, periodic, persistent, recurring, regular, repeated.

recusant *n.* person who refuses to submit or comply.

recycle *v.* convert (waste material) for reuse.

red *adj.* (**redder, reddest**) **1** of or like the colour of blood. **2** (of hair) reddish-brown. **3** Communist. ● *n.* **1** red colour or thing. **2** Communist. ▫ **in the red** with a debit balance. **red carpet** privileged treatment for an important visitor. **red-handed** *adj.* in the act of crime. **red herring** misleading clue or diversion. **red-hot** *adj.* **1** glowing red from heat. **2** (of news) completely new. **red-letter day** day of a very joyful occurrence. **red light 1** signal to stop. **2** danger signal. **red tape** excessive formalities in official transactions. **reddish** *adj.*, **redness** *n.*

redbrick *adj.* (of universities) founded in the 19th century or later.

redcurrant *n.* **1** small edible red berry. **2** bush bearing this.

redden *v.* make or become red.

■ blush, colour, flush, go red.

redeem *v.* **1** buy back. **2** convert (tokens etc.) into goods or cash. **3** reclaim. **4** free from sin. **5** make up for (faults). ▫ **redemption** *n.*, **redemptive** *adj.*

redeploy *v.* send to a new place or task. ▫ **redeployment** *n.*

redhead *n.* person with red hair.

redirect *v.* direct or send to another place. ▫ **redirection** *n.*

redolent *adj.* **1** smelling strongly. **2** reminiscent. ▫ **redolence** *n.*

■ **1** aromatic, odoriferous, odorous, perfumed, scented, sweet-smelling. **2** evocative, reminiscent, suggestive.

redouble *v.* increase or intensify.

redoubtable *adj.* formidable.

redress *v.* set right. ● *n.* reparation.

■ *v.* correct, put right, rectify, remedy. ● *n.* compensation, recompense, reparation, restitution.

reduce *v.* **1** make or become smaller or less. **2** make lower in rank. **3** slim. **4** convert into a simpler or more general form. ▫ **reduce to** bring into a specified state. **reduction** *n.*, **reducible** *adj.*

■ **1** abridge, curtail, cut, shorten; abate, decrease, diminish, lessen; ease, moderate, mitigate, tone down. **2** degrade, demote, downgrade. ▫ **reduce to** bring to, drive to, force to.

redundant *adj.* **1** superfluous. **2** no longer needed at work. ▫ **redundancy** *n.*

■ **1** inessential, superfluous, unnecessary **2** jobless, out of work, unemployed.

redwood *n.* **1** very tall evergreen Californian tree. **2** its wood.

re-echo *v.* **1** echo. **2** echo repeatedly. **3** resound.

reed *n.* **1** water or marsh plant with tall hollow stems. **2** its stem. **3** vibrating part of certain wind instruments.

reedy *adj.* (of the voice) having a thin high tone. ▫ **reediness** *n.*

reef *n.* **1** ridge of rock or sand etc reaching to or near the surface of water **2** part of a sail that can be drawn in when there is a high wind. ● *v.* shorten (a sail). ▫ **reef-knot** *n.* symmetrical double knot.

reefer *n.* thick double-breasted jacket.

reek *n.* strong usu. unpleasant smell. ● *v.* smell strongly.

reel *n.* **1** cylinder on which something is wound. **2** lively Scottish or folk dance ● *v.* **1** wind on or off a reel. **2** stagger ▫ **reel off** recite rapidly.

■ *v.* **1** coil, turn, twine, twist, wind. **2** lurch stagger, stumble, sway.

refectory *n.* dining room of a monastery or college etc.

refer *v.* (**referred**) **refer to 1** mention. **2** direct to an authority or specialist. **3** turn to for information. **4** be relevant to.

■ **1** allude to, bring up, mention, speak of touch on. **2** direct to, pass on to, send to. **3** check, look at, study; ask, consult, speak to. **4** apply to, concern, pertain to, relate to.

referee *n.* **1** umpire, esp. in football and boxing. **2** person to whom disputes are referred for decision. **3** person willing to testify to the character or ability of one applying for a job. ● *v.* (**refereed**) act as referee (for).

■ *n.* **1,2** arbitrator, judge, umpire.

reference *n.* **1** act of referring. **2** mention. **3** testimonial. **4** person willing to testify to another's character, ability, etc **5** source of information. ▫ **in** or **with reference to** in connection with, about **reference book** book providing information. **reference library** one con taining books that can be consulted but not taken away.

■ 2 allusion, mention. 3 endorsement, recommendation, testimonial.

referendum *n.* (*pl.* **-ums**) referring of a question to the people for decision by a general vote.

referral *n.* referring.

refill *v.* /reéfil/ fill again. ● *n.* /reéfil/ **1** second or later filling. **2** material used for this.

refine *v.* **1** remove impurities or defects from. **2** make elegant or cultured. □ **refined** *adj.*

■ 1 cleanse, decontaminate, purify. 2 civilize, cultivate, improve, perfect, polish.

refinement *n.* **1** refining. **2** elegance of behaviour. **3** improvement added.

■ 1 development, enhancement, improvement, perfection; cleansing, purification, refining. 2 breeding, culture, elegance, gentility, polish, sophistication, urbanity. 3 addition, alteration, improvement, modification.

refinery *n.* establishment where crude substances are refined.

reflate *v.* restore (a financial system) after deflation. □ **reflation** *n.*, **reflationary** *adj.*

reflect *v.* **1** throw back (light, heat, or sound). **2** show an image of. **3** have as a cause or source. **4** think deeply. **5** bring (credit or discredit). □ **reflection** *n.*

■ 2 mirror. 3 demonstrate, exemplify, exhibit, illustrate, point to, reveal, show. 4 brood, cogitate, deliberate, meditate, muse, ponder, ruminate, think.

reflective *adj.* **1** reflecting. **2** thoughtful.

■ 2 contemplative, meditative, pensive, ruminative, thoughtful.

reflector *n.* thing that reflects light or heat.

reflex *n.* reflex action. □ **reflex action** involuntary or instinctive movement in response to a stimulus. **reflex angle** angle of more than 180°.

reflexive *adj.* & *n.* (word or form) showing that the action of the verb is performed on its subject (e.g. *he washed himself*).

reflux *n.* flowing back.

reform *v.* **1** improve by removing faults. **2** (cause to) give up bad behaviour. ● *n.* reforming. □ **reformation** *n.*, **reformer** *n.*

■ *v.* 1 ameliorate, amend, better, correct, emend, improve, make better, put right, rectify, revise. 2 go straight, mend one's ways.

reformatory *adj.* reforming. ● *n.* institution to which offenders are sent to be reformed.

refract *v.* bend (a ray of light) where it enters water or glass etc. obliquely. □ **refraction** *n.*, **refractor** *n.*, **refractive** *adj.*

refractory *adj.* **1** resisting control or discipline. **2** resistant to treatment or heat.

refrain¹ *n.* **1** recurring lines of a song. **2** music for these.

refrain² *v.* **refrain from** keep oneself from doing something.

■ abstain from, avoid, eschew, forbear from; cease, desist from, discontinue, *US* quit, stop.

refresh *v.* **1** restore the vigour of by food, drink, or rest. **2** stimulate (a person's memory).

■ 1 fortify, invigorate, pep up, perk up, restore, revive.

refreshing *adj.* **1** restoring vigour, cooling. **2** welcome and interesting because of its novelty. □ **refreshingly** *adv.*

■ 1 enlivening, fortifying, invigorating, restorative, reviving, stimulating; bracing, cool, crisp.

refreshment *n.* **1** process of refreshing. **2** (*pl.*) food and drink.

■ 2 (**refreshments**) drinks, eatables, food, *sl.* grub, nibbles, snacks.

refrigerate *v.* make extremely cold, esp. in order to preserve. □ **refrigerant** *n.*, **refrigeration** *n.*

refrigerator *n.* cabinet or room in which food is stored at a very low temperature.

refuge *n.* shelter from pursuit or danger.

■ asylum, haven, hiding place, retreat, sanctuary, shelter.

refugee *n.* person who has left home and seeks refuge (e.g. from war or persecution) elsewhere.

refund *v.* /rifúnd/ pay back. ● *n.* /reéfund/ repayment, money refunded.

■ *v.* pay back, reimburse, repay.

refurbish *v.* brighten up, redecorate. □ **refurbishment** *n.*

■ decorate, do up, redecorate, renovate, revamp.

refuse[1] /rifyo͞oz/ *v.* say or show that one is unwilling to accept or do (what is asked or required). ▫ **refusal** *n.*

■ decline, *colloq.* pass up, turn down; deny, deprive of, withhold.

refuse[2] /réfyo͞oss/ *n.* waste material.

■ debris, dross, garbage, litter, rubbish, trash, waste.

refute *v.* prove falsity or error of. ▫ **refutation** *n.*

■ *colloq.* debunk, discredit, disprove, prove wrong.

regain *v.* **1** obtain again after loss. **2** reach again.

regal *adj.* like or fit for a monarch. ▫ **regally** *adv.*, **regality** *n.*

■ dignified, kingly, majestic, noble, princely, queenly, royal.

regale *v.* feed or entertain well.

regalia *n.pl.* emblems of royalty or rank.

regard *v.* **1** look steadily at. **2** consider to be. ● *n.* **1** attention, care. **2** respect. **3** (*pl.*) kindly greetings conveyed in a message. ▫ **as regards, with regard to** regarding.

■ *v.* **1** *poetic* behold, contemplate, gaze at, look at, observe, survey, view. **2** consider, judge, perceive, think of, view. ● *n.* **1** attention, care, concern, heed, notice, thought. **2** consideration, deference, esteem, honour, respect, reverence. **3** (**regards**) best wishes, compliments, greetings, salutations.

regarding *prep.* with reference to.

regardless *adv.* without paying attention. ▫ **regardless of** despite.

regatta *n.* boat or yacht races organized as a sporting event.

regency *n.* **1** rule by a regent. **2** period of this.

regenerate *v.* give new life or vigour to. ▫ **regeneration** *n.*, **regenerative** *adj.*

regent *n.* person appointed to rule while the monarch is a child, ill, or absent.

reggae *n.* W. Indian style of music, with a strong beat.

regicide *n.* killing or killer of a king. ▫ **regicidal** *adj.*

regime /rayzhe͞em/ *n.* method or system of government.

regimen *n.* **1** prescribed course of treatment etc. **2** way of life.

regiment *n.* **1** permanent unit of an army. **2** large array or number of things. ● *v.* organize rigidly. ▫ **regimentation** *n.*

regimental *adj.* of an army regiment.

region *n.* **1** part of a surface, space, or body. **2** administrative division of a country. ▫ **in the region of** approximately. **regional** *adj.*

■ **2** area, district, division, locality, part, province, quarter, sector, territory, zone.

register *n.* **1** official list. **2** range of a voice or musical instrument. ● *v.* **1** enter in a register. **2** record esp. in writing. **3** notice and remember. **4** indicate. **5** make an impression. ▫ **register office** place where records of births, marriages, and deaths are kept and civil marriages are performed. **registration** *n.*

■ *n.* **1** catalogue, directory, inventory, list, record, roll. ● *v.* **1** enter, note *or* put down; check in. **2** log, note, record, write down. **3** mark, note, notice, take note of. **4** indicate, measure, record; display, exhibit, express, show.

registrar *n.* **1** official responsible for keeping written records. **2** hospital doctor ranking just below specialist.

registry *n.* **1** registration. **2** place where written records are kept. ▫ **registry office** register office.

regress *v.* relapse to an earlier or more primitive state. ▫ **regression** *n.*, **regressive** *adj.*

regret *n.* feeling of sorrow about a loss, or of annoyance or repentance. ● *v.* (**regretted**) feel regret about. ▫ **regretful** *adj.*, **regretfully** *adv.*

■ *n.* compunction, contrition, guilt, penitence, remorse, repentance, sadness, sorrow; (**regrets**) qualms, second thoughts. ● *v.* bemoan, be sorry about, feel remorse for, lament, rue.

regrettable *adj.* that is to be regretted. ▫ **regrettably** *adv.*

■ awful, deplorable, dreadful, lamentable, sad, terrible, unfortunate.

regular *adj.* **1** acting, occurring, or done in a uniform manner or at a fixed time or interval. **2** conforming to a rule or habit. **3** even, symmetrical. **4** forming a country's permanent armed forces. ● *n.* **1** regular soldier etc. **2** (*colloq.*) regular customer etc. ▫ **regularly** *adv.*, **regularity** *n.*

■ *adj.* **1** even, rhythmical, steady, uniform; fixed, ordered, orderly, systematic. **2** accustomed, customary, habitual, normal, ordinary, routine, standard, traditional, usual; correct, legal, official, proper. **3** even, symmetrical, well-proportioned.

regularize *v.* make regular, lawful, or correct. ▫ **regularization** *n.*

regulate *v.* **1** control by rules. **2** adjust to work correctly or according to one's requirements. □ **regulator** *n.*

■ **1** administer, control, direct, govern, manage, monitor. **2** adjust, control, modify, modulate.

regulation *n.* **1** process of regulating. **2** rule.

■ **1** control, management; adjustment, modulation. **2** decree, directive, edict, law, order, rule, statute.

regurgitate *v.* **1** bring (swallowed food) up again to the mouth. **2** reproduce (information etc.). □ **regurgitation** *n.*

rehabilitate *v.* restore to a normal life or good condition. □ **rehabilitation** *n.*

rehash *v.* /reéhásh/ put (old material) into a new form. ● *n.* /reéhash/ **1** rehashing. **2** thing made of rehashed material.

rehearse *v.* practise beforehand. □ **rehearsal** *n.*

rehouse *v.* provide with new accommodation.

reign *n.* (period of) sovereignty or rule. ● *v.* **1** rule as king or queen. **2** prevail.

■ *v.* **1** govern, rule. **2** be prevalent, hold sway, predominate, prevail.

reimburse *v.* repay (a person), refund. □ **reimbursement** *n.*

■ pay back, refund, repay.

rein *n.* (also *pl.*) **1** long strap fastened to a bridle, used to guide or check a horse. **2** means of control. ● *v.* check or control with reins.

reincarnation *n.* rebirth of the soul in another body after death of the first. □ **reincarnate** *v.*

reindeer *n.* (*pl.* **reindeer**) deer of Arctic regions, with large antlers.

reinforce *v.* strengthen with additional people, material, or quantity. □ **reinforcement** *n.*

■ bolster, buttress, fortify, shore up, strengthen, support.

reinstate *v.* restore to a previous position. □ **reinstatement** *n.*

reiterate *v.* say or do again or repeatedly. □ **reiteration** *n.*

reject *v.* /rijékt/ **1** refuse to accept. **2** rebuff. ● *n.* /reéjekt/ person or thing rejected. □ **rejection** *n.*

■ *v.* **1** decline, refuse, say no to, turn down, veto; discount, dismiss, disregard; discard, scrap, throw away. **2** brush off, rebuff, shun, snub, spurn.

rejoice *v.* feel or show great joy.

■ be happy *or* delighted, celebrate, exult, glory.

rejoin *v.* **1** retort. **2** join again.

rejoinder *n.* answer, retort.

■ answer, reply, response, retort.

rejuvenate *v.* restore youthful appearance or vigour to. □ **rejuvenation** *n.*, **rejuvenator** *n.*

relapse *v.* fall back (into worse state after improvement). ● *n.* relapsing.

■ *v.* backslide, decline, deteriorate, lapse, regress, retrogress.

relate *v.* **1** narrate. **2** establish a relation between. □ **relate to 1** have a connection with. **2** feel sympathetic to.

■ **1** describe, narrate, recite, recount, tell. **2** associate, connect, link, tie. □ **relate to 1** apply to, be relevant to, concern, have a bearing on, pertain to, refer to. **2** empathize with, identify with, sympathize with, understand.

related *adj.* having a common descent or origin.

relation *n.* **1** similarity connecting people or things. **2** relative. **3** narrating. **4** (*pl.*) dealings with others. □ **in relation to** with reference to.

■ **1** association, connection, correspondence, interconnection, link, relationship, tie.

relationship *n.* **1** state of being related. **2** connection, relation. **3** emotional (esp. sexual) association between two people.

■ **3** affair, liaison, love affair, romance.

relative *adj.* considered in relation to something else. ● *n.* person related to another by descent or marriage. □ **relatively** *adv.*

■ *n.* (**relatives**) family, flesh and blood, folk, kindred, kinsfolk, kith and kin, relations.

relativity *n.* **1** being relative. **2** Einstein's theory of the universe, showing that all motion is relative and treating time as a fourth dimension related to space.

relax *v.* **1** make or become less tight, tense, or strict. **2** rest from work, indulge in recreation. □ **relaxation** *n.*

■ **1** loosen, release, slacken; calm down, unbend, *colloq.* unwind; moderate, modify, soften, temper, tone down. □ **relaxation** amusement, entertainment, fun, pleasure, recreation, rest.

relaxed *adj.* not worried or tense.

■ calm, carefree, cool, easygoing, happy-go-lucky, insouciant, nonchalant, peaceful, serene, tranquil.

relay *n.* /reélay/ **1** fresh set of workers relieving others. **2** relay race. **3** relayed message or transmission. **4** device relaying things or activating an electrical circuit. ● *v.* /riláy/ (**relayed**) receive and pass on or retransmit. □ **relay race** race between teams in which each person in turn covers a part of the total distance.

release *v.* **1** set free. **2** remove from a fixed position. **3** make (information, a film or recording) available to the public. ● *n.* **1** releasing. **2** handle or catch etc. that unfastens something. **3** record, film, etc. released.

■ *v.* **1** discharge, free, let go, liberate, loose, set free. **3** broadcast, circulate, disseminate, issue, make public, publish.

relegate *v.* consign to a less important position or group. □ **relegation** *n.*

■ degrade, demote, downgrade.

relent *v.* become less severe or more lenient. □ **relentless** *adj.*, **relentlessly** *adv.*

■ capitulate, give way, melt, soften, yield.

relevant *adj.* related to the matter in hand. □ **relevance** *n.*

■ applicable, apposite, appropriate, germane, pertinent, to the point.

reliable *adj.* able to be relied on. □ **reliably** *adv.*, **reliability** *n.*

■ dependable, infallible, safe, trusted, trustworthy, unfailing; honest, reputable, responsible.

reliance *n.* trust, confidence. □ **reliant** *adj.*

relic *n.* **1** thing that survives from earlier times. **2** (*pl.*) remains.

relief *n.* **1** ease given by reduction or removal of pain or anxiety etc. **2** thing that breaks up monotony. **3** assistance to those in need. **4** person replacing one who is on duty. **5** carving etc. in which the design projects from a surface. **6** similar effect given by colour or shading. □ **relief road** road by which traffic can avoid a congested area.

■ **1** abatement, alleviation, easing; comfort, consolation, solace. **3** aid, assistance, charity, help, support.

relieve *v.* **1** give or bring relief to. **2** release from a task or duty. □ **relieve oneself** urinate or defecate.

■ **1** alleviate, diminish, ease, lessen, moderate, reduce, soothe.

religion *n.* **1** belief in and worship of a superhuman controlling power. **2** system of this. **3** influence compared to religious faith.

■ **2** belief, creed, faith; denomination, sect.

religious *adj.* **1** of religion. **2** devout, pious. **3** very conscientious. □ **religiously** *adv.*

■ **2** devout, God-fearing, holy, pious. **3** conscientious, meticulous, painstaking, scrupulous.

relinquish *v.* give up, cease from. □ **relinquishment** *n.*

■ abandon, cede, concede, forfeit, give up, surrender, waive, yield.

reliquary *n.* receptacle for relic(s) of a holy person.

relish *n.* **1** great enjoyment of something. **2** appetizing flavour, thing giving this. ● *v.* enjoy greatly.

■ *n.* **1** delight, eagerness, enjoyment, enthusiasm, gusto, joy, pleasure, zest. ● *v.* appreciate, delight in, enjoy, love, revel in, take pleasure in.

relocate *v.* move to a different place. □ **relocation** *n.*

reluctant *adj.* unwilling, grudging one's consent. □ **reluctantly** *adv.*, **reluctance** *n.*

■ disinclined, hesitant, loath, unwilling.

rely *v.* **rely on** trust confidently, depend on for help etc.

■ bank on, be sure of, count on, depend on, have faith in, trust.

remain *v.* **1** stay. **2** be left or left behind. **3** continue in the same condition.

■ **3** carry on as, continue to be, go on as, keep, stay.

remainder *n.* **1** remaining people, things, or part. **2** quantity left after subtraction or division. ● *v.* dispose of unsold copies of (a book) at a reduced price.

■ *n.* **1** excess, residue, rest, surplus. **2** balance, difference.

remains *n.pl.* **1** what remains, surviving parts. **2** dead body.

■ **1** debris, fragments, leavings, leftovers, remnants, scraps. **2** body, cadaver, corpse, *sl.* stiff.

remand *v.* send back (a prisoner) into custody while further evidence is sought. ◻ **on remand** remanded.

remark *n.* spoken or written comment. ● *v.* **1** make a remark, say. **2** notice.

■ *n.* comment, observation, statement, utterance. ● *v.* **1** comment, declare, mention, note, observe, say. **2** note, notice, observe, perceive.

remarkable *adj.* worth noticing, unusual. ◻ **remarkably** *adv.*

■ amazing, astonishing, exceptional, extraordinary, impressive, incredible, marvellous, outstanding, singular, uncommon, unusual, wonderful.

remedy *n.* thing that cures or relieves a disease or puts right a matter. ● *v.* be a remedy for, put right. ◻ **remedial** *adj.*

■ *n.* cure, medicine, therapy, treatment. ● *v.* correct, cure, fix, heal, make good, put right, rectify, repair.

remember *v.* keep in one's mind and recall at will. ◻ **remembrance** *n.*

■ recall, recollect; reminisce.

remind *v.* cause to remember.

reminder *n.* thing that reminds someone, letter sent as this.

■ cue; keepsake, memento, souvenir.

reminisce *v.* think or talk about past events.

reminiscence *n.* **1** reminiscing. **2** (usu. *pl.*) account of what one remembers.

reminiscent *adj.* having characteristics that remind one (of something).

■ evocative, redolent, suggestive.

remiss *adj.* negligent.

■ careless, heedless, inattentive, lax, negligent, slack, thoughtless.

remission *n.* **1** remitting of a debt or penalty. **2** reduction of force or intensity.

remit *v.* /rimít/ (**remitted**) **1** cancel (a debt or punishment). **2** make or become less intense. **3** send (money etc.). **4** refer (a matter for decision) to an authority. ● *n.* /reémit/ terms of reference.

remittance *n.* **1** sending of money. **2** money sent.

remnant *n.* **1** small remaining quantity. **2** surviving trace.

■ **1** (**remnants**) debris, fragments, leftovers, remains, scraps. **2** relic, trace, vestige.

remonstrate *v.* make a protest. ◻ **remonstrance** *n.*

■ complain, expostulate, *colloq.* kick up, make a fuss, protest.

remorse *n.* deep regret for one's wrongdoing. ◻ **remorseful** *adj.*, **remorsefully** *adv.*

■ contrition, guilt, penitence, regret, repentance, self-reproach.

remorseless *adj.* without compassion. ◻ **remorselessly** *adv.*

■ callous, cruel, hard-hearted, merciless, pitiless, ruthless.

remote *adj.* **1** far away in place or time, not close. **2** slight. ◻ **remotely** *adv.*, **remoteness** *n.*

■ **1** distant, far-away, far-off, outlying; isolated, lonely, out of the way. **2** faint, slight.

remove *v.* **1** take off or away. **2** dismiss from office. **3** get rid of. ● *n.* degree of remoteness or difference. ◻ **removable** *adj.*, **remover** *n.*, **removal** *n.*

■ *v.* **1** doff, take off; detach, disconnect, separate, undo, unfasten; move, take away. **2** discharge, dismiss, fire, *colloq.* sack. **3** eliminate, eradicate, get rid of, root out; delete, efface, erase, rub out, wipe out.

remunerate *v.* pay or reward for services. ◻ **remuneration** *n.*

■ ◻ **remuneration** emolument, fee, pay, payment, salary, wages.

remunerative *adj.* giving good remuneration, profitable.

■ gainful, lucrative, profitable.

Renaissance *n.* **1** revival of art and learning in Europe in the 14th–16th centuries. **2** (**renaissance**) any similar revival.

renal *adj.* of the kidneys.

rend *v.* (**rent**) tear.

render *v.* **1** give, esp. in return. **2** cause to become. **3** perform. **4** submit (a bill etc.). **5** translate. **6** melt down (fat).

rendezvous /róndivōō/ *n.* (*pl.* **-vous**) prearranged meeting or meeting place. ● *v.* meet at a rendezvous.

■ *n.* appointment, assignation, *colloq.* date, meeting; venue.

rendition *n.* way something is rendered or performed.

renegade *n.* person who deserts from a group or cause etc.

renege /rináyg/ *v.* fail to keep a promise or agreement.

renew *v.* **1** revive. **2** replace. **3** resume. **4** repeat. ▫ **renewal** *n.*

■ **1** refresh, reinvigorate, restore, revive. **2** replace; refill, replenish. **3** restart, resume, return to, take up. **4** reiterate, repeat, restate.

rennet *n.* substance used to curdle milk in making cheese.

renounce *v.* **1** give up formally. **2** reject. ▫ **renouncement** *n.*

■ **1** forgo, give up, relinquish, surrender. **2** deny, disown, reject, repudiate, spurn.

renovate *v.* repair, restore to good condition. ▫ **renovation** *n.*, **renovator** *n.*

■ decorate, do up, modernize, overhaul, redecorate, refurbish, repair, revamp.

renown *n.* fame. ▫ **renowned** *adj.*

■ celebrity, distinction, eminence, fame, glory, note, prominence. ▫ **renowned** celebrated, distinguished, eminent, famous, illustrious, noted, well-known.

rent[1] *see* **rend**. *n.* torn place.

rent[2] *n.* periodical payment for use of land, rooms, machinery, etc. ● *v.* pay or receive rent for.

■ *v.* hire (out), lease (out), let (out); charter.

rental *n.* **1** rent. **2** renting.

renunciation *n.* renouncing.

rep[1] *n.* (*colloq.*) business firm's travelling representative.

rep[2] *n.* (*colloq.*) repertory.

repair *v.* **1** put into good condition after damage or wear. **2** make amends for. ● *n.* **1** process of repairing. **2** repaired place. **3** condition for use. ▫ **repairer** *n.*

■ *v.* **1** fix, mend, overhaul, patch up, put right, renovate, restore, service. ● *n.* **3** condition, form, shape, working order.

reparation *n.* **1** making amends. **2** compensation.

repartee *n.* **1** witty reply. **2** exchange of witty remarks.

repast *n.* (*formal*) a meal.

repatriate *v.* send or bring back (a person) to his or her own country. ▫ **repatriation** *n.*

repay *v.* (**repaid**) pay back. ▫ **repayment** *n.*, **repayable** *adj.*

■ pay back, recompense, refund, reimburse, settle up with.

repeal *v.* withdraw (a law) officially. ● *n.* repealing of a law.

■ *v.* abolish, abrogate, annul, cancel, rescind, revoke.

repeat *v.* **1** say, do, produce, or occur again. **2** tell (a thing told to oneself) to another person. ● *n.* **1** repeating. **2** thing repeated. ▫ **repeatedly** *adv.*

■ *v.* **1** *colloq.* recap, recapitulate, reiterate, restate, say again; echo; duplicate, reproduce; recur.

repel *v.* (**repelled**) **1** drive away. **2** be impossible for (a substance) to penetrate. **3** be repulsive to. ▫ **repellent** *adj.* & *n.*

■ **1** drive back, fend *or* fight *or* stave *or* ward off, repulse. **3** disgust, revolt, sicken.

repent *v.* feel regret about (what one has done or failed to do). ▫ **repentance** *n.*, **repentant** *adj.*

repercussion *n.* **1** indirect effect or reaction. **2** recoil.

■ **1** consequence, effect, outcome, result, upshot.

repertoire /réppərtwaar/ *n.* stock of songs, plays, etc., that a person or company is prepared to perform,

repertory *n.* **1** repertoire. **2** theatrical performances of various plays for short periods by one company (**repertory company**).

repetition *n.* **1** repeating. **2** instance of this. ▫ **repetitious** *adj.*, **repetitive** *adj.*

repine *v.* fret, be discontented.

replace *v.* **1** put back in its place. **2** take the place of. **3** be or find a substitute for. ▫ **replacement** *n.*

■ **1** put back, return. **2** follow, succeed, supersede, supplant, take the place of.

replay *v.* /reéplάy/ play again. ● *n.* /reéplay/ replaying.

replenish *v.* **1** refill. **2** renew (a supply etc.). ▫ **replenishment** *n.*

replete *adj.* **1** well-fed. **2** filled or well-supplied. ▫ **repletion** *n.*

■ **1** full, sated, satiated, well-fed.

replica *n.* exact copy.

replicate *v.* make a replica of. ▫ **replication** *n.*

reply *v.* & *n.* answer.

■ *v.* answer, rejoin, respond, retort. ● *n.* answer, rejoinder, response, retort, riposte.

report *v.* **1** give an account of. **2** tell as news. **3** make a formal complaint about. **4** present oneself on arrival. ● *n.* **1** spoken or written account. **2** written statement about a pupil's work etc. **3** rumour. **4** explosive sound.

■ *v.* **1** describe, give an account of, recount, relate, tell of. **2** announce, broadcast. ● *n.* **1** account, description, record, statement; article, bulletin, feature, item, piece, story. **4** bang, blast, boom, explosion, shot.

reporter *n.* person employed to report news etc. for publication or broadcasting.

■ correspondent, hack, journalist.

repose *n.* **1** rest, sleep. **2** tranquillity. ● *v.* rest, lie.

■ *n.* **1** inactivity, peace, relaxation, rest; sleep, slumber. **2** calm, peace, quiet, stillness, tranquillity.

repository *n.* storage place.

repossess *v.* take back something for which payments have not been made. □ **repossession** *n.*

reprehend *v.* rebuke.

reprehensible *adj.* deserving rebuke. □ **reprehensibly** *adv.*

■ blameworthy, disgraceful, deplorable, indefensible, inexcusable, shameful, unforgivable.

represent *v.* **1** be an example or embodiment of. **2** show in a picture or play etc. **3** describe or declare (to be). **4** act on behalf of. □ **representation** *n.*

■ **1** embody, epitomize, exemplify, stand for, symbolize, typify. **2** depict, portray, show.

representative *adj.* typical of a group or class. ● *n.* **1** person's or firm's agent. **2** person chosen to represent others.

■ *adj.* characteristic, illustrative, typical. ● *n.* **1** agent, *colloq.* rep, salesman, saleswoman. **2** agent, delegate, emissary, envoy, spokesperson.

repress *v.* suppress, keep (emotions) from finding an outlet. □ **repression** *n.*, **repressive** *adj.*

■ contain, control, hold back, restrain, stifle, suppress.

reprieve *n.* **1** postponement or cancellation of punishment (esp. death sentence). **2** respite. ● *v.* give a reprieve to.

reprimand *v.* & *n.* rebuke.

■ *v.* berate, chastise, chide, rebuke, reproach, scold, take to task, *colloq.* tell off, *colloq.* tick off, upbraid.

reprint *v.* /reeprint/ print again. ● *n.* /reeprint/ book reprinted.

reprisal *n.* act of retaliation.

reproach *v.* express disapproval to (a person) for a fault or offence. ● *n.* **1** act or instance of reproaching. **2** (cause of) discredit. □ **reproachful** *adj.*, **reproachfully** *adv.*

■ *v.* admonish, blame, castigate, censure, criticize, rebuke, reprehend, reprimand, reprove, take to task, upbraid.

reprobate *n.* immoral or unprincipled person.

reproduce *v.* **1** produce again. **2** produce a copy of. **3** produce further members of the same species. □ **reproduction** *n.*, **reproductive** *adj.*

■ **1,2** copy, duplicate, replicate. **3** breed, multiply, procreate.

reproof *n.* expression of condemnation for a fault or offence.

■ admonition, castigation, censure, condemnation, criticism, reproach.

reprove *v.* give a reproof to.

reptile *n.* member of the class of cold-blooded animals with a backbone and rough or scaly skin. □ **reptilian** *adj.* & *n.*

republic *n.* country in which the supreme power is held by the people or their representatives.

republican *adj.* of or advocating a republic. ● *n.* person advocating republican government.

repudiate *v.* reject or disown utterly, deny. □ **repudiation** *n.*

■ deny, disown, reject, renounce.

repugnant *adj.* distasteful, repulsive. □ **repugnance** *n.*

repulse *v.* **1** drive back (an attacking force). **2** reject, rebuff. ● *n.* **1** driving back. **2** rebuff.

■ *v.* **1** drive back, fend *or* fight *or* stave *or* ward off, repel.

repulsion *n.* **1** repelling. **2** strong feeling of distaste, revulsion.

repulsive *adj.* **1** arousing disgust. **2** able to repel. □ **repulsively** *adv.*, **repulsiveness** *n.*

■ **1** abhorrent, disgusting, distasteful, foul, hateful, horrible, loathsome, nasty, obnoxious, odious, offensive, repellent, repugnant, revolting, sickening, unpleasant, unsavoury, vile.

reputable *adj.* having a good reputation, respected.

reputation *n.* what is generally believed about a person or thing.

repute *n.* reputation.

reputed *adj.* said or thought to be. □ **reputedly** *adv.*

■ alleged, believed, considered, rumoured, supposed, thought.

request *n.* **1** asking for something. **2** thing asked for. ● *v.* make a request (for or of).

■ *n.* appeal, application, call, demand, entreaty, petition, plea. ● *v.* apply for, ask for, beg, beseech, demand, entreat, plead for, solicit.

requiem /rékwi-em/ *n.* **1** special mass for the repose of the soul(s) of the dead. **2** music for this.

require *v.* **1** need, depend on for success etc. **2** order, oblige. □ **requirement** *n.*

■ **1** demand, entail, necessitate, need; be short of, lack, want. **2** command, compel, force, instruct, make, oblige, order; demand, insist (on).

requisite *adj.* required, necessary. ● *n.* thing needed.

requisition *n.* formal written demand, order laying claim to use of property or materials. ● *v.* demand or order by this.

■ *v.* demand, order, request; appropriate, commandeer, confiscate, expropriate, seize, take possession of.

requite *v.* make a return for.

resale *n.* sale to another person of something one has bought.

rescind *v.* repeal or cancel.

rescue *v.* save from danger or capture etc. ● *n.* rescuing. □ **rescuer** *n.*

■ *v.* deliver, free, liberate, release, set free; salvage, save.

research *n.* study and investigation, esp. to discover new facts. ● *v.* perform research (into). □ **researcher** *n.*

■ *v.* analyse, examine, inquire into, look into, investigate, probe, study.

resemble *v.* be like. □ **resemblance** *n.*

■ be similar to, correspond to, look like, take after.

resent *v.* feel displeased and indignant about. □ **resentment** *n.*, **resentful** *adj.*, **resentfully** *adv.*

■ □ **resentful** angry, annoyed, bitter, displeased, grudging, envious, jealous, indignant, irritated, *colloq.* peeved.

reservation *n.* **1** reserving. **2** reserved accommodation etc. **3** doubt. **4** land set aside, esp. for occupation by American Indians.

■ **1,2** booking. **3** doubt, misgiving, qualm, scruple; hesitation, reluctance.

reserve *v.* **1** put aside for future or special use. **2** order or set aside for a particular person. **3** retain, esp. a power. ● *n.* **1** thing(s) reserved, extra stock available. **2** tendency to avoid showing feelings or friendliness. **3** (also *pl.*) forces outside the regular armed services. **4** land set aside for special use, esp. as habitat. **5** substitute player in team games. **6** lowest acceptable price for an item to be auctioned. □ **in reserve** unused, available.

■ *v.* **1** conserve, keep back, preserve, retain, save, set aside. **2** book, order; save. ● *n.* **1** cache, fund, hoard, stock, stockpile, store, supply. **2** caution, self-control, self-restraint; quietness, reticence, shyness, taciturnity. □ **in reserve** available, on hand, *colloq.* on tap, ready.

reserved *adj.* (of a person) showing reserve of manner.

■ aloof, cool, distant, formal, reticent, standoffish, taciturn, uncommunicative, unforthcoming, unresponsive, unsociable, withdrawn.

reservist *n.* member of a reserve force.

reservoir /rézzərvwaar/ *n.* **1** natural or artificial lake that is a source or store of water to a town etc. **2** container for a supply of fluid.

reshuffle *v.* interchange, reorganize. ● *n.* reshuffling.

reside *v.* dwell permanently.

residence *n.* **1** residing. **2** one's dwelling. □ **in residence** living or working in a specified place.

■ **2** abode, domicile, dwelling, home, house, lodging, quarters.

resident *adj.* residing, in residence. ● *n.* **1** permanent inhabitant. **2** (at a hotel) person staying overnight.

residential *adj.* **1** containing dwellings. **2** of or based on residence.

residual *adj.* left over as a residue. □ **residually** *adv.*

residue *n.* what is left over.

■ excess, remainder, rest, surplus; dregs, lees, remains.

resign *v.* give up (one's job, claim, etc.). □ **resign oneself to** accept (a situation etc.) reluctantly. **resignation** *n.*

■ □ **resign oneself to** accept, accommodate oneself to, come to terms with, face up to, reconcile oneself to.

resigned *adj.* having resigned oneself. □ **resignedly** *adv.*

resilient *adj.* **1** springy. **2** readily recovering from shock etc. □ **resiliently** *adv.*, **resilience** *n.*

resin *n.* **1** sticky substance from plants and certain trees. **2** similar substance made synthetically, used in plastics. □ **resinous** *adj.*

resist *v.* **1** oppose strongly or forcibly. **2** withstand the action or effect of. **3** abstain from (pleasure etc.). □ **resistance** *n.*, **resistant** *adj.*, **resistible** *adj.*

■ **1** combat, fight (against), oppose, strive *or* struggle against. **2** block, check, curb, hinder, impede, stem, stop, thwart; be proof against, withstand. **3** decline, forgo, *colloq.* pass up, refuse, turn down.

resistivity *n.* resistance to the passage of electric current.

resistor *n.* device having resistance to the passage of electric current.

resolute *adj.* showing great determination. □ **resolutely** *adv.*, **resoluteness** *n.*

■ adamant, decided, determined, dogged, firm, purposeful, resolved, steadfast, unshakeable, unswerving, unwavering.

resolution *n.* **1** firm intention. **2** great determination, resolve. **3** formal statement of a committee's opinion. **4** resolving.

resolve *v.* **1** decide firmly. **2** solve or settle (a problem or doubts). **3** separate into constituent parts. ● *n.* great determination.

■ *v.* **1** decide, determine, make up one's mind. **2** settle, solve, sort out. ● *n.* determination, doggedness, firmness, perseverance, persistence, purpose, resoluteness, resolution, steadfastness, will-power.

resonant *adj.* **1** resounding, echoing. **2** reinforcing sound, esp. by vibration. □ **resonance** *n.*

resonate *v.* produce or show resonance. □ **resonator** *n.*

resort *v.* **resort to** adopt as an expedient. ● *n.* **1** expedient. **2** popular holiday place.

resound *v.* **1** fill a place or be filled with sound. **2** echo.

■ boom, echo, resonate, reverberate, ring.

resource *n.* **1** something to which one can turn for help. **2** ingenuity. **3** (*pl.*) available assets.

■ **3** (**resources**) assets, capital, cash, finances, funds, means, money, wealth, *colloq.* wherewithal.

resourceful *adj.* clever at finding ways of doing things. □ **resourcefully** *adv.*, **resourcefulness** *n.*

respect *n.* **1** admiration or esteem. **2** consideration. **3** particular aspect. **4** (*pl.*) polite greetings. ● *v.* **1** feel or show respect for. **2** agree to recognize. □ **with respect to** about, concerning. **respectful** *adj.*, **respectfully** *adv.*

■ *n.* **1** admiration, esteem, regard. **2** civility, consideration, courtesy, deference, politeness, reverence; attention, care, concern, heed, regard, thought. **3** aspect, characteristic, detail, feature, particular, point, quality. **4** (**respects**) best wishes, compliments, greetings, regards, salutations. ● *v.* **1** admire, esteem, have a high opinion of, honour, look up to, revere, think highly of, value. **2** fulfil, honour, keep. □ **with respect to** about, apropos, as regards, concerning, re, regarding, with reference to, with regard to.

respectable *adj.* **1** worthy of respect. **2** considerable. □ **respectably** *adv.*, **respectability** *n.*

■ **1** decent, estimable, honourable, respected, upright, worthy; decorous, genteel, proper, seemly. **2** appreciable, considerable, large, significant, sizeable, substantial.

respective *adj.* belonging to each as an individual. □ **respectively** *adv.*

respiration *n.* breathing.

respirator *n.* **1** device worn over the nose and mouth to purify air before it is inhaled. **2** device for giving artificial respiration.

respiratory *adj.* of respiration.

respire *v.* breathe.

respite *n.* **1** interval of rest or relief. **2** permitted delay.

■ **1** break, breather, interlude, intermission, interval, pause, rest.

resplendent *adj.* brilliant with colour or decorations. □ **resplendently** *adv.*

respond *v.* **1** answer. **2** react.

■ **1** answer, rejoin, reply, retort.

respondent *n.* defendant in a lawsuit, esp. in a divorce case.

response *n.* **1** answer. **2** act, feeling, or movement caused by a stimulus etc.

■ **1** answer, rejoinder, reply, retort, riposte.

responsibility *n.* **1** being responsible. **2** thing for which one is responsible.

■ **1** accountability, liability. **2** burden, charge, duty, job, role, task.

responsible *adj.* **1** liable to be blamed for loss or failure etc. **2** being the cause of or to blame for. **3** reliable, trustworthy. **4** involving important duties. □ **responsibly** *adv.*

■ **1** accountable, answerable, liable. **2** at fault, culpable, guilty, in the wrong, to blame. **3** dependable, honest, reliable, reputable, sensible, trustworthy.

responsive *adj.* **1** responding well to an influence. **2** sympathetic. □ **responsiveness** *n.*

rest[1] *v.* **1** be still or asleep. **2** (cause or allow to) cease from tiring activity. **3** place or be placed for support. **4** rely. **5** (of a matter) be left without further discussion. ● *n.* **1** (period of) inactivity or sleep. **2** prop or support for an object.

■ *v.* **1** doze, repose, sleep, slumber, snooze. **2** relax, slow down, take a break, *colloq.* unwind. **3** lay, place, prop, put; be supported, lean, lie. ● *n.* **1** break, breather, interlude, intermission, lull, pause, respite; ease, relaxation, repose; doze, nap, siesta, sleep, snooze.

rest[2] *v.* remain in a specified state. ● *n.* the remaining part, the others.

■ *n.* balance, difference, remainder, residue, surplus; leftovers, remains, remnants.

restaurant *n.* place where meals can be bought and eaten.

restaurateur *n.* restaurant-keeper.

restful *adj.* giving rest, relaxing. □ **restfully** *adv.*, **restfulness** *n.*

■ calm, peaceful, quiet, relaxing, serene, tranquil.

restitution *n.* **1** restoring of a thing to its proper owner or original state. **2** compensation.

■ **2** amends, compensation, indemnification, recompense, redress, reparation.

restive *adj.* restless, impatient. □ **restively** *adv.*, **restiveness** *n.*

restless *adj.* unable to rest or be still. □ **restlessly** *adv.*, **restlessness** *n.*

■ agitated, edgy, fidgety, fretful, ill at ease, *colloq.* jittery, jumpy, nervous, restive, uneasy.

restoration *n.* **1** restoring. **2** restored thing.

restorative *adj.* restoring health or strength. ● *n.* restorative food, medicine, or treatment.

restore *v.* **1** bring back to its original state (e.g. by repairing), or to good health or vigour. **2** put back in a former position. □ **restorer** *n.*

■ **1** do up, rebuild, refurbish, renew, renovate, repair; cure, heal, make better; fortify, invigorate, refresh, revive.

restrain *v.* **1** hold back from movement or action. **2** keep under control. □ **restraint** *n.*

■ **1** check, control, curb, hinder, hold back, impede, prevent, stop. **2** contain, control, hold back, repress, stifle, suppress.

restrict *v.* put a limit on, subject to limitations. □ **restriction** *n.*, **restrictive** *adj.*

■ bound, circumscribe, confine, delimit, limit.

result *n.* **1** product of an activity, operation, or calculation. **2** score, marks, or name of the winner in a sports event or competition. ● *v.* occur or have as a result.

■ *n.* **1** consequence, effect, outcome, repercussion, upshot. ● *v.* come about, develop, emerge, happen, occur.

resultant *adj.* occurring as a result.

resume *v.* **1** get or take again. **2** begin again after stopping. □ **resumption** *n.*

résumé /rézyoomay/ *n.* summary.

resurface *v.* **1** put a new surface on. **2** return to the surface.

resurgence *n.* revival after destruction or disappearance. □ **resurgent** *adj.*

resurrect *v.* bring back into use.

resurrection *n.* **1** rising from the dead. **2** revival after disuse.

resuscitate *v.* revive. □ **resuscitation** *n.*

retail *n.* selling of goods to the public. ● *adj.* & *adv.* in the retail trade. ● *v.* sell or be sold in the retail trade. □ **retailer** *n.*

retain *v.* **1** keep, esp. in one's possession or in use. **2** keep in mind.

■ **1** keep, preserve, reserve, save.

retainer *n.* fee paid to retain services.

retaliate *v.* repay an injury or insult etc. by inflicting one in return. □ **retaliation** *n.*, **retaliatory** *adj.*

retard *v.* cause delay to. □ **retardation** *n.*

■ delay, hold up, set back, slow down; hamper, hinder, impede.

retarded *adj.* backward in mental or physical development.

retch *v.* strain one's throat as if vomiting.

retention *n.* retaining.

retentive *adj.* able to retain things.

rethink *v.* (**rethought**) reconsider, plan again and differently.

reticent *adj.* not revealing one's thoughts. □ **reticence** *n.*

■ diffident, quiet, reserved, retiring, secretive, shy, silent, taciturn, uncommunicative, unforthcoming.

retina *n.* (*pl.* **-as**) membrane at the back of the eyeball, sensitive to light.

retinue *n.* attendants accompanying an important person.

■ attendants, entourage, escort, suite, train.

retire *v.* **1** give up one's regular work because of age. **2** cause (an employee) to do this. **3** withdraw, retreat. **4** go to bed. □ **retirement** *n.*

■ **2** pension off. **3** depart, go (away), leave, move away, retreat, withdraw.

retiring *adj.* shy, avoiding society.

■ bashful, diffident, meek, self-effacing, shy, timid, unassuming.

retort[1] *v.* make (as) a witty or angry reply. ● *n.* reply of this kind.

■ *n.* answer, reaction, rejoinder, reply, response, riposte.

retort[2] *n.* **1** vessel with a bent neck, used in distilling. **2** vessel used in making gas or steel.

retouch *v.* touch up (a picture or photograph).

retrace *v.* go back over or repeat (a route).

retract *v.* withdraw. □ **retraction** *n.*, **retractor** *n.*, **retractable** *adj.*

■ draw back, pull back; cancel, deny, disavow, disclaim, recant, revoke, take back, withdraw.

retractile *adj.* retractable.

retreat *v.* withdraw, esp. after defeat or when faced with difficulty. ● *n.* **1** retreating, withdrawal. **2** military signal for this. **3** place of shelter or seclusion.

■ *v.* decamp, flee, move back, pull out, retire, withdraw. ● *n.* **1** flight, retirement, withdrawal. **3** asylum, haven, refuge, sanctuary, shelter.

retrench *v.* reduce expenses. □ **retrenchment** *n.*

retrial *n.* trial of a lawsuit or defendant again.

retribution *n.* punishment for crime or evil.

■ retaliation, revenge, vengeance; punishment.

retrieve *v.* **1** regain possession of. **2** bring back. **3** set right (an error etc.). □ **retrieval** *n.*

■ **1** get back, reclaim, recover, redeem, regain, repossess. **2** bring back, fetch, get.

retriever *n.* dog of a breed used to retrieve game.

retroactive *adj.* operating retrospectively. □ **retroactively** *adv.*

retrograde *adj.* **1** going backwards. **2** reverting to an inferior state.

retrogress *v.* **1** move backwards. **2** deteriorate. □ **retrogression** *n.*, **retrogressive** *adj.*

retrospect *n.* **in retrospect** when one looks back on a past event.

retrospective *adj.* **1** looking back on the past. **2** (of a law etc.) applying to the past as well as the future. □ **retrospectively** *adv.*

retroverted *adj.* turned backwards. □ **retroversion** *n.*

retry *v.* (**-tried**) try (a lawsuit or defendant) again.

return *v.* **1** come or go back. **2** bring, give, put, or send back. ● *n.* **1** returning. **2** profit. **3** return ticket. **4** return match. **5** formal report submitted by order. □ **return match** second match between the same opponents. **return ticket** ticket for a journey to a place and back again.

■ *v.* **1** come back, reappear; go *or* turn back; recur, reoccur. **2** bring back, put back, replace, restore; give back, refund, reimburse, repay. ● *n.* **1** reappearance, recurrence, re-emergence, renewal; replacement, restoration. **2** income, profit, revenue, yield.

reunion *n.* gathering of people who were formerly associated.

reunite *v.* bring or come together again.

rev (*colloq.*) *n.* revolution of an engine. ● *v.* (**revved**) **1** cause (an engine) to run quickly. **2** (of an engine) revolve.

Rev. *abbr.* Reverend.

revalue *v.* put a new (esp. higher) value on. □ **revaluation** *n.*

revamp *v.* renovate, give a new appearance to.

reveal *v.* **1** make visible by uncovering. **2** make known.

■ **1** display, expose, lay bare, show, uncover, unveil. **2** disclose, divulge, give away, *sl.* let on, make known.

reveille /riválli/ *n.* military waking-signal.

revel *v.* (**revelled**) make merry. ● *n.* (usu. *pl.*) lively festivities, merrymaking. □ **revel in** take great delight in. **reveller** *n.*, **revelry** *n.*

■ *v.* celebrate, make merry. ● *n.* (**revels**) celebrations, festivity, fun, jollification, merrymaking, revelry. □ **revel in** delight in, enjoy, glory in, relish, savour, take pleasure in, wallow in.

revelation *n.* **1** revealing. **2** (surprising) thing revealed.

■ **2** admission, confession, declaration, disclosure, leak.

revenge *n.* **1** injury inflicted in return for what one has suffered. **2** opportunity to defeat a victorious opponent. ● *v.* avenge.

■ *n.* **1** reprisal, retaliation, retribution, vengeance.

revenue *n.* **1** income esp. from taxes etc. **2** department collecting this.

reverberate *v.* echo, resound. □ **reverberation** *n.*

revere *v.* feel deep respect or religious veneration for.

■ admire, esteem, have a high opinion of, honour, idolize, look up to, respect, venerate, worship.

reverence *n.* feeling of awe and respect or veneration.

■ admiration, awe, esteem, honour, respect, veneration.

reverend *adj.* **1** deserving to be treated with respect. **2** (**Reverend**) title of a member of the clergy.

reverent *adj.* feeling or showing reverence. □ **reverently** *adv.*

reverie *n.* daydream.

revers /riveer/ *n.* (*pl.* **revers**) turned-back front edge at the neck of a jacket or bodice.

reverse *adj.* **1** opposite in character or order. **2** upside down. ● *v.* **1** turn the other way round, upside down, or inside out. **2** convert to the opposite. **3** annul (a decree etc.). **4** move backwards or in the opposite direction. ● *n.* **1** reverse or opposite side or effect. **2** piece of misfortune. **3** reverse motion or gear. □ **reversal** *n.*, **reversible** *adj.*

■ *adj.* **1** contrary, converse, inverse, opposite. ● *v.* **1** invert, overturn, turn over *or* upside down; change, exchange, interchange, transpose. **3** annul, cancel, negate, nullify, override, overturn, rescind, revoke. ● *n.* **1** antithesis, contrary, converse, opposite.

revert *v.* **1** return to a former condition or habit. **2** return to a subject in talk etc. **3** (of property etc.) pass to another holder when its present holder relinquishes it. □ **reversion** *n.*

review *n.* **1** general survey of events or a subject. **2** reconsideration. **3** critical report of a book or play etc. **4** ceremonial inspection of troops etc. ● *v.* make or write a review of. □ **reviewer** *n.*

■ *n.* **1** analysis, assessment, examination, study, survey. **2** reappraisal, reassessment, reconsideration, re-examination, rethinking. **3** commentary, criticism, notice, *colloq.* write-up. ● *v.* analyse, assess, comment on, consider, evaluate, judge, look at, study, survey; reassess, reconsider.

revile *v.* criticize angrily in abusive language.

revise *v.* **1** re-examine and alter or correct. **2** study again (work already learnt) in preparation for an exam. □ **reviser** *n.*, **revision** *n.*, **revisory** *adj.*

■ **1** alter, amend, change, correct, edit, emend, modify, update.

revivalist *n.* person who seeks to promote religious fervour. □ **revivalism** *n.*, **revivalistic** *adj.*

revive *v.* come or bring back to life, consciousness, or vigour, or into use. □ **revival** *n.*

■ awake, come to, wake up; resuscitate, revive; invigorate, perk up, refresh, restore; re-establish, renew, resurrect.

revoke *v.* withdraw (a decree or licence etc.). □ **revocable** *adj.*, **revocation** *n.*

■ annul, cancel, declare null and void, invalidate, nullify, quash, repeal, rescind, withdraw.

revolt *v.* **1** take part in a rebellion. **2** cause strong disgust in. ● *n.* **1** act or state of rebelling. **2** sense of disgust.

■ *v.* **1** mutiny, rebel, rise (up). **2** appal, disgust, horrify, offend, repel, shock, sicken. ● *n.* **1** coup, insurrection, mutiny, rebellion, revolution, uprising.

revolting *adj.* **1** in revolt. **2** causing disgust.

■ **2** abhorrent, disgusting, dreadful, execrable, hateful, loathsome, objectionable, obnoxious, odious, offensive, repellent, repulsive, unpleasant, vile.

revolution *n.* **1** revolving, single complete orbit or rotation. **2** complete change of method or conditions. **3** substitution of a new system of government, esp. by force.

■ **1** rotation, spin, turn, twirl; circuit, lap, orbit. **2** change, reorganization, transformation, upheaval.

revolutionary *adj.* **1** involving a great change. **2** of political revolution. ● *n.* person who begins or supports a political revolution.

■ *adj.* **1** innovative, new, novel, original; avant-garde, progressive.

revolutionize *v.* alter completely.

revolve *v.* **1** turn round. **2** move in an orbit.

■ circle, gyrate, loop, orbit, rotate, spin, spiral, turn, twirl, whirl.

revolver *n.* a kind of pistol.

revue *n.* entertainment consisting of a series of items.

revulsion *n.* **1** strong disgust. **2** sudden violent change of feeling.

■ **1** abhorrence, antipathy, detestation, disgust, hatred, loathing, odium, repugnance.

reward *n.* something given or received in return for a service or merit. ● *v.* give a reward to.

rewire *v.* renew the electrical wiring of.

rhapsodize *v.* talk or write about something ecstatically.

rhapsody *n.* **1** ecstatic statement. **2** romantic musical composition. □ **rhapsodic** *adj.*

rheostat *n.* device for varying the resistance to electric current.

rhesus *n.* small Indian monkey used in biological experiments. □ **rhesus factor** substance usu. present in human blood (**rhesus-positive** having this; **rhesus-negative** not having it).

rhetoric *n.* **1** art of using words impressively. **2** impressive language.

rhetorical *adj.* expressed so as to sound impressive. □ **rhetorical question** one used for dramatic effect, not seeking an answer. **rhetorically** *adv.*

rheumatic *adj.* of or affected with rheumatism. □ **rheumaticky** *adj.*

rheumatism *n.* disease causing pain in the joints, muscles, or fibrous tissue.

rheumatoid *adj.* having the character of rheumatism.

rhinestone *n.* imitation diamond.

rhino *n.* (*pl.* **rhino** or **-os**) (*colloq.*) rhinoceros.

rhinoceros *n.* (*pl.* **-oses**) large thick-skinned animal with one horn or two horns on its nose.

rhizome *n.* root-like stem producing roots and shoots.

rhododendron *n.* evergreen shrub with clusters of flowers.

rhomboid *adj.* like a rhombus. ● *n.* rhomboid figure.

rhombus *n.* quadrilateral with opposite sides and angles equal (and not right angles).

rhubarb *n.* plant with red leaf-stalks that are used like fruit.

rhyme *n.* **1** similarity of sound between words or syllables. **2** word providing a rhyme to another. **3** poem with line-endings that rhyme. ● *v.* form a rhyme.

rhythm *n.* pattern produced by emphasis and duration of notes in music or of syllables in words, or by regular movements or events. □ **rhythmic** *adj.*, **rhythmical** *adj.*, **rhythmically** *adv.*

■ accent, beat, cadence, lilt, measure, metre, tempo, time.

rib *n.* **1** one of the curved bones round the chest. **2** structural part resembling this. **3** pattern of raised lines in knitting. ● *v.* (**ribbed**) **1** knit as rib. **2** (*colloq.*) tease.

ribald *adj.* irreverent, coarsely humorous. □ **ribaldry** *n.*

riband *n.* ribbon.

ribbon *n.* **1** narrow band of silky material. **2** strip resembling this.

ribonucleic acid substance controlling protein synthesis in cells.

rice *n.* **1** cereal plant grown in marshes in hot countries, with seeds used as food. **2** these seeds.

rich *adj.* **1** having much wealth. **2** splendid, costly, valuable. **3** abundant. **4** fertile. **5** (of colour, sound, or smell) pleasantly deep and strong. **6** containing a large proportion of something (e.g. fat, fuel). □ **richness** *n.*

■ **1** affluent, moneyed, opulent, prosperous, wealthy, *colloq.* well-heeled, well-off, well-to-do. **2** gorgeous, luxurious, magnificent, palatial, splendid, sumptuous; costly, expensive, precious, priceless, valuable. **3** abundant, ample, bountiful, copious, plentiful, profuse. **4** fecund, fertile, fruitful, productive, prolific. **5** dark, deep, intense, strong; full, low, mellow, resonant, sonorous; aromatic, fragrant.

riches *n.pl.* wealth.

> ■ affluence, money, opulence, prosperity, wealth.

richly *adv.* **1** in a rich way. **2** fully, thoroughly.

rick[1] *n.* built stack of hay etc.

rick[2] *n.* slight sprain or strain. ● *v.* sprain or strain slightly.

rickety *adj.* shaky, insecure.

> ■ flimsy, insubstantial, ramshackle, shaky, tumbledown.

rickshaw *n.* two-wheeled hooded vehicle used in the Far East, drawn by one or more people.

ricochet /ríkkəshay/ *n.* & *v.* (**ricocheted**) rebound from a surface after striking it with a glancing blow.

rid *v.* (**rid, ridding**) free from something unpleasant or unwanted. □ **get rid of** cause to go away or free oneself of.

> ■ □ **get rid of** discard, dispense with, dispose of, *sl.* ditch, do away with, jettison, throw away; destroy, eliminate, eradicate, remove.

ridden *see* **ride**. *adj.* full of.

riddle[1] *n.* **1** question etc. designed to test ingenuity, esp. for amusement. **2** something puzzling or mysterious.

> ■ **1** conundrum, poser, problem, puzzle. **2** enigma, mystery.

riddle[2] *n.* coarse sieve. ● *v.* **1** pass through a riddle. **2** permeate thoroughly.

ride *v.* (**rode, ridden**) **1** sit on and be carried by (a horse or bicycle etc.). **2** travel in a vehicle. **3** float (on). ● *n.* **1** spell of riding. **2** journey in a vehicle. **3** track for riding on.

> ■ *n.* **2** drive, excursion, jaunt, journey, outing, run, spin, trip.

rider *n.* **1** one who rides a horse etc. **2** additional statement.

ridge *n.* **1** narrow raised strip. **2** line where two upward slopes meet. **3** elongated region of high barometric pressure.

ridicule *n.* making or being made to seem ridiculous. ● *v.* subject to ridicule, make fun of.

> ■ *n.* derision, mockery, taunting. ● *v.* deride, gibe (at), jeer at, laugh at, make fun of, mock, poke fun at, scoff at, taunt.

ridiculous *adj.* deserving to be laughed at, unreasonable. □ **ridiculously** *adv.*

> ■ absurd, daft, foolish, idiotic, insane, laughable, ludicrous, mad, nonsensical, preposterous, risible, silly, stupid, unreasonable.

rife *adj.* occurring frequently, widespread. □ **rife with** full of.

riff *n.* short repeated phrase in jazz etc.

riffle *v.* flick through (pages etc.).

riff-raff *n.* disreputable people.

rifle *n.* a gun with a long rifled barrel. ● *v.* **1** search and rob. **2** cut spiral grooves in (a gun barrel).

rift *n.* **1** cleft in earth or rock. **2** crack, split. **3** breach in friendly relations. □ **rift-valley** *n.* steep-sided valley formed by subsidence.

> ■ **1,2** cleft, crack, crevasse, crevice, fissure, split. **3** breach, break, estrangement, rupture, split.

rig[1] *v.* (**rigged**) **1** provide with clothes or equipment. **2** fit (a ship) with spars, ropes, etc. **3** set up, esp. in a makeshift way. ● *n.* **1** way a ship's masts and sails etc. are arranged. **2** apparatus for drilling an oil well etc. □ **rig-out** *n.* (*colloq.*) outfit.

rig[2] *v.* (**rigged**) manage or control fraudulently.

rigging *n.* ropes etc. used to support a ship's masts and sails.

right *adj.* **1** morally good. **2** proper. **3** correct, true. **4** in a good condition. **5** of or on the east side of a person facing north. ● *n.* **1** what is just. **2** something one is entitled to. **3** right hand, foot, or side. **4** people supporting more conservative policies than others in their group. ● *v.* **1** set right. **2** restore to a correct or upright position. ● *adv.* **1** directly. **2** exactly, correctly. **3** completely. **4** on or towards the right-hand side. □ **in the right** having truth or justice on one's side. **right angle** angle of 90°. **right away** immediately. **right-hand man, woman** indispensable assistant. **right-handed** *adj.* using the right hand. **right of way 1** right to pass over another's land. **2** path subject to this. **3** right to proceed while another vehicle must wait. **rightly** *adv.*

> ■ *adj.* **1** ethical, fair, good, honest, honourable, just, moral, proper, righteous. **2** apposite, appropriate, correct, fitting, proper, suitable. **3** accurate, correct, exact, precise, true. ● *n.* **1** equity, fairness, honesty, justice, morality, rightfulness, truth, virtue. **2** entitlement, freedom, liberty, licence, prerogative, privilege. ● *v.* **1** amend, correct, cure, fix, put right, rectify, remedy, repair. ● *adv.* **1** directly, immediately, straight. **2** exactly, just, precisely; accurately, correctly, properly, well.

□ **right away** at once, immediately, instantly, now, straight away, without delay.

righteous /ríchəss/ *adj.* **1** doing what is morally right, making a show of this. **2** morally justifiable. □ **righteously** *adv.*, **righteousness** *n.*

rightful *adj.* proper, legal. □ **rightfully** *adv.*, **rightfulness** *n.*

■ correct, lawful, legal, legitimate, proper, true.

rigid *adj.* **1** stiff. **2** strict, inflexible. □ **rigidly** *adv.*, **rigidity** *n.*

■ **1** firm, hard, inflexible, stiff, strong, unbendable. **2** harsh, inflexible, intransigent, strict, unyielding.

rigmarole *n.* **1** long statement. **2** complicated procedure.

rigor mortis *n.* stiffening of the body after death.

rigorous *adj.* **1** strict, severe. **2** exact, accurate. □ **rigorously** *adv.*, **rigorousness** *n.*

rigour *n.* **1** strictness, severity. **2** harsh conditions.

rile *v.* (*colloq.*) annoy.

rim *n.* edge or border of something more or less circular. □ **rimmed** *adj.*

rind *n.* tough outer layer on fruit, cheese, bacon, etc.

ring[1] *n.* **1** outline of a circle. **2** circular metal band usu. worn on a finger. **3** enclosure where a performance or activity takes place. **4** group of people acting together dishonestly. ● *v.* **1** put a ring on or round. **2** surround.

■ *n.* **1** circle, disc, round. ● *v.* **2** bound, circle, encircle, encompass, surround.

ring[2] *v.* (**rang, rung**) **1** give out a loud clear resonant sound. **2** cause (a bell) to do this. **3** signal by ringing. **4** telephone. **5** be filled with sound. ● *n.* **1** act or sound of ringing. **2** specified tone or feeling of a statement etc. **3** (*colloq.*) telephone call. □ **ring off** end a telephone call. **ring the changes** vary things. **ring up** make a telephone call.

■ *v.* **1** chime, peal, toll. **4** call, *colloq.* phone, ring up, telephone. **5** echo, resonate, resound, reverberate.

ringleader *n.* person who leads others in wrongdoing or riot etc.

ringlet *n.* long curly lock of hair.

ringside *n.* area beside a boxing ring. □ **ringside seat** position from which one has a clear view of the scene of action.

ringworm *n.* skin disease producing round scaly patches.

rink *n.* skating-rink.

rinse *v.* **1** wash lightly. **2** wash out soap etc. from. ● *n.* **1** process of rinsing. **2** solution for tinting or conditioning hair.

riot *n.* **1** wild disturbance by a crowd of people. **2** profuse display. **3** (*colloq.*) very amusing person or thing. ● *v.* take part in a riot. □ **run riot 1** behave in an unruly way. **2** grow in an uncontrolled way.

■ *n.* **1** affray, disturbance, fracas, fray, tumult, uproar.

riotous *adj.* disorderly, unruly. □ **riotously** *adv.*

■ boisterous, disorderly, noisy, rowdy, tumultuous, uncontrollable, unruly, uproarious, wild.

rip *v.* (**ripped**) **1** tear apart. **2** make (a hole) by ripping. **3** become torn. **4** rush along. ● *n.* **1** act of ripping. **2** torn place. □ **rip-cord** *n.* cord for pulling to release a parachute. **rip off** (*colloq.*) **1** defraud. **2** steal. **rip-off** *n.* **ripper** *n.*

■ *v.* **1** pull apart, rend, split, tear. ● *n.* **2** hole, rent, split, tear.

ripe *adj.* **1** ready to be gathered and used. **2** (of age) advanced. **3** ready. □ **ripeness** *n.*

ripen *v.* make or become ripe.

riposte /ripóst/ *n.* quick counterstroke or retort. ● *v.* deliver a riposte.

ripple *n.* **1** small wave(s). **2** gentle sound that rises and falls. ● *v.* form ripples (in).

rise *v.* (**rose, risen**) **1** come, go, or extend upwards. **2** get up from lying or sitting, get out of bed. **3** increase, become higher. **4** rebel. **5** have its origin or source. ● *n.* **1** act or amount of rising, increase. **2** upward slope. **3** increase in wages. □ **give rise to** cause.

■ *v.* **1** arise, ascend, climb, come *or* go up, mount, soar; fly. **2** arise, get to one's feet, get up, stand up. **3** climb, escalate, increase, go up, rocket, shoot up. ● *n.* **1** ascent, climb; growth, escalation, increase, jump; gain, improvement, surge. **2** gradient, hill, incline, slope. □ **give rise to** bring about, cause, create, engender, lead to, occasion, produce, result in.

risible *adj.* ridiculous. □ **risibly** *adv.*, **risibility** *n.*

risk *n.* **1** possibility of meeting danger or suffering harm. **2** person or thing representing a source of risk. ● *v.* **1** expose to the chance of injury or loss. **2** accept the risk of.

■ *n.* **1** chance, hazard, threat; danger, peril. ● *v.* **1** endanger, hazard, imperil, jeopardize; chance, venture.

risky *adj.* (**-ier**, **-iest**) full of risk. □ **riskily** *adv.*, **riskiness** *n.*

risotto *n.* (*pl.* **-os**) savoury rice dish.

risqué /riskáy/ *adj.* slightly indecent.

rissole *n.* fried cake of minced meat.

rite *n.* ritual.

ritual *n.* series of actions used in a religious or other ceremony. ● *adj.* of or done as a ritual. □ **ritually** *adv.*, **ritualistic** *adj.*, **ritualism** *n.*

rival *n.* person or thing that competes with or can equal another. ● *v.* (**rivalled**) **1** be a rival of. **2** seem as good as. □ **rivalry** *n.*

■ *n.* adversary, antagonist, challenger, competitor, contender, opponent. ● *v.* **2** be on a par with, compare with, equal, match, measure up to.

river *n.* **1** large natural stream of water. **2** great flow.

rivet *n.* bolt for holding pieces of metal together, with its end pressed down to form a head when in place. ● *v.* (**riveted**) **1** fasten with a rivet. **2** attract and hold (the attention of).

■ *v.* **2** captivate, engross, enthral, entrance, fascinate, grip, hold spellbound, transfix.

rivulet *n.* small stream.

RNA *abbr.* ribonucleic acid.

roach *n.* (*pl.* **roach**) small freshwater fish of the carp family.

road *n.* **1** prepared track along which people and vehicles may travel. **2** way of reaching something. □ **on the road** travelling. **road-hog** *n.* reckless or inconsiderate driver. **road-metal** *n.* broken stone for making the foundation of a road or railway.

■ **1** avenue, boulevard, carriageway, highway, motorway, roadway, street, thoroughfare.

roadway *n.* road, esp. as distinct from a footpath beside it.

roadworks *n.pl.* construction or repair of roads.

roadworthy *adj.* (of a vehicle) fit to be used on a road. □ **roadworthiness** *n.*

roam *v.* & *n.* wander.

■ *v.* drift, range, rove, wander.

roan *n.* horse with a dark coat sprinkled with white hairs.

roar *n.* **1** long deep sound like that made by a lion. **2** loud laughter. ● *v.* **1** give a roar. **2** express in this way.

■ *v.* **2** bark, bawl, bellow, cry, shout, thunder, yell.

roaring *adj.* **1** noisy. **2** briskly active.

roast *v.* cook or be cooked by exposure to open heat or in an oven. ● *n.* roast joint of meat.

rob *v.* (**robbed**) **1** steal from. **2** deprive. □ **robber** *n.*, **robbery** *n.*

■ **1** burgle, loot, pillage, plunder, raid, ransack, rifle; hold up, mug. □ **robber** burglar, housebreaker, intruder, mugger, thief.

robe *n.* long loose esp. ceremonial garment. ● *v.* dress in a robe.

robin *n.* brown red-breasted bird.

robot *n.* **1** machine resembling and acting like a person. **2** piece of apparatus operated by remote control. □ **robotic** *adj.*

robotics *n.* study of robots and their design, operation, etc.

robust *adj.* strong, vigorous. □ **robustly** *adv.*, **robustness** *n.*

■ durable, hard-wearing, strong, sturdy, tough; brawny, burly, fit, hale, healthy, muscular, powerful, strapping, vigorous.

rock[1] *n.* **1** hard part of earth's crust, below the soil. **2** mass of this, large stone. **3** hard sugar sweet made in sticks. □ **on the rocks** (of a drink) served with ice cubes. **rock-bottom** *adj.* very low. **rock-cake** *n.* small cake with a rough surface.

■ **2** crag, tor, outcrop; boulder, stone.

rock[2] *v.* **1** move to and fro while supported. **2** disturb greatly by shock. ● *n.* **1** rocking movement. **2** a kind of modern music with a strong beat. □ **rock and roll** rock music with elements of blues.

■ *v.* **1** sway, swing; teeter, totter, wobble; lurch, pitch, roll, shake, toss.

rocker *n.* **1** thing that rocks. **2** pivoting switch.

rockery *n.* collection of rough stones with soil between them on which small plants are grown.

rocket *n.* **1** firework that shoots into the air when ignited and then explodes. **2** structure that flies by expelling burning gases. ● *v.* (**rocketed**) move rapidly upwards or away.

rocketry *n.* science or practice of rocket propulsion.

rocky[1] *adj.* (**-ier**, **-iest**) of, like, or full of rocks. □ **rockiness** *n.*

rocky[2] *adj.* (**-ier**, **-iest**) (*colloq.*) unsteady. □ **rockily** *adv.*, **rockiness** *n.*

rococo *adj.* & *n.* (of or in) an ornate style of decoration in Europe in the 18th century.

rod *n.* **1** slender straight round stick or metal bar. **2** fishing rod.

rode *see* **ride**.

rodent *n.* animal with strong front teeth for gnawing things.

rodeo *n.* (*pl.* **-os**) competition or exhibition of cowboys' skill.

roe[1] *n.* **1** mass of eggs in a female fish's ovary (**hard roe**). **2** male fish's milt (**soft roe**).

roe[2] *n.* (*pl.* **roe** or **roes**) a kind of small deer. □ **roebuck** *n.* male roe.

roentgen /rúntgən/ *n.* unit of ionizing radiation.

roger *int.* (in signalling) message received and understood.

rogue *n.* **1** dishonest, unprincipled, or mischievous person. **2** wild animal living apart from the herd. □ **roguery** *n.*

■ **1** blackguard, cheat, *colloq.* crook, knave, malefactor, miscreant, rascal, ruffian, scoundrel, trickster, villain, wretch.

roguish *adj.* mischievous, playful. □ **roguishly** *adv.*

role *n.* **1** actor's part. **2** person's or thing's function.

■ **1** character, part. **2** duty, function, job, responsibility, task.

roll *v.* **1** move (on a surface) on wheels or by turning over and over. **2** turn on an axis or over and over. **3** form into a cylindrical or spherical shape. **4** flatten with a roller. **5** move or pass steadily. **6** rock from side to side. **7** undulate. ● *n.* **1** act of rolling. **2** cylinder of flexible material turned over and over upon itself. **3** official list or register. **4** long deep sound. **5** undulation. **6** small individual loaf of bread. □ **be rolling (in money)** (*colloq.*) be wealthy. **roll-call** *n.* calling of a list of names to check that all are present. **rolled gold** thin coating of gold on another metal. **rolling-pin** *n.* roller for flattening dough. **rolling-stock** *n.* railway engines and carriages, wagons, etc. **rolling stone** person who does not settle in one place.

■ *v.* **1,2** trundle, wheel; revolve, rotate, spin, turn. **5** glide, go, move, pass. **6** heel, list, lurch, pitch, toss. ● *n.* **1** revolution, rotation, spin, turn. **3** directory, index, inventory, list, record, register, roster. **4** boom, reverberation, roar, rumble.

roller *n.* **1** cylinder rolled over things to flatten or spread them, or on which something is wound. **2** long swelling wave. □ **roller coaster** switchback at a fair etc. **roller skate** (*see* **skate**[2]).

rollicking *adj.* full of boisterous high spirits.

roly-poly *n.* pudding of suet pastry spread with jam, rolled up, and boiled. ● *adj.* plump, podgy.

Roman *adj.* & *n.* **1** (native, inhabitant) of Rome or of the ancient Roman republic or empire. **2** Roman Catholic. □ **Roman Catholic** (member) of the Church that acknowledges the Pope as its head. **Roman numerals** letters representing numbers (I = 1, V = 5, etc.).

roman *n.* plain upright type.

romance *n.* **1** love story, love affair resembling this. **2** imaginative story or literature. **3** romantic situation, event, or atmosphere. **4** picturesque exaggeration. ● *v.* distort the truth or invent imaginatively. □ **Romance languages** those descended from Latin.

■ *n.* **1** affair, intrigue, liaison, relationship. **2** epic, fantasy, legend, melodrama, saga, tale. **3** exoticism, glamour, mystery.

Romanesque *adj.* & *n.* (of or in) a style of art and architecture in Europe about 1050–1200.

romantic *adj.* **1** appealing to the emotions by its imaginative, heroic, or picturesque quality. **2** involving a love affair. ● *n.* romantic person. □ **romantically** *adv.*

■ *adj.* **1** fairy-tale, fanciful, idealized, idyllic, imaginary; exotic, glamorous, picturesque.

romanticism *n.* romantic style.

romanticize *v.* **1** make romantic. **2** indulge in romance.

Romany *adj.* & *n.* **1** gypsy. **2** (of) the gypsy language.

romp *v.* **1** play about in a lively way. **2** (*colloq.*) go along easily. ● *n.* spell of romping.

rondeau *n.* short poem with the opening words used as a refrain.

rondo *n.* (*pl.* **-os**) piece of music with a recurring theme.

roof *n.* (*pl.* **roofs**) upper covering of a building, car, cavity, etc. ● *v.* **1** cover with a roof. **2** be the roof of. □ **roofer** *n.*

rook[1] *n.* bird of the crow family.

rook[2] *n.* chess piece with a top shaped like battlements.

rookery *n.* colony of rooks.

room *n.* **1** space that is or could be occupied. **2** enclosed part of a building. **3** scope.

■ **1** area, capacity, space. **3** leeway, latitude, margin, scope.

roomy *adj.* (**-ier**, **-iest**) having plenty of space.

■ big, capacious, commodious, large, sizeable, spacious.

roost *n.* place where birds perch or rest. ● *v.* perch, esp. for sleep.

root[1] *n.* **1** part of a plant that grows into the earth and absorbs water and nourishment from the soil. **2** embedded part of hair, tooth, etc. **3** source, basis. **4** number in relation to another which it produces when multiplied by itself a specified number of times. **5** (*pl.*) emotional attachment to a place. ● *v.* **1** (cause to) take root. **2** cause to stand fixed and unmoving. □ **root out** or **up 1** drag or dig up by the roots. **2** get rid of. **take root 1** send down roots. **2** become established. **rootless** *adj.*

■ *n.* **1** radicle, rhizome, tuber. **3** base, basis, cause, fount, origin, source. □ **root out 2** destroy, do away with, eliminate, eradicate, get rid of, remove, stamp out.

root[2] *v.* **1** (of an animal) turn up ground with the snout or beak in search of food. **2** rummage, extract.

rope *n.* strong thick cord. ● *v.* **1** fasten or secure with rope. **2** fence off with rope(s). □ **know** or **show the ropes** know or show the procedure. **rope in** persuade to take part in.

rosary *n.* **1** set series of prayers. **2** string of beads for keeping count in this.

rose[1] *n.* **1** ornamental usu. fragrant flower. **2** bush or shrub bearing this. **3** deep pink colour.

rose[2] *see* **rise**.

rosé /rṓzay/ *n.* light pink wine.

rosemary *n.* shrub with fragrant leaves used to flavour food.

rosette *n.* round badge or ornament made of ribbons.

rosewood *n.* dark fragrant wood used for making furniture.

rosin *n.* a kind of resin.

roster *n.* & *v.* list showing people's turns of duty etc.

rostrum *n.* (*pl.* **-tra**) platform for one person.

rosy *adj.* (**-ier**, **-iest**) **1** deep pink. **2** promising, hopeful.

■ **2** auspicious, bright, encouraging, favourable, hopeful, optimistic, promising.

rot *v.* (**rotted**) **1** lose its original form by chemical action caused by bacteria or fungi etc. **2** cause to do this. **3** perish through lack of use. ● *n.* **1** rotting, rottenness. **2** (*sl.*) nonsense.

■ *v.* **1** decay, decompose, fester, go bad, go off, moulder, putrefy, spoil. ● *n.* **1** decay, decomposition, mould, putrefaction.

rota *n.* list of duties to be done or people to do them in rotation.

■ list, roster, schedule.

rotary *adj.* acting by rotating.

rotate *v.* **1** revolve. **2** arrange or occur or deal with in a recurrent series. □ **rotation** *n.*, **rotatory** *adj.*

■ **1** circle, go round, revolve, roll, spin, turn round.

rote *n.* **by rote 1** by memory without thought of the meaning. **2** by a fixed procedure.

rotor *n.* rotating part.

rotten *adj.* **1** rotted, breaking easily from age or use. **2** worthless, unpleasant. □ **rottenness** *n.*

■ **1** bad, decayed, decomposing, mouldy, off, putrescent, putrid.

Rottweiler *n.* dog of a large black and tan breed.

rotund *adj.* rounded, plump. □ **rotundity** *n.*

■ chubby, dumpy, fat, plump, podgy, pudgy, roly-poly, tubby.

rotunda *n.* circular domed building or hall.

rouble *n.* unit of money in Russia.

rouge *n.* reddish cosmetic colouring for the cheeks. ● *v.* colour with rouge.

rough *adj.* **1** having an uneven or irregular surface. **2** not gentle or careful, violent. **3** (of weather) stormy. **4** harsh in sound, taste, etc. **5** incomplete. **6** approximate. **7** unpleasant, severe. **8** lacking comfort, finish, etc. ● *adv.* in a rough way. ● *n.* **1** rough thing, person, or state. **2** rough ground. ● *v.* make rough. □ **rough-and-ready** *adj.* rough but effective. **rough-and-tumble** *n.* haphazard struggle. **rough diamond** person of good nature but lacking polished manners. **rough it** do without ordinary comforts.

rough out plan or sketch roughly. **roughly** *adv.*, **roughness** *n.*

■ *adj.* **1** bumpy, coarse, irregular, pitted, rocky, stony, uneven. **2** cruel, hard, harsh, severe, violent; disorderly, rowdy, tumultuous, uproarious, wild; blunt, brusque, ill-mannered, impolite, rude, uncouth, ungracious. **3** blustery, squally, stormy, tempestuous. **4** gruff, guttural, harsh, hoarse, husky, rasping, throaty. **5** incomplete, rudimentary, sketchy, uncompleted, unfinished. **6** approximate, estimated, general, imprecise, inexact, vague. **7** arduous, demanding, difficult, severe, tough, unpleasant. **8** basic, crude, primitive, rude, rudimentary, spartan.

roughage *n.* dietary fibre.

roughen *v.* make or become rough.

roughshod *adj.* **ride roughshod over** treat inconsiderately.

roulette *n.* gambling game played with a small ball on a revolving disc.

round *adj.* curved, circular, spherical, or cylindrical. ● *n.* **1** round object. **2** circular or recurring course, route, or series. **3** song for two or more voices that start at different times. **4** shot from a firearm, ammunition for this. **5** one section of a competition. ● *prep.* **1** so as to circle or enclose. **2** to or on the other side of. **3** in the area near a place. **4** in or to many parts of. ● *adv.* **1** in a circle or curve. **2** so as to face in a different direction. **3** from one person, place, etc. to another. **4** to a person's house etc. **5** by a circuitous route. **6** measuring or marking the edge of. ● *v.* **1** make or become round. **2** express (a number) approximately. **3** travel round (corner etc.). □ **in the round** with all sides visible. **round about 1** nearby. **2** approximately. **round of applause** outburst of clapping. **round off** complete. **round robin** statement signed by a number of people. **round the clock** continuously through day and night. **round trip 1** circular tour. **2** outward and return journey. **round up** gather into one place. **round-up** *n.* **roundness** *n.*

■ *n.* **1** circle, disc, ring; ball, globe, orb, sphere. **2** circuit, lap, orbit; course, cycle, series, sequence. ● *prep.* **1** about, around, encircling, enclosing. **3** about, around, in the vicinity of, near. □ **round off** close, complete, crown, end, finish. **round up** assemble, collect, gather, marshal.

roundabout *n.* **1** revolving platform at a funfair, with model horses etc. to ride on. **2** road junction with a circular island round which traffic has to pass in one direction. ● *adj.* indirect.

■ *adj.* circuitous, indirect, tortuous; circumlocutory, oblique.

roundel *n.* **1** small disc. **2** rondeau.

rounders *n.* team game played with bat and ball, in which players have to run round a circuit. □ **rounder** *n.* unit of scoring in this.

roundly *adv.* **1** thoroughly, severely. **2** in a rounded shape.

roundworm *n.* parasitic worm with a rounded body.

rouse *v.* **1** wake. **2** cause to become active or excited.

■ **1** arouse, awaken, wake up, waken. **2** arouse, excite, fire, galvanize, invigorate, move, stimulate, stir.

rousing *adj.* vigorous, stirring.

rout *n.* **1** utter defeat. **2** disorderly retreat. ● *v.* **1** defeat completely. **2** put to flight.

■ *v.* **1** beat, conquer, crush, defeat, get the better of, *colloq.* lick, overwhelm, prevail over, thrash, triumph over, trounce, vanquish.

route *n.* course or way from starting point to finishing point. □ **route march** training-march for troops.

■ course, direction, itinerary, journey, path, road, track, way.

routine *n.* **1** standard procedure. **2** set sequence of movements. ● *adj.* **1** in accordance with routine. **2** unvarying, mechanical. □ **routinely** *adv.*

■ *n.* **1** custom, *colloq.* drill, form, procedure, way. ● *adj.* **1** commonplace, everyday, normal, ordinary, regular, usual. **2** dull, humdrum, mechanical, mundane, run-of-the-mill, tedious, uninteresting, unvarying.

roux /roo/ *n.* mixture of fat and flour used in sauces.

rove *v.* wander. □ **rover** *n.*

row[1] /rō/ *n.* people or things in a line.

row[2] /rō/ *v.* **1** propel (a boat) by using oars. **2** carry in a boat that one rows. ● *n.* spell of rowing. □ **rowboat, rowing boat** *ns.*

row[3] /row/ *n.* **1** loud noise. **2** quarrel, angry argument. ● *v.* quarrel, argue angrily.

■ *n.* **1** din, clamour, commotion, hubbub, noise, racket, rumpus, uproar. **2** altercation, argument, disagreement, dispute, fight, quarrel, *colloq.* scrap, squabble, tiff, wrangle. ● *v.* argue, bicker, disagree, dispute, fight, quarrel, *colloq.* scrap, spar, squabble, wrangle.

rowan *n.* tree bearing hanging clusters of red berries.

rowdy *adj.* (**-ier**, **-iest**) noisy and disorderly. ● *n.* rowdy person. □ **rowdily** *adv.*, **rowdiness** *n.*

■ *adj.* boisterous, disorderly, noisy, obstreperous, unruly, uproarious, wild.

rowlock /róllək/ *n.* device on the side of a boat securing and forming a fulcrum for an oar.

royal *adj.* **1** of or suited to a king or queen. **2** of the family or in the service or under the patronage of royalty. **3** splendid, of great size. ● *n.* (*colloq.*) member of a royal family. □ **royal blue** bright blue. **royally** *adv.*

■ *adj.* **1** imperial, kingly, majestic, noble, princely, queenly, regal.

royalist *n.* person supporting or advocating monarchy.

royalty *n.* **1** being royal. **2** royal person(s). **3** payment to an author etc. for each copy or performance of his or her work, or to a patentee for use of his or her patent.

RSVP *abbr.* (French *répondez s'il vous plaît*) please reply.

rub *v.* (**rubbed**) **1** press against a surface and slide to and fro. **2** polish, clean, dry, or make sore etc. by rubbing. ● *n.* act or process of rubbing. □ **rub it in** emphasize or remind a person constantly of an unpleasant fact. **rub out** erase with a rubber.

■ *v.* **2** buff, burnish, polish, shine, wipe; scour, scrub; chafe, gall, scrape, scratch.

rubber *n.* **1** tough elastic substance made from the juice of certain plants or synthetically. **2** piece of this for rubbing out pencil or ink marks, eraser. **3** device for rubbing things. □ **rubber-stamp** *v.* approve automatically without consideration. **rubbery** *adj.*

rubberize *v.* treat or coat with rubber.

rubbish *n.* **1** waste or worthless material. **2** nonsense.

■ **1** debris, garbage, junk, litter, refuse, scrap, trash, waste. **2** balderdash, *sl.* bilge, *sl.* boloney, bunkum, double Dutch, drivel, gibberish, *colloq.* gobbledegook, mumbo-jumbo, nonsense, *sl.* poppycock, *sl.* rot, twaddle.

rubble *n.* waste or rough fragments of stone or brick etc.

rubella *n.* German measles.

rubric *n.* words put as a heading or note of explanation.

ruby *n.* **1** red gem. **2** deep red colour. ● *adj.* deep red.

ruche /roōsh/ *n.* fabric gathered as trimming. ● *v.* gather thus.

ruck *v.* & *n.* crease, wrinkle.

rucksack *n.* capacious bag carried on the back in hiking etc.

ructions *n.pl.* (*colloq.*) protests and noisy arguments, a row.

rudder *n.* flat piece hinged to the stern of a boat or aircraft, used for steering.

ruddy *adj.* (**-ier**, **-iest**) reddish. □ **ruddily** *adv.*, **ruddiness** *n.*

rude *adj.* **1** impolite, showing no respect. **2** primitive, roughly made. **3** indecent. □ **rudely** *adv.*, **rudeness** *n.*

■ **1** bad-mannered, discourteous, disrespectful, ill-mannered, impertinent, impolite, impudent, insolent, offensive, uncivil, uncouth, ungracious. **2** crude, primitive, rough, rudimentary, simple.

rudiment *n.* **1** rudimentary part. **2** (*pl.*) elementary principles.

■ **2** (**rudiments**) basics, elements, essentials, first principles, foundations, fundamentals, *sl.* nitty-gritty.

rudimentary *adj.* **1** basic, elementary. **2** incompletely developed.

■ **1** basic, elementary, fundamental, introductory, primary. **2** basic, crude, primitive, rude; imperfect, incomplete.

rue[1] *n.* shrub with bitter leaves.

rue[2] *v.* repent, regret.

rueful *adj.* showing or feeling good-humoured regret. □ **ruefully** *adv.*

ruff *n.* **1** pleated frill worn round the neck. **2** projecting or coloured ring of feathers or fur round a bird's or animal's neck. **3** bird of the sandpiper family.

ruffian *n.* violent lawless person.

■ hooligan, lout, rogue, thug, tough, vandal, *sl.* yob.

ruffle *v.* **1** disturb the calmness or smoothness (of). **2** annoy. ● *n.* gathered frill.

■ *v.* **1** dishevel, disorder, mess (up), rumple, tousle. **2** anger, annoy, displease, irritate, perturb, *colloq.* rile, upset.

rufous *adj.* reddish-brown.

rug *n.* **1** floor-mat. **2** thick woollen wrap or coverlet.

Rugby *n.* (in full **Rugby football**) game played with an oval ball which may be kicked or carried.

rugged *adj.* **1** rough, uneven, rocky. **2** sturdy, robust, tough. □ **ruggedly** *adv.*, **ruggedness** *n.*

rugger *n.* (*colloq.*) Rugby football.

ruin *n.* **1** destruction. **2** loss of one's fortune or prospects. **3** broken remains. **4** cause of ruin. ● *v.* **1** cause ruin to. **2** reduce to ruins. □ **ruination** *n.*

■ *n.* **1** destruction, devastation, ruination. **2** collapse, downfall, end, failure. ● *v.* **1** blight, dash, destroy, end, put an end to, shatter, undo, wreck; damage, mar, spoil. **2** destroy, devastate, lay waste to, obliterate, ravage, raze, wreck.

ruinous *adj.* **1** bringing ruin. **2** in ruins, ruined. □ **ruinously** *adv.*

rule *n.* **1** statement of what can or should be done in certain circumstances or in a game. **2** governing, control. **3** dominant custom. **4** ruler used by carpenters etc. ● *v.* **1** govern, keep under control. **2** give an authoritative decision. **3** draw (a line) using a ruler. □ **as a rule** usually. **rule of thumb** rough practical method of procedure. **rule out** exclude.

■ *n.* **1** decree, directive, edict, law, order, ordinance, regulation, ruling. **2** administration, authority, command, control, dominion, government, jurisdiction, leadership, management, sovereignty, sway. **3** convention, custom, norm, practice, tradition. ● *v.* **1** be in charge of, command, control, dominate, govern, hold sway over, reign over. **2** decide, decree, dictate, find, judge, pronounce. □ **as a rule** generally, in general, mostly, normally, on the whole, ordinarily, usually. **rule out** ban, bar, forbid, exclude, preclude, prohibit, proscribe, veto.

ruler *n.* **1** person who rules. **2** straight strip used in measuring or for drawing straight lines.

■ **1** chief, chieftain, commander, head, king, leader, monarch, queen, sovereign.

ruling *n.* authoritative decision.

rum[1] *n.* alcoholic spirit distilled from sugar cane or molasses.

rum[2] *adj.* (*colloq.*) strange, odd.

rumba *n.* ballroom dance of Cuban origin.

rumble[1] *v.* make a low continuous sound. ● *n.* rumbling sound.

rumble[2] *v.* (*sl.*) detect the true character of.

rumbustious *adj.* (*colloq.*) boisterous, uproarious.

ruminant *n.* animal that chews the cud. ● *adj.* ruminating.

ruminate *v.* **1** chew the cud. **2** meditate, ponder. □ **rumination** *n.*, **ruminative** *adj.*

■ **2** cogitate, deliberate, meditate, muse, ponder, reflect, think.

rummage *v.* & *n.* search by disarranging things. □ **rummage sale** jumble sale.

rummy *n.* card game played usually with two packs.

rumour *n.* information spread by talking but not certainly true. □ **be rumoured** be spread as a rumour.

■ gossip, hearsay, talk, tittle-tattle.

rump *n.* **1** buttocks. **2** bird's back near the tail.

rumple *v.* make or become crumpled.

■ crease, crinkle, crumple, crush, dishevel, tousle, wrinkle.

rumpus *n.* uproar, angry dispute.

■ commotion, din, disturbance, fracas, fray, furore, fuss, hullabaloo, *colloq.* kerfuffle, racket, row, stir, to-do, tumult, uproar.

run *v.* (**ran, run, running**) **1** move with quick steps and with always at least one foot off the ground. **2** go smoothly or swiftly. **3** spread, flow, exude liquid. **4** function. **5** manage, organize. **6** extend. **7** travel or convey from one point to another. **8** compete in a race. **9** be current or valid. **10** own and use (a vehicle etc.). ● *n.* **1** spell of running. **2** trip or journey. **3** continuous stretch or sequence. **4** enclosure where domestic animals can range. **5** ladder in fabric. **6** point scored in cricket or baseball. **7** permission to make unrestricted use of something. □ **in** or **out of the running** with a good or with no chance of winning. **in the long run** in the end, over a long period. **on the run** fleeing. **run across** happen to meet or find. **run a risk** take a risk. **run a temperature** be feverish. **run away** flee. **run down 1** reduce the numbers of. **2** knock down with a vehicle. **3** speak of in a slighting way. **4** discover after searching. **be run down** be weak or exhausted. **run-down** *n.* detailed analysis. **run into 1** collide with. **2** happen to meet. **run-of-the-mill** *adj.* ordinary. **run out** become used up. **run out of** have used up (one's stock). **run over** knock down or crush with a vehicle. **run up** allow (a bill) to mount. **run-up** *n.* period leading up to an event.

■ *v.* **1,2** jog, lope, sprint, trot; bolt, dart, hare, hurry, race, rush, speed, tear, zoom. **3** course, flood, flow, gush, pour, stream; seep, trickle. **4** be in working order, function, go, operate, work. **5** administer, control, direct, manage, organize, supervise. **6** extend, go, lead, reach, stretch. **7** convey, drive, take, transport. ● *n.* **2** drive, excursion, jaunt, journey, ride, spin, trip. **3** period, session, spell, stint, stretch. □ **run away** bolt, decamp, *sl.* do a bunk, escape, flee, fly, make off, *colloq.* skedaddle, take to one's heels. **run down 3** criticize, denigrate, deprecate, disparage, *sl.* knock, malign. **run-of-the-mill** average, commonplace, everyday, normal, ordinary, standard, unexceptional, usual.

rune *n.* letter of an early Germanic alphabet. □ **runic** *adj.*

rung[1] *n.* crosspiece of a ladder etc.

rung[2] *see* **ring**[2].

runner *n.* **1** person or animal that runs. **2** messenger. **3** creeping stem that roots. **4** groove, strip, or roller etc. for a thing to move on. **5** long narrow strip of carpet or ornamental cloth. □ **runner-up** *n.* one who finishes second in a competition.

runny *adj.* semi-liquid, tending to flow or exude fluid.

runt *n.* undersized person or animal.

runway *n.* surface on which aircraft may take off and land.

rupee *n.* unit of money in India, Pakistan, etc.

rupture *n.* **1** breaking, breach. **2** abdominal hernia. ● *v.* **1** burst, break. **2** cause hernia in.

■ *n.* **1** breach, break, estrangement, rift, split; crack, fissure, fracture.

rural *adj.* of, in, or like the country.

ruse *n.* deception, trick.

■ deception, device, ploy, stratagem, trick, wile.

rush[1] *n.* marsh plant with a slender pithy stem.

rush[2] *v.* **1** go or convey with great speed. **2** act hastily. **3** force into hasty action. **4** attack with a sudden assault. ● *n.* **1** rushing, instance of this. **2** period of great activity. □ **rush hour** one of the times of day when traffic is busiest.

■ *v.* **1,2** hasten, hurry, make haste; dart, dash, fly, race, scoot, speed, tear, whiz, zoom.

rusk *n.* a kind of biscuit.

russet *adj.* soft reddish-brown. ● *n.* **1** russet colour. **2** a kind of apple with a rough skin.

rust *n.* brownish corrosive coating formed on iron exposed to moisture. ● *adj.* reddish-brown. ● *v.* make or become rusty. □ **rustproof** *adj.* & *v.*

rustic *adj.* **1** of or like country life or people. **2** made of rough timber or untrimmed branches. □ **rusticity** *n.*

rustle *v.* **1** (cause to) make a sound like paper being crumpled. **2** (*US*) steal (horses or cattle). ● *n.* rustling sound. □ **rustler** *n.*

rusty *adj.* (**-ier, -iest**) **1** affected with rust. **2** rust-coloured. **3** having lost quality by lack of use. □ **rustiness** *n.*

rut[1] *n.* **1** deep track made by wheels. **2** habitual usu. dull course of life.

rut[2] *n.* periodic sexual excitement of a male deer, goat, etc. ● *v.* (**rutted**) be affected with this.

ruthless *adj.* having no pity. □ **ruthlessly** *adv.*, **ruthlessness** *n.*

■ cruel, hard-hearted, harsh, heartless, inhuman, merciless, pitiless, remorseless, unfeeling, unforgiving, unsympathetic.

rye *n.* **1** a kind of cereal. **2** whisky made from rye.

Ss

S. *abbr.* **1** south. **2** southern.

sabbath *n.* day of worship and rest from work (Saturday for Jews, Sunday for Christians).

sabbatical *n.* leave granted at intervals to a university professor etc. for study and travel.

sable *n.* **1** small Arctic mammal with dark fur. **2** its fur. ● *adj.* black.

sabotage *n.* wilful damage to machinery or materials, or disruption of work. ● *v.* commit sabotage on. ▫ **saboteur** *n.*

sabre *n.* curved sword.

sac *n.* bag-like part in an animal or plant.

saccharin *n.* very sweet substance used instead of sugar.

saccharine *adj.* intensely and unpleasantly sweet.

sachet /sáshay/ *n.* small bag or sealed pack.

sack¹ *n.* **1** large bag of strong coarse fabric. **2** (**the sack**) (*colloq.*) dismissal from one's employment. ● *v.* (*colloq.*) dismiss. ▫ **sackful** *n.*

sack² *v.* plunder (a captured town). ● *n.* this act or process.

> ■ *v.* loot, pillage, plunder, raid, ransack.

sackcloth *n.* (also **sacking**) coarse fabric for making sacks.

sacral /sáykrəl/ *adj.* of the sacrum.

sacrament *n.* any of the symbolic Christian religious ceremonies. ▫ **sacramental** *adj.*

sacred *adj.* **1** holy. **2** dedicated (to a person or purpose). **3** connected with religion. **4** sacrosanct. ▫ **sacred cow** idea etc. which its supporters will not allow to be criticized.

> ■ **1** blessed, holy, sanctified.

sacrifice *n.* **1** slaughter of a victim or presenting of a gift to win a god's favour. **2** this victim or gift. **3** giving up of a valued thing for the sake of something else. **4** thing given up, loss entailed. ● *v.* offer, kill, or give up as a sacrifice. ▫ **sacrificial** *adj.*

> ■ *v.* immolate; forgo, give up, relinquish, renounce, surrender.

sacrilege *n.* disrespect to a sacred thing. ▫ **sacrilegious** *adj.*

sacrosanct *adj.* reverenced or respected and not to be harmed.

sacrum /sáykrəm/ *n.* bone at the base of the spine.

sad *adj.* (**sadder**, **saddest**) **1** showing or causing sorrow. **2** regrettable. ▫ **sadly** *adv.*, **sadness** *n.*

> ■ **1** crestfallen, dejected, depressed, despondent, dispirited, downcast, downhearted, glum, heartbroken, melancholy, miserable, sorrowful, unhappy, woebegone, wretched; bleak, depressing, disheartening, dismal, gloomy, sombre. **2** awful, deplorable, lamentable, regrettable, terrible, unfortunate.

sadden *v.* make or become sad.

saddle *n.* **1** seat for a rider. **2** joint of meat consisting of the two loins. ● *v.* **1** put a saddle on (an animal). **2** burden with a task.

saddler *n.* person who makes or deals in saddles and harness.

sadism *n.* (sexual) pleasure from inflicting or watching cruelty. ▫ **sadist** *n.*, **sadistic** *adj.*, **sadistically** *adv.*

safari *n.* expedition to hunt or observe wild animals. ▫ **safari park** park where exotic wild animals are kept in the open for visitors to see.

safe *adj.* **1** free from risk or danger. **2** secure. **3** reliable. ● *adv.* safely. ● *n.* strong lockable cupboard for valuables. ▫ **safe conduct** immunity from arrest or harm. **safe deposit** building containing safes and strongrooms for hire. **safely** *adv.*

> ■ *adj.* **1** harmless, innocuous; unharmed, unhurt, uninjured, unscathed. **2** impenetrable, impregnable, secure. **3** reliable, secure, sound, sure.

safeguard *n.* means of protection. ● *v.* protect.

> ■ *v.* conserve, defend, guard, keep (safe), maintain, preserve, protect, save.

safety *n.* being safe, freedom from risk or danger. ▫ **safety pin** brooch-like pin with a guard protecting and securing the point. **safety-valve** *n.* **1** valve that opens automatically to relieve excessive pressure in a steam boiler. **2** harmless outlet for emotion.

saffron *n.* **1** orange-coloured stigmas of a crocus, used to colour and flavour food. **2** colour of these.

sag *v.* **(sagged)** droop or curve down in the middle under weight or pressure. ● *n.* sagging.

■ *v.* bend, dip, droop, hang (down), sink, slump.

saga *n.* long story.

sagacious *adj.* wise. □ **sagaciously** *adv.*, **sagacity** *n.*

■ astute, discerning, intelligent, judicious, perceptive, perspicacious, prudent, sage, sensible, shrewd, wise.

sage[1] *n.* herb with fragrant grey-green leaves.

sage[2] *adj.* wise. ● *n.* old and wise man. □ **sagely** *adv.*

sago *n.* starchy pith of the sago palm, used in puddings.

said *see* **say.**

sail *n.* **1** piece of fabric spread to catch the wind and drive a boat along. **2** journey by boat. **3** arm of a windmill. ● *v.* **1** travel by water. **2** start on a voyage. **3** control (a boat). **4** move smoothly.

■ *v.* **3** navigate, pilot, steer. **4** float, glide, slide, sweep.

sailboard *n.* board with a mast and sail, used in windsurfing. □ **sailboarder** *n.*, **sailboarding** *n.*

sailcloth *n.* **1** canvas for sails. **2** canvas-like dress material.

sailor *n.* member of a ship's crew. □ **bad** or **good sailor** person liable or not liable to seasickness.

■ **1** boatman, mariner, old salt, sea dog, seafarer, seaman.

saint *n.* **1** holy person, esp. one venerated by the RC or Orthodox Church. **2** very good, patient, or unselfish person. □ **sainthood** *n.*, **saintly** *adj.*, **saintliness** *n.*

■ □ **saintly** devout, God-fearing, holy, pious, religious, reverent, righteous, virtuous.

sake[1] *n.* **for the sake of** in order to please or honour (a person) or to get or keep (a thing).

sake[2] /saáki/ *n.* Japanese fermented liquor made from rice.

salacious *adj.* lewd, erotic. □ **salaciously** *adv.*, **salaciousness** *n.*, **salacity** *n.*

salad *n.* cold dish of (usu. raw) vegetables etc.

salamander *n.* lizard-like animal.

salami *n.* strongly flavoured sausage, eaten cold.

salaried *adj.* receiving a salary.

salary *n.* fixed regular payment to an employee.

sale *n.* **1** selling, exchange of a commodity for money. **2** event at which goods are sold. **3** disposal of stock at reduced prices. □ **for** or **on sale** offered for purchase.

saleable *adj.* fit to be sold, likely to find a purchaser.

salesman, saleswoman, salesperson *ns.* one employed to sell goods.

salesmanship *n.* skill at selling.

salient *adj.* projecting, most noticeable. ● *n.* projecting part.

saline *adj.* salty, containing salt(s). □ **salinity** *n.*

saliva *n.* colourless liquid that forms in the mouth.

salivary *adj.* of or producing saliva.

salivate *v.* produce saliva. □ **salivation** *n.*

sallow[1] *adj.* (of the complexion) yellowish.

sallow[2] *n.* low-growing willow.

sally *n.* **1** sudden swift attack. **2** lively or witty remark. ● *v.* **sally forth 1** rush out in attack. **2** set out on a journey.

■ *n.* **1** assault, attack, charge, foray, incursion, onslaught, raid, sortie.

salmon /sámmən/ *n.* (*pl.* **salmon**) **1** large fish with pinkish flesh. **2** salmon-pink. □ **salmon-pink** *adj.* & *n.* yellowish-pink.

salmonella *n.* bacterium causing food poisoning.

salon *n.* **1** elegant room for receiving guests. **2** place where a hairdresser, couturier, etc. receives clients.

saloon *n.* **1** public room, esp. on board ship. **2** saloon car. □ **saloon car** car with body closed off from luggage area.

salsify *n.* plant with a long fleshy root used as a vegetable.

salt *n.* **1** sodium chloride used to season and preserve food. **2** chemical compound of a metal and an acid. **3** (*pl.*) substance resembling salt in form, esp. a laxative. ● *adj.* **1** tasting of salt. **2** impregnated with salt. ● *v.* **1** season with salt. **2** preserve in salt. □ **old salt** experienced sailor. **salt away** (*colloq.*) put aside for the future. **salt-cellar** *n.* small container for salt used at meals. **salt marsh** marsh flooded by the sea at high tide. **take with a grain** or **pinch of salt** regard sceptic-

ally. **worth one's salt** competent. **salty** *adj.*, **saltiness** *n.*
salting *n.* salt marsh.
saltpetre *n.* salty white powder used in gunpowder, in medicine, and in preserving meat.
salubrious *adj.* health-giving.

■ beneficial, healthful, healthy, salutary, wholesome.

saluki *n.* (*pl.* **-is**) tall swift silky-coated dog.
salutary *adj.* producing a beneficial or wholesome effect.
salutation *n.* **1** word(s) or gesture of greeting. **2** expression of respect.
salute *n.* gesture of respect or greeting. ● *v.* make a salute to.

■ *v.* address, greet, hail; acclaim, extol, honour, pay tribute to, praise.

salvage *n.* **1** rescue of a ship or its cargo from loss at sea, or of property from fire etc. **2** saving and use of waste material. **3** items saved in this way. ● *v.* save from loss or for use as salvage.

■ *v.* recover, rescue, retrieve, save.

salvation *n.* saving, esp. from the consequences of sin.
salve *n.* **1** soothing ointment. **2** thing that soothes. ● *v.* soothe (conscience etc.).
salver *n.* small tray.
salvo *n.* (*pl.* **-oes**) **1** simultaneous firing of guns. **2** round of applause.
sal volatile /sál voláttili/ solution of ammonium carbonate used as a remedy for faintness.
samba *n.* ballroom dance of Brazilian origin.
same *adj.* **1** being of one kind, not changed or different. **2** previously mentioned. ● *pron.* the same person or thing. □ **sameness** *n.*

■ *adj.* **1** identical, selfsame, very; unchanging, unvarying.

samphire *n.* plant with edible leaves, growing by the sea.
sample *n.* **1** small part showing the quality of the whole. **2** specimen. ● *v.* test by taking a sample or getting an experience of.

■ *n.* **1** example, illustration, specimen. ● *v.* experience, taste, test, try.

sampler *n.* piece of embroidery worked in various stitches to show one's skill.
sanatorium *n.* (*pl.* **-ums**) **1** establishment for treating chronic diseases or convalescents. **2** room for sick people in a school.
sanctify *v.* make holy or sacred. □ **sanctification** *n.*
sanctimonious *adj.* making a show of righteousness or piety. □ **sanctimoniously** *adv.*, **sanctimoniousness** *n.*
sanction *n.* **1** permission, approval. **2** penalty imposed on a country or organization. ● *v.* give sanction to, authorize.

■ *n.* **1** agreement, approbation, approval, assent, authorization, blessing, consent, go-ahead, permission. ● *v.* agree to, allow, approve (of), assent to, authorize, consent to, endorse, give the go-ahead for, *colloq.* give the green light to, permit.

sanctity *n.* sacredness, holiness.
sanctuary *n.* **1** sacred place. **2** place where birds or wild animals are protected. **3** refuge.

■ **2** conservation area, reserve. **3** asylum, haven, refuge, retreat, shelter.

sanctum *n.* **1** holy place. **2** person's private room.
sand *n.* **1** very fine loose fragments of crushed rock. **2** (*pl.*) expanse of sand, sandbank. ● *v.* **1** sprinkle with sand. **2** smooth with sandpaper.
sandal *n.* light shoe with straps.
sandalwood *n.* a kind of scented wood.
sandbag *n.* bag filled with sand, used to protect a wall or building. ● *v.* (**sandbagged**) protect with sandbags.
sandbank *n.* underwater deposit of sand.
sandblast *v.* treat with a jet of sand driven by compressed air or steam.
sandcastle *n.* structure of sand, usu. made by a child.
sandpaper *n.* paper with a coating of sand or other abrasive substance, used for smoothing surfaces. ● *v.* smooth with this.
sandstone *n.* rock formed of compressed sand.
sandstorm *n.* desert storm of wind with blown sand.
sandwich *n.* **1** two or more slices of bread with a layer of filling between. **2** thing arranged like this. ● *v.* put between two others.
sandy *adj.* (**-ier, -iest**) **1** like sand. **2** covered with sand. **3** yellowish-red.
sane *adj.* **1** having a sound mind. **2** rational. □ **sanely** *adv.*

■ **1** lucid, normal, of sound mind, rational, right-minded.

sang *see* **sing**.

sangria *n.* Spanish drink of red wine, lemonade, and fruit.

sanguinary *adj.* **1** full of bloodshed. **2** bloodthirsty.

sanguine *adj.* optimistic.

■ cheerful, confident, full of hope, hopeful, optimistic, *colloq.* upbeat.

sanitary *adj.* **1** of hygiene or sanitation. **2** hygienic.

■ **2** aseptic, clean, disinfected, germ-free, hygienic, sterile.

sanitation *n.* arrangements to protect public health, esp. drainage and disposal of sewage.

sanitize *v.* make sanitary.

sanity *n.* condition of being sane.

sank *see* **sink**.

sap *n.* **1** vital liquid in plants. **2** (*sl.*) foolish person. ● *v.* (**sapped**) exhaust gradually. □ **sappy** *adj.*

■ *v.* debilitate, drain, enervate, exhaust, weaken, wear out.

sapling *n.* young tree.

sapphire *n.* **1** blue precious stone. **2** its colour. ● *adj.* bright blue.

saprophyte *n.* fungus or related plant living on decayed matter. □ **saprophytic** *adj.*

sarcasm *n.* ironically scornful remark(s). □ **sarcastic** *adj.*, **sarcastically** *adv.*

■ □ **sarcastic** biting, bitter, caustic, cutting, mordant, sardonic, scathing, sharp, spiteful, stinging.

sarcophagus *n.* (*pl.* **-gi**) stone coffin.

sardine *n.* young pilchard or similar small fish.

sardonic *adj.* humorous in a grim or sarcastic way. □ **sardonically** *adv.*

sargasso *n.* seaweed with berry-like air-vessels.

sari *n.* (*pl.* **-is**) length of cloth draped round the body, worn by Hindu women.

sarong *n.* strip of cloth worn tucked round the body.

sarsen *n.* sandstone boulder.

sartorial *adj.* **1** of tailoring. **2** of men's clothing.

sash[1] *n.* strip of cloth worn round the waist or over one shoulder.

sash[2] *n.* frame holding a pane of a window and sliding up and down in grooves.

sat *see* **sit**.

satanic *adj.* **1** of or like Satan. **2** devilish, hellish.

Satanism *n.* worship of Satan.

satchel *n.* bag for school books, hung over the shoulder(s).

sate *v.* satiate.

sateen *n.* closely woven cotton fabric resembling satin.

satellite *n.* **1** heavenly or artificial body revolving round a planet. **2** country that is subservient to another. □ **satellite dish** dish-shaped aerial for receiving broadcasts transmitted by satellite.

satiate /sáyshiayt/ *v.* satisfy fully, glut. □ **satiation** *n.*

■ glut, gorge, overfill, overindulge, sate, satisfy, surfeit.

satiety /sətí-iti/ *n.* condition of being satiated.

satin *n.* silky material that is glossy on one side. ● *adj.* smooth as satin. □ **satiny** *adj.*

satire *n.* **1** use of ridicule, irony, or sarcasm. **2** novel or play etc. that ridicules something. □ **satirical** *adj.*, **satirically** *adv.*

■ **1** caricature, irony, parody, ridicule. **2** burlesque, lampoon, parody, *colloq.* spoof, take-off.

satirize *v.* attack or describe with satire. □ **satirist** *n.*

■ burlesque, caricature, lampoon, make fun of, parody, ridicule, send up, take off.

satisfactory *adj.* satisfying, adequate. □ **satisfactorily** *adv.*

■ acceptable, adequate, all right, fair, good enough, *colloq.* OK, passable, tolerable.

satisfy *v.* **1** meet the expectations or wishes of. **2** meet (an appetite or want). **3** rid (a person) of an appetite or want. **4** make pleased or contented. **5** convince. □ **satisfaction** *n.*

■ **1** answer, comply with, fulfil, meet. **2** quench, sate, satiate, slake. **3** gratify, indulge, pander to. **4** appease, content, mollify, pacify, placate, please. **5** convince, persuade.

satsuma *n.* a kind of tangerine.

saturate *v.* **1** make thoroughly wet. **2** cause to absorb or accept as much as possible. □ **saturation** *n.*

saturnine *adj.* having a gloomy temperament or appearance.

satyr /sáttər/ *n.* woodland god in classical mythology, with a goat's ears, tail, and legs.

sauce *n.* **1** liquid preparation added to food to give flavour or richness. **2** (*sl.*) impudence.

saucepan *n.* metal cooking pot with a long handle.

saucer *n.* **1** curved dish on which a cup stands. **2** thing shaped like this.

saucy *adj.* (**-ier, -iest**) impudent. □ **saucily** *adv.*

■ cheeky, disrespectful, forward, impertinent, impudent, irreverent, pert, rude.

sauerkraut /sówərkrowt/ *n.* chopped pickled cabbage.

sauna *n.* Finnish-style steam bath.

saunter *v.* & *n.* stroll.

■ *v.* amble, stroll, walk, wander.

saurian *adj.* of or like a lizard.

sausage *n.* minced seasoned meat in a tubular case of thin skin.

savage *adj.* **1** fierce, cruel. **2** wild, primitive. ● *n.* savage person. ● *v.* maul savagely. □ **savagely** *adv.*, **savagery** *n.*

■ *adj.* **1** barbaric, bestial, bloodthirsty, brutal, cruel, ferocious, fierce, merciless, ruthless, vicious, violent. **2** primitive, uncivilized, untamed, wild.

savannah *n.* grassy plain in hot regions.

save *v.* **1** rescue. **2** protect. **3** avoid wasting. **4** keep for future use, put aside money in this way. **5** prevent the scoring of (a goal etc.). ● *n.* act of saving in football etc. □ **saver** *n.*

■ *v.* **1** deliver, release, rescue; recover, retrieve, salvage. **2** conserve, keep (safe), preserve, protect, safeguard. **4** accumulate, collect, garner, hoard, keep back, lay up, put by, reserve, *colloq.* salt away, set aside, stockpile, store up; economize.

saveloy *n.* highly seasoned sausage.

savings *n.pl.* money put aside for future use.

saviour *n.* person who rescues people from harm or danger.

savory *n.* spicy herb.

savour *n.* **1** flavour. **2** smell. ● *v.* appreciate, enjoy.

■ *n.* **1** flavour, tang, taste. ● *v.* appreciate, delight in, enjoy, luxuriate in, relish, revel in, take pleasure in.

savoury *adj.* **1** having an appetizing taste or smell. **2** salty or piquant, not sweet. ● *n.* savoury dish, esp. at the end of a meal.

savoy *n.* a kind of cabbage.

saw[1] *see* **see**[1].

saw[2] *n.* cutting tool with a zigzag edge. ● *v.* (**sawed, sawn**) **1** cut with a saw. **2** make a to-and-fro movement.

saw[3] *n.* old saying, maxim.

sawdust *n.* powdery fragments of wood, made in sawing timber.

sawfish *n.* large sea fish with a jagged blade-like snout.

sawmill *n.* mill where timber is cut into planks etc.

sawn *see* **saw**[2].

sax *n.* (*colloq.*) saxophone.

saxe blue *adj.* & *n.* greyish-blue.

saxifrage *n.* a kind of rock plant.

saxophone *n.* brass wind instrument with finger-operated keys. □ **saxophonist** *n.*

say *v.* (**said**) **1** utter. **2** express in words, state. **3** give as an opinion. **4** suppose as a possibility etc. ● *n.* **1** opportunity to express view. **2** power to decide.

■ *v.* **1** articulate, pronounce, speak, utter, voice. **2** communicate, express, put into words, verbalize; affirm, announce, assert, comment, declare, mention, observe, remark, state. **3** conjecture, estimate, guess, imagine, judge. ● *n.* **2** authority, *colloq.* clout, influence, power, voice, weight.

saying *n.* well-known phrase or proverb.

■ aphorism, axiom, expression, idiom, maxim, motto, phrase, proverb, saw.

scab *n.* **1** crust forming over a sore. **2** skin disease or plant disease causing similar roughness. **3** (*colloq.*, *derog.*) blackleg. □ **scabby** *adj.*

scabbard *n.* sheath of a sword etc.

scabies *n.* contagious skin disease causing itching.

scabious *n.* herbaceous plant with clustered flowers.

scabrous *adj.* **1** rough-surfaced. **2** indecent.

scaffold *n.* **1** platform for the execution of criminals. **2** scaffolding.

scaffolding *n.* temporary structure of poles, planks, etc. for building work.

scald *v.* **1** injure with hot liquid or steam. **2** cleanse with boiling water. ● *n.* injury by scalding.

scale[1] *n.* **1** each of the overlapping plates of horny membrane protecting the skin of many fish and reptiles. **2** thing resembling this. **3** incrustation caused by hard water or forming on teeth. ● *v.* **1** remove scale(s) from. **2** come off in scales. □ **scaly** *adj.*

scale[2] *n.* **1** pan of a balance. **2** (*pl.*) instrument for weighing things.

scale[3] *n.* **1** ordered series of units or qualities etc. for measuring or classifying things. **2** relationship between actual size and the size on a map or plan. **3**

fixed series of notes in a system of music. **4** relative size or extent. ● *v.* climb.

■ *n.* **2** proportion, ratio. ● *v.* ascend, clamber up, climb, go up, mount.

scallop *n.* **1** shellfish with hinged fan-shaped shells. **2** one shell of this. **3** (*pl.*) semicircular curves as an ornamental edging. □ **scalloped** *adj.*

scallywag *n.* rascal.

scalp *n.* skin of the head excluding the face. ● *v.* cut the scalp from.

scalpel *n.* surgeon's or painter's small straight knife.

scamp *n.* rascal.

scamper *v.* run hastily or in play. ● *n.* scampering run.

■ *v.* dash, hasten, hurry, rush, scoot, scramble, scurry, scuttle.

scampi *n.pl.* large prawns.

scan *v.* (**scanned**) **1** look at all parts of intently or quickly. **2** pass a radar or electronic beam over. **3** (of verse) have a regular rhythm. ● *n.* scanning. □ **scanner** *n.*

■ *v.* **1** examine, inspect, peruse, scrutinize, study, survey; flick *or* leaf *or* skim through, glance at.

scandal **1** something disgraceful. **2** malicious gossip. □ **scandalous** *adj.*, **scandalously** *adv.*

■ *n.* **1** disgrace, outrage, shame, sin. **2** dirt, gossip, rumour, tittle-tattle. □ **scandalous** deplorable, disgraceful, monstrous, outrageous, shameful, shocking, sinful, terrible, wicked.

scandalize *v.* shock, outrage.

■ appal, horrify, outrage, shock.

scandalmonger *n.* person who invents or spreads scandal.

Scandinavian *adj.* & *n.* (native) of Scandinavia.

scansion *n.* scanning of verse.

scant *adj.* not or barely enough.

scanty *adj.* (**-ier**, **-iest**) **1** small in amount or extent. **2** barely enough. □ **scantily** *adv.*, **scantiness** *n.*

■ insufficient, limited, meagre, minimal, scant, scarce, sparse.

scapegoat *n.* person bearing the blame due to others.

scapula *n.* (*pl.* **-lae**) shoulder blade. □ **scapular** *adj.*

scar *n.* mark where a wound has healed. ● *v.* (**scarred**) **1** mark with a scar. **2** form scar(s).

■ *n.* cicatrice. ● *v.* **1** blemish, disfigure, mar, mark.

scarab *n.* sacred beetle of ancient Egypt.

scarce *adj.* not enough to supply a demand, rare.

scarcely *adv.* **1** only just, almost not. **2** not, surely not.

■ **1** barely, hardly, (only) just.

scarcity *n.* lack or shortage.

■ dearth, deficiency, insufficiency, lack, need, paucity, shortage, want.

scare *v.* **1** frighten. **2** be frightened. ● *n.* sudden fright.

■ *v.* **1** alarm, dismay, frighten, intimidate, panic, petrify, *colloq.* put the wind up, shock, startle, terrify. ● *n.* fright, shock, start, turn.

scarecrow *n.* figure dressed in old clothes and set up to scare birds away from crops.

scaremonger *n.* alarmist. □ **scaremongering** *n.*

scarf *n.* (*pl.* **scarves**) piece or strip of material worn round the neck or over the head.

scarify *v.* **1** make slight cuts in. **2** criticize harshly.

scarlet *adj.* & *n.* brilliant red. □ **scarlet fever** infectious fever producing a scarlet rash.

scarp *n.* steep slope on a hillside.

scary *adj.* (**-ier**, **-iest**) (*colloq.*) frightening.

scathing *adj.* (of criticism) very severe.

■ biting, bitter, caustic, cutting, mordant, savage, scornful, sharp, stinging, vitriolic.

scatter *v.* **1** throw or put here and there. **2** go or send in different directions. ● *n.* small scattered amount.

■ *v.* **1** distribute, litter, spread, sprinkle, strew.

scatterbrain *n.* forgetful person. □ **scatterbrained** *adj.*

■ □ **scatterbrained** absent-minded, careless, disorganized, forgetful, *colloq.* scatty.

scatty *adj.* (**-ier**, **-iest**) (*colloq.*) lacking concentration.

scavenge *v.* **1** search for (usable objects) among rubbish etc. **2** (of animals) search for decaying flesh as food. □ **scavenger** *n.*

scenario *n.* (*pl.* **-os**) **1** script or summary of a film or play. **2** imagined sequence of events.

scene *n.* **1** place of an event. **2** display of temper or emotion. **3** view of a place. **4** incident. **5** piece of continuous action in a play or film. **6** stage scenery. **7** (*sl.*) area of activity. □ **behind the scenes** hidden from view.

■ **1** location, place, setting, site. **2** commotion, disturbance, fuss, tantrum, to-do. **3** landscape, panorama, prospect, view, vista.

scenery *n.* **1** general (esp. picturesque) appearance of a landscape. **2** structures used on a theatre stage to represent the scene of action.

scenic *adj.* picturesque.

scent *n.* **1** pleasant smell. **2** liquid perfume. **3** animal's trail perceptible to a hound's sense of smell. ● *v.* **1** discover by smell. **2** suspect the presence or existence of. **3** apply scent to, make fragrant.

■ *n.* **1** aroma, bouquet, fragrance, perfume, redolence. **2** eau-de-Cologne, perfume, toilet water.

sceptic /sképtik/ *n.* sceptical person.

sceptical /sképtik'l/ *adj.* unwilling to believe things. □ **sceptically** *adv.*, **scepticism** *n.*

■ cynical, distrustful, doubtful, dubious, incredulous, mistrustful, unconvinced.

sceptre *n.* ornamental rod carried as a symbol of sovereignty.

schedule *n.* programme or timetable of events. ● *v.* appoint in a schedule.

■ *n.* plan, programme, timetable. ● *v.* arrange, fix, organize, plan, programme, time.

schematic *adj.* in the form of a diagram. □ **schematically** *adv.*

schematize *v.* put into schematic form. □ **schematization** *n.*

scheme *n.* plan of work or action. ● *v.* make plans, plot. □ **schemer** *n.*

■ *n.* plan, programme, strategy; game, plot, ploy, *sl.* racket, ruse. ● *v.* conspire, intrigue, plan, plot.

scheming *adj.* cunning, deceitful.

■ artful, crafty, cunning, deceitful, devious, sly, treacherous, tricky, underhand, wily.

scherzo /skáirtsō/ *n.* (*pl.* **-os**) lively piece of music.

schism /síss'm/ *n.* division into opposing groups through difference in belief or opinion. □ **schismatic** *adj.* & *n.*

schist /shist/ *n.* rock with components in layers.

schizoid *adj.* like or suffering from schizophrenia. ● *n.* schizoid person.

schizophrenia *n.* mental disorder in which a person is unable to act or reason rationally. □ **schizophrenic** *adj.* & *n.*

scholar *n.* **1** learned person. **2** pupil. **3** holder of a scholarship. □ **scholarly** *adj.*

■ **1** academic, highbrow, intellectual, thinker. □ **scholarly** academic, erudite, highbrow, intellectual, learned, scholastic.

scholarship *n.* **1** grant of money towards education. **2** scholars' methods and achievements.

scholastic *adj.* **1** of schools or education. **2** academic.

school[1] *n.* shoal of fish or whales.

school[2] *n.* **1** institution for educating children or giving instruction. **2** group of artists etc. following the same principles. ● *v.* train, discipline. □ **schoolboy** *n.*, **schoolchild** *n.*, **schoolgirl** *n.*

■ *v.* coach, drill, educate, instruct, teach, train, tutor.

schoolteacher *n.* teacher in a school.

schooner *n.* **1** a kind of sailing ship. **2** measure for sherry etc.

science *n.* branch of knowledge requiring systematic study and method, esp. dealing with substances, life, and natural laws. □ **scientific** *adj.*, **scientifically** *adv.*

scientist *n.* expert in science(s).

scimitar *n.* short curved sword.

scintillating *adj.* lively, witty.

scion /sī́ən/ *n.* **1** shoot, esp. cut for grafting. **2** descendant.

scissors *n.pl.* cutting instrument with two pivoted blades.

sclerosis *n.* abnormal hardening of tissue.

scoff[1] *v.* speak contemptuously.

■ gibe, jeer, laugh, mock, sneer.

scoff[2] *v.* (*colloq.*) eat quickly.

scold *v.* rebuke (esp. a child). □ **scolding** *n.*

■ berate, chide, rebuke, reprimand, reprove, *colloq.* tell off, *colloq.* tick off, upbraid.

sconce *n.* ornamental bracket on a wall, holding a light.

scone /skon, skōn/ *n.* soft flat cake eaten buttered.

scoop *n.* **1** deep shovel-like tool. **2** a kind of ladle. **3** piece of news published by one newspaper before its rivals. ● *v.* **1**

lift or hollow with (or as if with) a scoop. **2** forestall with a news scoop.

scoot *v.* run, dart.

scooter *n.* **1** child's toy vehicle with a footboard and long steering-handle. **2** lightweight motor cycle.

scope *n.* **1** range of a subject etc. **2** opportunity.

■ **1** area, breadth, compass, extent, range, sphere. **2** freedom, latitude, leeway, opportunity, room, space.

scorch *v.* burn or become burnt on the surface.

scorching *adj.* extremely hot.

score *n.* **1** number of points gained in a game or competition. **2** set of twenty. **3** line or mark cut into something. **4** written or printed music. **5** music for a film or play. ● *v.* **1** gain (points etc.) in a game or competition. **2** keep a record of the score. **3** cut a line or mark into. **4** achieve (an advantage). **5** write or compose as a musical score. □ **score out** cross out. **scorer** *n.*

■ *n.* **1** tally, total. ● *v.* **1** notch up, win. **3** cut, incise, mark, nick, notch, scratch.

scorn *n.* strong contempt. ● *v.* **1** feel or show scorn for. **2** reject with scorn. □ **scornful** *adj.*, **scornfully** *adv.*, **scornfulness** *n.*

■ *n.* contempt, derision, disdain. ● *v.* **1** be contemptuous of, deride, despise, disdain, look down on, mock, ridicule, sneer at.

scorpion *n.* small animal of the spider group with lobster-like claws and a sting in its long tail.

Scot *n.* native of Scotland.

Scotch *adj.* Scottish. ● *n.* Scotch whisky.

scotch *v.* put an end to (a rumour).

scot-free *adj.* **1** unharmed, not punished. **2** free of charge.

Scots *adj.* Scottish. ● *n.* Scottish form of the English language. □ **Scotsman** *n.*, **Scotswoman** *n.*

Scottish *adj.* of Scotland or its people.

scoundrel *n.* dishonest or unprincipled person.

■ blackguard, good-for-nothing, knave, rascal, rat, rogue, scallywag, scamp, villain, wretch.

scour[1] *v.* **1** cleanse by rubbing. **2** clear out (a channel etc.) by flowing water. ● *n.* scouring. □ **scourer** *n.*

scour[2] *v.* search thoroughly.

scourge /skurj/ *n.* **1** whip. **2** great affliction. ● *v.* **1** flog. **2** afflict greatly.

■ *n.* **2** affliction, bane, blight, curse, misfortune, torment.

scout *n.* person sent to gather information, esp. about enemy movements etc. ● *v.* **1** act as scout. **2** search.

scowl *n.* sullen or angry frown. ● *v.* make a scowl.

■ *v.* frown, glare, glower, grimace.

scrabble *v.* scratch or search busily with hands, paws, etc.

scraggy *adj.* (**-ier**, **-iest**) lean and bony. □ **scragginess** *n.*

scram *v.* (**scrammed**) (*colloq.*) go away.

scramble *v.* **1** clamber. **2** move hastily or awkwardly. **3** mix indiscriminately. **4** cook (eggs) by heating and stirring. **5** make (a transmission) unintelligible except by means of a special receiver. **6** (of aircraft or crew) hurry to take off quickly. ● *n.* **1** scrambling walk or movement. **2** eager struggle. **3** motor cycle race over rough ground. □ **scrambler** *n.*

■ *v.* **1** clamber, climb. **2** dash, hurry, race, rush, scurry; blunder, flounder, struggle, stumble. **3** confuse, jumble, mix up, muddle. ● *n.* **2** race, rush, scrimmage, scrum, scuffle, struggle, tussle.

scrap[1] *n.* **1** fragment. **2** waste material. **3** discarded metal suitable for reprocessing. ● *v.* (**scrapped**) discard as useless.

■ *v.* **1** bit, fragment, piece, remnant, shred, snippet. **2** debris, junk, rubbish, waste. ● *v.* discard, dispose of, *sl.* ditch, dump, get rid of, jettison, throw away.

scrap[2] *n.* & *v.* (**scrapped**) (*colloq.*) fight, quarrel.

scrapbook *n.* book in which to keep newspaper cuttings etc.

scrape *v.* **1** clean, smooth, or damage by passing a hard edge across a surface. **2** pass (an edge) across in this way. **3** make the sound of scraping. **4** get along or through etc. with difficulty. **5** be very economical. ● *n.* **1** scraping movement or sound. **2** scraped place. **3** awkward situation. □ **scraper** *n.*

■ *v.* **1** bark, chafe, graze, scratch. ● *n.* **2** abrasion, graze, scratch. **3** difficulty, fix, *colloq.* jam, mess, *colloq.* pickle, predicament.

scrappy *adj.* (**-ier**, **-iest**) made up of scraps or disconnected elements.

scratch *v.* **1** scrape with the fingernails. **2** damage or wound superficially, esp.

with sharp object. **3** withdraw from a race or competition. ● *n.* **1** mark, wound, or sound made by scratching. **2** spell of scratching. ● *adj.* collected from what is available. □ **from scratch** from the very beginning or with no preparation. **up to scratch** up to the required standard. **scratchy** *adj.*

■ *n.* **1** abrasion, cut, graze, mark, nick, scrape. □ **up to scratch** adequate, competent, good enough, satisfactory.

scratchings *n.pl.* crisp residue of pork fat left after rendering lard.

scrawl *n.* **1** bad handwriting. **2** something written in this. ● *v.* write in a scrawl, scribble.

scrawny *adj.* (**-ier, -iest**) scraggy.

■ bony, emaciated, gaunt, lean, scraggy, skeletal, skinny, thin.

scream *v.* make a long piercing cry or sound. ● *n.* **1** screaming cry or sound. **2** (*sl.*) extremely amusing person or thing.

■ *v.* cry, howl, screech, shriek, squeal, wail, yowl.

scree *n.* mass of loose stones on a mountain side.

screech *n.* harsh high-pitched scream or sound. ● *v.* make or utter with a screech.

screed *n.* **1** tiresomely long letter etc. **2** thin layer of cement.

screen *n.* **1** structure used to conceal, protect, or divide something. **2** windscreen. **3** blank surface on which pictures, cinema films, or television transmissions etc. are projected. ● *v.* **1** shelter, conceal. **2** show (images etc.) on a screen. **3** examine for the presence or absence of a disease or quality etc.

■ *n.* **1** barrier, divider, partition; protection, shelter, shield. ● *v.* **1** camouflage, conceal, cover (up), disguise, hide, mask, protect, shelter, shield, veil.

screw *n.* **1** metal pin with a spiral ridge round its length, fastened by turning. **2** thing twisted to tighten or press something. **3** propeller. **4** act of screwing. ● *v.* **1** fasten or tighten with screw(s). **2** turn (a screw). **3** twist, become twisted. **4** oppress, extort. □ **screw up 1** summon up (one's courage etc.). **2** (*sl.*) bungle.

screwdriver *n.* tool for turning screws.

scribble *v.* **1** write hurriedly or carelessly. **2** make meaningless marks. ● *n.* something scribbled.

scribe *n.* **1** person who (before the invention of printing) made copies of writings. **2** (in New Testament times) professional religious scholar.

scrimmage *n.* confused struggle. ● *v.* engage in this.

scrimp *v.* skimp.

script *n.* **1** handwriting. **2** style of printed characters resembling this. **3** text of a play, film, or broadcast talk.

scripture *n.* **1** sacred writings. **2** (**the Scriptures**) those of the Christians or the Jews. □ **scriptural** *adj.*

scroll *n.* **1** roll of paper or parchment. **2** ornamental design in this shape. ● *v.* move (a display on a VDU screen) up or down as the screen is filled.

scrotum *n.* (*pl.* **-ta**) pouch of skin enclosing the testicles.

scrounge *v.* obtain by cadging. □ **scrounger** *n.*

■ beg, borrow, cadge, sponge.

scrub[1] *n.* **1** vegetation consisting of stunted trees and shrubs. **2** land covered with this.

scrub[2] *v.* (**scrubbed**) **1** rub hard esp. with something coarse or bristly. **2** (*colloq.*) cancel. ● *n.* process of scrubbing.

■ *v.* **1** clean, rub, scour.

scruff *n.* back of the neck.

scruffy *adj.* (**-ier, -iest**) (*colloq.*) shabby and untidy. □ **scruffily** *adv.*, **scruffiness** *n.*

scrum *n.* **1** scrummage. **2** confused struggle.

scrummage *n.* grouping of forwards in Rugby football to struggle for possession of the ball by pushing.

scrumptious *adj.* (*colloq.*) delicious.

scrunch *v.* & *n.* crunch.

scruple *n.* doubt about doing something produced by one's conscience. ● *v.* hesitate because of scruples.

■ *n.* compunction, doubt, hesitation, misgiving, qualm, reservation.

scrupulous *adj.* very conscientious or careful. □ **scrupulously** *adv.*, **scrupulousness** *n.*

■ careful, conscientious, meticulous, painstaking, precise, punctilious, thorough.

scrutinize *v.* make a scrutiny of.

■ examine, go over, inspect, look into, peruse, study; contemplate, look at, observe, regard, survey.

scrutiny *n.* careful look or examination.

scuba *n.* an aqualung (acronym from *s*elf-*c*ontained *u*nderwater *b*reathing *a*pparatus). □ **scuba-diving** *n.*

scud *v.* (**scudded**) move along fast and smoothly.
scuff *v.* **1** scrape or drag (one's feet) in walking. **2** mark or scrape by doing this.
scuffle *n.* confused struggle or fight. ● *v.* take part in a scuffle.
■ *n.* brawl, fight, *colloq.* scrap, scrimmage, struggle, tussle.
scull *n.* **1** one of a pair of small oars. **2** oar used to propel a boat from the stern. ● *v.* row with scull(s).
sculpt *v.* sculpture.
sculptor *n.* maker of sculptures.
sculpture *n.* **1** art of carving or modelling. **2** work made in this way. ● *v.* **1** represent in or decorate with sculpture. **2** be a sculptor. □ **sculptural** *adj.*
■ *n.* **2** bust, figure, statue, statuette. ● *v.* **1** carve, chisel, fashion, hew, model, sculpt.
scum *n.* **1** layer of impurities or froth etc. on the surface of a liquid. **2** worthless person(s).
scupper *n.* opening in a ship's side to drain water from the deck. ● *v.* **1** sink (a ship) deliberately. **2** (*sl.*) ruin.
scurf *n.* **1** flakes of dry skin, esp. from the scalp. **2** similar scaly matter. □ **scurfy** *adj.*
scurrilous *adj.* grossly or obscenely abusive. □ **scurrilously** *adv.*, **scurrility** *n.*
■ abusive, defamatory, derogatory, disparaging, insulting, opprobrious, offensive, slanderous, vituperative.
scurry *v.* run hurriedly, scamper. ● *n.* scurrying, rush.
■ *v.* dart, dash, hasten, hurry, race, run, rush, scamper, scoot, scramble, scuttle.
scurvy *n.* disease caused by lack of vitamin C in the diet.
scuttle[1] *n.* box or bucket for holding coal in a room.
scuttle[2] *v.* sink (a ship) by letting in water.
scuttle[3] *v.* & *n.* scurry.
scythe /sīth/ *n.* implement with a curved blade on a long handle, for cutting long grass or grain.
SE *abbr.* **1** south-east. **2** south-eastern.
sea *n.* **1** expanse of salt water surrounding the continents. **2** section of this. **3** large inland lake. **4** waves of the sea. **5** vast expanse. □ **at sea 1** in a ship on the sea. **2** perplexed. **sea dog** old sailor. **sea-green** *adj.* & *n.* bluish-green. **sea horse** small fish with a horse-like head. **sea level** level corresponding to the mean level of the sea's surface. **sea lion** large seal. **sea urchin** sea animal with a round spiky shell.
seaboard *n.* coast.
seafarer *n.* seafaring person.
seafaring *adj.* & *n.* working or travelling on the sea.
seafood *n.* fish or shellfish from the sea eaten as food.
seagoing *adj.* for sea voyages.
seagull *n.* gull.
seal[1] *n.* amphibious sea animal with thick fur or bristles.
seal[2] *n.* **1** engraved piece of metal used to stamp a design. **2** its impression. **3** action etc. serving to confirm or guarantee something. **4** paper sticker. **5** thing used to close an opening very tightly. ● *v.* **1** stick down. **2** close or coat so as to prevent penetration. **3** settle, decide. □ **seal off** prevent access to (an area).
sealant *n.* substance for coating a surface to seal it.
sealskin *n.* seal's skin or fur used as a clothing material.
seam *n.* **1** line where two edges join. **2** layer of coal etc. in the ground. ● *v.* join by a seam. □ **seamless** *adj.*
seaman *n.* **1** sailor. **2** person skilled in seafaring. □ **seamanship** *n.*
seamstress *n.* woman who sews.
seance /sáyonss/ *n.* spiritualist meeting.
seaplane *n.* aircraft designed to take off from and land on water.
sear *v.* scorch, burn.
search *v.* look, feel, or go over (a person or place etc.) in order to find something. ● *n.* process of searching. □ **searcher** *n.*
■ *v.* comb, hunt *or* rummage through, ransack, scour; *sl.* frisk. ● *n.* hunt, pursuit, quest.
searching *adj.* thorough.
searchlight *n.* **1** outdoor lamp with a powerful beam. **2** its beam.
seascape *n.* picture or view of the sea.
seasick *adj.* made sick by a ship's motion. □ **seasickness** *n.*
seaside *n.* coast, esp. as a place for holidays.
season *n.* **1** section of the year associated with a type of weather. **2** time when something takes place or is plentiful. ● *v.* **1** give extra flavour to (food). **2** dry or treat until ready for use. □ **season ticket** ticket valid for any number of journeys or performances in a specified period.
seasonable *adj.* **1** suitable for the season. **2** timely. □ **seasonably** *adv.*

seasonal *adj.* **1** of a season or seasons. **2** varying with the seasons. □ **seasonally** *adv.*

seasoned *adj.* experienced.

seasoning *n.* substance used to season food.

■ condiments, flavouring, spice.

seat *n.* **1** thing made or used for sitting on. **2** buttocks, part of a garment covering these. **3** place where something is based. **4** country mansion. **5** place as member of a committee or parliament etc. ● *v.* **1** cause to sit. **2** have seats for. □ **be seated** sit down. **seat belt** strap securing a person to a seat in a vehicle or aircraft.

■ *n.* **1** chair, *colloq.* pew, place. **3** base, centre, cradle, focus, heart, hub. ● *v.* **2** accommodate, have room for, hold.

seaward *adj.* & *adv.* towards the sea. □ **seawards** *adv.*

seaweed *n.* any plant that grows in the sea.

seaworthy *adj.* (of ships) fit for a sea voyage. □ **seaworthiness** *n.*

sebaceous *adj.* secreting an oily or greasy substance.

secateurs *n.pl.* clippers for pruning plants.

secede *v.* withdraw from membership. □ **secession** *n.*

seclude *v.* keep (a person) apart from others. □ **seclusion** *n.*

secluded *adj.* hidden from view, very quiet.

■ isolated, lonely, private, quiet, remote, sequestered.

second[1] /sékkənd/ *adj.* **1** next after the first. **2** additional. ● *n.* **1** second thing, class, etc. **2** sixtieth part of a minute of time or (in measuring angles) degree. **3** short time. **4** (*pl.*) goods of an inferior quality. ● *v.* state one's support of (a proposal) formally. □ **second-best** *adj.* **1** next to the best in quality. **2** inferior. **second-class** *adj.* & *adv.* next or inferior to first-class in quality etc. **second cousin** (*see* **cousin**). **second-hand** *adj.* bought after use by a previous owner. **second nature** habit or characteristic that has become automatic. **second-rate** *adj.* inferior in quality. **second sight** supposed power to foresee future events. **second thought(s)** revised opinion. **second wind** renewed capacity for effort.

■ *adj.* **1** following, next, subsequent. **2** additional, extra, further, other, supplementary. ● *n.* **3** bit, instant, minute, moment, split second, *colloq.* tick, while. ● *v.* approve, back, endorse, support. □ **second-rate** bad, inferior, poor, second-best, second-class, shoddy, substandard, tawdry.

second[2] /sikónd/ *v.* transfer temporarily to another job or department. □ **secondment** *n.*

secondary *adj.* coming after or derived from what is primary. ● *n.* secondary thing. □ **secondary colours** those obtained by mixing two primary colours. **secondary education, secondary school** that for children who have received primary education. **secondarily** *adv.*

■ *adj.* additional, ancillary, extra, minor, non-essential, peripheral, subordinate, subsidiary.

secondly *adv.* furthermore.

secret *adj.* kept from the knowledge of most people. ● *n.* **1** something secret. **2** mystery. □ **in secret** secretly. **secretly** *adv.*, **secrecy** *n.*

■ *adj.* concealed, hidden; clandestine, covert, furtive, surreptitious, undercover; classified, confidential, hush-hush, private; arcane, cryptic, mysterious.

secretariat *n.* administrative office or department.

secretary *n.* **1** person employed to deal with correspondence and routine office work. **2** official in charge of an organization's correspondence. **3** ambassador's or government minister's chief assistant. □ **Secretary-General** *n.* principal administrative officer. **secretarial** *adj.*

secrete *v.* **1** put into a place of concealment. **2** produce by secretion. □ **secretor** *n.*

■ **1** cache, conceal, hide, *colloq.* stash away. **2** discharge, emit, exude, ooze, produce.

secretion *n.* **1** process of secreting. **2** production of a substance within the body. **3** this substance.

secretive *adj.* making a secret of things. □ **secretively** *adv.*, **secretiveness** *n.*

■ close, reserved, reticent, silent, taciturn, uncommunicative, unforthcoming; conspiratorial, furtive, mysterious.

secretory *adj.* of physiological secretion.

sect *n.* group with beliefs that differ from those generally accepted.

■ cult, denomination, faith, order, persuasion, religion.

sectarian *adj.* **1** of a sect or sects. **2** narrow-mindedly promoting the interests of one's sect.

section *n.* **1** distinct part. **2** subdivision. **3** cross-section. ● *v.* divide into sections.

■ *n.* **1** part, portion, segment; stage. **2** branch, department, division, sector, subdivision.

sectional *adj.* **1** of a section or sections. **2** made in sections.

sector *n.* **1** part of an area. **2** branch of an activity. **3** section of a circular area between two lines drawn from centre to circumference.

■ **1** area, district, part, quarter, region, territory, zone.

secular *adj.* of worldly (not religious or spiritual) matters.

■ earthly, material, mundane, temporal, worldly; lay.

secure *adj.* **1** safe. **2** certain not to move or fail. ● *v.* **1** make secure. **2** fasten securely. **3** obtain. □ **securely** *adv.*

■ *adj.* **1** impregnable, inviolable, invulnerable, protected, safe; bolted, closed, fastened, locked, shut. **2** attached, fast, fastened, fixed; assured, certain, reliable, safe, sound, sure. ● *v.* **1** defend, guard, make *or* keep safe, protect, safeguard. **2** bolt, close, fasten, lock, shut. **3** acquire, come by, gain, get, obtain, procure, win.

security *n.* **1** safety, confidence. **2** safety or precaution against espionage, theft, or other danger. **3** thing serving as a pledge. **4** certificate showing ownership of financial stocks etc.

■ **1** protection, safety, shelter; assurance, certainty, confidence, conviction.

sedate[1] *adj.* calm and dignified. □ **sedately** *adv.*, **sedateness** *n.*

■ calm, collected, composed, cool, decorous, dignified, imperturbable, placid, serene, staid, tranquil, unruffled.

sedate[2] *v.* treat with sedatives. □ **sedation** *n.*

sedative *adj.* having a calming effect. ● *n.* sedative drug.

■ *n.* barbiturate, narcotic, opiate, tranquillizer.

sedentary *adj.* **1** seated. **2** (of work) done while sitting.

sedge *n.* grass-like plant(s) growing in marshes or by water.

sediment *n.* particles of solid matter in a liquid or deposited by water or wind. □ **sedimentary** *adj.*, **sedimentation** *n.*

■ deposit, dregs, grounds, lees, precipitate, residue, silt.

sedition *n.* words or actions inciting rebellion. □ **seditious** *adj.*, **seditiously** *adv.*

■ □ **seditious** insurgent, mutinous, rebellious, revolutionary, riotous, subversive.

seduce *v.* entice into sexual activity or wrongdoing. □ **seducer** *n.*, **seduction** *n.*, **seductive** *adj.*, **seductiveness** *n.*

■ □ **seductive** attractive, desirable, erotic, sexy, voluptuous; alluring, captivating, enticing, irresistible, tempting.

sedulous *adj.* diligent and persevering. □ **sedulously** *adv.*

see[1] *v.* (**saw, seen**) **1** perceive with the eye(s) or mind. **2** understand. **3** find out. **4** consider. **5** escort. **6** make sure. **7** interview, visit. **8** meet socially. **9** watch. **10** experience. □ **see to** attend to. **seeing that** in view of the fact that.

■ **1** catch sight of, descry, discern, distinguish, espy, make out, note, notice, observe, perceive, *colloq.* spot; envisage, foresee, foretell, imagine. **2** appreciate, comprehend, *colloq.* get, grasp, know, realize, understand. **3** ascertain, determine, discover, find out. **4** consider, deliberate, reflect, think. **5** accompany, escort, take. **6** ensure, make sure, mind. □ **see to** arrange, attend to, deal with, do, organize, prepare, sort out, take care of.

see[2] *n.* district of a bishop or archbishop.

seed *n.* **1** plant's fertilized ovule. **2** semen, milt. **3** origin. ● *v.* **1** produce seed. **2** sprinkle with seeds. **3** remove seeds from. □ **go** or **run to seed 1** cease flowering as seed develops. **2** become shabby or less efficient. **seedless** *adj.*

seedling *n.* very young plant growing from a seed.

seedy *adj.* (**-ier, -iest**) looking shabby and disreputable.

■ dilapidated, disreputable, shabby, sleazy, squalid.

seek *v.* (**sought**) **1** try to find or obtain. **2** try (to do something).

■ **1** hunt for, look for, search for, try to find. **2** aim, attempt, endeavour, strive, try.

seem *v.* appear to be or exist or be true. □ **seemingly** *adv.*

■ appear, feel, give an impression of, look, strike one as.

seemly *adj.* (**-ier, -iest**) in good taste, decorous. □ **seemliness** *n.*

■ decent, decorous, dignified, genteel, proper, respectable.

seen *see* **see**[1].

seep *v.* ooze slowly out or through. □ **seepage** *n.*

seer *n.* prophet.

seersucker *n.* fabric woven with a puckered surface.

see-saw *n.* **1** a long board balanced on a central support so that children sitting on each end can ride up and down alternately. **2** constantly repeated up-and-down change. ● *v.* **1** make this movement or change. **2** vacillate.

■ *v.* **2** alternate, change, fluctuate, oscillate, swing, vacillate, vary.

seethe *v.* **1** bubble as if boiling. **2** be very angry.

segment *n.* **1** part cut off or marked off or separable from others. **2** part of a circle or sphere cut off by a straight line or plane. □ **segmented** *adj.*

■ **1** division, fragment, part, piece, portion, section.

segregate *v.* put apart from others. □ **segregation** *n.*

seine /sayn/ *n.* large fishing net that hangs from floats.

seismic /sízmik/ *adj.* of earthquake(s).

seismograph *n.* instrument for recording earthquakes.

seismology *n.* study of earthquakes. □ **seismological** *adj.*, **seismologist** *n.*

seize *v.* **1** take hold of forcibly or suddenly. **2** take possession of by force or legal right. **3** affect suddenly. □ **seize on** make use of eagerly. **seize up** become stuck through overheating.

■ **1** clasp, clutch, grab, grasp, snatch, take hold of. **2** annex, capture, conquer, occupy, take possession of; appropriate, commandeer, confiscate, take; abduct, carry off, kidnap; apprehend, arrest.

seizure *n.* **1** seizing. **2** sudden violent attack of an illness.

■ **2** attack, fit, paroxysm, spasm, *colloq.* turn.

seldom *adv.* rarely, infrequently.

select *v.* pick out as best or most suitable. ● *adj.* **1** carefully chosen. **2** exclusive. □ **selector** *n.*

■ *v.* choose, decide on, go for, opt for, pick, plump for, single out; name, nominate. ● *adj.* **1** choice, exceptional, finest, first-class, first-rate, prime, superior.

selection *n.* **1** selecting. **2** thing(s) selected. **3** things from which to choose.

■ **2** choice, option; excerpt, extract, passage. **3** assortment, collection, range, variety.

selective *adj.* chosen or choosing carefully. □ **selectively** *adv.*, **selectivity** *n.*

■ *colloq.* choosy, discerning, discriminating, fastidious, fussy, particular, *colloq.* pernickety.

self *n.* (*pl.* **selves**) **1** person as an individual. **2** person's special nature. **3** one's own advantage or interests.

self- *pref.* of, to, or done by oneself or itself. □ **self-assertive** *adj.* assertive in promoting oneself, one's claims, etc. **self-assured** *adj.* self-confident. **self-centred** *adj.* thinking chiefly of oneself. **self-confident** *adj.* having confidence in oneself. **self-conscious** *adj.* shy, embarrassed. **self-contained** *adj.* complete in itself, having all the necessary facilities. **self-controlled** *adj.* able to control one's behaviour. **self-denial** *n.* deliberately going without things one would like to have. **self-determination** *n.* **1** free will. **2** nation's own choice of its form of government or allegiance etc. **self-evident** *adj.* clear without proof. **self-governing** *adj.* governing itself or oneself. **self-interest** *n.* one's own advantage. **self-made** *adj.* successful or rich by one's own efforts. **self-possessed** *adj.* calm and dignified. **self-respect** *n.* proper regard for oneself and one's own dignity and principles etc. **self-righteous** *adj.* smugly sure of one's own righteousness. **self-satisfied** *adj.* smugly pleased with oneself. **self-seeking** *adj.* & *n.* seeking to promote one's own interests rather than those of others. **self-service** *adj.* at which customers help themselves and pay a cashier for goods taken. **self-styled** *adj.* using a name or description one has adopted without right. **self-sufficient** *adj.* capable of supplying one's own needs. **self-willed** *adj.* obstinately doing what one wishes.

■ □ **self-centred** egocentric, egotistical, narcissistic, selfish. **self-conscious** awkward, diffident, embarrassed, ill at ease, shy, uncomfortable. **self-evident** clear, evident, incontrovertible, manifest, patent, plain, unmistakable. **self-possessed** calm, collected, composed, controlled, dignified, even-tempered, imperturbable, placid, self-controlled, serene, tranquil, *colloq.* unflappable, unperturbed. **self-righteous** complacent, goody-goody, pious, priggish, sanctimonious, self-satisfied, smug. **self-willed** determined, headstrong, intractable, intransigent, obstinate, pigheaded, stubborn, uncontrollable, unruly, wilful.

selfish *adj.* thinking only about one's own needs or wishes. □ **selfishly** *adv.*, **selfishness** *n.*

■ egocentric, egotistical, inconsiderate, self-centred, self-seeking, thoughtless.

selfless *adj.* unselfish. □ **selflessly** *adv.*

■ altruistic, considerate, thoughtful, unselfish.

selfsame *adj.* the very same.

sell *v.* (**sold**) **1** exchange (goods etc.) for money. **2** keep (goods) for sale. **3** promote sales of. **4** (of goods) be sold. **5** have a specified price. **6** persuade into accepting (an idea etc.). ● *n.* manner of selling. □ **sell off** dispose of by selling, esp. at a reduced price. **sell out 1** dispose of all one's stock etc. by selling. **2** betray. **sell up** sell one's house or business. **seller** *n.*

■ *v.* **1** *sl.* flog, vend. **2** carry, have in stock, keep, stock; deal in, handle, market, retail, trade in, traffic in. **3** advertise, market, *colloq.* plug, promote.

Sellotape *n.* [P.] adhesive usu. transparent tape.

selvedge *n.* (also **selvage**) edge of cloth woven to prevent fraying.

semantic *adj.* of meaning in language. □ **semantically** *adv.*

semantics *n.* study of meaning.

semaphore *n.* **1** system of signalling with the arms. **2** signalling device with mechanical arms.

semblance *n.* **1** outward appearance, show. **2** resemblance.

■ **1** air, appearance, aspect, exterior, façade, front, guise, look, manner, mask, show, veneer.

semen *n.* sperm-bearing fluid produced by male animals.

semi- *pref.* **1** half. **2** partly.

semibreve *n.* note in music, equal to two minims.

semicircle *n.* half of a circle. □ **semicircular** *adj.*

semicolon *n.* punctuation mark (;).

semiconductor *n.* substance that conducts electricity in certain conditions.

semi-detached *adj.* (of a house) joined to another on one side.

semifinal *n.* match or round preceding the final. □ **semifinalist** *n.*

seminal *adj.* **1** of seed or semen. **2** giving rise to new developments. □ **seminally** *adv.*

seminar *n.* small class for discussion and research.

seminary *n.* training college for priests or rabbis.

semiprecious *adj.* (of a gem) less valuable than a precious stone.

semiquaver *n.* note in music, equal to half a quaver.

Semite *n.* member of the group of races that includes Jews and Arabs. □ **Semitic** *adj.*

semitone *n.* half a tone in music.

semolina *n.* hard grains of wheat used for puddings.

senate *n.* **1** upper house of certain parliaments. **2** governing body of certain universities.

senator *n.* member of a senate.

send *v.* (**sent**) **1** order or cause to go to a certain destination. **2** send a message. **3** cause to move, go, or become. □ **send for** order to come or be brought. **send-off** *n.* friendly demonstration of goodwill at a person's departure. **send up** ridicule by imitating.

■ **1** communicate, consign, convey, dispatch, forward, mail, post, ship, transmit; broadcast, radio, telegraph. **3** discharge, launch, project, propel, shoot, throw. □ **send up** caricature, lampoon, make fun of, parody, ridicule, satirize, take off.

senescent *adj.* growing old. □ **senescence** *n.*

senile *adj.* weak in body or mind because of old age. □ **senility** *n.*

senior *adj.* **1** older. **2** higher in rank or authority. **3** for older children. ● *n.* **1** senior person. **2** member of a senior school. □ **senior citizen** elderly person, pensioner. **seniority** *n.*

senna *n.* dried pods or leaves of a tropical tree, used as a laxative.

sensation *n.* **1** feeling produced by stimulation of a sense organ or of the mind. **2** excited interest. **3** person or thing producing this.

■ **1** awareness, consciousness, feeling, impression, perception, sense. **2** commotion, furore, stir. **3** hit, success, *sl.* wow.

sensational *adj.* causing great excitement or admiration. □ **sensationally** *adv.*

■ amazing, astonishing, astounding, breathtaking, exciting, incredible, phenomenal, remarkable, spectacular, staggering, thrilling; *colloq.* fantastic, marvellous, *colloq.* stunning, superb, wonderful.

sensationalism *n.* use of or interest in sensational matters. □ **sensationalist** *n.*

sense *n.* **1** any of the special powers (sight, hearing, smell, taste, touch) by which a living thing becomes aware of the external world. **2** ability to perceive a thing, consciousness. **3** practical wisdom. **4** meaning. **5** purpose. **6** (*pl.*) sanity. ● *v.* perceive by sense(s) or by a mental impression. □ **make sense 1** have a meaning. **2** be a sensible idea. **make sense of** find a meaning in.

■ *n.* **1** faculty. **2** awareness, consciousness, feeling, impression, perception, sensation. **3** brain(s), common sense, intellect, intelligence, *colloq.* nous, understanding, wisdom, wit. **4** drift, gist, import, meaning, purport. ● *v.* become aware of, detect, discern, divine, feel, perceive, realize, suspect.

senseless *adj.* **1** foolish. **2** unconscious.

■ **1** absurd, crazy, daft, fatuous, foolish, idiotic, ludicrous, meaningless, nonsensical, pointless, ridiculous, silly, stupid.

sensibility *n.* sensitiveness.

sensible *adj.* **1** having or showing good sense. **2** aware. □ **sensibly** *adv.*

■ **1** down-to-earth, intelligent, judicious, level-headed, logical, practical, prudent, rational, realistic, sagacious, shrewd, wise.

sensitive *adj.* **1** easily hurt or damaged. **2** showing sympathetic understanding. **3** easily hurt or offended. **4** requiring tact. □ **sensitively** *adv.*, **sensitivity** *n.*

■ **1** delicate, tender. **2** perceptive, sympathetic, understanding. **3** emotional, highly-strung, temperamental, touchy. **4** awkward, delicate, difficult, *colloq.* sticky, ticklish.

sensitize *v.* make sensitive. □ **sensitization** *n.*, **sensitizer** *n.*

sensor *n.* device that responds to a certain stimulus.

sensory *adj.* **1** of the senses. **2** receiving and transmitting sensations.

sensual *adj.* **1** gratifying to the body. **2** indulging oneself with physical pleasures. □ **sensualism** *n.*, **sensually** *adv.*, **sensuality** *n.*

■ **1** carnal, erotic, sexual. **2** hedonistic, pleasure-loving, sybaritic, voluptuous.

sensuous *adj.* affecting the senses pleasantly. □ **sensuously** *adv.*, **sensuousness** *n.*

sent *see* **send.**

sentence *n.* **1** series of words making a single complete statement. **2** punishment awarded by a law court. ● *v.* declare sentence on.

sententious *adj.* dull and moralizing. □ **sententiously** *adv.*, **sententiousness** *n.*

sentient *adj.* capable of perceiving and feeling things. □ **sentiently** *adv.*, **sentience** *n.*

sentiment *n.* **1** mental feeling, opinion. **2** sentimentality.

■ **1** belief, feeling, opinion, thought, view. **2** sentimentalism, sentimentality.

sentimental *adj.* full of romantic or nostalgic feeling. □ **sentimentalism** *n.*, **sentimentally** *adv.*, **sentimentality** *n.*

■ emotional, maudlin, mawkish, mushy, nostalgic, romantic, *colloq.* soppy.

sentinel *n.* sentry.

sentry *n.* soldier posted to keep watch and guard something.

■ guard, lookout, sentinel, watchman.

sepal *n.* each of the leaf-like parts forming a calyx.

separable *adj.* able to be separated. □ **separability** *n.*

separate *adj.* /séppərət/ not joined or united with others. ● *v.* /séppərayt/ **1** divide. **2** set, move, or keep apart. **3** cease to live together as a married couple. □ **separately** *adv.*, **separation** *n.*, **separator** *n.*

■ *adj.* detached, disconnected, separated, unattached; different, disparate, distinct, individual, other, unrelated. ● *v.* **1** detach, disconnect, divide, partition, sever, uncouple; branch, diverge, fork, split, subdivide. **2** classify, sort; disband, disperse; keep apart, isolate, segregate. **3** part, split (up).

separatist *n.* person who favours separation from a larger (esp. political) unit. □ **separatism** *n.*

sepia *n.* **1** brown colouring matter. **2** rich reddish-brown.

sepsis *n.* septic condition.

septet *n.* **1** group of seven instruments or voices. **2** music for these.

septic *adj.* infected with harmful micro-organisms. □ **septic tank** tank in which sewage is liquefied by bacterial activity.

septicaemia /séptiséemiə/ *n.* blood poisoning.

septuagenarian *n.* person in his or her seventies.

sepulchre /séppəlkər/ *n.* tomb.

sequel *n.* **1** what follows, esp. as a result. **2** novel or film etc. continuing the story of an earlier one.

sequence *n.* **1** following of one thing after another. **2** set of things belonging next to each other in a particular order. **3** section dealing with one topic in a film.

> ■ **1,2** chain, course, cycle, order, pattern, series, set, string, succession, train.

sequential *adj.* forming a sequence. □ **sequentially** *adv.*

sequester *v.* **1** seclude. **2** confiscate.

sequestrate *v.* confiscate. □ **sequestration** *n.*

> ■ appropriate, commandeer, confiscate, expropriate, remove, seize, sequester, take away.

sequin *n.* circular spangle. □ **sequinned** *adj.*

sequoia /sikwóyə/ *n.* Californian tree growing to a great height.

seraglio /seraáliō/ *n.* (*pl.* **-os**) harem of a Muslim palace.

seraph *n.* (*pl.* **-im**) member of the highest order of angels. □ **seraphic** *adj.*

serenade *n.* music played by a lover to his lady. ● *v.* perform a serenade to.

serendipity *n.* making of pleasant discoveries by accident.

serene *adj.* calm and cheerful. □ **serenely** *adv.*, **serenity** *n.*

> ■ calm, collected, composed, peaceful, placid, relaxed, self-controlled, self-possessed, tranquil, unperturbed, unruffled, untroubled.

serf *n.* **1** medieval farm labourer forced to work for his landowner. **2** oppressed labourer. □ **serfdom** *n.*

serge *n.* strong twilled fabric.

sergeant /saárjənt/ *n.* **1** non-commissioned army officer ranking just above corporal. **2** police officer ranking just below inspector.

serial *n.* story presented in a series of instalments. ● *adj.* of or forming a series. □ **serially** *adv.*

serialize *v.* produce as a serial. □ **serialization** *n.*

series *n.* (*pl.* **series**) number of things of the same kind, or related to each other, occurring or arranged or produced in order.

> ■ chain, line, progression, row, sequence, set, string, succession, train.

serious *adj.* **1** solemn. **2** sincere. **3** important. **4** not slight. □ **seriously** *adv.*, **seriousness** *n.*

> ■ **1** grave, earnest, humourless, sober, solemn, sombre, stern, unsmiling. **2** earnest, honest, sincere. **3** crucial, grave, important, major, momentous, pressing, significant, urgent, weighty. **4** acute, critical, grave, life-threatening; awful, bad, desperate, dire, dreadful, severe, terrible.

sermon *n.* talk on a religious or moral subject, esp. during a religious service.

sermonize *v.* give a long moralizing talk. □ **sermonizer** *n.*

serpent *n.* large snake.

serpentine *adj.* twisting like a snake.

serrated *adj.* having a series of small projections. □ **serration** *n.*

serried /sérrid/ *adj.* arranged in a close series.

serum *n.* (*pl.* **sera**) **1** fluid left when blood has clotted, esp. used for inoculation. **2** watery fluid in animal tissue. □ **serous** *adj.*

servant *n.* person employed to do domestic work in a household or as an attendant.

serve *v.* **1** perform or provide services for. **2** be employed (in the army etc.). **3** be suitable (for). **4** present (food etc.) for others to consume. **5** (of food) be enough for. **6** attend to (customers). **7** set the ball in play at tennis etc. **8** deliver (a legal writ etc.) to (a person). ● *n.* service in tennis etc. □ **server** *n.*

> ■ *v.* **3** answer, be acceptable *or* suitable, do, fulfil, meet, satisfy, suffice; act, function. **4** dish up, present; wait on.

service *n.* **1** assistance, use. **2** working for an employer. **3** maintenance and repair of machinery. **4** religious ceremony or meeting. **5** department of people employed by a public organization. **6** system that performs work for customers or supplies public needs. **7** set of dishes etc. for serving a meal. **8** act of serving in tennis etc. **9** game in which one serves. **10** (*pl.*) armed forces. ● *v.* **1** maintain and repair (machinery). **2** supply with service(s). **3** pay the interest on (a loan). □ **service area** area beside a motorway where petrol and refreshments etc. are available. **service flat** flat where domestic service is provided. **service road** road giving access to houses etc. but not for use by through traffic. **service station** garage selling petrol etc.

■ *n.* **1** aid, assistance, help, support; advantage, benefit, use, utility. **3** maintenance, overhaul, repair, servicing. **4** ceremonial, ceremony, rite, ritual.

serviceable *adj.* **1** useful, usable. **2** hard-wearing.

■ **1** functional, practical, useful, utilitarian; operative, usable.

serviceman, **servicewoman** *ns.* member of the armed services.

serviette *n.* table napkin.

servile *adj.* **1** of or like a slave. **2** excessively submissive. □ **servilely** *adv.*, **servility** *n.*

■ **2** deferential, fawning, flattering, grovelling, obsequious, submissive, subservient, sycophantic, unctuous.

serving *n.* quantity of food served to one person.

servitude *n.* condition of being forced to work for others.

servo- *pref.* power-assisted.

sesame /séssəmi/ *n.* **1** tropical plant with seeds that yield oil or are used as food. **2** its seeds.

session *n.* **1** meeting(s) for discussing or deciding something. **2** period spent in an activity. **3** academic year in certain universities.

■ **2** bout, period, run, spell, stint, time.

set[1] *v.* (**set, setting**) **1** put, place, or fix in position or readiness. **2** make or become hard or firm. **3** fix or appoint (a date etc.). **4** assign as something to be done. **5** put into a specified state, cause to do something. **6** represent a story, play, etc. as happening in a certain time or place. **7** be brought towards or below the horizon by earth's movement. ● *n.* **1** way a thing sets or is set. **2** scenery or stage for a play or film. **3** = **sett**. ● *adj.* **1** fixed, determined. **2** ready. □ **be set on** be determined about. **set about 1** begin (a task). **2** attack. **set back 1** halt or slow the progress of. **2** (*colloq.*) cost (a person) a specified amount. **set-back** *n.* setting back of progress. **set eyes on** catch sight of. **set fire to** cause to burn. **set forth** set off. **set in** become established. **set off 1** begin a journey. **2** cause to begin. **3** cause to explode. **4** improve the appearance of by contrast. **set out 1** exhibit. **2** begin a journey. **3** declare. **set piece** formal or elaborate construction. **set sail** begin a voyage. **set square** right-angled triangular drawing instrument. **set up** establish, build. **set-up** *n.* structure of an organization.

■ *v.* **1** deposit, lay, place, pose, position, put, situate, stand, *colloq.* stick; arrange, prepare; adjust, correct, regulate. **2** clot, coagulate, congeal, harden, *colloq.* jell, solidify, thicken. **3** appoint, arrange, decide, designate, determine, establish, fix, name, ordain, schedule, settle, specify. ● *adj.* **1** arranged, determined, established, fixed, prescribed; accustomed, customary, normal, regular, standard, usual. **2** prepared, ready. □ **set about 1** begin, buckle down to, embark on, start, tackle. **2** assail, assault, attack, beat up, *sl.* go for, *colloq.* lay into. **set back 1** delay, hold up, retard, slow up *or* down. **set-back** difficulty, hindrance, hitch, hold-up, problem, snag. **set fire to** fire, ignite, kindle, light, set alight, set on fire. **set off 1** be on one's way, depart, go, leave, *sl.* push off, sally forth, set forth, set out. **2** initiate, prompt, stimulate, trigger. **3** blow up, detonate, explode, touch off. **set out 1** arrange, display, exhibit, lay out, spread out. **set up** create, establish, found, institute, organize, start; assemble, build, construct, erect, put together *or* up.

set[2] *n.* **1** people or things grouped as similar or forming a unit. **2** games forming part of a match in tennis etc. **3** radio or television receiver.

■ **1** circle, clique, coterie, faction, gang; assortment, batch, collection, group, lot, series.

sett *n.* badger's burrow.

settee *n.* sofa.

setter *n.* **1** person or thing that sets. **2** dog of a long-haired breed.

setting *n.* **1** way or place in which a thing is set. **2** surroundings, environment. **3** cutlery etc. for one person.

■ **1** location, place, position, situation, spot. **2** environment, habitat, milieu, surroundings.

settle[1] *n.* wooden seat with a high back and arms.

settle[2] *v.* **1** make or become comfortably positioned. **2** make one's home. **3** occupy (a previously unoccupied area). **4** make or become calm or orderly. **5** arrange or agree finally or satisfactorily, deal with.

6 come to rest. **7** sink, subside. **8** pay (a bill etc.). **9** bestow legally. □ **settle on** choose, decide on. **settle up** pay what is owing. **settler** *n.*

■ **2** set up home, take up residence. **3** colonize, people, populate. **4** calm, lull, pacify, quiet, quieten; become quiet, quieten down. **5** arrange, choose, decide on, determine, establish, finalize, fix, select, set; patch up, put right, reconcile, resolve; organize, put in order, sort out, straighten out. **6** alight, come to rest, land.

settlement *n.* **1** settling. **2** agreement or arrangement. **3** amount or property settled legally on a person. **4** place occupied by settlers.

■ **2** agreement, arrangement, bargain, contract, deal, pact, understanding.

seven *adj.* & *n.* one more than six. □ **seventh** *adj.* & *n.*

seventeen *adj.* & *n.* one more than sixteen. □ **seventeenth** *adj.* & *n.*

seventy *adj.* & *n.* seven times ten. □ **seventieth** *adj.* & *n.*

sever *v.* cut or break off. □ **severance** *n.*

■ break off, cut off, disconnect, split off, sunder.

several *adj.* a few, more than two but not many. ● *pron.* several people or things.

severe *adj.* **1** strict, harsh. **2** extreme, intense. **3** serious, critical. **4** plain in style. □ **severely** *adv.*, **severity** *n.*

■ **1** draconian, harsh, rigorous, stiff, strict, tough; austere, dour, grim, forbidding, stern, stony. **2** acute, bitter, extreme, fierce, harsh, intense, keen, sharp, violent. **3** critical, dangerous, grave, perilous, serious.

sew /sō/ *v.* (**sewed**, **sewn** or **sewed**) **1** fasten by passing thread through material, using a threaded needle or an awl etc. **2** make or fasten (a thing) by sewing. □ **sewing** *n.*

■ baste, darn, hem, stitch, tack.

sewage /sóō-ij/ *n.* liquid waste drained from houses etc. for disposal.

sewer[1] /sōər/ *n.* one who sews.

sewer[2] /sóōər/ *n.* drain for carrying sewage.

sewerage *n.* system of sewers.

sewn *see* **sew**.

sex *n.* **1** either of the two main groups (*male* and *female*) into which living things are placed according to their reproductive functions. **2** fact of belonging to one of these. **3** sexual feelings or impulses. **4** sexual intercourse. ● *v.* judge the sex of.

sexagenarian *n.* person in his or her sixties.

sexism *n.* prejudice or discrimination against people (esp. women) because of their sex. □ **sexist** *adj.* & *n.*

sexless *adj.* **1** lacking sex, neuter. **2** not involving sexual feelings.

sextant *n.* instrument for finding one's position by measuring the height of the sun etc.

sextet *n.* **1** group of six instruments or voices. **2** music for these.

sexton *n.* official in charge of a church and churchyard.

sextuplet *n.* one of six children born at one birth.

sexual *adj.* **1** of sex or the sexes. **2** (of reproduction) occurring by fusion of male and female cells. □ **sexual intercourse** insertion of the penis into the vagina. **sexually** *adv.*, **sexuality** *n.*

sexy *adj.* (**-ier**, **-iest**) sexually attractive or stimulating. □ **sexily** *adv.*, **sexiness** *n.*

■ arousing, desirable, erotic, seductive, sensual, voluptuous.

shabby *adj.* (**-ier**, **-iest**) **1** dilapidated, worn. **2** poorly dressed. **3** unfair, dishonourable. □ **shabbily** *adv.*, **shabbiness** *n.*

■ **1** dilapidated, in disrepair, ramshackle, seedy, squalid; frayed, ragged, *colloq.* scruffy, tattered, tatty, threadbare, worn. **3** contemptible, despicable, dishonourable, mean, shameful, unfair.

shack *n.* roughly built hut.

■ cabin, hovel, hut, shanty.

shackle *n.* one of a pair of metal rings joined by a chain, for fastening a prisoner's wrists or ankles. ● *v.* **1** put shackles on. **2** impede, restrict.

shade *n.* **1** comparative darkness, place sheltered from the sun. **2** screen or cover used to block or moderate light. **3** small amount. **4** degree or depth of colour. **5** differing variety. ● *v.* **1** block the rays of. **2** give shade to. **3** darken (parts of a drawing etc.). **4** pass gradually into another colour or variety.

■ *n.* **2** awning, blind, cover, curtain, screen. **3** hint, suggestion, tinge, touch, trace. **4** hue, tinge, tint, tone. ● *v.* **1,2** cover, protect, screen, shield.

shadow *n.* **1** shade. **2** patch of this where a body blocks light rays. **3** person's inseparable companion. **4** slight trace. **5**

gloom. ● *v.* **1** cast shadow over. **2** follow and watch secretly. □ **shadow-boxing** *n.* boxing against an imaginary opponent. **shadower** *n.*, **shadowy** *adj.*

■ *v.* **2** follow, pursue, stalk, *colloq.* tail, trail.

shady *adj.* (**-ier**, **-iest**) **1** giving or situated in shade. **2** disreputable, not completely honest. □ **shadily** *adv.*, **shadiness** *n.*

shaft *n.* **1** arrow, spear. **2** long slender straight part of a thing. **3** long bar or axle. **4** vertical or sloping passage or opening.

shag *n.* **1** shaggy mass. **2** strong coarse tobacco. **3** cormorant.

shaggy *adj.* (**-ier**, **-iest**) **1** having long rough hair or fibre. **2** (of hair etc.) rough and thick. □ **shaggy-dog story** lengthy anecdote. **shagginess** *n.*

shagreen *n.* **1** untanned granulated leather. **2** sharkskin.

shake *v.* (**shook**, **shaken**) **1** move quickly up and down or to and fro, tremble, vibrate. **2** shock. **3** make less firm. **4** (of the voice) become uneven. **5** (*colloq.*) shake hands. ● *n.* shaking, being shaken. □ **shake hands** clasp right hands in greeting, parting, or agreement. **shake up 1** mix by shaking. **2** rouse from lethargy, shock. **shake-up** *n.* upheaval, reorganization.

■ *v.* **1** agitate; rock, sway, swing; quiver, shiver, tremble, vibrate. **2** disconcert, distress, disturb, fluster, perturb, ruffle, shock, trouble, unsettle, upset.

shaky *adj.* (**-ier**, **-iest**) **1** shaking, unsteady. **2** unreliable. □ **shakily** *adv.*, **shakiness** *n.*

■ **1** flimsy, insubstantial, rickety, unstable, unsteady; shaking, trembling, wobbly. **2** tenuous, uncertain, unreliable, weak.

shale *n.* slate-like stone. □ **shaly** *adj.*

shall *v.aux.* used with *I* and *we* to express future tense, and with other words in promises or statements of obligation.

shallot *n.* onion-like plant.

shallow *adj.* **1** of little depth. **2** superficial. ● *n.* shallow place. □ **shallowness** *n.*

sham *n.* **1** pretence. **2** person or thing that is not genuine. ● *adj.* pretended, not genuine. ● *v.* (**shammed**) pretend (to be).

■ *adj.* affected, artificial, bogus, fake, false, feigned, fraudulent, insincere, *colloq.* phoney, simulated. ● *v.* fake, feign, pretend, simulate.

shamble *v.* & *n.* walk or run in a shuffling or lazy way.

shambles *n.* scene or condition of great bloodshed or disorder.

shame *n.* **1** painful mental feeling aroused by something dishonourable or ridiculous. **2** ability to feel this. **3** something regrettable. **4** person or thing causing shame. ● *v.* **1** bring shame on, make ashamed. **2** compel by arousing shame. □ **shameful** *adj.*, **shamefully** *adv.*, **shameless** *adj.*, **shamelessly** *adv.*

■ *n.* **1** chagrin, embarrassment, humiliation, mortification; discredit, disgrace, dishonour, disrepute, ignominy, obloquy, opprobrium. **3** pity. ● *v.* **1** embarrass, humiliate, mortify; discredit, disgrace, dishonour.

shamefaced *adj.* showing shame.

■ abashed, ashamed, chastened, embarrassed, mortified.

shammy *n.* chamois leather.

shampoo *n.* **1** liquid used to wash hair. **2** preparation for cleaning upholstery etc. **3** process of shampooing. ● *v.* wash or clean with shampoo.

shamrock *n.* clover-like plant.

shandy *n.* mixed drink of beer and ginger beer or lemonade.

shank *n.* **1** leg, esp. from knee to ankle. **2** thing's shaft or stem.

shantung *n.* soft Chinese silk.

shanty[1] *n.* shack. □ **shanty town** area with makeshift housing.

shanty[2] *n.* sailors' traditional song.

shape *n.* **1** area or form with a definite outline. **2** form, condition. **3** orderly arrangement. ● *v.* **1** give shape to. **2** influence. □ **shapeless** *adj.*

■ *n.* **1** configuration, contour, figure, form, outline, profile, silhouette. **2** appearance, guise, form; condition, health, state. ● *v.* **1** create, fashion, form, make, model, mould. **2** affect, control, determine, govern, influence.

shapely *adj.* (**-ier**, **-iest**) having a pleasant shape. □ **shapeliness** *n.*

shard *n.* broken piece of pottery.

share *n.* **1** part or amount of something divided between several people. **2** one of the equal parts forming a business company's capital and entitling the holder to a proportion of the profits. ● *v.* give, have, or use a share (of), divide.

□ **share-out** *n.* division into shares. **shareholder** *n.*, **sharer** *n.*

> ■ *n.* **1** allocation, *colloq.* cut, part, percentage, portion, quota, ration. ● *v.* apportion, distribute, divide, split.

shark *n.* **1** large voracious sea fish. **2** (*colloq.*) extortioner, swindler.

sharkskin *n.* fabric with a slightly lustrous textured weave.

sharp *adj.* **1** having a fine edge or point capable of cutting. **2** severe, harsh. **3** mentally alert. **4** abrupt, not gradual. **5** pungent. **6** well-defined. **7** unscrupulous. **8** above the correct pitch in music. ● *adv.* **1** punctually. **2** suddenly. **3** at a sharp angle. ● *n.* (sign indicating) music note raised by a semitone. □ **sharp practice** barely honest dealing. **sharply** *adv.*, **sharpness** *n.*

> ■ *adj.* **1** keen, pointed, sharpened. **2** acute, intense, severe, strong, violent; bitter, cutting, harsh, hurtful, scathing. **3** astute, bright, clever, intelligent, perceptive, shrewd, smart. **4** abrupt, sheer, steep, sudden. **5** piquant, pungent, tangy, tart. **6** clear, distinct, vivid, well-defined.

sharpen *v.* make or become sharp or sharper. □ **sharpener** *n.*

> ■ grind, hone, whet.

sharpshooter *n.* marksman.

shatter *v.* **1** break violently into small pieces. **2** destroy utterly. **3** severely upset.

> ■ **1** break, shiver, smash, splinter. **2** dash, destroy, put an end to, ruin, spoil, wreck. **3** crush, devastate, overwhelm.

shave *v.* **1** scrape (growing hair) off the skin. **2** clear (the chin etc.) of hair in this way. **3** cut thin slices from. **4** graze gently in passing. ● *n.* shaving. □ **shaver** *n.*

shaven *adj.* shaved.

shaving *n.* thin strip of wood etc. shaved off.

shawl *n.* large piece of soft fabric worn round the shoulders or wrapped round a baby.

she *pron.* female (or thing personified as female) previously mentioned. ● *n.* female animal.

sheaf *n.* (*pl.* **sheaves**) **1** bundle of papers. **2** tied bundle of cornstalks.

shear *v.* (**shorn** or **sheared**) cut or trim with shears or other sharp device. □ **shearer** *n.*

shears *n.pl.* large cutting instrument shaped like scissors.

sheath /sheeth/ *n.* close-fitting cover, esp. for a blade or tool.

sheathe /sheeth/ *v.* **1** put into a case. **2** encase in a covering.

shed[1] *n.* building for storing things, or for use as a workshop.

shed[2] *v.* (**shed, shedding**) **1** lose by a natural falling off. **2** take off. **3** allow to fall or flow.

sheen *n.* gloss, lustre.

sheep *n.* (*pl.* **sheep**) grass-eating animal with a thick fleecy coat.

sheepdog *n.* dog trained to guard and herd sheep.

sheepish *adj.* bashful, embarrassed. □ **sheepishly** *adv.*, **sheepishness** *n.*

sheepskin *n.* sheep's skin with the fleece on.

sheer[1] *adj.* **1** pure, not mixed or qualified. **2** very steep. **3** (of fabric) diaphanous. ● *adv.* directly, straight up or down. □ **sheerly** *adv.*, **sheerness** *n.*

> ■ **1** absolute, complete, downright, out and out, pure, unmitigated, utter. **2** abrupt, precipitous, sharp, steep. **3** diaphanous, flimsy, thin, transparent.

sheer[2] *v.* swerve from a course.

sheet *n.* **1** piece of cotton used in pairs as inner bedclothes. **2** large thin piece of glass, metal, paper, etc. **3** expanse of water, flame, etc. **4** rope securing the lower corner of a sail. □ **sheet anchor 1** large anchor for emergency use. **2** thing on which one relies.

sheikh /shayk/ *n.* Arab ruler. □ **sheikhdom** *n.* his territory.

shekel *n.* **1** unit of money in Israel. **2** (*pl.*, *colloq.*) money, riches.

shelf *n.* (*pl.* **shelves**) **1** board or slab fastened horizontally for things to be placed on. **2** thing resembling this, ledge. □ **shelf-life** *n.* time for which a stored thing remains usable.

shell *n.* **1** hard outer covering of eggs, nut kernels, and of animals such as snails and tortoises. **2** firm framework or covering. **3** metal case filled with explosive, for firing from a large gun. ● *v.* **1** remove the shell(s) of. **2** fire explosive shells at. □ **shell-shock** *n.* nervous breakdown from exposure to battle conditions.

shellac *n.* resinous substance used in varnish. ● *v.* (**shellacked**) coat with this.

shellfish *n.* water animal that has a shell.

shelter *n.* **1** structure that shields against danger, wind, rain, etc. **2** refuge.

3 shielded condition. ● *v.* **1** provide with shelter. **2** find or take shelter.

■ *n.* **2** haven, refuge, sanctuary. **3** cover, protection, safety, security. ● *v.* **1** conceal, guard, harbour, hide, protect, screen, shield.

shelve *v.* **1** put aside, esp. temporarily. **2** slope.

■ **1** defer, hold in abeyance, postpone, put off.

shelving *n.* **1** shelves. **2** material for these.

shepherd *n.* person who tends sheep. ● *v.* guide (people). □ **shepherd's pie** pie of minced meat topped with mashed potato.

■ *v.* conduct, escort, guide, herd, steer, usher.

sherbet *n.* **1** fizzy sweet drink. **2** powder from which this is made.

sheriff *n.* **1** Crown's chief executive officer in a county. **2** chief judge of a district in Scotland. **3** (*US*) chief law-enforcing officer of a county.

Sherpa *n.* member of a Himalayan people of Nepal and Tibet.

sherry *n.* strong wine orig. from southern Spain.

shibboleth *n.* old slogan or principle still considered essential by some members of a party.

shield *n.* **1** piece of armour carried on the arm to protect the body. **2** trophy in the form of this. **3** thing giving protection. ● *v.* protect, screen.

■ *n.* **3** defence, protection, safeguard. ● *v.* defend, guard, protect, screen, shelter.

shift *v.* **1** change or move from one position to another. **2** remove. **3** (*sl.*) move quickly. ● *n.* **1** change of place or form etc. **2** set of workers who start work when another set finishes. **3** time for which they work.

■ *v.* **1** budge, change position, move; carry, transfer. ● *n.* **1** alteration, change, movement, swing, switch, transfer.

shiftless *adj.* lazy and inefficient.

shifty *adj.* (**-ier**, **-iest**) seeming untrustworthy.

Shiite /shee-īt/ *n.* & *adj.* (member) of a Muslim sect opposed to the Sunni.

shilly-shally *v.* be unable to make up one's mind firmly.

shimmer *v.* & *n.* (shine with) a soft quivering light.

shin *n.* **1** front of the leg below the knee. **2** lower foreleg. ● *v.* (**shinned**) **shin up** climb.

shindig *n.* (*colloq.*) din, brawl.

shine *v.* (**shone**) **1** give out or reflect light, be bright. **2** cause to shine. **3** excel. **4** (**shined**) polish. ● *n.* **1** brightness. **2** high polish.

■ *v.* **1** beam, blaze, gleam, glint, glisten, glitter, glow, shimmer, sparkle, twinkle. **4** buff, burnish, polish. ● *n.* **2** gleam, gloss, lustre, patina, polish, sheen.

shingle[1] *n.* wooden roof tile.

shingle[2] *n.* **1** small rounded pebbles. **2** stretch of these, esp. on a shore. □ **shingly** *adj.*

shingles *n.* disease with a rash of small blisters.

Shinto *n.* Japanese religion revering ancestors and nature-spirits. □ **Shintoism** *n.*

shiny *adj.* (**-ier**, **-iest**) shining, glossy. □ **shininess** *n.*

ship *n.* large sea-going vessel. ● *v.* (**shipped**) **1** put or take on board a ship. **2** transport. □ **shipper** *n.*

■ *n.* boat, craft, vessel. ● *v.* **2** carry, convey, ferry, transport.

shipmate *n.* fellow member of a ship's crew.

shipment *n.* **1** shipping of goods. **2** consignment shipped.

shipping *n.* ships collectively.

shipshape *adv.* & *adj.* in good order, tidy.

shipwreck *n.* destruction of a ship by storm or striking rock etc. □ **shipwrecked** *adj.*

shipyard *n.* establishment where ships are built.

shire *n.* county. □ **shire-horse** *n.* horse of a heavy powerful breed.

shirk *v.* avoid (duty or work etc.) selfishly. □ **shirker** *n.*

■ avoid, dodge, duck, evade, get out of.

shirt *n.* lightweight garment for the upper part of the body.

shirty *adj.* (*colloq.*) annoyed.

shiver[1] *v.* tremble slightly esp. with cold or fear. ● *n.* shivering movement. □ **shivery** *adj.*

■ *v.* quake, quiver, shake, shudder, tremble.

shiver[2] *v.* shatter.

shoal[1] *n.* great number of fish swimming together. ● *v.* form shoals.

shoal² *n.* **1** shallow place. **2** underwater sandbank.

shock¹ *n.* **1** effect of a violent impact or shake. **2** sudden violent effect on the mind or emotions. **3** acute weakness caused by injury, pain, or mental shock. **4** effect of a sudden discharge of electricity through the body. ● *v.* astonish, horrify, outrage.

■ *n.* **2** blow, bombshell, jolt, surprise; fright, scare, start, *colloq.* turn. ● *v.* astonish, astound, dismay, dumbfound, shake, shatter, stagger, stun, take aback; alarm, appal, devastate, horrify, outrage, revolt, scandalize, sicken, upset.

shock² *n.* bushy mass of hair.

shocker *n.* (*colloq.*) shocking person or thing.

shocking *adj.* **1** causing great indignation, disgust, etc. **2** (*colloq.*) very bad.

■ **1** appalling, deplorable, disgusting, disgraceful, dreadful, horrible, horrifying, monstrous, outrageous, revolting, scandalous, shameful, sickening.

shod *see* **shoe**.

shoddy *adj.* (**-ier, -iest**) of poor quality. □ **shoddily** *adv.*, **shoddiness** *n.*

■ inferior, poor, second-rate, substandard, tawdry, trashy.

shoe *n.* **1** outer covering for a person's foot. **2** horseshoe. **3** part of a brake that presses against a wheel. ● *v.* (**shod, shoeing**) fit with a shoe or shoes. □ **shoe-tree** *n.* shaped block for keeping a shoe in shape.

shoehorn *n.* curved implement for easing one's heel into a shoe.

shoelace *n.* a cord for lacing up shoes.

shoestring *n.* **1** shoelace. **2** (*colloq.*) barely adequate amount of capital.

shone *see* **shine**.

shoo *int.* sound uttered to frighten animals away. ● *v.* (**shooed**) drive away by this.

shook *see* **shake**.

shoot *v.* (**shot**) **1** fire (a gun etc., or a missile). **2** kill or wound with a missile from a gun etc. **3** hunt with a gun for sport. **4** send out or move swiftly. **5** (of a plant) put forth buds or shoots. **6** take a shot at goal. **7** photograph, film. ● *n.* young branch or new growth of a plant. □ **shooting star** small meteor seen to move quickly. **shooting stick** walking stick with a small folding seat in the handle. **shoot up 1** rise suddenly. **2** grow rapidly. **shooter** *n.*

■ *v.* **1** fire, open fire; launch. **4** dart, dash, fly, hurry, race, rush, speed, tear, zoom. **5** germinate, grow, sprout. ● *n.* offshoot, scion, sprout. □ **shoot up 1** escalate, go up, increase, rocket.

shop *n.* **1** building or room where goods or services are sold to the public. **2** workshop. **3** one's own work as a subject of conversation. ● *v.* (**shopped**) **1** go into a shop or shops to buy things. **2** (*sl.*) inform against. □ **shop around** look for the best bargain. **shop-floor** *n.* workers as distinct from management or senior union officials. **shop-soiled** *adj.* soiled from being on display in a shop. **shop steward** trade union official elected by fellow workers as their spokesperson.

shoplifter *n.* person who steals goods that are displayed in a shop. □ **shoplifting** *n.*

shopper *n.* **1** person who shops. **2** bag for holding shopping.

shopping *n.* **1** buying goods in shops. **2** goods bought.

shore¹ *n.* land along the edge of a sea or lake.

■ beach, coast, littoral, seashore, seaside, strand.

shore² *v.* prop or support with a length of timber.

shorn *see* **shear**.

short *adj.* **1** measuring little from end to end in space or time, or from head to foot. **2** (having) insufficient. **3** concise, brief. **4** curt. **5** (of pastry) crisp and easily crumbled. ● *adv.* abruptly. ● *n.* **1** (*colloq.*) drink of spirits. **2** short circuit. **3** (*pl.*) trousers that do not reach the knee. ● *v.* (*colloq.*) short-circuit. □ **short-change** *v.* cheat, esp. by giving insufficient change. **short circuit** fault in an electrical circuit when current flows by a shorter route than the normal one. **short-circuit** *v.* **1** cause a short circuit in. **2** bypass. **short cut** quicker route or method. **short-handed** *adj.* having insufficient workers. **short-list** *v.* put on a short list from which a final choice will be made. **short-sighted** *adj.* **1** able to see clearly only what is close. **2** lacking foresight. **short ton** (*see* **ton**). **short wave** radio wave of frequency greater than 3 MHz.

■ *adj.* **1** diminutive, little, slight, small; brief, ephemeral, fleeting, fugitive, transient, transitory. **2** deficient, lacking, wanting; inadequate, insufficient, low, scant, scanty, unplentiful. **3** brief, concise, laconic, pithy, succinct, terse. **4** abrupt, blunt, brusque, curt, rude, terse.

shortage *n.* lack, insufficiency.
■ dearth, deficiency, insufficiency, lack, paucity, scarcity, want.

shortbread *n.* rich sweet biscuit.

shortcake *n.* shortbread.

shortcoming *n.* failure to reach a required standard, fault.
■ defect, failing, fault, flaw, imperfection, weakness, weak spot.

shorten *v.* make or become shorter.
■ abbreviate, abridge, cut, edit, truncate; curtail, cut down, decrease, reduce.

shortfall *n.* deficit.

shorthand *n.* method of writing rapidly with special symbols.

shortly *adv.* **1** after a short time. **2** in a few words. **3** curtly.

shot *see* **shoot**. *n.* **1** firing of a gun etc. **2** sound of this. **3** person of specified skill in shooting. **4** missile(s) for a cannon or gun etc. **5** attempt. **6** photograph. **7** heavy ball thrown as a sport. **8** attempt to hit something or reach a target. **9** stroke in certain ball games. **10** injection. **11** (*colloq.*) dram of spirits. □ **like a shot** (*colloq.*) without hesitation.
■ **2** bang, blast, crack, report. **5** attempt, go, *colloq.* stab, try. **6** photo, snap, snapshot.

shotgun *n.* gun for firing small shot at close range.

should *v.aux.* used to express duty or obligation, possible or expected future event, or (with *I* and *we*) a polite statement or a conditional or indefinite clause.

shoulder *n.* **1** part of the body where the arm or foreleg is attached. **2** animal's upper foreleg as a joint of meat. ● *v.* **1** push with one's shoulder. **2** take (blame or responsibility) on oneself. □ **shoulder blade** large flat bone of the shoulder.

shout *n.* loud cry or utterance. ● *v.* **1** utter a shout. **2** call loudly. □ **shout down** silence by shouting.
■ *v.* bawl, bellow, call, cry, roar, scream, thunder, yell.

shove *n.* rough push. ● *v.* **1** push roughly. **2** (*colloq.*) put.

shovel *n.* **1** spade-like tool for scooping earth etc. **2** mechanical scoop. ● *v.* (**shovelled**) **1** shift or clear with or as if with a shovel. **2** scoop roughly.

shoveller *n.* duck with a broad shovel-like beak.

show *v.* (**showed, shown**) **1** allow or cause to be seen, offer for inspection or viewing. **2** demonstrate, point out, prove. **3** conduct. **4** present an image of. **5** cause to understand. **6** be able to be seen. ● *n.* **1** display. **2** public exhibition or performance. **3** outward appearance. □ **show business** entertainment profession. **show off 1** display well or proudly or ostentatiously. **2** try to impress people. **show of hands** raising of hands in voting. **show-piece** *n.* excellent specimen used for exhibition. **show up 1** make or be clearly visible. **2** reveal (a fault etc.). **3** (*colloq.*) arrive.
■ *v.* **1** display, exhibit, present. **2** demonstrate, display, evince, indicate, manifest; expose, lay bare, reveal, uncover; confirm, establish, illustrate, point out, prove, verify. **3** accompany, conduct, escort, guide, lead, usher. **4** depict, illustrate, portray, represent. ● *n.* **1** array, display, exhibition, spectacle. **2** exhibition, exposition, fair; musical, performance, play, production. **3** appearance, pretence, semblance, veneer. □ **show off 1** display, flaunt, parade.

showdown *n.* confrontation that settles an argument.

shower *n.* **1** brief fall of rain or of snow, stones, etc. **2** sudden influx of letters or gifts etc. **3** device or cabinet in which water is sprayed on a person's body. **4** wash in this. ● *v.* **1** send or come in a shower. **2** take a shower. □ **showery** *adj.*
■ *v.* **1** hail, pelt, pour, rain; bestow, heap, lavish.

showerproof *adj.* (of fabric) able to keep out slight rain. ● *v.* make showerproof.

showjumping *n.* competitive sport of riding horses to jump over obstacles. □ **showjumper** *n.*

showman *n.* **1** proprietor of a circus etc. **2** person skilled at presenting entertainment, goods, etc. □ **showmanship** *n.*

shown *see* **show**.

showroom *n.* room where goods are displayed for inspection.

showy *adj.* (**-ier, -iest**) making a good display, brilliant, gaudy. □ **showily** *adv.*, **showiness** *n.*
■ baroque, fancy, flamboyant, flashy, florid, gaudy, ornate, ostentatious, rococo.

shrank *see* **shrink**.

shrapnel *n.* pieces of metal scattered from an exploding bomb.

shred *n.* **1** small strip torn or cut from something. **2** small amount. ● *v.* (**shredded**) tear or cut into shreds. □ **shredder** *n.*

■ *n.* **1** bit, fragment, piece, scrap. **2** bit, iota, jot, speck, trace.

shrew *n.* small mouse-like animal.

shrewd *adj.* showing sound judgement, clever. □ **shrewdly** *adv.*, **shrewdness** *n.*

■ acute, astute, canny, clever, intelligent, perceptive, perspicacious, sharp, smart, wise.

shriek *n.* shrill cry or scream. ● *v.* utter (with) a shriek.

■ *v.* cry, scream, screech, shout, squeal.

shrike *n.* bird with a strong hooked beak.

shrill *adj.* piercing and high-pitched in sound. □ **shrilly** *adv.*, **shrillness** *n.*

■ high-pitched, penetrating, piercing, piping, squeaky.

shrimp *n.* **1** small edible shellfish. **2** (*colloq.*) very small person.

shrine *n.* sacred or revered place.

shrink *v.* (**shrank**, **shrunk**) **1** make or become smaller. **2** draw back to avoid something. ● *n.* (*sl.*) psychiatrist.

shrinkage *n.* **1** shrinking of textile fabric. **2** loss by theft or wastage.

shrivel *v.* (**shrivelled**) shrink and wrinkle from great heat or cold or lack of moisture.

shroud *n.* **1** cloth wrapping a dead body for burial. **2** thing that conceals. **3** one of the ropes supporting a ship's mast. ● *v.* **1** wrap in a shroud. **2** conceal.

shrub *n.* woody plant smaller than a tree. □ **shrubby** *adj.*

shrubbery *n.* area planted with shrubs.

shrug *v.* (**shrugged**) raise (one's shoulders) as a gesture of indifference, doubt, or helplessness. ● *n.* this movement.

shrunk *see* **shrink**.

shrunken *adj.* having shrunk.

shudder *v.* shiver or shake violently. ● *n.* this movement.

■ *v.* quake, quiver, shake, shiver, tremble.

shuffle *v.* **1** walk without lifting one's feet clear of the ground. **2** rearrange, jumble. ● *n.* **1** shuffling movement or walk. **2** rearrangement.

■ *v.* **2** disarrange, jumble, mix (up), muddle, rearrange.

shun *v.* (**shunned**) avoid.

■ avoid, give a wide berth to, keep away from, steer clear of.

shunt *v.* **1** move (a train) to a side track. **2** divert.

shush *int.* & *v.* (*colloq.*) hush.

shut *v.* (**shut, shutting**) **1** move (a door or window etc.) into position to block an opening. **2** be moved in this way. **3** (make a business etc.) close for trade. **4** bring or fold parts of (a thing) together. **5** trap or exclude by shutting something. □ **shut down** stop or cease working or business. **shut-down** *n.* this process. **shut-eye** *n.* (*colloq.*) sleep. **shut up 1** shut securely. **2** (*colloq.*) stop or cease talking or making a noise.

■ **1** bolt, close, fasten, latch, lock, secure, shut up.

shutter *n.* **1** screen that can be closed over a window. **2** device that opens and closes the aperture of a camera. □ **shuttered** *adj.*

shuttle *n.* **1** device carrying the weft-thread in weaving. **2** vehicle used in a shuttle service. **3** spacecraft for repeated use. ● *v.* move, travel, or send to and fro. □ **shuttle service** transport service making frequent journeys to and fro.

shuttlecock *n.* small cone-shaped feathered object struck to and fro in badminton.

shy[1] *adj.* timid and lacking self-confidence. ● *v.* jump or move suddenly in alarm. □ **shyly** *adv.*, **shyness** *n.*

■ *adj.* bashful, diffident, inhibited, introverted, reserved, reticent, self-conscious, self-effacing, timid, withdrawn.

shy[2] *v.* & *n.* throw.

SI *abbr.* Système International, the international metric system.

Siamese *adj.* of Siam, former name of Thailand. □ **Siamese cat** cat with pale fur and darker face, paws, and tail. **Siamese twins** twins whose bodies are joined at birth.

sibling *n.* brother or sister.

sibyl *n.* pagan prophetess.

sic *adv.* used or spelt in the way quoted.

Sicilian *adj.* & *n.* (native) of Sicily.

sick *adj.* **1** unwell. **2** vomiting, likely to vomit. **3** disgusted. **4** finding amusement in misfortune or morbid subjects

■ **1** ailing, ill, in bad health, indisposed, infirm, not well, off colour, poorly, sickly, under the weather, unwell. **2** ill, nauseous, queasy.

sicken *v.* **1** become ill. **2** disgust. □ **be sickening for** be in the first stages of (a disease).

■ **2** appal, disgust, outrage, repel, repulse, revolt, shock.

sickle *n.* curved blade used for cutting corn etc.

sickly *adj.* (**-ier**, **-iest**) **1** unhealthy. **2** causing sickness. **3** faint, pale. □ **sickliness** *n.*

■ **1** delicate, feeble, frail, infirm, peaky, unhealthy, weak.

sickness *n.* **1** illness. **2** vomiting.

■ **1** bad health, ill health, infirmity; ailment, complaint, condition, disease, disorder, indisposition.

side *n.* **1** surface of an object, esp. one that is not the top, bottom, front, back, or end. **2** bounding line of a plane or solid figure. **3** either of the two halves into which something is divided. **4** either of the two surfaces of something flat and thin. **5** region next to or farther from a person or thing. **6** part near an edge. **7** aspect of a problem etc. **8** one of two opposing groups or teams etc. **9** slope of a hill. ● *adj.* **1** at or on the side. **2** subordinate. ● *v.* join forces (with a person) in a dispute. □ **on the side 1** as a sideline. **2** as a surreptitious activity. **side by side** close together. **side effect** secondary (usu. bad) effect. **side-saddle** *n.* saddle on which a woman rider sits with both legs on the same side of the horse. *adv.* sitting in this way.

■ *n.* **2** facet, plane, surface. **6** border, boundary, brink, edge, margin, perimeter, rim, verge. **7** aspect, element, facet, feature. **8** faction, movement, party; squad, team.

sideboard *n.* **1** piece of dining-room furniture with drawers and cupboards for china etc. **2** (*pl., colloq.*) sideburns.

sideburns *n.pl.* short whiskers on the cheek.

sidelight *n.* **1** incidental information. **2** either of two small lights on the front of a vehicle.

sideline *n.* **1** thing done in addition to one's main activity. **2** (*pl.*) lines bounding the sides of a football pitch etc., place for spectators. **3** (*pl.*) position etc. apart from the main action.

sidelong *adj.* & *adv.* sideways.

sidereal /sīdēriəl/ *adj.* of or measured by the stars.

sideshow *n.* small show forming part of a large one.

sidestep *v.* (**-stepped**) **1** avoid by stepping sideways. **2** evade.

■ avoid, circumvent, dodge, duck, evade, steer clear of.

sidetrack *v.* divert.

■ deflect, distract, divert, turn aside.

sidewalk *n.* (*US*) pavement.

sideways *adv.* & *adj.* **1** to or from one side. **2** with one side forward.

siding *n.* short track by the side of a railway, used in shunting.

sidle *v.* advance in a timid, furtive, or cringing way.

siege *n.* surrounding and blockading of a place by armed forces, in order to capture it.

sienna *n.* **1** a kind of clay used as a pigment. **2** its colour of reddish- or yellowish-brown.

sierra *n.* chain of mountains with jagged peaks, esp. in Spain or Spanish America.

siesta *n.* afternoon nap or rest, esp. in hot countries.

sieve /siv/ *n.* utensil with a mesh through which liquids or fine particles can pass. ● *v.* put through a sieve.

sift *v.* **1** sieve. **2** examine carefully and select or analyse.

sigh *n.* long deep breath given out audibly in sadness, tiredness, relief, etc. ● *v.* give or express with a sigh.

sight *n.* **1** ability to see. **2** seeing, being seen. **3** thing seen or worth seeing. **4** unsightly thing. **5** device looked through to aim or observe with a gun or telescope etc. ● *v.* **1** get a sight of. **2** aim or observe with a gunsight etc. □ **at** or **on sight** as soon as seen. **catch sight of** see suddenly or for a moment. **sight-read** *v.* play or sing (music) without preliminary study of the score.

■ *n.* **1** eyesight, vision. **2** glimpse, look, view. **3** display, scene, spectacle; marvel, rarity, wonder. **4** eyesore, fright. □ **catch sight of** descry, discern, espy, glimpse, notice, observe, perceive, see, *colloq.* spot, spy.

sightless *adj.* blind.

sightseeing *n.* visiting places of interest. □ **sightseer** *n.*

■ □ **sightseer** holiday-maker, tourist, traveller, tripper, visitor.

sign *n.* **1** mark or symbol used to represent something. **2** thing perceived that suggests the existence of something. **3**

board, notice, etc. displayed. **4** action or gesture conveying information etc. **5** any of the twelve divisions of the zodiac. ● *v.* **1** make a sign. **2** write (one's name) on a document. **3** convey or engage or acknowledge by this.

■ *n.* **1** emblem, insignia, logo, mark, symbol, trade mark. **2** clue, hint, indication, manifestation, mark, proof, suggestion, token, trace; augury, omen, portent, presage, warning. **3** notice, placard, poster. **4** gesticulation, gesture, motion, signal.

signal *n.* **1** sign or gesture giving information or a command. **2** object placed to give notice or warning. **3** event which causes immediate activity. **4** sequence of electrical impulses or radio waves transmitted or received. ● *v.* **(signalled)** **1** make a signal or signals. **2** communicate with or announce thus. ● *adj.* noteworthy. □ **signal-box** *n.* small railway building with signalling apparatus. **signally** *adv.*

■ *n.* **1** cue, gesticulation, gesture, motion, movement, sign. ● *v.* **1** beckon, gesticulate, gesture, indicate, motion, wave.

signalman *n.* person responsible for operating railway signals.

signatory *n.* one of the parties who sign an agreement.

signature *n.* **1** person's name or initials written when signing something. **2** indication of key or tempo at the beginning of a musical score. □ **signature tune** tune used to announce a particular performer or programme.

signet ring finger ring with an engraved design.

significance *n.* **1** meaning. **2** importance. □ **significant** *adj.*, **significantly** *adv.*

■ **1** essence, gist, meaning, purport, sense, signification. **2** consequence, import, importance, moment, relevance, value, weight.

signification *n.* meaning.

signify *v.* **1** be a sign or symbol of. **2** have as a meaning. **3** make known. **4** matter.

■ **1** betoken, denote, indicate, represent, stand for, symbolize. **2** connote, denote, indicate, mean. **3** communicate, convey, express, make known. **4** be important, count, matter.

signpost *n.* post showing the direction of certain places.

Sikh /seek/ *n.* member of an Indian religion believing in one God. □ **Sikhism** *n.*

silage *n.* green fodder stored and fermented in a silo.

silence *n.* absence of sound or of speaking. ● *v.* make silent.

■ *n.* hush, peace, quiet, stillness; reticence, taciturnity. ● *v.* hush, quiet, quieten, *colloq.* shush, *colloq.* shut up.

silencer *n.* device for reducing sound.

silent *adj.* without sound, not speaking. □ **silently** *adv.*

■ hushed, noiseless, quiescent, soundless; *colloq.* mum, mute, quiet, tongue-tied.

silhouette /siloo-ét/ *n.* dark shadow or outline seen against a light background. ● *v.* show as a silhouette.

■ *n.* contour, form, outline, profile, shadow, shape.

silica *n.* compound of silicon occurring as quartz and in sandstone etc. □ **siliceous** *adj.*

silicate *n.* compound of silicon.

silicon *n.* chemical substance found in the earth's crust in its compound forms. □ **silicon chip** silicon microchip.

silicone *n.* organic compound of silicon, used in paint, varnish, and lubricants.

silicosis *n.* lung disease caused by inhaling dust that contains silica.

silk *n.* **1** fine strong soft fibre produced by silkworms. **2** thread or cloth made from it or resembling this. □ **silky** *adj.*

silken *adj.* like silk.

silkworm *n.* caterpillar which spins its cocoon of silk.

sill *n.* strip of stone, wood, or metal at the base of a doorway or window opening.

silly *adj.* **(-ier, -iest)** **1** lacking good sense, foolish, unwise. **2** feeble-minded. □ **silliness** *n.*

■ **1** absurd, asinine, *sl.* batty, crazy, daft, foolish, idiotic, illogical, imbecilic, inane, irrational, laughable, ludicrous, mad, nonsensical, *sl.* potty, preposterous, ridiculous, risible, senseless, stupid, unreasonable, unwise, witless; babyish, childish, frivolous, giddy, immature, infantile, juvenile, puerile.

silo *n.* (*pl.* **-os**) **1** pit or airtight structure for holding silage. **2** pit or tower for storing grain, cement, or radioactive waste. **3** underground place where a missile is kept ready for firing.

silt *n.* sediment deposited by water in a channel or harbour etc. ● *v.* block or become blocked with silt.

silver *n.* **1** white precious metal. **2** articles made of this. **3** coins made of an alloy resembling it. **4** household cutlery. **5** colour of silver. ● *adj.* made of or coloured like silver. □ **silver jubilee, wedding** 25th anniversary.

silverfish *n.* small wingless insect with a fish-like body.

silverside *n.* joint of beef cut from the haunch, below topside.

silvery *adj.* **1** like silver. **2** having a clear gentle ringing sound.

simian *adj.* monkey-like, ape-like. ● *n.* monkey, ape.

similar *adj.* resembling but not the same, alike. □ **similarly** *adv.*, **similarity** *n.*

■ akin, alike, analogous, comparable, equivalent, like.

simile /símmili/ *n.* figure of speech in which one thing is compared to another.

similitude *n.* similarity.

simmer *v.* **1** boil very gently. **2** be in a state of barely suppressed anger or excitement. □ **simmer down** become less excited.

simper *v.* smile in an affected way. ● *n.* affected smile.

simple *adj.* **1** not difficult. **2** not showy or elaborate, plain. **3** foolish, inexperienced. **4** of one element or kind. □ **simply** *adv.*, **simplicity** *n.*

■ **1** basic, easy, effortless, elementary, straightforward, uncomplicated, understandable. **2** austere, basic, plain, spartan, unadorned, unostentatious, unpretentious; homely, modest, ordinary.

simpleton *n.* foolish person.

simplify *v.* make simple, make easy to do or understand. □ **simplification** *n.*

simplistic *adj.* over-simplified. □ **simplistically** *adv.*

simulate *v.* **1** pretend. **2** imitate the form or condition of. □ **simulation** *n.*, **simulator** *n.*

■ **1** affect, fake, feign, pretend, sham. **2** imitate, replicate, reproduce.

simultaneous *adj.* occurring at the same time. □ **simultaneously** *adv.*, **simultaneity** *n.*

■ □ **simultaneously** at once, at the same time, in unison, together.

sin *n.* **1** breaking of a religious or moral law. **2** act which does this. ● *v.* (**sinned**) commit a sin.

■ *n.* **1** corruption, depravity, evil, immorality, iniquity, misconduct, wickedness, wrongdoing, vice. **2** crime, fault, misdeed, misdemeanour, offence, transgression, wrong. ● *v.* do wrong, err, offend, transgress.

since *prep.* **1** after. **2** from (a specified time) until now, within that period. ● *conj.* **1** from the time that. **2** because. ● *adv.* since that time or event.

sincere *adj.* without pretence or deceit. □ **sincerely** *adv.*, **sincerity** *n.*

■ candid, earnest, frank, honest, genuine, heartfelt, real, straightforward, true, truthful, unfeigned, wholehearted.

sine *n.* ratio of the length of one side of a right-angled triangle to the hypotenuse.

sinecure *n.* position of profit with no work attached.

sinew *n.* **1** tough fibrous tissue joining muscle to bone, tendon. **2** (*pl.*) muscles, strength. □ **sinewy** *adj.*

sinful *adj.* full of sin, wicked. □ **sinfully** *adv.*, **sinfulness** *n.*

■ bad, corrupt, depraved, evil, immoral, iniquitous, vile, wicked.

sing *v.* (**sang, sung**) **1** make musical sounds with the voice. **2** perform (a song). **3** make a humming sound. □ **singer** *n.*

singe /sinj/ *v.* (**singeing**) **1** burn slightly. **2** burn the ends or edges of. ● *n.* slight burn.

single *adj.* **1** one only, not double or multiple. **2** designed for one person or thing. **3** unmarried. **4** (of a ticket) valid for an outward journey only. ● *n.* **1** one person or thing. **2** room etc. for one person. **3** single ticket. **4** pop record with one piece of music on each side. **5** (usu. *pl.*) game with one player on each side. ● *v.* **single out** choose or distinguish from others. □ **single figures** numbers from 1 to 9. **single-handed** *adj.* & *adv.* without help. **single-minded** *adj.* with one's mind set on a single purpose. **single parent** person bringing up a child or children without a partner. **singly** *adv.*

■ *adj.* **1** distinct, individual, particular, separate; lone, sole, solitary. ● *v.* choose, decide on, go for, opt for, pick, select. □ **single-minded** determined, dogged, persevering, resolute, tireless, unwavering.

singlet *n.* sleeveless vest.

singsong *adj.* with a monotonous rise and fall of the voice. ● *n.* **1** singsong manner. **2** informal singing by a group of people.

singular *n.* form of a noun or verb used in referring to one person or thing. ● *adj.* **1** of this form. **2** uncommon, extraordinary. □ **singularly** *adv.*, **singularity** *n.*

> ■ *adj.* **2** exceptional, extraordinary, outstanding, rare, remarkable, uncommon, unparalleled, unusual; bizarre, curious, eccentric, odd, peculiar, strange.

sinister *adj.* **1** suggestive of evil. **2** involving wickedness.

> ■ **1** dark, forbidding, menacing, ominous, threatening. **2** criminal, evil, nefarious, wicked.

sink *v.* (**sank**, **sunk**) **1** fall or come gradually downwards. **2** make or become submerged. **3** lose strength or value. **4** dig. **5** send (ball) into a pocket at billiards or hole at golf etc. **6** invest (money). ● *n.* fixed basin with a drainage pipe. □ **sink in** become understood. **sinking fund** money set aside regularly for repayment of a debt etc.

> ■ *v.* **1** descend, droop, drop, fall, go down, sag, slump. **2** founder, go under, submerge; scupper, scuttle. **3** decline, degenerate, deteriorate, fade, languish, weaken, worsen; decrease, dwindle. **4** bore, dig, drill, excavate.

sinker *n.* weight used to sink a fishing line etc.

sinner *n.* person who sins.

> ■ evildoer, offender, malefactor, miscreant, reprobate, transgressor, wrongdoer.

sinuous *adj.* curving, undulating.

sinus /sínəss/ *n.* (*pl.* **-uses**) cavity in bone or tissue, esp. that connecting with the nostrils.

sip *v.* (**sipped**) drink in small mouthfuls. ● *n.* amount sipped.

siphon *n.* **1** bent pipe or tube used for transferring liquid by utilizing atmospheric pressure. **2** bottle from which soda water etc. is forced out by pressure of gas. ● *v.* **1** flow or draw out through a siphon. **2** take from a source.

sir *n.* **1** polite form of address to a man. **2** (**Sir**) title of a knight or baronet.

sire *n.* animal's male parent. ● *v.* be the father of.

siren *n.* **1** device that makes a loud prolonged sound as a signal. **2** dangerously fascinating woman.

sirloin *n.* upper (best) part of loin of beef.

sirocco *n.* (*pl.* **-os**) hot wind that reaches Italy from Africa.

sisal /sís'l/ *n.* fibre made from a tropical plant.

sissy *n.* weak or timid person. ● *adj.* characteristic of a sissy.

sister *n.* **1** daughter of the same parents as another person. **2** woman who is a fellow member of a group or Church etc. **3** nun. **4** senior female nurse. □ **sister-in-law** *n.* (*pl.* **sisters-in-law**) **1** husband's or wife's sister. **2** brother's wife. **sisterly** *adj.*

sisterhood *n.* **1** relationship of sisters. **2** order of nuns. **3** group of women with common aims.

sit *v.* (**sat**, **sitting**) **1** take or be in a position with the body resting on the buttocks. **2** cause to sit. **3** pose for a portrait. **4** perch. **5** (of animals) rest with legs bent and body on the ground. **6** (of birds) remain on the nest to hatch eggs. **7** be situated. **8** be a candidate (for). **9** (of a committee etc.) hold a session.

sitar *n.* guitar-like Indian musical instrument.

sitcom *n.* (*colloq.*) situation comedy.

site *n.* place where something is, was, or is to be located. ● *v.* locate, provide with a site.

> ■ *n.* location, place, position, spot. ● *v.* locate, place, position, put, situate.

sitter *n.* **1** person sitting. **2** babysitter.

sitting *see* **sit**. *n.* **1** time during which a person or assembly etc. sits. **2** clutch of eggs. □ **sitting room** room in which to sit and relax. **sitting tenant** one already in occupation.

situate *v.* place or put in a certain position. □ **be situated** be in a certain position.

> ■ locate, place, position, put, site.

situation *n.* **1** place (with its surroundings) occupied by something. **2** set of circumstances. **3** position of employment. □ **situation comedy** broadcast comedy involving the same characters in a series of episodes. **situational** *adj.*

> ■ **1** location, position, setting, site, spot. **2** case, circumstances, state of affairs. **3** appointment, job, place, position, post.

six *adj.* & *n.* one more than five.

sixteen *adj* & *n.* one more than fifteen. □ **sixteenth** *adj.* & *n.*

sixth *adj.* & *n.* next after fifth. □ **sixth sense** supposed intuitive faculty. **sixthly** *adv.*

sixty *adj.* & *n.* six times ten. □ **sixtieth** *adj.* & *n.*

size[1] *n.* **1** relative bigness, extent. **2** one of a series of standard measurements in which things are made and sold. ● *v.* group according to size. □ **size up 1** estimate the size of. **2** (*colloq.*) form a judgement of.

■ *n.* **1** area, dimensions, expanse, extent, magnitude, measurements, proportions. □ **size up 2** appraise, assess, evaluate, judge, weigh up.

size[2] *n.* gluey solution used to glaze paper or stiffen textiles etc. ● *v.* treat with size.

sizeable *adj.* fairly large.

sizzle *v.* make a hissing sound like that of frying.

skate[1] *n.* (*pl.* **skate**) edible marine flatfish.

skate[2] *n.* boot with a blade or (**roller skate**) wheels attached, for gliding over ice or a hard surface. ● *v.* move on skates. □ **skate over** make only a passing reference to. **skater** *n.*

skateboard *n.* small board with wheels for riding on while standing. ● *v.* ride on a skateboard.

skedaddle *v.* (*colloq.*) run away.

skein /skayn/ *n.* **1** loosely coiled bundle of yarn. **2** flock of wild geese etc. in flight.

skeletal *adj.* of or like a skeleton.

skeleton *n.* **1** hard supporting structure of an animal body. **2** any supporting structure, framework. □ **skeleton service, staff** one reduced to a minimum. **skeleton key** key made so as to fit many locks.

sketch *n.* **1** rough drawing or painting. **2** brief account. **3** short usu. comic play. ● *v.* make a sketch or sketches (of).

■ *n.* **1** draft, outline, rough. ● *v.* draft, draw, outline, rough out.

sketchy *adj.* (**-ier, -iest**) rough and not detailed or substantial. □ **sketchily** *adv.*, **sketchiness** *n.*

■ cursory, incomplete, patchy, perfunctory, rough, superficial.

skew *adj.* slanting, askew. ● *v.* **1** make skew. **2** turn or twist round.

skewbald *adj.* (of an animal) with irregular patches of white and another colour.

skewer *n.* pin to hold meat or pieces of food together while cooking. ● *v.* pierce with a skewer.

ski *n.* (*pl.* **-is**) one of a pair of long narrow strips of wood etc. fixed under the feet for travelling over snow. ● *v.* (**ski'd, skiing**) travel on skis. □ **skier** *n.*

skid *v.* (**skidded**) (of a vehicle) slide uncontrollably. ● *n.* skidding movement. □ **skid-pan** *n.* surface used for practising control of skidding vehicles.

skiff *n.* small light rowing boat.

skilful *adj.* having or showing great skill. □ **skilfully** *adv.*

■ able, accomplished, adept, adroit, capable, competent, deft, dexterous, expert, handy, masterly, practised, proficient, skilled, talented.

skill *n.* ability to do something well. □ **skilled** *adj.*

■ ability, aptitude, competence, expertise, know-how, proficiency, prowess, talent, workmanship.

skim *v.* (**skimmed**) **1** take (matter) from the surface of (liquid). **2** glide. **3** read quickly. □ **skimmed milk** milk with the cream removed.

skimp *v.* supply or use rather less than what is necessary.

skimpy *adj.* (**-ier, -iest**) scanty. □ **skimpily** *adv.*, **skimpiness** *n.*

skin *n.* **1** outer covering of the human or other animal body. **2** material made from animal skin. **3** complexion. **4** outer layer. **5** skin-like film on liquid. ● *v.* (**skinned**) strip skin from. □ **skin diving** sport of swimming under water with flippers and breathing apparatus.

■ *n.* **1** epidermis. **2** fleece, hide, pelt. **4** coating, covering, film; peel rind, shell. ● *v.* bark, graze, scrape; flay, strip.

skinflint *n.* miserly person.

skinny *adj.* (**-ier, -iest**) very thin.

■ bony, emaciated, gaunt, scraggy, scrawny, thin.

skint *adj.* (*sl.*) with no money left.

skip[1] *v.* (**skipped**) **1** move lightly, esp. taking two steps with each foot in turn. **2** jump with a skipping rope. **3** (*colloq.*) omit. **4** (*colloq.*) go away hastily or secretly. ● *n.* skipping movement.

■ *v.* **1** bound, caper, cavort, frisk, gambol, leap, prance, spring. **3** disregard, leave out, omit, overlook, pass over.

skip[2] *n.* large container for builders' rubbish etc.

skipper *n.* & *v.* captain.

skipping rope rope turned over the head and under the feet while jumping.

skirmish *n.* minor fight or conflict. ● *v.* take part in a skirmish.

■ *n.* brawl, conflict, fight, fray, *colloq.* scrap, struggle, tussle.

skirt *n.* **1** woman's garment hanging from the waist. **2** this part of a garment. **3** similar part. **4** cut of beef from the lower flank. ● *v.* go or be along the edge of.

skirting (board) *n.* narrow board round the bottom of the wall of a room.

skit *n.* short parody.

skittish *adj.* frisky.

skittle *n.* one of the wooden pins set up to be bowled down with a ball in the game of **skittles**.

skua *n.* large seagull.

skulduggery *n.* (*colloq.*) trickery.

skulk *v.* loiter stealthily.

skull *n.* **1** bony framework of the head. **2** representation of this.

skullcap *n.* small cap with no peak.

skunk *n.* **1** black bushy-tailed American animal able to spray an evil-smelling liquid. **2** (*sl.*) contemptible person.

sky *n.* region of the clouds or upper air. □ **sky-blue** *adj.* & *n.* bright clear blue.

skydiving *n.* parachuting in which the parachute is not opened until the last moment.

skylark *n.* lark that soars while singing. ● *v.* play mischievously.

skylight *n.* window set in the line of a roof or ceiling.

skyscraper *n.* very tall building.

slab *n.* broad flat piece of something solid.

slack *adj.* **1** not tight. **2** negligent. **3** not busy. ● *n.* slack part. ● *v.* **1** slacken. **2** be lazy about work. □ **slackness** *n.*

■ *adj.* **1** limp, loose. **2** careless, inattentive, lax, negligent, remiss.

slacken *v.* make or become slack.

slacks *n.pl.* casual trousers.

slag *n.* solid waste left when metal has been smelted. ● *v.* (**slagged**) (*sl.*) criticize, insult. □ **slag-heap** *n.* mound of waste matter.

slain *see* **slay**.

slake *v.* **1** satisfy (thirst). **2** combine (lime) with water.

slalom *n.* **1** ski race down a zigzag course. **2** obstacle race in canoes etc.

slam *v.* (**slammed**) **1** shut forcefully and noisily. **2** put or hit forcefully. **3** (*sl.*) criticize severely. ● *n.* slamming noise.

slander *n.* **1** false statement uttered maliciously that damages a person's reputation. **2** crime of uttering this. ● *v.* utter slander about. □ **slanderer** *n.*, **slanderous** *adj.*

■ *n.* calumny, defamation, denigration, obloquy, vilification. ● *v.* calumniate, defame, insult, libel, malign, traduce, vilify.

slang *n.* words or phrases or particular meanings of these used very informally for vividness or novelty. □ **slangy** *adj.*

slant *v.* **1** slope. **2** present (news etc.) from a particular point of view. ● *n.* **1** slope. **2** way news etc. is slanted, bias. □ **slantwise** *adv.*

■ *v.* **1** bank, incline, lean, list, slope, tilt. **2** bias, colour, distort, misrepresent, twist. ● *n.* **1** inclination, incline, gradient, pitch, slope, tilt. **2** angle, bias, perspective, viewpoint.

slap *v.* (**slapped**) **1** strike with the open hand or with something flat. **2** place forcefully or carelessly. ● *n.* slapping blow. ● *adv.* with a slap, directly. □ **slap-happy** *adj.* (*colloq.*) cheerfully casual. **slap-up** *adj.* (*colloq.*) lavish.

slapdash *adj.* hasty and careless.

■ careless, hasty, hurried, slipshod, sloppy.

slapstick *n.* boisterous comedy.

slash *v.* **1** gash. **2** make a sweeping stroke. **3** strike in this way. **4** reduce drastically. ● *n.* slashing stroke or cut.

■ *v.* **1, 3** cut, gash, knife, lacerate, slit. ● *n.* diagonal, line, stroke.

slat *n.* narrow strip of wood, metal, etc.

slate *n.* **1** rock that splits easily into flat blue-grey plates. **2** piece of this used as roofing-material or (formerly) for writing on. ● *v.* **1** cover with slates. **2** (*colloq.*) criticize severely.

slaughter *v.* **1** kill (animals) for food. **2** kill ruthlessly or in great numbers. ● *n.* this process.

■ *v.* **2** butcher, exterminate, kill, massacre, murder, put to death, slay. ● *n.* carnage, bloodshed, butchery, genocide, killing, massacre.

slaughterhouse *n.* place where animals are killed for food.

Slav *adj.* & *n.* (member) of any of the peoples of Europe who speak a Slavonic language.

slave *n.* **1** person who is owned by and must work for another. **2** victim of or to a dominating influence. **3** drudge. ● *v.*

work very hard. □ **slave-driver** *n.* hard taskmaster.

slaver *v.* have saliva flowing from the mouth.

slavery *n.* **1** existence or condition of slaves. **2** very hard work.

■ **1** bondage, captivity, servitude.

slavish *adj.* excessively submissive or imitative. □ **slavishly** *adv.*

■ fawning, obsequious, servile, submissive, sycophantic; unimaginative, unoriginal.

Slavonic *adj.* & *n.* (of) the group of languages including Russian and Polish.

slay *v.* (**slew, slain**) kill.

sleazy *adj.* (**-ier, -iest**) dirty, disreputable. □ **sleaziness** *n.*

sledge *n.* cart with runners instead of wheels, used on snow. ● *v.* travel or convey in a sledge.

sledgehammer *n.* large heavy hammer.

sleek *adj.* **1** smooth and glossy. **2** looking well fed and thriving. □ **sleekness** *n.*

sleep *n.* **1** natural condition of rest with unconsciousness and relaxation of muscles. **2** spell of this. ● *v.* (**slept**) **1** be or spend (time) in a state of sleep. **2** provide with sleeping accommodation.

■ *n.* **2** catnap, doze, nap, rest, siesta, snooze. ● *v.* **1** doze, repose, rest, slumber, snooze.

sleeper *n.* **1** one who sleeps. **2** beam on which the rails of a railway rest. **3** railway coach fitted for sleeping in. **4** ring worn in a pierced ear to keep the hole from closing.

sleeping bag padded bag for sleeping in.

sleepwalk *v.* walk about while asleep. □ **sleepwalker** *n.*

sleepy *adj.* (**-ier, -iest**) **1** feeling a desire to sleep. **2** quiet. □ **sleepily** *adv.*, **sleepiness** *n.*

■ **1** drowsy, somnolent, tired. **2** peaceful, quiet, tranquil.

sleet *n.* hail or snow and rain falling simultaneously. ● *v.* fall as sleet. □ **sleety** *adj.*

sleeve *n.* **1** part of a garment covering the arm. **2** tube-like cover. **3** cover for a record. □ **up one's sleeve** concealed but available.

sleigh /slay/ *n.* sledge, esp. drawn by horses.

sleight of hand /slīt/ skill in using the hands to perform conjuring tricks etc.

slender *adj.* **1** slim and graceful. **2** small in amount. □ **slenderness** *n.*

■ **1** slim, svelte, thin, willowy. **2** little, meagre, scanty, slight, slim, small.

slept *see* **sleep**.

sleuth /sloōth/ *n.* detective.

slew[1] *v.* turn or swing round.

slew[2] *see* **slay**.

slice *n.* **1** thin flat piece (or a wedge) cut off an item of food. **2** portion. **3** implement for lifting or serving food. **4** slicing stroke. ● *v.* **1** cut, esp. into slices. **2** strike (a ball) so that it spins away obliquely. □ **slicer** *n.*

■ *n.* **1** bit, helping, piece, portion, serving, slab, wedge. **2** part, portion, share.

slick *adj.* **1** skilful, efficient. **2** smooth. ● *n.* patch of oil on the sea. ● *v.* make sleek.

slide *v.* (**slid**) **1** (cause to) move along a smooth surface touching it always with the same part. **2** move or pass smoothly. ● *n.* **1** act of sliding. **2** smooth slope or surface for sliding. **3** sliding part. **4** piece of glass for holding an object under a microscope. **5** picture for projecting onto a screen. **6** hinged clip for holding hair in place. □ **sliding scale** scale of fees or taxes etc. that varies according to the variation of some standard.

■ *v.* glide, skid, skim, slip, slither.

slight *adj.* **1** not much, not great, not thorough. **2** slender, small. ● *v.* treat disrespectfully, ignore. ● *n.* act of slighting. □ **slightly** *adv.*, **slightness** *n.*

■ *adj.* **1** inconsequential, insignificant, minimal, minor, negligible, small, trifling; faint, remote, slim, small. **2** diminutive, petite, slender, slim, small, thin. ● *v.* affront, insult, offend; ignore, rebuff, scorn, snub, spurn. ● *n.* affront, insult, slur, snub.

slim *adj.* (**slimmer, slimmest**) **1** of small girth or thickness. **2** small, slight, insufficient. ● *v.* (**slimmed**) make (oneself) slimmer by dieting, exercise, etc. □ **slimmer** *n.*, **slimness** *n.*

■ *adj.* **1** lean, skinny, slender, slight, thin.

slime *n.* unpleasant thick slippery liquid substance. □ **slimy** *adj.*, **slimily** *adv.*, **sliminess** *n.*

sling *n.* **1** strap or bandage etc. looped round an object to support or lift it. **2** looped strap used to throw a stone etc. ● *v.* (**slung**) **1** suspend, lift, or hurl with a sling. **2** (*colloq.*) throw.

slink *v.* (**slunk**) move in a stealthy or shamefaced way.

■ creep, skulk, sneak, steal.

slinky *adj.* smooth and sinuous.

slip[1] *v.* (**slipped**) **1** slide accidentally. **2** lose one's balance thus. **3** go or put smoothly. **4** escape hold or capture. ● *n.* **1** act of slipping. **2** casual mistake. **3** petticoat. **4** liquid containing clay for coating pottery. □ **give a person the slip** escape from or avoid him or her. **slip by** (of time) pass rapidly. **slipped disc** disc of cartilage between vertebrae that has become displaced and causes pain. **slip-road** *n.* road for entering or leaving a motorway. **slip up** (*colloq.*) make a mistake. **slip-up** *n.*

■ *v.* **1** glide, skid, slide, slither. **2** fall, stumble, trip, tumble. ● *n.* **2** blunder, *colloq.* boob, error, gaffe, mistake, oversight, *colloq.* slip-up.

slip[2] *n.* small piece of paper.

slipper *n.* light loose shoe for indoor wear.

slippery *adj.* **1** smooth or wet and difficult to hold or causing slipping. **2** (of a person) not trustworthy. □ **slipperiness** *n.*

slipshod *adj.* done or doing things carelessly.

■ careless, disorganized, haphazard, slapdash, sloppy.

slipstream *n.* current of air driven backward as something is propelled forward.

slipway *n.* sloping structure on which boats are landed or ships built or repaired.

slit *n.* narrow straight cut or opening. ● *v.* (**slit, slitting**) **1** cut a slit in. **2** cut into strips.

■ *n.* aperture, cleft, crack, cut, fissure, opening, split, vent.

slither *v.* slide unsteadily.

sliver *n.* small thin strip.

slobber *v.* slaver, dribble.

sloe *n.* **1** blackthorn. **2** its small dark plum-like fruit.

slog *v.* (**slogged**) **1** hit hard. **2** work or walk hard and steadily. ● *n.* **1** hard hit. **2** spell of hard steady work or walking.

slogan *n.* word or phrase adopted as a motto or in advertising.

sloop *n.* small ship with one mast.

slop *v.* (**slopped**) spill, splash. ● *n.* **1** unappetizing liquid. **2** slopped liquid. **3** (*pl.*) liquid refuse.

slope *v.* lie or put at an angle from the horizontal or vertical. ● *n.* **1** sloping surface. **2** amount by which a thing slopes.

■ *v.* bank, lean, slant, tilt, tip. ● *n.* angle, gradient, inclination, incline, pitch, slant, tilt.

sloppy *adj.* (**-ier, -iest**) **1** wet, slushy. **2** slipshod. **3** weakly sentimental. □ **sloppily** *adv.*, **sloppiness** *n.*

slosh *v.* **1** (*colloq.*) splash, pour clumsily. **2** (*sl.*) hit. ● *n.* **1** (*colloq.*) splashing sound. **2** (*sl.*) blow.

sloshed *adj.* (*sl.*) drunk.

slot *n.* **1** narrow opening into or through which something is to be put. **2** position in a series or scheme. ● *v.* (**slotted**) **1** make slot(s) in. **2** put or fit into a slot. □ **slot machine** machine operated by inserting a coin into a slot.

■ *n.* **1** channel, cleft, groove, notch, slit. **2** niche, position, space, spot.

sloth /slōth/ *n.* **1** laziness. **2** slow-moving S. American animal.

slothful *adj.* lazy. □ **slothfully** *adv.*

■ idle, indolent, inert, lazy, lethargic, slow, sluggish, torpid.

slouch *v.* stand, sit, or move in a lazy awkward way. ● *n.* slouching movement or posture.

slough[1] /slow/ *n.* swamp, marsh.

slough[2] /sluf/ *v.* **1** shed (skin). **2** be shed in this way.

slovenly *adj.* careless and untidy. □ **slovenliness** *n.*

■ careless, messy, sloppy, unmethodical, untidy.

slow *adj.* **1** not quick or fast. **2** showing an earlier time than the correct one. **3** stupid. ● *adv.* slowly. ● *v.* reduce the speed (of). □ **slowly** *adv.*, **slowness** *n.*

■ *adj.* **1** easy, dilatory, gentle, gradual, leisurely, sluggish, steady, unhurried. ● *v.* brake, decelerate; delay, hinder, hold up, impede, retard, set back.

slowcoach *n.* person who is slow in his or her actions or work.

slow-worm *n.* legless lizard.

sludge *n.* thick mud.

slug[1] *n.* **1** small slimy animal like a snail without a shell. **2** small lump of metal. **3** bullet.

slug[2] *v.* (**slugged**) (*US*) hit hard.

sluggard *n.* slow or lazy person.

sluggish *adj.* slow-moving, not lively. ▫ **sluggishly** *adv.*, **sluggishness** *n.*

▪ inactive, indolent, languid, lethargic, listless, slow, torpid.

sluice /slōoss/ *n.* **1** sliding gate controlling a flow of water. **2** this water. **3** channel carrying off water.

slum *n.* squalid house or district

slumber *v.* & *n.* sleep.

slump *n.* sudden great fall in prices or demand. ● *v.* **1** undergo a slump. **2** sit or fall limply.

▪ *n.* crash, depression, economic decline, recession. ● *v.* **1** decline, drop, fall, nose-dive, plummet, plunge. **2** collapse, drop, fall, flop.

slung *see* **sling**.

slunk *see* **slink**.

slur *v.* (**slurred**) **1** utter with each letter or sound running into the next. **2** pass lightly over (a fact). ● *n.* **1** slurred sound. **2** curved line marking notes to be slurred in music. **3** discredit.

slurp *v.* & *n.* (make) a noisy sucking sound.

slurry *n.* thin semi-liquid cement, mud, manure, etc.

slush *n.* **1** partly melted snow on the ground. **2** silly sentimental talk or writing. ▫ **slushy** *adj.*

slut *n.* slovenly or promiscuous woman. ▫ **sluttish** *adj.*

sly *adj.* **1** unpleasantly cunning and secret. **2** mischievous and knowing. ▫ **on the sly** secretly. **slyly** *adv.*, **slyness** *n.*

▪ **1** artful, crafty, cunning, deceitful, devious, scheming, tricky, underhand, wily.

smack[1] *n.* **1** slap. **2** loud kiss. ● *v.* **1** slap, hit hard. **2** close and part (lips) noisily.

smack[2] *n.* & *v.* (have) a slight flavour or trace.

smack[3] *n.* single-masted boat.

small *adj.* **1** not large or great. **2** unimportant. **3** petty. ● *n.* **1** narrowest part (of the back). **2** (*pl.*, *colloq.*) underwear. ● *adv.* in a small way. ▫ **small hours** period soon after midnight. **small-minded** *adj.* narrow or selfish in outlook. **small talk** social conversation on unimportant subjects. **small-time** *adj.* unimportant. **smallness** *n.*

▪ *adj.* **1** diminutive, little, miniature, petite, slight, *colloq.* teeny, tiny, *Sc.* wee; humble, modest, unpretentious. **2** inconsequential, insignificant, negligible, slight, trifling, trivial, unimportant. ▫ **small-minded** insular, narrow-minded, parochial; mean, petty, selfish, ungenerous.

smallholding *n.* small farm. ▫ **smallholder** *n.*

smallpox *n.* disease with pustules that often leave bad scars.

smarmy *adj.* (**-ier**, **-iest**) (*colloq.*) ingratiating, obsequious. ▫ **smarmily** *adv.*, **smarminess** *n.*

smart *adj.* **1** neat and elegant. **2** clever. **3** fashionable. **4** brisk. ● *v.* & *n.* (feel) a stinging pain. ▫ **smartly** *adv.*, **smartness** *n.*

▪ *adj.* **1** dapper, elegant, neat, *sl.* snazzy, spruce, trim, well-groomed, well-turned-out. **2** able, brainy, bright, clever, gifted, intelligent, quick, sharp. **3** chic, elegant, fashionable, *colloq.* posh, stylish, *colloq.* swish. **4** brisk, energetic, lively, quick, vigorous.

smarten *v.* make or become smarter.

smash *v.* **1** break noisily into pieces. **2** strike forcefully. **3** crash. **4** ruin. ● *n.* **1** act or sound of smashing. **2** collision.

▪ *v.* **1** break, shatter, shiver, splinter. ● *n.* **2** accident, collision, crash, pile-up.

smashing *adj.* (*colloq.*) excellent.

smattering *n.* slight knowledge.

smear *v.* **1** spread with a greasy or dirty substance. **2** try to damage the reputation of. ● *n.* **1** mark made by smearing. **2** slander. ▫ **smeary** *adj.*

▪ *v.* **1** dirty, smudge, soil, stain, streak. **2** blacken, calumniate, defame, malign, slander, sully, tarnish, vilify.

smell *n.* **1** ability to perceive things with the sense organs of the nose. **2** quality perceived in this way. **3** unpleasant quality of this kind. **4** act of smelling. ● *v.* (**smelt** or **smelled**) **1** perceive the smell of. **2** give off a smell. ▫ **smelly** *adj.*

▪ *n.* **2** aroma, bouquet, fragrance, perfume, redolence, savour, scent. **3** odour, *sl.* pong, reek, stench, stink. ▫ **smelly** fetid, foul-smelling, putrid, rancid, rank, reeking, stinking.

smelt[1] *see* **smell**.

smelt[2] *v.* **1** heat and melt (ore) to extract metal. **2** obtain (metal) thus.

smelt[3] *n.* small fish related to the salmon.

smile *n.* facial expression indicating pleasure or amusement, with lips stretched and their ends upturned. ● *v.* **1** give a smile. **2** look favourably.

smirch *v.* & *n.* discredit.

smirk *n.* self-satisfied smile. ● *v.* give a smirk.
smite *v.* (**smote, smitten**) (*old use*) **1** hit hard. **2** affect suddenly.
smith *n.* **1** person who makes things in metal. **2** blacksmith.
smithereens *n.pl.* small fragments.
smithy *n.* blacksmith's workshop.
smitten *see* **smite**.
smock *n.* loose overall.
smog *n.* dense smoky fog.
smoke *n.* **1** visible vapour given off by a burning substance. **2** spell of smoking tobacco. **3** (*sl.*) cigarette, cigar. ● *v.* **1** give out smoke or steam. **2** preserve with smoke. **3** draw smoke from (a cigarette, cigar, or pipe) into the mouth. **4** do this as a habit. □ **smoky** *adj.*
smokeless *adj.* with little or no smoke.
smoker *n.* person who smokes tobacco as a habit.
smokescreen *n.* thing intended to disguise or conceal activities.
smooth *adj.* **1** having an even surface with no projections. **2** moving evenly without bumping. **3** free from difficulties or problems. **4** polite but perhaps insincere. **5** not harsh in sound or taste. ● *v.* make smooth. □ **smoothly** *adv.*, **smoothness** *n.*

■ *adj.* **1** even, flat, level, plane; glossy, satiny, shiny, silken, silky, sleek. **3** easy, effortless, steady. **4** glib, slick, suave, urbane.

smote *see* **smite**.
smother *v.* **1** suffocate, stifle. **2** cover thickly. **3** suppress.

■ **1** asphyxiate, stifle, suffocate. **2** cover, heap, pile. **3** hide, muffle, repress, stifle, suppress.

smoulder *v.* **1** burn slowly with smoke but no flame. **2** burn inwardly with concealed anger etc.
smudge *n.* dirty or blurred mark. ● *v.* **1** make a smudge on or with. **2** become smudged. **3** blur. □ **smudgy** *adj.*, **smudginess** *n.*

■ *n.* blot, blotch, mark, smear, splodge, splotch, spot.

smug *adj.* (**smugger, smuggest**) self-satisfied. □ **smugly** *adv.*, **smugness** *n.*

■ complacent, self-righteous, self-satisfied.

smuggle *v.* **1** convey secretly. **2** bring (goods) illegally into or out of a country, esp. without paying customs duties. □ **smuggler** *n.*

smut *n.* **1** small flake of soot. **2** small black mark. **3** indecent talk, pictures, or stories. □ **smutty** *adj.*
snack *n.* small or casual meal. □ **snack bar** place where snacks are sold.
snaffle *n.* horse's bit without a curb. ● *v.* (*colloq.*) take for oneself.
snag *n.* **1** unexpected drawback. **2** jagged projection. **3** tear caused by this. ● *v.* (**snagged**) catch or tear on a snag.

■ *n.* **1** catch, complication, difficulty, hitch, problem, set-back, stumbling block.

snail *n.* soft-bodied animal with a shell. □ **snail's pace** very slow pace.
snake *n.* reptile with a long narrow body and no legs. ● *v.* move in a winding course. □ **snaky** *adj.*

■ *v.* bend, curve, meander, turn, twist, wind, zigzag.

snakeskin *n.* leather made from snakes' skins.
snap *v.* (**snapped**) **1** (cause to) make a sharp cracking sound. **2** break suddenly. **3** speak with sudden irritation. **4** bite at suddenly. **5** move smartly. **6** take a snapshot of. ● *n.* **1** act or sound of snapping. **2** snapshot. **3** small crisp biscuit. ● *adj.* done without forethought. □ **snap up** take eagerly.
snapdragon *n.* plant with flowers that have a mouth-like opening.
snapper *n.* any of several edible sea fish.
snappy *adj.* (**-ier, -iest**) **1** irritable. **2** brisk. **3** neat and elegant. □ **snappily** *adv.*, **snappiness** *n.*
snapshot *n.* photograph taken informally or casually.
snare *n.* trap, usu. with a noose. ● *v.* trap in a snare.

■ *v.* capture, catch, ensnare, trap.

snarl *v.* **1** growl angrily with teeth bared. **2** speak or utter in a bad-tempered way. **3** become entangled. ● *n.* **1** act or sound of snarling. **2** tangle.
snarl-up *n.* **1** traffic jam. **2** muddle.
snatch *v.* **1** seize quickly or eagerly. **2** steal, kidnap. ● *n.* **1** act of snatching. **2** short or brief part.

■ *v.* **1** clasp, clutch, grab, grasp, seize, take hold of. **2** *sl.* nab, make off with, seize, steal, take; abduct, capture, carry off, kidnap.

snazzy *adj.* (**-ier, -iest**) (*sl.*) stylish.
sneak *v.* **1** go, convey, or (*sl.*) steal furtively. **2** (*sl.*) tell tales. ● *n.* (*sl.*) tell-tale.

■ *v.* **1** creep, pussyfoot, sidle, slink, steal, tiptoe.

sneaking *adj.* persistent but not openly acknowledged.

sneer *n.* scornful expression or remark. ● *v.* show contempt by a sneer.

■ *v.* **be scornful, jeer, mock, scoff.**

sneeze *n.* sudden audible involuntary expulsion of air through the nose. ● *v.* give a sneeze.

snicker *v.* & *n.* snigger.

snide *adj.* sneering slyly.

sniff *v.* **1** draw air audibly through the nose. **2** draw in as one breathes. **3** try the smell of. ● *n.* act or sound of sniffing. □ **sniffer** *n.*

sniffle *v.* sniff slightly or repeatedly. ● *n.* this act or sound.

snifter *n.* (*sl.*) small drink of alcoholic liquor.

snigger *v.* & *n.* (give) a sly giggle.

■ *v.* **giggle, laugh, snicker, titter.**

snip *v.* (**snipped**) cut with scissors or shears in small quick strokes. ● *n.* **1** act or sound of snipping. **2** (*sl.*) bargain.

snipe *n.* (*pl.* **snipe**) wading bird with a long straight bill. ● *v.* **1** fire shots from a hiding place. **2** make sly critical remarks. □ **sniper** *n.*

snippet *n.* small piece.

snivel *v.* (**snivelled**) cry in a miserable whining way.

■ **cry, grizzle, weep, whimper, whine,** *colloq.* **whinge.**

snob *n.* person with an exaggerated respect for social position, wealth, or certain tastes and who despises those he or she considers inferior. □ **snobbery** *n.*, **snobbish** *adj.*

■ □ **snobbish condescending, hoity-hoity, pretentious,** *colloq.* **snooty,** *colloq.* **stuck-up, supercilious, superior,** *colloq.* **uppity.**

snooker *n.* game played on a baize-covered table with 15 red and 6 other coloured balls.

snoop *v.* (*colloq.*) pry. □ **snooper** *n.*

snooty *adj.* (*colloq.*) haughty and contemptuous. □ **snootily** *adv.*

snooze *n.* & *v.* nap.

snore *n.* snorting or grunting sound made during sleep. ● *v.* make such sounds. □ **snorer** *n.*

snorkel *n.* tube by which an underwater swimmer can breathe. ● *v.* (**snorkelled**) swim with a snorkel.

snort *n.* sound made by forcing breath through the nose, esp. in indignation. ● *v.* **1** make a snort. **2** (*sl.*) inhale (a powdered drug).

snout *n.* animal's long projecting nose or nose and jaws.

snow *n.* **1** frozen atmospheric vapour falling to earth in white flakes. **2** fall or layer of snow. ● *v.* fall as or like snow. □ **snowed under 1** covered with snow. **2** overwhelmed with work etc. **snowstorm** *n.* **snowy** *adj.*

snowball *n.* snow pressed into a compact mass for throwing. ● *v.* increase in size or intensity.

snowdrift *n.* mass of snow piled up by the wind.

snowdrop *n.* plant with white flowers blooming in spring.

snowman *n.* figure made of snow.

snowplough *n.* device for clearing roads of snow.

snub[1] *v.* (**snubbed**) treat (a person) contemptuously with sharp words or a lack of politeness. ● *n.* treatment of this kind.

■ *v.* **cold-shoulder, cut, ignore, put down, rebuff, reject, slight, spurn.**

snub[2] *adj.* (of the nose) short and stumpy. □ **snub-nosed** *adj.*

snuff[1] *n.* powdered tobacco for sniffing up the nostrils.

snuff[2] *v.* put out (a candle). □ **snuff it** (*sl.*) die.

snuffle *v.* breathe with a noisy sniff. ● *n.* snuffling sound.

snug *adj.* (**snugger, snuggest**) **1** cosy. **2** close-fitting. □ **snugly** *adv.*

■ **1 comfortable, cosy, homely, warm, welcoming.**

snuggle *v.* nestle, cuddle.

so *adv.* & *conj.* **1** to the extent or in the manner or with the result indicated. **2** very. **3** for that reason. **4** also. □ **so-and-so** *n.* **1** person or thing that need not be named. **2** (*colloq.*) disliked person. **so-called** *adj.* called (wrongly) by that name. **so that** in order that.

soak *v.* **1** make or become thoroughly wet. **2** (of liquid) penetrate. **3** absorb. ● *n.* **1** soaking. **2** (*sl.*) heavy drinker.

■ *v.* **1 drench, saturate, wet; douse, immerse, souse, steep, submerge. 2 penetrate, permeate.**

soap *n.* **1** substance used in washing things, made of fat or oil and an alkali. **2** (*colloq.*) soap opera. ● *v.* apply soap to. □ **soap opera** television or radio serial dealing with domestic issues.

soapstone *n.* steatite.

soapsuds *n.pl.* froth of soapy water.

soapy *adj.* **1** of or like soap. **2** containing or smeared with soap. **3** unctuous. □ **soapiness** *n.*

soar *v.* rise high, esp. in flight.

■ ascend, fly, rise, wing; escalate, increase, rocket, shoot up.

sob *n.* uneven drawing of breath when weeping or gasping. ● *v.* (**sobbed**) weep, breathe, or utter with sobs.

■ *v.* bawl, blubber, cry, howl, wail, weep, whimper.

sober *adj.* **1** not drunk. **2** serious. **3** (of colour) not bright. ● *v.* make or become sober. □ **soberly** *adv.*, **sobriety** *n.*

■ *adj.* **2** dignified, earnest, grave, restrained, sedate, self-controlled, serious, solemn, staid, steady, temperate.

soccer *n.* football played with a spherical ball not to be handled in play except by the goalkeeper.

sociable *adj.* **1** fond of company. **2** characterized by friendly companionship. □ **sociably** *adv.*, **sociability** *n.*

■ affable, companionable, congenial, convivial, friendly, genial, gregarious, outgoing, social.

social *adj.* **1** living in a community. **2** of society or its organization. **3** sociable. ● *n.* social gathering. □ **social security** state assistance for those who lack economic security. **social services** welfare services provided by the state. **social worker** person trained to help people with social problems. **socially** *adv.*

socialism *n.* political and economic theory that resources, industries, and transport should be owned and managed by the state. □ **socialist** *n.*, **socialistic** *adj.*

socialite *n.* person prominent in fashionable society.

socialize *v.* behave sociably.

society *n.* **1** organized community. **2** system of living in this. **3** people of the higher social classes. **4** mixing with other people. **5** group organized for a common purpose.

■ **1** civilization, community, culture. **4** companionship, company, fellowship. **5** alliance, association, circle, club, group, guild, organization, union.

sociology *n.* study of human society or of social problems. □ **sociological** *adj.*, **sociologist** *n.*

sock¹ *n.* **1** knitted covering for the foot. **2** loose insole.

sock² (*colloq.*) *v.* hit forcefully. ● *n.* forceful blow.

socket *n.* hollow into which something fits. □ **socketed** *adj.*

sod *n.* **1** turf. **2** a piece of this.

soda *n.* **1** compound of sodium. **2** soda water. □ **soda water** water made fizzy by being charged with carbon dioxide under pressure.

sodden *adj.* made very wet.

sodium *n.* soft silver-white metallic element.

sofa *n.* long upholstered seat with a back and raised ends.

soft *adj.* **1** not hard, firm, or rough. **2** not loud or bright. **3** gentle. **4** (too) sympathetic and kind. **5** weak, foolish. **6** (of drinks) non-alcoholic. **7** (of drugs) not highly addictive. □ **soft fruit** small stoneless fruit. **soft option** easy alternative. **soft-pedal** *v.* refrain from emphasizing. **soft spot** feeling of affection. **softly** *adv.*, **softness** *n.*

■ **1** ductile, flexible, malleable, plastic, pliant; fleecy, fluffy, furry, silky, smooth, velvety. **2** hushed, low, muffled, muted, quiet, subdued; light, pale, pastel. **3** gentle, light, mild, moderate, pleasant. **4** easygoing, indulgent, lax, lenient, liberal, permissive. □ **soft spot** affection, fondness, liking, love, partiality, penchant, predilection, weakness.

soften *v.* make or become soft or softer. □ **softener** *n.*

■ alleviate, assuage, cushion, diminish, ease, lessen, mitigate, moderate, reduce, temper; deaden, muffle, mute, tone down; appease, mollify, pacify, placate.

software *n.* computer programs or tapes containing these.

softwood *n.* soft wood of coniferous trees.

soggy *adj.* (**-ier, -iest**) sodden, moist and heavy. □ **sogginess** *n.*

soil¹ *n.* **1** loose earth. **2** ground as territory.

■ **1** dirt, earth, ground, loam.

soil² *v.* make or become dirty.

■ dirty, foul, muddy, stain.

sojourn *n.* temporary stay. ● *v.* stay temporarily.

solace *v.* & *n.* comfort in distress.

■ *n.* comfort, consolation, relief, support.

solar *adj.* **1** of or from the sun. **2** reckoned by the sun. □ **solar cell** device converting solar radiation into electricity. **solar plexus 1** network of nerves at the pit of the stomach. **2** this area. **solar system** sun with the planets etc. that revolve round it.

sold *see* **sell**.

solder *n.* soft alloy used to cement metal parts together. ● *v.* join with solder. □ **soldering iron** tool for melting and applying solder.

soldier *n.* member of an army. ● *v.* serve as a soldier. □ **soldier on** (*colloq.*) persevere doggedly. **soldierly** *adj.*

sole[1] *n.* **1** under-surface of a foot. **2** part of a shoe etc. covering this. ● *v.* put a sole on.

sole[2] *n.* edible flatfish.

sole[3] *adj.* **1** one and only. **2** belonging exclusively to one person or group. □ **solely** *adv.*

■ **1 lone, only, single, solitary.**

solemn *adj.* **1** not smiling or cheerful. **2** formal and dignified. □ **solemnly** *adv.*, **solemnity** *n.*

■ **1 earnest, grave, serious, sober, sombre, unsmiling. 2 august, ceremonial, ceremonious, dignified, formal, ritual, stately.**

solemnize *v.* **1** celebrate (a festival etc.). **2** perform with formal rites. □ **solemnization** *n.*

solenoid *n.* coil of wire magnetized by electric current.

sol-fa *n.* system of syllables (*doh, ray, me,* etc.) representing the notes of a musical scale.

solicit *v.* seek to obtain by asking (for). □ **solicitation** *n.*

■ **appeal for, ask for, beg, beseech, entreat, importune, petition, request.**

solicitor *n.* lawyer who advises clients and instructs barristers.

solicitous *adj.* anxious about a person's welfare or comfort. □ **solicitously** *adv.*, **solicitude** *n.*

solid *adj.* **1** keeping its shape, firm, not liquid or gas. **2** without holes or spaces. **3** of the same substance throughout. **4** three-dimensional. **5** strong. **6** sound and reliable. **7** continuous. ● *n.* solid substance or body or food. □ **solidly** *adv.*, **solidity** *n.*

■ **1 firm, hard, rigid. 2 compact, dense. 3 pure, sterling. 5 firm, robust, stable, strong, sturdy. 6 dependable, reliable, sound, stalwart, staunch, steadfast; cogent, convincing, persuasive, valid. 7 continuous, unbroken, undivided, uninterrupted.**

solidarity *n.* unity resulting from common aims or interests etc.

solidify *v.* make or become solid. □ **solidification** *n.*

■ **clot, coagulate, congeal, harden, *colloq.* jell, set.**

soliloquize *v.* utter a soliloquy.

soliloquy *n.* speech made aloud to oneself.

solitaire *n.* **1** gem set by itself. **2** game for one person played on a board with pegs.

solitary *adj.* **1** single. **2** not frequented. **3** lonely. ● *n.* recluse.

■ ***adj.* 1 lone, single, sole. 2 desolate, isolated, lonely, out of the way, remote, secluded, unfrequented. 3 forlorn, friendless, lonely, lonesome.**

solitude *n.* being solitary.

solo *n.* (*pl.* **-os**) **1** music for a single performer. **2** unaccompanied performance etc. ● *adj.* & *adv.* unaccompanied, alone.

soloist *n.* performer of a solo.

solstice *n.* either of the times (about 21 June and 22 Dec.) or points reached when the sun is furthest from the equator.

soluble *adj.* **1** able to be dissolved. **2** able to be solved. □ **solubility** *n.*

solution *n.* **1** process of solving a problem etc. **2** answer found. **3** liquid containing something dissolved. **4** process of dissolving.

solve *v.* find the answer to. □ **solvable** *adj.*

■ **crack, decipher, figure out, resolve, sort out, work out.**

solvent *adj.* **1** having enough money to pay one's debts etc. **2** able to dissolve another substance. ● *n.* liquid used for dissolving something. □ **solvency** *n.*

sombre *adj.* dark, gloomy.

■ **black, dark, murky, overcast; dismal, gloomy, lugubrious, melancholy, morose, sad, serious, solemn.**

sombrero *n.* (*pl.* **-os**) man's hat with a very wide brim.

some *adj.* **1** unspecified quantity or number of. **2** unknown, unnamed. **3** considerable quantity. **4** approximately. ● *pron.* some people or things.

somebody *n.* & *pron.* **1** unspecified person. **2** person of importance.

somehow *adv.* in an unspecified or unexplained manner.

someone *n.* & *pron.* somebody.

somersault *n.* & *v.* leap or roll turning one's body upside down and over.

something *n.* & *pron.* **1** unspecified thing or extent. **2** important or praiseworthy thing. □ **something like** approximately.

sometime *adj.* former. ● *adv.* at some unspecified time.

sometimes *adv.* at some times but not all the time.

■ from time to time, now and again, now and then, occasionally, once in a while, periodically.

somewhat *adv.* to some extent.

■ fairly, moderately, pretty, quite, rather, relatively, to some extent.

somewhere *adv.* at, in, or to an unspecified place.

somnambulist *n.* sleepwalker. □ **somnambulism** *n.*

somnolent *adj.* **1** sleepy. **2** inducing sleep. □ **somnolence** *n.*

son *n.* male in relation to his parents. □ **son-in-law** *n.* (*pl.* **sons-in-law**) daughter's husband.

sonar *n.* device for detecting objects under water by reflection of sound waves.

sonata *n.* musical composition for one instrument or two, usu. in several movements.

song *n.* **1** music for singing. **2** singing. □ **going for a song** (*colloq.*) being sold very cheaply.

songbird *n.* bird with a musical cry.

songster *n.* **1** singer. **2** songbird.

sonic *adj.* of sound waves.

sonnet *n.* type of poem of 14 lines.

sonorous *adj.* resonant. □ **sonorously** *adv.*, **sonority** *n.*

■ deep, full, loud, resonant, resounding, ringing.

soon *adv.* **1** in a short time. **2** early. **3** readily. □ **as soon as 1** at the moment that. **2** as early as. **sooner or later** at some time, eventually.

■ **1** *old use* anon, before long, by and by, directly, in a minute *or* moment, in the near future, presently, shortly. **2** early, fast, quickly, speedily. **3** gladly, happily, readily, willingly.

soot *n.* black powdery substance in smoke. □ **sooty** *adj.*

soothe *v.* **1** calm. **2** ease (pain etc.). □ **soothing** *adj.*

■ **1** appease, calm, lull, mollify, pacify, placate, quieten, settle. **2** allay, alleviate, ease, lessen, palliate, reduce, relieve.

soothsayer *n.* prophet.

sop *n.* concession to pacify a troublesome person. ● *v.* (**sopped**) soak up (liquid).

sophisticated *adj.* **1** cultured and refined. **2** complicated, elaborate. □ **sophistication** *n.*

■ **1** cultured, discriminating, elegant, polished, refined, suave, urbane. **2** complex, complicated, elaborate, intricate.

sophistry *n.* (also **sophism**) clever but perhaps misleading reasoning. □ **sophist** *n.*

soporific *adj.* tending to cause sleep. ● *n.* soporific drug etc.

sopping *adj.* very wet, drenched.

soppy *adj.* (**-ier**, **-iest**) (*colloq.*) sentimental in a sickly way.

soprano *n.* (*pl.* **-os**) highest female or boy's singing voice.

sorbet /sórbay/ *n.* flavoured water ice.

sorcerer *n.* magician. □ **sorcery** *n.*

■ enchanter, enchantress, magician, witch, wizard.

sordid *adj.* **1** dirty, squalid. **2** (of motives etc.) not honourable. □ **sordidly** *adv.*, **sordidness** *n.*

■ **1** dirty, filthy, foul, seedy, sleazy, squalid. **2** base, corrupt, despicable, dishonourable, ignoble, mean, shameful, vile.

sore *adj.* **1** causing or suffering pain from injury or disease. **2** hurt and angry. ● *n.* **1** sore place. **2** source of distress or annoyance. □ **soreness** *n.*

■ *adj.* **1** inflamed, painful, raw, tender. **2** aggrieved, angry, annoyed, irked, irritated, *colloq.* peeved, upset, vexed.

sorely *adv.* very much, severely.

sorghum *n.* tropical cereal plant.

sorrel[1] *n.* sharp-tasting herb.

sorrel[2] *adj.* reddish-brown.

sorrow *n.* **1** mental suffering caused by loss or disappointment etc. **2** thing causing this. ● *v.* feel sorrow, grieve. □ **sorrowful** *adj.*, **sorrowfully** *adv.*

■ *n.* **1** anguish, despair, distress, grief, heartache, misery, sadness, unhappiness, woe. **2** affliction, trial, tribulation, trouble, woe, worry. ● *v.* grieve, lament, mourn.

sorry *adj.* (**-ier**, **-iest**) **1** feeling pity, regret, or sympathy. **2** wretched.

■ **1** apologetic, conscience-stricken, contrite, penitent, regretful, remorseful, repentant, rueful; distressed, sad, sorrowful, unhappy. **2** miserable, pathetic, pitiful, wretched.

sort *n.* particular kind or variety. ● *v.* arrange systematically. □ **sort out 1** select from a group. **2** resolve. **3** disentangle, put into order.

■ *n.* brand, class, kind, make, type, variety; breed, genus, species. ● *v.* arrange, categorize, classify, group, order, organize, sort out, systematize. □ **sort out 1** choose, decide on, pick, select, single out. **2** fix, put right, resolve, settle, solve.

sortie *n.* **1** sally by troops from a besieged place. **2** flight of an aircraft on a military operation.

SOS *n.* **1** international code-signal of distress. **2** urgent appeal for help.

sot *n.* habitual drunkard.

soufflé *n.* light dish made with beaten egg white.

sough /sow, suf/ *n.* & *v.* (make) a moaning or whispering sound as of wind in trees.

sought *see* **seek**.

soul *n.* **1** person's spiritual or immortal element. **2** mental, moral, or emotional nature. **3** personification. **4** person. **5** (also **soul music**) type of black American music.

■ **1,2** psyche, spirit. **3** embodiment, epitome, essence, personification, quintessence. **4** human being, individual, mortal, person.

soulful *adj.* showing deep feeling, emotional. □ **soulfully** *adv.*

soulless *adj.* **1** lacking sensitivity or noble qualities. **2** dull.

sound[1] *n.* **1** vibrations of air detectable by the ear. **2** thing that can be heard. **3** mental impression produced by a piece of news, description, etc. ● *v.* **1** produce or cause to produce sound. **2** seem when heard. **3** utter, pronounce. □ **sound barrier** high resistance of air to objects moving at speeds near that of sound. **sound off** (*colloq.*) express one's opinions loudly.

■ *v.* **1** resonate, resound, reverberate, ring; blow, play. **2** appear, give the impression of, seem, strike one as.

sound[2] *adj.* **1** not diseased or damaged. **2** financially secure. **3** correct, well-founded. **4** thorough. □ **soundly** *adv.*, **soundness** *n.*

■ **1** fit, hale, healthy, undiseased, uninjured; intact, undamaged, unscathed, whole. **2** good, reliable, safe, secure. **3** correct, judicious, sensible, valid, well-founded, wise.

sound[3] *v.* **1** test the depth of (a river or sea etc.). **2** examine with a probe.

sound[4] *n.* strait.

sounding board board to reflect sound or increase resonance.

soundproof *adj.* impervious to sound. ● *v.* make soundproof.

soup *n.* liquid food made from meat or vegetables etc. ● *v.* **soup up** (*colloq.*) increase the power of (an engine etc.). □ **soup-kitchen** *n.* place supplying free soup to the needy. **soupy** *adj.*

sour *adj.* **1** tasting sharp. **2** not fresh, tasting or smelling stale. **3** bad-tempered. ● *v.* make or become sour. □ **sourly** *adv.*, **sourness** *n.*

■ *adj.* **1** acid, sharp, tart, vinegary. **2** bad, curdled, off, rancid, spoiled. **3** bad-tempered, bitter, churlish, crabby, crusty, embittered, jaundiced, surly.

source *n.* **1** place from which something comes or is obtained. **2** person or book etc. supplying information. **3** river's starting point.

■ **1** cause, origin, root, start.

sourpuss *n.* (*colloq.*) bad-tempered person.

souse *v.* **1** steep in pickle. **2** drench.

south *n.* **1** point or direction to the right of a person facing east. **2** southern part. ● *adj.* **1** in the south. **2** (of wind) from the south. ● *adv.* towards the south. □ **south-east** *n.* point or direction midway between south and east. **south-easterly** *adj.* & *n.*, **south-eastern** *adj.* **south-west** *n.* point or direction midway between south and west. **south-westerly** *adj.* & *n.*, **south-western** *adj.*

southerly *adj.* towards or blowing from the south.

southern *adj.* of or in the south.

southerner *n.* native of the south.

southernmost *adj.* furthest south.

southpaw *n.* (*colloq.*) left-handed person.

southward *adj.* towards the south. □ **southwards** *adv.*

souvenir *n.* thing serving as a reminder of an incident or place visited.

■ keepsake, memento, reminder, token.

sou'wester *n.* waterproof hat with a broad flap at the back.

sovereign *n.* king or queen who is the supreme ruler of a country. ● *adj.* **1** supreme. **2** (of a state) independent. □ **sovereignty** *n.*

■ *n.* emperor, empress, king, monarch, potentate, queen, ruler. ● *adj.* **1** absolute, royal, supreme. **2** autonomous, free, independent, self-governing.

sow[1] /sō/ *v.* (**sowed, sown** or **sowed**) **1** plant or scatter (seed) for growth. **2** plant seed in. **3** implant (ideas etc.). □ **sower** *n.*

sow[2] /sow/ *n.* adult female pig.

soy *n.* soya bean.

soya *n.* plant from whose seed (**soya bean**) an edible oil and flour are obtained.

sozzled *adj.* (*colloq.*) drunk.

spa *n.* place with a therapeutic mineral spring.

space *n.* **1** continuous expanse in which things exist and move. **2** portion of this. **3** empty area or extent. **4** universe beyond earth's atmosphere. **5** interval. ● *v.* arrange with spaces between.

■ *n.* **1** area, expanse, extent. **2** gap, interval, opening. **3** area, capacity, room. **5** duration, hiatus, interval, lapse, period, time, while.

spacecraft *n.* (*pl.* **-craft**) vehicle for travelling in outer space.

spaceship *n.* spacecraft.

spacious *adj.* providing much space, roomy. □ **spaciousness** *n.*

■ capacious, extensive, large, roomy, sizeable.

spade[1] *n.* tool for digging, with a broad metal blade on a handle.

spade[2] *n.* playing card of the suit marked with black figures shaped like an inverted heart with a small stem.

spadework *n.* hard preparatory work.

spaghetti *n.* pasta in long thin strands.

span *n.* **1** extent from end to end. **2** part between the uprights of a bridge etc. ● *v.* (**spanned**) extend across.

■ *n.* **1** extent, reach, sweep; duration, interval, period, stretch, time. ● *v.* bridge, cross, pass over, traverse.

spangle *n.* small piece of glittering material ornamenting a dress etc. ● *v.* cover with spangles or sparkling objects.

Spaniard *n.* native of Spain.

spaniel *n.* a kind of dog with drooping ears and a silky coat.

Spanish *adj.* & *n.* (language) of Spain.

spank *v.* slap on the buttocks.

spanner *n.* tool for gripping and turning the nut on a screw etc.

spar[1] *n.* strong pole used as a ship's mast, yard, or boom.

spar[2] *v.* (**sparred**) **1** box, esp. for practice. **2** quarrel, argue.

spare *v.* **1** refrain from hurting or harming, not inflict. **2** afford to give. **3** use with restraint. ● *adj.* **1** additional to what is usually needed or used, kept in reserve. **2** thin. ● *n.* extra thing kept in reserve. □ **to spare** additional to what is needed.

■ *v.* **1** let off, pardon, reprieve; protect, save. **2** afford, do without, give up, manage without, part with, relinquish. ● *adj.* **1** additional, auxiliary, extra, other, supplementary.

sparing *adj.* economical, not generous or wasteful.

■ careful, economical, frugal, prudent, thrifty.

spark *n.* **1** fiery particle. **2** flash of light produced by an electrical discharge. **3** particle (of energy, genius, etc.). ● *v.* give off spark(s). □ **spark(ing) plug** device for making a spark in an internal-combustion engine.

sparkle *v.* **1** shine with flashes of light. **2** be lively or witty. **3** (of wine etc.) effervesce. ● *n.* sparkling light.

■ *v.* **1** blaze, flash, flicker, gleam, glint, glitter, glow, shimmer, shine, twinkle, wink.

sparkler *n.* sparking firework.

sparrow *n.* small brownish-grey bird.

sparrowhawk *n.* small hawk.

sparse *adj.* thinly scattered. □ **sparsely** *adv.*, **sparseness** *n.*

spartan *adj.* (of conditions) simple and sometimes harsh.

spasm *n.* **1** strong involuntary contraction of a muscle. **2** sudden brief spell of activity or emotion etc.

■ **1** convulsion, paroxysm.

spasmodic *adj.* of or occurring in spasms. □ **spasmodically** *adv.*

■ fitful, intermittent, irregular, occasional, periodic, sporadic, uneven.

spastic *adj.* affected by cerebral palsy which causes jerky, involuntary movements. ● *n.* person with this condition. □ **spasticity** *n.*

spat[1] *see* **spit**[1].

spat[2] *n.* short gaiter.

spate *n.* sudden flood.

spatial *adj.* of or existing in space. □ **spatially** *adv.*

spatter *v.* scatter or fall in small drops (on). ● *n.* **1** splash(es). **2** sound of spattering.

■ *v.* shower, splash, splatter, spray.

spatula *n.* broad-bladed implement used esp. by artists and in cookery.

spawn *n.* eggs of fish, frogs, etc. ● *v.* **1** deposit spawn. **2** generate.

spay *v.* sterilize (a female animal) by removing the ovaries.

speak *v.* (**spoke**, **spoken**) **1** utter (words) in an ordinary voice. **2** converse. **3** express by speaking. **4** know (a language).

> ■ **1** enunciate, pronounce, say, state, utter, vocalize, voice. **2** chat, communicate, converse, talk. **3** communicate, express, make known, verbalize.

speaker *n.* **1** person who speaks, esp. in public. **2** loudspeaker.

spear *n.* **1** weapon for hurling, with a long shaft and pointed tip. **2** pointed stem. ● *v.* pierce with or as if with a spear.

spearhead *n.* foremost part of an advancing force. ● *v.* be the spearhead of.

spearmint *n.* a kind of mint.

spec *n.* **on spec** (*colloq.*) as a speculation.

special *adj.* **1** distinctive, specific. **2** for a particular purpose. **3** exceptional. □ **specially** *adv.*

> ■ **1** characteristic, distinctive, individual, particular, peculiar, specific, unique. **2** exclusive, specialized. **3** exceptional, extraordinary, rare, remarkable, singular, uncommon, unusual; especial, extra, particular.

specialist *n.* expert in a particular branch of a subject.

> ■ authority, connoisseur, expert, professional.

speciality *n.* special quality, product, or activity.

specialize *v.* **1** be or become a specialist. **2** adapt for a particular purpose. □ **specialization** *n.*

species *n.* (*pl.* **species**) group of similar animals or plants which can interbreed.

specific *adj.* **1** particular. **2** exact, not vague. ● *n.* specific aspect. □ **specifically** *adv.*

> ■ *adj.* **1** distinctive, individual, particular, peculiar, unique. **2** clear, definite, exact, precise, unambiguous, unequivocal.

specification *n.* **1** specifying. **2** details describing a thing to be made or done.

specify *v.* **1** mention definitely. **2** include in specifications.

> ■ **1** designate, detail, enumerate, indicate, itemize, list, spell out, state, stipulate; determine, establish, fix, settle.

specimen *n.* part or individual taken as an example or for examination or testing.

> ■ example, illustration, instance, model, representation, sample.

specious *adj.* seeming good or sound but lacking real merit. □ **speciously** *adv.*

speck *n.* small spot or particle.

> ■ bit, dot, fleck, particle, pinch, spot.

speckle *n.* small spot, esp. as a natural marking. □ **speckled** *adj.*

spectacle *n.* **1** lavish public show. **2** impressive sight. **3** ridiculous sight. **4** (*pl.*) pair of lenses set in a frame, worn to assist sight.

> ■ **1** ceremony, display, pageant, parade, performance, show, sight, spectacular.

spectacular *adj.* impressive. ● *n.* spectacular performance or production. □ **spectacularly** *adv.*

> ■ *adj.* breathtaking, dramatic, exciting, impressive, sensational, *colloq.* stunning, thrilling.

spectator *n.* person who watches a show, game, or incident.

spectral *adj.* **1** of or like a spectre. **2** of the spectrum.

spectre *n.* **1** ghost. **2** haunting fear.

spectrum *n.* (*pl.* **-tra**) **1** bands of colour or sound forming a series according to their wavelengths. **2** entire range of ideas etc.

speculate *v.* **1** form opinions by guessing. **2** buy in the hope of making a profit. □ **speculation** *n.*, **speculator** *n.*, **speculative** *adj.*

> ■ **1** cogitate, conjecture, meditate, muse, ponder, reflect, ruminate, theorize, think, wonder.

speculum *n.* medical instrument for looking into bodily cavities.

sped *see* **speed**.

speech *n.* **1** act, power, or manner of speaking. **2** spoken communication, esp. to an audience. **3** language, dialect.

> ■ **1** diction, elocution, enunciation, pronunciation. **2** address, discourse, lecture, oration, talk; monologue, soliloquy. **3** dialect, jargon, language, parlance, slang.

speechless *adj.* unable to speak because of great emotion.

speed *n.* **1** rate of time at which something moves or operates. **2** rapidity. ● *v.* **1** (**sped**) move, pass, or send quickly. **2**

(**speeded**) travel at an illegal speed. □ **speed up** accelerate.

■ *n.* **1** pace, rate, tempo, velocity. **2** haste, promptness, rapidity, swiftness. ● *v.* **1** dash, fly, race, rush, shoot, tear, zoom. □ **speed up** accelerate, go faster, pick up speed; expedite, hasten, hurry, precipitate, quicken.

speedboat *n.* fast motor boat.

speedometer *n.* device in a vehicle, showing its speed.

speedway *n.* **1** arena for motor cycle racing. **2** (*US*) road for fast traffic.

speedy *adj.* (**-ier**, **-iest**) rapid. □ **speedily** *adv.*, **speediness** *n.*

■ expeditious, fast, prompt, quick, rapid, snappy, swift.

spell[1] *n.* **1** words supposed to have magic power. **2** their influence. **3** fascination.

■ **1** charm, incantation.

spell[2] *v.* (**spelt**) **1** give in correct order the letters that form (a word). **2** produce as a result. □ **spell out** state explicitly. **speller** *n.*

spell[3] *n.* period of time, weather, or activity.

■ bout, interval, patch, period, session, stint, time.

spellbound *adj.* entranced.

spelt *see* **spell**[2].

spend *v.* (**spent**) **1** pay out (money) in buying something. **2** pass (time etc.). **3** use up. □ **spender** *n.*

■ **1** disburse, expend, lay out, pay. **2** fill, kill, occupy, pass, use, while away. **3** consume, exhaust, expend, go through, use up.

spendthrift *n.* wasteful spender.

spent *see* **spend**.

sperm *n.* (*pl.* **sperms** or **sperm**) **1** male reproductive cell. **2** semen. □ **sperm whale** large whale.

spermatozoon /spérmətōzṓ-on/ *n.* (*pl.* **-zoa**) fertilizing cell of a male organism.

spermicidal *adj.* killing sperm.

spew *v.* **1** vomit. **2** cast out in a stream.

sphagnum *n.* moss growing on bogs.

sphere *n.* **1** perfectly round solid geometric figure or object. **2** field of action or influence etc.

■ **1** ball, globe, orb, round. **2** area, domain, field, preserve, province, speciality, subject, territory.

spherical *adj.* shaped like a sphere.

sphincter *n.* ring of muscle controlling an opening in the body.

sphinx *n.* **1** ancient Egyptian statue with a lion's body and human or ram's head. **2** enigmatic person.

spice *n.* **1** flavouring substance with a strong taste or smell. **2** thing that adds zest. ● *v.* flavour with spice. □ **spicy** *adj.*

spider *n.* small animal with a segmented body and eight legs. □ **spidery** *adj.*

spigot *n.* plug stopping the vent-hole of a cask or controlling the flow of a tap.

spike *n.* **1** pointed thing. **2** pointed piece of metal. ● *v.* **1** put spikes on. **2** pierce or fasten with a spike. **3** (*colloq.*) add alcohol to (drink). □ **spiky** *adj.*

■ *n.* **2** barb, nail, pin, point, prong, skewer, spine, spit, stake, tine.

spill[1] *n.* thin strip of wood or paper for transferring flame.

spill[2] *v.* (**spilt**) **1** cause or allow to run over the edge of a container. **2** become spilt. ● *n.* **1** spilling or being spilt. **2** fall. □ **spillage** *n.*

spin *v.* (**spun**, **spinning**) **1** turn rapidly on its axis. **2** draw out and twist into threads. **3** make (yarn etc.) thus. ● *n.* **1** spinning movement. **2** short drive for pleasure. □ **spin-off** *n.* incidental benefit. **spin out** prolong.

■ *v.* **1** circle, go round, gyrate, revolve, rotate, turn round, twirl. ● *n.* **1** revolution, rotation, turn, twirl. **2** drive, excursion, jaunt, outing, ride, run, trip. □ **spin out** draw out, extend, prolong, protract, stretch out.

spinach *n.* vegetable with dark-green leaves.

spinal *adj.* of the spine.

spindle *n.* **1** rod on which thread is wound in spinning. **2** revolving pin or axis.

spindly *adj.* long or tall and thin.

spindrift *n.* sea spray.

spine *n.* **1** backbone. **2** needle-like projection. **3** part of a book where the pages are hinged. □ **spine-chilling** terrifying.

spineless *adj.* lacking determination.

spinet *n.* small harpsichord.

spinnaker *n.* large extra sail on a racing yacht.

spinneret *n.* thread-producing organ in a spider, silkworm, etc.

spinney *n.* (*pl.* **-eys**) thicket.

spinster *n.* unmarried woman.

spiny *adj.* full of spines, prickly.

spiral *adj.* forming a continuous curve round a central point or axis. ● *n.* **1** spiral line or thing. **2** continuous increase or decrease in two or more quantities alternately. ● *v.* (**spiralled**) move in a spiral course. □ **spirally** *adv.*

■ *n.* **1** coil, convolution, corkscrew, curl, helix, whirl, whorl. ● *v.* coil, curl, snake, twist, wind.

spire *n.* tall pointed structure esp. on a church tower.

spirit *n.* **1** mind or animating principle as distinct from body. **2** soul. **3** ghost. **4** person's nature. **5** liveliness, boldness. **6** state of mind or mood. **7** real meaning. **8** distilled extract. **9** (*pl.*) person's feeling of cheerfulness or depression. **10** (*pl.*) strong distilled alcoholic drink. ● *v.* carry off swiftly and mysteriously. □ **spirit level** device used to test horizontality.

■ *n.* **3** apparition, ghost, phantom, spectre, *colloq.* spook, wraith. **4** character, disposition, nature, temper, temperament. **5** energy, life, liveliness, vitality, vivacity; boldness, courage, *colloq.* guts, mettle, pluck, valour. **6** atmosphere, mood, morale. **7** essence, meaning, purport, sense, tenor.

spirited *adj.* lively, bold. □ **spiritedly** *adv.*

■ bold, brave, courageous, energetic, enthusiastic, lively, sprightly, vigorous, vivacious.

spiritual *adj.* **1** of the human spirit or soul. **2** of the Church or religion. ● *n.* religious folk song of black Americans. □ **spiritually** *adv.*, **spirituality** *n.*

spiritualism *n.* attempted communication with spirits of the dead. □ **spiritualist** *n.*, **spiritualistic** *adj.*

spirituous *adj.* strongly alcoholic.

spit[1] *v.* (**spat** or **spit**, **spitting**) **1** eject from the mouth. **2** eject saliva. **3** (of rain) fall lightly. ● *n.* **1** spittle. **2** act of spitting.

spit[2] *n.* **1** spike holding meat while it is roasted. **2** strip of land projecting into the sea.

spite *n.* ill will, malice. ● *v.* hurt or annoy from spite. □ **in spite of** not being prevented by. **spiteful** *adj.*, **spitefully** *adv.*

■ *n.* animosity, bitterness, ill will, malevolence, malice, rancour, resentment, spitefulness, venom. ● *v.* annoy, irritate, upset, vex. □ **in spite of** despite, notwithstanding, regardless of. **spiteful** *colloq.* bitchy, hurtful, malevolent, malicious, mean, nasty, unkind, venomous, vindictive.

spitfire *n.* fiery-tempered person.

spittle *n.* saliva.

splash *v.* **1** cause (liquid) to fly about in drops. **2** move or fall or wet with such drops. **3** decorate with irregular patches of colour etc. **4** display in large print. **5** spend (money) freely. ● *n.* **1** act, mark, or sound of splashing. **2** patch of colour or light. **3** striking display. □ **splashy** *adj.*

■ *v.* **1,2** *colloq.* slosh, spatter, splatter, spray, sprinkle.

splatter *v.* & *n.* splash, spatter.

splay *v.* **1** spread apart. **2** slant outwards or inwards. ● *adj.* splayed.

spleen *n.* abdominal organ regulating the proper condition of the blood. □ **splenic** *adj.*

splendid *adj.* **1** brilliant, very impressive. **2** excellent. □ **splendidly** *adv.*

■ **1** beautiful, brilliant, dazzling, fine, glorious, gorgeous, grand, impressive, luxurious, magnificent, resplendent, spectacular, sumptuous, superb. **2** excellent, first-class, first-rate, marvellous, outstanding, *colloq.* super, *colloq.* terrific, *colloq.* tremendous, wonderful.

splendour *n.* splendid appearance.

splenetic *adj.* bad-tempered.

splice *v.* join by interweaving or overlapping the ends.

splint *n.* rigid framework preventing a limb etc. from movement, e.g. while a broken bone heals. ● *v.* secure with a splint.

splinter *n.* thin sharp piece of broken wood etc. ● *v.* break into splinters. □ **splinter group** small group that has broken away from a larger one.

split *v.* (**split**, **splitting**) **1** break or come apart, esp. lengthwise. **2** divide, share. ● *n.* **1** splitting (up). **2** split thing or place. **3** (*pl.*) acrobatic position with legs stretched fully apart. □ **split second** very brief moment. **split up** end a relationship.

■ *v.* **1** break, cleave, crack, fracture, rupture, separate. **2** apportion, distribute, divide, halve, share; bifurcate, branch, fork. ● *n.* **2** break, cleft, crack, cranny, crevice, fissure, fracture, rift, rupture, slit. □ **split up** divorce, part, separate.

splodge *v.* & *n.* splotch.

splotch *v.* & *n.* splash, blotch.

■ *n.* blot, blotch, mark, smudge, splash, splodge, spot.

splurge *n.* ostentatious display, esp. of wealth. ● *v.* make a splurge, spend money freely.

splutter *v.* **1** make a rapid series of spitting sounds. **2** speak or utter incoherently. ● *n.* spluttering sound.

spoil *v.* (**spoilt** or **spoiled**) **1** make useless or unsatisfactory. **2** harm the character of (a person) by being indulgent. **3** become unfit for use. ● *n.* (also *pl.*) plunder.

■ *v.* **1** damage, disfigure, harm, impair, mar; destroy, mess up, queer, ruin, wreck. **2** coddle, indulge, mollycoddle, pamper. **3** decompose, go bad, go off, moulder, rot. ● *n.* (**spoils**) booty, loot, plunder, *sl.* swag.

spoilsport *n.* person who spoils others' enjoyment.

spoke[1] *n.* any of the bars connecting the hub to the rim of a wheel.

spoke[2], **spoken** *see* **speak**.

spokesperson *n.* person who speaks on behalf of a group.

spoliation *n.* pillaging.

sponge *n.* **1** water animal with a porous structure. **2** its skeleton, or a similar substance, esp. used for washing, cleaning, or padding. **3** sponge cake. ● *v.* **1** wipe or wash with a sponge. **2** live off the generosity of others. □ **sponge cake, pudding** one with a light open texture. **spongeable** *adj.*, **spongy** *adj.*

sponger *n.* person who sponges on others.

sponsor *n.* **1** person who gives to charity in return for another's activity. **2** godparent. **3** one who provides funds for a broadcast, sporting event, etc. ● *v.* act as sponsor for. □ **sponsorship** *n.*

■ *n.* **3** backer, patron, supporter. ● *v.* back, finance, fund, subsidize, support, underwrite.

spontaneous *adj.* **1** acting, done, or occurring without external cause. **2** instinctive. □ **spontaneously** *adv.*, **spontaneity** *n.*

■ **2** automatic, impulsive, instinctive, involuntary, reflex, unconscious.

spoof *n.* (*colloq.*) hoax, parody.

spook *n.* (*colloq.*) ghost. □ **spooky** *adj.*

spool *n.* reel on which something is wound. ● *v.* wind on a spool.

spoon *n.* **1** utensil with a rounded bowl and a handle, for eating, serving, or stirring food. **2** amount it contains. ● *v.* take with a spoon. □ **spoonful** *n.*

spoonfeed *v.* (**-fed**) **1** feed from a spoon. **2** give excessive help to.

spoor *n.* track or scent left by an animal.

sporadic *adj.* occurring here and there or now and again. □ **sporadically** *adv.*

■ fitful, intermittent, occasional, periodic, spasmodic.

spore *n.* one of the tiny reproductive cells of fungi, ferns, etc.

sporran *n.* pouch worn hanging in front of a kilt.

sport *n.* **1** athletic (esp. outdoor) activity. **2** game(s), pastime(s). **3** (*colloq.*) sportsmanlike person. ● *v.* **1** play, amuse oneself. **2** wear. □ **sports car** open low-built fast car. **sports jacket** man's jacket for informal wear.

■ *n.* **2** activity, game, pastime.

sporting *adj.* **1** of or interested in sport. **2** like a sportsman. □ **sporting chance** reasonable chance of success.

sportive *adj.* playful. □ **sportively** *adv.*, **sportiveness** *n.*

sportsman, sportswoman *ns.* **1** person engaging in sport. **2** fair and generous person. □ **sportsmanlike** *adj.*, **sportsmanship** *n.*

spot *n.* **1** round mark or stain. **2** pimple. **3** place. **4** small amount. **5** drop. **6** spotlight. ● *v.* (**spotted**) **1** mark with a spot or spots. **2** (*colloq.*) notice. **3** rain slightly. **4** watch for and take note of. □ **on the spot 1** without delay or change of place. **2** (*colloq.*) compelled to take action or justify oneself. **spot check** random check. **spotter** *n.*

■ *n.* **1** blot, blotch, dot, fleck, mark, patch, smudge, speck, speckle, splodge, splotch, stain. **2** blackhead, pimple, pustule. **3** location, locality, place, position, setting, site, situation, venue. **4** bit, dash, jot, morsel, piece, pinch, scrap, touch, trace. **5** bead, blob, drip, drop, droplet, globule. ● *v.* **1** blot, fleck, mark, smudge, spatter, splash, stain. **2** catch sight of, descry, discern, glimpse, make out, notice, observe, see, spy.

spotless *adj.* free from stain or blemish. □ **spotlessly** *adv.*, **spotlessness** *n.*

■ clean, immaculate, unstained, untarnished; faultless, flawless, impeccable, pure, unblemished, unsullied, untainted.

spotlight *n.* lamp or its beam directed on a small area. ● *v.* (**-lit** or **-lighted**) **1** direct a spotlight on. **2** draw attention to.

spotty *adj.* marked with spots.

spouse *n.* husband or wife.

spout *n.* **1** projecting tube through which liquid is poured or conveyed. **2** jet of liquid. ● *v.* **1** come or send out forcefully as a jet of liquid. **2** utter or speak lengthily.

■ *v.* **1** flow, gush, pour, shoot, spew, spurt, squirt, stream; discharge, disgorge, eject, emit. **2** declaim, expatiate, pontificate, ramble on, rant, talk.

sprain *v.* injure by wrenching violently. ● *n.* this injury

sprang *see* **spring**.

sprat *n.* small herring-like fish

sprawl *v.* **1** sit, lie, or fall with arms and legs spread loosely. **2** spread untidily. ● *n.* sprawling attitude or arrangement.

spray[1] *n.* **1** branch with its leaves and flowers. **2** bunch of cut flowers. **3** ornament in similar form.

spray[2] *n.* **1** liquid dispersed in very small drops. **2** liquid for spraying. **3** device for spraying liquid. ● *v.* **1** come or send out as spray. **2** wet with liquid thus. □ **spray-gun** *n.* device for spraying paint etc. **sprayer** *n.*

■ *n.* **1** mist, spindrift; sprinkle, sprinkling. **3** aerosol, atomizer, sprayer, spray-gun, sprinkler. ● *v.* **1** gush, spout, spurt. **2** shower, spatter, sprinkle, water.

spread *v.* (**spread**) **1** open out. **2** become longer or wider. **3** apply as a layer. **4** make or become widely known, felt, etc. **5** distribute, become distributed. ● *n.* **1** spreading. **2** (*colloq.*) lavish meal. **3** thing's range. **4** paste for spreading on bread. □ **spread-eagled** *adj.* with arms and legs spread.

■ *v.* **1** fan out, lay out, open out, unfold, unfurl, unroll. **2** broaden, enlarge, expand, extend, grow, increase, mushroom, proliferate, sprawl, widen. **3** apply, cover, paint, plaster, rub, smear. **4** air, announce, broadcast, circulate, disseminate, make known *or* public, pronounce, promulgate, publicize, publish. **5** distribute, scatter, sow, strew. ● *n.* **1** broadening, enlargement, expansion, extension, growth, increase, proliferation, widening; broadcasting, circulation, dissemination, promulgation. **2** banquet, feast, meal, *formal* repast. **3** extent, range, reach, scope, span, sweep.

spreadsheet *n.* computer program for manipulating esp. tabulated numerical data.

spree *n.* (*colloq.*) lively outing.

sprig *n.* twig, shoot.

sprightly *adj.* (**-ier**, **-iest**) lively, full of energy. □ **sprightliness** *n.*

■ active, animated, brisk, energetic, frisky, full of life, jaunty, lively, perky, spirited, spry, vigorous, vivacious.

spring *v.* (**sprang**, **sprung**) **1** jump, move rapidly. **2** issue, arise. **3** produce or cause to operate suddenly. ● *n.* **1** act of springing, jump. **2** device that reverts to its original position after being compressed, tightened, or stretched. **3** elasticity. **4** place where water or oil flows naturally from the ground. **5** season between winter and summer. □ **spring-clean** *v.* clean (one's home etc.) thoroughly. **spring tide** tide when there is the largest rise and fall of water.

■ *v.* **1** bounce, bound, dart, jump, leap, skip, vault. **2** arise, begin, derive, descend, issue, originate, proceed, start, stem. ● *n.* **1** bounce, bound, jump, leap, skip, vault. **3** bounce, elasticity, flexibility, resilience, springiness. **4** fountain, geyser, spa.

springboard *n.* flexible board for leaping or diving from.

springbok *n.* S. African gazelle.

springtime *n.* season of spring.

springy *adj.* (**-ier**, **-iest**) able to spring back easily after being squeezed, tightened, or stretched. □ **springiness** *n.*

■ elastic, flexible, pliable, resilient, stretchy, whippy.

sprinkle *v.* scatter or fall in drops or particles on (a surface). ● *n.* light shower. □ **sprinkler** *n.*

■ *v.* dust, dredge, powder, scatter, spray, strew.

sprinkling *n.* small sparse number or amount.

sprint *v.* & *n.* run or swim etc. at full speed. □ **sprinter** *n.*

sprite *n.* elf, fairy, or goblin.

sprocket *n.* projection engaging with links on a chain etc.

sprout *v.* **1** begin to grow or appear. **2** put forth. ● *n.* **1** plant's shoot. **2** Brussels sprout.

■ *v.* **1** bud, germinate; develop, grow, shoot up, spring up. **2** grow, put forth.

spruce[1] *adj.* neat, smart. ● *v.* smarten. □ **sprucely** *adv.*, **spruceness** *n.*

■ *n.* dapper, neat, smart, tidy, trim, well-groomed, well-turned-out. ● *v.* neaten, smarten, straighten, tidy, *colloq.* titivate.

spruce[2] *n.* a kind of fir.

sprung *see* **spring**.

spry *adj.* lively, sprightly. □ **spryly** *adv.*, **spryness** *n.*

spud *n.* **1** narrow spade. **2** (*sl.*) potato.

spume *n.* froth.

spun *see* **spin**.

spur *n.* **1** pricking-device worn on a rider's heel. **2** stimulus, incentive. **3** projection. ● *v.* (**spurred**) **1** urge on (a horse) with one's spurs. **2** incite, stimulate. □ **on the spur of the moment** on impulse.

> ■ *n.* **2** goad, impetus, incentive, incitement, inducement, prod, provocation, stimulation, stimulus. ● *v.* **2** drive, egg on, encourage, goad, incite, inspire, motivate, prod, prompt, provoke, stimulate, urge.

spurious *adj.* not genuine or authentic. □ **spuriously** *adv.*

> ■ artificial, bogus, counterfeit, fake, false, feigned, forged, fraudulent, imitation, mock, *colloq.* phoney, sham, simulated.

spurn *v.* reject contemptuously.

> ■ brush off, cold-shoulder, rebuff, reject, repulse, scorn, snub.

spurt *v.* **1** (cause to) gush out in a jet or stream. **2** increase speed suddenly. ● *n.* **1** sudden gush. **2** short sudden effort or increase in speed.

sputter *v.* & *n.* splutter.

sputum *n.* expectorated matter.

spy *n.* person who secretly watches or gathers information. ● *v.* **1** catch sight of. **2** be a spy. □ **spy on** watch secretly.

> ■ *v.* **1** catch sight of, descry, discern, espy, glimpse, make out, see, *colloq.* spot. □ **spy on** follow, keep under observation *or* surveillance, observe, shadow, *colloq.* tail, watch.

sq. *abbr.* square.

squabble *v.* quarrel pettily or noisily. ● *n.* quarrel of this kind.

squad *n.* small group working or being trained together.

squadron *n.* **1** division of a cavalry unit or of an airforce. **2** detachment of warships.

squalid *adj.* **1** dirty and unpleasant. **2** morally degrading. □ **squalidly** *adv.*, **squalor** *n.*

> ■ **1** *sl.* crummy, dirty, disgusting, filthy, foul, grimy, grubby, insanitary, seedy, sleazy, sordid, unpleasant. **2** degrading, dishonourable, shameful, sordid.

squall *n.* sudden storm or wind. □ **squally** *adj.*

squander *v.* spend wastefully.

> ■ dissipate, fritter away, misspend, splurge, waste.

square *n.* **1** geometric figure with four equal sides and four right angles. **2** area or object shaped like this. **3** product of a number multiplied by itself. ● *adj.* **1** of square shape. **2** right-angled. **3** of or using units expressing the measure of an area. **4** properly arranged. **5** equal, not owed or owing anything. **6** fair and honest. **7** direct. ● *adv.* squarely, directly. ● *v.* **1** make right-angled. **2** mark with squares. **3** place evenly. **4** multiply by itself. **5** settle (an account etc.). **6** make or be consistent. **7** (*colloq.*) bribe. □ **square dance** dance in which four couples face inwards from four sides. **square meal** substantial meal. **square root** number of which a given number is the square. **square up to 1** face in a fighting attitude. **2** face resolutely. **squarely** *adv.*

> ■ *n.* **2** piazza, *colloq.* quad, quadrangle. ● *adj.* **4** arranged, in order, organized, settled, straight, straightened out. **5** equal, even, level, level-pegging, on a par; quits. **6** above-board, decent, equitable, ethical, fair, honest, just, on the level, straightforward, upright. ● *v.* **5** discharge, pay, settle (up). **6** make consistent, reconcile; agree, be consistent, correspond, match, tally.

squash *v.* **1** crush, squeeze or become squeezed flat or into pulp. **2** crowd. **3** suppress. **4** silence with a crushing reply. ● *n.* **1** crowd of people squashed together. **2** fruit-flavoured soft drink. **3** (also **squash rackets**) game played with rackets and a small ball in a closed court. □ **squashy** *adj.*

> ■ *v.* **1** compress, crush, flatten, mash, pulp, squeeze. **2** cram, crowd, jam, pack, squeeze. **3** crush, put an end to, put down, quash, quell, stamp out, suppress.

squat *v.* (**squatted**) **1** sit on one's heels. **2** (*colloq.*) sit. **3** be a squatter. ● *n.* **1** squatting posture. **2** place occupied by squatters. ● *adj.* dumpy.

squatter *n.* person who takes unauthorized possession of unoccupied premises.

squawk *n.* loud harsh cry. ● *v.* make or utter with a squawk.

squeak *n.* short high-pitched cry or sound. ● *v.* make or utter with a squeak. □ **narrow squeak** narrow escape. **squeaky** *adj.*

squeal *n.* long shrill cry or sound. ● *v.* **1** make or utter with a squeal. **2** (*sl.*) become an informer.

squeamish *adj.* **1** easily nauseated or disgusted. **2** excessively fastidious. □ **squeamishly** *adv.*, **squeamishness** *n.*

■ **1** fastidious, queasy. **2** dainty, fastidious, finicky, fussy, particular, *colloq.* pernickety.

squeeze *v.* **1** exert pressure on. **2** extract moisture in this way. **3** force into or through, force one's way, crowd. **4** obtain (money etc.) by extortion. **5** bring pressure to bear on. ● *n.* **1** squeezing. **2** affectionate clasp or hug. **3** small quantity produced by squeezing. **4** crowd, crush. **5** restrictions on borrowing.

■ *v.* **1** compress, constrict, crush, nip, pinch, press, squash, wring. **2** express, extract. **3** cram, crowd, force, jam, pack, ram, squash, stuff, wedge. **4** bleed, extort, screw, wrest. **5** lean on, pressure, pressurize. ● *n.* **2** clasp, cuddle, embrace, hug. **3** bit, dash, drop, spot, touch.

squelch *v.* & *n.* sound like someone treading in thick mud.

squib *n.* small exploding firework.

squid *n.* sea creature with ten arms round its mouth.

squiggle *n.* short curly line. □ **squiggly** *adj.*

squint *v.* **1** have the eyes turned in different directions. **2** look sideways or through a small opening. ● *n.* **1** squinting condition. **2** sideways glance.

squire *n.* country gentleman, esp. landowner.

squirm *v.* **1** wriggle. **2** feel embarrassment. ● *n.* wriggle.

■ *v.* **1** fidget, twist, wiggle, wriggle, writhe. **2** be embarrassed *or* uncomfortable, writhe.

squirrel *n.* small tree-climbing animal with a bushy tail.

squirt *v.* **1** send out (liquid) or be sent out in a jet. **2** wet in this way. ● *n.* jet of liquid.

squish *v.* & *n.* (move with) a soft squelching sound. □ **squishy** *adj.*

St *abbr.* Saint.

St. *abbr.* Street.

stab *v.* (**stabbed**) **1** pierce or wound with something pointed. **2** aim a blow with such a weapon. ● *n.* **1** act or result of stabbing. **2** sharply painful sensation. **3** (*colloq.*) attempt.

■ *v.* **1** gore, impale, knife, lance, skewer, spear, spike, transfix. **2** jab, lunge, poke, thrust.

stabilize *v.* make or become stable. □ **stabilization** *n.*, **stabilizer** *n.*

stable[1] *adj.* firmly fixed or established, not easily shaken or destroyed. □ **stably** *adv.*, **stability** *n.*

■ anchored, fast, firm, fixed, immovable, moored, secure, steady; enduring, established, lasting, solid, steadfast, strong, unchanging, unwavering.

stable[2] *n.* **1** building in which horses are kept. **2** establishment for training racehorses. **3** horses, people, etc. from the same establishment. ● *v.* put or keep in a stable.

staccato *adj.* & *adv.* in a sharp disconnected manner.

stack *n.* **1** orderly pile or heap. **2** (*colloq.*) large quantity. **3** tall chimney. **4** storage section of a library. ● *v.* **1** arrange in a stack or stacks. **2** arrange (cards) secretly for cheating. **3** cause (aircraft) to fly at different levels while waiting to land.

■ *n.* **1** accumulation, collection, heap, mass, mound, mountain, pile, stockpile. **2** (**stacks**) a lot, *colloq.* heaps, *colloq.* loads, lots, *colloq.* piles, plenty, quantities. ● *v.* **1** accumulate, amass, collect, gather, heap, pile (up).

stadium *n.* sports ground with tiered seats for spectators.

staff *n.* **1** stick used as a weapon, support, or symbol of authority. **2** the people employed by an organization. **3** (*pl.* **staves**) set of five horizontal lines on which music is written. ● *v.* provide with a staff of people.

■ *n.* **1** baton, cane, club, crook, mace, pike, pole, rod, sceptre, standard, stave, stick, truncheon. **2** employees, personnel, workforce.

stag *n.* fully grown male deer. □ **stag beetle** beetle with branched projecting mouthparts. **stag-night** *n.* all-male party for a man about to marry.

stage *n.* **1** raised floor or platform. **2** one on which plays etc. are performed. **3** (**the stage**) theatrical profession. **4** division of or point reached in a process or journey. ● *v.* **1** present on the stage. **2** arrange and carry out. □ **go on the stage** become an actor or actress. **stage fright**

nervousness on facing an audience. **stage whisper** one meant to be overheard.

■ *n.* **1** dais, platform, podium, rostrum. **4** juncture, period, phase, point, step; lap, leg, part, section. ● *v.* **1** mount, present, produce, put on. **2** arrange, engineer, orchestrate, organize.

stagger *v.* **1** move or go unsteadily. **2** shock deeply. **3** arrange so as not to coincide exactly. ● *n.* staggering movement.

■ *v.* **1** dodder, falter, lurch, reel, stumble, sway, teeter, totter, wobble. **2** amaze, astonish, astound, dumbfound, *colloq.* flabbergast, overwhelm, shake, shock, startle, stun, stupefy, surprise, take aback.

staggering *adj.* astonishing.

stagnant *adj.* **1** not flowing, still and stale. **2** inactive, sluggish. □ **stagnancy** *n.*

■ **1** motionless, still, unmoving; contaminated, dirty, foul, polluted, stale. **2** inactive, quiet, sluggish, static, torpid.

stagnate *v.* be or become stagnant. □ **stagnation** *n.*

■ decline, degenerate, deteriorate, *colloq.* go to pot, spoil, vegetate.

staid *adj.* steady and serious.

■ quiet, restrained, sedate, serious, sober, solemn, steady.

stain *v.* **1** be or become discoloured. **2** blemish. **3** colour with a penetrating pigment. ● *n.* **1** mark caused by staining. **2** blemish. **3** substance for staining things.

■ *v.* **1** blotch, discolour, mark, smudge, soil, spot, tinge. **2** blemish, blot, mar, spoil, sully, taint, tarnish. **3** colour, dye, paint. ● *n.* **1** blot, blotch, discoloration, mark, smudge, spot, splodge, splotch. **2** blemish, blot, taint. **3** colour, colourant, dye, paint, pigment, tint.

stainless *adj.* free from stains. □ **stainless steel** steel alloy not liable to rust or tarnish.

stair *n.* **1** one of a set of fixed indoor steps. **2** (*pl.*) such a set.

staircase *n.* stairs and their supporting structure.

stairway *n.* staircase.

stake *n.* **1** pointed stick or post for driving into the ground. **2** money etc. wagered. **3** share or interest in an enterprise etc. ● *v.* **1** fasten, support, or mark with a stake or stakes. **2** wager. □ **at stake** being risked. **stake a claim** claim a right to something. **stake out** place under surveillance. **stake-out** *n.*

■ *n.* **1** picket, pole, post, stave, stick, upright. **2** ante, bet, wager. **3** concern, interest, investment, share. ● *v.* **1** fasten, hitch, lash, leash, picket, secure, tether, tie (up). **2** bet, gamble, wager; chance, hazard, risk, venture.

stalactite *n.* deposit of calcium carbonate hanging like an icicle.

stalagmite *n.* deposit of calcium carbonate standing like a pillar.

stale *adj.* **1** not fresh. **2** unpleasant or uninteresting from lack of freshness. **3** spoilt by too much practice. ● *v.* make or become stale. □ **staleness** *n.*

■ *adj.* **1** limp, mouldy, musty, old, sour, spoiled, wilted, withered; close, fusty, stuffy. **2** banal, clichéd, *colloq.* corny, hackneyed, platitudinous, stereotyped, trite, unoriginal.

stalemate *n.* **1** drawn position in chess. **2** drawn contest. **3** deadlock. ● *v.* bring to such a state.

stalk[1] *n.* stem or similar supporting part.

stalk[2] *v.* **1** walk in a stately or haughty manner. **2** track or pursue stealthily. □ **stalker** *n.*

■ **1** flounce, march, stride, strut, walk. **2** follow, hunt, pursue, shadow, *colloq.* tail, track, trail.

stall *n.* **1** stable, shelter for cows. **2** compartment in this. **3** ground floor seat in a theatre. **4** booth or stand where goods are displayed for sale. **5** stalling of an aircraft. ● *v.* **1** place or keep in a stall. **2** (of an engine) stop suddenly through lack of power. **3** (of an aircraft) begin to drop because the speed is too low. **4** cause to stall. **5** play for time when being questioned.

■ *n.* **1** byre, cowshed, shed, stable. **4** booth, cubicle, kiosk, stand. ● *v.* **5** delay, hesitate, hedge, play for time, prevaricate, stonewall, temporize.

stallion *n.* uncastrated male horse.

stalwart *adj.* **1** strong, sturdy. **2** dependable and loyal. ● *n.* stalwart person.

■ *adj.* **1** hale, hardy, healthy, hearty, lusty, muscular, robust, stout, strapping, strong, tough, vigorous. **2** dependable, faithful, firm, loyal, reliable, staunch, steadfast, true, unswerving, unwavering.

stamen *n.* pollen-bearing part of a flower.

stamina *n.* ability to withstand long physical or mental strain.

stammer *v.* speak with involuntary pauses or repetitions of a syllable. ● *n.* this act or tendency.

stamp *v.* **1** bring (one's foot) down heavily on the ground. **2** flatten or crush in this way. **3** walk with heavy steps. **4** press so as to cut or leave a mark or pattern. **5** fix a postage stamp to. **6** give a specified character to. ● *n.* **1** act or sound of stamping. **2** instrument for stamping a mark etc., this mark. **3** small adhesive label for affixing to an envelope or document to show the amount paid as postage or a fee etc. **4** characteristic quality. □ **stamping ground** usual haunt. **stamp out** destroy, suppress by force.

■ *v.* **2** crush, flatten, press, squash, tramp, trample. **3** clump, stomp, tramp. **4** emboss, engrave, impress, imprint, mark, print. **6** brand, categorize, characterize, designate, label, mark. ● *n.* **4** characteristic(s), hallmark, mark, quality. □ **stamp out** crush, destroy, eliminate, put an end to, put down, quash, quell, repress, squash, suppress, wipe out.

stampede *n.* sudden rush of animals or people. ● *v.* (cause to) take part in a stampede.

stance *n.* manner of standing.

■ attitude, pose, position, posture, stand.

stanch *v.* stop or slow down the flow of (blood) from a wound.

stanchion *n.* upright post or support.

stand *v.* (**stood**) **1** have, take, or keep a stationary upright position. **2** be situated. **3** place, set upright. **4** stay firm or valid. **5** offer oneself for election. **6** endure. **7** provide at one's own expense. ● *n.* **1** stationary condition. **2** position or attitude taken up. **3** resistance to attack. **4** rack, pedestal. **5** raised structure with seats at a sports ground etc. **6** stall for goods. □ **stand a chance** have a chance of success. **stand by 1** look on without interfering. **2** stand ready for action. **3** support in a difficulty. **4** keep to (a promise etc.). **stand-by** *adj.* & *n.* (person or thing) available as a substitute. **stand down** withdraw. **stand for 1** represent. **2** (*colloq.*) tolerate. **stand in** deputize. **stand-in** *n.* deputy, substitute. **stand one's ground** not yield. **stand to reason** be logical. **stand up 1** come to or place in a standing position. **2** be valid. **3** (*colloq.*) fail to keep an appointment with. **stand up for** speak in defence of. **stand up to 1** resist courageously. **2** be resistant to (wear, use, etc.).

■ *v.* **1** arise, get to one's feet, get up, rise, stand up. **3** place, position, prop, put, rest, set. **4** apply, hold, obtain, prevail, remain in force *or* valid. **6** abide, accept, allow, bear, brook, endure, put up with, *colloq.* stick, stomach, *colloq.* stand for, take, tolerate; experience, undergo, suffer, survive, weather, withstand. **7** buy, treat to. ● *n.* **4** frame, rack, rest; pedestal, plinth; easel, tripod. **6** booth, kiosk, stall. □ **stand by 3** back, be loyal to, defend, stand up for, *colloq.* stick up for, support, take the side of, uphold. **4** abide by, adhere to, stick to. **stand for 1** betoken, denote, indicate, mean, represent, signify, symbolize. **stand-in** deputy, locum, replacement, substitute, surrogate, understudy. **stand up to 1** brave, challenge, confront, face (up to), resist.

standard *n.* **1** thing serving as a basis, example, or principle to which others (should) conform, or by which others are judged. **2** average quality. **3** required level of quality or proficiency. **4** distinctive flag. ● *adj.* **1** serving as or conforming to a standard. **2** of normal or usual quality. □ **standard lamp** household lamp set on a tall support.

■ *n.* **1** benchmark, criterion, gauge, measure, touchstone, yardstick; archetype, exemplar, model, paradigm, pattern, type; guideline, precept, principle. **3** grade, level, rating. **4** banner, ensign, flag, pennant, pennon. ● *adj.* **2** average, conventional, customary, normal, ordinary, regular, stock, usual.

standardize *v.* cause to conform to a standard. □ **standardization** *n.*

standing *n.* **1** status. **2** duration.

■ **1** footing, place, position, rank, station, status; eminence, prominence, reputation, repute.

standoffish *adj.* cold or distant in manner.

■ aloof, cold, cool, distant, frosty, reserved, unapproachable, unfriendly, unsociable, withdrawn.

standpipe *n.* vertical pipe for fluid to rise in, esp. for attachment to a water main.

standpoint *n.* point of view.

■ angle, attitude, opinion, outlook, perspective, point of view, view, viewpoint.

standstill *n.* stoppage, inability to proceed.

> ■ halt, stop, stoppage; deadlock, impasse, stalemate.

stank *see* **stink**.

stanza *n.* verse of poetry.

staphylococcus *n.* (*pl.* **-ci**) pus-producing bacterium. □ **staphylococcal** *adj.*

staple[1] *n.* U-shaped piece of wire for fastening papers together, fixing netting to a post, etc. ● *v.* secure with staple(s). □ **stapler** *n.*

staple[2] *adj.* & *n.* principal or standard (food or product etc.).

star *n.* **1** heavenly body appearing as a point of light. **2** asterisk. **3** star-shaped mark indicating a category of excellence. **4** famous actor or performer etc. ● *v.* (**starred**) **1** put an asterisk beside (an item). **2** present or perform as a star actor.

> ■ *n.* **1** comet, lodestar, nova, pole star, shooting star. **2** asterisk, pentagram. **4** celebrity, luminary, personality.

starboard *n.* right-hand side of a ship or aircraft.

starch *n.* **1** white carbohydrate. **2** preparation for stiffening fabrics. **3** stiffness of manner. ● *v.* stiffen with starch. □ **starchy** *adj.*

stardom *n.* being a star actor etc.

stare *v.* gaze fixedly, esp. in surprise, horror, etc. ● *n.* staring gaze.

> ■ *v.* gape, *colloq.* gawp, gaze, glare, goggle.

starfish *n.* star-shaped sea creature.

stark *n.* **1** desolate, bare. **2** sharply evident. **3** downright. ● *adv.* completely. □ **starkly** *adv.*, **starkness** *n.*

> ■ *adj.* **1** austere, bare, barren, bleak, cheerless, desolate, dreary, empty, grim. **2** clear, conspicuous, evident, obvious, overt, patent, plain, unmistakable. **3** absolute, complete, downright, out and out, outright, perfect, pure, sheer, thorough, total, unmitigated, utter.

starling *n.* noisy bird with glossy black speckled feathers.

starry *adj.* **1** full of stars. **2** starlike. □ **starry-eyed** *adj.* (*colloq.*) enthusiastic but impractical.

start *v.* **1** begin, cause to begin. **2** (cause to) begin operating. **3** begin a journey. **4** make a sudden movement, esp. from pain or surprise. ● *n.* **1** beginning. **2** place where a race etc. starts. **3** advantage gained or allowed in starting. **4** sudden movement of pain or surprise. □ **starter** *n.*

> ■ *v.* **1** begin, commence, embark on, get going, get under way, open, set about, set in motion; establish, found, inaugurate, initiate, institute, launch, originate, pioneer, set up. **2** activate, crank up, switch on, turn on. **3** depart, go, leave, move off, set off *or* out. **4** blench, draw back, flinch, jump, recoil, shrink back. ● *n.* **1** beginning, commencement, dawn, dawning, foundation, founding, genesis, inauguration, inception, initiation, institution, kick-off, launch, onset, opening, origin, outset, threshold. **3** advantage, lead. **4** jump, jolt, twitch.

startle *v.* shock, surprise.

> ■ alarm, astonish, catch unawares, frighten, jolt, scare, shock, surprise, take aback.

starve *v.* **1** die or suffer acutely from lack of food. **2** cause to do this. **3** force by starvation. **4** (*colloq.*) feel very hungry. □ **starvation** *n.*

stash *v.* (*colloq.*) **1** conceal in a safe place. **2** hoard.

state *n.* **1** mode of being, with regard to characteristics or circumstances. **2** (*colloq.*) excited or agitated condition of mind. **3** grand imposing style. **4** (often **State**) political community under one government or forming part of a federation. **5** civil government. ● *adj.* **1** of or involving the state. **2** ceremonial. ● *v.* **1** express in words. **2** specify.

> ■ *n.* **1** circumstance(s), condition(s), position, situation; form, order, repair, shape, trim. **2** *colloq.* flap, panic, *colloq.* stew, *colloq.* tizzy. **3** glory, grandeur, magnificence, pomp, splendour. **4** commonwealth, country, land, nation, realm, republic. ● *adj.* **1** federal, governmental, national. **2** ceremonial, dignified, formal, majestic, regal, royal, stately. ● *v.* **1** affirm, announce, assert, communicate, declare, express, maintain, make known, proclaim, profess, say, utter, voice. **2** define, designate, fix, set, specify.

stateless *adj.* not a citizen or subject of any country.

stately *adj.* (**-ier**, **-iest**) dignified, grand. □ **stately home** large historic house, esp. one open to the public. **stateliness** *n.*

> ■ august, ceremonial, dignified, grand, formal, impressive, majestic, noble, regal, solemn.

statement *n.* **1** process of stating. **2** thing stated. **3** formal account of facts. **4** written report of a financial account.

■ **1,2** affirmation, allegation, announcement, assertion, declaration, proclamation, profession, utterance. **3** account, affidavit, deposition, report, testimony.

stateroom *n.* **1** room used on ceremonial occasions. **2** private cabin on a passenger ship.

statesman, **stateswoman** *ns.* experienced and respected politician. □ **statesmanship** *n.*

static *adj.* **1** of force acting by weight without motion. **2** stationary, not moving. ● *n.* **1** electrical disturbances in the air, causing interference in telecommunications. **2** (also **static electricity**) electricity present in a body, not flowing as current.

■ *adj.* **2** immobile, immovable, inert, invariable, motionless, stagnant, stationary, still, unchanging, unmoving, unvarying.

station *n.* **1** place where a person or thing stands or is stationed. **2** place where a public service or specialized activity is based. **3** broadcasting channel. **4** stopping place on a public transport route. **5** status. ● *v.* put at or in a certain place for a purpose. □ **station wagon** (*US*) estate car.

■ *n.* **1** location, place, position, post, site, situation, spot; base, centre, headquarters. **4** *US* depot, terminal, terminus; stop. **5** level, position, rank, standing, status. ● *v.* locate, place, position, post, put, set, site, situate, stand.

stationary *adj.* **1** not moving. **2** not movable.

stationer *n.* dealer in stationery.

stationery *n.* writing paper, envelopes, labels, etc.

statistic *n.* **1** item of information expressed in numbers. **2** (*pl.*) science of collecting and interpreting numerical information. □ **statistical** *adj.*, **statistically** *adv.*

statistician *n.* expert in statistics.

statue *n.* sculptured, cast, or moulded figure.

■ bust, colossus, figure, figurine, image, likeness, model, representation, sculpture, statuette.

statuesque *adj.* like a statue, esp. in dignity or beauty.

statuette *n.* small statue.

stature *n.* **1** bodily height. **2** greatness gained by ability or achievement.

status *n.* (*pl.* **-uses**) **1** person's position or rank in relation to others. **2** high rank or prestige. □ **status quo** existing previous state of affairs.

■ **1** footing, level, place, position, rank, standing, station. **2** eminence, importance, pre-eminence, prestige, prominence, reputation, repute, stature.

statute *n.* law passed by Parliament or a similar body.

statutory *adj.* fixed, done, or required by statute.

staunch *adj.* firm in opinion or loyalty. □ **staunchly** *adv.*

■ constant, dependable, faithful, firm, loyal, reliable, stalwart, steadfast, steady, true, trusted, trustworthy, trusty, unswerving, unwavering.

stave *n.* **1** one of the strips of wood forming the side of a cask or tub. **2** staff in music. ● *v.* (**stove** or **staved**) dent, break a hole in. □ **stave off** (**staved**) ward off.

stay *v.* **1** continue in the same place or state. **2** dwell temporarily. **3** postpone. **4** show endurance. ● *n.* **1** period of staying. **2** postponement. □ **stay the course** be able to reach the end of it. **staying power** endurance.

■ *v.* **1** hang about, linger, loiter, remain, stop, wait; continue, keep. **2** sojourn, visit; dwell, live, lodge. **3** defer, delay, postpone, put off, suspend. **4** endure, last. ● *n.* **1** sojourn, visit. **2** deferral, delay, postponement, suspension.

stead *n.* **in a person's** or **thing's stead** as a substitute. **stand in good stead** be useful to.

steadfast *adj.* firm and not changing or yielding. □ **steadfastly** *adv.*

■ constant, dependable, devoted, faithful, firm, loyal, stalwart, staunch, steady, true, unswerving, unwavering; determined, indefatigable, persevering, resolute, resolved, single-minded, tireless, unflagging.

steady *adj.* (**-ier**, **-iest**) **1** not shaking. **2** regular, uniform. **3** dependable, not excitable. **4** constant, persistent. ● *adv.* steadily. ● *v.* make or become steady. □ **steadily** *adv.*, **steadiness** *n.*

■ *adj.* **1** balanced, fast, firm, poised, secure, settled, stable. **2** consistent, constant, even, invariable, regular, uniform, unchanging, unvarying. **3** dependable, imperturbable, level-headed, reliable, sedate, sensible, serious, sober, staid, temperate, *colloq.* unflappable. **4** ceaseless, constant, continual, continuous, endless, incessant, never-ending, non-stop, perpetual, persistent, unbroken, unceasing, unremitting. ● *v.* balance, secure, stabilize; calm, compose, control, settle.

steak *n.* slice of meat (esp. beef) or fish, usu. grilled or fried.

steal *v.* (**stole, stolen**) **1** take dishonestly. **2** move stealthily. ● *n.* (*colloq.*) bargain, easy task. □ **steal the show** outshine other performers.

■ *v.* **1** appropriate, embezzle, filch, *sl.* knock off, lift, make off with, misappropriate, *sl.* nick, pilfer, *sl.* pinch, pocket, purloin, *colloq.* snaffle, *colloq.* swipe, thieve, *sl.* whip; copy, pirate, plagiarize. **2** creep, pussyfoot, sidle, slink, sneak, tiptoe.

stealth *n.* secrecy, secret behaviour.

stealthy *adj.* (**-ier, -iest**) done or moving with stealth. □ **stealthily** *adv.*, **stealthiness** *n.*

■ clandestine, covert, furtive, secret, surreptitious, undercover.

steam *n.* **1** gas into which water is changed by boiling. **2** this as motive power. **3** energy, power. ● *v.* **1** give out steam. **2** cook or treat by steam. **3** move by the power of steam. **4** cover or become covered by steam. □ **steam engine** engine or locomotive driven by steam. **steamy** *adj.*

steamer *n.* **1** steam-driven ship. **2** container in which things are cooked or heated by steam.

steamroller *n.* heavy engine with a large roller, used in road-making.

steatite /stéeətīt/ *n.* greyish talc that feels smooth and soapy.

steel *n.* **1** strong alloy of iron and carbon. **2** steel rod for sharpening knives. ● *v.* make resolute. □ **steel wool** fine shavings of steel used as an abrasive. **steely** *adj.*, **steeliness** *n.*

steep[1] *v.* soak or bathe in liquid.

■ bathe, douse, drench, immerse, saturate, soak, souse, submerge, wet; marinade.

steep[2] *adj.* **1** sloping sharply not gradually. **2** (*colloq.*, of price) unreasonably high. □ **steeply** *adv.*, **steepness** *n.*

■ **1** abrupt, precipitous, sharp, sheer. **2** dear, excessive, exorbitant, extortionate, high, stiff, unreasonable.

steeple *n.* tall tower with a spire, rising above a church roof.

steeplechase *n.* race for horses or athletes with fences to jump.

steeplejack *n.* repairer of tall chimneys etc.

steer[1] *n.* bullock.

steer[2] *v.* direct the course of, guide by mechanism. □ **steer clear of** avoid.

■ direct, guide, navigate, pilot. □ **steer clear of** avoid, circumvent, dodge, give a wide berth to, keep away from, shun.

steersman *n.* person who steers a ship.

stellar *adj.* of a star or stars.

stem[1] *n.* **1** supporting usu. cylindrical part, esp. of a plant. **2** main usu. unchanging part of a noun or verb. ● *v.* (**stemmed**) **stem from** have as its source.

stem[2] *v.* (**stemmed**) **1** restrain the flow of. **2** dam.

stench *n.* foul smell.

stencil *n.* **1** sheet of card etc. with a cut-out design, painted over to reproduce this on the surface below. **2** design reproduced thus. ● *v.* (**stencilled**) produce or ornament by this.

stenographer *n.* shorthand writer.

stentorian *adj.* (of a voice) extremely loud.

step *v.* (**stepped**) **1** lift and set down a foot or alternate feet. **2** move a short distance thus. **3** progress. ● *n.* **1** movement of a foot and leg in stepping. **2** distance covered thus. **3** short distance. **4** pattern of steps in dancing. **5** one of a series of actions. **6** level surface for placing the foot on in climbing. **7** stage in a scale. **8** (*pl.*) stepladder. □ **in step 1** stepping in time with others. **2** conforming. **mind** or **watch one's step** take care. **step by step** gradually. **step in 1** intervene. **2** enter. **step up** increase.

■ *v.* **1** pace, stride, tread, walk. ● *n.* **1,2** footstep, pace, stride. **5** action, initiative, measure, move. **6** rung, stair, tread; (**steps**) staircase, stairs, stairway. **7** level, stage. □ **step up** augment, escalate, increase, intensify, raise, redouble, strengthen.

step- *pref.* related by re-marriage of a parent, as **stepfather**, **stepmother**, **stepson**, etc.

stepladder *n.* short ladder with a supporting framework.

steppe *n.* grassy plain, esp. in south-east Europe and Siberia.

stepping-stone *n.* **1** raised stone for stepping on in crossing a stream etc. **2** means of progress.

stereo *n.* (*pl.* **-os**) **1** stereophonic sound or record player etc. **2** stereoscopic effect.

stereophonic *adj.* using two transmission channels so as to give the effect of naturally distributed sound.

stereoscopic *adj.* giving a three-dimensional effect.

stereotype *n.* standardized conventional idea or character etc. ● *v.* standardize, cause to conform to a type.

sterile *adj.* **1** barren. **2** free from micro-organisms. □ **sterility** *n.*

> ■ **1** arid, barren, infertile, unfruitful, unproductive. **2** antiseptic, aseptic, clean, disinfected, germ-free, sanitary, sterilized.

sterilize *v.* **1** make sterile. **2** deprive of the power of reproduction. □ **sterilization** *n.*, **sterilizer** *n.*

> ■ **1** clean, cleanse, disinfect, fumigate, pasteurize, purify.

sterling *n.* British money. ● *adj.* **1** of standard purity. **2** excellent.

stern[1] *adj.* not kind or cheerful. □ **sternly** *adv.*, **sternness** *n.*

> ■ austere, dour, forbidding, grave, grim, serious, solemn, sombre, stony, unsmiling; hard, harsh, rigorous, severe, strict, stringent, tough.

stern[2] *n.* rear of a ship or aircraft.

sternum *n.* breastbone.

steroid *n.* any of a group of organic compounds that includes certain hormones.

stertorous *adj.* making a snoring or rasping sound.

stethoscope *n.* instrument for listening to heart, lungs, etc.

stetson *n.* hat with a wide brim and high crown.

stevedore *n.* docker.

stew *v.* **1** cook by simmering in a closed vessel. **2** (*colloq.*) swelter. ● *n.* **1** dish made by stewing. **2** (*colloq.*) state of great anxiety.

steward *n.* **1** person employed to manage an estate etc. **2** passengers' attendant on a ship, aircraft, or train. **3** official at a race meeting or show etc. □ **stewardess** *n.* female steward on a ship etc.

stick[1] *n.* **1** thin piece of wood. **2** thing shaped like this for use as a support, weapon, etc. **3** implement used to propel the ball in hockey, polo, etc. **4** (*colloq.*) criticism.

> ■ **1** branch, stalk, switch, twig. **2** crook, pole, rod, staff, stake, wand; cane, walking stick; baton, club, cudgel, truncheon.

stick[2] *v.* (**stuck**) **1** thrust (a thing) into something. **2** (*colloq.*) put. **3** fix or be fixed by glue or suction etc. **4** jam. **5** (*colloq.*) remain in a specified place. **6** (*colloq.*) endure. □ **stick-in-the-mud** *n.* person who will not adopt new ideas etc. **stick out 1** stand above the surrounding surface. **2** be conspicuous. **stick to 1** remain faithful to. **2** keep to (a subject or position etc.). **stick together** (*colloq.*) remain united. **stick to one's guns** not yield. **stick up for** (*colloq.*) stand up for.

> ■ **1** dig, jab, plunge, poke, push, thrust. **2** deposit, dump, park, place, put, set, *colloq.* shove. **3** affix, attach, cement, fasten, fix, glue, gum, paste, solder. **4** jam, lodge, wedge. **5** linger, remain, stay. **6** abide, bear, endure, put up with, stand, stomach, take, tolerate. □ **stick-in-the-mud** conservative, fogey, *sl.* fuddy-duddy. **stick out 1** beetle (out), bulge (out), jut (out), poke out, project, protrude. **stick up for** defend, stand by, stand up for, support, take the side of.

sticker *n.* adhesive label or sign.

sticking plaster adhesive fabric for covering small cuts.

stickleback *n.* small fish with sharp spines on its back.

stickler *n.* **stickler for** person who insists on something.

sticky *adj.* (**-ier, -iest**) **1** sticking to what is touched. **2** humid. **3** (*colloq.*) difficult, unpleasant. □ **stickily** *adv.*, **stickiness** *n.*

> ■ **1** adhesive, gummed; gluey, glutinous, *colloq.* gooey, gummy, tacky, viscid, viscous. **2** clammy, close, damp, humid, muggy, oppressive, steamy, sultry.

stiff *adj.* **1** not bending or moving easily. **2** difficult. **3** formal in manner. **4** (of wind) blowing briskly. **5** (of a drink etc.) strong. **6** (of a penalty or price) severe. ● *n.* (*sl.*) corpse. □ **stiff-necked** *adj.* **1** obstinate. **2** haughty. **stiffly** *adv.*, **stiffness** *n.*

> ■ *adj.* **1** brittle, firm, hard, inelastic, inflexible, rigid, stiffened, unyielding; stretched, taut, tense. **2** arduous, challenging, difficult, exhausting, fatiguing, hard, laborious, onerous, tiring, tough. **3** chilly, cool, distant, formal, prim, standoffish, starchy, unfriendly. **4** brisk, forceful, fresh, gusty, strong, vigorous. **5** alcoholic, potent, powerful, strong. **6** drastic, harsh, punitive, severe, strict, stringent, tough; excessive, exorbitant, extortionate, high, *colloq.* steep, unreasonable.

stiffen *v.* make or become stiff. ◻ **stiffener** *n.*

■ clot, coagulate, congeal, harden, set, solidify, thicken; strengthen, tauten, tighten.

stifle *v.* **1** feel or cause to feel unable to breathe. **2** suppress.

■ **1** asphyxiate, choke, smother, suffocate. **2** curb, hold back, muffle, repress, restrain, silence, suppress, withhold; crush, destroy, quash, quell, stamp out.

stigma *n.* (*pl.* **-as**) **1** shame, disgrace. **2** part of a pistil.

■ **1** blot, discredit, disgrace, dishonour, shame, slur, smirch, taint.

stigmatize *v.* describe as unworthy or disgraceful.

stile *n.* steps or bars for people to climb over a fence.

stiletto *n.* (*pl.* **-os**) dagger with a narrow blade. ◻ **stiletto heel** long tapering heel of a shoe.

still[1] *adj.* **1** with little or no motion or sound. **2** (of drinks) not fizzy. ● *n.* **1** silence and calm. **2** photograph taken from a cinema film. ● *adv.* **1** without moving. **2** then or now as before. **3** nevertheless. **4** in a greater amount or degree. ◻ **still life** picture of inanimate objects. **stillness** *n.*

■ *adj.* **1** calm, immobile, inactive, inert, motionless, peaceful, placid, quiescent, serene, static, stationary, stock-still, tranquil, undisturbed, unmoving; noiseless, quiet, silent, soundless. ● *n.* **1** calm, calmness, hush, quiet, peace, peacefulness, quietness, serenity, stillness, tranquillity.

still[2] *n.* distilling apparatus.

stillborn *adj.* born dead.

stilted *adj.* stiffly formal.

stilts *n.pl.* **1** pair of poles with footrests for walking at a distance above the ground. **2** piles or posts supporting a building.

stimulant *adj.* stimulating. ● *n.* stimulating drug or drink.

stimulate *v.* **1** make more active. **2** apply a stimulus to. ◻ **stimulation** *n.*, **stimulator** *n.*, **stimulative** *adj.*

■ animate, arouse, encourage, excite, fire, galvanize, goad, inflame, inspire, kindle, prompt, provoke, rouse, spur, stir up, whet, whip up.

stimulus *n.* (*pl.* **-li**) something that rouses a person or thing to activity or energy.

■ encouragement, goad, impetus, incentive, incitement, inducement, inspiration, prod, spur.

sting *n.* **1** sharp wounding part or organ of an insect or plant etc. **2** wound made thus. **3** its infliction. **4** sharp bodily or mental pain. ● *v.* (**stung**) **1** wound or affect with a sting. **2** feel or cause sharp pain. **3** (*sl.*) overcharge, extort money from.

stingy *adj.* (**-ier**, **-iest**) spending, giving, or given grudgingly or in small amounts. ◻ **stingily** *adv.*, **stinginess** *n.*

■ cheese-paring, close, mean, miserly, niggardly, parsimonious, penny-pinching, *colloq.* tight, tight-fisted.

stink *n.* **1** offensive smell. **2** (*colloq.*) row or fuss. ● *v.* (**stank** or **stunk**) **1** give off a stink. **2** (*colloq.*) seem very unpleasant or dishonest.

stinker *n.* (*sl.*) **1** very objectionable person. **2** very difficult task.

stinking *adj.* **1** that stinks. **2** (*sl.*) very objectionable.

stint *v.* restrict to a small allowance. ● *n.* allotted amount of work.

stipend /stípend/ *n.* salary.

stipendiary *adj.* receiving a stipend.

stipple *v.* paint, draw, or engrave in small dots. ● *n.* this process or effect.

stipulate *v.* demand or insist (on) as part of an agreement. ◻ **stipulation** *n.*

■ demand, insist on, prescribe, require, specify. ◻ **stipulation** condition, demand, prerequisite, proviso, requirement, requisite, specification.

stir *v.* (**stirred**) **1** move. **2** mix (a substance) by moving a spoon etc. round in it. **3** stimulate, excite. ● *n.* **1** act or process of stirring. **2** commotion, excitement. ◻ **stir up 1** mix thoroughly. **2** incite. **3** arouse, stimulate.

■ *v.* **1** disturb, move, ruffle, rustle; budge, shift. **2** beat, blend, mix, stir up, whip, whisk. **3** animate, arouse, excite, galvanize, kindle, inspire, quicken, rouse, stimulate. ● *n.* **2** ado, bustle, commotion, disturbance, excitement, flurry, fuss, hullabaloo, hubbub, *colloq.* kerfuffle, to-do.

stirrup *n.* support for a rider's foot, hanging from the saddle.

stitch *n.* **1** single movement of a thread in and out of fabric in sewing, or of a needle or hook in knitting or crochet. **2** loop made thus. **3** method of making a stitch. **4** sudden pain in the side. ● *v.* **1** sew. **2** join or close with stitches. □ **in stitches** (*colloq.*) laughing uncontrollably.

stoat *n.* animal of the weasel family.

stock *n.* **1** amount of something available. **2** livestock. **3** lineage. **4** business company's capital, portion of this held by an investor. **5** standing or status. **6** liquid made by stewing bones, meat, fish, or vegetables. **7** plant with fragrant flowers. **8** plant into which a graft is inserted. **9** handle of a rifle. **10** cravat. **11** (*pl.*) framework on which a ship rests during construction. **12** (*pl.*) wooden frame with holes for a seated person's legs, used like the pillory. ● *adj.* **1** stocked and regularly available. **2** commonly used, hackneyed. ● *v.* **1** keep in stock. **2** provide with a supply. □ **in stock** available immediately for sale. **stock-car** *n.* car used in racing where deliberate bumping is allowed. **stock exchange** stock market. **stock-in-trade** *n.* all the requisites for carrying on a trade or business. **stock market 1** institution for buying and selling stocks and shares. **2** transactions of this. **stock-still** *adj.* motionless. **stocktaking** *n.* making an inventory of stock.

■ *n.* **1** accumulation, cache, hoard, quantity, reserve, stockpile, store, supply; commodities, goods, merchandise, wares. **2** animals, cattle, fatstock, livestock. **3** ancestry, blood, descent, family, line, lineage, origins, parentage, pedigree. **4** assets, capital, funds. **5** reputation, standing, status. ● *adj.* **1** ordinary, regular, standard. **2** banal, clichéd, commonplace, conventional, *colloq.* corny, customary, hackneyed, predictable, routine, set, standard, stereotyped, trite, unimaginative, unoriginal, usual. ● *v.* **1** carry, handle, have, keep, sell. **2** equip, furnish, provide, supply.

stockade *n.* protective fence.

stockbroker *n.* person who buys and sells shares for clients.

stockinet *n.* fine machine-knitted fabric used for underwear etc.

stocking *n.* close-fitting covering for the foot and leg.

stockist *n.* firm that stocks certain goods.

stockpile *n.* accumulated stock of goods etc. kept in reserve. ● *v.* accumulate a stockpile of.

stocky *adj.* (**-ier**, **-iest**) short and solidly built. □ **stockily** *adv.*

■ beefy, burly, chunky, dumpy, sturdy, thickset.

stodge *n.* (*colloq.*) stodgy food.

stodgy *adj.* (**-ier**, **-iest**) **1** (of food) heavy and filling. **2** dull.

stoic /stṓ-ik/ *n.* stoical person.

stoical *adj.* calm and uncomplaining. □ **stoically** *adv.*, **stoicism** *n.*

■ calm, forbearing, impassive, imperturbable, long-suffering, patient, philosophical, phlegmatic, resigned, stolid, uncomplaining, *colloq.* unflappable.

stoke *v.* tend and put fuel on (a fire etc.). □ **stoker** *n.*

stole[1] *n.* woman's wide scarf-like garment.

stole[2], **stolen** *see* **steal**.

stolid *adj.* not excitable. □ **stolidly** *adv.*, **stolidity** *n.*

stomach *n.* **1** internal organ in which the first part of digestion occurs. **2** abdomen. **3** appetite, inclination. ● *v.* endure, tolerate.

■ *n.* **2** abdomen, belly, insides, *colloq.* tummy; paunch, pot-belly. **3** appetite, hunger, inclination, liking, relish, taste. ● *v.* abide, accept, bear, brook, endure, put up with, stand, *colloq.* stick, swallow, take, tolerate.

stomp *v.* tread heavily.

stone *n.* **1** piece of rock. **2** stones or rock as a substance or material. **3** gem. **4** hard substance formed in the bladder or kidney etc. **5** hard case round the kernel of certain fruits. **6** (*pl.* **stone**) unit of weight, 14 lb. ● *adj.* made of stone. ● *v.* **1** pelt with stones. **2** remove stones from (fruit). □ **Stone Age** prehistoric period when weapons and tools were made of stone.

stonemason *n.* person who shapes stone or builds in stone.

stonewall *v.* give noncommittal replies.

stoneware *n.* heavy kind of pottery.

stony *adj.* (**-ier**, **-iest**) **1** full of stones. **2** unfeeling, unresponsive. □ **stonily** *adv.*

■ **1** gravelly, pebbly, rocky, shingly. **2** callous, cold, hard, hard-hearted, heartless, merciless, pitiless, uncaring, unfeeling, unresponsive.

stood *see* **stand**.

stooge *n.* **1** comedian's assistant. **2** person whose actions are controlled by another.

stool *n.* **1** movable seat without arms or raised back. **2** footstool. **3** (*pl.*) faeces. □ **stool-pigeon** *n.* decoy, esp. to trap a criminal.

stoop *v.* **1** bend forwards and down. **2** condescend. **3** descend to (something shameful). ● *n.* stooping posture.

■ *v.* **1** bend, bow, duck, lean. **2,3** condescend, deign, demean oneself, descend, lower oneself, sink.

stop *v.* (**stopped**) **1** put an end to movement, progress, or operation (of). **2** discontinue (an action etc.). **3** effectively hinder or prevent. **4** come to an end. **5** cease from motion etc. **6** not permit or supply as usual. **7** close by plugging or obstructing. ● *n.* **1** stopping, being stopped. **2** place where a train or bus etc. stops regularly. **3** thing that stops or regulates motion. **4** row of organ pipes providing tones of one quality, knob etc. controlling these. □ **stop press** late news inserted in a newspaper after printing has begun.

■ *v.* **1** bring to a halt, bring to an end, put an end to, terminate, wind up. **2** cease, conclude, desist from, discontinue, end, finish, give up, *colloq.* knock off, *colloq.* lay off, refrain from, *US* quit. **3** bar, frustrate, hamper, hinder, impede, prevent, restrain, thwart; arrest, check, stanch, stem; intercept, waylay. **4** be over, come to an end, draw to a close, end, finish, fizzle out. **5** come to a halt *or* standstill, draw up, halt, pause, pull up. **6** discontinue, interrupt, suspend, withhold. **7** block, clog, fill, jam, obstruct, plug, seal, stuff. ● *n.* **1** cessation, close, conclusion, end, finish, halt, standstill, stoppage, termination. **2** depot, station, terminal, terminus.

stopcock *n.* valve regulating the flow in a pipe etc.

stopgap *n.* temporary substitute.

stoppage *n.* **1** stopping. **2** obstruction.

stopper *n.* plug for closing a bottle etc.

stopwatch *n.* watch that can be instantly started and stopped.

storage *n.* **1** storing. **2** space for this.

store *n.* **1** supply of something available for use. **2** large shop. **3** storehouse. ● *v.* **1** collect and keep for future use. **2** deposit in a warehouse. □ **in store 1** being stored. **2** destined to happen. **set store by** value greatly.

■ *n.* **1** accumulation, cache, collection, hoard, mine, quantity, reserve, reservoir, stock, stockpile, supply. **2** department store, hypermarket, shop, supermarket. **3** depository, repository, storehouse, warehouse. ● *v.* **1** accumulate, amass, hoard, collect, lay up, keep, put by, *colloq.* salt away, *colloq.* stash (away), stockpile.

storehouse *n.* place where things are stored.

storeroom *n.* room used for storing things.

storey *n.* (*pl.* **-eys**) each horizontal section of a building.

stork *n.* large wading bird.

storm *n.* **1** disturbance of the atmosphere with strong winds and usu. rain or snow. **2** shower (of missiles etc.). **3** outbreak (of anger or abuse etc.). ● *v.* **1** rage, be violent. **2** attack or capture suddenly. □ **stormy** *adj.*

■ *n.* **1** squall, tempest; cyclone, hurricane, mistral, tornado, typhoon, whirlwind; cloudburst, downpour, thunderstorm; blizzard, snowstorm. **2** barrage, bombardment, hail, shower, torrent, volley. **3** eruption, explosion, outbreak, outburst. ● *v.* **1** bluster, explode, fume, rage, rant, rave, roar, thunder. **2** assault, attack, charge, raid, rush; capture, overrun, take. □ **stormy** blustery, choppy, gusty, rough, squally, tempestuous, thundery, wild, windy.

story *n.* account of real or imaginary events.

■ account, anecdote, narrative, tale, *colloq.* yarn; allegory, epic, fable, fairy story *or* tale, folk-tale, legend, myth, parable, romance, saga; article, composition, piece; detective story, mystery, thriller, *colloq.* whodunit.

stout *adj.* **1** thick and strong. **2** fat. **3** brave and resolute. ● *n.* strong dark beer. □ **stoutly** *adv.*, **stoutness** *n.*

■ *adj.* **1** solid, strong, sturdy, thick. **2** bulky, burly, corpulent, fat, overweight, plump, portly, rotund, stocky, tubby. **3** bold, brave, courageous, dauntless, fearless, gallant, intrepid, plucky, resolute, valiant, valorous; determined, firm, resolute, staunch, steadfast.

stove[1] *n.* **1** apparatus containing an oven. **2** closed apparatus used for heating rooms etc.

stove[2] *see* **stave**.

stow *v.* place in a receptacle for storage. □ **stow away** conceal oneself as a stowaway.

stowaway *n.* person who hides on a ship etc. so as to travel free of charge.

straddle *v.* **1** sit or stand (across) with legs wide apart. **2** stand or place (things) in a line across.

strafe *v.* attack with gunfire.

straggle *v.* **1** grow or spread untidily. **2** trail behind others in a march, race, etc. □ **straggler** *n.*, **straggly** *adj.*

straight *adj.* **1** extending or moving in one direction, not curved or bent. **2** correctly or tidily arranged, level. **3** in unbroken succession. **4** honest, frank. **5** without additions. ● *adv.* **1** in a straight line. **2** in the right direction. **3** without delay. **4** frankly. ● *n.* straight part. □ **go straight** live honestly after being a criminal. **straight away** without delay. **straight face** not smiling. **straight fight** contest between only two candidates. **straight off** (*colloq.*) immediately. **straightness** *n.*

■ *adj.* **1** direct, linear, unbending, undeviating, unswerving. **2** in order, neat, orderly, organized, shipshape, tidy; even, flat, level, horizontal; erect, perpendicular, upright, vertical. **3** consecutive, successive, unbroken. **4** candid, direct, forthright, frank, honest, straightforward; decent, equitable, fair, honourable, just, *colloq.* on the level, trustworthy, upright. **5** neat, pure, unadulterated, undiluted, unmixed. □ **straight away** at once, directly, immediately, instantly, post-haste, right away, *colloq.* straight off, without delay.

straighten *v.* make or become straight.

■ unbend, uncurl, unravel, untwist; arrange, neaten, put in order, tidy (up).

straightforward *adj.* **1** honest, frank. **2** without complications. □ **straightforwardly** *adv.*

■ **1** candid, direct, forthright, frank, honest, plain, straight, truthful. **2** easy, elementary, simple, uncomplicated.

strain[1] *n.* **1** lineage. **2** variety or breed of animals etc. **3** slight or inherited tendency.

strain[2] *v.* **1** make taut. **2** injure by excessive stretching or over-exertion. **3** make an intense effort (with). **4** sieve to separate solids from liquid. ● *n.* **1** straining, force exerted thus. **2** injury or exhaustion caused by straining. **3** severe demand on strength or resources. **4** part of a tune or piece of music. □ **strainer** *n.*

■ *v.* **1** stretch, tauten, tense, tighten. **2** damage, hurt, injure, rick, sprain, twist, wrench; overburden, overtax, push, stretch, tax, try. **3** exert oneself, labour, strive, struggle. **4** filter, riddle, sieve, sift. ● *n.* **1** force, pressure, stress, tension. **2** injury, rick, sprain, twist, wrench. **3** burden(s), demand(s), pressure, stress. **4** air, melody, tune.

strained *adj.* (of manner etc.) tense, not natural or relaxed.

■ artificial, awkward, forced, laboured, stiff, tense, unnatural.

strait *n.* **1** (also *pl.*) narrow stretch of water connecting two seas. **2** (*pl.*) difficult state of affairs. □ **strait-jacket** *n.* strong garment put round a violent person to restrain their arms. **strait-laced** *adj.* puritanical.

straitened *adj.* (of conditions) poverty-stricken.

strand[1] *n.* **1** single thread. **2** each of those twisted to form a cable or yarn etc. **3** lock of hair.

■ **1** fibre, filament, thread. **3** lock, tress, wisp.

strand[2] *n.* shore. ● *v.* **1** run aground. **2** leave in difficulties.

strange *adj.* **1** unusual, odd. **2** unfamiliar, alien. **3** unaccustomed. □ **strangely** *adv.*, **strangeness** *n.*

■ **1** abnormal, atypical, bizarre, curious, eccentric, exceptional, extraordinary, funny, inexplicable, kinky, mysterious, odd, offbeat, outlandish, out of the ordinary, peculiar, quaint, queer, singular, surprising, unaccountable, uncanny, uncommon, unheard-of, unnatural, unusual, weird. **2** alien, exotic, foreign, unfamiliar. **3** unaccustomed, unused.

stranger *n.* person new to a place, company, etc.

■ alien, foreigner, newcomer, outsider, visitor.

strangle *v.* **1** kill or be killed by squeezing the throat. **2** restrict the growth or utterance of. □ **strangler** *n.*

■ **1** choke, garrotte, throttle. **2** curb, gag, repress, restrict, stifle, suppress.

stranglehold *n.* strangling grip.

strangulation *n.* strangling.

strap *n.* strip of leather or other flexible material for holding things together or in place, or supporting something. ● *v.* (**strapped**) secure with strap(s). □ **strapped for** (*colloq.*) short of.

strapping *adj.* tall and robust.

strata *see* **stratum**.

stratagem *n.* cunning plan or scheme.

■ artifice, device, dodge, manoeuvre, plan, plot, ploy, ruse, scheme, subterfuge, tactic, trick, wile.

strategic *adj.* **1** of strategy. **2** (of weapons) very long-range. □ **strategically** *adv.*

strategist *n.* expert in strategy.

strategy *n.* **1** planning and directing of a campaign or war. **2** plan, policy.

■ **2** blueprint, plan, policy, programme, scheme.

stratify *v.* arrange in strata. □ **stratification** *n.*

stratosphere *n.* layer of the atmosphere about 10–60 km above the earth's surface.

stratum *n.* (*pl.* **strata**) one of a series of layers or levels.

straw *n.* **1** dry cut stalks of corn etc. **2** single piece of this. **3** thin tube for sucking liquid through. □ **straw poll** unofficial poll as a test of general feeling.

strawberry *n.* soft juicy edible red fruit with yellow seeds on the surface. □ **strawberry mark** red birthmark.

stray *v.* **1** leave one's group or proper place aimlessly, wander. **2** deviate from a subject. ● *adj.* **1** having strayed. **2** isolated, occasional. ● *n.* stray animal.

■ *v.* **1** drift, go astray, meander, range, roam, rove, straggle, wander. **2** deviate, digress, get off the subject, get sidetracked, go off at a tangent, ramble, wander. ● *adj.* **1** homeless, lost, strayed, vagrant, wandering. **2** accidental, chance, haphazard, isolated, lone, occasional, odd, random.

streak *n.* **1** thin line or band of a colour or substance different from its surroundings. **2** element, strain. **3** spell, series. ● *v.* **1** mark with streaks. **2** move very rapidly. **3** (*colloq.*) run naked in a public place. □ **streaker** *n.*, **streaky** *adj.*

■ *n.* **1** band, bar, line, mark, striation, strip, stripe. **2** element, strain, trace. **3** patch, period, run, series, spell, stretch. ● *v.* **1** line, mark, stripe. **2** dart, dash, flash, fly, hurtle, race, run, rush, scoot, shoot, speed, sprint, tear, whiz, zip, zoom.

stream *n.* **1** small river. **2** flow of liquid, things, or people. **3** direction of this. **4** section into which schoolchildren of the same level of ability are placed. ● *v.* **1** flow. **2** run with liquid. **3** float or wave at full length. **4** arrange (schoolchildren) in streams. □ **on stream** in active operation or production.

■ *n.* **1** beck, brook, *Sc.* burn, *US* creek, river, rivulet, tributary, watercourse. **2** cascade, current, deluge, flood, flow, gush, outpouring, rush, spate, spurt, surge, torrent. ● *v.* **1** course, flow, glide, run, slide; cascade, flood, gush, pour, shoot, spill, spout, spurt, surge. **3** blow, flap, float, flutter, waft, wave.

streamer *n.* **1** long narrow flag. **2** strip of ribbon or paper etc. attached at one or both ends.

streamline *v.* **1** give a smooth even shape that offers least resistance to movement through water or air. **2** make more efficient by simplifying.

street *n.* public road in a town or village lined with buildings. □ **street credibility** familiarity with a fashionable urban subculture.

■ avenue, boulevard, byway, crescent, cul-de-sac, lane, road, thoroughfare.

strength *n.* **1** quality of being strong. **2** its intensity. **3** advantageous skill or quality. **4** total number of people making up a group. □ **on the strength of** relying on as a basis or support.

■ **1** brawn, durability, endurance, force, might, mightiness, muscle, power, resilience, robustness, sinews, stamina, sturdiness, toughness, vigour; courage, firmness, fortitude, *colloq.* grit, *colloq.* guts, nerve, perseverance, pertinacity, resoluteness, resolution, tenacity, will-power; concentration, intensity, potency; cogency, efficacy, persuasiveness, weight. **3** advantage, asset, strong point, virtue.

strengthen *v.* make or become stronger.

■ bolster, brace, buttress, fortify, reinforce, shore up, toughen; heighten, increase, intensify, step up; boost, encourage, hearten, invigorate, rejuvenate, restore, revive; back up, confirm, corroborate, substantiate, support.

strenuous *adj.* making or requiring great effort. □ **strenuously** *adv.*, **strenuousness** *n.*

■ active, determined, dogged, energetic, indefatigable, tenacious, tireless, unflagging, vigorous, zealous; arduous, demanding, difficult, hard, exhausting, gruelling, laborious, taxing, tiring, toilsome, tough.

streptococcus *n.* (*pl.* **-ci**) bacterium causing serious infections.

stress *n.* **1** emphasis. **2** extra force used on a sound in speech or music. **3** pressure or tension exerted on an object **4** physical or mental strain. ● *v.* lay stress on.

■ *n.* **1** emphasis, importance, weight. **2** accent, accentuation, emphasis. **3** pressure, strain, tension. **4** anxiety, distress, pressure, strain, tension, trauma, worry. ● *v.* accent, accentuate, dwell on, emphasize, feature, highlight, point up, spotlight, underline.

stretch *v.* **1** pull out tightly or to a greater extent. **2** be able or tend to become stretched. **3** be continuous. **4** thrust out one's limbs. **5** strain. **6** exaggerate. ● *n.* **1** stretching. **2** ability to be stretched. **3** continuous expanse or period. ● *adj.* able to be stretched. □ **stretch a point** agree to something not normally allowed. **stretch out 1** extend (a limb etc.). **2** prolong. **3** relax by lying at full length. **stretchy** *adj.*

■ *v.* **1** draw *or* pull out, elongate, lengthen, widen; make taut, tense, tighten. **3** extend, range, reach, spread. **5** overburden, overtax, push, strain, tax. **6** exaggerate, overstate. ● *n.* **3** area, expanse, extent, range, reach, spread, sweep, tract; period, run, spell, stint. □ **stretch out 1** extend, hold out, outstretch, reach out. **2** draw out, extend, lengthen, prolong, protract, spin out. **3** lie (down), lounge, recline, sprawl.

stretcher *n.* framework for carrying a sick or injured person in a lying position.

strew *v.* (**strewed, strewn** or **strewed**) **1** scatter over a surface. **2** cover with scattered things.

striation *n.* each of a series of lines or grooves.

stricken *adj.* afflicted by an illness, shock, or grief.

strict *adj.* **1** precisely limited or defined. **2** without exception, complete. **3** requiring complete obedience or exactitude. □ **strictly** *adv.*, **strictness** *n.*

■ **1** accurate, exact, literal, precise, rigid. **2** absolute, complete, perfect, total, utter. **3** authoritarian, hard, harsh, inflexible, rigorous, severe, stern, stringent, tough, uncompromising.

stricture *n.* **1** severe criticism. **2** abnormal constriction.

stride *v.* (**strode, stridden**) walk with long steps. ● *n.* **1** single long step. **2** manner of striding. **3** (usu. in *pl.*) progress.

strident *adj.* loud and harsh. □ **stridently** *adv.*, **stridency** *n.*

■ discordant, grating, harsh, jarring, loud, noisy, rasping, raucous, screeching, shrill, unmelodious, unmusical.

strife *n.* quarrelling, conflict.

■ animosity, antagonism, arguing, bickering, conflict, disagreement, discord, disharmony, dissension, enmity, friction, hostility, ill will, quarrelling, squabbling.

strike *v.* (**struck**) **1** hit. **2** come or bring sharply into contact with. **3** attack suddenly. **4** (of a disease) afflict. **5** ignite (a match) by friction. **6** agree on (a bargain). **7** indicate (the hour) or be indicated by a sound. **8** find (oil etc.) by drilling. **9** occur to the mind of, produce a mental impression on. **10** take down (a flag or tent etc.). **11** stop work in protest. **12** assume (an attitude) dramatically. ● *n.* **1** act or instance of striking. **2** attack. **3** workers' refusal to work as a protest. □ **on strike** taking part in an industrial strike. **strike home** deal an effective blow. **strike out** cross out. **strike up 1** begin playing or singing. **2** start (a friendship etc.) casually.

■ *v.* **1** bash, *sl.* belt, *sl.* biff, *sl.* clobber, clout, cuff, hit, punch, rap, slap, smack, *old use* smite, spank, thump, thwack, *sl.* wallop, *colloq.* whack; batter, beat, hammer, pound. **2** bang into, bump into, collide with, crash into, dash against, hit, knock, run into. **3** assault, attack, storm. **4** affect, afflict, attack, hit. **5** ignite, light. **6** agree (on), conclude, make, reach, settle (on). **7** chime, sound. **9** come to, dawn on, occur to. **10** let down, lower, take down; dismantle. **11** go on strike, stop work, walk out. **12** affect, assume, display. ● *n.* **2** assault, attack, blitz, incursion, invasion, offensive, onslaught, raid, sortie. □ **strike out** blot out, cross out, delete, erase, obliterate, rub out, score out.

strikebound *adj.* immobilized by a workers' strike.

striker *n.* **1** person or thing that strikes. **2** worker who is on strike. **3** footballer whose main function is to try to score goals.

striking *adj.* **1** sure to be noticed. **2** impressive. □ **strikingly** *adv.*

■ **1** conspicuous, marked, noticeable, salient, prominent, unmistakable. **2** imposing, impressive, magnificent, marvellous, outstanding, remarkable, splendid, stunning, superb, wonderful.

string *n.* **1** narrow cord. **2** stretched piece of catgut or wire etc. in a musical instrument, vibrated to produce tones. **3** set of things strung together. **4** series of people or things. **5** (*pl.*) conditions insisted upon. **6** (*pl.*) stringed instruments in an orchestra etc. ● *v.* (**strung**) **1** fit or fasten with string(s). **2** thread on a string. □ **pull strings** use one's influence. **string along** (*colloq.*) **1** deceive. **2** go along (with). **string out** spread out on a line. **string up 1** hang up on string. **2** kill by hanging. **3** make tense.

■ *n.* **1** cord, thread, twine; lace, thong, tie; lead, leash. **4** chain, column, concatenation, file, line, procession, queue, row, sequence, series, succession, train.

stringent *adj.* strict, with firm restrictions. □ **stringently** *adv.*, **stringency** *n.*

stringy *adj.* **1** like string. **2** fibrous, tough.

strip[1] *v.* (**stripped**) **1** remove (clothes, coverings, or parts etc.). **2** pull or tear away (from). **3** undress. **4** deprive, e.g. of property or titles. □ **stripper** *n.*

■ **1** unclothe, undress; excoriate, flay, peel, skin; defoliate; dismantle, take apart, take to pieces. **2** peel off, remove, take off. **3** disrobe, get undressed, remove *or* take off one's clothes, undress (oneself). **4** deprive, dispossess, divest.

strip[2] *n.* long narrow piece or area. □ **comic strip, strip cartoon** sequence of cartoons. **strip light** tubular fluorescent lamp.

■ band, bar, belt, ribbon, sliver, stripe, swath.

stripe *n.* **1** long narrow band on a surface, differing in colour or texture from its surroundings. **2** chevron on a sleeve, indicating rank. □ **striped** *adj.*, **stripy** *adj.*

■ **1** band, bar, line, streak, striation, strip.

stripling *n.* a youth.

striptease *n.* entertainment in which a performer gradually undresses.

strive *v.* (**strove, striven**) **1** make great efforts. **2** struggle or contend.

■ **1** attempt, endeavour, exert oneself, make an effort, seek, strain, struggle, take pains, try, work. **2** battle, compete, contend, fight, struggle.

strobe *n.* (*colloq.*) stroboscope.

stroboscope *n.* apparatus for producing a rapidly flashing bright light. □ **stroboscopic** *adj.*

strode *see* **stride**.

stroke[1] *n.* **1** act of striking something. **2** single movement, action, or effort. **3** particular sequence of movements (e.g. in swimming). **4** mark made by a movement of a pen or paintbrush etc. **5** sound made by a clock striking. **6** sudden loss of ability to feel and move, caused by rupture or blockage of the brain artery.

■ **1** blow, hit, knock, rap, slap, smack, *colloq.* swipe, tap, thump, thwack, *sl.* wallop, *colloq.* whack.

stroke[2] *v.* pass the hand gently along the surface of. ● *n.* act of stroking.

■ *v.* caress, fondle, pat, pet.

stroll *v.* & *n.* walk in a leisurely way. □ **stroller** *n.*

■ *v.* & *n.* amble, meander, saunter, ramble, walk, wander.

strong *adj.* **1** capable of exerting or resisting great power. **2** powerful in numbers, resources, etc. **3** (of emotions, opinons, etc.) firmly held. **4** forceful or powerful in effect. **5** having a great effect on the senses, intense. **6** concentrated, containing much alcohol. **7** having a specified number of members. ● *adv.* strongly. □ **strong language** forcible language, swearing. **strong-minded** *adj.* determined. **strongly** *adv.*

■ *adj.* **1** durable, hard-wearing, indestructible, solid, stout, sturdy, substantial, tough, unbreakable; brawny, burly, hardy, husky, lusty, mighty, muscular, powerful, resilient, robust, sinewy, stalwart, strapping, vigorous, wiry; fit, hale, healthy, well. **2** formidable, great, mighty, powerful, redoubtable. **3** decided, definite, deep, earnest, fervent, fierce, firm, heartfelt, intense, keen, passionate, profound, unshakeable, unwavering, vehement. **4** forceful, influential, powerful, vigorous; cogent, compelling, convincing, persuasive, potent, weighty. **5** bright, dazzling, glaring, intense, vivid; acrid, piquant, pungent, sharp, spicy. **6** concentrated, undiluted; alcoholic, stiff.

stronghold *n.* **1** fortified place. **2** centre of support for a cause.

strongroom *n.* room designed for safe storage of valuables.

strontium *n.* silver-white metallic element. □ **strontium 90** its radioactive isotope.

strove *see* **strive**.

struck *see* **strike**.

structure *n.* **1** constructed unit. **2** way a thing is constructed, organized, etc. □ **structural** *adj.*, **structurally** *adv.*

■ **1** building, construction, edifice. **2** arrangement, composition, configuration, construction, design, form, formation, make-up, organization, shape.

struggle *v.* **1** make vigorous efforts to get free. **2** make great efforts under difficulties. **3** fight strenuously. **4** make one's way with difficulty. ● *n.* **1** act or spell of struggling. **2** vigorous effort. □ **struggle along** or **on** manage to survive in spite of difficulties.

■ *v.* **1, 2** attempt, endeavour, exert oneself, labour, strain, strive, toil, try, wrestle. **3** battle, contend, fight, grapple, wrestle; scuffle, tussle. **4** flounder, scramble, toil. ● *n.* **1** battle, competition, contest, fight, scrimmage, scuffle, tussle. **2** effort, endeavour, exertion, travail.

strum *v.* (**strummed**) play unskilfully or monotonously on (a musical instrument). ● *n.* sound made by strumming.

strung *see* **string**.

strut *n.* **1** bar of wood or metal supporting something. **2** strutting walk. ● *v.* (**strutted**) walk in a pompous self-satisfied way.

strychnine /strîkneen/ *n.* bitter highly poisonous substance.

stub *n.* **1** short stump. **2** counterfoil of a cheque or receipt etc. ● *v.* (**stubbed**) strike (one's toe) against a hard object. □ **stub out** extinguish (a cigarette) by pressure.

stubble *n.* **1** cut stalks of corn left in the ground after harvest. **2** short stiff hair, esp. on an unshaven face. □ **stubbly** *adj.*

stubborn *adj.* obstinate. □ **stubbornly** *adv.*, **stubbornness** *n.*

■ adamant, *colloq.* bloody-minded, headstrong, inflexible, intractable, intransigent, obdurate, obstinate, persistent, pertinacious, pigheaded, recalcitrant, refractory, self-willed, stiff-necked, uncompromising, uncooperative, unyielding, wilful.

stubby *adj.* (**-ier**, **-iest**) short and thick. □ **stubbiness** *n.*

stucco *n.* plaster or cement for coating walls or moulding into decorations. □ **stuccoed** *adj.*

stuck *see* **stick**[2]. *adj.* unable to move. □ **stuck-up** *adj.* (*colloq.*) **1** conceited. **2** snobbish.

stud[1] *n.* **1** projecting nail-head or similar knob on a surface. **2** device for fastening e.g. a detachable shirt-collar. ● *v.* (**studded**) **1** decorate with studs or precious stones. **2** strengthen with studs.

stud[2] *n.* **1** horses kept for breeding. **2** place keeping these.

student *n.* person engaged in studying something, esp. at a college or university.

■ fresher, undergraduate; pupil, scholar, schoolboy, schoolchild, schoolgirl; apprentice, trainee.

studied *adj.* deliberate and artificial.

■ affected, calculated, conscious, contrived, feigned, forced, intentional, purposeful, wilful.

studio *n.* (*pl.* **-os**) **1** workroom of a painter, photographer, etc. **2** place for making films, recordings, or broadcasts. □ **studio flat** one-room flat with a kitchen and bathroom.

studious *adj.* **1** spending much time in study. **2** deliberate and careful. □ **studiously** *adv.*, **studiousness** *n.*

■ **1** academic, bookish, scholarly. **2** assiduous, careful, deliberate, diligent, industrious, meticulous, painstaking, sedulous, thorough.

study *n.* **1** process of acquiring information etc., esp. from books. **2** work presenting the results of studying. **3** investigation of a subject. **4** musical composition designed to develop a player's skill. **5** preliminary drawing. **6** room used for studying. ● *v.* **1** give one's attention to acquiring knowledge of (a subject). **2** examine attentively.

■ *n.* **1** cramming, learning, reading, research, *colloq.* swotting, work. **2** dissertation, essay, paper, thesis. **3** analysis, examination, exploration, investigation, review, survey. **5** drawing, sketch. **6** den, library. ● *v.* **1** learn (about), read; cram, revise, *colloq.* swot. **2** analyse, examine, explore, go over, inquire into, inspect, investigate, look into, monitor, observe, scrutinize, survey; look at, peruse, pore over, read, scan.

stuff *n.* **1** material. **2** unnamed things, belongings, subjects, etc. ● *v.* **1** pack tightly. **2** force or cram (a thing). **3** eat greedily.

■ *n.* **1** fabric, material, matter, substance. **2** articles, objects, things; accessories, accoutrements, belongings, *sl.* clobber, effects, equipment, gear, kit, paraphernalia, possessions, property, tackle, trappings. ● *v.* **1** fill, pack, pad. **2** cram, force, jam, press, ram, shove, squash, squeeze, thrust, wedge. **3** gobble, gorge, overeat, scoff.

stuffing *n.* **1** padding used to fill something. **2** mixture put inside poultry etc., before cooking.

stuffy *adj.* (**-ier, -iest**) **1** lacking fresh air or ventilation. **2** dull. **3** (*colloq.*) prim, narrow-minded. □ **stuffily** *adv.*, **stuffiness** *n.*

■ **1** airless, close, frowzy, fuggy, fusty, oppressive, stale, stifling, suffocating, unventilated. **2** boring, dreary, dull, tedious, uninteresting. **3** conventional, *sl.* fuddy-duddy, narrow-minded, old-fashioned, priggish, prim, prudish, strait-laced.

stultify *v.* impair, make ineffective. □ **stultification** *n.*

stumble *v.* **1** trip and lose one's balance. **2** walk with frequent stumbles. **3** make mistakes in speaking etc. ● *n.* act of stumbling. □ **stumbling block** obstacle, difficulty.

■ *v.* **1** fall, lose one's balance, slip, trip. **2** falter, lurch, stagger, teeter, totter. **3** blunder, falter, stammer, stutter. □ **stumbling block** bar, barrier, hindrance, hurdle, impediment, obstacle, obstruction; catch, difficulty, drawback, hitch, snag.

stump *n.* **1** base of a tree left in the ground when the rest has gone. **2** similar remnant of something cut, broken, or worn down. **3** one of the uprights of a wicket in cricket. ● *v.* **1** walk stiffly or noisily. **2** (*colloq.*) baffle. □ **stump up** (*colloq.*) pay over (money required).

stumpy *adj.* (**-ier, -iest**) short and thick. □ **stumpiness** *n.*

stun *v.* (**stunned**) **1** knock senseless. **2** astound.

■ **1** daze, knock out, stupefy. **2** amaze, astonish, astound, bowl over, dumbfound, *colloq.* flabbergast, overcome, overwhelm, shock, stagger, stupefy, transfix.

stung *see* **sting**.

stunk *see* **stink**.

stunner *n.* (*colloq.*) stunning person or thing.

stunning *adj.* (*colloq.*) very impressive or attractive. □ **stunningly** *adv.*

stunt[1] *v.* hinder the growth or development of.

■ hamper, hinder, impede, inhibit, retard, slow (down).

stunt[2] *n.* something unusual or difficult done as a performance.

stupefy *v.* **1** dull the wits or senses of. **2** stun. □ **stupefaction** *n.*

stupendous *adj.* **1** amazing. **2** exceedingly great. □ **stupendously** *adv.*

■ **1** amazing, astonishing, astounding, breathtaking, extraordinary, marvellous, miraculous, phenomenal, remarkable, staggering, wonderful. **2** colossal, enormous, gigantic, huge, immense, massive, prodigious.

stupid *adj.* **1** not clever or intelligent. **2** showing lack of good judgement. **3** in a stupor. □ **stupidly** *adv.*, **stupidity** *n.*

■ **1** bovine, brainless, dense, *colloq.* dim, doltish, dopey, dull, *colloq.* dumb, foolish, obtuse, *colloq.* thick, unintelligent, witless. **2** absurd, asinine, *sl.* barmy, *sl.* batty, *colloq.* crack-brained, crazy, daft, fatuous, foolhardy, foolish, hare-brained, idiotic, ill-advised, imprudent, inane, insane, laughable, ludicrous, lunatic, mad, mindless, nonsensical, ridiculous, risible, senseless, silly, unthinking, unwise, wild. **3** dazed, insensible, stunned, stupefied, unconscious.

stupor *n.* dazed almost unconscious condition.

sturdy *adj.* (**-ier, -iest**) **1** strongly built, hardy, **2** vigorous and determined. □ **sturdily** *adv.*, **sturdiness** *n.*

■ **1** durable, solid, sound, strong, stout, substantial, tough, well-made; brawny, burly, hardy, husky, muscular, powerful, robust, stocky, strapping. **2** determined, firm, resolute, stalwart, staunch, steadfast, unswerving, unwavering, vigorous.

sturgeon *n.* (*pl.* **sturgeon**) large shark-like fish yielding caviare.

stutter *v.* & *n.* stammer.

sty[1] *n.* pigsty.

sty[2] *n.* inflamed swelling on the edge of the eyelid.

style *n.* **1** manner of writing, speaking, or doing something. **2** design. **3** elegance. **4** fashion in dress etc. ● *v.* design, shape, or arrange, esp. fashionably. □ **in style** elegantly, luxuriously.

■ *n.* **1** approach, manner, method, technique, way; phraseology, wording. **2** cut, design, genre, kind, make, manner, sort, type, variety. **3** chic, elegance, panache, polish, sophistication, stylishness, taste, tastefulness. **4** craze, fad, fashion, mode, trend, vogue. ● *v.* arrange, cut, design, make, shape, tailor.

stylish *adj.* fashionable, elegant. □ **stylishly** *adv.*, **stylishness** *n.*

■ chic, dapper, dressy, elegant, fashionable, modish, smart, snappy, *sl.* snazzy, *colloq.* trendy.

stylist *n.* **1** person who has a good style. **2** person who styles things.

stylistic *adj.* of literary or artistic style. □ **stylistically** *adv.*

stylized *adj.* painted, drawn, etc. in a conventional style.

stylus *n.* (*pl.* **-uses** or **-li**) needle-like device for cutting or following a groove in a record.

stymie *v.* (**stymieing**) thwart.

styptic *adj.* checking bleeding by causing blood vessels to contract.

suave /swaav/ *adj.* smooth-mannered, polite, sophisticated. □ **suavely** *adv.*, **suavity** *n.*

■ charming, courteous, civilized, cultivated, debonair, gracious, polished, polite, smooth-mannered, sophisticated, urbane.

sub *n.* (*colloq.*) **1** submarine. **2** subscription. **3** substitute.

sub- *pref.* **1** under. **2** subordinate.

subaltern *n.* army officer below the rank of captain.

subatomic *adj.* **1** smaller than an atom. **2** occurring in an atom.

subcommittee *n.* committee formed from some members of a main committee.

subconscious *adj.* & *n.* (of) our own mental activities of which we are not aware. □ **subconsciously** *adv.*

■ *adj.* hidden, latent, repressed, subliminal, suppressed, unconscious; instinctive, intuitive.

subcontinent *n.* large land mass forming part of a continent.

subcontract *v.* give or accept a contract to carry out all or part of another contract. □ **subcontractor** *n.*

subculture *n.* culture within a larger one.

subcutaneous *adj.* under the skin.

subdivide *v.* divide (a part) into smaller parts. □ **subdivision** *n.*

subdue *v.* **1** bring under control. **2** make quieter or less intense.

■ **1** conquer, control, crush, defeat, gain the upper hand over, get the better of, master, overpower, put down, quash, quell, repress, subjugate, suppress, tame, triumph over, vanquish. **2** moderate, mute, quieten, soften, temper, tone down.

subhuman *adj.* **1** less than human. **2** not fully human.

subject[1] /súbjikt/ *adj.* not politically independent. ● *n.* **1** person subject to a particular political rule or ruler. **2** person or thing being discussed or studied. **3** word(s) in a sentence that name who or what does the action of the verb. □ **subject-matter** *n.* matter treated in a book or speech etc. **subject to 1** conditional on. **2** liable or exposed to. **subjection** *n.*

■ *n.* **1** citizen, national. **2** guinea-pig, patient; subject-matter, topic; area, course of study, discipline, field. □ **subject to 1** conditional on, contingent on, dependent on. **2** at the mercy of, exposed to, liable to, open to, prone to, susceptible to, vulnerable to.

subject[2] /səbjékt/ *v.* subjugate. □ **subject to** cause to undergo.

subjective *adj.* dependent on personal taste or views etc. □ **subjectively** *adv.*

■ idiosyncratic, individual, personal; biased, partial, prejudiced.

subjugate *v.* conquer, bring into subjection. □ **subjugation** *n.*

■ conquer, crush, enslave, master, overpower, quash, subdue, subject, suppress, vanquish.

sublet *v.* (**sublet, subletting**) let (rooms etc. that one holds by lease) to a tenant.

sublimate *v.* divert the energy of (an emotion or impulse) into a culturally higher activity. □ **sublimation** *n.*

sublime *adj.* of the most admirable kind, causing awe and reverence. □ **sublimely** *adv.*, **sublimity** *n.*

■ awe-inspiring, exalted, glorious, great, lofty, magnificent, majestic, noble, splendid, transcendent.

subliminal *adj.* below the level of conscious awareness.

sub-machine-gun *n.* lightweight machine-gun held in the hand.

submarine *adj.* under the surface of the sea. ● *n.* vessel that can operate under water.

submerge *v.* **1** put or go below the surface of water or other liquid. **2** flood. □ **submersion** *n.*

■ **1** dip, douse, dunk, immerse, soak, steep; descend, dive, plummet, plunge, sink. **2** bury, deluge, drown, engulf, flood, inundate, overwhelm, swamp.

submersible *adj.* able to submerge. ● *n.* submersible craft.

submission *n.* **1** submitting. **2** thing submitted. **3** obedience.

■ **1** acquiescence, capitulation, concession, giving in, surrender, yielding. **2** application, presentation, proposal, proposition, suggestion, tender. **3** compliance, deference, docility, meekness, obedience, passivity, submissiveness, timidity, tractability.

submissive *adj.* humble, obedient. □ **submissively** *adv.*, **submissiveness** *n.*

■ acquiescent, amenable, biddable, compliant, deferential, docile, humble, malleable, meek, obedient, passive, timid, tractable, unassertive, yielding; ingratiating, obsequious, servile, slavish, subservient, sycophantic.

submit *v.* (**submitted**) **1** yield to authority or control, surrender. **2** present for consideration. □ **submit to** subject to a process.

■ **1** bow, capitulate, cave in, give in *or* up *or* way, knuckle under, succumb, surrender, throw in the towel, yield; accede, agree, comply, concede, consent. **2** advance, offer, present, proffer, propose, put forward; hand *or* send in.

subordinate *adj.* /səbórdinət/ of lesser importance or rank. ● *n.* /səbórdinət/ person working under another's authority. ● *v.* /səbórdinayt/ make or treat as subordinate. □ **subordination** *n.*

■ *adj.* minor, secondary, subsidiary; inferior, junior, lesser, lower. ● *n.* aide, assistant, junior; inferior, *derog.* minion, servant, underling.

suborn *v.* induce by bribery to commit perjury or other crime.

subpoena /səbpéénə/ *n.* writ ordering a person to appear in a law court. ● *v.* (**subpoenaed**) serve a subpoena on.

subscribe *v.* **subscribe to 1** pay (a specified sum) for membership of an organization or receipt of a publication. **2** agree with an opinion etc. □ **subscriber** *n.*

subscription *n.* **1** sum of money contributed. **2** membership fee. **3** process of subscribing.

subsequent *adj.* occurring after. □ **subsequently** *adv.*

■ ensuing, following, future, later, next, succeeding, successive; consequent, resultant, resulting.

subservient *adj.* **1** subordinate. **2** servile. □ **subserviently** *adv.*, **subservience** *n.*

■ **1** secondary, subordinate, subsidiary. **2** fawning, ingratiating, obsequious, servile, *colloq.* smarmy, submissive, sycophantic, toadying, unctuous.

subside *v.* **1** become less intense. **2** (of water) sink to a lower or normal level. **3** (of ground) cave in. □ **subsidence** *n.*

■ **1** abate, calm (down), decrease, die (down), diminish, dwindle, ebb, lessen, *colloq.* let up, moderate, wane. **2** drop, go down, recede, sink. **3** cave in, collapse, give way.

subsidiary *adj.* **1** of secondary importance. **2** (of a business company) controlled by another. ● *n.* subsidiary thing.

■ *adj.* **1** ancillary, auxiliary, lesser, minor, secondary, subordinate.

subsidize *v.* pay a subsidy to or for. □ **subsidization** *n.*

■ back, finance, fund, invest in, sponsor, support, underwrite.

subsidy *n.* money given to support an industry etc. or to keep prices down.

subsist *v.* keep oneself alive, exist. □ **subsistence** *n.*

subsoil *n.* soil lying immediately below the surface layer.

subsonic *adj.* of or flying at speeds less than that of sound.

substance *n.* **1** particular kind of matter with more or less uniform properties. **2** essence of something spoken or written. **3** reality, solidity.

■ **1** material, matter, stuff. **2** core, essence, gist, heart, nub, pith, quintessence; import, meaning, point, purport, significance, signification. **3** actuality, concreteness, corporeality, reality, solidity.

substantial *adj.* **1** of solid material or structure. **2** of considerable amount, intensity, or validity. **3** wealthy. **4** in essentials. □ **substantially** *adv.*

■ **1** durable, solid, sound, strong, stout, sturdy, well-built. **2** ample, big, considerable, generous, handsome, large, respectable, sizeable; important, material, significant, valuable, worthwhile. 3 affluent, moneyed, prosperous, rich, successful, wealthy.

substantiate *v.* support with evidence. □ **substantiation** *n.*

■ attest, authenticate, back up, confirm, corroborate, prove, support, validate, verify.

substitute *n.* person or thing that acts or serves in place of another. ● *v.* **substitute for** act or cause to act as a substitute. □ **substitution** *n.*

■ *n.* alternative, deputy, locum, proxy, relief, replacement, reserve, stand-by, stand-in, surrogate, understudy. ● *v.* deputize for, fill in for, relieve, stand in for, take the place of; exchange for, replace with.

substructure *n.* underlying or supporting structure.

subsume *v.* bring or include under a particular classification.

subtenant *n.* person to whom a room etc. is sublet.

subterfuge *n.* trick used to avoid blame or defeat etc.

■ artifice, device, dodge, manoeuvre, ploy, ruse, scheme, stratagem, trick, wile.

subterranean *adj.* underground.

subtitle *n.* **1** subordinate title. **2** caption on a cinema film. ● *v.* provide with subtitle(s).

subtle /sútt'l/ *adj.* **1** hard to detect or describe. **2** making fine distinctions. **3** ingenious. □ **subtly** *adv.*, **subtlety** *n.*

■ **1** delicate, elusive, faint, slight, understated; fine, nice. **2** acute, astute, discerning, discriminating, perceptive, sensitive, shrewd. **3** clever, cunning, ingenious, sophisticated.

subtotal *n.* total of part of a group of figures.

subtract *v.* remove (a part, quantity, or number) from a greater one. □ **subtraction** *n.*

subtropical *adj.* of regions bordering on the tropics.

suburb *n.* residential area outside the central part of a town. □ **suburban** *adj.*, **suburbanite** *n.*

suburbia *n.* suburbs and their inhabitants.

subvention *n.* subsidy.

subvert *v.* destroy the authority of (a political system etc.). □ **subversion** *n.*, **subversive** *adj.*

■ destroy, overthrow, overturn, ruin, sabotage, undermine, wreck.

subway *n.* **1** underground passage. **2** (*US*) underground railway.

succeed *v.* **1** be successful. **2** take the place previously filled by. **3** come next in order.

■ **1** *colloq.* arrive, be a success, be successful, do well, flourish, get on, make good, make it, make the grade, prevail, prosper, thrive; be effective, *colloq.* do the trick, work. **2** follow, replace, supplant, take over from, take the place of.

success *n.* **1** favourable outcome. **2** attainment of one's aims, or of wealth, fame, etc. **3** successful person or thing.

successful *adj.* having success. □ **successfully** *adv.*

■ triumphant, victorious, winning; celebrated, eminent, famed, famous, renowned, well-known; fruitful, lucrative, money-making, profitable, remunerative; affluent, flourishing, prosperous, rich, thriving, wealthy, well-to-do.

succession *n.* **1** following in order. **2** series of people or things following each other. **3** succeeding to a throne or other position. □ **in succession** one after another.

■ **1,2** chain, course, cycle, line, order, progression, round, run, sequence, series, string. □ **in succession** consecutively, in a row, in turn, one after another, *colloq.* on the trot, successively.

successive *adj.* following in succession. □ **successively** *adv.*

successor *n.* person who succeeds another.

succinct /səksíngkt/ *adj.* brief, concise. □ **succinctly** *adv.*

■ brief, compact, concise, condensed, pithy, short, terse.

succour *v.* & *n.* help.

succulent *adj.* **1** juicy. **2** (of plants) having thick fleshy leaves. ● *n.* succulent plant. □ **succulence** *n.*

succumb *v.* give way to something overpowering.

■ bow, capitulate, give in, give way, submit, surrender, yield,

such *adj.* **1** of the same or that kind or degree. **2** so great or intense. ● *pron.*

that. □ **such-and-such** *adj.* particular but not now specified.

suchlike *adj.* of the same kind.

suck *v.* **1** draw (liquid or air etc.) into the mouth. **2** draw liquid from in this way. ● *n.* act or process of sucking. □ **suck in** absorb, involve (a person). **suck up to** (*colloq.*) treat sycophantically.

sucker *n.* **1** organ or device that can adhere to a surface by suction. **2** (*sl.*) person who is easily deceived.

suckle *v.* feed at the breast.

suckling *n.* unweaned child or animal.

sucrose *n.* sugar.

suction *n.* **1** sucking. **2** production of a partial vacuum so that external atmospheric pressure forces fluid etc. into the vacant space or causes adhesion.

sudden *adj.* happening or done quickly or without warning. □ **all of a sudden** suddenly. **suddenly** *adv.*, **suddenness** *n.*

■ surprising, unannounced, unanticipated, unexpected, unforeseen; abrupt, hurried, immediate, meteoric, quick, rapid, swift; hasty, impetuous, impulsive, precipitate, rash, snap. □ **suddenly** in a flash, in an instant, in a trice, instantly; abruptly, quickly, rapidly, speedily, swiftly; all of a sudden, out of the blue, unexpectedly, without warning.

suds *n.pl.* soapsuds.

sue *v.* (**suing**) take legal proceedings against.

suede /swayd/ *n.* leather with a velvety nap on one side.

suet *n.* hard fat round an animal's kidneys, used in cooking.

suffer *v.* **1** feel pain, discomfort, grief, etc. **2** undergo or be subjected to (pain, loss, damage, etc.). **3** tolerate. □ **suffering** *n.*

■ **1** agonize, be in pain *or* distress, hurt. **2** endure, experience, feel, go through, sustain, undergo. **3** abide, bear, endure, put up with, stand, stomach, take, tolerate. □ **suffering** agony, anguish, distress, grief, hardship, misery, pain, sorrow, torment, torture, tribulation, unhappiness, woe

sufferance *n.* **on sufferance** tolerated but only grudgingly.

suffice *v.* be enough (for).

■ answer, be enough *or* sufficient, do, serve.

sufficient *adj.* enough. □ **sufficiently** *adv.*, **sufficiency** *n.*

■ adequate, ample, enough.

suffix *n.* (*pl.* **-ixes**) letter(s) added at the end of a word to make another word.

suffocate *v.* **1** kill by stopping the breathing. **2** cause discomfort to by making breathing difficult. **3** be suffocated. □ **suffocation** *n.*

■ **1** asphyxiate, choke, smother, stifle.

suffrage *n.* right to vote in political elections.

suffuse *v.* spread throughout or over. □ **suffusion** *n.*

■ cover, flood, permeate, pervade, saturate, spread through *or* over.

sugar *n.* sweet crystalline substance obtained from the juices of various plants. □ **sugar beet** white beet from which sugar is obtained. **sugar cane** tall tropical plant from which sugar is obtained. **sugary** *adj.*

suggest *v.* **1** bring to mind. **2** propose (theory, plan, etc.).

■ **1** call to mind, evoke; hint (at), imply, indicate, insinuate, intimate. **2** advance, move, present, propound, propose, put forward, recommend; advise, counsel, urge.

suggestible *adj.* easily influenced. □ **suggestibility** *n.*

suggestion *n.* **1** suggesting. **2** thing suggested. **3** slight trace.

■ **2** piece of advice, proposal, proposition, recommendation, tip; implication, insinuation, intimation. **3** hint, shade, suspicion, tinge, touch, trace.

suggestive *adj.* **1** conveying a suggestion. **2** suggesting something indecent. □ **suggestively** *adv.*

suicidal *adj.* **1** of or involving suicide. **2** liable to commit suicide. □ **suicidally** *adv.*

suicide *n.* **1** intentional killing of oneself. **2** person who commits suicide. **3** act destructive to one's own interests. □ **commit suicide** kill oneself intentionally.

suit /so͞ot/ *n.* **1** set of clothing, esp. jacket and trousers or skirt. **2** any of the four sets into which a pack of cards is divided. **3** lawsuit. ● *v.* **1** make or be suitable or convenient for. **2** give a pleasing appearance upon.

■ *n.* **1** ensemble, outfit. **3** action, case, cause, lawsuit, proceedings. ● *v.* **1** accommodate, adapt, adjust, fit, make suitable, tailor; be acceptable (to), be suitable *or* convenient (for), please, satisfy; agree with. **2** become, befit, look good on.

suitable *adj.* right for the purpose or occasion. □ **suitably** *adv.*, **suitability** *n.*

■ apposite, appropriate, apt, becoming, befitting, fit, fitting, proper, right, seemly; acceptable, convenient, opportune, satisfactory; eligible, qualified.

suitcase *n.* rectangular case for carrying clothes.

suite /sweet/ *n.* **1** set of rooms or furniture. **2** retinue. **3** set of musical pieces.

sulk *v.* be sullen because of resentment or bad temper. ● *n.* (also **the sulks**) fit of sulking. □ **sulky** *adj.*, **sulkily** *adv.*

sullen *adj.* **1** gloomy and unresponsive. **2** dark and dismal. □ **sullenly** *adv.*, **sullenness** *n.*

■ **1** bad-tempered, brooding, gloomy, grumpy, ill-humoured, moody, morose, resentful, sulking, sulky, uncommunicative, unresponsive. **2** dark, dismal, gloomy, leaden, murky, sombre.

sully *v.* stain, blemish.

sulphate *n.* salt of sulphuric acid.

sulphide *n.* compound of sulphur and an element or radical.

sulphite *n.* salt of sulphurous acid.

sulphur *n.* pale yellow non-metallic element. □ **sulphurous** *adj.*

sulphuric acid strong corrosive acid.

sultan *n.* Muslim sovereign.

sultana *n.* **1** seedless raisin. **2** sultan's wife, mother, or daughter.

sultanate *n.* sultan's territory.

sultry *adj.* (**-ier, -iest**) **1** hot and humid. **2** (of a woman) passionate and sensual. □ **sultriness** *n.*

■ **1** close, hot, humid, muggy, oppressive, steamy, sticky, stifling, suffocating, sweltering.

sum *n.* **1** total. **2** amount of money. **3** problem in arithmetic. □ **sum total** total. **sum up 1** give the total of. **2** summarize. **3** form an opinion of.

■ **1** aggregate, sum total, total, whole. **2** amount, quantity. □ **sum up 1** add up, calculate, reckon, total, *colloq.* tot up. **2** encapsulate, *colloq.* recap, recapitulate, summarize. **3** assess, evaluate, form an opinion of, *colloq.* size up.

summarize *v.* make or be a summary of. □ **summarization** *n.*

■ abridge, condense, encapsulate, give a résumé *or* synopsis of, outline, précis, *colloq.* recap, recapitulate, sum up.

summary *n.* statement giving the main points of something. ● *adj.* brief, without details or formalities. □ **summarily** *adv.*

■ *n.* abridgement, abstract, digest, encapsulation, outline, précis, *colloq.* recap, recapitulation, résumé, synopsis. ● *adj.* abrupt, brief, hasty, hurried, perfunctory, quick, rapid, short, sudden.

summation *n.* **1** adding up. **2** summarizing.

summer *n.* warmest season of the year. □ **summer house** small building in a garden giving shade in summer. **summertime** *n.* summer. **summer time** period from March to October when clocks are advanced one hour. **summery** *adj.*

summit *n.* **1** highest point. **2** top of a mountain. **3** conference between heads of states.

■ **1** acme, climax, culmination, height, peak, pinnacle, zenith. **2** apex, crest, crown, peak, tip, top.

summon *v.* **1** send for (a person). **2** order to appear in a law court. **3** call together. □ **summon up** gather together (courage etc.).

■ **1** call upon, send for. **2** subpoena, summons. **3** assemble, call, convene, convoke, gather together, rally.

summons *n.* **1** command summoning a person. **2** written order to appear in a law court. ● *v.* serve with a summons.

sump *n.* **1** reservoir of oil in a petrol engine. **2** hole or low area into which liquid drains.

sumptuous *adj.* splendid and costly-looking. □ **sumptuously** *adv.*, **sumptuousness** *n.*

■ de luxe, gorgeous, luxurious, magnificent, opulent, palatial, plush, plushy, rich, splendid.

sun *n.* **1** star around which the earth travels. **2** light or warmth from this. **3** any fixed star. ● *v.* (**sunned**) expose to the sun.

sunbathe *v.* expose one's body to the sun. □ **sunbather** *n.*

sunbeam *n.* ray of sun.

sunburn *n.* inflammation from exposure to sun. □ **sunburnt** *adj.*

sundae /súnday/ *n.* dish of ice cream and fruit, nuts, syrup, etc.

Sunday school school for religious instruction of Christian children, held on Sundays.

sunder *v.* break or tear apart.

sundial *n.* device that shows the time by means of a shadow on a scaled dial.

sundown *n.* sunset.

sundry *adj.* various. ● *n.pl.* (**sundries**) various small items. □ **all and sundry** everyone.

■ *adj.* assorted, different, diverse, miscellaneous, mixed, various.

sunflower *n.* tall plant with large yellow flowers.

sung *see* **sing**.

sunk *see* **sink**.

sunken *adj.* lying below the level of the surrounding surface.

Sunni *n.* & *adj.* (*pl.* **Sunni** or **-is**) (person) belonging to a Muslim sect opposed to Shiites.

sunny *adj.* (**-ier**, **-iest**) **1** full of sunshine. **2** cheerful. □ **sunnily** *adv.*

■ **1** bright, clear, cloudless, fine, sunlit, sunshiny. **2** bright, bubbly, buoyant, cheerful, cheery, gay, happy, light-hearted, merry.

sunrise *n.* rising of the sun.

sunset *n.* **1** setting of the sun. **2** coloured sky associated with this.

sunshade *n.* **1** parasol. **2** awning.

sunshine *n.* direct sunlight. □ **sunshiny** *adj.*

sunspot *n.* **1** dark patch observed on the sun's surface. **2** (*colloq.*) place with a sunny climate.

sunstroke *n.* illness caused by too much exposure to sun.

super *adj.* (*colloq.*) excellent.

superb *adj.* of the most impressive or splendid kind. □ **superbly** *adv.*

■ *colloq.* divine, excellent, exceptional, exquisite, *colloq.* fabulous, *colloq.* fantastic, fine, first-class, first-rate, glorious, gorgeous, *colloq.* great, *colloq.* heavenly, impressive, magnificent, marvellous, outstanding, peerless, sensational, *colloq.* smashing, splendid, stupendous, *colloq.* super, superlative, *colloq.* wonderful.

supercharge *v.* increase the power of (an engine) by a device that forces extra air or fuel into it. □ **supercharger** *n.*

supercilious *adj.* haughty and superior. □ **superciliously** *adv.*, **superciliousness** *n.*

■ arrogant, condescending, contemptuous, disdainful, haughty, hoity-toity, lofty, patronizing, scornful, snobbish, *colloq.* snooty, *colloq.* stuck-up, superior.

superficial *adj.* **1** of or on the surface, not deep or penetrating. **2** having no depth of character. □ **superficially** *adv.*, **superficiality** *n.*

■ **1** exterior, external, outside, surface; cursory, hasty, hurried, perfunctory, quick, rapid, swift. **2** frivolous, shallow, trivial.

superfluous *adj.* more than is required. □ **superfluously** *adv.*, **superfluity** *n.*

■ excess, excessive, extra, redundant, surplus, unnecessary, unneeded; gratuitous, needless.

superhuman *adj.* exceeeding ordinary human capacity or power.

superimpose *v.* place on top of something else. □ **superimposition** *n.*

superintend *v.* supervise. □ **superintendence** *n.*

superintendent *n.* **1** supervisor. **2** police officer next above inspector.

superior *adj.* **1** higher in position or rank. **2** above average in quality etc. **3** showing that one feels wiser or better etc. than others. ● *n.* person or thing of higher rank, ability, or quality. □ **superiority** *n.*

■ *adj.* **1** higher, higher-ranking, senior. **2** better, choice, *colloq.* classier, excellent, exceptional, first-class, first-rate, outstanding, select, superlative, supreme. **3** condescending, disdainful, haughty, hoity-toity, lofty, patronizing, *colloq.* snooty, *colloq.* stuck-up, supercilious.

superlative *adj.* **1** of the highest quality. **2** of the grammatical form expressing 'most'. ● *n.* superlative form.

■ *adj.* **1** exceptional, first-class, first-rate, incomparable, matchless, outstanding, peerless, perfect, superb, superior, supreme, unequalled, unparalleled, unrivalled, unsurpassed.

superman *n.* man of superhuman powers.

supermarket *n.* large self-service shop.

supernatural *adj.* of or involving a power above the forces of nature. □ **supernaturally** *adv.*

■ magical, mysterious, mystic(al), occult, preternatural, psychic, uncanny, unearthly, unnatural, weird; ghostly, spectral.

supernumerary *adj.* & *n.* extra.

superpower *n.* extremely powerful nation.

superscript *adj.* written or printed above.

supersede *v.* take the place of.

■ displace, replace, succeed, supplant, take over from.

supersonic *adj.* of or flying at speeds greater than that of sound. ▫ **supersonically** *adv.*

superstition *n.* **1** belief in magical and similar influences. **2** idea or practice based on this. **3** widely held but wrong idea. ▫ **superstitious** *adj.*

superstore *n.* supermarket.

superstructure *n.* structure that rests on something else.

supervene *v.* occur as an interruption or a change. ▫ **supervention** *n.*

supervise *v.* direct and inspect. ▫ **supervision** *n.*, **supervisor** *n.*, **supervisory** *adj.*

■ be in charge of, control, direct, handle, head, manage, oversee, run, superintend, watch (over). ▫ **supervisor** *colloq.* boss, chief, director, foreman, *colloq.* gaffer, head, manager, overseer, superintendent, superior.

supine /sóopīn/ *adj.* **1** lying face upwards. **2** indolent.

supper *n.* evening meal, last meal of the day.

supplant *v.* oust and take the place of. ▫ **supplanter** *n.*

■ displace, eject, oust, remove, replace, supersede, take the place of, unseat.

supple *adj.* bending easily. ▫ **supply** *adv.*, **suppleness** *n.*

■ bendable, flexible, pliable, pliant, whippy; athletic, graceful, lissom, lithe, willowy.

supplement *n.* thing added as an extra part. ● *v.* provide or be a supplement to.

■ *n.* addition, adjunct; extra, surcharge; addendum, appendix, codicil, insert, postscript, rider. ● *v.* add to, augment, increase, top up.

supplementary *adj.* serving as a supplement.

■ accessory, added, additional, ancillary, extra, further.

suppliant /súpliənt/ *n.* & *adj.* (person) asking humbly for something.

supplicate *v.* make a humble petition to or for. ▫ **supplication** *n.*

supply *v.* **1** give or provide with, make available. **2** satisfy (a need). ● *n.* **1** supplying. **2** stock, amount provided or available.

■ *v.* **1** afford, contribute, donate, give, furnish, present, provide; equip, fix up, kit out, provision. **2** fulfil, meet, satisfy. ● *n.* **1** providing, provision, supplying. **2** cache, fund, hoard, reserve, reservoir, stock, stockpile, store.

support *v.* **1** bear the weight of. **2** keep from falling, sinking, or failing. **3** provide for. **4** help by one's approval or sympathy, or by giving money. **5** give corroboration to. **6** speak in favour of. **7** be a regular customer or a fan of. ● *n.* **1** act or instance of supporting. **2** person or thing that supports. ▫ **supporter** *n.*, **supportive** *adj.*

■ *v.* **1** bear, carry, hold (up), sustain, take. **2** bolster, brace, buttress, fortify, prop (up), reinforce, shore up, strengthen. **3** keep, look after, maintain, provide for, take care of. **4** aid, assist, buoy up, encourage, fortify, help, sustain; back (up), champion, stand by, stand up for, *colloq.* stick up for, uphold; finance, sponsor, subsidize, underwrite. **5** authenticate, back up, confirm, corroborate, endorse, ratify, substantiate, validate, verify. **6** advocate, recommend, second, speak in favour of. ● *n.* **1** aid, assistance, backing, encouragement, fortification, help, succour. **2** beam, brace, buttress, column, foundation, frame, guy, joist, pillar, post, prop, stanchion, strut, substructure, truss; keep, maintenance, subsistence, upkeep; finances, funding; comfort, mainstay, tower of strength. ▫ **supporter** adherent, admirer, advocate, assistant, backer, benefactor, champion, defender, devotee, enthusiast, helper, fan, follower, patron, promoter, sponsor, upholder. **supportive** caring, encouraging, helpful, sympathetic, understanding.

suppose *v.* **1** be inclined to think. **2** assume to be true. **3** consider as a proposal. ▫ **be supposed** be expected or required.

■ **1** believe, conjecture, fancy, guess, imagine, reckon, think. **2** assume, presuppose, presume, surmise, take for granted. **3** postulate, theorize.

supposedly *adv.* according to supposition.

supposition *n.* **1** process of supposing. **2** what is supposed.

■ **1** assumption, inference, postulation. **2** assumption, belief, conjecture, hypothesis, guess, inference, surmise, theory.

suppository *n.* solid medicinal substance placed in the rectum or vagina and left to melt.

suppress *v.* **1** put an end to the activity or existence of. **2** keep from being seen, heard, known, etc. □ **suppression** *n.*, **suppressor** *n.*

■ **1** *colloq.* crack down on, crush, extinguish, overcome, overpower, put an end to, put down, quash, quell, squash, stamp out, subdue. **2** contain, control, curb, repress, restrain, silence, smother, stifle, strangle, withhold; conceal, cover up, hide, keep secret; censor.

suppurate *v.* form pus, fester. □ **suppuration** *n.*

supra- *pref.* above, over.

supreme *adj.* highest in authority, rank, or quality. □ **supremely** *adv.*, **supremacy** *n.*

■ chief, first, foremost, greatest, highest, principal, sovereign, top, topmost, uppermost; best, incomparable, inimitable, matchless, outstanding, peerless, perfect, superlative, unequalled, unparalleled, unsurpassed.

supremo *n.* (*pl.* **-os**) supreme leader.

surcharge *n.* additional charge. ● *v.* make a surcharge on or to.

sure *adj.* **1** without doubt or uncertainty. **2** reliable, unfailing. ● *adv.* (*colloq.*) certainly. □ **make sure** act so as to be certain. **sure-footed** *adj.* never slipping or stumbling. **sureness** *n.*

■ *adj.* **1** assured, certain, confident, convinced, definite, positive. **2** accurate, dependable, foolproof, infallible, reliable, unerring, unfailing; inescapable, inevitable, unavoidable.

surely *adv.* **1** in a sure manner. **2** (used for emphasis) that must be right. **3** (as an answer) certainly.

surety *n.* **1** guarantee. **2** guarantor of a person's promise.

surf *n.* foam of breaking waves. ● *v.* engage in surfing.

surface *n.* **1** outside or outward appearance of something. **2** any side of an object. **3** top of a liquid or of the ground. ● *adj.* of or on the surface. ● *v.* **1** put a specified surface on. **2** come or bring to the surface. **3** become visible or known.

■ *n.* **1** façade, face, exterior, outside, top. **2** face, side. **3** meniscus. ● *adj.* exterior, external, outside, outward, superficial. ● *v.* **1** coat, top; concrete, pave, tarmac. **2, 3** appear, come to light, come up, crop up, emerge, materialize.

surfboard *n.* narrow board for riding over surf.

surfeit /súrfit/ *n.* too much, esp. of food or drink. ● *v.* **1** overfeed. **2** satiate.

■ *n.* excess, glut, over-abundance, superfluity, surplus. ● *v.* **1** gorge, overfeed, stuff. **2** glut, sate, satiate.

surfing *n.* sport of riding on a surfboard.

surge *v.* move forward in or like waves. ● *n.* **1** surging movement. **2** sudden occurrence or increase.

■ *v.* billow, eddy, heave, roll, swell; flow, gush, pour, rush, stream. ● *n.* **1** billowing, eddying, heaving, rolling, swell; gush, flood, rush, stream. **2** increase, rise, upsurge.

surgeon *n.* doctor qualified to perform surgical operations.

surgery *n.* **1** treatment by cutting or manipulation of affected parts of the body. **2** place where or times when a doctor or dentist or an MP etc. is available for consultation. □ **surgical** *adj.*, **surgically** *adv.*

surly *adj.* (**-ier**, **-iest**) bad-tempered and unfriendly. □ **surliness** *n.*

■ bad-tempered, cantankerous, churlish, crabby, cross, crotchety, crusty, disagreeable, grumpy, ill-tempered, irascible, irritable, rude, sour, splenetic, sulky, sullen, testy, unpleasant, unfriendly.

surmise *n.* conjecture. ● *v.* infer doubtfully.

■ *v.* assume, conclude, conjecture, deduce, fancy, feel, gather, guess, imagine, infer, presume, speculate, suppose, suspect, understand.

surmount *v.* overcome (a difficulty or an obstacle). □ **surmountable** *adj.*

surname *n.* family name.

surpass *v.* outdo, be better than.

■ beat, better, cap, eclipse, exceed, excel, outclass, outdo, outshine, outstrip, overshadow, top, transcend.

surplus *n.* amount left over after what is needed has been used.

■ excess, glut, superfluity, surfeit; balance, leftovers, remainder, residue, rest.

surprise *n.* **1** emotion aroused by something sudden or unexpected. **2** thing causing this. ● *v.* **1** cause to feel surprise. **2** come upon or attack unexpectedly. □ **take by surprise** affect with surprise.

■ *n.* **1** astonishment, incredulity, shock, stupefaction, wonder. **2** blow, bombshell, eye-opener, jolt, shock, *colloq.* shocker. ● *v.* **1** amaze, astonish, astound, bowl over, dumbfound, *colloq.* flabbergast, shock, stagger, startle, stun, stupefy, take aback, take by surprise. **2** ambush, pounce on, swoop on; catch napping, catch red-handed, catch unawares.

surrealism *n.* style of art and literature seeking to express what is in the subconscious mind. □ **surrealist** *n.*, **surrealistic** *adj.*

surrender *v.* **1** hand over, give into another's power or control, esp. under compulsion. **2** give oneself up. ● *n.* surrendering.

■ *v.* **1** cede, concede, forgo, forsake, give up, hand over, let go (of), part with, relinquish, renounce, yield. **2** admit defeat, bow, capitulate, cave in, give oneself up, submit, throw in the towel.

surreptitious *adj.* acting or done stealthily. □ **surreptitiously** *adv.*

■ clandestine, covert, furtive, secret, sly, stealthy, underhand.

surrogate *n.* deputy. □ **surrogate mother** woman who bears a child on behalf of another. **surrogacy** *n.*

surround *v.* come, place, or be all round, encircle. ● *n.* border.

■ *v.* circle, encircle, enclose, encompass, envelop, girdle, hedge in, hem in, ring; beset, besiege.

surroundings *n.pl.* things or conditions around a person or place.

■ background, conditions, element, environment, habitat, milieu, setting.

surveillance *n.* close observation.

survey *v.* /sərváy/ **1** look at and take a general view of. **2** examine the condition of (a building). **3** determine the area and features of (a piece of land). **4** investigate the behaviour, opinions, etc. of (a group of people). ● *n.* /súrvay/ **1** general view or consideration etc. **2** report produced by surveying.

■ *v.* **1** contemplate, examine, inspect, look at, observe, scan, scrutinize, view; appraise, assess, consider, evaluate, investigate, review, look into, study. **3** map out, measure, plot; explore, reconnoitre. **4** canvass, interview, question. ● *n.* **1** appraisal, assessment, consideration, contemplation, evaluation, examination, inspection, investigation, review, scrutiny, study.

surveyor *n.* person who surveys land or buildings, esp. professionally.

survival *n.* **1** surviving. **2** relic.

survive *v.* **1** continue to live or exist. **2** remain alive or in existence after. **3** remain alive in spite of (a danger, accident, etc.). □ **survivor** *n.*

■ **1** continue, endure, exist, keep going, last, live (on), remain, persist, subsist; pull through; cope, get by, make do, manage, struggle along *or* on. **2** outlast, outlive. **3** live through, stand, weather, withstand.

susceptible *adj.* impressionable. □ **susceptible to** likely to be affected by. **susceptibility** *n.*

■ credulous, gullible, impressionable, naive, suggestible.

sushi *n.* Japanese dish of flavoured balls of cold rice usu. garnished with fish.

suspect *v.* /səspékt/ **1** feel that something may exist or be true. **2** mistrust. **3** feel to be guilty but have no proof. ● *n.* /súspekt/ person suspected of a crime etc. ● *adj.* /súspekt/ suspected, open to suspicion.

■ *v.* **1** be inclined to think, believe, guess, fancy, feel, have a feeling, imagine, sense, suppose, think. **2** be suspicious of, disbelieve, distrust, doubt, have doubts *or* misgivings about, mistrust. ● *adj.* doubtful, dubious, fishy, questionable, shady, suspicious.

suspend *v.* **1** hang up. **2** keep from falling or sinking in air or liquid. **3** stop temporarily. **4** debar temporarily from a position, privilege, etc.

■ **1** hang (up), sling. **3** adjourn, defer, delay, hold in abeyance, interrupt, postpone, shelve.

suspender *n.* attachment to hold up a sock or stocking by its top.

suspense *n.* anxious uncertainty while awaiting an event etc.

■ anxiety, anxiousness, apprehension, excitement, expectation, expectancy, nervousness, tension, uncertainty.

suspension *n.* **1** suspending. **2** means by which a vehicle is supported on its axles. □ **suspension bridge** bridge suspended from cables that pass over supports at each end.

suspicion *n.* **1** suspecting. **2** unconfirmed belief. **3** mistrust. **4** slight trace.

■ **2** belief, feeling, guess, hunch, idea, impression, inkling, notion, premonition, presentiment. **3** distrust, doubt, misgiving, mistrust, scepticism, uncertainty, wariness.

4 hint, shade, suggestion, tinge, touch, trace.

suspicious *adj.* feeling or causing suspicion. ◻ **suspiciously** *adv.*

■ disbelieving, distrustful, doubtful, incredulous, mistrustful, sceptical, wary; dubious, fishy, shady, shifty, suspect.

sustain *v.* **1** support. **2** keep alive. **3** keep (a sound or effort) going continuously. **4** undergo. **5** endure. **6** uphold the validity of.

■ **1** bear, carry, hold, support, take. **2** feed, keep (alive), maintain, nourish. **3** continue, keep going, keep up, maintain. **4** be subjected to, experience, suffer, undergo. **5** endure, stand, tolerate, weather, withstand. **6** ratify, support, uphold, validate.

sustenance *n.* food, nourishment.

■ comestibles, eatables, food, foodstuffs, *sl.* grub, nourishment, nutriment, provisions, viands.

suture /so͞ochər/ *n.* **1** surgical stitching of a wound. **2** stitch or thread used in this. ● *v.* stitch (a wound).

svelte *adj.* slender and graceful.

SW *abbr.* **1** south-west. **2** south-western.

swab *n.* **1** mop or pad for cleansing, drying, or absorbing things. **2** specimen of a secretion taken with this. ● *v.* (**swabbed**) cleanse with a swab.

swaddle *v.* swathe in wraps or warm garments.

swag *n.* (*sl.*) loot.

swagger *v.* walk or behave self-importantly. ● *n.* this gait or manner.

■ *v.* parade, strut; boast, brag, show off, *colloq.* swank.

Swahili *n.* Bantu language widely used in E. Africa.

swallow[1] *v.* **1** cause or allow to go down one's throat. **2** work throat muscles in doing this. **3** take in and engulf or absorb. **4** accept. ● *n.* **1** act of swallowing. **2** amount swallowed.

■ *v.* **1** consume, devour, eat, ingest; bolt, gobble, guzzle, scoff, wolf; *colloq.* down, drink, gulp, imbibe, quaff, swill. **3** absorb, assimilate, consume, engulf, envelop. **4** accept, believe, credit, fall for.

swallow[2] *n.* small migratory bird with a forked tail.

swam *see* **swim**.

swamp *n.* marsh. ● *v.* **1** flood, drench or submerge in water. **2** overwhelm with numbers or quantity. ◻ **swampy** *adj.*

■ *n.* bog, fen, marsh, mire, morass, quagmire, slough. ● *v.* **1** deluge, flood, immerse, inundate, submerge; drench, soak. **2** deluge, engulf, flood, inundate, overload, overwhelm, snow under.

swan *n.* large usu. white waterbird with a long slender neck.

swank (*colloq.*) *n.* **1** boastful person or behaviour. **2** ostentation. ● *v.* behave with swank.

swansong *n.* person's last performance or achievement etc.

swap *n.* & *v.* (**swapped**) exchange.

swarm[1] *n.* large cluster of people, insects, etc. ● *v.* move in a swarm. ◻ **swarm with** be crowded or infested with.

■ *n.* army, drove, flock, herd, horde, host, mass, multitude, pack, throng. ● *v.* crowd, flock, pour, stream, surge, throng. ◻ **swarm with** abound in, be full of, be infested with, be overrun with, teem with.

swarm[2] *v.* **swarm up** climb by gripping with arms and legs.

swarthy *adj.* (**-ier**, **-iest**) having a dark complexion.

swashbuckling *adj.* & *n.* swaggering boldly. ◻ **swashbuckler** *n.*

swastika *n.* symbol formed by a cross with ends bent at right angles.

swat *v.* (**swatted**) hit hard with something flat. ◻ **swatter** *n.*

swatch *n.* sample(s) of cloth etc.

swath /swawth/ *n.* (*pl.* **swaths**) strip cut in one sweep or passage by a scythe or mowing-machine.

swathe *v.* wrap with layers of coverings.

■ bandage, envelop, muffle (up), shroud, swaddle, wrap.

sway *v.* **1** (cause to) lean unsteadily from side to side. **2** influence the opinions of. **3** waver in one's opinion. ● *n.* **1** swaying movement. **2** rule or control.

■ *n.* **1** bend, lean, move to and fro *or* from side to side, roll, swing; lurch, reel, rock, totter, wobble. **2** convince, influence, persuade, prevail on, talk into, win over; incline, move, swing. **3** fluctuate, oscillate, vacillate, waver. ● *n.* **2** authority, command, control, dominion, influence, jurisdiction, leadership, mastery, power, sovereignty.

swear *v.* (**swore**, **sworn**) **1** state or promise on oath. **2** state emphatically. **3**

use a swear word. □ **swear by** have great confidence in. **swear word** profane or indecent word.

■ **1** give one's word, pledge, promise, undertake, vow. **2** assert, avow, declare, insist, testify. **3** blaspheme, curse, *colloq.* cuss, utter profanities. □ **swear by** believe in, count on, have confidence *or* faith in, rely on, trust (in).

sweat *n.* **1** moisture given off by the body through the pores. **2** state of sweating. **3** moisture forming in drops on a surface. ● *v.* **1** exude sweat or as sweat. **2** be in a state of great anxiety. **3** work long and hard. □ **sweat-band** *n.* band of material worn to absorb or wipe away sweat. **sweated labour** labour of workers with poor pay and conditions. **sweaty** *adj.*

sweater *n.* jumper, pullover.

sweatshirt *n.* cotton sweater with sleeves.

sweatshop *n.* place employing sweated labour.

Swede *n.* native of Sweden.

swede *n.* large variety of turnip.

Swedish *adj.* & *n.* (language) of Sweden.

sweep *v.* (**swept**) **1** clear away with a broom or brush. **2** clean or clear (a surface) thus. **3** go smoothly and swiftly or majestically. **4** extend in a continuous line. ● *n.* **1** sweeping movement or line. **2** act of sweeping. **3** range or scope. **4** chimney sweep. **5** sweepstake. □ **sweep the board** win all the prizes.

sweeping *adj.* **1** wide in range or effect. **2** taking no account of particular cases.

■ **1** broad, comprehensive, exhaustive, extensive, general, radical, thorough, universal, wholesale, wide-ranging. **2** broad, generalized, inexact, oversimplified.

sweepstake *n.* form of gambling in which the money staked is divided among those who have drawn numbered tickets for the winners.

sweet *adj.* **1** tasting as if containing sugar. **2** fragrant. **3** melodious. **4** pleasant. **5** (*colloq.*) charming. ● *n.* **1** small shaped piece of sweet substance. **2** sweet dish forming one course of a meal. **3** beloved person. □ **sweet pea** climbing plant with fragrant flowers. **sweet tooth** liking for sweet things. **sweetly** *adv.*, **sweetness** *n.*

■ *adj.* **1** honeyed, sugared, sugary, sweetened, syrupy. **2** aromatic, balmy, fragrant, perfumed, scented. **3** dulcet, euphonious, harmonious, melodious, musical, silvery, tuneful. **4** agreeable, amiable, considerate, easygoing, friendly, genial, kind, nice, pleasant, thoughtful, warm. **5** appealing, attractive, charming, *colloq.* cute, endearing, lovable, winning, winsome. ● *n.* **1** (**sweets**) *US* candy, confectionery. **2** dessert, pudding.

sweetbread *n.* animal's thymus gland or pancreas used as food.

sweeten *v.* make or become sweet or sweeter.

sweetener *n.* **1** thing that sweetens. **2** (*colloq.*) bribe.

sweetheart *n.* either of a pair of people in love with each other.

swell *v.* (**swelled**, **swollen** or **swelled**) **1** (cause to) become larger from pressure within. **2** make or become greater in amount or intensity. ● *n.* **1** act or state of swelling. **2** heaving of the sea. **3** a crescendo. **4** (*colloq.*) fashionable or stylish person.

■ *v.* **1** balloon, billow, blow up, bulge, dilate, inflate, puff up *or* out. **2** escalate, expand, grow, increase, mount, multiply, mushroom, rise, snowball; heighten, intensify, raise, step up. ● *n.* **1** enlargement, escalation, expansion, growth, increase, rise. **2** billowing, heaving, rolling, surging.

swelling *n.* abnormally swollen place, esp. on the body.

■ bulge, bump, distension, excrescence, lump, node, nodule, protrusion, protuberance, tumescence, tumour; abscess, blister, boil, carbuncle, pustule, spot.

swelter *v.* be uncomfortably hot.

swept *see* **sweep**.

swerve *v.* turn aside from a straight course. ● *n.* swerving movement or direction.

■ *v.* change direction, deviate, diverge, sheer off, skew, swing, turn (aside), veer.

swift *adj.* quick, rapid. ● *n.* swift-flying long-winged bird. □ **swiftly** *adv.*, **swiftness** *n.*

■ *adj.* brisk, fast, fleet, *colloq.* nippy, quick, rapid, speedy; meteoric, sudden; immediate, instant, instantaneous, prompt.

swill *v.* **1** wash, rinse. **2** (of water) pour. **3** drink greedily. ● *n.* **1** rinse. **2** sloppy food fed to pigs.

swim *v.* (**swam**, **swum**) **1** travel through water by movements of the body. **2** be covered with liquid. **3** seem to be whirling or waving. **4** be dizzy. ● *n.* act or period of swimming. □ **swimming bath**,

pool artificial pool for swimming in. **swimmer** *n.*

swimmingly *adv.* with easy unobstructed progress.

swindle *v.* cheat in a business transaction. ● *n.* piece of swindling. ▫ **swindler** *n.*

■ *v.* bilk, cheat, *sl.* con, deceive, defraud, *colloq.* do down, double-cross, *colloq.* fiddle, fleece, hoodwink, *colloq.* rip off, sting, trick. ● *n. sl.* con, confidence trick, *colloq.* fiddle, fraud, *colloq.* rip-off, scam, trick.

swine *n.* **1** (*pl.* **swine**) pig. **2** (*colloq.*) (*pl.* **swine** or **-s**) hated person or thing.

swing *v.* (**swung**) **1** move to and fro while supported. **2** turn in a curve. **3** (cause to) change from one mood or opinion to another. ● *n.* **1** act, movement, or extent of swinging. **2** hanging seat for swinging in. **3** jazz with the time of the melody varied. ▫ **in full swing** with activity at its greatest. **swing-bridge** *n.* bridge that can be swung aside for ships to pass. **swing-wing** *n.* aircraft wing that can be moved to slant backwards. **swinger** *n.*

■ *v.* **1** dangle, move to and fro *or* back and forth, oscillate, rock, sway. **2** spin, turn, veer, wheel. **3** change, fluctuate, oscillate, see-saw; incline, move, sway. ● *n.* **1** fluctuation, oscillation; change, movement, shift, switch, variation.

swingeing *adj.* **1** forcible. **2** huge in amount or scope.

swipe (*colloq.*) *v.* **1** hit with a swinging blow. **2** snatch, steal. ● *n.* swinging blow.

swirl *v.* & *n.* whirl, flow with a whirling movement.

swish *v.* move with a hissing sound. ● *n.* swishing sound. ● *adj.* (*colloq.*) smart, fashionable.

Swiss *adj.* & *n.* (native) of Switzerland. ▫ **Swiss roll** thin flat sponge cake spread with jam etc. and rolled up.

switch *n.* **1** device operated to turn electric current on or off. **2** flexible stick or rod, whip. **3** tress of hair tied at one end. **4** shift in opinion or method etc. ● *v.* **1** (cause to) change, esp. suddenly. **2** reverse the positions of, exchange. ▫ **switch on** or **off** turn (an electrical device) on or off.

■ *n.* **4** alteration, change, reversal, shift, U-turn. ● *v.* **1** change, divert, shift, redirect, transfer. **2** change, exchange, reverse, substitute, swap, transpose.

switchback *n.* **1** railway used for amusement at a fair etc., with alternate steep ascents and descents. **2** road with similar slopes.

switchboard *n.* panel of switches for making telephone connections or operating electric circuits.

swivel *n.* link or pivot enabling one part to revolve without turning another. ● *v.* (**swivelled**) turn on or as if on a swivel.

swollen *see* **swell**.

swoop *v.* **1** make a sudden downward rush. **2** attack suddenly. ● *n.* swooping movement or attack.

swop *v.* & *n.* = **swap**.

sword /sord/ *n.* weapon with a long blade and a hilt. ▫ **swordsman** *n.*

■ claymore, cutlass, rapier, sabre, scimitar.

swordfish *n.* sea fish with a long sword-like upper jaw.

swore *see* **swear**.

sworn *see* **swear**. *adj.* open and determined, esp. in enmity.

swot (*colloq.*) *v.* (**swotted**) study hard. ● *n.* person who studies hard.

swum *see* **swim**.

swung *see* **swing**.

sybarite *n.* person who is excessively fond of comfort and luxury. ▫ **sybaritic** *adj.*

sycamore *n.* large tree of the maple family.

sycophant *n.* person who tries to win favour by flattery. ▫ **sycophantic** *adj.*

■ ▫ **sycophantic** fawning, flattering, ingratiating, obsequious, servile, slavish, *colloq.* smarmy, subservient, toadying, unctuous.

syllable *n.* unit of sound in a word. ▫ **syllabic** *adj.*

syllabub *n.* dish of whipped cream flavoured with wine.

syllabus *n.* (*pl.* **-buses**) statement of the subjects to be covered by a course of study.

sylph *n.* slender girl or woman.

symbiosis *n.* (*pl.* **-oses**) relationship of different organisms living in close association. ▫ **symbiotic** *adj.*

symbol *n.* **1** thing regarded as suggesting something. **2** mark or sign with a special meaning.

■ **1** emblem, figure, metaphor, representation, sign, token. **2** badge, emblem, insignia, hallmark, logo, mark, trade mark; character, pictograph.

symbolic *adj.* (also **symbolical**) of, using, or used as a symbol. ◻ **symbolically** *adv.*

■ allegorical, emblematic, figurative, metaphorical, representative.

symbolism *n.* use of symbols to express things. ◻ **symbolist** *n.*

symbolize *v.* **1** be a symbol of. **2** represent by means of a symbol.

■ **1** betoken, connote, denote, epitomize, exemplify, express, indicate, mean, represent, stand for, typify.

symmetry *n.* state of having parts that correspond in size, shape, and position on either side of a dividing line or round a centre. ◻ **symmetrical** *adj.*, **symmetrically** *adv.*

sympathetic *adj.* **1** feeling, showing, or resulting from sympathy. **2** likeable. ◻ **sympathetically** *adv.*

■ **1** caring, compassionate, concerned, considerate, kind, kind-hearted, kindly, solicitous, supportive, understanding; like-minded. **2** agreeable, companionable, congenial, friendly, likeable, nice, pleasant.

sympathize *v.* **sympathize with** feel or express sympathy for. ◻ **sympathizer** *n.*

■ commiserate with, condole with, feel sorry for; empathize with, identify with, relate to, understand.

sympathy *n.* **1** ability to share the feelings of others. **2** (*sing.* or *pl.*) feeling or expression of sorrow or pity. **3** liking for each other.

■ **1** empathy, fellow-feeling. **2** commiseration, compassion, concern, pity, solicitousness, solicitude; condolences. **3** affinity, compatibility, harmony, liking, rapport.

symphony *n.* long elaborate musical composition for a full orchestra. ◻ **symphonic** *adj.*

symposium *n.* (*pl.* **-ia**) meeting for discussing a particular subject.

symptom *n.* sign of the existence of a condition.

symptomatic *adj.* serving as a symptom.

synagogue *n.* building for public Jewish worship.

synchromesh *n.* device that makes gear wheels revolve at the same speed.

synchronize *v.* **1** (cause to) occur or operate at the same time. **2** cause (clocks etc.) to show the same time. ◻ **synchronization** *n.*

synchronous *adj.* occurring or existing at the same time.

syncopate *v.* change the beats or accents in (music). ◻ **syncopation** *n.*

syndicate *n.* /síndikət/ association of people or firms to carry out a business undertaking. ● *v.* /síndikayt/ **1** combine into a syndicate. **2** arrange publication in many newspapers etc. simultaneously. ◻ **syndication** *n.*

■ *n.* alliance, association, bloc, cartel, combine, confederation, consortium, group, league, trust, union.

syndrome *n.* combination of signs, symptoms, etc. characteristic of a specified condition.

synonym *n.* word or phrase meaning the same as another in the same language. ◻ **synonymous** *adj.*

synopsis *n.* (*pl.* **-opses**) summary, brief general survey.

syntax *n.* grammatical arrangement of words. ◻ **syntactic** *adj.*

synthesis *n.* (*pl.* **-theses**) **1** process or result of combining. **2** artificial production of a substance that occurs naturally.

synthesize *v.* make by synthesis.

synthesizer *n.* electronic musical instrument able to produce a great variety of sounds.

synthetic *adj.* made by synthesis. ● *n.* synthetic substance or fabric. ◻ **synthetically** *adv.*

syphilis *n.* a venereal disease. ◻ **syphilitic** *adj.*

syringe *n.* device for drawing and injecting liquid. ● *v.* wash out or spray with a syringe.

syrup *n.* **1** thick sweet liquid. **2** water sweetened with sugar. ◻ **syrupy** *adj.*

system *n.* **1** set of connected things that form a whole or work together. **2** animal body as a whole. **3** (**the system**) traditional practices, methods, and rules existing in a society, insitution, etc. **4** method of classification, notation, or measurement. **5** orderliness.

■ **1** arrangement, network, organization, set-up, structure. **4** approach, method, practice, procedure, process, scheme, technique, way. **5** method, order, orderliness.

systematic *adj.* methodical. □ **systematically** *adv.*

■ businesslike, efficient, methodical, orderly, organized, planned, systematized, well-organized.

systematize *v.* arrange according to a carefully organized system. □ **systematization** *n.*

systemic *adj.* of or affecting the body as a whole.

Tt

tab *n.* small projecting flap or strip. □ **keep tabs on** (*colloq.*) keep under observation.

tabard *n.* short sleeveless tunic-like garment.

tabby *n.* cat with grey or brown fur and dark stripes.

table *n.* **1** piece of furniture with a flat top supported on one or more legs. **2** list of facts or figures arranged in columns. ● *v.* submit (a motion or report) for discussion. □ **at table** taking a meal. **table tennis** game played with bats and a light hollow ball on a table.

■ *n.* **2** chart, graph, index, inventory, list.

tableau /táblō/ *n.* (*pl.* **-eaux**) silent motionless group arranged to represent a scene.

table d'hôte /taab'l dōt/ (meal) served at a fixed inclusive price.

tableland *n.* plateau of land.

tablespoon *n.* **1** large spoon for serving food. **2** amount held by this. □ **tablespoonful** *n.*

tablet *n.* **1** slab bearing an inscription etc. **2** measured amount of a drug compressed into a solid form.

■ **2** capsule, lozenge, pill.

tabloid *n.* small-sized newspaper, often sensational in style.

taboo *n.* ban or prohibition made by religion or social custom. ● *adj.* prohibited by a taboo.

tabular *adj.* arranged in tables.

tabulate *v.* arrange in tabular form. □ **tabulation** *n.*

tachograph *n.* device in a motor vehicle to record speed and travel time.

tacit *adj.* implied or understood without being put into words. □ **tacitly** *adv.*

■ implicit, implied, silent, undeclared, unspoken.

taciturn *adj.* saying very little. □ **taciturnity** *n.*

■ quiet, reserved, reticent, uncommunicative, unforthcoming, untalkative.

tack[1] *n.* **1** small broad-headed nail. **2** course of action or policy. **3** long temporary stitch. **4** sailing ship's oblique course. ● *v.* **1** nail with tack(s). **2** stitch with tacks. **3** sail a zigzag course. □ **tack on** add as an extra thing.

■ *n.* **1** drawing-pin, nail, pin. **2** approach, course, direction, line, method, policy, procedure, way. ● *v.* **1** nail, pin. **2** baste, sew, stitch. □ **tack on** add (on), append, attach, tag on.

tack[2] *n.* harness, saddles, etc.

tackle *n.* **1** equipment for a task or sport. **2** set of ropes and pulleys for lifting etc. **3** act of tackling in football etc. ● *v.* **1** try to deal with or overcome (an opponent or problem etc.). **2** intercept (an opponent who has the ball in football etc.).

■ *n.* **1** accoutrements, apparatus, *sl.* clobber, equipment, gear, paraphernalia, tools. ● *v.* **1** address (oneself to), apply oneself to, face up to, have a go at, take on.

tacky *adj.* (of paint etc.) sticky, not quite dry. □ **tackiness** *n.*

tact *n.* skill in avoiding offence or in winning goodwill. □ **tactful** *adj.*, **tactless** *adj.*

■ consideration, delicacy, diplomacy, discretion, finesse, politeness, sensitivity, thoughtfulness. □ **tactful** considerate, delicate, diplomatic, discreet, judicious, polite, sensitive, thoughtful. **tactless** blunt, clumsy, gauche, hurtful, impolite, impolitic, inappropriate, inconsiderate, indelicate, indiscreet, insensitive, thoughtless, undiplomatic.

tactic *n.* **1** (usu. *pl.*) plan or means of achieving something, skilful device(s). **2** (*pl.*) the skilful arrangement and use of military forces to win a battle.

■ **1** device, manoeuvre, plan, ploy, ruse, scheme, stratagem; **(tactics)** policy, strategy.

tactical *adj.* **1** of tactics. **2** (of weapons) for use in a battle or at close quarters. □ **tactically** *adv.*

tactician *n.* expert in tactics.

tactile *adj.* of or using the sense of touch. □ **tactility** *n.*

tadpole *n.* larva of a frog or toad etc. at the stage when it has gills and a tail.

taffeta *n.* shiny silk-like fabric.

tag *n.* **1** label. **2** much-used phrase or quotation. **3** metal point on a shoelace etc. ● *v.* (**tagged**) label. □ **tag on** attach, add.

tail *n.* **1** animal's hindmost part, esp. when extending beyond its body. **2** rear, hanging, or inferior part. **3** (*colloq.*) person tailing another. **4** (*pl.*) reverse of a coin, turned upwards after being tossed. ● *v.* (*colloq.*) follow closely. □ **tail-end** *n.* very last part. **tail-light** *n.* light at the back of a motor vehicle or train etc. **tail off 1** gradually diminish. **2** end inconclusively.

tailback *n.* queue of traffic extending back from an obstruction.

tailboard *n.* hinged or removable back of a lorry etc.

tailcoat *n.* man's coat with the skirt tapering and divided at the back.

tailgate *n.* rear door in a motor vehicle.

tailor *n.* maker of men's clothes, esp. to order. ● *v.* **1** make (clothes) as a tailor. **2** make or adapt for a special purpose. □ **tailor-made** *adj.*

tailplane *n.* horizontal part of an aeroplane's tail.

tailspin *n.* aircraft's spinning dive.

taint *n.* trace of decay, infection, or other bad quality. ● *v.* affect with a taint.

> ■ *n.* blemish, blot, flaw, mark, stain. ● *v.* blemish, blot, damage, harm, smear, stain, sully, tarnish.

take *v.* (**took, taken**) **1** get possession of. **2** capture. **3** make use of. **4** cause to come or go with one. **5** carry. **6** remove. **7** accept, endure. **8** study or teach (a subject). **9** need. **10** make a photograph (of). ● *n.* **1** amount taken or caught. **2** instance of photographing a scene for a cinema film. □ **be taken by** or **with** find attractive. **be taken ill** become ill. **take after** resemble (a parent etc.). **take-away** *n.* & *adj.* **1** (cooked meal) bought at a restaurant etc. for eating elsewhere. **2** (place) selling this. **take back** withdraw (a statement). **take in 1** deceive, cheat. **2** include. **3** make (a garment etc.) smaller. **4** understand. **take off 1** remove (clothing etc.). **2** mimic humorously. **3** become airborne. **take-off** *n.* **1** humorous mimicry. **2** process of becoming airborne. **take on 1** undertake. **2** engage (an employee). **3** accept as an opponent. **4** acquire. **take oneself off** depart. **take one's time** not hurry. **take out** remove. **take over** take control of. **takeover** *n.* **take part** share in an activity. **take place** occur. **take sides** support one side or another. **take to 1** develop a liking or ability for. **2** adopt as a habit or custom. **3** go to as a refuge. **take up 1** take as a hobby or cause. **2** occupy (time or space). **3** resume. **4** accept (an offer). **5** interrupt or question (a speaker). **take up with** begin to associate with. **taker** *n.*

> ■ *v.* **1** acquire, get (hold of), lay one's hands on, obtain, pick up, procure; grab, grasp, grip, seize, snatch. **2** abduct, apprehend, arrest, capture, catch, kidnap, seize, take prisoner. **3** book, engage, hire, rent, reserve; travel by, use; buy, subscribe to. **4** accompany, conduct, escort, go with, guide, lead. **5** contain, hold, support; bring, carry, convey, transport. **6** appropriate, *sl.* nick, *sl.* pinch, pocket, purloin, remove, *colloq.* snaffle, steal. **7** abide, accept, bear, brook, cope with, endure, put up with, stand, *colloq.* stick, stomach, tolerate, withstand. **8** learn, read, study; teach. **9** demand, necessitate, need, require. □ **take in 1** cheat, *sl.* con, deceive, dupe, fool, hoodwink, mislead, trick. **2** comprise, cover, encompass, include, incorporate. **take off 1** doff, remove, strip off. **2** caricature, imitate, mimic, parody, satirize, send up.

taking *adj.* attractive, captivating. ● *n.pl.* money taken in business.

talc *n.* **1** a kind of smooth mineral. **2** talcum powder.

talcum *n.* talc. □ **talcum powder** talc powdered and usu. perfumed for use on the skin.

tale *n.* **1** narrative, story. **2** report spread by gossip.

> ■ **1** account, anecdote, narration, narrative, story, *colloq.* yarn; fable, legend, saga.

talent *n.* special ability.

> ■ ability, aptitude, bent, flair, genius, gift, knack.

talented *adj.* having talent.

> ■ able, accomplished, adept, bright, capable, clever, gifted, good, proficient, skilful, skilled.

talisman *n.* (*pl.* **-mans**) object supposed to bring good luck.

> ■ amulet, charm, mascot.

talk *v.* **1** convey or exchange ideas by spoken words. **2** use (a specified language) in talking. ● *n.* **1** talking, conversation. **2** style of speech. **3** informal lecture. **4** rumour. □ **talk down to** speak condescendingly to. **talk into** persuade by talking. **talk out of** dissuade by talking. **talk over** discuss. **talking-to** *n.* reproof. **talker** *n.*

■ *v.* **1** chat, chatter, converse, *colloq.* natter, speak. ● *n.* **1** chat, conversation, dialogue, discourse, discussion, *colloq.* natter, tête-à-tête. **2** dialect, jargon, language, speech. **3** address, discourse, lecture, speech. **4** gossip, hearsay, rumour, tittle-tattle. □ **talk over** confer about, consider, debate, deliberate about *or* over, discuss.

talkative *adj.* fond of talking.

■ chatty, garrulous, loquacious, voluble.

tall *adj.* of great or specified height. □ **tall order** difficult task. **tall story** (*colloq.*) one that is hard to believe. **tallness** *n.*

■ big, giant, high, lofty, towering.

tallboy *n.* tall chest of drawers.

tallow *n.* animal fat used to make candles, lubricants, etc.

tally *n.* total of a debt or score. ● *v.* correspond.

■ *n.* score, sum, total. ● *n.* accord, agree, coincide, correspond, match, square.

Talmud *n.* body of Jewish law and tradition. □ **Talmudic** *adj.*

talon *n.* bird's large claw.

tambourine *n.* percussion instrument with jingling discs.

tame *adj.* **1** (of animals) domesticated, not wild or shy. **2** docile. **3** not exciting. ● *v.* make tame or manageable. □ **tamely** *adv.*, **tameness** *n.*

■ *adj.* **1** domesticated, house-trained. **2** compliant, docile, gentle, meek, mild, obedient, submissive, tractable. **3** bland, boring, dreary, dull, humdrum, mundane, unexciting, uninteresting. ● *v.* domesticate, train; calm, control, master, subdue.

Tamil *n.* member or language of a people of southern India and Sri Lanka.

tamp *v.* pack down tightly.

tamper *v.* **tamper with** meddle or interfere with.

tampon *n.* plug of absorbent material inserted into the body.

tan *v.* (**tanned**) **1** convert (hide) into leather. **2** make or become brown by exposure to sun. **3** (*sl.*) thrash. ● *n.* **1** yellowish-brown. **2** brown colour in sun-tanned skin. ● *adj.* yellowish-brown.

tandem *n.* bicycle for two people one behind another. ● *adv.* one behind another. □ **in tandem** arranged in this way.

tandoor *n.* Indian etc. clay oven.

tandoori *n.* food cooked in a tandoor.

tang *n.* strong taste, flavour, or smell. □ **tangy** *adj.*

■ flavour, piquancy, savour, sharpness, taste.

tangent *n.* straight line that touches the outside of a curve without intersecting it. □ **go off at a tangent** diverge suddenly from a line of thought etc. **tangential** *adj.*

tangerine *n.* **1** a kind of small orange. **2** its colour.

tangible *adj.* **1** able to be perceived by touch. **2** clear and definite, real. □ **tangibly** *adv.*, **tangibility** *n.*

■ concrete, definite, material, palpable, physical, real.

tangle *v.* twist into a confused mass, entangle. ● *n.* tangled mass or condition. □ **tangle with** become involved in conflict with.

tango *n.* (*pl.* **-os**) ballroom dance with gliding steps. ● *v.* dance a tango.

tank *n.* **1** large container for liquid or gas. **2** armoured fighting vehicle moving on Caterpillar tracks.

tankard *n.* large one-handled usu. metal drinking vessel.

tanker *n.* ship, aircraft, or vehicle for carrying liquid in bulk.

tanner *n.* person who tans hides.

tannery *n.* place where hides are tanned into leather.

tannic acid tannin.

tannin *n.* substance used in tanning and dyeing.

tantalize *v.* torment by the sight of something desired but kept out of reach or withheld.

■ frustrate, provoke, taunt, tease, tempt, torment.

tantamount *adj.* equivalent.

■ comparable, equal, equivalent.

tantra *n.* any of a class of Hindu or Buddhist mystical or magical writings.

tantrum *n.* outburst of bad temper.

tap[1] *n.* **1** tubular plug with a device for allowing liquid to flow through. **2** connection for tapping a telephone. ● *v.* (**tapped**) **1** obtain supplies etc. or information from. **2** fit a listening device in (a telephone circuit). **3** fit a tap into. **4** draw off through a tap or incision. □ **on tap** (*colloq.*) available for use. **tap root** plant's chief root.

tap[2] *v.* (**tapped**) knock gently. ● *n.* **1** light blow. **2** sound of this. □ **tap-dance** *n.*

dance in which the feet tap an elaborate rhythm.

■ *n.* & *v.* hit, knock, pat, rap.

tape *n.* **1** narrow strip of material for tying, fastening, or labelling things. **2** tape recording. **3** magnetic tape. **4** tape-measure. ● *v.* **1** record on magnetic tape. **2** tie or fasten with tape. □ **have a thing taped** (*colloq.*) understand fully. **tape-measure** *n.* strip of tape etc. marked for measuring length. **tape recorder** apparatus for recording and reproducing sounds on magnetic tape. **tape recording** recording made on magnetic tape.

taper *n.* **1** thin candle. **2** narrowing. ● *v.* make or become gradually narrower. □ **taper off** diminish.

tapestry *n.* textile fabric woven or embroidered ornamentally.

tapeworm *n.* tape-like worm living as a parasite in intestines.

tapioca *n.* starchy foodstuff prepared from cassava.

tapir /táypeer/ *n.* small pig-like animal with a long snout.

tappet *n.* projection used in machinery to tap against something.

tar *n.* **1** thick dark liquid distilled from coal etc. **2** similar substance formed by burning tobacco. ● *v.* (**tarred**) coat with tar.

tarantella *n.* rapid whirling dance.

tarantula *n.* large black hairy spider.

tardy *adj.* (**-ier**, **-iest**) **1** slow. **2** late. □ **tardily** *adv.*, **tardiness** *n.*

■ **1** dilatory, slow, sluggish. **2** belated, delayed, late, overdue, unpunctual.

tare¹ *n.* a kind of vetch.

tare² *n.* allowance for the weight of the container or vehicle weighed with the goods it holds.

target *n.* **1** object or mark to be hit in shooting etc. **2** object of criticism. **3** objective. ● *v.* (**targeted**) aim at (as) a target.

■ *n.* **2** butt, object, victim. **3** aim, ambition, end, goal, object, objective.

tariff *n.* **1** list of fixed charges. **2** duty to be paid.

■ **1** charges, price-list. **2** duty, excise, levy, tax, toll.

Tarmac *n.* **1** [P.] broken stone or slag mixed with tar. **2** (**tarmac**) area surfaced with this. □ **tarmacked** *adj.*

tarnish *v.* **1** (cause to) lose lustre. **2** blemish (a reputation). ● *n.* **1** loss of lustre. **2** blemish.

■ *v.* **1** discolour, dull, stain. **2** blemish, damage, disgrace, mar, spoil, stain, sully, taint.

tarot /tárrō/ *n.* pack of 78 cards mainly used for fortune-telling.

tarpaulin *n.* waterproof canvas.

tarragon *n.* aromatic herb.

tarsus *n.* (*pl.* **-si**) set of small bones forming the ankle.

tart¹ *adj.* acid in taste or manner. □ **tartly** *adv.*, **tartness** *n.*

■ acid, sharp, sour, tangy, vinegary; astringent, biting, caustic, cutting, harsh, mordant.

tart² *n.* **1** pie or pastry flan with sweet filling. **2** (*sl.*) prostitute. ● *v.* **tart up** (*colloq.*) dress gaudily, smarten up.

tartan *n.* **1** pattern (orig. of a Scottish clan) with coloured stripes crossing at right angles. **2** cloth with this.

Tartar *n.* **1** member of a group of Central Asian peoples. **2** bad-tempered or difficult person.

tartar *n.* **1** hard deposit forming on teeth. **2** deposit formed by fermentation in a wine cask.

tartare sauce sauce of mayonnaise, chopped gherkins, etc.

tartlet *n.* small tart.

task *n.* piece of work to be done. □ **take to task** rebuke. **task force** group organized for a special task.

■ assignment, business, charge, chore, duty, errand, job, piece of work, responsibility. □ **take to task** berate, chastise, chide, criticize, rebuke, reprimand, reproach, scold, *colloq.* tell off, *colloq.* tick off, upbraid.

taskmaster *n.* person who imposes a task, esp. regularly or severely.

tassel *n.* ornamental bunch of hanging threads. □ **tasselled** *adj.*

taste *n.* **1** sensation caused in the tongue by things placed upon it. **2** ability to perceive this. **3** small quantity (of food or drink). **4** liking. **5** ability to perceive what is beautiful or fitting. **6** slight experience. ● *v.* **1** discover or test the flavour of. **2** have a certain flavour. **3** experience.

■ *n.* **1** flavour, savour. **3** bit, dash, drop, hint, pinch, *colloq.* spot, touch; morsel, nibble, titbit; nip, sip. **4** appetite, desire, fancy, fondness, liking, partiality, penchant, preference. **5** discernment, discrimination, judgement, refinement, style.

tasteful *adj.* showing good taste. □ **tastefully** *adv.*, **tastefulness** *n.*

■ aesthetic, artistic, attractive, charming, elegant, graceful, stylish; decorous, proper, seemly.

tasteless *adj.* **1** having no flavour. **2** showing poor taste. □ **tastelessly** *adv.*, **tastelessness** *n.*

■ **1** bland, flavourless, insipid, vapid, watery. **2** flashy, garish, gaudy, loud, tawdry.

tasty *adj.* (**-ier**, **-iest**) having a strong flavour, appetizing.

■ appetizing, delicious, luscious, savoury, *colloq.* scrumptious, *colloq.* yummy.

tat *n.* (*colloq.*) tatty thing(s).

tattered *adj.* ragged.

■ ragged, ripped, shabby, tatty, threadbare, torn, worn-out.

tatters *n.pl.* torn pieces.

tattle *v.* chatter idly, reveal information thus. ● *n.* idle chatter.

■ *v.* chatter, gossip, *colloq.* natter, prattle, talk; blab, *sl.* squeal, tittle-tattle. ● *n.* chatter, gossip, small talk, talk.

tattoo[1] *n.* **1** military display or pageant. **2** tapping sound.

tattoo[2] *v.* **1** mark (skin) by puncturing it and inserting pigments. **2** make (a pattern) thus. ● *n.* tattooed pattern.

tatty *adj.* (**-ier**, **-iest**) **1** ragged, shabby and untidy. **2** tawdry. □ **tattily** *adv.*, **tattiness** *n.*

■ **1** old, ragged, *colloq.* scruffy, shabby, tattered, threadbare, untidy, worn-out.

taught *see* **teach**.

taunt *v.* jeer at provocatively. ● *n.* taunting remark.

■ *v.* gibe at, goad, jeer at, make fun of, mock, tease, torment. ● *n.* barb, dig, gibe, jeer.

taut *adj.* stretched tightly.

■ firm, rigid, stiff, stretched, tense, tight.

tauten *v.* make or become taut.

tautology *n.* repetition of the same thing in different words. □ **tautological** *adj.*, **tautologous** *adj.*

tavern *n.* (*old use*) inn, pub.

tawdry *adj.* (**-ier**, **-iest**) showy but without real value.

■ flashy, garish, gaudy, loud, showy, tasteless, tatty.

tawny *adj.* orange-brown.

tax *n.* **1** money to be paid to a government. **2** thing that makes a heavy demand. ● *v.* **1** impose a tax on. **2** make heavy demands on. □ **taxation** *n.*, **taxable** *adj.*

■ *n.* **1** customs, duty, excise, levy, tariff. **2** burden, demand, drain, pressure, strain. ● *v.* **2** burden, overload, strain, stretch.

taxi *n.* (*pl.* **-is**) car with driver which may be hired. ● *v.* (**taxied**, **taxiing**) (of an aircraft) move along ground under its own power. □ **taxi-cab** *n.* taxi.

taxidermy *n.* process of preparing, stuffing, and mounting the skins of animals in lifelike form. □ **taxidermist** *n.*

taxonomy *n.* classification of organisms. □ **taxonomical** *adj.*, **taxonomist** *n.*

taxpayer *n.* person who pays tax.

tea *n.* **1** dried leaves of a tropical evergreen shrub. **2** hot drink made by infusing these (or other substances) in boiling water. **3** afternoon or evening meal at which tea is drunk. □ **tea bag** small porous bag holding tea for infusion. **tea chest** wooden box in which tea is exported. **tea cloth** tea towel. **tea leaf** leaf of tea, esp. after infusion. **tea rose** rose with scent like tea. **tea towel** towel for drying washed crockery etc.

teacake *n.* bun for serving toasted and buttered.

teach *v.* (**taught**) impart information or skill to (a person) or about (a subject). □ **teachable** *adj.*, **teacher** *n.*

■ coach, drill, educate, instruct, school, train, tutor. □ **teacher** coach, instructor, lecturer, schoolmaster, schoolmistress, professor, trainer, tutor.

teak *n.* hard durable wood.

teal *n.* (*pl.* **teal**) a kind of duck.

team *n.* **1** set of players. **2** set of people or animals working together. ● *v.* **team up** combine into a team or set.

■ *n.* **1** side, squad. **2** band, corps, crew, gang, group, pair. ● *v.* combine, gang up, join together, link up, unify, unite.

teamwork *n.* combined effort, cooperation.

teapot *n.* vessel with a spout, in which tea is made.

tear[1] /tair/ *v.* (**tore**, **torn**) **1** pull forcibly apart or away or to pieces. **2** make (a hole etc.) thus. **3** become torn. **4** move or travel hurriedly. ● *n.* hole etc. torn.

■ *v.* **1** pull apart, rend, rip. **4** bolt, bound, dart, dash, fly, hasten, hurry, race, run, rush, scoot, shoot, speed, sprint, zoom. ● *n.* hole, rent, rip, split.

tear[2] /teer/ *n.* drop of liquid forming in and falling from the eye. □ **in tears** with tears flowing. **tear gas** gas causing severe irritation of the eyes.

tearaway *n.* unruly young person.

tearful *adj.* shedding or ready to shed tears. □ **tearfully** *adv.*

tease *v.* **1** try to provoke in a playful or unkind way. **2** pick into separate strands. ● *n.* person fond of teasing others.

■ *v.* **1** bait, chaff, goad, make fun of, provoke, *sl.* rag, *colloq.* rib, taunt, torment.

teasel *n.* plant with bristly heads.

teaset *n.* set of cups and plates etc. for serving tea.

teashop *n.* shop where tea is served to the public.

teaspoon *n.* **1** small spoon for stirring tea etc. **2** amount held by this. □ **teaspoonful** *n.*

teat *n.* **1** nipple on a breast or udder. **2** rubber nipple for sucking milk from a bottle.

technical *adj.* **1** of the mechanical arts and applied sciences. **2** using technical terms. **3** of a particular subject or craft etc. **4** in a strict legal sense. □ **technically** *adv.*, **technicality** *n.*

technician *n.* **1** expert in the techniques of a subject or craft. **2** skilled mechanic.

technique *n.* method of performing or doing something.

■ art, artistry, craft, craftsmanship, expertise, knack, skill; approach, manner, method, mode, style, system, way.

technocracy *n.* rule by technical experts. □ **technocrat** *n.*

technology *n.* **1** study of mechanical arts and applied sciences. **2** these subjects. **3** their application in industry etc. □ **technological** *adj.*, **technologically** *adv.*, **technologist** *n.*

teddy bear toy bear.

tedious *adj.* tiresome because of length, slowness, or dullness. □ **tediously** *adv.*, **tediousness** *n.*, **tedium** *n.*

■ boring, deadly, dreary, dull, flat, humdrum, interminable, long-winded, monotonous, repetitive, tiresome, uninspiring, uninteresting, wearisome.

tee *n.* **1** cleared space from which a golf ball is driven at the start of play. **2** small heap of sand or piece of wood for supporting this ball. ● *v.* **(teed) tee off** make the first stroke in golf.

teem[1] *v.* be present in large numbers. □ **teem with** be full of.

■ □ **teem with** abound in, be abundant in, be full of, be overflowing with, swarm with.

teem[2] *v.* (of water or rain) pour.

teenager *n.* person in his or her teens.

teens *n.pl.* years of age from 13 to 19. □ **teenage** *adj.*, **teenaged** *adj.*

teeny *adj.* (**-ier**, **-iest**) (*colloq.*) tiny.

tee shirt = **T-shirt**.

teeter *v.* stand or move unsteadily.

■ rock, sway, totter, wobble.

teeth *see* **tooth**.

teethe *v.* (of a baby) have its first teeth appear through the gums. □ **teething troubles** problems in the early stages of an enterprise.

teetotal *adj.* abstaining completely from alcohol. □ **teetotaller** *n.*, **teetotalism** *n.*

telecommunication *n.* **1** communication by telephone, radio, etc. **2** (*pl.*) technology for this.

telegram *n.* message sent by telegraph.

telegraph *n.* system or apparatus for sending messages, esp. by electrical impulses along wires. ● *v.* communicate in this way.

telegraphist *n.* person employed in telegraphy.

telegraphy *n.* communication by telegraph. □ **telegraphic** *adj.*, **telegraphically** *adv.*

telemeter *n.* apparatus for recording and transmitting the readings of an instrument at a distance. □ **telemetry** *n.*

telepathy *n.* communication between minds other than by the senses. □ **telepathic** *adj.*, **telepathist** *n.*

telephone *n.* device for transmitting speech by wire or radio. ● *v.* send (a message) to (a person) by telephone. □ **telephonic** *adj.*, **telephonically** *adv.*, **telephony** *n.*

telephonist *n.* operator of a telephone switchboard.

telephoto lens lens producing a large image of a distant object for photography.

teleprinter *n.* telegraph instrument for sending and receiving typewritten messages.

telesales *n.pl.* selling by telephone.

telescope *n.* optical instrument for making distant objects appear larger. ● *v.* **1** make or become shorter by sliding each section inside the next. **2** compress or become compressed forcibly. □ **telescopic** *adj.*, **telescopically** *adv.*

teletext *n.* service transmitting written information to subscribers' television screens.

televise *v.* transmit by television.

television *n.* **1** system for reproducing on a screen a view of scenes etc. by radio transmission. **2** televised programmes. **3** (in full **television set**) apparatus for receiving these. □ **televisual** *adj.*

telex *n.* system of telegraphy using teleprinters and public transmission lines. ● *v.* send (a message) to (a person) by telex.

tell *v.* (**told**) **1** make known in words. **2** give information to. **3** reveal a secret. **4** predict. **5** distinguish. **6** direct, order. **7** produce an effect. □ **tell apart** distinguish between. **tell off** (*colloq.*) reprimand. **tell-tale** *n.* person who tells tales. **tell tales** reveal secrets.

■ **1** narrate, recount, relate. **2** announce, communicate, declare, disclose, divulge, impart, make known, say, state, utter. **3** blab, *sl.* squeal, tattle. **4** forecast, foresee, foretell, predict. **5** differentiate, distinguish. **6** bid, command, direct, instruct, order. □ **tell off** *colloq.* bawl out, berate, chastise, chide, rebuke, reprimand, reproach, scold, take to task, *colloq.* tick off, upbraid.

teller *n.* **1** narrator. **2** person appointed to count votes. **3** bank cashier.

telling *adj.* having a noticeable effect.

■ considerable, marked, noticeable, significant, striking.

telly *n.* (*colloq.*) television.

temerity *n.* **1** impudence. **2** rashness.

temp *n.* (*colloq.*) temporary employee.

temper *n.* **1** mental disposition, mood. **2** fit of anger. **3** calmness under provocation. ● *v.* **1** bring (metal or clay) to the required hardness or consistency. **2** moderate the effects of.

■ *n.* **1** character, disposition, make-up, nature, personality, temperament; frame of mind, humour, mood. **2** fury, rage, tantrum. **3** calm, calmness, composure, *sl.* cool, equanimity, self-control. ● *v.* **1** anneal, harden, strengthen, toughen. **2** assuage, cushion, moderate, modify, soften, tone down.

tempera *n.* method of painting using colours mixed with egg.

temperament *n.* person's nature and character.

■ character, disposition, make-up, nature, personality, temper

temperamental *adj.* **1** of or relating to temperament. **2** excitable or moody. □ **temperamentally** *adv.*

■ **2** changeable, emotional, excitable, highly-strung, mercurial, moody, sensitive, touchy, volatile.

temperance *n.* **1** self-restraint. **2** total abstinence from alcohol.

temperate *adj.* **1** self-restrained, avoiding excess. **2** (of climate) without extremes. □ **temperately** *adv.*

■ **1** disciplined, equable, even-tempered, moderate, restrained, self-controlled, sensible.

temperature *n.* **1** degree of heat or cold. **2** body temperature above normal.

tempest *n.* violent storm.

tempestuous *adj.* stormy.

■ stormy, turbulent, wild.

template *n.* pattern or gauge, esp. for cutting shapes.

temple[1] *n.* building dedicated to the presence or service of god(s).

temple[2] *n.* flat part between forehead and ear.

tempo *n.* (*pl.* **-os** or **-i**) **1** time, speed, or rhythm of a piece of music. **2** speed, pace.

■ **1** beat, cadence, measure, metre, rhythm, stress, tempo, time. **2** pace, rate, speed.

temporal *adj.* **1** earthly. **2** of or denoting time. **3** of the temple(s) of the head.

■ **1** earthly, human, material, mortal, physical, secular, worldly.

temporary *adj.* lasting for a limited time. □ **temporarily** *adv.*

■ interim, makeshift, provisional.

temporize *v.* avoid committing oneself in order to gain time.

tempt *v.* **1** persuade or try to persuade by the prospect of pleasure or advantage. **2** arouse a desire in. □ **temptation** *n.*, **tempter** *n.*, **temptress** *n.*

■ **1** coax, entice, inveigle, persuade, woo. **2** allure, attract, captivate, lure, seduce. □ **temptation** allure, appeal, attractiveness, charm, fascination, pull, seductiveness.

ten *adj.* & *n.* one more than nine.

tenable *adj.* able to be defended or held. □ **tenability** *n.*

■ arguable, defensible, justifiable, plausible, reasonable, supportable, viable, workable.

tenacious *adj.* holding or sticking firmly. □ **tenaciously** *adv.*, **tenacity** *n.*

■ determined, dogged, persistent, pertinacious, resolute, single-minded, steadfast, unswerving.

tenancy *n.* use of land or a building etc. as a tenant.

tenant *n.* person who rents land or a building etc. from a landlord.

tench *n.* (*pl.* **tench**) fish of the carp family.

tend[1] *v.* take care of.

■ care for, look after, mind, minister to, see to, take care of.

tend[2] *v.* have a specified tendency.

tendency *n.* **1** way a person or thing is likely to be or behave. **2** thing's direction.

■ **1** bent, inclination, leaning, predisposition, proclivity, propensity, susceptibility, trend.

tendentious *adj.* biased, not impartial. □ **tendentiously** *adv.*

tender[1] *adj.* **1** not tough or hard. **2** delicate. **3** painful when touched. **4** compassionate. **5** loving, gentle. □ **tenderly** *adv.*, **tenderness** *n.*

■ **2** delicate, fragile, sensitive, weak. **3** painful, raw, sensitive, sore. **4** compassionate, kind, soft-hearted, sympathetic. **5** affectionate, caring, fond, loving, warm; delicate, gentle, light, soft.

tender[2] *v.* **1** offer formally. **2** make a tender for. ● *n.* formal offer to supply goods or carry out work at a stated price. □ **legal tender** currency that must be accepted in payment.

■ *v.* **1** offer, proffer, submit. ● *n.* bid, offer, quotation.

tender[3] *n.* **1** vessel or vehicle conveying goods or passengers to and from a larger one. **2** truck attached to a steam locomotive and carrying fuel and water etc.

tendon *n.* strip of strong tissue connecting a muscle to a bone etc.

tendril *n.* **1** thread-like part by which a climbing plant clings. **2** slender curl of hair etc.

tenement *n.* large house let in portions to tenants.

tenet *n.* firm belief or principle.

■ axiom, belief, canon, conviction, creed, doctrine, maxim, precept, principle, theory.

tenfold *adj.* & *adv.* ten times as much or as many.

tennis *n.* ball game played with rackets over a net, with a soft ball on an open court (**lawn tennis**) or with a hard ball in a walled court (**real tennis**).

tenon *n.* projection shaped to fit into a mortise.

tenor *n.* **1** general meaning. **2** highest ordinary male singing voice. ● *adj.* of tenor pitch.

tense[1] *n.* any of the forms of a verb that indicate the time of the action.

tense[2] *adj.* **1** stretched tightly. **2** nervous, anxious. ● *v.* make or become tense. □ **tensely** *adv.*, **tenseness** *n.*

■ *adj.* **1** firm, rigid, stiff, stretched, taut, tight. **2** agitated, anxious, concerned, edgy, fretful, *colloq.* jittery, jumpy, keyed up, nervous, on tenterhooks, uneasy, *colloq.* uptight, worried; *colloq.* fraught, stressful, worrying.

tensile *adj.* **1** of tension. **2** capable of being stretched.

tension *n.* **1** stretching. **2** tenseness, esp. of feelings. **3** effect produced by forces pulling against each other. **4** electromotive force.

■ **1** force, pressure, pull, strain, stress, tightness. **2** anxiety, apprehension, nervousness, strain, stress, suspense, tenseness.

tent *n.* portable shelter or dwelling made of canvas etc.

tentacle *n.* slender flexible part of certain animals, used for feeling or grasping things.

tentative *adj.* **1** hesitant. **2** done as a trial. □ **tentatively** *adv.*

■ **1** cautious, hesitant, uncertain, unsure. **2** experimental, exploratory, provisional.

tenterhooks *n.pl.* **on tenterhooks** in suspense because of uncertainty.

tenth *adj.* & *n.* next after ninth. □ **tenthly** *adv.*

tenuous *adj.* **1** very slight. **2** very thin. □ **tenuousness** *n.*

■ **1** doubtful, dubious, flimsy, insubstantial, shaky, slight, weak.

tenure *n.* holding of office or of land or accommodation etc.

tepee /teépee/ *n.* conical tent used by N. American Indians.

tepid *adj.* lukewarm.

tequila *n.* Mexican liquor made from the sap of an agave plant.

tercentenary *n.* 300th anniversary.

term *n.* **1** fixed or limited period. **2** word or phrase. **3** (*pl.*) conditions offered or accepted. **4** (*pl.*) relations between people. **5** period of weeks during which a school etc. is open or in which a law court holds sessions. **6** each quantity or expression in a mathematical series or ratio etc. □ **come to terms with** reconcile oneself to (a difficulty etc.).

■ **1** duration, period, span, spell, stretch, time. **2** denomination, designation, name, phrase, title, word. **3** (**terms**) conditions, prerequisites, provisions, provisos, requirements, stipulations. □ **come to terms with** accept, face up to, reconcile oneself to, resign oneself to.

termagant *n.* bullying woman.

terminal *adj.* **1** of or forming an end. **2** of or undergoing the last stage of a fatal disease. ● *n.* **1** point of input or output to a computer etc. **2** terminus. **3** building where air passengers arrive and depart. **4** point of connection in an electric circuit. □ **terminally** *adv.*

terminate *v.* end. □ **termination** *n.*

■ bring to an end, conclude, discontinue, end, finish, halt, put an end to, stop, wind up. □ **termination** cessation, close, conclusion, discontinuation, ending.

terminology *n.* technical terms of a subject. □ **terminological** *adj.*

terminus *n.* (*pl.* **-ni**) **1** end. **2** last stopping place on a rail or bus route.

termite *n.* small insect that is very destructive to timber.

tern *n.* seabird with long wings.

terrace *n.* **1** raised level place. **2** paved area beside a house. **3** row of houses joined by party walls.

terracotta *n.* **1** brownish-red unglazed pottery. **2** its colour.

terra firma dry land, the ground.

terrain *n.* land with regard to its natural features.

■ country, ground, land, territory.

terrapin *n.* freshwater tortoise.

terrestrial *adj.* **1** of the earth. **2** of or living on land.

terrible *adj.* **1** appalling, distressing. **2** very bad. □ **terribly** *adv.*

■ **1** appalling, awful, bad, dire, disastrous, distressing, dreadful, fearful, frightful, ghastly, grave, horrible, serious, severe. **2** hopeless, incompetent, inept, inexpert, unskilful.

terrier *n.* small active dog.

terrific *adj.* **1** (*colloq.*) huge, excellent. **2** causing terror. □ **terrifically** *adv.*

terrify *v.* fill with terror.

■ alarm, dismay, frighten, horrify, petrify, scare, terrorize.

terrine *n.* **1** pâté or similar food. **2** earthenware dish for this.

territorial *adj.* of territory. □ **Territorial Army** a volunteer reserve force.

territory *n.* **1** land under the control of a person, state, etc. **2** sphere of action or thought.

■ **1** area, district, dominion, land, province, region, sector, zone.

terror *n.* **1** extreme fear. **2** terrifying person or thing. **3** (*colloq.*) troublesome person or thing.

■ **1** alarm, dread, fear, fright, horror, panic, trepidation.

terrorism *n.* use of violence and intimidation. □ **terrorist** *n.*

terrorize *v.* **1** fill with terror. **2** coerce by terrorism.

■ **1** frighten, petrify, scare, terrify. **2** browbeat, bully, coerce, cow, intimidate, threaten.

terry *n.* looped cotton fabric used for towels etc.

terse *adj.* **1** concise. **2** curt. □ **tersely** *adv.*, **terseness** *n.*

■ **1** brief, concise, laconic, pithy, short, succinct, to the point. **2** abrupt, blunt, brusque, curt, short.

tertiary /térshəri/ *adj.* next after secondary.

tessellated *adj.* resembling mosaic.

test *n.* **1** something done to discover a person's or thing's qualities or abilities etc. **2** examination (esp. in a school) on a limited subject. **3** test match. ● *v.* subject to a test. □ **test match** one of a series of international cricket or Rugby football matches. **test-tube** *n.* tube of thin glass with one end closed, used in laboratories. **tester** *n.*

■ *n.* **1** appraisal, assessment, check, evaluation, examination, trial. **2** exam, examination. ● *v.* appraise, assess, check, evaluate, examine, screen, try (out).

testament *n.* **1** a will. **2** written statement of beliefs. **3** (**Testament**) main division of the Christian Bible.

testate *adj.* having left a valid will at death. □ **testacy** *n.*
testator *n.* person who has made a will.
testes *see* **testis**.
testicle *n.* male organ that secretes sperm-bearing fluid.
testify *v.* **1** give evidence. **2** be evidence of.

■ **1** affirm, assert, attest, avow, declare, state, swear.

testimonial *n.* **1** formal statement testifying to character, abilities, etc. **2** gift showing appreciation.
testimony *n.* **1** declaration (esp. under oath). **2** supporting evidence.

■ **1** attestation, declaration, deposition, evidence, statement, submission.

testis *n.* (*pl.* **testes**) testicle.
testosterone *n.* male sex hormone.
testy *adj.* irritable. □ **testily** *adv.*

■ bad-tempered, crabby, cross, crotchety, disagreeable, fractious, grumpy, irritable, peevish, petulant, prickly, quarrelsome, snappy, splenetic, surly.

tetanus *n.* bacterial disease causing painful muscular spasms.
tête-à-tête /táytaatáyt/ *n.* private conversation, esp. between two people. ● *adj.* & *adv.* together in private.
tether *n.* rope etc. fastening an animal so that it can graze. ● *v.* fasten with a tether. □ **at the end of one's tether** having reached the limit of one's endurance.
tetrahedron *n.* (*pl.* **-dra**) solid with four sides.
Teutonic *adj.* of Germanic peoples or their languages.
text *n.* **1** main body of a book as distinct from illustrations etc. **2** sentence from Scripture used as the subject of a sermon. □ **textual** *adj.*
textbook *n.* book of information for use in studying a subject.
textile *n.* woven or machine-knitted fabric. ● *adj.* of textiles.

■ *n.* & *adj.* cloth, fabric, material.

texture *n.* way a fabric etc. feels to the touch. □ **textural** *adj.*
textured *adj.* having a noticeable texture.
than *conj.* used to introduce the second element in a comparison.
thank *v.* express gratitude to. □ **thank you** polite expression of thanks. **thanks** *n.pl.* **1** expressions of gratitude. **2** (*colloq.*) thank you.
thankful *adj.* feeling or expressing gratitude. □ **thankfully** *adv.*

■ appreciative, beholden, grateful, indebted, obliged.

thankless *adj.* not likely to win thanks. □ **thanklessness** *n.*
thanksgiving *n.* expression of gratitude, esp. to God.
that *adj.* & *pron.* (*pl.* **those**) **1** the (person or thing) referred to. **2** further or less obvious (one) of two. ● *adv.* to such an extent. ● *rel.pron.* used to introduce a defining clause. ● *conj.* introducing a dependent clause.
thatch *n.* roof made of straw or reeds etc. ● *v.* roof with thatch. □ **thatcher** *n.*
thaw *v.* **1** make or become unfrozen. **2** become less cool or less formal in manner. ● *n.* thawing, weather that thaws ice etc.
the *adj.* **1** applied to a noun standing for a specific person or thing, or one or all of a kind, or used to emphasize excellence or importance. **2** (of prices) per.
theatre *n.* **1** place for the performance of plays etc. **2** plays and acting. **3** lecture hall with seats in tiers. **4** room where surgical operations are performed.

■ **2** acting, drama, show business, the stage.

theatrical *adj.* **1** of or for the theatre. **2** exaggerated for effect. ● *n.pl.* theatrical (esp. amateur) performances. □ **theatrically** *adv.*, **theatricality** *n.*

■ *adj.* **1** dramatic, histrionic, stage. **2** affected, camp, exaggerated, *sl.* hammed, histrionic, overacted, overdone.

thee *pron.* (*old use*) objective case of *thou.*
theft *n.* stealing.

■ burglary, embezzlement, misappropriation, pilfering, robbery, shoplifting, stealing, thievery, thieving.

their *adj.*, **theirs** *poss.pron.* belonging to them.
theism /thée-iz'm/ *n.* belief that the universe was created by a god. □ **theist** *n.*, **theistic** *adj.*
them *pron.* objective case of *they.*
theme *n.* **1** subject being discussed. **2** melody which is repeated. □ **theme park** park with amusements organized round one theme. **thematic** *adj.*

■ **1** subject, subject-matter, text, thesis, topic.

themselves *pron.* emphatic and reflexive form of *they* and *them.*

then *adv.* **1** at that time. **2** next, and also. **3** in that case. ● *adj.* & *n.* (of) that time.
thence *adv.* from that place or source.
thenceforth *adv.* from then on.
theocracy *n.* form of government by a divine being or by priests. □ **theocratic** *adj.*
theodolite *n.* surveying instrument for measuring angles.
theology *n.* study or system of religion. □ **theological** *adj.*, **theologian** *n.*
theorem *n.* mathematical statement to be proved by reasoning.
theoretical *adj.* based on theory only. □ **theoretically** *adv.*

■ hypothetical, putative, speculative, unproven.

theorist *n.* person who theorizes.
theorize *v.* form theories.

■ conjecture, guess, speculate.

theory *n.* **1** set of ideas formulated to explain something. **2** opinion, supposition. **3** statement of the principles of a subject.

■ **2** conjecture, hypothesis, idea, opinion, supposition, thesis, view.

theosophy *n.* system of philosophy that aims at direct intuitive knowledge of God. □ **theosophical** *adj.*
therapeutic /thérrəpyóótik/ *adj.* of or for the cure of disease. □ **therapeutically** *adv.*

■ beneficial, healing, health-giving, medicinal, restorative.

therapist *n.* specialist in therapy.
therapy *n.* medical or healing treatment.
there *adv.* **1** in, at, or to that place. **2** at that point. **3** in that matter. ● *n.* that place. ● *int.* exclamation of satisfaction or consolation.
thereabouts *adv.* near there.
thereafter *adv.* after that.
thereby *adv.* by that means.
therefore *adv.* for that reason.

■ as a result, consequently, for that reason, hence, so, thus.

therein *adv.* in that place.
thereof *adv.* of that.
thereto *adv.* to that.
thereupon *adv.* in consequence of that, because of that.
thermal *adj.* of or using heat. ● *n.* rising current of hot air.
thermodynamics *n.* science of the relationship between heat and other forms of energy.
thermometer *n.* instrument for measuring heat.
thermonuclear *adj.* of or using nuclear reactions that occur only at very high temperatures.
Thermos *n.* [P.] vacuum flask.
thermostat *n.* device that regulates temperature automatically. □ **thermostatic** *adj.*, **thermostatically** *adv.*
thesaurus *n.* (*pl.* **-ri**) dictionary of synonyms.
these *see* **this**.
thesis *n.* (*pl.* **theses**) **1** theory put forward and supported by reasoning. **2** lengthy essay submitted for a university degree.

■ **1** argument, case, contention, premiss, proposition, theory. **2** dissertation, essay, paper.

they *pron.* people or things mentioned or unspecified.
thick *adj.* **1** of great or specified distance between opposite surfaces. **2** densely covered or filled. **3** fairly stiff in consistency. **4** (*colloq.*) stupid. **5** (*colloq.*) intimate. ● *adv.* thickly. ● *n.* busiest part. □ **thick-skinned** *adj.* not sensitive to criticism or snubs. **thickly** *adv.*, **thickness** *n.*

■ *adj.* **1** broad, wide. **2** chock-a-block, crowded, dense, full, impenetrable, packed. **3** clotted, gelatinous, glutinous, viscid, viscous. ● *n.* centre, core, heart.

thicken *v.* make or become thicker.

■ clot, coagulate, congeal, *colloq.* jell, set, solidify, stiffen.

thicket *n.* close group of shrubs and small trees etc.

■ coppice, copse, covert, grove, spinney.

thickset *adj.* **1** set or growing close together. **2** stocky, burly.
thief *n.* (*pl.* **thieves**) one who steals. □ **thievish** *adj.*, **thievery** *n.*

■ burglar, housebreaker, pickpocket, robber, shoplifter.

thieve *v.* steal, be a thief.

■ filch, lift, misappropriate, *sl.* nick, pilfer, *sl.* pinch, pocket, purloin, *colloq.* rip off, steal, *colloq.* swipe.

thigh *n.* upper part of the leg, between hip and knee.
thimble *n.* hard cap worn to protect the end of the finger in sewing.
thin *adj.* (**thinner, thinnest**) **1** not thick. **2** lean, not plump. **3** not plentiful. **4** lacking substance, weak. ● *adv.* thinly. ● *v.*

(**thinned**) make or become thinner. □ **thinly** *adv.*, **thinness** *n.*

■ *adj.* **1** fine, narrow, slender; delicate, diaphanous, filmy, fine, gauzy. **2** lean, scraggy, scrawny, skinny, slender, slight, slim, spindly. **3** inadequate, insufficient, meagre, scanty, scarce, sparse. **4** insubstantial, feeble, flimsy, lame, tenuous; watery, weak.

thine *adj.* & *poss.pron.* (*old use*) belonging to thee.

thing *n.* **1** any unspecified object or item. **2** any fact, event, quality, action, etc. **3** (*pl.*) belongings, utensils, circumstances.

■ **1** article, item, object, whatnot. **2** detail, fact, piece of information; event, happening, incident, occurrence; characteristic, feature, quality; act, action, deed. **3** (**things**) belongings, *sl.* clobber, effects, goods, possessions, property; apparatus, equipment, gear, tools, utensils; circumstances, conditions, matters.

think *v.* (**thought**) **1** exercise the mind, form ideas. **2** form or have as an idea or opinion. ● *n.* (*colloq.*) act of thinking. □ **think about** or **of** consider. **think better of it** change one's mind after thought. **think-tank** *n.* group providing ideas and advice on national or commercial problems. **think up** (*colloq.*) devise. **thinker** *n.*

■ *v.* **1** cogitate, contemplate, deliberate, meditate, muse, ponder, reflect, ruminate. **2** believe, consider, deem, imagine, judge, reckon, regard (as), suppose. □ **think about** consider, contemplate, have in mind, intend, propose.

third *adj.* next after second. ● *n.* **1** third thing, class, etc. **2** one of three equal parts. □ **third party** another person etc. besides the two principals. **third-rate** *adj.* very inferior in quality. **Third World** developing countries of Asia, Africa, and Latin America. **thirdly** *adv.*

thirst *n.* **1** feeling caused by a desire to drink. **2** strong desire. ● *v.* feel a thirst. □ **thirst for** have a strong desire for. **thirsty** *adj.*, **thirstily** *adv.*

■ *n.* **2** appetite, craving, desire, hankering, hunger, itch, longing, yearning, *colloq.* yen. □ **thirst for** crave, desire, hanker after, hunger for, long for, lust after, want, wish for, yearn for.

thirteen *adj.* & *n.* one more than twelve. □ **thirteenth** *adj.* & *n.*

thirty *adj.* & *n.* three times ten. □ **thirtieth** *adj.* & *n.*

this *adj.* & *pron.* (*pl.* **these**) the (person or thing) near, present, or mentioned.

thistle *n.* prickly plant. □ **thistledown** *n.* very light fluff on thistle seeds. **thistly** *adj.*

thong *n.* strip of leather used as a fastening or lash etc.

thorax *n.* part of the body between head or neck and abdomen. □ **thoracic** *adj.*

thorn *n.* **1** small sharp pointed projection on a plant. **2** thorn-bearing tree or shrub. □ **thorny** *adj.*

■ **1** prickle, spine. **2** bramble, brier. □ **thorny** prickly, scratchy, sharp, spiky, spiny.

thorough *adj.* **1** complete in every way. **2** done with great care. **3** absolute. □ **thoroughly** *adv.*, **thoroughness** *n.*

■ **1** complete, comprehensive, detailed, exhaustive, extensive, far-reaching, full, sweeping, wide-ranging. **2** assiduous, careful, methodical, meticulous, painstaking, scrupulous. **3** absolute, complete, downright, out and out, perfect, pure, sheer, total, unmitigated, utter.

thoroughbred *n.* & *adj.* (horse etc.) bred of pure or pedigree stock.

thoroughfare *n.* public way open at both ends.

those *see* **that**.

thou *pron.* (*old use*) you.

though *conj.* in spite of the fact that, even supposing. ● *adv.* (*colloq.*) however.

thought *see* **think**. *n.* **1** process, power, or way of thinking. **2** idea etc. produced by thinking. **3** intention. **4** consideration.

■ **1** intelligence, reason, reasoning; cogitation, consideration, contemplation, meditation, reflection, rumination. **2** *US* brainstorm, brainwave, idea. **3** design, expectation, intention, hope, plan. **4** attention, concern, consideration, regard, solicitude, thoughtfulness.

thoughtful *adj.* **1** thinking deeply. **2** thought out carefully. **3** considerate. □ **thoughtfully** *adv.*, **thoughtfulness** *n.*

■ **1** contemplative, introspective, meditative, pensive, reflective. **2** intelligent, reasoned. **3** caring, considerate, helpful, kind, neighbourly, solicitous, unselfish.

thoughtless *adj.* **1** careless. **2** inconsiderate. □ **thoughtlessly** *adv.*, **thoughtlessness** *n.*

■ **1** careless, heedless, inattentive, negligent, remiss, unthinking. **2** impolite, inconsiderate, insensitive, rude, tactless, undiplomatic.

thousand *adj.* & *n.* ten hundred. □ **thousandth** *adj.* & *n.*

thrash *v.* **1** beat, esp. with a stick or whip. **2** defeat thoroughly. **3** thresh. **4** make flailing movements. □ **thrash out** discuss thoroughly.

■ **1** bash, batter, beat, *sl.* clobber, flog, hit, lash, *colloq.* lay into, strike, thwack, *sl.* wallop, *colloq.* whack, whip. **2** beat, conquer, crush, defeat, destroy, get the better of, *colloq.* lick, overcome, overpower, rout, subdue, triumph over, trounce, vanquish.

thread *n.* **1** thin length of spun cotton or wool etc. **2** thing compared to this. **3** spiral ridge of a screw. ● *v.* **1** pass a thread through. **2** pass (a strip or thread etc.) through or round something. **3** make (one's way) through a crowd etc.

■ *n.* **1** fibre, filament, (piece of) yarn, strand.

threadbare *adj.* **1** with nap worn and threads visible. **2** shabbily dressed.

■ **1** ragged, shabby, tattered, tatty, worn-out. **2** *colloq.* scruffy, shabby, tatty, untidy.

threadworm *n.* small thread-like parasitic worm.

threat *n.* **1** expression of intention to punish, hurt, or harm. **2** person or thing thought likely to bring harm or danger.

■ **1** intimidation, menace, warning. **2** danger, hazard, peril, risk.

threaten *v.* make or be a threat (to).

■ browbeat, bully, intimidate, menace, terrorize; endanger, imperil, jeopardize, put in jeopardy.

three *adj.* & *n.* one more than two. □ **three-dimensional** *adj.* having or appearing to have length, breadth, and depth.

threefold *adj.* & *adv.* three times as much or as many.

threesome *n.* group of three people.

thresh *v.* **1** beat out (grain) from husks of corn. **2** make flailing movements.

threshold *n.* **1** piece of wood or stone forming the bottom of a doorway. **2** point of entry. **3** beginning.

■ **2** doorway, entrance. **3** beginning, brink, commencement, dawn, outset, start, verge.

threw *see* **throw**.

thrice *adv.* (*old use*) three times.

thrift *n.* **1** economical management of resources. **2** plant with pink flowers. □ **thrifty** *adj.*

■ **1** economy, frugality, husbandry, parsimony, prudence. □ **thrifty** careful, economical, frugal, parsimonious, provident, prudent, sparing.

thrill *n.* wave of feeling or excitement. ● *v.* feel or cause to feel a thrill.

■ *n.* buzz, *colloq.* kick. ● *v.* delight, excite, stimulate, stir, titillate.

thriller *n.* exciting story or play etc., esp. involving crime.

thrive *v.* (**throve** or **thrived**, **thrived** or **thriven**) **1** grow or develop well. **2** prosper.

■ **1** blossom, burgeon, develop, flourish, grow. **2** boom, do well, flourish, prosper, succeed.

throat *n.* **1** front of the neck. **2** passage from mouth to oesophagus or lungs. **3** narrow passage.

throaty *adj.* uttered deep in the throat, hoarse. □ **throatily** *adv.*

■ deep, gruff, guttural, hoarse, husky.

throb *v.* (**throbbed**) **1** (of the heart or pulse) beat with more than usual force. **2** vibrate or sound with a persistent rhythm. ● *n.* throbbing beat or sound.

■ *v.* **1** beat, palpitate, pound, pulsate, pulse. **2** beat, vibrate.

throes *n.pl.* severe pangs of pain. □ **in the throes of** struggling with the task of.

thrombosis *n.* formation of a clot of blood in a blood vessel or organ of the body.

throne *n.* **1** ceremonial seat for a monarch, bishop, etc. **2** sovereign power.

throng *n.* crowded mass of people. ● *v.* **1** move or press in a throng. **2** fill with a throng.

■ *n.* crowd, crush, drove, flock, gathering, herd, horde, host, mob, multitude, pack, swarm. ● *v.* **1** assemble, collect, congregate, crowd, flock, gather, herd, mass, press, swarm.

throttle *n.* **1** valve controlling the flow of fuel or steam etc. to an engine. **2** lever controlling this. ● *v.* strangle. □ **throttle back** or **down** reduce an engine's speed by means of the throttle.

through *prep.* & *adv.* **1** from end to end or side to side (of), entering at one point and coming out at another. **2** from beginning to end (of). **3** by the agency, means, or fault of. **4** so as to have finished, so as to have passed (an exam). **5** so as to be connected by telephone (to). ● *adj.* **1** going through. **2** passing without stopping.

throughout *prep.* & *adv.* right through, from end to end (of).

throughput *n.* amount of material processed.

throve *see* **thrive**.

throw *v.* (**threw**, **thrown**) **1** send with some force through the air. **2** cause to fall. **3** (*colloq.*) disconcert. **4** put (clothes etc.) on or off hastily. **5** shape (pottery) on a wheel. **6** cause to be in a certain state. **7** operate (a switch or lever). **8** have (a fit or tantrum). **9** (*colloq.*) give (a party). ● *n.* **1** act of throwing. **2** distance something is thrown. □ **throw away 1** part with as useless or unwanted. **2** fail to make use of. **throw-away** *adj.* to be thrown away after use. **throw in the towel** admit defeat or failure. **throw out 1** discard. **2** reject. **throw up 1** bring to notice. **2** resign from. **3** vomit. **4** raise, erect. **thrower** *n.*

■ *v.* **1** *sl.* bung, cast, *colloq.* chuck, fling, *colloq.* heave, hurl, lob, pitch, shy, *colloq.* sling, toss. **3** confound, confuse, discomfit, disconcert, floor, *colloq.* flummox, fluster, *colloq.* rattle, *colloq.* stump, unnerve. □ **throw away 1** discard, dispose of, *sl.* ditch, dump, get rid of, jettison, scrap, throw out, toss out. **2** fritter away, squander, waste.

thrush[1] *n.* kind of songbird.

thrush[2] *n.* fungal infection of the mouth, throat, or vagina.

thrust *v.* (**thrust**) **1** push forcibly. **2** make a forward stroke with a sword etc. ● *n.* thrusting movement or force.

■ *v.* **1** drive, force, push, ram, shove. **2** lunge, plunge, stab.

thud *n.* dull low sound like that of a blow. ● *v.* (**thudded**) make or fall with a thud.

thug *n.* vicious ruffian. □ **thuggish** *adj.*, **thuggery** *n.*

■ hoodlum, hooligan, lout, ruffian, tough, *sl.* yob.

thumb *n.* short thick finger set apart from the other four. ● *v.* **1** touch or turn (pages etc.) with the thumbs. **2** request (a lift) by signalling with one's thumb. □ **under the thumb of** completely under the influence of.

thump *v.* strike or knock heavily (esp. with the fist), thud. ● *n.* **1** heavy blow. **2** sound of thumping.

■ *v.* bash, beat, *sl.* clobber, clout, hit, pound, punch, strike, thwack, *sl.* wallop, *colloq.* whack.

thunder *n.* **1** loud noise that accompanies lightning. **2** similar sound. ● *v.* **1** sound with or like thunder. **2** utter loudly. **3** make a forceful attack in words. □ **steal a person's thunder** forestall him or her. **thundery** *adj.*

■ *v.* **1** boom, resonate, resound, reverberate, roar, rumble. **2** bawl, bellow, roar, shout, yell.

thunderbolt *n.* **1** imaginary missile thought of as sent to earth with a lightning flash. **2** startling formidable event or statement.

thunderclap *n.* clap of thunder.

thunderous *adj.* like thunder.

■ booming, deafening, loud, noisy, roaring.

thunderstorm *n.* storm accompanied by thunder.

thunderstruck *adj.* amazed.

thus *adv.* **1** in this way. **2** as a result of this. **3** to this extent.

thwack *v.* strike with a heavy blow. ● *n.* this blow or sound.

thwart *v.* prevent from doing what is intended or from being accomplished. ● *n.* oarsman's bench across a boat.

■ *v.* baulk, foil, frustrate, hinder, impede, *sl.* scupper, stop, stymie.

thy *adj.* (*old use*) belonging to thee.

thyme /tīm/ *n.* herb with fragrant leaves.

thymus *n.* ductless gland near the base of the neck.

thyroid gland large ductless gland in the neck.

thyself *pron.* (*old use*) emphatic and reflexive form of *thou* and *thee*.

tiara *n.* woman's jewelled semi-circular headdress.

tibia *n.* (*pl.* **-ae**) shin-bone.

tic *n.* involuntary muscular twitch.

tick[1] *n.* **1** regular clicking sound, esp. made by a clock or watch. **2** (*colloq.*) moment. **3** small mark placed against an item in a list etc., esp. to show that it is correct. ● *v.* **1** (of a clock etc.) make a series of ticks. **2** mark with a tick. □ **tick off** (*colloq.*) reprimand. **tick over** (of an engine) idle. **tick-tack** *n.* semaphore signalling by racecourse bookmakers.

tick[2] *n.* parasitic insect.

tick[3] *n.* (*colloq.*) financial credit.

ticket *n.* **1** marked piece of card or paper entitling the holder to a certain right (e.g. to travel by train etc.). **2** certificate of qualification as a ship's master or pilot etc. **3** label. **4** notification of a traffic offence. **5** list of candidates for office. **6** (**the ticket**) (*colloq.*) the correct or desirable thing. ● *v.* (**ticketed**) put a ticket on.

tickle *v.* **1** touch or stroke lightly so as to cause a slight tingling sensation. **2** feel this sensation. **3** amuse, please. ● *n.* act or sensation of tickling.

ticklish *adj.* **1** sensitive to tickling. **2** (of a problem) requiring careful handling.

■ **2** awkward, delicate, difficult, sensitive, *colloq.* sticky, tricky.

tidal *adj.* of or affected by tides.

tiddler *n.* (*colloq.*) small fish, esp. stickleback or minnow.

tiddly *adj.* (*colloq.*) slightly drunk.

tiddly-winks *n.pl.* game of flicking small counters (**tiddly-winks**) into a receptacle.

tide *n.* **1** sea's regular rise and fall. **2** trend of feeling or events etc. □ **tide over** help temporarily.

tidings *n.pl.* news.

tidy *adj.* (**-ier, -iest**) neat and orderly. ● *v.* make tidy. □ **tidily** *adv.*, **tidiness** *n.*

■ *adj.* neat, orderly, shipshape, smart, spruce, straight, trim. ● *v.* arrange, neaten, put in order, smarten (up), sort (out), spruce (up), straighten.

tie *v.* (**tying**) **1** attach or fasten with cord etc. **2** form into a knot or bow. **3** connect. **4** make the same score as another competitor. **5** restrict, limit. ● *n.* **1** cord etc. used for tying something. **2** thing that unites or restricts. **3** equality of score between competitors. **4** sports match between two of a set of teams or players. **5** strip of material worn below the collar and knotted at the front of the neck. □ **tie-break** *n.* means of deciding the winner when competitors have tied. **tie-clip, -pin** *ns.* ornamental clip or pin for holding a necktie in place. **tie in** link or (of information etc.) be connected with something else. **tie up 1** fasten with cord etc. **2** make (money etc.) not readily available for use. **3** occupy fully. **tie-up** *n.* connection, link.

■ *v.* **1** bind, fasten, hitch, lash, make fast, moor, rope, secure, tether. **2** knot. **3** associate, connect, link. **4** be even *or* level, be neck and neck, draw, finish equal. **5** confine, curb, limit, restrain, restrict. ● *n.* **1** band, cord, ligature, string. **2** association, bond, connection, link, relationship. **3** draw, stalemate.

tied *adj.* **1** (of a pub) bound to supply only one brewer's beer. **2** (of a house) for occupation only by a person working for its owner.

tier /teer/ *n.* any of a series of rows or ranks or units of a structure placed one above the other.

■ level, line, rank, row, storey.

tiff *n.* petty quarrel.

■ altercation, argument, disagreement, quarrel, row, *colloq.* scrap, squabble.

tiger *n.* large striped animal of the cat family. □ **tiger lily** orange lily with dark spots.

tight *adj.* **1** held or fastened firmly, fitting closely. **2** tense, not slack. **3** (*colloq.*) stingy. **4** (*colloq.*) drunk. **5** (of money etc.) severely restricted. ● *adv.* tightly. □ **tight corner** difficult situation. **tight-fisted** *adj.* stingy. **tightly** *adv.*, **tightness** *n.*

■ *adj.* **1** fast, firm, fixed, secure; close-fitting. **2** firm, rigid, stiff, stretched, taut, tense.

tighten *v.* make or become tighter.

tightrope *n.* tightly stretched rope on which acrobats perform.

tights *n.pl.* garment covering the legs and lower part of the body.

tigress *n.* female tiger.

tile *n.* thin slab of baked clay etc. for roofing, paving, etc. ● *v.* cover with tiles.

till[1] *v.* prepare and use (land) for growing crops. □ **tillage** *n.*

■ cultivate, farm, plough, work.

till[2] *prep.* & *conj.* up to (a specified time).

till[3] *n.* receptacle for money in a shop etc.

tiller *n.* bar by which a rudder is turned.

tilt *v.* move into a sloping position. ● *n.* sloping position. □ **at full tilt** at full speed or force.

■ *v.* angle, careen, heel over, incline, lean, list, slant, slope.

tilth *n.* **1** tillage. **2** tilled soil.

timber *n.* **1** wood prepared for use in building or carpentry. **2** trees suitable for this. **3** wooden beam used in constructing a house or ship.

■ **1** beams, lumber, planks, wood.

timbered *adj.* **1** constructed of timber or with a timber framework. **2** (of land) wooded.

timbre /támbər/ *n.* characteristic quality of the sound of a voice or instrument.

time *n.* **1** all the years of the past, present, and future. **2** portion or point of this. **3** occasion, instance. **4** rhythm in music. **5** (*pl.*) contemporary circumstances. **6** (*pl.*, in multiplication or comparison) taken a number of times. ● *v.* **1** choose the time for. **2** measure the time taken by. □ **behind the times** out of date. **for the time being** until another arrangement is made. **from time to time** at intervals. **in time 1** not late. **2** eventually. **on time** punctual(ly). **time-and-motion** *adj.* concerned with measuring the efficiency of effort. **time bomb** bomb that can be set to explode after an interval. **time-honoured** *adj.* respected because of antiquity, traditional. **time-lag** *n.* interval between two connected events. **time-share** *n.* share in a property that allows use by several joint owners at agreed different times. **time switch** one operating automatically at a set time. **time zone** region (between parallels of longitude) where a common standard time is used.

■ *n.* **2** age, day(s), epoch, era, period; interval, period, spell, stretch; hour, instant, juncture, moment, point. **3** chance, occasion, opportunity; instance, occurrence. **4** beat, measure, rhythm, stress, tempo. **5** (**times**) circumstances, conditions, things. □ **behind the times** antiquated, old, old-fashioned, outdated, outmoded, out of date. **time-honoured** accepted, conventional, customary, established, habitual, traditional, usual.

timeless *adj.* not affected by the passage of time. □ **timelessness** *n.*

■ ageless, eternal, everlasting, immortal, immutable, unchangeable, unchanging, undying.

timely *adj.* occurring at just the right time. □ **timeliness** *n.*

■ convenient, opportune, seasonable, well-timed.

timepiece *n.* clock or watch.

timetable *n.* list showing the times at which certain events take place.

timid *adj.* easily alarmed, shy. □ **timidly** *adv.*, **timidity** *n.*

■ bashful, coy, diffident, faint-hearted, fearful, frightened, meek, mousy, retiring, self-conscious, scared, shy, timorous, unassuming.

timorous *adj.* timid. □ **timorously** *adv.*, **timorousness** *n.*

timpani *n.pl.* kettledrums. □ **timpanist** *n.*

tin *n.* **1** silvery-white metal. **2** metal box or other container, one in which food is sealed for preservation. ● *v.* (**tinned**) **1** coat with tin. **2** seal (food) into a tin.

tincture *n.* **1** solution of a medicinal substance in alcohol. **2** slight tinge.

tinder *n.* any dry substance that catches fire easily.

tine *n.* prong or point of a fork, harrow, or antler.

tinge *v.* (**tingeing**) **1** colour slightly. **2** give a slight trace of an element or quality to. ● *n.* slight colouring or trace.

tingle *v.* have a slight pricking or stinging sensation. ● *n.* this sensation.

tinker *n.* **1** travelling mender of pots and pans. **2** (*colloq.*) mischievous person or animal. ● *v.* work at something casually trying to repair or improve it.

tinkle *n.* series of short light ringing sounds. ● *v.* make or cause to make a tinkle.

tinsel *n.* glittering decorative metallic strips or threads.

tint *n.* variety or slight trace of a colour. ● *v.* colour slightly.

■ *n.* colour, hue, shade, tincture, tinge, tone. ● *v.* colour, tinge.

tiny *adj.* (**-ier, -iest**) very small.

■ diminutive, little, miniature, minuscule, minute, small, *colloq.* teeny, *colloq.* wee.

tip[1] *n.* end, esp. of something small or tapering. ● *v.* (**tipped**) provide with a tip.

■ *n.* cap, end, head, point.

tip[2] *v.* (**tipped**) **1** tilt, topple. **2** discharge (a thing's contents) by tilting. **3** name as a likely winner. **4** make a small present of money to, esp. in acknowledgement of services. ● *n.* **1** small money present. **2** useful piece of advice. **3** place where rubbish etc. is tipped. □ **tip off** give a warning or hint to. **tip-off** *n.* such a warning etc. **tipper** *n.*

■ *v.* **1** incline, lean, list, slant, slope, tilt; knock over, overturn, topple (over), upend, upset. ● *n.* **1** gratuity. **2** piece of advice, recommendation, suggestion, tip-off. **3** dump, rubbish heap. □ **tip off** advise, alert, caution, forewarn, notify, warn.

tipple *v.* drink (wine or spirits etc.) repeatedly. ● *n.* (*colloq.*) alcoholic or other drink.

tipster *n.* person who gives tips about racehorses etc.

tipsy *adj.* slightly drunk.

tiptoe *v.* (**tiptoeing**) walk very quietly or carefully.

tiptop *adj.* (*colloq.*) first-rate.

tirade *n.* long angry piece of criticism or denunciation.

tire *v.* make or become tired.

■ drain, exhaust, fatigue, *sl.* knacker, wear out, weary.

tired *adj.* feeling a desire to sleep or rest. □ **tired of** having had enough of.

■ all in, dead beat, drained, exhausted, *colloq.* fagged, fatigued, *sl.* knackered, weary, *colloq.* whacked, worn-out.

tireless *adj.* not tiring easily. □ **tirelessly** *adv.*, **tirelessness** *n.*

■ energetic, hard-working, indefatigable, industrious, tenacious, unflagging, untiring, vigorous.

tiresome *adj.* **1** annoying. **2** tedious.

■ **1** annoying, bothersome, exasperating, *colloq.* infernal, infuriating, irksome, irritating, maddening, trying, vexatious. **2** boring, deadly, dreary, dull, tedious, uninteresting, wearisome.

tissue *n.* **1** substance forming an animal or plant body. **2** tissue-paper. **3** piece of soft absorbent paper used as a handkerchief etc. □ **tissue-paper** *n.* thin soft paper used for packing things.

tit *n.* any of several small birds.

titanic *adj.* gigantic.

■ colossal, enormous, gargantuan, giant, gigantic, huge, immense, mammoth, massive, monumental, vast.

titanium *n.* dark-grey metal.

titbit *n.* choice bit of food or item of information.

■ delicacy, *colloq.* goody, treat.

tithe *n.* one-tenth of income or produce formerly paid to the Church.

titillate *v.* excite or stimulate pleasantly. □ **titillation** *n.*

titivate *v.* (*colloq.*) smarten up, put finishing touches to. □ **titivation** *n.*

title *n.* **1** name of a book, poem, or picture etc. **2** word denoting rank or office, or used in speaking of or to the holder. **3** championship in sport. **4** legal right to ownership of property. □ **title-deed** *n.* legal document proving a person's title to a property. **title role** part in a play etc. from which the title is taken.

titled *adj.* having a title of nobility.

titmouse *n.* (*pl.* **-mice**) = **tit**.

titter *n.* high-pitched giggle. ● *v.* give a titter.

■ *n.* & *v.* giggle, laugh, snicker, snigger.

tittle-tattle *v.* & *n.* gossip.

■ *n.* gossip, hearsay, rumour, talk.

titular *adj.* **1** of a title. **2** having the title of ruler etc. but no real authority.

tizzy *n.* (*colloq.*) state of nervous agitation or confusion.

TNT *abbr.* trinitrotoluene, a powerful explosive.

to *prep.* **1** towards. **2** as far as. **3** as compared with, in respect of. **4** for (a person or thing) to hold or possess or be affected by. **5** (with a verb) forming an infinitive, or expressing purpose or consequence etc.; used alone when the infinitive is understood. ● *adv.* **1** to a closed position. **2** into a state of consciousness or activity. □ **to and fro** backwards and forwards. **to-do** *n.* fuss.

toad *n.* frog-like animal living chiefly on land. □ **toad-in-the-hole** *n.* sausages baked in batter.

toadflax *n.* wild plant with yellow or purple flowers.

toadstool *n.* fungus (usu. poisonous) with a round top on a stalk.

toady *n.* sycophant. ● *v.* behave sycophantically.

toast *n.* **1** toasted bread. **2** person or thing in whose honour a company is requested to drink. **3** this request or instance of drinking. ● *v.* **1** brown by heating. **2** express good wishes to by drinking.

toaster *n.* electrical device for toasting bread.

tobacco *n.* **1** plant with leaves that are used for smoking or snuff. **2** its prepared leaves.

tobacconist *n.* shopkeeper who sells cigarettes etc.

toboggan *n.* small sledge used for sliding downhill. ● *v.* ride on a toboggan.

toby jug mug or jug in the form of a seated old man.

tocsin *n.* **1** bell rung as an alarm signal. **2** signal of disaster.

today *adv.* & *n.* **1** (on) this present day. **2** (at) the present time.

toddle *v.* (of a young child) walk with short unsteady steps.

toddler *n.* child who has only recently learnt to walk.

toddy *n.* sweetened drink of spirits and hot water.

toe *n.* **1** any of the divisions (five in humans) of the front part of the foot. **2** part of a shoe or stocking covering the toes. ● *v.* touch with the toe(s). □ **on one's toes** alert or eager. **toe-hold** *n.* slight foothold. **toe the line** conform, obey orders.

toff *n.* (*sl.*) distinguished or well-dressed person.

toffee *n.* sweet made with heated butter and sugar. □ **toffee-apple** *n.* toffee-coated apple on a stick.

tog *n.* **1** unit for measuring the warmth of duvets. **2** (*pl.*, *colloq.*) clothes. ● *v.* (**togged**) **tog out** or **up** (*colloq.*) dress.

toga *n.* loose outer garment worn by men in ancient Rome.

together *adv.* **1** simultaneously. **2** in or into company or conjunction. **3** towards each other.

■ **1** at once, at the same time, in unison, simultaneously.

toggle *n.* **1** short piece of wood etc. passed through a loop as a fastening device. **2** switch that turns a function on and off alternately.

toil *v.* work or move laboriously. ● *n.* **1** laborious work. **2** (*pl.*) net, snare. □ **toilsome** *adj.*

■ *v.* drudge, exert oneself, labour, slave (away), slog (away), work. ● *n.* **1** donkey work, drudgery, *colloq.* fag, *sl.* graft, grind, labour, slog, travail, work.

toilet *n.* **1** lavatory. **2** process of dressing and grooming oneself. □ **toilet water** light perfume.

toiletries *n.pl.* articles used in washing and grooming oneself.

token *n.* **1** sign, symbol. **2** keepsake. **3** voucher that can be exchanged for goods. **4** disc used as money in a slot machine etc. ● *adj.* **1** symbolic. **2** perfunctory.

■ *n.* **1** badge, mark, sign, symbol. **2** keepsake, memento, reminder, souvenir. **3** coupon, voucher. **4** counter, disc. ● *adj.* **1** emblematic, symbolic. **2** nominal, perfunctory, slight, superficial.

tokenism *n.* granting of minimal concessions.

told *see* **tell**. **all told** counting everything or everyone.

tolerable *adj.* **1** endurable. **2** passable. □ **tolerably** *adv.*

■ **1** acceptable, bearable, endurable, supportable. **2** adequate, average, fair, mediocre, *colloq.* OK, passable, satisfactory.

tolerance *n.* **1** willingness to tolerate. **2** permitted variation. □ **tolerant** *adj.*, **tolerantly** *adv.*

■ **1** broad-mindedness, charity, forbearance, indulgence, lenience, patience, understanding. **2** allowance, play, variation. □ **tolerant** broad-minded, charitable, forbearing, indulgent, lenient, liberal, patient, permissive, understanding, unprejudiced.

tolerate *v.* **1** permit without protest or interference. **2** endure. □ **toleration** *n.*

■ **1** accept, allow, consent to, permit, sanction, *colloq.* stand for. **2** abide, bear, brook, endure, live with, put up with, stand, *colloq.* stick, stomach, take.

toll[1] *n.* **1** tax paid for the use of a public road etc. **2** loss or damage caused by a disaster etc. □ **toll-gate** *n.* barrier preventing passage until a toll is paid.

toll[2] *v.* ring with slow strokes, esp. to mark a death. ● *n.* stroke of a tolling bell.

tom *n.* (in full **tom-cat**) male cat.

tomahawk *n.* light axe used by N. American Indians.

tomato *n.* (*pl.* **-oes**) red fruit used as a vegetable.

tomb *n.* grave or other place of burial.

■ crypt, grave, mausoleum, sepulchre, vault.

tombola *n.* lottery resembling bingo.

tomboy *n.* girl who enjoys rough and noisy activities.

tombstone *n.* memorial stone set up over a grave.

tome *n.* large book.

tomorrow *adv.* & *n.* **1** (on) the day after today. **2** (in) the near future.

tom-tom *n.* drum beaten with the hands.

ton *n.* **1** measure of weight, either 2240 lb (**long ton**) or 2000 lb (**short ton**) or 1000 kg (**metric ton**). **2** unit of volume in shipping.

tone *n.* **1** musical or vocal sound, esp. with reference to its pitch, quality, and strength. **2** full interval between one note and the next in an octave. **3** shade of colour. **4** general character. **5** proper firmness of muscles etc. ● *v.* **1** give tone to. **2** harmonize in colour. □ **tone-deaf** *adj.* unable to perceive differences of musical pitch. **tone down** make less intense. **tonal** *adj.*, **tonally** *adv.*, **tonality** *n.*

■ *n.* **1** note, sound; inflection, modulation, pitch, quality, timbre. **3** colour, hue, shade, tincture, tinge, tint. **4** atmosphere, character, feeling, mood, spirit. □ **tone down** moderate, modify, reduce, soften, temper.

toneless *adj.* without positive tone, not expressive. □ **tonelessly** *adv.*

tongs *n.pl.* instrument with two arms used for grasping things.

tongue *n.* **1** muscular organ in the mouth, used in tasting and speaking. **2**

tongue of an ox etc. as food. **3** language. **4** projecting strip. **5** tapering jet of flame. ▫ **tongue-tied** *adj.* too shy to speak. **tongue-twister** *n.* sequence of words difficult to pronounce quickly and correctly. **tongue-in-cheek** *adj.* with sly sarcasm.

tonic *n.* **1** medicine etc. with an invigorating effect. **2** keynote in music. **3** tonic water. ● *adj.* invigorating. ▫ **tonic sol-fa** (*see* **sol-fa**). **tonic water** mineral water, esp. flavoured with quinine.

■ *n.* **1** restorative, stimulant. ● *adj.* fortifying, invigorating, refreshing, restorative, reviving, stimulating, strengthening.

tonight *adv.* & *n.* (on) the present or approaching evening or night.

tonnage *n.* **1** ship's carrying capacity expressed in tons. **2** charge per ton for carrying cargo.

tonne *n.* metric ton, 1000 kg.

tonsil *n.* either of two small organs near the root of the tongue.

tonsillitis *n.* inflammation of the tonsils.

tonsure *n.* **1** shaving the top or all of the head as a religious symbol. **2** this shaven area. ▫ **tonsured** *adj.*

too *adv.* **1** to a greater extent than is desirable. **2** also. **3** (*colloq.*) very.

■ **1** excessively, overly, unduly. **2** additionally, also, as well, besides, furthermore, in addition.

took *see* **take**.

tool *n.* **1** thing used for working on something. **2** person used by another for his or her own purposes. ● *v.* **1** shape or ornament with a tool. **2** equip with tools.

■ *n.* **1** appliance, implement, instrument, utensil. **2** cat's-paw, dupe, pawn, puppet, stooge.

toot *n.* short sound produced by a horn or whistle etc. ● *v.* make or cause to make a toot.

tooth *n.* (*pl.* **teeth**) **1** each of the white bony structures in the jaws, used in biting and chewing. **2** tooth-like part or projection. ▫ **toothed** *adj.*

toothpaste *n.* paste for cleaning the teeth.

toothpick *n.* small pointed instrument for removing bits of food from between the teeth.

toothy *adj.* having many or large teeth.

top[1] *n.* **1** highest point or part or position. **2** thing forming the upper part or covering. **3** upper surface. **4** utmost degree or intensity. **5** garment for the upper part of the body. ● *adj.* highest in position or rank etc. ● *v.* **(topped) 1** provide or be a top for. **2** be higher or better than. **3** add as a final thing. **4** reach the top of. ▫ **on top of** in addition to. **top dog** (*colloq.*) master, victor. **top dress** apply fertilizer on the top of (soil). **top hat** man's tall hat worn with formal dress. **top-heavy** *adj.* heavy at the top and liable to fall over. **top-notch** *adj.* (*colloq.*) first-rate. **top secret** of utmost secrecy. **top up 1** complete (an amount). **2** fill up (something half empty).

■ *n.* **1** apex, crest, crown, head, peak, pinnacle, point, summit, tip, vertex; acme, climax, culmination, height, zenith. **2** cap, cover, covering, lid. ● *adj.* highest, topmost, uppermost; best, finest, first, foremost, greatest, leading, pre-eminent, supreme. ● *v.* **2** beat, better, cap, exceed, outdo, outstrip, surpass, transcend. **3** decorate, garnish.

top[2] *n.* toy that spins on its point when set in motion.

topaz *n.* semiprecious stone of various colours, esp. yellow.

topcoat *n.* **1** overcoat. **2** final coat of paint etc.

topiary *adj.* & *n.* (of) the art of clipping shrubs etc. into ornamental shapes.

topic *n.* subject of a discussion or written work.

■ issue, subject, text, theme.

topical *adj.* having reference to current events. ▫ **topically** *adv.*, **topicality** *n.*

topknot *n.* tuft, crest, or bow etc. on top of the head.

topless *adj.* leaving or having the breasts bare.

topmost *adj.* highest.

topography *n.* local geography, position of the rivers, roads, buildings, etc., of a place or district. ▫ **topographical** *adj.*

topper *n.* (*colloq.*) top hat.

topple *v.* **1** be unsteady and fall. **2** cause to do this.

■ **1** fall, keel over, overbalance, tumble. **2** knock over, overturn, up-end, upset.

topside *n.* beef from the upper part of the haunch.

topsoil *n.* top layer of the soil.

topsy-turvy *adv.* & *adj.* **1** upside down. **2** in or into great disorder.

■ *adj.* **2** chaotic, disordered, disorderly, disorganized, higgledy-piggledy, jumbled, messy, untidy.

tor *n.* hill or rocky peak.

torch *n.* **1** small hand-held electric lamp. **2** burning piece of wood etc. carried as a light. □ **torchlight** *n.*

tore *see* **tear**[1].

toreador *n.* fighter (esp. on horseback) in a bullfight.

torment *n.* /tórment/ **1** severe suffering. **2** cause of this. ● *v.* /tormént/ **1** subject to torment. **2** tease, annoy. □ **tormentor** *n.*

■ *n.* **1** agony, anguish, distress, misery, pain, suffering, torture. **2** affliction, bane, blight, bother, curse, nuisance, scourge. ● *v.* **1** afflict, bedevil, distress, rack, trouble; abuse, ill-treat, maltreat, mistreat, torture. **2** bait, chaff, goad, make fun of, provoke, *sl.* rag, tease, taunt; *colloq.* aggravate, annoy, bother, *sl.* bug, drive mad, infuriate, madden, nettle, pester, *colloq.* rile.

torn *see* **tear**[1].

tornado *n.* (*pl.* **-oes**) violent destructive whirlwind.

torpedo *n.* (*pl.* **-oes**) explosive underwater missile. ● *v.* attack or destroy with a torpedo.

torpid *adj.* sluggish and inactive. □ **torpidly** *adv.*, **torpidity** *n.*

■ apathetic, inactive, languid, lethargic, listless, slow, sluggish.

torpor *n.* sluggish condition.

■ apathy, inactivity, indolence, inertia, languor, lassitude, lethargy, listlessness, sluggishness.

torque *n.* force that produces rotation.

torrent *n.* **1** rushing stream or flow. **2** downpour. □ **torrential** *adj.*

torrid *adj.* intensely hot.

torsion *n.* twisting, spiral twist.

torso *n.* (*pl.* **-os**) trunk of the human body.

tort *n.* any private or civil wrong (other than breach of contract) for which damages may be claimed.

tortoise *n.* slow-moving reptile with a hard shell.

tortoiseshell *n.* mottled yellowish-brown shell of certain turtles, used for making combs etc. □ **tortoiseshell cat** one with mottled colouring.

tortuous *adj.* full of twists and turns. □ **tortuously** *adv.*

■ circuitous, convoluted, curved, meandering, serpentine, twisting, winding, zigzag.

torture *n.* **1** severe pain. **2** infliction of this as a punishment or means of coercion. ● *v.* inflict torture upon. □ **torturer** *n.*

■ *n.* **1** agony, anguish, pain, suffering, torment.

toss *v.* **1** throw lightly. **2** throw up (a coin) to settle a question by the way it falls. **3** roll about from side to side. **4** coat (food) by gently shaking it in dressing etc. ● *n.* tossing. □ **toss off 1** drink rapidly. **2** compose rapidly. **toss-up** *n.* **1** tossing of a coin. **2** even chance.

■ *v.* **1** *sl.* bung, cast, *colloq.* chuck, pitch, *colloq.* sling, throw. **2** flick, flip. **3** pitch, rock, roll.

tot[1] *n.* **1** small child. **2** small quantity of spirits.

tot[2] *v.* (**totted**) **tot up** (*colloq.*) add up.

total *adj.* **1** including everything or everyone. **2** complete. ● *n.* total amount. ● *v.* (**totalled**) **1** reckon the total of. **2** amount to. □ **totally** *adv.*, **totality** *n.*

■ *adj.* **1** complete, entire, full, gross, overall, whole. **2** absolute, complete, out and out, perfect, pure, thorough, unmitigated, unqualified, utter. ● *n.* amount, sum. ● *v.* **1** add (up), count (up), reckon, sum up, *colloq.* tot up. **2** add up to, amount to, come to, equal, make.

totalitarian *adj.* of a regime in which no rival parties or loyalties are permitted. □ **totalitarianism** *n.*

totalizator *n.* device that automatically registers bets, so that the total amount can be divided among the winners.

tote[1] *n.* (*sl.*) totalizator.

tote[2] *v.* (*US*) carry. □ **tote bag** large capacious bag.

totem *n.* tribal emblem among N. American Indians. □ **totem-pole** *n.* pole carved or painted with totems.

totter *v.* walk or rock unsteadily. ● *n.* tottering walk or movement. □ **tottery** *adj.*

■ *v.* dodder, rock, stagger, sway, teeter, wobble.

toucan *n.* tropical American bird with an immense beak.

touch *v.* **1** be, come, or bring into contact. **2** feel or stroke. **3** press or strike lightly. **4** reach. **5** match. **6** rouse sympathy in. **7** (*sl.*) persuade to give or lend money. ● *n.* **1** act, fact, or manner of touching. **2** ability to perceive things through touching them. **3** style of workmanship. **4** slight trace. **5** detail. □ **in touch** in communication. **touch-**

-and-go *adj.* uncertain as regards result. **touch down 1** touch the ball on the ground behind the goal line in Rugby football. **2** (of an aircraft) land. **touchdown** *n.* **touchline** *n.* side limit of a football field. **touch off 1** cause to explode. **2** start (a process). **touch on** mention briefly. **touch up** improve by making small additions.

■ *v.* **1** abut, be against, border on, brush against. **2** feel, finger, handle; caress, fondle, stroke. **3** brush, graze, knock, tap. **4** attain, make, reach, rise to. **5** be in the same league as, be on a par with, compare with, equal, match, rival. **6** affect, move. ● *n.* **1** brush, caress; knock, pat, tap. **2** feeling, sensation. **3** manner, method, style, technique. **4** bit, dash, hint, jot, piece, pinch, speck, spot, suggestion, trace. □ **touch on** allude to, bring up, broach, mention, raise, refer to, speak of.

touché /tōōsháy/ *int.* acknowledgement of a hit in fencing, or of a valid criticism.

touching *adj.* moving, pathetic. ● *prep.* concerning.

■ *adj.* emotional, emotive, moving, pathetic, poignant, stirring. ● *prep.* about, apropos, concerning, re, regarding, with respect to.

touchstone *n.* standard or criterion.

■ benchmark, criterion, measure, norm, standard, yardstick.

touchy *adj.* (**-ier, -iest**) easily offended. □ **touchiness** *n.*

■ highly-strung, moody, sensitive, temperamental, volatile.

tough *adj.* **1** hard to break, cut, or chew. **2** hardy. **3** unyielding. **4** difficult. **5** (*colloq.*, of luck) hard. ● *n.* rough violent person. □ **toughness** *n.*

■ *adj.* **1** durable, hard-wearing, long-lasting, stout, strong, sturdy; chewy, fibrous, gristly, hard, leathery, stringy. **2** hardy, robust, stalwart, strong, sturdy. **3** adamant, hard, harsh, inflexible, severe, stern, strict, unyielding. **4** arduous, challenging, demanding, difficult, exacting, onerous, taxing; baffling, knotty, perplexing, puzzling.

toughen *v.* make or become tough or tougher.

toupee /tōōpay/ *n.* small wig.

tour *n.* journey through a place, visiting things of interest or giving performances. ● *v.* make a tour (of). □ **on tour** touring.

■ *n.* excursion, expedition, jaunt, journey, outing, trip.

tour de force feat of strength or skill.

tourism *n.* organized touring or other services for tourists.

tourist *n.* person travelling or visiting a place for recreation.

■ holiday-maker, sightseer, traveller, tripper, visitor, voyager.

tournament *n.* contest of skill involving a series of matches.

tousle *v.* make (hair etc.) untidy by ruffling.

tout *v.* pester people to buy. ● *n.* **1** person who touts. **2** person who sells tickets for popular events at high prices.

tow[1] *n.* coarse fibres of flax etc.

tow[2] *v.* pull along behind one. ● *n.* act of towing. □ **tow-path** *n.* path beside a canal or river, orig. for horses towing barges.

■ *v.* drag, draw, haul, lug, pull, trail, tug.

toward *prep.* towards.

towards *prep.* **1** in the direction of. **2** in relation to. **3** as a contribution to. **4** near.

towel *n.* absorbent cloth for drying oneself or wiping things dry. ● *v.* (**towelled**) rub with a towel.

towelling *n.* fabric for towels.

tower *n.* tall structure, esp. as part of a church or castle etc. ● *v.* be very tall. □ **tower block** tall building with many storeys. **tower of strength** source of strong reliable support.

town *n.* **1** collection of buildings (larger than a village). **2** its inhabitants. **3** central business and shopping area. □ **go to town** (*colloq.*) do something lavishly or with enthusiasm. **town hall** building containing local government offices etc. **townsman** *n.*, **townswoman** *n.*

■ **1** borough, city, conurbation, metropolis, municipality.

township *n.* **1** small town. **2** (*S. Afr.*) urban area formerly set aside for black people.

toxaemia /tokséemiə/ *n.* **1** blood poisoning. **2** abnormally high blood pressure in pregnancy.

toxic *adj.* **1** of or caused by poison. **2** poisonous. □ **toxicity** *n.*

■ **2** dangerous, deadly, harmful, lethal, noxious, poisonous.

toxicology *n.* study of poisons. □ **toxicologist** *n.*

toxin *n.* poisonous substance, esp. formed in the body.

toy *n.* **1** thing to play with. **2** (*attrib.*, of a dog) of a diminutive variety. ● *v.* **toy with 1** handle idly. **2** deal with (a thing) without seriousness. □ **toy boy** (*colloq.*) woman's much younger male lover.

trace[1] *n.* **1** track or mark left behind. **2** sign of what has existed or occurred. **3** very small quantity. ● *v.* **1** follow or discover by observing marks or other evidence. **2** outline. **3** copy by using tracing-paper or carbon paper. □ **trace element** one required only in minute amounts. **traceable** *adj.*, **tracer** *n.*

■ *n.* **1** footprint, mark, print, spoor, track, trail. **2** clue, evidence, hint, indication, mark, sign, vestige. **3** bit, dash, drop, hint, jot, pinch, speck, spot, suggestion, touch. ● *v.* **1** detect, discover, find, unearth; follow, pursue, shadow, stalk, *colloq.* tail, track, trail. **2** delineate, map out, outline, sketch (out).

trace[2] *n.* each of the two side-straps by which a horse draws a vehicle. □ **kick over the traces** become insubordinate.

tracery *n.* **1** openwork pattern in stone. **2** lacelike pattern.

trachea /trəkée̍ə/ *n.* windpipe.

tracheotomy *n.* surgical opening into the trachea.

tracing *n.* copy of a map or drawing etc. made by tracing it. □ **tracing-paper** *n.* transparent paper used in this.

track *n.* **1** mark(s) left by a moving person or thing. **2** course. **3** path, rough road. **4** railway line. **5** particular section on a record or recording tape. **6** continuous band round the wheels of a tractor etc. ● *v.* follow the track of. □ **keep** or **lose track of** keep or fail to keep oneself informed about. **track down** find by tracking. **track suit** loose warm suit worn for exercising etc. **tracker** *n.*

■ *n.* **1** footprint, mark, print, spoor, trace, trail. **2** course, racecourse, racetrack. **3** bridle path, footpath, path, route, trail. **4** line, rails, railway. ● *v.* follow, pursue, shadow, stalk, *colloq.* tail, trace, trail. □ **track down** detect, dig up, discover, ferret out, find, locate, trace, turn up, uncover, unearth.

tract[1] *n.* **1** stretch of land. **2** system of connected parts of the body, along which something passes.

tract[2] *n.* pamphlet with a short essay, esp. on a religious subject.

tractable *adj.* easy to deal with or control. □ **tractability** *n.*

■ amenable, biddable, compliant, docile, manageable, yielding.

traction *n.* **1** pulling. **2** grip of wheels on the ground. **3** use of weights etc. to exert a steady pull on an injured limb etc.

tractor *n.* powerful motor vehicle for pulling heavy equipment.

trade *n.* **1** exchange of goods for money or other goods. **2** business or customers of a particular kind. **3** skilled job. ● *v.* **1** engage in trade, buy and sell. **2** exchange (goods) in trading. □ **trade in** give (a used article) as partial payment for a new one. **trade-in** *n.* **trade mark** company's registered emblem or name etc. used to identify its goods. **trade off** exchange as a compromise. **trade union** (*pl.* **trade unions**) organized association of employees formed to further their common interests. **trade-unionist** *n.* member of a trade union. **trade wind** constant wind blowing towards the equator. **trader** *n.*

■ *n.* **1** barter, business, buying and selling, commerce, dealings, traffic. **2** business, clientele, custom, customers, patrons, shoppers. **3** business, calling, career, craft, job, line, profession, vocation, work. ● *v.* **1** buy and sell, do business, have dealings, traffic. **2** barter, change, exchange, interchange, swap, switch.

tradesman *n.* **1** craftsman. **2** shopkeeper. **3** person who delivers goods to private houses.

■ **1** artisan, craftsman. **2** dealer, merchant, retailer, shopkeeper, trader.

tradition *n.* **1** belief or custom handed down from one generation to another. **2** long-established procedure. □ **traditional** *adj.*, **traditionally** *adv.*

■ **1** convention, custom, habit, institution, practice, usage, way; belief, folklore, lore. □ **traditional** conventional, customary, established, habitual, normal, regular, routine, time-honoured, usual.

traditionalist *n.* person who upholds traditional beliefs etc. □ **traditionalism** *n.*

traduce *v.* slander. □ **traducement** *n.*

traffic *n.* **1** vehicles, ships, or aircraft moving along a route. **2** trading. ● *v.* (**trafficked**) trade. □ **traffic warden** official who controls the movement and parking of road vehicles. **trafficker** *n.*

tragedian *n.* **1** writer of tragedies. **2** actor in tragedy.

tragedienne *n.* actress in tragedy.

tragedy *n.* **1** event causing great sadness. **2** serious drama with unhappy events or a sad ending.

■ **1** calamity, cataclysm, catastrophe, disaster, misfortune.

tragic *adj.* **1** causing great sadness. **2** of or in tragedy. □ **tragically** *adv.*

■ **1** appalling, awful, calamitous, catastrophic, disastrous, dreadful, fatal, fearful, horrible, terrible.

tragicomedy *n.* drama of mixed tragic and comic events.

trail *v.* **1** drag behind, esp. on the ground. **2** hang loosely. **3** lag, straggle. **4** track. **5** move wearily. ● *n.* **1** thing that trails. **2** track, trace. **3** beaten path. **4** line of people or things following something.

■ *v.* **1** drag, pull, tow. **2** dangle, hang. **3** dawdle, lag behind, linger, loiter, straggle. **4** follow, pursue, shadow, stalk, *colloq.* tail, trace, track. ● *n.* **2** footprints, marks, prints, spoor, tracks, traces. **3** footpath, path, route, track.

trailer *n.* **1** truck etc. designed to be hauled by a vehicle. **2** short extract from a film etc., shown in advance to advertise it.

train *n.* **1** railway engine with linked carriages or trucks. **2** retinue. **3** sequence of things. **4** part of a long robe that trails behind the wearer. ● *v.* **1** bring or come to a desired standard of efficiency or condition or behaviour etc. by instruction and practice. **2** aim (a gun etc.). **3** cause (a plant) to grow in the required direction. □ **in train** in preparation.

■ *n.* **2** entourage, escort, retinue, suite. **3** chain, line, sequence, series, string, succession. ● *v.* **1** coach, discipline, drill, educate, instruct, school, teach, tutor; exercise, practise, work out.

trainee *n.* person being trained.

trainer *n.* **1** person who trains horses or athletes etc. **2** (*pl.*) soft sports or running shoes.

traipse *v.* (*colloq.*) trudge.

trait *n.* characteristic.

■ attribute, characteristic, feature, idiosyncrasy, peculiarity, property, quality, quirk.

traitor *n.* person who behaves disloyally, esp. to his or her country. □ **traitorous** *adj.*

trajectory *n.* path of a projectile.

tram *n.* public passenger vehicle running on rails laid in the road.

tramcar *n.* tram.

tramlines *n.pl.* **1** rails on which a tram runs. **2** (*colloq.*) pair of parallel sidelines in tennis etc.

trammel *v.* (**trammelled**) hamper, restrain.

tramp *v.* **1** walk heavily. **2** go on foot across (an area). **3** trample. ● *n.* **1** sound of heavy footsteps. **2** long walk. **3** vagrant. **4** (*sl.*) promiscuous woman.

■ *v.* **1** clump, plod, stamp, stomp, *colloq.* traipse, trudge. **2** hike, march, slog, trek, trudge, walk. ● *n.* **2** hike, march, slog, trek, trudge, walk. **3** *US sl.* bum, down-and-out, vagabond, vagrant.

trample *v.* tread repeatedly, crush or harm by treading.

■ stamp, step, tramp, tread; crush, flatten, press, squash.

trampoline *n.* sheet attached by springs to a frame, used for jumping on in acrobatic leaps. ● *v.* use a trampoline.

trance *n.* sleep-like or dreamy state.

tranquil *adj.* calm and undisturbed. □ **tranquilly** *adv.*, **tranquillity** *n.*

■ calm, peaceful, placid, quiet, relaxed, serene, still, untroubled.

tranquillize *v.* make calm.

■ sedate; calm, pacify, quiet, quieten, soothe.

tranquillizer *n.* drug used to relieve anxiety and tension.

transact *v.* perform or carry out (business etc.). □ **transaction** *n.*

transatlantic *adj.* **1** on or from the other side of the Atlantic. **2** crossing the Atlantic.

transcend *v.* **1** go beyond the range of (experience, belief, etc.). **2** surpass. □ **transcendent** *adj.*, **transcendence** *n.*

■ **2** exceed, excel, outdo, outshine, outstrip, surpass, top.

transcendental *adj.* **1** transcendent. **2** abstract, obscure, visionary.

transcontinental *adj.* crossing or extending across a continent.

transcribe *v.* **1** copy in writing. **2** produce in written form. **3** arrange (music) for a different instrument etc. □ **transcription** *n.*

transcript *n.* written copy.

transducer *n.* device that receives waves or other variations from one system and conveys related ones to another.

transept *n.* part lying at right angles to the nave in a church.

transfer *v.* /transfér/ (**transferred**) convey from one place, person, or application to another. ● *n.* /tránsfer/ **1** transferring. **2** document transferring property or a right. **3** design for transferring from one surface to another. ▫ **transference** *n.*, **transferable** *adj.*

■ *v.* carry, convey, move, remove, send, shift, transport; hand over, make over, pass on.

transfigure *v.* make nobler or more beautiful in appearance. ▫ **transfiguration** *n.*

transfix *v.* **1** pierce through, impale. **2** make motionless with fear or astonishment.

■ **2** freeze, paralyse, stun; captivate, enthral, fascinate, hold spellbound, hypnotize, mesmerize, rivet.

transform *v.* **1** change greatly in appearance or character. **2** change the voltage of (electric current). ▫ **transformation** *n.*, **transformer** *n.*

■ **1** alter, change, convert, metamorphose, revolutionize, transfigure, transmute. ▫ **transformation** alteration, change, conversion, metamorphosis, transfiguration, transmutation.

transfuse *v.* **1** give a transfusion of or to. **2** permeate, imbue.

transfusion *n.* injection of blood or other fluid into a blood vessel.

transgress *v.* break (a rule or law). ▫ **transgression** *n.*, **transgressor** *n.*

■ break, contravene, defy, disobey, flout, infringe, violate. ▫ **transgression** error, fault, misdeed, misdemeanour, offence, sin, wrongdoing, violation.

transient *adj.* passing away quickly. ▫ **transience** *n.*

■ brief, ephemeral, fleeting, fugitive, momentary, short, temporary, transitory.

transistor *n.* **1** very small semiconductor device which controls the flow of an electric current. **2** portable radio set using transistors. ▫ **transistorized** *adj.*

transit *n.* process of going or conveying across, over, or through.

transition *n.* process of changing from one state or style etc. to another. ▫ **transitional** *adj.*

■ alteration, change, conversion, development, evolution, metamorphosis, shift, transformation.

transitory *adj.* brief, fleeting.

translate *v.* **1** express in another language or other words. **2** be able to be translated. **3** transfer. ▫ **translation** *n.*, **translator** *n.*

transliterate *v.* convert to the letters of another alphabet. ▫ **transliteration** *n.*

translucent *adj.* allowing light to pass through but not transparent. ▫ **translucence** *n.*

transmigrate *v.* (of the soul) pass into another body after a person's death. ▫ **transmigration** *n.*

transmission *n.* **1** transmitting. **2** broadcast. **3** gear transmitting power from engine to axle.

transmit *v.* (**transmitted**) **1** pass on from one person, place, or thing to another. **2** send out (a signal or programme etc.) by cable or radio waves. ▫ **transmissible** *adj.*, **transmitter** *n.*

■ **1** convey, deliver, relay, send, transfer; communicate, pass on, spread. **2** broadcast, relay.

transmute *v.* change in form or substance. ▫ **transmutation** *n.*

transom *n.* **1** horizontal bar across the top of a door or window. **2** fanlight.

transparency *n.* **1** being transparent. **2** photographic slide.

transparent *adj.* **1** able to be seen through. **2** easily understood. ▫ **transparently** *adv.*

■ **1** clear, crystalline, limpid, pellucid, translucent. **2** apparent, clear, evident, manifest, obvious, plain, unmistakable.

transpire *v.* **1** become known. **2** (of plants) give off (vapour) from leaves etc. ▫ **transpiration** *n.*

■ **1** become known, be revealed, come to light, emerge, turn out.

transplant *v.* /tranzplaánt/ **1** plant elsewhere. **2** transfer (living tissue). ● *n.* /tránzplaant/ **1** transplanting of tissue. **2** thing transplanted. ▫ **transplantation** *n.*

transport *v.* /tranzpórt/ convey from one place to another. ● *n.* /tránzport/ **1** process of transporting. **2** means of conveyance. **3** (*pl.*) condition of strong emotion. ▫ **transportation** *n.*, **transporter** *n.*

■ *v.* bear, carry, *sl.* cart, convey, ferry, haul, move, send, ship. ● *n.* **1** carriage, conveyance, haulage, moving, shipping, shipment.

transpose *v.* **1** cause (two or more things) to change places. **2** change the position of. **3** put (music) into a different key. □ **transposition** *n.*

■ **1** exchange, interchange, rearrange, reverse, swap, switch.

transsexual *n.* & *adj.* (person) having the physical characteristics of one sex and psychological characteristics of the other. □ **transsexualism** *n.*

transuranic *adj.* having atoms heavier than those of uranium.

transverse *adj.* crosswise.

transvestism *n.* dressing in clothing of the opposite sex. □ **transvestite** *n.*

trap *n.* **1** device for capturing an animal. **2** scheme for tricking or catching a person. **3** trapdoor. **4** compartment from which a dog is released in racing. **5** curved section of a pipe holding liquid to prevent gases from coming upwards. **6** two-wheeled horse-drawn carriage. **7** (*sl.*) mouth. ● *v.* (**trapped**) catch in or by means of a trap.

■ *n.* **1** snare. **2** device, ploy, ruse, stratagem, trick, wile. ● *v.* capture, catch, ensnare, net, snare; deceive, dupe, fool, trick; confine, imprison.

trapdoor *n.* door in a floor, ceiling, or roof.

trapeze *n.* a kind of swing on which acrobatics are performed.

trapezium *n.* **1** quadrilateral with only two opposite sides parallel. **2** (*US*) trapezoid.

trapezoid *n.* **1** quadrilateral with no sides parallel. **2** (*US*) trapezium.

trapper *n.* person who traps animals, esp. for furs.

trappings *n.pl.* accessories.

■ accessories, accoutrements, apparatus, equipment, finery, gear, paraphernalia, trimmings.

trash *n.* worthless stuff. □ **trashy** *adj.*

trauma *n.* **1** wound, injury. **2** emotional shock producing a lasting effect. □ **traumatic** *adj.*

■ □ **traumatic** distressing, disturbing, harrowing, painful, shocking, upsetting.

travail *n.* & *v.* labour.

travel *v.* (**travelled**) **1** go from one place to another. **2** journey along or through. ● *n.* travelling, esp. abroad. □ **traveller** *n.*

■ *v.* **1** go, journey, take a trip, tour, voyage. ● *n.* globe-trotting, touring, tourism. □ **traveller** globe-trotter, holiday-maker, sightseer, tourist, voyager.

traverse *v.* travel, lie, or extend across. ● *n.* **1** thing that lies across another. **2** lateral movement. □ **traversal** *n.*

■ *v.* range, rove, travel, wander; bridge, cross, pass over, span.

travesty *n.* absurd or inferior imitation. ● *v.* make or be a travesty of.

trawl *n.* large wide-mouthed fishing net. ● *v.* fish with a trawl.

trawler *n.* boat used in trawling.

tray *n.* **1** board with a rim for dishes etc. **2** shallow box for papers etc.

treacherous *adj.* **1** showing treachery. **2** not to be relied on, dangerous. □ **treacherously** *adv.*

■ **1** disloyal, false, perfidious, traitorous, treasonous. **2** dangerous, hazardous, perilous, precarious, unreliable, unsafe.

treachery *n.* **1** betrayal of a person or cause. **2** act of disloyalty.

■ **1** betrayal, disloyalty, faithlessness, perfidy, treason.

treacle *n.* thick sticky liquid produced when sugar is refined. □ **treacly** *adj.*

tread *v.* (**trod, trodden**) **1** walk, step. **2** walk on, press or crush with the feet. ● *n.* **1** manner or sound of walking. **2** horizontal surface of a stair. **3** part of a tyre that touches the ground. □ **tread water** keep upright in water by making treading movements.

■ *v.* **1** go, step, walk. **2** crush, flatten, press, squash, stamp on, step on, tramp on, trample.

treadle *n.* lever worked by the foot to drive a wheel.

treadmill *n.* **1** mill-wheel formerly turned by people treading on steps round its edge. **2** tiring monotonous routine work.

treason *n.* treachery towards one's country. □ **treasonous** *adj.*

treasonable *adj.* involving treason.

treasure *n.* **1** collection of precious metals or gems. **2** highly valued object or person. ● *v.* **1** value highly. **2** store as precious. □ **treasure trove** treasure of unknown ownership, found hidden.

■ *n.* **1** fortune, riches, valuables, wealth. **2** darling, delight, gem, jewel, joy. ● *v.* **1** cherish, hold dear, prize, value.

treasurer *n.* person in charge of the funds of an institution.

treasury *n.* **1** place where treasure is kept. **2** department managing a country's revenue.

treat *v.* **1** act or behave towards or deal with in a specified way. **2** give medical treatment to. **3** buy something for (a person) in order to give pleasure. **4** subject to a chemical or other process. ● *n.* something special that gives pleasure.

■ *v.* **1** deal with, handle, manage; consider, regard, view. **2** dose, medicate, minister to, tend. **3** entertain, pay for, stand, take out. ● *n.* gift, present; delicacy, *colloq.* goody, luxury, titbit.

treatise *n.* written work dealing with one subject.

■ dissertation, essay, paper.

treatment *n.* **1** manner of dealing with a person or thing. **2** something done to relieve illness etc.

■ **1** care, handling, management. **2** healing, therapy; cure, remedy.

treaty *n.* formal agreement made, esp. between countries.

■ agreement, compact, entente, pact, settlement.

treble *adj.* **1** three times as much or as many. **2** (of a voice) high-pitched, soprano. ● *n.* **1** treble quantity or thing. **2** treble voice, person with this. □ **trebly** *adv.*

tree *n.* perennial plant with a single thick woody stem.

trefoil *n.* **1** plant with three leaflets (e.g. clover). **2** thing shaped like this.

trek *n.* long arduous journey. ● *v.* (**trekked**) make a trek.

■ *n.* & *v.* hike, march, slog, tramp, trudge, walk.

trellis *n.* light framework of crossing strips of wood etc.

tremble *v.* **1** shake, quiver. **2** feel very anxious. ● *n.* trembling movement. □ **trembly** *adj.*

■ *v.* **1** quake, quaver, quiver, shake, shiver, shudder.

tremendous *adj.* **1** immense. **2** (*colloq.*) excellent. □ **tremendously** *adv.*

■ **1** big, colossal, enormous, gigantic, huge, immense, large, massive, vast.

tremor *n.* **1** slight trembling movement. **2** thrill of fear etc.

tremulous *adj.* trembling, quivering. □ **tremulously** *adv.*

trench *n.* deep ditch. □ **trench coat** belted double-breasted raincoat.

trenchant *adj.* (of comments etc.) strong and effective.

■ effective, incisive, keen, penetrating, pointed.

trend *n.* **1** general tendency. **2** fashion. □ **trend-setter** *n.* person who leads the way in fashion etc.

■ **1** bent, bias, inclination, leaning, tendency. **2** craze, fad, fashion, mode, style, vogue.

trendy *adj.* (**-ier, -iest**) (*colloq.*) following the latest fashion. □ **trendily** *adv.*, **trendiness** *n.*

trepidation *n.* fear, anxiety.

trespass *v.* **1** enter land or property unlawfully. **2** intrude. ● *n.* act of trespassing. □ **trespasser** *n.*

tress *n.* lock of hair.

trestle *n.* one of a set of supports on which a board is rested to form a table. □ **trestle-table** *n.*

trews *n.pl.* close-fitting usu. tartan trousers.

tri- *pref.* three times, triple.

triad *n.* **1** group of three. **2** Chinese usu. criminal secret society.

trial *n.* **1** examination in a law court by a judge. **2** process of testing qualities or performance. **3** trying person or thing. □ **on trial** undergoing a trial.

■ **1** case, hearing, lawsuit, proceedings. **2** check, *colloq.* dry run, dummy run, test, try-out. **3** bane, bother, nuisance, pest, worry; affliction, difficulty, hardship, misfortune, ordeal, tribulation, trouble.

triangle *n.* **1** geometric figure with three sides and three angles. **2** triangular steel rod used as a percussion instrument.

triangular *adj.* **1** shaped like a triangle. **2** involving three people.

triangulation *n.* measurement or mapping of an area by means of a network of triangles.

tribe *n.* related group of families living as a community. □ **tribal** *adj.*, **tribesman** *n.*

tribulation *n.* great affliction.

tribunal *n.* board of officials appointed to adjudicate on a particular problem.

tributary *n.* & *adj.* (stream) flowing into a larger stream or a lake.

tribute *n.* **1** something said or done as a mark of respect. **2** payment that one country or ruler was formerly obliged to pay to another.

■ **1** accolade, commendation, compliment, honour, praise.

trice *n.* **in a trice** in an instant.

trichology *n.* study of hair and its diseases. □ **trichologist** *n.*

trick *n.* **1** something done to deceive or outwit someone. **2** technique, knack. **3** mannerism. **4** mischievous act. ● *v.* deceive or persuade by a trick. □ **do the trick** (*colloq.*) achieve what is required.

■ *n.* **1** artifice, *sl.* con, deceit, deception, device, ploy, ruse, stratagem, subterfuge, trap, wile. **2** art, knack, skill, technique. **3** characteristic, idiosyncrasy, mannerism, peculiarity, quirk, trait. **4** gag, hoax, (practical) joke, prank. ● *v.* cheat, *sl.* con, deceive, double-cross, dupe, fool, hoodwink, mislead, pull the wool over someone's eyes, swindle, take in.

trickery *n.* use of tricks, deception.

■ artfulness, chicanery, deceit, deception, double-dealing, duplicity, fraud, guile, hocus-pocus, *colloq.* jiggery-pokery, *colloq.* skulduggery.

trickle *v.* **1** (cause to) flow in a thin stream. **2** come or go gradually. ● *n.* trickling flow.

■ *v.* **1** dribble, drip, drop, flow, leak, ooze, seep.

tricky *adj.* (**-ier**, **-iest**) **1** crafty, deceitful. **2** requiring careful handling. □ **trickiness** *n.*

■ **1** artful, crafty, cunning, deceitful, devious, dishonest, double-dealing, duplicitous, guileful, scheming, shifty, slippery, sly, wily. **2** awkward, delicate, difficult, sensitive, ticklish.

tricolour *n.* flag with three colours in stripes.

tricycle *n.* three-wheeled pedal-driven vehicle. □ **tricyclist** *n.*

trident *n.* three-pronged spear.

triennial *adj.* **1** happening every third year. **2** lasting three years.

trier *n.* person who tries hard.

trifle *n.* **1** thing of only slight value or importance. **2** very small amount. **3** sweet dish of sponge cake and jelly etc. topped with custard and cream. ● *v.* **trifle with** treat frivolously.

trifling *adj.* trivial.

■ *sl.* footling, frivolous, inconsequential, insignificant, minor, petty, trivial, unimportant.

trigger *n.* lever for releasing a spring, esp. to fire a gun. ● *v.* (also **trigger off**) set in action, cause. □ **trigger-happy** *adj.* apt to shoot on slight provocation.

trigonometry *n.* branch of mathematics dealing with the relationship of sides and angles of triangles etc.

trike *n.* (*colloq.*) tricycle.

trilateral *adj.* having three sides or three participants.

trilby *n.* man's soft felt hat.

trill *n.* vibrating sound, esp. in music or singing. ● *v.* sound or sing with a trill.

trillion *n.* **1** a million million million. **2** a million million.

trilobite *n.* a kind of fossil crustacean.

trilogy *n.* group of three related books, plays, etc.

trim *adj.* (**trimmer**, **trimmest**) neat and orderly. ● *v.* (**trimmed**) **1** reduce or neaten by cutting. **2** ornament. **3** make (a boat or aircraft) evenly balanced by distributing its load. ● *n.* **1** ornamentation. **2** colour or type of upholstery etc. in a car. **3** trimming of hair etc.

■ *adj.* neat, orderly, shipshape, smart, spruce, straight, tidy. ● *v.* **1** bob, clip, crop, cut, dock, lop, pare, prune, shear, snip. **2** adorn, beautify, deck, decorate, dress, embellish, garnish, ornament. ● *n.* **1** adornment, decoration, embellishment, ornamentation.

trimaran *n.* vessel like a catamaran, with three hulls.

trimming *n.* **1** thing added as a decoration. **2** (*pl.*) pieces cut off when something is trimmed.

trinity *n.* **1** group of three. **2** (**the Trinity**) the three members of the Christian deity (Father, Son, Holy Spirit) as constituting one God.

trinket *n.* small fancy article or piece of jewellery.

trio *n.* (*pl.* **-os**) **1** group or set of three. **2** music for three instruments or voices.

trip *v.* (**tripped**) **1** go lightly and quickly. **2** (cause to) catch one's foot and lose balance. **3** (cause to) make a blunder. **4** release (a switch etc.) so as to operate a mechanism. ● *n.* **1** journey or excursion, esp. for pleasure. **2** stumble. **3** (*colloq.*) visionary experience caused by a drug. **4** device for tripping a mechanism.

■ *v.* **1** caper, cavort, dance, frisk, frolic, gambol, skip. **2** fall, slip, stumble, tumble. ● *n.* **1** drive, excursion, expedition, jaunt, journey, outing, tour.

tripartite *adj.* consisting of three parts.

tripe *n.* **1** stomach of an ox etc. as food. **2** (*sl.*) nonsense.

triple *adj.* **1** having three parts or members. **2** three times as much or as many. ● *v.* increase by three times its amount.

triplet *n.* **1** one of three children born at one birth. **2** set of three.

triplex *adj.* triple, threefold.

triplicate *adj.* & *n.* existing in three examples. □ **in triplicate** as three identical copies.

tripod *n.* three-legged stand.

tripper *n.* person who goes on a pleasure trip.

triptych /tríptik/ *n.* picture or carving with three panels fixed or hinged side by side.

trite *adj.* hackneyed.

■ banal, clichéd, *colloq.* corny, hackneyed, platitudinous, stale, unimaginative, unoriginal.

triumph *n.* **1** fact of being successful or victorious. **2** great success. **3** joy at this. ● *v.* be successful or victorious, rejoice at this. □ **triumphant** *adj.*, **triumphantly** *adv.*

■ *n.* **2** conquest, success, victory, win. **3** elation, exultation, joy, jubilation, rejoicing. ● *v.* be victorious, prevail, succeed, win. □ **triumphant** conquering, successful, victorious, winning.

triumphal *adj.* celebrating or commemorating a triumph.

triumvirate *n.* government or control by a board of three.

trivet *n.* metal stand for a kettle or hot dish etc.

trivia *n.pl.* trivial things.

trivial *adj.* of only small value or importance. □ **trivially** *adv.*, **triviality** *n.*

■ *sl.* footling, frivolous, inconsequential, insignificant, minor, petty, trifling, unimportant.

trod, trodden *see* **tread**.

troglodyte *n.* cave dweller.

troll *n.* giant or dwarf in Scandinavian mythology.

trolley *n.* (*pl.* **-eys**) **1** platform on wheels for transporting goods. **2** small cart. **3** small table on wheels.

trollop *n.* promiscuous or slovenly woman.

trombone *n.* large brass wind instrument with a sliding tube.

troop *n.* **1** company of people or animals. **2** cavalry or artillery unit. ● *v.* go as a troop or in great numbers.

■ *n.* **1** band, company, crowd, drove, flock, gang, group, herd, horde, throng. ● *v.* file, march, walk.

trooper *n.* **1** soldier in a cavalry or armoured unit. **2** (*US*) member of a state police force.

trophy *n.* **1** memento of any success. **2** object awarded as a prize.

■ **1** keepsake, memento, reminder, souvenir, token. **2** award, cup, medal, prize.

tropic *n.* **1** line of latitude 23° 27' north or south of the equator. **2** (*pl.*) region between these, with a hot climate. □ **tropical** *adj.*

troposphere *n.* layer of the atmosphere between the earth's surface and the stratosphere.

trot *n.* **1** running action of a horse etc. **2** moderate running pace. ● *v.* (**trotted**) go or cause to go at a trot. □ **on the trot** (*colloq.*) **1** continually busy. **2** in succession. **trot out** (*colloq.*) produce, repeat.

troth *n.* promise, fidelity.

trotter *n.* animal's foot as food.

troubadour *n.* medieval romantic poet.

trouble *n.* **1** difficulty, distress, misfortune. **2** cause of this. **3** conflict. **4** illness. **5** exertion. ● *v.* **1** cause trouble to. **2** make or be worried. **3** exert oneself.

■ *n.* **1** anxiety, concern, difficulty, distress, *colloq.* hassle, inconvenience, misfortune, unpleasantness, woe, worry. **2** bother, inconvenience, irritation, nuisance, pest. **3** conflict, discord, disorder, disturbance, fighting, strife, turmoil, unrest, violence. **4** disease, illness, sickness; ache, pain. **5** bother, care, effort, exertion, pains. ● *v.* **1** annoy, bother, harass, *colloq.* hassle, nag, pester, *colloq.* plague, torment, vex; discommode, impose on, inconvenience, put out. **2** agitate, alarm, disquiet, perturb, unsettle, upset, worry. **3** be concerned, bother, care, concern oneself, exert oneself, take pains.

troubleshooter *n.* person employed to deal with faults or problems.

troublesome *adj.* causing trouble, annoying.

■ awkward, burdensome, difficult, inconvenient; annoying, bothersome, irksome, irritating, tiresome, vexatious, vexing, worrying.

trough *n.* **1** long open receptacle, esp. for animals' food or water. **2** channel or hollow like this. **3** region of low atmospheric pressure.

trounce *v.* defeat heavily.

■ beat, conquer, defeat, *colloq.* lick, overwhelm, thrash, vanquish.

troupe /tro͞op/ *n.* company of actors or other performers.

trouper *n.* **1** member of a troupe. **2** staunch colleague.

trousers *n.pl.* two-legged outer garment reaching from the waist usu. to the ankles.

trousseau /tro͞ossō/ *n.* (*pl.* **-eaux**) bride's collection of clothing etc. for her marriage.

trout *n.* (*pl.* **trout**) freshwater fish related to the salmon.

trowel *n.* **1** small garden tool for digging. **2** similar tool for spreading mortar etc.

troy weight system of weights used for precious metals.

truant *n.* pupil who stays away from school without permission. □ **play truant** stay away as a truant. **truancy** *n.*

truce *n.* agreement to cease hostilities temporarily.

■ armistice, ceasefire, peace.

truck *n.* **1** open container on wheels for transporting loads. **2** open railway wagon. **3** lorry.

trucker *n.* lorry driver.

truculent *adj.* defiant and aggressive. □ **truculently** *adv.*, **truculence** *n.*

■ aggressive, antagonistic, bad-tempered, belligerent, defiant, ill-tempered, quarrelsome, surly.

trudge *v.* walk laboriously. ● *n.* laborious walk.

■ *v.* plod, slog, *colloq.* traipse, tramp, trek.

true *adj.* **1** in accordance with fact, genuine. **2** exact, accurate. **3** loyal, faithful. ● *adv.* truly, accurately.

■ *adj.* **1** authentic, bona fide, genuine, legitimate, real, rightful, veritable; accurate, correct, factual, faithful, literal, truthful, veracious. **2** accurate, correct, exact, precise, proper. **3** constant, dependable, devoted, faithful, firm, loyal, staunch, steadfast, trusted, trustworthy.

truffle *n.* **1** rich-flavoured underground fungus valued as a delicacy. **2** soft chocolate sweet.

truism *n.* self-evident or hackneyed statement.

truly *adv.* **1** truthfully. **2** genuinely. **3** faithfully.

trump *n.* **1** playing card of a suit temporarily ranking above others. **2** (*colloq.*) person who behaves in a helpful or useful way. ● *v.* **trump up** invent fraudulently. □ **turn up trumps** (*colloq.*) be successful or helpful.

trumpet *n.* **1** metal wind instrument with a flared tube. **2** thing shaped like this. ● *v.* (**trumpeted**) **1** proclaim loudly. **2** (of an elephant) make a loud sound with its trunk. □ **trumpeter** *n.*

truncate *v.* shorten by cutting off the end. □ **truncation** *n.*

truncheon *n.* short thick stick carried as a weapon.

trundle *v.* roll along, move along heavily on wheels.

trunk *n.* **1** tree's main stem. **2** body apart from head and limbs. **3** large box with a hinged lid, for transporting or storing clothes etc. **4** elephant's long flexible nose. **5** (*US*) boot of a car. **6** (*pl.*) shorts for swimming etc. □ **trunk road** important main road.

■ **1** bole, stem. **2** torso. **3** box, chest, coffer, crate, tea chest.

truss *n.* **1** framework supporting a roof etc. **2** device worn to support a hernia. ● *v.* tie up securely.

trust *n.* **1** firm belief in the reliability, truth, or strength etc. of a person or thing. **2** confident expectation. **3** responsibility, care. **4** association of business firms, formed to defeat competition. **5** property legally entrusted to someone. ● *v.* **1** have or place trust in. **2** entrust. **3** hope earnestly. □ **in trust** held as a trust. **on trust** accepted without investigation. **trustful** *adj.*, **trustfully** *adv.*

■ *n.* **1** belief, certainty, confidence, credence, faith, reliance. **2** belief, conviction, expectation. **3** care, charge, custody, guardianship, keeping, protection. **4** cartel, corporation, group, syndicate. ● *v.* **1** bank on, believe in, be sure of, count on, depend on, have faith in, rely on. **2** commit, consign, delegate, entrust, give, hand over.

trustee *n.* **1** person who administers property held as a trust. **2** one of a group managing the business affairs of an institution.

trustworthy *adj.* worthy of trust.

■ dependable, faithful, honest, loyal, principled, reliable, responsible, steady, true, trusty.

trusty *adj.* trustworthy.

truth *n.* **1** quality of being true. **2** something that is true.

■ **1** accuracy, correctness, genuineness, truthfulness, veracity. **2** actuality, fact(s), reality.

truthful *adj.* **1** habitually telling the truth. **2** true. □ **truthfully** *adv.*, **truthfulness** *n.*

■ **1** candid, frank, honest, reliable, sincere, veracious. **2** accurate, correct, factual, true, veracious.

try *v.* **1** attempt. **2** test, esp. by use. **3** be a strain on. **4** hold a trial of. ● *n.* **1** attempt. **2** touchdown in Rugby football, entitling the player's side to a kick at goal. □ **try on** put (a garment) on to see if it fits. **try out** test by use. **try-out** *n.*

■ *v.* **1** attempt, endeavour, essay, have a go, make an effort, seek, strive, venture. **2** appraise, assess, evaluate, test, sample. **3** strain, tax. **4** adjudge, adjudicate, judge. ● *n.* **1** attempt, bid, effort, endeavour, go, shot, *colloq.* stab.

trying *adj.* **1** annoying. **2** difficult.

■ **1** annoying, bothersome, exasperating, infuriating, irksome, irritating, maddening, tiresome, vexatious. **2** demanding, difficult, hard, stressful, taxing, tiring, tough.

tsar /zaar/ *n.* title of the former emperor of Russia.

tsetse *n.* African fly that transmits disease by its bite.

T-shirt *n.* short-sleeved casual cotton top.

T-square *n.* large T-shaped ruler used in technical drawing.

tub *n.* open usu. round container.

tuba *n.* large low-pitched brass wind instrument.

tubby *adj.* (**-ier**, **-iest**) short and fat. □ **tubbiness** *n.*

■ chubby, dumpy, plump, podgy, portly, pudgy, roly-poly, rotund.

tube *n.* **1** long hollow cylinder. **2** thing shaped like this. **3** cathode-ray tube.

tuber *n.* short thick rounded root or underground stem from which shoots will grow.

tubercle *n.* small rounded swelling.

tubercular *adj.* of or affected with tuberculosis.

tuberculosis *n.* infectious wasting disease, esp. affecting lungs.

tuberous *adj.* **1** of or like a tuber. **2** bearing tubers.

tubing *n.* **1** tubes. **2** a length of tube.

tubular *adj.* tube-shaped.

tuck *n.* flat fold stitched in a garment etc. ● *v.* **1** put into or under something so as to be concealed or held in place. **2** cover or put away compactly. □ **tuck in** (*colloq.*) eat heartily. **tuck shop** shop selling sweets etc. to schoolchildren.

tuft *n.* bunch of threads, grass, or hair etc. held or growing together at the base. □ **tufted** *adj.*

■ bunch, clump, knot.

tug *v.* (**tugged**) **1** pull vigorously. **2** tow. ● *n.* **1** vigorous pull. **2** small powerful boat for towing others. □ **tug of war** contest of strength in which two teams pull opposite ways on a rope.

■ *v.* **1** jerk, pull, tweak, *colloq.* yank. **2** haul, pull, tow.

tuition *n.* teaching or instruction, esp. of an individual or small group.

■ education, instruction, schooling, teaching, tutelage.

tulip *n.* garden plant with a cup-shaped flower. □ **tulip-tree** *n.* tree with tulip-like flowers.

tulle /tyool/ *n.* fine silky net fabric.

tumble *v.* **1** (cause to) fall. **2** roll, push, or move in a disorderly way. **3** perform somersaults etc. **4** rumple. ● *n.* **1** fall. **2** untidy state. □ **tumble-drier** *n.* machine for drying washing in a heated rotating drum. **tumble to** (*colloq.*) grasp the meaning of.

■ *v.* **1** collapse, *sl.* come a cropper, drop, fall, overbalance, stumble, topple (over). **4** disarrange, disorder, mess (up), ruffle, rumple, tousle. ● *n.* **1** *sl.* cropper, fall, spill, stumble. **2** clutter, disorder, jumble, mess, muddle.

tumbledown *adj.* dilapidated.

■ decrepit, derelict, dilapidated, ramshackle, rickety, ruined.

tumbler *n.* **1** drinking glass with no handle or foot. **2** pivoted piece in a lock etc. **3** acrobat.

tumescent *adj.* swelling. □ **tumescence** *n.*

tummy *n.* (*colloq.*) stomach.

tumour *n.* abnormal mass of new tissue growing in or on the body.

tumult *n.* **1** uproar. **2** conflict of emotions. □ **tumultuous** *adj.*

■ **1** commotion, disorder, disturbance, fracas, pandemonium, rumpus, turmoil, unrest, uproar. **2** confusion, turmoil, upheaval. □ **tumultuous** boisterous, disorderly, excited, riotous, rowdy, *colloq.* rumbustious, turbulent, unruly, uproarious.

tun *n.* **1** large cask. **2** fermenting-vat.

tuna *n.* (*pl.* **tuna**) **1** tunny. **2** its flesh as food.

tundra *n.* vast level Arctic regions with permafrost.

tune *n.* melody. ● *v.* **1** put (a musical instrument) in tune. **2** set (a radio) to the desired wavelength. **3** adjust (an engine) to run smoothly. □ **in tune 1** playing or singing at the correct musical pitch. **2** harmonious. **out of tune** not in tune. **tune up** bring musical instruments to the correct or uniform pitch.

■ *n.* air, melody, motif, song, strain, theme.

tuneful *adj.* melodious.

■ dulcet, euphonious, harmonious, mellifluous, melodious, musical, sweet-sounding.

tungsten *n.* heavy grey metal.

tunic *n.* **1** close-fitting jacket worn as part of a uniform. **2** loose garment reaching to the knees.

tunnel *n.* underground passage. ● *v.* (**tunnelled**) make a tunnel (through), make (one's way) in this manner.

■ *v.* bore, burrow, dig, excavate.

tunny *n.* large edible sea fish.

turban *n.* **1** Muslim or Sikh man's headdress of cloth wound round the head. **2** woman's hat resembling this.

turbid *adj.* **1** (of liquids) muddy, not clear. **2** disordered. □ **turbidity** *n.*

turbine *n.* machine or motor driven by a wheel that is turned by a flow of water or gas.

turbo- *pref.* **1** using a turbine. **2** driven by such engines.

turbot *n.* large flat edible sea fish.

turbulent *adj.* **1** in a state of commotion or unrest. **2** moving unevenly. □ **turbulently** *adv.*, **turbulence** *n.*

■ **1** confused, disorderly, restless, riotous, tumultuous, wild.

tureen *n.* deep covered dish from which soup is served.

turf *n.* (*pl.* **turfs** or **turves**) **1** short grass and the soil just below it. **2** piece of this. **3** (**the turf**) horse racing. ● *v.* lay (ground) with turf. □ **turf accountant** bookmaker. **turf out** (*colloq.*) throw out.

turgid *adj.* **1** swollen and not flexible. **2** (of language) pompous. □ **turgidly** *adv.*, **turgidity** *n.*

■ **2** bombastic, flowery, grandiloquent, pompous, pretentious.

Turk *n.* native of Turkey.

turkey *n.* (*pl.* **-eys**) **1** large bird reared for its flesh. **2** this as food.

Turkish *adj.* & *n.* (language) of Turkey. □ **Turkish bath** hot air or steam bath followed by massage etc. **Turkish delight** sweet consisting of flavoured gelatine coated in powdered sugar.

turmeric *n.* bright yellow spice from the root of an Asian plant.

turmoil *n.* state of great disturbance or confusion.

■ chaos, confusion, disorder, havoc, mayhem, pandemonium, tumult, unrest, uproar.

turn *v.* **1** move round a point or axis. **2** take or give a new direction (to). **3** aim. **4** invert, reverse. **5** pass (a certain hour or age). **6** change in form or appearance etc. **7** make or become sour. **8** give an elegant form to. **9** shape in a lathe. ● *n.* **1** process of turning. **2** change of direction or condition etc. **3** bend in a road. **4** service of a specified kind. **5** opportunity or obligation coming in succession. **6** short performance in an entertainment. **7** (*colloq.*) attack of illness, momentary nervous shock. □ **in turn** in succession. **out of turn 1** before or after one's proper turn. **2** indiscreetly, presumptuously. **to a turn** so as to be cooked perfectly. **turn against** make or become hostile to. **turn down 1** reject. **2** fold down. **3** reduce the volume or flow of. **turn in 1** hand in. **2** (*colloq.*) go to bed. **turn off 1** stop the operation of. **2** (*colloq.*) cause to lose interest. **turn on 1** start the operation of. **2** (*colloq.*) excite sexually. **turn out 1** expel. **2** turn off (a light etc.). **3** equip, dress. **4** produce by work. **5** appear. **6** prove to be the case eventually. **7** empty and search or clean. **turn-out** *n.* **1** number of people attending a gathering. **2** outfit. **3** process of turning out a room etc. **turn the tables** reverse a situation and put oneself in a superior position. **turn up 1** discover. **2** be found. **3** make one's appearance. **4** increase the volume or flow of. **turn-up** *n.* turned-up part, esp. at the lower end of a trouser leg.

■ *v.* **1** circle, gyrate, orbit, revolve, roll, rotate, spin, swivel, whirl. **2** bear, head, move, swerve, swing, veer, wheel; bend, curve, meander, snake, twist, wind. **3** aim, direct, point, train. **6** adapt, alter, change, modify, refashion, remodel, reshape, transform; become, get. **7** curdle, go bad, go off, sour, spoil. **8** construct, create, fashion, form. ● *n.* **1** revolution, rotation, spin, twirl, whirl. **2** alteration, change, shift; direction, drift, trend. **3** bend, corner, curve, turning, twist. **4** act, deed, service. **5** chance, go, opportunity. **6** act, performance, routine. **7** attack, seizure; fright, scare, shock, start, surprise. □ **turn down 1** decline, refuse, reject, *colloq.* pass up. **turn out 1** eject, evict, expel, *colloq.* kick out, oust, throw out, *colloq.* turf out. **2** extinguish, switch off, turn off. **turn up 1** come across, dig up, discover, find, uncover, unearth. **2** appear, come to light. **3** appear, arrive, put in an appearance, *colloq.* show up.

turncoat *n.* person who changes his or her principles.

turner *n.* person who works with a lathe.

turning *n.* place where one road meets another, forming a corner. □ **turning point** point at which a decisive change takes place.

turnip *n.* (plant with) a round white root used as a vegetable.

turnover *n.* **1** pasty. **2** amount of money taken in a business. **3** rate of replacement.

turnstile *n.* revolving barrier for admitting people one at a time.

turntable *n.* circular revolving platform.

turpentine *n.* oil used for thinning paint and as a solvent.

turpitude *n.* wickedness.

turps *n.* (*colloq.*) turpentine.

turquoise *n.* **1** blue-green precious stone. **2** its colour.

turret *n.* small tower-like structure. □ **turreted** *adj.*

turtle *n.* sea creature like a tortoise. □ **turn turtle** capsize. **turtle-dove** *n.* wild dove noted for its soft cooing. **turtle-neck** *n.* high round close-fitting neckline.

tusk *n.* long pointed tooth outside the mouth of certain animals.

tussle *v.* & *n.* struggle, conflict.

■ *n.* conflict, fight, *colloq.* scrap, scuffle, skirmish, struggle.

tussock *n.* tuft or clump of grass.

tutelage *n.* **1** guardianship. **2** tuition.

tutor *n.* private or university teacher. ● *v.* act as tutor (to).

■ *n.* lecturer, professor, teacher. ● *v.* coach, educate, instruct, school, teach, train.

tutorial *adj.* of a tutor. ● *n.* student's session with a tutor.

tut-tut *int.* exclamation of annoyance, impatience, or rebuke.

tutu *n.* dancer's short skirt made of layers of frills.

tuxedo /tukseédō/ *n.* (*pl.* **-os**) (*US*) dinner jacket.

twaddle *n.* nonsense.

twang *n.* **1** sharp ringing sound like that made by a tense wire when plucked. **2** nasal intonation. ● *v.* make or cause to make a twang.

tweak *v.* & *n.* (pinch, twist, or pull with) a sharp jerk.

twee *adj.* affectedly dainty or quaint.

tweed *n.* **1** thick woollen fabric. **2** (*pl.*) clothes made of tweed. □ **tweedy** *adj.*

tweet *n.* & *v.* chirp.

tweeter *n.* loudspeaker for reproducing high-frequency signals.

tweezers *n.pl.* small pincers for handling very small things.

twelve *adj.* & *n.* one more than eleven. □ **twelfth** *adj.* & *n.*

twenty *adj.* & *n.* twice ten. □ **twentieth** *adj.* & *n.*

twerp *n.* (*sl.*) stupid or objectionable person.

twice *adv.* **1** two times. **2** in double amount or degree.

twiddle *v.* twist idly about. ● *n.* act of twiddling. □ **twiddle one's thumbs** have nothing to do.

twig[1] *n.* small shoot issuing from a branch or stem.

twig[2] *v.* (**twigged**) (*colloq.*) realize, grasp the meaning of.

twilight *n.* **1** light from the sky after sunset. **2** period of this.

■ **2** dusk, evening, gloaming, sundown, sunset.

twill *n.* fabric woven so that parallel diagonal lines are produced. □ **twilled** *adj.*

twin *n.* **1** one of two children or animals born at one birth. **2** one of a pair that are exactly alike. ● *adj.* being a twin or twins. ● *v.* (**twinned**) combine as a pair. □ **twin towns** two towns that establish special social and cultural links.

twine *n.* strong thread or string. ● *v.* twist, wind.

■ *n.* cord, string, thread. ● *v.* braid, coil, entwine, plait, twist, wind.

twinge *n.* slight or brief pang.

twinkle *v.* shine with a light that flickers rapidly. ● *n.* twinkling light.

■ *v.* glint, glisten, glitter, shimmer, shine, sparkle.

twirl *v.* twist lightly or rapidly. ● *n.* **1** twirling movement. **2** twirled mark. □ **twirly** *adj.*

■ *v.* gyrate, pirouette, revolve, rotate, spin, turn, twist, whirl.

twist *v.* **1** wind (strands etc.) round each other, esp. to form a single cord. **2** make by doing this. **3** bend, turn. **4** contort, wrench. **5** distort. **6** wriggle. ● *n.* **1** process of twisting. **2** thing formed by twisting. **3** bend, turn. **4** unexpected development etc.

■ *v.* **1,2** braid, coil, entwine, plait, twine, weave, wind. **3** bend, curve, meander, snake, wind, zigzag. **4** contort, distort, screw up; rick, sprain, wrench. **5** distort, falsify, garble, misrepresent, pervert, slant. **6** squirm, wiggle, worm, wriggle, writhe. ● *n.* **2** braid, plait; coil, curl, helix, spiral, whorl. **3** bend, corner, curve, turn, zigzag.

twit *n.* (*sl.*) foolish person.

twitch *v.* **1** quiver or contract spasmodically. **2** pull with a light jerk. ● *n.* twitching movement.

twitter *v.* **1** make light chirping sounds. **2** talk rapidly in an anxious or nervous way. ● *n.* twittering.

■ *v.* **1** cheep, chirp, tweet, warble.

two *adj.* & *n.* one more than one. □ **be in two minds** be undecided. **two-dimensional** *adj.* having or appearing to have length and breadth but no depth. **two-faced** *adj.* insincere, deceitful.

■ □ **two-faced** deceitful, double-dealing, duplicitous, hypocritical, insincere, untrustworthy.

twofold *adj.* & *adv.* twice as much or as many.

twosome *n.* group of two people.

tycoon *n.* magnate.

tying *see* **tie**.

tympanum *n.* (*pl.* **-na**) **1** eardrum. **2** space between the lintel and the arch above a door.

type *n.* **1** kind, class. **2** typical example or instance. **3** set of characters used in printing. **4** (*colloq.*) person of specified character. ● *v.* **1** write with a typewriter. **2** classify according to type.

■ *n.* **1** category, class, genre, genus, group, kind, make, order, set, sort, species, variety. **2** archetype, epitome, exemplar, model, paradigm. **3** lettering, fount.

typecast *v.* cast (an actor) repeatedly in similar roles.

typesetter *n.* person or machine that sets type for printing.

typescript *n.* typewritten document.

typewriter *n.* machine for producing print-like characters on paper, by pressing keys. □ **typewritten** *adj.*

typhoid *n.* **typhoid fever** serious infectious feverish disease.

typhoon *n.* violent hurricane.

typhus *n.* infectious feverish disease transmitted by parasites.

typical *adj.* having the distinctive qualities of a particular type of person or thing. □ **typically** *adv.*

■ characteristic, classic, representative, standard; average, conventional, normal, ordinary, regular, usual.

typify *v.* be a representative specimen of.

■ characterize, epitomize, exemplify, illustrate, personify, represent, symbolize.

typist *n.* person who types.

typography *n.* art or style of printing. □ **typographical** *adj.*

tyrannize *v.* rule as or like a tyrant.

■ browbeat, bully, domineer over, intimidate, oppress, subjugate, terrorize, torment.

tyranny *n.* **1** government by a tyrant. **2** tyrannical use of power. □ **tyrannical** *adj.*, **tyrannically** *adv.*

■ **1** authoritarianism, autocracy, despotism, dictatorship, totalitarianism. □ **tyrannical** authoritarian, autocratic, despotic, dictatorial, domineering, imperious, overbearing.

tyrant *n.* ruler or other person who uses power in a harsh or oppressive way.

■ autocrat, despot, dictator; bully, martinet, slave-driver.

tyre *n.* covering round the rim of a wheel to absorb shocks.

Uu

ubiquitous *adj.* found everywhere. □ **ubiquity** *n.*
udder *n.* bag-like milk-producing organ of a cow, goat, etc.
ugly *adj.* (**-ier**, **-iest**) **1** unpleasant to look at or hear. **2** threatening, hostile. □ **ugliness** *n.*

■ **1** grotesque, hideous, *US* homely, plain, unattractive, unsightly, unprepossessing. **2** hostile, menacing, nasty, ominous, sinister, threatening, unpleasant.

UHF *abbr.* ultra-high frequency.
UK *abbr.* United Kingdom.
ukulele /yōōkəláyli/ *n.* small four-stringed guitar.
ulcer *n.* open sore. □ **ulcerous** *adj.*
ulcerated *adj.* affected with ulcer(s). □ **ulceration** *n.*
ulna *n.* (*pl.* **-ae**) thinner long bone of the forearm. □ **ulnar** *adj.*
ulterior *adj.* beyond what is obvious or admitted.

■ concealed, covert, hidden, secret, undisclosed.

ultimate *adj.* **1** last, final. **2** fundamental. □ **ultimately** *adv.*

■ **1** closing, concluding, eventual, final, last, terminating. **2** basic, elemental, essential, fundamental, primary, underlying.

ultimatum *n.* (*pl.* **-ums**) final demand, with a threat of hostile action if this is rejected.
ultra- *pref.* **1** beyond. **2** extremely.
ultra-high *adj.* (of frequency) between 300 and 3000 MHz.
ultramarine *adj.* & *n.* deep bright blue.
ultrasonic *adj.* above the range of normal human hearing.
ultrasound *n.* ultrasonic waves.
ultraviolet *adj.* of or using radiation with a wavelength shorter than that of visible light rays.
ululate *v.* howl, wail. □ **ululation** *n.*
umber *n.* natural brownish colouring matter.
umbilical *adj.* of the navel. □ **umbilical cord** flexible tube connecting the placenta to the navel of a foetus.
umbra *n.* (*pl.* **-ae**) area of total shadow cast by the moon or earth in an eclipse.
umbrage *n.* offence taken.
umbrella *n.* circle of fabric on a folding framework of spokes attached to a central stick, used as protection against rain.
umpire *n.* person appointed to supervise a game or contest etc. and see that rules are observed. ● *v.* act as umpire in.

■ *n.* arbitrator, judge, referee.

umpteen *adj.* (*sl.*) very many. □ **umpteenth** *adj.*
UN *abbr.* United Nations.
un- *pref.* **1** not. **2** reversing the action indicated by a verb, e.g. *unlock*. (The number of words with this prefix is almost unlimited and many of those whose meaning is obvious are not listed below.)
unaccountable *adj.* **1** unable to be explained. **2** not having to account for one's actions etc. □ **unaccountably** *adv.*

■ **1** incomprehensible, inexplicable, mysterious, odd, strange.

unadulterated *adj.* pure, complete.
unalloyed *adj.* pure.
unanimous *adj.* **1** all agreeing. **2** agreed by all. □ **unanimously** *adv.*, **unanimity** *n.*
unarmed *adj.* without weapons.
unassuming *adj.* not arrogant.
unavoidable *adj.* unable to be avoided. □ **unavoidably** *adv.*

■ certain, ineluctable, inescapable, inevitable.

unawares *adv.* **1** unexpectedly. **2** without noticing.

■ **1** abruptly, by surprise, off (one's) guard, suddenly, unexpectedly. **2** by accident *or* mistake, inadvertently, unconsciously, unintentionally, unwittingly.

unbalanced *adj.* **1** not balanced. **2** mentally unsound. **3** biased.
unbeknown *adj.* unknown.
unbend *v.* (**unbent**) **1** change from a bent position. **2** become relaxed or affable.
unbending *adj.* inflexible, refusing to alter one's demands.
unbidden *adj.* not commanded or invited.
unbounded *adj.* without limits.

■ boundless, endless, infinite, limitless, unlimited.

unbridled *adj.* unrestrained.

unburden *v.* **unburden oneself** reveal one's thoughts and feelings.

uncalled-for *adj.* given or done impertinently or unjustifiably.

uncanny *adj.* (**-ier**, **-iest**) **1** strange and rather frightening. **2** extraordinary. □ **uncannily** *adv.*

■ **1** eerie, frightening, ghostly, mysterious, *colloq.* spooky, strange, unearthly, weird. **2** astonishing, extraordinary, incredible, remarkable.

unceremonious *adj.* without proper formality or dignity.

uncertain *adj.* **1** not known or not knowing certainly. **2** not to be depended on. □ **uncertainly** *adv.*, **uncertainty** *n.*

■ **1** ambiguous, indefinite, questionable, touch-and-go, undetermined, unforeseeable; ambivalent, doubtful, dubious, indecisive, in two minds, irresolute, undecided, unsure, vague. **2** changeable, unpredictable, unreliable, unsettled, variable.

uncle *n.* brother or brother-in-law of one's father or mother.

unclean *adj.* **1** not clean. **2** ritually impure.

uncommon *adj.* unusual.

uncompromising *adj.* not allowing or not seeking compromise, inflexible.

■ dogged, inflexible, intransigent, resolute, stubborn, unyielding.

unconcern *n.* lack of concern.

unconditional *adj.* not subject to conditions. □ **unconditionally** *adv.*

■ absolute, complete, full, total, unequivocal, unqualified.

unconscionable *adj.* **1** unscrupulous. **2** contrary to what one's conscience feels is right.

unconscious *adj.* **1** not conscious. **2** not aware. **3** done without conscious intention. □ **unconsciously** *adv.*, **unconsciousness** *n.*

■ **1** comatose, insensible, knocked out, out. **2** heedless, oblivious, unaware. **3** inadvertent, instinctive, involuntary, subconscious, unintentional, unpremeditated, unthinking.

unconsidered *adj.* disregarded.

uncork *v.* pull the cork from.

uncouple *v.* disconnect (things joined by a coupling).

uncouth *adj.* uncultured, rough.

uncover *v.* **1** remove a covering from. **2** reveal, expose.

■ **2** disclose, discover, expose, lay bare, make known, reveal, unmask, unveil.

unction *n.* **1** anointing with oil, esp. as a religious rite. **2** excessive politeness.

unctuous *adj.* **1** unpleasantly flattering. **2** greasy. □ **unctuously** *adv.*, **unctuousness** *n.*

■ **1** fawning, flattering, obsequious, servile, *colloq.* smarmy, sycophantic.

undeniable *adj.* undoubtedly true. □ **undeniably** *adv.*

under *prep.* **1** in or to a position or rank etc. lower than. **2** less than. **3** governed or controlled by. **4** subjected to. **5** in accordance with. **6** known or classified as. ● *adv.* **1** in or to a lower position or subordinate condition. **2** in or into a state of unconsciousness. **3** below a certain quantity, rank, age, etc. □ **under age 1** not old enough, esp. for some legal right. **2** not yet of adult status. **under way** making progress.

under- *pref.* **1** below. **2** lower, subordinate. **3** insufficiently.

underarm *adj.* & *adv.* **1** in the armpit. **2** with the hand brought forwards and upwards.

undercarriage *n.* **1** aircraft's landing wheels and their supports. **2** supporting framework of a vehicle.

underclass *n.* social class below mainstream society.

undercliff *n.* terrace or lower cliff formed by a landslip.

underclothes *n.pl.* (also **underclothing**) underwear.

undercoat *n.* layer of paint used under a finishing coat.

undercover *adj.* done or doing things secretly.

undercurrent *n.* **1** current flowing below a surface. **2** underlying feeling, influence, etc.

■ **1** undertow. **2** atmosphere, feeling, overtone, sense, suggestion, undertone.

undercut *v.* (**undercut**) **1** cut away the part below. **2** sell or work for a lower price than.

underdog *n.* person etc. in an inferior or subordinate position.

underdone *adj.* not thoroughly cooked.

underestimate *v.* & *n.* (make) too low an estimate (of).

■ *v.* belittle, minimize, set little store by, underrate, undervalue.

underfelt *n.* felt for laying under a carpet.

underfoot *adv.* **1** on the ground. **2** under one's feet.

undergo *v.* (**-went, -gone**) experience, endure.

■ bear, be subjected to, endure, experience, go *or* live through, put up with, stand, suffer, weather, withstand.

undergraduate *n.* person studying for a first degree.

underground *adj.* **1** under the surface of the ground. **2** secret. ● *n.* underground railway.

■ *adj.* **1** buried, subterranean. **2** clandestine, covert, secret, undercover. ● *n. US* subway.

undergrowth *n.* thick growth of shrubs and bushes under trees.

underhand *adj.* **1** secret, sly. **2** underarm.

■ **1** clandestine, covert, cunning, devious, dishonest, furtive, secret, sly, surreptitious.

underlay *n.* material laid under another as a support.

underlie *v.* (**-lay, -lain, -lying**) be the basis of. □ **underlying** *adj.*

underline *v.* **1** draw a line under. **2** emphasize.

■ **2** accentuate, draw attention to, emphasize, highlight, point up, stress.

underling *n.* subordinate.

undermanned *adj.* having too few staff or crew etc.

undermine *v.* **1** weaken gradually. **2** weaken the foundations of.

■ **1** damage, debilitate, harm, ruin, spoil, subvert, weaken, wreck.

underneath *prep.* & *adv.* below or on the inside of (a thing).

underpants *n.pl.* undergarment for the lower part of the torso.

underpass *n.* road passing under another.

■ subway, tunnel.

underpin *v.* (**-pinned**) support, strengthen from beneath.

underprivileged *adj.* not having the normal standard of living or rights in a community.

underrate *v.* underestimate.

underseal *v.* coat the lower surface of (a vehicle) with a protective layer. ● *n.* this coating.

undersell *v.* (**-sold**) sell at a lower price than.

undershoot *v.* (**-shot**) land short of (a runway etc.).

undersigned *adj.* who has or have signed this document.

underskirt *n.* petticoat.

understand *v.* (**-stood**) **1** see the meaning or importance of. **2** be sympathetically aware of the character or nature of. **3** infer, take as implied. □ **understandable** *adj.*

■ **1** apprehend, *colloq.* catch on to, comprehend, *colloq.* cotton on to, fathom, follow, *colloq.* get, grasp, make out, perceive, realize, recognize, see, take in, *colloq.* twig. **2** accept, appreciate, empathize with, sympathize with. **3** assume, conclude, deduce, gather, infer, presume, suppose, surmise.

understanding *adj.* showing insight or sympathy. ● *n.* **1** ability to understand. **2** judgement of a situation. **3** sympathetic insight. **4** agreement. **5** thing agreed.

■ *adj.* compassionate, considerate, kind, sensitive, supportive, sympathetic, thoughtful. ● *n.* **1** brain(s), intellect, intelligence, mind, sense. **2** interpretation, judgement, opinion, perception, view. **3** compassion, empathy, feeling, insight, sensitivity, sympathy. **4** accord, agreement, concord, harmony. **5** agreement, arrangement, bargain, compact, contract, deal, pact, settlement.

understate *v.* **1** express in restrained terms. **2** represent as less than it really is. □ **understatement** *n.*

understudy *n.* actor ready to take another's role when required. ● *v.* be an understudy for.

undertake *v.* (**-took, -taken**) agree or promise (to do something).

■ agree, consent, guarantee, pledge, promise, swear, vow.

undertaker *n.* one whose business is to organize funerals.

undertaking *n.* **1** work etc. undertaken. **2** promise, guarantee.

■ **1** endeavour, enterprise, job, project, task, venture. **2** agreement, commitment, guarantee, pledge, promise, vow.

undertone *n.* underlying quality or feeling.

■ atmosphere, feeling, overtone, sense, suggestion, undercurrent.

undertow *n.* undercurrent moving in the opposite direction to the surface current.

underwear *n.* garments worn under indoor clothing.

■ lingerie, *colloq.* smalls, underclothes, *colloq.* undies.

underwent *see* **undergo**.

underworld *n.* **1** (in mythology) abode of spirits of the dead, under the earth. **2** part of society habitually involved in crime.

underwrite *v.* (**-wrote**, **-written**) **1** accept liability under (an insurance policy). **2** undertake to finance. □ **underwriter** *n.*

undesirable *adj.* not desirable, objectionable. □ **undesirably** *adv.*

■ disagreeable, objectionable, obnoxious, offensive, repugnant, unacceptable, unpleasant.

undies *n.pl.* (*colloq.*) women's underwear.

undo *v.* (**-did**, **-done**) **1** unfasten. **2** annul, cancel. **3** ruin.

■ **1** loosen, open, unfasten, untie; unbolt, unlock; uncover, unwrap. **2** annul, cancel, nullify, reverse. **3** destroy, ruin, spoil, wreck.

undone *adj.* **1** unfastened. **2** not done.

undoubted *adj.* not disputed. □ **undoubtedly** *adv.*

undress *v.* take clothes off.

undue *adj.* excessive. □ **unduly** *adv.*

■ excessive, extreme, immoderate, unjustifiable, unreasonable.

undulate *v.* have or cause to have a wavy movement or appearance. □ **undulation** *n.*

undying *adj.* everlasting.

■ endless, eternal, everlasting, immortal, perpetual, unending.

unearth *v.* **1** uncover or bring out from the ground. **2** find by searching.

■ **1** dig up, disinter, excavate, exhume, uncover. **2** come across, discover, find, turn up, uncover.

unearthly *adj.* **1** not of this earth. **2** mysterious and frightening. **3** (*colloq.*) very early.

■ **1** celestial, heavenly. **2** preternatural, supernatural; creepy, eerie, frightening, ghostly, mysterious, *colloq.* scary, *colloq.* spooky, strange, uncanny, weird.

uneasy *adj.* disturbed or uncomfortable in body or mind. □ **uneasily** *adv.*, **uneasiness** *n.*

■ anxious, apprehensive, disturbed, ill at ease, nervous, restive, restless, tense, troubled, worried.

uneatable *adj.* not fit to be eaten.

uneconomic *adj.* not profitable.

unemployable *adj.* not fit for paid employment.

unemployed *adj.* **1** without a paid job. **2** not in use. □ **unemployment** *n.*

■ **1** jobless, laid off, *colloq.* on the dole, out of work. **2** idle, inactive, unused.

unequivocal *adj.* clear, unambiguous. □ **unequivocally** *adv.*

■ categorical, clear, direct, explicit, plain, unambiguous, unmistakable.

unerring *adj.* making no mistake.

uneven *adj.* **1** not level. **2** of variable quality. □ **unevenly** *adv.*, **unevenness** *n.*

■ **1** bumpy, rough; lopsided, unbalanced, unequal. **2** erratic, inconsistent, patchy, variable.

unexceptionable *adj.* with which no fault can be found.

unfailing *adj.* **1** constant. **2** reliable.

unfair *adj.* not impartial, not in accordance with justice. □ **unfairly** *adv.*, **unfairness** *n.*

■ biased, bigoted, inequitable, one-sided, prejudiced, unjust.

unfaithful *adj.* **1** not loyal. **2** having committed adultery. □ **unfaithfully** *adv.*, **unfaithfulness** *n.*

■ **1** disloyal, faithless, false, fickle, inconstant, perfidious, unreliable.

unfeeling *adj.* lacking sensitivity, callous. □ **unfeelingly** *adv.*

■ callous, hard, hard-hearted, harsh, heartless, insensitive, ruthless, unkind, unsympathetic.

unfit *adj.* **1** unsuitable. **2** not in perfect physical condition. ● *v.* (**unfitted**) make unsuitable.

unflappable *adj.* (*colloq.*) remaining calm in a crisis.

unfold *v.* **1** open, spread out. **2** become known.

■ **1** open, spread (out), stretch out, uncoil, unfurl, unwind.

unforgettable *adj.* impossible to forget. □ **unforgettably** *adv.*

unfortunate *adj.* **1** having bad luck. **2** regrettable. □ **unfortunately** *adv.*

■ **1** hapless, luckless, poor, unlucky, wretched. **2** deplorable, lamentable, regrettable.

unfounded *adj.* with no basis.

■ baseless, groundless, unjustified, unproven, unsupported, unwarranted.

unfrock *v.* dismiss (a priest) from the priesthood.

unfurl *v.* unroll, spread out.

ungainly *adj.* awkward-looking, not graceful. □ **ungainliness** *n.*

■ awkward, clumsy, gauche, gawky, maladroit, ungraceful.

ungodly *adj.* **1** impious, wicked. **2** (*colloq.*) outrageous.

ungovernable *adj.* uncontrollable.

ungracious *adj.* not courteous or kindly. □ **ungraciously** *adv.*

■ churlish, discourteous, gauche, ill-mannered, impolite, rude, uncivil, unkind.

unguarded *adj.* **1** not guarded. **2** incautious.

■ **2** careless, ill-considered, imprudent, incautious, indiscreet, thoughtless, unthinking, unwise.

unguent /úngɡwənt/ *n.* ointment, lubricant.

ungulate *adj.* & *n.* hoofed (animal).

unhappy *adj.* (**-ier**, **-iest**) **1** not happy, sad. **2** unfortunate. **3** unsuitable. □ **unhappily** *adv.*, **unhappiness** *n.*

■ **1** blue, crestfallen, dejected, depressed, despondent, disconsolate, dispirited, downcast, downhearted, fed up, forlorn, glum, heavy-hearted, in low spirits, in the doldrums, melancholy, miserable, sad, sorrowful, woebegone. **2** hapless, luckless, unfortunate, unlucky. **3** inappropriate, infelicitous, unfortunate, unsuitable.

unhealthy *adj.* (**-ier**, **-iest**) **1** not healthy. **2** harmful to health. □ **unhealthily** *adv.*

■ **1** ailing, ill, indisposed, poorly, sick, sickly, unwell. **2** insalubrious, insanitary, unhygienic, unwholesome.

unheard-of *adj.* unprecedented.

unhinge *v.* cause to become mentally unbalanced.

unholy *adj.* (**-ier**, **-iest**) **1** wicked, irreverent. **2** (*colloq.*) very great.

unicorn *n.* mythical horse-like animal with one straight horn.

uniform *n.* distinctive clothing worn by members of the same organization etc. ● *adj.* always the same. □ **uniformly** *adv.*, **uniformity** *n.*

■ *adj.* consistent, constant, even, regular, unchanging, unvarying.

unify *v.* unite. □ **unification** *n.*

unilateral *adj.* done by or affecting one person or group etc. and not another. □ **unilaterally** *adv.*

unimpeachable *adj.* completely trustworthy.

uninviting *adj.* unattractive, repellent.

■ disagreeable, offensive, *colloq.* off-putting, repellent, repulsive, unappealing, unappetizing, unattractive, unpleasant, unsavoury.

union *n.* **1** uniting, being united. **2** a whole formed by uniting parts. **3** trade union (*see* **trade**). □ **Union Jack** national flag of the UK.

■ **1** amalgamation, coalition, combination, fusion, merger, synthesis, unification. **2** alliance, association, coalition, confederation, consortium, federation, partnership, syndicate.

unionist *n.* **1** member of a trade union. **2** supporter of trade unions. **3** one who favours union.

unionize *v.* organize into or cause to join a union. □ **unionization** *n.*

unique *adj.* **1** being the only one of its kind. **2** unequalled. □ **uniquely** *adv.*

■ **2** incomparable, inimitable, matchless, peerless, unequalled, unparalleled.

unisex *adj.* designed in a style suitable for people of either sex.

unison *n.* **in unison** **1** all together. **2** sounding or singing together.

■ at once, at the same time, simultaneously, together.

unit *n.* **1** individual thing, person, or group, esp. as part of a complex whole. **2** fixed quantity used as a standard of measurement. □ **unit trust** investment company paying dividends based on the average return from the various securities which they hold.

■ **1** component, constituent, element, item, part, piece.

unitary *adj.* of a unit or units.

unite *v.* **1** join together, make or become one. **2** act together, cooperate.

■ **1** amalgamate, coalesce, combine, integrate, join, merge, synthesize. **2** ally, cooperate, join forces, team up.

unity *n.* **1** state of being one or a unit. **2** agreement. **3** complex whole.

■ **2** accord, agreement, concord, consensus, harmony, solidarity, unanimity.

universal *adj.* of, for, or done by all. □ **universally** *adv.*

■ common, general, global, widespread, worldwide.

universe *n.* all existing things, including the earth and its creatures and all the heavenly bodies.

university *n.* educational institution for advanced learning.

unkempt *adj.* untidy, neglected.

■ dishevelled, messy, *colloq.* scruffy, untidy.

unkind *adj.* harsh, hurtful. □ **unkindly** *adv.*, **unkindness** *n.*

■ callous, cruel, hard, hard-hearted, heartless, hurtful, inconsiderate, insensitive, malicious, mean, nasty, spiteful, thoughtless, uncaring, uncharitable, unfeeling, unsympathetic.

unknown *adj.* not known. ● *n.* unknown person, thing, or place.

unleaded *adj.* (of petrol etc.) without added lead.

unleash *v.* release, let loose.

unless *conj.* except when, except on condition that.

unlettered *adj.* illiterate.

unlike *adj.* not like. ● *prep.* differently from.

unlikely *adj.* not likely to happen or be true or be successful.

■ doubtful, dubious, far-fetched, implausible, improbable.

unlisted *adj.* not included in a (published) list.

unload *v.* **1** remove a load or cargo (from). **2** get rid of. **3** remove the ammunition from (a gun).

unlock *v.* **1** release the lock of. **2** release by unlocking.

unlooked-for *adj.* unexpected.

unmanned *adj.* operated without a crew.

unmask *v.* **1** remove a mask (from). **2** expose the true character of.

unmentionable *adj.* not fit to be spoken of.

unmistakable *adj.* clear, obvious, plain. □ **unmistakably** *adv.*

■ clear, incontrovertible, indisputable, manifest, obvious, patent, plain, undeniable.

unmitigated *adj.* not modified, absolute.

■ absolute, complete, downright, out and out, perfect, pure, sheer, thorough, total, unadulterated, unqualified, utter.

unmoved *adj.* not moved, not persuaded, not affected by emotion.

unnatural *adj.* different from what is normal or expected, not natural. □ **unnaturally** *adv.*

■ abnormal, bizarre, extraordinary, odd, outlandish, peculiar, strange, uncanny, uncharacteristic, unusual, weird; affected, artificial, false, feigned, forced, insincere, *colloq.* phoney, studied.

unnecessary *adj.* **1** not necessary. **2** more than is necessary. □ **unnecessarily** *adv.*

■ **1** dispensable, expendable, inessential, needless, unneeded. **2** excessive, superfluous, undue.

unnerve *v.* deprive of confidence.

■ discomfit, disconcert, dismay, disturb, fluster, perturb, *colloq.* rattle, ruffle, shake, *colloq.* throw, unsettle, upset.

unnumbered *adj.* **1** not marked with a number. **2** countless.

unobtrusive *adj.* not making oneself or itself noticed. □ **unobtrusively** *adv.*

■ discreet, inconspicuous, low-key, unassuming, unpretentious.

unpack *v.* **1** open and remove the contents of (a suitcase etc.). **2** take out from its packaging.

unparalleled *adj.* never yet equalled.

■ incomparable, inimitable, matchless, peerless, unequalled, unique, unmatched, unprecedented, unrivalled, unsurpassed.

unpick *v.* undo the stitching of.

unplaced *adj.* not placed as one of the first three in a race etc.

unpleasant *adj.* not pleasant. □ **unpleasantly** *adv.*, **unpleasantness** *n.*

■ *colloq.* beastly, disagreeable, distasteful, *colloq.* horrible, horrid, nasty, objectionable, offensive, repellent, unsavoury.

unpopular *adj.* not popular, disliked. □ **unpopularity** *n.*

unprecedented *adj.* having no precedent, unparalleled.

unprepared *adj.* **1** not prepared beforehand. **2** not ready or not equipped to do something.

unpretentious *adj.* not pretentious, not showy or pompous.

■ humble, modest, plain, unassuming, unostentatious.

unprincipled *adj.* without good moral principles, unscrupulous.

unprintable *adj.* too indecent or libellous etc. to be printed.

unprofessional *adj.* contrary to the standards of behaviour for members of a profession. □ **unprofessionally** *adv.*

unqualified *adj.* **1** not qualified. **2** unmitigated.

unquestionable *adj.* too clear to be doubted. □ **unquestionably** *adv.*

■ certain, incontestable, incontrovertible, indubitable, irrefutable, undeniable, undisputed, undoubted, unmistakable.

unravel *v.* (**unravelled**) **1** disentangle. **2** undo (knitted fabric). **3** become unravelled.

unreasonable *adj.* **1** not reasonable. **2** excessive. □ **unreasonably** *adv.*

■ **1** absurd, foolish, illogical, irrational, ludicrous, preposterous, ridiculous. **2** excessive, exorbitant, extravagant, immoderate, inordinate, outrageous, unjustifiable.

unrelieved *adj.* monotonously uniform.

unremitting *adj.* incessant.

unrequited *adj.* (of love) not returned.

unreservedly *adv.* without reservation, completely.

unrest *n.* disturbance, turmoil.

■ disquiet, disturbance, strife, trouble, turbulence, turmoil.

unrivalled *adj.* having no equal.

unroll *v.* open after being rolled.

unruly *adj.* not easy to control, disorderly. □ **unruliness** *n.*

■ disobedient, disorderly, insubordinate, mutinous, obstreperous, rebellious, recalcitrant, riotous, rowdy, uncontrollable, undisciplined, ungovernable, unmanageable, wayward, wild.

unsaid *adj.* not spoken or expressed.

unsavoury *adj.* **1** disagreeable to the taste or smell. **2** morally offensive.

■ **1** unappetizing, unpalatable. **2** disgusting, distasteful, nasty, objectionable, offensive, repugnant, repulsive, revolting.

unscathed *adj.* without suffering any injury.

■ safe, intact, unharmed, unhurt, uninjured.

unscramble *v.* **1** sort out. **2** make (a scrambled transmission) intelligible.

unscrew *v.* **1** loosen (a screw etc.). **2** unfasten by removing screw(s).

unscripted *adj.* without a prepared script.

unscrupulous *adj.* not prevented by scruples of conscience.

■ amoral, corrupt, *colloq.* crooked, dishonest, immoral, unethical, unprincipled.

unseat *v.* **1** dislodge (a rider). **2** remove from a parliamentary seat.

unselfish *adj.* considering others' needs before one's own. □ **unselfishly** *adv.*, **unselfishness** *n.*

■ altruistic, charitable, considerate, generous, kind, selfless, thoughtful.

unsettle *v.* make uneasy, disturb.

■ agitate, discomfit, disquiet, disturb, perturb, *colloq.* rattle, ruffle, shake, *colloq.* throw, unnerve, upset, worry.

unsettled *adj.* (of weather) changeable.

unshakeable *adj.* firm.

unsightly *adj.* not pleasant to look at, ugly. □ **unsightliness** *n.*

■ hideous, plain, ugly, unattractive, unlovely, unprepossessing.

unskilled *adj.* not having or needing skill or special training.

unsocial *adj.* **1** not sociable. **2** not conforming to normal social practices.

unsolicited *adj.* not requested.

unsophisticated *adj.* simple and natural or naive.

■ artless, callow, childlike, green, inexperienced, ingenuous, innocent, naive, natural, simple, unworldly.

unsound *adj.* not sound or strong. □ **of unsound mind** insane.

unsparing *adj.* giving lavishly.

■ generous, lavish, liberal, munificent, open-handed.

unspeakable *adj.* too bad to be described in words.

unstable *adj.* **1** not stable. **2** likely to change suddenly.

■ **2** capricious, changeable, unpredictable, unreliable, volatile.

unstuck *adj.* detached after being stuck on or together. □ **come unstuck** (*colloq.*) suffer disaster, fail.

unstudied *adj.* natural in manner.

unsullied *adj.* not sullied, pure.

unswerving *adj.* **1** not turning aside. **2** unchanging.

■ **2** constant, firm, steadfast, unchanging, unwavering.

unthinkable *adj.* too bad or too unlikely to be thought about.

unthinking *adj.* **1** thoughtless. **2** inadvertent.

■ **1** careless, inconsiderate, rude, tactless, thoughtless. **2** inadvertent, involuntary, unintentional.

untidy *adj.* (**-ier**, **-iest**) not tidy. ▫ **untidily** *adv.*, **untidiness** *n.*

■ chaotic, disordered, disorderly, disorganized, higgledy-piggledy, messy, topsy-turvy; dishevelled, *colloq.* scruffy, unkempt.

untie *v.* **1** unfasten. **2** release from being tied up.

■ **1** loose, loosen, open, undo, unfasten. **2** free, release, unchain, unfetter, unleash.

until *prep.* & *conj.* = **till**[2].

untimely *adj.* **1** inopportune. **2** premature.

■ **1** ill-timed, inappropriate, inconvenient, inopportune, unseasonable, unsuitable. **2** early, premature.

untold *adj.* **1** not told. **2** too much or too many to be counted.

■ **2** countless, innumerable, many, numerous, *sl.* umpteen; immeasurable, incalculable, unlimited.

untouchable *adj.* that may not be touched. ● *n.* member of the lowest Hindu social group.

untoward *adj.* unexpected and inconvenient.

untruth *n.* **1** untrue statement, lie. **2** lack of truth. ▫ **untruthful** *adj.*, **untruthfully** *adv.*

■ **1** fabrication, falsehood, fib, lie, *sl.* whopper.

unusual *adj.* **1** not usual. **2** remarkable, rare. ▫ **unusually** *adv.*

■ **1** abnormal, atypical, bizarre, curious, different, odd, out of the ordinary, peculiar, strange, surprising, uncommon, unconventional, unexpected. **2** exceptional, extraordinary, rare, remarkable, singular.

unutterable *adj.* beyond description. ▫ **unutterably** *adv.*

unvarnished *adj.* **1** not varnished. **2** plain, straightforward.

unveil *v.* **1** remove a veil (from). **2** remove concealing drapery from. **3** disclose, make publicly known.

■ **3** disclose, divulge, expose, lay bare, make known, reveal, uncover, unmask.

unversed *adj.* **unversed in** not experienced in.

unwarranted *adj.* unjustified.

■ indefensible, inexcusable, uncalled-for, unjustified.

unwell *adj.* not in good health.

■ ailing, ill, indisposed, off-colour, poorly, sick, sickly, under the weather, unhealthy.

unwieldy *adj.* awkward to move or control because of its size, shape, etc. ▫ **unwieldiness** *n.*

■ awkward, bulky, clumsy, cumbersome, unmanageable.

unwind *v.* (**unwound**) **1** draw out or become drawn out from being wound. **2** (*colloq.*) relax.

unwise *adj.* not wise, foolish. ▫ **unwisely** *adv.*

■ foolhardy, foolish, ill-advised, ill-judged, imprudent, incautious, misguided, rash, reckless, senseless, short-sighted, silly, stupid.

unwitting *adj.* **1** unaware. **2** unintentional. ▫ **unwittingly** *adv.*

unwonted *adj.* not customary, not usual. ▫ **unwontedly** *adv.*

unworldly *adj.* **1** spiritual. **2** naive. ▫ **unworldliness** *n.*

unworthy *adj.* of little or no value. ▫ **unworthy of** unsuitable to the character of (a person or thing).

■ inferior, mediocre, second-rate, substandard. ▫ **unworthy of** below, beneath, inappropriate to, inconsistent with, unbecoming to, unsuitable for.

unwritten *adj.* **1** not written down. **2** based on custom not statute.

up *adv.* **1** to, in, or at a higher place or state etc. **2** to a vertical position. **3** completed or completely. **4** as far as a stated place, time, or amount. **5** out of bed. **6** (*colloq.*) amiss, happening. ● *prep.* **1** upwards along or through or into. **2** at a higher part of. ● *adj.* **1** directed upwards. **2** travelling towards a central place. ● *v.* (**upped**) (*colloq.*) **1** raise. **2** get up (and do something). ▫ **time is up** time is finished. **up in** (*colloq.*) knowledgeable about. **ups and downs** alternate good and bad fortune. **up to** **1** occupied with, doing. **2** required as a duty or obligation from. **3** capable of.

upbeat *n.* unaccented beat. ● *adj.* (*colloq.*) optimistic, cheerful.

upbraid *v.* reproach.

■ admonish, berate, chastise, chide, criticize, rebuke, reprimand, reproach, reprove, scold, take to task, *colloq.* tell off.

upbringing *n.* training and education during childhood.

update *v.* bring up to date.

up-end *v.* set or rise up on end.

upgrade *v.* raise to higher grade.

upheaval *n.* **1** sudden heaving upwards. **2** violent disturbance.

■ **2** chaos, confusion, disorder, disruption, disturbance, havoc, turmoil.

uphill *adj.* & *adv.* going or sloping upwards.

uphold *v.* (**upheld**) support.

■ defend, maintain, preserve, protect, stand by, support, sustain.

upholster *v.* put fabric covering, padding, etc. on (furniture).

upholstery *n.* **1** upholstering. **2** material used in this.

upkeep *n.* **1** keeping (a thing) in good condition and repair. **2** cost of this.

■ **1** maintenance. **2** expenses, overheads, running costs.

upland *n.* & *adj.* (of) higher or inland parts of a country.

uplift *v.* raise. ● *n.* **1** being raised. **2** mentally elevating influence.

upon *prep.* on.

upper *adj.* higher in place, position, or rank. ● *n.* part of a shoe above the sole. □ **upper case** capital letters. **upper class** highest class of society. **upper crust** (*colloq.*) aristocracy. **upper hand** dominance.

■ *adj.* higher, superior. □ **upper class** aristocracy, elite, nobility, *colloq.* upper crust. **upper hand** advantage, authority, command, control, dominance, power, superiority, sway.

uppermost *adj.* & *adv.* in, on, or to the top or most prominent position.

■ *adj.* highest, top, topmost; first, foremost, most important, paramount, preeminent, principal.

uppity *adj.* (*colloq.*) having too high an opinion of oneself.

upright *adj.* **1** in a vertical position. **2** honest or honourable. ● *n.* vertical part or support.

■ *adj.* **1** erect, perpendicular, straight, upstanding, vertical. **2** decent, ethical, good, honest, honourable, incorruptible, law-abiding, moral, principled, righteous, virtuous. ● *n.* column, pillar, pole, post, stanchion.

uprising *n.* rebellion.

■ *coup d'état*, insurrection, mutiny, rebellion, revolt, revolution.

uproar *n.* outburst of noise and excitement or anger.

■ clamour, commotion, din, fracas, hubbub, hullabaloo, noise, pandemonium, racket, rumpus, tumult.

uproarious *adj.* noisy, with loud laughter. □ **uproariously** *adv.*

■ clamorous, deafening, noisy, riotous, rowdy, tumultuous, wild.

uproot *v.* **1** pull out of the ground together with its roots. **2** force to leave an established place.

upset *v.* /úpsét/ (**upset, upsetting**) **1** overturn. **2** disrupt. **3** disturb the temper, composure, or digestion of. ● *n.* /úpset/ upsetting, being upset.

■ *v.* **1** knock over, overturn, tip over, topple, up-end, upturn. **2** disrupt, disturb, mess up, ruin, spoil, thwart, wreck. **3** annoy, dismay, distress, disturb, grieve, hurt, offend, pain, perturb, put out, trouble, unsettle, worry.

upshot *n.* outcome.

■ conclusion, consequence, effect, outcome, repercussion, result.

upside down **1** with the upper part underneath. **2** in great disorder.

upstage *adv.* & *adj.* nearer the back of a theatre stage. ● *v.* divert attention from, outshine.

upstairs *adv.* & *adj.* to or on a higher floor.

upstanding *adj.* **1** strong and healthy. **2** standing up.

upstart *n.* person newly risen to a high position, esp. one who behaves arrogantly.

upstream *adj.* & *adv.* in the direction from which a stream flows.

upsurge *n.* upward surge, rise.

upswing *n.* upward movement or trend.

uptake *n.* **quick on the uptake** (*colloq.*) quick to understand.

uptight *adj.* (*colloq.*) nervously tense or angry.

upturn *v.* /úptúrn/ turn up or upwards or upside down. ● *n.* /úpturn/ **1** upheaval. **2** upward trend, improvement.

upward *adj.* moving or leading up.

upwards *adv.* towards a higher place etc.

uranium *n.* heavy grey metal used as a source of nuclear energy.

urban *adj.* of a city or town.

urbane *adj.* having smooth manners. □ **urbanely** *adv.*, **urbanity** *n.*

■ charming, civilized, courteous, cultivated, debonair, gracious, polished, sophisticated, suave, well-mannered.

urbanize *v.* change (a place) into an urban area. □ **urbanization** *n.*

urchin *n.* mischievous child.

Urdu *n.* language related to Hindi.

ureter /yooréetər/ *n.* duct from the kidney to the bladder.

urethra /yooréethrə/ *n.* duct carrying urine from the body.

urge *v.* **1** impel, drive forcibly. **2** encourage or entreat earnestly or persistently. **3** advocate emphatically. ● *n.* feeling or desire urging a person to do something.

■ *v.* **1** drive, force, hasten, hustle, impel, push; egg on, goad, incite, prod, prompt, spur. **2** beg, beseech, encourage, entreat, exhort, implore, plead with, press. **3** advise, advocate, counsel, recommend. ● *n.* craving, desire, longing, yearning, *colloq.* yen.

urgent *adj.* needing or calling for immediate attention or action. □ **urgently** *adv.*, **urgency** *n.*

■ acute, critical, desperate, imperative, important, necessary, pressing, serious, vital.

urinal *n.* receptacle or structure for receiving urine.

urinate *v.* discharge urine from the body. □ **urination** *n.*

urine *n.* waste liquid which collects in the bladder and is discharged from the body. □ **urinary** *adj.*

urn *n.* **1** a kind of vase, esp. for a cremated person's ashes. **2** large metal container with a tap, for keeping water etc. hot.

ursine *adj.* of or like a bear.

us *pron.* objective case of *we.*

US, **USA** *abbrs.* United States (of America).

usable *adj.* able or fit to be used.

usage *n.* **1** manner of using or treating something. **2** customary practice.

■ **1** handling, management, treatment, use. **2** convention, custom, habit, practice, tradition.

use *v.* /yōōz/ **1** cause to act or to serve for a purpose. **2** treat. **3** exploit selfishly. ● *n.* /yōōss/ **1** using, being used. **2** power of using. **3** purpose for which a thing is used.

■ *v.* **1** employ, exercise, make use of, utilize. **2** act *or* behave towards, treat. **3** abuse, capitalize on, cash in on, exploit, make use of, misuse, profit from, take advantage of. ● *n.* **1** application, employment, usage, utilization. **2** function, power. **3** advantage, benefit, gain, good, profit, service; object, point, purpose.

used /yōōzd/ *adj.* second-hand. □ **used to** /yōōsst/ *v.* was accustomed to (do). *adj.* familiar with by practice or habit.

useful *adj.* **1** fit for a practical purpose. **2** able to produce good results. □ **usefully** *adv.*, **usefulness** *n.*

■ **1** convenient, functional, handy, helpful, practical, serviceable, utilitarian. **2** advantageous, beneficial, constructive, fruitful, productive, salutary, valuable, worthwhile.

useless *adj.* not usable, not useful. □ **uselessly** *adv.*, **uselessness** *n.*

■ fruitless, futile, impractical, ineffective, ineffectual, pointless, purposeless, unproductive, vain.

user *n.* one who uses something. □ **user-friendly** *adj.* easy for a user to understand and operate.

usher *n.* person who shows people to their seats in a public hall etc. ● *v.* lead, escort.

■ *v.* accompany, conduct, escort, guide, lead, shepherd.

usherette *n.* woman who ushers people to seats in a cinema etc.

usual *adj.* such as happens or is done or used etc. in many or most instances. □ **usually** *adv.*

■ accustomed, common, conventional, customary, established, everyday, familiar, habitual, normal, ordinary, regular, routine, set, standard, traditional, typical. □ **usually** as a rule, customarily, for the most part, generally, in general, mainly, mostly, normally.

usurp *v.* take (power, position, or right) wrongfully or by force. □ **usurpation** *n.*, **usurper** *n.*

usury *n.* lending of money at excessively high rates of interest. □ **usurer** *n.*

utensil *n.* instrument or container, esp. for domestic use.

■ appliance, implement, instrument, tool.

uterus *n.* womb. ◻ **uterine** *adj.*

utilitarian *adj.* useful rather than decorative or luxurious.

■ functional, practical, serviceable, useful.

utilitarianism *n.* theory that actions are justified if they benefit the majority.

utility *n.* **1** usefulness. **2** useful thing. ● *adj.* severely practical.

utilize *v.* use, find a use for. ◻ **utilization** *n.*

utmost *adj.* & *n.* furthest, greatest, or extreme (point or degree etc.).

Utopia *n.* imaginary place or state where all is perfect. ◻ **Utopian** *adj.*

utter[1] *adj.* complete, absolute. ◻ **utterly** *adv.*

■ absolute, complete, downright, out and out, outright, pure, sheer, thorough, total, unmitigated, unqualified.

utter[2] *v.* **1** make (a sound or words) with the mouth or voice. **2** speak. ◻ **utterance** *n.*

■ **1** emit, give. **2** articulate, express, pronounce, speak, voice.

uttermost *adj.* & *n.* = **utmost**.

U-turn *n.* **1** driving of a vehicle in a U-shaped course to reverse its direction. **2** reversal of policy or opinion.

uvula /yo͞ovyoolər/ *n.* small fleshy projection hanging at the back of the throat. ◻ **uvular** *adj.*

uxorious *adj.* excessively fond of one's wife.

V *abbr.* volt(s).

vac *n.* (*colloq.*) vacation.

vacancy *n.* **1** state of being vacant. **2** vacant place, post, etc.

vacant *adj.* **1** unoccupied. **2** showing no interest. □ **vacantly** *adv.*

■ **1** empty, free, uninhabited, unoccupied. **2** blank, emotionless, expressionless, vacuous.

vacate *v.* cease to occupy.

■ abandon, desert, evacuate, leave, quit, withdraw from.

vacation *n.* **1** interval between terms in universities and law courts. **2** (*US*) holiday. **3** vacating of a place etc.

vaccinate *n.* inoculate with a vaccine. □ **vaccination** *n.*

vaccine /vákseen/ *n.* preparation that gives immunity from an infection.

vacillate *v.* keep changing one's mind. □ **vacillation** *n.*

vacuous *adj.* **1** inane. **2** expressionless, vacant. □ **vacuously** *adv.*, **vacuity** *n.*

■ **1** foolish, inane, senseless, silly, stupid, unintelligent.

vacuum *n.* (*pl.* **-cua** or **-cuums**) space from which air has been removed. ● *v.* (*colloq.*) clean with a vacuum cleaner. □ **vacuum cleaner** electrical apparatus that takes up dust by suction. **vacuum flask** container for keeping liquids hot or cold. **vacuum-packed** *adj.* sealed after removal of air.

vagabond *n.* wanderer or vagrant

■ gypsy, rolling stone, rover, tramp, vagrant, wanderer, wayfarer.

vagary *n.* capricious act or idea, fluctuation.

■ caprice, fancy, quirk, whim.

vagina *n.* passage leading from the vulva to the womb. □ **vaginal** *adj.*

vagrant *n.* person without a settled home. □ **vagrancy** *n.*

vague *adj.* **1** not clearly explained or perceived. **2** (of a person or mind) inexact in thought, expression, etc. □ **vaguely** *adv.*, **vagueness** *n.*

■ **1** ambiguous, generalized, hazy, imprecise, unclear, woolly; amorphous, blurred, dim, faint, fuzzy, indistinct, obscure, nebulous, shadowy, shapeless. **2** hesitant, uncertain, undecided, unsure; absent-minded, dreamy.

vain *adj.* **1** conceited. **2** useless, futile. □ **in vain** uselessly. **vainly** *adv.*

■ **1** cocky, conceited, egotistical, haughty, narcissistic, proud, self-important, *colloq.* stuck-up, vainglorious. **2** fruitless, futile, unsuccessful, useless.

vainglory *n.* great vanity. □ **vainglorious** *adj.*

valance *n.* short curtain or hanging frill.

vale *n.* valley.

valediction *n.* farewell. □ **valedictory** *adj.*

valency *n.* (also **valence**) combining power of an atom as compared with that of the hydrogen atom.

valentine *n.* **1** person to whom one sends a romantic greetings card on St Valentine's Day (14 Feb.). **2** this card.

valet *n.* man's personal attendant. ● *v.* (**valeted**) act as valet to.

valetudinarian *n.* person of poor health or unduly anxious about health.

valiant *adj.* brave. □ **valiantly** *adv.*

■ bold, brave, courageous, daring, fearless, heroic, indomitable, intrepid, plucky, spirited, valorous.

valid *adj.* **1** having legal force, usable. **2** logical. □ **validity** *n.*

■ **2** logical, rational, reasonable, sound, well-founded.

validate *v.* make valid, confirm. □ **validation** *n.*

valley *n.* (*pl.* **-eys**) low area between hills.

■ coomb, dale, dingle, glen, vale.

valour *n.* bravery. □ **valorous** *adj.*

■ boldness, bravery, courage, daring, fearlessness, fortitude, *colloq.* guts, nerve, spirit.

valuable *adj.* of great value or worth. ● *n.pl.* valuable things.

■ *adj.* expensive, invaluable, precious, priceless, treasured; advantageous, beneficial, helpful, useful, worthwhile.

valuation *n.* estimation or estimate of a thing's worth.

value *n.* **1** amount of money or other commodity etc. considered equivalent to something else. **2** usefulness, importance. **3** (*pl.*) moral principles. ● *v.* **1** estimate the value of. **2** consider to be of great worth. □ **value added tax** tax on the amount by which a thing's value has been increased at each stage of its production. **value judgement** subjective estimate of quality etc.

■ *n.* **1** cost, price, worth. **2** advantage, benefit, importance, merit, usefulness. **3** (**values**) ethics, morals, philosophy, principles. ● *v.* **1** assess, estimate, evaluate, price. **2** appreciate, prize, rate highly, respect, treasure.

valuer *n.* person who estimates values professionally.

valve *n.* **1** device controlling flow through a pipe. **2** structure allowing blood to flow in one direction only. **3** each half of the hinged shell of an oyster etc. □ **valvular** *adj.*

vamoose *v.* (*US sl.*) depart hurriedly.

vamp *n.* upper front part of a boot or shoe. ● *v.* improvise (esp. a musical accompaniment).

vampire *n.* ghost or reanimated body supposed to suck blood.

van[1] *n.* **1** covered vehicle for transporting goods etc. **2** railway carriage for luggage or goods.

van[2] *n.* vanguard, forefront.

vandal *n.* person who damages things wilfully. □ **vandalism** *n.*

■ hoodlum, hooligan, lout, ruffian, *sl.* yob.

vandalize *v.* damage wilfully.

vanguard *n.* foremost part of an advancing army etc.

vanilla *n.* a kind of flavouring, esp. obtained from the pods of a tropical orchid.

vanish *v.* disappear completely.

■ become invisible, disappear, dissolve, evaporate, fade (away), melt (away).

vanity *n.* **1** conceit. **2** futility. □ **vanity case** woman's small case for carrying cosmetics etc.

■ **1** conceit, egotism, haughtiness, narcissism, pride, self-admiration, self-love, vainglory. **2** futility, pointlessness, uselessness, worthlessness.

vanquish *v.* conquer.

■ beat, conquer, defeat, get the better of, *colloq.* lick, overcome, subjugate, thrash, triumph over.

vantage *n.* advantage (esp. as a score in tennis). □ **vantage point** position giving a good view.

vapid *adj.* insipid, uninteresting. □ **vapidly** *adv.*, **vapidity** *n.*

■ bland, dull, insipid, unexciting, uninteresting, wishy-washy.

vaporize *v.* convert or be converted into vapour. □ **vaporization** *n.*

vapour *n.* moisture suspended in air, into which certain liquids or solids are converted by heating. □ **vaporous** *adj.*

variable *adj.* varying. ● *n.* thing that varies. □ **variability** *n.*

■ *adj.* changeable, erratic, fluctuating, inconsistent, inconstant, mutable, protean, uncertain, unpredictable, unstable, varying.

variance *n.* **at variance** disagreeing.

variant *adj.* differing. ● *n.* variant form or spelling etc.

variation *n.* **1** varying, extent of this. **2** variant. **3** repetition of a melody in a different form.

■ **1** alteration, change, difference, modification; choice, diversity, variety.

varicose *adj.* (of veins) permanently swollen.

varied *adj.* of different sorts.

■ assorted, diverse, heterogeneous, miscellaneous, mixed.

variegated *adj.* having irregular patches of colours. □ **variegation** *n.*

variety *n.* **1** quality of not being the same. **2** quantity of different things. **3** sort or kind. **4** entertainment with a series of short performances.

■ **1** contrast, difference, diversity, heterogeneity, variation. **2** array, assortment, collection, miscellany, mixture, multiplicity, number, range, selection. **3** brand, category, class, genre, group, kind, make, sort, species, strain, type.

various *adj.* **1** of several kinds. **2** several. □ **variously** *adv.*

■ **1** assorted, different, diverse, miscellaneous, varying. **2** many, numerous, several, sundry.

varnish *n.* liquid that dries to form a shiny transparent coating. ● *v.* coat with varnish.

vary *v.* make, be, or become different.

■ adjust, alter, change, modify; deviate, differ, diverge; alternate, fluctuate, shift, swing.

vascular *adj.* of vessels or ducts for conveying blood or sap.

vase *n.* decorative jar, esp. for holding cut flowers.

vasectomy *n.* surgical removal of part of the ducts that carry semen from the testicles, esp. as a method of birth control.

vassal *n.* humble subordinate.

vast *adj.* very great in area or size. □ **vastly** *adv.*, **vastness** *n.*

■ **colossal, elephantine, enormous, extensive, gigantic, great, huge, immense, large, mammoth, massive, monumental, prodigious, tremendous.**

VAT *abbr.* value added tax.

vat *n.* large tank for liquids.

vault[1] *n.* **1** arched roof. **2** cellar used as a storage place. **3** burial chamber. □ **vaulted** *adj.*

vault[2] *v.* & *n.* jump, esp. with the help of the hands or a pole.

vaunt *v.* & *n.* boast.

VDU *abbr.* visual display unit.

veal *n.* calf's flesh as food.

vector *n.* **1** thing (e.g. velocity) that has both magnitude and direction. **2** carrier of disease.

veer *v.* change direction.

vegan /véegən/ *n.* person who eats no meat or animal products.

vegetable *n.* plant grown for food. ● *adj.* of or from plants.

vegetarian *n.* person who eats no meat or fish. □ **vegetarianism** *n.*

vegetate *v.* live an uneventful life.

vegetation *n.* plants collectively.

vehement *adj.* showing strong feeling. □ **vehemently** *adv.*, **vehemence** *n.*

■ **ardent, eager, enthusiastic, fervent, impassioned, intense, keen, passionate, zealous.**

vehicle *n.* **1** conveyance for transporting passengers or goods on land or in space. **2** means by which something is expressed or displayed. □ **vehicular** *adj.*

veil *n.* piece of fine net or other fabric worn to protect or conceal the face. ● *v.* cover with or as if with a veil.

■ *v.* **camouflage, cloak, conceal, cover, hide, mask, shroud.**

vein *n.* **1** any of the blood vessels conveying blood towards the heart. **2** a similar structure in an insect's wing etc. **3** narrow layer in rock etc. **4** mood. □ **veined** *adj.*

■ **3 lode, seam, stratum. 4 humour, mood, spirit, temper.**

vellum *n.* **1** fine parchment. **2** smooth writing paper.

velocity *n.* speed.

■ **celerity, quickness, rapidity, speed, swiftness.**

velour /vəlóor/ *n.* heavy plush-like fabric.

velvet *n.* woven fabric with thick short pile on one side. □ **velvet glove** outward gentleness concealing inflexibility. **velvety** *adj.*

velveteen *n.* cotton velvet.

venal *adj.* **1** able to be bribed. **2** involving bribery. □ **venality** *n.*

vend *v.* sell, offer for sale.

vendetta *n.* feud.

■ **conflict, dispute, feud, quarrel.**

vending machine slot machine selling small articles.

vendor *n.* seller.

veneer *n.* **1** thin covering layer of fine wood. **2** superficial show of a quality. ● *v.* cover with a veneer.

■ *n.* **2 appearance, façade, guise, mask, semblance, show.**

venerate *v.* respect deeply, esp. because of great age. □ **veneration** *n.*, **venerable** *adj.*

■ **admire, esteem, honour, look up to, respect, revere, worship.**

venereal *adj.* (of infections) contracted by sexual intercourse with an infected person.

Venetian *adj.* & *n.* (native) of Venice. □ **Venetian blind** window blind of adjustable horizontal slats.

vengeance *n.* punishment inflicted for a wrong. □ **with a vengeance** in an extreme degree.

■ **retaliation, retribution, revenge.**

vengeful *adj.* seeking vengeance.

venial *adj.* (of a sin) pardonable, not serious. □ **veniality** *n.*

venison *n.* deer's flesh as food.

venom *n.* **1** poisonous fluid of snakes etc. **2** bitter feeling or language. □ **venomous** *adj.*

■ **1 poison, toxin. 2 animosity, bitterness, enmity, hate, hatred, hostility, ill will, malevolence, malice, rancour, spite.**

vent[1] *n.* slit at the lower edge of the back or side of a coat.

vent² *n.* opening allowing gas or liquid to pass through. ● *v.* give vent to. □ **give vent to** give an outlet to (feelings).

■ *n.* aperture, hole, opening, outlet; chimney, duct, flue, funnel. □ **give vent to** air, articulate, express, release, voice.

ventilate *v.* cause air to circulate freely in. □ **ventilation** *n.*

■ aerate, air, freshen.

ventilator *n.* device for ventilating a room etc.

ventral *adj.* of or on the abdomen.

ventricle *n.* cavity, esp. in the heart or brain. □ **ventricular** *adj.*

ventriloquism *n.* skill of speaking without moving the lips. □ **ventriloquist** *n.*

venture *n.* undertaking that involves risk. ● *v.* **1** dare. **2** dare to go or say. □ **venturesome** *adj.*

■ *n.* endeavour, enterprise, gamble, undertaking. ● *v.* **1** dare, endeavour, presume, try. **2** chance, hazard, risk.

venue *n.* appointed place for a meeting etc.

veracious *adj.* **1** truthful. **2** true. □ **veraciously** *adv.*, **veracity** *n.*

■ **1** candid, frank, honest, sincere, truthful. **2** accurate, correct, factual, true, truthful.

veranda *n.* roofed terrace.

verb *n.* word indicating action or occurrence or being.

verbal *adj.* **1** of or in words. **2** spoken. **3** of a verb. □ **verbally** *adv.*

verbalize *v.* **1** express in words. **2** be verbose. □ **verbalization** *n.*

verbatim /verbáytim/ *adv.* & *adj.* in exactly the same words.

verbiage *n.* excessive number of words.

verbose *adj.* using more words than are needed. □ **verbosely** *adv.*, **verbosity** *n.*

■ long-winded, wordy.

verdant *adj.* (of grass etc.) green.

verdict *n.* **1** decision reached by a jury. **2** decision or opinion reached after testing something.

■ conclusion, decision, finding, judgement, opinion, ruling.

verdigris *n.* green deposit forming on copper or brass.

verdure *n.* green vegetation.

verge¹ *n.* **1** extreme edge. **2** grass edging of a road etc. □ **on the verge of** very near to.

■ **1** border, brim, brink, edge, lip, rim, side.

verge² *v.* **verge on** border on.

verger *n.* church caretaker.

verify *v.* check the truth or correctness of. □ **verification** *n.*

■ authenticate, check, confirm, corroborate, make sure of, prove, substantiate, validate.

verisimilitude *n.* appearance of being true.

veritable *adj.* real, rightly named.

■ actual, authentic, genuine, proper, real, true.

vermicelli *n.* pasta made in slender threads.

vermiform *adj.* worm-like in shape.

vermilion *adj.* & *n.* bright red.

vermin *n.* (*pl.* **vermin**) animal or insect regarded as a pest.

verminous *adj.* infested with vermin.

vermouth *n.* wine flavoured with herbs.

vernacular *n.* ordinary language of a country or district.

vernal *adj.* of or occurring in spring.

veronica *n.* a kind of flowering herb or shrub.

verruca /vəro͞okə/ *n.* (*pl.* **-ae**) wart, esp. on the foot.

versatile *adj.* able to do or be used for many different things. □ **versatility** *n.*

■ adaptable, flexible, resourceful; handy, multi-purpose.

verse *n.* **1** metrical (not prose) composition. **2** group of lines forming a unit in a poem or hymn. **3** numbered division of a Bible chapter.

versed *adj.* **versed in** experienced in.

versify *v.* **1** express in verse. **2** compose verse. □ **versification** *n.*

version *n.* **1** particular account of a matter. **2** special or variant form. **3** translation.

■ **1** account, description, report, story. **2** design, form, kind, model, style, type, variant.

versus *prep.* against.

vertebra *n.* (*pl.* **-brae**) segment of the backbone. □ **vertebral** *adj.*

vertebrate *n.* & *adj.* (animal) having a backbone.

vertex *n.* (*pl.* **-tices**) highest point of a hill etc., apex.

■ apex, crest, crown, peak, pinnacle, point, summit, top.

vertical *adj.* perpendicular to the horizontal, upright. ● *n.* vertical line or position. ◻ **vertically** *adv.*

■ *adj.* erect, on end, perpendicular, upright.

vertigo *n.* dizziness.

■ dizziness, giddiness, light-headedness.

verve *n.* enthusiasm, vigour.

■ animation, *colloq.* bounce, dynamism, élan, enthusiasm, gusto, life, liveliness, pep, spirit, vigour, *colloq.* vim, vitality, vivacity, zeal, zest, zip.

very *adv.* in a high degree, extremely. ● *adj.* **1** actual, truly such. **2** mere. ◻ **very best, worst, etc.** absolute best, worst, etc. **very high frequency** frequency in the range 30–300 MHz. **very well** expression of consent.

■ *adv.* especially, exceedingly, exceptionally, extremely, greatly, highly, most, particularly, really, thoroughly, totally, truly, unusually, utterly. ● *adj.* **1** actual, exact, precise; same, selfsame.

vesicle *n.* **1** sac, esp. containing liquid. **2** blister.

vessel *n.* **1** receptacle, esp. for liquid. **2** boat, ship. **3** tube-like structure conveying blood or other fluid in the body of an animal or plant.

■ **1** container, receptacle. **2** craft, boat, ship.

vest *n.* **1** undergarment covering the trunk. **2** (*US*) waistcoat. ● *v.* confer or furnish with (power) as a firm or legal right. ◻ **vested interest** advantageous right held by a person or group.

vestibule *n.* **1** entrance hall. **2** porch.

vestige *n.* **1** small remaining bit. **2** very small amount. ◻ **vestigial** *adj.*, **vestigially** *adv.*

■ **1** evidence, mark, sign, suggestion, trace. **2** fragment, iota, particle, scrap, shred, speck; hint, suggestion, suspicion.

vestment *n.* ceremonial garment, esp. of clergy or a church choir.

vet (*colloq.*) *n.* veterinary surgeon. ● *v.* (**vetted**) examine critically for faults etc.

vetch *n.* plant of the pea family used as fodder for cattle.

veteran *n.* person with long experience, esp. in the armed forces.

veterinarian *n.* veterinary surgeon.

veterinary *adj.* of or for the treatment of diseases and disorders of animals. ◻ **veterinary surgeon** person skilled in such treatment.

veto *n.* (*pl.* **-oes**) **1** rejection of something proposed. **2** right to make this. ● *v.* reject by a veto.

■ *n.* **1** ban, embargo, interdict, prohibition. ● *v.* ban, disallow, forbid, outlaw, prevent, prohibit, proscribe, quash, reject, stop, turn down.

vex *v.* annoy. ◻ **vexed question** problem that is much discussed. **vexation** *n.*, **vexatious** *adj.*

■ *colloq.* aggravate, anger, annoy, drive mad, enrage, exasperate, gall, incense, infuriate, irk, irritate, madden, *colloq.* rile.

VHF *abbr.* very high frequency.

via *prep.* by way of, through.

viable *adj.* **1** practicable. **2** capable of living or surviving. ◻ **viability** *n.*

■ **1** feasible, possible, practicable, practical, workable.

viaduct *n.* long bridge over a valley.

vial *n.* small bottle.

viands *n.pl.* articles of food.

vibrant *adj.* **1** vibrating, resonant. **2** thrilling with energy.

■ **2** alive, animated, dynamic, energetic, lively, spirited, vivacious.

vibraphone *n.* percussion instrument like a xylophone but with a vibrating effect.

vibrate *v.* **1** move rapidly and continuously to and fro. **2** sound with rapid slight variation of pitch. ◻ **vibrator** *n.*, **vibratory** *adj.*

■ **1** oscillate, pulsate, pulse, quiver, shake, throb, tremble.

vibration *n.* **1** vibrating. **2** (*pl.*) atmosphere etc. felt intuitively.

vicar *n.* member of the clergy in charge of a parish.

vicarage *n.* house of a vicar.

vicarious *adj.* **1** experienced indirectly. **2** acting or done etc. for another. ◻ **vicariously** *adv.*

vice[1] *n.* **1** great wickedness. **2** criminal and immoral practices.

■ corruption, degeneracy, depravity, evil, immorality, iniquity, sin, turpitude, venality, viciousness, villainy, wickedness, wrongdoing.

vice[2] *n.* instrument with two jaws for holding things firmly.

vice- *pref.* **1** substitute or deputy for. **2** next in rank to.

vice-chancellor *n.* chief administrator of a university.

viceroy *n.* person governing a colony etc. as the sovereign's representative. □ **viceregal** *adj.*

vice versa with terms the other way round.

vicinity *n.* surrounding district. □ **in the vicinity (of)** near.

■ area, district, environs, locality, neighbourhood, precincts, region.

vicious *adj.* **1** bad-tempered, spiteful. **2** violent and dangerous. □ **vicious circle** bad situation producing effects that intensify its original cause. **viciously** *adv.*

■ **1** bad-tempered, *colloq.* bitchy, bitter, malicious, malignant, mean, nasty, spiteful, venomous, vitriolic. **2** aggressive, barbaric, brutal, cruel, dangerous, ferocious, fierce, savage, violent.

vicissitude *n.* change of circumstances or luck.

victim *n.* **1** person injured, killed, or made to suffer. **2** creature sacrificed to a god etc.

■ **1** martyr; casualty, fatality.

victimize *v.* single out to suffer ill-treatment. □ **victimization** *n.*

victor *n.* winner.

■ champion, conquerer, winner.

Victorian *n.* & *adj.* (person) of the reign of Queen Victoria (1837–1901).

victorious *adj.* having gained victory.

■ conquering, successful, triumphant, winning.

victory *n.* success achieved by gaining mastery over opponent(s) or having the highest score.

vicuña /vikyōōnə/ *n.* **1** S. American animal related to the llama. **2** soft cloth made from its wool.

video *n.* (*pl.* **-os**) **1** recording or broadcasting of pictures. **2** apparatus for this. **3** videotape. ● *v.* make a video of.

videotape *n.* magnetic tape suitable for recording television pictures and sound. ● *v.* record on this.

videotex *n.* (also **videotext**) electronic information system, esp. teletext or viewdata.

vie *v.* (**vying**) carry on a rivalry, compete.

■ compete, contend, contest, strive, struggle.

view *n.* **1** range of vision. **2** what is seen. **3** mental attitude, opinion. ● *v.* **1** look at. **2** consider. **3** watch television. □ **in view of** considering. **on view** displayed for inspection. **with a view to** with the hope or intention of. **viewer** *n.*

■ *n.* **1** sight, vision. **2** aspect, outlook, panorama, prospect, scene, vista. **3** attitude, belief, conviction, feeling, opinion, point of view, standpoint, viewpoint. ● *v.* **1** *poetic* behold, contemplate, gaze at, look at, observe, regard, survey, watch. **2** consider, deem, judge, regard, think of.

viewdata *n.* news and information service from a computer source to which a television screen is connected by a telephone link.

viewfinder *n.* device on a camera showing the extent of the area being photographed.

viewpoint *n.* point of view.

vigil *n.* period of staying awake to keep watch or pray.

vigilant *adj.* watchful. □ **vigilantly** *adv.*, **vigilance** *n.*

■ alert, attentive, awake, careful, circumspect, heedful, observant, on one's guard, on one's toes, on the lookout, wary, watchful, *colloq.* wide awake.

vigilante /vijilánti/ *n.* member of a self-appointed group trying to prevent crime etc.

vignette /veenyét/ *n.* short written description.

vigour *n.* active physical or mental strength, forcefulness. □ **vigorous** *adj.*, **vigorously** *adv.*

■ animation, drive, dynamism, energy, enthusiasm, force, gusto, liveliness, power, spirit, strength, verve, *colloq.* vim, zeal, zest.

Viking *n.* ancient Scandinavian trader and pirate.

vile *adj.* **1** extremely disgusting. **2** wicked. □ **vilely** *adv.*, **vileness** *n.*

■ **1** disgusting, distasteful, foul, loathsome, nasty, objectionable, obnoxious, offensive, repellent, repugnant, repulsive, sickening. **2** bad, base, contemptible, corrupt, degenerate, depraved, immoral, iniquitous, sinful, villainous, wicked.

vilify *v.* say evil things about. □ **vilification** *n.*, **vilifier** *n.*

■ abuse, calumniate, defame, insult, libel, malign, run down, slander.

villa *n.* **1** holiday home, esp. abroad. **2** house in a suburban district.

village *n.* group of houses etc. in a country district. □ **villager** *n.*

villain *n.* wicked person. □ **villainous** *adj.*, **villainy** *n.*

■ blackguard, criminal, *colloq.* crook, evildoer, malefactor, miscreant, rogue, scoundrel, wretch, wrongdoer.

vim *n.* (*colloq.*) vigour.
vinaigrette *n.* salad dressing of oil and vinegar.
vindicate *v.* **1** clear of blame. **2** justify. □ **vindication** *n.*

■ **1** absolve, acquit, clear, exonerate. **2** justify, prove, support.

vindictive *adj.* showing a desire for vengeance. □ **vindictively** *adv.*, **vindictiveness** *n.*

■ avenging, revengeful, vengeful; malicious, spiteful, unforgiving.

vine *n.* climbing plant whose fruit is the grape.
vinegar *n.* sour liquid made from wine, malt, etc., by fermentation. □ **vinegary** *adj.*
vineyard *n.* plantation of vines for wine-making.
vintage *n.* **1** wine from a season's grapes, esp. when of high quality. **2** date of origin or existence. ● *adj.* of high quality, esp. from a past period.
vintner *n.* wine-merchant.
vinyl /vínil/ *n.* a kind of plastic.
viola[1] /vióla/ *n.* instrument like a violin but of lower pitch.
viola[2] /víələ/ *n.* plant of the genus to which violets belong.
violate *v.* **1** break (an oath or treaty etc.). **2** treat (a sacred place) irreverently. **3** disturb. **4** rape. □ **violation** *n.*, **violator** *n.*

■ **1** break, contravene, disobey, disregard, flout, infringe, transgress. **2** desecrate, profane.

violent *adj.* **1** involving great force or intensity. **2** using excessive physical force. □ **violently** *adv.*, **violence** *n.*

■ **1** acute, extreme, forceful, furious, intense, mighty, powerful, severe, strong, uncontrollable, vehement, wild. **2** barbaric, brutal, cruel, ferocious, fierce, savage, vicious.

violet *n.* **1** small plant, often with purple flowers. **2** bluish-purple colour. ● *adj.* bluish-purple.
violin *n.* musical instrument with four strings of treble pitch, played with a bow. □ **violinist** *n.*
violoncello /víələnchéllō/ *n.* (*pl.* **-os**) cello.
viper *n.* small poisonous snake.
virago *n.* (*pl.* **-os**) aggressive woman.
viral /víral/ *adj.* of a virus.
virgin *n.* **1** person who has never had sexual intercourse. **2** (**the Virgin**) Mary, mother of Christ. ● *adj.* **1** virginal. **2** not yet used. □ **virginal** *adj.*, **virginity** *n.*
virile *adj.* having masculine strength or procreative power. □ **virility** *n.*
virology *n.* study of viruses. □ **virologist** *n.*
virtual *adj.* being so in effect though not in name. □ **virtually** *adv.*

■ □ **virtually** all but, almost, as good as, more or less, nearly, practically.

virtue *n.* **1** moral excellence, goodness. **2** chastity. **3** good characteristic. □ **by** or **in virtue of** because of.

■ **1** decency, goodness, honesty, honour, integrity, morality, probity, righteousness. **2** chastity, innocence, purity, virginity. **3** asset, good point, strength.

virtuoso *n.* (*pl.* **-si**) expert performer. □ **virtuosity** *n.*

■ expert, genius, maestro, master, wizard.

virtuous *adj.* morally good. □ **virtuously** *adv.*

■ decent, good, honest, honourable, moral, noble, principled, respectable, righteous, upright; chaste, innocent, pure, virginal.

virulent *adj.* **1** (of poison or disease) extremely strong or violent. **2** bitterly hostile. □ **virulently** *adv.*, **virulence** *n.*

■ **1** dangerous, deadly, harmful, life-threatening. **2** acrimonious, bitter, hostile, malicious, malignant, nasty, spiteful, venomous, vicious, vitriolic.

virus *n.* (*pl.* **-uses**) **1** minute organism able to cause disease. **2** (*colloq.*) such a disease. **3** destructive code hidden in a computer program.
visa *n.* official mark on a passport, permitting the holder to enter a specified country.
visage *n.* person's face.
vis-à-vis /veézaavée/ *adv.* & *prep.* **1** with regard to. **2** as compared with.
viscera /víssərə/ *n.pl.* internal organs of the body. □ **visceral** *adj.*
viscid /víssid/ *adj.* viscous.
viscose *n.* **1** viscous cellulose. **2** fabric made from this.
viscount /víkownt/ *n.* nobleman ranking between earl and baron.
viscountess *n.* **1** woman with the rank of viscount. **2** viscount's wife or widow.

viscous *adj.* thick and gluey. ▫ **viscosity** *n.*

visibility *n.* **1** state of being visible. **2** range of vision or clarity.

visible *adj.* able to be seen or noticed. ▫ **visibly** *adv.*

▪ apparent, conspicuous, discernible, evident, manifest, obvious, noticeable, perceivable, perceptible, plain.

vision *n.* **1** ability to see, sight. **2** thing seen in the imagination or a dream. **3** foresight. **4** person of unusual beauty.

▪ **1** eyesight, sight. **2** dream, fantasy; apparition, chimera, ghost, hallucination, mirage, phantom, spectre. **3** foresight, imagination.

visionary *adj.* **1** fanciful. **2** not practical. ● *n.* person with visionary ideas.

▪ *adj.* **1** fanciful, idealistic, romantic, unrealistic. ● *n.* dreamer, idealist, romantic.

visit *v.* **1** go or come to see. **2** stay temporarily with or at. ● *n.* act of visiting. ▫ **visitor** *n.*

▪ *v.* **1** call on, drop in on, look up, pay a visit to. ● *n.* call, sojourn, stay. ▫ **visitor** caller, guest.

visitation *n.* **1** official visit. **2** trouble regarded as divine punishment.

visor /vízər/ *n.* **1** movable front part of a helmet, covering the face. **2** shading device at the top of a vehicle's windscreen.

vista *n.* extensive view, esp. seen through a long opening.

visual *adj.* of or used in seeing. ▫ **visual display unit** device displaying a computer output or input on a screen. **visually** *adv.*

visualize *v.* form a mental picture of. ▫ **visualization** *n.*

▪ conceive of, envisage, imagine, picture, think of.

vital *adj.* **1** essential to life. **2** essential to a thing's existence or success. **3** full of vitality. ● *n.pl.* vital organs of the body. ▫ **vital statistics 1** those relating to population figures. **2** (*colloq.*) measurements of a woman's bust, waist, and hips. **vitally** *adv.*

▪ *adj.* **1,2** critical, crucial, essential, fundamental, imperative, indispensable, key, necessary. **3** alive, animated, energetic, lively, spirited, vibrant, vivacious.

vitality *n.* liveliness, persistent energy.

▪ animation, *colloq.* bounce, dynamism, élan, energy, go, gusto, life, liveliness, pep, spirit, verve, vigour, *colloq.* vim, vivacity, zeal, zest, zip.

vitamin *n.* any of the organic substances present in food and essential to nutrition.

vitaminize *v.* add vitamins to.

vitiate /víshiayt/ *v.* make imperfect or ineffective.

▪ harm, impair, mar, ruin, spoil.

viticulture *n.* vine-growing.

vitreous *adj.* having a glass-like texture or finish.

vitrify *v.* change into a glassy substance. ▫ **vitrifaction** *n.*

vitriol *n.* **1** sulphuric acid. **2** caustic speech. ▫ **vitriolic** *adj.*

▪ ▫ **vitriolic** bitter, caustic, cruel, hostile, hurtful, malicious, savage, scathing, vicious, virulent.

vituperate *v.* criticize abusively. ▫ **vituperation** *n.*, **vituperative** *adj.*

vivacious *adj.* lively, high-spirited. ▫ **vivaciously** *adv.*, **vivacity** *n.*

▪ animated, bubbly, cheerful, ebullient, energetic, exuberant, high-spirited, lively, perky, spirited, sprightly, spry.

vivarium *n.* (*pl.* **-ia**) place for keeping living animals etc. in natural conditions.

vivid *adj.* **1** bright and strong. **2** clear. **3** (of imagination) lively. ▫ **vividly** *adv.*, **vividness** *n.*

▪ **1** bright, brilliant, colourful, intense, strong. **2** clear, detailed, graphic, realistic. **3** active, creative, fertile, inventive, lively, prolific.

vivify *v.* put life into.

viviparous *adj.* bringing forth young alive, not egg-laying.

vivisection *n.* performance of experiments on living animals.

vixen *n.* female fox.

viz. *adv.* namely.

vocabulary *n.* **1** list of words with their meanings. **2** words known or used by a person or group.

vocal *adj.* of, for, or uttered by the voice. ● *n.* piece of sung music. ▫ **vocally** *adv.*

vocalist *n.* singer.

vocalize *v.* utter. ▫ **vocalization** *n.*

vocation *n.* **1** strong desire or feeling of fitness for a certain career. **2** trade, profession. ▫ **vocational** *adj.*

▪ **1** calling. **2** business, career, employment, job, line, occupation, profession, trade, work.

vociferate *v.* say loudly, shout. □ **vociferation** *n.*

■ bawl, bellow, roar, scream, shout, thunder, yell.

vociferous *adj.* making a great outcry. □ **vociferously** *adv.*

■ clamorous, insistent, loud, noisy, obstreperous.

vodka *n.* alcoholic spirit distilled chiefly from rye.

vogue *n.* current fashion. □ **in vogue** in fashion.

■ craze, fad, fashion, mode, trend. □ **in vogue** chic, fashionable, in, in fashion, *colloq.* trendy.

voice *n.* **1** sounds formed in the larynx and uttered by the mouth. **2** expressed opinion. **3** right to express an opinion. ● *v.* express, vocalize. □ **voice-over** *n.* narration in a film etc. without a picture of the speaker.

■ *n.* **2** belief, feeling, opinion, view. **3** say, vote. ● *v.* air, articulate, communicate, declare, enunciate, give vent to, make known, put into words, state, utter, verbalize.

void *adj.* **1** empty. **2** not valid. ● *n.* empty space, emptiness. ● *v.* **1** make void. **2** excrete.

■ *adj.* **1** empty, unfilled, unoccupied, vacant. **2** invalid, null and void. ● *n.* emptiness, vacuum. ● *v.* **1** annul, cancel, invalidate, nullify, quash.

voile *n.* very thin dress fabric.

volatile *adj.* **1** evaporating rapidly. **2** changing quickly in mood. □ **volatility** *n.*

■ **2** capricious, changeable, emotional, fickle, inconstant, mercurial, moody, temperamental, unpredictable, unstable.

vol-au-vent /vóllōvon/ *n.* puff pastry case with a savoury filling.

volcano *n.* (*pl.* **-oes**) mountain with a vent through which lava is expelled. □ **volcanic** *adj.*

vole *n.* small rodent.

volition *n.* use of one's own will in making a decision etc.

■ choice, choosing, discretion, (free) will, preference.

volley *n.* (*pl.* **-eys**) **1** simultaneous discharge of missiles etc. **2** outburst of questions or other words. **3** return of the ball in tennis etc. before it touches the ground. ● *v.* send in a volley.

■ *n.* **1** barrage, bombardment, cannonade, salvo. **2** hail, shower, storm, stream, torrent.

volleyball *n.* game for two teams of six people sending a large ball by hand over a net.

volt *n.* unit of electromotive force.

voltage *n.* electromotive force expressed in volts.

volte-face /voltfaass/ *n.* complete change of attitude to something.

voluble *adj.* speaking or spoken with a great flow of words. □ **volubly** *adv.*, **volubility** *n.*

■ chatty, garrulous, loquacious, talkative; fluent, glib.

volume *n.* **1** book. **2** amount of space occupied or contained. **3** amount. **4** strength of sound.

■ **1** book, tome. **2** bulk, capacity, content, mass, size. **3** amount, quantity, sum total.

voluminous *adj.* **1** having great volume, bulky. **2** prolific.

■ **1** ample, big, bulky, capacious, enormous, great, large, roomy, spacious, substantial.

voluntary *adj.* **1** done, given, or acting by choice. **2** working or done without payment. **3** maintained by voluntary contributions. □ **voluntarily** *adv.*

■ **1** discretionary, elective, optional. **2** unpaid.

volunteer *n.* **1** person who offers to do something. **2** one who enrols voluntarily for military service. ● *v.* **1** undertake or offer voluntarily. **2** be a volunteer.

voluptuary *n.* person fond of luxury etc.

voluptuous *adj.* **1** full of or fond of sensual pleasures. **2** having a full attractive figure. □ **voluptuously** *adv.*, **voluptuousness** *n.*

■ **1** hedonistic, pleasure-loving, sensual, sensuous, sybaritic. **2** attractive, *colloq.* curvaceous, desirable, seductive, sexy, shapely.

vomit *v.* (**vomited**) eject (matter) from the stomach through the mouth. ● *n.* vomited matter.

voodoo *n.* form of religion based on witchcraft. □ **voodooism** *n.*

voracious *adj.* **1** greedy. **2** insatiable. □ **voraciously** *adv.*, **voracity** *n.*

■ **1** gluttonous, greedy, *colloq.* gutsy, insatiable, ravenous. **2** ardent, avid, eager, enthusiastic, insatiable, passionate.

vortex *n.* (*pl.* **-exes** or **-ices**) **1** whirlpool. **2** whirlwind.

vote *n.* **1** formal expression of one's opinion or choice on a matter under discussion. **2** choice etc. expressed thus. **3** right to vote. ● *v.* give or decide by a vote. □ **vote for** choose by a vote. **voter** *n.*

■ *n.* **1** ballot, election, plebiscite, poll, referendum. **2** choice, preference, selection. **3** franchise, suffrage. □ **vote for** choose, elect, pick, opt for, select.

votive *adj.* given to fulfil a vow.

vouch *v.* **vouch for** guarantee the accuracy or reliability etc. of.

■ answer for, attest to, bear witness to, certify, confirm, guarantee.

voucher *n.* **1** document exchangeable for certain goods or services. **2** a kind of receipt.

vow *n.* solemn promise, esp. to a deity or saint. ● *v.* make a vow.

■ *n.* oath, pledge, promise. ● *v.* pledge, promise, swear, undertake.

vowel *n.* **1** speech sound made without audible stopping of the breath. **2** letter(s) representing this.

voyage *n.* journey made by water or in space. ● *v.* make a voyage. □ **voyager** *n.*

■ *n.* cruise, journey, passage, trip. ● *v.* cruise, journey, sail, travel.

voyeur /vwaayőr/ *n.* person who gets sexual pleasure from watching others having sex or undressing.

vulcanite *n.* hard black vulcanized rubber.

vulcanize *v.* strengthen (rubber) by treating with sulphur. □ **vulcanization** *n.*

vulgar *adj.* lacking refinement or good taste. □ **vulgar fraction** one represented by numbers above and below a line. **vulgarly** *adv.*, **vulgarity** *n.*

■ boorish, coarse, ill-mannered, uncouth; cheap, flashy, garish, gaudy, tasteless, tawdry; crude, indecent, indelicate, obscene, rude, unseemly.

vulgarian *n.* vulgar (esp. rich) person.

vulgarism *n.* vulgar word(s) etc.

vulnerable *adj.* able to be hurt or injured. □ **vulnerability** *n.*

■ defenceless, exposed, unguarded, unprotected, weak.

vulture *n.* large carrion-eating bird of prey.

vulva *n.* external parts of the female genital organs.

vying *see* **vie**.

Ww

W. *abbr.* **1** west. **2** western.
wad *n.* **1** pad of soft material. **2** bunch of papers or banknotes. ● *v.* (**wadded**) pad.
wadding *n.* padding.
waddle *v.* & *n.* walk with short steps and a swaying movement.
wade *v.* walk through water or mud. □ **wade through** proceed slowly and laboriously through (work etc.).
wader *n.* **1** long-legged waterbird. **2** (*pl.*) high waterproof boots.
wafer *n.* **1** thin light biscuit. **2** small thin slice. □ **wafery** *adj.*
waffle[1] (*colloq.*) *n.* vague wordy talk or writing. ● *v.* talk or write waffle.
waffle[2] *n.* small cake of batter cooked in a **waffle-iron**.
waft *v.* carry or travel lightly through air or over water. ● *n.* wafted odour.

■ *v.* blow, drift, float, glide.

wag *v.* (**wagged**) shake briskly to and fro. ● *n.* **1** wagging movement. **2** humorous person.
wage[1] *v.* engage in (war).
wage[2] *n.* (also **wages**) employee's regular pay.

■ emolument, pay, remuneration, salary, stipend.

wager *n.* & *v.* bet.

■ *n.* bet, *colloq.* flutter, gamble, stake. ● *v.* bet, gamble, risk, stake, venture.

waggle *v.* & *n.* wag.
wagon *n.* (also **waggon**) **1** four-wheeled goods vehicle pulled by horses or oxen. **2** open railway truck.
wagtail *n.* small bird with a long tail that wags up and down.
waif *n.* homeless child.
wail *v.* & *n.* **1** (utter) a long sad cry. **2** lament.

■ *v.* **1** cry, howl, lament, sob, weep.

wainscot *n.* (also **wainscoting**) wooden panelling in a room.
waist *n.* **1** part of the human body between ribs and hips. **2** narrow middle part.
waistcoat *n.* close-fitting waist-length sleeveless jacket.
waistline *n.* outline or size of the waist.
wait *v.* **1** defer an action or departure until a specified time or event occurs. **2** be postponed. **3** act as waiter or waitress. ● *n.* act or period of waiting. □ **wait on** hand food and drink to (people) at a meal.

■ *v.* **1** *sl.* hang on, linger, pause, remain, stay. **2** be deferred, be postponed, be put off. ● *n.* delay, hold-up, interval, pause.

waiter, waitress *ns.* person employed to wait on customers in a restaurant etc.
waive *v.* refrain from using or insisting on. □ **waiver** *n.*

■ forgo, give up, relinquish, renounce; disregard, overlook.

wake[1] *v.* (**woke** or **waked, woken** or **waked**) **1** cease or cause to cease sleeping. **2** evoke. ● *n.* (*Ir.*) **1** watch by a corpse before burial. **2** attendant lamentations and merrymaking. □ **wake up** **1** wake. **2** make or become alert.

■ *v.* **1** awake, awaken, rouse, waken, wake up. **2** arouse, evoke, stir up. ● *n.* **1** vigil. □ **wake up 2** animate, enliven, galvanize, rouse, stimulate, stir.

wake[2] *n.* track left on water's surface by a ship etc. □ **in the wake of 1** behind. **2** following.
wakeful *adj.* **1** unable to sleep. **2** sleepless. □ **wakefulness** *n.*
waken *v.* wake.
walk *v.* **1** progress by setting down one foot and then lifting the other(s) in turn. **2** travel (over) in this way. **3** accompany in walking. ● *n.* **1** journey on foot. **2** manner or style of walking. **3** place or route for walking. □ **walk of life** occupation. **walk out 1** depart suddenly and angrily. **2** go on strike. **walk-out** *n.* **walk out on** desert. **walkover** *n.* easy victory.

■ *v.* **1** amble, hike, march, pace, pad, perambulate, plod, ramble, saunter, stalk, step, stride, stroll, strut, trudge. **3** accompany, escort, take. ● *n.* **1** amble, constitutional, hike, perambulation, ramble, saunter, stroll. **2** gait. **3** footpath, lane, path, route, track, trail; beat, circuit, route. □ **walk out on** abandon, desert, forsake, leave.

walkabout *n.* informal stroll among a crowd by royalty etc.

walkie-talkie *n.* small portable radio transmitter and receiver.

walking stick stick carried or used as a support when walking.

wall *n.* **1** continuous upright structure forming one side of a building, room, or area. **2** thing like this in form or function. ● *v.* surround or enclose with a wall.

■ *n.* **2** barrier, fence, partition, screen; impediment, obstacle.

wallaby *n.* small species of kangaroo.

wallet *n.* small folding case for banknotes or documents.

wallflower *n.* garden plant with fragrant flowers.

wallop (*sl.*) *v.* (**walloped**) thrash, hit hard. ● *n.* heavy blow.

wallow *v.* roll in mud or water etc. ● *n.* act of wallowing. □ **wallow in** take unrestrained pleasure in.

■ □ **wallow in** bask in, delight in, glory in, indulge (oneself) in, luxuriate in, revel in, savour.

wallpaper *n.* paper for covering the interior walls of rooms.

wally *n.* (*sl.*) stupid person.

walnut *n.* **1** nut containing a wrinkled edible kernel. **2** tree bearing this. **3** its wood.

walrus *n.* large seal-like Arctic animal with long tusks.

waltz *n.* **1** ballroom dance. **2** music for this. ● *v.* **1** dance a waltz. **2** (*colloq.*) move gaily or casually.

wan *adj.* pallid. □ **wanly** *adv.*

■ anaemic, ashen, colourless, pale, pallid, pasty, peaky, washed out, white.

wand *n.* slender rod, esp. associated with the working of magic.

wander *v.* **1** go from place to place aimlessly. **2** stray from a path etc. **3** digress. ● *n.* act of wandering. □ **wanderer** *n.*

■ *v.* **1** drift, meander, *colloq.* mooch, ramble, range, roam, rove, stray. **3** deviate, digress, drift, go off at a tangent, ramble. ● *n.* amble, ramble, saunter, stroll.

wanderlust *n.* strong desire to travel.

wane *v.* **1** decrease in vigour or importance. **2** (of the moon) show a decreasing bright area after being full. □ **on the wane** waning.

■ **1** abate, decline, decrease, diminish, dwindle, ebb, fall, lessen, subside, weaken.

wangle *v.* (*colloq.*) contrive to obtain (a favour etc.).

want *v.* **1** desire. **2** need. **3** lack. **4** be without. ● *n.* **1** desire. **2** need. **3** lack.

■ *v.* **1** crave, desire, fancy, hanker after, *colloq.* have a yen for, wish for. **2** need, require. **3** be deficient in, be short of, lack. ● *n.* **1** craving, desire, fancy, hankering, wish, *colloq.* yen. **2** need, poverty. **3** dearth, deficiency, insufficiency, lack, need, scarcity, shortage.

wanted *adj.* (of a suspected criminal) sought by the police.

wanting *adj.* lacking, deficient.

■ deficient, disappointing, imperfect, inadequate, insufficient, lacking, unsatisfactory.

wanton *adj.* **1** random, arbitrary. **2** licentious.

■ **1** arbitrary, gratuitous, indiscriminate, motiveless, random, senseless, unjustified, unnecessary, unprovoked. **2** dissolute, immoral, licentious, profligate, promiscuous.

wapiti /wóppiti/ *n.* large N. American deer.

war *n.* **1** (a period of) fighting (esp. between countries). **2** open hostility. **3** organized campaign of action. ● *v.* (**warred**) make war. □ **at war** engaged in a war.

■ *n.* **1** battle, combat, conflict, fighting, hostilities, warfare. **3** battle, campaign, crusade, drive, fight, struggle. ● *v.* do battle, fight, strive, struggle, wage war.

warble *v.* sing, esp. with a gentle trilling note. ● *n.* warbling sound.

ward *n.* **1** room with beds for patients in a hospital. **2** division of a city or town, electing a councillor to represent it. **3** person (esp. a child) under the care of a guardian or law court. ● *v.* **ward off** keep away, repel.

■ *n.* **2** district, division, quarter, sector, zone. **3** charge, dependant, protégé. ● *v.* fend off, fight off, forestall, hold at bay, keep away, repel, stave off.

warden *n.* **1** official with supervisory duties. **2** churchwarden.

warder *n.* prison officer.

wardrobe *n.* **1** large cupboard for storing clothes. **2** stock of clothes.

ware *n.* **1** manufactured goods of the kind specified. **2** (*pl.*) articles offered for sale.

■ **2 (wares)** commodities, goods, merchandise, stock.

warehouse *n.* building for storing goods or furniture.

warfare *n.* making war, fighting.

warhead *n.* explosive head of a missile.

warlike *adj.* **1** fond of making war, aggressive. **2** of or for war.

■ **1** aggressive, bellicose, belligerent, combative, militaristic, pugnacious. **2** martial, military.

warm *adj.* **1** moderately hot. **2** providing warmth. **3** enthusiastic, hearty. **4** kindly and affectionate. ● *v.* make or become warm. □ **warm-blooded** *adj.* having blood that remains warm permanently. **warm to** become cordial towards (a person) or more animated about (a task). **warm up 1** warm. **2** reheat. **3** prepare for exercise etc. by practice beforehand. **4** make or become more lively. **warmly** *adv.*, **warmness** *n.*

■ *adj.* **1** balmy; lukewarm, tepid. **3** ardent, eager, earnest, enthusiastic, hearty, sincere, wholehearted. **4** affable, amiable, cheerful, cordial, friendly, genial, hospitable, kindly, pleasant; affectionate, caring, loving, tender.

warmonger *n.* person who seeks to bring about war.

warmth *n.* warmness.

warn *v.* **1** inform about a present or future danger or difficulty etc. **2** advise about action in this. □ **warn off** tell (a person) to keep away or to avoid (a thing).

■ alert, apprise, forewarn, inform, notify, tip off; advise, caution, counsel.

warning *n.* thing that serves to warn a person.

■ notification, tip-off; advice, caution, counsel; augury, indication, omen, portent, sign, signal.

warp *v.* **1** make or become bent by uneven shrinkage or expansion. **2** pervert. ● *n.* **1** warped condition. **2** lengthwise threads in a loom.

warrant *n.* **1** thing that authorizes an action etc. **2** written authorization. ● *v.* **1** justify. **2** guarantee.

■ *n.* **1** authorization, certification, permission, sanction; guarantee, warranty. **2** licence, permit, writ. ● *v.* **1** excuse, justify. **2** answer for, attest to, certify, guarantee, vouch for.

warranty *n.* guarantee.

warren *n.* **1** series of burrows where rabbits live. **2** labyrinthine building or district.

warrior *n.* person who fights in a battle.

wart *n.* small hard abnormal growth. □ **warthog** *n.* African wild pig with wart-like growths on its face. **warty** *adj.*

wary *adj.* (**-ier, -iest**) cautious, on one's guard. □ **warily** *adv.*, **wariness** *n.*

■ apprehensive, careful, cautious, chary, circumspect, heedful, on one's guard, prudent, vigilant, watchful.

wash *v.* **1** cleanse with water or other liquid. **2** wash oneself or clothes etc. **3** be washable. **4** flow past, against, or over. **5** carry by flowing. **6** coat thinly with paint. **7** (*colloq.*, of reasoning) be valid or credible. ● *n.* **1** process of washing or being washed. **2** clothes etc. to be washed. **3** disturbed water or air behind a moving ship or aircraft etc. **4** thin coating of paint. □ **washed out 1** pallid. **2** faded. **washed up** (*sl.*) defeated, having failed. **wash one's hands of** refuse to take responsibility for. **wash out 1** make (a sport) impossible by heavy rainfall. **2** (*colloq.*) cancel. **wash-out** *n.* (*colloq.*) complete failure. **wash up 1** wash (dishes etc.) after use. **2** cast up on the shore. **washable** *adj.*

■ *v.* **1** clean, cleanse, mop, scour, scrub, shampoo, sponge, wipe. **2** bath, shower; launder. ● *n.* **1** ablutions, bath, shampoo, shower; scour, scrub, wipe. **3** backwash, wake. □ **washed out 1** ashen, colourless, pale, pallid, pasty, peaky, wan, white. **wash-out** disaster, failure, fiasco, *sl.* flop.

washbasin *n.* bowl for washing one's hands and face.

washer *n.* ring of rubber or metal etc. placed between two surfaces to give tightness.

washing *n.* clothes etc. to be washed. □ **washing-up** *n.* **1** dishes etc. for washing after use. **2** process of washing these.

washroom *n.* (*US*) room with a lavatory.

wasp *n.* stinging insect with a black and yellow striped body.

waspish *adj.* irritable. □ **waspishly** *adv.*

■ bad-tempered, crabby, crotchety, grumpy, irritable, peevish, snappy, splenetic, surly, testy.

wastage *n.* **1** loss or diminution by waste. **2** loss of employees by retirement or resignation.

waste *v.* **1** use or be used extravagantly or without adequate result. **2** fail to use. **3** make or become gradually weaker. ● *adj.* **1** thrown away because not wanted. **2** (of land) unfit for use. ● *n.* **1** act or instance of wasting. **2** waste material or food etc. **3** waste land. **4** waste pipe. □ **waste pipe** pipe carrying off used or superfluous water or steam.

■ *v.* **1** dissipate, fritter away, misuse, squander. **2** lose, miss. **3** debilitate, disable, enervate, enfeeble, weaken; atrophy, wither. ● *adj.* **1** superfluous, unused, unwanted, useless, worthless. **2** barren, desert, uncultivated, unproductive. ● *n.* **1** dissipation, misuse; extravagance, improvidence, prodigality, profligacy, wastefulness. **2** debris, dross, garbage, litter, refuse, rubbish, trash.

wasteful *adj.* extravagant. □ **wastefully** *adv.*, **wastefulness** *n.*

■ extravagant, improvident, lavish, overindulgent, prodigal, profligate, spendthrift, thriftless, uneconomical.

waster *n.* **1** wasteful person. **2** good-for-nothing person.

watch *v.* **1** keep under observation. **2** be in an alert state. **3** pay attention to. **4** exercise protective care. ● *n.* **1** act of watching, constant observation or attention. **2** sailor's period of duty, people on duty in this. **3** small portable device indicating the time. □ **on the watch** waiting alertly. **watch out** be on one's guard. **watch-tower** *n.* tower from which observation can be kept. **watcher** *n.*

■ *v.* **1** contemplate, eye, gaze at, look at, observe, regard, view. **2** be one one's guard, be on the watch *or* lookout, look out. **3** note, observe, pay attention to, take heed of. **4** look after, mind, supervise, take care of, tend. ● *n.* **1** lookout; observation, surveillance.

watchdog *n.* **1** dog kept to guard property. **2** guardian of people's rights etc.

watchful *adj.* watching closely. □ **watchfully** *adv.*, **watchfulness** *n.*

■ alert, attentive, heedful, observant, on one's toes, on the lookout, wary, *colloq.* wide awake.

watchmaker *n.* person who makes or repairs watches.

watchman *n.* man employed to guard a building etc.

watchword *n.* word or phrase expressing a group's principles.

water *n.* **1** colourless odourless tasteless liquid that is a compound of hydrogen and oxygen. **2** this as supplied for domestic use. **3** lake, sea. **4** watery secretion, urine. **5** level of the tide. ● *v.* **1** sprinkle, supply, or dilute with water. **2** secrete tears or saliva. □ **by water** in a boat etc. **water-bed** *n.* mattress of rubber etc. filled with water. **water-biscuit** *n.* thin unsweetened biscuit. **water-butt** *n.* barrel used to catch rainwater. **water-cannon** *n.* device giving a powerful jet of water to dispel a crowd etc. **water-closet** *n.* lavatory flushed by water. **water colour** **1** artists' paint mixed with water (not oil). **2** painting done with this. **water down** **1** dilute. **2** make less forceful. **water ice** frozen flavoured water. **water lily** plant with broad floating leaves. **water main** main pipe in a water supply system. **water-meadow** *n.* meadow that is flooded periodically by a stream. **water melon** melon with red pulp and watery juice. **water-mill** *n.* mill worked by a waterwheel. **water-pistol** *n.* toy pistol that shoots a jet of water. **water polo** ball game played by teams of swimmers. **water-power** *n.* power obtained from flowing or falling water. **water-rat** *n.* small rodent living beside a lake or stream. **water-skiing** *n.* sport of skimming over water on flat boards while towed by a motor boat. **water-table** *n.* level below which the ground is saturated with water. **water-wings** *n.pl.* floats worn on the shoulders by a person learning to swim.

■ *v.* **1** damp, dampen, douse, drench, flood, hose, irrigate, moisten, saturate, soak, splash, spray, sprinkle, wet. **2** run, stream. □ **water down** **1** dilute, weaken. **2** moderate, modify, soften, tone down.

watercourse *n.* **1** stream or artificial waterway. **2** its channel.

watercress *n.* a kind of cress that grows in streams and ponds.

waterfall *n.* stream that falls from a height.

■ cascade, cataract, falls.

waterfront *n.* part of a town that borders on a river, lake, or sea.

watering-can *n.* container with a spout for watering plants.

watering place **1** pool where animals drink. **2** spa, seaside resort.

waterline *n.* line along which the surface of water touches a ship's side.

waterlogged *adj.* saturated with water.

watermark *n.* manufacturer's design in paper, visible when the paper is held against light.

waterproof *adj.* unable to be penetrated by water. ● *n.* waterproof garment. ● *v.* make waterproof.

watershed *n.* **1** line of high land separating two river systems. **2** turning point in the course of events.

waterspout *n.* column of water between sea and cloud, formed by a whirlwind.

watertight *adj.* **1** made or fastened so that water cannot get in or out. **2** impossible to disprove.

waterway *n.* navigable channel.

waterwheel *n.* wheel turned by water to drive machinery.

waterworks *n.* establishment with machinery etc. for supplying water to a district.

watery *adj.* **1** of or like water. **2** containing too much water. **3** (of colour) pale. □ **wateriness** *n.*

■ **1** aqueous. **2** diluted, runny, thin, watered down, weak.

watt *n.* unit of electric power.

wattage *n.* amount of electric power, expressed in watts.

wattle[1] *n.* interwoven sticks used as material for fences, walls, etc.

wattle[2] *n.* fold of skin hanging from the neck of a turkey etc.

wave *n.* **1** moving ridge of water. **2** wave-like curve(s), e.g. in hair. **3** temporary increase of an influence or condition. **4** act of waving. **5** advancing group. **6** wave-like motion by which heat, light, sound, or electricity etc. is spread. **7** single curve in this. ● *v.* **1** move loosely to and fro or up and down. **2** move (one's arm etc.) thus as a signal. **3** give or have a wavy course or appearance.

■ *n.* **1** billow, breaker, ripple, roller, wavelet, white horse. **2** curl, kink. **3** surge, swell, upsurge. **4** gesticulation, gesture, sign, signal. ● *v.* **1** billow, flap, flutter, ripple, undulate. **2** gesticulate, gesture, sign, signal.

waveband *n.* range of wavelengths.

wavelength *n.* distance between corresponding points in a sound wave or electromagnetic wave.

wavelet *n.* small wave.

waver *v.* **1** be or become unsteady. **2** show hesitation or uncertainty. □ **waverer** *n.*

■ **1** rock, shake, sway, teeter, wobble. **2** be in two minds, be uncertain, dither, shilly-shally, vacillate.

wavy *adj.* (**-ier**, **-iest**) full of wave-like curves.

wax[1] *n.* **1** beeswax. **2** any of various similar soft substances. **3** polish containing this. ● *v.* coat, polish, or treat with wax. □ **waxy** *adj.*

wax[2] *v.* **1** increase in vigour or importance. **2** (of the moon) show an increasing bright area until becoming full.

waxwing *n.* small bird with red tips on some of its wing-feathers.

waxwork *n.* wax model, esp. of a person.

way *n.* **1** path, road, street, etc. **2** route, direction. **3** space free of obstacles so that people etc. can pass. **4** progress. **5** aspect. **6** method, style, manner. **7** manner of behaving, idiosyncrasy. **8** chosen or desired course of action. **9** distance in space or time. ● *adv.* (*colloq.*) far. □ **by the way** incidentally, as an irrelevant comment. **by way of 1** by going through. **2** serving as. **in a way** to a limited extent, in some respects. **in the way** forming an obstacle or hindrance. **on one's way** in the process of travelling or approaching. **on the way 1** on one's way. **2** (of a baby) conceived but not yet born. **way of life** normal pattern of life of a person or group.

■ *n.* **1** path, road, street, thoroughfare, track. **2** course, direction, route. **4** headway, progress. **5** aspect, detail, feature, particular, point, respect. **6** approach, course of action, method, procedure, technique; fashion, manner, mode, style. **7** character, disposition, nature, personality, temperament; habit, idiosyncrasy, mannerism, peculiarity.

wayfarer *n.* traveller.

waylay *v.* (**-laid**) lie in wait for.

wayside *n.* side of a road or path.

wayward *adj.* childishly self-willed, capricious. □ **waywardness** *n.*

■ capricious, difficult, disobedient, headstrong, perverse, refractory, self-willed, uncooperative, wilful.

we *pron.* **1** used by a person referring to himself or herself and another or others. **2** used instead of 'I' in newspaper editorials and by a royal person in formal proclamations.

weak *adj.* **1** lacking strength, power, or resolution. **2** not convincing. **3** much diluted. □ **weak-kneed** *adj.* lacking determination. **weakly** *adv.*

■ **1** debilitated, delicate, enervated, exhausted, feeble, frail, groggy, incapacitated, infirm, sickly, worn-out; defenceless, helpless, powerless, vulnerable; cowardly, impotent, ineffectual, irresolute, namby-pamby, powerless, pusillanimous, spineless, unassertive, weak-kneed. **2** feeble, flimsy, hollow, insubstantial, lame, poor, unconvincing. **3** diluted, thin, watery.

weaken *v.* make or become weaker.

■ debilitate, diminish, enervate, enfeeble, exhaust, lessen, lower, reduce, sap, undermine, vitiate; dilute, water down; abate, ebb, decline, dwindle, fade, flag, subside, wane.

weakling *n.* feeble person or animal.

weakness *n.* **1** state of being weak. **2** weak point, fault. **3** self-indulgent liking.

■ **1** debility, feebleness, fragility, frailty, infirmity, vulnerability; impotence, powerlessness, timidity. **2** Achilles heel, defect, deficiency, failing, fault, flaw, foible, imperfection, shortcoming. **3** appetite, fondness, liking, partiality, penchant, predilection, soft spot, taste.

weal *n.* ridge raised on flesh esp. by the stroke of a rod or whip.

wealth *n.* **1** money and valuable possessions. **2** possession of these. **3** great quantity.

■ **1** assets, money, riches. **2** affluence, opulence, prosperity. **3** abundance, profusion.

wealthy *adj.* (**-ier, -iest**) having wealth, rich. □ **wealthiness** *n.*

■ affluent, moneyed, prosperous, rich, *colloq.* well-heeled, well off, well-to-do.

wean *v.* **1** accustom (a baby) to take food other than milk. **2** cause to give up something gradually.

weapon *n.* **1** thing designed or used for inflicting harm or damage. **2** means of coercing someone.

wear *v.* (**wore, worn**) **1** have on the body as clothing or ornament. **2** damage or become damaged by prolonged use. **3** endure continued use. ● *n.* **1** wearing, being worn. **2** clothing. **3** capacity to endure being used. □ **wear down** overcome (opposition etc.) by persistence. **wear off** pass off gradually. **wear out 1** use or be used until no longer usable. **2** tire or become tired out. **wearer** *n.*, **wearable** *adj.*

■ *v.* **1** be dressed *or* clothed in, sport. **2** damage, erode, fray, rub away, wash away. **3** endure, last, survive. ● *n.* **2** apparel, attire, clothes, clothing, dress, garb. □ **wear off** decrease, diminish, fade, subside, wane. **wear out 2** drain, exhaust, fatigue, *sl.* knacker, tire, weary.

wearisome *adj.* tedious.

■ boring, dreary, monotonous, tedious, tiresome, uninteresting.

weary *adj.* (**-ier, -iest**) **1** very tired. **2** tiring. ● *v.* make or become weary. □ **wearily** *adv.*, **weariness** *n.*

■ *adj.* **1** all in, dead beat, drained, exhausted, *colloq.* fagged, fatigued, *sl.* knackered, tired, *colloq.* whacked, worn-out.

weasel *n.* small ferocious reddish-brown flesh-eating mammal.

weather *n.* state of the atmosphere with reference to sunshine, rain, wind, etc. ● *v.* **1** (cause to) be changed by exposure to the weather. **2** come safely through (a storm). □ **under the weather** feeling unwell or depressed. **weather-beaten** *adj.* bronzed or worn by exposure to weather. **weather vane** weathercock.

■ *v.* **2** endure, live through, survive, withstand. □ **under the weather** ailing, ill, indisposed, infirm, not well, off colour, poorly, sick, unwell; depressed, dispirited.

weatherboard *n.* sloping board for keeping out rain.

weathercock *n.* revolving pointer to show the direction of the wind.

weave[1] *v.* (**wove, woven**) **1** make (fabric etc.) by passing crosswise threads or strips under and over lengthwise ones. **2** form (thread etc.) into fabric thus. **3** compose (a story etc.). ● *n.* style or pattern of weaving.

weave[2] *v.* move in an intricate course to avoid obstacles.

weaver *n.* **1** person who weaves. **2** tropical bird that builds an intricately woven nest.

web *n.* **1** network of fine strands made by a spider etc. **2** skin filling the spaces between the toes of ducks, frogs, etc. □ **web-footed** *adj.* having toes joined by web.

webbing *n.* strong band(s) of woven fabric used in upholstery.

wed *v.* (**wedded**) **1** marry. **2** unite.

■ **1** get married, espouse, marry. **2** ally, combine, join, unite.

wedding *n.* marriage ceremony and festivities.

■ marriage, nuptials.

wedge *n.* piece of solid substance thick at one end and tapering to a thin edge at the other. ● *v.* **1** force apart or fix firmly with a wedge. **2** crowd tightly.

■ *v.* **2** cram, crowd, jam, pack, ram, squeeze, stuff.

wedlock *n.* married state.

wee *adj.* **1** (*Sc.*) little. **2** (*colloq.*) tiny.

weed *n.* **1** wild plant growing where it is not wanted. **2** thin weak-looking person. ● *v.* uproot and remove weeds (from). □ **weed out** remove as inferior or undesirable. **weedy** *adj.*

week *n.* **1** period of seven successive days, esp. from Monday to Sunday or Sunday to Saturday. **2** the weekdays of this. **3** working period during a week.

weekday *n.* day other than Sunday and usu. Saturday.

weekend *n.* Saturday and Sunday.

weekly *adj.* & *adv.* (produced or occurring) once a week. ● *n.* weekly periodical.

weep *v.* (**wept**) **1** shed (tears). **2** shed or ooze (moisture) in drops. ● *n.* spell of weeping. □ **weeper** *n.*, **weepy** *adj.*

■ *v.* **1** bawl, *sl.* blub, blubber, cry, grizzle, lament, shed tears, snivel, sob, wail, whimper.

weeping *adj.* (of a tree) having drooping branches.

weevil *n.* small beetle that feeds on grain, nuts, tree-bark, etc.

weft *n.* crosswise threads in weaving.

weigh *v.* **1** measure the weight of. **2** have a specified weight. **3** consider the relative importance of. **4** have influence. □ **weigh anchor** raise the anchor and start a voyage. **weigh down 1** bring or keep down by its weight. **2** depress, oppress. **weigh on** make a person worry. **weigh up** form an estimate of.

■ **3** assess, consider, contemplate, estimate, evaluate, gauge, judge, mull over, ponder, weigh up. **4** be important, carry weight, count, matter. □ **weigh down 2** burden, depress, oppress, strain, trouble, worry.

weighbridge *n.* weighing machine with a plate set in a road etc. for weighing vehicles.

weight *n.* **1** object's mass numerically expressed using a recognized scale of units, heaviness. **2** unit or system of units used thus. **3** piece of metal of known weight used in weighing. **4** heavy object. **5** burden. **6** influence. ● *v.* **1** attach a weight to. **2** hold down with a weight. **3** bias. □ **weightless** *adj.*

■ *n.* **5** burden, cross, load, millstone, onus. **6** authority, *colloq.* clout, influence, power; consequence, importance, significance, value.

weighting *n.* extra pay given in special cases.

weighty *adj.* (**-ier**, **-iest**) **1** heavy. **2** important. **3** influential.

■ **1** heavy, hefty, massive, ponderous. **2** consequential, important, momentous, serious, significant. **3** authoritative, cogent, convincing, forceful, influential, persuasive, powerful.

weir *n.* **1** small dam built so that some of a stream's water flows over it. **2** waterfall formed thus.

weird *adj.* uncanny, bizarre. □ **weirdly** *adv.*, **weirdness** *n.*

■ bizarre, curious, freakish, odd, outlandish, peculiar, *colloq.* spooky, strange, uncanny.

welcome *adj.* **1** received with pleasure. **2** freely permitted. ● *int.* greeting expressing pleasure. ● *n.* kindly greeting or reception. ● *v.* receive with a welcome.

■ *adj.* **1** agreeable, gratifying, pleasant, pleasing. ● *n.* greeting, reception. ● *v.* greet, hail, meet, receive.

weld *v.* **1** unite or fuse (pieces of metal) by heating or pressure. **2** unite into a whole. ● *n.* welded joint. □ **welder** *n.*

welfare *n.* **1** well-being. **2** practical care for the health, safety, etc. of a particular group. □ **welfare state** country with highly developed social services.

■ **1** benefit, good, happiness, health, well-being.

well[1] *n.* **1** shaft sunk to obtain water or oil etc. **2** enclosed shaft-like space. ● *v.* rise or spring.

■ *v.* flow, rise, spring, surge.

well[2] *adv.* (**better**, **best**) **1** in a satisfactory way. **2** with distinction or talent. **3** thoroughly, carefully. **4** favourably, kindly. **5** probably. **6** to a considerable extent. ● *adj.* **1** in good health. **2** satisfactory. ● *int.* expressing surprise, relief, or resignation etc., or said when one is hesitating. □ **as well 1** in addition. **2** desirable, reasonably. **as well as** in addition to. **well-adjusted** *adj.* mentally and emotionally stable. **well-appointed** *adj.* well-equipped. **well-being** *n.* good health, happiness, and prosperity. **well-disposed** *adj.* having kindly or favourable feelings. **well-founded** *adj.* based on good evidence. **well-heeled** *adj.* (*colloq.*) wealthy. **well-meaning, -meant** *adjs.* acting or done with good intentions. **well off 1** in a fortunate situation. **2** rich. **well-read** *adj.* having read much literature. **well-spoken** *adj.* speaking in a polite and correct way. **well-to-do** *adj.* prosperous.

■ *adv.* **1** correctly, nicely, properly, satisfactorily. **2** ably, competently, expertly, proficiently, skilfully. **3** carefully, exhaustively, thoroughly. **4** approvingly, favourably, highly, warmly; considerately, hospitably, kindly, thoughtfully. ● *adj.* **1** fit, hale, healthy, in good health, *colloq.* in the pink, strong, thriving. **2** all right, fine, good, in order, *colloq.* OK, satisfactory. □ **as well 1** additionally, also, besides, furthermore, in addition, moreover, too. **well off 1** blessed, fortunate, lucky. **2** affluent, moneyed, prosperous, rich, wealthy, *colloq.* well-heeled, well-to-do.

wellington *n.* boot of rubber or other waterproof material.

well-nigh *adv.* almost.

Welsh *adj.* & *n.* (language) of Wales. □ **Welsh rabbit** or **rarebit** melted cheese on toast. **Welshman** *n.*, **Welshwoman** *n.*

welsh *v.* **1** avoid paying one's debts. **2** break an agreement.

welt *n.* **1** leather rim attaching the top of a boot or shoe to the sole. **2** ribbed or strengthened border of a knitted garment. **3** weal. ● *v.* **1** provide with a welt. **2** raise weals on, thrash.

welter *n.* **1** turmoil. **2** disorderly mixture.

■ **1** chaos, confusion, disorder, turmoil. **2** jumble, hotchpotch, mess, muddle, tangle.

wen *n.* benign tumour on the skin.

wend *v.* **wend one's way** go.

went *see* **go**.

wept *see* **weep**.

werewolf /weerwoolf/ *n.* (*pl.* **-wolves**) (in myths) person who at times turns into a wolf.

west *n.* **1** point on the horizon where the sun sets. **2** direction in which this lies. **3** western part. ● *adj.* **1** in the west. **2** (of wind) from the west. ● *adv.* towards the west. □ **go west** (*sl.*) be destroyed, lost, or killed.

westerly *adj.* towards or blowing from the west.

western *adj.* of or in the west. ● *n.* film or novel about cowboys in western N. America.

westerner *n.* native or inhabitant of the west.

westernize *v.* make (an oriental country etc.) more like a western one in ideas and institutions. □ **westernization** *n.*

westernmost *adj.* furthest west.

westward *adj.* towards the west. □ **westwards** *adv.*

wet *adj.* (**wetter, wettest**) **1** soaked or covered with water or other liquid. **2** rainy. **3** not dry. **4** (*colloq.*) lacking vitality. ● *v.* (**wetted**) make wet. ● *n.* **1** moisture, water. **2** wet weather. □ **wet blanket** gloomy person. **wet-nurse** *n.* woman employed to suckle another's child. *v.* **1** act as wet-nurse to. **2** coddle as if helpless. **wetsuit** *n.* porous garment worn by a skin diver etc. **wetly** *adv.*, **wetness** *n.*

■ *adj.* **1** damp, drenched, dripping, moist, soaked, sodden, soggy, sopping, waterlogged, watery; clammy, dank, humid. **2** rainy, showery. **3** sticky, tacky. ● *v.* damp, dampen, douse, drench, moisten, saturate, soak, spray, sprinkle, water.

wether *n.* castrated ram.

whack *v.* & *n.* (*colloq.*) hit. □ **do one's whack** (*sl.*) do one's share.

whacked *adj.* (*colloq.*) tired out.

whale *n.* very large sea mammal.

whalebone *n.* horny substance from the upper jaw of whales, formerly used as stiffening.

whaler *n.* **1** whaling ship. **2** seaman hunting whales.

whaling *n.* hunting whales.

wham *int.* & *n.* sound of a forcible impact.

wharf *n.* (*pl.* **wharfs**) landing-stage where ships load and unload.

what *adj.* **1** asking for a statement of amount, number, or kind. **2** how great or remarkable. **3** the or any that. ● *pron.* **1** what thing(s). **2** what did you say? ● *adv.* to what extent or degree. ● *int.* exclamation of surprise. □ **what for** why?

whatever *adj.* of any kind or number. ● *pron.* **1** anything or everything. **2** no matter what.

whatnot *n.* something trivial or indefinite.

whatsoever *adj.* & *pron.* whatever.

wheat *n.* **1** grain from which flour is made. **2** plant producing this.

wheatear *n.* a kind of small bird.

wheaten *adj.* made from wheat.

wheedle *v.* coax.

■ cajole, charm, coax, entice, inveigle, persuade.

wheel *n.* **1** disc or circular frame that revolves on a shaft passing through its centre. **2** wheel-like motion. ● *v.* **1** push or pull (a cart or bicycle etc.) along. **2** turn. **3** move in circles or curves. □ **at the wheel 1** driving a vehicle, directing a ship. **2** in control of affairs. **wheel and deal** engage in scheming to exert influence. **wheel-clamp** *v.* fix a clamp on (an illegally parked car etc.).

■ *v.* **2** spin, swing, swivel, turn, veer, whirl. **3** circle, gyrate.

wheelbarrow *n.* open container for moving small loads, with a wheel at one end.

wheelbase *n.* distance between a vehicle's front and rear axles.

wheelchair *n.* chair on wheels for a person who cannot walk.

wheeze *v.* breathe with a hoarse whistling sound. ● *n.* this sound. □ **wheezy** *adj.*

whelk *n.* spiral-shelled mollusc.

whelp *n.* young dog, puppy. ● *v.* give birth to (puppies).

when *adv.* **1** at what time, on what occasion. **2** at which time. ● *conj.* **1** at the time that. **2** whenever. **3** as soon as. **4** although. ● *pron.* what or which time.

whence *adv.* & *conj.* **1** from where. **2** from which.

whenever *conj.* & *adv.* **1** at whatever time. **2** every time that.

where *adv.* & *conj.* **1** at or in which place or circumstances. **2** in what respect. **3** from what place or source. **4** to what place. ● *pron.* what place.

whereabouts *adv.* in or near what place. ● *n.* a person's or thing's approximate location.

whereas *adv.* **1** since it is the fact that. **2** but in contrast.

whereby *adv.* by which.

whereupon *adv.* after which, and then.

wherever *adv.* at or to whatever place.

wherewithal *n.* (*colloq.*) (money etc.) needed for a purpose.

wherry *n.* **1** light rowing boat. **2** large light barge.

whet *v.* (**whetted**) **1** sharpen by rubbing against a stone etc. **2** stimulate (appetite or interest).

■ **1** hone, sharpen. **2** arouse, awaken, excite, fire, kindle, stimulate, stir.

whether *conj.* introducing an alternative possibility.

whetstone *n.* shaped hard stone used for sharpening tools.

whey *n.* watery liquid left when milk forms curds.

which *adj.* & *pron.* **1** what particular one(s) of a set. **2** and that. ● *rel.pron.* thing or animal referred to.

whichever *adj.* & *pron.* any which, that or those which.

whiff *n.* puff of air or odour etc.

■ breath, puff, scent, waft.

while *n.* **1** period of time. **2** time spent in doing something. ● *conj.* **1** during the time that, as long as. **2** although. ● *v.* **while away** pass (time) in a leisurely or interesting way.

whilst *conj.* while.

whim *n.* sudden fancy.

■ caprice, fancy, impulse, notion, vagary.

whimper *v.* make feeble crying sounds. ● *n.* whimpering sound.

■ *v.* cry, grizzle, sniffle, snivel.

whimsical *adj.* **1** impulsive and playful. **2** fanciful, quaint. □ **whimsically** *adv.*, **whimsicality** *n.*

■ **1** capricious, impulsive, mischievous, playful. **2** curious, fanciful, fantastic, odd, quaint, unusual.

whine *v.* **1** make a long high complaining cry or a similar shrill sound. **2** complain or utter with a whine. ● *n.* whining sound or complaint. □ **whiner** *n.*

whinge *v.* (*colloq.*) whine, complain.

whinny *n.* gentle or joyful neigh. ● *v.* utter a whinny.

whip *n.* **1** cord or strip of leather on a handle, used for striking a person or animal. **2** food made with whipped cream etc. ● *v.* (**whipped**) **1** strike or urge on with a whip. **2** beat into a froth. **3** move or take suddenly. **4** (*sl.*) steal. □ **have the whip hand** have control. **whip-round** *n.* appeal for contributions from a group. **whip up** incite.

■ *n.* **1** lash, scourge, thong. ● *v.* **1** beat, flagellate, flay, flog, lash, scourge, thrash. **2** beat, stir, whisk. **3** dart, dash, fly, hurry, race, run, rush, scamper, scoot, scurry; jerk, pull, snatch, whisk, *colloq.* yank.

□ **whip up** arouse, excite, incite, kindle, rouse, stir up, work up.

whipcord *n.* **1** cord of tightly twisted strands. **2** twilled fabric with prominent ridges.

whiplash *n.* **1** lash of a whip. **2** jerk.

whippet *n.* small dog resembling a greyhound, used for racing.

whipping-boy *n.* scapegoat.

whippy *adj.* flexible, springy.

whirl *v.* **1** swing or spin round and round. **2** convey or go rapidly in a vehicle. ● *n.* **1** whirling movement. **2** confused state. **3** bustling activity.

■ *v.* **1** circle, gyrate, revolve, rotate, spin, swirl, turn, twirl. ● *n.* **1** pirouette, spin, swirl, turn, twirl.

whirlpool *n.* current of water whirling in a circle.

■ eddy, vortex.

whirlwind *n.* mass of air whirling rapidly about a central point.

■ cyclone, hurricane, tornado, typhoon.

whirr *n.* continuous buzzing or vibrating sound. ● *v.* make this sound.

■ *n.* & *v.* buzz, drone, hum, murmur, purr.

whisk *v.* **1** convey or go rapidly. **2** brush away lightly. **3** beat into a froth. ● *n.* **1** whisking movement. **2** instrument for beating eggs etc. **3** bunch of bristles etc. for brushing or flicking things.

■ *v.* **1** dart, dash, hasten, hurry, rush, scurry; jerk, pull, snatch, whip, *colloq.* yank. **2** brush, sweep. **3** beat, stir, whip.

whisker *n.* **1** long hair-like bristle near the mouth of a cat etc. **2** (*pl.*) hair growing on a man's cheek. □ **whiskered** *adj.*, **whiskery** *adj.*

whiskey *n.* Irish whisky.

whisky *n.* spirit distilled from malted grain (esp. barley).

whisper *v.* **1** speak or utter softly, not using the vocal cords. **2** rustle. ● *n.* **1** whispering sound or speech or remark. **2** rumour.

whist *n.* card game usu. for two pairs of players.

whistle *n.* **1** shrill sound made by blowing through a narrow opening between the lips. **2** similar sound. **3** instrument for producing this. ● *v.* **1** make this sound. **2** signal or produce (a tune) in this way. □ **whistler** *n.*

Whit *adj.* of or close to **Whit Sunday**, seventh Sunday after Easter.

white *adj.* **1** of the colour of milk or fresh snow. **2** having a light-coloured skin. **3** pale from illness or fear etc. ● *n.* **1** white colour or thing. **2** transparent substance round egg yolk. **3** member of a light-skinned race. □ **white ant** termite. **white-collar** *adj.* (of worker or work) non-manual, clerical. **white elephant** useless possession. **white hope** person expected to achieve much. **white horses** white-crested waves on sea. **white-hot** *adj.* (of metal) glowing white after heating. **white lie** harmless lie. **white sale** sale of household linen. **white spirit** light petroleum used as a solvent. **white wine** wine of yellow colour. **whiteness** *n.*, **whitish** *adj.*

■ *adj.* **1** chalky, ivory, milky, snowy; silver. **3** anaemic, ashen, pale, pallid, pasty, wan.

whitebait *n.* (*pl.* **whitebait**) small silvery-white fish.

whiten *v.* make or become white or whiter.

■ blanch, bleach, fade, lighten.

whitewash *n.* **1** liquid containing quicklime or powdered chalk, used for painting walls or ceilings etc. **2** means of glossing over mistakes. ● *v.* **1** paint with whitewash. **2** gloss over mistakes in.

■ *v.* **2** camouflage, conceal, cover (up), disguise, gloss over, hide.

whither *adv.* (*old use*) to what place.

whiting *n.* (*pl.* **whiting**) small edible sea fish.

whitlow *n.* small abscess under or affecting a nail.

Whitsun *n.* Whit Sunday and the days close to it. □ **Whitsuntide** *n.*

whittle *v.* trim (wood) by cutting thin slices from the surface. □ **whittle down** reduce by removing various amounts.

whiz *v.* (**whizzed**) **1** make a sound like something moving at great speed through air. **2** move very quickly. ● *n.* whizzing sound. □ **whiz-kid** *n.* (*colloq.*) brilliant or successful young person.

who *pron.* **1** what or which person(s)? **2** the particular person(s).

whodunit *n.* (*colloq.*) detective or mystery story or play etc.

whoever *pron.* any or every person who, no matter who.

whole *adj.* **1** with no part removed or left out. **2** not injured or broken. ● *n.* **1** full amount, all parts or members. **2** complete system made up of parts. □ **on the whole** considering everything.

wholehearted *adj.* without doubts or reservations. **whole number** number consisting of one or more units with no fractions.

■ *adj.* **1** complete, entire, full, total. **2** undamaged, unharmed, unhurt, uninjured, unscathed; complete, entire, intact, unabridged, uncut. □ **wholehearted** ardent, earnest, fervent, genuine, hearty, sincere, true; complete, heartfelt, unconditional, unqualified, unreserved.

wholemeal *adj.* made from the whole grain of wheat etc.

wholesale *n.* selling of goods in large quantities to be retailed by others. ● *adj.* & *adv.* **1** in the wholesale trade. **2** on a large scale. □ **wholesaler** *n.*

wholesome *adj.* good for health or well-being.

■ healthful, healthy, nourishing, nutritious, salubrious.

wholly *adv.* entirely.

■ absolutely, completely, entirely, fully, quite, thoroughly, totally.

whom *pron.* objective case of *who.*

whoop *v.* utter a loud cry of excitement. ● *n.* this cry.

■ *v.* cheer, cry, shout, shriek, yell.

whooping cough infectious disease esp. of children, with a violent convulsive cough.

whopper *n.* (*sl.*) **1** something very large. **2** great lie.

whore *n.* prostitute.

whorl *n.* **1** one turn of a spiral. **2** circle of ridges in a fingerprint. **3** ring of leaves or petals.

whose *pron.* **1** of whom. **2** of which.

whosoever *pron.* whoever.

why *adv.* **1** for what reason or purpose? **2** on account of which. ● *int.* exclamation of surprised discovery or recognition.

wick *n.* length of thread in a candle or lamp etc., by which the flame is kept supplied with melted grease or fuel.

wicked *adj.* **1** given to or involving immorality. **2** playfully malicious. **3** (*colloq.*) very bad. □ **wickedly** *adv.*, **wickedness** *n.*

■ **1** amoral, bad, base, corrupt, criminal, degenerate, depraved, evil, heinous, immoral, iniquitous, nefarious, scandalous, shameful, sinful, vile, villainous, wrong. **2** devilish, mischievous, naughty, sly, roguish.

wicker *n.* osiers or thin canes interwoven to make furniture or baskets etc. □ **wickerwork** *n.*

wicket *n.* **1** set of three stumps and two bails used in cricket. **2** part of a cricket ground between or near the two wickets.

wide *adj.* **1** measuring much from side to side. **2** having a specified width. **3** extending far. **4** fully opened. **5** far from the target. ● *adv.* widely. □ **wide awake** **1** fully awake. **2** (*colloq.*) alert. **widely** *adv.*, **wideness** *n.*

■ *adj.* **1** broad, extensive, large, spacious, vast. **3** broad, comprehensive, extensive, sweeping, wide-ranging, widespread.

widen *v.* make or become wider.

■ broaden, enlarge, expand, increase; dilate, distend.

widespread *adj.* found or distributed over a wide area.

■ common, extensive, general, prevalent, sweeping.

widgeon *n.* wild duck.

widow *n.* woman whose husband has died and who has not remarried. □ **widowhood** *n.*

widowed *adj.* made a widow or widower.

widower *n.* man whose wife has died and who has not remarried.

width *n.* **1** distance from side to side. **2** wideness. **3** piece of material of full width.

wield *v.* **1** hold and use (a tool etc.) with the hands. **2** have and use (power).

■ **1** employ, ply, use; brandish, flourish, wave. **2** exercise, exert.

wife *n.* (*pl.* **wives**) married woman in relation to her husband. □ **wifely** *adj.*

wig *n.* artificial covering of hair worn on the head.

wiggle *v.* move repeatedly from side to side. ● *n.* act of wiggling.

wigwam *n.* N. American Indian's conical tent.

wild *adj.* **1** not domesticated, tame, or cultivated. **2** not civilized. **3** disorderly. **4** stormy. **5** full of strong unrestrained feeling. **6** extremely foolish. **7** random. ● *adv.* in a wild manner. ● *n.* (usu. *pl.*) desolate uninhabited place. □ **wild-goose chase** useless quest. **wildly** *adv.*, **wildness** *n.*

■ *adj.* **1** feral, undomesticated, untamed; natural, uncultivated. **2** primitive, savage, uncivilized. **3** chaotic, disorderly, riotous, rowdy, uncontrollable, uncontrolled, undisciplined, unmanageable, unrestrained, unruly, uproarious, tumultuous. **4** stormy, tempestuous, turbulent. **5** excited, feverish, frantic, frenzied, hysterical; *colloq.* crazy, enthusiastic, mad, passionate. **6** absurd, crazy, foolhardy, foolish, ill-advised, impractical, imprudent, silly. **7** arbitrary, haphazard, random.

wildcat *adj.* **1** reckless. **2** (of strikes) sudden and unofficial.

wildebeest *n.* gnu.

wilderness *n.* wild uncultivated area.

wildfire *n.* **spread like wildfire** spread very fast.

wildfowl *n.* birds hunted as game.

wildlife *n.* wild animals and plants.

wile *n.* piece of trickery.

■ artifice, device, ploy, ruse, stratagem, subterfuge, trap, trick.

wilful *adj.* **1** intentional, not accidental. **2** self-willed. □ **wilfully** *adv.*, **wilfulness** *n.*

■ **1** conscious, deliberate, intentional, premeditated. **2** headstrong, obdurate, obstinate, pigheaded, recalcitrant, refractory, self-willed, stubborn, unyielding.

will[1] *v.aux.* used with *I* and *we* to express promises or obligations, and with other words to express a future tense.

will[2] *n.* **1** mental faculty by which a person decides upon and controls his or her actions. **2** determination. **3** desire. **4** person's attitude in wishing good or bad to others. **5** written directions made by a person for disposal of their property after their death. ● *v.* **1** try to cause by will-power. **2** intend unconditionally. **3** bequeath by a will. □ **at will** whenever one pleases. **will-power** *n.* control exercised by one's will.

■ *n.* **2** determination, drive, purpose, resolution, resolve, will-power. **3** desire, wish. **4** attitude, disposition, feeling(s). **5** testament. ● *v.* **1** compel, force, make. **2** command, desire, intend, ordain, order, require, want, wish. **3** bequeath, hand down, leave, make over, pass on.

willing *adj.* **1** desiring to do what is required, not objecting. **2** given or done readily. □ **willingly** *adv.*, **willingness** *n.*

■ **1** acquiescent, agreeable, amenable, complaisant, compliant, eager, enthusiastic, game, pleased, prepared, ready.

will-o'-the-wisp *n.* **1** phosphorescent light seen on marshy ground. **2** hope or aim that can never be fulfilled.

willow *n.* **1** tree or shrub with flexible branches. **2** its wood.

willowy *adj.* **1** full of willows. **2** slender and supple.

■ **2** flexible, lissom, lithe, slender, slim, supple, svelte.

willy-nilly *adv.* whether one desires it or not.

wilt *v.* **1** lose freshness, droop. **2** lose one's energy.

■ **1** droop, wither. **2** flag, grow tired, languish, tire, weaken.

wily /wīli/ *adj.* (**-ier**, **-iest**) full of wiles, cunning. □ **wiliness** *n.*

■ artful, canny, clever, crafty, cunning, deceitful, designing, devious, guileful, machiavellian, scheming, shrewd, sly.

wimp *n.* (*sl.*) feeble or ineffective person.

win *v.* (**won**, **winning**) **1** be victorious (in). **2** obtain as the result of a contest etc., or by effort. ● *n.* victory, esp. in a game. □ **win over** gain the favour or support of.

■ *v.* **1** be victorious, come *or* finish first, prevail, succeed, triumph. **2** achieve, attain, carry off, earn, gain, get, obtain, receive, secure. ● *n.* conquest, success, triumph, victory. □ **win over** charm, get round, influence, persuade, sway.

wince *v.* make a slight movement from pain or embarrassment etc. ● *n.* this movement.

■ *v.* blench, cringe, flinch.

winceyette *n.* cotton fabric with a soft downy surface.

winch *n.* machine for hoisting or pulling things by a cable that winds round a revolving drum. ● *v.* hoist or pull with a winch.

wind[1] /wind/ *n.* **1** current of air. **2** gas in the stomach or intestines. **3** useless or boastful talk. **4** breath as needed in exertion or speech etc. **5** orchestra's wind instruments. ● *v.* cause to be out of breath. □ **get wind of** hear a hint or rumour of. **in the wind** happening or about to happen. **put the wind up** (*colloq.*) frighten. **take the wind out of a person's sails** frustrate a person by anticipating him or her. **wind-break** *n.* screen shielding something from the wind. **wind-chill** *n.* cooling effect of the wind. **wind instrument** musical instrument sounded by a current of air, esp. by the player's breath. **wind-sock** *n.* canvas cylinder flown at an airfield to show the direction of the wind. **wind-tunnel** *n.*

enclosed tunnel in which winds can be created for testing things.

■ *n.* **1** breeze, draught, gust, *poetic* zephyr. **3** bluster, boasting, claptrap, *colloq.* gas, *colloq.* hot air.

wind² /wīnd/ *v.* (**wound**) **1** move or go in a curving or spiral course. **2** wrap closely around something or round upon itself. **3** move by turning a windlass or handle etc. **4** wind up (a clock etc.). □ **wind up 1** set or keep (a clock etc.) going by tightening its spring. **2** bring or come to an end. **3** settle the affairs of and close (a business company). **winder** *n.*

■ **1** bend, curve, meander, snake, turn, twist, zigzag. **2** enfold, envelop, wrap; coil, reel, turn, twine, twist. □ **wind up 2** bring to an end, close, conclude, end, finish, terminate. **3** dissolve, liquidate.

windbag *n.* (*colloq.*) person who talks lengthily.

windfall *n.* **1** fruit blown off a tree by the wind. **2** unexpected gain, esp. a sum of money.

windlass *n.* winch-like device using a rope or chain that winds round a horizontal roller.

windmill *n.* mill worked by the action of wind on projecting parts that radiate from a shaft.

window *n.* **1** opening in a wall etc. to admit light and often air, usu. filled with glass. **2** this glass. **3** space for display of goods behind the window of a shop. □ **window-box** *n.* trough fixed outside a window, for growing flowers etc. **window-dressing** *n.* **1** arranging a display of goods in a shop window. **2** presentation of facts so as to give a favourable impression. **window-shopping** *n.* looking at displayed goods without buying.

windpipe *n.* air passage from the throat to the bronchial tubes.

windscreen *n.* glass in the window at the front of a vehicle.

windsurfing *n.* sport of surfing on a board to which a sail is fixed.

windswept *adj.* exposed to strong winds.

windward *adj.* situated in the direction from which the wind blows. ● *n.* this side or region.

wine *n.* **1** fermented grape-juice as an alcoholic drink. **2** fermented drink made from other fruits or plants. **3** dark red. ● *v.* **1** drink wine. **2** entertain with wine. □ **wine bar** bar or small restaurant serving wine as the main drink.

wing *n.* **1** each of a pair of projecting parts by which a bird or insect etc. is able to fly. **2** wing-like part of an aircraft. **3** projecting part. **4** bodywork above the wheel of a car. **5** either end of a battle array. **6** player at either end of the forward line in football or hockey etc., side part of playing area in these games. **7** extreme section of a political party. **8** (*pl.*) sides of a theatre stage. ● *v.* **1** fly, travel by wings. **2** wound slightly in the wing or arm. □ **on the wing** flying. **take wing** fly away. **under one's wing** under one's protection. **wing-collar** *n.* high stiff collar with turned-down corners.

winged *adj.* having wings.

winger *n.* wing player in football etc.

wink *v.* **1** blink one eye as a signal. **2** shine with a light that flashes or twinkles. ● *n.* act of winking.

■ *v.* **2** flash, glint, glitter, shine, sparkle, twinkle.

winker *n.* flashing indicator.

winkle *n.* edible sea snail. ● *v.* **winkle out** extract, prise out.

winner *n.* **1** person or thing that wins. **2** something successful.

■ **1** champion, conquerer, victor.

winning *adj.* charming, attractive. ● *n.pl.* money won. □ **winning post** post marking the end of a race.

■ *adj.* appealing, attractive, captivating, charming, delightful, enchanting, endearing, engaging, pleasing, winsome.

winnow *v.* fan or toss (grain) to free it of chaff.

winsome *adj.* charming.

winter *n.* coldest season of the year. ● *v.* spend the winter. □ **wintry** *adj.*

winy *adj.* wine-flavoured.

wipe *v.* **1** clean, dry, or remove by rubbing. **2** spread thinly on a surface. ● *n.* act of wiping. □ **wipe out** destroy completely.

■ □ **wipe out** annihilate, destroy, eliminate, eradicate, exterminate, get rid of, obliterate, stamp out.

wiper *n.* device that automatically wipes rain etc. from a windscreen.

wire *n.* **1** strand of metal. **2** length of this used for fencing, conducting electric current, etc. ● *v.* provide, fasten, or strengthen with wire(s). □ **wire-haired** *adj.* having stiff wiry hair.

wireworm *n.* destructive larva of a beetle.

wiring *n.* system of electric wires in a building, vehicle, etc.

wiry *adj.* (**-ier**, **-iest**) **1** like wire. **2** lean but strong. □ **wiriness** *n.*

wisdom *n.* **1** being wise, soundness of judgement. **2** wise sayings. □ **wisdom tooth** hindmost molar tooth, not usu. cut before the age of 20.

■ **1** astuteness, discernment, insight, intelligence, judgement, perspicacity, sagacity, sense, shrewdness, understanding.

wise *adj.* **1** showing soundness of judgement. **2** having knowledge. □ **wisely** *adv.*

■ **1** astute, discerning, intelligent, perspicacious, sagacious, sage, shrewd; advisable, judicious, prudent, sensible. **2** erudite, intelligent, knowledgeable, learned, well-educated, well-read.

wiseacre *n.* person who pretends to have great wisdom.

wisecrack (*colloq.*) *n.* witty remark. ● *v.* make a wisecrack.

wish *n.* **1** desire, request, or aspiration. **2** expression of this. ● *v.* **1** have or express as a wish. **2** express one's hopes for. **3** (*colloq.*) foist.

■ *n.* **1** aspiration, craving, desire, fancy, hankering, longing, need, request, want, yearning, *colloq.* yen.

wishbone *n.* forked bone between a bird's neck and breast.

wishful *adj.* desiring. □ **wishful thinking** belief founded on wishes not facts.

wishy-washy *adj.* weak in colour, character, etc.

■ bland, insipid, vapid; feeble, ineffectual, namby-pamby, spineless, weak-kneed, *colloq.* wet.

wisp *n.* **1** small separate bunch. **2** small streak of smoke etc. □ **wispy** *adj.*, **wispiness** *n.*

wisteria *n.* climbing shrub with hanging clusters of flowers.

wistful *adj.* full of sad or vague longing. □ **wistfully** *adv.*, **wistfulness** *n.*

■ forlorn, melancholy, mournful, pensive, sad, sorrowful.

wit *n.* **1** amusing ingenuity in expressing words or ideas. **2** person who has this. **3** intelligence. □ **at one's wits' end** worried and not knowing what to do.

■ **1** drollery, humour, wittiness. **2** comedian, comedienne, comic, wag. **3** brains, intelligence, *colloq.* nous, sense, understanding, wisdom.

witch *n.* **1** person (esp. a woman) who practises witchcraft. **2** bewitching woman. □ **witch-doctor** *n.* tribal magician. **witch hazel 1** N. American shrub. **2** astringent lotion made from its bark. **witch-hunt** *n.* persecution of people thought to hold unorthodox or unpopular views.

■ **1** enchantress, sorceress.

witchcraft *n.* practice of magic.

■ enchantment, magic, necromancy, sorcery, wizardry.

with *prep.* **1** in the company of, among. **2** having, characterized by. **3** by means of. **4** of the same opinion as. **5** at the same time as. **6** because of. **7** under the conditions of. **8** by addition or possession of. **9** in regard to, towards.

withdraw *v.* (**withdrew**, **withdrawn**) **1** take back or away. **2** remove (deposited money) from a bank etc. **3** cancel (a statement). **4** go away from a place or from company. □ **withdrawal** *n.*

■ **1** draw back, pull back, retract; extract, remove, take out. **3** annul, nullify, rescind, retract, revoke, take back. **4** depart, leave, retire, retreat.

withdrawn *adj.* (of a person) unsociable.

■ aloof, distant, introverted, reserved, standoffish, unfriendly, unsociable.

wither *v.* **1** shrivel, lose freshness or vitality. **2** subdue by scorn.

■ **1** droop, shrivel, wilt.

withhold *v.* (**withheld**) **1** refuse to give. **2** restrain.

■ **1** deny, deprive of, refuse. **2** control, curb, hold back, repress, restrain, suppress.

within *prep.* **1** inside. **2** not beyond the limit or scope of. **3** in a time no longer than. ● *adv.* inside.

without *prep.* **1** not having. **2** in the absence of. **3** with no action of. ● *adv.* outside.

withstand *v.* (**withstood**) endure successfully.

■ bear, brave, cope with, endure, handle, stand, *colloq.* stick, survive, take, tolerate, weather.

withy *n.* tough flexible shoot, esp. of willow.

witless *adj.* foolish.

witness *n.* **1** person who sees or hears something. **2** one who gives evidence in a law court. **3** one who confirms another's signature. **4** thing that serves as evidence. ● *v.* be a witness of. □ **bear witness to** attest the truth of.

■ *n.* **1** bystander, eyewitness, observer, onlooker, spectator. ● *v.* observe, see, watch. □ **bear witness to** attest (to), confirm, prove, show, testify (to).

witticism *n.* witty remark.

witty *adj.* (**-ier**, **-iest**) full of wit. □ **wittily** *adv.*, **wittiness** *n.*

■ amusing, comical, droll, funny, humorous, scintillating, sparkling.

wives *see* **wife**.

wizard *n.* **1** male witch, magician. **2** person with amazing abilities. □ **wizardry** *n.*

■ **1** enchanter, magician, sorcerer. **2** expert, genius, maestro, master, mastermind, virtuoso, *colloq.* whiz-kid.

wizened /wízz'nd/ *adj.* full of wrinkles, shrivelled with age.

woad *n.* **1** blue dye obtained from a plant. **2** this plant.

wobble *v.* **1** stand or move unsteadily. **2** quiver. ● *n.* **1** wobbling movement. **2** quiver. □ **wobbly** *adj.*

■ *v.* **1** dodder, rock, sway, teeter, totter, waver. **2** quaver, quiver, shake, tremble.

wodge *n.* (*colloq.*) chunk, wedge.

woe *n.* **1** sorrow, distress. **2** trouble causing this, misfortune. □ **woeful** *adj.*, **woefully** *adv.*

■ **1** anguish, distress, grief, heartache, misery, sadness, sorrow, suffering, unhappiness. **2** affliction, calamity, hardship, misfortune, problem, trial, tribulation, trouble.

woebegone *adj.* looking unhappy.

■ crestfallen, dejected, disconsolate, dispirited, doleful, downhearted, forlorn, gloomy, glum, melancholy, miserable, sad, sorrowful, unhappy.

wok *n.* bowl-shaped frying pan used esp. in Chinese cookery.

woke, woken *see* **wake**[1].

wold *n.* (esp. in *pl.*) area of open upland country.

wolf *n.* (*pl.* **wolves**) **1** wild animal of the dog family. **2** (*sl.*) aggressive male flirt. ● *v.* eat quickly and greedily. □ **cry wolf** raise false alarms. **wolf-whistle** *n.* man's admiring whistle at an attractive woman. **wolfish** *adj.*

wolfram *n.* tungsten (ore).

wolverine *n.* N. American animal of the weasel family.

woman *n.* (*pl.* **women**) **1** adult female person. **2** women in general.

womanhood *n.* state of being a woman.

womanize *v.* (of a man) be promiscuous. □ **womanizer** *n.*

womankind *n.* women in general.

womanly *adj.* having qualities considered characteristic of a woman. □ **womanliness** *n.*

womb *n.* hollow organ in female mammals in which the young develop before birth.

wombat *n.* small bear-like Australian animal.

women *see* **woman**.

womenfolk *n.* **1** women in general. **2** women of one's family.

won *see* **win**.

wonder *n.* **1** feeling of surprise and admiration, curiosity, or bewilderment. **2** remarkable thing. ● *v.* **1** feel wonder or surprise. **2** be curious to know.

■ *n.* **1** admiration, amazement, astonishment, awe, curiosity, fascination, surprise, wonderment. **2** curiosity, marvel, miracle, phenomenon, prodigy, spectacle. ● *v.* **1** be amazed *or* awed, marvel. **2** conjecture, meditate, muse, ponder, speculate, think.

wonderful *adj.* very good or remarkable. □ **wonderfully** *adv.*

■ *colloq.* fabulous, *colloq.* fantastic, *colloq.* great, *colloq.* smashing, splendid, *colloq.* super, *colloq.* terrific, *colloq.* tremendous; amazing, astounding, excellent, extraordinary, first-class, incredible, magnificent, marvellous, outstanding, remarkable, superb.

wonderland *n.* place full of wonderful things.

wonderment *n.* feeling of wonder.

wont *adj.* accustomed. ● *n.* custom, habit.

woo *v.* **1** court. **2** try to achieve, obtain, or coax.

wood *n.* **1** tough fibrous substance of a tree. **2** this cut for use. **3** (also *pl.*) trees growing fairly densely over an area of ground. □ **out of the wood(s)** clear of danger or difficulty.

■ **2** lumber, timber. **3** coppice, copse, forest, grove, spinney, thicket, woodland.

woodbine *n.* wild honeysuckle.

woodcock *n.* a kind of game bird.

woodcut *n.* **1** engraving made on wood. **2** picture made from this.

wooded *adj.* covered with trees.

wooden *adj.* **1** made of wood. **2** expressionless. □ **woodenly** *adv.*

woodland *n.* wooded country.

woodlouse *n.* (*pl.* **-lice**) small crustacean with many legs, living in decaying wood etc.

woodpecker *n.* bird that taps tree trunks to find insects.

woodpigeon *n.* a kind of large pigeon.

woodwind *n.* wind instruments originally made of wood.

woodwork *n.* **1** art or practice of making things from wood. **2** wooden things or fittings.

woodworm *n.* larva of a kind of beetle that bores in wood.

woody *adj.* **1** like or consisting of wood. **2** full of woods.

woof *n.* dog's gruff bark. ● *v.* make this sound.

woofer *n.* loudspeaker for reproducing low-frequency signals.

wool *n.* **1** soft hair from sheep or goats etc. **2** yarn or fabric made from this. ◻ **pull the wool over someone's eyes** deceive him or her.

woollen *adj.* made of wool. ● *n.pl.* woollen cloth or clothing.

woolly *adj.* (**-ier, -iest**) **1** covered with wool. **2** like wool, woollen. **3** vague. ● *n.* (*colloq.*) woollen garment. ◻ **woolliness** *n.*

word *n.* **1** sound(s) expressing a meaning independently and forming a basic element of speech. **2** this represented by letters or symbols. **3** thing said. **4** talk. **5** message, news. **6** promise or assurance. **7** command. ● *v.* express in words. ◻ **word of mouth** spoken (not written) words. **word-perfect** *adj.* having memorized every word perfectly. **word processor** computer programmed for storing, correcting, and printing out text entered from a keyboard.

> ■ *n.* **3** comment, declaration, expression, observation, remark, statement, utterance. **4** chat, conversation, discussion, talk, tête-à-tête. **5** information, intelligence, news; bulletin, communiqué, message, report. **6** assurance, guarantee, oath, pledge, promise, undertaking, vow. **7** command, direction, order, signal. ● *v.* couch, express, phrase, put.

wording *n.* form of words used.

wordy *adj.* verbose.

wore *see* **wear**.

work *n.* **1** use of bodily or mental power in order to do or make something. **2** task to be undertaken. **3** thing done or produced by work. **4** things made of certain materials or with certain tools. **5** literary or musical composition. **6** employment. **7** (*pl.*, usu. treated as *sing.*) factory. **8** (*pl.*) operations of building etc. **9** (*pl.*) operative parts of a machine. **10** (usu. *pl.*) defensive structure. ● *v.* **1** perform work. **2** make efforts. **3** be employed. **4** operate, esp. effectively. **5** bring about, accomplish. **6** shape, knead, or hammer etc. into a desired form or consistency. **7** make (a way) or pass gradually by effort. **8** become (loose etc.) through repeated stress or pressure. **9** be in motion. **10** ferment. **11** (**the works**) (*sl.*) all that is available. ◻ **work off** get rid of by activity. **work out 1** find or solve by calculation. **2** plan the details of. **3** have a specified result. **4** take exercise. **work to rule** cause delay by over-strict observance of rules, as a form of protest. **work up 1** bring gradually to a more developed state. **2** excite progressively. **3** advance (to a climax).

> ■ *n.* **1** donkey work, drudgery, effort, exertion, *sl.* graft, grind, industry, labour, slog, toil, travail. **5** composition, piece. **6** employment, occupation, profession, trade. ● *v.* **1** beaver (away), drudge, exert oneself, *sl.* graft, labour, *colloq.* plug (away), slave (away), slog (away), toil. **2** exert oneself, make an effort, strive. **4** be in working order, function, go, operate, run; be effective, succeed. **5** accomplish, achieve, bring about, cause, create, effect, produce. ◻ **work out 1** calculate, figure out, solve. **2** develop, devise, draw up, formulate, plan, prepare, put together. **3** go, pan out, turn out. **4** exercise, train.

workable *adj.* able to be done or used successfully.

> ■ feasible, practicable, practical, viable.

workaday *adj.* ordinary, everyday, practical.

worker *n.* **1** person who works. **2** member of the working class. **3** neuter bee or ant that does the work of the hive or colony.

workforce *n.* workers engaged or available, number of these.

workhouse *n.* former public institution where people unable to support themselves were housed.

working *adj.* engaged in work. ● *n.* excavation(s) made in mining, tunnelling, etc. ◻ **in working order** (esp. of a machine) able to function properly. **working class** class of people who are employed for wages, esp. in manual or industrial work. **working knowledge** knowledge adequate to work with.

workman *n.* man employed to do manual labour.

workmanlike *adj.* showing practical skill.

workmanship *n.* skill in working or in a thing produced.

■ art, artistry, craftsmanship, skill, skilfulness.

workout *n.* session of physical exercising.

workshop *n.* room or building in which manual work or manufacture etc. is carried on.

workstation *n.* **1** computer terminal and keyboard. **2** desk with this. **3** location of a stage in a manufacturing process.

world *n.* **1** the earth. **2** all that concerns or all who belong to a specified class, time, or sphere of activity. **3** all that exists. **4** very great amount.

■ **1** earth, globe. **2** community, domain, field, realm, sphere; age, epoch, era, period, time. **3** cosmos, universe.

worldly *adj.* of or concerned with earthly life or material gains, not spiritual. □ **worldliness** *n.*

■ carnal, earthly, human, material, mortal, mundane, physical, secular, temporal.

worldwide *adj.* extending through the whole world.

■ global, universal, world.

worm *n.* **1** animal with a long soft body and no backbone or limbs. **2** (*pl.*) internal parasites. **3** insignificant or contemptible person. **4** spiral part of a screw. ● *v.* **1** make one's way with twisting movements. **2** insinuate oneself. **3** obtain by crafty persistence. **4** rid of parasitic worms. □ **worm-cast** *n.* pile of earth cast up by an earthworm. **wormy** *adj.*

wormeaten *adj.* full of holes made by insect larvae.

wormwood *n.* woody plant with a bitter flavour.

worn *see* **wear**. *adj.* **1** damaged or altered by use or wear. **2** looking exhausted. □ **worn-out** *adj.*

■ **1** old, ragged, *colloq.* scruffy, shabby, tattered, tatty, threadbare. **2** drawn, haggard.

worried *adj.* feeling or showing worry.

■ anxious, apprehensive, fearful, fretful, nervous, tense, troubled, uneasy, *colloq.* uptight.

worry *v.* **1** give way to anxiety. **2** make anxious. **3** be troublesome to. **4** seize with the teeth and shake or pull about. ● *n.* **1** worried state, mental uneasiness. **2** thing causing this. □ **worrier** *n.*

■ *v.* **1** agonize, brood, fret. **2** concern, disquiet, distress, disturb, trouble, unnerve, unsettle. **3** annoy, bother, harass, *colloq.* hassle, nag, pester, trouble. ● *n.* **1** anxiety, apprehension, concern, disquiet, distress, nervousness, unease, uneasiness. **2** bother, burden, care, concern, headache, problem, trouble.

worse *adj.* & *adv.* **1** more bad, more badly, more evil or ill. **2** less good. ● *n.* something worse.

worsen *v.* make or become worse.

■ aggravate, exacerbate, make worse; decline, degenerate, deteriorate, get worse.

worship *n.* **1** reverence and respect paid to a deity. **2** adoration of or devotion to a person or thing. ● *v.* (**worshipped**) **1** honour as a deity. **2** idolize, treat with adoration. **3** take part in an act of worship. □ **worshipper** *n.*

■ *n.* **1** glorification, homage, respect, reverence, veneration. **2** adoration, devotion, idolization, love. ● *v.* **1** glorify, honour, praise, pray to, revere, venerate. **2** adore, idolize, love.

worst *adj.* & *adv.* most bad, most badly, least good. ● *n.* worst part, feature, event, etc. ● *v.* defeat. □ **get the worst of** be defeated in.

■ *v.* beat, conquer, defeat, get the better of, overcome, rout, subdue, trounce, vanquish.

worsted /wóostid/ *n.* smooth woollen yarn or fabric.

worth *adj.* **1** having a specified value. **2** deserving. **3** possessing as wealth. ● *n.* **1** value, merit, usefulness. **2** amount that a specified sum will buy. □ **for all one is worth** (*colloq.*) with all one's energy. **worth while** or **worth one's while** worth the time or effort needed.

■ *n.* **1** advantage, benefit, good, importance, merit, usefulness, value.

worthless *adj.* without merit or value. □ **worthlessness** *n.*

■ meaningless, paltry, valueless; fruitless, futile, pointless, vain; cheap, tawdry, trashy.

worthwhile *adj.* worth while.

■ advantageous, beneficial, fruitful, gainful, good, helpful, profitable, useful, valuable.

worthy *adj.* (**-ier, -iest**) **1** having great merit. **2** deserving. ● *n.* worthy person. □ **worthily** *adv.*, **worthiness** *n.*

■ *adj.* **1** admirable, commendable, creditable, estimable, honourable, laudable, meritorious, praiseworthy. ● *n.* dignitary, luminary.

would *v.aux.* used in senses corresponding to *will*[1] in the past tense, conditional statements, questions, polite requests and statements, and to express probability or something that happens from time to time. □ **would-be** *adj.* desiring or pretending to be.

wound[1] /woond/ *n.* **1** injury done to tissue by violence. **2** injury to feelings. ● *v.* inflict a wound upon.

■ *n.* **1** cut, injury, laceration, lesion, trauma. **2** affront, injury, insult, slight. ● *v.* harm, hurt, injure, maim; distress, grieve, offend, pain, upset.

wound[2] /wownd/ *see* **wind**[2].

wove, woven *see* **weave**[1].

wow *int.* exclamation of astonishment. ● *n.* (*sl.*) sensational success.

wrack *n.* a type of seaweed.

wraith *n.* ghost, spectral apparition of a living person.

■ apparition, ghost, phantom, spectre, spirit, *colloq.* spook.

wrangle *v.* argue or quarrel noisily. ● *n.* noisy argument.

■ *v.* argue, bicker, disagree, dispute, fall out, fight, quarrel, row, squabble. ● *n.* altercation, argument, disagreement, dispute, fight, quarrel, row, squabble, tiff.

wrap *v.* (**wrapped**) arrange (a soft or flexible covering) round (a person or thing). ● *n.* shawl. □ **be wrapped up in** have one's attention deeply occupied by.

■ *v.* enfold, envelop, shroud, swaddle, swathe. ● *n.* cape, cloak, mantle, shawl, stole.

wrapper *n.* cover of paper etc. wrapped round something.

wrapping *n.* material for wrapping things.

wrasse /rass/ *n.* brightly coloured sea fish.

wrath *n.* anger, indignation. □ **wrathful** *adj.*, **wrathfully** *adv.*

■ anger, annoyance, fury, indignation, ire, irritation, rage, vexation.

wreak *v.* inflict (vengeance etc.).

wreath /reeth/ *n.* (*pl.* **-ths**) flowers or leaves etc. fastened into a ring, used as a decoration or placed on a grave etc.

wreathe /reeth/ *v.* **1** encircle. **2** twist into a wreath. **3** wind, curve.

wreck *n.* **1** destruction, esp. of a ship by storms or accident. **2** ship that has suffered this. **3** something ruined or dilapidated. **4** person whose health or spirits have been destroyed. ● *v.* **1** cause the wreck of. **2** ruin. □ **wrecker** *n.*

■ *v.* **2** destroy, devastate, put paid to, ruin, shatter, spoil.

wreckage *n.* remains of something wrecked.

■ debris, flotsam, fragments, remains, rubble.

wren *n.* very small bird.

wrench *v.* **1** twist or pull violently round. **2** damage or pull by twisting. ● *n.* **1** violent twisting pull. **2** pain caused by parting. **3** adjustable spanner-like tool.

■ *v.* **1** jerk, pull, rip, tear, tug, twist, wrest, *colloq.* yank. **2** rick, sprain, twist.

wrest *v.* **1** wrench away. **2** obtain by force or effort.

wrestle *v.* **1** fight (esp. as a sport) by grappling with and trying to throw an opponent to the ground. **2** struggle to deal with.

■ **2** battle, fight, grapple, struggle.

wretch *n.* **1** unfortunate or pitiable person. **2** rascal.

■ **2** blackguard, knave, rascal, rat, rogue, scallywag, scoundrel, villain.

wretched *adj.* **1** miserable, unhappy. **2** of bad quality. **3** contemptible. □ **wretchedly** *adv.*, **wretchedness** *n.*

■ **1** dejected, desolate, despondent, forlorn, melancholy, miserable, sad, sorrowful, unhappy, woebegone, woeful; pathetic, pitiful, sorry, unfortunate. **2** inferior, mean, miserable, poor, shabby, squalid. **3** contemptible, despicable, shameful, vile.

wriggle *v.* move with short twisting movements. ● *n.* wriggling movement.

■ *v.* fidget, squirm, twist, wiggle, worm, writhe.

wring *v.* (**wrung**) **1** twist and squeeze, esp. to remove liquid. **2** squeeze firmly or forcibly. **3** obtain with effort or difficulty.

wrinkle *n.* **1** crease in skin or other flexible surface. **2** (*colloq.*) useful hint or device. ● *v.* form wrinkles (in). □ **wrinkly** *adj.*

■ *n.* **1** crease, crinkle, fold, furrow, line, pucker.

wrist *n.* **1** joint connecting hand and forearm. **2** part of a garment covering this. □ **wrist-watch** *n.* watch worn on the wrist.

wristlet *n.* band or bracelet etc. worn round the wrist.

writ *n.* formal written authoritative command.

write *v.* (**wrote, written, writing**) **1** make letters or other symbols on a surface, esp. with a pen or pencil. **2** compose in written form, esp. for publication. **3** be an author. **4** write and send a letter. □ **write down** record in writing. **write off** recognize as lost. **write-off** *n.* something written off, esp. a vehicle too damaged to be worth repairing. **write up** write an account of. **write-up** *n.* (*colloq.*) published account of something, review.

■ **1** inscribe, scrawl, scribble. **2** compose, draft, pen. □ **write down** jot down, log, note, put down, record, register.

writer *n.* person who writes, author. □ **writer's cramp** cramp in the muscles of the hand.

■ author, columnist, dramatist, journalist, novelist, playwright, poet, scriptwriter.

writhe *v.* **1** twist one's body about, as in pain. **2** wriggle. **3** suffer because of embarrassment, squirm.

writing *n.* **1** handwriting. **2** literary work. □ **in writing** in written form. **writing paper** paper for writing (esp. letters) on.

written *see* **write**.

wrong *adj.* **1** incorrect, not true. **2** contrary to morality or law. **3** not what is required or desirable. **4** not in a satisfactory condition. ● *adv.* in a wrong manner or direction, mistakenly. ● *n.* **1** what is wrong, wrong action etc. **2** injustice. ● *v.* treat unjustly. □ **in the wrong** not having truth or justice on one's side. **wrongly** *adv.*, **wrongness** *n.*

■ *adj.* **1** erroneous, fallacious, false, inaccurate, incorrect, inexact, mistaken, untrue. **2** amoral, bad, evil, immoral, sinful, unethical, unconscionable, wicked; criminal, dishonest, illegal, illicit, unlawful. **3** improper, imprudent, inappropriate, incorrect, misguided, unacceptable, undesirable, unsuitable. **4** amiss, awry, defective, faulty, not working, out of order. ● *adv.* amiss, awry, badly, imperfectly, improperly, inappropriately, incorrectly, wrongly. ● *n.* **1** crime, misdeed, misdemeanour, offence, transgression; inequality, iniquity, unfairness, unjustness. **2** bad turn, injury, injustice. ● *v.* abuse, harm, ill-treat, injure, maltreat. □ **in the wrong** at fault, blameworthy, culpable, guilty, responsible.

wrongdoer *n.* person who acts illegally or immorally. □ **wrongdoing** *n.*

■ criminal, *colloq.* crook, lawbreaker, malefactor, miscreant, offender, transgressor, villain.

wrongful *adj.* contrary to what is right or legal. □ **wrongfully** *adv.*

■ injurious, unfair, unjust, unjustified; illegal, unlawful.

wrote *see* **write**.

wrought /rawt/ *adj.* (of metals) shaped by hammering. □ **wrought iron** pure form of iron used for decorative work.

wrung *see* **wring**.

wry *adj.* **1** (of the face) contorted in disgust or disappointment. **2** (of humour) dry, mocking. □ **wryly** *adv.*, **wryness** *n.*

■ **1** contorted, distorted, twisted. **2** droll, dry, ironic, sarcastic, sardonic.

wych elm elm with broad leaves and spreading branches.

xenophobia /zénnəfóbiə/ *n.* strong dislike or distrust of foreigners.
Xmas *n.* (*colloq.*) Christmas.
X-ray *n.* photograph or examination made by a kind of electromagnetic radiation (**X-rays**) that can penetrate solids. ● *v.* photograph, examine, or treat by X-rays.
xylophone *n.* musical instrument with flat wooden bars struck with small hammers.

yacht /yot/ *n.* **1** light sailing vessel for racing. **2** larger vessel for private pleasure trips. ● *v.* race or cruise in a yacht. □ **yachtsman** *n.*, **yachtswoman** *n.*

yak *n.* long-haired Asian ox.

yam *n.* **1** tropical climbing plant. **2** its edible tuber. **3** sweet potato.

yank (*colloq.*) *v.* pull sharply. ● *n.* sharp pull.

yap *n.* shrill bark. ● *v.* (**yapped**) bark shrilly.

yard[1] *n.* **1** measure of length, = 3 feet (0.9144 metre). **2** pole slung from a mast to support a sail.

yard[2] *n.* piece of enclosed ground, esp. attached to a building.

yardage *n.* length measured in yards.

yardstick *n.* standard of comparison.

■ benchmark, criterion, gauge, measure, standard, touchstone.

yarmulke /yaárməlkə/ *n.* skullcap worn by Jewish men.

yarn *n.* **1** any spun thread. **2** (*colloq.*) tale.

yarrow *n.* perennial plant with strong-smelling flowers.

yashmak *n.* veil worn by Muslim women in certain countries.

yaw *v.* (of a ship or aircraft) fail to hold a straight course. ● *n.* yawing.

yawl *n.* a kind of fishing boat or sailing boat.

yawn *v.* **1** open the mouth wide and draw in breath, as when sleepy or bored. **2** have a wide opening. ● *n.* act of yawning.

yaws *n.* tropical skin disease.

yd *abbr.* yard.

year *n.* **1** time taken by the earth to orbit the sun (about 365¼ days). **2** period from 1 Jan. to 31 Dec. inclusive. **3** consecutive period of twelve months. **4** (*pl.*) age.

yearling *n.* animal between 1 and 2 years old.

yearly *adj.* happening, published, or payable once a year. ● *adv.* annually.

yearn *v.* **yearn for** feel great longing for. □ **yearning** *n.*

■ crave, desire, hunger for, long for, pine for, thirst for, want, wish for.

yeast *n.* fungus used to cause fermentation in making beer and wine and as a raising agent.

yell *v.* & *n.* shout.

■ *v.* bawl, bellow, cry, roar, shout.

yellow *adj.* **1** of the colour of buttercups and ripe lemons. **2** (*colloq.*) cowardly. ● *n.* yellow colour or thing. ● *v.* turn yellow. □ **yellowish** *adj.*

yellowhammer *n.* bird of the finch family with a yellow head, neck, and breast.

yelp *n.* shrill yell or bark. ● *v.* utter a yelp.

yen[1] *n.* (*pl.* **yen**) unit of money in Japan.

yen[2] *n.* (*colloq.*) longing, yearning.

yes *adv.* & *n.* expression of agreement or consent, or of reply to a summons etc. □ **yes-man** *n.* weakly acquiescent person.

yesterday *adv.* & *n.* **1** (on) the day before today. **2** (in) the recent past.

yet *adv.* **1** up to this or that time, still. **2** besides. **3** eventually. **4** even. **5** nevertheless. ● *conj.* nevertheless, in spite of that.

yeti *n.* (*pl.* **-is**) large human-like or bear-like animal said to exist in the Himalayas.

yew *n.* **1** evergreen tree with dark needle-like leaves. **2** its wood.

Yiddish *n.* language used by Jews from eastern Europe.

yield *v.* **1** give as fruit, gain, or result. **2** surrender. **3** allow (victory, right of way, etc.) to another. **4** bend or break under pressure. ● *n.* amount yielded or produced.

■ *v.* **1** give, produce, supply; earn, generate, net, pay. **2** capitulate, cave in, concede, give in, submit, succumb, surrender, throw in the towel. **3** abandon, cede, give up, relinquish, surrender.

yielding *adj.* **1** submissive, compliant. **2** able to bend.

yippee *int.* exclamation of delight or excitement.

yob *n.* (*sl.*) lout, hooligan. □ **yobbish** *adj.*

yodel *v.* (**yodelled**) sing with a quickly alternating change of pitch. ● *n.* yodelling cry. □ **yodeller** *n.*

yoga *n.* Hindu system of meditation and self-control.

yoghurt *n.* food made of milk that has been thickened by the action of certain bacteria.

yoke *n.* **1** wooden crosspiece fastened over the necks of two oxen pulling a plough etc. **2** piece of wood shaped to fit a person's shoulders and hold a load slung from each end. **3** top part of a garment. **4** oppression. ● *v.* **1** harness with a yoke. **2** unite.

■ *v.* **2** connect, couple, join, link, tie, unite.

yokel *n.* country fellow, bumpkin.

yolk *n.* yellow inner part of an egg.

yonder *adj.* & *adv.* over there.

yonks *adv.* (*sl.*) a long time.

yore *n.* **of yore** long ago.

Yorkshire pudding baked batter pudding eaten with meat.

you *pron.* **1** person(s) addressed. **2** one, anyone, everyone.

young *adj.* **1** having lived or existed for only a short time. **2** youthful. **3** having little experience. ● *n.* offspring of animals.

■ *adj.* **1,2** immature, juvenile, youthful. **3** callow, green, immature, inexperienced, innocent, naive, unsophisticated. ● *n.* brood, litter, offspring, progeny.

youngster *n.* child.

■ adolescent, boy, child, girl, juvenile, *colloq.* kid, lad, *Sc. & N. Engl.* lass, minor, teenager, youth.

your *adj.*, **yours** *poss.pron.* belonging to you.

yourself *pron.* (*pl.* **yourselves**) emphatic and reflexive form of *you*.

youth *n.* (*pl.* **youths**) **1** state or period of being young. **2** young man. **3** young people. □ **youth club** club with leisure activities for young people. **youth hostel** hostel providing cheap accommodation for young travellers.

■ **1** adolescence, boyhood, childhood, girlhood, minority, puberty, teens. **3** adolescents, children, juveniles, *colloq.* kids, young people.

youthful *adj.* **1** young. **2** having the characteristics of youth. □ **youthfulness** *n.*

yowl *v.* & *n.* howl.

yucca *n.* tall plant with white flowers and spiky leaves.

yule *n.* (in full **yule-tide**) (*old use*) Christmas festival.

yummy *adj.* (*colloq.*) delicious.

yuppie *n.* (*colloq.*) young urban professional person.

Zz

zany *adj.* (**-ier**, **-iest**) crazily funny. ● *n.* zany person.

zeal *n.* enthusiasm, hearty and persistent effort.

■ ardour, eagerness, enthusiasm, fervour, gusto, passion, zest.

zealot /zéllət/ *n.* fanatic.

■ bigot, extremist, fanatic, maniac.

zealous /zélləss/ *adj.* full of zeal. □ **zealously** *adv.*

■ ardent, eager, earnest, energetic, enthusiastic, fervent, fervid, impassioned, keen, passionate.

zebra *n.* African horse-like animal with black and white stripes. □ **zebra crossing** striped pedestrian crossing.

Zen *n.* form of Buddhism.

zenith *n.* **1** the part of the sky that is directly overhead. **2** highest point.

■ **2** apex, climax, height, peak, pinnacle, summit, top.

zephyr *n.* (*poetic*) soft gentle wind.

zero *n.* (*pl.* **-os**) **1** nought, the figure 0. **2** nil. **3** point marked 0 on a graduated scale, temperature corresponding to this. □ **zero hour** hour at which something is timed to begin. **zero in on 1** aim at or for. **2** focus attention on.

■ **1,2** cipher, nil, nothing, nought. □ **zero in on 1** take aim at, target. **2** concentrate on, fix on, focus on.

zest *n.* **1** keen enjoyment or interest. **2** orange or lemon peel as flavouring. □ **zestful** *adj.*

■ **1** appetite, eagerness, enjoyment, enthusiasm, gusto, interest, relish, zeal.

zigzag *n.* line or course turning right and left alternately at sharp angles. ● *adj.* & *adv.* as or in a zigzag. ● *v.* (**zigzagged**) move in a zigzag.

zinc *n.* bluish-white metal.

zing (*colloq.*) *n.* vigour, zip. ● *v.* move swiftly or shrilly.

Zionism *n.* movement that campaigned for a Jewish homeland in Palestine. □ **Zionist** *n.*

zip *n.* **1** short sharp sound. **2** vigour, liveliness. **3** zip-fastener. ● *v.* (**zipped**) **1** fasten with a zip-fastener. **2** move with vigour or at high speed. □ **zip-fastener** *n.* fastening device with teeth that interlock when brought together by a sliding tab.

zipper *n.* zip-fastener.

zircon *n.* bluish-white gem cut from a translucent mineral.

zirconium *n.* grey metallic element.

zither *n.* stringed instrument played with the fingers.

zodiac *n.* (in astrology) band of the sky divided into twelve equal parts (**signs of the zodiac**) each named from a constellation. □ **zodiacal** *adj.*

zombie *n.* **1** (in voodoo) corpse said to have been revived by witchcraft. **2** (*colloq.*) person who seems to have no mind or will.

zone *n.* area with particular features, purpose, or use. ● *v.* divide into zones. □ **zonal** *adj.*

■ *n.* area, belt, district, quarter, region, sector.

zoo *n.* place where wild animals are kept for exhibition and study.

zoology *n.* study of animals. □ **zoological** *adj.*, **zoologist** *n.*

zoom *v.* **1** move quickly, esp. with a buzzing sound. **2** rise sharply. **3** (in photography) make a distant object appear gradually closer by means of a **zoom lens**.

■ **1** dart, dash, fly, race, rush, shoot, speed, streak, tear, whiz.

zucchini /zookéeni/ *n.* (*pl.* **-i** or **-is**) courgette.

Zulu *n.* member or language of a Bantu people of S. Africa.

zygote *n.* cell formed by the union of two gametes.